KIRK-OTHMER

ENCYCLOPEDIA OF CHEMICAL TECHNOLOGY

THIRD EDITION

VOLUME 7

COPPER ALLOYS
TO
DISTILLATION

A WILEY-INTERSCIENCE PUBLICATION

John Wiley & Sons

NEW YORK • CHICHESTER • BRISBANE • TORONTO

Copyright © 1979 by John Wiley & Sons, Inc.

All rights reserved. Published simultaneously in Canada.

Reproduction or translation of any part of this work beyond that permitted by Sections 107 or 108 of the 1976 United States Copyright Act without the permission of the copyright owner is unlawful. Requests for permission or further information should be addressed to the Permissions Department, John Wiley & Sons, Inc.

Library of Congress Cataloging in Publication Data:

Main entry under title:
   Encyclopedia of chemical technology.

  At head of title: Kirk-Othmer.
  "A Wiley-Interscience publication."
  Includes bibliographies.
  1. Chemistry, Technical—Dictionaries. I. Kirk, Raymond Eller, 1890–1957. II. Othmer, Donald Frederick, 1904–   III. Grayson, Martin. IV. Eckroth, David. V. Title: Kirk-Othmer encyclopedia of chemical technology.

TP9.E685 1978   660'.03   77-15820
ISBN 0-471-02043-5

Printed in the United States of America

CONTENTS

Copper alloys, 1
Copper compounds, 97
Cork, 110
Corrosion and corrosion inhibitors, 113
Cosmetics, 143
Cotton, 176
Coumarin, 196
Crotonaldehyde, 207
Crotonic acid, 218
Cryogenics, 227
Crystallization, 243
Cumene, 286
Cyanamides, 291
Cyanides, 307
Cyanine dyes, 335
Cyanocarbons, 359
Cyanoethylation, 370
Cyanohydrins, 385
Cyanuric and isocyanuric acids, 397
Cyclohexanol and cyclohexanone, 410
Cyclopentadiene and dicyclopentadiene, 417

Defoamers, 430
Deformation recording media, 448
Dental materials, 461
Dentifrices, 522
Design of experiments, 526
Deuterium and tritium, 539
Dialysis, 564
Diamines and higher amines, aliphatic, 580
Diatomite, 603
Dicarboxylic acids, 614
Dietary fiber, 628
Diffusion separation methods, 639
Digital displays, 724
Dimensional analysis, 752
Dimer acids, 768
Diphenyl and terphenyls, 782
Disinfectants and antiseptics, 793
Dispersants, 833
Distillation, 849

EDITORIAL STAFF FOR VOLUME 7

Executive Editor: **Martin Grayson**
Associate Editor: **David Eckroth**
Production Supervisor: **Michalina Bickford**
Editors: **Galen J. Bushey**   **Caroline L. Eastman**   **Anna Klingsberg**
   Lorraine van Nes

CONTRIBUTORS TO VOLUME 7

Siegfried Aftergut, *General Electric Company, Schenectady, New York,* Deformation recording media

A. F. Alciatore, *Grefco, Inc., New Orleans, Louisiana,* Diatomite

B. A. Kottes Andrews, *United States Department of Agriculture, New Orleans, Louisiana,* Cotton

William F. Baxter, Jr., *Eastman Chemical Products, Inc., Kingsport, Tennessee,* Crotonaldehyde; Crotonic acid

B. F. Brown, *The American University, Washington, D. C.,* Corrosion and corrosion inhibitors

Joseph V. Burakevich, *FMC Corporation, Princeton, New Jersey,* Cyanuric and isocyanuric acids

Wai-Kai Chen, *Ohio University, Athens, Ohio,* Dimensional analysis

Michael S. Cholod, *Rohm and Haas Company, Philadelphia, Pennsylvania,* Cyanohydrins

Ines V. de Gruy, *United States Department of Agriculture, New Orleans, Louisiana,* Cotton

M. Fefer, *Exxon Chemical Company, Houston, Texas,* Cyclopentadiene and dicyclopentadiene

William B. Fisher, *Allied Chemical Corp., Petersburg, Virginia,* Cyclohexanol and cyclohexanone

R. T. Foley, *The American University, Washington, D. C.,* Corrosion and corrosion inhibitors

Ted Gammon, *Diamond Shamrock Chemical Co., Morristown, New Jersey,* Defoamers

Richard M. Goodman, *American Cyanamid Co., Stamford, Connecticut,* Dispersants

CONTRIBUTORS TO VOLUME 7

William Gump, *Consultant, Upper Montclair, New Jersey,* Disinfectants and antiseptics
Earl R. Hafslund, *E. I. du Pont de Nemours & Co., Inc., Wilmington, Delaware,* Distillation
Gerald J. Hahn, *General Electric Company, Schenectady, New York,* Design of experiments
Robert W. Haisty, *Texas Instruments, Inc., Dallas, Texas,* Digital displays
Robert J. Harper, Jr., *United States Department of Agriculture, New Orleans, Louisiana,* Cyanoethylation
R. L. Hoglund, *Union Carbide Nuclear Division, Oak Ridge, Tennessee,* Diffusion separation methods
R. H. Hutchison, *Armstrong Cork Co., Lancaster, Pennsylvania,* Cork
William R. Jenks, *E. I. du Pont de Nemours & Co., Inc., Memphis, Tennessee,* Hydrogen cyanide; Alkali metal cyanides; Alkaline earth metal cyanides all under Cyanides
Joseph J. Katz, *Argonne National Laboratory, Argonne, Illinois,* Deuterium and tritium
Roger N. Kust, *Kennecott Copper Corporation, Lexington, Massachusetts,* Copper compounds
Harry Isacoff, *International Flavors and Fragrances Inc., New York, New York,* Cosmetics
Edward C. Leonard, *Humko Sheffield, Memphis, Tennessee,* Dimer acids
Edward F. Leonard, *Columbia University, New York, New York,* Dialysis
Betty Lewis, *Cornell University, Ithaca, New York,* Dietary fiber
Irwin Lichtman, *Diamond Shamrock Chemical Co., Morristown, New Jersey,* Defoamers
Donald R. May, *Cyanamid Canada Inc., Niagara Falls, Ontario, Canada,* Cyanamides
Walter C. Meuly, *Rhodia Inc., New Brunswick, New Jersey,* Coumarin
Paul Morgan, *Consultant, Westchester, Pennsylvania,* Dicarboxylic acids
J. W. Mullin, *University College, London, London, England,* Crystallization
E. L. Neu, *Grefco Inc., Los Angeles, California,* Diatomite
C. L. Newton, *Air Products and Chemicals, Inc., Allentown, Pennsylvania,* Cryogenics
G. C. Paffenberger, *American Dental Association, Health Foundation Research Unit, National Bureau of Standards, Washington, D. C.,* Dental materials
P. Andrew Penz, *Texas Instruments Inc., Dallas, Texas,* Digital displays
Anthony F. Posteraro, *College of Dentistry, New York University, New York, New York,* Dentifrices
Ralph E. Ricksecker, *Chase Brass & Copper Co., Inc., Cleveland, Ohio,* Wrought copper and wrought copper alloys; Cast copper alloys both under Copper alloys
N. W. Rupp, *American Dental Association, Health Foundation Research Unit, National Bureau of Standards, Washington, D. C.,* Dental materials
J. Shacter, *Union Carbide Nuclear Division, Oak Ridge, Tennessee,* Diffusion separation methods
P. B. Simmons, *Dow Chemical U.S.A., Midland, Michigan,* Diphenyl and terphenyls
A. B. Small, *Exxon Chemical Company, Baton Rouge, Louisiana,* Cyclopentadiene and dicyclopentadiene
R. D. Spitz, *Dow Chemical U.S.A., Freeport, Texas,* Diamines and higher amines, aliphatic
David M. Sturmer, *Eastman Kodak Company, Rochester, New York,* Cyanine dyes
Kishin H. Surtani, *Texas Instruments Inc., Dallas, Texas,* Digital displays
Q. E. Thompson, *Monsanto Co., St. Louis, Missouri,* Diphenyl and terphenyls
Jan F. Van Peppen, *Allied Chemical Corp., Petersburg, Virginia,* Cyclohexanol and cyclohexanone
E. Von Halle, *Union Carbide Nuclear Division, Oak Ridge, Tennessee,* Diffusion separation methods

Dennis J. Ward, *UOP Inc., Des Plaines, Illinois,* Cumene

W. C. Weaver, *Dow Chemical U.S.A., Midland, Michigan,* Diphenyl and terphenyls

O. W. Webster, *E. I. du Pont de Nemours & Co., Inc., Wilmington, Delaware,* Cyanocarbons

NOTE ON CHEMICAL ABSTRACTS SERVICE REGISTRY NUMBERS AND NOMENCLATURE

Chemical Abstracts Service (CAS) Registry Numbers are unique numerical identifiers assigned to substances recorded in the CAS Registry System. They appear in brackets in the *Chemical Abstracts* (CA) substance and formula indexes following the names of compounds. A single compound may have many synonyms in the chemical literature. A simple compound like phenethylamine can be named β-phenylethylamine or, as in *Chemical Abstracts,* benzeneethanamine. The usefulness of the Encyclopedia depends on accessibility through the most common correct name of a substance. Because of this diversity in nomenclature careful attention has been given the problem in order to assist the reader as much as possible, especially in locating the systematic CA index name by means of the Registry Number. For this purpose, the reader may refer to the CAS Registry Handbook-Number Section which lists in numerical order the Registry Number with the Chemical Abstracts index name and the molecular formula; eg, **458-88-8,** Piperidine, 2-propyl-, (S)-, $C_8H_{17}N$; in the Encyclopedia this compound would be found under its common name, coniine [*458-88-8*]. The Registry Number is a valuable link for the reader in retrieving additional published information on substances and also as a point of access for such on-line data bases as Chemline, Medline, and Toxline.

In all cases, the CAS Registry Numbers have been given for title compounds in articles and for all compounds in the index. All specific substances indexed in *Chemical Abstracts* since 1965 are included in the CAS Registry System as are a large number of substances derived from a variety of reference works. The CAS Registry System identifies a substance on the basis of an unambiguous computer-language description of its molecular structure including stereochemical detail. The Registry Number is a machine-checkable number (like a Social Security number) assigned in sequential order to each substance as it enters the registry system. The value of the number lies in the fact that it is a concise and unique means of substance identification, which is independent of, and therefore bridges, many systems of chemical nomenclature. For polymers, one Registry Number is used for the entire family; eg, polyoxyethylene (20)sorbitan monolaurate has the same number as all of its polyoxyethylene homologues.

Registry numbers for each substance will be provided in the third edition index (eg, Alkaloids shows the Registry Number of all alkaloids (title compounds) in a table in the article as well, but the intermediates will have their Registry Numbers shown only in the index). Articles such as Absorption, Adsorptive separation, Air conditioning, Air pollution, Air pollution control methods have no Registry Numbers in the text.

Cross-references have been inserted in the index for many common names and for some systematic names. Trademark names appear in the index. Names that are incorrect, misleading or ambiguous are avoided. Formulas are given very frequently in the text to help in identifying compounds. The spelling and form used, even for industrial names, follow American chemical usage, but not always the usage of *Chemical Abstracts* (eg, *coniine* is used instead of *(S)-2-propylpiperidine*, *aniline* instead of *benzenamine*, and *acrylic acid* instead of *2-propenoic acid*).

There are variations in representation of rings in different disciplines. The dye industry does not designate aromaticity or double bonds in rings. All double bonds and aromaticity will be shown in the *Encyclopedia* as a matter of course. For example, tetralin has an aromatic ring and a saturated ring and its structure will appear in the

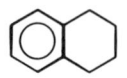

Encyclopedia with its common name, Registry Number enclosed in brackets, and parenthetical CA index name, ie, tetralin [*119-64-2*] (1,2,3,4-tetrahydronaphthalene). With names and structural formulas, and especially with CAS Registry Numbers the aim is to help the reader have a concise means of substance identification.

CONVERSION FACTORS, ABBREVIATIONS, AND UNIT SYMBOLS

SI Units (Adopted 1960)

A new system of measurement, the International System of Units (abbreviated SI), is being implemented throughout the world. This system is a modernized version of the MKSA (meter, kilogram, second, ampere) system, and its details are published and controlled by an international treaty organization (The International Bureau of Weights and Measures) (1).

SI units are divided into three classes:

BASE UNITS

length	meter[†] (m)
mass[‡]	kilogram (kg)
time	second (s)
electric current	ampere (A)
thermodynamic temperature[§]	kelvin (K)
amount of substance	mole (mol)
luminous intensity	candela (cd)

[†] The spellings "metre" and "litre" are preferred by ASTM; however "-er" will be used in the Encyclopedia.
[‡] "Weight" is the commonly used term for "mass".
[§] Wide use is made of "Celsius temperature" (t) defined by

$$t = T - T_0$$

where T is the thermodynamic temperature, expressed in kelvins, and $T_0 = 273.15$ K by definition. A temperature interval may be expressed in degrees Celsius as well as in kelvins.

FACTORS, ABBREVIATIONS, AND SYMBOLS

SUPPLEMENTARY UNITS

plane angle	radian (rad)
solid angle	steradian (sr)

DERIVED UNITS AND OTHER ACCEPTABLE UNITS

These units are formed by combining base units, supplementary units, and other derived units (2–4). Those derived units having special names and symbols are marked with an asterisk in the list below:

Quantity	Unit	Symbol	Acceptable equivalent
*absorbed dose	gray	Gy	J/kg
acceleration	meter per second squared	m/s^2	
*activity (of ionizing radiation source)	becquerel	Bq	1/s
area	square kilometer	km^2	
	square hectometer	hm^2	ha (hectare)
	square meter	m^2	
*capacitance	farad	F	C/V
concentration (of amount of substance)	mole per cubic meter	mol/m^3	
*conductance	siemens	S	A/V
current density	ampere per square meter	A/m^2	
density, mass density	kilogram per cubic meter	kg/m^3	g/L; mg/cm^3
dipole moment (quantity)	coulomb meter	C·m	
*electric charge, quantity of electricity	coulomb	C	A·s
electric charge density	coulomb per cubic meter	C/m^3	
electric field strength	volt per meter	V/m	
electric flux density	coulomb per square meter	C/m^2	
*electric potential, potential difference, electromotive force	volt	V	W/A
*electric resistance	ohm	Ω	V/A
*energy, work, quantity of heat	megajoule	MJ	
	kilojoule	kJ	
	joule	J	N·m
	electron volt[†]	eV[†]	
	kilowatt-hour[†]	kW·h[†]	

[†] This non-SI unit is recognized by the CIPM as having to be retained because of practical importance or use in specialized fields (1).

Quantity	Unit	Symbol	Acceptable equivalent
energy density	joule per cubic meter	J/m^3	
*force	kilonewton	kN	
	newton	N	kg·m/s^2
*frequency	megahertz	MHz	
	hertz	Hz	1/s
heat capacity, entropy	joule per kelvin	J/K	
heat capacity (specific), specific entropy	joule per kilogram kelvin	J/(kg·K)	
heat transfer coefficient	watt per square meter kelvin	W/(m^2·K)	
*illuminance	lux	lx	lm/m^2
*inductance	henry	H	Wb/A
linear density	kilogram per meter	kg/m	
luminance	candela per square meter	cd/m^2	
*luminous flux	lumen	lm	cd·sr
magnetic field strength	ampere per meter	A/m	
*magnetic flux	weber	Wb	V·s
*magnetic flux density	tesla	T	Wb/m^2
molar energy	joule per mole	J/mol	
molar entropy, molar heat capacity	joule per mole kelvin	J/(mol·K)	
moment of force, torque	newton meter	N·m	
momentum	kilogram meter per second	kg·m/s	
permeability	henry per meter	H/m	
permittivity	farad per meter	F/m	
*power, heat flow rate, radiant flux	kilowatt	kW	
	watt	W	J/s
power density, heat flux density, irradiance	watt per square meter	W/m^2	
*pressure, stress	megapascal	MPa	
	kilopascal	kPa	
	pascal	Pa	N/m^2
sound level	decibel	dB	
specific energy	joule per kilogram	J/kg	
specific volume	cubic meter per kilogram	m^3/kg	
surface tension	newton per meter	N/m	
thermal conductivity	watt per meter kelvin	W/(m·K)	
velocity	meter per second	m/s	
	kilometer per hour	km/h	
viscosity, dynamic	pascal second	Pa·s	
	millipascal second	mPa·s	
viscosity, kinematic	square meter per second	m^2/s	

xvi FACTORS, ABBREVIATIONS, AND SYMBOLS

Quantity	Unit	Symbol	Acceptable equivalent
	square millimeter per second	mm^2/s	
volume	cubic meter	m^3	
	cubic decimeter	dm^3	L(liter) (5)
	cubic centimeter	cm^3	mL
wave number	1 per meter	m^{-1}	
	1 per centimeter	cm^{-1}	

In addition, there are 16 prefixes used to indicate order of magnitude, as follows:

Multiplication factor	Prefix	Symbol	Note
10^{18}	exa	E	
10^{15}	peta	P	
10^{12}	tera	T	
10^9	giga	G	
10^6	mega	M	
10^3	kilo	k	
10^2	hecto	h[a]	[a] Although hecto, deka, deci, and centi are SI prefixes, their use should be avoided except for SI unit-multiples for area and volume and nontechnical use of centimeter, as for body and clothing measurement.
10	deka	da[a]	
10^{-1}	deci	d[a]	
10^{-2}	centi	c[a]	
10^{-3}	milli	m	
10^{-6}	micro	μ	
10^{-9}	nano	n	
10^{-12}	pico	p	
10^{-15}	femto	f	
10^{-18}	atto	a	

For a complete description of SI and its use the reader is referred to ASTM E 380 (4) and the article Units and Conversion Factors which will appear in a later volume of the *Encyclopedia*.

A representative list of conversion factors from non-SI to SI units is presented herewith. Factors are given to four significant figures. Exact relationships are followed by a dagger. A more complete list is given in ASTM E 380-76(4) and ANSI Z210.1-1976 (6).

Conversion Factors to SI Units

To convert from	To	Multiply by
acre	square meter (m^2)	4.047 × 10^3
angstrom	meter (m)	1.0 × 10^{-10}†
are	square meter (m^2)	1.0 × 10^{2}†
astronomical unit	meter (m)	1.496 × 10^{11}
atmosphere	pascal (Pa)	1.013 × 10^5
bar	pascal (Pa)	1.0 × 10^{5}†

† Exact.

To convert from	To	Multiply by
barrel (42 U.S. liquid gallons)	cubic meter (m^3)	0.1590
Bohr magneton μ_β	J/T	9.274×10^{-24}
Btu (International Table)	joule (J)	1.055×10^3
Btu (mean)	joule (J)	1.056×10^3
Btu (thermochemical)	joule (J)	1.054×10^3
bushel	cubic meter (m^3)	3.524×10^{-2}
calorie (International Table)	joule (J)	4.187
calorie (mean)	joule (J)	4.190
calorie (thermochemical)	joule (J)	4.184†
centipoise	pascal second (Pa·s)	1.0×10^{-3}†
centistoke	square millimeter per second (mm^2/s)	1.0†
cfm (cubic foot per minute)	cubic meter per second (m^3/s)	4.72×10^{-4}
cubic inch	cubic meter (m^3)	1.639×10^{-5}
cubic foot	cubic meter (m^3)	2.832×10^{-2}
cubic yard	cubic meter (m^3)	0.7646
curie	becquerel (Bq)	3.70×10^{10}†
debye	coulomb·meter (C·m)	3.336×10^{-30}
degree (angle)	radian (rad)	1.745×10^{-2}
denier (international)	kilogram per meter (kg/m)	1.111×10^{-7}
	tex‡	0.1111
dram (apothecaries')	kilogram (kg)	3.888×10^{-3}
dram (avoirdupois)	kilogram (kg)	1.772×10^{-3}
dram (U.S. fluid)	cubic meter (m^3)	3.697×10^{-6}
dyne	newton (N)	1.0×10^{-5}†
dyne/cm	newton per meter (N/m)	1.00×10^{-3}†
electron volt	joule (J)	1.602×10^{-19}
erg	joule (J)	1.0×10^{-7}†
fathom	meter (m)	1.829
fluid ounce (U.S.)	cubic meter (m^3)	2.957×10^{-5}
foot	meter (m)	0.3048†
footcandle	lux (lx)	10.76
furlong	meter (m)	2.012×10^{-2}
gal	meter per second squared (m/s^2)	1.0×10^{-2}†
gallon (U.S. dry)	cubic meter (m^3)	4.405×10^{-3}
gallon (U.S. liquid)	cubic meter (m^3)	3.785×10^{-3}
gallon per minute (gpm)	cubic meter per second (m^3/s)	6.308×10^{-5}
	cubic meter per hour (m^3/h)	0.2271
gauss	tesla (T)	1.0×10^{-4}
gilbert	ampere (A)	0.7958
gill (U.S.)	cubic meter (m^3)	1.183×10^{-4}
grad	radian	1.571×10^{-2}
grain	kilogram (kg)	6.480×10^{-5}
gram force per denier	newton per tex (N/tex)	8.826×10^{-2}
hectare	square meter (m^2)	1.0×10^4†
horsepower (550 ft·lbf/s)	watt (W)	7.457×10^2

† Exact.
‡ See footnote on p. xiv.

FACTORS, ABBREVIATIONS, AND SYMBOLS

To convert from	To	Multiply by
horsepower (boiler)	watt (W)	9.810×10^3
horsepower (electric)	watt (W)	$7.46 \times 10^{2\dagger}$
hundredweight (long)	kilogram (kg)	50.80
hundredweight (short)	kilogram (kg)	45.36
inch	meter (m)	$2.54 \times 10^{-2\dagger}$
inch of mercury (32°F)	pascal (Pa)	3.386×10^3
inch of water (39.2°F)	pascal (Pa)	2.491×10^2
kilogram force	newton (N)	9.807
kilowatt hour	megajoule (MJ)	$3.6^\dagger$
kip	newton (N)	4.48×10^3
knot (international)	meter per second (m/s)	0.5144
lambert	candela per square meter (cd/m^2)	3.183×10^3
league (British nautical)	meter (m)	5.559×10^3
league (statute)	meter (m)	4.828×10^3
light year	meter (m)	9.461×10^{15}
liter (for fluids only)	cubic meter (m^3)	$1.0 \times 10^{-3\dagger}$
maxwell	weber (Wb)	$1.0 \times 10^{-8\dagger}$
micron	meter (m)	$1.0 \times 10^{-6\dagger}$
mil	meter (m)	$2.54 \times 10^{-5\dagger}$
mile (U.S. nautical)	meter (m)	$1.852 \times 10^{3\dagger}$
mile (statute)	meter (m)	1.609×10^3
mile per hour	meter per second (m/s)	0.4470
millibar	pascal (Pa)	1.0×10^2
millimeter of mercury (0°C)	pascal (Pa)	$1.333 \times 10^{2\dagger}$
minute (angular)	radian	2.909×10^{-4}
myriagram	kilogram (kg)	10
myriameter	kilometer (km)	10
oersted	ampere per meter (A/m)	79.58
ounce (avoirdupois)	kilogram (kg)	2.835×10^{-2}
ounce (troy)	kilogram (kg)	3.110×10^{-2}
ounce (U.S. fluid)	cubic meter (m^3)	2.957×10^{-5}
ounce-force	newton (N)	0.2780
peck (U.S.)	cubic meter (m^3)	8.810×10^{-3}
pennyweight	kilogram (kg)	1.555×10^{-3}
pint (U.S. dry)	cubic meter (m^3)	5.506×10^{-4}
pint (U.S. liquid)	cubic meter (m^3)	4.732×10^{-4}
poise (absolute viscosity)	pascal second (Pa·s)	$0.10^\dagger$
pound (avoirdupois)	kilogram (kg)	0.4536
pound (troy)	kilogram (kg)	0.3732
poundal	newton (N)	0.1383
pound-force	newton (N)	4.448
pound per square inch (psi)	pascal (Pa)	6.895×10^3
quart (U.S. dry)	cubic meter (m^3)	1.101×10^{-3}
quart (U.S. liquid)	cubic meter (m^3)	9.464×10^{-4}
quintal	kilogram (kg)	$1.0 \times 10^{2\dagger}$

† Exact.

To convert from	To	Multiply by
rad	gray (Gy)	1.0×10^{-2}†
rod	meter (m)	5.029
roentgen	coulomb per kilogram (C/kg)	2.58×10^{-4}
second (angle)	radian (rad)	4.848×10^{-6}
section	square meter (m^2)	2.590×10^6
slug	kilogram (kg)	14.59
spherical candle power	lumen (lm)	12.57
square inch	square meter (m^2)	6.452×10^{-4}
square foot	square meter (m^2)	9.290×10^{-2}
square mile	square meter (m^2)	2.590×10^6
square yard	square meter (m^2)	0.8361
stere	cubic meter (m^3)	1.0†
stokes (kinematic viscosity)	square meter per second (m^2/s)	1.0×10^{-4}†
tex	kilogram per meter (kg/m)	1.0×10^{-6}†
ton (long, 2240 pounds)	kilogram (kg)	1.016×10^3
ton (metric)	kilogram (kg)	1.0×10^3†
ton (short, 2000 pounds)	kilogram (kg)	9.072×10^2
torr	pascal (Pa)	1.333×10^2
unit pole	weber (Wb)	1.257×10^{-7}
yard	meter (m)	0.9144†

† Exact.

Abbreviations and Unit Symbols

Following is a list of commonly used abbreviations and unit symbols appropriate for use in the *Encyclopedia*. In general they agree with those listed in *American National Standard Abbreviations for Use on Drawings and in Text* (*ANSI Y1.1*) (6) and *American National Standard Letter Symbols for Units in Science and Technology* (*ANSI Y10*) (6). Also included is a list of acronyms for a number of private and government organizations as well as common industrial solvents, polymers, and other chemicals.

Rules for Writing Unit Symbols (4):

1. Unit symbols should be printed in upright letters (roman) regardless of the type style used in the surrounding text.

2. Unit symbols are unaltered in the plural.

3. Unit symbols are not followed by a period except when used as the end of a sentence.

4. Letter unit symbols are generally written in lower-case (eg, cd for candela) unless the unit name has been derived from a proper name, in which case the first letter of the symbol is capitalized (W,Pa). Prefix and unit symbols retain their prescribed form regardless of the surrounding typography.

5. In the complete expression for a quantity, a space should be left between the numerical value and the unit symbol. For example, write 2.37 lm, *not* 2.37lm, and 35 mm, *not* 35mm. When the quantity is used in an adjectival sense, a hyphen is often used, for example, 35-mm film. *Exception:* No space is left between the numerical value and the symbols for degree, minute, and second of plane angle, and degree Celsius.

6. No space is used between the prefix and unit symbols (eg, kg).

7. Symbols, not abbreviations, should be used for units. For example, use "A," not "amp," for ampere.

8. When multiplying unit symbols, use a raised dot:

$$N \cdot m \text{ for newton meter}$$

In the case of W·h, the dot may be omitted, thus:

$$Wh$$

An exception to this practice is made for computer printouts, automatic typewriter work, etc, where the raised dot is not possible, and a dot on the line may be used.

9. When dividing unit symbols use one of the following forms:

$$m/s \text{ or } m \cdot s^{-1} \text{ or } \frac{m}{s}$$

In no case should more than one slash be used in the same expression unless parentheses are inserted to avoid ambiguity. For example, write:

$$J/(mol \cdot K) \text{ or } J \cdot mol^{-1} \cdot K^{-1} \text{ or } (J/mol)/K$$

but *not*

$$J/mol/K$$

10. Do not mix symbols and unit names in the same expression. Write:

$$\text{joules per kilogram } or \text{ J/kg } or \text{ J} \cdot \text{kg}^{-1}$$

but *not*

$$\text{joules/kilogram } nor \text{ joules/kg } nor \text{ joules} \cdot \text{kg}^{-1}$$

ABBREVIATIONS AND UNITS

A	ampere	AIChE	American Institute of Chemical Engineers
A	anion (eg, HA)		
a	atto (prefix for 10^{-18})	AIP	American Institute of Physics
AATCC	American Association of Textile Chemists and Colorists	alc	alcohol(ic)
		Alk	alkyl
		alk	alkaline (not alkali)
ABS	acrylonitrile–butadiene–styrene	amt	amount
		amu	atomic mass unit
abs	absolute	ANSI	American National Standards Institute
ac	alternating current, *n*.		
a-c	alternating current, *adj*.	AO	atomic orbital
ac-	alicyclic	APHA	American Public Health Association
ACGIH	American Conference of Governmental Industrial Hygienists	API	American Petroleum Institute
ACS	American Chemical Society	aq	aqueous
AGA	American Gas Association	Ar	aryl
Ah	ampere hour	*ar*-	aromatic

as-	asymmetric(al)	**d**	differential operator
ASH-RAE	American Society of Heating, Refrigerating, and Air Conditioning Engineers	*d-*	dextro-, dextrorotatory
		da	deka (prefix for 10^1)
		dB	decibel
		dc	direct current, *n.*
ASM	American Society for Metals	d-c	direct current, *adj.*
ASME	American Society of Mechanical Engineers	dec	decompose
		detd	determined
ASTM	American Society for Testing and Materials	detn	determination
		dia	diameter
at no.	atomic number	dil	dilute
at wt	atomic weight	*dl-*; DL-	racemic
av(g)	average	DMF	dimethylformamide
bbl	barrel	DMG	dimethyl glyoxime
bcc	body-centered cubic	DOE	Department of Energy
Bé	Baumé	DOT	Department of Transportation
bid	twice daily	dp	dew point; degree of polymerization
BOD	biochemical (biological) oxygen demand		
		dstl(d)	distill(ed)
bp	boiling point	dta	differential thermal analysis
Bq	becquerel	(*E*)-	entgegen; opposed
C	coulomb	ϵ	dielectric constant (unitless number)
°C	degree Celsius		
C-	denoting attachment to carbon	e	electron
		ECU	electrochemical unit
c	centi (prefix for 10^{-2})	ed.	edited, edition, editor
ca	circa (approximately)	ED	effective dose
cd	candela; current density; circular dichroism	EDTA	ethylenediamine tetraacetic acid
CFR	Code of Federal Regulations	emf	electromotive force
cgs	centimeter–gram–second	emu	electromagnetic unit
CI	Color Index	eng	engineering
cis-	isomer in which substituted groups are on same side of double bond between C atoms	EPA	Environmental Protection Agency
		epr	electron paramagnetic resonance
		eq.	equation
cl	carload	esp	especially
cm	centimeter	esr	electron-spin resonance
cmil	circular mil	est(d)	estimate(d)
cmpd	compound	estn	estimation
COA	coenzyme A	esu	electrostatic unit
COD	chemical oxygen demand	exp	experiment, experimental
coml	commercial(ly)	ext(d)	extract(ed)
cp	chemically pure	F	farad (capacitance)
cph	close-packed hexagonal	f	femto (prefix for 10^{-15})
CPSC	Consumer Product Safety Commission	FAO	Food and Agriculture Organization (United Nations)
D-	denoting configurational relationship		

FACTORS, ABBREVIATIONS, AND SYMBOLS

fcc	face-centered cubic	IUPAC	International Union of Pure and Applied Chemistry
FDA	Food and Drug Administration	IV	iodine value
FEA	Federal Energy Administration	J	joule
		K	kelvin
fob	free on board	k	kilo (prefix for 10^3)
fp	freezing point	kg	kilogram
FPC	Federal Power Commission	L	denoting configurational relationship
frz	freezing		
G	giga (prefix for 10^9)	L	liter (for fluids only) (5)
g	gram	l-	*levo-*, levorotatory
(g)	gas, only as in H_2O(g)	(l)	liquid, only as in NH_3(l)
g	gravitational acceleration	LC$_{50}$	conc lethal to 50% of the animals tested
gem-	geminal		
glc	gas-liquid chromatography	LCAO	linear combination of atomic orbitals
g-mol wt; gmw	gram-molecular weight	lcl	less than carload lots
		LD$_{50}$	dose lethal to 50% of the animals tested
grd	ground		
Gy	gray	liq	liquid
H	henry	lm	lumen
h	hour; hecto (prefix for 10^2)	ln	logarithm (natural)
ha	hectare	LNG	liquefied natural gas
HB	Brinell hardness number	log	logarithm (common)
Hb	hemoglobin	LPG	liquefied petroleum gas
HK	Knoop hardness number	ltl	less than truckload lots
HRC	Rockwell hardness (C scale)	lx	lux
HV	Vickers hardness number	M	mega (prefix for 10^6); metal (as in MA)
hyd	hydrated, hydrous		
hyg.	hygroscopic	*M*	molar
Hz	hertz	m	meter; milli (prefix for 10^{-3})
^{i}(eg, Pr^i)	iso (eg, isopropyl)		
i-	inactive (eg, i-methionine)	*m*	molal
IACS	International Annealed Copper Standard	*m*-	meta
		max	maximum
ibp	initial boiling point	MCA	Manufacturing Chemists' Association
ICC	Interstate Commerce Commission	MEK	methyl ethyl ketone
ICT	International Critical Table	meq	milliequivalent
ID	inside diameter; infective dose	mfd	manufactured
		mfg	manufacturing
		mfr	manufacturer
IPS	iron pipe size	MIBC	methylisobutyl carbinol
IPT	Institute of Petroleum Technologists	MIBK	methyl isobutyl ketone
		MIC	minimum inhibiting concentration
ir	infrared		
ISO	International Organization for Standardization	min	minute; minimum
		mL	milliliter

MLD	minimum lethal dose	NRI	New Ring Index
MO	molecular orbital	NSF	National Science Foundation
mo	month	NTA	nitrilotriacetic acid
mol	mole	NTSB	National Transportation Safety Board
mol wt	molecular weight		
mom	momentum	O-	denoting attachment to oxygen
mp	melting point		
MR	molar refraction	o-	ortho
ms	mass spectrum	OD	outside diameter
mxt	mixture	OPEC	Organization of Petroleum Exporting Countries
μ	micro (prefix for 10^{-6})		
N	newton (force)	OSHA	Occupational Safety and Health Administration
N	normal (concentration)		
N-	denoting attachment to nitrogen	owf	on weight of fiber
		Ω	ohm
n (as n_D^{20})	index of refraction (for 20°C and sodium light)	P	peta (prefix for 10^{15})
		p	pico (prefix for 10^{-12})
		p-	para
n (as Bu^n), n-	normal (straight-chain structure)	p.	page
		Pa	pascal (pressure)
		pd	potential difference
n	nano (prefix for 10^{-9})	pH	negative logarithm of the effective hydrogen ion concentration
na	not available		
NAS	National Academy of Sciences		
		pmr	proton magnetic resonance
NASA	National Aeronautics and Space Administration	pos	positive
		pp.	pages
nat	natural	ppb	parts per billion
NBS	National Bureau of Standards	ppm	parts per million
		ppt(d)	precipitate(d)
neg	negative	pptn	precipitation
NF	*National Formulary*	Pr (no.)	foreign prototype (number)
NIH	National Institutes of Health	pt	point; part
		PVC	poly(vinyl chloride)
NIOSH	National Institute of Occupational Safety and Health	pwd	powder
		qv	quod vide (which see)
		R	univalent hydrocarbon radical
nmr	nuclear magnetic resonance		
NND	New and Nonofficial Drugs (AMA)	(R)-	rectus (clockwise configuration)
no.	number	rad	radian; radius
NOI-(BN)	not otherwise indexed (by name)	rds	rate determining step
		ref.	reference
NOS	not otherwise specified	rf	radio frequency, *n.*
nqr	nuclear quadrople resonance	r-f	radio frequency, *adj.*
NRC	Nuclear Regulatory Commission; National Research Council	rh	relative humidity
		RI	Ring Index
		RT	room temperature

xxiv FACTORS, ABBREVIATIONS, AND SYMBOLS

s (eg, Bu^s);		T	tera (prefix for 10^{12}); tesla (magnetic flux density)
sec-	secondary (eg, secondary butyl)	t	metric ton (tonne); temperature
S	siemens	TAPPI	Technical Association of the Pulp and Paper Industry
(S)-	sinister (counterclockwise configuration)	tex	tex (linear density)
S-	denoting attachment to sulfur	THF	tetrahydrofuran
		tlc	thin layer chromatography
s-	symmetric(al)	TLV	threshold limit value
s	second	trans-	isomer in which substituted groups are on opposite sides of double bond between C atoms
(s)	solid, only as in H$_2$O(s)		
SAE	Society of Automotive Engineers		
SAN	styrene–acrylonitrile	TSCA	Toxic Substance Control Act
sat(d)	saturate(d)		
satn	saturation	Twad	Twaddell
SCF	self-consistent field	UL	Underwriters' Laboratory
Sch	Schultz number	USDA	United States Department of Agriculture
SFs	Saybolt Furol seconds		
SI	Le Système International d'Unités (International System of Units)	USP	*United States Pharmacopeia*
		uv	ultraviolet
sl sol	slightly soluble	V	volt (emf)
sol	soluble	var	variable
soln	solution	*vic-*	vicinal
soly	solubility	vol	volume (not volatile)
sp	specific; species	vs	versus
sp gr	specific gravity	v sol	very soluble
sr	steradian	W	watt
std	standard	Wb	Weber
STP	standard temperature and pressure (0°C and 101.3 kPa)	Wh	watt hour
		WHO	World Health Organization (United Nations)
SUs	Saybolt Universal seconds	wk	week
syn	synthetic	yr	year
^{t} (eg, Bu^t), *t-*,		(Z)-	zusammen; together
tert-	tertiary (eg, tertiary butyl)		

Non-SI (Unacceptable and Obsolete) Units   *Use*

Å	angstrom	nm
at	atmosphere, technical	Pa
atm	atmosphere, standard	Pa
b	barn	cm^2
bar†	bar	Pa

† Do not use bar (10^5Pa) or millibar (10^2Pa) because they are not SI units, and are accepted internationally only for a limited time in special fields because of existing usage.

Non-SI. (Unacceptable and Obsolete) Units   *Use*

bhp	brake horsepower	W
Btu	British thermal unit	J
bu	bushel	m^3; L
cal	calorie	J
cfm	cubic foot per minute	m^3/s
Ci	curie	Bq
cSt	centistokes	mm^2/s
c/s	cycle per second	Hz
cu	cubic	exponential form
D	debye	C·m
den	denier	tex
dr	dram	kg
dyn	dyne	N
erg	erg	J
eu	entropy unit	J/K
°F	degree Fahrenheit	°C; K
fc	footcandle	lx
fl	footlambert	lx
fl oz	fluid ounce	m^3; L
ft	foot	m
ft·lbf	foot pound-force	J
gf den	gram-force per denier	N/tex
G	gauss	T
Gal	gal	m/s^2
gal	gallon	m^3; L
Gb	gilbert	A
gpm	gallon per minute	(m^3/s); (m^3/h)
gr	grain	kg
hp	horsepower	W
ihp	indicated horsepower	W
in.	inch	m
in. Hg	inch of mercury	Pa
in. H_2O	inch of water	Pa
in.·lbf	inch pound-force	J
kcal	kilogram-calorie	J
kgf	kilogram-force	N
kilo	for kilogram	kg
L	lambert	lx
lb	pound	kg
lbf	pound-force	N
mho	mho	S
mi	mile	m
MM	million	M
mm Hg	millimeter of mercury	Pa
mμ	millimicron	nm
mph	miles per hour	km/h
μ	micron	μm
Oe	oersted	A/m
oz	ounce	kg
ozf	ounce-force	N
η	poise	Pa·s
P	poise	Pa·s
ph	phot	lx
psi	pounds-force per square inch	Pa
psia	pounds-force per square inch absolute	Pa
psig	pounds-force per square inch gage	Pa

qt	quart	m³; L
°R	degree Rankine	K
rd	rad	Gy
sb	stilb	lx
SCF	standard cubic foot	m³
sq	square	exponential form
thm	therm	J
yd	yard	m

BIBLIOGRAPHY

1. The International Bureau of Weights and Measures, BIPM, (Parc de Saint-Cloud, France) is described on page 22 of Ref. 4. This bureau operates under the exclusive supervision of the International Committee of Weights and Measures (CIPM).
2. *Metric Editorial Guide (ANMC-75-1)*, American National Metric Council, 1625 Massachusetts Ave. N.W., Washington, D.C. 20036, 1975.
3. *SI Units and Recommendations for the Use of Their Multiples and of Certain Other Units (ISO 1000-1973)*, American National Standards Institute, 1430 Broadway, New York, N.Y. 10018, 1973.
4. Based on *ASTM E 380-76 (Standard for Metric Practice)*, American Society for Testing and Materials, 1916 Race Street, Philadelphia, Pa. 19103, 1976.
5. *Fed. Regist.*, Dec. 10, 1976 (41 FR 36414).
6. For ANSI address, see Ref. 3.

R. P. LUKENS
American Society for Testing and Materials

C *continued*

COPPER ALLOYS

Wrought copper and wrought copper alloys,   1
Cast copper alloys,   69

WROUGHT COPPER AND WROUGHT COPPER ALLOYS

Copper classification of important alloys,   2
Properties,   3
Precipitation-hardening copper alloys,   34
Copper-zinc alloy brasses,   39
Corrosion of copper and copper-base alloys,   48
Behavior of copper alloys in various media,   60
Alloy selection,   67
Economic aspects,   67

Copper and its alloys form a group of materials of major commercial importance and are produced in large quantities. Properties influencing the general application of wrought copper and copper alloys are their electrical and thermal conductivity, both hot and cold formability, machinability, the ease with which they may be joined by welding, brazing, and soldering, fatigue characteristics, generally good to excellent corrosion resistance, and the fact that the materials are nonmagnetic. Copper and its alloys have a wide range of warm, pleasing colors. They extend through a variety of red, pink, yellow, and white shades. Gold is the only other metal having such distinctive coloring. The copper alloys are all easily finished by buffing, scratch brushing, electroplating, chemical coloring, or by applying a clear protective coating system.

A significant percentage of copper produced is used as the basis of a group of commercially important alloys, marketed as a variety of high copper content alloys, so-called engineering alloys including a number of precipitation or age-hardenable alloys, the brasses or the copper–zinc series with or without other designated alloying elements, nickel silvers, cupro–nickels, and the bronzes. Bronzes are copper alloys in which the major alloying element is at least one other than zinc or nickel. Originally, the term bronze described copper alloys with tin as the sole or principal alloying element. The term bronze is seldom used by itself, but rather with a descriptive adjective. The four main classes of wrought bronzes are: copper–tin–phosphorus alloys known as the phosphor bronzes; copper–tin–lead–phosphorus alloys called the leaded phosphor bronzes; copper–aluminum (with or without other modifying elements) alloys termed aluminum bronzes; and the copper–silicon (with or without other modifying elements) alloys identified as the silicon bronzes.

The alloys are prepared by melting the various elements and casting the resulting liquid into appropriate mold shapes to produce bars or cakes and billets to be processed into plate, sheet or strip, rod, wire, pipe, or tube by rolling, extruding, piercing, tube reducing, or drawing. The rather coarse dendritic structure characteristic of cast alloys is refined to a controlled uniform structure during working and annealing. The refinement of the grain enhances mechanical properties of the wrought alloy.

Copper Classifications of Important Alloys

Electrolytic and fire-refined copper relate to the method of refining. The former is refined by electrolytic deposition as a cathode in the final step of the refining process, and the latter by the use of a furnace process.

Tough-pitch copper, deoxidized copper, and oxygen-free copper relate to the characteristic of the copper during melt preparation and casting (see also Copper).

Electrolytic tough pitch (ETP) (the exact state or quality of well reduced and refined copper) has long been the standard type of wrought copper used in the manufacture of rod, wire, sheet, and strip products. ETP copper contains small amounts of oxygen (0.02–0.05%) present as cuprous oxide which only slightly affects the mechanical properties; the material is entirely satisfactory for most purposes. It is fabricated by most of the usual metal-working processes and is used in numerous applications, eg, automobile radiators, sheet metal for buildings, screen cloth, printing rolls, vats, kettles, chemical process equipment, gaskets, rivets, and soldering coppers.

Deoxidized copper is produced by introducing an element that reacts with oxygen (CuO) present in a molten bath of copper. Copper ready for deoxidation is usually in the tough-pitch condition. It may be taken directly from the refining furnace or it may be obtained by melting solid copper either in the form of virgin or scrap material. Elements used as deoxidizers are phosphorus, calcium, silicon, lithium, beryllium, aluminum, magnesium, zinc, barium, strontium, boron, and carbon. The excess amount of deoxidizer allowed to remain in the copper depends upon its intended end use, and the effect the deoxidizing element has upon the properties of the alloy.

Phosphorus is the most common deoxidizer although some copper alloys are deoxidized with lithium, boron, or calcium. Most phosphorus-deoxidized copper contains residual phosphorus in concentrations of 0.01–0.04% and is marketed as Deoxidized High Residual Phosphorus (DHP) copper.

Up to 0.10% phosphorus does not significantly impair the mechanical properties of copper. The electrical and thermal conductivities are seriously affected. If phosphorus is used to deoxidize copper intended for maximum electrical conductivity, the amount must be carefully controlled so that very little remains in the metal (0.005–0.009%). To ensure complete deoxidation, some minimum residual phosphorus (0.01%) must remain in the solidified alloy because of the possibility of the existence of a copper–phosphorus–oxygen equilibrium at low phosphorus concentrations. In general, the oxygen in tough-pitch copper is less harmful than the residual deoxidizing elements.

Deoxidized copper can pick up as much as 0.01% oxygen in traveling from the melting furnace to the mold through a launder with gas burners firing over the molten copper unless proper precautions are taken. Figure 1 shows the degree of copper oxide found in the following three kinds of copper: (1) Copper saturated with copper oxide (not a commercial product) containing about 0.35% oxygen; (2) Tough-pitch copper containing 0.03% oxygen; and (3) Oxygen-free copper containing 0.005% oxygen.

A variety of copper known as oxygen-free high conductivity or OFHC, was introduced in 1953. OFHC is the registered trademark for the oxygen-free high conductivity copper produced exclusively by American Metals Climax. OFHC is not a deoxidized copper, but is made by melting select copper cathodes in a furnace constructed to exclude air and using a protective gas atmosphere over the molten copper. The copper is protected from any oxidizing condition during melting, casting, and solidification. The dry, protective gas is maintained in all launders, boxes, and molds during the pouring operation to ensure an oxide-free casting. Water cooled molds of either the static or direct chill continuous type are employed.

Deoxidized copper is widely used for pipe and tube, eg, in plumbing, gas and oil lines, refrigeration, condensers, heat exchangers, and brewery and distilling tubes. Both deoxidized (DHP) and oxygen-free copper are used where the copper must be heated in reducing atmospheres such as hydrogen to prevent reactions between the copper oxide and the reducing gas which can damage the alloy.

The oxygen-free copper is used for electrical conductors, lead-in wires and anodes for vacuum tubes, transistor components, glass to metal seals, and many other purposes.

Electrolytic tough-pitch (ETP) copper is used in the architectural, automotive, electrical, hardware fields as well as the chemical industry for processing equipment such as kettles, pans, and vats.

Properties

The coppers are soft with relatively low tensile strength and high elongation, and are capable of high reduction in area by cold-working processes, even in the as-cast condition. They are not as greatly increased in hardness by a given reduction in cross-sectional area by cold-working as are alloys of lower copper content. Owing to the initial softness and low rate of work-hardening under deformation, copper is rated as having a high working capacity. It is very well suited to forming and shaping by both hot and cold metal-working operations (see Copper). For example, phosphorus deoxidized copper can be cold-reduced as much as 99% without the application of intermediate annealings.

The low rate of work-hardening is probably accounted for by copper's fcc crystallization. Plastic deformation takes place by a shearing action along certain planes within a crystal or grain. The shearing action is discontinuous, occurring by distinct movement on certain planes through the structure. The movement is easiest along

4 COPPER ALLOYS (WROUGHT)

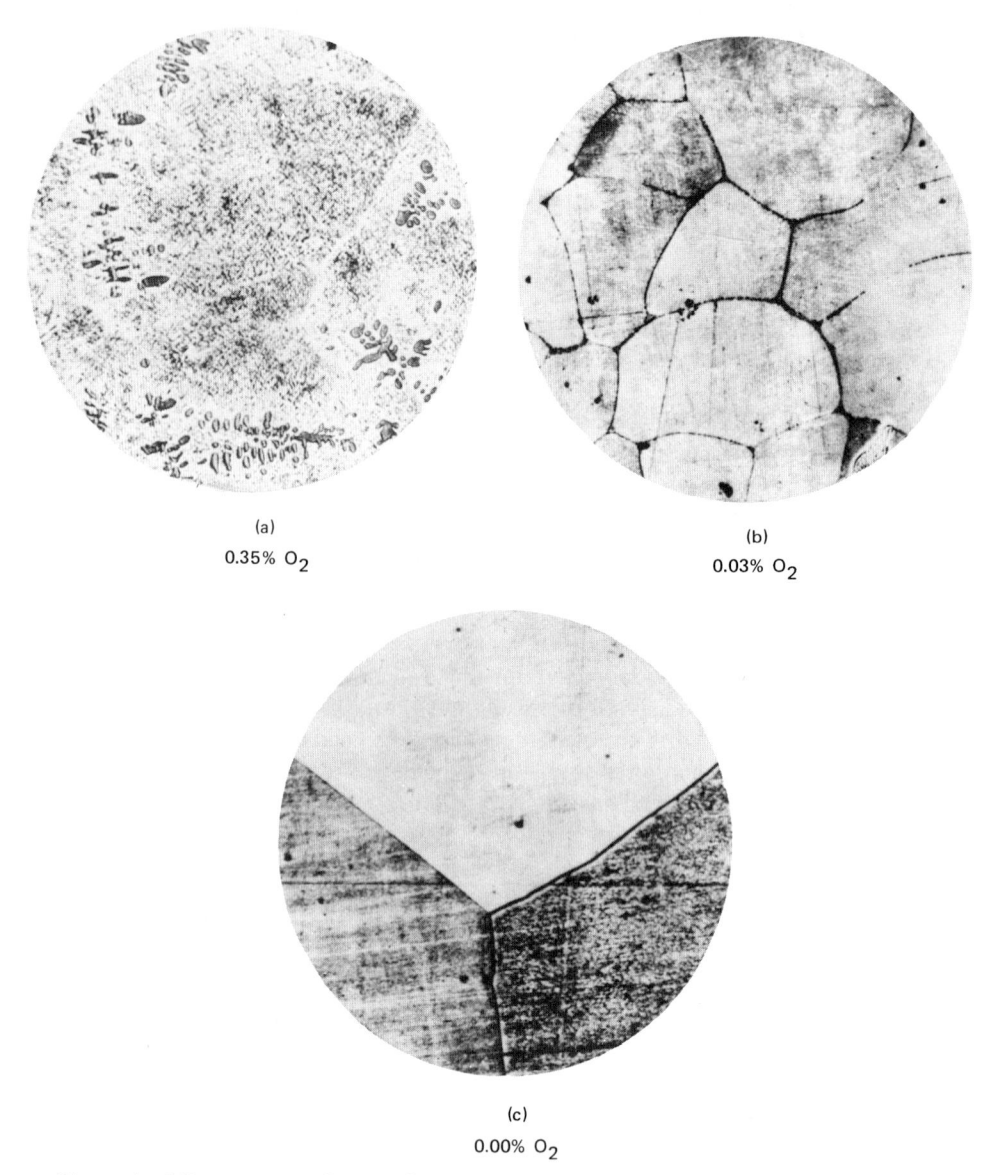

(a) 0.35% O$_2$

(b) 0.03% O$_2$

(c) 0.00% O$_2$

Figure 1. Microstructure of oxygen-bearing copper compared with oxygen-free copper. Photomicrographs (50×) represent (a) set copper showing structure of coarse crystals of cuprous oxide in matrix of the eutectic; (b) tough-pitch copper showing the presence of the eutectic in the grain boundaries; and (c) oxygen-free copper.

the planes having the greatest atomic density. In the fcc lattices of copper the (111) planes slip most readily. The direction of movement is also the direction corresponding to the greatest atomic density.

Copper has no allotropic modifications; thus, it has no critical temperature at which changes occur in the crystal pattern. Heat-treatment operations used on other alloys that depend upon allotropic changes have no application to copper.

Annealing (Recovery, Recrystallization, and Softening). Cold-work distorts the copper structure, and results in increased hardness and strength, and decreased ductility and electrical conductivity. Heating the cold-worked material gradually restores these properties to values characteristic of soft or annealed copper. The properties do not recover simultaneously. In copper the changes take place at a temperature that more nearly accompanies the appearance of recrystallization. The process causing marked changes in the structure of a deformed or cold-worked metal during heating is a rearrangement of the dislocations into a lower energy level without a very marked decrease in the total number of dislocations. This process is sometimes called polygonization. The evidence of polygonization and subboundary movement in copper is only relevant to the recrystallization characteristics of the metal which has been cold-deformed to reductions less than those needed for the formation of the cubic texture. Severe cold deformation leads to an annealing phenomenon in which stacking faults seem to assume an important function.

Recrystallization takes place at a temperature somewhat higher than that of recovery which results from rearrangement of atoms of the solid in an entirely new set of crystals. These crystals grow from a few small nuclei until they meet each other, absorbing the intervening unrecrystallized material. The new crystals have a much lower dislocation content than the deformed material. The driving force for recrystallization is the decrease in free energy resulting from the decrease in dislocation content.

The recrystallization and grain-growth characteristics of the coppers depend upon the degree of deformation or the residual strain energy in the crystal lattice, temperature, and the time the material remains at that temperature. The recrystallization temperature usually decreases for material that has been severely deformed, and grain size attained for a given time at recrystallization temperature is smaller than that for material deformed to a lesser extent.

The grain-growth and softening characteristics of silver-bearing or lake copper, electrolytic tough-pitch, phosphorized and oxygen-free copper have been investigated. The electrolytic copper (0.03% oxygen) exhibits the lowest recrystallization temperature for a given condition of prior cold-work and heating characteristics. This is caused by the effect of oxygen upon the minor impurities present. Phosphorus- and silver-bearing copper exhibit higher softening characteristics, because of the solid solution effect involving phosphorus and silver.

Figure 2 defines some mechanical properties and grain size characteristics for four types of copper. The shape in which the tensile property curves is typical for the recovery, recrystallizations, and grain-growth phenomena.

Refined copper containing either traces of impurities or elements added in low concentrations illustrates the solid solution effect of increasing softening temperatures.

Cadmium added in small amounts (0.05–0.15%) to deoxidized copper results in a dilute alloy having superior resistance to softening at the temperatures and for the short times involved in soldering operations. Resistance to softening under these conditions is retained even after very severe cold deformation. The severe deformation results in higher strength material, and since the alloy does not readily soften, a higher strength product such as an automobile radiator can be produced. The application of the cadmium alloy allows the use of higher melting point and higher strength solders.

6 COPPER ALLOYS (WROUGHT)

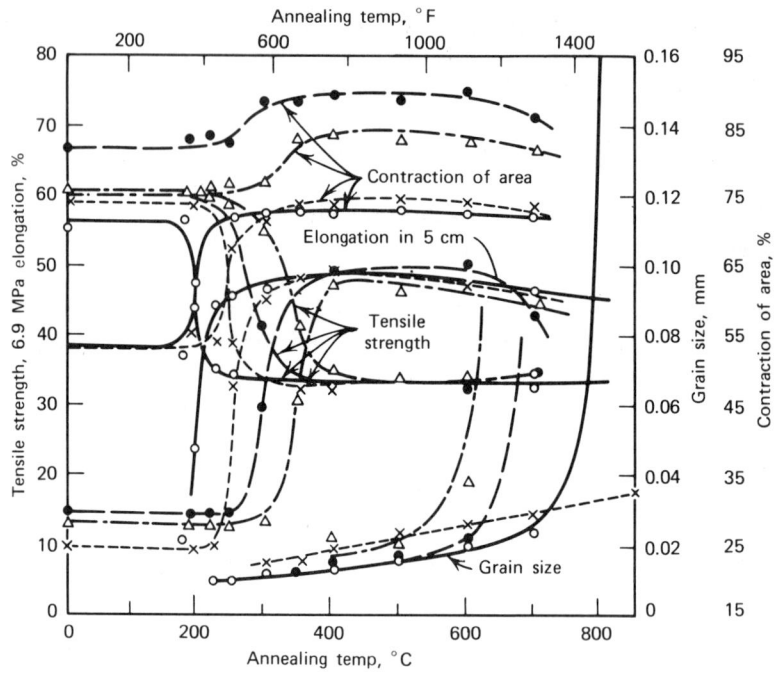

Figure 2. Effect of annealing temperature upon several properties and grain size of four types of copper. Samples tested at room temperature were wires previously cold drawn 62.5% reduction in area; annealing time is 1 h (1). - - - × = Lake tough-pitch; ———○ = electrolytic toughpitch; — — —△ = phosphorized; and — —● = OFHC. To convert MPa to psi, multiply by 145.

Figure 3 shows the relationship of softening characteristics of three severely cold-worked copper alloys.

Silver is added to copper to increase its resistance to softening at elevated temperature and improves its creep characteristics compared to ETP copper. Figure 4 illustrates some creep characteristics of silver-bearing copper compared to ETP copper. The silver-bearing copper performs better at 225°C at high stress levels than does ETP copper at 130°C. The higher temperature tests on ETP copper show values that do not fall on this curve. Figure 4 also shows the relationship of cold-work and grain size on the creep resistance of silver-bearing copper. Increased cold-work or grain size raise the creep resistance.

Fire-refined copper of high purity containing trace amounts of nickel, selenium, and tellurium also illustrates the influence of dilute solid solutions on increasing softening characteristics. Figure 5 shows the softening characteristics in terms of tensile strength for ETP, silver-bearing copper, and fire-refined copper.

Electrical and Thermal Properties. Copper is widely applied commerically because of its relatively high electrical and thermal conductivity. It is second only to silver.

Electrical conductivity is expressed as a percentage of the International Annealed Copper Standard (IACS), adopted in 1913 to represent average electrical conductivity of high-grade copper. However, refining methods have improved enough since 1913 so that electrical conductivity of over 101%, and occasionally as high as 102% of IACS are not unusual. The standard is expressed in terms of mass resistivity, namely the resistivity of a wire one meter long weighing one gram. This standard resistance is 0.15328 $(\Omega \cdot g)/m^2$ (875.2 $(\Omega \cdot lb)/mi^2$) at 20°C.

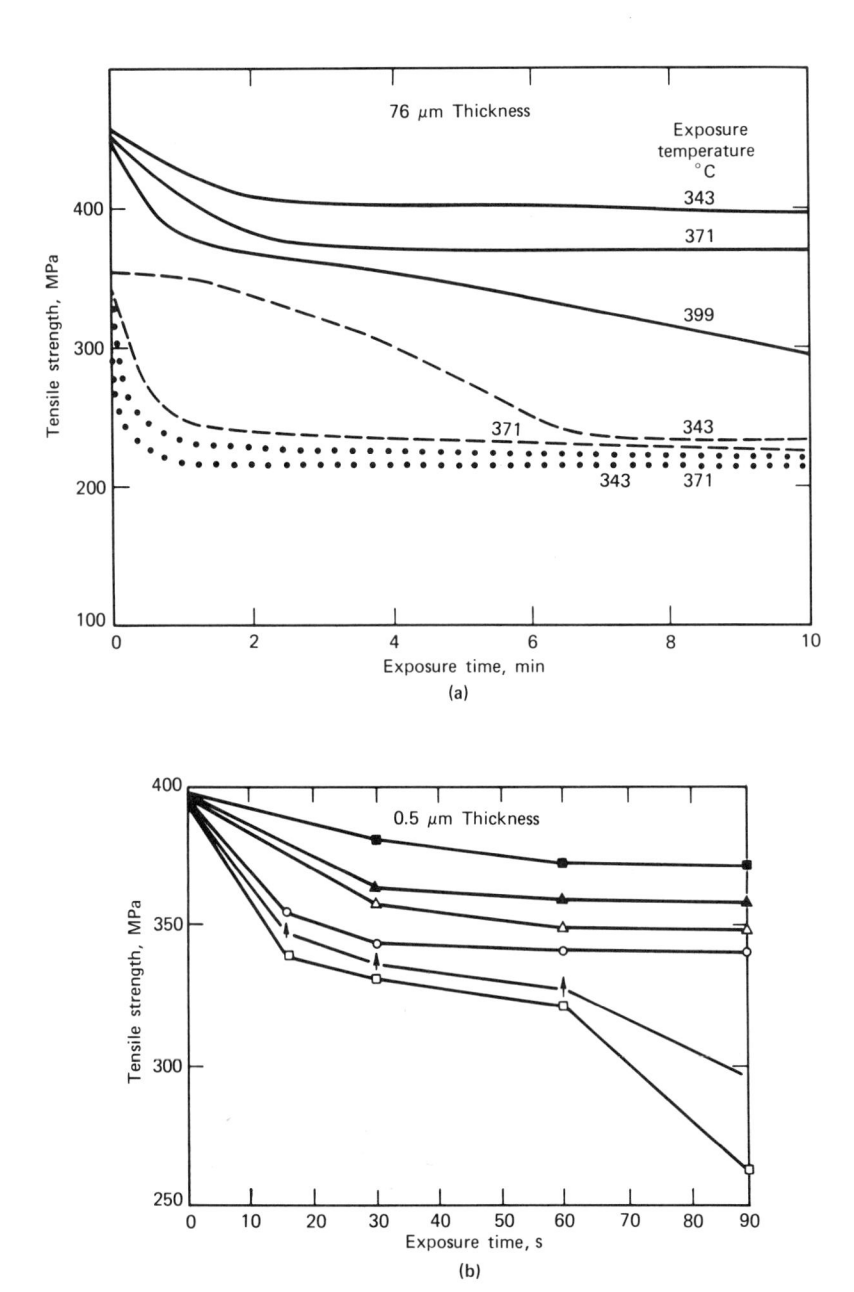

Figure 3. Softening characteristics of severely cold-rolled deoxidized cadmium copper, silver-bearing copper, and electrolytic tough-pitch copper with relatively short exposure times. (a) Comparison of softening resistance typical curves where ——— = 99.9% Cu, 0.07 Cd; - - - = 0.28 kg silver-bearing copper; and . . . = electrolytic tough-pitch copper. (b) Softening resistance 99.9% Cu, 0.07% Cd where exposure temperature °C is ○ = 482; † = 454; □ = 468; ■ = 343; ▲ = 371; and △ = 399. To convert MPa to psi, multiply by 145. Courtesy of T. S. Howald, Chase Brass and Copper Company, Inc.

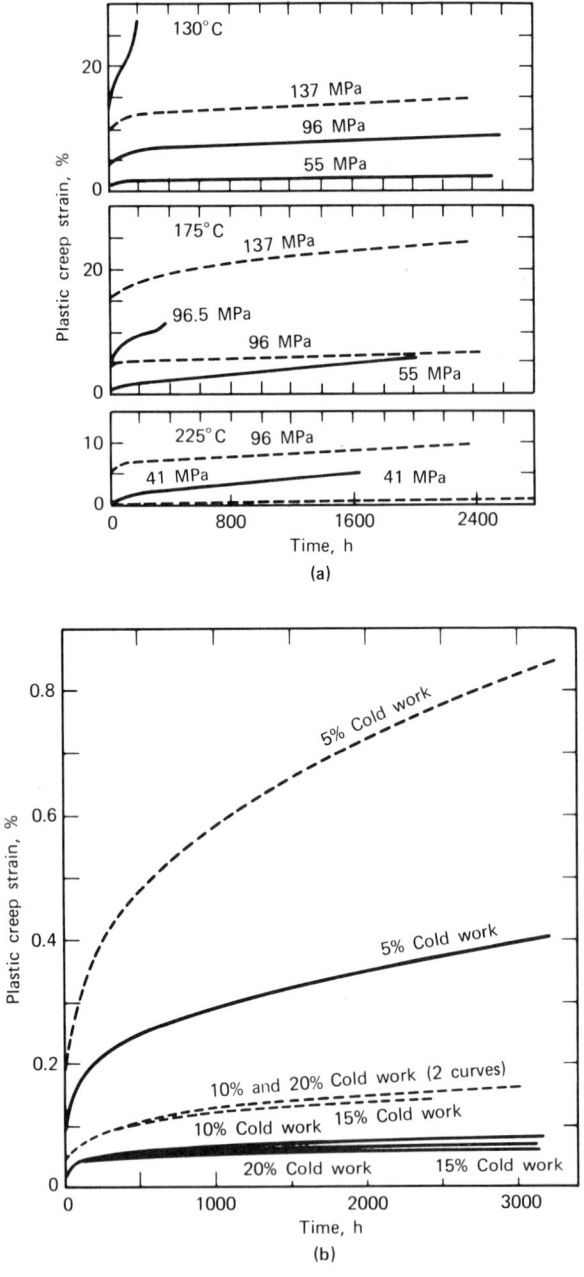

Figure 4. Creep characteristics. (a) Creep tests on annealed tough-pitch coppers showing the effect of the addition of 0.1% silver where ——— = tough-pitch and - - - = tough-pitch + silver. To convert MPa to psi, multiply by 145. (b) Effect of cold-work and grain size on the creep of silver-bearing tough-pitch coppers stressed at 96.5 MPa (14,000 psi) at 175°C where ——— = large grain-size and - - - = small grain-size. (c) Effect of cold-work on the creep of silver-bearing tough-pitch coppers stressed at 96.5 MPa (14,000 psi) at 175°C where material cold-worked by tensile overstrain is - - - ▲ - - - = fine-grain (0.025 mm dia) and — × — = coarse-grain (0.045 mm dia), and drawn material 0.25 m thick is - - - ● - - - = 0.030–0.035 mm dia (2). Courtesy of *The Journal of the Institute of Metals*.

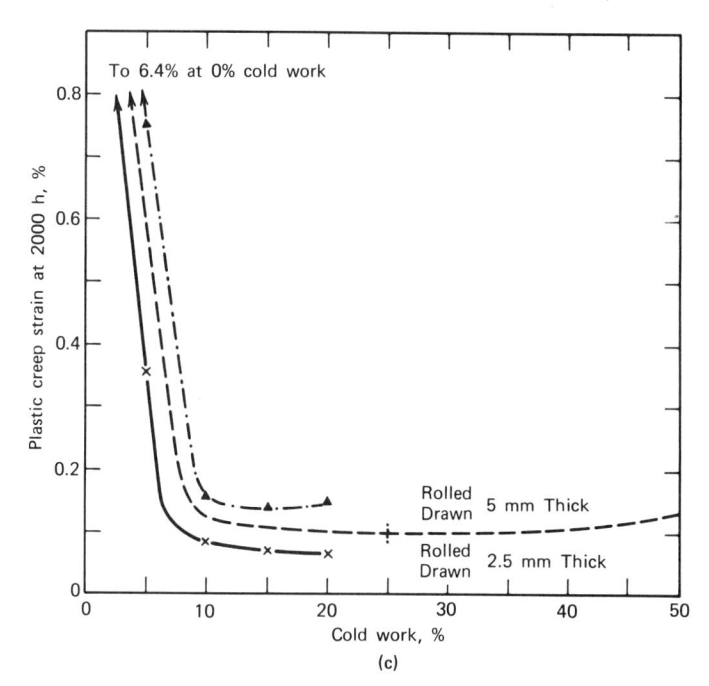

Figure 4. (*continued*)

High purity fire-refined, chemically refined, OFHC, ETP and deoxidized low residual phosphorus (DLP) copper are all known as high conductivity coppers. These materials meet ASTM standards and other specification requirements for electrical and thermal conductivities.

Deoxidized copper having high residual phosphorus (0.01–0.04% P) DHP is known as low conductivity copper (80–90% IACS).

The electrical conductivity of annealed copper is not affected by grain size, annealing temperature, or crystallographic orientation.

Cold-work does decrease electrical conductivity of copper.

Elements held in solid solution in copper have a marked effect upon electrical conductivity, particularly when used in quantities added by intent for alloying. Small concentrations of impurity elements may have a significant detrimental effect. Figure 6 defines the effect of 15 elements upon the electrical conductivity of otherwise pure copper. The presence of oxygen along with the impurity element often forms compounds with the impurity removing it from solid solution and thereby nullifying the effect of the element on the conductivity of the copper. The effect of this type of reaction on conductivity is also shown in Figure 6.

Elements that are insoluble in copper or form insoluble intermetallic compounds do not significantly effect electrical conductivity. Similarly, heat treatment (aging—see Precipitation hardening alloys) which causes precipitation of elements or intermetallic compounds from solid solution also causes the electrical conductivity to increase. The measure of electrical conductivity is used to judge when proper aging has been accomplished. Conversely, if elements are present that can go into solid solution with copper, the electrical conductivity will decrease.

The thermal conductivity of copper with an electrical conductivity of 100% IACS

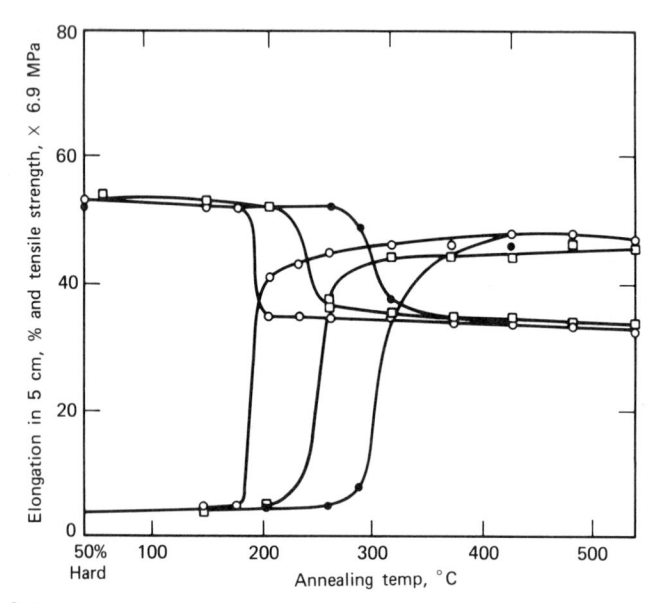

Figure 5. Softening characteristics of three types of tough-pitch (oxygen-bearing) copper. Samples were 0.8 mm thick strip, previously annealed at 538°C cold-rolled 50%, and finally annealed one hour at temperature shown (3). Tough-pitch (oxygen) coppers are O = electrolytic; ● = lake (0.044% Ag); and □ = fire-refined (0.04 Ni, 0.016 Te and Se). To convert MPa to psi, multiply by 145.

is 391 W/(m·K) 20°C. The ratio of thermal conductivity and the product of electrical conductivity times absolute temperature is approximately constant.

Since thermal conductivity is not easily measured, electrical conductivity is much more frequently employed.

Oxidation. *Surface Oxidation.* When heated in air, copper develops a cuprous oxide film that exhibits a succession of interference tints as a function of film thickness. The colors associated with oxide film thicknesses are (21–23):

Color	Film thickness, nm
dark brown	37–38
very dark purple	45–46
violet	48–48.5
dark blue	50–52
yellow	98–94
orange	112–120
red	124–126

Black cupric oxide forms over the cuprous oxide layer as film thickness increases beyond the interference color range.

If the velocity of oxidation is controlled by the rate of oxygen meeting the copper, the attack on the metal can be expressed by the Tammann, Pilling, Bedworth parabolic equation (24), $dy/dt = k/y$, y = thickness of film, t = time, and k = oxidation constant.

The oxidation of copper has been studied from temperatures of −183°C up to the melting point. At −183°C, copper rapidly reacts with a very small amount of oxygen to a thickness of about 2.4 nm that may represent a monolayer of absorbed oxygen;

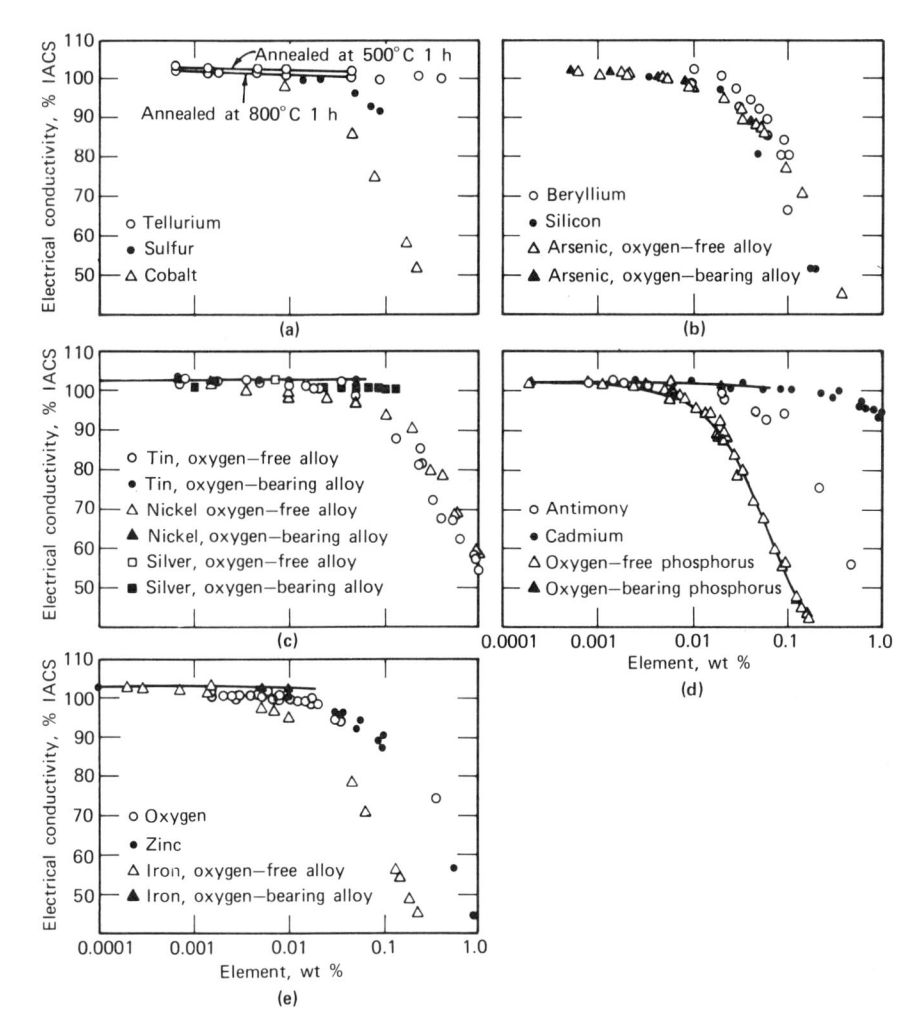

Figure 6. The effect of low concentrations of a number of elements on the electrical conductivity of copper: tellurium (4); cobalt (5–6); beryllium (7–8); silicon (9); arsenic (10–11); tin (12–13); silver (14); phosphorus (15–16); zinc (17–18); and oxygen (19).

after this further oxidation is very small. The oxidation rate increases with temperature as is shown in Figure 7.

Internal Oxidation. Dilute copper alloys of base metals exhibiting a relatively high solubility and diffusivity for atomic oxygen are subject to internal oxidation. Dilute copper alloys containing Al, Zn, Cd, Be, P, Ni, Si, and other elements show this type of oxidation.

The oxygen diffusion into copper is dependent upon a second element such as a low concentration of silicon or phosphorus. The oxygen reacts with the phosphorus beneath the copper surface. If the concentration of the second element is high, new atoms of the element tend to diffuse to the surface to meet the oxygen thereby keeping the internal oxidation to a shallow depth. Time and temperature control the rate of reaction. Absolutely pure copper cannot be internally oxidized because oxygen then reacts with the copper atoms to produce a surface layer of copper oxide.

12   COPPER ALLOYS (WROUGHT)

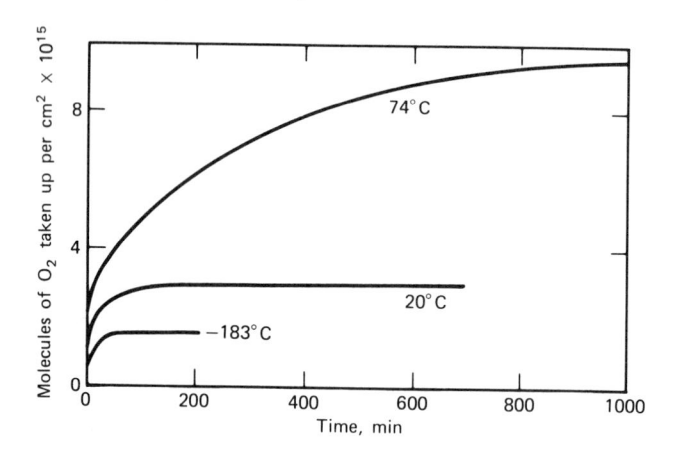

Figure 7. Oxidation rate for copper at low and moderate temperatures.

Directional Properties. An effect caused by cold-rolling (work) is the possible development of directional properties. Such directionality can be mechanical or crystallographic.

Mechanical directionality, sometimes called fibering usually results from the attenuation in the working direction of second phases, such as cuprous oxide in the case of electrolytic tough-pitch copper or the attenuation of grains alone in the single phase material.

The mechanical properties such as tensile strength and elongation vary when measured in different planes in respect to the rolling direction as fibering develops. Therefore, it is said that the strip has developed directional properties.

The more severely copper is rolled, the greater the directional property, although some directionality must be present even at light reductions.

The preferred orientation developed by cold-work frequently changes on annealing or recrystallization to a very different directionality.

Other Effects of Cold-Work. Cold-work increases the tensile strength and decreases ductility. However, it is possible to cold-reduce the cross section of copper castings by 99% without causing mechanical failure.

The mechanical properties are relatively unaffected by any of the variables in the working process.

Some mechanical properties of copper related to degree of cold-work in quantitative detail are shown in Figure 8.

Copper strip is used in forming operations; the temper (degree of cold-rolling) has slight effect upon the cupping limit, but it does have a serious effect upon stretch forming, ie, Erichsen indentations, stretch formed collars, and others.

The compositions of a number of commercial copper alloys are given in Tables 1–6.

Effect of Other Elements in Copper. The elements considered are those present as impurities, residual deoxidizing elements, or deliberate additions intended to enhance a particular property.

Arsenic, antimony, bismuth, iron, lead, cadmium, cobalt, nickel, silver, sulfur, selenium, and tellurium may be classified as impurities. Although oxygen is intentionally controlled, this element may be classified as an impurity, but perhaps it is

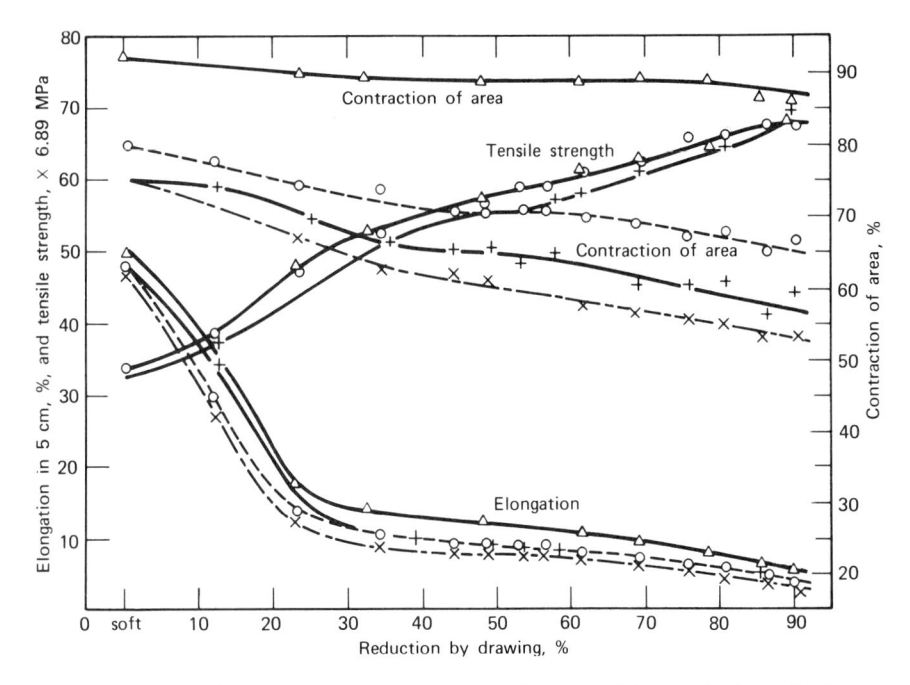

Figure 8. The mechanical properties of copper as a function of degree of cold-work where × - - - = lake tough-pitch, + —— = electrolytic tough-pitch, ○ - - - - = phosphorized, and ▲ —— = OFHC (1,3). To convert MPa to psi, multiply by 145.

more correctly listed as an alloying element. Some of the elements present as impurities in certain instances are added on other occasions as alloying elements. Examples are silver, tellurium, or lead.

The elements phosphorus, lithium, silicon, boron, and calcium can be used as deoxidizing elements and are allowed to remain only in low concentrations. Elements such as aluminum, magnesium, manganese, barium, and strontium might be used as deoxidizers, but they are not used to any significant extent commercially.

Arsenic, beryllium, cadmium, cobalt, chromium, lead, nickel, phosphorus, silicon, silver, tellurium, tin, zinc, and zirconium are added intentionally.

The elements present are shown in Table 7.

Arsenic. Arsenic occurs naturally in some copper ores and may be permitted to remain after the refining process or added in concentrations of ca 0.3%. The element is sometimes added up to 0.5% and marketed under the name of arsenical copper for use as a heat exchanger or as condenser tubes (see Heat exchange technology). Arsenic increases tensile strength slightly in the cold-worked condition and raises the recrystallization temperature. Silver added to arsenical copper also aids in raising the recrystallization point.

Arsenic has a significant adverse effect upon electrical conductivity even in the presence of oxygen (see Fig. 6).

The presence of oxygen improves the casting characteristics of arsenical copper. Conversely, arsenic improves the working properties of oxygen-bearing copper. The beneficial effect of arsenic on the cold-working properties of oxygen-bearing copper is explained by influence on the structure of the alloy. Cuprous oxide normally occurs

Table 1. Composition of Copper Alloys

Name	UNS designation[a]		Composition, wt %
electrolytic tough-pitch	C 11000	Cu	99.9+
		O	0.04
oxygen-free	C 10200	Cu	99.95+
low phosphorus deoxidized	C 12000	Cu	99.9
		P	0.008
high phosphorus deoxidized	C 12200	Cu	99.9
		P	0.02
arsenical	C 14200	Cu	99.68
		As	0.30
		P	0.02
silver-bearing	C 11300, C 11400, C 11500, C 11600	Cu	99.9 min
		Ag	(27, 34, 55, or 86) $\times 10^{-3}$
			(8, 10, 16, or 25 troy oz/ton)
cadmium	C 14300	Cu	99.9
		Cd	0.07
tellurium	C 14500	Cu	99.50
		Te	0.50
		P	0.008
sulfur	C 14700	Cu	99.6
		S	0.4
zirconium	C 15000	Cu	99.8 min
		Zr	0.15
cadmium	C 16200	Cu	99.0
		Cd	1.0
beryllium	C 17000	Cu	98.3
		Be	1.7
chromium	C 18400	Cu	99.2
		Cr	0.8
leaded	C 18700	Cu	99.0
		Pb	1.0

[a] UNS = Unified Numbering System (25).

Table 2. Tin Bronzes

Name	UNS designation	Copper	Tin	Phosphorus	Lead	Zinc
phosphor bronze 5%, grade A	C 51000	95	4.75	0.25		
leaded phosphor bronze, grade B	C 53400	94	4.75	0.25	1.0	
phosphor bronze 8%, grade C	C 52100	92	7.75	0.25		
phosphor bronze 10%, grade D	C 52400	90	9.85	0.15		
444 bronze	C 54600	88.4	4	0.1	4	3.5

as an eutectic surrounding the ductile copper crystals. As the arsenic concentration increases from about 0.1%, the eutectic is replaced by relatively large isolated globules located in the individual grains. The new constituent is probably a reaction product of arsenic plus cuprous oxide. It may be copper arsenate. The constituent is optically active and forms interference figures when examined under polarized light. Similar optical activity has been reported for copper–oxygen–antimony systems.

Table 3. Aluminum Bronzes

Name	UNS designation	Composition, wt %		
		Copper	Aluminum	Iron
aluminum bronze 5%	C 60800	95	5	
aluminum bronze D	C 61400	91	7	2
	C 61900	86.5	9.5	4

Table 4. Silicon Bronzes

Name	UNS designation	Composition, wt %			
		Copper	Silicon	Zinc	Aluminum
olympic bronze, type A	C 65100	96	3	1	
olympic bronze, type B	C 65500	97.5	1.5	1	
aluminum silicon bronze	C 64200	91	2		7

Table 5. Copper–Nickel Alloys

Name	UNS designation	Composition, wt %				
		Copper	Nickel	Zinc	Iron	Manganese
nickel silver, 12%	C 75700	64	12	24		
nickel silver, 18%	C 75200	65	18	17		
cupro–nickel, 10%	C 70600	88.5	10		1.3	0.25
cupro–nickel, 30%	C 71500	69	30		0.5	0.5

Table 6. Copper–Iron–Phosphorus Alloys

UNS designation	Composition, wt %					
	Copper	Iron	Phosphorus	Zinc	Cobalt	Tin
C 19200	98.97	1.00	0.03			
C 19400	97.43	2.40	0.04	0.13		
C 19500	97.00	1.50	0.10		0.80	0.60
C 19600	98.65	1.00	0.30	0.35		

Antimony. Antimony is usually found only in very low concentration in copper. It forms a solid solution with oxygen-free copper and, therefore, has an effect upon electrical conductivity, but to a somewhat lesser extent than arsenic. Antimony also forms globules in the presence of cuprous oxide and the resulting compound is optically active.

Copper containing up to 0.5% Sb in the presence of 0.02–0.03% oxygen can be successfully hot-rolled. Alloys containing both arsenic and antimony in concentrations of 0.05–0.5% for each element in the presence of 0.02% oxygen can be hot-worked. Antimony increases the tensile strength and fatigue properties as compared with plain arsenical copper, and the endurance limit increases. Antimony, even in oxygen bearing-copper, increases the recrystallization temperature.

COPPER ALLOYS (WROUGHT)

Table 7. Elements Present in Copper Alloys

Impurities	Deoxidizers	Controlled alloying elements	
arsenic	phosphorus	aluminum	silicon
antimony	lithium	arsenic	silver
bismuth	silicon	beryllium	sulfur
iron	boron	cadmium	tellurium
cobalt	calcium	cobalt	tin
nickel		chromium	zinc
silver		lead	zirconium
sulfur		nickel	
selenium		oxygen	
tellurium		phosphorus	

Bismuth. Bismuth is almost insoluble in copper, 0.02% at 980°C. Bismuth appears as intergranular films in oxygen-free copper. Since it has a low melting point (271°C) it promotes hot-shortness. Since the major portion of all copper strip is produced by hot-rolling, even small amounts of bismuth cannot be tolerated. The presence of phosphorus tends to repress the hot-shortness caused by bismuth.

Reese and Conda (26) found that with annealing at 900°C the unit effect of bismuth on the half hard temperature was 2.5°C per ppm in the range of 1.6–3 ppm. Like arsenic and antimony, bismuth reacts with cuprous oxide causing a progressive coalescence of the copper–copper oxide eutectic structure with increasing bismuth concentrations until globules of a new constituent are formed. With sufficiently high oxygen concentrations in bismuth-bearing copper, no bismuth films are found under the microscope. The presence of oxygen in the copper–bismuth system can be expected to improve the hot-rolling properties.

Bismuth also effects the cold-working properties by the tendency to make the material brittle. The presence of lead decreases the adverse affects of bismuth probably by alloying with it, thereby destroying its tendency to form grain boundary films.

Arsenic and antimony in the presence of oxygen act to decrease the hot-shortness of bismuth-bearing copper probably as the result of redistribution of the bismuth in the microstructure. Complex oxides with antimony and bismuth may be formed preventing the segregation of bismuth in the grain boundaries.

Sulfur, Selenium, and Tellurium. These three elements do not induce hot- or cold-shortness and have little effect upon tensile strength, although there may be some loss in ductility.

Although these elements are usually present only in very small concentrations in refined copper, they increase the ease with which copper may be machined when added in quantities up to 1%. All three form compounds with copper and in a qualitative way form similar constitution diagrams (see Fig. 9).

The solid solubility of each element is low. Based upon electrical resistivity measurements the solubilities are as follows:

Temperature, °C	Sulfur, wt %	Selenium, wt %	Tellurium, wt %
600	0.0005	0.001	0.0003
800	0.002	0.015	0.0075

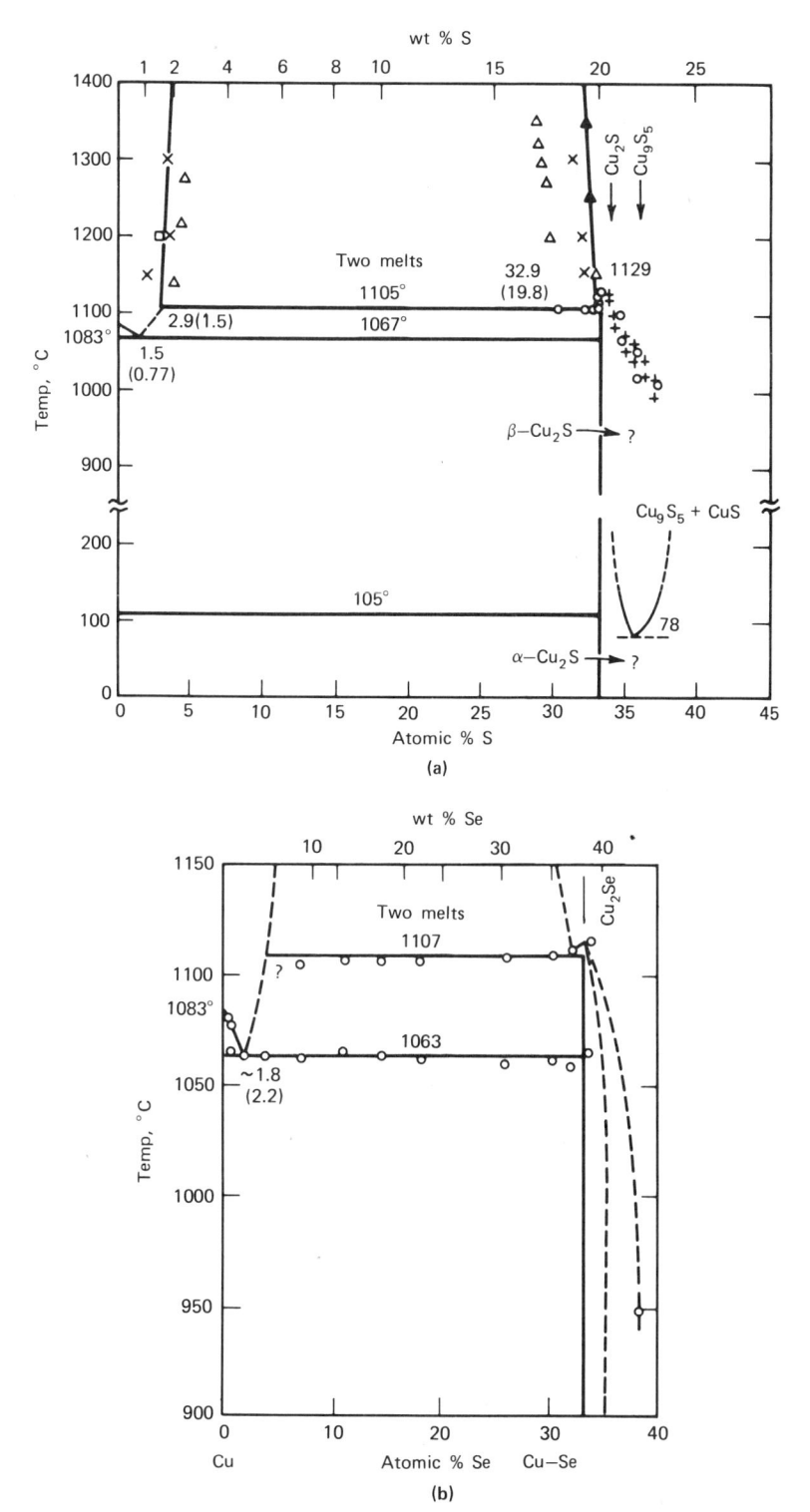

Figure 9. Constitution diagrams (27). (**a**) Copper–sulfur system. (**b**) Copper–selenium system. (**c**) Copper–tellurium system.

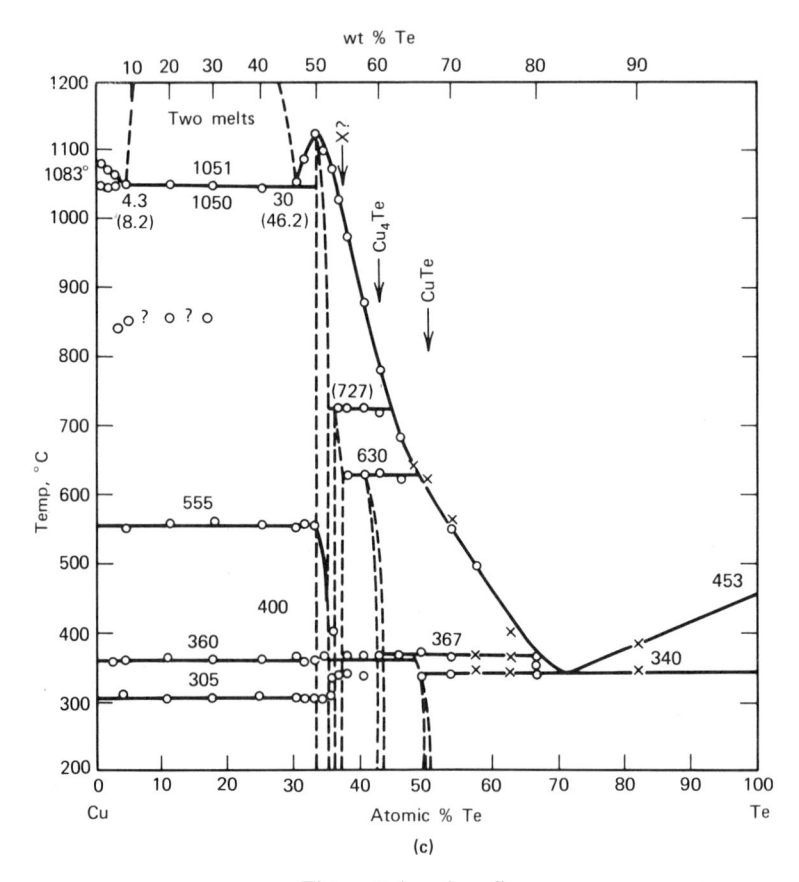

Figure 9. (*continued*)

Sulfur has a significant effect on the recrystallization temperature of copper at low concentrations up to 10 ppm. The effect decreases at higher concentrations because of the limited solid solubility. Part of the effect could be caused by the formation of insoluble Cu_2S. The recrystallization temperature increases to ca 0.7°C per ppm of sulfur in the 6–25 ppm range (28).

Selenium has a dominating effect upon the softening of copper (29). At one ppm selenium causes a reduction in spring elongation of 40–60 mm, compared to sulfur's effect of 20 mm.

The unit effect in terms of softening temperature is ca 12°C per ppm in the 0–5 ppm range for annealing at 850°C (30).

There is a rather high unit effect of 6°C per ppm of tellurium (26). Oxygen does not seem to change the behavior of tellurium.

Copper alloys containing sulfur or tellurium are sold as free-machining copper. Pure copper is soft and tough, making it difficult to machine. When copper is cut, the long stringy curls that are formed slow the cutting rate, foul the tool and increase heat production, and give copper its low machinability rating. Compared to free-cutting brass (61% Cu, 3% Pb, 36% Zn) as a standard (rating of 100), copper has a rating of 20%. Sulfur copper, 99.7% Cu, 0.3% S, and tellurium copper, 99.5% Cu, 0.5% Te, both have a machinability rating of 85%. The mechanical properties are similar to those of copper.

Lead is also added to copper to improve machinability but the alloy has a tendency to crack. Tellurium copper is relatively free from fire-cracking and lends itself well to hot-forging operations. The copper telluride phase in the alloy is relatively brittle and of very low strength, causing the chips to break instead of curl. The brittle phase does lead to tool wear, thus carbide-tipped tools are needed in automatic machines for cutting tellurium copper.

Lead. Lead is practically insoluble in copper and occurs as globules in the grain boundaries throughout the structure, and can lead to hot-shortness. Generally, lead contents much in excess of 0.02% in copper cannot be tolerated in metal for hot-rolling, and concentrations of less than 0.01% lead are used in copper intended to be pierced into tubes. These operations exert tensile forces upon the metal being worked and cracks result. However, copper containing at least 1% lead can be successfully hot-extruded into either rod or tube forms, because the stresses are mostly compressive in extrusion.

The presence of oxygen with low lead concentrations restrains hot-shortness in hot rod-rolling (31). There is probably a reaction of cuprous oxide and lead to form a mixed oxide.

Reported allowable impurity limits can show considerable variation since various other factors must also be considered such as temperature and its control, finish of roll surfaces, shape, relation of thickness of stock to roll diameter, variable cast grain structure, segregation, inaccuracy in analysis, and the possible presence of other impurities. The effect of lead on hot-rolling should be expressed as a concentration range over which hot-rolling becomes increasingly difficult, rather than as a definite limiting concentration above which failures always result.

The variable results reported concerning the effect lead has on softening temperature is probably attributable to the presence of other impurity elements, such as oxygen, sulfur, and selenium.

Lead is deliberately added to copper to be fabricated into rod by hot-extrusion. The function of the lead particle is to increase the brittleness of the alloy, allowing it to break off readily into short chips when the material is machined. Although even small lead additions produce this effect, the addition of 1% reduces necessary tool pressures and improves tool life. The machinability of leaded copper rod is rated at 80% of that for free-cutting brass.

Iron. The very small quantity of iron normal to commercial copper has no significant effect upon the mechanical properties. In oxygen-free copper, the presence of 0.001–0.005% iron causes a loss of electrical conductivity amounting to 0.8% IACS for every 0.001% iron product. The presence of cuprous oxide completely nullifies this effect by forming an insoluble iron oxide.

In a series of copper–iron alloys up to 2% iron both strengthens and hardens copper without significantly reducing ductility (32).

Iron in oxygen-free copper increases the softening temperature of the copper and in the presence of phosphorus the effect is compounded. When both iron and phosphorus are present, the grain structure of the annealed copper can become very nonuniform.

Copper–iron–phosphorus alloys of the age-hardenable type are obtained with relatively small quantities of iron and phosphorus. The optimum electrical conductivity results when iron and phosphorus are present in a ratio of 3.6 to 1 which corresponds to the compound Fe_2P. Since high electrical conductivity is attainable in such alloys, the solid solubility of iron phosphorus compound must be very low at temperatures used in normal annealing.

Cobalt. The effects of cobalt upon the properties of copper are similar to those of iron. Cobalt is precipitated from solid solution in oxygen-bearing copper, thereby nullifying its effect upon electrical conductivity. Cobalt has a significant effect upon electrical conductivity in oxygen-free copper (see Fig. 6), but only a slight effect on softening temperature. At 100 ppm cobalt the softening temperature is raised ca 25°C (33).

Nickel. Nickel in oxygen-free copper lowers electrical conductivity about 0.09% IACS per 0.001% for nickel concentrations up to 0.05%.

The reaction between nickel and cuprous oxide in copper is reversible (33). The quantity of the nickel oxide formed depends on the concentration of the reactants and the time at the reaction temperature. Small additions of nickel slightly increase the elongation, and concentrations up to 1.0% steadily increase the fatigue characteristics (31).

Nickel in concentrations up to 0.05% has no effect upon the recrystallization temperature.

The existence of a continuous series of solid solutions of copper and nickel in all concentrations has been confirmed by both microscopic and x-ray examination. All alloys of copper nickel have melting points greater than that of copper; all alloys have the fcc crystal structure.

The constitution diagram is very simple, ie, there are no compounds and no eutectic mixtures.

Generally, nickel slightly hardens copper and increases its strength without reducing its ductility. Nickel has a striking effect upon the color of copper, ie, only 10% nickel produces an alloy with a pink cast whereas higher nickel content alloys are white in appearance.

Three copper-base nickel alloys have gained commercial importance. These alloys contain 10, 20, and 30% nickel-dissolved copper with small amounts of manganese and iron to enhance casting qualities and corrosion resistance. These alloys are known as the cupro–nickels and are well suited for application in the chemical industry, and as condenser and heat exchanger tubes because of their generally high corrosion resistance and particular resistance to impingement attack. The alloys are superior to copper and other common copper-base alloys in resisting attack by acid solutions. They have high resistance toward stress-corrosion type failure but their behavior to sulfur corrosion appears to be erratic. Generally, their behavior in regard to sulfur is not as good as that of the brasses, such as the admiralty alloys.

Carbon, sulfur, and phosphorus as impurities in these alloys can have marked effect upon the fabricating and welding characteristics and are commercially controlled.

A variety of cupro–nickel alloys have been used in coinage.

A series of alloys containing varying amounts of copper, nickel, and zinc are known as the nickel silvers. The alloys find use for hardware and optical goods as well as for hollow ware, costume jewelry, and slide fasteners.

Phosphorus. Phosphorus forms copper-rich solid solutions with copper, and has a strong effect upon electrical conductivity in the absence of cuprous oxide. Phosphorus does increase the softening temperature in the oxygen-free condition. It slightly increases both the tensile and fatigue strength without decreasing ductility significantly. The constitution diagram suggests that copper–phosphorus alloys can be hardened by precipitation. Although an alloy of ca 1% P can be hardened slightly by precipitation-hardening techniques, the strength of the alloy is not significantly affected.

Phosphorus readily reacts with cuprous oxide and has become the most important element used commercially as a deoxidizer in melt shops. Two commercially important phosphorus-deoxidized coppers are available. They are known as high conductivity low-residual-phosphorus DLP copper (0.003% P) and low conductivity high-residual-phosphorus DHP copper (0.02% P). The electrical conductivity of these alloys is 101% and 80–90% IACS, respectively.

DLP copper is used for bus bars, electrical conductors, and tubular bus whereas DHP copper is used in air conditioning, gas lines, heater lines, burner tubes, plumbing pipe, and evaporator, condenser, and heat exchanger tubes. Some of these uses require flanging, flaring, or spinning which could be adversely affected by the copper–copper oxide eutectic dispersion that is characteristic of tough-pitch copper. The phosphorized coppers are to be used for applications demanding heating under reducing conditions, such as can be encountered during brazing or welding operations. The low-conductivity alloy (0.015–0.040% P) is preferred because, with lower phosphorus concentrations, it is possible to obtain an equilibrium condition between oxygen and phosphorus in the copper alloy. Under certain conditions it is possible to damage (embrittle) the structure of the high-conductivity alloy by heating it in reducing gases.

Copper-containing phosphorus has been used as an anode material in electroplating systems to help control sludge formation in the plating bath.

Silicon. Silicon can also be used as a deoxidizer and residual low concentrations have little effect upon the mechanical properties. Small additions of silicon have a pronounced effect on electrical conductivity in oxygen-free copper. These effects are nullified in the presence of cuprous oxide. Silicon forms solid solutions in the manner described in the constitution diagram shown in Figure 10. Silicon is added to copper to form the silicon bronze alloys which also usually contain one of four other elements, ie, manganese, zinc, tin, or iron, in concentrations of 0.80–1.50%.

Although there are a number of silicon bronze alloys available, they differ from each other principally in the silicon content and the nature of the third element. There has been a general division of the alloys into two types, known as Grade A and B, containing 3% and 1.5% Si, respectively.

Silicon is a very potent hardener of copper, and the addition of 3% Si practically doubles the tensile strength of copper. The tensile strength of hard-drawn wire may be as high as 1 GPa (145,000 psi), indicating that these alloys work-harden at a higher rate than do most coppers and other copper alloys.

Silicon has a significant effect upon electrical conductivity in oxygen-free copper and has about the same quantitative effect as arsenic and beryllium. The conductivities are 7 and 12% IACS for Grade A and Grade B materials, respectively.

The silicon bronze alloys are readily welded by commercial methods including gas, carbon, arc, metal arc, resistance, and the MIG (metal electrode inert gas) and TIG (tungsten electrode inert gas) systems. Much of the use of these alloys for structures and equipment is caused by their ease of manufacture by welding. The 1.5% Si

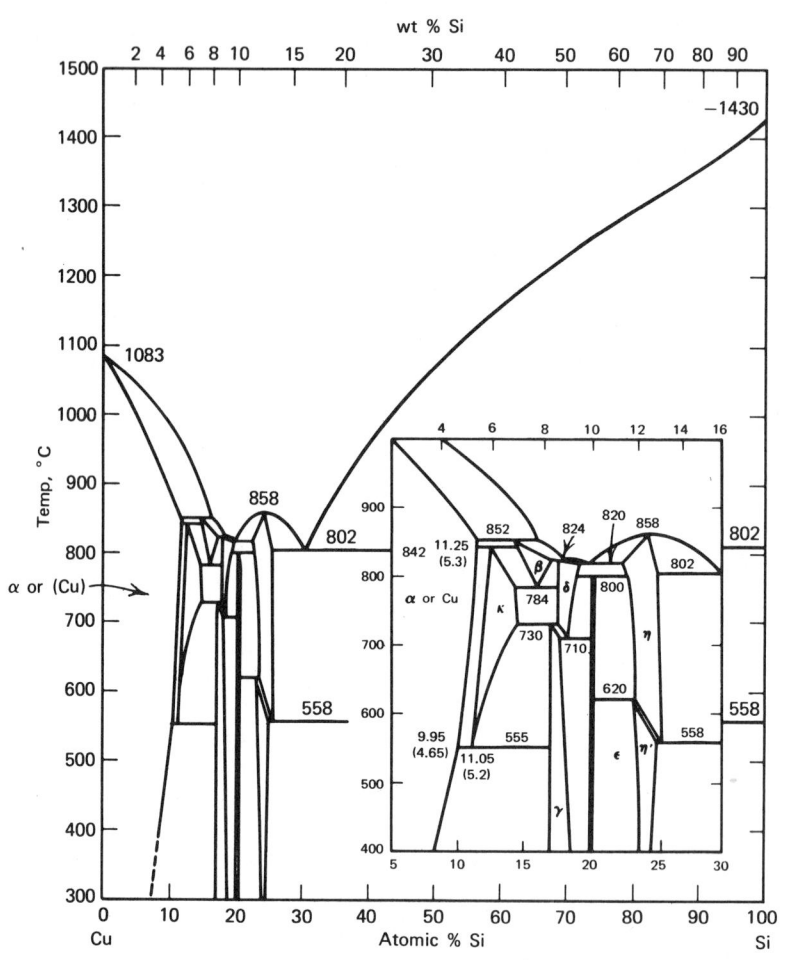

Figure 10. Constitution diagram for copper–silicon system (27). The atomic % and the wt % are shown respectively, eg, 11.25 atomic % and 5.3 wt % Si.

alloys are of lower strength, but are commonly used in the cold-worked state for such articles as cold-leaded bolts that have attractively high strength. Both grades have excellent hot-working properties and can be readily rolled, forged, and extruded. Lead must be controlled as an impurity. The melts are apt to dissolve gases; thus, careful steps must be taken to prevent this or gross porosity may be present in the castings. Covering melts of these alloys with unburned charcoal has been known to lead to worthless castings. Annealing both grades is done at ca 425–760°C depending upon the degree of softening required. If annealing is done in air, the scale formed is very refractory and is very difficult to remove even by wire brushing. A pickling solution of 10% H_2SO_4, 20% HNO_3, and 5% HF has been used effectively. Annealing atmospheres containing quantities of CO as low as 2% can also oxidize the silicon bronze alloys depending upon temperature and exposure time. Dry nitrogen atmospheres will protect surfaces during annealing.

Silicon bronzes are used for construction of tanks, chemical equipment, hydraulic lines, bolts, screws, nuts, bearing plates, and shafting.

Oxygen. Oxygen in the form of cuprous oxide is always present in tough-pitch coppers, and under ordinary conditions does not interfere with the use of this material. Since the great bulk of rolled copper is oxygen-bearing, there is an interest in what kind of effect the presence of cuprous oxide has upon hot-rolling and what the effect is of impurities in the presence of oxygen. Little information can be found in the literature, perhaps because oxygen in concentrations normal to tough-pitch copper (0.03–0.08%) does not interfere with the hot-rolling operation. The probable limit to which this element may be tolerated is six to seven times its usual concentration.

Copper containing up to 0.36% O can be hot rod-rolled successfully from 2.5 cm diameter to 2.2 cm diameter (31).

The variation of certain properties with increasing oxygen content in rolled and annealed condition is shown in Table 8.

Table 8. The Effect of Varying Oxygen Contents Upon the Properties of Rolled and Annealed Copper

Oxygen concentration, %	Tensile strength, MPa[a]	Contraction in area, %	Density	Impact test energy absorbed, J[b]	Electrical conductivity, % IACS
0.015	193	75	8.912	277.9	102.1
0.09	203	65	8.882	255.3	100.6
0.36	299	38	8.76	89.5	96.2

[a] To convert MPa to psi, multiply by 145.
[b] To convert J to ft·lbf, divide by 1.356.

The change in tensile strength, elongation, contraction in area, impact value, and mass conductivity is slight for values of oxygen below 0.10%. Therefore, the element in the range of 0.03–0.08%, which is normal for tough-pitch copper, usually does not have much effect upon the mechanical properties of commercial varieties of copper. If significant oxide segregation is present in the copper microstructure, mechanical properties and drawability may be adversely affected.

Oxygen copper does tend to work-harden at a slightly higher rate, and is somewhat less ductile than the oxygen-free or deoxidized varieties.

Some investigators have suggested that small amounts of oxygen actually increase its electrical conductivity. Such an effect is produced by the reaction of cuprous oxide with impurities, causing their partial or complete precipitation from solid solution as oxides, thereby increasing the purity of the base metal.

Hydrogen Damage (Embrittlement). All copper alloys that contain appreciable amounts of cuprous oxide such as the tough-pitch or DLP coppers are susceptible to damage called embrittlement when heated under reducing conditions, especially if hydrogen is the reducing agent.

When oxygen-bearing copper is heated in hydrogen or hydrogen-bearing gases, the hydrogen easily diffuses into the metal and reacts with the copper oxide to form steam. Under substantial pressure, steam produces holes and fissures and the ductility of the structure is decreased; the condition is generally known as embrittlement. Tough-pitch copper heated to 595°C in hydrogen has been damaged to a depth of 0.9 mm after one hour and to a depth of 2.5 mm after 10 h exposure. The effect of the hydrogen, copper oxide reaction on the microstructure of wrought copper is illustrated in Figure 11. Although both the hydrogen and the steam formed would occupy equal

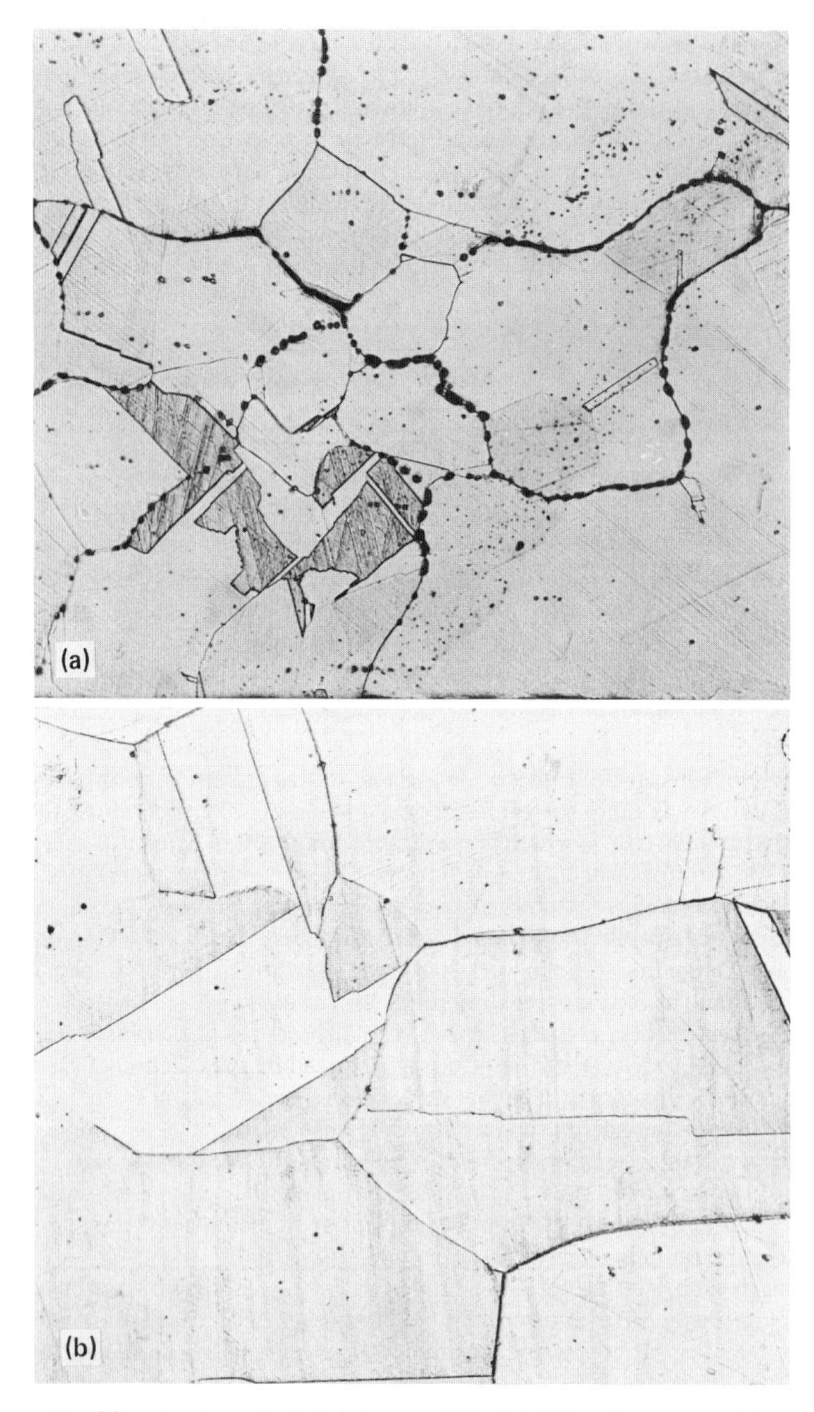

Figure 11. Microstructure of embrittled copper (200×). (**a**) Note the rupture of the metal at the grain boundaries. (**b**) Normal tough-pitch copper. Courtesy of T. S. Howald, Chase Brass and Copper Company, Inc.

volumes if both were present as a gas, it is thought that the hydrogen is dissolved in the copper until the steam is formed.

As is the case in most chemical reactions, time and temperature, as well as availability of a reducing gas such as hydrogen, are important factors. For copper containing as much as 0.06% O, the rate of reaction decreases markedly below 540°C and is negligible below 400°C.

Oxygen-free or deoxidized copper is not susceptible to embrittlement upon heating in hydrogen atmospheres unless it has first been heated in an oxidizing atmosphere. During the oxidizing cycle oxygen diffuses into the copper, depositing a subscale composed of the oxide of the residual deoxidizer or other low concentration elements. Upon subsequent treatment with hot hydrogen, copper becomes ruptured along the grain boundaries to a depth sharply limited by the previous penetration of oxygen as defined by the inner boundary of the subscale.

Foreign oxides embedded in copper are reduced by hydrogen at elevated temperatures and a type of hydrogen embrittlement can result from this cause alone (34). The oxides known to be reduced by hydrogen, such as those of manganese, nickel, phosphorus, tin, and zinc can produce embrittlement. The oxides of lithium, magnesium, and zirconium are only slightly affected. Intermediate in their effect are oxides of arsenic, bismuth, boron, calcium, chromium, iron, lead, antimony, and titanium. Embrittlement can take place in coppers containing arsenic, antimony, or bismuth. Volatile hydrides such as arsine, stibine, or bismuthine may occur under sufficient pressure to cause the alloy to rupture (34).

Most commercially produced copper is annealed in bright or controlled gas atmospheres consisting largely of N_2, but also containing 2–5% of combustibles such as a combination of CO and H_2. Commercial annealing in these atmospheres containing 2% H_2 is regularly done at temperatures as high as 510°C for as much as 3 h without embrittlement taking place.

When tough-pitch copper is to be brazed or welded, the possibility of hydrogen damage or embrittlement must be anticipated; if hydrogen atmospheres must be used, oxygen-free or deoxidized coppers should be selected.

Silver. Silver in varying concentrations is one of the most common impurities found in copper. It is common practice for specifications governing copper alloys to count silver as copper in the chemical analysis. The usual mechanical properties of copper are not significantly modified by the presence of silver, but the presence of this element does have significant influence upon the softening temperature of copper.

Silver-bearing copper is commercially available with silver contents of 0.25–1 kg/t. The greatest increment of influence is noted for concentrations of silver up to 0.25 kg/t. The increase in softening temperature continues with increasing silver concentrations, but it does so at a slower rate. The higher silver amounts are specified when the maximum softening temperature of the copper is required.

Resistance to creep (the phenomenon of extension under loads lower than the yield strength of the material) has been shown to increase for silver contents of 0.79 kg/t. Increases in both cold-work and grain size increases the resistance to creep.

The silver-bearing coppers have been used where the mechanical properties of strain-hardened copper may be impaired in soldering or tinning operations or in commutator sections subjected to heat owing to surges in electric loads.

Tin. Tin has a significant effect upon electrical conductivity of oxygen-free copper. Small amounts of oxygen in the copper precipitate tin quantitatively.

Tin in concentrations up to 0.10% has no effect on the softening incidence, and also has no effect on grain growth at low and moderate temperatures.

Tin forms a complex series of alloys with copper. Tin bronzes are alloys of copper and tin. The commercial wrought bronzes usually contain 11% or less of tin. It is common practice to deoxidize the bronze melts with phosphorus leaving a residual of 0.03–0.04%; these alloys are known as the phosphor bronzes.

Alloys containing about 5% and more of tin are subject to inverse segregation of a tin-rich, lower-melting point phase known as delta.

Two general groups of alloys are produced, ie, alloys containing 0.5–2.0% Sn, and alloys containing 4–11% Sn. Tin is added to copper in concentrations up to 2% to produce alloys of relatively high strength combined with relatively high electrical conductivity not obtained in most other high-strength alloys. For example, an alloy containing 1.3% tin has an electrical conductivity of 43% IACS.

Two common types of bronzes contain 5% and 8% Sn. The alloys have excellent cold-working properties. However, the tin bronzes are not considered to be good hot-working alloys, since they are only slightly plastic within the narrow temperature range of 620–670°C.

Some alloys contain lead to improve machinability.

In general, this series of tin alloys combines higher strength with good fatigue characteristics. Phosphor bronzes of Grades A (5% Sn), UNS C 51000, C (8% Sn) UNS C 52100, and D (10% Sn), UNS C 52400, are used as welding rods, springs, diaphragms, contact points, and applications requiring good resistance to corrosion and fatigue in combination with high tensile properties.

Cadmium. Cadmium may be added to copper to increase its strength while having only a minimum influence upon electrical conductivity. It is soluble in copper to the extent of about 2.8% at the eutectic temperature of 557°C. Under equilibrium conditions, the solid solubility decreases as the temperature is lowered, but in practice amounts as high as 1% Cd apparently remain in solution. These characteristics indicate that copper cadmium alloys are precipitation hardenable; however, the increase in strength is commercially insignificant.

Cadmium is added to copper when resistance is desired. In some respects small additions of cadmium may be superior to silver for increasing the softening temperature. The dilute copper cadmium alloys can be severely cold-worked and still retain a relatively elevated softening temperature (see Fig. 3).

Many elements may be added to increase the tensile strength of copper in the annealed condition and to produce even greater strengths in the strain-hardened state. However, the great majority of these elements so greatly decrease electrical conductivity that the resultant alloy is undesirable as an electrical conductor. Cadmium does produce the desired effect of increasing hardness and has a minimum effect upon electrical conductivity. A commercially important alloy containing a nominal 1% cadmium can have a tensile strength of 606–689 MPa (88,000–100,000 psi) for severely drawn wire. The alloy has an electrical conductivity of 90% IACS. The copper cadmium alloys possess a better combination of strength and electrical conductivity than does pure copper.

The 1% alloy is used for electric blanket elements, heating pad elements, connectors, cable wrap, trolley wire, and switchgear parts.

Aluminum. Alloys of copper and aluminum fall into two commercially important types, ie, the single-phase alpha solid-solution alloys and the alpha–beta alloys. Under equilibrium conditions 9.8% Al is soluble in copper before the beta phase forms, but in commercial nonequilibrium conditions, alloys containing much in excess of 7.5% Al exhibit two phases. The commercially important aluminum bronzes usually contain 4–10% Al with or without other metals.

Additions of iron, manganese, silicon, or nickel are frequently made to increase strength and hardness.

The alpha aluminum bronzes have excellent cold-working and good hot-working properties and can be readily hot-forged, rolled, and extruded. The best hot-working range is 790–900°C.

The hot plasticity range increases with aluminum content while the cold-work rate significantly increases also. The alpha–beta alloys can be hot-forged into intricate shapes. Their hot-working behavior compares with those for alpha–beta brasses, but they can be cold-worked relatively lightly. Generally, these aluminum bronzes are produced in the hot-rolled or extruded condition. These alloys possess high tensile strength coupled with low ductility. The alpha–beta alloys respond to heat treatment with a general improvement of mechanical properties. Quenching is done in water from temperatures of 815–870°C and reannealing between 370–540°C depending upon size of section and alloy.

The annealing characteristics of the alpha alloys are similar to those of the alpha brasses. Annealing can be performed within the range of 430–760°C, depending upon the required properties. All of the aluminum alloys have good resistance to oxidation at elevated temperatures and are better in that respect than other copper-base alloys. The resistance toward oxidation increases with the aluminum content. The resistance of aluminum bronze to corrosion is largely caused by the formation of aluminum oxide on exposed surfaces. Aluminum oxide is resistant to most acid solutions and, therefore, renders these alloys very resistant toward acid attack. Since aluminum and its oxides are soluble in alkalies, aluminum bronze alloys offer considerably less resistance to the strong alkalies than other copper-base alloys. Under certain conditions of corrosion, the alpha–beta aluminum bronzes are susceptible to a form of corrosion called dealuminumization which is analogous to dezincification in the brasses.

Table 9 lists some mechanical properties for three aluminum bronze alloys. The aluminum bronze alloys are useful in applications requiring high tensile properties

Table 9. Mechanical Properties of Aluminum Bronzes

Name	UNS designation	Temper	Tensile strength, MPa[c]	Yield strength, MPa[c]	Elongation %, 5 cm
aluminum bronze 5%	C 60800	soft	413.4	186	55
aluminum bronze D	C 61400	soft	523.7–565	227.4–310.1	45–40
	C 61900	soft	634	338	60
		hard[a]	931	829[b]	2
		spring[a]	999.2	931[b]	3

[a] After heat treatment for 1 h at 232°C.
[b] 0.2% Offset.
[c] To convert MPa to psi, multiply by 145.

and wear resistance in combination with good behavior in corrosive environments for such parts as propeller blades, bolts, pump rods, slide liners, wear strips, valve guides, gear shifter forks, bearings, bushings, and condenser tubes.

Zinc. Zinc is seldom present in copper as an impurity, but is intentionally added to form a series of industrially important alloys called the brasses. These alloys contain from a few percent of zinc to about 40% and are treated in detail below. The presence of small quantities in oxygen-free copper does not produce either hot- or cold-shortness and has little effect upon the mechanical properties. The effect on electrical and thermal properties is somewhat greater than that for either silver or cadmium.

Other Properties of Wrought Copper Alloys. *Modulus of Elasticity.* Generally, copper-base alloys do not exhibit a true proportionality between stress and strain. The moduli of elasticity for almost all copper alloys are quite similar, and the values in tension suitable for ordinary calculations are listed in Table 10 for various alloy groups.

Table 10. Modulus of Elasticity in Tension

Alloy groups	Modulus, GPa[a]
leaded brasses, 3% Pb 58–62% Cu	96
other leaded brasses and nonleaded brasses	
58–62% Cu	103
66–80% Cu	110
80–100% Cu	117
tin bronzes (phosphorus bronzes) including leaded alloys	
88–90% Cu	103
90–95% Cu	110
95–100% Cu	117
cupro–nickels	
90% Cu	124
70% Cu	152
silicon bronzes	
3% Si, 1% Zn	103
1.5% Si, 1% Zn	117
precipitation-hardenable alloys	
chromium copper, 0.8% Cr	131
97.5% Cu, 2% Ni, 0.5% Si	124
98.68% Cu, 1.1% Ni, 0.22% P	124
beryllium copper, 98.1% Cu, 1.9% Be	131

[a] To convert GPa to psi, multiply by 145,000.

Modulus of Rigidity (Shear). Maximum shear stresses of wrought copper alloys in torsion or simple shear are related to the tensile strength of the material involved. The modulus of rigidity has a value of about 60% for the soft annealed condition. Actual possible distortion of rod in torsion, as expressed in number of turns per centimeter, varies with rod diameter, temper, and ductility.

The modulus of rigidity is sometimes termed the modulus of elasticity in shear. For copper alloys this modulus is generally expressed as 0.38 times the modulus of elasticity in tension.

Endurance (Fatigue). Fatigue failure is fracture that occurs as a result of many applications of a stress which, if applied only once, would not cause failure. The number of times the stress must be applied to cause failure is related to the magnitude of the stress. The number of applications of stress before failure increases as the stress decreases. The fatigue limit does not appear to be related to the ductility. Time is unimportant; the number of applications for a given stress is not affected by the frequency of the applied stress. Stress concentrations have a marked effect; the maximum stress that can be applied for a given life is very significantly reduced if a severe stress raiser is present.

The fatigue strength as measured by rotating-beam methods for a large number of reversals is reported for brass, 65% Cu, 35% Zn, from ca 96 MPa (14,000 psi) as annealed to ca 138 MPa (20,000 psi) as rolled 8 or 10 B & S numbers (60–68% reduction in thickness) hard.

Corresponding values for 70% Cu, 30% Zn sheet are only slightly higher and, as copper content increases to 85% Cu, 15% Zn, the value increases to ca 165 MPa (24,000 psi) for material rolled 8 B & S numbers hard. The endurance limit for fatigue strength for copper ranges from ca 69 MPa (10,000 psi) as annealed to ca 138 MPa (20,000 psi) as severely cold-worked.

Alloying does have an effect upon this property of copper. Hard-rolled or drawn silicon copper 96% Cu, 3% Si, 1% Zn, and the phosphor bronzes have increased fatigue resistance as compared to the copper–zinc alloys. The fatigue strength for silicon copper of 8 B & S number hardness is ca 186 MPa (27,000 psi) and that for the phosphor bronzes is slightly higher for the same temper. Beryllium copper has excellent fatigue resistance and reportedly has limiting values of 227–351 MPa (33,000–51,000 psi) dependent upon alloy composition and heat-treatment.

Magnetic Properties. Copper alloys are generally considered to be nonmagnetic. Under certain conditions, however, they show definite magnetic properties. These ferromagnetic characteristics are most often associated with the iron concentration of the alloys and the mode of its occurrence. The magnetic susceptibility of copper alloys depends not only on the amount of iron present, but also whether the iron is in solid solution. Iron in solid solution does not result in significant ferromagnetic properties. The iron precipitated from solid solution may be in either the magnetic or nonmagnetic form. If in the magnetic form, its effect varies with the particle size and distribution of the precipitate (see Magnetic materials).

Copper alloys with low iron contents are diamagnetic with negative susceptibility and a permeability of less than one. They are repelled from the poles of a magnet and tend to move toward a weaker field.

The magnetic properties of copper alloys depend on complex relationships; they depend not only upon iron content, but also on elements other than iron, and on the previous mechanical and thermal history of the alloy. The same material could have different values for permeability and susceptibility with different treatments. Table 11 lists susceptibility and permeability values for some commonly used alloys. The values listed are the maximum likely to be experienced in these alloys. The tests were made on samples prepared in the laboratory with iron contents of near the maximum allowed by specification and given various treatments to produce maximum values.

Directionality in Cold-Rolled and Annealed Copper Alloys. When the polycrystalline brasses are plastically deformed, the individual grains tend to rotate into a common orientation. The results are described in terms of an ideal orientation such as (110)

COPPER ALLOYS (WROUGHT)

Table 11. Maximum Magnetic Properties

Alloy	Iron, wt %	Maximum specific (mass) susceptibility, 10^6 cgs units	Maximum permeability, 10^6 H/m[a]
commercial bronze	0.039	0.00	1.256597
cartridge brass	0.05	0.945	1.256724
Muntz metal	0.07	15.8	1.258687
leaded commercial bronze	0.112	59.3	1.264840
free-cutting brass	0.35	474	1.320181
architectural bronze	0.15	139	1.275194
naval brass	0.087	59.1	1.264450
phosphor bronze, 5%	0.085	6.6	1.257523
phosphor bronze, 8%	0.07	5.83	1.257416
nickel silver 65–18	0.25	1.91	1.256858
telnic bronze	0.17	1.36	1.256789
aluminum–silicon bronze	0.38	76	1.266637
high-silicon bronze	0.29	91	1.269163
silnic bronze	0.13	0.840	1.256715

[a] To convert H/m to cgs units, multiply by 79.58×10^4.

[112] for a rolling texture; this means that a plane in the form of (110) lies parallel to the rolling plane, and a plane of the [112] type is parallel to the rolling direction. In practice, there will be a greater or lesser degree of scatter about the ideal orientation. For pure copper, deformation texture is observed that could be a mixture of the two textures (110) [112] and (112) [111].

The deformation texture is altered to another texture upon annealing. The formation of a new annealing texture (100) [001] is related to the deformation texture by standard lattice rotation. Elements that enter into solid solution with copper gradually change the annealing texture to that represented by orientation (113) [112]. This is demonstrated by the alpha brasses. The amount of solute element necessary to bring about this transformation depends upon the degree of misfit of the solute atom in the solvent lattice, the valency, and the temperature of deformation. Nickel, which is completely soluble in copper, does not substantially change the annealing texture. Other elements change the texture in a systematic manner and effect, in a very rough way, the same changes at the same relative concentration (relative concentration is concentration divided by solubility limit).

The orientations found in the brasses do not lend themselves to easy identification micrographically as is the case with copper. The orientations developed on annealing are sensitive within the commercial annealing temperatures.

Annealing. Work-hardened brasses can be softened by annealing or heating at 425–760°C, depending upon the zinc content, the desired degree of softening, and prior amount of plastic deformation (cold-work). The lower temperatures produce relatively harder, stiffer, and less ductile material. The best criterion for expressing the degree of annealing is grain size which accurately determines the properties of the material in the annealed state for most alloys. In general, small grain sizes (0.005 mm) are used where excellent surface, considerable stiffness, and good ductility are desired. Increased grain size up to 0.035 mm for high-copper alloys, or 0.070 mm for cartridge brass, 70% Cu, 30% Zn, might be required when subsequent forming operations are expected to be severe.

Figure 12 illustrates the variation of some mechanical properties in relationship to composition and grain size. The basic strengthening effect of zinc is evident. Copper itself exhibits little change in strength within the usual grain size range, but the difference in strength owing to grain size change in the brasses first increases as copper content decreases to about 80%, and then remains almost constant to about 65% Cu. These curves representing changes in yield strength, elongation, and hardness related to various grain sizes indicate these property relationships to be similar to those for tensile strength.

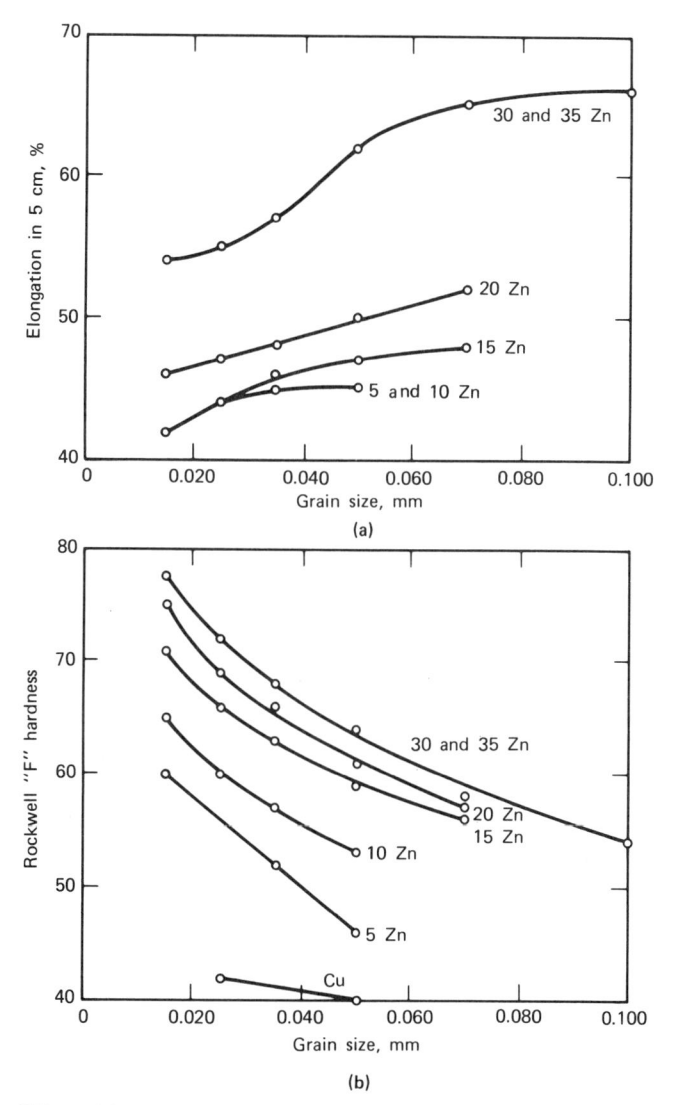

Figure 12. Effect of degree of anneal as measured by grain size upon various properties of copper and copper–zinc alloys (35). (**a**) Effect of grain size on elongation of annealed 1-mm brass strip. (**b**) Effect of grain size on Rockwell hardness of annealed 1-mm strip of copper and brass. (**c**) Tensile strength of 1-mm strip of copper and brasses (Cu–Zn) as annealed to designated grain sizes. (**d**) Effect of grain size on yield strength of annealed 1-mm brass strip 5% Zn, UNS C 2100; 10% Zn, UNS C 2200; 15% Zn, UNS C 2300; 20% Zn, UNS C 2400; 30% Zn, UNS C 2600. Courtesy of W. M. Baldwin, Jr., Chase Brass and Copper Company, Inc.

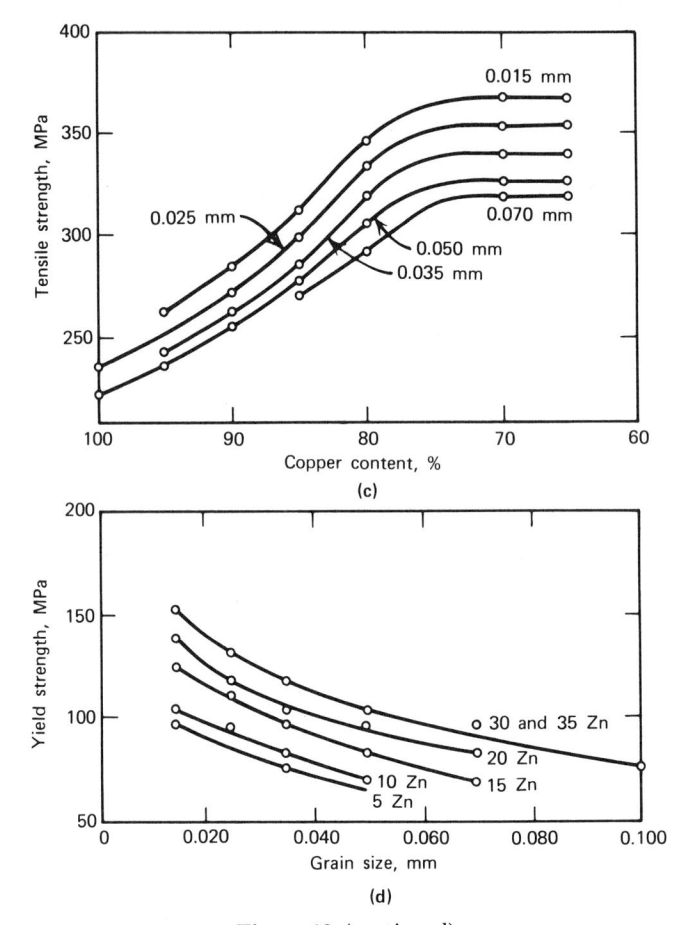

Figure 12. (*continued*)

Other properties of brass affected significantly by the degree of annealing are creep, stress-corrosion cracking, and fatigue strength. For annealed admiralty brass (71% Cu, 1% Sn, 28% Zn) the same creep strength was found for two grain sizes, 0.018 and 0.0035 mm, at 149°C, but at higher temperatures up to 260°C, the material of larger grain sizes was superior (35). The fatigue strength of some brasses decreases as grain size increases. A more pronounced effect was obtained for 85% Cu, 15% Zn alloy. It was shown that the fatigue strength, and the ratio of fatigue strength to tensile strength were greater for the smaller grain size. The relationship was found to prevail even after the material annealed to the two different grain sizes had been rolled 21%. For reductions of 37% and 60%, the ratio of fatigue strength to tensile strength was the same regardless of previous grain size (36).

The recrystallization and grain growth characteristics of the alpha brasses are generally governed by four factors, ie, the degree of cold-work before the annealing, the effect of previous grain size, the time and temperature of annealing, and the alloy composition.

Recrystallization of 70% Cu, 30% Zn alloy can be distinguished at successively lower temperatures as the degree of cold-work increases, and the first recrystallized grains become smaller and smaller. Considerable spread in small final grain size results when the previous grain size is large. The effect is less pronounced for large final grain sizes.

Grain size of brass increases with length of annealing time. Evaluation of the time factor has been difficult in commercial practice, because in most annealing the greatest part of the heating cycle is consumed in heating the material to the annealing temperature rather than heating at temperature for any prolonged time. In actual practice, the temperature attained is of much greater importance than the length of time held at temperature. The average grain size may be approximately doubled by a twenty-five fold increase in annealing time at a given temperature. This relationship holds for isothermal annealing at ca 455–705°C, and for annealing times from 5 min to 25 h.

A guide for attaining given grain sizes for several alloys is shown in Table 12.

Controlled atmospheres are frequently used during the annealing of the coppers and copper-base alloys. Gas obtained from the controlled combustion of natural gas is important as a protective or bright atmosphere and consists mostly of nitrogen and with up to about 4% combustibles. The combustibles are ca 50% CO and 50% H_2. Care must be exercised to keep sulfur out of such atmospheres to control discoloration of the copper-rich alloys. Cracked ammonia is also a suitable protective atmosphere. Best results are obtained when these gases contain little water vapor.

The brasses present a more difficult bright annealing problem in usual atmospheres. Zinc has an appreciable vapor pressure at normal annealing temperatures and distills from the alloy. Zinc is oxidized in furnace atmospheres containing carbon dioxide and water vapor, and the color and surface finish is adversely affected. Such surfaces have a haze and are usually pickled (immersed in 10% sulfuric acid) after annealing.

The annealing of brass is complicated by the presence of small quantities of impurities that have been inadvertently introduced by the use of contaminated scrap. Continual vigilance must be exercised in melt shops to avoid casting material containing undesirable elements. Impurities in general repress recrystallization or lead to nonuniform grain growth or both. Elements known to have such an effect either singly or in combination include iron, aluminum, lead, manganese, nickel, phosphorus, and silicon.

Table 12. Approximate Annealing Temperature Ranges for Annealing of Fabricated Parts

Alloy	Average grain size, mm	Metal temp[a], °C
copper	0.025	400–482
copper	0.020	343–400
commercial bronze,	0.040	496–635
90% Cu, 10% Zn	0.025	413–566
red brass, 85% Cu,	0.040	482–609
15% Zn	0.025	400–482
cartridge brass, 70%	0.040	454–593
Cu, 30% Zn	0.025	391–482
yellow brass, 66% Cu,	0.040	454–593
34% Zn	0.025	391–454
phosphor bronze A,	0.030	566–649
94.8% Cu, 5% Sn, 0.2% P	0.015	482–566
nickel silver, 65% Cu,	0.035	593–904
17% Zn, 18% Ni	0.020	482–621

[a] Variable factors such as size of load, distribution of parts, and shape make it difficult to suggest times for annealing.

Fire-Cracking. Under certain conditions, the brasses are susceptible to failure known as fire-cracking during the annealing process. It is a phenomenon that occurs when differentially stressed metal is suddenly subjected to high temperature. Fire-cracking occurs infrequently in copper alloy strip because the deformations produced by rolling are relatively uniform, much more so than those produced in tube and rod drawing.

Light reductions can favor fire-cracking. Coarse columnar crystals present a restricted number of slip systems to deformation and concentrations of stress are accordingly set up at the grain boundaries. Fire-cracking can develop at these points. If the alloy is prone to segregation or has low melting or brittle intercrystalline films (lead and others), the tendency to fire-crack is accentuated.

Scalping performed on castings, hot-rolled, or lightly cold-rolled brasses can lead to fire-cracking.

Copper alloys that are worked and annealed in a manner to develop a fine or small grain size are relatively immune to fire-cracking. The finer grain is more compatible under strain and no severe stress concentrations are developed at the infinitely longer grain boundaries per unit cross section.

Effect of Other Elements on Copper–Zinc Alloys. Other elements present in the copper–zinc alloy system fall into one or more of the three following classifications: impurities; deoxidizers (which are usually not necessary); and elements intentionally added for a specific reason.

Impurities such as lead, bismuth, and antimony are sometimes encountered and cause serious trouble in hot-working operations. Phosphorus and iron have a significant effect upon recrystallization temperature, especially if both are present. Selenium, tellurium, sulfur, cobalt, arsenic, silver, silicon, chromium, and cadmium in amounts normally found as impurities do not interfere with hot-rolling or extrusion, but may effect the recrystallization temperature.

Precipitation-Hardening Copper Alloys

Alloys that can be precipitation-hardened have the common characteristic of a decreasing solid solubility of some phase, element, or intermetallic compound with decreasing temperature. This is the sole requirement for precipitation or age-hardening.

Precipitation is a decomposition of a solid solution leading to a new phase of different composition; namely, the precipitation of the new phase from the solid solution that comprises the matrix. In such alloy systems, cooling at an appropriately rapid rate (quenching) from a temperature well within the all-alpha field will preserve the alloy as a single solid solution possessing relatively low hardness, strength, and electrical conductivity.

A second heat treatment (aging) at a lower temperature will cause precipitation of the unstable phase. The process of precipitation is usually accompanied by an increase in strength, hardness, and electrical conductivity.

The interaction of a dislocation with a foreign particle offers an explanation for the hardening mechanism. A precipitated particle has a different elastic constant than that of the matrix, causing a stress state when a dislocation approaches it; this may give rise to several effects. One is that it should be more difficult to drive a dislocation through a crystal that contains a large number of such areas; therefore, the stress required to cause strain would be higher than for the homogenous solid solution. Secondly, if the precipitated particles offered enough resistance to the passage of dislocations, the dislocations would pass around rather than through them leaving closed loops around the particles. As each successive dislocation moves across the crystal, another loop is formed. The space between each obstacle decreases making it increasingly more difficult for dislocations to be forced through. The maximum effect is achieved when each particle is big enough to restrict the dislocations, and are as close together as possible.

In some alloy combinations, only a very slight increase in hardness follows aging, but often the increase is very significant. Probably the number of fine particles per unit volume is a prominent determining factor in the hardening effect; the smallest conceivable precipitate and the densest population are the most effective form for any given amount of precipitation. One universal feature of this process prolonged aging causes continued coalescence of the particles and results in relatively few particles of larger size, their number and distribution being insufficient to cause great hardness. Alloys in this condition are said to be overaged and are softer than properly aged material of the same composition. Typical aging curves are shown in Figure 13 for a Cu–Ni–Si alloy where the precipitant is Ni_3Si. Aging is a function of temperature and time as indicated by change in hardness. The age-hardening curve for a given temperature passes through a maximum that is reached sooner with higher aging temperatures. The hardness gradually decreases as a result of overaging. Cold-work also causes maximum hardness to be reached at a shorter time for a given temperature. The eventual softening is a consequence of the approach of the alloy to the equilibrium condition with increasing time and temperature. A greatly overaged alloy is very similar in hardness to an annealed alloy.

It may be expected that the initial nuclei will be small and grow with time and that new nuclei will continue to form and grow. The nuclei should be distributed at random throughout the system. They are visible if their size exceeds the limit of resolution for the microscope employed. Chromium precipitates in copper; these precipitates have not been detected at magnifications of 1000 diameters, whereas precipitates of Ni_3Si are visible at 500 diameters.

In addition to composition and heat-treating conditions, the properties obtained in certain alloys are significantly affected by cold-working after solution treatment.

Regardless of the degree of hardness obtained after aging, any effective precipitation invariably results in some increase in electrical conductivity. In some alloy systems, maximum conductivity can be realized only after exceedingly long aging treatments that result in overaging the material and a loss in hardness. The beneficial effect of aging on electrical conductivity is ascribed to the fact that any given element or compound has a lesser effect upon this property in the dispersed form than in the solid-solution state.

Nineteen elements form copper-base binary alloys having the type of alpha solid solution that is a requirement for precipitation hardening (37). Many of these alloys do not show a marked or reasonably useful degree of hardening. The commercially important age-hardening copper alloys contain at least one of these nineteen elements. Table 13 lists the elements with solubility data.

Beryllium. The copper–beryllium constitution diagram shows a decreasing solid solubility of beryllium in copper with decreasing temperature in the alpha region, thus fulfilling the requirements for all precipitation-hardenable alloys.

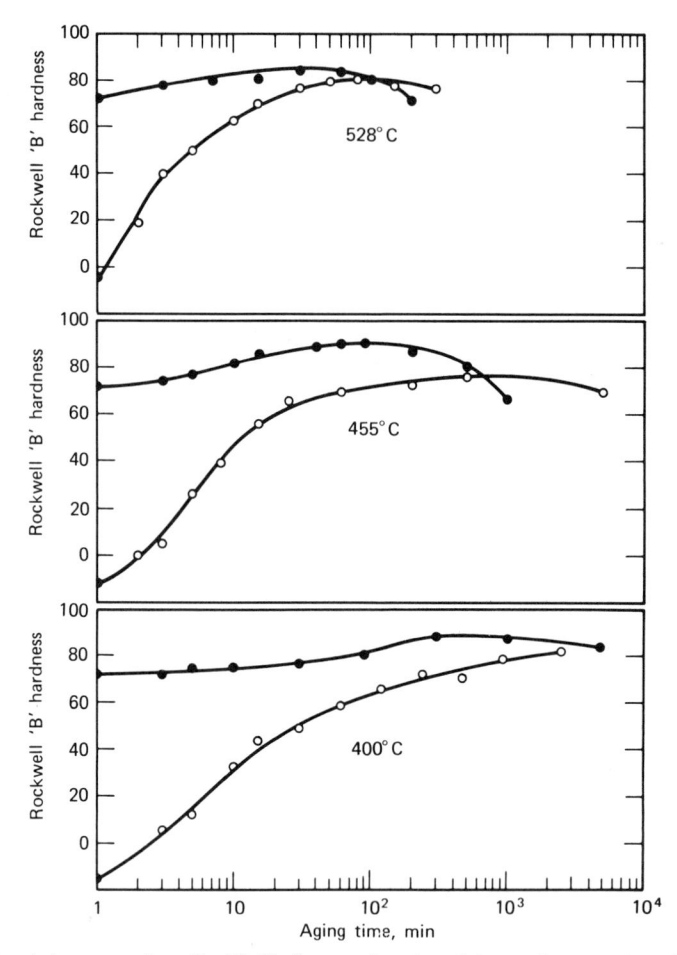

Figure 13. Aging curves for a Cu–Ni–Si alloy as a function of time and temperature for the quenched, and drawn and quenched conditions. ● = 75% Reduction before aging and ○ = 0% reduction before aging. These data can be plotted showing the hardness as a function of the reciprocal aging temperature (1/T(K)), corresponding to the general equation, Rate = 1/time = exp (−Q/RT) for diffusion controlled processes. The apparent activation energy Q is 182 kJ (43.6 kcal) for the undeformed state and 101 kJ (24.2 kcal) when aging is preceded by a 75% reduction of area. Courtesy of V. F. Nole, Chase Brass and Copper Company, Inc.

Alloys containing 2.0–2.5% Be are most important because the maximum mechanical properties can be readily achieved. Alloys containing as much as 4% Be may be hardened by heat treatment; however, the alloy tends to be brittle probably as a result of elements present that do not enter into the hardening reaction.

Ordinarily the alloys are heat-treated in the range of 760–810°C to homogenize or to dissolve the beryllium in the copper, and then quenched in water to retain the solid solution at room temperature. The soft as-quenched condition may be followed by an aging operation. Harder and stronger material can be produced by quenching, cold-working, and aging rather than by quenching and aging alone. The aging operation is usually performed at 288–343°C for suitable periods, depending upon the mass of the material involved.

Table 13. Maximum Alpha Solid Solubility

Alloy system	Approximate soly, wt %	Temp, °C
copper–aluminum[a]		
copper–antimony	11.3	630
copper–arsenic	8.0	680
copper–beryllium	2.4	870
copper–cadmium	1.2	500
copper–chromium	1.25	1076
copper–cobalt	5	1100
copper–iron	3.8	1080
copper–gallium	21.9	620
copper–indium	19.2	574
copper–magnesium	3.0	730
copper–phosphorus	1.7	700
copper–silicon	5.3	850
copper–silver	7.9	780
copper–tin	16	540
copper–thorium		
copper–titanium	5	895
copper–vanadium		
copper–francium	1	980

[a] Hardening of copper–aluminum is not caused by varying alpha solubility.

A third element is commonly added. Cobalt is added for high temperature stability, and nickel is used to refine the grain structure.

The tensile strength of material in the quenched and aged condition is ca 1.25 GPa (180,000 psi), whereas material quenched, rolled to 11% reduction in thickness, and then aged has a tensile strength of ca 1.4 GPa (200,000 psi), 2.8% elongation, and an electrical conductivity of less than 50% IACS.

The high order of the endurance limit of these alloys accounts for their many industrial applications including, springs, diaphragms, bearing plates, and other applications requiring high strength, resistance to shock and fatigue.

Chromium. Chromium is intentionally added in amounts of up to ca 1%, forming a heat-treatable alloy. The alloy has fairly high strength and exceptionally high electrical conductivity after suitable heat treatment. A chromium content of 0.5% is desirable to obtain high strength, hardness, and good electrical conductivity, but alloys containing less than 0.3% Cr can yield satisfactory properties by proper manipulation. Chromium values much over 0.8% do not greatly enhance the mechanical properties. Phosphorus, silicon, or lithium have been used as deoxidizing elements, and of the three elements, residual lithium has the least deleterious affect on electrical conductivity.

Homogenization is usually performed from 927–982°C followed by a water quench and aging at 482°C for a suitable time, dependent upon the mass of material involved. Work loads are standardized as to size to make for more consistent heat treatment. Quenching is done from a high temperature salt bath and aging in an atmosphere controlled furnace. Quenching has also been carried out after Cu–Cr alloy rod has been heated in an induction furnace.

Cold-work is sometimes superimposed on the precipitation hardened condition to enhance strength and hardness. Often, cold-working of heat-treated alloys is not possible because they are too brittle. Copper–chromium alloys are an exception. Tensile strength properties of 517–551 MPa (75,000–80,000 psi) in the heat-treated worked condition are not unusual. These alloys have a minimum electrical conductivity in this condition of 80% IACS. Chrome copper has a fair fatigue resistance, and an endurance limit of 172–207 MPa (25,000–30,000 psi). The material also has a wear resistance of about three times that of hard-drawn copper.

The copper–chromium alloys have gained considerable standing in industry, and are used for resistance-welding electrodes, structural members for electrical switchgear, current-carrying members, and springs. These alloys are well suited for the fabrication of resistance-welding electrodes because they have excellent thermal and electrical conductivities and sufficiently high strength, hardness, and resistance to flow at slightly elevated temperature to withstand the combination of pressure and temperature to which they are subjected.

Zirconium. Copper–zirconium is included with this group of materials because it responds to heat treatment; however, its strength is primarily developed through strain hardening or the application of cold-work. Heat treatment primarily restores high electrical conductivity and ductility, while increasing surface hardness.

A tensile strength of 482 MPa (70,000 psi) at room temperature coupled with a conductivity of 88% IACS can be developed. Copper–zirconium is especially well suited where strength must be retained along with high conductivity at elevated temperatures.

The recommended fabricating procedure for processing of copper–zirconium rod is to anneal in solution at 900–925°C, preferably using a bright or slightly reducing atmosphere (to prevent internal oxidation) followed by a water quench. Cold-work is then applied to obtain the desired strength and, finally, the material is aged at 370–425°C dependent upon strength and conductivity requirements.

The alloy can be used in the production of resistance-welding wheels and tips, stud bases for power transmitters and rectifiers, and commutators and electrical switchgear for use at higher temperatures.

Copper, Nickel, and Silicon or Phosphorus. Precipitation hardening in the copper–beryllium and copper–chromium alloys involves the precipitation of a single element. The precipitation of compounds can also be responsible for the heat-treatable behavior of certain copper alloy systems. Two such systems are the copper–nickel–silicon and copper–nickel–phosphorus alloys.

Copper–nickel–silicon alloys respond to heat treatment because of the precipitation of a nickel–silicon phase. The typical composition for such a copper alloy is 2.0% Ni, 0.6% Si.

Tensile strength of 724 MPa (105,000 psi) can be developed coupled with a yield strength of 606 MPa (88,000 psi), 20% elongation in 5 cm, a contraction of area of 48%, and an electrical conductivity of 32% IACS. The alloy may be furnished in a so called cold-forming temper that is suitable for severe cold-forming application performed after the solution annealing. Such material can be heat-treated after forming to achieve the precipitation hardened condition. This alloy has properties for applications similar to phosphor bronze, silver bronze, and related alloys.

Copper–nickel–phosphorus alloy is precipitation-hardenable because of the separation of a nickel phosphorus phase with decreasing temperature.

Wire drawn from an alloy containing 98.65% Cu, 1.1% Ni, and 0.25% P has a tensile strength in the soft solution-annealed condition of 262 MPa (38,000 psi), and an electrical conductivity of 32% IACS. After aging, the strength increases to 448 MPa (65,000 psi) and the conductivity of 68% IACS is attained. Cold-work can be applied to the heat-treated condition, and for severely drawn wire, a tensile strength of about 827 MPa (120,000 psi) can be obtained. The material can be useful for springs, clips, high strength electrical conductors, bolts, nails, and screws. The alloy offers a combination of high strength, high electrical and thermal conductivities, high resistance to fatigue and creep, and good formability.

Copper, Iron, and Phosphorus. Copper–iron–phosphorus alloy UNS C 19500 is precipitation-hardenable because of the formation of an iron-phosphorus phase, probably Fe_2P, with decreasing temperature. The alloy has moderate strength and moderate electric conductivity suitable for many electrical spring applications. The tensile strength in the annealed precipitation-hardened condition is 551 MPa (80,000 psi) and, when precipitation hardened and rolled to a super spring condition, the tensile strength is ca 669 MPa (97,000 psi).

Iron in the range of 1 to 2% in combination with 0.03–0.3% of phosphorus significantly increases the hardness and strength of copper although a moderate loss of electrical conductivity is experienced. Where higher strength copper is required the decrease in conductivity is not considered to be too severe.

Several copper–iron–phosphorus alloys have been developed, one of which is furnished in the precipitation hardened condition. The alloys find a variety of uses such as automotive hydraulic brake lines, flexible hose, electrical terminals, circuit-breaker components, contact springs and welded condenser tube. The alloys are listed in Table 6.

The capability for being either hot- or cold-formed or welded is excellent. All the alloys can be joined easily by soldering, brazing, oxyacetylene welding and by butt-resistance welding.

Electrical conductivity of iron–phosphorus alloys is 50–65% IACS, depending upon chemical composition. The mechanical strengths of the alloys of UNS numbers C 19200 and C 19400 are ca 310 MPa (45,000 psi) in the annealed condition and ca 531 MPa (77,000) for extra spring temper in strip form. Alloy UNS C 19500 in strip form has a tensile strength of 551 MPa (80,000 psi) in the precipitation heat-treated condition and 641 MPa (93,000 psi) when precipitation hardened and cold-rolled to a spring temper. The tensile strength of electrolytic tough-pitch copper alloy UNS C 11000 is ca 234 MPa (34,000 psi) in the annealed condition and 393 MPa (57,000 psi) in the extra spring temper.

Copper–Zinc Alloy Brasses

Copper and zinc melted together in various proportions produce one of the most useful groups of alloys known as the brasses. Six distinct phases are formed in the complete range of possible compositions. The relationship between composition and phases, alpha, beta, gamma, delta, epsilon, and eta are graphically shown in Figure 14. Alloys having less than ca 58% Cu have limited commercial application because they become increasingly difficult to fabricate. The alloys containing more than 58% Cu are characterized by their high ductility and malleability, good strength, pleasing color, and excellent resistance to most corrosive media. More complex alloy systems containing various other elements in addition to zinc are produced commercially; they

COPPER ALLOYS (WROUGHT)

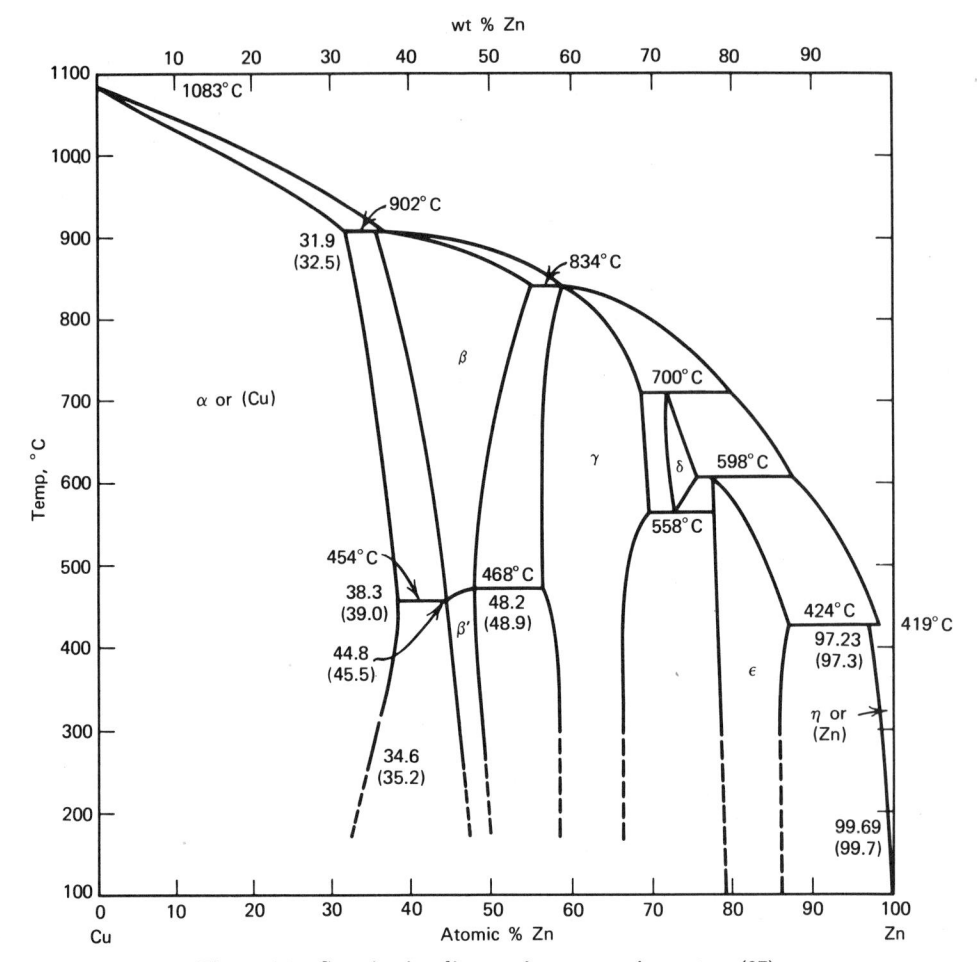

Figure 14. Constitution diagram for copper–zinc system (27).

are treated below. As the copper content decreases from 100%, the brasses vary in color from rich bronze (90% Cu, 10% Zn) to golden (35% Cu, 15% Zn), to yellow (70% Cu, 30% Zn), to reddish-yellow (60% Cu, 40% Zn). Table 14 defines the chemical composition of the principal commercial binary alloys.

All the commercially important brass alloys are cold-workable, some to a greater extent than others. The alloys may be hot-worked at certain optimum temperatures. The hot-working range is somewhat limited by a loss in ductility that occurs to a varying degree for the various alloys in the system. The degree of the property change is quantitatively indicated in Figure 15, which shows the effect of temperature on some of the mechanical properties for one of the most important of all the alloys in this binary series, cartridge brass (70% Cu, 30% Zn). Although the ductility as measured by contraction of area is at its lowest value at ca 315°C, enough ductility remains in the alloy to allow it to be rolled at this temperature without failure occurring. However, if a part were to be drawn from a blank at this temperature, trouble could be expected. The condition is accentuated by the presence of small amounts of insoluble elements (eg, lead) or compounds that are above their own melting point at the temperature

Table 14. Brasses

Name	UNS designation	Composition, wt % Copper	Zinc
Muntz metal	C 28000	62	38
yellow brass	C 26800	66	34
cartridge brass, 70%	C 26000	70	30
low brass, 80%	C 24000	80	20
Fourdrinier brass	C 23500	83	17
red brass, 85%	C 23000	85	15
jewelry bronze	C 22600	87.5	12.5
commercial bronze, 90%	C 22000	90	10
gilding, 95%	C 21000	95	5

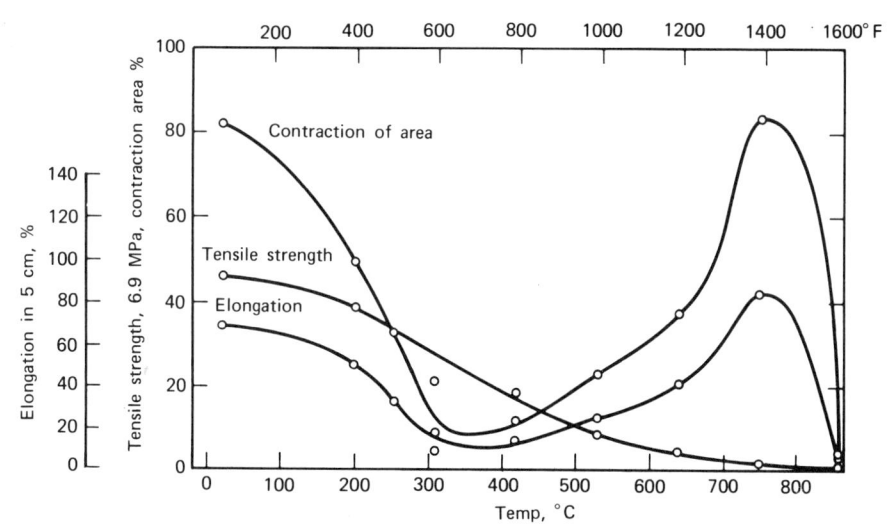

Figure 15. Effect of temperature on some properties of cartridge brass. Diameter of test section, 0.90 cm; approximate testing speed, 1.25 cm/(cm·min); composition, 70.48% Cu, 0.020% Pb, less than 0.005% Fe, and 20.50% Zn. Courtesy of H. L. Burghoff, Chase Brass and Copper Company, Inc.

of hot deformation. Because of the presence of lead, the hot-piercing process for the production of seamless tube must be done at a carefully controlled temperature. The hot-piercing operation demands the utmost from the metal being worked upon because of the nature of the forces developed in the process.

Two broad classifications of copper zinc alloys are commercially important: those containing 65–99% Cu consist of a single phase and are known as the alpha brasses; and those containing ca 55–64% Cu contain two phases, and are known as the alpha–beta brasses.

Properties of the brasses vary with condition or temper and composition. The properties can be made to vary with the degree of deformation, the deformation temperature, and the time and temperature of annealing.

Cold-working the brasses has a marked effect upon their electrical conductivity. The conductivity of brass alloys of compositions 60–100% Cu decrease regularly as the percentage reduction increases to 84% (see Fig. 16) (35). There are slight changes in conductivity for compositions of 90–100% Cu, for a given cold reduction, more abrupt changes for 80–90% Cu, and relatively less change for alloys containing less than 80% Cu.

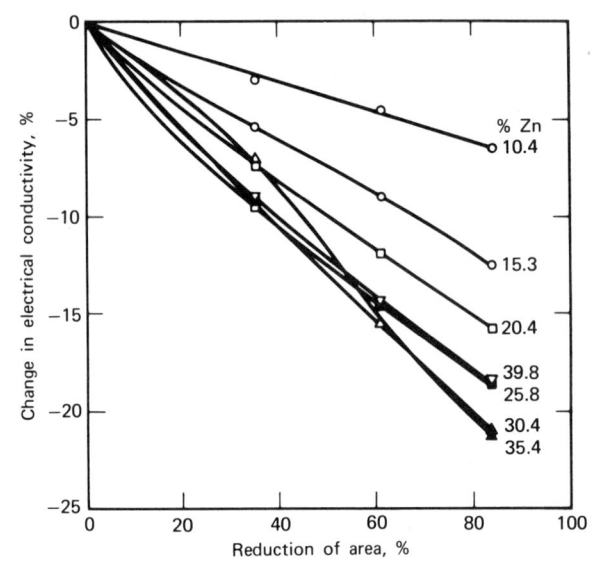

Figure 16. Changes in electrical conductivity of copper–zinc alloys as annealed and drawn (35).

Other properties of copper-zinc alloys such as color and susceptibility to stress corrosion can change rapidly in the 80–90% copper range.

Toughness as measured by impact strength for annealed material is considered to be high for the entire nonleaded-brass series, increasing with zinc content to a maximum value of 30% Zn. It decreases thereafter, with a marked reduction with the appearance of the beta phase. Cold-working reduces toughness, although not to the point of brittleness; it is most effective in the alpha solid solution range rather than in the alpha–beta range. Table 15 illustrates these facts quantitatively. Toughness and strength are actually increased with no evidence of brittleness as temperature decreases to −130°C. At elevated temperatures, property values generally decrease, with 204°C being a safe upper limit for most copper and copper alloys.

Table 15. Toughness of Copper and Some Copper-Base Alloys

Name	Composition, wt %	Microstructure	Temper	Tensile impact energy[a], J
high-conductivity phosphorus- deoxidized copper	99.95% Cu, 0.01% P	all alpha	annealed drawn 60%	50.2 10.8
cartridge brass	70% Cu, 30% Zn	all alpha	annealed drawn 60%	119.3 17.6
Muntz metal	59% Cu, 41% Zn	alpha and beta	annealed drawn 60%	89.5 38.0

[a] For 2.6 mm diameter wires (specimen 10.4 cm between grips). To convert J to cal divide by 4.184.

Alloys containing only small amounts of zinc (less than ca 2%) are sometimes deoxidized with other elements such as phosphorus. Zinc is a powerful deoxidizer with respect to copper; it reacts readily with oxygen in the melt to form zinc oxide. Zinc oxide is insoluble in molten brass as well as in the solid state and is skimmed off before the melt is poured.

Elements added intentionally to improve special properties for particular reasons include lead, tellurium, tin, iron, manganese, phosphorus, aluminum, and nickel. A brass may be modified by the addition of one or more of these elements. The effects of these various elements are treated individually below. Tables 16 and 17 show compositions of some commercial brasses containing additional elements.

Table 16. Lead-Bearing Brasses

Name	UNS designation	Copper	Zinc	Lead
architectural bronze	C 38500	57.0	40.0	3.0
bright extruded bronze[a]	C 38000	58.5	38.7	2.5
forging brass	C 37700	60.0	38.0	2
free-cutting brass	C 36000	62	34.75	3.25
high-leaded brass	C 35300	63.5	34.6	1.9
high-leaded brass (tube)	C 33200	66.5	31.9	1.6
medium-leaded brass	C 34000	64.5	34.5	1
low-leaded brass (tube)	C 33000	66.5	33	0.5
leaded commercial bronze	C 31400	89	9	2
nickel-leaded commercial bronze[b]	C 31600	90.25	6.9	1.75

[a] 0.3% Aluminum.
[b] Also contains 1.00% Ni, 0.10% P.

Table 17. Special Brasses and Nickel Silvers

Name	UNS designation	Copper	Zinc	Miscellaneous
manganese bronze A	C 67500	58.5	39.25	1% tin, 1% iron, 0.25% manganese
naval brass	C 46400	60.5	38.75	0.75% tin
leaded naval brass	C 48200	60.5	36.75	0.75% tin, 2% lead
inhibited admiralty		71.5	27.5	1% tin
type B	C 44300			0.035% arsenic
type C	C 44400			0.035% antimony
type D	C 44500			0.04% phosphorus
inhibited aluminum brass	C 68700	77.5	20.5	2% aluminum, 0.035% arsenic
low-tin commercial bronze		90	9.5	0.5% tin
nickel silver 65-10	C 74500	65	25	10% nickel
nickel silver 65-18	C 75200	65	17	18% nickel
nickel silver 65-15	C 75400	65	20	15% nickel
nickel silver 65-12	C 75700	65	23	12% nickel
	C 79400	62.5	18.5	18% nickel, 1% lead

The following discussion concerns the effects of various impurities experienced in normal mill and possibly subsequent customer processing when contaminated alloys are encountered.

Bismuth. This element is not added to copper or its alloys intentionally. It may find its way into copper alloys by way of contaminated scrap. Bismuth is insoluble in the brasses and appears as intergranular films that are conducive to hot- and even cold-shortness. It has a very low melting point (271°C). In alloys containing lead, bismuth is substantially less harmful, and it is suspected that the bismuth is soluble in the lead phase and does not form films in the grain boundaries. A bearing alloy containing 88% Cu, 4% Sn, and 4% Zn containing more than 0.10% bismuth was processed into strip with no problems (see Bearing materials).

Antimony. Antimony is found in some copper to a slight extent, and may also find its way into the brasses from certain brass scrap to which it has been added intentionally or through recycling of fired cartridge cases containing antimony sulfide in the priming charge. Antimony in the brasses may cause hot-shortness. The antimony content that may be tolerated in hot-rolling of brasses is not defined clearly. A wide spread of values has been reported in the literature. The confusion that exists on the subject may be caused by the segregation in castings and the presence of antimony oxide. The maximum tolerance limit for antimony in brasses increases as the zinc content increases and closely parallels the solubility of antimony in brass. Antimony contents in admiralty alloy 70% Cu, 1% Sn, 29% Zn up to ca 0.04% can be hot-extruded into tubing provided the temperature and extrusion speed are controlled. Antimony contents much over 0.05% in the same alloy become very difficult to extrude. Short circumferential cracks develop on the outside of the extruded shell and are referred to as cross checks.

Antimony as well as arsenic and phosphorus is intentionally added to admiralty alloy in concentrations of about 0.03% for the purpose of inhibiting the type of corrosion known as dezincification. The admiralty alloy finds extensive use as condenser tubing, and the inhibited variety is remarkably resistant to failure from dezincification. Many metric tons of inhibited alloy have been in service in heat exchanger equipment without a single failure from dezincification being reported (see Heat exchange technology).

Arsenic. Arsenic does not affect the hot-shortness of the brasses in the amounts usually present as an impurity. The element is intentionally added to admiralty alloy as an inhibitor of dezincification.

Phosphorus. This element does not lead to either hot- or cold-shortness in the brasses. The element present in concentrations of 0.04% in 70% Cu, 30% Zn cartridge brass slightly increases tensile strength and somewhat lowers the elongation. Phosphorus has a very strong effect upon annealing characteristics and significantly increases recrystallization temperature, and leads to nonuniform grain size of the recrystallized material. For cartridge brass of low iron content (less than 0.003%), the initial softening of phosphorus-containing alloys was significantly retarded when only 0.002% phosphorus was present. Even for low temperature annealing, the grain size expected was retarded to about half. Mill experience has demonstrated that amounts of phosphorus as low as 0.005% combined with 0.02–0.03% iron in cartridge brass reduced grain size from an expected 0.100 to 0.055 mm. In the presence of iron, low concentrations of phosphorus have a very strong influence in restricting grain growth at all usual annealing temperatures. Combinations of phosphorus and nickel that may form intermetallic compounds also have a significant effect upon annealing characteristics and grain growth. These effects are more pronounced than is the case if the same concentration of impurity is present alone.

Phosphorus is added intentionally to admiralty alloy as an inhibitor of dezincification, and precautions must be taken in annealing.

Lead. The solubility of lead in the brasses, as in copper, is very slight. Therefore, it appears as a separate phase and, because it has a low melting point in respect to the brass alloy system, it is the last to freeze, is squeezed into the grain boundaries of the alloy, and appears in the form of globules. Small quantities of lead (0.02–0.03%) in the alpha brasses seriously limit hot-rolling procedures. The lead in the grain boundaries causes hot-shortness, whereas in beta brass, it is more uniformly distributed throughout the structure and alloys containing a high amount of beta phase (53% Cu, 32% Zn) and as much as 1% lead have been successfully hot-rolled. Hot-shortness is a relative phenomenon that can be partly controlled by regulating the physical conditions surrounding the hot-working operation. Brasses containing up to 4% lead are successfully hot-extruded commercially. Leaded alloys are cold-worked readily using ordinary reductions. Lead is intentionally added to brasses in nominal concentrations of 0.5–3.5% to enhance their machining properties. In this respect, its effect in concentrations of 3% is pronounced and lead is an essential element in brasses intended for engraving and automatic screw-machine operations. Lead is also desirable in alloys intended for shearing and blanking operations since it promotes clean smooth edges after these operations.

The chip form ordinarily obtained by cutting a leaded brass is short and brittle rather than long, curly, and ductile. The chips tend to free themselves very rapidly from the cutting tool, and leaded alloys are therefore more suitable for machining. The rating of 100% machinability is assigned to Free Cutting Brass, an alloy containing 61.5% Cu, 3% Pb, and 35.5% Zn. The machinability rating of all copper-base alloys is measured in relation to this alloy.

Although such operations as staking, knurling, and roll-threading can be performed on rod stack, the lead content, conditions of the temper of the alloy, and the mechanical attributes of the metal-working process must be carefully controlled.

Annealing of leaded-brasses must be done with care, because the sudden exposure of cold-worked parts to high temperature may result in failure by fire-cracking since these alloys are more susceptible to this type of failure than the nonleaded brasses. Where fire-cracking is a problem, a more gradual exposure to high heat is suggested. Lead has no significant effect upon softening but does retard grain growth.

Tellurium. The addition of tellurium, selenium, or sulfur offers the possibility of improving the machining characteristics of the brass alloys. Tellurium forms insoluble copper telluride and appears as a separate phase in the structure of the alloy, and as in the case with lead, accounts for a freely separating brittle chip.

Copper containing 0.5% Te is almost as free-machining (90%) as free cutting brass; it is free from the tendency to fire crack, and can be hot-worked quite well. Tellurium has the same general effect of improving the machinability of various alpha brasses as it has with copper however, concentrations as high as 1.8% do not improve machinability in the alpha–beta alloys.

Manganese. Manganese does not adversely affect hot-shortness in the brasses. Alloys containing as much as 5% Mn in the 85% Cu range have been successfully hot-rolled. Cartridge brass, 70% Cu, 30% Zn, containing 1% Mn has been commercially hot-rolled.

Alloys containing manganese have commercial application because they can be resistance-welded by using either spot or wheel techniques. The reason for this improvement over ordinary brasses is that manganese decreases the electrical conductivity of the base alloy. For example, 5% Mn added to red brass, 85% Cu, 15% Zn, reduces the conductivity from 37% to 7 or 8% IACS. The alloy has a conductivity comparable to 3% silicon bronze and can be spot- or seam-welded with ease.

Manganese-bearing brasses have a brownish surface oxide. However, these alloys can be bright-annealed and can be cleaned in the usual 10% H_2SO_4 pickle solutions.

Up to 0.13% manganese has an insignificant effect upon initial softening and recrystallization (38). Grain size is restricted in direct proportion to the manganese concentration; however, for smaller grain sizes the effect is small. Cartridge brass containing 0.13% manganese annealed at 510°C was found to have a grain size of 0.035 mm instead of an expected 0.045 mm grain size. Manganese additions of 1% to cartridge brass causes an increase in the annealing temperature of ca 10°C compared with that needed to produce a grain size of 0.050 mm in commercially pure material.

Tin. Tin is intentionally added to the copper–zinc system of alloys in varying concentrations to improve their corrosion resistance, and to increase their strength and hardness. Tin tends to lighten the color of the alloy.

The addition of 0.75% Sn to Muntz metal (60% Cu, 40% Zn) results in the well known naval brass. When prepared from high purity materials and by fabricating with processing schedules designed to produce a very uniform grain structure, an improvement in resistance toward corrosion fatigue is effected. The alloy is used primarily for marine applications, where comparatively high strength is needed coupled with good corrosion resistance. Leaded varieties containing about 2% Pb have improved machinability and are commercially available. The leaded alloys are not suited for extreme bending or upsetting operations.

All of the tin brasses, with the exception of the leaded variety, have excellent hot-working properties and reasonably good cold-working properties. However, the leaded as well as the nonleaded naval brasses are commercially hot-extruded.

The addition of 1% Sn to cartridge brass 70% Cu, 30% Zn, results in an alloy known as admiralty metal. Admiralty was developed in 1890 by the British navy as a condenser tube material which was a more corrosion-resistant alloy than cartridge brass. It has satisfactory corrosion resistance in seawater, although it does tend to dezincify. Inhibited admiralty alloys have been developed and have very remarkable resistance toward dezincification. The inhibited alloys contain small amounts (0.01–0.05%) of phosphorus, antimony, or arsenic.

Admiralty is essentially a cold-working alloy, although it can be hot-extruded into tube or hot-rolled into strip.

The addition of 1–2% Sn to brasses of 85–90% copper content produces greater resistance to nonoxidizing acid attack in these alloys, particularly as encountered in acid mine water. These alloys also have an attractive color and have been used as watch case material. A 90% Cu, 9.5% Zn alloy with 0.5% Sn added has given good service as a bearing material intended to carry light to moderate loads.

Another bearing alloy containing 88% Cu, 4% Sn, 4% Pb, and 4% Zn has long been used as a bearing and bushing material in automotive and aircraft piston engines.

Tin as an impurity in concentrations up to 0.10% in cartridge brass has no effect on the softening incidence, and also has no effect on grain growth at low and moderate annealing temperatures.

Aluminum. Aluminum is added in amounts of 2% to the 70% Cu, 28% Zn composition to produce an aluminum-brass condenser tube alloy for marine and coastal power plant installations.

Aluminum tends to form a self-healing film of aluminum oxide on the surface of the tube, thereby accounting for the high resistance of the alloy to impingement corrosion.

Aluminum has the property of imparting a bright hue to the copper-base alloys. The color effect has been used to good advantage in architectural bronze containing nominally 58.5% Cu, 38.7% Zn, 2.5% Pb, and 0.3% Al. The alloy finds use in sections, butts, extruded shapes, and hinges.

Aluminum up to 0.13% has an effect upon initial softening and recrystallization in 70% Cu, 30% Zn alloy.

Nickel. The alloys of copper, nickel, and zinc are called nickel silvers or German silvers. The commercially important alloys contain 60–65% Cu, 7–30% Ni, and the balance Zn. The series of alloys form two distinct types, ie, the all-alpha structure and the alpha–beta structure. The composition-phase relations are shown in Figure 17. The alloys containing 65% or more of nickel and copper, and 35% or less of zinc consist of the single alpha phase; the alloys containing 55–60% nickel and copper consist of the alpha and beta phases. The all alpha alloys have fcc crystal structure and possess excellent cold-working properties. Therefore, they are used more frequently for those applications requiring ductility in the cold condition. The single-phase alloys have been used for articles to be plated, such as tableware and hollow ware. One of the most important wrought nickel silvers is the 18% nickel silver containing 65% Cu, 18% Ni, 17% Zn. About 1% Pb may be added to this alloy to improve machinability where this property is required.

The two-phase alloys containing the beta phase rapidly work-harden and are not well suited to cold-forming operations. Their hot-working properties are excellent and several nickel-silver alloys are produced by hot extrusion and hot forging. Occasionally, lead is added to these alloys to improve machinability.

Nickel has a very pronounced whitening effect on copper and copper alloys. An alloy with 5% Ni has a yellowish-white color, with 10% Ni a white color with a tinge of yellow, and with 15% Ni and higher the color is white. Nickel is added to copper–zinc

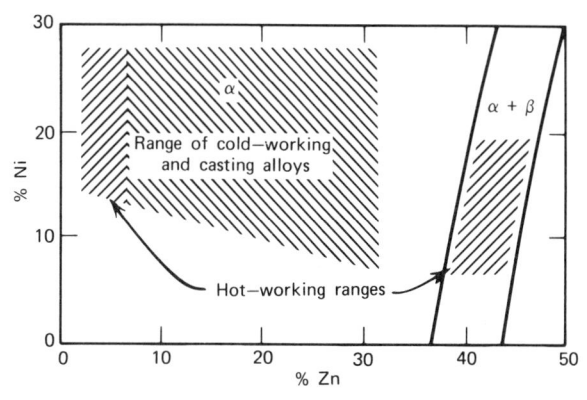

Figure 17. Composition ranges of nickel silvers indicating phase boundaries and hot-work ranges (4).

alloys primarily because of its influence upon the color of the resulting alloys. The nickel silvers are used as a base for silver-plated ware, costume jewelry, hollow ware, screen cloth, welding rod, architectural applications and ornamental work, marine fittings, and for many other applications in which desirable color combinations are required.

Iron. Iron is usually not added to the brass system of alloys except in the few alloys already noted. Iron has long been recognized as being detrimental to the control of grain size in annealed structures. The solubility of iron in brass is variable and depends upon the annealing temperature. The effects of iron on grain growth are related to whether the iron is present in solid solution or as a second phase that hinders grain growth. The iron solubilities at several annealing temperatures are as follows:

Fe solubility, %	Temp, °C
0.3	700
0.15	600
ca 0.05	500

Since iron has a deleterious effect on the annealing of brass, care must be used in the melt shop so that the element is not introduced into the molten alloy. The usual source of iron in the melt shop is free iron in the form of scrap.

A combination of iron and phosphorus has a pronounced effect upon softening and grain growth leading to nonuniform grain sizes.

Corrosion of Copper and Copper-Base Alloys

Copper and its alloys have been used extensively for many years in a variety of corrosive environments. Copper itself possesses very high resistance to the effects of the atmosphere, fresh and seawater, alkaline solutions (except those containing ammonia), and many organic chemicals. The behavior of copper in acids depends mainly on the severity of existing oxidizing conditions. Many salt solutions are handled successfully. Sulfur and sulfides attack copper and high-copper alloys vigorously, and can lead to premature failure. The brasses with high zinc content offer much better resistance toward sulfur-bearing environments than does copper, but they too may be seriously damaged by sulfur-bearing materials (see Corrosion and corrosion inhibitors).

Since the commercially important coppers and copper-base alloys comprise a wide range of chemical compositions, they vary in corrosion resistance. Some of the alloys compare favorably with copper, others have better resistance and still others have less resistance than copper in a specific environment. The alloys may offer improved mechanical properties, and in a specific application, it may be necessary to choose an alloy because of desired mechanical properties at the sacrifice of certain corrosion-resistant properties. For example, manganese bronze may be selected because of its high tensile strength, although the alloy may dezincify. The damage from dezincification in relationship to cross section and strength of the base metal is insignificant.

Many of the alloying elements added to copper improve corrosion-resistance of the parent metal and enhance its mechanical properties.

Extensive successful application of copper in many service areas has included the following broad classifications:

(1) Atmospheric exposures, eg, roofing and related uses, hardware, building fronts, grill work, hand rails, butts, hinges, extruded shapes, forgings, lock bodies, door knobs, and plates.

(2) Fresh water supply lines where superior resistance to corrosion by various types of water is important, and for resistance to corrosion by soil although the materials should not be buried in cinder-fill or marshland where they can come into direct contact with sulfur-bearing conditions.

(3) Seawater applications; usually for water supply lines, heat exchangers, and condensers; also shafting, marine hardware, and valve stems.

(4) Heat exchangers and condensers, including marine equipment and liquid or gas-to-air exchangers.

(5) Industrial and chemical plant equipment involving exposure to a wide variety of conditions.

Copper roofing, in addition to forming a protection from the elements, develops a pleasing green patina consisting essentially of green basic copper sulfate. The basic copper sulfate constitutes a protective layer, and thus enhances the natural corrosion resistance of copper.

The various grades of copper, such as electrolytic tough-pitch, fire-refined tough-pitch, and the deoxidized alloys vary slightly in some physical and mechanical properties; however, these variations do not extend significantly to their corrosion resistance.

The coppers and their alloys may be damaged by the various types of corrosion listed in Figure 18, depending upon service conditions. Most of the listed forms of attack are not peculiar to copper, they occur generally with other materials.

Many factors may be involved in a given environment that influence the rate of reaction such as stress, velocity, effects of galvanic coupling, concentration cells, effect of surface condition, and contamination of the surrounding media. If damage results from pitting, intergranular corrosion, or selective leaching as in dezincification, a measure of corrosion rate based on change in weight may be misleading. For these forms of corrosion, measurement in terms of loss in mechanical strength is more meaningful. It is convenient to classify corrosion by the appearance of the corroded material.

General Thinning. General thinning is an attack that results in the uniform or overall solution of the material with little or no localized penetration. It is the least damaging type of corrosion, because of the lack of localized attack. This form of corrosion is perhaps the only one for which weight loss data correlated with time can be used to estimate penetration with significant accuracy.

Corrosion is general or uniform when it results from long contact with the environment where the rate of attack is very low, as is the case of copper in contact with materials such as fresh, brackish, or salt waters, many types of soil, neutral, alkaline, or acidic salts, organic acids, and sugar juices. Many other materials may produce a uniform, but faster rate of corrosion. Oxidizing acids, sulfur-bearing compounds, and ammonical solutions, where the very soluble copper ammonium complex ion is formed, and cyanide may cause this general thinning type of attack.

Galvanic Corrosion. When dissimilar metals are immersed in an electrically conductive solution, an electropotential difference usually results between the metals. If the metals are coupled electrically, this potential difference produces electron flow. Corrosion of less resistant metal is usually increased and the attack on more resistant

Figure 18. Typical appearances of some manifestations of corrosion attack. (**a**) Impingement attack. Flattened section of red brass illustrating characteristic horseshoe shaped pits associated with impingement attack. High velocity and turbulence of the fluid caused the pitting (3×). (**b**) Deposit attack. Pits are a result of oxygen concentration cells formed under buildup of corrosion product or debris; note typical spherical appearance (3×). (**c**) Galvanic corrosion showing the coupling of a 3.8 cm steel nipple with two sections of red brass pipe. The red brass was not attacked, but the steel was severely corroded (3×). (**d**) Layer type dezincification showing the copper-rich phase in the top half of the illustration, as compared to the normal alloy structure beneath the dezincified layer (75×). (**e**) Plug type dezincification; note the copper-rich rosette surrounded by the normal alloy structure (75×). (**f**) Intercrystalline failure typical of stress-corrosion cracking; failure resulted from combined action of corrosion (contact with mercury) and presence of a stress differential (100×). (**g**) Transcrystalline cracking typical of corrosion fatigue; note the roughened surface

from which the cracks originated. Failure resulted from the presence of corrosive media and cyclic stresses probably caused by vibrations in the equipment (250×). (**h**) Corrosion fatigue external appearance of failed condenser tube (1×). (**i**) Intercrystalline corrosion in admiralty tube (70.37% Cu, 1.15% Sn, 0.04% As, 28.43% Zn). The attack has penetrated a number of grains from the surface, a form of corrosion not usually encountered in copper base alloys (75×). Courtesy of R. E. Ricksecker and V. F. Nole, Chase Brass and Copper Company, Inc.

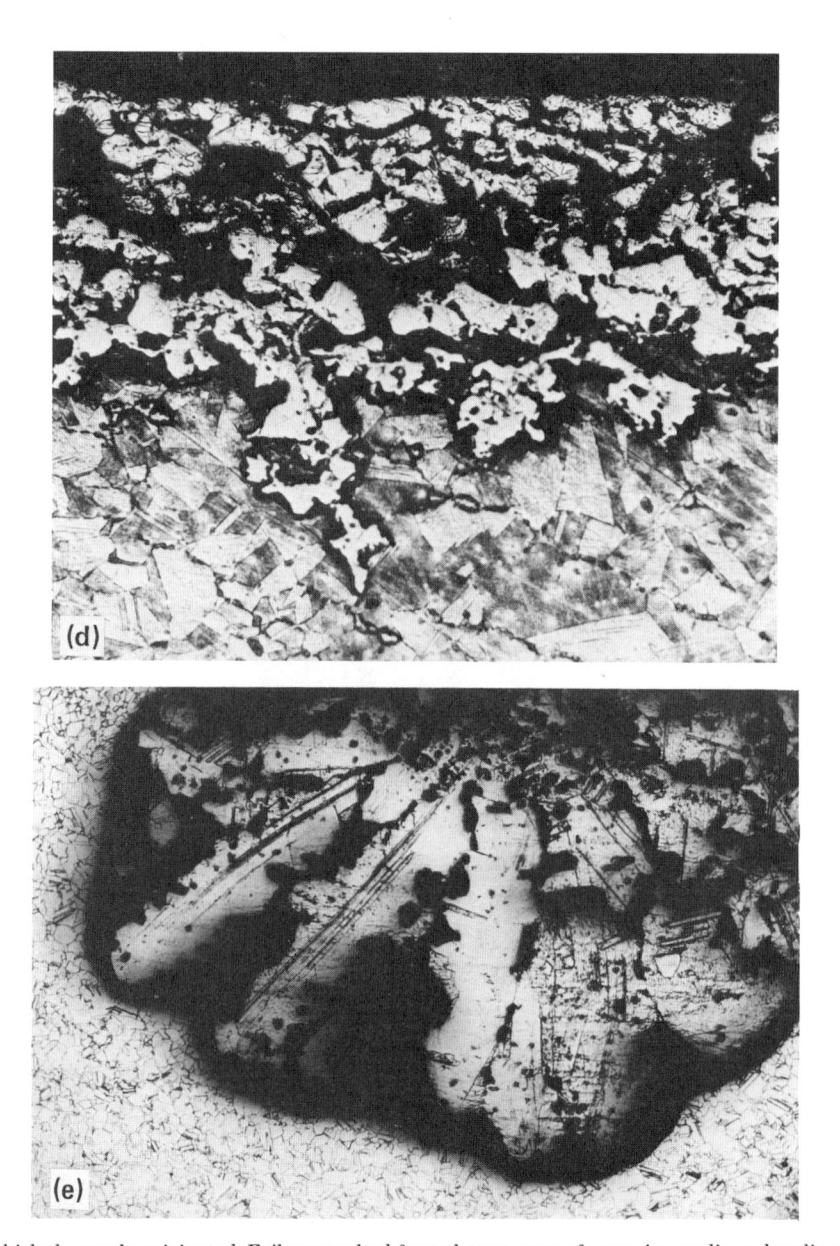

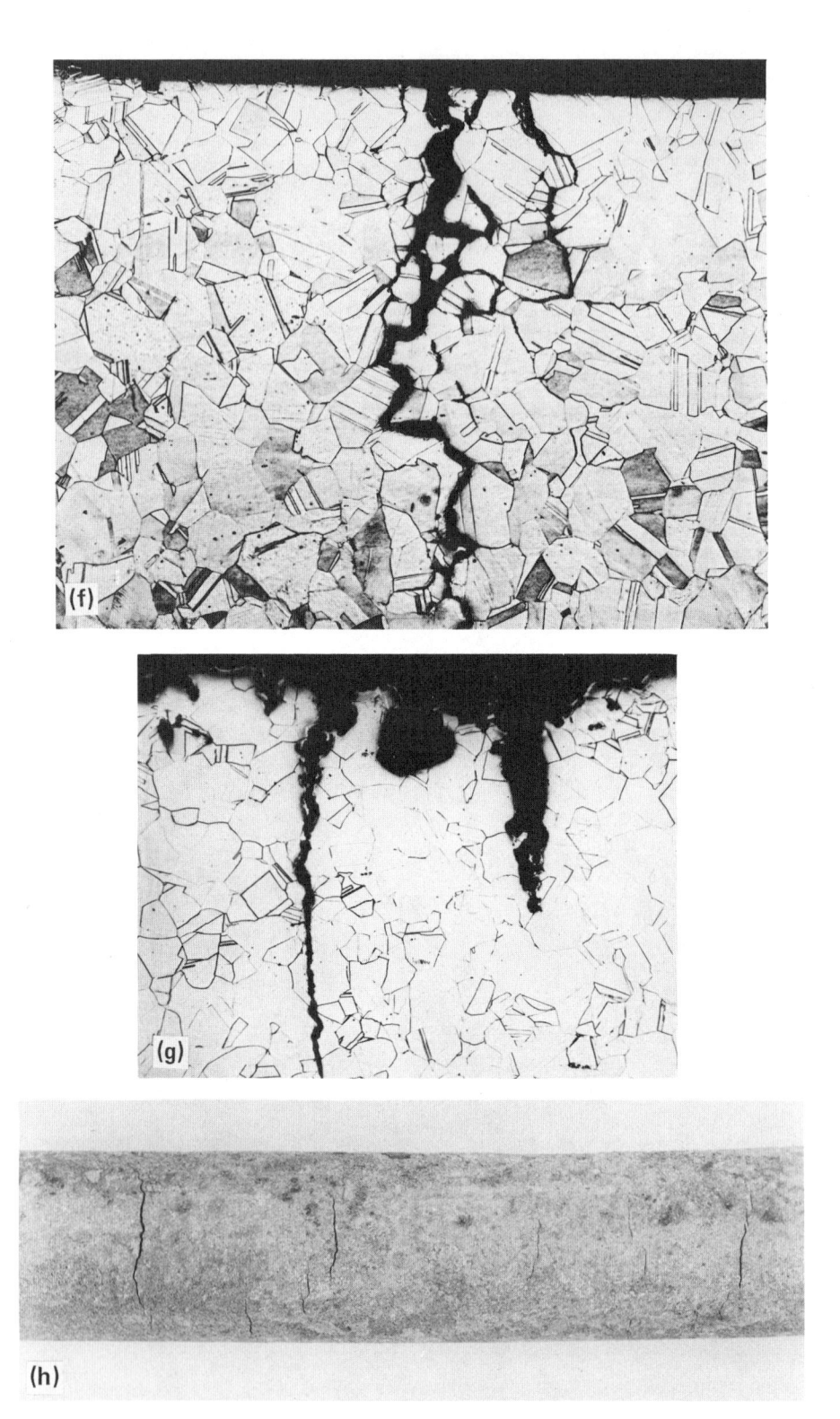

Figure 18. (*continued*)

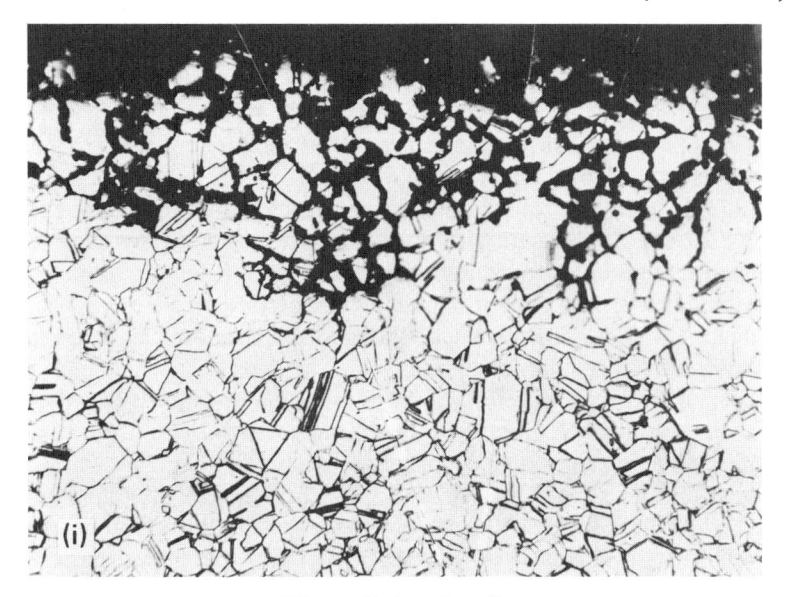

Figure 18. (*continued*)

material is decreased, as compared to the behavior of the same metals when they are not in electrical contact. The less resistant material becomes anodic and the more resistant metal cathodic. Copper and its alloys are cathodic to the commonly encountered materials such as iron, steel, zinc, and aluminum. Coupling with more common grades of stainless steel may lead to variable behavior depending upon exposure conditions.

The conditions necessary for galvanic action are good electrical contact between dissimilar metals, immersion of the couple in a conductive solution, and the presence of oxygen in the electrolyte to act as a depolarizer.

The potential differences between coupled copper and copper alloys is small. The additions of zinc and aluminum to copper move the potentials toward the anodic end of the range, and additions of tin or nickel move the potential toward the cathode end. Galvanic corrosion between coupled copper alloys in an electrolyte is seldom a significant problem, because the potential difference is insignificant.

Pitting. The copper alloys undergo corrosion of the pitting type when exposed to certain conditions of service as do all the commercial metals and alloys. Occasionally pitting is general over the entire surface, giving the metal an irregular and rough appearance, or the pits are concentrated in a specific area and exist in a variety of shapes and sizes.

Pitting attack occurs in a number of forms and these are referred to as deposit attack, waterline attack, crevice attack, impingement attack, and fretting corrosion.

Impingement Corrosion. Impingement attack sometimes called erosion corrosion occurs where gases, vapors, or liquids impinge on metal surfaces at high velocities. The erosive action removes protective films from localized areas on the metal surface, thereby forming differential cells and localized pitting of anodic points of the cell.

Fretting Corrosion. A classic case of fretting corrosion sometimes occurs during the shipment of bundles of free-cutting hexagonal, octagonal, or rectangular brass rod having flat faces. A relief annealing applied to the rod just prior to shipment helps remove traces of process lubricant from the rod faces, enhancing metal to metal contact, and increases the incidence of fretting.

Fretting has been controlled, and in some cases eliminated, in the shipment of brass rod by leaving a lubricant with low viscosity on the surface to reduce friction between interfaces and also to exclude oxygen.

Dezincification. Some copper–zinc alloys are subject to corrosion by a process known as dezincification. Zinc is selectively lost from the alloy, leaving a porous mass of copper having poor mechanical strength.

Corrosion of this type manifests itself in two general ways: uniform or layer-type dezincification and local or plug type. The uniform attack seems to be prevalent in alloys of high zinc content, such as cartridge brass (70% Cu, 30% Zn), and plain or uninhibited admiralty alloy (71% Cu, 1% Sn, 28% Zn). The local or plug type of attack appears to be associated with the more severe service conditions which usually occur at higher operating temperatures. The alpha–beta structural alloys may undergo a selective type dezincification where the beta grains are attacked first, and then the alpha grains may become involved.

Dezincification may occur when uninhibited brasses are in contact with stagnant or slowly moving waters or solutions that have high oxygen and carbon dioxide content.

The resistance toward dezincification in the brasses changes abruptly at ca 80% Cu, 20% Zn. Alloys ranging in composition from 95% Cu, 5% Zn, to 80% Cu, 20% Zn, have very high resistance to dezincification damage. The use of red brass, 85% Cu, 15% Zn, has eliminated dezincification in plumbing pipe systems. Alloys containing more than 20% zinc may dezincify under corrosive conditions, and Muntz metal has the least resistance of the commercial brass alloys.

Tin has some inhibiting effectiveness, especially in cast alloys. Naval brass and manganese bronze, which are alpha–beta brasses containing about 1% tin, are widely used for naval equipment, and have reasonably good resistance to dezincification. Phosphorus, arsenic, and antimony added in small amounts (0.02–0.05%) to admiralty alloy (all alpha phase) are very effective inhibitors of dezincification.

Inhibitors do not prevent dezincification of the beta phase, and hence are not entirely effective in preventing the attack of the alpha–beta alloys.

Dezincification need not be a serious problem since red brass, commercial bronze, and inhibited aluminum brass can be used successfully where this form of corrosion has been experienced or is anticipated.

Parting. Selective solution similar to dezincification occurs in other alloys and is sometimes known as parting. Dealuminification occurs in some copper–aluminum alloys. Decobaltification of cobalt alloys has been reported and denickelification of copper nickel alloys may occur under special conditions.

Intercrystalline Stress-Corrosion Cracking. Stress-corrosion cracking, sometimes referred to as season cracking, is encountered in copper alloys in varying degrees depending upon alloy composition, level of stress in the article, and the enviroment. The failure is most often seen as a ragged crack or system of cracks paralleling the working direction of the deformation needed to form the part. Cracks may follow a zig-zag pattern and eventually form a system of cracks allowing pieces to fall out from the corroding member. Microscopic examination has shown the cracks to be largely intercrystalline in form, although some transcrystalline tearing may also be observed.

Environment. One of two specific classifications of compounds is needed in the environment to cause stress-corrosion cracking; one is ammonia and ammonium compounds, and the other is mercury and mercury compounds. Both oxygen or air and moisture are also needed. Other compounds, such as carbon dioxide, are thought to accelerate cracking in ammoniacal atmospheres. Moisture films on metal surfaces will dissolve significant quantities of ammonia from atmospheres that are low in ammonia concentration. Ammonium hydroxide solutions thus formed can have a very significant amount of ammonia. Cracks experienced in brass exposed to ammonia are usually branching, and they create a broad network pattern of intergranular cracks.

Mercury and mercury compounds can also cause cracking by forming an amalgam on the surface of the brass article and by allowing the stresses to produce cracking. Mercury rarely comes in contact with brass alloys in practice; however, this happens sometimes in heat exchanger or condenser systems when abrupt changes in operating pressure inadvertently cause mercury to enter the equipment from pressure-measuring devices. Failure induced in a mercury environment appear more in a zig-zag line of intergranular cracks rather than as a network, as in the case of ammonia exposures.

Stress. Although test results may indicate that a finished part is free from susceptibility to stress-corrosion cracking, such an indication does not absolutely assure freedom from failure.

The susceptibility is a function of stress magnitude. Stresses near the yield stress of the material are usually required to cause stress-corrosion cracking, although failures have occurred at lower stresses. Depending upon alloy, environment, and grain size, values as low or lower than 6.9 MPa (1000 psi) have been related to failure in ammoniacal atmospheres, whereas stress values of 55–69 MPa (8000–10,000 psi) have been associated with articles that cracked in exposure to mercury.

Alloy Composition and Structure. Usage indicates that brasses containing ≤15% zinc are practically immune to stress-corrosion cracking. Phosphorus-deoxidized copper or tough-pitch copper rarely crack even under the most severe conditions. The alloys containing 40% zinc are highly susceptible to stress-corrosion cracking, but only slightly more so than alloys containing somewhat more than 20% zinc. Elements added to the brasses such as phosphorus, arsenic, magnesium, tellurium, tin, beryllium, and manganese may decrease the tendency toward stress-corrosion cracking under certain conditions.

The effects of structure on cracking are not sharply defined, since they are interrelated with alloy composition and stress factors. In general, the rate of cracking increases as grain size increases.

Control. Several means are available for overcoming failure by stress-corrosion cracking:

(1) Selection of copper alloys that have high resistance to this type of failure; the alloys in the copper–zinc series with ≤20% zinc have very high resistance.

(2) Reduction of the stress differential to safe levels by stress-relief annealing can be accomplished without significantly changing the hardness of the material.

(3) Improvement of the environment such as, eg, replacing ammonia boiler water treatment with other water treatment procedures. Ammonia does build up to alarming and most damaging concentrations in certain areas of condensers and after-coolers although the addition of ammonia to the boiler water system is thought to be very small.

COPPER ALLOYS (WROUGHT)

(4) Use of coatings such as electroplates, lacquers, or paint that separate the material from the environment. Such coatings must be maintained and must be impermeable (see Coatings, resistant).

Residual and assembly stresses may be eliminated by a recrystallizing annealing after forming. However, annealing is not always possible because the rigidity in strain hardening may be a service requirement that would be lost. A thermal treatment referred to as a relief anneal (although annealing as such is not involved), may be applied to reduce the stresses present in a given part. The treatment consists of heating the brass for a relatively short time at a temperature below the recrystallization point. Safe times and temperatures vary with alloy composition, severity of cold deformation, and the prevailing stresses.

Typical stress-relieving times and temperature for a few alloys are given in Table 18. The exact thermal treatment for a specific part should be established by examining the article for residual stresses through the application of the mercurous nitrate or ammonia test. The thermal treatment and time for a given size furnace load should be adjusted until the test is satisfactorily passed. Parts treated in layers may not reach the desired temperature in the center of the load in the same time as do the parts at or near the top or sides.

Table 18. Stress Relief for Alloys

Alloy	UNS designation	Temp, °C	Time, h
commercial bronze	C 22000	204	1
cartridge brass	C 26000	260	1
Muntz metal	C 28000	191	½
admiralty	C 44300, C 44400, C 44500	302	1
cupro-nickel, 30%	C 71500	427	1
phosphor bronze, 5% or 10%	C 51000, C 52400	191	1
silicon bronze, 3%	C 65500	371	1

Two tests have been devised to indicate the liability of a given part to stress-corrode by cracking. The simplest method is the standard test that is made by immersing the clean test piece in an acidified solution of mercurous nitrate (10.7 g of $HgNO_3 \cdot H_2O$ and 10 mL HNO_3/L soln) for a definite time and then examining for cracks (ASTM Specification B154-73). If the part passes the test, its susceptibility to season crack is fairly low.

Another test consists of exposing the specimen to a gaseous atmosphere of ammonia, moisture, and oxygen or air for a period of as much as 24 h before examining for cracks. Materials passing this test are considered to be safe for commercial application and relatively free from residual stresses.

The mercurous nitrate and ammonia in each of the tests remove the surface of the test material and allow unbalanced stress conditions present in the article being tested to manifest themselves in the form of cracks. Specimens that do not crack after exposure to either the mercurous nitrate or ammonia are free from the tendency to stress-corrosion cracking in service unless the stress patterns have been dangerously altered during use. The ammonia test is more sensitive than mercurous nitrate.

Intercrystalline Corrosion. Allied to stress-corrosion cracking is still another form of intercrystalline attack that is referred to as intercrystalline corrosion. The metal is penetrated along its grain boundaries and stress is not a factor. Microscopic examination reveals that the attack extends to a depth of several grains from the surface which distinguishes it from plain surface roughening. Intercrystalline corrosion is seldom encountered except in handling high pressure steam. Muntz metal, the admiralties, aluminum brass, and silicon–bronze are apparently the alloys most susceptible.

Corrosion Fatigue. A combination of corrosion, usually of the pitting type, and cyclic stresses applied externally may result in corrosion fatigue cracking. Cracking of this type is characterized by its localized nature and abruptness of the failure with very little distortion of the part as a whole. Condenser tubing fails by corrosion-fatigue cracking within a matter of months, whereas the expected life for the corrosive condition alone is usually years.

Copper and copper alloys offer good resistance toward failure by corrosion fatigue in many applications that involve repeated stressing and corrosion. These applications include such parts as springs, switches, diaphragms, bellows of the syphon type, pipes and tubing in gasoline, and oil lines of aircraft and automobiles, tubes for condensers and heat exchangers, and Fourdrinier wire in the paper industry. Some of the alloys frequently used for service under conditions favorable for the occurrence of corrosion-fatigue cracking include beryllium copper, phosphor bronze, aluminum bronze, and cupro–nickel.

Fouling. As a group, the copper alloys tend to resist fouling by marine organisms, although the degree of resistance varies among the various alloys. Alloys with greater than 85% copper (excepting the aluminum bronzes), offer the greatest resistance to fouling and are selected when this property is of prime importance. They function by slowly releasing copper ions into the aqueous environment that poisons the barnacles and other organisms. The copper or copper alloy must be allowed to corrode freely at a rate of ca 25 μm/yr (1 mpy). Some observations of fouling resistance in cupro–nickels has been made in cases where the instantaneous corrosion rate is on the order of 2.5 μm/yr (0.1 mpy) which implies that the corrosion product film itself is resistant to barnacle and other creature attachment. The corrosion rate of the copper alloys is so low that they can only protect themselves from fouling, and do not release enough copper ions into the seawater to protect adjacent noncopper surfaces (see Coatings, marine).

Effect of Alloy Composition. *Copper–Zinc Alloys (Brasses).* The most widely used group of copper alloys are the copper zinc series with zinc content as high as 42%. The resistance of these alloys to corrosion by aqueous solutions does not change significantly until the zinc content exceeds about 15%; then dezincification can occur. Stagnant or slowly moving salt solutions, brackish or seawater, or mildly nonoxidizing acidic solutions may lead to dezincification in the susceptible alloys.

The occurrence of stress-corrosion cracking of the copper zinc alloys follows much the same pattern as dezincification, that is, the alloys of higher zinc concentrations are the most susceptible. A marked increase in resistance toward stress-corrosion cracking occurs above 80% Cu, although the increase in this instance is not as pronounced as for increased resistance toward dezincification. Red brass (85% Cu, 15% Zn) offers good resistance to this phenomenon, and season cracking of copper is virtually unknown.

Where exposure to sulfur compounds is involved, alloys containing the highest amounts of zinc have the best resistance. The inhibited admiralty alloys have been used extensively for heat exchangers in oil refineries where corrosion from sulfur compounds and contaminated water may be very severe. There is no abrupt change in sulfur attack in the 80–85% copper range. However, red brass does have better resistance toward corrosion by sulfur attack than copper, but the difference is not significant enough to recommend red brass over copper for handling sulfur-bearing materials of the sulfide type.

Lead, tellurium, silicon, and manganese are added to the brasses to enhance some mechanical property, and have little effect upon corrosion resistance.

Tin-Bearing Brasses. Tin added to certain brasses significantly increases the resistance toward acidic media as well the tendency to dezincify. The addition of 0.75% to Muntz metal (60% Cu, 40% Zn) results in the familiar naval brass alloys, all of which have improved resistance to damage by dezincification, although under severe conditions dezincification does occur. The beneficial effects of tin in manganese bronze (58% Cu, 1% Sn, 1% Fe, 0.1% Mn, 39.9% Zn) are probably offset by the presence of iron. The addition of 1% Sn to cartridge brass forms admiralty alloy with improved resistance toward dezincification, but does not prevent damage from this source under severe conditions.

Aluminum Bronze and Brass. Copper alloys containing from 5–12% aluminum have excellent resistance to impingement attack and also to high temperature oxidation. Aluminum bronze is used for beater bars and blades in wood pulp machines because of its high resistance to mechanical abrasion as well as to attack by sulfite solutions.

Aluminum additions to brasses, as well as to copper, tend to form self-healing films on the surface of the alloy. The film contains aluminum oxide which markedly increases the alloy's resistance toward impingement damage. The films are fairly resistant to acid conditions, but as expected dissolve readily in alkalies.

Muntz metal containing 2% Al is used as a propeller bronze. Aluminum added in concentrations ≤2% to cartridge brass (70% Cu, 30% Zn) produces aluminum brass, a condenser tube alloy used in marine and coastal power plant installations. The alloy has significantly greater resistance to impingement than the admiralty alloys. Handling clean seawater, admiralty has impingement resistance to flows up to 1.8 m/s, whereas aluminum brass has resistance to flows up 2.4 m/s.

Aluminum brass is susceptible to dezincification unless the alloy is inhibited by the addition of a small amount (0.05%) of arsenic.

Phosphorus, Arsenic, and Antimony. The additions of very small amounts of any one of these three elements (0.02–0.10%) to the alpha-structured brass alloys containing more than 2.0% zinc eliminates or greatly reduces the incidence of damage by dezincification. Hundreds of thousands of metric tons of admiralty alloy and aluminum brass have been safeguarded over years of service by the addition of one of these elements with occurrences of only superficial dezincification.

Phosphor Bronzes. The phosphor bronze alloys, each of which is copper-containing up to 10% Sn and deoxidized to contain 0.25% P, are comparable in respect to the coppers in general corrosion resistance. The tin does improve the resistance to most nonoxidizing acids except hydrochloric and the higher tin alloys have been reported to have high impingement-attack resistance. The phosphor bronze alloys have good resistance toward damage by stress-corrosion cracking. The behavior of these alloys toward sulfur is about the same as that for copper.

Copper–Silicon Alloys. Silicon added to copper enhances its resistance in acidic environments, probably because of oxide films containing SiO_2 on the surface of the copper. Silicon does not appear to alter the general behavior of copper in corrosive environments. There is some indication that 3% silicon bronze is corroded intergranularly by high temperature steam.

Cupro–Nickel Alloys. Of the three alloys of copper with nickel additions (10%, 20%, and 30% Ni), the 10% and 30% cupro–nickels are more common. All three alloys contain additions of both iron and manganese, and perhaps offer the best general resistance to aqueous solutions of all the commercially important wrought copper alloys. The 30% cupro–nickel has the best general-corrosion resistance of all, although 10% cupro–nickel may be an economical choice. The resistance to impingement attack is superior to that for aluminum brass and is of prime importance because these alloys allow greater cooling-water velocities in heat exchanger equipment that can lead to better unit operating efficiency. These alloys have very high resistance to stress-corrosion cracking. The behavior in sulfurous conditions is perhaps as good as that for the admiralty alloys.

Nickel Silvers. The two most common nickel silver alloys contain 65% or 55% Cu, 18% Ni, and the remainder zinc. These alloys have good resistance to corrosion by fresh and salt water. The resistance to corrosion in contact with saline solutions is better than that of the brasses of the same copper content, because the relatively large amount of nickel inhibits dezincification.

Protective Coatings. The relatively high general resistance of copper and its alloys to corrosion in many kinds of environments is a result of the protective reaction products developed by converting the surface of the metal or alloy into one or more compounds. The application of various protective coatings may be used to enhance the natural resistance of the materials.

The coatings may be either metallic or organic in nature (see Coatings, industrial; Paint; Metallic coatings). Coatings must be able to resist corrosion adequately, and success in service may depend upon impermeability, continuity, and adhesion to the base metal. Metallic coatings may make the electropotential relationship of coating to base metal important, especially at uncoated or cut edges.

Tin, lead, and solder are used as metallic coatings and are usually applied by hot dipping, although electroplating is also used. Tin arrests corrosion caused by sulfur in rubber insulation in contact with copper wire conductors and cable. Lead coated copper roofing material is resistant in contact with flue gases or other products that contain dilute sulfuric acid. Tin or lead-coated copper is used for atmospheric exposure. These coatings serve primarily for architectural effects, since the atmospheric resistance of bare copper is excellent.

Chromium plate is used mostly for decoration, improvement of wear, or for reflectivity. Chromium plates are porous and, for protection, should be applied over other plates such as nickel.

Clear lacquer is sometimes applied to preserve a bright copper or alloy color for decorative reasons. Incralac developed by the International Copper Research Association is an excellent coating for such purposes. Recently, it has been effectively applied to copper roof installations to maintain the warm color of copper.

Behavior of Copper Alloys in Various Media

Corrosion-accelerating rate factors may vary from one situation to another, accordingly it is advisable to conduct preliminary service or field tests under actual operating conditions before purchasing large quantities of any material. Service data from similar situations are also most valuable. All rate data given below are to be considered as guides only, since various conditions can accelerate attack very appreciably.

Atmospheric Exposure. Copper and copper alloys give years of excellent service and resistance toward corrosion by industrial, marine, and rural atmospheres. The most widely used copper materials for atmospheric exposure are copper, commercial bronze, red brass, architectural bronze, and the nickel silvers. Copper is a satisfactory material for roofing, flashings, gutters, and downspouts.

Soils. Copper has a high resistance to corrosion by soils. Data compiled by the National Bureau of Standards (Circular 579) compare the behavior of copper, zinc, lead, and iron in the following four types of soils, considered to be representative of most soils in the United States:

(1) Well-aerated acid soils low in soluble salts (Cecil clay).
(2) Poorly aerated soils (Lake Charles clay).
(3) Alkaline soils high in soluble salts (Docas clay).
(4) Soils high in sulfides (rifle peat with sulfide).

For exposures of more than 14 yr, copper had the lowest corrosion rate in all four soils.

The highest rate of corrosion for copper occurred in sulfide soils. Tough-pitch copper, deoxidized copper, silicon bronzes, and red brasses are corroded by soils containing cinder filled with high concentrations of sulfides, chlorides, or hydrogen ions.

Where local soil conditions are unusually corrosive, it may be advisable to employ some means of protection, eg, cathodic protection, neutralizing backfill (such as limestone), or protective coating or wrapping.

Fresh Water Exposure. Copper is used extensively for handling fresh water. Compared to steel, construction costs are lower for copper joined with any of several types of fittings.

The greatest single application of copper tubing is in hot and cold distribution lines, although significant quantities have been used in heating lines including radiant heating for homes, for drainage, and fire safety systems.

Minerals in water combine with dissolved carbon dioxide and oxygen and react with copper to form protective films. The rate of corrosion in most cases is low (5–25 μm/yr or 0.2–1.0 mpy).

The admiralty alloys, aluminum brass, and cupro–nickels all have high resistance to fresh water and, under conditions found in condensers and heat exchangers, may have corrosion rates of <25 μm/yr (1 mpy) under proper operating conditions of the equipment.

Steam. Copper and copper alloys are resistant to attack by pure steam (qv), but if much carbon dioxide, oxygen, or ammonia is present, the condensate becomes corrosive.

The cupro–nickels are preferred for the high temperatures and pressures found in boiler feed-water systems.

Steam Condensate. With proper treatment, steam condensate rendered relatively free of noncondensable gases is relatively noncorrosive to copper and copper alloys. Such conditions can exist in power generating stations. Corrosion rates in many such exposures are <2.5 μm/yr (0.1 mpy). Condensates contaminated with oil, possible where reciprocating engine condensate is involved, are virtually noncorrosive to copper alloys because of the oil film formed on the metal surface.

The rate of attack is significantly increased by condensates containing carbon dioxide or oxygen, or both. Condensates containing 5 ppm of oxygen and 15 ppm of carbon dioxide have been known to have a corrosion rate of 0.18–0.35 mm/yr (7–14 mpy) in contact with copper alloys.

The amount of ammonia present in a condensing system is a very significant factor influencing the life of copper alloys. The practice of treating boiler water with ammonia has led to contamination and subsequent corrosion or cracking in the condensing and heat exchanger equipment.

Salt Water. Copper alloys find use in handling seawater in ships and tide water power stations. Copper is usually less resistant than the inhibited admiralty alloys, aluminum brass, or cupro–nickel alloys. The superior behavior of these alloys is the result of their insolubility in seawater, and their ability to form films resistant to impingement attack.

The corrosion rate of copper and its alloys in relatively quiet seawater is usually <50 μm/yr (2 mpy). The cupro–nickels and aluminum brass often experience corrosion rates <25 μm/yr (1 mpy). The instantaneous corrosion rate of copper alloys tends to decrease as the duration of the exposure increases. Localized pitting with penetration may increase by a factor of 5 under deposits or debris, or adjacent to crevices. Local dezincification or dealloying of uninhibited high zinc brasses and certain grades of aluminum bronzes may occur at rates >250 μm/yr (10 mpy).

Acids. Copper is widely used for industrial equipment handling acid solutions. In general, copper alloys are used in contact with nonoxidizing acids such as acetic, sulfuric, and phosphoric if the concentration of oxidizing agents such as dissolved oxygen (air), or ferric or bichromate ions is low.

Acids that are oxidizing agents themselves, eg, nitric, sulfurous and hot concentrated sulfuric acids, and carry oxidizing materials such as ferric salts, bichromate, or permanganate ions, cannot be handled in copper alloy equipment.

For dilute solutions of acids <1%, the corrosion rate in contact with copper is relatively low (usually <250 μm/yr or 10 mpy).

The nonoxidizing acids except hydrochloric acid that contain as much air as is experienced in quiet contact with air are weakly corrosive. The rates may be of the order of less than 25–250 μm/yr (1–10 mpy). Air-saturated solutions of nonoxidizing acids are likely to be strongly corrosive, having a rate of attack from perhaps 200–1250 μm/yr (8–50 mpy) in contact with copper alloys. The rate is higher for hydrochloric acid. The actual rate of corrosion depends on concentration, temperature, rate of movement, and other factors difficult to classify.

Nonoxidizing acids containing almost no air have virtually no corrosive effect on copper alloys. Corrosion attack rates for 1.2 N sulfuric, hydrochloric, and acetic acids in absence of air are <3.8 μm/yr (0.15 mpy).

COPPER ALLOYS (WROUGHT)

Comparison corrosion rates for 3 acids are:

Acid	Corrosion rate, μm/yr (mpy)
nitric, 32%	2.3×10^5 (9,000)
hydrochloric, conc	760 (30)
sulfuric, 17%	100 (4)

Under similar conditions phosphoric, acetic, tartaric, formic, oxalic, malic, and similar acids are expected to react in a manner comparable to sulfuric acid.

Copper and copper alloys are used in the form of heat-exchanger tube, pipe, and fittings handling phosphoric acid, although some of the rates may be relatively high. The concentration of phosphoric acid seems to have less effect on the rate of attack than does the concentration of impurities. The impure phosphoric acid produced by the sulfuric acid process may contain markedly higher concentrations of ferric sulfate, sulfite, chloride, and fluoride ions than acid made by the electric furnace process. These impurities can increase the corrosion rate by a factor of 150.

Hydrochloric Acid. Hydrochloric acid is one of the most corrosive of the nonoxidizing acids in contact with copper alloys, and is successfully handled in dilute solutions (see Hydrogen chloride). The corrosion rate of cupro–nickel in 2 N HCl at 24°C is 2.3–7.6 mm/yr (90–300 mpy), depending upon the degree of aeration and other factors.

Hydrofluoric Acid. Hydrofluoric acid is less corrosive than hydrochloric acid and can be successfully handled by 30% cupro–nickel (see Fluorine compounds, inorganic).

Acetic Acid and Acetic Anhydride. Copper and copper alloys are used successfully in commercial processes involving exposure to acetic acid (qv) and related chemical compounds or in the manufacture of this acid. Copper has been successfully employed in contact with solutions containing, acetic acid, acetic anhydride, ketones, and esters. Various exposures indicate corrosion rates of 5–125 μm/yr (0.2–5 mpy) for copper dependent upon conditions of operation.

Fatty Acids. Oleic and stearic acids attack the copper alloys under severe service conditions to form metallic soaps (see Carboxylic acids). The rate of attack is somewhat higher than those for other organic acids such as acetic or citric acids. Corrosion rates are:

Alloy	Fatty acid	Corrosion rate, μm/yr (mpy)
cupro–nickel, 20%	mixture plus 1.2% H_2SO_4	66–178 (2.6–7.0)
tin bronze	oleic	51 (2)
aluminum brass		51 (2)
cartridge brass		508 (20)
copper	stearic	508–1270 (20–50)
silicon bronze		508–1270 (20–50)
cartridge brass		508–1270 (20–50)

Alkalies. Copper and its alloys are resistant to alkaline solutions, except those containing ammonium hydroxide or compounds that hydrolyze to ammonium hydroxide or cyanides. Ammonium hydroxide forms soluble complex copper cations, and cyanides form soluble complex copper anions.

The rate of attack for the copper–zinc series of alloys exposed to alkalies other than ammonia and cyanide is estimated to be ca 50–500 μm/yr (2–20 mpy) at room temperature and in quiet solutions, but increases to ca 50–1800 (2–70 mpy) when solutions are aerated and boiling.

The copper–tin alloys corrode at <250 μm/yr (10 mpy) in 1–2 N NaOH at room temperature, and aeration appears to have an insignificant effect. Cupro–nickel 30%, corrodes at a rate <5 μm/yr (0.2 mpy) in 1–2 N NaOH at room temperature; the rate is ca 25 μm/yr (1 mpy) for boiling temperatures, and aeration usually produces no significant effect.

Ammonium Hydroxide. Strong ammonium hydroxide solutions attack copper and copper alloys very rapidly compared to the rate of metallic hydroxides. However, in some applications the rate of attack on copper exposed to dilute solutions of ammonium hydroxide is low. Copper specimens submerged in 0.01 N NH_4OH at room temperature for a week showed a corrosion rate of 65 μm/yr (2.5 mpy). Quiet 2 N solutions of NH_4OH at room temperature attack the following alloys at the rates shown:

Material	Corrosion rate mm/yr (mpy)
copper–zinc alloys	1.8–6.6 (70–260)
copper–silicon alloys	0.8–5.1 (30–200)
copper–tin alloys	1.3–2.5 (50–100)
cupro–nickel alloys	0.3–0.5 (10–20)

Alloys containing more than 15% zinc, exposed in a stressed condition to ammonium hydroxide, are susceptible to stress-corrosion cracking.

Salts. The copper alloys are useful in equipment exposed to salt solutions of various kinds, particularly those that are nearly neutral upon hydrolysis. Among these salts are sodium and potassium nitrate, sulfate, and chloride. The chlorides are usually more corrosive than the other salts, especially with strong agitation and aeration.

The nonoxidizing salts, such as the alums and certain metal chlorides such as magnesium and calcium that hydrolyze to produce an acidic pH, behave essentially the same as the dilute solutions of the corresponding acids.

The corrosion rate of copper in sodium chloride refrigeration systems containing sodium bichromate as an inhibitor at <−15°C is <25 μm/yr (1 mpy).

Alkaline salts, such as sodium silicate, sodium phosphate, and sodium carbonate attack copper alloys at varying rates at room temperature, but in general the rates are considered to be low. Alkali cyanides are aggressive and attack copper at a high rate because of the formation of the soluble complex copper anion.

Oxidizing Salts. Copper and copper alloys corrode rapidly in contact with all but dilute solutions of oxidizing salts. Aqueous solutions of sodium or potassium bichromate can be safely handled in copper, but an addition of a highly ionized acid such as sulfuric acid may increase the rate of attack several hundred times because the bichromate ions act as oxidizing agents in acid solution.

Solutions containing ferric, mercuric, or stannic ions in sufficient concentration have corroded copper alloys at the rate of 76–760 μm/d.

Mercury or silver nitrates corrode copper alloys rapidly with a displacement of mercury or silver onto the copper alloy surface. The rate is dependent upon concentration, temperature, and acidity. A mercury film on stressed brasses containing more than 15% zinc may cause intercrystalline cracking or stress-corrosion cracking.

Organic Compounds. Copper and copper alloys are resistant to a range of organic compounds including the amines, alkanolamines, esters, glycols, ethers, ketones, alcohols, aldehydes, naphtha, and gasoline.

64 COPPER ALLOYS (WROUGHT)

The corrosion rate of copper in contact with various alkanolamines and various amine solutions is low; however, the attack rate is significantly increased if these compounds are present, especially at high temperatures.

Gasoline. Gasoline (qv), naphtha, and other related hydrocarbons in pure form essentially do not attack copper alloys. The manufacture of hydrocarbon materials produces streams that are likely to be contaminated with one or more materials such as water, sulfides, acids, gases, or a variety of organic compounds. The contaminants increase the corrosive attack on copper and its alloys. Corrosion rates for the inhibited admiralties and the cupro–nickels are low and these two alloy groups are successfully used for gasoline refining.

Creosote. Copper and copper alloys are generally suitable for handling creosote, although there is some attack on high-zinc brasses. Copper, cartridge brass, red brass, tin bronze, and silicon bronze are estimated to corrode at <500 µm/yr (20 mpy) when exposed to creosote at 24°C (see Alkylphenols; Tar and pitch).

Sugar. Copper has been successfully used for vacuum pan heating coils, evaporators and juice extractors in the preparation of both cane and beet sugar. Inhibited admiralty, aluminum brass, aluminum bronze, and the cupro–nickels are also used for tubes in juice heaters and evaporators (see Sugar).

Beer. Copper has been used extensively in brewing beer and in one installation copper kettle walls have thinned from an original 16 mm to 9.5 mm in 30 years of service. Steam coils require most frequent replacement of the brewing equipment and have an expected life of 15–20 years. The service life of other copper items exposed to process flows is 30–40 years (see Beer).

Sulfur. Sulfur compounds such as H_2S, Na_2K, or K_2S react freely with copper to form CuS. Reaction rates depend upon alloy composition, temperature, and the presence of water. The alloys of highest resistance are those of higher zinc content. Muntz metal (60% Cu, 40% Zn) exhibits good corrosion resistance under conditions that cause the complete destruction of red brass (85% Cu, 15% Zn).

Inhibited admiralty alloy is also an excellent material for use in heat-exchangers and condensers handling sulfur-bearing petroleum products. These alloys offer good resistance toward sulfur attack and excellent resistance to the water portion of the system.

Table 19. Accelerated Corrosion Tests of Some Copper Alloys In a Sulfur-Bearing Medium [a]

Exposure temperature, °C		360	316	285	254
Exposure time, d		115	26–34	24–34	24–34
Alloy	Composition, %	\multicolumn{4}{c}{Loss of tensile strength, %}			
silicon bronze 3%	96 Cu, 3 Si, 1 Zn		100	100	100
red brass	85 Cu, 15 Zn	100	100	100	100
inhibited aluminum brass	77.5 Cu, 2 Al 20.5 Zn, 0.035 As		7	16	10
inhibited admiralty	71.5 Cu, 1 Sn, 27.5 Zn, 0.035 Sb	16.5	6	4	2.5
naval brass	60.5 Cu, 0.75 Sn, 38.75 Zn			1.5	2
Muntz metal	60 Cu, 40 Zn	12		0–1	1.5

[a] Specimens exposed in a high pressure oil refinery fractionating tower. 1.4% Sulfur present in the media.

Table 20. Alloy Classification According to Composition

Group 1 coppers oxygen free	Group 3 alloys containing less than 80% copper
electrolytic tough-pitch	low brass 80%
deoxidized low phosphorus	cartridge brass 70%
deoxidized high phosphorus	yellow brass 65%
silver bearing	Muntz metal 60%
cadmium bearing	67 Cu, 32.5 Zn, 0.5 Pb
tellurium	64 Cu, 35.5 Zn, 0.5 Pb
leaded	63 Cu, 37 Zn
sulfur	62 Cu, 35.5 Zn, 2.5 Pb
zirconium	free cutting brass
arsenical	60 Cu, 0.5 Pb nickel silver
beryllium	60 Cu, 2.0 Pb nickel silver
chromium	57 Cu, 3.0 Pb nickel silver
phosphor bronze, 1.25% Sn	naval brass
	manganese brass

Group 2 alloys containing more than 80% copper	Group 4 special brasses
gilding 95%	inhibited brasses
commercial bronze	inhibited aluminum brass
87 Cu, 13 Zn	inhibited admiralty brass
red brass 85%	
leaded commercial bronze	

Group 5 phosphor bronzes	Group 8 cupro–nickel alloys
95 Cu, 92 Cu, 90 Cu, 5 Sn, 8 Sn, 10 Sn	cupro–nickel (10% Ni)
88 Cu, 444 alloy 4 Sn, 4 Pb, 4 Zn	cupro–nickel (20% Ni)
88 Cu, 6 Sn, 1.5 Pb, 4.5 Zn	cupro–nickel (30% Ni)
87 Cu, 8 Sn, 1.0 Pb, 4.0 Zn	
85 Cu, 5 Sn, 9 Pb, 1 Zn	Group 9 nickel–silvers
80 Cu, 10 Sn, 10 Pb	55% Cu, 18% Ni
83 Cu, 7 Sn, 7 Pb, 3 Zn	65% Cu, 18% Ni
	65% Cu, 12% Ni

Group 6 aluminum bronze
7 Al, 3 Si
7 Al, 2 Fe
5 Al

Group 7 copper–silicon alloys
3 Si
1.5 Si

Alloys containing the highest zinc content offer the best resistance toward corrosion by sulfur.

Strip coupons of various copper alloys were exposed in various locations in a high pressure fractionating column processing oil containing 1.4% sulfur. The temperatures for specimen sites varied from 254–360°C and the time for tests ranged from 24–115 d. Results of the accelerated test using loss in tensile strength as the criterion for the rate of attack are shown in Table 19. Resistance to corrosion by sulfur attack is a function of zinc content starting with a minimum zinc concentration of 20%.

COPPER ALLOYS (WROUGHT)

Table 21. Prices (avg) of Copper and Several Copper Alloys, $/kg

Copper and alloys	Price In 1966	Price In 1976
electrolytic copper	0.79	1.51
copper wire	1.07	1.98
UNS C 11000 sheet copper		2.76
UNS C 11000 drawn copper rods		2.97
UNS C 12200 seamless copper tubing		2.93
UNS C 26000 sheet brass	1.35	2.49
UNS C 26000 brass wire	1.35	2.71
UNS C 33000 brass rod	1.11	1.92
UNS C 36000 seamless brass tube	1.49	2.83

Gases. *Carbon Monoxide and Carbon Dioxide.* Carbon monoxide and carbon dioxide in dry form are usually inert to copper and its alloys. When moisture is present some corrosion takes place, depending on temperature and the amount of water present.

Some alloy steels are attacked by carbon monoxide and high pressure equipment used for handling this gas may be lined with copper or its alloys.

Sulfur Dioxide. Gases containing sulfur dioxide attack copper in a manner similar to that of oxygen. The dry gas does not attack copper, but the moist gas reacts to produce a mixture of copper oxide and sulfide. Phosphor bronze and aluminum bronze offer fair resistance to paper mill vapors containing moist sulfur dioxide.

Hydrogen Sulfide. Moist hydrogen sulfide gas reacts with copper alloys to form copper sulfide. Hot, wet hydrogen sulfide corrodes Muntz metal, naval brass, and admiralty alloy at 50–76 μm/yr (2–3 mpy), whereas the rate for copper and red brass for the same condition may be as high as 1.3–1.7 mm/yr (50–65 mpy).

The Halogens. Fluorine, chlorine, and bromine and their hydrogen compounds are noncorrosive to copper in their dry state. They are aggressive when moisture is present. The rate of attack is similar to those experienced handling hydrochloric and hydrofluoric acid.

Hydrogen. When oxygen-bearing copper is heated in hydrogen or hydrogen-bearing gases, the hydrogen diffuses into the copper and reacts with the oxide present to form steam and damage the grain boundaries of the metal; the condition is known as embrittlement. Tough-pitch copper containing 0.03% oxygen heated at 593°C has been embrittled to a depth of 0.38 mm in 3 min and 2.5 mm in 15 h.

Oxygen. Copper and copper alloy tubes are used to convey oxygen at room temperature in hospital oxygen service systems.

Scaling results when copper is used at high temperature in air or oxygen. The oxide film increases linearly in thickness with the logarithm of time for temperatures up to 100°C. There is an irregular increase of scaling rate with increasing temperature, and pressure is increased rapidly to 1.6 kPa (12 mm Hg). Above 20 kPa (150 mm Hg) the rate of increase is steady.

Alloy Selection

The coppers and copper alloys are grouped according to composition in Table 20. Generally, the members of each group behave similarly in the same corrosive media, although operating experience may prove the superiority of one alloy within a group.

Economic Aspects

Prices of copper and copper alloys are shown in Table 21.

BIBLIOGRAPHY

"Copper Alloys, Wrought" in *ECT* 1st ed., Vol. 4, pp. 431–458, by R. E. Ricksecker, Chase Brass & Copper Co.; "Copper Alloys (Wrought)" in *ECT* 2nd ed., Vol. 6, pp. 181–244, by R. E. Ricksecker, Chase Brass & Copper Co.

1. W. R. Webster, J. L. Cristie, and R. S. Pratt, *Trans. A.I.M.E.* **104,** 166 (1933).
2. N. D. Benson, J. McKeown, and D. N. Mends, *J. Inst. Met.* **LXXX,** (1951–1952).
3. J. C. Bradley, *Trans. A.I.M.E.* **75,** 210 (1927).
4. C. S. Smith, *Trans. A.I.M.E.* **128,** 325 (1938).
5. J. S. Smart, Jr. and A. A. Smith, Jr., *Trans. A.I.M.E.* **147,** 48 (1942).
6. *Ann. Phys. N.Y.* **15,** 219 (1932).
7. S. Skowronski quoted by L. L. Wyman, *General Electric Review*, 1934, p. 123.
8. S. Mosing and L. W. Haase, *Wiss. Veroeff. Simmens Konzern* **7,** 321 (1928).
9. C. S. Smith and E. W. Palmer, Jr., *Trans. A.I.M.E.* **117,** 225 (1935).
10. J. S. Smart, Jr. and A. A. Smith, Jr., *A.I.M.E. Technical Publication*, 1945, p. 1807.
11. D. Hanson and C. B. Maryat, Jr., *Inst. Met.* **37,** 121 (1927).
12. C. S. Smith, *Trans. A.I.M.E.* **93,** 176 (1934).
13. N. B. Pilling and G. P. Holliwell, *A.I.M.E. Technical Publication*, 1926, p. 1548D.
14. H. C. Kenny and G. L. Craig, *Trans. A.I.M.E.* **111,** 196 (1934).
15. S. Skowronski in Hofman and Hayward, eds., *Metallurgy of Copper*, McGraw Hill Book Company, Inc., New York, 1924.
16. D. Hanson, S. L. Archbutt, and G. Ford, Jr., *Inst. Met.* **43,** 41 (1930).
17. C. S. Smith, *Trans. A.I.M.E.* **89,** 84 (1930).
18. J. W. Borough, *Trans. A.I.M.E.* **221,** 1274 (1961).
19. F. L. Antisell, *Trans. A.I.M.E.* **64,** 435 (1921).
20. N. B. Pilling and G. P. Halliwell, *Min. Metall.*, (Feb. 1926).
21. H. A. Miley, *J. Am. Chem. Soc.* **59,** 2626 (1937).
22. F. A. Constable, *Proc. R. Soc. London* **A115,** 520 (1927).
23. *Ibid.*, **A117,** 376 (1928).
24. N. B. Pilling and R. E. Bedworth, *J. Inst. Met.* **29,** 529 (1923).
25. *Unified Numbering System for Metals and Alloys*, SAE HS 1086a or ASTM DS-56A, 2nd ed., 1977.
26. D. A. Reese and L. W. Condra, *Wire J.* **2,** 42 (1969).
27. M. Hansen and K. Anderko, *Constitution of Binary Alloys*, McGraw Hill Book Company, Inc., New York, 1958.
28. S. Lundquist and S. Carlen, *Erzmetall. Z. Erzbergbau. Metall.* **9,** 145 (1956)..
29. K. E. McKay and G. Armstrong-Smith, *Inst. Min. Metall. Trans.* **75 C,** 269 (1966).
30. J. S. Smart, Jr. and A. A. Smith, Jr., *Trans. A.I.M.E.* **152,** 103 (1943).
31. S. L. Archbutt and W. E. Prytherch, *Effect of Impurities in Copper, Research Monograph No. 4*, British Non-Ferrous Research Assoc., London, 1937.
32. D. Hanson and G. W. Ford, *Communications of the National Physical Laboratory to the British Non-Ferrous Metals Research Assoc.*, Feb. 1925.
33. J. S. Smart and A. A. Smith, Jr., *Met. Tech. Am. Inst. Min. Metall. Eng.* **1434,** 1 (1942).

34. F. N. Rhimes and W. A. Anderson, *Trans. A.I.M.E.* **143,** 312 (1941).
35. D. K. Crampton, H. L. Burghoff, and J. I. Stacy, *Trans. A.I.M.E.* **143,** 228 (1941).
36. H. L. Burghoff and A. I. Blank, *Am. Soc. Test. Mater. Proc.* **47** (1947).
37. D. K. Crampton, *Age Hardening of Metals,* American Society for Metals, Cleveland, Ohio, 1940.
38. H. L. Burghoff, *Grain Control in Industrial Metallurgy,* American Society of Metals, Cleveland, Ohio, 1948.

General References

P. F. George and A. Cohen, "Performance of Copper Alloys in the Desalting Environment," paper presented at the CDA-ASM Conference on Copper, Cleveland, Ohio, Oct. 16–19, 1972.
M. W. Covington, "Heat Treating Copper and Copper Alloys," *Met. Prog.* (May 1974).
V. Sawicki, "Solderability and Contact Resistance of Some Copper Base Alloys," paper presented at the CDA-ASM Conference on Copper, Cleveland, Ohio, Oct. 16–19, 1972.
J. Crane and E. Shapiro, "Physical Metallurgy and Properties of Copper Alloys 195," paper presented at the CDA-ASM Conference on Copper, Cleveland, Ohio, Oct. 16–19, 1972.
J. C. Almago and G. W. Rowe, *Bridge Die Extrusion of Copper Alloys,* second annual report to INCRA, New York, Oct. 1972.
W. Hotz and co-workers, "Extrusion Defects in Copper-Zinc Alloy Rods," *Z. Metallkd.* **62,** 182 (1971).
D. J. Pedder, E. D. Boyes, and G. C. Smith, "Field-Ion Microscopy of Internally Oxidized Copper Alloys," *Met. Sci.* **10,** 437 (1976).
F. R. Fickett, *A Preliminary Investigation of the Behavior of High Purity Copper In High Magnetic Fields,* annual report to INCRA, New York, August 1974, 34 pp.
P. Sibo, B. Galloria, and C. H. P. Lupis, "The Surface Tension of Liquid Silver–Copper Alloys," *Met. Trans* **8B,** 691 (1977).
M. M. Shea and N. S. Staloff, "Plastic Deformation of Polycrystalline Binary and Ternary Beta Brass Alloys," *Met. Trans* **5,** 755 (1974).
B. Lengyel and L. L. Culver, "Properties of Materials Extruded By Orthodox Hydrostatic Extrusion" *J. Inst. Met.* **97,** 97 (April 1969).
G. J. Sellers and A. C. Anderson, "Calorimetry Below 1K: The Specific Heat of Copper," *Rev. Sci. Instrum.* **45,** 1256 (1974).
Y. T. Hsu and B. O'Reilly, "Impurity Effects In High-Conductivity Copper," *J. Met.* **29**(12), 21 (1977).
C. R. Nanda and G. H. Geiger, "The Kinetics of Deoxidation of Copper and Copper Alloys of Carbon Monoxide," *Met. Trans* **2,** 1101 (1971).
J. Gruenwald, L. Slominski, and A. Lassdau, "Some Physical Properties of Chemical Copper" *Galvano-Organo,* **43,** 256 (1974).
D. E. Taylor and R. B. Waterhouse, "An Electrochemical Investigation of Fretting Corrosion of A Number of Pure Metals in 0.5 M Sodium Chloride," *Corros. Sci.* **14**(2), 111 (1974).

RALPH E. RICKSECKER
Chase Brass & Copper Co., Inc.

CAST COPPER ALLOYS

Copper alloy castings are used for their generally superior corrosion resistance, high electrical and thermal conductivities, and good bearing and wear qualities. Some of the alloys are heat-treatable and couple high strength with good electrical and thermal conductivity. Irregular and complex external and internal shapes can be produced by various casting methods. The production of the same configurations by other methods may be mechanically impractical or too costly.

The chemical compositions of casting alloys have greater latitudes than those of comparable wrought alloys, because the cold- and hot-working properties of castings are not important since castings are usually not subject to mechanical forming operations. However, the chemistry of the heat-treatable alloys must be carefully controlled to attain the proper hardening effects. Furthermore, compositions that do not lend themselves to forming operations are available in cast form for specific applications. An example is in the bearing field for alloys containing up to 40% lead (see Bearing materials). Wrought methods for making the same product are unsatisfactory because the intermediate annealing of the alloy causes the lead to separate from the matrix, and then from the material itself. The tolerance for impurities is normally greater in castings than in wrought alloys because the former are not mechanically formed.

On the basis of consumption, red brass (85% Cu, 5% Sn, 5% Zn, 5% Pb) alloy, Unified Numbering System C 83600 (UNS C 83600), is the most important of the cast alloys.

Properties and Characteristics

The cast copper alloys can be classified into two main groups: the *single-phase alloys* characterized by moderate strength, high ductility (except for the leaded varieties), moderate hardness and good impact strength; and the *polyphase alloys* having high strength, moderate ductility, and moderate impact strength.

Single-Phase Alloys. The copper–tin–zinc–lead alloys, the tin–bronzes, and the leaded tin–bronzes have a narrow range of properties, namely 234–303 MPa (34,000–44,000 psi) tensile strength, 103–152 MPa (15,000–22,000 psi) yield strength and more than 20% elongation and contraction in area. The leaded alloys are lower in all the properties cited above. The fatigue strength of these alloys is, in general, in proportion to the tensile strength. Most of the alloys have a fatigue strength of about one third times the tensile strength at 100 million cycles.

The alloys containing nickel or aluminum have high strength and the impact data exhibits considerable scatter. However, satisfactory toughness for practical application is obtained. High lead contents leads to lower impact values. Generally the all-alpha single-phase also structure exhibits a high degree of ductility. The structure may be strengthened by the addition of elements that forms solid solutions with copper.

There is a relatively large difference between yield strength and tensile strength. Most of the alloys have a tensile strength two or three times greater than the yield strength, illustrating the great ductility of these alloys and also the degree that they can be cold-worked.

Polyphase Alloys. The two-phase alloys have a rather wide range of properties resulting from variations within the structure. If the second phase is distributed in critical depression, the hardness and strength are at a maximum, and the ductility is at a moderate level. Tensile strength may be 415–825 MPa (60,000–120,000 psi), yield strength of 170–585 MPa (25,000–85,000 psi), and elongation 10–40%.

The effect of a second phase is demonstrated in the copper–aluminum system where increasing aluminum concentration causes the alloy system to change to a polyphase alloy. By obtaining a fine dispersion of the γ_2 phase, the yield strength is increased from 225 to 480 MPa (33,000–70,000 psi). The alpha–beta aluminum alloys respond to heat treatment with a general improvement of mechanical properties. Heat treatment is accomplished by heating to 815–870°C, quenching in water, and reannealing at 370–535°C, dependent upon the size and section of the casting. Different combinations of strength, hardness, and ductility can be attained. Nickel in aluminum–bronze alloy is in solid solution with the matrix and helps to refine the γ_2 precipitate.

The desired balance of ductility and strength can be obtained in age-hardenable alloys, such as beryllium copper, by controlling the amount of precipitate. For higher strength, aging is conducted to provide a critical size dispersion. Greater amounts of precipitate are obtained by increasing the beryllium content of the alloy.

Copper–chromium alloys are also precipitation hardenable. The precipitate can be chromium itself or chromium silicide, and if the conductivity is critical the chromium–silicon ratio should be held at 10:1 so that appreciable amounts of either element are not left in solid solution in the copper *after* aging. Lithium used as a deoxidizer instead of silicon can lead to more favorable conditions in respect to electrical conductivity. For a discussion of the principle of age- or precipitation-hardening copper alloys, see Copper alloys, wrought.

The manganese bronzes form an interesting series in the alpha–beta area of the copper–zinc system. Their mechanical properties are dependent upon the relative amounts of the two phases and their distribution. Since the copper contents of the manganese bronze alloys are relatively close together, they do not by themselves account for the presence of varying quantities of beta phase present. The differences in beta content are accounted for by the effect of other elements upon the location of the phase boundary between the alpha phase and the alpha–beta phase in the copper–zinc system. Aluminum, iron, and manganese all tend to shift the phase

Table 1. Boundary Shift for Manganese–Bronze Alloys

	Alloy composition[a], %						
Cu	Sn	Pb[b]	Fe	Al	Mn	Shift	New boundary, % Zn
58–63	0.5–1.5	0.8–1.5	0.5[c]	0.5[c]		0	38
56–62	0.5–1.5	0.5–1.3	0.8–1.5	0.25–1.0	0.1–0.5	−1.98	36
55–60	1.0[c]	0.3	0.4–2.0	0.5–1.5	1.5[c]	−2.80	35
60–68	0.1[c]	0.1[c]	2.0–4.0	3.0–7.5	2.5–5.0	−22.32	16

[a] Remainder is zinc in all cases.
[b] Lead is insoluble and therefore does not enter into the calculation.
[c] Maximum.

boundary from the value of 38% zinc. In the shift of the phase boundary from 38% zinc, the shift is given as $-0.87 \times \% $ Mn, $+1.0 \times \% $ Fe, and $-4 \times \% $ Al. The boundary shift for the manganese–bronze alloys may be computed and is shown in Table 1.

The increase in strength in this alloy series is caused by increased amounts of beta phase in the structure. The silicon brasses show similar hardening effects accompanying a second phase.

Typical mechanical properties and electrical conductivity for various cast alloys are shown in Table 2.

Stresses and Stress Relieving. Nonuniform cooling leads to unbalanced residual stress patterns in the casting. Exposure of stressed castings to environments containing ammonia, ammoniacal compounds or mercury can cause cracking of the alloys. Castings stressed to a level of ca 80 MPa (12,000 psi) and greater may crack in mercury environments and stress levels of only a few MPa may produce cracks when exposed to ammonia. The coppers and copper–nickel–iron alloys usually have very high resistance to stress corrosion cracking and stress relieving is not considered to be needed.

Stress relieving to safe stress levels can be accomplished by a thermal treatment, sometimes called stress-relief annealing even though annealing may not be involved.

The following alloys are all effectively stress-relieved by heating at 260°C for 24 min/cm of section: high copper alloys, red brass, semired brass, yellow brass, manganese–bronze, silicon–bronze, tin–bronze, and nickel–bronze.

The aluminum–bronze alloys are treated at 316°C for 24 min/cm of section.

If the casting is loaded or stressed externally in service, prior thermal stress treatment is of little value. Stress–corrosion cracking does not distinguish between residual or applied stresses (see Copper alloys, wrought).

Machinability. The cast copper alloys can be placed in three groups relative to machinability, and are rated in the same general manner as the wrought copper alloys. Group one contains the leaded alloys and is considered to be free-machining. Lead causes chip breakage during machining operations, permitting higher cutting speeds, decreased tool wear, and improved surface finish.

Alloys in group two are polyphase alloys with a second phase generally harder than the matrix which can cause brittleness and some chip breakage. The group contains leaded tin–bronze, silicon–bronze, high tin–bronze, aluminum–bronze, and manganese–bronze.

Group three, the most difficult to machine, is made up of high-strength manganese and aluminum–bronze high in iron or nickel content. Machinability ratings are listed by alloy in Table 3.

Electrical and Thermal Conductivities. Electrical conductivity is customarily expressed as a percentage of The International Annealed Copper Standards (IACS) adopted in 1913 to represent the average electrical conductivity of high-grade copper.

Elements in solid solution with copper often have a marked effect upon both the electrical and thermal conductivity of the alloy. Alloying elements present in significant concentrations as well as low concentrations of deoxidized elements decrease both properties. The wrought alloys have higher conductivities than the comparable casting alloys, since the wrought alloys usually have lower concentrations and fewer alloying elements. Sound copper castings requiring at least an 85% electrical conductivity IACS

Table 2. Properties of Cast Copper Alloys[a]

Common name	UNS designation	0.5% yield strength, MPa[b]	Compressive yield strength, MPa[b]	Tensile strength, MPa[a]	Elongation in 5 cm, %	Brinell hardness, 500 kg load	Electrical conductivity, % IACS	Thermal conductivity at 20°C, W/(m·K)[c]
copper	C 80100	62		172	40	44	100	391
chromium–copper								
as cast	C 81500	83		214	35	63	45	315
precipitation hardened		296		379	18	110	82	
beryllium–copper	C 81700	469	551[d]	620	8	217[e]	48	188
ASTM B22								
high strength yellow brass	C 86300	124	413[f]	820	83	225[e]	8.0	35.4
gun metal	C 90500	138–158	276[g]	310	25	75	11.0	74
tin–bronze 84:16	C 91100	172		214	2	135[e]	8.5	
tin–bronze 81:19	C 91300	207		214		170[e]	7.0	
high leaded tin–bronze	C 93700	124	90[f]	214	20	60	10.0	47
ASTM B61								
steam–bronze	C 92200	137	262[d]	276	20	65	14	69
ASTM B62								
leaded red brass	C 83600	103	96[f]	241	32	62	15	73
ASTM B66								
phosphor–bronze	C 94400	110	303[d]	221	18	55	10	52
high leaded tin–bronze	C 93800	110	83[f]	208	16	55	11.5	52
medium bronze	C 94500	83	248[d]	172	12	50	10	52
high leaded tin–bronze	C 94300	90	76[f]	186	10	48	9	62
ASTM B148								
aluminum–bronze 9A	C 95200	172–208	186–214[f]	482–600	22–38	110–140	11	50
aluminum–bronze 9B								
as cast	C 95300	208–241	110–138[f]	482–586	20–35	110–160	13	62
as heat treated	C 95300	276–379	241–310[f]	551–655	12–16	160–225[e]		
aluminum–bronze 9C								
as cast	C 95400	208–283	689[d]	517–655	12–20	150–185[e]	13	59
as heat treated	C 95400	310–358	827[d]	620–689	6–15	190–235[e]		

aluminum–bronze 9D							
as cast	C 95500	276–345	827[d]	620–724	7–20	175–210[e]	8.5	41
as heat treated	C 95500	413–551	1034[d]	758–855	5–12	215–260[e]		
aluminum–silicon–bronze	C 95600	234		517	18	140[e]	8.5	38
manganese–aluminum–bronze	C 95700	310	1034[d]	655	26	180[e]	3.0	12
nickel–aluminum–bronze	C 95800	262	689[f]	655	25	159[e]	7.0	35
ASTM B176								
die casting yellow brass	C 85800	207[h]		379	15		20	
die cast silicon–brass	C 87800	345[h]		586	25		6.7	28
silicon–yellow brass	C 87900	241[h]		482	25		15.0	
ASTM B584								
leaded red brass	C 83600	103	96	255	32	60	15.0	71.9
	C 83800	83–117	76–83	207–262	15–27	50–60	15.0	73
leaded semired brass	C 84400	90–117		200–269	18–30	50–60	18.0	73
	C 84800	103	90	234	30	55–59	16.4	73
leaded yellow brass	C 85200	83–96	55–69[f]	241–276	25–40	40–50	15–22	83.9
commercial no. 1								
yellow brass	C 85400	76–103	62[f]	208–262	20–35	40–60	18–25	88
yellow brass	C 85700	96–138		276–310	15–40	50–75	20–26	83.9
high strength yellow brass	C 86200	331	345[f]	655	20	180[e]	7.5	35.4
	C 86300	572[h]	413[f]	820	18	225[e]	8.0	35.4
leaded high strength yellow brass	C 86400	172[h]	158[f]	448	20	105[e]	19.0	88
high strength yellow brass	C 86500	193[h]	165[f]	489	30	130[e]	22	86
leaded high strength yellow brass	C 86700	289		586	20	155[f]	16.7	
silicon–bronze	C 87200	172	124[f]	379	30	85	6.0	28.4
silicon–brass	C 87400	165		379	30	70	6.7	27.7
silicon–brass	C 87500	207	183[f]	462	21	115	6.7	27.7
tin–bronze	C 90300	145	90[f]	310	30	70	12	74.7
gun metal	C 90500	152	276[g]	310	25	75	11	74.7
leaded tin–bronze	C 92300	138	241[d]	276	25	70	12	74.7

Table 2 (*continued*)

Common name	UNS designation	0.5% yield strength, MPa[b]	Compressive yield strength, MPa[b]	Tensile strength, MPa[a]	Elongation in 5 cm, %	Brinell hardness, 500 kg load	Electrical conductivity, % IACS	Thermal conductivity at 20°C, W/(m·K)[c]
high leaded tin–bronze								
83:7-7-3	C 93200	117–148	317[d]	207–262	12–20	67	12	58.1
85:5:9:1	C 93500	83–103	90[f]	193–241	20–35	55–65	15	71
80:10:10	C 93700	124	122[f]	241	20	60	10	46.9
78:7:15	C 93800	96–138	90–100[f]	172–227	10–18	50–60	11.5	52
70:5:25	C 94300	76–103	83–96[f]	158–207	7–16	72–55	9	62.6
nickel–tin–bronze								
as cast	C 94700	138–158		310–345	25–35	85	12	53.9
as heat treated	C 94700	345–413		517–551	5–10	180[e]		
leaded nickel–tin–bronze								
as cast	C 94800	138–158		310–345	20–35	80	12	38.6
as heat treated		207		413	8	120		
	C 94900	103		262	15			
copper–nickel 90:10	C 96200	172		310	20		11	45
copper–nickel 70:30	C 96400	221		413	20	140[e]	5	29
12% nickel–silver	C 97300	117		241	20	55	5.7	28.5
15% nickel–silver	C 97400	117		262	20	70	5.5	27.3
20% nickel–silver	C 97600	117		262	20	70	5	22
25% nickel–silver	C 97800	165		276	20	80	4.5	25.4

[a] Mechanical property data were developed from separately cast test bars, values shown are based on technical literature.
[b] To convert MPa to psi, multiply by 145.
[c] To convert W/(m·K) to Btu·ft²/(ft·h·°F), multiply by 0.578.
[d] 0.1 cm set/cm.
[e] 3000 kg load.
[f] 0.001 cm set/cm.
[g] 0.01 cm set/cm.
[h] Off set of 0.2%.

Table 3. Machinability Ratings for Cast Copper Alloy [a]

Common name	UNS designation	Machinability rating, %
Group 1: free-cutting alloys		
leaded red brass	C 83600	84
	C 83800	90
leaded semired brass	C 84400	90
	C 84800	90
leaded yellow brass	C 85200	80
commercial no. 1 yellow brass	C 85400	80
yellow brass	C 85700	80
die casting yellow brass	C 85800	80
high leaded tin–bronze	C 93200	70
	C 93500	70
bushing–bronze	C 93700	80
high leaded tin–bronze	C 93800	80
	C 94300	80
phosphor–bronze	C 94400	80
medium bronze	C 94500	80
12% nickel–silver	C 97300	70
Group 2: moderately machinable alloys		
leaded high strength yellow brass	C 86400	65
	C 86700	55
silicon–bronze	C 87200	40
silicon–brass	C 87400	50
silicon–brass	C 87500	50
die cast silicon–brass	C 87800	40
steam–bronze	C 92200	40
leaded tin–bronze	C 92300	40
leaded nickel–tin–bronze		
as cast	C 94800	50
as heat treated	C 94800	40
aluminum–bronze 9A	C 95200	50
aluminum–bronze 9B	C 95300	55
aluminum–bronze 9C	C 95400	60
aluminum–bronze 9D	C 95500	50
aluminum–silicon–bronze	C 95600	60
manganese–aluminum–bronze	C 95700	50
nickel–aluminum–bronze	C 95800	50
15% nickel–silver	C 97400	60
20% nickel–silver	C 97600	65
25% nickel–silver	C 97800	60
Group 3: difficult-to-machine alloys		
copper	C 80100	10
chrome–copper	C 81500	20
beryllium–copper	C 81700	30
high strength yellow brass	C 86200	30
	C 86300	8
	C 86500	25
tin–bronze	C 90300	30
gun metal	C 90500	30
tin–bronze 84:16	C 91100	10
tin–bronze 81:19	C 91300	10
nickel–tin–bronze		
as cast	C 94700	30
as heat treated	C 94700	20
copper–nickel 90:10	C 96200	10
copper–nickel 70:30	C 96400	20

[a] Ratings are compared to free-cutting brass—60% Cu, 3% Pb, 37% Zn—which has a rating of 100%.

require care in the preparation of their melts. The ordinary deoxidizers, such as silicon, tin, zinc, aluminum, or phosphorus, cannot be used because small residual concentrations markedly lower the conductivity. Calcium boride or lithium help to produce sound castings with better conductivities. Quantitative effects of various elements on the electrical conductivity of copper can be found under Copper alloys, wrought.

The thermal conductivity of copper having an electrical conductivity of 100% IACS is 391 W/(m·K) at 20°C.

The electrical conductivity is comparatively easy to measure whereas thermal conductivity is not.

The ratio (Wiedemann-Franz ratio) of thermal conductivity and the product of electrical conductivity times absolute temperature is approximately constant.

The electrical conductivity values for the important cast alloys are listed in Table 2. Figure 1 schematically shows the electrical conductivity of the cast copper-base alloys compared to various other cast metals and alloys.

The equation $Y = 4.184 + 3.93x$ gives an approximation of thermal conductivity in relationship to electrical conductivity, where Y = W/(m·K) at 20°C; x = % IACS at 20°C.

Bearing and Wear Properties. Copper alloys have been used as bearing materials (qv) because of their combination of moderate to high strength, very good corrosion resistance, wear resistance, and self-lubricating characteristics. The bearing alloys containing copper can be placed into three groups: phosphor–bronze alloys; copper–tin–lead alloys; and aluminum–bronze and silicon–bronze alloys.

The *phosphor–bronze alloys* contain Cu, Sn or Cu, Sn, Pb, and have a residual phosphorus concentration of a few hundredths to 1%. Nickel can be added to refine the grain structure and it is claimed to disperse the lead phase. The copper–tin bearings have high wear resistance, high hardness, and moderately high strength.

An alloy containing 11% Sn, formerly referred to as gear bronze, is used as its name implies for making gears. Bronze alloys containing 18–20% Sn are successfully used for high loads and slow movement. A maximum load of 17 MPa (2500 psi) is permissible for the 17% Sn alloy. These alloys contain high amounts of phosphorus to improve hardness. Zinc in these alloys leads to seizing and galling, and therefore, is limited to a maximum of 0.25%.

The *copper–tin–lead alloys* are softer and are used as bearing material for lighter loads, below 5.5 MPa, moving at moderate speeds. These materials include alloys, UNS C 93700 (80% Cu, 10% Sn, 10% Pb), UNS C 93200 (83% Cu, 7% Sn, 7% Pb) and UNS C 93500 (85% Cu, 5% Sn, 9% Pb). Alloy C 93700 is an excellent general bearing material widely used in machine tools, in electrical and railroad equipment. Alloys C 93500 and C 93700 are slightly lower in cost and are used in maintenance service. Alloys UNS C 93800 (78% Cu, 7% Sn, 15% Pb) and UNS C 94300 (70% Cu, 5.0% Sn, 25.0% Pb), containing higher quantities of lead, are used where high loads are encountered under poor operating conditions, ie, poor or no lubrication, corrosive environment, or dirty applications.

Aluminum–bronze alloys containing 8–9% Al are widely used for bushings and bearings in light or high speed applications. Alloys containing 11% Al, both in the as-cast and heat-treated condition, are well suited for heavy service, such as valve guides, rolling mill bearings, screwdown nuts, cams and wear plates. As the aluminum content

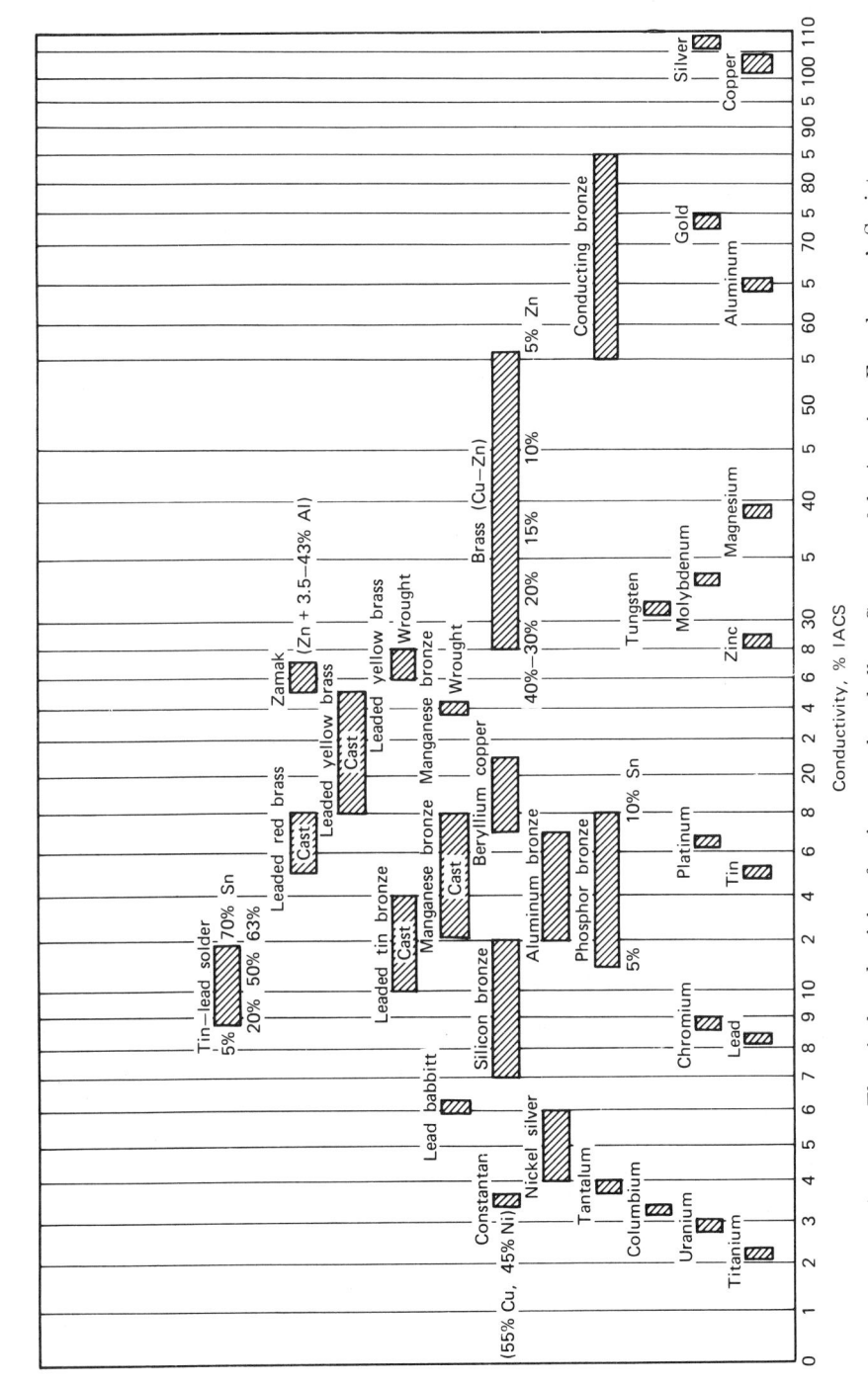

Figure 1. Electrical conductivity of various metals and alloys. Courtesy of the American Foundrymen's Society.

77

increases above 11%, the material becomes harder and stronger but the ductility decreases. Alloys containing more than 13% Al have a Brinell hardness of 300 or greater and the material is brittle.

Additions of manganese and silicon are frequently made to increase strength and hardness of the aluminum–bronze alloys. These elements form hard intermetallic compounds that are imbedded in the softer matrix producing a condition that enhances the bearing and wear resistance qualities.

The aluminum–bronzes, generally, have excellent bearing properties when used against steel. They will operate efficiently with a minimum of lubrication, and resist galling and scoring of the mating surface.

Joining Characteristics. The cast copper alloys may be joined by welding (qv), brazing, and soldering techniques with varying degrees of ease and success (see Solders and brazing alloys).

Table 4 lists a number of copper-base casting alloys and indicates the ease with which they can be joined by various methods.

Table 4. Joining Characteristics of Some Cast Copper Alloys[a]

Common name	UNS designation	Oxy-acetylene weldability	Carbon arc weldability	Metal arc weldability	Brazeability	Solderability
copper	C 80100	NR	F	NR	E	E
chromium–copper	C 81500	NR	F	NR	G	G
beryllium–copper	C 81700	NR	NR	NR	G	G
leaded red brass	C 83600	NR	NR	F	G[b]	E
	C 83800					
	C 84400					
leaded semired brass	C 84800	NR	NR	F	G[b]	E
steam–bronze	C 92200	NR	NR	NR	E[b]	E
leaded tin–bronze	C 92300	NR	NR	NR	G[b]	E
high leaded tin–bronze	C 93200, C 93500, C 93700	NR	NR	NR	G[b]	G
	C 93800, C 94300, C 94500	NR	NR	NR	P[b]	G
phosphor–bronze	C 94400	NR	NR	NR	G[b]	G
leaded high strength yellow brass	C 86400	P	P	P	F	F
high strength yellow brass	C 86200	G	NR	G	P	P
	C 86300	P	P	G	P	P
	C 86500	P	P	P	F	F
aluminum–bronze 9A	C 95200					
aluminum–bronze 9B	C 95300	NR	F	G	G	G
aluminum–bronze 9C	C 95400					
aluminum–bronze 9D	C 95500	NR	P	G	F	G
silicon–bronze	C 87200	G	P	F	F	NR
	C 87400					
silicon–brass	C 87500	F	NR	NR	F	NR
nickel–silver	C 97300, C 97400, C 97600, C 97800	NR	NR	NR	E	E

[a] E, excellent; G, good; F, fair; P, poor; NR, not recommended.
[b] Strain must be avoided during brazing and cooling to control cracking.

Mechanical Properties. Most alloys containing tin, lead, or zinc have moderate tensile and yield strength, and high elongation. Higher tensile or yield strengths are available through the use of aluminum and manganese bronzes, silicon brasses and bronzes, and some nickel alloys. Some of these alloys, such as beryllium–copper and chromium–copper and some of aluminum–bronze, are heat-treatable to attain maximum tensile strengths. Mechanical and physical properties for copper-base casting alloys are summarized in Table 2.

The entire group of leaded alloys are the easiest to cut and machine.

Most cast alloys can be joined by welding, brazing, and soldering (Table 4).

Bearing and wear-resistant alloys may be divided into three groups: phosphor-bronze (Cu–Sn); copper–tin–lead (low zinc); manganese–, aluminum–, and silicon-bronze alloys.

Copper–tin bearings have high resistance to wear, high hardness and moderately high strength. The copper–tin–lead alloys are used where a softer material is needed where moving parts are at moderate speeds and the loads are light. High strength manganese–bronze alloys have high tensile strength, hardness, and resistance to shock. General uses are for slow motion and relatively high compression applications.

Production Methods

Precautions. Care should be taken not to impair the quality of the metal as a result of the melting operation. The best practice is to select scrap of known good quality. The use of miscellaneous scrap may not be economical, despite the lower cost of such material; however, it is not common practice to make much use of virgin metals.

Ingots produced by secondary smelters and refiners and made to specifications are a good source of melting stock. Scrap, such as sprues and gates, and turnings from the foundry's own castings are very acceptable melting materials.

The melt may be exposed to gaseous and solid impurities unless suitable precautions are taken. The most deleterious of these contaminants are hydrogen or water vapor which may produce a porous structure in the cast alloy and ruin its quality. Oily scrap may also cause gaseous contamination of the melt.

If an open flame can come in contact with the melt, care should be taken to ensure a slightly oxidizing flame, preferably containing 0.1–0.2% excess oxygen.

Electric furnaces do not utilize gas as fuel and it is possible to operate with a neutral atmosphere. Care must be taken not to introduce water with the charge. Damp charcoal and lamp black are sources of water contamination; patched and newly lined furnaces or ladles must be thoroughly dried before use. Some foundries use electrical strip heaters to slowly remove most of the moisture and finally use torch flames to complete drying and fusing the surface of the refractory. Rapid drying may lead to cracked refractory.

Most molding methods can be used for the production of copper alloys. The choice of molding methods is based upon cost, surface finish requirements, and tolerances.

Sand Casting. Sand casting is the most popular method of molding and most copper alloys can be so cast. It is the cheapest method and has a low pattern cost. Care must be taken to obtain the desired surfaces. Dimensional tolerances are greater than those obtained using other methods of casting.

80 COPPER ALLOYS (CAST)

Permanent-Mold Castings. The so-called permanent mold is made of metal and the method has the optional use of a core. The high cost of the mold dictates that the number of castings be minimized for the process to be economical. Some advantages compared to sand molding are higher production rates, finer grain size and, therefore, improved mechanical properties of the casting, closer dimensional tolerances, lower machining costs, and higher yield. Permanent molding is generally limited to the tin, aluminum, silicon, and manganese bronzes, and to the yellow brasses. Generally, tolerance control is ca ±0.010 cm/cm, although closer control is possible.

Die Casting. Die casting is a process by which molten alloy is formed in a die under pressure into substantially finished, usually complex parts. Automatic or semiautomatic cyclic operations are normally employed. The die assembly consists of two halves with cores, and frequently with other parts to shape the inside and outside of the casting as required. The die-casting machine consists of a press, a special melting furnace, an injection mechanism, and necessary controls. The die halves are fixed to the opposing platens of the press so that they match when the press is closed. The other moving parts, including the injection mechanism designed to introduce molten metal at high velocity and under pressure into the die cavity, are integrated with the press so as to operate automatically in proper sequence as the press opens and closes.

The proper coordination and regulation of such factors as speed and pressure of the injection plunger, temperatures of both the dies and the metal, and gating and venting influence the quality of die castings, especially with regard to soundness of structure and surface finish.

Die casting is a special form of permanent-mold casting. Cost is an important factor in selecting this process. Yellow brass (UNS C 85700) is the largest volume alloy to be die cast, although more permanent mold alloys may be used. The method is used when a difficult cored section or relatively large surface area compared to weight is required. Cast sections having tolerances of 0.070 cm/cm of thickness are not difficult to obtain. Die casting is best suited for large numbers of small, intricate pieces.

Plaster Casting. Plaster casting is a refinement of sand molding. It uses a match-plate system, cope and drag assembly, cores, and gravity pouring.

Some features inherent in the plaster investment casting are the following: (*1*) the process helps to reduce cracking and the product stresses are low because of the yielding quality of the mold material. (*2*) The mold is free of volatile material that may lead to unsoundness. (*3*) Pouring time for filling the mold, and metal temperature are not critical; therefore, the pouring time can be increased up to about ten times that needed in sand casting; consequently, small orifice inlets to the mold cavity can be used and risers can be proportioned accurately to perform their function of supplying metal exactly where it is needed to control microshrinkage; (*4*) directional solidification can be controlled because the mold has relatively high insulating properties; also, small orifices can be employed at or near the bottom of cavities making it possible to control the flow of the melt and thereby establish a temperature differential between the feeder and the metal in remote sections of the mold.

The process can be largely automatic and continuous in operation.

The alloys best suited for this type of molding are aluminum bronzes, yellow brass, manganese bronze (UNS C 86200), low nickel bronze (UNS C 94700) and silicon bronze (UNS C 87200). Lead should be held to a minimum since it reacts with the calcium sulfate in the plaster, discoloring the surface of the casting.

Centrifugal Casting. This process consists of pouring the melt into a rotating mold which may be in either a horizontal or vertical position. Most alloys can be cast by this method. Size of castings is not a constraint. Castings from a hundred grams to 25 metric tons may be produced. Clean castings can be produced because dross and nonmetallics, which are lighter than the base metal, are thrown toward the surface where they can be removed by subsequent machining. However, machining and scrap cost must be considered. Soundness is enhanced by high pressures and consequent effective feeding during the solidification process.

Mechanical properties of centrifugally cast copper alloys vary with the alloy and mold material. Centrifugal castings made in sand molds where relatively slow cooling prevails have poorer mechanical properties than castings made in metal or graphite molds. However, even centrifuged castings made in sand molds have higher tensile strength than the same alloy and configuration cast statically in conventional sand molds.

The combination of handling molten metal and rotating equipment emphasizes the need for exercising safe operating practice and caution in the casting process.

Precision Investment Casting. Precision investment casting, also known as the *cire perdue* or *lost wax* process, is a special process for making small castings to very close dimensional tolerances. The process consists of making a pattern of such materials as wax, plastics, fusible alloy or frozen mercury. A suitable molding or investment compound such as ethyl silicate is cast around the pattern, cured, and the invested pattern melted out to form the finished mold. After the metal is cast, usually under pressure, the mold is broken and removed to free the castings, which usually only require the removal of gates for finishing. The investment casting method has been used to mass produce parts weighing 5–4500 g.

An older but similar process used in statuary founding involves modeling in wax over a core, and then covering with plaster. The wax is removed by melting, and a metal shell is cast in its place between the core and the plaster mold. The main advantage of this procedure is the ability to make intricate shapes without machining while attaining close dimensional tolerances. Size is a limitation and may range up to ca 8 cm maximum dimension with a maximum 4 cm section in thickness.

Shell Molding. Shell molding employs synthetic resin to make shells that can be used with or without cores. Both nonferrous and ferrous alloys can be cast using this method. Considerable equipment and accurate patterns are involved, but the process is feasible if high production rates and fine surface finish are required.

Continuous Casting. Continuous casting methods are being applied to the production of copper-base alloys. The molten metal is continuously poured into the top of the water-cooled lubricated mold, and the solid cast shape is continuously withdrawn mechanically from the bottom of the mold. The process is continuous as long as molten metal is available and the mold does not wear out. It may be made semicontinuous by casting a few meters in length. The process is then stopped, the casting is removed from the system, and after suitable mold preparation, the process can be restarted.

Where more than one strand of casting is required, a multihole water-cooled die can be fixed to the bottom of a holding furnace and the metal withdrawn as a solid strand from the bottom of the furnace, either in a horizontal or vertical plane.

The process requires considerable equipment, and skills in casting must be acquired. The process is useful where soundness and high volume of parts are needed. The process is also used to produce shapes, such as billets, bars, and cakes, as starting materials for the wrought product industry.

COPPER ALLOYS (CAST)

Comparison of Casting Methods. Table 5 compares several factors of various casting procedures. The factors are rated A through E; letter A is the most advantageous.

Table 5. Comparison of Casting Methods[a]

Factor	Sand	Die	Investment	Permanent mold	Plaster	Centrifugal	Continuous
tolerance	C	A	A	B	A	D	C
surface	D	B	A	C	A	D	C
thickness of section	D	B	D	C	A	A	A
structural density	A	C	A	B	A	A	A
pattern cost	A	E	B	C	B	A	A
ease of getting into production	A	D	B	C	B	D	D
production rate	B	A	C	C	D	B	C
cost per piece	B	A	C	C	D	A	C
flexibility as to alloy	A	C	B	D	B	A	A
limitation of size	A	D	C	B	C	A	A
reproducibility of successive cast quality	B	D	A	C	A	A	B
pressure tightness	B	D	A	C	A	A	A
low stress	B	D	C	C	A	C	C

[a] Comparisons are all relative based on letters A through E; A is most advantageous.

Alloy Identification and Chemistry

The nominal chemical composition and identification of the most important copper casting alloys are listed in Table 6. The alloys in Table 6 are identified by name and the Unified Numbering System. Alloy names are shown for information only; the use of names is not recommended.

Effect of Various Alloying Elements. Mechanical properties of the cast copper alloys are a function of alloying elements and their concentration. The specific effects of a number of these alloying elements are as follows:

Zinc. Zinc is added to copper as a predominating alloying element in concentrations of 5–40%, forming the alloy series known as the brasses. Zinc increases the tensile strength at a significant rate up to the concentration of ca 20%, whereas the tensile strength increases only slightly more for additions of zinc of 20–40%. The Rockwell F scale hardness is substantially an increasing straight line function with zinc additions of 5–35%.

The higher zinc concentrations in the high-strength yellow brass alloys form a duplex alpha–beta structure.

Zinc up to 5% is added to tin–bronze alloys to tighten the structure and to act as a deoxidizer, thereby aiding in producing sound castings for pressure applications. Zinc can impart a freedom from gas porosity because as the melt is heated until the zinc boils, the zinc vapor sweeps the melt free of gas.

Zinc is not considered a very harmful impurity in most alloys. The cast copper–zinc alloys are described as red brasses and leaded red brasses, semired, silicon, yellow, and high strength yellow brasses. The red brasses contain zinc as the principal alloying element along with some tin and lead. The semired brasses contain less copper than

Table 6. Nominal Composition of Some Casting Alloys

Common name	UNS designation	Cu	Sn	Pb	Zn	Fe	Ni	Al	Mn	Others
copper	C 80100	99.95								0.05[a]
chrome–copper	C 81500[b]	98.0[c]								1.0 Cr
beryllium–copper	C 81700[b]	94.25[c]								1.0 Ag
										0.4 Be
										0.9 Co
										0.9 Ni
ASTM B22										
high strength yellow brass	C 86300	63.0			25.0	3.0		6.0	3.0	
gun metal	C 90500	88.0	10.0		2.0					
tin–bronze 84:16	C 91100	84.0	16.0							
tin–bronze 81:19	C 91300	81.0	19.0							
high leaded tin–bronze	C 93700	80.0	10.0	10.0						
ASTM B61										
steam–bronze	C 92200	88.0	6.0	1.5	4.5					
ASTM B62										
leaded red brass	C 83600	85.0	5.0	5.0	5.0					
ASTM B66										
phosphorus–bronze	C 94400	81.0	8.0	11.0						0.35 P
high leaded tin–bronze	C 93800	78.0	7.0	15.0						
medium bronze	C 94500	73.0	7.0	20.0						
high leaded tin–bronze	C 94300	70.0	5.0	25.0						
ASTM B67										
journal bronze	C 94100	70.0	5.5	18.0	3.0[a]					
ASTM B148										
aluminum–bronze 9A	C 95200	88.0				3.0		9.0		
aluminum–bronze 9B	C 95300	89.0				1.0		10.0		
aluminum–bronze 9C	C 95400	85.0				4.0		11.0		
aluminum–bronze 9D	C 95500	81.0				4.0	4.0	11.0		
aluminum–silicon–bronze	C 95600	91.0						7.0		2.0 Si
manganese–aluminum–bronze	C 95700	75.0				3.0	2.0	8.0	12.0	
nickel–aluminum–bronze	C 95800	81.0				4.0	5.0	9.0	1.0	
ASTM B176										
die casting yellow brass	C 85800	58.0	1.0	1.0	40.0					
die cast silicon–brass	C 87800	82.0			14.0					4.0 Si
silicon–yellow brass	C 87900	65.0			34.0					1.0 Si
ASTM B584										
leaded red brass	C 83800	83.0	4.0	6.0	7.0					
leaded semired brass	C 84400	81.0	3.0	7.0	9.0					
	C 84800	76.0	3.0	6.0	15.0					
leaded yellow brass	C 85200	72.0	1.0	3.0	24.0					
commercial no. 1 yellow brass	C 85400	67.0	1.0	3.0	29.0					
yellow brass	C 85700	63.0	1.0	1.0	34.7			0.3		
high strength yellow brass	C 86200[d]	64.0			26.0	3.0		4.0	3.0	
leaded high strength yellow brass	C 86400	59.0		1.0	40.0	2.0[a]		1.5[a]	1.5[a]	
high strength yellow brass	C 86500	58.0	0.5		39.5	1.0		1.0	1.5[a]	
leaded high strength yellow brass	C 86700	58.0		1.0	41.0	3.0[a]		3.0[a]	3.5[a]	

84 COPPER ALLOYS (CAST)

Table 6 (*continued*)

Common name	UNS designation	Cu	Sn	Pb	Zn	Fe	Ni	Al	Mn	Others
silicon–bronze	C 87200	89.0[c]	1.0[a]	0.5[a]	5.0[a]	2.5[a]		1.5[a]	1.5[a]	4.0 Si
silicon–brass	C 87400	83.0			14.0					3.0 Si
silicon–brass	C 87500	82.0			14.0					4.0 Si
tin–bronze	C 90300	88.0	8.0		4.0					
leaded tin–bronze	C 92300	87.0	8.0	1.0[a]	4.0					
high leaded tin–bronze	C 93200	83.0	7.0	7.0	3.0					
	C 93500	85.0	5.0	9.0	1.0					
nickel–tin–bronze	C 94700	88.0	5.0		2.0		5.0			
leaded nickel–tin–bronze	C 94800	87.0	5.0	1.0[a]	2.5[a]		5.0			
	C 94900	80.0	5.0	5.0	5.0		5.0			
12% nickel–silver	C 97300	56.0	2.0	10.0	20.0		12.0			
15% nickel–silver	C 97400	59.0	3.0	5.0	16.0		17.0			
20% nickel–silver	C 97600	64.0	4.0	4.0	8.0		20.0			
25% nickel–silver	C 97800	66.0	5.0	2.0	2.0		25.0			
copper–nickel 90:10	C 96200	87.5				1.5	10.0		0.9 Mn	
copper–nickel 70:30	C 96400	68.0				0.7	30.0		0.8 Mn	

[a] Maximum.
[b] Responds to heat treatment.
[c] Minimum
[d] Several compositions are available that will meet mechanical property specifications.

the red brasses. Yellow brasses contain zinc as the principal alloying element accompanied by small amounts of tin and lead or other designated elements. High strength yellow brasses are alloys that contain zinc with smaller amounts of iron, aluminum, nickel, and lead.

Tin. Tin added to copper in concentrations of 5–20% forms the tin–bronze alloy series. Leaded tin–bronze is also produced. Usually a deoxidizer is also added to the melt for the purpose of producing a clean structure. Phosphorus is the usual element employed as a deoxidizer. Zinc may also be used, but is not effective as a deoxidizer until it is present in concentrations of ca 2%. Delta tin eutectoid develops in some castings containing as little as 6–8% tin because of nonequilibrium freezing conditions. The copper–tin constitution diagram under equilibrium condition indicates delta tin should not form until the tin concentration exceeds 16% at 520°C. Tin imparts strength and hardness to copper-base alloys, making them tough and wear resistant. Worm gears are often made from tin bronze containing 8% or more of tin. Tin also enhances the corrosion resistance of the copper-base alloys in nonoxidizing media.

Small amounts of tin (3–5%) are added to leaded red brass and semired brasses to increase the strength and hardness of alloys. Alloy UNS C 94700—88% Cu, 5% Sn, 5% Ni, and 2% Zn—deoxidized with phosphorus, is heat treatable to provide high strength.

Tin is usually not regarded as an impurity except in high tensile strength manganese–bronze where it is limited to 0.2% maximum. Tin lowers both the tensile strength and ductility of the alloy.

Lead. Lead is added to copper in amounts up to 40%. Lead is insoluble in the copper-base alloys, and because of its low melting point, it is found distributed in the grain boundaries of the casting. Since it imparts a certain degree of brittleness to the structure, it enhances machining operations by causing the alloy to break into chips as cutting tools are thrust into the matrix. Nonleaded alloys usually machine with the formation of long curls rather than chips and the surface is not as smooth. Additions of lead up to 1.5% significantly improve machinability without a serious decrease in tensile strength. Lead concentrations of 5–25% greatly increase machinability and, when present in amounts equal to or more than the tin content of the alloy, the material is used for bearing applications where resistance to both wear and friction are required.

Lead added to copper in amounts of about 35–40% forms a useful bearing alloy.

Lead is considered to be undesirable in high strength manganese–bronze, silicon–bronze, and silicon–brass. It affects the surface of silicon–brass and silicon–bronze, causing a noticeable darkening and pockmarking.

Silicon. Silicon added to copper forms alloys of high strength and toughness along with improved corrosion resistance, particularly in acidic media. Silicon acts as a deoxidizing agent and as such is added in small amounts. Silicon is a very harmful impurity in leaded tin bronze alloys because it contributes to lead sweat and unsoundness.

Aluminum. Aluminum added as the predominating alloying element to copper forms a series of high strength alloys called aluminum–bronzes. Aluminum forms solid solutions with copper up to about 9.5%. High strength yellow brasses contain aluminum in varying amounts. Some of the aluminum-bearing alloys form a second phase (gamma) and are heat-treatable.

Aluminum present in leaded tin–bronze alloys promote unsoundness.

Iron. Iron added to copper alloys adds strength to the silicon, aluminum and manganese bronzes. It combines with aluminum or manganese to form hard intermetallic compounds. The hard compounds embedded in the matrix, increase the alloy's wear resistance qualities. The undissolved iron in the alloy leads to nonuniform hardness and interferes with machining.

Phosphorus. Phosphorus is used principally as a deoxidizer in copper and high copper alloys. The alloy should contain a minimum residual of 0.02% phosphorus to ensure complete deoxidization. Lesser amounts of residual phosphorus can form an equilibrium system with copper and oxygen. Such castings are subject to damage (embrittlement) when heated in hydrogen atmosphere, depending upon temperature and hydrogen concentration of the environment.

Boron. Boron is a commercial deoxidizer.

Manganese. Manganese is added as an alloying element in high strength brasses where it forms compounds with other elements such as iron and aluminum. Manganese may also be used as a deoxidizer.

Nickel. Nickel added to copper markedly whitens the resulting alloy. The element added to copper at the rate of 10–30% produces the so-called cupro–nickel alloys which have very high corrosion resistance. Iron, up to a nominal 1.4%, added along with the nickel significantly enhances the resistance toward cavitation or impingement corrosion. Added to the bronzes, nickel refines the cast grain structure and adds toughness. Nickel improves strength and corrosion resistance. An alloy series containing 10–25% nickel along with tin, lead, and zinc as principal alloy elements is known as the nickel–silvers.

Nickel is also added to some of the high tin–gear bronze alloys to enhance wear properties. It does not have deleterious effects as an impurity and many specifications allow about 1%.

Beryllium. Beryllium added to copper forms a series of age or precipitation-hardenable alloys. These heat treatable alloys are the strongest of all known copper-base alloys (see Age-Hardenable Alloys under Copper alloys, wrought).

Chromium. Chromium also forms heat-treatable copper alloys. These alloys, in the heat-treated condition, have a Brinell hardness of about 120 and an electrical conductivity of about 80% IACS.

Arsenic, Antimony, and Phosphorus. Arsenic, antimony, and phosphorus can be added in small quantities (0.05%) to the all alpha-phase brass alloys containing less than 80% copper to inhibit the dezincification type of corrosion.

Economic Aspects

Selection. Copper alloy castings are used for their generally superior corrosion resistance, high electrical and thermal conductivities, and good bearing and wear qualities.

Irregular external and internal shapes can be produced by various casting methods that are impractical, impossible, or too costly to produce using other methods.

The choice of an alloy for any casting usually depends upon four factors: metal cost, castability, properties and final cost.

Metal cost is a minor consideration if only a few castings are made. The factor becomes of prime importance when large amounts of metal are required, or the casting is needed in large quantities, or is a competitive item or when metal is a major factor in the final cost. Final cost must be considered because the initial advantage of a lower metal cost may be eliminated by increasing cost resulting from an inherent property of the alloy. One such property that is very real to the foundryman is castability.

Castability is not to be confused with fluidity which is the ability of a molten alloy to fill a mold cavity in every detail. Castability is the ease with which an alloy responds to ordinary foundry practice without undue attention to gating, risering, melting, sand conditions, and any other factors involved in producing sound castings.

Cost. The design of a copper alloy casting requires a decision by the foundry or design engineer concerning the method of production for making various internal cavities. There are no general rules that apply to the question of cored vs coreless designs. A cost analysis is used to determine the more economical method of producing the casting, although frequently the choice can be based on experience.

Health and Safety Factors

During the melting and pouring, certain metals such as zinc volatilize, enter the atmosphere, and immediately oxidize to solid particulates forming a smoke. Some of these fumes or smokes can be hazardous to health, dependent upon the chemical composition of the particulate, its concentration in the off gas, and the time of exposure. More and more melt and casting shops are significantly controlling the atmospheric conditions in working areas by the use of adequate ventilating hoods, ducts, and exhaust fans in strategic locations. Rather than being exhausted into the atmosphere outside the operating plant, the particulates can be captured in bag-filtering and automatic collection systems (see Air pollution control methods). Care must be exercised in the design and operation of such devices to prevent hot gases or sparks from igniting the bag-filter material. Fires in the collecting system can be very costly. Gas temperatures are controlled by the length of ducts between the heat source and the filter, and also by the volume of gas being moved through the system. Often a cyclone dust collector is placed in the system just before the baghouse for the purpose of collecting sparks from smoldering or burning carbonaceous material to prevent them from contacting the filters in the baghouse. Masking of workers may also be practiced.

Local or national agencies have set limits on quantities of particulate thought permissible in casting shop atmospheres and also in the exhaust from collection systems. Agencies concerned in the United States are the EPA, NIOSH, and OSHA.

Particulates collected in foundry atmosphere have been found to contain compounds of zinc, copper, lead, iron, aluminum, magnesium, and silicon, as well as carbon and oily material. The specific composition and concentration of any smoke is dependent upon the composition of the material being melted as well as the melting and pouring practice.

Foundry Practice

Copper and High Copper Alloys. Copper castings contain a minimum of 99.3% copper. High copper alloys contain more than 94% copper. The coppers are usually specified to have high electrical conductivity, and therefore must be of high purity (see Electrical and Thermal Properties under Copper alloys, wrought). It is usually necessary to start with ingots of fire-refined or electrolytic-grade copper. High-grade or number one copper scrap of known chemistry may also be used.

Copper is susceptible to gassing and must therefore be protected from direct flame contact during melting. Copper melts can be contaminated with either hydrogen or oxygen. Hydrogen is soluble in the melt and oxygen forms copper oxide. The melts should be protected from air by the use of melt covers such as *dry* charcoal or carbon beads. The cover decreases the area of the melt exposed to air and also reduces the amount of fuming.

Copper oxide can also cause porosity in the finished casting, by combining with hydrogen formed by the dissociation of water in the mold material to form steam within the melt causing holes during solidification. The following reaction may take place.

$$CuO + H_2 \rightarrow Cu + H_2O_{(g)} \uparrow$$

If no cover is used, the copper should be melted under a slightly oxidizing atmosphere to reduce possible hydrogen solution, and the oxides floating upon the melt will tend to become the cover.

The copper melt heated to a temperature 30–55°C above its melting point should be deoxidized using calcium boride, lithium, or copper–phosphorus 15% alloy. The deoxidizer is plunged beneath the melt surface. A shrink test may be made by pouring a test piece ca 4 cm dia by 10–13 cm long into a mold of the same type intended to be used for the cast pieces. If the test piece shrinks, the melt is considered to be ready to pour. Experience will be a good guide.

If the test piece contains many holes, gas is indicated and the melt may be flushed with nitrogen to control the situation. Small amounts of deoxidizer should be added to the gas-free melt until shrinking of the test piece occurs.

88 COPPER ALLOYS (CAST)

Cores should be made with a minimum amount of binder or volatile compounds, and should also be readily collapsible because copper is susceptible to hot tearing.

Deoxidized copper is a high shrinkage material, and castings must be suitably risered to provide proper feeding. Exothermic compounds (iron oxide and magnesium) may be added to the top of the riser to keep the copper hot and molten to aid in feeding. Insulation and exothermic riser sleeves may also be used.

Contamination of the melt with elements that will remain in solid solution must be avoided, and excess deoxidizer must be controlled to attain high electrical conductivity castings. Properties are shown in Table 7.

Table 7. Properties of Copper and High Copper Alloys

Property	Copper C 80100	Chromium–copper C 81500	Beryllium–copper C 81700
melting point, °C	1064–1083	1075–1085	1029–1068
pouring temperature, °C			
light castings	1149–1204	1149–1204	1132–1188
heavy castings	1110–1166	1110–1165	1093–1149
specific gravity	8.94	8.82	8.75
density (lb/in^3)	(0.323)	(0.319)	(0.316)
dross generation	very low	high	low
gassing	high	moderate	moderate
cast yield	low	low	moderate
pattern maker's shrinkage cm/m	1.56	2.08	1.56

The coppers and high copper alloys can be successfully cast using the centrifugal, continuous, investment, permanent, plaster, and sand molding methods.

Uses. The following uses are typical for the copper group.

Alloy	Uses
UNS C 80100	electrical and thermal conductors, corrosion and oxidation resistant
UNS C 81500 (chromium–copper alloy)	electrical and thermal conductors used structurally where strength and hardness are required
UNS C 81700 (beryllium–copper alloy)	electrical and thermal conductors used structurally where high strength and hardness are required

Red Brass Alloys. This group of alloys includes the leaded red brasses and leaded semired brasses. Caution should be exercised to prevent gas absorption by flame impingement or the melting of oily scrap, or metal loss through excessive oxidation of the melt surface. The melt must be poured as soon as it reaches the proper temperature to prevent excessive zinc volatilization. The melt should be finally deoxidized and cast at ca 1065–1230°C as measured with a pyrometer. Fluxing is usually not needed if clean material has been melted.

The deoxidizer (copper–phosphorus 15% alloy) is added just before pouring in a quantity calculated to result in less than 0.01% residual phosphorus. The suggested addition is ca 1 g/kg of melt. Zinc-containing scrap should be melted first, and then any ingot copper along with the necessary zinc to replace any that had burned out during the melting. The amount of makeup zinc to be added varies with practice and should be judged by experience. Care should be taken not to heat the melt more than 83°C above the pouring temperature. Gating is very important to compensate for the good shrinkage that will be obtained. A mold facing wash (graphite) is used if good surfaces are needed although this is not a common practice.

Table 8. Properties of Red Brass Alloys

	Leaded red brass		Leaded semired brass	
Property	C 83600	C 83800	C 84400	C 84800
melting point, °C	854–1010	843–1004	843–1060	832–954
pouring temperature, °C				
light castings	1149–1288	1149–1266	1149–1260	1149–1260
heavy castings	1066–1177	1066–1177	1066–1177	1066–1177
specific gravity	8.7–8.88	8.6–8.7	8.6–8.8	8.55–8.70
density (lb/in^3)	(0.314–0.321)	(0.311–0.314)	(0.311–0.318)	(0.309–0.314)
dross generation	low	low	moderate	moderate
gassing	moderate	moderate	moderate	moderate
cast yield	high	high	high	high
pattern maker's shrinkage, cm/m	1.56	1.56	1.56	1.56

Properties of the red brass alloys are given in Table 8.

The members of the red brass group are cast using the centrifugal, continuous, investment, and sand molding methods. Their general tensile strengths vary from 170 to 210 MPa (25,000–30,000 psi) minimum as cast in sand molds.

Uses. The general uses of the red brass alloys are as follows:

Alloy	Uses
UNS C 83600 (leaded red brass)	fittings, plumbing goods, small gears
UNS C 83800 (leaded red brass)	low pressure valves and fittings, general hardware, plumbing goods
UNS C 84400 (leaded semired brass)	general hardware, ornamental casting, plumbing goods, low pressure valves, and fittings
UNS C 84800 (leaded semired brass)	cocks and faucets, general hardware, low pressure valves, and fittings

Tin–Bronze Alloys. Tin–bronze alloys are successfully melted in any type of foundry furnace. Best results are obtained by protecting the melt from either direct flame contact or excessive oxidation. Melting should be carried out in a slightly oxidizing atmosphere. Rapid melting is desired to reduce the opportunity of gas absorption in the melt. The melt should be somewhat superheated. Fracture tests exhibit an open grain and discoloration when the melt has been exposed to reducing conditions. Oily scrap should be avoided. Pouring temperatures range from 1010 to 1260°C and should be measured with a pyrometer. Pouring should take place at a temperature that gives the best results for a particular gating or risering system. The temperature should be high enough to avoid internal shrinkage that can lead to voids because gating and risering may not always completely eliminate such internal unsoundness.

Alloy scrap should be melted first and then have any copper added to the melt.

It is general practice to deoxidize the melt before pouring. The alloys may be degassed by superheating the alloys ca 40°C to boil the zinc, and allowing it to quietly air cool before pouring. Zinc must be added to replace the amount boiled off. Flushing the melt with dry nitrogen (oil pumped) will also degas the melt and lost zinc again must be replaced.

Gating and risering of castings of alloys is very important and careful consideration should be exercised for each configuration. Sound castings are made by following fundamental foundry techniques.

Some properties of the copper–tin–lead alloys are found in Table 9.

The members of the tin–bronze alloy group are cast using the centrifugal, continuous, permanent, plaster, and sand molding methods. The leaded tin–bronze alloys have minimum tensile strengths of 234–248 MPa (34,000–36,000 psi) as cast in sand molds, whereas the minimum tensile property for the high leaded tin bronze alloys are 138–207 MPa. The values are based on measurement of test bars cast in sand molds.

Uses. General uses for the tin bronzes are the following:

Alloy	Uses
leaded tin–bronze	
UNS C 92200	valves, fittings, pressure parts
UNS C 92300	valves, fittings for high pressure steam
high leaded tin–bronze	
UNS C 93200	bearings and bushings
UNS C 93500	small bearings, bushings, backing for babbitt-lined bearings
UNS C 93700	bearings, high speeds and high pressures
UNS C 93800	bearings, general service, moderate pressure, acid mine water exposure
UNS C 94300	bearings, high speeds and light loads

Manganese–Bronze Alloys. The melt should be heated to the highest temperature without causing excessive zinc boil or flare, and pouring temperature is important enough to be measured. Zinc losses change the composition of the casting. Manganese–bronze alloys are dross formers and therefore tend to develop their own melt covers. However, if open flame melting is practiced, a flux cover is recommended. Since the alloys form dross (oxidation of zinc, aluminum, and manganese) readily, choke gates, strainer cores, and pouring basins should be arranged to cause the melt to enter the mold with as little turbulence as possible.

The manganese–bronze alloys have high shrink characteristics and special attention must be given to gating and risering systems for individual castings. Hot tops and insulated sleeves are often used in casting manganese–bronze because of an insulating effect on the riser, thus promoting better feeding to the casting.

Good castings can be produced if the composition is controlled and good melting, gating, and risering practice is followed.

See Table 10 for some properties and characteristics of manganese–bronze alloys.

The alloys in this series have been cast using centrifugal, continuous die, investment, permanent, plaster, and sand molding systems. Minimum tensile strength for the leaded high strength yellow brass is 413 MPa (60,000 psi) for sand cast test bars,

Table 9. Properties of Tin–Bronze Alloys

Property	Gun metal C 90500	Tin–bronze 84:16 C 91100	Tin–bronze 81:19 C 91300	Steam–bronze C 92200	Leaded tin–bronze C 92300	C 93200	C 93500	C 93700	C 93800	C 94300
								High leaded tin-bronze		
melting point, °C	854–999	818–960	818–889	826–988	854–999	954–982	854–999	762–929	854–954	899–927
pouring temperature, °C										
light castings	1093–1260	1093–1232	1093–1204	1093–1260	1149–1260	1093–1232	1038–1204	1093–1232	1093–1232	1093–1204
heavy castings	1038–1149	1038–1121	1038–1093	1038–1149	1049–1149	1038–1121	1038–1204	1010–1149	1038–1149	1010–1093
specific gravity	8.7			8.6–8.8	8.7–8.87	8.85–8.9	8.87	8.9–9.1	9.1–9.4	9.2–9.5
density (lb/in³)	(0.315)			(0.311–0.318)	(0.314–0.320)	(0.320–0.322)	(0.320–0.320)	(0.321–0.329)	(0.329–0.340)	(0.332–0.343)
dross generation	low	low	low	low	low	low	low	low	low	low
gassing	moderate	moderate to high	moderate to high	moderate	moderate	moderate	moderate	moderate	moderate	moderate to high
cast yield	moderate	moderate	moderate	moderate	moderate	high	high	high	high	high
pattern maker's shrinkage, cm/m	1.56	1.56	1.56	1.56–1.82	1.56–1.82	1.56–1.82	1.56	1.04–1.56	1.56	1.56

92 COPPER ALLOYS (CAST)

Table 10. Properties of Manganese–Bronze Alloys

Property	Leaded high strength yellow brass (leaded manganese–bronze) C 86400	C 86200	High strength yellow brass (manganese–bronze) C 86500	C 86300
melting point, °C	860–880	899–927	862–880	885–923
pouring temperature, °C				
light castings	1038–1121	1066–1177	1038–1093	1066–1177
heavy castings	954–1038	982–1066	954–1038	982–1066
specific gravity	8.0–8.4	7.9	8.0–8.5	7.7–8.0
density (lb/in³)	(0.289–0.303)	(0.288)	(0.289–0.309)	(0.278–0.289)
dross generation	high	high	high	high
gassing	low	low	low	low
cast yield	low	low	low	low
pattern maker's shrinkage, cm/m	1.82–2.08	2.08–2.60	1.82–2.34	2.08–2.60

and the minimum tensile strength for the high strength yellow brasses varies from 448 to 758 MPa (65,000–110,000 psi).

Uses. The general uses for the manganese alloy group are as follows:

Alloy	Uses
manganese and leaded manganese–bronze	
UNS C 86200	marine fittings, gears, gun mounts,
UNS C 86200	bearings, bushings
UNS C 86400	free machining manganese bronze, valve stems, marine fittings, light duty gears

Aluminum–Bronze Alloys. The melting precautions are basically the same as those for the other copper-base alloys. The danger of gas absorption is not as great as with the tin–bronze alloys because of the protective film of aluminum oxide on the surface of the melt. Nevertheless, the alloys will dissolve gas, and therefore, precautions must be taken to prevent this. Although pouring temperatures are not as critical, temperature control is needed. The pouring range is 1065–1260°C dependent upon casting size and section; ladle surfaces should be kept well skimmed during pouring, and the melt should be exposed to air as little as possible. The shortest distance from the pouring ladle to the mold should be used to hold turbulence to an absolute minimum. Every precaution to eliminate dross should be taken.

The gating and risering system for cast aluminum bronze is extremely important, and must be arranged to introduce the metal quietly at the lowest portion of the mold. The alloys shrink well, hence the gating and risering must be well adapted to the particular casting. See Table 11 for properties of these alloys.

The aluminum–bronze alloys have been successfully cast in the centrifugal, continuous, permanent, plaster, and sand molding methods.

Dependent upon alloy the minimum tensile strengths of sand cast test bars are 448–620 MPa (65,000–90,000 psi).

Table 11. Properties of Aluminum–Bronze Alloys

Property	Aluminum–bronze			
	C 95200	C 95300	C 95400	C 95500
melting point, °C	1038–1043	1038–1052	1027–1038	1038–1054
pouring temperature, °C				
light castings	1121–1204	1121–1204	1260	1232
heavy castings	1093–1149	1066–1149	1777	1066
specific gravity	7.3–7.5	7.3–7.65	7.50	7.49–7.66
density (lb/in^3)	(0.279)	(0.264–0.276)	(0.270)	(0.275)
dross generation	high	high	high	high
gassing	moderate	moderate	moderate	moderate
pattern maker's shrinkage, cm/m	2.60–2.86	1.82–2.60	1.82–2.60	2.08

Uses. Some general use for the aluminum bronze follow:

Alloy	Uses
aluminum–bronze	
UNS C 95200	acid resistant pumps, pump rods, bushings, gears
UNS C 95300	pickling baskets, gears, marine equipment
UNS C 95400	bearings, gears, valve seats, valve guides, pickling hooks
UNS C 95500	valve guides and seats, corrosion resistant parts, bushings, gears, worms

Silicon–Bronze and Silicon–Brass Alloys. Best casting results are obtained with these alloys if the melting is accomplished using slightly oxidizing conditions. Rapid heating is required to the melting point to control dissolved gases in the melt. Superheating the melt to ca 85°C above the pouring temperature is desired. Undisturbed cooling to the pouring temperature allows the dross to float on the melt surface. Both melting and pouring temperatures are important and should be measured. Pouring temperatures for these alloys and 1040–1175°C depending upon size and section thickness of the casting. Prolonged holding of the melt at pouring temperature should be avoided. Molds should be ready to receive the melt as soon as it is at proper temperature. Melting oily scrap should be avoided. Silicon is a powerful deoxidizing elements and can easily be lost from the melt in the form of dross.

Covers for the melt are recommended, except for charcoal because silicon alloys absorb gas from green charcoal leading to unsoundness upon solidification. Preburned charcoal has been successfully used as a cover and is applied hot directly from the preburner.

Fluxing and deoxidation is not necessary because of the strong affinity silicon has for oxygen.

Silicon-bearing scrap must be kept segregated from other scrap, because silicon as an impurity in other alloys can promote a very coarse dendritic structure and weak porous castings.

94 COPPER ALLOYS (CAST)

If gas has been absorbed, some degassing treatment is necessary. The melt may be flushed with dry nitrogen or a degassing flux may be used. The copper–silicon alloys shrink less than the manganese–bronze and aluminum–bronze, but more than the tin–bronze. Good gating and risering systems are required.

Properties are listed in Table 12.

The alloys in the copper silicon group have been cast using centrifugal, investment, die, permanent, plaster, and sand molding methods. The minimum tensile strengths for sand cast test bars are 310–413 MPa (45,000–60,000 psi).

Table 12. Properties of Silicon–Bronze and Silicon–Brass Alloys

Property	Silicon–bronze C 87200	Silicon–brass C 87400	Silicon–brass C 87500
melting point, °C	860–971	821–916	821–917
pouring temperature, °C			
light castings	1093–1177	1093–1177	1093–1177
heavy castings	1038–1066	1038–1066	1038–1066
specific gravity	8.30–8.44	8.30–8.44	8.30–8.44
density (lb/in^3)	(0.300–0.305)	(0.300–0.305)	(0.300–0.305)
dross generation	low	low	low
gassing	high	moderate to high	moderate to high
casting yield	moderate	moderate	moderate
pattern maker's shrinkage, cm/m	1.82	1.82	1.82

Uses. Some general uses for the copper-silicon alloys are as follows:

Alloy	Uses
silicon–bronze UNS C 87200	bearings, pumps, valve parts, marine fittings, corrosion-resistant castings
silicon brass UNS C 87400	bearings, gears, impellers, valve stems, clamps
UNS C 87500	small propellers, valve stems, gears, bearings

Copper–Nickel and Leaded Nickel Bronze and Brass Alloys. High pouring temperatures cause most casting troubles with this group. The temperatures make a sand mold of low permeability practically unusable because the steam generated may exceed the venting power of the sand. Gases dissolve more readily at the higher temperatures and should be removed before pouring the melt. Degasifiers are used as a matter of standard practice. The copper–nickel alloys should be melted under slightly oxidizing conditions. Scrap must be clean. Cutting oils or compounds, or other organic matter must be removed before melting to control dissolved gases in the melt that can lead to inferior castings.

Time must be allowed for the degasifier to be effective, and the treated melt should be tested with a so-called pitch test to determine the condition of the melt.

The International Copper Research Association has recently developed an accurate and rapid method for evaluating the gas content of melts. The method consists of solidification of a sample of the melt in a vacuum and measuring the degree of swelling in the solidified sample.

Melting and pouring temperatures must be controlled by measurement and the melt poured at 1095–1425°C, depending upon the alloy, casting size, and section thickness. Gating and risering are especially important and careful consideration of the particular casting must be made.

Properties of the copper–nickel alloys are listed in Table 13.

Table 13. Properties of Copper–Nickel Alloys, and Leaded Nickel–Brass and Bronze

Property	Copper–nickel 90:10, C 96200	Copper–nickel 70:30, C 96400	Leaded nickel–brass nickel–silver 12%, C 97300	Leaded nickel-bronze Nickel–silver 20%, C 97600	Leaded nickel-bronze Nickel–silver 25%, C 97800
melting point, °C	1099–1149	1171–1238	1010–1040	1108–1143	1140–1180
pouring temperature, °C					
light castings	1316–1427	1371–1483	1204–1316	1260–1427	1316–1427
heavy castings	1204–1316	1316–1399	1993–1204	1232–1316	1260–1316
specific gravity	8.94	8.94	8.9–8.95	8.8–8.9	8.8–8.9
density (lb/in^3)	(0.323)	(0.323)	(0.318–0.322)	(0.318–0.322)	(0.318–0.322)
dross generation	low	moderate	high	moderate to high	moderate to high
gassing	high	high	moderate	moderate to high	moderate to high
cast yield	low	low	moderate	moderate	moderate
pattern maker's shrinkage, cm/m	1.56	1.82	1.04–1.56	1.56	1.56

The alloys in the copper–nickel group have been successfully cast using the centrifugal, investment, permanent, and sand molding methods. The minimum tensile strengths on test bars cast in sand mold are 207–310 MPa (30,000–45,000 psi).

Uses. The following uses are considered typical for leaded nickel brass and bronze.

Alloy	Uses
copper–nickel–iron	
UNS C 96200 (90:10 copper:nickel)	corrosion-resistant marine (sea water) applications
UNS C 96400 (70:30 copper:nickel)	corrosion-resistant marine elbows, flanges, valves, and pumps
leaded nickel–brass	
UNS C 97300 (12% nickel–silver)	hardware fittings, valves, statuary, ornamental castings
leaded nickel–bronze	
UNS 97600 (20% nickel–silver)	marine castings, ornamental castings, sanitary castings, valves, pumps

leaded nickel–bronze
UNS C 97800
(25% nickel–silver)

ornamental hardware, sanitary castings, valves, seats, musical instrument parts

BIBLIOGRAPHY

"Cast Copper Alloys" under "Copper Alloys" in *ECT* 1st ed., Vol. 4, pp. 458–467, by G. P. Halliwell, H. Kramer & Co.; "Cast Copper Alloys" under "Copper Alloys" in *ECT* 2nd ed., Vol. 6, pp. 244–265, by R. E. Ricksecker, Chase Brass & Copper Company.

General References

Cast Metals Handbook, American Foundrymen's Society, Des Plaines, Ill., 1957.
G. J. Cook, *Engineered Castings,* McGraw-Hill Book Co., New York, 1961.
J. L. Morris, *Metal Castings,* Prentice-Hall Inc., Englewood Cliffs, N.J., 1957.
Brass and Bronze Ingot Institute Manual, ASTM Standard, Non-ferrous Metals, ASTM, Philadelphia, Pa., 1951.
Symposium on Principles of Gating, American Foundrymen's Association, Des Plaines, Ill., 1951.
R. A. Flinn, *Copper and Brass and Bronze Castings—Their Structures, Properties and Applications,* The Non-Ferrous Founders' Society, Cleveland, Ohio, 1961.
Casting Design Handbook, American Society for Metals, Metals Park, Ohio, 1962.
P. R. Beeley, *Foundry Technology,* Halsted Press, a division of John Wiley & Sons, Inc., New York, 1972.
R. W. Heine and P. C. Rosenthal, *Principles of Metal Casting,* McGraw Hill Book Co., Inc., New York, 1955.
Analysis of Casting Defects, American Foundrymen's Association, Chicago, Ill., 1947.
J. G. Sylvia, *Cast Metals Technology,* American Foundrymen's Society Training and Research Institute, Addison-Wesley Publishing Co., Reading, Mass., 1972.
Metals Handbook, 8th ed., Vols. 1, 3, 5, and 6, American Society for Metals, Metals Park, Ohio, 1961.
Standards Book 5, American Society for Testing Materials, Philadelphia, Pa., 1977; various specifications.
Unified Numbering System for Metals and Alloys, SAE HS 1086a or ASTM DS-56a, 2nd ed., 1977.
C. R. Loper, P. D. Bhawalker, and L. M. Hepler, "Effects of Section Size in the Mechanical Properties of Cu Alloy Castings," *Trans. Am. Foundrymen's Soc.* 1970.
H. Ichikawa, *J. Jpn Copper Brass Res. Assn.* 8(1), 20 (1969).
I. D. Simpson and N. Standish, *Metallography,* **10**(4) 433 (Oct. 1977).
C. J. Evans, "Continuous Casting of Bronze," *Foundry Trade J.* 143 (Dec. 22, 1977).
A. W. Hudd, "Development of Continuous Casting of Copper Slab and Billet at IMI Refiners Limited," *Metall. Mater. Technol.* **9**(11), (1977).
"Applications of Cast Copper Base Alloys," in *Copper-Base Alloys Foundry Practice,* 3rd ed., American Foundrymen's Society, Des Plaines, Ill., 1965, Chapt. 22.
"High Leaded Tin Bronzes," in *Copper-Base Alloys Foundry Practice,* 3rd ed., American Foundrymen's Society, Des Plaines, Ill., 1965, Chapt. 17.

RALPH E. RICKSECKER
Chase Brass & Copper Co., Inc.

COPPER COMPOUNDS

The element copper, atomic weight of 63.54, is in Group IB of the periodic table. Its electronic configuration is $1s^2 2s^2 2p^6 3s^2 3p^6 3d^{10} 4s$, and consequently, it cannot be considered a transition element in the strict sense. However, because much of the chemistry of copper is that of divalent copper, which has an incomplete $3d$ shell, copper along with its compounds is usually discussed with the transition metals. Although the single outer s electron would suggest a chemistry similar to the alkali metals, the relatively high penetration of the d electrons decreases their shielding effect and, consequently, the $4s$ electron is rather tightly held. This is reflected in a high value for the first ionization energy, 745 kJ/mol (178 kcal/mol), and a somewhat inert nature. Copper displays four oxidation states. Copper(II) and copper(I) compounds are most common, copper(III) compounds are few but possibly becoming more significant; several compounds of copper(0) have also been claimed. At present, only copper(II) and copper(I) compounds are industrially important (see also Copper; Copper alloys).

The relative stabilities of the metal and its several oxidation states can be seen from the oxidation potentials relating the various states.

$$Cu \rightarrow Cu^+ + e \qquad E^0_{1/2} = -0.521 \text{ V}$$
$$Cu \rightarrow Cu^{2+} + 2e \qquad E^0_{1/2} = -0.337 \text{ V}$$
$$Cu^+ \rightarrow Cu^{2+} + e \qquad E^0_{1/2} = -0.153 \text{ V}$$
$$Cu^+ \rightarrow Cu^{3+} + 2e \qquad E^0_{1/2} < -1.8 \text{ V}$$

The negative potentials indicate that the metallic state of copper is preferred. The dissolution of metallic copper requires a mild oxidizing agent or a strong complexing agent that stabilizes one of the positive-valent ions. For example, nitric acid dissolves copper with the evolution of nitric oxide.

$$3\,Cu + 8\,HNO_3 \rightarrow 3\,Cu(NO_3)_2 + 2\,NO + 4\,H_2O$$

Sulfuric acid dissolves copper only under oxidizing conditions, such as high acid concentrations and relatively high temperatures. Sulfur dioxide is usually obtained.

$$Cu + 2\,H_2SO_4 \rightarrow CuSO_4 + SO_2 + 2\,H_2O$$

Cyanide forms very strong complexes with copper(I) and hence dissolves copper in aqueous solution with the evolution of hydrogen.

$$2\,Cu + 8\,KCN + 2\,H_2O \rightarrow 2\,K_3Cu(CN)_4 + 2\,KOH + H_2$$

The copper(I) ion is unstable in aqueous solution, tending to disproportionate to copper(II) ions and copper metal unless a stabilizing ligand is present. From the half reactions listed above, the potential for the following reaction is +0.37 V, which is equivalent to −35.5 kJ (−8.5 kcal).

$$2\,Cu^+ \text{ (aq)} \rightleftharpoons Cu + Cu^{2+} \text{ (aq)}$$

This leads to an equilibrium constant of about 1.8×10^6. Consequently, only very small concentrations of Cu^+ can exist in aqueous solutions and the only copper(I) compounds that are stable in water are the insoluble ones such as Cu_2S [22205-45-4], CuCN [544-92-3], and the copper(I) halides. However, in the presence of ligands with larger formation constants for Cu(I) than Cu(II), the copper(I) state can be stabilized. Examples of such ligands are ammonia and chloride ion. In aqueous ammoniacal solutions containing the $Cu(NH_3)_4^{2+}$ species, the following reaction proceeds rapidly and the blue color of the $Cu(NH_3)_4^{2+}$ species is completely discharged in the presence of excess copper metal.

$$Cu(NH_3)_4^{2+} + Cu \rightarrow 2\, Cu(NH_3)_2^+$$
[16828-95-8]   [16089-31-9]

Similarly, when copper is added to a solution of copper(II) chloride in strong hydrochloric acid, the complex copper(I) chlorides, $CuCl_2^-$ [15697-16-2], $CuCl_3^{2-}$ [29931-61-1], and $CuCl_4^{3-}$ [15444-92-5] are formed.

Until recently, few copper(III) compounds were known. They are difficult to prepare, and are only of academic interest. Alkali and alkaline earth cuprates(III) have been made by heating mixtures of copper(II) oxide with an alkali or alkaline earth oxide in oxygen. The compounds formed, eg, sodium cuprate(III) [12174-73-1], $NaCuO_2$, are unstable and decompose in water or moist air. In simple aqueous systems the Cu(III) species are stable only for a short time (1–2). As in the case of Cu(I), the Cu(III) species can be stabilized by the presence of certain complexing agents. For example, the dicarbamylurea complex of Cu(II),

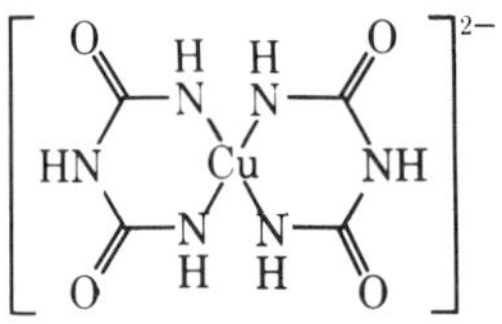

[57674-47-2]

can be oxidized to $Cu(NHCONHCONH)_2^-$ with oxidants such as potassium persulfate, $K_2S_2O_8$ (3). Periodates and tellurates of Cu(III) have also been made.

Recently, it has been discovered (4) that deprotonated amides, such as tetraglycine, can stabilize Cu(III) sufficiently to allow relatively easy oxidation of Cu(II) to the Cu(III) state. The standard half-cell potential for the following reaction has been measured to be −0.631 V.

$$CuL^{2-} \rightarrow CuL^- + e$$

where L is the tetraglycine ligand. This is particularly important to biochemists for it allows a 2-electron oxidation step from Cu(I) to Cu(III) in some enzymatic reactions, hitherto explained by a free-radical mechanism.

Copper is widely distributed throughout the earth; its crustal abundance is estimated at between 55 and 65 ppm. It occurs primarily in sulfidic minerals such as chalcopyrite [1308-56-1], $CuFeS_2$, and bornite [12517-64-5], Cu_5FeS_4, which are both of considerable economic value. Deposits of other sulfides, such as chalcocite [21112-20-9], Cu_2S, and covellite [19138-68-2], CuS, are becoming increasingly important, particularly in the central African copper region. The basic carbonates malachite [1319-53-5], $Cu_2CO_3(OH)_2$, and azurite [1319-45-5], $Cu_3(CO_3)_2(OH)_2$, the oxide cuprite [1308-76-5], Cu_2O, and the silicate chrysocolla [14567-86-3], $CuSiO_3\cdot 2H_2O$, are also important. A potentially large source of copper occurs in ocean manganese nodule deposits. In some areas, the ocean nodules average over 1.2% copper in addition to other valuable metals (see Ocean raw materials). Large deposits of native copper have been known since antiquity, but these have been largely depleted. Ores are usually processed to make copper metal or copper sulfate. Most industrially important copper compounds are made from copper metal.

Copper is one of the trace metals essential to life; over 18 cuproproteins have been identified as participating in various metabolic cycles. Particularly important are those copper-bearing enzymes participating in oxygen transfer such as mammalian dopamine hydroxylase (5) and galactose oxidase (6). Naturally occurring sulfidic copper deposits serve as a main source of copper for plant growth. Techniques have been developed for copper prospecting by analyzing plants for copper. Above average quantities of copper could indicate the occurrence of a copper deposit near the earth's surface (7).

On the other hand, copper in quantities larger than necessary can be poisonous, especially to lower life forms such as fungi. For many years, copper salts have been used as fungicides (qv). Copper sulfate has been used in particular to kill fungi of the genus *Tilletia* which causes the disease bunt in cereal grasses such as wheat and rye. Many other copper compounds have been developed for use as fungicides, the best known being a Bordeaux mixture (a basic copper(II) hydroxide formed from copper sulfate solution and slaked lime). Paris green (see below) is an effective insecticide (see Insect control technology).

Copper metal and copper compounds find many applications as catalysts. Both organic and inorganic oxidations are frequently accelerated by the presence of copper salts. This property is exemplified in the well-known Deacon process for the manufacture of chlorine (see Alkali and chlorine products). In this process chlorine is displaced from hydrogen chloride by oxygen at high temperature. Increasing the temperature of the reaction increases the rate but shifts the equilibrium away from chlorine and toward oxygen. The use of copper(II) chloride as a catalyst allows the reaction to proceed at a reasonable rate at about 400°C favoring the production of chlorine. In the classical Kjeldahl determination of organic nitrogen, copper(II) sulfate catalyzes the destructive oxidation of the sample by sulfuric acid.

Industrially important copper compounds and their uses are too numerous to allow a complete discussion. Uses with over 100 references in the period 1962–1977 were deemed important and are noted in Table 1. In some instances, such as cloud seeding and pollution control, the timeliness was considered important. Some compounds that had only a few references, such as copper(I) hydride, are included because their potential had not yet been fully realized. A summary of the most important copper compounds and their uses is given in Table 1.

Acetates. *Copper(I) acetate* [598-54-9], cuprous acetate, CH_3COOCu, mol wt 122.6, usually occurs in the form of white needles. It is stable when dry, but decomposes slowly in water. It is obtained by the slow addition of a solution of hydroxylamine hydrogen sulfate to a hot mixture of ammoniacal copper(II) acetate solution buffered with a large amount of ammonium acetate. The copper(I) acetate is precipitated by an excess of acetic acid.

Solutions of copper(I) acetate have been used to absorb olefins, to separate γ-globulins, and as a catalyst in the oxidation of phenols to poly(phenylene ethers) and diphenoquinones in the presence of tertiary amines (8). It displays some fire-retardant capability, and has a potential application as an oxidant in hydrogen–air fuel cells (9).

Table 1. Uses of Important Copper Compounds

Compound	Agriculture (insecticides, fungicides, herbicides)	Analytical chemistry	Antifouling paints	Catalyst	Cloud seeding	Corrosion inhibitors	Electrolysis and electroplating	Electronics	Fabrics, textiles	Flameproofing	Fuel oil treatment, additives
copper(I) acetate				X							
copper(II) acetate	X	X		X			X		X	X	
copper(I) acetylide				X							
copper(II) arsenate											
copper(I) bromide	X			X							
copper(II) bromide				X				X			
copper(II) carbonate, basic	X		X	X					X		
copper(I) chloride	X	X		X		X		X	X		X
copper(II) chloride	X	X		X		X		X	X	X	X
copper(II) chloride hydroxide	X			X			X				
copper(II) chromate(VI)				X			X		X		
copper(II) ferrate				X				X			
copper(II) fluoroborate				X			X	X	X		
copper(II) gluconate											
copper(II) hydroxide	X						X	X	X		
copper(I) iodide		X		X	X		X	X			
copper(II) nitrate	X	X				X	X	X	X		X
copper(II) oxide	X	X	X	X	X	X	X	X	X	X	X
copper(II) oxide	X	X	X	X	X			X	X		X
copper(II) soaps	X	X	X					X	X		X
copper(II) sulfate	X	X				X	X	X	X	X	

Table 1 (continued)

Compound	Glass, ceramics, cement	Medical, food and drugs	Metallurgy	Mining and metals	Nylon	Organic reactions	Paper, paper products	Pigments, dyes	Pollution control (catalyst)	Printing, photocopying	Pyrotechnics, rocketry	Wood preservative
copper(I) acetate						X						
copper(II) acetate		X			X	X		X		X		
copper(I) acetylide						X						
copper(II) arsenate						X						X
copper(I) bromide			X		X	X						
copper(II) bromide	X		X			X				X		
copper(II) carbonate, basic						X						
copper(I) chloride	X		X	X	X	X				X		
copper(II) chloride	X	X	X	X	X	X	X	X		X	X	
copper(II) chloride hydroxide	X											
copper(II) chromate(VI)						X			X		X	X
copper(II) ferrate										X		
copper(II) fluoroborate												X
copper(II) gluconate		X			X	X				X		
copper(II) hydroxide		X			X	X	X			X		
copper(I) iodide										X		
copper(II) nitrate	X	X	X	X	X	X	X			X	X	
copper(II) oxide	X		X	X	X	X	X		X	X	X	X
copper(II) oxide	X	X	X	X	X	X	X		X	X	X	X
copper(II) soaps					X	X		X	X	X		X
copper(II) sulfate	X	X	X	X	X	X	X	X	X	X	X	X

Copper(II) acetate monohydrate [142-71-2], cupric acetate monohydrate, $(CH_3COO)_2Cu \cdot H_2O$, mol wt 199.6, melts at 115°C and decomposes at 240°C. It forms dark green to blue green monoclinic crystals and tends to effloresce in dry air. It is soluble in water and other hydroxylic solvents such as alcohols and glycerols.

Copper(II) acetate is usually made by the reaction of copper(II) oxide with hot acetic acid. This type of reaction is conveniently carried out by refluxing acetic acid vapors over copper shot in the presence of air. The air oxidizes the copper and the refluxing acetic acid returns the oxidized copper to the boiler.

Copper(II) acetate is an effective catalyst in many polymerization processes involving styrene, acrylonitrile, and vinyl pyridines, among others. It is also used as a preservative for cellulosic materials, a stabilizer for polyurethanes and nylon, and as a pigment and a pigment intermediate.

Basic copper(II) acetate [52503-63-6], $Cu(C_2H_3O_2)_2 \cdot CuO \cdot 6H_2O$ (variable formula), also known as green or blue verdigris, is obtained by neutralizing acetic acid solutions of copper(II) acetate. It is used in the manufacture of Paris green, as a pigment in oil- and water-based paints, and a pigment intermediate (see Pigments).

Copper(I) Acetylide. Ammoniacal solutions of Cu(I) react directly with acetylene to form a flocculant, voluminous precipitate of copper(I) acetylide monohydrate [1117-94-8], $CuC{\equiv}CCu \cdot H_2O$. Upon drying *in vacuo* at 110°C, anhydrous copper(I) acetylide [1117-94-8] is obtained, an unstable, red powder that oxidizes in air to Cu_2O, carbon, and water. It appears, however, to be more stable than most other heavy metal acetylides, particularly if kept moist.

Copper(I) acetylide has been used as a catalyst in the manufacture of 2-propyn-1-ol (10) and acrylonitrile, and in promoting the reaction of acetylene with formaldehyde (see Acetylene-derived chemicals). It has also been used as an intermediate in the preparation of pure copper powder.

Arsenic Compounds. *Copper(II) acetoarsenite* [12002-03-8], $3Cu(AsO_2)_2 \cdot Cu(C_2H_3COO)_2$, known as Paris green, is a highly poisonous, emerald green powder. It is made by refluxing arsenic(III) oxide, As_2O_3, in a hot solution of 8% acetic acid to which copper oxide has been added, or by the interaction of a cupric sulfate solution with a solution of sodium carbonate and arsenic(III) oxide. Copper(II) arsenite is formed and allowed to settle. Addition of cold, dilute acetic acid gives the acetoarsenite:

$4\ CuSO_4 + 2\ CH_3COOH + 3\ As_2O_3 + 4\ Na_2CO_3 \rightarrow$

$3Cu(AsO_2)_2 \cdot Cu(C_2H_3O_2)_2 + 4\ Na_2SO_4 + H_2O + 4\ CO_2$

It is used primarily as an insecticide, wood preservative, and paint pigment. The intermediate copper(II) arsenite can also be used as an insecticide and a pigment.

Copper(II) arsenate [10103-61-4], $Cu_3(AsO_4)_2$, varies in color from light blue to blue green, depending on the amount of water of crystallization. It is prepared by the reaction of copper sulfate with sodium arsenate or arsenic acid followed by neutralization with sodium carbonate or sodium hydroxide. Like other arsenates, it shows a marked tendency to form complexes and double salts. Copper(II) arsenate is used as an insecticide, fungicide, and wood preservative, and possibly as an emulsifying agent (11). An extensive study of copper arsenate has been made (12).

Copper(II) Carbonate. Green basic copper(II) carbonate [12069-69-1], $Cu_2CO_3(OH)_2$, occurs in nature as the mineral malachite. A simple, neutral carbonate of copper(II) does not appear to have been isolated, most likely because of the relative insolubilities of copper(II) hydroxide and oxide. These compounds tend to precipitate at a pH so low that only extremely small amounts of the carbonate ion can exist, the bicarbonate species being the major one present.

Basic copper carbonate is prepared by the addition of an alkali carbonate, such as soda ash, to a solution of copper(II) nitrate. The compound has fungicidal and herbicidal properties. It is used in the manufacture of antifouling paint (13), and as a catalyst in hydrogenations and in the curing of rubber. It is frequently a source of copper in the manufacture of other copper salts.

Another basic carbonate of copper(II) is $Cu_3(CO_3)_2(OH)_2$ [12070-39-2] which occurs in nature as the mineral azurite (also called chessylite). Generally elevated pressures of carbon dioxide are required for its formation. It has been used as an artist's pigment called azurite; however, it deteriorates in moist air and gradually changes into malachite and becomes green.

Copper(II) Chromate(VI). Reddish-brown copper(II) chromate(VI) [13548-42-0], $CuCrO_4$, begins to decompose at 400°C, losing oxygen to form the blue–black copper(II) chromate(III) [12018-10-9]. The neutral salt can be prepared directly by heating copper(II) oxide and chromium(VI) oxide together. Precipitation from aqueous solutions generally gives a basic salt of variable composition.

Copper(II) chromate(VI) is used as a fungicide, in electrolytic treatment of metals, in weatherproofing of textiles (14), in epoxy adhesives, and as a wood preservative. The reduced salt copper(II) chromate(III) can be used as a mordant in textile dyeing, to catalyze ammonia oxidation (15), detergent manufacture, hydrogenation, oxidation of exhaust gases for pollution control (16), and the burning of solid fuel propellants.

Copper(II) Ferrate(III). Copper(II) ferrate(III) [12018-79-0], $CuFe_2O_3$, is formed when CuO and Fe_2O_3 are heated together at about 900°C. This reaction often takes place in the roasting of copper ores, where it is undesirable. Copper(II) ferrate(III) can also be made by precipitating the hydroxides of the two metals in stoichiometric proportions, drying the mixture, and sintering it at about 900°C. The sinter is then cooled and ground. Because of its spinel-type structure, it has magnetic properties. It is used in the manufacture of piezomagnets (17), and magnetic tapes (qv) and as a catalyst in the methylation of cresol and phenol (18). It may be employed as a catalyst for the production of hydrogen (qv) from carbon monoxide and water (19), and as an exhaust catalyst (20) (see Exhaust control).

Copper(II) Gluconate. Copper(II) gluconate [527-09-3], $Cu(C_6H_{11}O_7)_2$, forms light blue to blue–green water soluble crystals. It is prepared by the reaction of gluconic acid with basic copper(II) carbonate (21). Copper(II) gluconate is used in the treatment of arthritis (22) and as a deodorant for metabolic products of food (23). This latter property makes the compound useful in breath fresheners. It is also used as a dietary supplement to provide readily available copper (see Mineral nutrients).

Halogen Compounds. Copper(I) bromide [7787-70-4], cuprous bromide, CuBr, forms white tetrahedral crystals, mp 504°C, bp 1345°C. It oxidizes slowly in air acquiring a green color, and is only slightly soluble in water; it dissolves in solutions of halogen acids.

Copper(I) bromide is prepared by reducing a copper(II) sulfate solution containing potassium bromide with SO_2 or other standard reducing agents, or with metallic copper.

It has applications as a catalyst, eg, in the oxidative polymerization of xylenol (24), as an accelerator in the vulcanization of butyl rubber resins (25), and in photographic emulsions.

Copper(II) bromide [7789-45-9], cupric bromide, CuBr$_2$, mp 498°C, is obtained from warm solutions as black, deliquescent crystals. It is highly soluble in water (ca 1.2 g in 1 mL at 25°C). Copper(II) bromide is prepared by dissolving copper(II) oxide in hydrobromic acid. It can be used in cutting solutions for machining steel, as a stabilizer for acylated polyformaldehyde, as a bromination reagent, for removing lead from gasoline, as a catalyst in a variety of organic syntheses, in photothermographic copying, and for the extraction of sulfur from oil (26).

Copper(I) chloride [7758-89-6], CuCl, mol wt 99.0, mp 422°C, bp 1366°C, is a white to gray–white crystalline powder that turns green on exposure to air and brown on exposure to light. It is stable in dry air but oxidizes and hydrolyzes in moist air forming a variety of basic copper(II) chlorides. It is soluble in hydrochloric acid and in ammoniacal solutions.

Copper(I) chloride can be prepared by a variety of methods. Because it is frequently used in solution, it is convenient to reduce a concentrated solution of copper(II) chloride in strong hydrochloric acid with copper metal; the complex chlorides $CuCl_2^-$ and $CuCl_3^{2-}$ are also formed. Copper(I) chloride can also be precipitated from a copper sulfate solution containing sodium chloride by passing SO$_2$ gas through the solution. A pure product is obtained by melting impure CuCl under chlorine and adding a small amount of copper metal, followed by cooling and grinding to a powder (27). A continuous process based on the direct reaction of chlorine and copper metal has been developed (28).

Copper(I) chloride has a wide range of uses; eg, battery electrolyte, textile bleach, lubricant, ceramic decolorizer, soldering fluxes, stabilizer in nylon manufacture, catalyst in oxidation of olefins, and antioxidant in cellulose solutions. It is considered poisonous.

Copper(II) chloride [7447-39-4], CuCl$_2$, mp 630°C, bp 993°C (dec to CuCl), usually exists as the blue–green crystalline copper(II) chloride dihydrate [10125-13-0], CuCl$_2$.H$_2$O. In addition to high water solubility (1 g dihydrate in 1 mL H$_2$O at 25°C), copper(II) chloride is soluble in alcohols, and slightly soluble in acetone and ethylacetate. The anhydrous salt is obtained from the dihydrate by drying at 120°C. A stream of dry HCl gas should be passed over the salt to prevent hydrolysis.

Copper(II) chloride solutions can be easily prepared by direct metathesis using, eg, copper(II) sulfate and barium chloride. Direct action of chlorine on metallic copper in aqueous environment is very slow but can be accelerated by refluxing the solution over copper metal at a low pH (29) to prevent the formation of a basic copper(II) chloride. The anhydrous salt is prepared by the direct reaction of FeCl$_3$ with CuS (30) or from the dihydrate in dry HCl above 120°C.

Copper(II) chloride is widely used as a catalyst in organic chemical synthesis, particularly in chlorination reactions. It is also used as a mordant in textile dyeing, as a pigment, and as a wood preservative. Concentrated CuCl$_2$ solutions absorb large quantities of NO gas and hence are of interest in pollution control (31–32). Copper(II) chloride also removes sulfur (33) and lead compounds from gasoline (34) and oils.

Copper(II) chloride hydroxide [16004-08-3]. Basic copper(II) chlorides of a wide range of compositions can be made by precipitation at controlled pH. The compositions Cu$_2$OCl$_2$, copper(II) chloride oxide [12167-76-9] (2:2:1), and Cu$_4$O$_3$Cl$_2$ [12356-86-4] are frequently reported in the literature. These compounds are employed in crop protection, electronics, metallurgy, and as catalysts.

Copper(II) fluoroborate [14735-84-3], CuBF$_4$, is prepared by neutralizing fluoroboric acid, HBF$_4$, with copper(II) hydroxide or basic carbonate, followed by concentration of the solution to achieve crystallization. This salt is used as an additive in electroplating baths and as a catalyst in the epoxy resin curing of textiles. It has also had application in the preparation of electrically conducting coatings, as a wood preservative, and as a catalyst in the cold curing of cycloaliphatic epoxy resins.

Copper(I) iodide [7681-65-4], CuI, mp 606°C, bp 1290°C, is a pure-white crystalline powder; it occurs in nature as the rare mineral marshite [24401-69-2] (red–brown crystals). It is insoluble in water but dissolves in aqueous solutions of ammonia, alkali cyanides, thiosulfates, and iodides but decomposes in solutions of oxidizing acids. It is generally stable in air and is less photosensitive than the chloride or bromide.

There are numerous good procedures for making copper(I) iodide (35). A simple one is the reaction of a solution of a copper(II) salt such as copper(II) sulfate with potassium iodide at a slightly acidic pH. The iodide is partially oxidized by Cu(II) giving iodine and copper(I) iodide:

$$2\,Cu^{2+} + 4\,KI \rightarrow 2\,CuI + I_2 + 4\,K^+$$

This reaction forms the basis of the excellent iodometric titration analysis for copper which uses a sodium thiosulfate solution to titrate the liberated iodine. To obtain pure copper(I) iodide, an additional reducing agent such as sulfur dioxide or sodium sulfite is recommended (36). This prevents the iodine being adsorbed on the precipitated iodide.

Copper(I) iodide is used in the manufacture of photographic emulsions (37) and conductive transparent films (38). It is used as a catalyst in a wide variety of organic reactions such as the polymerization of butadiene (qv) with olefins. Copper(I) iodide is effective as a promoter for ice nucleation in cloud seeding (39–40).

Copper(I) Hydride. Copper(I) hydride [135717-00-5], CuH, has a formula weight of 64.5 but the composition is variable from CuH$_{0.6}$ to CuH. It is light red–brown in color and begins to decompose into the elements above 60°C and almost explosively above 100°C.

Reduction of an acidic copper(II) sulfate solution with hypophosphorous acid at 65°C precipitates copper(I) hydride (41). It can also be prepared by treating a solution of copper(I) iodide in a polar anhydrous solvent such as pyridine with lithium aluminum hydride in ether. After the reaction is complete the soluble CuH is precipitated by adding excess ether (42). The highest purity product is obtained by slowly heating a solution of copper(II) sulfate and monosodium hypophosphite to 50°C in a 100 g/L sulfuric acid medium (43).

Copper (I) hydride is a reducing reagent for some organic reactions (see Hydrides).

Copper(II) Nitrate. Copper(II) nitrate [3251-23-8] is commercially available in the form of copper(II) nitrate trihydrate [10031-43-3], mp 114.5°C. Both the trihydrate and hexahydrate [13478-38-1] are blue and deliquescent and highly soluble in water. Dehydration of these salts does not give the anhydrous compound because hydrolysis leads to a basic copper(II) nitrate [12158-75-7], Cu$_2$(NO$_3$)(OH)$_3$. Various methods are available for making the anhydrous salt (44). For example, sublimation of copper(II) nitrate under vacuum from an intimate mixture of copper(II) bromide and silver nitrate at about 200°C (45) yields well-formed blue–green crystals.

Solutions of copper(II) nitrate are obtained by dissolving copper(II) oxide or basic carbonate in dilute nitric acid. The hydrated salts can then be crystallized.

Copper(II) nitrate has wide applications, including as a ceramic color, a mordant in dyeing, a catalyst in solid rocket fuel, a nitrating agent, and as a wood and cellulosic material preservative. It has been employed in the flotation of cinnabar (46), as a drilling mud dispersant, to reduce carcinogenic gases in tobacco smoke (47), in sarcoma inhibition (48), and as a corrosion inhibitor.

Oxides and Hydroxide. Copper(I) oxide [1317-39-1], Cu_2O, mp 1235°C, dec above 1800°C, occurs in nature as the brown–red mineral cuprite. Copper(I) oxide is stable at a high temperature and can be made by thermally decomposing copper(II) oxide above 1030°C and then cooling in an inert atmosphere. In general, copper(I) oxide is prepared by the reduction of a copper(II) salt solution with a mild reducing agent such as metallic copper or hydroxylamine followed by precipitation with sodium hydroxide. In a commercial process (49), an ammoniacal solution of copper(II) basic carbonate is reduced with scrap copper. Sodium hydroxide is carefully added to precipitate the copper(I) as copper(I) hydroxide but not as copper(II). The slurry of copper(I) hydroxide is then treated with more sodium hydroxide and heated, whereupon the copper(I) oxide is formed. Copper(I) oxide is used as a catalyst, eg, in the chlorination of ethylene (50), as a fungicide, and as an antioxidant in lubricants. It is used in the purification of helium and is particularly useful for carbon monoxide absorption.

Copper(II) oxide [1317-38-0], CuO, mp 1326°C, occurs naturally as the mineral tenorite [1317-92-6] and the mineral melaconite [1317-92-6] formed by the weathering of copper sulfide ores. It can be prepared by oxidation of copper turnings at 800°C in air or oxygen or ignition of copper(II) nitrate or the basic carbonate. Copper(II) hydroxide is easily converted to the oxide by heating.

Like most copper compounds, copper(II) oxide is a fungicide and herbicide (see Herbicides). It is used in antifouling marine paint (see Coatings, marine) and as an ice nucleating agent in cloud seeding. Copper(II) oxide is used as a heat collecting surface in solar energy devices (51) because thin, black coatings of CuO are nearly opaque to short wavelength light, but nearly transparent to infrared radiation. Copper(II) oxide reduces tar in tobacco smoke (52) and is used as catalyst in ammonia manufacture and for the oxidation of exhaust gases from internal combustion engines (53) (see Exhaust control, automotive).

Copper(II) hydroxide [20427-59-2], $Cu(OH)_2$, is obtained by addition of sodium hydroxide to an ammoniacal solution of a copper(II) salt at ambient temperature. Usually a blue gel is formed that can be dried under vacuum to a powder of variable water content. However, copper(II) hydroxide is unstable with respect to copper(II) oxide and decomposes to the black oxide on slight warming. There is some indication that the conversion is catalyzed by the presence of colloidal copper(II) oxide. The method of preparation also effects the degree of metastability. A preferred method for making a stabilized form of the hydroxide is to electrolyze a copper anode in an electrolyte containing sodium sulfate and trisodium phosphate (54). Stabilized copper(II) hydroxide is highly useful as a fungicide and in antifouling marine paints. It is also used in the cuprammonium process for rayon (qv) manufacture and as a nylon stabilizer (see Polyamides).

Copper(II) Soaps. Many fatty acids form soaps with copper that have a variety of uses. Generally, the fatty acid reacts with copper(II) oxide or basic carbonate or an alkali salt of the acid is treated with a copper(II) sulfate solution. Copper(II) oleate [1120-44-1] and copper(II) stearate [660-60-6] are the ones most frequently mentioned in the literature but the valerate [15432-57-2], linoleate [7721-15-5], and octanoate [20543-04-8] have also been made. Copper(II) oleate coalesces mercury droplets and improves fuel oil combustion (55) by reducing the smoke and fumes of burning oil. It is used as a textile fungicide and in antifouling paints. Copper(II) stearate is used in coatings for xerographic plates (56) and in heat-sensitive coatings for photoduplication (see Electrophotography), as a color stabilizer in metal-containing dyes, and as a catalyst in a variety of organic reactions.

Copper(II) Sulfate. Copper(II) sulfate [7758-98-7], $CuSO_4$, as the pentahydrate [7758-99-8], $CuSO_4 \cdot 5H_2O$, is the most important copper salt in terms of the amount produced. It usually occurs in the form of blue triclinic crystals (blue vitriol). In nature it is found as the mineral chalcanthite. The pentahydrate can be dehydrated to intermediate hydrates and the anhydrous salt. The anhydrous salt, which also occurs naturally as the mineral hydrocyanite [14567-54-5], is white. Copper(II) sulfate solutions are usually obtained in the processing of copper ores and most of the copper metal produced is either electrorefined or recovered electrolytically from acidic copper(II) sulfate solutions. The pentahydrate is usually made by dissolving scrap copper in hot, concentrated sulfuric acid with the generation of sulfur dioxide, or by the air oxidation of scrap copper in dilute sulfuric acid. Large quantities of copper(II) sulfate are used in agriculture as a fungicide and algicide, a source of copper in animal nutrition, and as fertilizer. It is also a primary source from which other copper compounds can be derived.

Table 2. United States Copper Sulfate Production, 1967–1976

Year	Thousands of metric tons	¢/kg
1967	36.4	40.8
1969	45.9	46.1
1971	31.4	50.3
1973	39.4	55.6
1974	38.2	82.0
1975	32.3	77.6
1976	29.1	

Table 3. United States Consumption of a Selection of Copper Compounds, 1975[a]

Compound	Metric tons
copper(II) carbonate	45.4
copper(II) naphthenate	408.2
copper(II) oleate	90.7
copper(II) oxychloride sulfate	498.9
copper(I) oxide	181.4

[a] Ref. 57.

Economic Aspects

Copper(II) sulfate is by far the most important copper compound in terms of volume of production. Table 2, based on data obtained from the U.S. Bureau of Mines, gives the copper sulfate production in the United States between 1967 and 1976 and the prices per kilogram. Also listed are the average prices for primary copper for the same period.

Consumption figures for 1975 for other copper(II) compounds are summarized in Table 3.

BIBLIOGRAPHY

"Copper Compounds" in *ECT* 1st ed., Vol. 4, pp. 467–479, by L. G. Utter and S. B. Tuwiner, Phelps Dodge Corporation; "Copper Compounds" in *ECT* 2nd ed., Vol. 6, pp. 265–280, by E. A. Winter, Mary Jane Montesinos, and J. E. Singley, Tennessee Corporation.

1. J. S. Magee and R. H. Wood, *Can. J. Chem.* **43,** 1234 (1965).
2. D. Meyerstein, *Inorg. Chem.* **10,** 638 (1971).
3. J. J. Baur and J. J. Steggenda, *Chem. Commun.*, 85 (1967).
4. D. W. Margerum and co-workers, *J. Am. Chem. Soc.* **97,** 6894 (1975).
5. M. Goldstein and co-workers, *Fed. Proc.* **24,** 604 (1965).
6. J. Peisach, P. Aisen, and W. E. Blumberg, eds., *Biochemistry of Copper,* Academic Press, Inc., New York, 1966.
7. D. L. Poskotin and M. V. Lyukimova, *Geokhimiya,* 603 (1963).
8. Fr. Pat. 1,322,152 (Mar. 29, 1963), A. S. Hay (to Compagnie Francais Thompson-Houston).
9. A. Dravnieks, *Nature (London)* **199,** 1182 (1963).
10. U.S. Pat. 3,218,362 (Nov. 16, 1965), G. L. Moore (to Cumberland Chemical Corp.).
11. S. N. Srivastava, *Rheol. Acta* **2,** 210 (1962).
12. R. Mas, *Ann. Chim.* **4,** 459 (1946).
13. Belg. Pat. 650,131 (Jan. 6, 1965), (to Deutsche Akademie der Wissenschaften zu Berlin).
14. N. D. Bhandavi and co-workers, *Ind. J. Technol.* **1,** 86 (1963).
15. N. M. Morozov, L. I. Luk'yanova, and M. I. Tomkin, *Kinet. Katal.* **7,** 172 (1966).
16. U.S. Pat. 3,230,034 (Jan. 18, 1966), A. B. Stiles (to E. I. du Pont de Nemours & Co., Inc.).
17. U.S. Pat. 3,100,194 (Aug. 6, 1963), C. M. Van der Burgt and E. Put (to North American Philips Co., Inc.).
18. U.S. Pat. 3,716,589 (Feb. 13, 1973), T. Kotanigawa, M. Yamamato, and K. Shimokawa (to Agency of Industrial Sciences and Technology).
19. Ger. Pat. 2,036,059 (Apr. 8, 1971), A. Sugier (to Institut Francais du Petrole, des Carburants et Lubrifiants).
20. C. A. Leech, III and L. E. Campbell, *Adv. Chem. Ser.* **143,** 161 (1975).
21. Jpn. Pat. 2859 (April 8, 1963), G. Suzuki and co-workers (to Dainippon Pharmaceutical Co., Ltd.).
22. H. Picard, *6th Int. Congr. Therap. Strasbourg,* 461 (1959).
23. Fr. Pat. 1,394,875 (Apr. 9, 1965), C. Erwin.
24. Jpn. Pat. 73 30,156 (Sept. 18, 1973), S. Izawa and co-workers (to Ashai Dow Ltd.).
25. D. Khristov, Cv. Benchev, and N. Nenov, *Kautsch. Gummi Kunstst.* **28,** 260 (1975).
26. Jpn. Pat. 75 31,561 (Oct. 13, 1975), K. Aomura, T. Yotsuyanagi, and T. Uedate (to Mitsubishi Petrochemical Co., Ltd.).
27. A. Kajima, Jpn. Pat. 155,854 (April 16, 1943).
28. Fr. Pat. 2,009,852 (Feb. 13, 1970), (to Deutsche Gold und Selber-Scheideanstalt vorm. Roessler).
29. B. Popyankov, M. Nishev, and D. Krusteva, *Khim Ind. (Sofia),* 66 (1968).
30. M. Spasic and D. Vucurovic, *Tehnika (Belgrade)* **23,** 1962 (1968).
31. Fr. Pat. 1,388,242 (Feb. 5, 1965), T. Kitagawa (to Kobe Steel Works Ltd.).
32. Jpn. Pat. 74 92,226 (Aug. 14, 1974), K. Ohki and co-workers (to Asahi Chemical Industry Co., Ltd.).
33. U.S. Pat. 3,907,666 (Sept. 23, 1975), S. Chun, H. Hamilton, and A. Montagna (to Gulf Research & Development Co.).

34. U.S. Pat. 3,893,912 (July 8, 1975), A. Zimmerman (to Exxon Research & Engineering Co.).
35. G. B. Kaufman and R. P. Pinnel, *Inorg. Syn.* **6,** 3 (1960).
36. Neth. Pat. 6,701,808 (Aug. 8, 1967), N. V. Kodak.
37. Brit. Pat. 1,222,374 (Feb. 10, 1971), E. A. Sutherns (to Kodak Ltd.); J. Franco and N. K. Patel, *Ann. Chim. (Rome)* **65,** 99 (1975).
38. Ger. Pat. 1,800,653 (May 8, 1969), J. B. Wells (to Rank Xerox Ltd.).
39. L. Krustanov, *Dokl. Bolg. Akad. Nauk.* **25,** 1515 (1972).
40. R. E. Passavelli, Jr., H. Chessin, and B. Vonnegert, *Science* **181,** 549 (1973).
41. O. Neunhoeffer and F. Nerdel, *J. Prokt. Chem.* **144,** 63 (1935).
42. E. Wiberg and W. Henle, *Z. Naturforsch.* **7b,** 250 (1952); J. A. Dilts and D. F. Shriver, *J. Am. Chem. Soc.* **90,** 5769 (1968).
43. V. I. Mikheeva, N. N. Mal'tseva, and I. M. Kuvshinnikov, *Zh. Neorg. Khim.* **11,** 2001 (1966).
44. C. C. Addison and N. Logan in H. J. Emeleus and A. G. Sharpe, eds., *Advances in Inorganic Chemistry and Radiochemistry,* Vol. 6, Academic Press, Inc., New York, 1964.
45. R. N. Kust, unpublished.
46. V. A. Glembotskii, P. M. Solozhenkin, and L. L. Ogneva, *Azv. Akad. Nauk Tadzh. SSR Otd. Siz-Tekhn. Khim. Nauk,* 64 (1965).
47. E. L. Wynder and D. Hoffmann, *J. Am. Med. Assoc.* **192,** 88 (1965).
48. Y. Kimura and S. Makino, *Gann* **54,** 155 (1963).
49. U.S. Pat. 2,474,497 (June 28, 1949), P. J. Rowe (to Lake Chemical Co.).
50. Belg. Pat. 616,762 (Oct. 22, 1962), (to Distillers Co., Ltd.).
51. E. Salam and F. Daniels, *Bull. Inst. Desert Egypte* **8,** 131 (1958).
52. Fr. Pat. 1,402,087 (June 11, 1965), (to Oxy-Catalyst, Inc.).
53. Jpn. Pat. 75 131,668 (Oct. 17, 1975), S. Adachi, T. Miyakoshi, and M. Hattori (to TDK Electronics Co., Ltd.).
54. U.S. Pat. 3,194,749 (July 13, 1965), W. H. Furness (to Kennecott Copper Corp.).
55. Fr. Pat. 1,381,305 (Dec. 11, 1964), L. Alliot, M. Auclair, and R. Gay (to Esso Research & Engineering Co.).
56. Neth. Pat. 6,514,028 (May 2, 1966), (to Rank-Xerox Ltd.).
57. *Chemical Economics Handbook,* Stanford Research Institute, Menlo Park, Calif., Oct. 1976, pp. 573.5007G-J.

General References

F. A. Cotton and G. Wilkinson, *Advanced Inorganic Chemistry,* 3rd ed., Interscience Publishers, a division of John Wiley & Sons, Inc., New York, 1972, pp. 905–922.

P. Pascal, *Nouveau Traité de Chimie Minérale,* Vol. 3, Masson et Cie., Editeurs, Paris, Fr., 1957, pp. 155–421.

H. Remy, *Treatise on Inorganic Chemistry,* Vol. II, Elsevier, Amsterdam, The Netherlands, 1956, pp. 377–391.

A. G. Massey in A. F. Trotman-Dickenson, ed., *Comprehensive Inorganic Chemistry,* Vol. 3, Pergamon Press, Oxford, Eng., 1973, pp. 1–78.

W. E. Hatfield and R. Whyman in R. L. Carlin, ed., *Transition Metal Chemistry,* Vol. 5, Marcel Dekker, Inc., New York, 1969, pp. 47–179.

ROGER N. KUST
Kennecott Copper Corporation

CORK

Cork [*61789-98-8*] is one of the few naturally grown closed-cell foams that has never been duplicated with synthetic material. It has been an item of commerce for at least 2500 years and was used by the early Greeks for shoes, floats, and stoppers.

As referred to in this article, cork is the outer bark of the cork oak, *Quercus Suber*, which grows mainly near the Mediterranean Sea on the Iberian Peninsula and along the shores of North Africa. The cork oak is unique in that the outer bark of cork, or the phellem, can be stripped from most of the tree trunk and major limbs without harming the tree. Furthermore, the cork bark undergoes rejuvenation and can be stripped repeatedly at specific intervals. The harvest starts when a new tree reaches maturity (20–25 yr) and the first, or virgin cork stripping is made. Successive strippings as a rule are made every nine years. The stripped cork bark is steamed, cleaned, graded, and baled for marketing. The yield per tree ranges from 15–100 kg. Mature trees of normal age between 50 and 200 years measure 1.5–4.0 m in trunk circumference and 10–20 m in height. The maintenance of trees and harvest of cork are now controlled to various degrees by the governments of the growing regions. Many countries have tagged and registered each cork oak and have established laws to protect the trees from being cut for firewood, lumber, or other use.

Structure and Physical Properties

In contrast to the bark of other trees, cork is nonfibrous. It is composed of tiny, closely packed cells that are tetrakaidecahedral, or 14-sided; six of the faces are quadrilateral and eight are hexagonal. This shape provides optimum packing of the cells without extra void space, which accounts for the excellent gasketing and flotation characteristics of cork. The cells are in the range of 0.025–0.050 mm in the longest dimension. The cell walls and resinous binder are highly resistant to water, most organic liquids, and all but strong acid and alkali solutions. Cork has a specific gravity of 0.1–0.3 which provides superior flotation and thermal-insulation properties. A fresh-cut surface of cork exhibits a high coefficient of friction even when subjected to water and oil.

Chemical Properties

Considerable work has been done over the past 200 years to study the chemical composition of cork. The various origins and growing conditions of natural cork cause variations in composition so that it is reasonable that a broad, general analysis must be used to represent the composition. A representative analysis is as follows:

Content	wt %
fatty acids	30
miscellaneous organics	17
lignin	16
other acids	13
ceroids	10
tannins	4
glycerol	4
cellulose	3
inorganic ash	3

The first two groups of chemicals are composed of the following:

Common name	Systematic name	Formula	CAS Registry No.
Fatty acids			
phellonic acid	22-hydroxydocosanoic acid	$C_{22}H_{44}O_3$	[506-45-6]
phellogenic acid	docosanedioic acid	$C_{22}H_{42}O_4$	[505-56-6]
phloionic acid	9,10-dihydroxyoctadecanedioic acid	$C_{18}H_{34}O_6$	[23843-52-9]
phloionolic acid	9,10,18-trihydroxyoctadecanoic acid	$C_{18}H_{36}O_5$	[583-86-8]
suberonic acid	a mixture		
suberolic acid	a mixture		
cordicinic acid	a mixture		
Miscellaneous organics			
resorcinol		$C_6H_6O_2$	[108-46-3]
hydroquinone		$C_6H_6O_2$	[123-31-9]
gallic acid		$C_7H_6O_5$	[149-91-7]
salicylic acid		$C_7H_6O_3$	[69-72-7]
glycerol		$C_3H_8O_3$	[56-81-5]
phloroglucinol		$C_6H_6O_3$	[108-73-6]
friedelin		$C_{30}H_{50}O$	[559-74-0]
oxalic acid		$C_2H_2O_4$	[144-62-7]
sterols		a mixture	

Processing

While enroute from the point of harvest to the manufacturing plant, the cork bark is visually classified by thickness, amount of noncork structure, color, and density. The better grades are used for cork-stopper production to be sold for bottling still wine, champagne, other alcoholic beverages, and perfumes. In spite of the development of plastic stoppers and closures for many of these applications, the volume of cork stoppers produced annually exceeds 5 billion. Natural stoppers are cut with the cylindrical axis normal to the bark thickness. Large, thin stoppers are cut parallel to the annual rings and laminates are made for large cork stoppers. In contrast to the old art of hand-cutting stoppers, most of the stopper production is now done by automatic sawing, punching, turning, and sorting equipment. The finished stoppers are cleaned, bleached, sterilized, and bulk-packaged for shipment from Spain, Portugal, and North Africa to all parts of the world. The 50% of the cork from stopper production that remains as waste is granulated to provide furnish for cork composition and corkboard thermal insulation (see Insulation, thermal).

Cork Composites

Today most of the cork harvested from trees ends up in cork compositions of many types. The cork-stopper waste and poorer grades of cork bark are ground, cleaned, and classified into various particle ranges. Binders, plasticizers, and cutting aids are mixed with the ground cork, and the composition is molded or extruded into blocks, bars, sheets, tubes, rods, and other shapes. The final products are cut from these stocks and finished for market. Compositions range from ground cork with 5% protein binder to ground cork with 70% rubber or plastic binder. Cork compositions are used in a wide variety of gasket and packing applications because of their inherent chemical resistance, true compressibility, and low specific gravity. Cork compositions are also used for such products as handles for fishing rods and sports rackets, coasters, table pads, bulletin boards, and wall covering. A unique use for composition cork was developed by the aerospace industry. Certain compositions have been used to provide ablative insulation covering for rocket-engine casings and fuel tanks (see Ablative materials). Other assemblies used cork compositions to protect Apollo vehicles on the way to and from the moon.

Economic Aspects

The evolution of cork from its position as a relatively low-cost raw material to its current position in economic competition with a variety of synthetic substitute materials has caused a gradual decline in world consumption of cork since 1960. Pressure from increasing labor and shipping costs and higher tariffs induced an average cork price escalation of 300% during the period of 1950–1976. This trend is reflected in the decline in the United States' total imports of both manufactured and unmanufactured cork through the past several decades, as shown:

Year	U.S. imports, metric tons
1941	154,000
1950	135,000
1961	53,000
1976	20,000

Currently, most of the baled cork imported to the United States is either scrap corkwood to be processed or granulated cork. The largest volume of imported manufactured cork is corkboard insulation in block and sheet form which is also used for such applications as decorative paneling, lamps, and furniture.

The major consumers of cork are now Western and Eastern European and Asiatic countries. Portugal and Spain supply most of the Western European and North and South American markets, and North Africa supplies the rest of the importing countries. Total world consumption of cork for 1976 was estimated at 300,000 metric tons valued in raw material cost at $130 million.

BIBLIOGRAPHY

"Cork" in *ECT* 1st ed., Vol. 4, pp. 480–487, by G. B. Cooke, "Cork" in *ECT* 2nd ed., Vol. 6, pp. 281–288, by G. B. Cooke, Essex Community College.

General References

E. Palmgren, *Cork Production and International Cork Trade,* International Institute of Agriculture, F.A.O., 1947.
J. Marcos de Lanuga, *The Cork of Quercus Suber, Inst. For. Invest. Exper. Madrid* **35**(82), 1964.
P. Pla Casadevall, *El Suro,* Universitat Politecnica de Barcelona, 1976.
Report FT 246 Annual 1976, U.S. Imports for Consumption and General Imports, 1976.

R. H. HUTCHINSON
Armstrong Cork Co.

CORROSION. See Corrosion and corrosion inhibitors.

CORROSION AND CORROSION INHIBITORS

Corrosion is defined as the destructive attack of a metal by the environment, by chemical or electrochemical processes. Corrosion does not include the wasting of a metal by mechanical means such as the erosion of a metal structure by sand in the desert; but it does include conjoint mechanical and chemical action to produce early failure of a load-carrying metal structure. Stress-corrosion cracking, for example, is produced by the conjoint action of mechanical stress and specific chemical agents. The usual connotation of corrosion is one of an undesirable phenomenon; however, in technology there are many cases of useful corrosion. Among these are the formation of decorative coatings on metal items and the controlled corrosion of zinc in a primary cell to produce electrical energy (see Batteries).

In this article corrosion is treated as an undesirable wastage of metal and metal structures and consideration is given to its prevention.

Manifestations of Corrosion

The most common form in which corrosion appears is uniform attack. The entire metal surface is covered with the corrosion product. The rusting of iron in any humid atmosphere and the tarnishing of copper and silver alloys in sulfur-containing environments are examples. High temperature oxidation, or dry oxidation, is usually uniform in character.

An especially pernicious type of attack is called pitting corrosion. In such cases a pit or a number of pits may cause considerable damage to a metal structure, or may indeed penetrate the structure without the metal exhibiting any appreciable loss in weight. Pitting corrosion is usually seen on metals that are normally passivated with an oxide film. Aluminum alloys, such as those used in household storm window frames, exhibit this type of attack. Stainless steels of the nickel–chromium type exposed to quiet seawater will pit.

114   CORROSION AND CORROSION INHIBITORS

Several of the conjoint types of corrosion attack appear as cracking failures. In stress-corrosion cracking a metal part exposed simultaneously to a constant tensile stress and a specific corroding agent will crack intergranularly (between the metal grains) or transgranularly (across the metal grains). When the stress is cyclic rather than constant the failure is termed corrosion fatigue. Stress-corrosion cracking has assumed such importance in modern technology that this subject is discussed in more detail below.

Corrosion may appear in the form of intergranular attack. Often intergranular attack is the result of stress corrosion, but can also occur because the grain boundary and grain proper have different tendencies to corrode, ie, different electrical potentials. Intergranular attack becomes serious because it results in a loss in strength or ductility of the metal.

A further type of attack has been described as dezincification. Certain alloys of importance in commerce such as the Cu–Zn alloys contain a reactive metal (Zn) and a more noble metal (Cu). Upon exposure to a particularly corrosive environment the brass becomes covered with a reddish film suggesting that the zinc has gone into the solution and the copper remained behind. Actually, both elements dissolve and the copper is redeposited to form the reddish appearing surface.

Copper–aluminum alloys are also subjected to the same type of attack as are gold–silver alloys. In the latter case the reaction has been called parting.

Most of the observed corrosion occurs as one or more of these five cases. There are some special forms of corrosion such as the dissolution of a metal or alloy in a hot fused salt bath which are difficult to categorize.

Origin of Corrosion

The most common origin of corrosion is the basic thermodynamic tendency for metal to react as expressed in terms of the free energy of reaction (see Thermodynamics). However, in most cases of interest in engineering there are factors imposed on this basic tendency that accelerate the corrosion rate.

Some free energies of formation of compounds important in corrosion are given in Table 1 (1). The free energy of formation of the compound is the free energy for the reaction of the elemental metal with the other chemical species also in its elemental state, eg,

$$2\,\text{Fe} + \frac{3}{2}\text{O}_2 \rightarrow \text{Fe}_2\text{O}_3 \quad \Delta G° = -741.0 \text{ kJ/mol} \,(-177.1 \text{ kcal/mol})$$

The negative free energy of formation indicates a tendency for the metal to react, that is, the oxide is stable. A positive $\Delta G°$ indicates that the elemental metal is stable; note the positive $\Delta G°$ for Au_2O_3 (Table 1).

Although a large free energy of formation indicates that the formation of the compound is favored thermodynamically, the specific reaction may not go readily because of a high activation energy for the reaction. This is illustrated by the corrosion of lithium metal by air and its component gases. A comparison of free energy of formations of compounds that should be obtained from the exposure of the metal to several gases in air indicates that lithium should react readily with dry oxygen and dry carbon dioxide to form Li_2O and Li_2CO_3. However, a detailed study of the reactivity of Li metal with gases (2) showed that there was no detectable reaction with Li under 250°C. Water vapor is required for the reaction of Li with CO_2, probably as:

Table 1. Standard Free Energy of Formation of Some Corrosion Product Compounds, 25°C[a]

Compound	$\Delta G°$, kJ/mol[b]
$AlCl_3$	−636.8
$Al_2O_3 \cdot H_2O$	−1820.0
CdO	−225.1
CaO	−604.2
CuO	−127.2
Cu_2O	−146.4
Au_2O_3	+163.2
Fe_2O_3	−741.0
Fe_3O_4	−1014.2
$Fe(OH)_2$	−483.5
$Fe(OH)_3$	−694.5
MgO	−569.6
NiO	−216.7
Ag_2O	−10.8
ZnO	−318.2
$Zn(OH)_2$	−554.8
$ZnCO_3$	−731.4
Cr_2O_3	−1046.8

[a] Ref. 1.
[b] To convert J to cal, divide by 4.184.

$$2\,Li + H_2O + CO_2 \rightarrow Li_2CO_3 + H_2$$

in spite of the large free energy decrease for Li_2CO_3 formation.

A measure of reaction tendency (free energy) is the electrode potential. A series of electrode potentials of common elements is given in Table 2 (3). The negative potential indicates the strong tendency for the metal to oxidize (corrode). The standard electrode potential refers to the metal in a solution containing its ions at unit activity; obviously the potential will be modified by the environment. For this reason special

Table 2. Standard Electrode (Reduction) Potentials, 25°C[a]

Electrode	$E°$ (vs standard hydrogen electrode)
$Li^+ + e \rightarrow Li$	−3.045
$Zn^{2+} + 2\,e \rightarrow Zn$	−0.763
$Fe^{2+} + 2\,e \rightarrow Fe$	−0.440
$Cd^{2+} + 2\,e \rightarrow Cd$	−0.403
$Co^{2+} + 2\,e \rightarrow Co$	−0.277
$Ni^{2+} + 2\,e \rightarrow Ni$	−0.250
$Sn^{2+} + 2\,e \rightarrow Sn$	−0.140
$Pb^{2+} + 2\,e \rightarrow Pb$	−0.126
$Sn^{4+} + 2\,e \rightarrow Sn^{2+}$	+0.15
$Cu^{2+} + e \rightarrow Cu^+$	+0.153
$Cu^{2+} + 2\,e \rightarrow Cu$	+0.337
$Fe^{3+} + e \rightarrow Fe^{2+}$	+0.771
$Ag^+ + e \rightarrow Ag$	+0.7991
$½\,Br_2 + e \rightarrow Br^-$	+1.0652
$½\,Cl_2 + e \rightarrow Cl^-$	+1.3595

[a] Ref. 3.

116 CORROSION AND CORROSION INHIBITORS

tables of electrode potentials are required to measure the proper galvanic relationships existing between metals and alloys in the specific environment. This has been done for seawater and Figure 1 shows such a series (4).

A further factor altering the quantitative significance of the tabulated standard electrode potential as a measure of corrosion tendency is polarization or the alteration

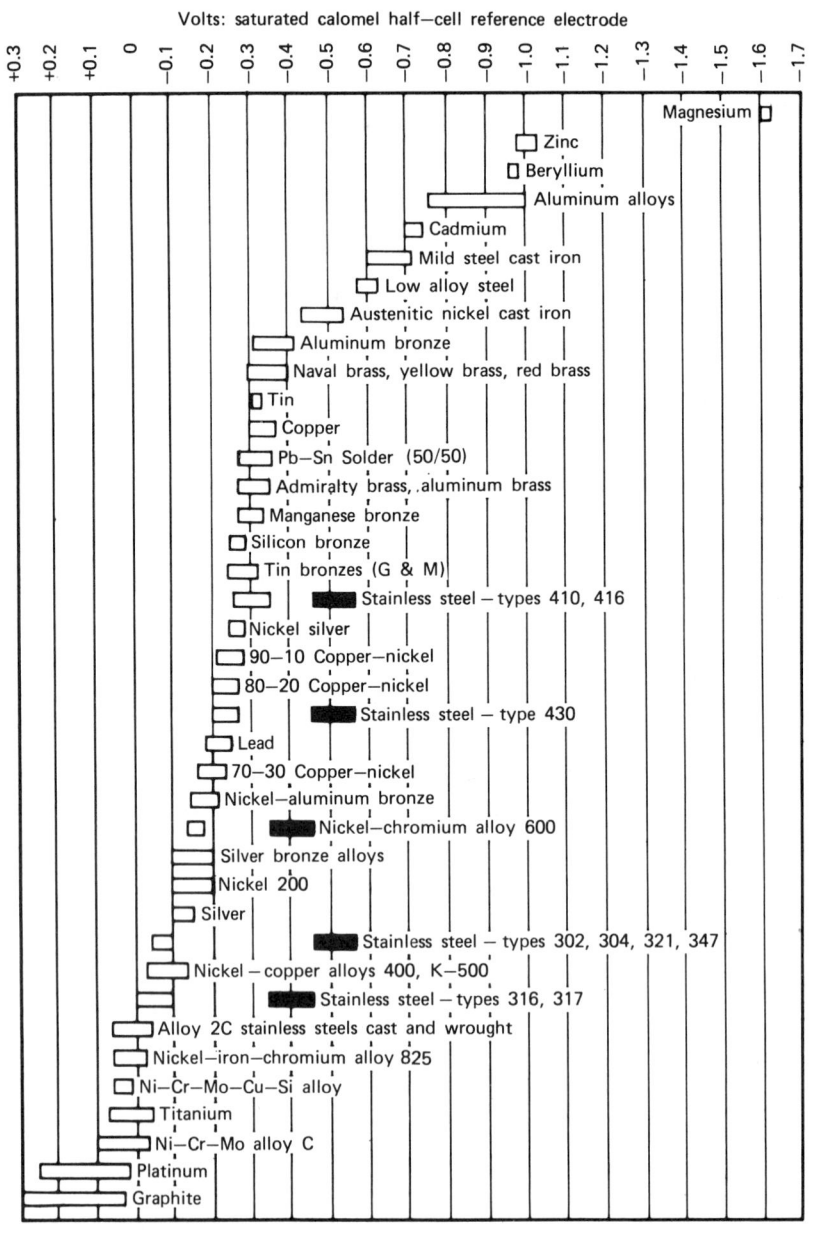

Figure 1. Galvanic series in seawater (4). Alloys are listed in the order of the potential they exhibit in flowing seawater. Certain alloys indicated by the symbol ■, in low-velocity or poorly aerated water, and at shielded areas, may become active and exhibit a potential near −0.5 V.

of the potential as current flows. This is discussed below when corrosion is considered as an electrochemical process.

In the absence of these complicating effects, it is possible to calculate the electromotive force of the galvanic cell from the expression $\Delta G = -nFE°$. Here ΔG is the free energy change for the reaction in kJ (cal), $E°$ is the emf of the cell in volts, n the number of electrons involved in the oxidation, and F the Faraday, 96.49 kJ/V (23,061 cal/V).

Imposed on this basic thermodynamic tendency for metals to seek their lowest energy level are other sources that lead to accelerated corrosion. One of the most common types of corrosion is galvanic corrosion, often labeled dissimilar metal corrosion. Recalling the table of standard potentials, it is apparent that if, for example, Cu ($E° = +0.337$ V) is connected with Fe ($E° = -0.440$ V) a potential difference between the two will be immediately established and a current will flow in a battery with an Fe anode and a Cu cathode. A galvanic cell of this type will operate if an iron pipe is connected to a copper pipe in a moist environment such as the soil. Many of the early unsuccessful applications of aluminum alloys were caused by the use of adjacent copper or iron bolts, screws, or fasteners which accelerated the corrosion of the contiguous aluminum sections.

A major source of corrosion in the field arises from concentration cells. These are of two types as illustrated in Figure 2. In Figure 2(**a**) a salt concentration cell is depicted in which the common metal, Cu, is in contact with two concentrations of the same salt, $CuSO_4$. Corrosion is accelerated at the site of low salt concentration. In Figure 2(**b**) an oxygen concentration cell or a differential aeration cell is depicted. Here corrosion is accelerated at the low O_2 concentration site because the high concentration site is cathodic.

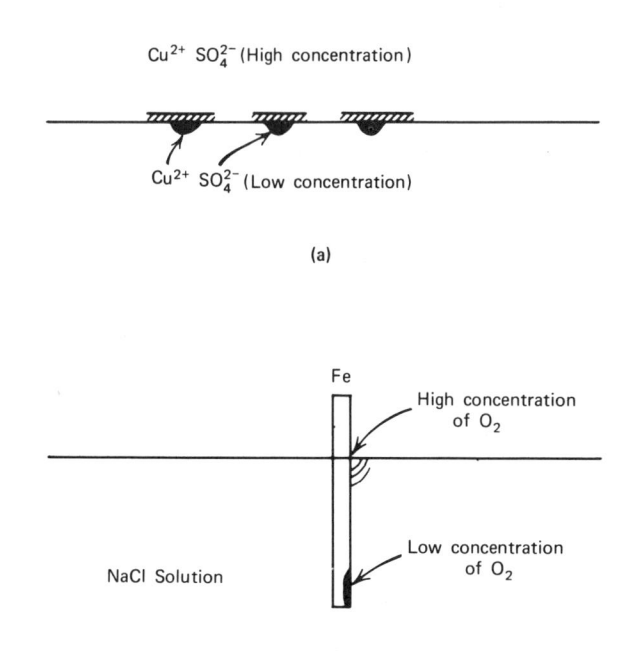

Figure 2. Concentration cells. (**a**) A salt concentration cell. (**b**) A differential aeration cell.

Concentration cells (salt or oxygen) are related to the phenomenon of crevice corrosion. Because of the geometry of a structure, one site may be covered or have restricted movement of solution adjacent to it. This leads to changes in local solution chemistry; that is, sections of the metal surface are exposed to widely different environments. In a crevice or a pit in seawater with a pH of 8.2, it was observed that the solution can achieve an acid pH (~3.5) which leads to rapid attack in the sheltered area. Filiform corrosion occurs under organic coatings in the form of threads, again caused by changes in local solution chemistry by the preferential availability of oxygen. Corrosion can also be accelerated by differential temperature cells. For example, given the cell:

$$Cu\ (at\ 60°C)/CuSO_4/Cu\ (at\ 20°C)$$

the lower temperature Cu site becomes anodic (corrodes) and the high temperature Cu site cathodic (copper plates out). All metal couples, eg, Cu/Cu^{2+}, Fe/Fe^{2+}, and Ag/Ag^+ have temperature coefficients which, under proper conditions, can yield thermogalvanic corrosion.

The acceleration of corrosion by the conjoint action of mechanical forces with chemical ones is mentioned above. This takes several forms dependent on the type of mechanical forces. Fretting corrosion arises from the relative slippage of surfaces in contact with each other. Often the slippage arises through vibration. Fretting corrosion has been observed between the strands of a steel elevator cable; eg, it is characterized by the presence of loose, reddish iron oxide (hematite).

Corrosion fatigue occurs when a metal is stressed in a cyclic manner in the presence of a corrosion environment. A metal that will survive a certain number of cycles at a specified stress in air, will fail at much fewer cycles in a corrosive environment.

Thermodynamic Basis. The thermodynamic data pertinent to the corrosion of metals in aqueous solutions have been systematically assembled by Pourbaix in a form that has become known as a Pourbaix diagram (5). The data include the potential and pH dependence of metal, metal oxide, and metal hydroxide reactions and, in some cases, complex ions. Along with the specific data for a given metal, the potential and pH dependence of the hydrogen and oxygen reaction are superimposed on the diagram. The Pourbaix diagram for the iron–water system is shown in Figure 3. The hydrogen reaction line is (a), the line representing the reduction reaction:

$$2\ H^+ + 2\ e \rightarrow H_2$$

and the oxygen line is (b), ie, the line representing the oxidation reaction:

$$2\ H_2O \rightarrow O_2 + 4\ H^+ + 4\ e$$

If the potential is moved below the dashed line (more negative) at a fixed pH, H_2 will be evolved. At a fixed pH, movement of the potential above the dashed line (b) will cause O_2 to be evolved. Thus, the region between the dashed line indicates the region of thermodynamic stability of H_2O.

The diagrams are constructed from data for reactions that are: (*1*) potential dependent and pH independent; (*2*) pH dependent and potential independent; (*3*) dependent on both pH and potential; and (*4*) dependent on neither pH nor potential. On this basis a diagram that summarizes an enormous amount of data can be divided into three major areas (see Fig. 4). The first is the region of immunity. In this region the pH is such, and the potential is sufficiently negative so that metallic iron is ther-

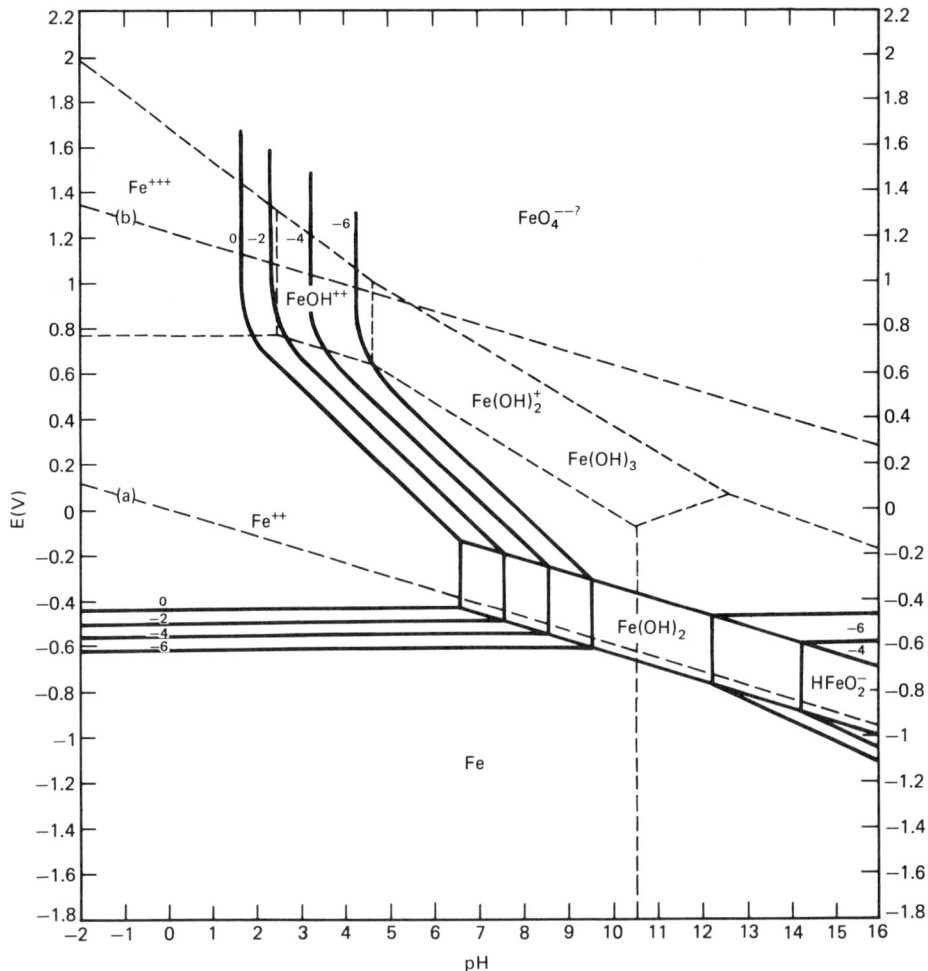

Figure 3. Potential-pH equilibrium diagram for the system iron–water, at 25°C (considering as solid substances only Fe, Fe(OH)$_2$, and Fe(OH)$_3$) (5).

modynamically stable and will not corrode. The second region is the corrosion area. In this region of pH and potential, Fe^{2+} and Fe^{3+} ions are stable. The third region is one of passivation. The metallic form of iron does corrode but forms insoluble hydroxides or oxides such as Fe(OH)$_3$ and Fe$_2$O$_3$ which prevent further dissolution.

On the Pourbaix diagram for Fe a corrosion area is designated at a pH of about 14 and a potential of −1.0 V. In this region the iron hydroxide oxide species, HFeO$_2$, which is soluble, is stable. This illustrates one of the limitations of the diagram. In certain environments the presence of certain anions causes the oxide film on the metal surface to dissolve as a complex ion, eg, Cu(NH$_3$)$_4$$^{2+}$. Owing to the large number of such complexing species, it would be impractical to construct all the pertinent diagrams. Thus, in the presence of chloride ions, the corrosion–passivation line for Fe would be moved to the right. Another complicating feature is the existence of a local chemistry adjacent to the corroding metal that differs from the solution chemistry. Thus, cathodically generated OH$^-$ will cause the pH to go up, and H$^+$ generated by

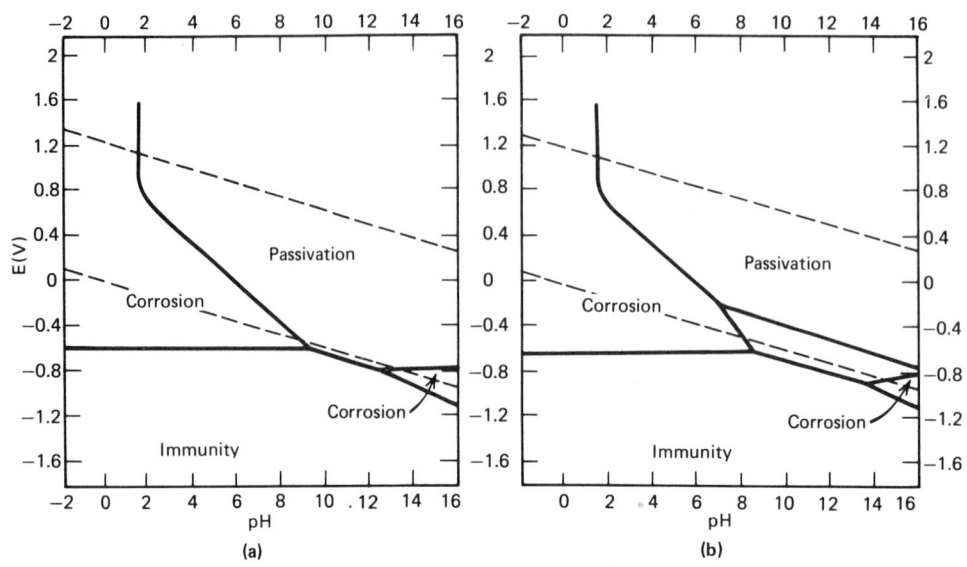

Figure 4. Theoretical condition of corrosion, immunity, and passivation of iron (5). (a) Assuming passivation by a film of Fe_2O_3. (b) Assuming passivation by films of Fe_2O_3 and Fe_3O_4.

the dissolution of the active metal will cause the pH to go down. Generally, however, the Pourbaix diagram represents well the basic thermodynamic stability of the metal–water system and should be considered the starting point for any corrosion study.

Electrochemical Basis. Early in the history of corrosion, it was discovered that corrosion is an electrochemical process rather than a strictly chemical reaction (see also Electrochemical processing). It was visualized that a corroding metal surface is comprised of a large number of local anodes and a large number of local cathodes whose sites may actually shift as the corrosion reaction ensues. At the anode site, of course, the metal is being oxidized as:

$$M \rightarrow M^{2+} + 2\,e$$

(for a divalent cation). In neutral solution at the cathode site oxygen (air) is being reduced as:

$$O_2 + 2\,H_2O + 4\,e \rightarrow 4\,OH^-$$

In acid solutions H^+ ions are being reduced. As current is drawn from the local cell, the anode potential moves in the positive direction and the cathode potential in the negative direction so that the metal surface assumes a uniform potential overall. This situation is depicted in Figure 5 (6). In the lower left the reversible or equilibrium potential for the $M \rightleftharpoons M^{2+}$ couple is indicated. The exchange current density is also indicated; this is the current for the forward reaction $M \rightarrow M^{2+}$, as well as the reverse reaction $M^{2+} \rightarrow M$, which are equal at this equilibrium potential. The magnitude of exchange current density for an electrode reaction is approximately a measure of the tendency for the reaction to go, or its reversibility. Thus, the exchange current density for the reduction of H_2 on platinized Pt is 10^{-3} A/cm^2 and that on Hg is 10^{-13} A/cm^2, meaning that it would be possible to construct a reversible hydrogen electrode on

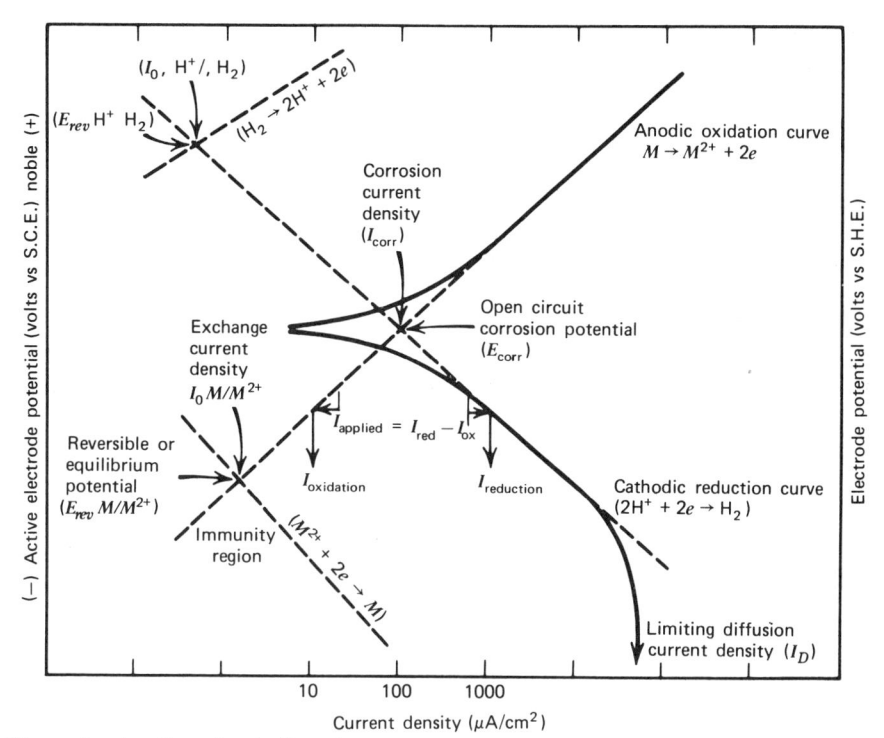

Figure 5. Anodic and cathodic polarization curves (6). SCE = standard calomel electrode; SHE = standard hydrogen electrode.

platinized platinum but not on mercury. In the upper left of Figure 5, the potential for the local cathode reaction is in this case the reduction of H^+. As current is drawn from this local cell (as current density moves to the right), the potentials of the two local cell reactions move to the open circuit corrosion potential E_{corr}, the intersection of the two polarization curves. The current at this point is labeled the corrosion current, i_{corr}.

If now, at this point, an external current is applied in the anodic direction, that is, the metallic electrode is required to assume a more positive potential, the potential-current density behavior moves up the anodic oxidation curve. On the other hand, polarization in the negative direction by this externally applied current will move the electrode down the cathode reduction curve. It should be noted that these last two plots are linear when the potential is plotted against the logarithm of the current density rather than the current density directly. The solid curves in Figure 5 represent polarization curves.

In the polarization curve (the cathodic curve) in the lower right part of the diagram, starting with the open circuit corrosion potential, the potential moves in the negative direction. The overpotential, η, is the difference between the actual potential E and the corrosion potential:

$$\eta = E - E_{corr}$$

The solid curve up to fairly negative potentials is represented by the Tafel plot.

$$\eta = a + \beta \log i$$

The applied current at any overpotential is the difference between the reduction current and the oxidation current, ie, $i_{applied} = i_{red} - i_{ox}$.

The Tafel equation was first observed empirically but is now interpreted in terms of modern electrode kinetics. The constant a incorporates the exchange current density for the reaction and the magnitude of β is indicative of the mechanism of the rate determining step in the electrode process.

The lower right of Figure 5 shows that the logarithm of the current density, instead of increasing in a linear manner, reaches a limiting diffusion current density i_D. Up to this point there is an ample supply of reducible hydrogen ions. However, owing to acceleration of the reduction reaction by the very negative potential, every hydrogen ion that reaches the electrode is reduced and the reduction current is limited by the diffusion of new hydrogen ions through the solution to the electrode surface. Whereas the current density, or the rate of reaction up to this point, has been limited by the electrochemical activation process (Tafel behavior), at this point the process becomes limited by concentration polarization.

Extrapolation of the η-log i curve, the Tafel plot, back to the corrosion potential E_{corr} yields a linear dotted line, or if the applied current causes the overpotential, η, to be only about 5–10 mV, the η-i relationship, rather than the logarithmic one, is linear.

This relationship has been used by Stern and Geary (7) to develop the linear polarization method for measuring corrosion rates. Their equation:

$$i_{corr} = \frac{i_{applied}}{2.3\,\eta} \frac{\beta_c \beta_a}{\beta_c + \beta_a}$$

includes β_c and β_a, the Tafel constants for the cathodic reaction (eg, hydrogen reduction) and the anodic reaction (eg, Fe oxidation), respectively. If reasonable values for these constants can be found or estimated, the slope of the $i_{applied} - \eta$ curve at E_{corr} can be used as a rapid method of measuring corrosion.

If sufficient cathodic current is applied, the potential of the metal will move below the potential of the reversible or equilibrium potential for the reaction (Fig. 5), $M = M^{2+} + 2\,e$. Therefore, the tendency for the metal will be to plate out rather than dissolve and the metal will be cathodically protected. The current can be applied by an external source, such as a rectifier, or a sacrificial anode, such as magnesium, which is employed for the protection of iron pipelines (qv). In practice, the situation might not be this simple since inadequate control of the potential might lead to hydrogen evolution and absorption by the metal leading to hydrogen embrittlement (note hydrogen evolution line on Pourbaix diagram). It is also critical to establish the proper ohmic resistance of the circuit.

The anodic oxidation curve (upper right, Fig. 5) indicates continual dissolution with increasing positive potential. In the Pourbaix diagram (Fig. 3), it is evident that movement of potential in the positive direction can promote the formation of insoluble hydroxides or oxides.

Some metals under certain environmental conditions, notably in the absence of chloride ion, can be anodically protected. If the metal is potentiostatically maintained in the proper potential range, the metal will resist corrosion indefinitely. By applying an anodic potential, stainless steel reactors have been protected against corrosion in concentrated sulfuric acid in the chemical industry.

Environmental Effects. Completely apart from metallurgical effects, the environment has a significant effect on corrosion rate. This means that even if a metal were completely pure and completely homogeneous, eg, a pure single crystal, it would corrode if dictated by thermodynamic and electrochemical considerations. The important environmental factors are the oxygen concentration in water or the atmosphere, the pH of the electrolyte or the temperature, and concentrations of various salts in solution in contact with the metal. The relative importance of some of these effects may be compared in the corrosion of iron.

An awareness of the important role of oxygen was developed in the 1920s. In his classical drop experiments (8), Evans showed that the corrosion of iron or steel by drops of electrolytes depends on electrochemical action between the central unaerated area, which becomes anodic and suffers attack, and the peripheral aerated portion which becomes cathodic and remains unattacked. He concluded that the velocity of iron dissolution depends on whether the product of the anodic reaction is soluble or insoluble.

Vernon (9) showed a linear relationship between rate of iron corrosion and oxygen pressure from 0–2.5 MPa (0–25 atm). Iron corrodes more slowly in concentrated chloride solution because oxygen solubility is less. Müller (10) concluded that oxygen: (1) acts as a cathode depolarizer causing bivalent Fe to go into solution; (2) oxidizes the dissolved ferrous hydroxide to rust; and (3) restores the oxide film.

For the corrosion of steel in solutions of LiCl, LiBr, and LiI, Paramonova and Balezin (11) concluded that the main effects were caused by oxygen solubility in the electrolytes rather than the nature of the anion. The data in Table 3 are typical results where the gas atmosphere is the only variable.

Table 3. The Solution Corrosion Rate of Steel-3 in Lithium Halides Under Conditions of Differential Aeration[a], g/(m²·h)

Substance	Control	Aeration with O_2	Aeration with H_2
LiCl	0.032	1.014	0.008
LiBr	0.033	0.768	0.012
LiI	0.037	1.32	0.008

[a] Concentration, 0.24 N; time, 48 h; temperature, 20°C; gas flow rate, 3 L/h. Ref. 11.

Table 4 shows the oxygen solubilities (oxygen content of aerated solution) for three solutions (12). At higher concentrations (greater than 0.1 N) a linear relationship exists between the observed weight loss of electrolytic iron and oxygen solubility in the electrolyte.

Table 4. Oxygen Solubility as a Function of Salt Concentration[a]

| Salt concentration, M | Solubility of O_2 in cm³/L |||
	KCl	NaCl	Na_2SO_4
0.0	6.13	6.13	6.13
0.5	5.23	5.20	4.87
1.0	4.51	4.46	3.98
2.0	3.41	3.19	2.67
3.0	2.55	2.31	

[a] Ref. 12.

The concentration dependence of corrosion in the steady state condition in KCl, NaCl, and LiCl solutions is shown in Figure 6 (13). In all three cases there is a maximum in corrosion rate; with NaCl it is at ca 0.5 N. With regard to temperature, the corrosion rate of iron reaches a maximum at ca 70°C. The increased rate of chemical reaction achieved with increased temperature is balanced by a decreased cathodic depolarization reaction owing to the inverse solubility of oxygen in the solution.

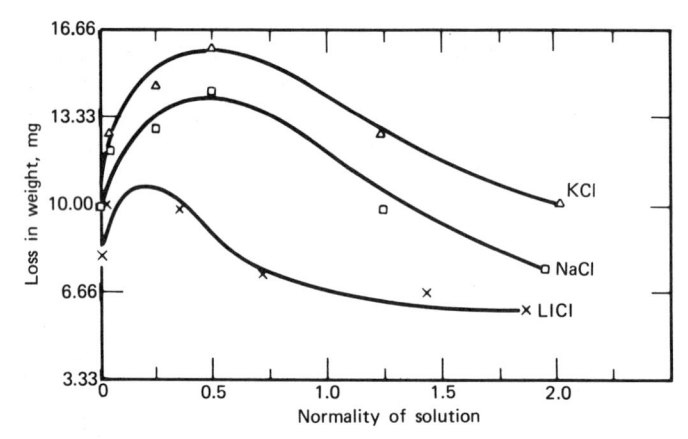

Figure 6. Corrosion-concentration curves for alkali chlorides (13).

The corrosion rate of iron in aerated water is also a function of pH and generally follows the pattern in Figure 7 (14). At 4–10 pH, the rate is controlled by the availability of oxygen as described above. In more acid solutions (lower pH) the rate is accelerated with the reduction of hydrogen ion replacing the reduction of oxygen as the rate controlling cathodic reaction. Some other metals follow approximately the same behavior with the exception of those metals that dissolve to form amphoteric ions. Zinc forms

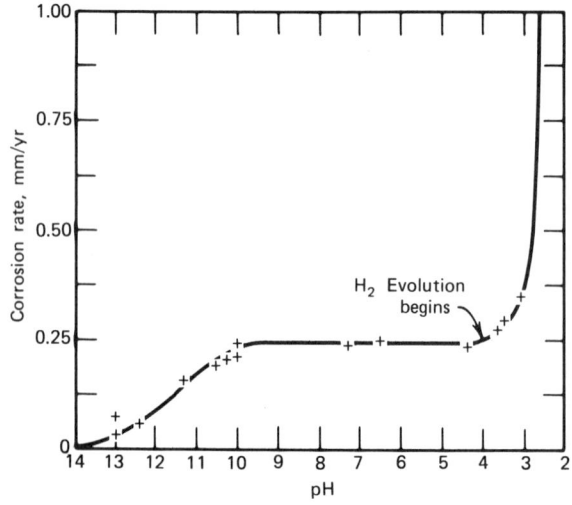

Figure 7. Effect of pH on corrosion of iron in aerated water at room temperature (14). To convert mm/yr to mpy, divide by 0.0254.

the zincate ion, ZnO_2^{2-}, which causes Zn to corrode excessively above a pH of 12; whereas Al forms the aluminate ion, AlO_2^-, which increases the dissolution rate above a pH of about 8.

The role that the chloride ion plays in the corrosion of metals is an important consideration for environmental effects because chlorides are ubiquitous in nature. A review of the role of chloride ion in iron corrosion (15) led to the classification of the theoretical analysis into the following five categories.

Oxide Film Properties. The chloride ion functions through its property of penetrating oxide films that otherwise are protective. The consideration that chloride ion has the property of dispersing normally protective films of colloidal nature must be added to this more traditional viewpoint.

Adsorption. The chloride ion is adsorbed preferential to some passivating species.

Field Effect. Chloride ions are adsorbed on the metal surface or on the thin oxide film to produce a strong electric field that can draw ions from the metal.

Catalytic. The halide ion catalyzes the reaction by forming some sort of intermediate bridging structure.

Complex Formation. The halide ion forms a surface complex with iron, the stability of which determines the corrosion kinetics.

Very often the environment is reflected in the composition of corrosion products, eg, the composition of the green patina formed on copper roofs over a period of years. The determination of the chemical composition of this green patina was one of the first systematic corrosion studies ever made (see Copper). The composition of patina varied considerably depending on the location of the structure and the approximate composition varied with the atmosphere as shown in Table 5.

Metallurgical Factors. The primary determining factor in corrosion behavior of metals and alloys is usually the chemical composition as reported by standard chemical analysis, eg, the amount of chromium in a stainless steel. There are, however, metallurgical factors not normally included in such analyses that may play roles varying from trivial to highly important, depending on the alloy, the nature of the metallurgical

Table 5. Composition of Green Patina on Copper from Different Locations[a]

Location of structure	Age of structure (yr)	Composition of green patina, %
urban	30	$CuSO_4$ (49.8)
		$CuCO_3$ (14.6)
		$Cu(OH)_2$ (9.6)
rural	300	$CuSO_4$ (25.6)
		$CuCO_3$ (1.4)
		$Cu(OH)_2$ (58.5)
marine	13	$CuSO_4$ (2.5)
		$CuCO_3$ (12.8)
		$CuCl_2$ (26.7)
		$Cu(OH)_2$ (52.5)
urban–marine	38	$CuSO_4$ (29.7)
		$CuCl_2$ (4.6)
		$Cu(OH)_2$ (61.5)

[a] Refs. 16–17.

factor, and the form of corrosion. These metallurgical factors include crystallography, grain size and shape, grain heterogeneity, second phases, impurity inclusions, and residual stress owing to cold-work.

The technologically important structural metals and alloys are polycrystalline aggregates, each grain of which is a single crystal. Grain size and orientation may vary over a wide range with a negligibly small effect on practical corrosion considerations. Grain shape may likewise vary greatly depending on the alloy and processing history. Figure 8 shows schematically the shapes of grains in a wrought high-strength aluminum alloy. These elongated, flattened grains can be produced by rolling, extruding, or forging, and their shape and orientation with respect to sustained tensile stress are of primary importance to stress-corrosion cracking, as is discussed further below.

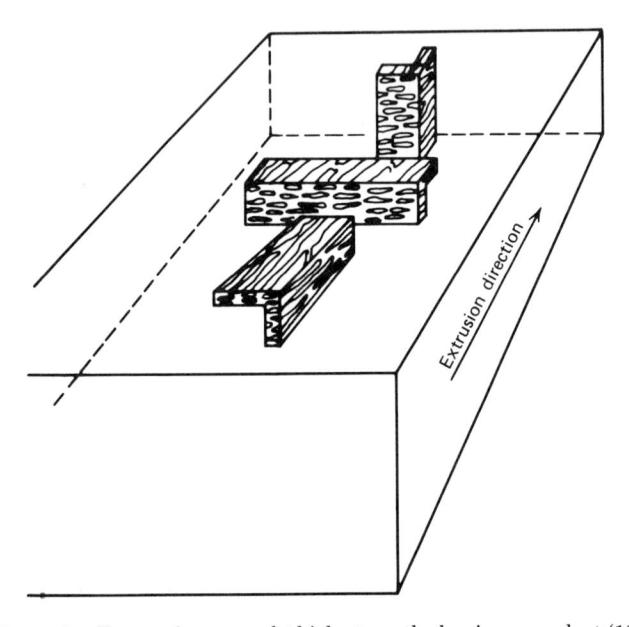

Figure 8. Texture in a wrought high-strength aluminum product (18).

Individual grains of an alloy, particularly in the as-cast condition, may exhibit inhomogeneity in composition from grain interior toward the grain boundary, a condition called coring; this can produce different electrochemical characteristics on a grain-diameter scale that can promote disintegration of the alloy by corrosion of a continuous anodic phase. This problem can be of great practical importance to wrought stainless steels and nickel alloys, as is discussed further under those headings below. Second phases such as ferrite grains in an otherwise austenitic stainless steel and beta grains in an otherwise alpha brass can be of considerable practical importance in some alloy systems and some forms of corrosion (see also below on stainless steels and copper alloys). Residual stresses from cold-working or other sources can cause differences in the reversible electrode potential of the metal, but these differences, where measured or calculated, have been found to be in the microvolt range, whereas the difference in potential of the bare metal and oxide-coated metal can be in the hundreds of millivolts. Thus, residual stresses from cold-work are of trivial effect on most forms of corrosion. One important exception is stress-corrosion cracking, in which residual stresses can be the determining factor.

Some of the important metallurgical factors in the more technologically important alloy families are as follows:

Stainless Steels. The most common and serious metallurgical factor affecting the corrosion resistance of stainless steels is termed sensitization. This condition is caused by the precipitation of chromium-rich carbides in the grain boundaries, giving rise to chromium-depleted grain-boundary media; the anodic areas tend to dissolve giving intergranular corrosion. Sensitization also makes stainless steel more prone to stress-corrosion cracking in chloride environments at elevated temperature. The precipitation responsible for sensitization occurs in the range of roughly 425–900°C and may take place during cooling from forging temperature, during welding, or during annealing when, for example, a stainless steel component has been attached to a carbon steel pressure vessel that must be stress-relieved thermally. There are several measures available to mitigate sensitization, ie, low carbon grades are available, such as AISI 304L, etc, which have a lesser proclivity toward sensitization. Stabilized grades such as AISI 321 and 347 are available that contain titanium and niobium, respectively, to tie up the carbon. Sensitization can be detected chemically (eg, using ASTM A262 methods) employing coupons that have been processed with the structure or component of interest.

Sulfide inclusions have been identified as the origins of stress-corrosion cracks in stainless steel structures under some conditions. Conceivably, in critical components in a marginal situation, it might be useful to specify premium-grade low sulfur steel, and certainly to avoid grades that have been resulfurized for improved machinability (see Steel).

Copper Alloys. The metallurgical factor of greatest importance in the corrosion response of copper alloys is residual stresses left from forming which can cause stress-corrosion cracking. The other metallurgical factor that can be of importance is the presence of the beta phase. Whether this phase appears, and its proportion if it does appear, are dependent on both composition and processing history. Alpha-phase brasses are subject to a form of corrosion known as dezincification in such environments as tap water, seawater, and some moist soils. This form of attack can be greatly mitigated by alloying with Sb, As, or P. But these inhibitors are ineffective against dezincification of beta grains (see Copper and copper alloys).

Aluminum Alloys. Wrought high-strength aluminum alloys, whether the products are rolled, forged, or extruded, tend to be highly textured because of the manner in which secondary phases or inclusions are strung out. This processing causes the grains of the primary alpha phase to be flattened and elongated, as is shown schematically in Figure 8 (18). This metallurgical texture is of comparatively minor consequence to most forms of corrosion, but it can promote exfoliation corrosion (intergranular corrosion leading to the leafing-off of uncorroded grain-bodies). And the texture is of first importance to the stress-corrosion cracking technology of high-strength aluminum alloys (see Aluminum and aluminum alloys).

Nickel Alloys. The nickel-base alloys containing chromium for passivity are of great importance to the chemical industry. They can become sensitized similarly to the austenitic stainless steels and become vulnerable to intergranular corrosion. A standard accelerated method to verify whether a given lot of material is susceptible to this form of attack or not is to expose it to boiling acidified ferric sulfate for 24–120 h (depending on the alloy). The procedure for conducting the test and evaluating the results is available as a standard method (ASTM G28).

The National Association of Corrosion Engineers has compiled (19) a comprehensive listing of the performance of many metals and alloys in environments including chemical processes, water, soil, and high temperature (see Nickel and nickel alloys).

Stress-Corrosion Cracking. Stress-corrosion cracking (SCC) is a fracturing process caused by the conjoint action of tensile stress and corrosion. The stress may be design operating stress, or residual stresses from welding, heat treatment, fit-up, cold-forming, or combinations of these. A large proportion of service failures are caused by stresses other than design stress. The corrodent responsible for cracking need not be present in either large quantities of high concentrations, nor need it be severely corrosive; in fact most stress-corrosion cracking in practice occurs when the alloy is almost but not totally inert to the environment. It was once supposed that there was a specific species in the environment that was responsible for SCC in a given alloy. This rule of specificity, which appears in the older literature, is now known to be invalid. Several or even many species may be responsible for SCC in a given alloy. Only alloys and not pure metals are susceptible to SCC in a practical sense.

The path of SCC may either be between the grains of the alloy (intergranular mode), or across the grains of the alloy (transgranular mode). Small changes in the environment can cause transgranular cracking to become intergranular, or vice versa. Relatively small changes in the alloy can also produce a change in the fracture mode. Likewise the crack may be single and unbranched, or cracking may have multiple origins and may also be branched. Thus, neither intergranularity nor transgranularity of crack path is indicative of or excludes SCC as the cause of a given service crack. Similarly, cracking of a nonbranching character does not exclude SCC as a possible cause of a service crack. But multiple branching of a crack, particularly if there is also multiple nucleation of families of branched cracks, is usually diagnostic evidence of SCC in a service failure. The conditions that can cause SCC may also cause corrosion fatigue if the stress is cyclic. Usually it is possible to distinguish between SCC and corrosion fatigue by electron fractography (20).

It is possible for overload fracture, fatigue, and presumably corrosion fatigue to occur as slant fractures (roughly 45 degrees to the maximum principal stress), but stress corrosion cracks occur only perpendicular to the tensile stress that causes them, that is, they occur by the opening mode, as in the opening of a book. Most stress corrosion cracks have the appearance of macroscopic brittleness.

The damage caused by SCC can be grossly out of proportion to the amount of corrosion that has occurred. For example, SCC can initiate fatigue cracks in aircraft fuselage panels, initiate brittle fracture of high-strength alloy components, or perforate a condenser tube wall permitting cooling water to contaminate a boiler.

It is generally agreed that there is no single theory that can account for all SCC. Most of the theories can be fitted into the following categories: (1) *Mechano–electrochemical.* Stress opens the crack to admit corrodent that reacts with the metal at the crack tip. (2) *Film rupture.* A brittle corrosion-product film forms, breaks at the crack tip, and the exposed bare metal reacts to form a film, and the cycle repeats. (3) *Embrittlement.* The metal near the crack tip becomes embrittled, as by adsorbing hydrogen, and permits the crack to advance. (4) *Adsorption.* The energy required to generate fracture surfaces is reduced by the adsorption of specific species. (5) *Periodic electrochemical–mechanical.* The fractures are generated partly by electrochemical dissolution which leaves behind resistant "posts"; these posts later rupture by a purely mechanical process. Extended discussions of theories are given in reference 21. None of these theories has been developed to the point of being useful to the technologist at this time, and present day SCC control measures are based either on service experience or on characterization using macroscopic specimens.

Evaluation of Stress-Corrosion Cracking. Many different types of macroscopic specimens are in current use to evaluate SCC in various systems (22). The smooth (as opposed to precracked) specimens in common use are shown in Figure 9. The Brinell impression, the Erichsen cup, and the U-bend tests are not suitable for quantifying the stress which varies at different points in a given specimen from above the yield strength all the way to zero. Such tests are used to identify combinations of alloys and environments that produce or do not produce SCC. With the beam, tensile, or C-ring specimens, it is possible to quantify the stress factor, and using such specimens one can proceed to search for a threshold stress below which cracking is not observed in an arbitrary exposure period (see also under Corrosion testing). Much of the engineering practice now in use to mitigate SCC is based on such data. The U-bend, beam, tensile, and C-ring specimens, as well as a number of environments, have been standardized either as Methods or as Recommended Practices by ASTM; new standards are added from time to time which are available in the List of Subjects in Part 10 of the latest *ASTM Annual Book of Standards*.

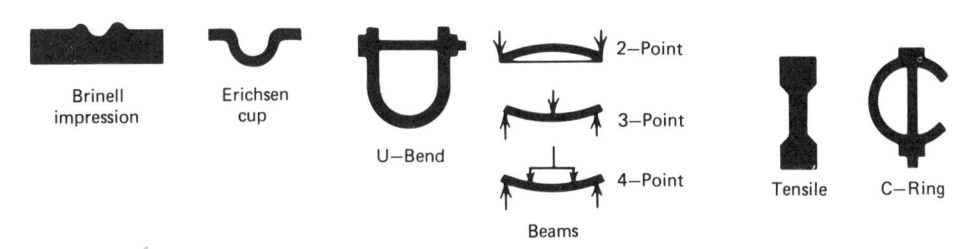

Figure 9. Smooth stress-corrosion cracking specimens (22).

In addition to tests using the specimens of Figure 9, there are two other fundamentally different tests for characterizing the SCC response of alloys. One of these tests employs a precracked specimen, four types of which are shown in Figure 10. The rationale behind such a test is that in a large or complex structure one must address the possibility that it will go into service with one or more crack-like flaws, and for prudence one would like to know the response of a cracked, stressed body in the presence of the service environment. In the presence of a crack, a nominal stress is a fiction, but the stress field at a crack tip can be quantified by the methods of linear elastic fracture mechanics. The methods for determining the stress intensity parameter

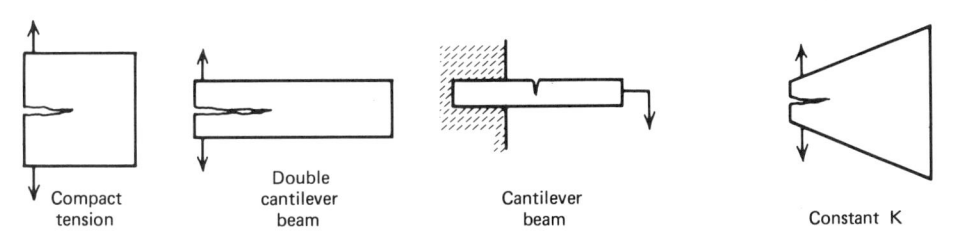

Figure 10. Principal fracture mechanics of specimens that have been used in stress-corrosion cracking studies (22).

as a function of load, specimen geometry, and crack geometry for various specimens are given in reference 23. In at least some systems there is a threshold stress intensity designated K_{Iscc} above which SCC growth is observed and below which it has not been detected. The value of such a parameter is that it affords, through the Irwin equation, a means to estimate the combinations of flaw size, flaw shape, remote stress level, and yield strength that will be expected to produce SCC in the environment in which K_{Iscc} was measured. The relationship between these terms is shown in the following equation:

$$K^2 = \frac{1.2 \pi \sigma^2 a}{\phi^2 - 0.212 \left(\frac{\sigma}{\sigma_y}\right)^2}$$

where a is the depth of a surface crack, σ is the stress, σ_y is the yield strength, and ϕ is a factor for the shape of the crack. If the length of the crack is $2b$, then ϕ^2 has the values for various crack shapes listed in Table 6.

Table 6. Values of ϕ^2

$\dfrac{a}{b}$	ϕ^2
0 (very long, thin crack)	1.00
0.25	1.14
0.5	1.46
0.75	1.89
1.0 (semicircular)	2.46

The value of a parameter such as K_{Iscc} is that it enables one to start making a rationale for the flaw detection that must be assured to avoid SCC in various field situations.

The second fundamentally different method for assessing SCC characteristics is known variously as the constant extension rate test, the constant strain rate test, or the slow strain rate test. In this test the ductility and the fracture strength of the specimen, which is often a miniature tensile specimen, is affected by the amount of SCC, if any, that has preceded terminal fracture. Thus, the ductility or strength of the specimen reflects the amount of SCC (see reference 24 for examples). By conducting such a test over a range of potentials, for example, it is possible to identify potential ranges in which SCC is a hazard for the system under study.

The phenomenon of SCC involves stress, an alloy, and a chemical (or electrochemical) environment. All the measures that are possible to mitigate the problem fit into these three categories. Typically, in practice one cannot exercise absolute control over any given category, because of various constraints, and it is common engineering practice to operate on factors in two or even all three categories to enhance the degree of mitigation.

Stress-Corrosion Cracking Mitigation by Alloy Families. It is most convenient to treat the SCC mitigative measures by alloy families rather than by component because the hazardous environmental species responsible for SCC and the measures taken to avoid SCC tend to group themselves by alloy families (22).

Copper Alloys. Ammoniacal SCC is the most common form of SCC in copper-base alloys. In addition to ammonia or its derivatives, this form of cracking occurs only in the presence of oxygen (or presumably an oxidizer) and water, which may be in the form of a thin film of condensate or a deliquescent layer on surface corrosion products. Typically, the stresses that cause ammoniacal SCC in service are not design stresses but are usually forming stresses. Forming stresses may be reduced or removed by an annealing treatment, suggested temperatures for which are available for various copper alloys (25). An old procedure for determining whether residual stresses have been reduced is the acidified mercurous nitrate test (ASTM B154-71). But this is not a fully conservative test, and items that pass this test uncracked may subsequently crack in an ammoniacal environment. The various alloys differ markedly in their susceptibility to ammoniacal SCC.

An example of multiple precautions that may be taken in engineering practice is the large modern power plant main condenser. The tubing may experience an ammoniacal environment on the steam side because of hydrazine used to scavenge oxygen from the boiler water. The oxygen level in a well-operated condenser is very low (except in the air removal section). Note above that oxygen is necessary to cause ammoniacal SCC, and low oxygen in the steam might be a sufficient measure. However, the available data are not sufficiently quantitative, nor is our knowledge of the composition of the steam environment; hence, in addition to the low level of oxygen, it is common practice to tube the condensers with cupro–nickel which has a low degree of susceptibility; furthermore the tubing is annealed with precautions to minimize the reintroduction of stress during handling and fabricating.

For small heat exchangers, whose failure through SCC would cause high economic losses either directly or through cost of down time, some engineers prefer to anneal the entire fabricated heat exchanger to ensure minimal fabrication residual stresses (see Heat exchange technology).

Atmospheric air heavily polluted with SO_2 can also cause cracking of some copper alloys. Sulfamic acid in the concentrations used for cleaning is capable of causing SCC in brass condenser and heat exchanger tubes. Nitric acid fumes have been observed to crack stressed brass. Aluminum bronze D (C61400) is susceptible to SCC in live steam. Mercury can cause cracking in a wide variety of copper alloys, including cupro–nickels that have very low susceptibility to ammoniacal SCC. Other diverse species such as tartrate, acetate, citrate, and other anions have been found capable of causing cracking in the laboratory, but service failures attributable to these species have been rare, if indeed they have occurred.

Aluminum Alloys. The aluminum alloys that have caused the most problems because of SCC in service are the 2000-series and 7000-series high strength alloys. These alloys in the wrought mill product forms are highly textured (see Fig. 8), and the SCC behavior differs greatly according to the direction of tensile stress with respect to the texture directions. The alloys are most vulnerable to SCC if stressed parallel to the short transverse grain direction (parallel to the thinnest dimension of the grain), most resistant if stressed only parallel to the longest grain dimension, and show intermediate susceptibility if stressed parallel to the long transverse direction. In practice, it is the short transverse-direction stresses that cause SCC problems; hence prudent practice is to avoid designs in which high sustained stresses are imposed across the short transverse direction. For example, one avoids an interference-fit fastener, or a taper pin fastener oriented to stress a part across the vulnerable texture unless the alloy is inherently of low susceptibility to SCC. Table 7 summarizes alloys and tempers in various categories of susceptibility as judged from specimens with short-transverse (most vulnerable) orientation.

In addition to the possibility of selecting alloys with the minimum SCC susceptibility while retaining other properties as needed, there are other steps possible to reduce the probability of SCC: (1) Avoid designs that permit water to accumulate in contact with the aluminum. (2) Avoid conditions in which salts, especially chlorides, can concentrate in contact with the aluminum. (3) Use an alloy clad with an anodic coating where available and otherwise acceptable.

High Strength Steels. Steels that owe their strength to heat-treatment, whether they are martensitic, precipitation hardening, or maraging, and whether stainless or not, are susceptible to SCC in aqueous environments, including water vapor. The primary factor in determining the degree of susceptibility of a given steel is its strength. There is no sharply defined threshold strength that defines a threshold susceptibility, but above 1200 MPa (ca 175,000 psi) yield strength the problem becomes of increasing concern, until at 1400 MPa (ca 200,000 psi) the problem is properly termed acute. This does not mean that steels cannot be used at such strength levels or even higher since they are so used. But if they are used at these strengths, great care must be exercised in design that any tensile stresses (or bending stresses) are small, or that moisture is excluded from the surface of the steel.

Table 7. Categories of Aqueous Susceptibility of Commercial Wrought Aluminum Alloys in Plate Form—Short Transverse Orientation[a]

Susceptibility category	Alloy	Temper[b]
very low	1100	all
	3003, 3004, 3005	all
	5000, 5050, 5052, 5154	all
	5454, 6063	
	5086	O, H32, H34
	6061, 6262	O, T6
	Alclad: 2014, 2219, 6061, 7075	all
low	2219	T6, T8
	5086	H36
	5083, 5456	controlled
	6061	T4
	6161, 5351	all
	6066, 6070, 6071	T6
	2021	T8
	7049, 7050, 7075	T73
moderate	2024, 2124	T8
	7050, 7175	T736
appreciable	7049, 7075, 7178	T6
	2024, 2219	T3, T4
	2014, 7075, 7079, 7178	T6
	5083, 5086, 5456	sensitized
	7005, 7039	T5, T6

[a] Ref. 26.
[b] Standard mill designation.

Cadmium plating is a useful cathodic protection measure that does not risk over-protection, but hydrogen either must not be codeposited with cadmium to a dangerous degree, or if codeposited (as is the case with the cyanide plating bath), it must be safely redistributed by thermal treatment before the high-strength steel component is stressed; otherwise the steel may experience hydrogen-embrittlement cracking which can be as serious a problem as SCC, if not worse.

Stainless Steels. Austenitic stainless steels undergo SCC when stressed in hot aqueous environments containing chloride ion. The oxygen level of the environment can be important, probably through its effect in establishing the electrode potential of the steel. The total matrix of combinations of stress level, chloride ion and oxygen concentrations, and temperature that will cause SCC has not been worked out. At about yield strength stress and about 290°C, 1 ppm Cl$^-$ and 1 ppm O$_2$ are approximately the minimum levels that initiate SCC. At room temperature, chloride SCC seldom occurs in austenitic stainless steels except when they are heavily sensitized. If they are not heavily sensitized, there is not usually a problem except at elevated temperatures. There is no well defined threshold temperature for vulnerability to SCC in stainless steels, but above about 60°C it becomes of increasing concern.

The relative susceptibilities of the common grades of austenitic stainless steel to chloride SCC do not differ greatly, although the steels that can and do become sensitized are decidedly inferior in this condition. The high purity ferritic grades of stainless steel, which have become commercially available only recently, offer appreciable improvement in SCC resistance compared with the austenitic grades, but they are not immune to cracking; care must also be exercised to avoid the ductile-to-brittle transition problem and the potential embrittlement by sigma-phase formation in elevated temperature service. The standard methods for avoiding chloride SCC in austenitic stainless steels include avoiding introducing fabrication stresses, minimizing chloride ion level, and minimizing oxygen concentration in the environment. Additionally, where feasible, it is possible to mitigate against SCC in these alloys by cathodic protection, such as can be provided by coupling to iron or to lead.

Corrosion Resistant Materials

A large number of alloys has been developed with varying degrees of corrosion resistance in response to various environmental needs. At the lower end of the alloying scale are the low alloy steels. These are iron-base alloys containing from 0.5–3.0% of Ni, Cr, Mo, or Cu and controlled amounts of P, N, and S with the exact composition varying with the manufacturer. The corrosion resistance of the alloy is based on the protective nature of the surface film, which in turn is based on the physical and chemical properties of the oxide film. As a rule, this alloying will reduce the rate of rusting by 50% over the first few years of atmosphere exposure. These low alloy steels have been used outdoors without protection.

The stainless steels contain appreciable Cr, Ni, or both. The straight chrome steels, types 410, 416, and 430, contain about 12, 13, and 16% Cr, respectively. The chrome–nickel steels include type 302 (18% Cr and 9% Ni), type 304 (19% Cr and 10% Ni), and type 316 (19% Cr and 12% Ni). Additionally, type 316 contains 2–3% Mo which greatly improves the resistance to crevice corrosion in seawater as well as general corrosion resistance. All of the stainless steels offer exceptional improvement in all sorts of atmospheric conditions. They depend for their unusual corrosion resistance on the formation of a passive film, and for this reason are susceptible to pitting. Type 304 stainless, the so-called 18-8 alloy, has very good resistance to moving seawater but does pit in stagnant seawater owing to the development of concentration cells (see Figure 2(**a**) and (**b**)).

Several copper alloys are exceptionally resistant to certain atmospheres. The Cu–Ni alloys, 90% Cu, 10% Ni, and 70% Cu, 30% Ni have outstanding resistance to corrosion in seawater, assuming the Fe content (0.5–1%) is properly controlled. In this application the alloy must resist fouling as well as corrosion. Monel alloy (66.5% Ni, 31.5% Cu) is widely used in marine service where strength is also required. Several copper alloys, silicon bronzes, aluminum bronzes, and manganese bronzes are resistant to severe atmospheric corrosion conditions, and because of their high strength are used as bolts, clamps or load-carrying parts in outdoor environments. Specific alloys have been found suitable for very corrosive environments. For example, Carpenter 20 (20% Cr, 29% Ni, 2.5% Mo, and 3.5% Cu) has outstanding resistance to concentrated sulfuric acid.

For most environments studies have been reported with quantitative data describing the corrosion rate of various materials including a number of corrosion resistant alloys. For example, Table 8 gives weight losses suffered by various corrosion resistant alloys in 28% phosphoric acid, 20–22% sulfuric acid, and 1–15% fluoride (27).

Table 8. Plant Test in Sulfuric Acid Dilution with Recirculated Phosphoric Acid[a]

Materials	Corrosion rate, μm/yr (mpy)	Concentration cell depth, μm(mils)
Carpenter stainless no. 20Cb	28 (1.1)	
Aloyco 20	38 (1.5)	
Incoloy alloy 825	69 (2.7)	127 (5)
Hastelloy alloy C	71 (2.8)	
Illium "G"	76 (3.0)	
Inconel alloy 718[b]	81 (3.2)[c]	
Worthite	109 (4.3)	76 (3)
Incoloy alloy 901[d]	155 (6.1)	127 (5)
Illium "R"	175, 363 (6.9, 14.3)	
Monel alloy K-500[e]	787 (31)	[f]
Monel alloy 400	965 (38)	
Hastelloy alloy B	1980 (78)	
stainless type 317	>3800 (>150) (s)	
stainless type 316	>4600 (>180) (s)	

[a] Phosphoric acid (wet-process) 28% (20% P_2O_5) sulfuric acid 20–22%, fluoride approx 1–1.5%, probably as hydrofluosilicic acid; temperature 82–110°C, average 93°C; and duration of test 42 days, moderate aeration, agitation by convection only.
[b] Composition 52.5% Ni, 18.6% Cr, 18.5% Fe, 5.0% Cb, and 3.1% Mo.
[c] Pitted to a maximum depth of 127 μm (5 mils).
[d] Composition 42.7% Ni, 34.0% Fe, 13.5% Cr, 6.2% Mo, and 2.5% Ti.
[e] Composition 65.0% Ni, 29.5% Cu, 2.8% Al, and 1.0% Fe.
[f] Pitted to a maximum depth of 76 μm (3 mils).
[g] Stress-corrosion cracking around markings.

Corrosion Inhibitors

Inhibitors are defined as chemical substances that, when added in small amounts to the environment in which the metal would corrode, will retard or entirely prevent this corrosion. In this sense, sodium sulfite and hydrazine added to water to remove oxygen, or silica gel to remove water from the atmosphere may be called inhibitors. However, in this article the term inhibitors is restricted to those materials that must interact with the metal surface to prevent corrosion.

The subject of inhibitors from theoretical and applied viewpoint is well covered in a monograph, *Corrosion Inhibitors* (28).

Inhibitors may be discussed as (*1*) inorganic inhibitors, (*2*) organic inhibitors, and (*3*) vapor-phase inhibitors, without any implication that inorganic inhibitors function by a mechanism difference from organic or vapor-phase inhibitors.

Inorganic Inhibitors. Considering first inorganic inhibitors, a subclassification is usually made on the basis of the functioning of the inhibitor with or without oxygen. Inhibitors that can function without oxygen are sometimes called passivators (30). These compounds include chromate and nitrate. They themselves are readily reduced and are able to oxidize the metal surface, usually iron, to form a passive oxide film. Other inorganic compounds required oxygen. These include sodium phosphates, silicates, and borates.

Inhibitors may also be classified in terms of their mechanism, that is, whether they function by influencing the anodic or cathodic side of the electrochemical corrosion cell, although there is not general agreement with regard to a given inhibitor functioning as an anodic or cathodic inhibitor under all conditions of pH, oxygen content, and temperature. However, chromates, nitrites, silicates, phosphates, and borates are usually considered to be anodic inhibitors and those cations that react with the cathodically generated hydroxide to form an insoluble compound, such as Mg^{2+}, Cu^{2+}, Zn^{2+}, Cd^{2+}, Mn^{2+}, and Ni^{2+}, are considered to be cathodic inhibitors; thus, calcium polyphosphate [7758-87-4] may be viewed as a cathodic inhibitor. The differentiation is made by the direction in which the potential moves upon the addition of the inhibitor to the system. An anodic inhibitor will cause the potential to move in the positive direction, the cathodic inhibitor will move the potential in the negative direction, that is, towards the equilibrium potential of the anodic reaction.

To inhibit corrosion in cooling waters, polyphosphates, nitrites, and chromates have been used, although in recent years the use of the latter has been discouraged even in closed systems because of environmental considerations. In municipal water supplies low concentrations (eg, 10–200 ppm) of polyphosphate and silicate have been employed (see Water, municipal water treatment). In some hot water systems borax [12447-40-4] has been used, and in common with some other systems, an adjustment in pH to the neutral range has been required (see Water, industrial water treatment). The antifreeze (qv) mixtures used in automobile cooling systems present a difficult problem because within the system several metals, iron, copper, lead–tin solders, aluminum, as well as rubber are in contact, encouraging galvanic corrosion. The successful formulations are proprietary, but some contain borates, polyphosphates, and mercaptobenzothiazole [1321-08-0] (an organic inhibitor). In acid solutions as, for example, those used in the pickling of steel, organic compounds have been used most successfully (see Metal surface treatments).

The concentrations required for inhibition are dependent on such factors as the presence or absence of chloride ion, the temperature, and the movement of the corroding solution; however, usually the effective concentration for inorganic inhibition falls in the range of several hundred ppm. Oxidizing inhibitors usually function at considerably lower concentrations than nonoxidizing inhibitors. It should be noted that, with respect to the oxidizing anions such as chromate, nitrite, molybdate, and tungstate there exists a critical concentration (31). These along with the concentrations to achieve inhibition are shown in Figure 11 (32).

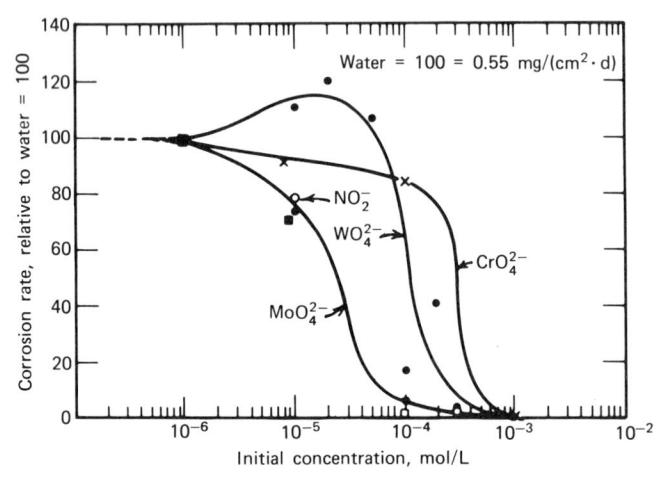

Figure 11. Concentrations of several inhibitors required to achieve inhibition (32).

Organic Inhibitors. A large number of organic compounds have been used as organic inhibitors; one review (33) lists 141 basic structures that have been utilized. Organic inhibitors and their effectiveness have been systematically discussed (34) on the basis of the type of bonding the organic molecule achieves with the metal. It is generally recognized that to be effective the compound must be adsorbed, but the type of adsorption bond varies with the chemical configuration of the molecule. The main types of adsorption involve electrostatic adsorption, chemisorption, and π-bond (delocalized electron) adsorption. Foroulis (34) cites examples of each type of adsorption. Inhibition by electrostatic adsorption is illustrated by aniline [62-53-3] and substituted anilines, pyridine [25275-41-6], butylamine [109-73-9], benzoic acid [65-85-0], and substituted benzoic acids and compounds such as benzenesulfonic acid [98-11-3].

Some of the types of compounds that function through electrostatic adsorption may also function by a chemisorption process. Chemisorption is most evident with nitrogen or sulfur heterocycles. Benzotriazole [27556-51-0] and tolytriazole [29385-43-1], both effective inhibitors of copper corrosion, are believed to operate through chemisorption as does 0.1 M butylamine which is effective in inhibiting the corrosion of iron in concentrated perchloric acid. However a single compound may apparently utilize different mechanisms to be effective.

The interaction of delocalized electrons (π-bond orbital interaction) with the metallic surface may be quite effective. Data on the corrosion inhibition of 1020 carbon steel in 2.8 N HCl at 65°C, given in Table 9 (35), illustrates this effect. As the structure goes from the single to the double to the triple bond the opportunity for π-bond interaction with the metal increases. However, as indicated by the last compound, such factors as steric interference may decrease the efficiency of inhibition.

Table 9. Corrosion Inhibition Data with 1020 Carbon Steel[a]

Inhibitor[b]	CAS Registry No.	Corrosion rate, metal loss in mg/(dm²·d)
blank		>48,900
CH₃CH₂CH₂OH	[71-23-8]	>48,900
CH₂=CHCH₂OH	[107-18-6]	13,200
HC≡CCH₂OH	[107-19-7]	146
HC≡CCCH₂CH₃ with CH₃ and OH substituents	[77-75-8]	1,956

[a] In 2.8 N HCl, 65°C.
[b] 0.4 wt %.

Generally, the concentration of organic inhibitor is substantially higher than that required for inorganic inhibitors such as chromates. The corrosion of iron in 6 N HCl at 30°C is inhibited by nonamethyleneimine [4396-27-4]; 1% addition reduces the corrosion current by an order of magnitude (36). Typical pickling inhibitors include quinoline [91-22-5] and substituted quinolines, thiourea [62-56-6], dihexylamine [143-16-8], and tolualdehyde [1334-78-9]. A number of organic inhibitors have been effective in reducing corrosion of tin plate in citric acid. These include diphenylurea [102-07-8], carbon disulfide [75-15-0], and allylthiourea [109-57-9] (37). In boiler treatment, so-called filming amines, such as octadecylamine [123-30-1] and hexadecylamine [143-27-1] are effective in retarding corrosion by carbonic acid.

Vapor-Phase Inhibitors. Vapor-phase inhibitors are volatile compounds containing one or more functional groups capable of inhibiting corrosion. The principle is to saturate with the volatile compound the vapor in which the metal object resides. The compound is adsorbed and, in the presence of atmospheric moisture, dissociates to develop functional groups on the surface that retard corrosion. The surface of the metal does not have to be prepared in any special way, as is the case with electroplating, and the inhibitor will function even if the surface is oxidized or rusted before application. The inhibitor will not remove the rust but will prevent further rusting.

The operation of volatile inhibitors has been related to electrochemical theory by pointing out that the inhibitors operate by altering the electrochemical kinetics in a fashion similar to aqueous corrosion (30). The inhibitor must have three properties to be effective: (1) it must contain certain functional groups to provide inhibition; (2) it must have a vapor pressure above a minimum value; and (3) it must be adsorbed on the metal surface. Some of the classes of compounds that have furnished successful volatile inhibitors have been tabulated (38). These include: (1) amine salts with nitrous or chromic acids; (2) amine salts with carbonic, carbamic, acetic, and substituted or unsubstituted benzoic acids; (3) organic esters of nitrous, phthalic, or carbonic acids; (4) primary, secondary, and tertiary aliphatic amines; (5) cycloaliphatic and aromatic amines; (6) polymethylene amines; (7) mixtures of nitrites with urea [57-13-6], urotropine [100-97-0], and ethanolamines; (8) nitrobenzene [98-95-3] and 1-nitronaphthalene [86-57-7].

Dicyclohexylamine nitrite [3129-91-7] has been used commercially for many years and a number of volatile compounds are available commercially as powders or tablets

138   CORROSION AND CORROSION INHIBITORS

to be included in a package containing metallic parts to be protected. The system must be closed to retain the volatile compound, but this is not a significant problem and objects as large as the interior of an ocean-going tanker have been treated by this technique.

A large number of commercial inhibitors and inhibitor formulations are available; some of these are listed along with their intended application in Table 10.

Table 10. Typical Commercial Inhibitors

Type of inhibitor	Proprietary name	Application
chromate based (10–20 ppm as CrO_4)	CWT 102[a]	for open, recirculating cooling system
nonchromate, phosphate based, 10–12 ppm as PO_4	Drewgard 180[a]	for open, recirculating cooling system
chromate (200–500 ppm as CrO_4)	DEWT-L[a]	closed, recirculating cooling system
nonchromate, organic based	Drewgard 100[a]	closed, recirculating cooling system
phosphate (1–10 ppm as PO_4)	Drewgard 120[a]	once-through cooling system
silicate (8–10 ppm as SiO_2)	CIL[a]	once-through cooling system
sodium molybdate (3 ppm)	Molybdate Corrosion Inhibitor[b]	for ferrous and nonferrous metals, cooling and heating systems, hydraulic fluids
organic corrosion inhibitor (formulation of alkylthiophosphate, phosphate esters, and zinc salts)	Tol-Aeromer[c] ACW-11, ACW-15, ACW-16,	for recirculating cooling systems, heat exchangers
chromate, phosphate, and zinc salts	TCW-14[c], TCW-15	corrosion and fouling in piping and heat exchange recirculating cooling water systems
organic corrosion inhibitor (organic phosphate ester)	Tol-Aeromer[c] ACW-18	for recirculating cooling systems
organic corrosion inhibitor (no phosphorus compounds) (2000–4000 ppm)	Tol-Aeromer[c] ACW-22	for closed water systems
buffered organic corrosion and deposit inhibitor	Tol-Aeromer[c], ACW-61	for closed systems, with ferrous and nonferrous metals
chromate and zinc salts (100 ppm)	TCW-10[c], TCW-11, TCW-12	for corrosion and fouling in recirculating cooling water systems
vapor phase inhibitor	NI-22790[d]	for packaging, shipping, and storage of ferrous and nonferrous parts
vapor phase inhibitor	Cor-tab[d], CT-25, CT-50, CT-10	these are vapor phase corrosion inhibitors in tablet form
fatty acid–fatty amine	Cortron R-66[e]	for oil flow lines
quaternary ammonium chloride	Cortron RU-14[e]	for gas lines
fatty imidazoline–fatty acid salt	Cortron R-2258[e]	for oil wells

[a] Marketed by Drew Chemical Corporation, Boonton, N.J.
[b] Marketed by Climax Molybdenum Co., Greenwich, Conn.
[c] Marketed by Tretolite Division, Petrolite Corp., St. Louis, Mo.
[d] Marketed by Northern Instrument Co., St. Paul, Minnesota.
[e] Marketed by Champion Chemicals, Inc., Houston, Texas. (Other commercial suppliers of corrosion inhibitors or treatments are Betz Laboratories, Trevose, Pa.; Nalco Chemical Co., Oak Brook, Illinois; Calgon Corporation, Pittsburgh, Pa.; and Mogul Division of the Dexter Corporation, Chagrin Falls, Ohio.)

Coatings for Protection Against Corrosion

The coatings useful for the protection of metals against corrosion may be characterized by the temperature at which they are applied, whether or not they require electrical current for deposition, or whether or not they convert the original surface metal to another chemical compound involving the same metal (see Coating processes; Coatings).

Hot Dip Coatings. Hot dip coatings are produced by dipping the metal (usually steel) in the molten metal, zinc, or aluminum (see Metal surface treatments). Galvanized coatings are produced by immersing the steel in molten zinc to produce a coating about 0.06–0.08 mm thick. Galvanizing will protect steel in rural atmospheres and in marine atmospheres for 10 years or more, the protective character of the coating depending on the nature of the zinc compounds (oxide, hydroxide, carbonate) that form as well as the sacrificial nature of the zinc itself (see Table 2).

In areas that are polluted by sulfur-containing gases, for example, industrial areas where coal is burned producing atmospheres containing SO_2, SO_3, or H_2S, the protective life of zinc coatings is greatly reduced, eg, to 3–4 years for a 0.05 mm coating. In such environments and for protection against oxidation at elevated temperatures, aluminized (hot dip aluminum) coatings are superior to zinc. Aluminum, according to the electromotive force series, should also be sacrificial to steel but, owing to the severe anodic polarization, this galvanic cell does not operate effectively and the protective ability of the aluminized coating is established by the protective character of the oxide film on aluminum.

Hot dip lead coatings have been employed to protect electrical utility hardware exposed in areas with heavy sulfur gas contamination.

Electrodeposited Coatings. Electrodeposited coatings are usually much thinner than dip coatings (0.0025–0.012 mm) depending on the service life expected. Cadmium and zinc deposits are used on steel as sacrificial coatings. Nickel plating is widely used for protection and decoration in the automobile industry. The steel is plated first with a thin deposit of copper to improve the adherence of the nickel, then with a layer of nickel ranging in thickness from 6–12 μm, and finally with a very thin coating of chromium (0.5 μm) to maintain a bright finish in atmospheres that would normally tarnish nickel (see Electroplating).

Most of the metals that are cathodic to iron such as copper and silver, are not used to protect iron outdoors. These electroplating coatings have pin holes or holidays and even relatively thick deposits fail upon several months exposure outdoors.

In recent years a number of alloy plating systems have been developed, eg, tin–zinc, tin–nickel, and silver–cadmium to improve the corrosion resistance of single metal coatings. The plating procedures and properties of electrodeposited metals are discussed in depth in ref. 39 (see Metallic coatings).

Other Coatings. Conversion coatings may either be chemical or electrochemical. Such coatings may be protective or offer a basis for a more protective coating themselves. Phosphate coatings produced by the chemical action of phosphoric acid–manganese or –zinc solutions on steel form an iron phosphate [*10045-86-0*] film that improves paint adherence. Chromate coatings formed chemically on zinc or aluminum offer corrosion resistance in mild environments. Anodizing of aluminum involves the electrochemical buildup of an aluminum oxide layer by making the aluminum the anode in a dilute solution of sulfuric acid. Empirical relationships have been developed relating the electrolytic voltage applied to the anodizing cell to the thickness of anodic layer achieved in the process (see Aluminum).

Sprayed coatings have been found to be very useful in the field, for example, in the zinc or aluminum coating of permanent structures such as bridges. In this technique an aluminum or zinc wire is passed through a metallizing gun and the metal droplets are sprayed onto the metal structure. These coatings can be up to 0.1 mm thick.

Many organic coatings have been developed for corrosion protection of metals and usually specific coatings are developed for given applications. For example, Munger has an extensive discussion of vinyl and epoxy coatings for marine service in LaQue's monograph *Marine Corrosion* (4) (see Coatings, resistant; Coatings, marine).

Corrosion Testing

Testing of the corrosion resistance of metals has reached a high degree of sophistication as experimenters have become aware of the different behaviors of materials in varied environments, as well as the need to place the tests on a sound statistical basis (40).

Corrosion tests may be conveniently divided into two types: (a) laboratory tests which are usually intended to be accelerated tests; and (b) field and service tests which are designed to reproduce actual usage conditions.

Laboratory Tests. Several special types of laboratory tests have been developed in response to different problems. An important test deals with the measurement of galvanic corrosion, the corrosion of an active metal in contact with a more noble metal. Such tests recognize the existence of a practical emf series (see Fig. 1).

A number of laboratory tests have been based on polarization curves (see Fig. 5). These are usually short-time tests and some are based on the linear–current relationship observed near the corrosion potential of the metal (see under Electrochemical Basis).

Some of the newer technologies have required the development of specialized tests. The selection of materials for desalination plants has led to the development of hot-brine loops for screening materials. In such tests heat exchanger and evaporator materials can be evaluated directly. Nuclear reactors (qv) using either pressurized water or boiling water present unusual problems because of the high temperatures and pressures required and the need to maintain high purity in the coolant. Both static autoclaves and autoclaves wherein it is possible to replenish the solution are used at temperatures up to 400°C and pressures up to 24 MPa (3500 psig). Recirculatory loops are also used.

Nuclear breeder reactors operating at high temperature use liquid metals, alkali metals, and alloys of alkali metals as coolants. Loops have been constructed to confront two major problems in these systems wherein a metal like lithium is not the corroded metal but the corrodent. These problems include the dissolution of the container material by the alkali metal in one part of the system and the deposition of corrosion product in another part. The second severe problem involves the corrosion of container materials by intergranular penetration.

Field Tests. Atmospheric field tests are located in those geographical sites intended to fairly represent typical corrosion environments. For example, the ASTM Committee A-5 on Corrosion of Iron and Steel established the following sites for test racks in 1926: Key West, Fla., as a tropical marine site; State College, Pa., as a rural site; Sandy Hook, N.J., as an industrial marine site; Pittsburgh, Pa., as an industrial site; and Altoona, Pa., as another industrial site.

These sites varied in respect to the severity of corrosion to iron by a factor of 4 and to zinc by a factor of 9, but not in the same order. Measurements of weight loss and changes in appearance have been made for time durations of up to 20 years.

Extensive field tests are conducted at several seawater sites; one of the most active being the International Nickel seawater laboratory at Harbor Island, North Carolina, and the atmospheric test racks at Kure Beach, North Carolina. A distinction is made between total immersion tests, tests in the tidal zone, and tests in the splash zone. Seawater tests include experiments simulating impingement attack, rotating disks, jets, venturis, and pumps.

The National Bureau of Standards has maintained an active program of testing metals buried in soils. Here such properties as soil resistivity, acidity, water content, oxidation–reduction capability, and particularly biological activity must be considered.

Virtually all chemical plants have extensive in-plant test programs (see also Maintenance; Plant safety).

It is often the practice to install specimens or spools containing a large number of specimens directly in the actual reactor system. These samples are then exposed to the exact conditions as the process equipment of direct interest to the corrosion engineer.

BIBLIOGRAPHY

"Corrosion Inhibitors" in *ECT* 2nd ed., Vol. 6, pp. 317–346, by Charles C. Nathan.

1. *Selected Values of Chemical Thermodynamic Properties,* Circ. 500, National Bureau of Standards, Feb. 1, 1952.
2. M. M. Markowitz and D. A. Baryta, *J. Chem. Eng. Data* **7,** 586 (1962).
3. W. M. Latimer, *The Oxidation States of the Elements and Their Potentials in Aqueous Solutions,* Prentice-Hall, New York, 1952.
4. F. L. LaQue, *Marine Corrosion, Causes and Prevention,* John Wiley & Sons, Inc., New York, 1975.
5. M. Pourbaix, *Atlas of Electrochemical Equilibria in Aqueous Solutions,* Pergamon Press, London, 1966, pp. 313–314.
6. S. W. Dean, Jr., W. D. France, Jr., and S. J. Ketcham in W. H. Ailor, ed., *Handbook on Corrosion Testing and Evaluation,* John Wiley & Sons, Inc., New York, 1971, p. 176.
7. M. Stern and A. L. Geary, *J. Electrochem. Soc.* **104,** 56 (1957).
8. U. R. Evans, *J. Soc. Chem. Ind.* **43,** 315–22T (1924).
9. W. H. J. Vernon, *J. Sci. Instrum.* **22,** 226 (1945).
10. W. J. Müller, *Trans. Electrochem. Soc.* **76,** 167 (1939).
11. R. A. Paramonova and S. A. Balezin, *Zh. Prikl. Khim.* **38,** 84 (1965).
12. U. R. Evans and T. P. Hoar, *Proc. R. Soc. London A* **137,** 343 (1932).
13. C. W. Borgmann, *Ind. Eng. Chem.* **29,** 814 (1937).
14. W. Whitman, R. Russell, and V. Atteri, *Ind. Eng. Chem.* **16,** 665 (1924).
15. R. T. Foley, *Corrosion* **26,** 58 (1970).
16. W. H. J. Vernon and L. Whitby, *J. Inst. Met.* **42,** 181 (1929); **44,** 389 (1930).
17. W. H. J. Vernon, *J. Inst. Met.* **49,** 153 (1932).
18. H. S. Campbell in W. H. Ailor, ed., *Handbook on Corrosion Testing and Evaluation,* John Wiley & Sons, Inc., New York, 1971, p. 5.
19. N. E. Hamner, *Corrosion Data Survey—Metals,* National Association of Corrosion Engineers, Houston, Texas, 1974.
20. H. E. Boyer, ed., *Metals Handbook,* 8th ed., Vol. 10, ASM, 1975.
21. J. C. Scully, ed., *The Theory of Stress Corrosion Cracking in Alloys,* NATO Scientific Affairs Divisions, Brussels, 1972.
22. B. F. Brown, *Stress Corrosion Cracking Control Measures,* NBS Monograph 156, National Bureau of Standards, 1977.
23. H. R. Smith and D. E. Piper in B. F. Brown, ed., *Stress Corrosion Cracking in High Strength Steels and in Titanium and Aluminum Alloys,* Naval Research Lab., 1972, p. 17.

24. J. H. Payer, W. E. Berry, and W. K. Boyd in H. L. Craig, Jr., ed., *STP 610,* ASTM, 1976, p. 82.
25. *Metals Handbook,* 8th ed., Vol. 1, ASM 1961, p. 1001.
26. E. H. Spuhler and C. L. Burton, *Avoiding Stress Corrosion Cracking in High Strength Aluminum Alloy Structures,* ALCOA Green Letter, Aluminum Company of American, Apr. 1970.
27. *Corrosion Resistance of Nickel-Containing Alloys in Phosphoric Acid,* International Nickel Co. Corrosion Engineering Bulletin CEB-4, International Nickel Co., New York, 1966, p. 15.
28. C. C. Nathan, *Corrosion Inhibitors,* National Association of Corrosion Engineers, Houston, Texas, 1973.
29. H. H. Uhlig, *Corrosion and Corrosion Control,* John Wiley & Sons, Inc., New York, 1963, p. 224.
30. I. L. Rosenfeld, B. P. Persiantseva, and P. B. Terentiev, *Corrosion* **20,** 222t (1964).
31. M. J. Pryor and M. Cohen, *J. Electrochem. Soc.* **100,** 203 (1953).
32. W. Robertson, *J. Electrochem. Soc.* **98,** 94 (1951).
33. O. L. Riggs, Jr., in ref. 28, pp. 7–27.
34. Z. A. Foroulis, *Proceedings of Symposium on the Coupling of Basic and Applied Corrosion Research,* National Association of Corrosion Engineers, Houston, Texas, 1969, p. 24.
35. G. L. Foster, B. D. Oakes, and C. H. Kuceda, *Ind. Eng. Chem.* **7,** 825 (1959).
36. N. Hackerman, R. M. Hurd, and R. R. Armand, *Corrosion,* **18,** 37t (1962).
37. P. W. Board and R. V. Holland, *Brit. Corrosion J.* **3,** 31 (1968); **4,** 162 (1969).
38. G. Trabanelli and F. Zucchi, *Paper No. 82,* presented at the Meeting of the National Association of Corrosion Engineers, Houston, Texas, March 22, 1976.
39. F. A. Lowenheim, ed., *Modern Electroplating,* Wiley-Interscience, New York, 1974.
40. W. H. Ailor, *Handbook on Corrosion Testing and Evaluation,* John Wiley & Sons, Inc., New York, 1971.

R. T. Foley
B. F. Brown
The American University

CORUNDUM. See Abrasives; Aluminum compounds.

COSMETICS

Cosmetics are the product of cosmetic chemistry, a science that combines the skills of specialists in chemistry, physics, biology, and medicine (1–3). The professional societies of cosmetic chemistry are the Society of Cosmetic Chemists and the International Federation of Societies of Cosmetic Chemists (IFSCC). *The Journal of the Society of Cosmetic Chemists* publishes the papers presented at the meetings of both groups. The Cosmetic, Toiletry, and Fragrance Association (CTFA) keeps its member companies informed on legislation that affects the industry. CTFA finances the extensive test programs for safety of color additives and cosmetic ingredients. The CTFA's *Cosmetic Ingredient Dictionary* was established as an industry standard reference by one of the scientific panels of the CTFA (4).

Regulation of the Cosmetics Industry

In 1938 the *Federal Food and Drug Act of 1906* was revised to regulate cosmetics and cosmetic devices. Cosmetic products shipped in interstate commerce are also subject to the regulations of the FDA, FTC, and other federal agencies. The FDA regulations are, in part, contained in *Federal Food, Drug, and Cosmetic Act, as amended October 1976* (5). Additional regulations and information letters are added by all the regulatory agencies as required and the cosmetic industry, like the drug industry, must comply with manufacturing practices outlined by the regulatory agencies. In 1978 the Interagency Regulatory Liaison Group was formed consisting of the EPA, Consumer Product Safety Commission, OSHA, and the FDA. Any regulatory problems that affect the cosmetic industry can be passed to the liaison group.

Over-the-counter (OTC) panels are product and ingredient review panels of industry and regulatory agency scientists working with the FDA. The establishment of these panels and the reports of their deliberations are expected to aid the FDA in preparing suitable regulations for OTC ingredients and products. OTC panels of interest to the cosmetic industry are those concerned with antiperspirants, sunscreens, dental products, and antimicrobials. The OTC panel on sunscreens prepared and submitted a report to the FDA (6) (see Sunscreens). The panel on antiperspirants has submitted a draft monograph.

Economic Aspects

The cosmetic industry attracts a large portion of consumer spending each year. As shown in Table 1, reprinted from *Product Marketing* (the national news publication for cosmetics, toiletries, fragrances, and drugs) (7), the percentage increase in sales from 1973 to 1977 varies from a low of 13.8% (foot products) to a high of 69.4% (hand preparations). Hair preparations (qv) is the leader in the consumer dollar share as shown by a 65.1% increase in a single year (from 1976–1977).

Makeup preparations and accessories accounted for $1\frac{1}{2}$ billion dollars in sales in 1977—an increase of $271,900,000 in a single year, owing largely to the introduction of blushers and lip glosses which amounted to over $200,000,000 in retail sales.

Table 1. Consumer Expenditures on Health and Beauty Aids

Item	Retail sales, thousands of dollars 1973	1976	1977	Change, %, 1977 vs 1973	1976
Packaged medications[a]	3,134,040	3,653,860	4,358,970	39.1	19.3
vitamin concentrates[b]	434,350	475,650	519,580	19.6	9.2
cough and cold items[b]	696,650	809,500	843,520	21.1	4.4
laxatives and other elimination aids[c,d]	218,580	257,500	273,000	24.9	6.1
internal analgesics[a,b,e,f]	658,520	794,040	832,490	26.4	4.8
external antiseptics	76,900	82,170	83,880	9.1	2.1
external analgesics	135,560	145,890	147,700	9.0	1.2
antacids[a,e,g]	127,180	145,690	459,290	base changed	
contraceptives[f]	12,420	11,400	66,820		
other packaged medications (non-Rx)[a]	773,880	932,020	1,132,600	46.4	21.5
Prescriptions	7,084,430	8,949,370	9,617,500	35.8	7.5
Foot products[h]	118,590	136,090	134,940	13.8	0.8
foot powder	8,340	9,780	9,940		
salves and ointments	35,370	39,780	37,550		
pads and plasters	53,300	62,820	63,350		
arch supports, appliances, and insoles	21,580	23,710	24,100		
Baby needs and medicaments	261,380	360,000	403,320	54.3	12.0
milk modifiers and formulas	160,090	238,510	257,350		
baby powder	44,500	56,100	57,090		
oils and lotions	27,490	34,940	38,320		
infant suppositories	4,530	4,670	4,600		
diaper rash preparations	24,770	25,780	26,510		
disposable towelettes	19,450				
Dieting aids	146,450	136,040	131,510	10.2	3.3
Feminine hygiene[i]	502,810	556,750	595,220	18.4	6.9
sanitary napkins	207,250	211,690	219,670		
tampons	147,590	212,430	227,380		
sanitary belts	13,010	11,210	10,000		
sanitary panties	19,790	14,910	14,600		
liquid douches	13,020	40,130	41,630		
powdered douches	12,370	10,240	9,920		
other hygiene medications (non-Rx)	59,180	56,140	57,310		
feminine deodorants	30,600	17,270	14,710		
Total for drugs and medications listed	11,123,380	13,664,000	15,241,460	37.0	11.5
Oral hygiene products[a]	945,330	1,142,600	1,300,800	37.6	13.8
toothpaste	419,800	541,030	571,680		
toothpowder	4,090	3,800	3,570		
denture cleansers	38,680	52,560	121,020		
denture adhesives	46,090	56,470	91,920		
denture brushes	6,710	7,840	7,480		
toothbrushes, nonelectric	98,050	107,660	127,170		
electric toothbrushes	15,380	15,250	20,050		
oral lavages	25,670	27,840	38,480		
dental floss	3,640	4,660	5,020		
mouthwashes and gargles	252,750	278,310	268,140		
breath fresheners	34,470	47,180	46,270		
Hair preparations[b,j]	1,405,820	1,066,150	1,759,910	25.2	65.1
regular shampoos	365,020	544,230	580,880		
medicated dandruff shampoos	124,530	136,740	142,000		

Table 1 (*continued*)

Item	Retail sales, thousands of dollars 1973	1976	1977	Change, %, 1977 vs 1973	1976
color shampoos	11,600	12,690	9,500		
hair color rinses	46,420	49,630	46,050		
hair tints and dyes	208,730	285,520	318,220		
hair medications	13,140	25,580	13,360		
women's hair dressings	102,920	136,970	155,670		
hair straighteners and other dressings[b]	26,280	32,170	35,560		
cream rinse	68,350	85,150	105,460		
depilatories	15,450	18,220	18,480		
waveset preparations	13,350	14,650	14,030		
women's hairspray	263,830	279,160	180,570		
home permanent kits	52,340	55,720	57,230		
men's hair dressings, aerosol	35,160	25,730	13,430		
men's hair dressings, nonaerosol	58,700	57,650	69,470		
Cosmetics and accessories[b]	1,231,640	1,620,160	1,924,900	56.3	18.8
face creams	260,730	371,180	404,040		
makeup cake bases	9,930	9,620	9,180		
makeup lotion bases	41,840	43,550	44,480		
makeup cream bases	18,640	22,300	23,410		
pressed cake powder and compact[k]	140,620	162,160	100,360		
loose face powder and powder puffs	42,410	39,490	37,930		
blush makeup, brush[b]			30,520		
blush makeup, nonbrush[b]			42,400		
rouge	10,470	10,060	9,250		
face lotions and astringents	48,460	63,500	70,250		
liquid facial cleaners	64,100	81,180	85,370		
talcum and body powder	60,120	59,860	59,690		
lipsticks	318,540	458,450	552,400		
lipglosses			135,800		
mascara	45,380	52,430	56,110		
eyebrow pencil	20,780	25,300	29,650		
eyeshadow	32,420	41,390	43,620		
other	117,200	179,690	190,440		
Shaving preparations	473,910	632,040	706,680	49.1	11.8
aerosol shave cream	92,400	114,570	124,150		
other shave cream	10,140	9,660	11,270		
shaving soaps and sticks	5,690	5,790	6,040		
after-shave lotions	129,680	159,570	179,860		
preshave products	18,440	21,080	23,170		
men's cologne	103,710	181,690	188,160		
men's talcum	3,310	3,410	3,430		
styptics	1,660	2,120	2,200		
men's packaged toiletry sets	108,880	134,150	168,400		
Women's fragrances[l]	518,360	616,300	621,140	19.8	0.8
perfumes	83,020	107,030	113,390		
toilet water and cologne	363,570	427,210	426,690		
bubble bath	35,280	44,490	44,900		
other bath products	32,640	33,190	31,530		
purse atomizers	3,850	4,380	4,630		

Table 1 (*continued*)

Item	Retail sales, thousands of dollars			Change, %, 1977 vs	
	1973	1976	1977	1973	1976
Hand preparations	217,830	300,760	369,000	69.4	22.7
hand lotions[m]	66,740	87,720	107,230		
hand creams	18,910	19,770	21,440		
nail polish and enamel	104,830	155,990	198,240		
nail enamel removers	20,870	29,640	33,850		
cuticle softeners	6,480	7,640	8,240		
Personal cleanliness items[n]	934,690	1,186,570	1,334,870	42.8	12.5
deodorant soaps	205,460	293,470	381,590		
other bath soaps	176,140	284,700	340,680		
cream deodorants	24,500	27,980	27,490		
liquid and squeeze container deodorants	12,990	12,640	12,910		
stick deodorants	8,970	14,650	20,200		
roll-on deodorants	38,270	121,870	182,090		
pad deodorants	1,530	1,550	1,590		
feminine spray deodorants	30,600	17,270	14,460		
other aerosol deodorants	434,990	411,450	352,770		
powder deodorants	600	990	1,090		
Total for cosmetics and toiletries	5,727,580	6,564,580	8,017,300	40.0	22.1
Total for health and beauty aids	16,820,360	20,211,310	23,258,760	38.3	15.1

[a] Data base revised to correct for prior over or understatement.
[b] Category revised in 1977 to show additional detail.
[c] Includes epsom salts.
[d] Includes dioctyl sodium sulfosuccinate (DSS) products as well as those combined with laxatives.
[e] Effervescent compounds transferred to antacid category.
[f] Transferred from feminine needs category.
[g] Includes sodium bicarbonate.
[h] Athlete's foot products found under: packaged medication, other.
[i] Menstrual pain relievers and contraceptive categories transferred to packaged medication section.
[j] Excludes products sold for professional use in beauty parlors and barber shops, and medicated shampoos; includes baby shampoos and dandruff shampoos.
[k] 1976 figure includes blush makeup.
[l] Excludes door-to-door sales.
[m] Includes products promoted for hand and body.
[n] Excludes medicated soaps.

Cosmetic Emulsions

Cosmetic lotions and creams are emulsions of water-based and oil-based phases. An emulsion is a two-phase system consisting of two incompletely miscible liquids, the internal, or discontinuous, phase dispersed as finite globules in the other. Special designations have been devised for oil and water emulsions to indicate which is the dispersed and which the continuous phase. Oil-in-water (o/w) emulsions have oil as the dispersed phase in water as the continuous phase. In water-in-oil (w/o) emulsions, water is dispersed in oil, which is the external (continuous) phase (8–9) (see Emulsions).

Properties of Emulsions. The properties that are most apparent, and thus are usually most important, are: ease of dilution, viscosity, color, and stability. For a given type of emulsification equipment, these properties depend upon: (*1*) the properties of the continuous phase, (*2*) the ratio of the external to the internal phase, (*3*) the particle size of the emulsion, (*4*) the relationship of the continuous phase to the particles (including ionic charges), and (*5*) the properties of the discontinuous phase. In any given emulsion, the properties depend upon which liquid constitutes the external phase, ie, whether the emulsion is o/w or w/o. The resulting emulsion type is controlled by: (*1*) the emulsifier: type, and amount, (*2*) the ratio of ingredients, and (*3*) the order of addition of ingredients during mixing.

The dispersibility (solubility) of an emulsion is determined by the continuous phase; thus if the continuous phase is water-soluble, the emulsion can be diluted with water; conversely, if the continuous phase is oil-soluble, the emulsion can be diluted with oil.

The ease with which an emulsion can be diluted can be increased by decreasing the viscosity of the emulsion. The viscosity of an emulsion when the continuous phase is in excess is essentially the viscosity of the continuous phase. As the proportion of internal phase increases, the viscosity of the emulsion increases to the point that the emulsion is no longer fluid. When the volume of the internal phase exceeds the volume of the external phase, the emulsion particles become crowded and the apparent viscosity is partially structural viscosity.

An emulsion is stable as long as the particles of the internal phase do not coalesce. The stability of an emulsion depends upon: (*1*) the particle size; (*2*) the difference in density of the two phases; (*3*) the viscosity of the continuous phase and of the completed emulsion; (*4*) the charges on the particles; (*5*) the nature, effectiveness, and amount of the emulsifier used; and (*6*) conditions of storage, including temperature variation, agitation and vibration, and dilution or evaporation during storage or use. The stability of an emulsion is affected by almost all factors involved in its formulation and preparation. In formulas containing sizable amounts of emulsifier, stability is predominantly a function of the type and concentration of emulsifier.

Emulsifiers. Emulsifiers can be classified as ionic or nonionic according to their behavior. An ionic emulsifier is composed of an organic lipophilic group (L) and a hydrophilic group (H). The hydrophilic–lipophilic balance (HLB) is often used to characterize emulsifiers and related surfactant materials. The ionic types may be further divided into anionic and cationic, depending upon the nature of the ion-active group. The lipophilic portion of the molecule is usually considered to be the surface-active portion.

Nonionic emulsifiers are completely covalent and show no apparent tendency to ionize. They can, therefore, be combined with other nonionic surface-active agents and with either anionic or cationic agents as well. The nonionic emulsifiers are likewise less susceptible to the action of electrolytes than the anionic surface-active agents. The solubility of an emulsifier is of the greatest importance in the preparation of emulsifiable concentrates.

Emulsifiers, being surface-active agents, lower surface and interfacial tensions and increase the tendency of their solution to spread.

O/w emulsifying agents produce emulsions in which the continuous phase is hydrophilic; hence, such emulsions are generally dispersible in water and will conduct electricity. The surfactants that are capable of producing such emulsions usually have an HLB of more than 6.0 (preferably 7), the hydrophilic portion of their molecules being predominant. (Between HLB 5 and 7 many surfactants will function as either w/o or o/w emulsifiers, depending on how they are used.)

O/w emulsifiers		HLB
P.E.G. 300 distearate	nonionic	7.3
sorbitan monolaurate	nonionic	8.6
P.E.G. 400 distearate	nonionic	9.3
triethanolamine stearate	anionic	12.0
P.E.G. 6000 monolaurate	nonionic	19.2

W/o emulsifiers produce emulsions in which the continuous phase is lipophilic in character (oil, wax, fat, etc). Such emulsions are not generally dispersible in water and do not conduct electricity. The surfactants capable of producing such emulsions usually have an HLB of less than 6.0 and preferably below 5. The lipophilic portion of their molecules is predominant.

W/o emulsifiers		HLB
lanolin alcohols	nonionic	ca 1.0
ethylene glycol monostearate S/E	nonionic	2.0
propylene glycol monostearate S/E	anionic	3.2
sorbitan monooleate	nonionic	4.3
P.E.G. 200 dilaurate	nonionic	6.0

Cosmetic Creams

Materials used in creams may be prepared in o/w or in w/o emulsions. The esthetic effect and degree of emolliency depend to a great extent on the emulsion type as well as on the emulsion composition. O/w emulsions produce a cooling effect on application to the skin owing to water evaporation. W/o emulsions do not produce this effect since water evaporation is slowed by the film of the oil in the continuous phase (10).

The classical example of a cream was the USP Unguentum Aquae Rosae which was prepared from 3.0% beeswax, 11.8% spermaceti, 40.2% sweet almond oil, and 45.0% rose water. In 1890 the formula was changed to 12.1% beeswax, 12.6% spermaceti, 55.4% sweet almond oil, 0.5% borax, and 19.4% rose water.

This was the basic formula for the familiar cold cream that is now made with mineral oil instead of almond oil. Its occlusive action aided in rehydration of the corneum when allowed to remain on the skin for an appreciable length of time. Because the solvent action of mineral oil tends to remove skin surface lipids when the cream is applied for a short period of time, partial replacement with a vegetable oil is needed. These emulsions are w/o, the emulsifier is sodium cerotate formed by reaction of borax and free cerotic acid in the beeswax. If the water content is raised to approximately 45% or more the composition changes to an o/w emulsion.

Nonionic emulsifiers, such as glyceryl monostearate, propylene glycol and polyethylene glycol esters of fatty acids, sorbitol, and ethoxylated sorbitol esters of fatty acids, are used to prepare creams that have stability at acid pH as well as alkaline pH.

Anionic emulsifiers, such as the amine soaps prepared by reaction of fatty acids with various amines (eg, triethanolamine), are popular in preparing slightly alkaline creams. Most of these creams are of the o/w type. W/o creams can be prepared with anionic soaps such as calcium and magnesium soaps of fatty acids formed *in situ*.

Cationic emulsifiers are used in the preparation of emulsion systems to increase deposition of the emulsion on negatively charged surfaces such as skin and hair. A popular cream prepared with cationic emulsifiers has the following composition: 0.10% antioxidant, 3.00% cetyl alcohol, 3.00% stearyl alcohol, 3.00% dewaxed lanolin, 3.00% mineral oil, 0.15% *N*-(colaminoformylmethyl)pyridinium chloride, 1.20% *N*-(colaminoformylmethyl)pyridinium chloride stearate, 0.15% preservative [methyl and propyl paraben (5:1)], 4.00% isopropyl myristate, 6.00% propylene glycol, 76.05% distilled or deionized water, and 0.35% perfume.

Vanishing Cream. Vanishing cream can be considered to be an emulsion of a free fatty acid (usually stearic acid) in a nonalkaline medium. The basic ingredients are: 65–75% water, 15–20% stearic acid, 8–12% glycerol, 0.5–1.5% alkali (KOH), *qs* (as needed) preservative, and *qs* perfume. Of the stearic acid used, about 15–20% is saponified; the rest remains as free acid.

Manufacture. The oils, waxes, emulsifiers, and other oil-soluble components are heated to 75°C in a steam-jacketed kettle. The water-soluble components (alkalis, alkanolamines, polyhydric alcohols, and preservatives) are dissolved in the aqueous phase and heated to 75°C in another steam-jacketed kettle. To allow for evaporation of water during the heating and emulsification, about 3–5% excess water (based on formula weight) is added.

The procedure for preparing o/w and w/o emulsions is to add the warmed inner phase very slowly to the outer phase (also at 75°C), stirring constantly and homogenizing to assure efficient emulsification. Finely dispersed o/w emulsions can also be prepared by adding the aqueous phase to the oils. Initially the low concentration of water forms a w/o emulsion according to the phase–volume relationship. The slow addition and emulsification of the water increases the viscosity of the system while the oil phase expands to a maximum. At this point, the continuous oil phase breaks up into minute droplets as emulsion inversion occurs, characterized by a sudden decrease in viscosity. This emulsification technique proceeds smoothly at the critical inversion point in a well-balanced, low oil–wax system, but it frequently causes coagulation in high oil–wax emulsions. The conventional procedure of adding the inner phase to the outer is preferable for creams and lotions.

The rates of addition and mechanical agitation of the dispersed phase are critically important in determining the efficiency of emulsification. The product formed may vary from a completely dispersed inner phase in a well emulsified system, to a mixed emulsion in a poorly emulsified system, the latter owing to excessive rate of addition of inner phase and to inadequate stirring. This in turn affects the consistency, viscosity, and stability of creams and lotions.

Total stirring times and cooling rates are important to lotion viscosity, cream consistency, and emulsion stability. Experimental formulas are often developed in vessels that are not equipped with a heating and cooling jacket. Under these air-cooled conditions, longer stirring times are necessary. The transition to full-scale production in jacketed equipment introduces a variable in the physical factors contributing to emulsion preparation. If cooling is started too soon after emulsification is complete, crystallization of the higher melting waxes may occur.

The temperature at which the perfume oils are added to the cream or lotion is another factor contributing to emulsion instability. The addition of perfume to a w/o emulsion proceeds smoothly owing to its solubility in the external phase. In o/w systems, the oils must break through the continuous aqueous phase to be emulsified.

If the cream is to be hot-poured, it is stirred to 5°C above the congealing point, any required color solutions are added, and the cream held at that temperature with occasional stirring during the filling procedure. If cold-filling is preferred, the cream is stirred to 35°C, any color solutions are added, and filling proceeds at room temperature.

Cosmetic Lotions

The oils and waxes in lotions are identical to those of an emollient cream but they are present in lower concentration. An o/w emollient lotion usually contains more water than the corresponding cream; a w/o type may have the same water content, with oily components replacing part of the waxlike materials. These lotions are preferred for use during the day because they produce a lighter or less oily emollient film. However, they can be formulated to contain the same concentration of oil phase ordinarily used in creams. The sales appeal of emollient lotions derives partly from their convenience and partly from the greater variety of package design possible for liquid emulsions.

The formulation of lotions of all emulsifier and emulsion types is similar to that of emollient creams. The emulsion must be stable at elevated temperatures (45–50°C) for whatever period is deemed necessary and at room temperatures for a minimum of one year. In the freeze–thaw test of emulsion stability, the lotion is subjected to a temperature of −5°C for 24 h and is then allowed to return to room temperature, at which time it should be stable and pourable.

An example of an all-purpose hand, face, and body lotion prepared with an anionic emulsifier is: 4.00% stearic acid (triple-pressed), 1.50% lanolin anhydrous, 1.50% mineral oil, 1.00% cetyl alcohol, 0.80% triethanolamine, 0.25% preservative [methyl and propyl paraben (5:1)], 90.60% distilled or deionized water, and 0.35% perfume.

A representative all-purpose hand, face, and body lotion, prepared with cationic emulsifiers is: 1.00% cetyl alcohol, 0.50% stearyl alcohol, 1.00% lanolin anhydrous, 4.00% mineral oil, 0.10% *N*-(colaminoformylmethyl)pyridinium chloride, 0.80% *N*-(colaminoformylmethyl)pyridinium chloride stearate, 0.25% preservative [methyl and propyl paraben (5:1)], 6.00% propylene glycol, 64.80% distilled or deionized water, 0.125% sodium chloride, 0.125% sodium benzoate, 11.00% distilled or deionized water, 0.30% perfume, and 10.00% denatured alcohol no. 40.

Deodorants and Antiperspirants

Deodorant and antiperspirant products are marketed as aerosols, creams, gels, lotions, powders, soaps, and sticks (11–12).

The FDA product and ingredient review panel for over-the-counter antiperspirants (OTC panel) has assigned three classifications to antiperspirants (13): category 1, safe to use; category 2, does not meet requirements, remove offending materials; category 3, may be marketed while testing continues.

The initial proposal states that color and fragrance apparently have no effect on safety; buffers do not apparently make a product more effective; a product, to remain on the market, must be at least 20% effective on at least one half of the persons tested, with a confidence rate of 95%; special tests will be set up to evaluate claims of extra effectiveness (30% or more); products aimed at combatting problem and emotional perspiration will require special tests; if a new ingredient is used tests are required for skin irritation and compatibility with normal flora.

Active Ingredients. Aluminum chloride, once commonly used in deodorants and antiperspirants, may cause fabric damage and skin irritation because of its low pH. This has led to the development of various basic aluminum compounds. The most widely used compound is aluminum chlorhydroxide (ACH) (12). Other compounds that have been or are now used are: basic aluminum bromide, iodide, and nitrate, and basic aluminum hydroxychloride–zirconyl hydroxy oxychloride, with and without glycine. Zirconium salts, formerly used in antiperspirants, were banned for use in aerosol antiperspirants in 1977.

An example of an antiperspirant roll-on is: 33.60–40.60% methyl cellulose 65 HG (high gain), 400 mPa·s (= cP), 3% solution [96.9% distilled or deionized water, 3.0% methyl cellulose 65 HG, 400 mPa·s (= cP), 0.1% methyl paraben]; 30.00–36.00% aluminum chlorhydroxide complex, 50% solution; 16.25–29.25% alcohol SDA no. 40; 4.65% propylene glycol; and 2.50% solubilized perfume [solubilizer and perfume oil (4:1)].

Deodorant–Antiperspirant Sticks. Sodium stearate, the primary gelling agent for deodorant sticks, constitutes about 7–9 wt % of these products. The grade employed depends on the fatty acid used in making the stearate. Derivatives such as acetylated sucrose distearates also can be used but are more common in antiperspirant sticks to gel cetyl alcohol-based formulations (typical use level, 28%).

There are three main types of stick formulations for antiperspirants: hydroxyethyl stearamide, which produces clear-melting, homogeneous sticks; stearamides and cetyl alcohol dry-powder dispersions of ACH; and stearamides coupled with propoxylated alcohol. The stearamide wax content is 26–29%; the active concentration, 20–25%.

Volatile silicones, the siloxanes, are employed in the newer antiperspirant sticks at levels as high as 40–50%. They function as a processing aid, reducing cracking and crumbling of the stick. The tetrameric and pentameric cyclic silicones are preferred because of their volatility (see Silicon compounds). When the stick is applied to the skin, the silicones evaporate, leaving the ingredients on the skin in a nonsticky film. The silicones are useful in pump antiperspirants, acting as a lubricant for valves and a suspending agent for the active material. The compounds also act as skin lubricants and emollients.

A typical deodorant stick formula is: 75.35 wt % alcohol SDA no. 40; 17.00 wt % propylene glycol; 6.00 wt % sodium stearate, purified USP; 1.50 wt % perfume; and 0.15 wt % trichlorocarbanilide.

A typical antiperspirant stick formula is: 27.5 wt % stearic acid monoethanolamide; 24.0 wt % poly(dimethylsiloxane) (low viscosity, volatile); 20.0 wt % rehydrol powder; 15.0 wt % propoxylated myristyl alcohol; 9.0 wt % propylene glycol; 3.0 wt % distilled or deionized water; 1.0 wt % perfume; and 0.5 wt % myristyl lactate.

A typical dry compressed antiperspirant stick formula is: 65.75 wt % microcrystalline cellulose; 25.00 wt % ultrafine powdered aluminum chlorhydrate; 9.00 wt % talc; and 0.25 wt % magnesium stearate.

Cream and Lotion Antiperspirants. Most emulsions are nonionic systems because the nonionics have less tendency to be irritating and make more stable and uniform systems.

An example of a cream antiperspirant formula is: 61.5 wt % distilled or deionized water; 26.0 wt % aluminum chlorhydroxide complex, 50% solution; 7.0 wt % magnesium aluminum silicate; 4.0 wt % glyceryl monostearate, self emulsifying; 1.0 wt % imidazolidinyl urea (preservative); and 0.5 wt % perfume.

Sunscreens

The sun emits energy in a continuous band throughout the electromagnetic spectrum. The shorter wavelengths are absorbed in the upper atmosphere so that at sea level the radiation extends from a cutoff near 290 nm through the near ultraviolet to the conventional end of the ultraviolet range, which is near 400 nm. The intensity of the radiation varies nonlinearly throughout this range (14).

The production of erythema and the subsequent production of melanin pigment are both maximum with 296.7 nm radiation. As the wavelength increases, both responses fall rapidly, so that 10 $\mu W/cm^2$ of 307 nm radiation, 100 $\mu W/cm^2$ of 314 nm, 1000 μW of 330 nm, and 10,000 μW of 340 nm radiation are required to equal the effect of 1 $\mu W/cm^2$ of 296.7 nm radiation in the production of erythema.

A unit of erythemal flux, the E-viton, is equivalent to the erythema induced by 10 $\mu W/cm^2$ of 296.7 nm radiation. The response of the skin to an E-viton (or viton) is constant: irradiation by 10 vitons for one hour produces the same erythemal response as 5 vitons for two hours. The effects owing to irradiation by different wavelengths are additive. About twenty minutes exposure to midsummer sunlight (or 40 viton min) are needed to produce a minimum perceptible erythema (MPE) on normal Caucasian skin. Thus 1 E-viton acting for 40 min produces an MPE. 10 E-vitons acting 4 min or 40 E-vitons acting for 1 min will produce the same MPE.

With the realization that prolonged exposure to sunlight produces most of the aging effects on skin and the clear implication of exposure in at least three types of skin cancer (basal cell cancer, squamous cell cancer, melanoma), use of products that protect the skin from excessive exposure has become increasingly widespread.

Sunscreens are of two types: physical and chemical (15–16). Physical screening agents, such as titanium dioxide and zinc oxide, are opaque materials that block and scatter light, and thus act as mechanical barriers. Their action is nonselective over all wavelengths. Chemical screening agents, which act by absorbing uv light, offer selective protection against certain uv wave bands, depending on their absorption spectrum. Anthranilates, cinnamates, benzyl and homomenthyl salicylate, and *p*-aminobenzoic acid (PABA) and its ester derivatives belong in this category and have maximal absorption in the sunburn region. A useful sunscreen resists wash-off from swimming or sweating and does not come off during exercise or rubbing. Maintenance of significant protection following swimming or sweating would suggest that the chemical is substantive or can diffuse into the horny skin layer. PABA and its esters are moderately substantive and give appreciable protection after intensive sweating and brief periods of immersion in water (5 to 10 min).

OTC Report. Sunscreen products for over-the-counter (OTC) human use are reviewed in a report on recommendations to the FDA by the OTC panel on Topical Analgesics (6). Definitions and recommendations made by the panel are reviewed below.

Sun Protection Factor (SPF). The panel defined the ratio of the amount of energy required to produce a minimum erythemal dose (MED) to the amount of energy to produce the same MED without any treatment by the SPF. Examples given are: SPF

3 for nonsensitive skin, minimal protection; SPF 4 for normal sensitive skin, moderate protection; and SPF 6 for sensitive skin, extra protection.

Definitions. A sunscreen *sunburn preventive agent* is an active ingredient that absorbs 95% or more of light in the uv range of 290–320 nm. A sunscreen *suntanning agent* is an active ingredient that absorbs up to 85% of light in the uv range of 290–320 nm but transmits light at wavelengths longer than 320 nm. A sunscreen *opaque sunblock agent* is an opaque agent that reflects or scatters light in the uv and visible range at wavelengths of 290–777 nm.

Recommendations. The panel recommended the product designations based on SPF factors in Table 2.

Table 2. Sunscreen Product Designations Based on SPF Factors

Skin type	Sunburn and tanning history	Recommended SPF factor product category designation
1	burns easily, never tans	8 or more (max)
2	burns easily, minimal tan	6–7 (extra)
3	burns moderately, normal tan	4–5 (moderate)
4	burns minimally, tans well	2–3 (min)
5	burns rarely, tans profusely	2 (min)
6	never burns, deep pigment	none indicated

Safe and effective sunscreen ingredients are listed in Table 3.
For significant residual protection, PABA must be applied at least two hours prior

Table 3. Safe and Effective Sunscreen Ingredients

Ingredient	Trade or generic name
aminobenzoic acid	
2-ethoxyethyl *p*-methoxycinnamate	Cinoxate
hydroxyethylaminoethyl *p*-methoxycinnamate	
digalloyl trioleate	
2,2′-dihydroxy-4-methoxybenzophenone	dioxybenzone
ethyl 4-[bis(2-hydroxypropyl)amino]benzoate	
2-ethylhexyl 2-cyano-3,3-diphenylacrylate	
2-ethylhexyl *p*-methoxycinnamate	
2-ethylhexyl salicylate	
3,3,5-trimethylcyclohexyl salicylate	Homosalate
glyceryl aminobenzoate	
2-hydroxy-1,4-naphthoquinone	Lawsone
methyl anthranilate	
2-hydroxy-4-methoxybenzophenone	oxybenzone
pentyl 4-dimethylaminobenzoate	Padimate A
2-ethylhexyl 4-dimethylaminobenzoate	Padimate O
2-phenylbenzimidazole-5-sulfonic acid	
red petrolatum	
2-hydroxy-4-methoxybenzophenone-5-sulfonic acid	sulisobenzone
titanium dioxide	
2-[bis(2-hydroxyethyl)amino]ethyl salicylate	

to immersion. This probably reflects the time required for penetration into the stratum corneum. It is highly important, especially during the first few days of exposure to the sun, to reapply PABA-type sunscreens after swimming. With repeated daily applications there is a gradual build-up of the drug in the horny layer, resulting in significantly increased protection.

Sunscreens with much greater resistance to swimming can be formulated with acrylate polymers, which leave flexible films on the surface. A chemical sunscreen incorporated in the formulation will tend to be bound in the film and remain in place despite sweating and bathing.

Table 4 lists some effective physical and chemical sunscreens; information on various sunscreens is included in patents (17–18).

Table 4. Sunscreens

Chemical sunscreens	Trade name
2 ethoxyethyl *p*-methoxycinnamate	Giv-tan F (Givaudan)
menthyl anthranilate	
homomenthyl salicylate	
glyceryl *p*-aminobenzoate	Escalol 106 (Van Dyk)
isobutyl *p*-aminobenzoate	Cycloform (RSA Corp.)
isoamyl-*p*-dimethylaminobenzoate	Escalol 506 (Van Dyk)
2-hydroxy-4-methoxybenzophenone-5-sulfonic acid	Uvinul MS-40 (GAF)
2,2'-dihydroxy-4-methoxybenzophenone	Uvinul D-49 (GAF)
2-hydroxy-4-methoxybenzophenone	Uvinul M-40 (GAF)
p-aminobenzoic acid	
4-mono- and 4-bis(3-hydroxypropyl)amino isomers of ethyl benzoate	Amerscreen P (Amerchol)
2-ethylhexyl *p*-dimethylaminobenzoate	Escalol 507 (Van Dyk)

Natural oils (in order of effectiveness)
 (*1*) mink oil   (*6*) safflower oil
 (*2*) avocado oil   (*7*) peanut oil
 (*3*) sweet almond oil (*8*) jojoba oil
 (*4*) sesame oil   (*9*) coconut oil
 (*5*) persic oil   (*10*) olive oil

Physical sunscreens (opaque)
 titanium dioxide
 zinc oxide

Examples of sunscreen formulations are the following:

Clear Sunscreen Lotion: 74.50 wt % poly(alkylene glycol) (Ucon LB-625, Union Carbide); 20.00 wt % alcohol denatured no. 40; 5.00 wt % PABA (sunscreen); and 0.50 wt % perfume.

Opaque Sunscreen Lotion: 67.60 wt % distilled or deionized water; 10.00 wt % sesame oil (preserved, contains antioxidant); 10.00 wt % mineral oil (Saybolt viscosity at 37.8°C is 65/75); 5.00 wt % PABA (sunscreen); 3.00 wt % stearic acid, triple pressed; 2.90 wt % propylene glycol; 0.50 wt % cetyl alcohol; 0.50 wt % triethanolamine, 98 wt % solution; 0.25 wt % preservative [methyl and propyl paraben (5:1)]; and 0.25 wt % perfume.

Because of the extent of sunscreen application to the skin, any perfume used must be chosen with attention to its potential to irritate. Careful testing of its stability to heat and sunlight are also required (see also Uv absorbers).

Aerosols

Such cosmetic products as shaving creams, hair sprays, deodorants and antiperspirants, colognes, and sunscreens are now commonly packaged as aerosol sprays. In a bulletin released in 1966 the Aerosol Scientific Committee of the Chemical Specialties Manufacturers Association defined an aerosol product as a liquid, solid, gas, or a mixture that is discharged by a propellant force of liquified and/or nonliquified compressed gas, usually from a disposable-type container and through a valve (19–20).

The three main components of aerosol products are the propellants, solvents and the active ingredients. Previously, the liquified gas propellants have primarily been the chlorofluorocarbons (eg, Freons by DuPont) but these were banned in 1978 for use in all personal and household aerosol products, except for pharmaceuticals such as asthma spray.

Compressed gases (N_2, CO_2) and liquified hydrocarbons have become the primary propellants in revised formulations (21–22) (see Aerosols).

Make-Up Preparations

Color. Straight colorants are referred to in the trade as primary colors, primary dyes, and primaries. The principal primary colors are red, blue, and yellow. Practically every hue can be produced by blending red, yellow, and blue in proper proportions. Hue, in color, is a quality whereby one color differs from another. Brightness is a measure of the reflectance value of a dye. Strength is a measure of the tinting or coloring value of a dye (see Color).

Straight colorants include both primaries and lakes of such primary colorants made by extending the primary colorant on a substratum such as aluminum hydroxide, barium sulfate, a mixture of aluminum hydroxide and barium sulfate, aluminum benzoate, or zinc oxide. These lakes vary in pure dye content strength from 2–80%. The primaries contain over 80% pure dye. Primary colorants are frequently used in diluted forms. Mixtures of two or more primary colorants are sometimes referred to as primary mixtures. Mixtures of two primary colorants are referred to as secondary colorants, and mixtures of three primary colorants, as tertiary colorants.

The initials, FD&C, D&C, and external D&C, which are part of the name of straight colorants, indicate for what use the colorant has been certified. Colorants designated FD&C may be used for foods, drugs, and cosmetics. The name of a mixture usually tells nothing of its permitted use, but if there is a restriction in the use of the colorant, a statement to that effect is required on the product label (see Colorants for foods, drugs, and cosmetics).

For safety, colorants that have been certified for cosmetic use must contain not more than 0.002% of lead, not more than 0.0002% of arsenic, and not more than 0.003% of heavy metals other than lead and arsenic.

Most cosmetic colorants are certified coal-tar colorants (23–24). However, a few natural dyes are used in limited amounts to color foods, drugs, and cosmetics. The more important are alkanet, annatto, carotene, chlorophyll, cochineal, saffron, and henna (see Dyes, natural).

Inorganic Colorants. Except for the white pigments, such as titanium dioxide, zinc oxide, and talcum, the inorganic colorants are used in a number of cosmetic products. They usually have excellent light fastness and complete insolubility in solvents and aqueous solutions.

The naturally occurring colored minerals that depend for their color upon the presence of oxide of iron are known by such names as ochre, umber, sienna, etc. Because they are naturally occurring products, they possess certain inherent disadvantages. They show greater variation in color and tinting power than do manufactured pigments. The nature and amount of impurities they contain varies and may, from time to time, exclude their use in a domestic product.

The most important inorganic colorants or pigments are the iron oxides, chrome oxide greens, ultramarine pigment blues and pinks, and the carbon blacks. Iron oxide is sold in the form of yellow hydrated iron oxide, or ochre; brown iron oxide, red iron oxide, and black iron oxide. The yellow iron oxide is made by an alkaline precipitation from a ferrous salt, followed by an oxidation. the product is ca 85% Fe_2O_3, and is supplied in shades from light lemon yellow to a deep orange.

The brown iron oxides, umber, are made by treating a mixture of red, yellow, and black iron salt with alkali and partially oxidizing the precipitate. The brown oxide is a mixture of Fe_2O_3 and Fe_3O_4. The red iron oxides, sienna, are usually made by heating the precipitated yellow iron oxides. Light to dark red can be obtained depending on heating conditions, and the product is 96–98% Fe_2O_3.

The black iron oxide is formed by precipitation of Fe_3O_4 under carefully controlled conditions. The chromium oxide green is pure Cr_2O_3. The anhydrous pigment has a dull green shade; the hydrated pigment shows a bluish tint. This type of pigment is used in eye makeup.

Carbon blacks are of vegetable or natural gas origin. These blacks find wide cosmetic application, particularly in eye makeup.

The ultramarine blues and pinks are made from a mixture containing sulfur, sodium carbonate, charcoal, pitch or rosin, and several other ingredients carefully heated for a lengthy period. The finely pulverized pigment gives shades of ultramarine blue from the pink to the green, depending on its silica content.

Dye Stability. It is well known that many dyes are pH indicators. Such dyes change in hue with pH variations. Those that do not change much in hue can be affected in other important properties. It is, therefore, important to consider the pH conditions in a cosmetic product. In general, the inorganic colors are not as much affected by pH conditions as are the coal-tar dyes.

The hue of some dyes varies with the percentage of moisture present. The term bleeding is used to refer to the strengthening of the hue of a color in the presence of moisture, particularly in moist powders that are stored for a long period of time.

All water or organo-soluble dyes and most insoluble dyes and pigments are destroyed or chemically changed by the photochemical action of ultraviolet, visual light, infrared, and other short wavelength radiations. The degree of sensitivity varies with the dye and with the conditions.

The dyes change in tinctorial strength and hue, becoming progressively weaker, and usually duller, as the destruction of the dyestuff progresses. This is considered normal fading. The ability of the certified dyes to resist this chemical change owing to photochemical action has been charted for a number of the dyes and is available in tabular form (24).

The absence of moisture greatly retards photolytic effects. Absence of air, oxygen, hydrogen, and other oxidizing or reducing reagents also retards fading. The presence of titanium dioxide greatly accelerates it.

Judging Color Effects. Many factors affect the judgment of color, including the nature of the light source, the degree of transparency or opacity of the color, and eye strain or color memory. Ref. 25 discusses the factors involved in color measurement.

Face Powders. *Formulation.* Face powder, both the loose and the now-popular cake or pressed powder, is a blend of white pigments, tinted and perfumed. The function of a face powder is to impart a smooth, matte finish to the skin by masking any shine owing to the secretions of the sebaceous and sweat glands. To obtain this effect, the powder must be neither too transparent to mask the shine, nor too opaque, lest it give a masklike appearance. In addition, it must possess reasonable lasting properties; it must adhere to the skin and be reasonably resistant to the mixed secretions of the skin. Finally, through intimate contact of its perfume-laden particles over a warm and relatively large area, it should disseminate a pleasing odor, the psychological value of which is often underestimated.

No single substance possesses all the desired properties: covering power, slip, absorbency, adhesiveness, and bloom. Therefore, a modern face powder is a blend of several constituents, each one chosen for some specific quality. The materials most commonly used can be grouped in the following manner:

Covering Power. To cover skin defects, such as enlarged pores and skin shine. Examples: titanium dioxide, zinc oxide, kaolin, and magnesium oxide.

Slip. To assist in spreading and to give the characteristic smooth feeling. Examples: talc, zinc stearate, magnesium stearate, and starch.

Absorbency. To absorb sebaceous or oily secretions and perspiration, and thus reduce shine. Examples: precipitated chalk, magnesium carbonate, starch, and kaolin (see Clays).

Adherence. To improve clinging to the face. Examples: metallic soaps, such as magnesium and zinc stearates, and small quantities of oil or fatty materials in the powder base.

Bloom. To give a smooth, velvetlike appearance to the skin. A matte effect. Examples: chalk and starch.

Representative formulas for loose and pressed powders are:

Loose face powder	Wt %	
kaolin	3.0	
talc (Italian)	64.0	
talc (Sierra)		
magnesium stearate	1.5	base
magnesium carbonate	0.5	
zinc oxide	15.0	
corn starch	10.0	
D&C red #2 (lake) 20% in talc		
D&C red #3 (lake) 10% in talc		
D&C orange #4 (lake) 10% in talc	5.0	colorant dilutions
yellow iron oxide 20% in talc		
brown iron oxide 20% in talc		
perfume	1.0	

158 COSMETICS

Cake or pressed powder	Wt %	
kaolin	10.0	
zinc stearate	5.0	
zinc oxide	10.0	base
magnesium stearate	5.0	
talc (French)	61.4	
mineral oil light	2.0	
cetyl alcohol	1.0	binding agent
lanolin	0.3	
D&C orange #4 (lake) 10% in talc	2.7	
D&C red #2 (lake) 20% in talc	0.8	colorant dilutions
brown iron oxide 20% in talc	1.0	
perfume	0.8	

The function of face powder is to conceal shine and endow the skin with a new and better looking color or tint. The finer the particle size of face powder, the greater the diffusion of light and the better the bloom, with correspondingly less shine. The reason for this is that the light is reflected in many different angles, somewhat like reflection from the frosted surface of glass. Within certain limits, the smaller the powder particles, the higher the covering power and the better the tinting. Titanium dioxide, with its extremely fine particle size, provides especially good covering power. The percentage used is generally in the range of 5–8%. A common method of determining fineness of particles is the bite test, which consists of grating the ingredients between the teeth. The gritty threshold on grating chalk powder between the teeth has been placed at ca 12 μm. The materials used in a face powder should, therefore, be passed through at least a 50-μm (300-mesh) sieve, or, more properly, through a micronizing or micropulverizing machine.

The designation, light, medium, heavy, applied to a face powder denotes not weight, but covering power. A dry skin requires, in the absence of exceptional skin defects, a light powder; but since such a skin is lacking in grease, either the powder must possess increased powers of adherence, or a powder base must be used. A greasy skin, on the other hand, requires a powder of high opacity and grease resistance. Adherence is supplied by the skin itself, and in this case a powder containing sufficient grease-resistant constituents such as kaolin, together with a drying agent such as zinc oxide, may be used. The latter powder would be termed a heavy powder. Most women prefer a light to medium weight powder, since both are suitable to dry or normal skin.

Manufacture. The following is a procedure commonly used in manufacturing loose face powder: (*1*) All basic materials are screened into a Day spiral (ribbon) mixer, and blended. (2) The inorganic colorants and lakes or toners are diluted with either some of the mixed base material or with a portion of the talcum and screened. These diluted colorants are then added to the base. (3) The perfume, diluted and thoroughly mixed with a portion of the mixed base material or the talcum and screened, is added to the mixer. (*4*) The mixture is well blended and then transferred to a 50-μm (300-mesh) silk or stainless steel screen where it is screened several times and checked for color. An efficient and rapid method of screening consists of passing the material through a micronizer, using a suitable screen. (5) If the color check indicates that adjustments are necessary, the estimated amount of colorant dilutions are added and extra base and perfume to compensate for the additions are blended in. (6) When the color corresponds closely with the standard, the batch should again be thoroughly mixed in the spiral mixer and the entire batch passed through the 50-μm (300-mesh) screen or the Fitzpatrick comminuter or a micronizer. (7) The screened material can now be stored in a cool, dry place, or can be sent to the packaging room.

The manufacturing procedure for compact powder closely parallels that for loose powder, except that the moistening agent must be added slowly and mixed well in the base materials. If water is used to moisten the base materials, the water content of the mixed powder should be reduced by drying to ca 2.5–3%. Few manufacturers use water to moisten the powder because drying lengthens the process considerably and because water may cause some bleeding. The finished powder, having been passed by the control laboratory for shade, can be stored in the same way that the loose powder is stored.

Before the powder is placed in the pans for pressing it must be agitated because the inclusion of air, introduced during the mixing and screening process, may cause breakage of the compact powder unless it is vibrated before pressing.

Because vibration does not necessarily remove all included air, the process known as prepress is a part of the manufacturing procedure for compact powder. Prepress consists of applying moderate pressure to the material placed in the pans, and then releasing the pressure before passing the cakes on to the hydraulic press. At the hydraulic press sufficient pressure is applied to make a hard, compact powder cake.

Breakage of the compact cake is a very important factor for rejection. One firm tests the cake of compressed powder by dropping it from a height of ca 0.3 m onto a thin metal plate, cushioned by a few sheets of paper. The cake of compressed powder has to withstand more than ten drops from this height to pass the control test.

Product Stability. The most common reason that loose face powder loses its saleability is loss of perfume. The stability of the perfume in a particular face powder can be readily checked by using the method of minimum quantity to discover whether perfume and powder are reacting with each other. For this test one uses not more than 0.1% of perfume. A good perfume and suitable powder should maintain a slight but agreeable perfumed note for at least two days in a vessel open to the air. The usual failure is the disappearance of all of the perfume, or the appearance of a phenolic or medicated odor.

The stability of the pigment used is of great concern to every manufacturer of face powder. Not all shades can be produced with earth colors exclusively, and it is necessary to use some organic lakes and toners. These should be used sparingly, because they are not light stable for any length of time. A simple check of light stability is to place a puff on the smoothened surface of loose face powder and then put the sample into bright sunlight for about thirty minutes. Any appreciable bleaching of the unprotected area indicates possible trouble.

Zinc stearate may develop an odor on aging, and it is necessary to use a good grade of zinc stearate for face powder. Kaolin may develop a claylike odor when moistened, and magnesium carbonate may have a deleterious effect on the perfume. All materials used in the preparation of face powders must be carefully tested to prevent the inclusion of impurities. The same precautionary measures taken in preparing loose face powder must be observed in the manufacture of compact or pressed face powder. Additional care must be taken in the choice of ingredients since the compact face powders usually contain moistening ingredients. Off odors may also be introduced by the glues used in powder boxes, puffs or in gluing compressed powder to the metal pan.

Lipstick. Lipstick is a solid fatty-base product containing dissolved and suspended colorant materials. A good lipstick must possess a certain maximum and minimum of thixotropy; ie, it must soften enough to yield a smooth, even application with the minimum of pressure. It should not be necessary to apply a lipstick more often than every 4 to 6 hours. The applied film, no matter how thin, should to some extent be impervious to the mild abrasion encountered during eating, drinking, or smoking. The lipstick should be of such composition as to color only that portion of the lip to which it is applied, and not to bleed, streak or feather into the surrounding tissue of the mouth. It should be of a consistency that can be quickly and easily applied, and the perfume (usually present at 1–2%) should be carefully compounded so that it does not have an unpleasant taste. The colorants used in lipsticks are either insoluble lakes or oil-soluble dyes of the Eosin group, or both.

The bromo acids, bromo derivatives of fluorescein, are used to produce indelibility in the applied film. Two are most generally used, dibromo- and tetrabromofluorescein. Dibromofluorescein produces a yellow–red color; the tetra compound gives a more purple stain. Usually the two are used in combination, as neither produces desired results separately. From 1–5% of these dyes are used in lipsticks; but ordinarily, 2 or 3% is adequate. Other pigments are used in concentrations of 5–15%, 10% is an average.

A highly refined grade of castor oil (qv) is one of the most common ingredients in lipstick. It is used primarily to impart viscosity to the molded stick and, secondly, because it has a small solvent action on the bromo acids. Castor oil dissolves less than 0.3% of the bromo acids. Its solvent action is owing to the presence of free hydroxyl groups in the glycerides of ricinoleic acid, which make up a large percentage of castor oil.

In the so-called high-stain sticks, a solvent such as polyethylene glycol 400 is used. The polyethylene glycol is not soluble in castor oil. A mutual solvent or coupling agent is, therefore, necessary; propylene glycol monoricinoleate is used for this purpose. Isopropyl myristate, isopropyl palmitate, butyl stearate, and some of the higher esters of the mono- and dihydric alcohols are commonly used in lipsticks, both for their solvent action on the bromo acids, and to reduce the thickness of the castor oil film (26).

Lanolin and various derivatives are used for their emollient properties and for a degree of tackiness and drag. Acetylated lanolin is more soluble in castor oil than lanolin itself. Lanolin oils, solvent extractions of lanolin, eliminate much of the tack of lanolin. Lanolin helps hold the lipstick mass in a uniform homogenous mixture. Lanolin, used in the proper ratio, helps to prevent sweating of the solvent oils and confers some protection against abrupt temperature changes.

Various waxes are used to impart different characteristics, such as hardness, thixotropy, melting point, and ease of application. These waxes are used in amounts regulated by the nature of the other components making up the finished stick.

Carnauba wax is responsible for stiffening the stick. If used in the proper proportions, it imparts to the stick exactly the right degree of thixotropy. Its melting point is very high, much higher than any of the other natural waxes. Candelilla wax is used for the same reason, but to a lesser extent. Too much carnauba will cause a granular texture in the finished lipstick, and it is best to keep its concentration as low as possible, usually less than 5%.

Ceresin and ozokerite are both also good stiffening agents, with ceresin exhibiting the greater stiffening action. Since ceresin is merely a mixture of ozokerite and paraffin, the same effect can be obtained by using these two waxes in combination. Ceresin can have a variety of melting points, depending upon the ratio of ozokerite to paraffin, thus affecting the melting point of the lipstick. Ozokerite is used more often to raise the melting point. A combination of ozokerite and carnauba is most efficient.

Beeswax can be used to raise the melting point. It is a good binder, helping to produce a homogenous mass. In large amounts it causes granulation and a dulling effect (27–28). A typical lipstick base formulation is: 49.6 wt % castor oil; 26.3 wt % hydrogenated vegetable oil, a cocoa butter substitute; 11.6 wt % ozokerite 160; 5.0 wt % lanolin anhydrous; 4.5 wt % carnauba wax; 1.7 wt % cetyl alcohol; 1.0 wt % perfume; 0.2 wt % antioxidant; 0.1 wt % preservative, plus the necessary pigment pastes.

Manufacture. In manufacturing lipsticks, the colorants are mixed with part of the oil, usually in a Day Pony mixer, then passed through a 3-roll mill. The color and oil are ground a sufficient number of times to achieve a colorant distribution of 7 or better on a grind gage.

Gas that is adsorbed on the surface of pigment particles diffuses slowly into the surrounding lipstick mass. A marked improvement is noticed in a pigment dispersion that is placed under vacuum. Dispersion of pigment involves a complicated combination of surface phenomena, eg, adsorption, wetting, surface tension, interfacial tension, and formation of phase boundaries. Electrical charges must also be considered. Mechanical force alone cannot produce a satisfactory dispersion.

The base material is heated in a steam-jacketed kettle to 90°C and the ground colorants are added to the base material in the kettle. The batch is stirred with a high speed mixer for several hours. Any adjustment in shade is made by using the ground colorants. Base material is added in proportion to the amount of colorant used in the correction to keep the consistency approximately equal to that in the original formulation.

When the color match is satisfactory, the lipstick mass is drawn from the kettle and passed through a 74-μm (200-mesh) stainless steel screen. It is collected in a molding pan and slabbed for future molding. Prior to molding, the mass is melted with slow agitation in a melting kettle to remove entrapped air.

Temperature control is important throughout processing, with care being taken to avoid excessive temperature in the kettle. The warmed lipstick mass is poured into warm molds, and the molds are passed through a cooling chamber. When cool, the lipsticks are removed from the molds, inserted in the lipstick holders, and flamed by passing them rapidly through a bunsen gas flame. They are then ready for packaging and distribution.

Control Methods. Manufacturing control begins with a thorough check of raw materials. A test of heat resistance of the finished product is made by holding it for 24 h in a constant temperature box at a temperature of 55°C, to determine any tendency to droop or distort. Hardness may be tested using a penetrometer.

The tendency for the stick to sweat or bleed can be checked by putting filled lipstick containers in a desiccator containing water and kept at a constant temperature of 45°C.

Shade and intensity of stain are best judged on the skin by comparison with the standard. Skin tone and undertone are also checked on the skin. Mass tone is judged by comparison with a standard (29).

Mascara. Because of the proximity of these products to the eye, the law imposes strict limitations on the ingredients they may contain. The colorants are limited to natural dyes and inorganic and carbon pigments. The commercially available pigments used are the varying shades of the iron oxides, carbon blacks, and ultramarine blues.

Mascara, used to emphasize lashes, makes them appear longer since the ends of the lashes tend to be rather blondish, as compared to the rest of the hair length. The film applied should show good resistance to moisture and to gentle abrasion.

Cream Mascaras. In cream mascara the water has already been incorporated to provide a mascara ready for use without previous trituration. There are some advantages in this type of preparation since a much higher concentration of pigment can be incorporated than is possible in a cake. In addition, the optimum amount of water may be added rather than leaving this to the discretion of the consumer.

A typical cream mascara base is: 46.50 wt % hydrocarbon solvent (Soltrol no. 130); 33.25 wt % distilled or deionized water; 8.25 wt % caranuba wax no. 1 refined; 7.50 wt % beeswax, white; 4.00 wt % glyceryl monostearate S.E. (Arlacel C); 0.40 wt % polyoxyethylene (20) sorbitan stearate (Tween 60); 0.40 wt % borax .5H$_2$O; 0.10 wt % propyl p-hydroxybenzoate; and 0.05 wt % methyl p-hydroxybenzoate.

The pigments used in cream mascara are ground into the base material, and the fine grind is then mixed with the necessary water and humectants to form the cream.

Eye Shadows. The colors used in eye shadows are carbon blacks, ultramarine blue, and the various yellows, browns, and reds of the iron oxide pigments. There is usually a higher concentration of pigments in eye shadows than mascaras, sometimes as high as 25%. The base into which the colorants are ground usually consists of beeswax, ozokerite, mineral oil, lanolin, and petrolatum. The method of manufacture consists of melting the waxes and oils together, adding the colorants and titanium dioxide used to impart the desired opacity, and grinding in a three-roll mill.

Eye shadows that have a metallic luster are made by incorporating finely ground aluminum, bronze, silver, or gold powders. The completed formula is heated and stirred slowly to remove occluded air and is then poured into the container.

From time to time powder-shadow products, called Kohl, have appeared on the market. These are principally talcum-based with various inorganic colorants.

A prototype formula for an eye shadow is: 40.0 wt % mineral oil 65/75; 24.0 wt % ozokerite 80°C; 20.0 wt % chromic oxide hydrated (40% in petrolatum) 14.0 wt % petrolatum 45–50°C; 1.7 wt % paraffin; and 0.3 wt % propyl para-hydroxybenzoate.

Nail Products. *Nail Lacquer.* Nail lacquer or nail polish usually consists of a resin, plasticizer, solvents, and pigment (30–32). The resins can be nitrocellulose or dinitrocellulose, also known as pyroxylin. Nitrocellulose films tend to shrink, so that their surface adhesion may be only moderate. Other resins are, therefore, added to impart adhesion. Gloss and plasticizers are added to reduce shrinkage and make the film flexible.

The solvents used influence the ease of application of the lacquer, its rate of drying and hardening and the final characteristics of the film. The preferred solvents for nail lacquers are mixtures of low and medium boiling point alcohols, aromatic hydrocarbons, and aliphatic hydrocarbons.

Plasticizers may be added singly or in combinations of two or more, depending on the formulation. High molecular weight esters are used as well as castor oil. The esters can be dibutyl and dioctyl phthalate, triethyl citrate, and acetyl tributyl citrate.

The pigments used in nail lacquer formulations are carbon black, iron oxides, chromium oxides, ultramarines, metallic powders (gold, bronze, aluminum, copper), aluminum and calcium lakes of FD&C and D&C blue, red, yellow, and orange, and titanium dioxide. Transparent systems require the use of solvent-soluble colorants such as D&C red, green, yellow, and violet.

Pearl pigment in nail lacquer may be the natural pearl essence, guanine (2-amino-6-hydroxy purine), or a synthetic pearl that is usually a bismuth salt.

A typical nail lacquer formulation is: 30.5 wt % toluene; 12.8 wt % n-butyl acetate; 11.1 wt % n-amyl acetate; 11.1 wt % ethyl acetate; 10.0 wt % nitrocellulose; 10.0 wt % aryl sulfonamide–formaldehyde resin; 5.0 wt % isopropanol; 3.0 wt % bentonite; 2.5 wt % camphor (plasticizer); 2.5 wt % dibutyl phthalate; and 1.5 wt % ethanol.

Lacquer Removers. Lacquer removers may be made from any of a number of solvents. In most cases, any of the solvents used in the formulation of the enamels can be used as removers. The fastest acting seems to be acetone, but because of its high volatility, ethyl acetate has come into favor as well. Many of the lacquer removers contain an additive designed to leave a film on the nail, since the solvent itself has so high a degreasing tendency. Various oils and emollient compounds are added to the lacquer removers such as castor oil, lanolins, and lanolin derivatives.

A typical nail polish remover formula is: 40.0 wt % ethyl acetate; 30.0 wt % acetone; 19.0 wt % carbitol; 10.0 wt % dibutyl phthalate; 1.0 wt % sesame oil; and *qs* perfume.

Cuticle Removers and Softeners. Cuticle removers and cuticle softeners usually consist of a dilute solution of alkali in water with some glycerol or other humectant added to keep the water from evaporating too easily. Potassium hydroxide, the alkali most often used, results in a somewhat harsh preparation but is effective. Trisodium phosphate, triethanolamine, and some of the quaternary ammonium salts have been used as replacements for potassium hydroxides. These are not as effective as the alkali but do a fair job of softening the cuticle.

Hair Preparations

Shampoos. Present day shampoos usually contain a primary detergent which can be a fatty alcohol sulfate, ether sulfate, sarcosinate, or one of many other anionics (see Hair preparations; Surfactants). The primary detergent can also be amphoteric or nonionic. The secondary, or auxiliary, detergent is usually an alkanolamide, which controls the viscosity of the product and increases the quality and volume of the foam (33–39).

Soap Shampoo. Soap shampoos do not cleanse well in hard water areas. They are primarily aqueous solutions of soft soap combined with preservatives, sequestrants, color, and perfume.

Soapless Shampoo. Soapless shampoos are primarily based on aqueous solutions of sulfonated oils such as sulfonated castor oil and sulfonated olive oil combined with preservatives, color, and perfume.

Low pH Shampoos. Unlike many of the properties ascribed to shampoos, such as foaming, cleansing, luster, and manageability, pH is an intangible in that it cannot be observed or demonstrated during product usage, and the effects of low pH are not obvious to the user. It has been reported that mild aqueous acids cause antiswelling action on the cuticle scales of the hair. As the cuticle tightens, the hair gains luster because light is more efficiently reflected from the surface of the hair shaft. In the absence of external charges hair shows its highest strength and resiliency at pH 4.0–6.0. Some surfactants do not perform well in an acid medium. Fatty alcohol sulfates hydrolyze rapidly and decompose below pH 4.0. Alkanolamides generally react in a similar manner. However, combinations of triethanolamine, and sodium and ammonium lauryl sulfates with amine oxides can be adjusted to low pH with good stability. Examples of clear, stable shampoo formulations adjusted to pH 4.0 are: 54.0 wt % water; 40.0 wt % ammonium lauryl sulfate (30% solution); 6.0 wt % cocamidopropylamine oxide (30% solution); and qs (as needed) preservative, color, perfume, acid, etc.

Amphoteric Shampoos. Amphoteric surfactants are generally assumed to be less irritating to the eyes than other detergent or soap preparations. They are usually low foam materials, and it is customary to combine them with anionic surfactants. Amphoteric shampoos are based on the use of imidazoline, betaine, and sulfobetaine surfactants. Amphoteric-base shampoos are difficult to thicken and difficult to opacify. Amphoterics can be blended with equal parts of either high foaming anionic lauryl sulfates or the sulfo-succinate half esters. Amine oxides can also be used instead of the more irritating alkanolamides. The keratin substantivity of the betaine amphoterics is greatest at pH 5.5–6.5. The following shampoo formula, which has been adjusted to a pH of 5.8, illustrates the use of amphoterics: 17.0 wt % ammonium lauryl sulfate; 6.0 wt % Monomate CPA-40 (sulfo-succinate half ester); 2.0 wt % Monateric ISA-35 (isostearic imidazoline); and qs perfume, preservative, color, and water.

Dandruff Shampoos. Dandruff is the product of hyperkeratinization. The rate of keratinization has increased to the point that the scales become more visible. There are no known differences in the incidence of dandruff in men and women or among races. Dandruff shampoos contain ingredients that effectively control dandruff by allowing a normal turnover rate of epidermal cells. Ingredients used in some antidandruff shampoos are coal tar, quaternary ammonium compounds, resorcinol, salicylic acid, selenium sulfide, sulfur, undecylenic acid and derivatives and zinc pyrithione (zinc 2-mercaptopyridine 1-oxide; zinc omadine). Selenium sulfide and zinc pyrithione are cytostatic (reduce epidermal cell turnover rate). Zinc pyrithione is strongly adsorbed by the hair, with adsorption related to concentration, pH, temperature, product formulation, and time of exposure. Since these are medicated shampoos, formulations should be nonirritating and nonsensitizing.

A typical lotion cream shampoo is: 47.0 wt % distilled or deionized water; 42.1 wt % sodium lauryl sulfate, 28% active; 4.0 wt % ethylene glycol monostearate; 3.0 wt % coconut fatty acid diethanolamide; 3.0 wt % propylene glycol; 0.5 wt % perfume; and 0.4 wt % preservative.

A typical clear liquid shampoo is: 50.0 wt % triethanolamine lauryl sulfate, 65% active; 27.6 wt % distilled or deionized water; 10.0 wt % propylene glycol; 5.0 wt % lauric acid diethanolamide; 4.0 wt % myristic acid; 2.0 wt % oleyl alcohol; 0.8 wt % perfume; 0.4 wt % preservative; 0.1 wt % citric acid; and 0.1 wt % sequesterant Na_2.

A typical viscous liquid shampoo is: 60.0 wt % triethanolamine lauryl sulfate, 38% active; 21.1 wt % distilled or deionized water; 3.0 wt % myristic acid; 2.0 wt % oleyl alcohol; 2.0 wt % propylene glycol; 1.0 wt % perfume; 0.5 wt % poly(vinylpyrrolidinone) K-30; and 0.4 wt % preservative.

A typical antidandruff shampoo is: 65.3 wt % water; 25.5 wt % sodium lauryl sulfate; 5.0 wt % stearic acid; 2.1 wt % lauramine oxide; 1.0 wt % zinc pyrithione powder; 0.7 wt % sodium hydroxide pellets; 0.4 wt % sodium chloride; plus preservative, color, and perfume.

Shampoo Additives. Foam builders are principally fatty acid alkanolamides. Conditioning agents are amine oxides, fatty alcohols, lanolin derivatives, esters of fatty acids, silicones, cationic materials, and polyacrylamides (40–41). Opacifying agents are higher fatty alcohols, such as stearyl and cetyl, and glycol and glycerol esters of fatty acids. Sequestering agents prevent formation of insoluble calcium and magnesium soaps and prevent discoloration owing to iron contamination. Citric acid, tartaric acid and salts of EDTA are commonly used sequestrants. Viscosity builders are cellulose derivatives, alkanolamides, and carboxy-vinyl polymers. Preservatives are usually the hydroxybenzoates, formaldehyde, imidazolidinyl- urea compounds (Germall), sorbates, and 1-(3-chloroallyl)3,5,7-triazo-1-azoniaadamantane chloride (Dowicil).

Hair Straighteners. Though the thioglycolate straighteners are essentially the same as products used for hair waving, they differ in that the straighteners are appreciably more viscous to help keep the hair straight while it is being softened (see Hair preparations). After application, the hair is repeatedly combed, which may be a factor in scalp irritation. Since curly hair is quite resistant, the softening process takes a rather long time and the danger of hair damage and scalp irritation proportionately increases. When it reaches the desired straightness, the hair is rinsed and treated with a conventional neutralizer such as sodium bromate (42). A typical formula for a thioglycolate straightener is: 52.00–53.00 wt % distilled or deionized water; 22.00 wt % glyceryl monostearate, pure; 17.50 wt % ammonium thioglycolate (50% solution, 8.75% thio acid); ca 4.50 wt % ammonium hydroxide (28% solution, sufficient to product pH 9.2–9.4); 1.75 wt % Brij 35; 1.00 wt % mineral oil; 0.50–1.00 wt % perfume; and 0.15 wt % preservative [methyl p-hydroxybenzoate and propyl p-hydroxybenzoate (3:1)].

The Gant and Hersh patent relates to compositions and methods for straightening kinky hair or for curling straight hair (43).

Depilatories. Products used for the removal of unwanted hair by chemical agents are commonly called depilatories. The requirements of a depilatory are that it should be (1) nonirritating and innocuous, (2) efficient in action, removing hair within a 10 min period, and (3) be without odor or have the pleasant odor of the perfume.

The activity of the depilatory substance depends upon (1) the nature of the depilating or denaturing agent, (2) the duration of the action, (3) the pH of the cosmetic medium, and (4) the temperature of the reaction.

The depilatory most widely used today is based on calcium thioglycolate in a strongly alkaline medium, pH ca 12.3. It also contains a small amount of a wetting agent, a filler, enough slaked lime to attain the required pH, and alkali stable perfume. The commercial products are sold either as thin or very thick pastes, packaged in a jar or collapsible tube.

A typical formula for a depilatory is: 69.0 wt % base cream [60.10 wt % distilled water (solvent); 32.00 wt % calcium carbonate (filler); 6.15 wt % cetyl alcohol (bodying agent for cream); and 1.75 wt % product G-2135 (wetting agent and emulsifier)]; 15.0 wt % distilled water; 6.8 wt % calcium hydroxide (pH control); 5.4 wt % calcium thioglycolate (depilating agent); 3.4 wt % strontium hydroxide (pH control); and 0.4 wt % perfume. Perfumes used for depilatories have to be specially formulated to overcome the intense product sulfide odor and be stable in the high pH medium of 12.3.

In manufacture the following precautions must be observed: (1) avoidance of

166 COSMETICS

metal contamination from iron or copper or other heavy metals, pure stainless steel is satisfactory; (2) avoidance of unnecessary exposure to air as oxidation reduces product strength; (3) pH control at ca 12.3, which should be checked before packaging; (4) choice of proper packaging, pure tin is best, wax-lined lead tubes, glass, and polyethylene packaging are satisfactory.

Waxing. A very old method for removal of hair makes use of a material known as epilating wax. This wax is a mixture of rosin and beeswax, which is melted, applied to the area to be treated, and allowed to solidify, after which the wax mixture is peeled from the skin. During this peeling action, the hair is actually pulled out and removed from the skin along with the wax mixture.

Bath Products

When bathing, dirt, perspiration, and microorganisms are initially removed from the surface of the skin by the water and soap. The water also causes the outermost layers of the epidermis to swell slightly so that they are shed more rapidly. If the bath contains suitable additives, it also can show effects that extend beyond the hygienic aim and contribute cosmetically (44–46).

Bath Oils. The subjective response to the use of any bath oil is generally favorable. This is to be expected since bathing itself imparts a feeling of softness to the stratum corneum, and even the slightest deposition of an emollient adds to this effect. The addition of fragrances and fragrant oils to baths was recorded as early as the Roman era.

Bath oils fall into one of two major categories: (1) spreading or floating bath oils and (2) dispersible bath oils. Spreading bath oil is usually an anhydrous system that is oleaginous. The hydrophobic system floats on the surface of the bath water. The spreading action is facilitated by inclusion of small amounts (usually ca 5%) of an oil soluble surfactant. The lowering of the surface tension of bath water by the surfactant permits the oil to form a continuous film, rather than existing as individual droplets. Ideal surfactants should demonstrate a reasonably high hydrophilic–lipophilic balance (HLB) (ca 9) yet remain soluble in the oleaginous composition.

From a functional standpoint the floating layer of perfumed oil affords a pleasant fragrance and a desirable feel to the skin. The oleaginous film adheres to the body surface and imparts a hydrophobic barrier to the keratin.

The oleaginous composition may consist of one or more of the commonly accepted hydrophobic components: natural vegetable oils, animal-derived lipids, such as lanolin and mink oil, low melting fatty alcohols, such as oleyl and hexadecyl, low melting synthetic glycerides, such as glyceryl mono-oleate and glyceryl monolaurate, and mineral oils.

Dispersible bath oils contain a sufficient concentration of surface active agents to solubilize the oleaginous components through micellar solubilization, or to disperse them as the internal phase of an oil-in-water emulsion. The addition of either system to the bath water disperses oil throughout the bath water. Unlike the spreading bath oil, which maintains its oleaginous character in water, the dispersible bath oil loses some of its oleaginous character as the result of oil-in-water dispersibility. Dispersible bath oils leave a layer of dilute oil in a water dispersion on the skin. As the water evaporates from this, a hydrophobic residue remains, providing a moisture barrier on the skin surface.

The nature (HLB) and concentration of the surfactant component also dictates whether the dispersible bath oil turns milky on addition to water (blooming bath oil), whether it remains clear on dispersing in water, or whether the bath oil generates any foam (foaming bath oil). Surfactants with lower HLBs, such as polyoxyethylene oleyl ether (HLB 4.9), are oil soluble and yield oils that produce the characteristic blooming. Perfume is generally employed in a concentration of 3–20%, obviously depending on the nature and quantity of surfactant.

The dispersible bath oils are usually preferred by the consumer over the floating type because they are water dispersible (washable) whereas the latter may leave an oil slick in the tub and may create drain clogging problems.

Foaming Bath Oils and Foam Baths. These are based on suitable surfactants, either liquid or powder, which are blended with color, perfume, and foam stabilizers. Addition of warm or hot running water produces a fragrant foam.

Bath Salts. Two types of bath salts are available. The first is formulated with crystalline salts such as rock salt and epsom salt to which color and perfume are added. These are not water softeners. The second type is the water softening type based on sesquicarbonates, phosphates, and borates. Color and perfume are added and in some products, small percentages of fatty acid ester are included for nondrying effects.

Soaps

Soaps can be made from natural fats or from the isolated fatty acids (47–50). The natural fats, glycerides of various fatty acids, are used to make soap where cost is a consideration (see Soap).

Manufacture. Soap manufacture is generally by one of three methods. In the boiled process, the fats, oils, and alkalies are all boiled together until saponification is complete. The soap is salted out with or without varying percentages of glycerol left in the soap. This is the most widely used process. In the semiboiled process, exact chemical amounts of the fats, oils, and alkalis are mixed and heated to start the saponification, after which the reaction is allowed to complete itself. In the cold process, the fats, oils, and a calculated quantity (less than needed for complete saponification) of alkali are placed in a crutcher and gently heated and mixed. When saponification is underway the material is poured into soap frames where the reaction completes itself spontaneously. The glycerol which is the by-product is left in the soap.

Transparent soaps are prepared with sugar, alcohol and glycerol or a combination of these materials in the approximate proportions (51): sugar, 10–15 wt % of the soap; alcohol, 50% of the fatty acid present in the soap; glycerol, 50% of the fatty acid present in the soap.

Detergency or cleaning ability is usually determined by soiling cloth with a standard smudge or soiling mixture made according to the following formula: lamp or carbon black, 2.0 g; heavy mineral oil, 5.0 g; tallow, 3.0 g; carbon tetrachloride, 2000.0 mL.

It is well known that soaps and detergents remove the natural oil secretions (sebaceous matter) as well as the dirt. This drying effect can be partially overcome by superfatting the soap or detergent bar. The presence of superfatting agents reduces the cleansing efficiency thereby preventing complete degreasing of the skin surface. From a dermatological aspect, the superfatting of soaps should be advantageous. Apart from the dermatological aspect, the addition of superfatting agents to soap also makes the process of milling and plodding easier because the soap is rendered more plastic. It contributes to the quality of the final product by reducing the tendency of the soap cake to split or crack in use. Superfatted soaps have a superior velvety feel and produce an especially smooth lather.

Superfatting agents are: lecithin (qv) (softens and moisturizes skin); amine oxides and fatty acid amides (improve soap texture and lathering properties); lanolin (improves texture of soap); casein (improves lather and skin feel); and cocoa butter (emollient and moisturizer).

Deodorant Soaps. About half the soap market is claimed for deodorant soaps. These are usually quality soap bars containing either Irgasan DP-300 (Triclosan) or 3,4,4'-trichlorocarbanilide (Triclocarban). A patent was issued in 1972 for a germicidal detergent bar that contains poly(vinylpyrrolidinone)–iodine complex (52).

Shaving Preparations

Dry hair in the beard is usually quite compact and is resistant to cutting by a razor blade. Preliminary treatment with hot water and soap removes the sebum, a complex mixture of lipids secreted by the sebaceous glands, and absorbed by the beard, that presents a barrier to water and impedes hair softening. The use of a shaving soap or shaving cream provides a wet, lubricating blanket of water that reduces friction between the razor blade and the skin (53–54).

Shave Soap. Shaving soaps or shaving sticks are hard soaps, sodium salts of primarily saturated fatty acids, that are usually applied with a shaving brush.

Lather Shave Cream. Shave creams are formulated as lather shave and brushless shave. Lather shave creams are soft soaps formulated with potassium hydroxide salts of saturated and unsaturated fatty acids, and containing a small percentage of sodium hydroxide to regulate viscosity, body, and consistency. Added to the basic product are humectants such as glycerol and propylene glycol, emollients such as lanolin and lanolin derivatives, boric acid in small percentage to neutralize any excess alkali, and menthol for its cooling effect. The water content of a lather shave is about 35 to 40%, and the pH of the lather is about 10.

A typical formula for shaving cream is: 41.52 wt % water; 30.8 wt % stearic acid; 10.0 wt % glycerol; 7.7 wt % coconut fatty acids; 7.0 wt % potassium hydroxide; 1.0 wt % lanolin; 1.0 wt % boric acid; 0.57 wt % sodium hydroxide; 0.5 wt % sodium silicate; and qs perfume.

Brushless Shave Cream. Brushless shave creams are similar to vanishing creams, with an excess of free fatty acid, usually stearic, emulsified into a nonalkaline base of suitable viscosity or consistency. Its softening effect is not so rapid as that obtained with a lather shave. Brushless shave creams have a water content higher than that of lather shaves, and a pH of 6.0–6.5. Additives are similar to those used in lather shave formulations. A typical formulation is: 70.8 wt % water; 17.0 wt % stearic acid; 4.0 wt % glycerol; 4.0 wt % glyceryl monostearate; 2.0 wt % mineral oil; 1.2 wt % triethanolamine; 1.0 wt % lanolin; qs preservative; and qs perfume. All formulations contain preservatives and usually fragrance to cover the base odor of the soap and additives.

Aerosol Shave Creams. Aerosol shave creams in pressurized containers are convenient to use because the lather is easily obtained in a ready to use form from the can and is easily rinsed off the face. Formulations are either aqueous soap solutions or soap emulsions that are pressurized with a mixture of propane and isobutane. The product is expelled from the container through a valve. The expelled liquid expands to a foam as the propellant is released. Aerosol shave creams contain additives similar to those used in lather shave formulas.

Self-Heating Shave Creams. Self-heating shave creams are formed from an exothermic reaction that occurs when two components, one containing an oxidizing agent and the other a reductant, are expelled together from a dual compartment container (55) (see Aerosols).

Gel Shave Creams. Gel shave creams are emollient aqueous gels that are soap based and contain a volatile solvent such as pentane. Expelled as a soft gel from a pressure container, the shave cream is rubbed on the face, and the solvent is released. The released solvent expands as it volatilizes, and the gel becomes a foam.

Oral Products

Toothpastes. Toothpastes are suspensions of polishing agents and detergents in a suitable binder, generally packaged in a collapsible tube (56–57) (see Dentifrices). The Council on Dental Therapeutics of the American Dental Association has established definite criteria for approval of dentifrices. Some of the requirements are based on the degree of hardness, degree of abrasiveness, and particle size of the polishing agent. The use of medicinals or of drugs that should be used only under the supervision of a dentist are not approved by the Council for over-the-counter purchase by the consumer.

The principal ingredients in toothpastes or powders are: polishing agents (calcium carbonate, magnesium carbonate, di- or tricalcium phosphate, talc, etc); detergents (soap, anionic surface active agents); binders (excipients) and humectants (glycerol, propylene glycol, sorbitol, mucilages or gums, etc); sweeteners (saccharin, sorbitol); preservatives (benzoic acid, hydroxybenzoates); flavors (essential oils, etc); and water.

Cellulose derivatives, such as methyl cellulose, are sometimes used to maintain satisfactory consistency and stannous fluoride or sodium monofluorophosphate may be added as anticaries agents.

A prototype formula of toothpaste is: 32.200 wt % glycerol; 27.00 wt % calcium phosphate, dibasic 36.675 wt % water; 2.000 wt % sodium N-lauroyl sarcosinate; 1.00 wt % carrageenan TP-4; 0.500–1.000 wt % flavor; 0.400 wt % sodium monofluorophosphate; 0.125 wt % saccharin sodium USP; 0.100 wt % sodium benzoate USP.

A recently developed dentifrice contains synthetic, amorphous, porous silica xerogel as the polishing and cleansing ingredient (58). A formula example is: 39.0 wt % water; 34.76 wt % glycerol; 12.0 wt % silica Xerogel V; 6.0 wt % sodium lauryl sulfate (21% in glycerol); 5.0 wt % silica Aerogel 1; 1.5 wt % hydroxyethyl cellulose; 1.1 wt % flavor; 0.41 wt % stannous fluoride; 0.2 wt % saccharin; and 0.03 wt % FD&C Blue no. 1 (1% solution).

Preservatives

Changes and deterioration in cosmetic products can be brought about by changes in acidity or alkalinity, or as a result of hydrolysis, oxidation, heat and light, and bacterial or fungal contamination (59–62) (see Industrial antimicrobial agents). The source of bacterial or fungal contamination of cosmetic products can be the raw materials, water from dirty ion exchangers, contaminated equipment, or dirty bottles and cap liners. Even cosmetic products that are properly prepared require addition of preservatives to ensure adequate shelf life. In addition to all other quality assurance tests run by the laboratory, suitable microbiological tests should also be run on the finished product.

Characteristics of the ideal preservative or mixture of preservatives are: (1) activity of low concentration; (2) physical and chemical compatibility and effectiveness over a wide range of pH; (3) effectiveness against a wide range of microorganisms; (4) ready solubility and stability to heat and storage, and limited tendency to migrate into the nonaqueous phase; (5) lack of odor, color, toxicity, and irritation in the concentrations used.

Preservatives frequently used in cosmetic products are: p-hydroxybenzoates (parabens) (most effective in acid pH); dehydroacetic acid (active against fungi); sorbic acid and sorbates (active against fungi, pH-dependent); imidazolidinylurea (Germall 115) (synergistic with other preservatives); 2-bromo-2-nitro-1,3-propanediol (Bronopol) (active against gram-negative bacteria, releases formaldehyde): formaldehyde (strong antibacterial); and 1-(3-chloroallyl)-3,5,7-triaza-1-azoniaadamantane chloride (Dowicil 200 as cosmetic preservative) (formaldehyde donor).

Cosmetic formulations and packaging may interact with some preservatives. The nonionic emulsifiers commonly used in cosmetic formulations do not significantly interfere with preservative action when used in a concentration below 1%. In higher concentrations the emulsifier can form aggregates that extract preservatives from the oil–water phases, leaving a low concentration of preservative. Lipid-soluble preservatives may migrate into plastic container walls and into plastic cap liners. Some components of a cosmetic formulation can enhance preservative action. The ingredients of many perfume oils, eg, have antimicrobial properties. Synergisms of preservative mixtures are also possible. Mixtures of parabens are more effective than individual parabens, and mixtures of parabens and imidazolidinylurea are significantly more active and widen the range of microorganisms against which the cosmetic product is protected. Whatever preservative system is chosen should undergo microbial challenge testing in the final product (63–65). Methods of challenge testing have been outlined by the CFTA (61) (Table 5).

Antioxidants

An antioxidant is a substance capable of slowing the rate of oxidation of autoxidizable materials (66) (see Antioxidants). It should be soluble in both water and oil; be odorless, nontoxic, nonallergenic, and effective in low concentration; and be stable under conditions of use. Phenolic antioxidants are usually soluble in oil and generally insoluble in water. Propyl gallate has a solubility of about 1.8% in water. The addition of propyl gallate requires a chelating agent to prevent reactions with iron contaminants which will cause color changes. Calcium disodium EDTA is slightly soluble in oil. The most widely used antioxidants are butylated hydroxyanisole (BHA), butylated hydroxytoluene (BHT), PG and nordihydroguaiaretic acid (NDGA). A widely used antioxidant mixture contains 20% BHA, 6% PG, 4% citric acid, and 70% propylene glycol.

Table 5. Methods of Challenge Test Proposed by U.S.P. XVIII, CTFA, and SCC of Great Britain[a]

	U.S.P. XVIII	CTFA	SCC of Great Britain
organisms used	*Candida albicans* ATCC 10231 *Aspergillus niger* ATCC 16404 *Escherichia coli* ATCC 8739 *Staphylococcus aureus* ATCC 6538 *Pseudomonas aeruginosa* ATCC 9027	*S. aureus* ATCC 6538 *E. coli* *P. aeruginosa* 15442 or 13388 *C. albicans* *A. niger* 9642 *P. luteum* 9644 *Bacillus subtilis*	vigorous strains of factory or lab product contaminants *E. coli, A. aerogenes, Proteus morganii, P. aeruginosa, P. fluorescens, S. aureus, S. epidermidis, Streptococcus faecalis,* fungi, and yeast
inoculum	0.1 mL/20 mL; (1.25–5) × 10^5 cells/mL of product	10^6 cells/mL of product (20 mL)	check growth in unpreserved product, 0.1 mL/30 mL; 10^5–10^7 cells/mL of product
control	viability of cells in non-preserved product; 30–32°C	cells in nonpreserved product: 32–37°C for bacteria; 25–30°C for fungi	cells in unpreserved product; assay natural contaminants
sampling	at least two observations, 7 d apart, during 28-d test period	0, 1–2, 7, 14, 28 d	toothpaste: 0 1, 2, 3, 4 weeks, 28°C; shampoo: 0, 1, 2, 3, 4 weeks, 22, 30°C; creams and lotions: 0, 1, 2, 4, 8 weeks, 22, 32, 37°C; eye cosmetics: 0, 1, 2, 4 weeks, 30, 37°C; surface, saliva
effectiveness	vegetative cells: 0.1% survival at 2 sampling periods 7 d apart; *C. albicans* and *A. niger*: no significant increase during 28-d test period	7 d; rechallenge	toothpaste, decrease; shampoo, no count for 2 weeks, reinoculate; creams and lotions, low count for 3 weekly samplings; eye cosmetics, kill of *P. aeruginosa*, no fungi growth, kill of saliva bacteria

[a] Ref. 61.

Toxicity and Test Methods

Animal Studies. Even though true sensitization does not occur equally in animals and humans, the animal test provides some measure of protection by serving as a screen. Sensitization of animals makes a material suspect, and so great caution is exercised before using the material on humans. On the other hand, a negative sensitizing result on animals indicates that the material is probably not a strong sensitizer and further tests on humans offers little risk. The final tests must, however, be clinical trials on human subjects (67–68).

Preliminary animal tests are performed to determine whether the material has an acute or chronic toxicity, and, if so, the MLD_{50}. These tests involve feeding the material to the animals which are subsequently sacrificed for close examination of the organs, glands, and brain.

Percutaneous toxicity is determined by the application of the material to cleanly shaven necks and backs of animals over a regularly scheduled period. The condition of the skin is carefully checked and the animals are observed for any signs of systemic toxicity and their tissues are examined for pathological changes. The final tests must, however, be clinical trials on human subjects (67–68).

Human Safety Testing. Irrespective of the completeness of the program carried out with animals, initial trials on humans must be on a limited number of individuals (69). Evaluation of sensitizing potential should begin with 10 or 12 panelists. Full-scale tests must then be carried out before the product is released to market. Although 60 panelists are generally adequate as precursors to large-scale studies, forthcoming regulations are expected to require tests on 200.

In human safety trials of topically applied materials the principal concern in most cases is induced cutaneous reactions. If, however, animal studies indicate significant absorption of components of the formulation, clinical trials must also monitor such absorption by testing blood and urine which may reflect subclinical systemic toxic effects.

The cutaneous reactions that may result from application of cosmetics are of three types: primary irritation, contact sensitization, and reactions that develop following exposure to the sun. The potential of topically-applied materials to induce these reactions is evaluated by means of patch tests.

Sensitization Potential. One of the most critical evaluations is the determination of sensitizing potential. Two basic procedures have been most frequently employed, and numerous variations of these have been used. One of the basic procedures (70) has limited value since it detects only very strong sensitizers. Present consideration of the human patch test will primarily be limited to the repeated insult procedure and its various modifications. Two variations were originally developed—one by Shelanski and Shelanski and the other by Draize (68,71). Both of these procedures employ repeated applications made over a period of four or five weeks. They differ in that ten sensitizing applications are made in the Draize procedure and fifteen are made in the Shelanski procedure. They differ also in that challenge applications are made only to the original test sites in the Draize procedure, but are made to previously unpatched sites as well as the original test site in the Shelanski procedure.

A short review of these and other procedures involving human panelists has been presented by Kligman (72–73), who suggested the maximization procedure as an alternative. Various methods for achieving the objective of Kligman's maximization procedure have been employed, an interesting one being a Hill Top Research procedure in which the skin is abraded prior to sample application. The test materials are applied to abraded sites on the upper arm under patches. Reactions to the test applications are scored on Wednesday, Friday, and Monday following applications on the preceding Monday, Wednesday, and Friday. The patches are removed by the panelists 24 h after application. Nine applications are made in three successive weeks, and the challenge application is made on Monday of the sixth week. The challenge applications are made to previously unpatched sites. Abrasions of the test sites are made just prior to sample applications on test days 1, 4, 7, and 11, or preferably, prior to each application. This more frequent scheduling of abrading can be done when up to four materials are evaluated on abraded skin only. The abrasions, in the form of two concentric circles with diameters of 1.25 cm and 1 cm, are inflicted by a specially constructed apparatus. The skin areas are cleansed with isopropyl alcohol just prior to abrading. A separate sterile abrader is used for each panelist. All adhesive patches and absorbent pads applied over abraded areas are sterilized by exposure to ethylene oxide prior to use. The usual grading scale is utilized.

Patch test reaction scoring scale
- 0 no evidence of irritation
- 1 slight erythema
- 2 marked erythema
- 3 erythema and papules
- 4 edema, erythema may or may not be present
- 5 erythema, edema, and papules
- 6 vesicular eruption
- 7 strong reaction spreading beyond test site
- min reaction meets minimal requirements for grade assigned

Effect on superficial layers of skin
- A slight glazed appearance
- B marked glazing
- C glazing with peeling and cracking
- F glazing with fissures
- G film of dried serious exudate covering all or portions of the patch site
- H small petechial erosions and/or scabs

Another procedure employed is to strip the skin by repeated application and removal of adhesive tape from the test sites prior to sample applications.

Regardless of the procedure employed in tests on human panelists, minimal reactions, frequently evoked during the serial applications and at challenge, require expert judgment as to whether the material is a sensitizer or not. It is frequently necessary to conduct rechallenges in order to assess the significance of the observed reactions.

When properly applied, the repeated insult test is a severe test for the sensitizing potential of products. Many products that induce sensitization in a low percentage of panelists tested by this procedure can be used without untoward reactions. The proposed use must be considered. For example, a product that shows minimal sensitization potential and is to be used on the feet should not be marketed. On the other hand, a product that would be used infrequently under nonocclusive conditions could probably be safely marketed.

In addition to its value as a predictor of sensitizing potential, the repeated insult patch test furnishes valuable information on the primary irritant characteristics and cumulative primary irritant properties of test materials. The latter property is descriptively referred to frequently as skin fatigue.

Some assurance that products found to be minimal sensitizers can be safely marketed may be obtained by conducting product-use challenges of panelists who have been sensitized in the patch test procedure. If these reactive individuals do not respond to use applications, it is most unlikely that sensitization would be induced by normal use of the product. However, this reasoning cannot be applied to products used for whole-body applications. Whole-body exposures by sensitized panelists cannot be safely done, and limited area applications may be misleading.

Cosmetic Ingredient Review

The CTFA has set up a Cosmetic Ingredient Review (CIR), financed and directed by the Association but operating independently of the Association to evaluate the safety of cosmetic ingredients. Reviews will reflect information gathered from published scientific literature as well as data developed through test procedures.

To assign priority ranking, CIR began with the 2800 ingredients listed in CTFA's *Cosmetic Ingredient Dictionary,* deferred about 700 items that are under review by the FDA or other programs, and selected from the remainder those that are used in 25 or more cosmetic product formulations. Seven criteria then were used by the panel for priority weighting of the 189 substances remaining: frequency of occurrence, concentration in formulas, area of normal use, frequency of application, use by particular groups (eg, infants and elderly), suggestions of biological activity, and frequency of consumer complaints about product categories containing the ingredient being reviewed.

BIBLIOGRAPHY

"Cosmetics" in *ECT* 1st ed., Vol. 4, pp. 545–562, by Florence E. Wall, Consulting Chemist; "Cosmetics" in *ECT* 2nd ed., Vol. 6, pp. 346–375, by Harry Isacoff, International Flavors & Fragrances (U.S.).

1. F. E. Wall, *Origin and Development of Cosmetic Science and Technology,* Interscience Publishers, New York, 1957, Chapt. 2.
2. M. G. de Navarre, *International Encyclopedia of Cosmetic Material Trade Names,* Moore Publishing Co., New York, 1957.
3. L. A. Greenburg and D. Lester, *Handbook of Cosmetic Materials,* Interscience Publishers, New York, 1954.
4. *CTFA-Cosmetic Ingredient Dictionary,* 2nd ed., C.T.F. Association, Washington D.C., 1977.
5. *Federal Food, Drug, and Cosmetic Act, as amended Oct. 1976,* Food and Drug Administration, 1976, Washington, D.C.
6. *Fed. Reg.* **43**(166), 38206 (Aug. 25, 1978).
7. *Prod. Mark. Expend. Study 31st Ann. ed.,* (Aug. 1978).
8. P. Carter, *Am. Perfume Arom. Doc. Ed.* **1,** 233 (1960).
9. P. Carter, *J. Am. Perfum. Cosmet.* **77**(4), 10 (July 1962).
10. G. Barnett, *Emollient Creams and Lotions,* Vol. 1, Wiley-Interscience, 1972, pp. 27–104.
11. W. H. Mueller and R. P. Quatrale, "Antiperspirants and Deodorants," in M. G. de Navarre, ed., *Chemistry and Manufacture of Cosmetics,* 2nd ed., Vol. 3, Continental Press, Orlando, Fla., 1975, pp. 205–228.
12. A. B. G. Lansdown, *Soap Perfum. Cosmet.* **47,** 209 (May 1974).
13. *Society of Cosmetic Chemists, Annual Scientific Meeting, Nov. 30, 1978.*
14. S. I. Kreps, *J. Soc. Cosmet. Chem.* **14**(12), 625 (Dec. 1963).
15. S. I. Kreps, *J. Am. Perfum. Cosmet.* **78,** (Oct. 1963).
16. E. G. Klarman, *Suntan Preparations,* Interscience Publishers, New York, 1957, pp. 189–212.
17. U.S. Pat. 2,976,217 (Mar. 21, 1961), S. I. Kreps (to Van Dyk Co.).
18. U.S. Pat. 3,341,419 (Sept. 12, 1967), H. J. Eiermann and I. Rappaport (to Shulton).
19. *Glossary of Terms Used in the Aerosol Industry,* Chemical Specialties Manufacturers Assoc., New York, Aug. 25, 1955.
20. *Bulletin No. 211-66,* Chemical Specialties Manufacturers Assoc., New York, Dec. 27, 1966.
21. P. A. Sanders, *Principles of Aerosol Technology,* Van Nostrand Reinhold Company, New York, 1970.
22. M. A. Johnsen, W. E. Dorland, and E. K. Dorland, *The Aerosol Handbook* 1st ed., Wayne E. Dorland Co., New York, 1972.
23. W. H. Peacock, *The Application Properties of the Certified Coal-Tar Colors, Calco Technical Bulletin No. 715,* Calco Chemical Div., American Cyanamid Co., Wayne, N. J., Dec. 1944.

24. W. H. Peacock, *The Practical Art of Color Matching, Calco Technical Bulletin No. 573,* Calco Chemical Div., American Cyanamid Co., Wayne, N. J., 1948.
25. D. Nickerson, *Color Measurement,* U.S. Department of Agriculture, Washington, D.C., 1946.
26. U.S. Pat. 3,148,125 (Sept. 8, 1964), S. J. Strianse and M. Havass (to Yardley).
27. H. Hilfer, *Drug Cosmet. Ind.* **65,** 518 (1949).
28. P. G. Lauffer, *Lipsticks,* Interscience Publishers, New York, 1957, Chapt. 13.
29. H. Bishop, *J. Soc. Cosmet. Chem.* **5,** 2 (1954).
30. H. J. Wing, "Nail Preparations," in ref. 11, pp. 983–1010.
31. P. Alexander, *Soap Perfum. Chem.* **48,** 153 (Apr. 1975).
32. J. Peirano in E. Sagarin, ed., *Cosmetics, Science, and Technology,* Interscience Publishers, New York, 1957, p. 678.
33. J. R. Hart and E. F. Levy, *J. Soap Cosmet. Chem. Specialties,* **50,** (Aug. 1977).
34. A. J. Harris, K. C. James, and M. Powell, *J. Cosmet. Perf.* **90,** 23 (Oct. 1975).
35. F. V. Wells and I. I. Lubowe, *Cosmetics and the Skin,* Reinhold Publishing Corp., New York, 1964, pp. 397–415.
36. L. R. Smith and D. C. Zajac, *J. Household Personal Prod. Ind.* **13,** 34 (Mar. 1976).
37. L. R. Smith and M. Weinstein, *J. Household Personal Prod. Ind.,* 54 (Oct. 1977).
38. R. L. Goldemberg, "Amphoteric Shampoos," *Drug Cosmet. Ind.,* 26 (Jan. 1976).
39. W. R. Markland, *Low Eye Irritation Shampoo Systems, Norda Brief No. 479,* Norda Inc., East Hanover, N. J., Mar. 1977.
40. U.S. Pat. 3,313,734 (Apr. 11, 1967), E. W. Lang and H. W. McCune (to Procter & Gamble).
41. U.S. Pat. 3,001,949 (Sept. 26, 1961), K. R. Hansen (to Colgate Palmolive).
42. D. Y. Hsiung, "Hair Straightening," in ref. 11, pp. 1155–1166.
43. U.S. Pat. 2,787,274 (Apr. 2, 1957), V. A. Gant and H. I. Hersh (to Hersh).
44. E. A. Taylor, *Arch. Dermatol.* **87,** 137 (Mar. 1963).
45. *Soap Perfum. Cosmet.* **48,** 437 (Oct. 1975).
46. A. P. R. James, *J. Am. Geriatrics Soc.* **9**(5), 367 (May 1961).
47. E. Jungermann, *J. Cosmet. Toiletries* **91,** 50 (July 1976).
48. D. Osteroth, *J. Soap Cosmet. Chem. Specialties,* **50,** 29 (Apr. 1977).
49. E. T. Webb, *Soap Perfum. Cosmet.* **31,** 770 (Aug. 1958).
50. R. C. Reald, *Germicidal Detergent Bar, Norda Brief No. 447,* Norda Inc., East Hanover, N. J., Mar. 1973.
51. U.S. Pat. 2,820,768 (Jan. 21, 1958), L. E. G. H. Fromont (to Molenbeek St. Jean, Belgium).
52. U.S. Pat. 3,687,855 (Aug. 1972), A. Halpern (to Synergistics, Inc.).
53. U.S. Pat. 3,072,536 (Jan. 8, 1963), D. J. Pye (to The Dow Chemical Co.).
54. C. Jones, *Nucleus (Boston),* 3 (June 1976).
55. U.S. Pat. 3,341,418 (Sept. 12, 1967), R. E. Moses and P. W. Lucas (to Gillette).
56. S. D. Gershon, H. H. Pokras, and T. H. Rider, *Dentifrices,* Interscience Publishers, New York, 1967, Chapt. 15.
57. H. W. Zussman, *Proc. Sci. Sect. Toilet Goods Assoc. (CTFA)* (19), (1953).
58. U.S. Pat. 3,662,059 (May 9, 1972), W. Wiesner and M. Pader (to Lever Bros.).
59. M. S. Parker, *Soap Perfum. Cosmet.* **45,** 621 (1972).
60. F. V. Wells and I. I. Lubowe, *Cosmetics and the Skin,* Reinhold Publishing Corp., New York, 1964, pp. 586–598.
61. M. Yanagi, "Microbiology, The Chemistry and Manufacture of Cosmetics," in ref. 11, pp. 67–84.
62. W. E. Rosen and P. A. Berke, *J. Soc. Cosmet. Chem.* **24,** 663 (Sept. 1973).
63. I. R. Gucklhorn, *TGA Cosmet. J.* **I**(3), 15 (1969).
64. E. M. Owen, *TGA Cosmet. J.* **I**(3), 12 (1969).
65. A. Marinaro, *TGA Cosmet. J.* **I**(3), 16 (1969).
66. L. Chalmers, *Soap Perfum. Cosmet.* **44,** 29 (Jan. 1971).
67. M. J. Thomas and P. A. Majors, *J. Soc. Cosmet. Chem.* **24,** 140 (1973).
68. J. H. Draize, *Dermal Toxicity, Appraisal of the Safety of Chemicals in Foods, Drugs, and Cosmetics,* The Editorial Committee of the Association of Food and Drug Officals of the United States, Austin, Tex., 1959, p. 52.
69. R. A. Quisno, G. L. Quisno, and P. A. Majors, *J. Cosmet. Perfum.* **90,** 44 (Feb. 1975).
70. L. Schwartz and S. M. Peck, *Pub. Health Rep.* **59,** 2 (1944).
71. H. A. Shelanski and M. V. Shelanski, *Proc. Sci. Sect. Toilets Goods Assoc.* **19,** 46 (1953).
72. A. M. Kligman, *J. Invest. Dermatol.* **47,** 369 (1966).

73. B. Magnusson and A. M. Kligman, *J. Invest. Dermatol.* **52**, 268 (1964).

General References

M. G. de Navarre, *The Chemistry and Manufacture of Cosmetics,* 2nd ed., Continental Press, Orlando, Fla., Vols. 1 and 2, 1962; Vols. 3 and 4, 1975.
E. Sagarin, ed., *Cosmetics—Science and Technology,* Interscience Publishers, New York, 1957.
F. V. Wells and I. I. Lubowe, *Cosmetics and the Skin,* Reinhold Publishing Corp., New York, 1964.
R. G. Harry, "Modern Cosmeticology" in *Principles and Practice of Modern Cosmetics,* revised by J. B. Wilkinson, Chemical Publishing Co., Inc., New York, 1962.
R. G. Harry, "Cosmetic Materials" in *Principles and Practice of Modern Cosmetics,* revised by W. W. Myddleton, Chemical Publishing Co., Inc., New York, 1963.
M. S. Balsam and E. Sagarin, *Cosmetics, Science, and Technology,* 2nd ed., Vols. 1 and 2, Wiley-Interscience, New York, 1972.
W. R. Keithler, *The Formulation of Cosmetics and Cosmetic Specialties,* Drug and Cosmetic Industry, New York, 1956.
Official Methods of Analysis, 9th ed., Assoc. Offic. Agric. Chemists, Washington, D.C., 1960.

HARRY ISACOFF
International Flavors and Fragrances, Inc.

COTTON

The story of cotton predates recorded history, and although the actual origin of cotton is still unknown, there is evidence that it existed in Egypt as early as 12,000 BC. Its use in cloth in 3000 BC was indicated by archeological findings and recorded evidence exists of its cultivation in India as far back as 700 BC. In the fifth century BC, Herodotus wrote of trees growing wild in India bearing wool of a softness and beauty equivalent to that of the sheep; clothes made from this tree wool were described as garments of extraordinary perfection.

It is said that Alexander the Great introduced Indian cotton into Egypt in the 4th century BC, and from there it spread to Greece, Italy, and Spain. During the year 700 AD, China began growing cotton as a decorative plant, and 798 AD saw its introduction into Japan. Early explorers in Peru found cotton cloth on exhumed mummies that dated to 200 BC. Cotton was found in North America by Columbus in 1492. About three hundred years later the first cotton mill was built at Beverly, Mass., and in 1794 Eli Whitney was granted a patent for the invention of the cotton gin.

Cotton is the most important vegetable fiber used in spinning (see Fibers, vegetable). Its origin, breeding, morphology, and chemistry have been described in innumerable publications (1–4). It is a member of the Malvaceae or mallow family, a plant of the genus *Gossypium,* and is widely grown in warm climates the world over.

The average cotton plant is a herbaceous shrub having a normal height of 1.2–1.8 m, although some tree varieties reach a maximum height of 4.5–6.0 m. The most important species included in the genus *Gossypium* are *hirsutum, barbadense, arboreum,* and *herbaceum.*

G. hirsutum originated in Central America and is found in the Mayan culture of Mexico. It is the species that comprises all of the many varieties of American Upland cotton of commerce. *G. hirsutum,* a shrubby plant that reaches a maximum height of 1.8 m, is probably the origin of the green-seeded (ie, bearing green fuzz fibers) cotton of former times grown so extensively in the southern United States.

G. barbadense, originally from the Incan civilization, grows from black seeds and ranges in height from 1.8–4.5 m. It is the species with longest staple and includes Sea Island, Egyptian Giza strains, American–Egyptian, and Tanguis cottons; of these, Sea Island is the longest and silkiest of the commercial cottons.

G. arboreum, the tree wool of India, grows as tall as 4.5–6.0 m and includes both Indian and Asiatic varieties. Its seeds are covered with greenish-gray fuzz fibers below the white lint fibers.

G. herbaceum, the original cotton of India, averages 1.2–1.8 m. The fiber is grayish-white and grows from a seed encased in gray fuzz fibers. This species includes the short staple cottons of Asia and some of China, as well as most native Indian varieties. It is also grown commercially in Iran, Iraq, Turkey, and the USSR.

The most favorable growing conditions for cotton include a warm climate (17–27°C, mean temperature), where fairly moist and loamy, rather than rich, soil is an important factor; seed planted in dry soil produces fine and strong, but short, fibers of irregular lengths and shapes.

Under normal climatic conditions, cotton seeds germinate in 7–10 d. Flower buds (known as squares) appear in 35–45 d followed by open flowers 21–25 d later. After one day the cotton boll begins to grow rapidly if the flower has been fertilized. The mature boll opens 45–90 d after flowering, depending on variety and environmental conditions. Within the boll are 3–5 divisions called locks, each of which normally has 7–9 seeds that are covered both with lint and with fuzz fibers (Fig. 1). The fuzz fibers form a short, shrubby undergrowth beneath the lint hairs on the seed. Each seed

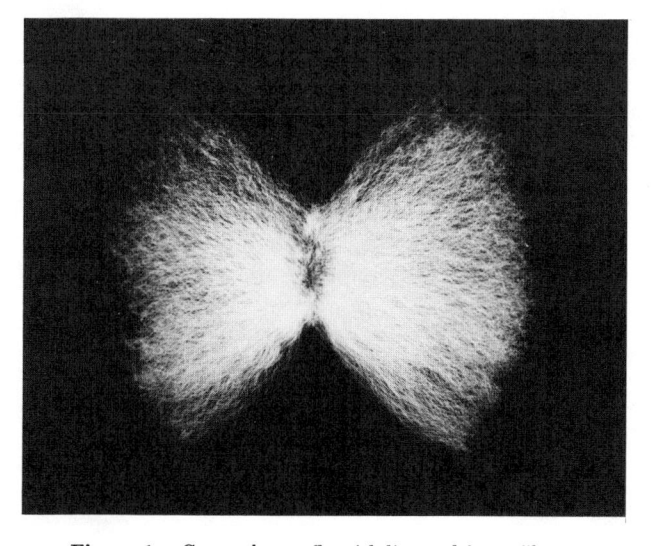

Figure 1. Cotton butterfly with lint and fuzzy fibers.

contains at least 10,000 fibers and there are close to 500,000 fibers in each boll. The usual range of planting time in the United States extends from the beginning of March to the end of May; harvest time is in the late summer or early fall.

The cotton fiber is a single cell that originates in the epidermis of the seed coat at about the time the flower opens; it first emerges on the broad, or chalazal end of the seed and progresses by degrees to the sharp, or micropylar end. As the boll matures, the fiber grows until it attains its maximum length, which averages about two thousand five hundred times its width (Fig. 2). During this time, the cell is composed of only a thin wall that is covered with a waxy, pectinaceous material and that encloses the protoplasm or plant juices. In ca 17–25 d after the flower opens, when the boll is half mature, the fiber virtually attains its full length, at which time it begins to deposit layers of cellulose on the inside of the thin casing, or primary wall. The pattern of deposition is such that one layer of cellulose is formed each day in a centripetal manner until the mature fiber has developed a thick secondary wall of cellulose, from the primary wall to the lumen, or central canal. The fiber now consists of three main parts: primary wall, secondary wall, and lumen. At the end of the growing period when the boll bursts open, the fibers dry out and collapse, forming shrivelled, twisted, flattened tubes.

The seed hairs of cultivated cottons are divided into two groups—fuzz and lint—that may be distinguished on the basis of such characteristics as length, width, pigmentation, and strength of adherence to the seed. The growth of fuzz fibers is much the same as that of lint, but they are usually about 0.33 cm long compared with the 2.5 cm average length of lint fibers, and are twice as thick, or about 32 μm (Fig. 3). Their color range is from greenish-brown to gray. After lint fibers have been ginned off the seed, the fuzz fibers or linters are left and must be removed by a machine similar to the regular cotton gin. Linters are an important source of cellulose for chemical purposes and are used in upholstery and batting.

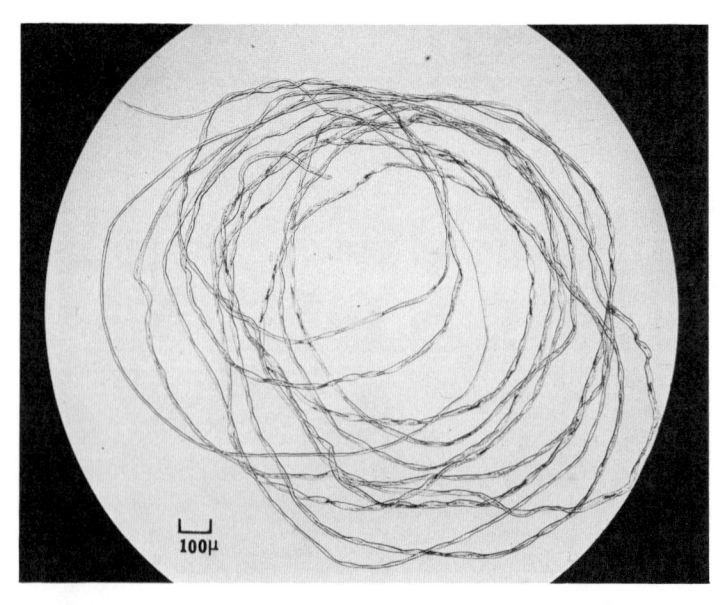

Figure 2. Single cotton fibers showing ratio of length to width.

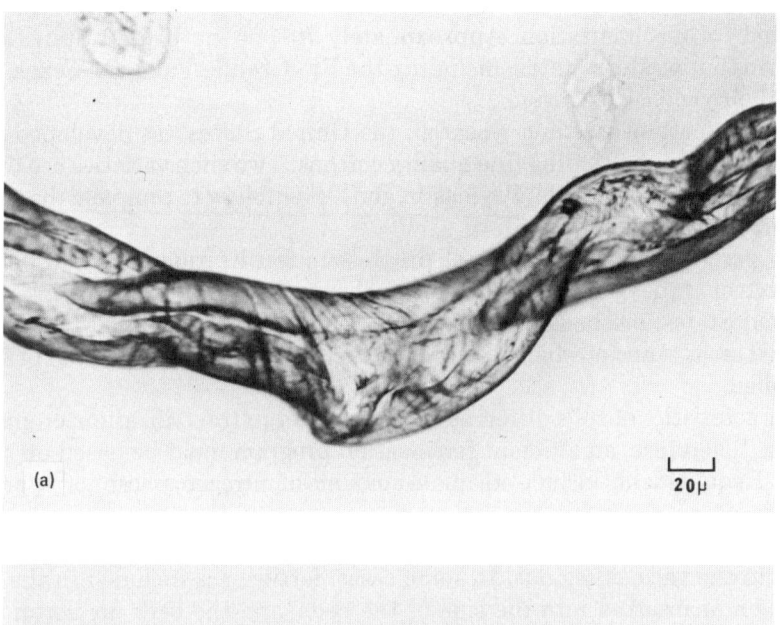

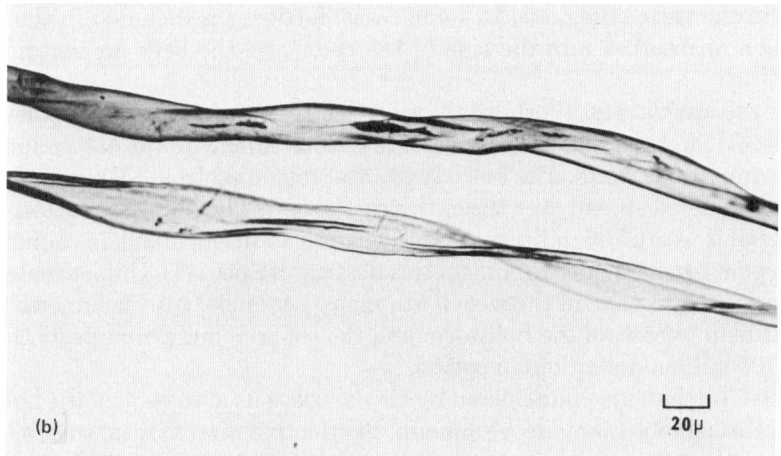

Figure 3. Longitudinal view of (a) fuzz fiber and (b) lint fibers.

Cultivation and Production

At present, the chief cotton-growing countries of the world are the Union of Soviet Socialist Republics (22%), the People's Republic of China (19%), the United States (11%), India (9%), Pakistan (4%), Brazil (4%), Turkey (4%), and Egypt (3%) (5). In the Western hemisphere, cotton is cultivated from the equator to about 37° N latitude and 32° S latitude; in the Eastern hemisphere, the limits extend to 47° N latitude and 30° S latitude.

The United States produces approximately 11% of the world's cotton. Although the actual size of the crop has changed very little, improved fertilizers, insecticides, and harvesting practices have increased the yield per hectare so much that only about half as much area of land is needed today as in 1930.

Cultivation of cotton differs markedly from one country to another, depending on methods of mechanization. Approximately 70% of the United States cotton is rain-grown, but western states, including the Rio Grande Valley of Texas, have increased their yield by irrigation.

Through cotton-breeding research, the United States has developed varieties that today remain the leading fine quality cottons. Two such varieties are Stoneville and Deltapine. Lankart, Acala, Paymastir, and Coker follow to complete the six leading cottons in the United States (6).

Fertilizers. Perhaps the greatest progress in fertilization of cotton fields is reflected in techniques of application. Throughout all of the cotton growing regions of the United States, methods of applying nutrients must be tailored to the needs of designated areas, some of which also require special attention in the types of fertilizers (qv) applied.

Characteristics of soils differ, as do practices of past fertilization, cropping, and irrigation. Therefore, an efficient fertilization program must be based on results of soil tests. Requirements include adequate amounts of nitrogen, phosphorus, potassium, and boron.

However, regardless of soil needs, methods of fertilizer application must be adapted to the various regions. In some cases, fertilizer is included in the seedbed preparation or inserted into the side of the beds, and the beds are rerun immediately.

Pests and Insecticides. Perhaps the most destructive pests of the cotton plant are the boll weevil and the bollworm, which are serious threats to the cotton industry in countries around the world. The boll weevil, after migrating from Mexico around 1892, within 30 years had spread over the entire cotton belt. The domestic cotton crop lost to the weevil is worth 200 million dollars a year; in addition, about 75 million dollars a year is spent for pesticides to control this destructive pest (7). Unfortunately, some insecticides used to control the weevil kill many beneficial insects, among them insects that help to control the bollworm and the tobacco budworm, pests that cause another 200 million dollar loss in cotton.

In 1916 calcium arsenate dusted by airplane was used to control the boll weevil; however, throughout many developments in effective insecticides, such as organophosphates, the boll weevil began to build up resistance to poisons that were formerly effective (see Insect control technology).

Several years ago, a pilot program to eradicate the boll weevil, carried out on an 8100-ha region of Mississippi, Alabama, and Louisiana, led to plans for beltwide eradication of this devastating pest (7). The plan of attack on the weevil has been mapped out, and it is believed that success can be realized by 1980.

One of the most destructive insects is the pink bollworm, which overwinters as diapausing (hibernating) larvae in the soil. After feeding on the late-blooming bolls, the larvae drop to the ground and hibernate for the winter, emerging as adults in the spring to lay eggs on the early cotton blooms. The eggs hatch and the new larvae bore into the fresh cotton bolls, go through molting stages, bore their way out, and drop to the ground. Throughout the growing season the cycle repeats itself, to the destruction of vast numbers of cotton plants in a single field.

A technique to combat this pest has been developed by agricultural research scientists in cooperation with the Arizona Agricultural Experiment Station in Phoenix. The number of overwintering pink bollworms is limited by reduction of their food supply late in the year. Growth regulators are used to prevent late boll formation; this has no effect on the existing bolls and causes no loss in harvesting, as most of the bolls fail to mature anyway. Termination of late-fruiting cotton is carried out by applying chemical formulations. Along with the chemical treatments, other eradication practices include early stalk shredding, early and deep tillage, and winter irrigation that drowns diapausing larvae (8).

To the problems involving the boll weevil and the pink bollworm must be added the efforts expended on research to eliminate other insects injurious to the cotton plant. Included in the list are aphids, leafhoppers, lygus bugs, mites, white flies, black fleahoppers, thrips, cutworms, and leafminers (9).

Harvesting. When the ripened boll bursts and the cotton dries and fluffs, it is ready for harvesting. Prior to the actual operation, however, the plants may be prepared by treatment with harvest-aid chemicals, classified as defoliants or desiccants. Defoliants usually cause an abcission or shedding of leaves earlier than normal, preventing further plant development. Desiccants, on the other hand, rapidly kill the plant by causing an immediate loss of water from the tissue. The dried leaves remain on the plant. Conditions of both plants and weather are of major importance in the success of these harvest-aid programs (see Herbicides; Plant-growth substances).

Until about the middle of this century, cotton was picked entirely by hand, and in many foreign countries this practice still exists. However, in the United States, mechanical harvesting essentially has replaced handpicking; in fact, over 99% of the cotton in the United States is harvested by machine.

The first type of mechanical harvester was the stripper, a machine that extracts from each plant all of the cotton in the field, regardless of maturity. As a result, large quantities of unopened bolls, stalks, and leaves are caught up by the stripper.

The spindle harvester straddles the cotton and extracts it from the bolls in one or two rows simultaneously. The lint from the open bolls is wound onto the spindle, from which it is removed and carried to a large container.

As a general rule, harvesting should take place when the relative humidity is 60% or less and cotton is not wet with dew. Storage of damp cotton on trailers affects the quality of the cotton and germination of the seed. Since the gin cannot process seed cotton at the same rate at which it is harvested, storage should be under optimum conditions.

Ginning. The greatest number of cotton gins in existence in the United States was in 1902; 30,948 gins, the majority on plantations, processed 10.6 million bales (of ca 217 kg each) of cotton (10). Since 1902 the number of gins has declined and the average number of bales handled has correspondingly increased. In 1975, 2856 active gins handled a crop of 8,151,223 bales for an average of 2854 bales per gin plant (11).

Mechanical harvesting systems have reduced the harvesting and ginning period from 4–5 mo to ca 6 wk of intensive operation (12). In 1957, the typical United States cotton gin had an average production rate of about 6 bales per hour; the present design capacity of new gins is in excess of 20 bales per hour.

Most of the United States gins are now operated as corporations or as cooperatives serving many cotton producers, in contrast to the plantation ownership pattern. Automatic devices do the work faster, more efficiently and more economically than hand labor. Examples of increased efficiency are (1) high-volume bulk seed cotton handling systems to get cotton into the gin, (2) reduced processing time for drying, ginning, and cleaning cotton, and (3) automated bale packaging devices.

Commercial Classification

Classing of cotton involves describing the quality of cotton in terms of grade and staple length in accordance with the official cotton standards of the United States. Classing in this way indicates the cotton's spinning utility and value to the buyers and sellers and provides information for evaluating efficiency of production, harvesting, ginning, and processing practices. About 95% of the cotton produced in the United States is classed by grade and staple length. This classification is often supplemented by inclusion of "Micronaire reading" to indicate the fineness of the cotton.

There are 37 grades for American Upland cotton within seven general color classifications in addition to below-grade cotton. Color (qv) is defined in terms of hue, lightness, and chroma. Color classifications are from white, actually a pale cream color, to light-spotted, spotted, tinged, yellow-stained, to light gray or gray, the last two indicating cotton that was left too long unpicked, with weather and possible fungal damage. There are ten grades for American–Egyptian cotton. In addition to color, leaf and ginning preparation are also used to establish grade. Leaf includes dried pieces of leaf, stem, stalk, seed coat, and other contaminants. Leaf is divided into two general groups based on size large leaf, and pin or pepper leaf. The penalty is greater for a large amount of fine trash because it is so difficult to remove. Ginning preparation describes the degree of smoothness and relative nappiness or neppiness of the ginned lint.

Classification of cotton in terms of staple length is in accordance with official United States standards and regulations. Current official standards for staple length include 14 standards for the range 2.1–3.2 cm (including most American Upland cotton), four standards covering 3.3–3.8 cm (including most American–Egyptian cotton), and four covering 3.8–4.4 cm (including most American Sea Island cotton). Cotton below 2.1 cm is classed as "below 13/16 in." (2.1 cm).

Physical Properties

Length is the most important dimension of the cotton fiber. The length of staple of any cotton is the length by measurement of a representative sample of fiber, without regard to quality or value, at 65% rh and 21°C (13). An experienced classifier determines the length by parallelizing a typical portion of fibers from pulls drawn from a sample.

Variations of length are peculiar to classes of cotton and range from less than 2.5 cm for short staple Upland varieties, and 2.6–2.8 cm for medium staple Uplands, to >2.85 cm for long staple (14).

Shorter staple varieties tend to be coarser, whereas the longer varieties such as Pima and Sea Island, which can measure up to 5.7 cm in length, are fine, silky, and soft.

Diameter of the fiber is an inherited characteristic that can be greatly influenced by soil and weather. This fiber dimension may best be observed in cross sections of bundles of fine, medium, or coarse fibers. Whereas the typical shape has been described as resembling that of the kidney bean, the shapes range from circular to elliptical to linear in most varieties (15).

Uniformity within a sample is largely dependent on variety and degree of maturity; hence diameter measurements must be calculated. Use of a bidiameter scale gives values for both the major and minor axes of the fiber cross sections. Other measurements from which diameter may be derived are wall thickness, lumen axes, or combinations of these.

Fineness is defined as a relative measure of size, diameter, linear density, or weight per unit length expressed as micrograms per inch (μg/in.) [0.394 μg/cm] (16), and is sometimes referred to as linear density. Today, however, the term Micronaire reading is accepted in the United States as an added criterion for marketing. It is a measure of the resistance of a plug of cotton to air flow and indicates the fineness of the cotton (16).

Strength of the cotton fiber has been attributed mostly, if not entirely, to the cellulose (qv) it contains and, in particular, to the molecular chain length and orientation of the cellulose (15) (see also Biopolymers). Fiber bundle testing and single fiber testing (17) have contributed much basic knowledge to this important property.

Directly associated with the strength of the fiber is the degree of maturity it has attained. Average maturity acceptable for mill usage is from 75–80%. When this percentage drops, an excess of immature fibers causes higher picker and card waste, formation of neps, and production of yarns that grade low in appearance.

Morphology

The cotton fiber is tapered for a short length at the tip, and along its entire length is twisted frequently, with the direction of twist reversing occasionally. These twists are referred to as convolutions and it is believed that they are important in spinning because they contribute to the natural interlocking of fibers in a yarn.

Cotton is essentially 95% cellulose (Table 1). The noncellulosic materials, consisting mostly of waxes, pectinaceous substances, and nitrogenous matter, are located to a large extent in the primary wall, with small amounts in the lumen.

Of the noncellulosic substances in cotton, protein normally occurs in the largest amounts. This protein is apparently the protoplasmic residue left behind on the gradual drying up of the living cell. It occurs almost entirely in the lumen. On the assumption that the cell content is approximately the same for each fiber, it has been suggested that the nitrogen content might be used as an indirect measure of wall thickening. However, no practical test of sufficient accuracy has been developed.

Table 1. Composition of Typical Cotton Fibers

Constituent	Composition, % of dry weight Typical	Range
cellulose	94.0	88.0–96.0
protein (% N × 6.25)[a]	1.3	1.1–1.9
pectic substances	1.2	0.7–1.2
ash	1.2	0.7–1.6
wax	0.6	0.4–1.0
total sugars	0.3	
pigment	trace	
others	1.4	

[a] Standard method of estimating percent protein from nitrogen content (% N).

Most of the pectin in the cotton fiber is in the primary wall. Total removal of the pectin substances is accomplished readily by scouring, which does not change greatly the properties of the cotton.

The wax of most cottons is a soft, low melting, complex mixture having some differences among varieties. The noncommercial, strongly pigmented green lint cotton may contain up to about 17% wax of high melting point. Because the wax becomes established in fibers, largely if not wholly, during the first phase of development, the wax content expressed as a percentage of the whole fiber mass decreases as the fiber maturity or degree of wall thickening increases. Removal of cotton wax by chemical treatment increases the friction between fibers and between fiber and metal.

Although the mature cotton fiber is considered as having a primary wall, a secondary wall, and a lumen, further examination of its structure shows a cuticle and a winding layer (Fig. 4). The cellulose of the primary wall exists as a woven network of microfibrils in and on which are deposited noncellulosic materials that form the cuticle. Just beneath the primary wall is the winding layer, which is also the first layer of the secondary wall. The winding layer appears to comprise a single layer of fibrillar bundles composed of microfibrils and oriented at an angle to the fiber axis. The main body of the fiber consists of cellulose fibrils packed tightly in a solid cylinder, which, under certain conditions of chemical swelling, can be induced to separate into more or less concentric layers. These layers seem to have a finer and more regular structure than does the winding layer, since the 20–50 secondary wall layers have cellulose microfibrils compactly parallelized along the axis of the fiber (see Cellulose).

The gross morphology of cotton, which refers to the relatively large structural elements above, is visible in the electron microscope. The microfibrillate structure includes pores, channels, and cavities that play an important role in the chemical modification of cotton. The arrangement of fibrils follows a spiral pattern and at times reverses itself; it is believed that regions of low strength along the length of the fiber occur within the reversal zones.

Microfibrils of the secondary wall are 10–40 nm in width, and these in turn are composed of elementary fibrils (crystallites) 3–6 nm wide.

Chemical modification of the cotton fiber must be achieved within the physical framework of this rather complicated architecture. Uniformity of reaction and distribution of reaction products are inevitably influenced by rates of diffusion, swelling and shrinking of the whole fiber, and by distension or contraction of the fiber's individual structural elements during finishing processes.

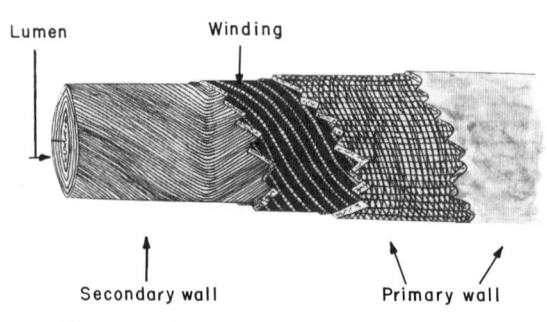

Figure 4. Schematic diagram for cotton fiber.

Chemical Properties

Although the cotton fiber, on drying, collapses from its round, never-dried shape and much of its water is removed, moisture is retained tenaciously in cotton. This moisture is expressed either as moisture content (amount of moisture as a percentage of original sample mass) or more commonly as moisture regain (amount of moisture as a percentage of oven-dry sample). Under ordinary atmospheric conditions, moisture regain is 7–11%. Because the cotton fiber has such a high cellulose content, the chemical properties are essentially those of the cellulose polymer. Indeed, the standard cellulose adopted by a committee of the ACS is simply a purified cotton. Like other forms of cellulose, the molecular chains of cotton cellulose consist of anhydroglucose units joined by 1–4 linkages (see Carbohydrates). These glycosidic linkages characterize the cotton cellulose as a polysaccharide, and are cleaved during hydrolysis, acetolysis, or oxidation resulting in shorter chains. If degradation is extensive enough, cellobiose or glucose derivatives are produced.

The cotton cellulose is a naturally occurring polymer; therefore, the molecular cellulose chains are of varying length. A measure of the average chain length can be obtained by determining the fluidity (or its reciprocal, the viscosity) of a solution of cotton. Solvents for this purpose include cuprammonium hydroxide solutions, phosphoric acid, nitric acid, quaternary ammonium bases, cadmium ethylenediamine hydroxide, and the one most used currently, cupriethylenediamine hydroxide (18). Low fluidities indicate cotton cellulose of high molecular weight; an increase in fluidity from that of untreated cotton usually signifies hydrolytic or oxidative damage to the cellulose. Oxidative damage also may be detected by increased adsorption of such dyes as methylene blue, by determination of copper number, or by titration of the resultant carboxyl group. If oxidation has proceeded to a great extent, marked alkali solubility results.

In addition to the length of cotton cellulose chains and to the chain length distribution, there is another factor that characterizes cotton cellulose. This is the degree of accessibility of the long, threadlike, polymeric molecules of cellulose within the elementary fibril of the cotton. Until the 1960s the fibrils were thought to divide into well-ordered crystalline and less organized amorphous regions. These dense crystalline regions, composed of highly hydrogen-bonded elementary fibrils, were believed to be the basis for the characteristic x-ray crystal diagram of the cellulose. The amorphous regions, thought to be composed of random, nonparallel polymeric aggregates, were responsible for the diffuse background in the x-ray crystal diagram. Thus relative amounts of crystalline and amorphous cellulose have been estimated from the x-ray diagram. Recent interpretations suggest that differences observed in the x-ray diagram of cellulose arise from structural defects that provide surfaces of varying accessibility, implying that cellulose is paracrystalline material (19–20). Also, whereas estimates formerly were made of crystalline and amorphous regions from rates and extent of reactions with hydrolytic and oxidative chemicals, it is now believed that these data merely indicate differences in accessibility (21).

These molecular chains that are associated into the elementary fibril are not parallel to the fiber axis; the spiral angle is approximately the same as that of the microfibrillar units of which the molecular chains form the ultimate unit. Furthermore, the helical configuration is irregular, due to frequent reversals in spiral angle. Chemical reactivity of the cellulose molecular chains within this complex physical structure is influenced by the location of the molecules on surfaces of the elementary fibril. Highly bonded surfaces result in inaccessible regions; surfaces under strain from spiral or reversal demands provide the most accessible regions.

Reactions. One of the earliest known modifications of cotton was mercerization. Traditionally, the process employed a cold concentrated sodium hydroxide treatment of yarn or woven fabric, followed by washing and a mild acetic acid neutralization. Maintaining the fabric under tension during the entire procedure was integral to the expected properties. The resultant mercerized cotton of commerce has improved luster and dyeability, and to a lesser extent, improved strength. A recent variation of this procedure substitutes hot sodium hydroxide that is allowed to cool while the cotton remains immersed in the caustic solution. More thorough initial penetration increases the efficiency of the mercerizing process. If the cotton is allowed to shrink freely during contact with mercerizing caustic, slack mercerization takes place; this technique produces a product with greatly increased stretch (stretch cotton) that has found application in both medical and apparel fields.

Effects similar to those from sodium hydroxide mercerization have been produced by exposure of the cotton to volatile primary amines or to ammonia. A procedure that utilizes liquid ammonia has found commercial adaptation (22). Improvements in luster and strength are similar to those from sodium hydroxide mercerization, but dyeability is not enhanced. A distinct difference exists chemically between cotton mercerized in sodium hydroxide and that mercerized in liquid ammonia. Under laboratory conditions of sodium hydroxide mercerization, Cellulose I of native cotton is converted to Cellulose II; under conditions of mercerization with liquid ammonia followed by nonaqueous quenching, Cellulose III is the product (23). Figure 5 shows the characteristic diffractograms (CuK_2 radiation) of native cellulose and cellulose mercerized with sodium hydroxide and liquid ammonia. With both treatments, there is increased accessibility, but differences in dye receptivity presumably result from differences in swelling loci between sodium hydroxide and liquid ammonia treatments.

The hydroxyl groups on the 2, 3, and 6 positions of the anhydroglucose residue are quite reactive (24), and by virtue of their functionality provide sites for much of the current modification of cotton cellulose to impart special properties. The two most common classes into which modifications fall include esterification and etherification of the cotton cellulose hydroxyls, as well as addition reactions with certain unsaturated compounds to produce a cellulose ether (see Cellulose derivatives, ethers).

Cellulose esters can be subdivided further into inorganic and organic esters (25). Of the three most common inorganic esters, cellulose nitrate, phosphate, and sulfate, only the cellulose sulfate is soluble in water. Cellulose sulfate attains water solubility at a degree of substitution (DS) of 3, indicating esterification of all three hydroxyls, whereas the sodium salt of cellulose sulfate is soluble in hot and cold water with a DS of only 0.33. Sodium cellulose sulfate is used in applications requiring suspension, thickening, stabilizing, and film-forming properties. A recent addition to cellulose-reactive dyestuffs is the class of phosphonic acid and phosphoric acid dyestuffs; these attach to cotton through esterification by the phosphonic acid or phosphoric acid group of the dyestuff (26). Organic esters of cotton cellulose, with two notable exceptions, are only of academic interest, although partial esterification of cotton by fatty acids has been reported to increase resiliency (27). A large class of cellulose-reactive dyestuffs in commercial use attach to the cellulose through an alkali-catalyzed esterification by the chlorotriazine moiety of the dyestuff:

$$\text{cellulose—OH} + \text{Cl—dyestuff} \rightarrow \text{cellulose—O—dyestuff} + \text{HCl}$$

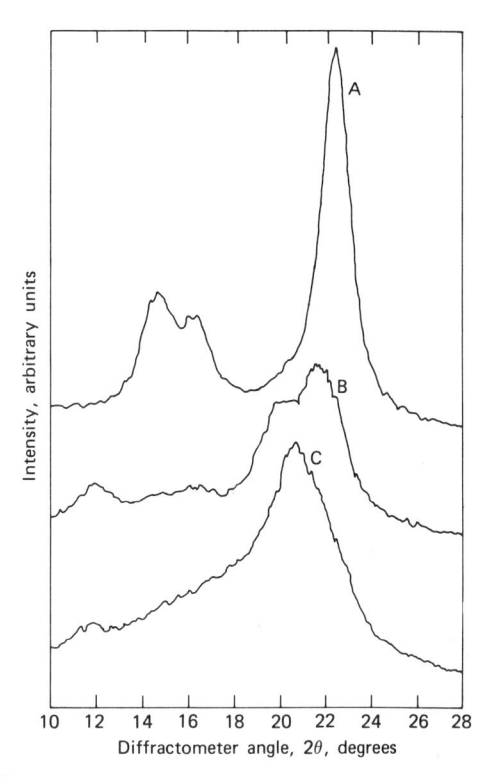

Figure 5. X-ray diffractograms: A, native; B, NaOH mercerized; and C, NH$_3$ mercerized cellulose (15).

Acetylation of cotton to an acetyl content slightly greater than 21% produces a material with greatly increased resistance to fungal and microbiological degradation, in addition to a tolerance of high temperatures not exhibited by native cotton; fibrous appearance and physical properties are unchanged by the acetylation. X-ray diffractograms indicate that, at this extent of substitution, only accessible regions of the cotton are involved in the acetylation (28). Differences in reaction rates between formation of inorganic and organic esters of cellulose depend on the availability to the reagents of the cellulose hydroxyls within the microstructure of cotton. Because only the most accessible regions can be reached by the organic reagents, organic esterification proceeds layer by layer. Spreading apart of the microstructure by swelling during inorganic esterification creates new accessible regions and allows essentially simultaneous reaction throughout the layers of the fibrillar structure (see Cellulose derivatives, esters).

By far the most important commercial modifications of cotton cellulose occur through etherification. For example, commercial modification of cotton to impart durable-press, smooth drying, or shrinkage resistance properties involves cross-linking adjacent cellulose chains through amidomethyl ether linkages. This cross-linking is commonly achieved via a pad–bake process. Although methylene, or oligomeric, cross-links from a pad–bake formaldehyde treatment have been produced, the resultant fabric exhibits severe strength loss. Most reagents for cross-linking cotton cellulose are di- or polyfunctional amidomethylol compounds (see Amino resins). The general formula for these compounds is shown with equations for synthesis of a methylol agent and its reaction with cellulose in Figure 6 (29). Commercially available cross-linking agents are dimethylolurea, dimethylolethyleneurea, dimethyloldihydroxyethyleneurea, dimethylolpropyleneurea, dimethylol alkylcarbamate, tetramethylolacetylenediurea, and methylolated melamine. The cross-linking proceeds via either Lewis or Bronsted catalysis (30) by a carbonium ion mechanism. Gross effects of the cross-linking are increased resiliency, manifested in wrinkle resistance, smooth drying properties, and greater shape holding properties, all important to cotton textiles, and conversely, reduced extensibility, strength, and moisture regain. These effects are observed at substitutions of 0.04–0.05 cross-links per anhydroglucose residue (31). The use of liquid ammonia treatment of cotton fabric followed by cross-linking attenuates the strength loss as well as an accompanying loss in abrasion resistance; a result of this unique combination of existing technologies is a reappearance of all-cotton fabrics in the woven shirting and high-fashion knitted fabric markets (32).

The microstructural effects of cross-linking via a pad–bake process can be seen in the response of the microfibrils and elementary fibrils to applied loads and to swelling agents. Resistance of cross-linked fibers to a methacrylate layer-expansion treatment that separates lamellae and reveals pore structure in untreated cotton is shown in Figure 7 (33). Although the ultimate chemical reaction—etherification of cellulose—is the same, the amount of moisture present in the fiber at the time of etherification influences both the response of the microstructure of the etherified fiber to swelling and solvation, and the physical properties of the cotton product (34). Thus resistance to layer-expansion treatment of the cotton fiber cross-linked with formaldehyde in a water medium that allows free fiber swelling lies between that of untreated fibers and pad–bake cross-linked fibers (35). Response to layer expansion by the cotton

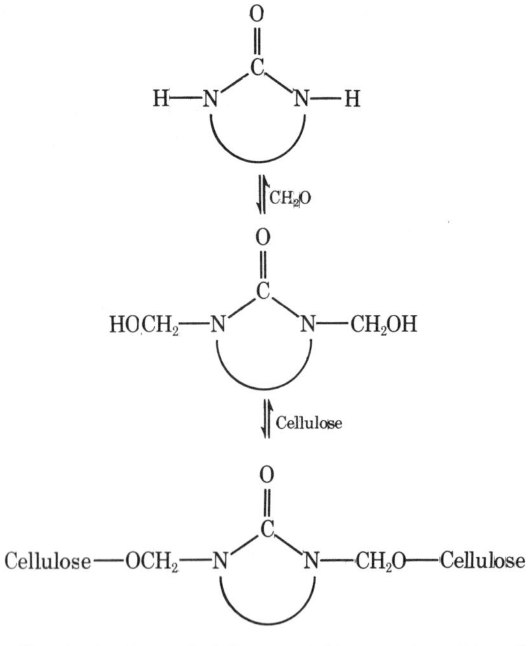

Figure 6. Synthesis of a methylol agent and its reaction with cellulose (19).

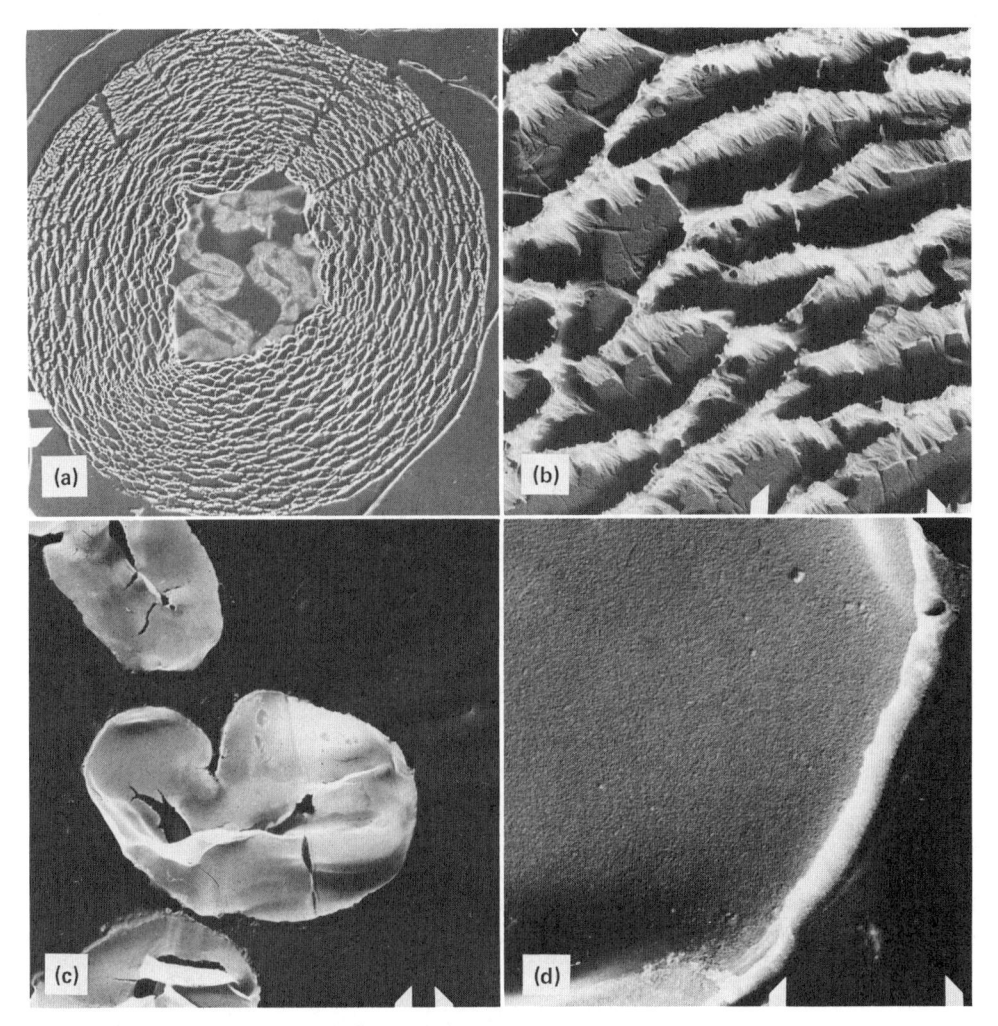

Figure 7. Methacrylate expansion patterns: (**a**) and (**b**), cross sections of unmodified cotton fiber at two levels of magnification; (**c**) and (**d**), expansion-resistant cross sections of cross-linked fiber (23). The gage marks on all four photographs indicate 1 μm.

fiber cross-linked with formaldehyde in the vapor phase, with moisture equivalent to normal regain, is similar to that by a conventionally cross-linked fiber (36). Cross-linking cotton fibers in a swollen state increases resiliency under wet but not under ambient conditions. As the amount of water present in the fiber at the time of reaction is decreased, wet resiliency decreases, and resiliency under ambient conditions increases.

Base-catalyzed reactions of cotton cellulose with either mono- or diepoxides to form cellulose ethers also result in fabrics with increased resiliency. Monoepoxides, believed to result only in cellulose hydroxyalkyl ethers or linear graft polymers (37), produce marked improvement in resiliency under wet conditions, but little improvement under ambient conditions. Difunctional epoxides, capable of cross-linking the cellulose, can produce increases in resiliency under both wet and ambient conditions (38). Microscopical examination of ultrathin cross sections from fabrics finished with both mono- and diepoxides indicates much less cross-linking in the former. Not only resiliency can be imparted through epoxide etherification; oil and water repellency can be imparted by reactions of monomeric perfluoro epoxides with cotton. Etherification of cotton with ethyleneimine has also provided the basis for imparting special properties to cotton, with the end product dependent on the attached group. Another base-catalyzed etherification of cotton is the cross-linking reaction between bis(hydroxyethyl) sulfone and cellulose; fabrics possessing increased resiliency under both wet and ambient conditions are obtained. The earliest application of sulfone cross-links to cotton textiles was the reaction of divinyl sulfone under alkaline conditions (39). However, the hazard of working with the vinyl compound led to modifications of the sulfone agent to replace the vinyl groups with more stable precursors such as the β-thiosulfatoethyl or β-sulfatoethyl (40) and β-hydroxyethyl groups (41). Etherification of cotton by divinyl sulfone and its precursors also forms the basis for another large class of fiber-reactive dyestuffs with the general formula: dyestuff—$SO_2CH=CH_2$ (42).

Other cotton cellulose ethers that have been prepared include carboxymethyl, carboxyethyl, hydroxyethyl, cyanoethyl, sulfoethyl, and aminoethyl (aminized cotton). Most of these ethers, with the exception of cyanoethylated and aminized cotton, are of interest in applications requiring solubility in water or alkali (26). In addition, ethers with pendant acid or basic groups have ion-exchange properties (43). Aminized cotton is of interest because it introduces into the cotton basic groups that provide sites for attachment of acid dyes. Simultaneous aminization of cotton and dyeing with an acid dyestuff marked the first successful attempt at dye attachment to cellulose through an ether linkage (44) (see Cellulose derivatives).

The flameproofing of cotton will not be discussed extensively here (see Flame-retardants; Flame-retardant textiles). Although certain cellulose esters such as the ammonium salt of phosphorylated cotton and cellulose phosphate are flame resistant, the attachment of most currently used durable polymeric flame-retardants for cotton is through an ether linkage to the cellulose at a relatively low degree of substitution (DS).

The development of water-repellent cellulose ethers has been reviewed by Marsh (45) (see Waterproofing). A typical example of a commercial etherification for waterproofing cotton is with stearamidomethylpyridinium chloride:

$$C_{17}H_{35}\overset{O}{\underset{}{\overset{\|}{C}}}-\underset{H}{\overset{}{N}}CH_2-\overset{+}{N}\bigcirc \; Cl^- + \text{cell}-OH \longrightarrow C_{17}H_{35}\overset{O}{\underset{}{\overset{\|}{C}}}-\underset{H}{\overset{}{N}}CH_2O-\text{cell} + HCl + N\bigcirc$$

N-Substituted, long chain alkyl monomethylol cyclic ureas have also been used to waterproof cotton through etherification. Other water repellent finishes for cotton are produced by cross-linked silicone films (46). In addition to the polymerization of the phosphorus-containing polymers on cotton to impart flame retardancy, and of silicone to impart water repellency, polyfluorinated polymers have been successfully applied to cotton to impart oil repellency. Chemical attachment to the cotton is not necessary for durability; oil repellency occurs because of the low surface energy of a fluorinated surface (47).

One of the earliest examples of etherification of cellulose by an unsaturated compound through vinyl addition is the cyanoethylation (qv) of cotton (48). This base-catalyzed reaction with acrylonitrile, a Michael addition, proceeds as follows:

$$CH_2=CHCN + cell\text{—}OH \rightarrow cell\text{—}OCH_2CH_2CN$$

For most textile uses, a DS less than 1 is desirable. Cyanoethylation can be used to impart a wide variety of properties to the cotton fabric such as rot resistance, heat and acid resistance, and receptivity to acid and acetate dyes. Acrylonitrile (qv) has also been radiation-polymerized onto cotton with a ^{60}Co source. Microscopical examination of ultrathin sections of the product shows the location of the polymer is within the fiber (49). Examination of the ir spectrum of cotton containing polymerized acrylonitrile indicates grafting does occur at the hydroxyl site of the cellulose (50). Another monomer grafted onto cellulose by irradiation is styrene (qv). Chemical properties, mechanisms, and textile properties of these graft polymers of cellulose are summarized in ref. 51. Graft polymerization onto cotton has also been induced by both chemical (52) and photochemical (53) initiation (see Radiation curing).

The effects of high energy radiation on cotton properties have also been investigated, eg, the effects of gamma radiation on cotton are described in refs. 54–56. Depolymerization of the cellulose occurs with increasing energy absorption; carbonyl formation, carboxyl formation, and chain cleavage occur in a ratio of 20:1:1. With these chemical changes, there is a corresponding increase in solubility in water and alkali and a decrease in fiber strength. In addition, the gamma-irradiated cotton was found to possess base ion-exchange properties. Irradiation of cotton with near ultraviolet light (325–400 nm) causes formation of cellulose free radicals and mild oxidative degradation of the cotton (57). Carbonyl and carboxyl contents of the cotton cellulose increase, and DP and tensile strength decrease, with increasing time of irradiation (58) (see Photochemical technology). The induction of cellulose free radicals by near uv irradiation forms the basis for photofinishing with vinyl monomers to produce graft polymers on the cotton:

$$cell\text{—}H \xrightarrow{h\nu} cell\cdot + H\cdot$$

$$cell\cdot + n\,M \rightarrow cell\text{—}(M)_{n-1}M\cdot$$

Another interesting reaction of cotton cellulose occurs with an ionized atmosphere. This is essentially a surface reaction. Glow discharge treatment of cotton yarn in air increase water absorbancy and strength (59), and surface-dependent properties of cotton fabric are drastically changed by exposure to low temperature–low pressure argon plasma generated by radio-frequency radiation (60). Because only a few extremely high energy electrons (10–15 eV) are generated, ambient temperature is maintained in the chamber. Light microscopy indicates a smoother surface, although scanning electron microscopy shows no change from native cotton. Spectral changes show some oxidation of the cotton, a decreased carbon to oxygen ratio. Free radicals similar to those from ^{60}Co radiation are formed. In addition, highly charged species are also formed, allowing such usually inert monomers as benzene to be polymerized onto the cotton; there is great capacity for bond cleavage. An increased rate of wetting and drying indicates potential end uses requiring absorbancy. The cohesiveness and fiber friction of cotton sliver was increased temporarily through air–trace chlorine corona treatments at 95°C and atmospheric pressure (61–62). With a 15 kV electrode voltage at a frequency of 2070 Hz, no chemical effects on the cotton could be noted. Dyeability, hand, and wettability were unaffected. The increase in cohesiveness was used to produce yarns with increased strength, abrasion resistance, and greater spinnability (63). Thus yarns of significantly lower twist can be produced with strength equal to, or higher than, untreated cotton yarns of higher twist.

Insolubilization of compounds within textiles parallels the history of man; the direct dyeing technique for cotton was highly advanced in the Bronze Age (see Dyes, natural). With the exception of fiber-reactive dyestuffs discussed earlier, all other cotton dyes—substantive, vat and sulfur—are insolubilized within the fiber, the latter two after an oxidizing step (see Dyes and dye intermediates). Insoluble metal oxides have been used to flameproof cotton. Certain zirconium compounds have been insolubilized on cotton to render the fabric microbial resistant (64) or mildew resistant (65) via a mineral dyeing process (see Textiles). Insolubilization, along with five other methods for imparting antimicrobial properties to cotton, are described in ref. 66. These methods can all be classified under one or more of the chemical reactions of cotton cited earlier; they include fiber reactions to form metastable bonds, grafting through thermosetting agents, formation of coordination compounds, ion-exchange methods, polymer formation with possible grafting, and a regeneration process. In the regeneration process, a compound is chemically bound to the cotton cellulose. This compound is capable of reversible reaction with antimicrobial species present in end-use conditions such as ozone, light, peroxide from water and ozone, or chemicals in detergents. As the antimicrobial species are consumed, newly present species recombine to regenerate the finish. Promising antimicrobial cotton products include halodeoxy cellulose fabrics (67) and cotton fabrics containing peroxide complexes of zirconyl acetate (68).

Economic Aspects

Marketing. There are several routes by which cotton fiber in the United States changes ownership from the grower to its final destination as a fiber, at the cotton mill. The grower may sell cotton directly to a spinning mill under a grower contract, or the grower may sell the cotton to a gin, or broker, or commission firm. Some growers, after ginning their cotton, may sell through a cooperative organization or may place the cotton in a depository as collateral under the Commodity Credit Corporation Loan Program, to be either withdrawn on repayment of the loan plus interest, or forfeited for sale by the government. These intermediate buyers then sell the cotton to shippers,

Table 2. World Production of Cotton, in 1000 Bales [a,b]

Area	1968	1973	1978
North America	10,272	16,938	17,618
South America	4,091	4,810	4,806
Western Europe	769	916	926
Eastern Europe and the U.S.S.R	9,500	11,505	12,670
Asia, Oceania, and Australia	18,355	23,365	24,049
Africa	4,759	5,879	5,155
Total	*47,746*	*63,413*	*65,224*

[a] Ref. 70.
[b] 1 bale = 216.8 kg.
[c] A year begins Aug. 1st of previous year to July 31st of year given.

who in turn sell to foreign mills or to domestic mills that have not purchased the cotton directly from the grower (69).

World Production and Prices. World production, consumption, and prices are shown in Tables 2 and 3.

Table 3. World Consumption and United States Prices of Cotton[a]

Factor	1968	1973	1978
consumption, t	11,471	13,105	11,452
price, $/kg	0.58–0.73	0.86–1.01	1.39–1.72

[a] Ref. 71.

Through the *Universal Cotton Standards of Agreement* (72), the official cotton standards of the United States for the grade of Upland cotton are recognized by 14 cotton associations and exchanges in 10 major consuming countries of Europe and Asia. The major differences between the United States system and those of other countries is the smaller number of grades established and the factors that influence quality, such as variety, moisture, and geographic background. In some instances, staple length is included as an attribute of grade, whereas in the United States, grade and staple length are separate criteria of the classification system.

Health and Safety Factors

Byssinosis. Byssinosis, also called mill fever or brown lung disease, is a pulmonary ailment, similar to bagassosis (see Bagasse) or silicosis, that is alleged to develop on repeated inhalation of cotton dust. A small percentage of textile workers exposed to cotton dust contract byssinosis. Other natural fibers whose dust afflicts workers with byssinosis are flax, sisal, soft hemp, and perhaps jute. Incidence of the disease can be controlled with adequate regulation of dust levels in work areas (see Air pollution control methods). OSHA has issued standards for control of cotton dust (73).

BIBLIOGRAPHY

"Cotton" in *ECT* 1st ed., Vol. 4, pp. 563–578, by Kyle Ward, Jr., R. B. Evans, Mary L. Rollins, Barkley Meadows, and Ines de Gruy, Southern Regional Research Laboratory, U.S. Department of Agriculture; "Cotton" in *ECT* 2nd ed., Vol. 6, pp. 376–412, by E. Lord, The Cotton Silk & Man-Made Fibres Research Association, Shirley Institute, Manchester.

1. W. L. Balls, *The Development and Properties of Raw Cotton*, A. & C. Black, Ltd., London, Eng., 1915.
2. C. B. Purves in E. Ott, H. M. Spurlin, and M. W. Graefflin, eds., *Cellulose and Cellulose Derivatives*, 2nd ed., Pt. 1, Interscience Publishers, Inc., New York, 1954, pp. 29–53.
3. R. D. Preston, *The Molecular Architecture of Plant Cell Walls*, John Wiley & Sons, Inc., New York, 1952.
4. H. B. Brown and J. O. Ware, *Cotton*, 3rd ed., McGraw-Hill Book Co., New York, 1958.
5. International Cotton Advisory Committee, *Cotton World Statistics* **30**(6, Pt. 2), 8 (Jan. 1977).
6. *Cotton Varieties Planted 1972–1976*, U.S. Agricultural Marketing Service, Memphis, Tenn., Sept. 1976.
7. G. A. Slater, *Cotton Int.* (*Memphis*) **43**, 90, 130, 138 (1976).

8. *Agric. Res.* **24**(12), 8 (June 1976).
9. *Cotton Int. (Memphis)* **43**, 55, 58 (1976).
10. U.S. Bur. of the Census, *Cotton Production in the United States; Crop of 1970,* U.S. Government Printing Office, Washington, D.C., 1971.
11. U.S. Bur. of the Census, *Cotton Ginnings in the United States; Crop of 1975,* U.S. Government Printing Office, Washington, D.C., 1976.
12. A. C. Griffin, private communication, Mar. 1977.
13. W. H. Fortenberry in D. S. Hamby, ed., *The American Cotton Handbook,* 3rd ed., Vol. 1, Interscience Publishers, Inc., New York, 1965, pp. 110–131.
14. *U.S. Cotton Handbook,* Cotton Council International and National Cotton Council of America, Washington, D.C., 1976.
15. M. L. Rollins in ref. 13, pp. 44–81.
16. L. A. Fiori and J. Compton in ref. 13, pp. 132–205.
17. J. N. Grant, *Text. Res. J.* **26**, 74 (1956).
18. E. Heuser, *The Chemistry of Cellulose,* John Wiley & Sons, Inc., New York, 1944, pp. 575–607.
19. S. Haworth and co-workers, *Carbohydr. Res.* **10**, 1 (1969).
20. R. Jeffries and co-workers, *Cellul. Chem. Technol.* **3**, 255 (1969).
21. R. Jeffries, J. G. Roberts, and R. N. Robinson, *Text. Res. J.* **38**, 234 (1968).
22. Brit. Pat. 1,084,612 (Sept. 27, 1967) (to Sentralinstitutt for Industriell Forskning and Norsk Tekstilforskningsinstitutt); Brit. Pat. 1,136,417 (Dec. 11, 1968), R. M. Gailey, (to J. and P. Coats, Ltd.).
23. T. A. Calamari, Jr., and co-workers, *Text. Chem. Color.* **3**, 235 (1971).
24. E. Heuser, *Text. Res. J.* **20**, 828 (1950); T. E. Timell, *Sven. Papperstidn.* **56**, 483 (1953).
25. R. M. Reinhardt and J. D. Reid, *Text. Res. J.* **27**, 59 (1957).
26. L. A. Graham and C. A. Suratt, *Am. Dyest. Rep.* **67**(7), 36 (1978).
27. J. B. McKelvey, R. J. Berni, and R. R. Benerito, *Text. Res. J.* **34**, 1102 (1964).
28. C. F. Goldthwait, E. M. Buras, Jr., and A. S. Cooper, Jr., *Text. Res. J.* **21**, 831 (1951).
29. J. G. Frick, Jr., *Chem. Technol.* **1**, 100 (1971).
30. A. G. Pierce, Jr., R. M. Reinhardt, and R. M. H. Kullman, *Text. Res. J.* **46**, 420 (1976).
31. J. G. Frick, Jr., B. A. Kottes Andrews, and J. D. Reid, *Text. Res. J.* **30**, 495 (1960).
32. S. A. Heap, *Colourage* **24**(7), 15 (1977).
33. M. L. Rollins and co-workers, *Norelco Rep.* **13**, 119, 132 (1966).
34. W. A. Reeves, R. M. Perkins, and L. H. Chance, *Text. Res. J.* **30**, 179 (1960).
35. S. P. Rowland, M. L. Rollins, and I. V. de Gruy, *J. Appl. Polym. Sci.* **10**, 1763 (1966).
36. A. M. Cannizzaro and co-workers, *Text. Res. J.* **40**, 1087 (1970).
37. J. B. McKelvey, B. G. Webre, and E. Klein, *Text. Res. J.* **29**, 918 (1959).
38. R. R. Benerito and co-workers, *J. Polym. Sci. Part A* **1**, 3407 (1963).
39. U.S. Pat. 2,524,399 (Oct. 3, 1950), D. L. Schoene and V. S. Chambers (to U.S. Rubber Co.).
40. G. C. Tesoro, P. Linden, and S. B. Sello, *Text. Res. J.* **31**, 283 (1961).
41. Can. Pat. 625,790 (Aug. 15, 1961), R. O. Steele (to Rohm and Haas Co.).
42. R. H. Peters, *Textile Chemistry III, The Physical Chemistry of Dyeing,* Elsevier Scientific Publishing Co., New York, 1975, pp. 624–629.
43. J. D. Guthrie, *Ind. Eng. Chem.* **44**, 2187 (1952).
44. J. D. Guthrie, *Am. Dyest. Rep.* **41**(1), P13, 30 (1952).
45. J. T. Marsh, *An Introduction to Textile Finishing,* 2nd ed., Chapman and Hall, London, Eng., 1966, pp. 458–494.
46. C. M. Welch, J. B. Bullock, and M. F. Margavio, *Text. Res. J.* **37**, 324 (1967).
47. E. J. Grajeck and W. H. Petersen, *Text. Res. J.* **32**, 320 (1962).
48. G. C. Daul, R. M. Reinhardt, and J. D. Reid, *Text. Res. J.* **25**, 246 (1955).
49. J. C. Arthur, Jr., and R. J. Demint, *Text. Res. J.* **30**, 505 (1960).
50. J. C. Arthur, Jr., and R. J. Demint, *Text. Res. J.* **31**, 988 (1961).
51. J. C. Arthur, Jr., and F. A. Blouin, *Am. Dyest. Rep.* **51**(26), 1024 (1962).
52. H. H. St. Mard, C. Hamalainen, and A. S. Cooper, Jr., *Am. Dyest. Rep.* **56**(5), 24 (1967).
53. A. H. Reine, N. A. Portnoy, and J. C. Arthur, Jr., *Text. Res. J.* **43**, 638 (1973).
54. F. A. Blouin and J. C. Arthur, Jr., *Text. Res. J.* **28**, 198 (1958).
55. J. C. Arthur, Jr., *Text. Res. J.* **28**, 204 (1958).
56. R. J. Demint and J. C. Arthur, Jr., *Text. Res. J.* **29**, 276 (1959).
57. J. C. Arthur, Jr., *U.S. Agric. Res. Serv. South. Reg. Rep.* **ARS-S-64**, 6 (1975).
58. J. C. Arthur, Jr., and O. Hinojosa, *Appl. Polym. Symp.* **26**, 147 (1975).

59. R. B. Stone and J. R. Barrett, Jr., *Text. Bull.* **88**(1), 65 (1962).
60. H. Z. Jung, T. L. Ward, and R. R. Benerito, *Text. Res. J.* **47,** 217 (1977).
61. W. J. Thorsen, *Text. Res. J.* **41,** 331 (1971).
62. *Ibid.,* p. 455.
63. D. P. Thibodeaux and H. R. Copeland, *Text. Chem. Process. Conf., 15th,* U.S.D.A. South. Reg. Res. Center, New Orleans, Louisiana, 1975, pp. 36–40.
64. C. J. Conner and co-workers, *Text. Res. J.* **34,** 347 (1954).
65. C. J. Conner and co-workers, *Text. Res. J.* **31,** 94 (1967).
66. D. D. Gagliardi, *Am. Dyest. Rep.* **51,** P49 (1962).
67. T. L. Vigo, D. J. Daigle, and C. M. Welch, *Text. Res. J.* **43,** 715 (1973).
68. T. L. Vigo, G. F. Danna, and C. M. Welch, *Text. Chem. Color.* **9,** 77 (1977).
69. R. S. Corkern, *Cotton Wool Situation (U.S.D.A. Economic Research Service)* **CWS-3,** 34 (Dec. 1975).
70. International Cotton Advisory Committee, *Cotton-World Statistics* **25**(6), 8 (1972); **31**(6), 8 (1978).
71. *Ibid.,* **22**(7), 14 (1968); **24**(6), 12 (1971); **28**(6), 34 (1975); **32**(6), 34 (1979).
72. *The Classification of Cotton Miscellaneous Publication No. 310,* U.S. Department of Agriculture, Washington, D.C., Jan. 1965, pp. 11–15.
73. *Fed. Reg.* **43**(122), 27350–27463 (June 23, 1978).

B. A. KOTTES ANDREWS
INES V. DE GRUY
United States Department of Agriculture

COTTONSEED. See Vegetable oils.

COUMARIN

Coumarin [91-64-5], 2H-1-benzopyran-2-one, $C_9H_6O_2$, is the odoriferous principle of the tonka bean (*Dipteryx odorata*), sweet clover (*Melilotus officinalis* and *alba*), and woodruff (*Asperula odorata*); widely distributed in the plant kingdom, it occurs in oil of lavender, oil of cassia, citrus oils, balsam Peru, and in some 60 other species of plants (1). In many cases, the plants are odorless, because coumarin occurs as a complex combined with sugars and acids. The resultant glucoside compounds can be split by acids, by natural enzymatic action or by uv irradiation. For instance, the odor of new-mown hay develops only after drying the cut grass.

Coumarin (1) was first isolated in 1820 by alcoholic extraction of tonka beans, which contain 1.5% coumarin (2). This method remained the source of coumarin until the synthetic methods, pioneered by Perkin in 1868, largely displaced coumarin from natural sources.

(1)

Coumarin is widely used in the perfumery, cosmetic, and related industries owing to its pleasant bitter–sweet and characteristic odor that recalls the fragrance of new-mown hay. It has other industrial uses unrelated to its odor qualities. Formerly it was an important food-flavoring material, often used in conjunction with vanillin (qv) for flavoring chocolates, candies, confections, and baked goods. Since 1954 the use in food has been suspended, because in a statement of general policy under the Federal Food, Drug, and Cosmetic Act, foods containing coumarin are regarded as adulterated (3). However, the consumption of coumarin has greatly increased during the past twenty years (see Flavors and spices; Perfumes).

Coumarin is the parent substance of a very large group of derivatives, many of which occur naturally and some of which are of economic importance.

Physical Properties

Coumarin is usually sold in the form of colorless shiny leaflets or rhombic prisms. It begins to sublime at 100°C. It is slightly water soluble and freely soluble in hot ethanol (see Table 1). The uv absorption spectrum of coumarin in alcohol shows a strong band at 265–275 nm (5).

Chemical Properties

Hydrolysis. The lactone is easily hydrolyzed by alkali to the corresponding salt of coumarinic acid, *cis-o*-hydroxycinnamic acid (2), eg, by treatment with dilute sodium hydroxide or by boiling with potassium carbonate. By contrast, heating coumarin with concentrated potassium hydroxide gives salts of *o*-coumaric acid, *trans-o*-hydroxycinnamic acid (3).

Table 1. Physical Properties of Coumarin

Property	Value
mol wt	146.14
mp, °C	70.6
bp at 101 kPa[a], °C	303
at 1.3 kPa[a]	154
at 0.67 kPa[a]	139
soly in H$_2$O	
at 25°C	0.25 g/100 mL
at 100°C	2 g/100 mL
soly in C$_2$H$_5$OH:H$_2$O[b]	g coumarin in 100 mL soln
vol % ethanol	at 20°C at 40°C
25	0.50 1.93
50	3.71 19.02
70	10.04 47.00

[a] To convert kPa to mm Hg, multiply by 7.5.
[b] Ref. 4.

(2)   (3)

Upon acidification with mineral acid, coumarinic acid readily reverts to coumarin. Fusion of coumarin with caustic alkali ruptures the double bond to yield salicylic and acetic acids.

Bisulfite Reaction. Coumarin combines readily with sodium bisulfite solutions across the 3,4-double bond to form stable, soluble sodium 3- or 4-hydrosulfonates. Since coumarin can be regenerated with acid, this method has been used for the purification of coumarin (6).

Halogenation. Treatment of coumarin in chloroform at room temperature with bromine yields coumarin-3,4-dibromide in good yield (7). Under more drastic conditions 3-bromo- and 3,6-dibromocoumarin are formed (8).

Sulfonation. Heating coumarin with fuming sulfuric acid effects ring substitution and at water bath temperature gives coumarin-6-sulfonic acid; at 150°C coumarin-3,6-disulfonic acid is formed.

Nitration. Fuming nitric acid forms mainly 6-nitrocoumarin along with small amounts of 8-nitrocoumarin. Coumarin resists the formation of dinitrocoumarin (8).

Hydrogenation. The 3,4-double bond of coumarin reacts readily with hydrogen. Hydrogenation with an active Raney catalyst at 40°C and 345 kPa (50 psi) yields 83% 3,4-dihydrocoumarin (9). Continued hydrogenation leads to the formation of the saturated octahydrocoumarin (10).

Reduction. Coumarin is reduced by sodium amalgam to melilotic acid (o-hydroxyhydrocinnamic acid).

Oxidation. Coumarin is oxidized with difficulty and is stable to chromic acid (11). However, oxidative biochemical attack occurs frequently at the 7-position.

Dimerization. Prolonged exposure to sunlight, or uv irradiation convert coumarin to a dimer, mp 263°C, involving the 3,4-double bond.

Methods of Preparation

The practical synthetic methods for coumarin use as starting materials either salicylaldehyde (or its equivalent in the form of side-chain chlorinated o-cresol) or phenol, from which the pyrone ring is elaborated. Some of the methods described below give poor yields of coumarin itself but often excellent yields of certain substituted coumarins.

From Salicylaldehyde. *Perkin Reaction.* William H. Perkin synthesized coumarin in 1868 (12) by heating the sodium salt of salicylaldehyde with acetic anhydride. Perkin later found that sodium acetate could serve as the base catalyst instead of sodium salicylaldehyde (12). The newer method permitted the extension of this reaction to nonphenolic aromatic aldehydes, such as benzaldehyde, leading to cinnamic acids. Named the Perkin reaction, this general method can be considered an example of Claisen condensation. Its mechanism has been studied from many viewpoints (13).

The reaction for the formation of coumarin (**1**) may be written (14):

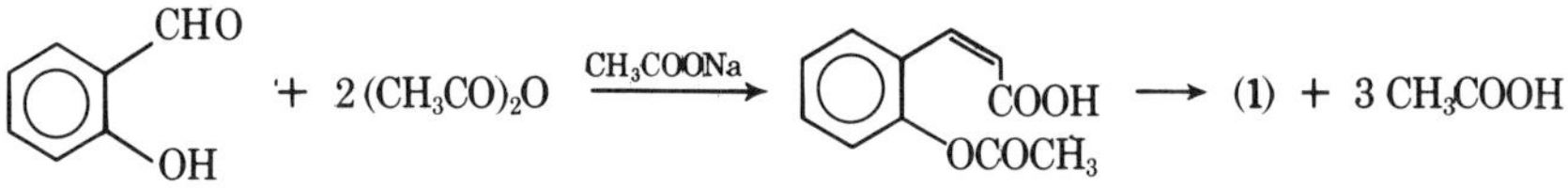

$$\text{salicylaldehyde} + 2\,(CH_3CO)_2O \xrightarrow{CH_3COONa} \text{cis-o-acetoxycinnamic acid} \longrightarrow (\mathbf{1}) + 3\,CH_3COOH$$

The reaction may proceed through the triacetyl derivative of salicylaldehyde, which has been shown to yield 70–80% coumarin (**1**) when heated to 180°C (15).

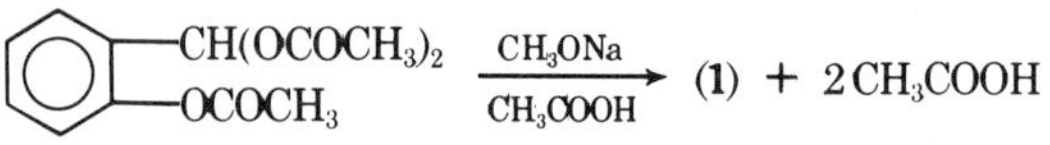

$$\text{triacetyl derivative} \xrightarrow[CH_3COOH]{CH_3ONa} (\mathbf{1}) + 2\,CH_3COOH$$

A 50% yield of coumarin has been reported (16) by heating 2-acetoxybenzaldehyde in the presence of sodium acetate.

The Perkin synthesis has been used extensively for the production of coumarin. Various reaction conditions using molar proportions of reactants per mole of salicylaldehyde are shown below. The yields of coumarin represent % of theoretical.

1.6 $(CH_3CO)_2O$, 2.4 CH_3CO_2Na; 180–195°C	27%	(17)
with trace iodine catalyst	70%	(17)
2.0 $(CH_3CO)_2O$, 2.0 CH_3CO_2Na; 180–195°C, $CoCl_2$ catalyst	64%	(18)
1.92 $(CH_3CO)_2O$, 0.3 CH_3CO_2Na; 160–180°C	52%	(19)
3.2 $(CH_3CO)_2O$, 0.8 K_2CO_3; 180–200°C	75%	(20)
1.9 $(CH_3CO)_2O$, trace CH_3CO_2Na; 160–190°C	60%	(16)
1.2 $(CH_3CO)_2O$, trace pyridine 160°C: (2-acetoxybenzaldehyde 97%) heating with trace CH_3CO_2Na at 208°C	50%	(16)

Knoevenagel Reaction. Knoevenagel discovered in 1896 (21) that the strenuous conditions (high temperature) of the Perkin reaction were not required, if acetic anhydride were replaced by acetic acid derivatives possessing an activated methylene group, and organic amines used as catalysts. Primary and secondary amines, ammonia, and their salts can be used, piperidine and pyridine are especially efficient. The condensation has been used primarily in the synthesis of many substituted coumarins but has also been employed for the production of coumarin (**1**), where it involves the extra step of the removal of the activating group. Knoevenagel obtained coumarin-3-carboxylic acid in nearly quantitative yield from malonic acid and salicylaldehyde (21).

$$\text{salicylaldehyde} + CH_2(COOR)_2 \xrightarrow{\text{organic base}} \text{coumarin-3-COOR} \xrightarrow{-CO_2} (1)$$

R = H, malonic acid
R = C$_2$H$_5$, ethyl malonate

Coumarin-3-carboxylic acid (22–23) can be decarboxylated to coumarin by heating to 290°C. Improved yields at lower temperature are obtained in the presence of mercuric salts (24). 3-Carbethoxycoumarin is decarboxylated to coumarin in 81% yield by heating the ester with concentrated aqueous metabisulfite followed by the addition of sulfuric acid and heating to 125°C (25).

Cyanoacetic acid and esters condensed with salicylaldehyde and followed by hydrolysis of the cyano group, also yield coumarin-3-carboxylic acid (**4**).

$$\text{salicylaldehyde} + \underset{\text{cyanoacetic acid}}{CH_2CN\text{-}COOH} \rightarrow \text{intermediate} \rightarrow \text{coumarin-3-COOH} \quad (4)$$

Raschig Method. Raschig discovered in 1909 (26) a practical method for making salicylaldehyde by dichlorinating the methyl group of esters of o-cresol (free o-cresol is ring-chlorinated), followed by hydrolysis of the benzal chloride intermediate. He also found (27) that coumarin can be obtained directly from the chloro compound by heating it with anhydrous sodium acetate. Processes based on chlorinated o-cresol esters have been used to a considerable extent in the commercial production of coumarin because of inexpensive starting materials such as the phosphate (27) and carbonate esters (28). Fusion of α,α-dichloro-o-cresyl carbonate with potassium acetate at 150–200°C yielded 71% coumarin (28). It was discovered (29) that alkali acetates in the fusion reaction could be replaced by acetic anhydride in the presence of metal catalysts, preferably 1% cobalt oxide, in which case acetyl chloride is formed and is recovered; yield of coumarin, 45% (29).

$$\text{(dichloro-o-cresyl carbonate)} + 3(CH_3CO)_2O \rightarrow 2\,\text{(salicylaldehyde acetate)} + 4\,CH_3COCl + CO_2$$
(5)

$$(5) + (CH_3CO)_2O \rightarrow (1) + 2\,CH_3COOH$$

From Phenol. ***Pechmann Condensation.*** Coumarin (**1**) and substituted coumarins can be obtained by heating phenols with malic acid in the presence of concentrated sulfuric acid, with elimination of CO (30). Beta-keto esters in place of malic acid (see Hydroxy carboxylic acids) yield coumarins substituted in the pyrone ring. These reactions are commonly known as the Pechmann reaction (31).

Maleic and fumaric acids can be used in place of malic acid (32–35) (see Maleic anhydride).

The condensation of ethyl acetoacetate with phenol is an example of a beta-keto ester reaction. It furnishes 4-methylcoumarin (6).

Phenyl acetoacetate obtained from phenol and diketene also yields 4-methylcoumarin (36).

Vinylation with Acrylic Esters. A novel method for producing coumarin is disclosed in recent patents (37). Phenol is vinylated with methyl acrylate in acidic medium and in the presence of air. After 6 h, 51% of coumarin and methyl o-coumarate were produced in a condensation favoring ortho over para condensation in a ratio of 2.4:1.0. The p-coumarate, also formed, does not yield coumarin, but the methyl o-coumarate (7) is converted into coumarin by continuously passing it through a reaction zone at 200–350°C. This cyclization can be carried out in the presence of coumarin (1) and methyl p-coumarate (38).

Another method (39) produces 3,4-dihydrocoumarins (8) by condensing phenols and methyl acrylate in the presence of aluminum chloride. Dehydrogenation with palladium on charcoal forms the corresponding coumarin in high yield. The method is applicable only to dihydroxybenzenes and produces 5-, 6-, or 7-hydroxydihydrocoumarins.

Purification and Shipment

Purification. In order to be suitable for perfumery purposes, the product of commerce must be of high purity since the human nose can detect minute amounts of odorous foreign material. Common methods of purification include fractional distillation under high vacuum and crystallization from suitable solvents, such as ethanol, and combinations of these methods. Treatment with lime has been proposed (40) to form the soluble calcium o-coumarate from which impurities can be removed by steam distillation and solvent extraction, followed by acidification. A later patent (6) takes advantage of the ability of coumarin to combine with bisulfite to form stable, water soluble solutions. Coumarin is recovered from the purified solution by acidification (evolution of SO_2), or by adding the equivalent amount of alkali (6).

Shipment. Coumarin is shipped in tin-lined containers or fiberboard drums. It is stable on storage and no special precautions in handling are required.

Economic Aspects

Data on the production, sales, and value of synthetic coumarin were published by the U.S. Tariff Commission until 1967 (Table 2). Later data are not available because the number of producers dropped to three or fewer.

Table 2. Production, Sales, and Value of Synthetic Coumarin in the United States

Year	Production, t	Quantity, t	Unit price, $/kg	Value, $
1940	112	99	5.14	508,000
1946	163	149	5.51	822,000
1956	354	299	6.68	2,000,000
1967	520	510	4.37	2,221,000
1977	>544	>544	9.92	>5,000,000

At the present time, Rhodia Inc. is a leading United States producer, and its parent company, Rhone-Poulenc, Paris, is the world's largest producer. The removal of coumarin in 1954 as an artificial flavor affected the growth curve of its market only temporarily. During the late 1960s the unit price dropped below that of the early 1960s.

Specifications

The official specifications of the Scientific Section of the Essential Oil Association of the United States are:

appearance and odor:	coumarin is a white crystalline solid with a sweet, fresh hay-like slightly spicy odor
congealing point:	68.0°C (min)
solubility in alcohol:	1 g is soluble in 8 mL 90% ethanol

Some manufacturers have additional internal specifications, such as melt appearance: water-white and free from turbidity; solidified appearance: practically water-white;

sulfuric acid test: almost colorless. In practice, the odor test is the most exacting. Although odor qualities cannot be quantified, comparison with a standard product reveals deviations that are not measurable by any instrumental method. Coumarin has very high odor strength. The concentration threshold at which it can be recognized by a trained observer is ca 0.01 ppm by weight in air and 0.4 ppm in water.

Health and Safety Factors (Toxicology)

Since 1954 coumarin has been classed by the FDA as a toxic substance, and foods containing coumarin are regarded as adulterated (3). The rather abrupt banning of coumarin as a food additive was not followed in European countries until some 20 years later. Sax lists the toxic hazards of coumarin as slight (41). Earlier references estimated the hazardous level for adults at 3.5 g, where it caused vomiting and weakness. The FDA action was based on liver toxicity found in rats. Extensive animal tests have since been carried out: acute oral LD_{50} in rats, 300–600 mg/kg, about twice the toxicity of methyl salicylate, and subacute toxicity, no effect in rat studies with 1000 ppm (0.1%) coumarin in the diet up to 2 yr (42). At 5000 ppm (0.5%) in the diet, one third of the rats surviving a 1½ yr study developed liver carcinomas (43). Biological attack of coumarin when fed to animals and man occurs preferentially at the 7-position and results in hydroxylation; another important metabolite is o-hydroxyphenylacetic acid, formed by ring opening and side chain oxidation (44). The metabolism of coumarin in man was reported in 1969 (45). After volunteers ingested 200 mg coumarin, urine samples after 24 h were analyzed. About 80% was excreted as 7-hydroxycoumarin and about 4% as o-hydroxyphenylacetic acid. By contrast, in rats only 1% is excreted as 7-hydroxycoumarin and 20% as o-hydroxyphenylacetic acid. Rabbits excrete mainly o-hydroxyphenylacetic acid. It is likely that liver damage is due to decreased activity of the enzyme glucose 6-phosphatase which is inhibited by o-hydroxyphenylacetic acid. It is doubtful whether evaluation of toxicity of coumarin to man, based on toxicity data for rats or rabbits, is justifiable. Coumarin is phytotoxic, eg, it is extremely toxic to wheat seedlings; it also has bacteriostatic action properties, eg, on *E. coli* (46).

When coumarin is administered to sweet clover shoots it is rapidly converted through reduction to o-hydroxyphenylpropionic acid (melilotic acid) (47). It was discovered early that sweet clover disease in cattle, caused by spoiled clover hay, could be simulated if coumarin were added to otherwise nontoxic spoiled alfalfa hay. Certain penicillin molds from soil convert coumarin into 4-hydroxycoumarin and traces of dicumarol (46).

Uses

Coumarin is one of the most widely used synthetic aroma chemicals, owing to its odor strength, tenacity, stability to alkali, and relatively cheap price. In perfumery it is used almost universally as a sweetener and fixative (see Perfumes). It is a dominant ingredient in fern (fougere), chypre and new-mown hay types. It is used extensively to enhance the character of fragrances based on natural essential oils such as lavender, citrus, rosemary, oakmoss, etc (see Oils, essential). It blends especially well with vanillin and heliotropin. Such combinations are used in numerous cosmetic products like colognes, lotions, bath powders, talcs, and especially soaps and detergents (see Cosmetics). Coumarin is used in tobacco to enhance its natural aroma (see Tobacco substitutes). A large field of application consists in the imparting of a pleasant odor to industrial products or to mask disagreeable odors (see Odor counteractants). It is a classical masking agent for iodoform as well as for phenolic and quinoline odors. The applications include paints (qv), printing inks (qv), insecticides, plastics, and synthetic rubbers (see Insect control technology; Rubber chemicals).

A substantial portion of coumarin is used by the electroplating (qv) industries. Addition of coumarin to the electroplating bath causes deposits of metals with reduced porosity and increased brightness, especially with nickel, zinc and cadmium.

Derivatives

Coumarin is widely distributed in the plant kingdom along with some fifty derivatives of coumarin, many of which were identified in the extended research by E. Spaeth (summary in ref. 8). Some products are physiologically active, but few are of economic importance. They include, for instance, alkyl, hydroxy, and methoxy derivatives as well as more complicated condensed systems. A large number of synthetic derivatives have been described, some of which are of commercial significance.

3,4-Dihydrocoumarin. 3,4-Dihydrocoumarin is obtained by hydrogenation of coumarin and is a coumarin metabolite as well. It has GRAS status (generally regarded as safe) and thus is permitted in imitation flavors, where its sweet caramel-like taste is used to fortify flavors such as vanilla, butter, rum, and caramel. The odor is similar to that of coumarin and it finds some use in fine perfumery. United States production in 1976 was 14 metric tons at a unit price of $15.72/kg. Ring-alkylated dihydrocoumarins that possess valuable musklike odor characteristics have been described (48).

6-Methylcoumarin. Obtained from p-cresol and fumaric acid in the presence of sulfuric acid, 6-methylcoumarin has GRAS status, possesses a delicate coumarin odor, and is used in imitation flavors and as a modifier in chypre and oriental perfumes (see Perfumes).

Umbelliferone, 7-Hydroxycoumarin. Umbelliferone is an important coumarin metabolite, since the 7-position is the preferred site of biochemical attack. It occurs naturally in gum resins of umbelliferae, such as galbanum, and it is readily manufactured from resorcinol and maleic or fumaric acid. Solutions show blue fluorescence and the product is used in sunscreen lotions and creams (see Cosmetics). Umbelliferone and β-methylumbelliferone are used as fluorescent brighteners (see Brighteners, fluorescent).

4-Hydroxycoumarin. 4-Hydroxycoumarin is a coumarin metabolite occurring in spoiled hay. Sweet clover disease, the hemorrhagic condition of cattle, is caused by the toxic agent dicumarol derived from 4-hydroxycoumarin, which inhibits the coagulant action of Vitamin K, (with whose chemical structure 4-hydroxycoumarins have some analogy) (see Vitamins) (49). Dicumarol has been introduced as an anticoagulant drug in the therapy of thrombotic diseases. A related product, Warfarin, is an important rodenticide that causes fatal hemorrhages. 4-Hydroxycoumarin, which has distinct analgesic properties, has been synthesized by treating acetyl methyl salicylate with metallic sodium (49). A synthesis of 4-hydroxycoumarins with substituents in the 3-position consists of subjecting o-hydroxyaryl alkyl ketones to a Kolbe-Schmitt type carboxylation step followed by lactonization (50).

With o-hydroxyacetophenone, the yield is over 30% 4-hydroxycoumarin.

Dicumarol. Dicumarol (9) is readily prepared by treating an aqueous or alcoholic solution of 4-hydroxycoumarin with an excess of 40% formaldehyde (49).

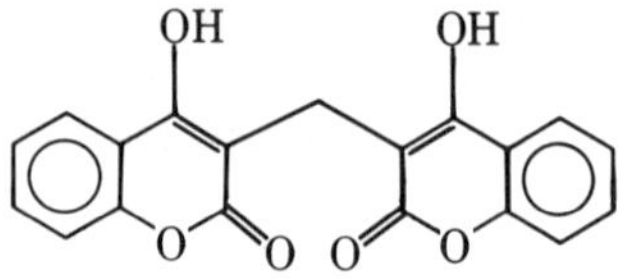

(9)

Link and co-workers described in detail the extraordinary sequence of 16 steps required to isolate dicumarol from spoiled sweet clover hay, constituting a 30,000-fold enrichment (51). Unlike heparin it may be administered orally (see Blood, coagulants and anticoagulants).

Warfarin. Compounds related to dicumarol, all 3-substituted 4-hydroxycoumarins, also possess hemorrhagic properties (52). Of these, Warfarin (10), 3-α-acetonylbenzyl)-4-hydroxycoumarin, is a highly effective rodenticide. It is produced by a Michael condensation of benzylidene acetone with 4-hydroxycoumarin (see Poisons, economic).

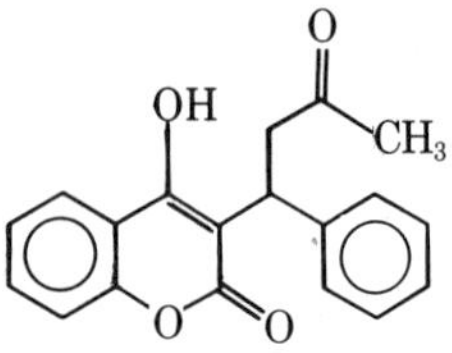

(10)

Phenprocoumon. Phenprocoumon (11), similar in structure to Warfarin, usually maintains the anticoagulant effect established by other anticoagulants (53–54) (see Blood coagulants and anticoagulants).

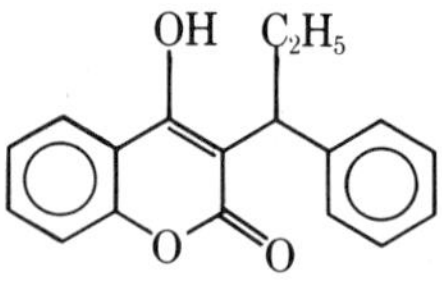

(11)

Coumarone, Coumarone–Indene Resins. Coumarone (benzofuran) (bp 171–172°C) can be obtained from coumarin by passing its vapor through an iron tube at 800°C. It occurs abundantly in a fraction of bituminous coal tar that also contains indene, bp 160–200°C. Polymerization with sulfuric acid or AlCl$_3$ catalysts converts the mixture to the important category of coumarone–indene resins, which were among the first synthetic resins in the United States (before 1920) and are used widely for floor tiles and in compounded rubbers (see Hydrocarbon resins).

BIBLIOGRAPHY

"Coumarin" in *ECT* 1st ed., Vol. 4, pp. 588–593, by Oliver DeGarmo, Monsanto Chemical Company; "Coumarin" in *ECT* 2nd ed., Vol. 6, pp. 425–433, by Oliver DeGarmo and Paula Raizman, Monsanto Company.

1. E. Spaeth, *Ber.* **70A,** 83 (1937).
2. Vogel, *Gilberts Annalen der Physik* **64,** 163 (1820).
3. *Fed. Reg.* **19,** 1239 (Mar. 5, 1954).
4. A. Seidell, *Solubilities of Organic Compounds,* Vol. 2, Van Nostrand Co., Inc., New York, 1941, p. 623.
5. E. Cingolani, *Gazz. Chim. Ital.* **84,** 825 (1954).
6. U.S. Pats. 1,945,182, 1,945,184 (Jan. 30, 1934), E. Clemmensen (to Monsanto Chemical Company); F. D. Dodge, *J. Am. Chem. Soc.* **52,** 1724 (1930); **38,** 446 (1916).
7. R. C. Fuson and co-workers, *Organic Syntheses Collective Volume 3,* John Wiley & Sons, Inc., New York, 1955, p. 209.
8. S. M. Sethna and co-workers, *Chem. Rev.* **36,** 27 (1945).
9. X. A. Dominguez and co-workers, *J. Org. Chem.* **26,** 1625 (1961).
10. De Bennerville and co-workers, *J. Am. Chem. Soc.* **62,** 283 (1940).
11. R. Parikh and co-workers, *J. Indian Chem. Soc.* **27,** 369 (1950).
12. W. H. Perkin, *J. Chem. Soc.* **21,** 53 (1868); **31,** 388 (1877).
13. J. Johnson, "The Perkin Reaction and Related Reactions" in *Organic Reactions,* Vol. 1, John Wiley & Sons, Inc., New York, 1942, p. 210.
14. F. Tiemann and H. Herzfeld, *Ber.* **10,** 283 (1877).
15. A. Marczewski, *Pol.* **42,** 431 (1960).
16. K. I. Bogacheva, *Tr. Vses. Nauchn. Issled. Sint. Nat. Dushistykh Veshchestv* (5), 30 (1961); *Chem. Abstr.* **57,** 9800 (1962).
17. H. Yanagisawa and co-workers, *J. Chem. Soc. Abstr. I* **120,** 682 (1921).
18. U.S. Pat. 2,204,008 (June 11, 1940), E. C. Britton and co-workers (to The Dow Chemical Company).
19. U.S. Pat. 3,631,067 (Dec. 28, 1971), R. J. Nankee and co-workers (to The Dow Chemical Company).
20. V. I. Isaguliants and co-workers, *J. Appl. Chem. (U.S.S.R.)* **11,** 946 (1938).
21. E. Knoevenagel, *Ber.* **29,** 172 (1896); **31,** 730 (1898); Ger. Pat. 161,171 (Dec. 1, 1903), E. Knoevenagel.
22. I. N. Bratus and co-workers in ref. 16, pp. 34–37.
23. E. C. Horning and co-workers in ref. 7, p. 165.
24. Ger. Pat. 440,341 (1924), (to I. G. Farbenindustrie).
25. U.S.S.R. Pat. 130,518 (Aug. 5, 1960), V. G. Voronin and co-workers; *Chem. Abstr.* **55,** 6500 (1960).
26. Ger. Pat. 233,631 (Jan. 27, 1909), F. Raschig.
27. Ger. Pat. 233,684 (April 27, 1909), F. Raschig.
28. U.S. Pat. 1,920,494 (Aug. 1, 1933), E. C. Britton and W. R. Reed (to The Dow Chemical Company).
29. U.S. Pat. 2,062,364 (Dec. 1, 1936), G. Krotchman and C. Collaud (to Givaudan Company).
30. H. Von Pechmann, *Ber.* **16,** 2119 (1883); **17,** 929 (1884).
31. S. M. Sethna and co-workers, *Organic Reactions,* Vol. 7, John Wiley & Sons, Inc., New York, 1953, pp. 1–58.
32. H. Simonis, *Ber.* **48,** 1584 (1915).
33. G. C. Bailey and co-workers, *Ind. Eng. Chem.* **13,** 905 (1921).
34. A. A. Shmuk, *Vsesoyuz Nauchn.-Issled. Inst. Tabach (140),* 45 (1939); *Chem. Abstr.* **34,** 5070 (1940).
35. V. G. Austerweil, *Compt. Rend.* **248,** 1810 (1959).
36. R. N. Lacey, *J. Chem. Soc.,* 854 (1954).
37. U.S. Pats. 3,859,311 (Jan. 5, 1975), 3,936,473 (Feb. 3, 1976), T. Symon and co-workers (to Universal Oil Products Company).
38. U.S. Pat. 3,888,883 (June 10, 1975), N. J. Christensen and co-workers (to Universal Oil Products Company).
39. A. K. Das Gupta and co-workers, *J. Chem. Soc. C,* 29 (1969).
40. U.S. Pat. 1,437,344 (Nov. 28, 1922), C. C. Loomis.
41. N. I. Sax, *Dangerous Properties of Industrial Materials,* 4th ed., Van Nostrand & Co., Inc., New York, 1975.
42. D. L. J. Opdyke, *Food Cosmet. Toxicol.* **12,** 385 (1974).
43. F. Baer and co-workers, *Med. Ernaehr.* **8,** 244 (1967).
44. M. Kaighen and co-workers, *J. Med. Pharm. Chem.* **3,** 25 (1961).
45. W. H. Shilling and co-workers, *Nature (London)* **221,** 664 (1969).
46. Y. Raoul, *Bull. Soc. Chem. Biol.* **29,** 518 (1947).
47. T. Kosuge and co-workers, *J. Biol. Chem.* **234,** 2133 (1959).

48. U.S. Pat. 3,144,467 (Aug. 11, 1964), W. J. Houlihan (to Universal Oil Products Company).
49. M. A. Stahman, C. F. Huebner, and K. P. Link, *J. Biol. Chem.* **138**, 21, 513, 529 (1941); Brit. Pat. 579,459 (Aug. 2, 1946), M. A. Stahman, C. F. Huebner, and K. P. Link.
50. P. Da Re and co-workers, *Ber.* **93**, 1085 (1960).
51. U.S. Pat. 2,601,204 (June 17, 1952), H. A. Campbell and co-workers (to Wisconsin Alumni Research Foundation).
52. U.S. Pat. 2,427,578 (Sept. 16, 1947), M. A. Stahman and co-workers (to Wisconsin Alumni Research Foundation).
53. C. H. Schroeder and K. P. Link, *J. Am. Chem. Soc.* **79**, 3291 (1957).
54. *AMA Drug Evaluations*, 3rd ed., Publishing Sciences Group, Inc., Littleton, Mass., 1977, p. 121.

WALTER C. MEULY
Rhodia Inc.

COUMARONE INDENE RESINS. See Hydrocarbon resins.

CRACKING. See Petroleum.

CREAM. See Milk products.

CRESIDINE, $CH_3C_6H_3(CH_3)NH_2$. See Amines—Aromatic amines—Aniline and its derivatives.

CRESOLS. See Alkylphenols.

CRESYLIC ACIDS, SYNTHETIC. See Alkylphenols.

CROTONALDEHYDE

Crotonaldehyde [4170-30-3] (2-butenal), $CH_3CH=CHCHO$, is a water-white liquid. Its strongly lacriminatory vapors make it a valuable warning agent. It can exist either as the *cis* or *trans* isomer. Commercial crotonaldehyde is more than 95% trans (1).

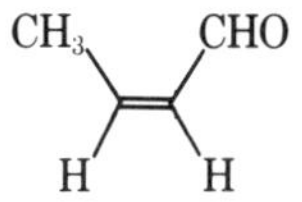

cis
[15798-64-8]

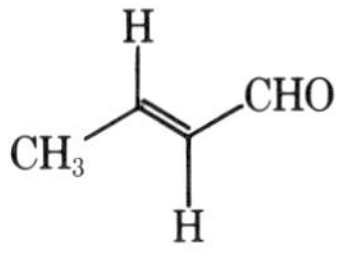

trans
[123-73-9]

Physical Properties

Because of the proximity of the aldehyde group and the ethylenic double bond, crotonaldehyde, like acrolein (qv), $CH_2=CHCHO$, is very reactive. Pure crotonaldehyde resinifies fairly readily, the amount and rate of polymerization depending upon temperature, type of container, etc. Since the products of resinification and oxidation vary from yellow to dark brown, it is difficult to store crotonaldehyde without its becoming discolored. Dicroton or dicrotonaldehyde is the best-known dimer of crotonaldehyde (2). One trimer is also known (3–4). Hydroquinone and water reportedly inhibit resinification and oxidation. Several physical properties of crotonaldehyde are given in Table 1.

Table 1. Physical Properties of Crotonaldehyde

density d_{20}^{20}, g/mL	0.853
melting point, °C	−69
boiling point, °C	102.2
refractive index $n_D^{17.3}$	1.4384
heat of combustion, kJ/mol[a]	2268
solubility, g crotonaldehyde/100 g H_2O	
20°C	18.1
5°C	19.2
solubility, g H_2O/100 g crotonaldehyde	
20°C	9.5
5°C	8.0
flash point, open cup °C	13
heat of vaporization, J/g[a]	515
specific heat, liquid, J/g[a]	3
vapor density (air = 1)	2.41
explosive limits in air, % by vol	2.91–15.5

[a] To convert J to cal, divide by 4.184.

Chemical Properties

Reactions. *Reduction.* Crotonaldehyde can be reduced selectively at the carbonyl group, at the carbon–carbon double bond, or at both groups. Reductive cyclization reactions also occur. Table 2 shows various reducing agents for crotonaldehyde, along with other pertinent data.

Oxidation. Autoxidation. The air oxidation of crotonaldehyde has been studied widely (40–45). When an organic solvent and acetates or crotonates of Mn, Co, and Cu (preferably 15% Co + 85% Cu salts) are used, crotonic anhydride and crotonic acid are formed. Yields are approximately 70% (41). In the presence of Mn catalyst, at 5–10°C in acetic acid (42), free oxygen converts crotonaldehyde to crotonic acid in yields of 40–80%. Water does not hinder the reaction (43, 46).

A by-product of crotonaldehyde production is the dimer 2,6-dimethyl-5,6-dihydro-3-formylpyran (dicroton). Air oxidation of this compound using $Mn(OOCCH_3)_2$ yields the corresponding carboxylic acid (40):

Table 2. Reducing Agents for Crotonaldehyde

Agent	Catalyst	Conditions	Yield, %	Product	Reference
Reduction of carbonyl group					
$LiBH_4$			70	crotyl alcohol[a]	5
$NaBH_4$			85	crotyl alcohol	6
KBH_4			55	crotyl alcohol	7
$LiAlH_4$			70	crotyl alcohol	8
H_2	colloidal Pt or Pd on boneblack			2-butene	9
H_2	$Co(Co)_4$	50–300°C		crotyl alcohol	10
H_2	Cd	20 MPa[b], 275°C		crotyl alcohol	11
Reduction of carbon–carbon double bond[c]					
H_2	Pd (amount and type of reduction depends on amount of catalyst)			butyraldehyde	12
H_2	Raney Ni	40–70°C	70–90	butyraldehyde	13–16
BuOH	Cu	300°C, 0.1 MPa[b]	70	butyraldehyde	17–18
H_2, BuOH	Cu	165°C	91	butyraldehyde	19
H_2	Ni–Cr–Mg or Ni–Cr–Al on pumice	112°C	84	butyraldehyde	20–21
Reduction of both groups[d]					
H_2	Raney Ni	140–150°C		butanol	13
H_2	Raney Cu			butanol	22
Mg in acetic acid				hydroxytetrahydrofuran	23
H_2	Cu or Cu–Ni on kieselguhr	180–200°C	95	butanol	24–26
H_2	vapor phase			butanol	27–29
H_2	CuO on inert support	150°C	100	butanol	30
H_2, CO			31	butanol	31

[a] Crotyl alcohol [6117-91-5] $CH_3CH=CHCH_2OH$ (32) has been offered as a development chemical. It is a clear liquid miscible in all proportions with water, ethyl alcohol, and acetone.
[b] To convert MPa to atm, divide by 0.101.
[c] See also references 33–36.
[d] See also references 37–39.

β-Dicarbonyl formation in the preflame reaction of several compounds including crotonaldehyde has been studied (47).

Bromine. At 200–600°C (48), oxidation of crotonaldehyde by bromine can be made to proceed almost exclusively at the carbonyl group, converting —CHO to —COBr. The reaction is exothermic and may be applied to compounds of the general structure RCH=CHCHO.

Halogenation. Direct chlorination of crotonaldehyde at room temperature results in the addition of two chlorine atoms; on warming to 50°C in the presence of water, the monohydrate, mp 78°C, is formed (49).

$$CH_3CH=CHCHO + Cl_2 + H_2O \rightarrow CH_3CHClCHClCHO \cdot H_2O$$

Bromination of cooled crotonaldehyde followed by boiling with an excess of 1% HCl in absolute ethanol for 46 h gives a 42% yield of 2-bromo-1,1,3-triethoxybutane. This compound dehydrobrominates on refluxing for 96 h with sodium ethoxide (50)

$$CH_3CH=CHCHO + Br_2 + C_2H_5OH \xrightarrow{HCl} CH_3CH(OC_2H_5)CHBrCH(OC_2H_5)_2 \rightarrow$$

$$CH_3C(OC_2H_5)=CHCH(OC_2H_5)_2 + HBr$$

The kinetics of the acid-catalyzed addition of Cl_2 and Br_2 to crotonaldehyde in acetic acid solution indicates a nucleophilic mechanism (51). Addition of water, however, reduces the nucleophilic rate. Sufficient water establishes the electrophilic reaction and further dilution increases the electrophilic rate.

The reaction of crotonaldehyde with phosphorus pentachloride proceeds as follows (52):

$$CH_3CH=CHCHO + PCl_5 \rightarrow CH_3CHClCH=CHCl + CH_3CH=CHCHCl_2$$
main product

Addition to C=C Double Bond. *Sulfur Compounds.* Thiols generally add to crotonaldehyde at the 2,3 positions (53).

The reaction with hydrogen sulfide gives predominantly the thio ether (54).

$$CH_3CH=CHCHO + H_2S \rightarrow HCOCH_2CH(CH_3)SCH(CH_3)CH_2CHO$$

Alcohols. Ethanol, in addition to forming the acetal (see below) adds to the carbon–carbon double bond (55).

$$C_2H_5OH + H_2SO_4 + CH_3CH=CHCHO \xrightarrow[16\ h,\ 50°C]{CO_2\ atmosphere} CH_3CH(OC_2H_5)CH_2CHO$$

$$\xrightarrow{neutralization} CH_3CH(OC_2H_5)CH_2CH(OC_2H_5)_2$$

Nitro Compounds. Nitromalonic ester (56) and nitroacetic ester (57), in the presence of a basic catalyst, add to crotonaldehyde. Thus nitromalonic ester leads to:

$$\begin{array}{c} NO_2 \\ | \\ C(COOR)_2 \\ | \\ CH_3CHCH_2CHO \end{array}$$

Diels-Alder Reaction. Crotonaldehyde undergoes the Diels-Alder reaction as both a diene and a dienophile (58–59).

Acetal Formation. Crotonaldehyde forms many acetals, usually under relatively mild conditions and with acid catalysts. Acetals with unsaturated alcohols undergo rearrangement to form 2-alkenyl crotonaldehydes (60).

Addition to the Carbonyl Group. *Amines.* With amines, crotonaldehyde undergoes characteristic reactions of aldehydes (61–64) (qv).

Organometallic Compounds. Metal alkyls react with the carbonyl group of crotonaldehyde to produce secondary alcohols after hydrolysis (see Organometallics).

A *Grignard reagent,* RMgCl gives, after hydrolysis, the expected product $CH_3CH=CHCH(OH)R$.

Nitroalkanes. Reactions of crotonaldehyde with nitroalkanes usually result in the secondary alcohol.

Sodium Bisulfite. $NaHSO_3$ adds to the carbonyl group, to the carbon–carbon double bond, or to both (65). The carbonyl bisulfite group is easily removed by treatment with alkali. The other bisulfite group is removed only with difficulty.

Formaldehyde. Formaldehyde reacts with crotonaldehyde or acetaldol to form the methylol derivative (66–68):

$$CH_3CH=CHCHO + H_2O \xrightarrow{OH^-} CH_3CHOHCH_2CHO \xrightarrow{HCHO} CH_3CHOHCHCHO$$
$$\text{acetaldol} \qquad \qquad |$$
$$CH_2OH$$

Other Aldehydes. Acetaldehyde reacts with crotonaldehyde to form, after hydrogenation, 1-hexanol, 1-octanol, and 1-decanol (69).

Ketene. Crotonaldehyde reacts with ketene (qv) in at least two ways, depending upon the catalyst used. In the presence of boric acid (70) or zinc salts (71), the β-lactone is formed.

$$CH_3CH=CHCHO + CH_2=C=O \longrightarrow$$

This lactone can be pyrolyzed to piperylene (70) or to sorbic acid (71).

With strong acids the acetoxy compound is formed (72):

$$CH_3CH=CHCHO + CH_2=C=O \longrightarrow CH_2=CHCH=CHOCCH_3$$

Phenols. Two molecules of a substituted phenol react with crotonaldehyde in acetic acid solution to give a derivative of 2-(substituted phenoxy)-4 methylchroman (1).

Autocondensation. The autocondensation of crotonaldehyde can be made to yield a variety of products by varying catalyst and conditions, eg, refluxing with aqueous hydrochloric acid yields a dimer:

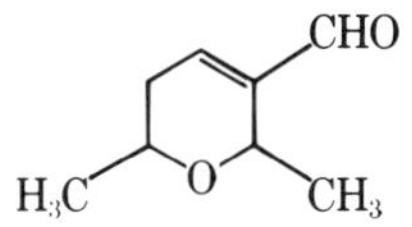

Polymers and Polymerization. The polymerization of crotonaldehyde has been studied extensively. Crotonaldehyde undergoes condensation polymerization and addition polymerization as well as reaction with numerous polymers.

Crotonaldehyde forms condensation polymers with ethylamine (4), ethyleneimine (73), and ammonium thiocyanate (74). It also reacts with such hydroxy compounds as resorcinol (75), phenol (76), sulfonated polyhydric phenols (77), alkylphenols (78), polyhydric alcohols (79–80) and poly(vinyl alcohol) (81).

Crotonaldehyde is polymerized by triethylamine to form a resin with film-forming properties (82). Other amines also catalyze this polymerization (83). Brittle, high-melting polymers are formed when crotonaldehyde is polymerized at pressures up to ca 1 GPa (10^4 atm) (84).

Synthesis and Manufacture

The most widely used method for the synthesis of crotonaldehyde is the aldol condensation of acetaldehyde, accompanied or followed by dehydration. Various catalysts and conditions are reported for this reaction (85–97).

$$2\ CH_3CHO \rightarrow CH_3CHOHCH_2CHO \rightarrow CH_3CH{=}CHCHO + H_2O$$
$$\text{acetaldol} \qquad \text{crotonaldehyde}$$

The equilibrium constant for this reaction, K = crotonaldehyde/acetaldol, is ≥0.61 at 40–50°C (98). Other reactions leading to crotonaldehyde are (99–102):

$$CH_3CHO + CH_2{=}C(OCH_3)_2 \rightarrow CH_3CH{=}CHCHO + 2\ CH_3OH + \text{resins}$$

$$CH_3CHO \xrightarrow{300\ °C} CH_3CH{=}CHCHO \xrightarrow{355\text{-}390\ °C} CH_2{=}CHCH{=}CH_2$$

$$\text{(ethylene oxide)} \xrightarrow[400\ °C]{MgO/Al_2O_3} CH_3CH{=}CHCHO + CH_3CHO + \text{mixed aldehydes}$$

$$CH{\equiv}CH + H_2O \xrightarrow[330\text{-}350\ °C]{ZnW\ on\ Al_2O_3} CH_3CH{=}CHCHO + CH_3CHO + CH_3COOH$$

Direct oxidation of hydrocarbons offers a possible synthesis of crotonaldehyde. In most cases only trace amounts are formed (see Hydrocarbon oxidation). However, the oxidation of 1,3-butadiene has been reported (103) to give conversions as high as 34% (Consortium process) (3,104–108).

$$CH_2{=}CH_2 + N_2O \xrightarrow{300°C,\ 51\ MPa\ (500\ atm)} CH_3CHO + CH_3CH{=}CHCHO$$

$$CH_3CH{=}CHCH_3 + O_2 \rightarrow CH_3CH{=}CHCHO + \text{mixed aldehydes}$$

Activated alumina in phosphomolybdic acid catalyzes the air oxidation of propane,

212 CROTONALDEHYDE

ethylene, acetaldehyde, acetic acid, propionic acid, or acetone to mixtures of maleic acid, acetic acid, crotonaldehyde, formaldehyde, and carbon dioxide (95).

Pyrolysis of various hydroxy compounds can lead to crotonaldehyde. Some examples are as follows (109–115):

$$CH_3CHOHCH_2CH_2OH \xrightarrow[400°C, 30 \text{ min}]{MgO \text{ or } ZnO} CH_3CH=CHCHO + CH_3COCH=CH_2$$
30% yield

$$CH_2=CHCHOHCH_2OH \xrightarrow[\text{reflux}]{HCl(aq)} CH_3CH=CHCHO + H_2O$$
7% yield

$$CH_2OHCH=CHCH_2OH \xrightarrow[200°C]{Al_2O_3} CH_3CH=CHCHO + H_2O$$

furan $\xrightarrow[\text{ground glass cat.}]{375°C} CH_3CH=CHCHO$

β-methyl-butyrolactone $\xrightarrow[300-320°C]{N_2} CH_3CH=CHCHO$ + resins
9% yield

Other reactions that produce crotonaldehyde are given in references 87 and 116–121.

Economic Aspects

Crotonaldehyde is currently produced and marketed in the United States. The late 1977 price of the product was ca $1.10/kg. Sales in 1977 of crotonaldehyde for the markets shown in Uses below are estimated to be less than 500 metric tons.

Analysis

Derivatives. Table 3 shows derivatives that can be used for the identification of crotonaldehyde.

Determination. Several methods have been developed for the quantitative analysis of crotonaldehyde. (1) Determination of crotonaldehyde and acetaldehyde in vinyl acetate (132–133). (2) Potentiometric titration with bromine in methanol. Methanol reacts to form the acetal of the aldehyde to prevent interference with bromine (134). In this way unsaturated aldehydes are determined in the presence of saturated al-

Table 3. Melting Points of Crotonaldehyde Derivatives

Derivative	Melting point, °C	Reference
oxime	119–120	122
semicarbazone	198–199	123
4-phenylsemicarbazone	179	124
thiosemicarbazone	167	125
phenylhydrazone	56	126
methone	193–194	127
2,4-dinitrophenylhydrazone	190	128–129
2,4-dinitrophenylsemicarbazone	230 (dec)	130–131

dehydes. (3) Acetaldehyde is determined in the presence of crotonaldehyde by heating the mixture with sodium bisulfite and then distilling acetaldehyde in the presence of sodium bicarbonate (135). (4) Iodometric method (136). (5) For other methods see references 137–139.

Neutralization. Crotonaldehyde may be neutralized in chemical process waste water by adjusting the water to a pH of 8 with alkali hydroxide, such as NaOH, and heating for 15–30 min at 80–100°C (140).

Health and Safety Factors

Crotonaldehyde is more toxic than its saturated analogue. Its most pronounced feature is its irritating effect on the nose, pharynx, and larynx. It damages the lungs of mice (141). The biochemical properties of crotonaldehyde and its derivatives have been the subject of considerable study. For example, crotonaldehyde has been used to treat certain plant virus diseases (142) and has been assayed for fungistatic activity against various organisms (143). It is a volatile active principle in tuberculin (144).

Crotonaldehyde is not mailable because of its lacrimatory properties and low flash point. In case of contact, immediately flush skin or eyes with plenty of water for at least 15 min; for eyes, get medical attention. Remove and wash clothing before reuse.

Uses

The largest use of crotonaldehyde has been in the manufacture of n-butanol. This process is still used commercially but has been displaced partially by the oxo process (qv). Currently, crotonaldehyde is widely used to produce sorbic acid (145). 3-Methoxybutanol has been made commercially by the reaction of methanol with crotonaldehyde followed by reduction. The aldehyde is also converted to crotonic acid (qv). Crotonaldehyde has been suggested as a fuel warning agent.

Despite its reactive nature, the aldehyde does not gain wide usage in the chemical trade. The following uses are suggested.

Polymers or polymerization. Crotonaldehyde reacts with wool (qv) to give a product which is less soluble in alkali (146); is a solvent for poly(vinyl chloride) (147); increases the strength of ordinary rubber (qv) when added in the form of resins by heating ketones and crotonaldehyde to high temperatures (148); the p-dodecylphenol derivative is a rubber tackifier (149); acts as a shortstop in vinyl chloride polymerization (150) (see Vinyl polymers); increases the yield of polymer formed in castor oil–olefin reactions (151) (see Castor oil); reacts with water-soluble fusible hydroxylated vinyl resins to render them insoluble (152); and forms resins that act as pickling inhibitors by condensing with amine thiocyanates (153). On addition to vinyl acetate polymerization, crotonaldehyde decreases the average degree of polymerization (dp) but narrows the distribution of dp (154–155). The styrene copolymer is a thermoplastic (156) and the m-cresol-copolymer may be used to treat nylon fibers for improved wet-fastness of colors (157). In the medical area, 1-vinyl-2-pyrrolidinone-crotonaldehyde copolymer is useful in cardiovascular surgery as an anticoagulant (see Blood, coagulants and anticoagulants) (158) and crotonaldehyde is used in tanning of valves for heterografts (159). The reaction of 2-sulfonilamide-4-methylpyrimidine with crotonaldehyde and sodium bisulfite yields solutions with chemotherapeutic (slow release) activity. Crotonaldehyde also forms plasticizers with terpene resins (160); an adhesive resin with resorcinol (161); and tanning materials with sulfonated phenols (162).

Miscellaneous. Various applications include the production of flavoring agents from crotonaldehyde and L-cystine (163) (see Flavors and spices); the tobacco flavoring agent, 2,4-diphenylcrotonaldehyde (164); the absorption of crotonaldehyde on hide (165–166); the use of crotonaldehyde as a hardening agent for gelatin (167); and as a stabilizer for tetraethyllead (168). Crotonaldehyde is also used in the preparation of surface-active agents (169–175), textile and paper sizes (176–177), a deodorizer for paper prepared from waste paper (178), and insecticidal compounds (179–180). Crotonaldehyde has some antimicrobial activity against spores, yeasts, and fungi (181–182) and bacteria in aqueous liquids used in petroleum processing (183). Crotonaldehyde–amine compounds improve the brightness of electroplated tin (184) and crotonaldehyde inhibits the acid corrosion of steel used in electroplating processes (185–186). Insecticides (179–180) and fertilizers (187) can be prepared from crotonaldehyde. In the processing of color photographs, the addition of crotonaldehyde to the bleach bath eliminates the hardening bath and rinse and increases the dye resistance to heat and light (188).

BIBLIOGRAPHY

"Crotonaldehyde" in *ECT* 1st ed., Vol. 4, pp. 608–612, by John Galaba, Niacet Chemicals Division, United States Vanadium Corporation; "Crotonaldehyde" in *ECT* 2nd ed., Vol. 6, pp. 445–464, by William M. Gearhart, Eastman Chemical Products, Inc.

1. D. W. Clayton and co-workers, *J. Chem. Soc.*, 581 (1953).
2. E. Späth, R. Lorenz, and E. Freund, *Monatsh.* **76,** 297 (1947); *Sitzungsber. Akad. Wiss. Wien Math. Naturwiss Kl. Abt. 2B* **155,** 297 (1946).
3. F. S. Bridson-Jones and co-workers, *J. Chem. Soc.*, 2299 (1951).
4. J. F. Carson and H. S. Olcott, *J. Am. Chem. Soc.*, **76,** 2257 (1954).
5. R. F. Nystrom, S. W. Chaikin, and W. G. Brown, *J. Am. Chem. Soc.*, **71,** 3245 (1949).
6. S. W. Chaikin and W. G. Brown, *J. Am. Chem. Soc.* **71,** 122 (1949).
7. R. Huls and Y. Simon, *Bull. Soc. R. Sci. Liege* **25,** 89 (1956).
8. R. F. Nystrom and W. G. Brown, *J. Am. Chem. Soc.* **69,** 1192 (1947).
9. Z. Csuros and I. Sello, *Hung. Acta Chim.* **1** (4/5), 27 (1949).
10. U.S. Pat. 2,614,107 (Oct. 14, 1952), I. Wender and M. Orchin (to U.S.A. by Secretary of the Interior).
11. Ger. Pat. 858,247 (Dec. 4, 1952), H. Brendlein (to Deutsche Gold- u. Silber-Scheideanstalt vorm. Roessler).
12. Z. Csuros, *Muegy. Kozl. Budapest,* 110 (1947).
13. K. Matsui, *J. Chem. Soc. Japan Pure Chem. Sect.* **64,** 1420 (1943).
14. Y. Orito, *Rept. Gov. Chem. Ind. Res. Inst. Tokyo* **47,** 63 (1952).
15. U.S. Pat. 2,501,708 (March 28, 1950), T. Bowley and J. A. Keeble (to Distillers Co., Ltd.).
16. Brit. Pat. 595,941 (Dec. 23, 1947), (to Distillers Co., Ltd.).
17. K. Satake, *Annu. Rep. Sci. Works Fac. Sci. Osaka Univ.* **2,** 17 (1954).
18. K. Satake and S. Akabori, *J. Chem. Soc. Japan, Pure Chem. Sect.* **70,** 84 (1949).
19. Brit. Pat. 656,860 (Sept. 5, 1951), (to Usines de Melle and Louis Alheritiere).
20. Jpn Pat. 1,271 (March 11, 1954), T. Fukumoto and co-workers, (to Ness Nippon Nitrogenous Fertilizers Co.).
21. Jpn Pat. 6,030 (Sept. 20, 1954), T. Fukumato and co-workers (to Ness Nippon Nitrogenous Fertilizers Co.).
22. J. Jadot and R. Braine, *Bull. Soc. R. Sci. Liege* **25,** 67 (1956).
23. C. Glacet, *Compt. Rend.* **222,** 501 (1946).

24. W. Reppe, *Reichsamt Wirtschaftsausbau Chem., Ber. Prüf-Nr. 36 U.S. Dept. Comm. Tech. Serv.* PB 56002, 35–46 (1940).
25. T. Moroe and J. Kito, *J. Chem. Soc. Japan, Ind. Chem. Sect.* **56,** 173 (1953).
26. Brit. Pat. 687,775 (Feb. 18, 1953), M. Erlenbach and A. Sieglitz.
27. Jpn. Pat. 1566 (April 15, 1953), I. Watabe (to Mitsubishi Chemical Industries Co.).
28. Brit. Pat. 649,554 (Jan. 31, 1951), C. E. Barraclough and co-workers, (to Distillers Co., Ltd.).
29. Jpn. Pat. 3,666 (1951), H. Arita (to Asaki Chemical Industries Co.).
30. Fr. Pat. 973,322 (Feb. 9, 1951), H. M. Guinot (to Usines de Melle).
31. I. Wender, R. Levine, and M. Orchin, *J. Am. Chem. Soc.* **72,** 4375 (1950).
32. *Crotyl Alcohol, Bulletin TDS X-156,* Eastman Chemical Products, Inc., 1963.
33. Z. Csuros, K. Zech, and I. Géczy, *Hung. Acta Chim.* **1,** 1 (1946).
34. Y. Suchiro, *J. Chem. Soc. Japan. Ind. Chem. Sect.* **51,** 77 (1948).
35. S. Tsutsumi H. Kayamori, and T. Kawamura, *J. Chem. Soc. Japan, Ind. Chem. Sect.* **54,** 27 (1951).
36. Brit. Pat. 736,074 (Aug. 31, 1955) (to Farbwerke Hoechst A.-G. vorm. Meister Lucius und Bruning).
37. G. P. Khomchenko and G. D. Vouchenko, *Vestn. Mosk. Univ. Ser. Fiz. Mat. Estestv. Nauk Ser 5* **10**(8), 91 (1955).
38. M. J. Wiemann and M. Paget, *Bull. Soc. Chim. Fr.,* 285 (1955).
39. Jpn. Pat. 158,671 (Sept. 2, 1943) (to Japan Celluloid Co.).
40. U.S. Pat. 2,378,996 (June 26, 1945), B. T. Freure (to Carbide and Carbon Chemicals Corp.).
41. U.S. Pat. 2,413,235 (Dec. 24, 1946), D. J. Kennedy (to Shawinigan Chemicals, Inc.).
42. U.S. Pat. 2,487,188 (Nov. 8, 1949), G. W. Seymour, B. B. White, and E. Barabash (to Celanese Corp. of America).
43. Brit. Pat. 595,170 (Nov. 27, 1947) (to Shawinigan Chemicals Ltd.).
44. Ger. Pat. 803,296 (April 2, 1951), H. G. Trieschmann (to Badische Anilin & Soda-Fabrik (I. G. Farbenindustrie A.G.)).
45. Ger. Pat. 870,846 (March 16, 1953), W. O. Herrmann and W. Heohnek (to Consortium für Elektrochemische Industrie G.m.b.H.).
46. Jpn. Pat. 75-112,314 (Sept. 3, 1975), I. Watansbe, K. Tanaka, and A. Aoshima (to Asahi Chem. Ind. Co., Ltd.).
47. M. R. Barusch and co-workers, *Ind. Eng. Chem.* **43,** 2766 (1951).
48. Brit. Pat. 626,772 (July 21, 1949), R. W. Tess, G. W. Hearne, and H. L. Yale (to N.V. de Bataafsche Petroleum Maatschappij).
49. Brit. Pat. 576,435 (April 3, 1946) (to Pennsylvania Salt Manufacturing Company).
50. F. M. Hammer and R. J. Rathbone, *J. Chem. Soc.,* 595 (1945).
51. P. B. D. De la Mare and P. W. Robertson, *J. Chem. Soc.,* 888 (1945).
52. L. J. Andrews, *J. Am. Chem. Soc.* **68,** 2584 (1946).
53. R. H. Hall and B. K. Howe, *J. Chem. Soc.,* 2723 (1949).
54. L. L. Gershbein and C. D. Hurd, *J. Am. Chem. Soc.* **69,** 241 (1947).
55. W. Flaig, *Reichsant Wirtschaftsausbau, Chem. Ber. Prüf.-Nr. 093, U.S. Dept. Commerce, Tech. Serv.* PB52020, 1073–1108 (1942).
56. U.S. Pats. 2,546,958 and 2,546,960 (April 3, 1951), D. T. Warner and O. A. Moe (to General Mills, Inc.).
57. U.S. Pat. 2,499,653 (June 10, 1952), O. A. Moe and D. T. Warner (to General Mills, Inc.).
58. W. Flaig, *Justus Liebigs Ann Chem.* **568,** 1 (1950).
59. U.S. Pat. 3,341,061 (Sept. 12, 1967), J. Mertzweiller (to Esso Research & Engineering Co.).
60. U.S. Pat. 2,501,144 (March 21, 1950), R. H. Saunders (to Hercules Powder Co.).
61. F. Nerdel and I. Huldschinsky, *Chem. Ber.* **86,** 1005 (1953).
62. C. W. Smith, D. G. Norton, and S. A. Ballard, *J. Am. Chem. Soc.* **75,** 3316 (1953).
63. A. Treibs and R. Derro, *Justus Liebigs Ann. Chem.* **589,** 176 (1954).
64. U.S. Pat. 2,408,127 (Sept. 24, 1946), G. W. Seymour and V. S. Salvin (to Celanese Corp. of America).
65. M. Hori, *J. Agr. Chem. Soc. Japan* **18,** 155 (1942).
66. R. Pummerer and co-workers, *Ann. Chem.* **583,** 161 (1953).
67. R. Pummerer, *Sitzungsber. Math. Naturwiss Kl. Bayer, Akad. Wiss. Muenchen* **1953,** 191 (1954).
68. Ger. Pat. 821,204 (Nov. 15, 1951), R. Pummerer, E. Person, and H. Rick (to Rudolf Pummerer).

69. Jpn. Pat. 2,217 (May 20, 1953), K. Matsuyo and S. Tsutsumi (to Nippon Synthetic Chemical Industries Co.).
70. U.S. Pat. 2,469,110 (May 3, 1949), H. J. Hagemeyer, Jr. (to Eastman Kodak Co.).
71. U.S. Pat. Appl. 252,194 (Aug 18, 1953), J. R. Caldwell.
72. U.S. Pat. 2,421,976 (June 10, 1947), A. H. Agett (to Eastman Kodak Co.).
73. J. B. Doughty, C. L. Lazzell, and A. R. Collett, *J. Am. Chem. Soc.* **72,** 2866 (1950).
74. U.S. Pat. 2,526,644 (Oct. 24, 1950), A. P. Dunlop and P. R. Stout (to Quaker Oats Co.).
75. J. Le Bras, *Deut. Kautschuk-Ges. Vortragstagung Hamburg,* 1956, 21 pp.
76. Brit. Pat. 747,093 (March 28, 1956), (to Farbenfabriken Bayer A.-G.).
77. Ger. Pat. 803,848 (April 12, 1951), R. Alles (to Badische Anilin- & Soda-Fabrik (I. G. Farbenindustrie A.G.)).
78. C. S. Marvel, R. J. Gander, and R. R. Chambers, *J. Polym. Sci.* **4,** 689 (1949).
79. U.S. Pat. 2,481,155 (Sept. 6, 1949), F. C. Schaefer (to American Cyanamid Co.).
80. U.S. Pat. 2,534,307 (Dec. 19, 1950), L. Shechter and J. M. Whelan, Jr. (to Union Carbide and Carbon Corp.).
81. U.S. Pat. 2,527,495 (Oct. 24, 1950), A. F. Fitzhugh (to Shawinigan Resins Corp.).
82. E. F. Degering and T. Stoudt *J. Polym. Sci.* **7,** 653 (1951).
83. S. Hünig, *Justus Liebigs Ann. Chem.* **569,** 198 (1950).
84. K. H. Klaassens and J. H. Gisolf, *J. Polym. Sci.* **10,** 149 (1953).
85. L. Canonica and co-workers *Atti Accad. Naz. Lincei Cl. Sci. Fis., Mat. Nat. Rend.* **14,** 814.
86. Brit. Pat. 704,854 (March 3, 1954) E. C. Craven and W. H. Gell (to Distillers Co. Ltd.).
87. G. Durr, *Compt. Rend.* **235,** 1314 (1952).
88. N. Hasebe, *Bull. Yamagata Univ.* **1,** 57 (1950).
89. P. Mastagli, *Compt. Rend.* **242,** 1031 (1956).
90. R. Nodzu and co-workers, *J. Chem. Soc. Japan, Ind. Chem. Sect.* **57,** 914 (1954).
91. H. H. Schlubach, V., Franzen, and E. Dahl, *Justus Liebigs Ann. Chem.* **587,** 124 (1954).
92. Swed. Pat. 124,737 (April 26, 1949), H. F. A. Topsoe and A. Nielson (to Stockholms Superfosfat Fabrik Aktiebolog).
93. Brit. Pat. 630,904 (Oct. 24, 1949) (to Usines de Melle).
94. Brit. Pat. 660,972 (Nov. 14, 1951), Badische Anilin- & Soda-Fabrik (I. G. Farbenindustrie A.G.).
95. Brit. Pat. 677,624 (Aug. 20, 1952), D. J. Hadley, R. Heap, and D. I. H. Jacobs (to Distillers Co. Ltd.).
96. U.S. Pat. 2,489,608 (Nov. 29, 1949), L. Alkeritière (to Usines de Melle, Société Anonyme).
97. U.S. Pat. 2,639,295 (May 19, 1953), H. J. Hagemeyer, Jr. (to Eastman Kodak Co.).
98. W. Langenbeck and R. Grochalski, *Z. Phys. Chem. Leipzig* **197,** 191 (1951).
99. S. M. McElvain, E. R. Degginger, and J. D. Behun, *J. Am. Chem. Soc.* **76,** 5736 (1954).
100. M. Ya. Kagan and co-workers *Bull. Acad. Sci. USSR Classe Sci. Chim.* 173 (1947).
101. M. S. Malinovskii and S. N. Baranov, *Sb. Statei Obshch. Shim., Akad. Nauk USSR* **2,** 1674 (1953).
102. Brit. Pat. 701,550 (Dec. 30, 1953), Chemische Werke Hulls G.m.b.H.).
103. J. Schmidt and co-workers, *Angew. Chem.* **71** (5), 176 (1959).
104. A. A. Dobrinskaya, M. B. Neiman, and N. K. Rudnevskii, *Zh. Fiz. Khim.* **27,** 1622–1630.
105. A. A. Dobrinskaya and M. B. Neiman, *Dokl. Akad. Nauk SSSR* **58,** 1969–1972.
106. Brit. Pat. 688,033 (Feb. 25, 1953), D. J. Hadley, R. Heap, and D. I. H. Jacobs (to Distillers Co. Ltd.).
107. U.S. Pat. 2,369,182 (Feb. 13, 1945), F. F. Rust (to Shell Development Co.).
108. U.S. Pat. 2,383,711 (Aug. 28, 1945), A. Clark and R. S. Shett (to Battelle Memorial Institute).
109. J. Lichtenberger and R. Lichtenberger, *Bull Soc. Chim. Fr.* 1002 (1948).
110. U.S. Pat. 2,620,357 (Dec. 2, 1952), E. Arundale and H. O. Moltern (to Standard Oil Development Co.).
111. W. E. Bissinger and co-workers *J. Am. Chem. Soc.* **69,** 2955 (1947).
112. N. O. Brace, *J. Am. Chem. Soc.* **77,** 4157 (1955).
113. C. Prevost, *Bull. Soc. Chim. Fr.* **11,** 218 (1944).
114. A. Vallette, *Justus Liebigs Ann. Chim.* **312,** 644 (1947).
115. C. L. Wilson, *J. Am. Chem. Soc.* **69,** 3002 (1947).
116. E. Späth, R. Lorenz, and E. Freund, *Chem. Ber.* **77B,** 354 (1944).
117. P. D. Bartlett and S. D. Ross, *J. Am. Chem. Soc.* **70,** 926 (1948).
118. M. S. Kharasch, B. M. Kuderna, and W. Urry, *J. Org. Chem.* **13,** 895 (1948).

119. A. Krattiger, *Bull. Soc. Chim. Fr.*, 222 (1953).
120. S. Olsen, *Acta Chem. Scand.* **4,** 901 (1950).
121. Ger. Pat. 844,441 (July 21, 1952), W. O. Herrmann and W. Hoenek (to Rudolf Decker & Hellmuth Holz).
122. M. Gouge, *Compt. Rend.* **225,** 637 (1947).
123. P. Grammaticakis, *Compt. Rend.* **226,** 189 (1948)
124. P. Grammaticakis, *Bull. Soc. Chim. Fr.*, 410 (1949).
125. *Ibid.*, 504 (1950).
126. *Ibid.*, 690 (1950).
127. E. C. Horning and M. G. Horning, *J. Org. Chem.* **11,** 95 (1946).
128. G. Matthiessen, *Arch. Pharm.* **284,** 62 (1951).
129. D. F. Meigh, *Nature* **170,** 579 (1952).
130. A. E. Gillam and D. G. Moss, *J. Chem. Soc.*, 1387 (1947).
131. J. L. McVeigh and J. D. Rose, *J. Chem. Soc.*, 713 (1945).
132. C. Capitani and E. Milani, *Chim. Ind. Milan* **37,** 177 (1955).
133. M. Yano and Matsumoto, *J. Chem. Soc. Japan Ind. Chem. Sect.* **55,** 498 (1952).
134. J. Kubias and S. Pilný, *Chem. Listy* **47,** 672 (1953).
135. E. Sjöstrom, *Acta Chem. Scand.* **7,** 1392 (1953).
136. R. I. Veksler, *Tr. Kom. Anal. Khim., Akad. Nauk SSSR* **3,** 369 (1951).
137. J. Mitchell Jr. and D. M. Smith, *Anal. Chem.* **22,** 746 (1950).
138. C. D. Willits and co-workers, *Anal. Chem.* **24,** 785 (1952).
139. B. Wurtzschmitt, *Z. Anal. Chem.* **128,** 549 (1948).
140. Ger. Pat. 2,427,088 (Dec. 19, 1974), E. Lashley (to Union Carbide Corp.).
141. E. Skog, *Acta Pharmacol. Toxicol.* **6,** 299 (1950).
142. U.S. Pat. 2,690,627 (Oct. 5, 1954), V. L. Nymon.
143. J. C. McGowan, P. W. Brian, and H. G. Hemming, *Ann. Appl. Biol.* **35,** 25 (1948).
144. S. Akabori and Y. Yamamura, *Proc. Jpn. Acad.* **26**(6), 37 (1950).
145. Ger. Pat. 2,203,712 (Aug. 16, 1973), H. Fernholz, H. Schmidt, and F. Wunder (to Farbwerke Hoechst A.-G.).
146. J. R. McPhee and M. Lipson, *Aust. J. Chem.* **7,** 387 (1954).
147. Fr. Pat. 982,926 (June 18, 1961) (to Soc. Anon. des Manufactures des Glaces et Produits Chimiques de Saint Gobain, Chauny et Cirey).
148. Belg. Pat. 451,209 (July, 1943) (to Knoll A. G. Chemische Fabriken).
149. Brit. Pat. 1,079,909 (Aug. 16, 1967), M. Jordan (to Schenectady Chemicals, Inc.).
150. U.S. Pat. 2,616,887 (Nov. 4, 1952), M. H. Danzig and L. F. Marous (to U.S. Rubber Co.).
151. U.S. Pat. 2,576,370 (Nov. 27, 1951), P.O. Tawney (to United States Rubber Co.).
152. U.S. Pat. 2,441,470 (May 11, 1948), T. S. Carswell (to Monsanto Chemical Co.).
153. U.S. Pat, 2,425,320 (Aug. 12, 1947), W. H. Hill (to Koppers Co., Inc.).
154. M. Matsumoto and H. Iwasaki, *Chem. High Polymers Tokyo* **7,** 402 (1950).
155. B. Takigawa, *J. Chem. Soc. Japan Ind. Chem. Sect.* **56,** 632 (1953).
156. Jap. Pat. 70-33, 417 (Oct. 27, 1970), T. Isojima (to Mitsubishi Rayon Co., Ltd.).
157. Ger. Pat. 1,963,994 (July 1, 1971), M. Meister and co-workers (to Farbenfabriken Boyer A.G.).
158. S. E. Vasyukov and co-workers, *Eksp. Khir. Anesteziol.* **14**(2), 51 (1969); *Chem. Abstr.* **71,** 62044s (1969).
159. T. Bass and co-workers *Pathol. Biol.* **22**(7), 593 (1974).
160. Fr. Pat. 961,575 (May 15, 1950), (to Vitex, S.A.).
161. U.S. Pat. 2,716,083 (Aug. 23, 1955), E. E. Tallis (to Courtaulds, Ltd.).
162. F. Stather and H. Nebe, *Gesammelte Abh. Dtsch. Lederinst. Freiberg* **7,** 34 (1951).
163. Y. Obata and T. Yamanishi, *J. Agr. Chem. Soc. Japan* **24,** 226 (1951).
164. Brit. Pat. 1,156,120 (June 25, 1964) (to R. J. Reynolds Tobacco Co.).
165. H. Nozaki and H. Hamada, *Bull. Natl. Inst. Agr. Sci. Japan Ser. G* **4,** 41 (1952).
166. U.S. Pat, 2,649,355 (Aug. 18, 1953), T. White and J. R. B. Hastings (to Forestal Land, Timber and Railways Company, Ltd.).
167. H. Yamaguchi and K. Aoki, *Konishiroku Rebyu* **7,** 1 (1956).
168. U.S. Pats. 2,660,591 and 2,660,592 (Nov. 24, 1953), G. Calingaert (to Ethyl Corporation).
169. Brit. Pat. 611,215 (Oct. 27, 1948) (to Ciba Ltd.).
170. Swiss Pat. 238,944 (Nov. 16, 1945) (to J. R. Geigy, A.G.).
171. Swiss Pat. 240,350 (May 1, 1946) (to J. R. Geigy, A.G.).

172. Swiss Pat. 243,597 (Jan. 16, 1947) (to Ciba Ltd.).
173. Swiss Pat. 246,668 (Oct. 16, 1947) (to Ciba Ltd.).
174. Swiss Pat. 249,001 (March 1, 1948) (to Ciba Ltd.) (addition to Swiss Pat. 243,597; see ref. 172).
175. Swiss Pat. 252,068 (Sept. 1, 1948) (to Ciba Ltd.).
176. F. Blikstad, *Tidsskr. Kjemi Bergves. Metall.* **9,** 168 (1949).
177. U.S. Pat. 2,415,039 (Jan. 28, 1947), J. B. Rust (to Montclair Research Corp.).
178. Ger. Pat. 1,938,634 (Feb. 11, 1971), T. Ploetz and E. Solich (to Leldmuehle A.-G.).
179. B. V. Travis and co-workers *J. Econ. Entomol.* **42,** 686 (1949).
180. U.S. Pat 2,690,988 (Oct. 5, 1954), R. H. Jones and G. E. Lukes (to Stauffer Chemical Co.).
181. W. Beilfuss and co-workers, *Zentralbl. Bakteriol. Parasitenkd. Infektionskr. Hyg. Abt. 1 Orig. Reihe A,* **234**(2), 271 (1976); *Chem. Abstr.* **84,** 160056k (1976).
182. L. Egyud, *Curr. Mod. Biol.* 1(1), 14 (1967).
183. U.S. Pat. 3,766,063 (Oct. 16, 1973), C. Blankenhorn and T. Telmann (to Shell Oil Co.).
184. Jap. Pat. 73-86,740 (Nov. 15, 1973), K. Aotani, K. Ishii, and J. Sakate.
185. N. I. Podobaer and A. G. Voskresenskii, *Zh. Prikl. Khim. Leningrad,* **43,** 834 (1970).
186. V. Zadorozhnyi, *Izv. Varonezh. Gos. Pedagog. Inst.* **94,** 12 (1969); *Chem. Abstr.* **75,** 14076q (1971).
187. Ger Pat. 2,238,800 (Feb. 21, 1974), E. Seifert (to Kolloidtechnik G.m.b.H.).
188. Ger. Pat. 2,062,032 (July 15, 1971), H. Ammano, H. Iwano, and K. Shirasu (to Fuji Photo Film Co., Ltd.).

WILLIAM F. BAXTER, JR.
Eastman Chemical Products, Inc.

CROTONIC ACID

Crotonic acid [3724-65-0] (*trans*-2-butenoic acid), CH₃CH=CHCOOH, is a white crystalline solid. It is used mainly in the preparation of synthetic resins, fungicides, surface coatings, plasticizers, and pharmaceuticals.

Crotonic acid was first prepared in 1863, but was not commercially available in North America until 1936.

Crotonic and isocrotonic acids are *cis–trans* stereoisomers:

(*trans*)-crotonic acid
[107-93-7]

(*cis*)-isocrotonic acid
[503-64-0]

Physical Properties

Crotonic acid crystals are monoclinic needles or prisms; the vapor pressure of the solid is 24 Pa (0.18 mm Hg) at 20°C and 880 Pa (6.6 mm Hg) at 70°C. Other physical

Table 1. Physical Properties of Crotonic and Isocrotonic Acids

Properties	Crotonic acid	Isocrotonic acid
mp, °C	71.4–71.7	15.5
bp, °C at 101.3 kPa[a]	184.7	169
at 1.7 kPa[a]	81	
at 2.0 kPa[a]		74
density, g/mL		
solid dens	1.018	
liquid dens	0.9648	
d_4^{20}		1.0265
n_D^t	1.4228 at 80°C	1.4456 at 20°C
heat of combustion, MJ/mol[b]	2.00	2.03
specific heat, J/g[b]		
(20–63°C)	3.03	
(80–95°C)	2.07	
latent heat of fusion, J/g[b]	150.7	
ionization content, K_a, at 25°C	2.0×10^{-5}	3.5×10^{-5}

[a] To convert kPa to mm Hg, multiply by 7.5.
[b] To convert J to cal, divide by 4.184.

properties are shown in Table 1. Solubility in water is shown below:

temp, °C	0	10	20	25	30	40	42
soly, g/100 water	4.15	5.46	7.61	9.4	12.2	65.6	126

At 25°C the solubility by wt of crotonic acid in ethyl alcohol is 52.5%; in toluene, 37.5%; and in acetone, 53.0%. The constant-boiling mixture with 96.86% water boils at 99.7°C. The eutectic mixture with isocrotonic acid contains 30% crotonic acid and melts at −3°C.

Isocrotonic acid (cis-2-butenoic acid) is miscible with water at 25°C. It is converted into the more stable form, crotonic acid, by warming with 60% sulfuric acid or a trace of hydrochloric acid or hydrogen bromide. In water or carbon disulfide solution the isomerization is catalyzed by traces of bromine or iodine in sunlight. Hydrogenation with palladium catalyst gives butyric acid. Isocrotonic acid may be prepared by the partial hydrogenation of tetrolic acid, $CH_3C{\equiv}CCOOH$, with palladium catalyst, and by the reduction of β-chloroisocrotonic acid [6213-90-7] with sodium amalgam. Although the sodium salt of isocrotonic acid is soluble in 4–5 parts of absolute ethyl alcohol, sodium crotonate [17342-77-7] is practically insoluble (soluble in 380 parts); this offers a method of separating the acids. Conversion to their amides has also been employed.

Chemical Properties

Crotonic acid contains the conjugated system —C=C—C=O, which influences the velocity and direction of many of its reactions (1–2). Esterification is much slower than for butyric acid and crotonic esters undergo hydrolysis and alcoholysis more slowly than the corresponding butyric esters. Ethyl crotonate is called a vinylog of ethyl acetate: the ability of ethyl acetate to condense with diethyl oxalate is relayed along the conjugated system:

$$C_2H_5OOCCOOC_2H_5 + CH_3CH{=}CHCOOC_2H_5 \rightarrow C_2H_5OOCCOCH_2CH{=}CHCOOC_2H_5 + C_2H_5OH$$

Isomerization and Polymerization. Crotonic acid in toluene solution has been converted into isocrotonic acid (about 4%) by uv radiation. When either acid is heated at 110–125°C for five to six months, the equilibrium mixture contains, in addition to both acids, a high-boiling liquid of twice their formula weight having one free carboxyl group and one double bond (3).

Oxidation. Concentrated nitric acid oxidizes crotonic acid to acetic and oxalic acids; chromic acid gives acetic acid and acetaldehyde; fusion with potassium hydroxide gives two moles of acetate. Ozonization yields 95% acetic and 5% formic acid. Crotonic acid is destructively oxidized by electrolysis; only 1.3% of the current consumed forms allenes. Crotonic acid is oxidized by the liver to acetoacetic acid. Crotonic acid vapor with excess air passed over vanadium pentoxide on aluminum or pumice at 400–500°C forms maleic anhydride. Peroxybenzoic acid oxidizes crotonic acid to DL-*erythro*-α,β-dihydroxybutyric acid [759-06-8], mp 81.5°C, and isocrotonic acid to the DL-*threo* form [5057-93-2], mp 74–75°C.

Silver chlorate in the presence of osmium tetroxide oxidizes crotonic acid to the threo form and isocrotonic to the erythro form (4).

Addition. Crotonic acid is hydrogenated to butyric acid in the presence of platinum, palladium, or nickel catalysts. Fluorine forms two isomeric α,β-difluorobutyric acids [62939-45-9], [62939-46-9], mp 81°C and bp$_{1.7 \text{ kPa (13 mm Hg)}}$ 110–112°C. Chlorine forms α,β-dichlorobutyric acid (crotonic acid dichloride [25620-56-8]), mp 63°C, which by partial isomerization with concentrated hydrochloric acid in a sealed tube at 100°C forms the higher-melting isomer, isocrotonic acid dichloride [26708-33-8], mp 78°C. Bromine addition is catalyzed by traces of water or hydrogen bromide. The addition of iodine is not quantitative. Hydrobromic acid is added to give β-bromobutyric acid [80-58-0], regardless of the solvent or presence of peroxides. In the presence of boiling 20% sulfuric acid or of 1–2N perchloric acid at 90–110°C, crotonic acid is hydrated in about 80% yield, forming an equilibrium mixture with β-hydroxybutyric acid [565-70-8] (5). Small quantities of β-chlorobutyric acid [4170-24-5] are formed, in addition to β-hydroxybutyric acid, when crotonic acid is warmed with 20% hydrochloric acid. Hypochlorous acid gives 74% α-chloro-β-hydroxy- [26708-34-9] and 26% β-chloro-α-hydroxybutyric acid [4250-23-1]. Ammonia under suitable conditions gives β-aminobutyric acid [541-48-0]; aniline and other amines can be added similarly. Crotonic acid heated with butadiene for 3 h at 150–170°C forms Δ^4-tetrahydro-*o*-toluic acid [10479-42-2], mp 68°C. The unsaturated carbon atoms in crotonic acid can also bridge the 9,10 positions in anthracene.

Manufacture

Crotonic acid is formed among numerous other compounds in the dry distillation of wood, and small quantities may be obtained from crude wood tars. It has been detected in a patch of infertile soil in Texas, and has been prepared from β-hydroxybutyric acid and its polymerization and dehydration products formed in the metabolism of several species of bacteria.

Crotonic acid is prepared commercially by the oxidation of crotonaldehyde (qv) with air or oxygen (6).

$$2 \text{ CH}_3\text{CH}=\text{CHCHO} + \text{O}_2 \rightarrow 2 \text{ CH}_3\text{CH}=\text{CHCOOH} + 536 \text{ kJ/mol (128 kcal/mol)}$$

It is believed that percrotonic acid [5813-77-4] is an intermediate product.

$$CH_3CH{=}CHCHO + O_2 \rightarrow CH_3CH{=}CHCO(O)_2H$$

$$\underset{\text{percrotonic acid}}{CH_3CH{=}CHCO(O)_2H} + \underset{\text{crotonaldehyde}}{CH_3CH{=}CHCHO} \rightarrow \underset{\text{crotonic acid}}{2\ CH_3CH{=}CHCOOH}$$

Acetic and formic acids are formed to the extent of a few percent by destructive oxidation. At higher temperatures some isocrotonic acid is also formed. Controlled oxidation in the laboratory with alkaline silver oxide at 15–20°C gives only the *trans* isomer (7).

The continuous process developed by the former Shawinigan Chemical Ltd. (now Gulf Oil Canada, Ltd.) uses an inert organic diluent such as benzene, methyl acetate, acetone, or methyl ethyl ketone in about equal volume with "wet" crotonaldehyde (about 10% water). The catalyst is a copper acetate–cobalt acetate mixture, ca 0.5–1 wt % of the total charge, at 35–50°C (8). Crotonaldehyde, diluent, and catalyst solution in the desired proportions are pumped into the oxidation kettle, and air is bubbled in, as the crude product is drawn off continuously. The diluent and unreacted crotonaldehyde are distilled and returned to the process; the crude crotonic acid is fractionated under reduced pressure and may be purified further by crystallization from water or other solvents (9).

Other catalysts and conditions for the oxidation of crotonaldehyde to crotonic acid have been proposed (10–11). Batch processes have been used in the Federal Republic of Germany (12–14).

In the laboratory, crotonic acid may be prepared in about 85% yield by condensing acetaldehyde in dry ether solution with malonic acid in the presence of pyridine (15). Other possible methods include the ready isomerization of vinylacetic acid, $CH_2{=}CHCH_2COOH$, by warming with 50% sulfuric acid.

Economic Aspects

It is estimated that 1977 usage was <2300 metric tons. The volume price in 1977 was $1.90/kg. Crotonic acid may be shipped by truck or railroad and is normally packaged in polyethylene-lined fiber drums.

Analytical Methods and Specifications

Crotonic acid may be isolated and identified as *S*-benzylthiouronium crotonate [62939-47-0], mp 162°C (16). In the absence of other unsaturated compounds the acid may be determined by quantitative bromination with an excess of 0.05 N bromate–bromide solution (5).

The acyl group in esters may be identified by refluxing the ester with benzylamine in the presence of 0.1 g ammonium chloride; crotonic esters form *N*-benzylcrotonamide [51944-67-3], mp 112.5–113.6°C (17). A typical specification for technical-grade crotonic acid is as follows:

appearance	free from foreign particles
solution color, APHA	25 ppm, max
melting point (dried sample)	70.0°C, min
assay (dried sample)	98.5%, min
water	1.0%, max
carbonyl, as C=O	0.5%, max

Health and Safety Factors

Crotonic acid (1% in the diet) and its butyl ester [7299-91-4] (0.5 mL/100 g food) failed to show any toxic effects in rats over a period of 41 d (18). The single-dose oral toxicity to rats, skin irritation, and eye injury from solutions or liquid have been evaluated for the acid and its ethyl ester (19). The acid should not be permitted to come in contact with the skin or eyes.

Crotonic acid is similar in strength to acrylic acid (qv) in its disinfecting action. It has a harmful effect on the growth of plants and delays the germination of seeds.

Uses

Crotonic acid, from ca 2.5% to equimolar proportions, forms copolymers with vinyl acetate; the 5% and 10% crotonic acid copolymers (Mowilith CT 5 and 10) were used in the German hat industry under the name Appretan H in preference to shellac (qv) (20). The vinyl acetate–crotonic acid copolymer is used as a hot-melt adhesive, especially in the book-bindery trade (21). These pressure–heat sensitive polymers are also used as tape coatings on natural and synthetic substrates (22–23).

Strippable wallpaper coatings may be made by using emulsions of crotonic acid–vinyl acetate copolymers [25609-89-6] (24); this type of adhesive is also used in paper–plywood bondings.

Crotonic acid copolymers are used extensively in the production and treatment of paper products (see Paper). From the initial flocculation (25), binder resins (26), and sizing resins to pigment coating compositions (27) for paper, crotonic acid copolymers are versatile components in today's paper industry. Epoxy resins (qv) prepared with crotonic acid are used in printing inks (28–29) as color intensifiers and binders (30). Further paper-related uses include components in film developers and electrostatic copying fluids (31).

With isobutylamine crotonic acid forms interpolymerization products suitable for sizing and wood sealing; with vinyl methyl ketone it forms products suitable for molding. When used as a modifier for drying oils, crotonic acid improves the gloss of surface coatings. Crotonic acid–olefin copolymers can be condensed with dimethylaminoethylamine or tetraethylenepentamine to prepare an ashless detergent useful in mineral lubricating oils (32).

4,6-Dinitro-2-(1-methylheptyl)phenyl crotonate [6119-92-2] is sold under the trade name of Karathane as a fungicide (33). Such crotonic acid-based fungicides are useful in treatment of powdery mildew on peaches (34), cucumbers (35), and apples (36). Additional fungicides based on crotonic acid control spider mites (37), regulate plant growth (38) (see Plant growth substances), control mildew on flowers (39), and aid in biological control of fruit pests as a component in insecticides (40). Insect attractants can also be prepared with crotonic acids that are highly specific for one species (41).

Electromagnetic steel can be treated with crotonic acid copolymers to reduce punching and improve corrosion resistance (42) or improve adhesion of polyolefins (43).

Water soluble resins can be formed by neutralizing crotonic acid–vinyl acetate copolymers. Such resins have a wide range of applications, but are used primarily for hairsprays (44–45), wave lotions (46), and cosmetic bases (47) (see Resins, water soluble).

In the textile industry, durable press resins for cotton fabrics can be prepared with crotonic acid derivatives (48). Sizes for polyester–cotton yarns used in weaving fabric in water jet looms can be prepared from crotonic acid compounds (49) and increase adhesion in later fabrication (50). Compounds prepared from polycarboxylic acid based on crotonic and suitable ether and hydroxy compounds are useful in imparting soil-release and antistatic properties to textile yarns (51). Special bleaches containing crotonic acid can be used on fibers of protein–acrylonitrile graft copolymers (52) (see Textiles).

The water solubility of crotonic acid copolymers makes them suitable in many pharmaceutical and health-related areas including suture coatings (53), tablet binders (54–55), and encapsulative coatings for tablets (56–57). Blood anticoagulants based on crotonic acid aid in cardiovascular surgery (58) (see Blood, coagulants and anticoagulants).

By selective copolymerization crotonic acid polymers can be used as binders for inorganic fiber such as rock wool and asbestos (qv) (59), in reverse osmosis (qv) membranes (60), and as flavoring agents for coffee (qv) (61).

Crotonic acid copolymers can be prepared by radiolysis-induced solid state polymerization to form unique polymers (62–63).

Crotonic acid is the starting material in the preparation of β- and α,β-substituted butyric acids and other derivatives, such as the amino acid DL-threonine (64).

Cellulose crotonates and cellulose acetate crotonates form clear, strong, flexible films suitable for coatings. Esters of crotonic acid with glycerol [26894-51-9], pentaerythritol [62939-44-7], and numerous complex alcohols are plasticizers for cellulose esters and synthetic resins. The amyl [25415-76-3], geranyl [56172-46-4], 2-methylallyl [27819-09-6], and 2-ethylhexyl esters are suitable for use in perfumes; the benzyl [65416-24-2], cyanomethyl [51977-58-3], and trimethylene glycol [69331-32-4] esters are insecticides (see Insect control technology). Crotonyl peroxide (see Peroxides) is used as a catalyst for the polymerization of vinyl- and vinylidene halides (10,65) (see Vinyl polymers).

Derivatives

Crotonamide [625-37-6]. Crotonamide (*trans*-2-butenamide), CH$_3$CH=CHCONH$_2$, forms white needles; mp 159–160°C (sublimes readily); 2.8% soluble in water, 0.085% in ethyl ether, and 0.084% in benzene at room temperature, more soluble in ethyl alcohol. Hydrolysis with dilute sodium hydroxide or dilute sulfuric acid gives crotonic acid. Crotonamide in acetone solution has been converted into its stereoisomer, isocrotonamide [31110-30-2], mp 102°C, to the extent of about 41% by uv radiation. Crotonamide is prepared from crotonic acid or its chloride and ammonia.

Crotonanilide [17645-30-6]. Crotonanilide (*trans*-2-butenanilide CH$_3$CH=CHCONHC$_6$H$_5$, crystallizes as needles from water and as prisms from dilute alcohol; mp 115–118°C; readily soluble in benzene, chloroform, ethyl alcohol, ethyl ether, and hot water; almost insoluble in water at room temperature. It is prepared from crotonyl chloride or crotonic anhydride and aniline. The *N*-methyl [7363-92-0], *N*-ethyl [56604-80-9], *N*-*n*-butyl [56604-84-3], and *N*-2-chloroallyl [69331-33-5] derivatives are insecticides.

Crotonic Anhydride [623-68-7]. Crotonic anhydride (*trans*-2-butenoic anhydride) (CH$_3$CH=CHCO)$_2$O, is a colorless liquid with a characteristic odor: bp 247°C, bp$_{1.6\text{ kPa (12 mm Hg)}}$ 114°C; d$_4^{20}$ 1.0397; n$_D^{20}$ 1.4745; infinitely soluble in ethyl ether; reacts with water and ethyl alcohol. It may be prepared by heating crotonyl chloride with sodium crotonate (80% yield) and also from crotonic acid with acetic anhydride or ketene. It is useful in the preparation of cellulose crotonate and other derivatives.

Crotonyl Chloride [625-35-4]. Crotonyl chloride CH$_3$CH=CHCOCl is a colorless liquid with a suffocating odor: bp$_{2.4\text{ kPa (18 mm Hg)}}$ 34–36°C; bp 124–125°C; d$_4^{20}$ 1.0905; n$_D^{18}$ 1.460. It is prepared from crotonic acid and chlorine-containing compounds, such as phosphorus trichloride or pentachloride, benzoyl chloride, or thionyl chloride. It is a reactive compound, useful in the preparation of esters and other derivatives of crotonic acid.

Esters. The esters of crotonic acid shown in Table 2 are all colorless liquids with pleasant odors.

Methyl crotonate is miscible with ethyl alcohol and ethyl ether but practically insoluble in water. The 20% methyl crotonate–80% vinyl acetate copolymer is a clear solid resin of high softening point, soluble in organic solvents such as benzene and methyl chloride. It is used in the stiffening of hats.

Ethyl crotonate, bp$_{2.3\text{ kPa (17 mm Hg)}}$ 45°C, is used as an organic intermediate and has been suggested as a solvent for cellulose esters and as a plasticizer for acrylic resins. The addition product with conjugated methyl linoleate is compatible with synthetic resins (66).

n-Butyl crotonate, bp 179.5°C, bp$_{3\text{ kPa (22 mm Hg)}}$ 79.0°C, is soluble in ethyl alcohol and ethyl ether but only slightly soluble in water.

Vinyl crotonate, bp$_{1.3\text{ kPa (9.5 mm Hg)}}$ 25°C; soly at 20°C in water, 4.3 wt %; soly of water in vinyl crotonate, 0.66 wt %; forms an azeotrope with water, bp 91°C, containing 75.8 wt % vinyl crotonate. It is prepared from crotonic acid and acetylene or vinyl acetate. It gives improved properties when used as a modifier or cross-linking agent in polystyrenes and other polymers.

Allyl crotonate, bp$_{9.3\text{ kPa (70 mm Hg)}}$ 88–89°C, modifies unsaturated alkyd-type resins to give improved properties for coatings and molded articles and may also be polymerized alone.

Table 2. Properties of Some Crotonic Acid Esters, CH$_3$CH=CHCOOR

Ester	CAS Registry No.	R	bp, °C	d$_4^{20}$	n$_D^{20}$
methyl	[18707-60-3]	—CH$_3$	119–120	0.9444	1.4242
ethyl	[10544-63-5]	—C$_2$H$_5$	136–138	0.91775	1.4245
n-propyl	[10352-87-1]	—CH$_2$CH$_2$CH$_3$	156–157	0.9059	1.4285
n-butyl	[7299-91-4]	—CH$_2$(CH$_2$)$_2$CH$_3$	178	0.8989	1.4325
vinyl	[3234-54-6]	—CH=CH$_2$	132.7$_{101\text{ kPa}}$[a]	0.9410	1.450
allyl	[50921-71-6]	—CH$_2$CH=CH$_2$	190–192$_{100\text{ kPa}}$[a]	0.9440	1.4465

2-ethylhexyl	[16931-00-3]	—CH$_2$CH(CH$_2$)$_3$CH$_3$
		              |
		           C$_2$H$_5$
			240	0.88	1.4438

[a] To convert kPa to mm Hg, multiply by 7.5.

BIBLIOGRAPHY

"Crotonic Acid" in *ECT* 1st ed., Vol. 4, pp. 613–619, by M. S. W. Small, Shawinigan Chemicals Limited; "Crotonic Acid" in *ECT* 2nd ed., Vol. 6, pp. 464–470, by William M. Gearhart, Eastman Chemical Products, Inc.

1. D. J. G. Ives, R. P. Linstead, and H. L. Riley, *J. Chem. Soc.*, 561 (1933); E. J. Boorman, R. P. Linstead, and H. N. Rydon, *J. Chem. Soc.*, 568 (1933); D. J. G. Ives, *J. Chem. Soc.*, 86 (1938).
2. A. I. Titov, *J. Gen. Chem. (USSR)* **16,** 1891 (1946).
3. E. L. Skau and B. Saxton, *J. Am. Chem. Soc.* **52,** 335 (1930).
4. G. Braun, *J. Am. Chem. Soc.* **51,** 228 (1929); **52,** 3185 (1930); **54,** 1133 (1932).
5. D. Pressman and H. J. Lucas, *J. Am. Chem. Soc.* **61,** 2271 (1939); **62,** 2078 (1940).
6. U.S. Pat. 2,945,058 (July 12, 1960), R. Watson and K. Finch (to Eastman Kodak Co.).
7. W. G. Young, *J. Am. Chem. Soc.* **54,** 2498 (1932).
8. Brit. Pat. 595,170 (Nov. 27, 1947), (to Shawinigan Chemicals Ltd.).
9. U.S. Pat. 2,413,235 (Dec. 24, 1946), D. J. Kennedy (to Shawinigan Chemicals Ltd.).
10. Brit. Pat. 612,346 (Nov. 11, 1948), (to British Celanese Ltd.).
11. U.S. Pat. 2,450,389 (Sept. 28, 1948), K. H. W. Tuerek (to Distillers Co., Ltd.).
12. R. L. Blackmore and co-workers, "Solvents and Plasticizers in Germany—Plasticizers Section," *BIOS, Final Rept. 1651,* 35 (1947).
13. W. Hunter, "German Acetylene Chemical Industry—Miscellaneous Organic Chemicals," *BIOS, Final Rept. 1053,* 120 (1947).
14. M. A. Matthews, "Manufacture of Crotonaldehyde and Crotonic Acid at I. G. Hoechst," *BIOS, Final Rept. 758,* (1946).
15. R. A. Letch and R. P. Linstead, *J. Chem. Soc.*, 454 (1932).
16. J. J. Donleavy, *J. Am. Chem. Soc.* **58,** 1004 (1936).
17. O. C. Dermer and J. King, *J. Org. Chem.* **8,** 168 (1943).
18. L. P. Dugal, *Report of Research Department on Acclimatization,* Univ. Laval, Quebec, July 16, 1947.
19. H. F. Smyth, Jr., and C. P. Carpenter, *J. Ind. Hyg. Toxicol.* **26,** 269 (1944).
20. S. J. Baum and R. D. Dunlop, "Polymerization of Vinyl Acetate," *Field Information Agency, Tech. Final Rept. 1102,* 19, 45 (1947); W. Baird, C. B. Brown, and G. R. Perdue, "Textile Auxiliary Products of I. G. Farbenindustrie: Application, Testing and Miscellaneous Information," *Brit. Intelligence Objectives Subcommittee (BIOS), Final Rept. 518,* 57 (1946); W. Starck, *U.S. Dept. Comm. Office Tech. Services Rept. PB-33051* (1939); J. Nerz and co-workers, *U.S. Dept. Comm. Office Tech. Services Rept. PBL-70174,* Frames 9665–9681 (1948); "Production and Properties of Polymers from Vinyl Acetate and Crotonic Acid," *U.S. Dept. Comm. Office Tech. Services Rept. PB-70248,* Frames 2401–2411 (1939).
21. Can. Pat. 984,084 (Feb. 17, 1976), W. N. Martin (to Consolidated Bathrust, Ltd.).
22. Jpn. Pat. 76-06.235 (Jan. 19, 1976), H. S. Horiki and K. Ito (to Nagoya Yukagaku Kogyo K.K.).
23. Ger. Pat. 2,302,473 (July 26, 1973), S. Ishihra and S. Kabayashi (to Showa Denko K.K.).
24. Jpn. Pat. 73-73,443 (Oct. 3, 1973), K. Azumaya.
25. U.S. Pat. 3,776,892 (Dec. 4, 1973), M. Bleyle (to W. R. Grace and Company).
26. Jpn. Pat. 72-42140 (Oct. 24, 1972), M. Fukushima and T. Ishibashi (to Nippon Synthetic Company, Ltd.).
27. Jpn. Pat. 74-21,424 (July 19, 1972), J. Fujyia, M. Ohtake, and S. Kato (to Denki Kagaku Kogyo K.K.).
28. Jpn. Pat. 70-93,229 (Oct. 23, 1972), T. Kurita, J. Ishiwata, and K. Futaki (to Mitsubishi Paper Mills, Ltd.).
29. Jpn. Pat. 73-55,280 (Aug. 3, 1973), T. Nishikubo, S. Ugai, and T. Ishijo (to Japan Oil Seal Industry Company, Ltd.).
30. W. Kiser, *Defazen-Dtsch. Farben Z.* **28,** 381 (1974).
31. U.S. Pat. 3,753,760 (Aug. 21, 1973), G. Kosel (to Phillip A. Hunt).
32. U.S. Pat. 3,483,125 (Dec. 9, 1969), T. Clough (to Sinclair Research, Inc.).
33. E. H. Fisher, ed., *Entoma,* 13th ed., The Entomological Society of America, Madison, Wis., 1960, p. 46.
34. G. Lorenzini, *Not. Mal. Piante* **90–91,** 331 (1974).
35. Jpn. Pat. 75-135,218 (Oct. 27, 1975), A. Kojima and co-workers (to Mitsui Toatsu Chemicals, Inc.).
36. D. Fieldgate, *Pesticide Science* **1**(5), 183 (1970).

37. W. Karg, *Nachr. Pflanzenschutzdienst DDR* **27**(8), 169 (1973).
38. U.S. Pat. 3,773,824 (June 14, 1971), J. Strong (to Mobil Oil Corporation).
39. M. Prutenfkaya, *Introd. Aklim. Rosl. UKR* **7**, 174 (1976).
40. R. C. Moore, *N. Haven Bull.*, 751 (1975).
41. U.S. Pat. 3,790,666 (Feb. 5, 1974), G. Eddy and co-workers (to USDA).
42. Jpn. Pat. 73-08,700 (Aug. 13, 1969), M. Kitsyma, M. Nakamura, and H. Okada (to Nippon Steel Corporation).
43. U.S. Pat. 3,466,207 (Sept. 9, 1969), G. Vincent and F. Saunders (to Dow Chemical Company).
44. Ger. Pat. 1,617,696 (July 26, 1973), G. Kalopissis (to Oreal S.A.).
45. Ger. Pat. 2,513,807 (Oct. 2, 1975), C. Papantoniou and J. Grognet (to Oreal S.A.).
46. Ger. Pat. 2,330,956 (Jan. 10, 1974), C. Papantoniou (to Oreal S.A.).
47. Span. Pat. 388,559 (Mar. 1, 1974), S. A. Pulcra.
48. W. E. Franklin, *Text. Chem. Color.* **8**(4), 63 (1976).
49. Jpn. Pat. 75-154,591 (Dec. 12, 1975), F. Suzaki and T. Hayakana (to Denki Kagaku Kogio K.K.).
50. Jpn. Pat. 74-30,239 (Aug. 10, 1974), K. Konishi and S. Toyooka (to Denki Kagaku Kogio K.K.).
51. Jpn. Pat. 72-45,639 (Nov. 17, 1972), H. Suzusho and K. Nohmi (to Kanebo Company, Ltd.).
52. Jpn. Pat. 73-03,356 (Jan. 30, 1973), K. Kakinuma and K. Nose (to Toyobo Company, Ltd.).
53. Ger. Pat. 2,600,174 (July 8, 1976), F. Mattei (to Ethicon, Inc.).
54. S. Niazi, R. El-Rashidy, and F. El-Khwas, *Drug. Dev. Commun.* **2**, 241 (1976).
55. F. El-Kahawas, M. Abd-el-Khalek, and R. El-Rashidy, *Pharm. Ind.* **38**, 648 (1976).
56. Ger. Pat. 2,364,104 (June 27, 1974), M. Trichot (to Société des Produits Chimiques de la Montagne Noire).
57. O. Langues and co-workers, *Ann. Pharm. Fr.* **33**, 235 (1975).
58. S. E. Vasyukov and co-workers, *Eksp. Khir. Anesteziol* **14**(2), 51 (1969).
59. Jpn. Pat. 74-435, (Apr. 18, 1974), F. Masuda and co-workers (to Denki Kagaku Kogoyo K.K.).
60. C. W. Saltonstall, *U.S. Office Salene Water, Research, Development Progress Review Report*, No. 434, 1969, pp. 154.
61. U.S. Pat. 3,978,241 (Aug. 31, 1976), M. Winter and co-workers (to Frimenich S.A.).
62. V. Dikin and V. Karpov, *Spektrosk. Polim. S. B. Pokl. Vses. Simp.*, 132 (1965).
63. I. Kietsu, A. Ito, and K. Hayashi, *J. Appl. Polym. Sci.* **17**, 3221 (1973).
64. K. Pfister, III and co-workers, *J. Am. Chem. Soc.* **71**, 1096 (1949).
65. Brit. Pat. 594,717 (Nov. 18, 1947), C. A. Brighton and J. J. P. Staundinger (to Distillers Co., Ltd.).
66. H. M. Teeter, C. R. Scholfield, and J. C. Cowan, *Oil Soap Egypt* **23**, 216 (1946).

WILLIAM F. BAXTER, JR.
Eastman Chemical Products, Inc.

CROWN ETHERS. See Catalysis, phase-transfer; Chelating agents.

CRYOGENICS

Cryogenics is the branch of physics that relates to the attainment and effects of very low temperatures (1–2). Cryogenic technology has contributed greatly to scientific research and has also achieved wide industrial use. Its economic benefits include: (*1*) the ability to store and ship large quantities of gases in the dense cryogenic liquid state rather than at high pressure; (*2*) the ability to produce low cost, high purity gases through fractional condensation and distillation; and (*3*) the ability to utilize low temperatures to refrigerate other materials or to alter their physical properties.

Currently, it is common practice to liquefy natural gas, helium, oxygen, hydrogen, and nitrogen for economical shipping and storage. Gases obtained at relatively low cost by cryogenic separation of air are used extensively in the electronics industry as high purity inerting gases (N_2, Ar), in steel industry BOF (Bessemer Oxygen Furnace) processes, and in municipal wastewater treatment (see Water) to increase processing capacity (O_2). Liquid nitrogen is used as a refrigerant in the food industry for efficient fast food freezing as well as for sustained food refrigeration during transporting; in several manufacturing processes in which molded rubber or metallic parts are frozen to facilitate deflashing; and for critical cryobiological applications for preservation of such materials as whole blood and semen. Table 1 gives cryogenic properties of several gases. Table 2 lists the 1977 U.S. production of the more common industrial gases.

Small cryogenic refrigerators provide low temperatures for enhancing the sensitivity of ir detectors required for spectroscopy and missile guidance systems (see Infrared technology). Other ultralow temperature refrigerators have recently been applied industrially for cryopumping to yield high pumping speeds and clean ultrahigh vacuum. The use of ultralow temperatures in superconductivity applications has great potential for electric power transmission, magnetic transportation systems, and magnets for generation of energy in fusion processes (see Superconducting materials; Magnetic materials).

Refrigeration Methods

Refrigeration for cryogenic temperatures is produced within a system by absorbing or extracting heat at low temperature levels and rejecting it at higher temperature levels to the surroundings (see Refrigeration). Three general methods of producing low temperature refrigeration for commercial applications are: (*1*) liquid vaporization; (*2*) Joule-Thomson (J-T) cycle; and (*3*) expansion in an engine with the performance of work. The first two are similar in that they both involve expansion of a fluid through an orifice or porous plug at constant enthalpy (heat content), and are both irreversible.

Liquid Vaporization. For this process, a fluid with the desired physical properties is compressed to the pressure at which it can be condensed by heat exchange with cooling water, ambient air, or another refrigerant that boils at a higher temperature. The pressure of the condensed liquid phase is reduced (isenthalpic expansion), resulting in formation of a flash vapor fraction which reduces the temperature and enthalpy of the remaining liquid. This liquid is then vaporized by heat exchange, thereby cooling the load stream. The vapor phase is recompressed to complete the cycle.

Table 1. Cryogenic Properties of Gases

	He	Ne	Ar	Kr	Xe	H$_2$	CH$_4$	NH$_3$	N$_2$	O$_2$	F$_2$
density (0°C, 101.3 kPa^a) kg/m^3 ^{b}	0.1784	0.9002	1.783	3.748	5.895	0.0899	0.718	0.770	1.251	1.429	1.70
boiling point, °C	−268.9	−245.9	−185.78	−151.8	−109.1	−252.8	−161.5	−33.35	−195.8	−183.0	−187
melting point, °C	(−272.2)	−248.7	−189.3	−169	−140	−259.2	−182.6	−77.7	−209.9	−218.4	−223
vapor density at bp, kg/m^3 ^{b}	16.0	9.50	5.89	8.30	9.71	1.33	1.80	0.891	4.61	4.74	
liquid density at bp, kg/m^3 ^{b}	125.0	1200	1390	2400	3100	70.0	424.0	682.1	804.0	1142	1512
vapor pressure (solid at mp), kPa^a		43.06	68.79	73.19	81.59	7.2	9.33	6.03	12.85	0.266	0.016
heat of vaporization at bp, J/g^c	23.96	87.0	163	108	96.3	452.2	577.8	1369	199	213	171
heat of fusion at mp, J/g^c	4.2	16.7	28.1	16.3	13.7	58.6	60.7	351.7	25.6	13.7	13.5
Cp (15°C, 101.3 kPa^a, J/(g·K)^d	5.23	Ca 1.05	0.523	Ca 0.25	Ca 0.17	14.2	2.21	2.19	1.04	0.913	0.754
Cp/Cv (15–20°C, 101.3 kPa^a (dimensionless)	1.66	1.64	1.67	1.68	1.66	1.41	1.31	1.31	1.40	1.40	
critical temperature, °C	−267.9	−228.7	−122.5	−63.7	16.6	−239.9	−82.5	132.4	−147.1	−118.4	−129.0
critical pressure, MPa^e	0.229	2.72	4.86	5.50	5.90	1.30	4.63	11.30	3.39	5.04	5.57

^{a} 101.3 kPa = 1 atm.
^{b} To convert kg/m^3 to lb/ft^3, multiply by 0.0624.
^{c} To convert J/g to Btu/lb, multiply by 0.429.
^{d} To convert J/(g·K) to Btu/(lb.°F), multiply by 0.239.
^{e} To convert MPa to atm, divide by 0.101.

Table 2. U.S. Production of Industrial Gases for 1977 [a]

	10^6 m^3/yr at STP	Percentage produced as liquid
argon	155	100
helium	18.4	70–75
hydrogen	2208	8–10
nitrogen	9148	28
oxygen	9463	18

[a] Ref. 3.

Joule-Thomson Cycle. This process makes use of the temperature reduction of a fluid upon pressure reduction (isenthalpic expansion) through an orifice. Joule-Thomson cooling must start with the compressed fluid precooled below its inversion temperature, ie, the temperature above which warming rather than cooling of the fluid occurs upon pressure reduction [ie, temperature increases following pressure reduction] at constant enthalpy. Lower temperatures are obtained with this cycle by incorporating a heat exchanger to cool the compressed, precooled fluid further with its returning expanded flow.

Expansion Engine Cycle. Expansion of a fluid that performs work approaches isentropic (reversible) expansion with a resultant reduction in fluid enthalpy and temperature. Both centrifugal and reciprocating expansion engines are used in present day cryogenic plants, but the trend has been toward development of a wider range of centrifugal machines because of their lower cost and greater reliability. Reciprocating expanders can handle relatively low volumetric flow rates and high expansion pressure ratios; centrifugal machines handle relatively large volumetric flow rates at low expansion ratios.

Applications

Air Separation. The practical application of cryogenics began with air separation. Today air separation is a very mature and highly competitive technology. The large capacity (up to 75,600 kg/h) air separation plants produce oxygen, nitrogen, argon, (approximately 21, 78, and 1% of air, respectively), and if desired, also recover the rare gases neon, krypton, and xenon present in trace amounts.

Figure 1 indicates the basic equipment and flow arrangement for a present day large tonnage air separation plant that uses a combination of expander and Joule-Thomson refrigeration (4). The air is compressed to a nominal 700 kPa (ca 100 psi) pressure in the compressor; the main exchanger precools the air against effluent product streams and extracts carbon dioxide and moisture through condensation and solidification; the expander provides the process refrigeration; high-pressure and low-pressure columns separate the oxygen and nitrogen primary components and concentrate the argon; and the crude argon column further separates the argon from oxygen.

The main switching heat exchanger is the heart of the basic air separation plant. As the air is cooled in the main exchanger, moisture and CO_2 are condensed and solidified. If not removed, they would eventually plug the circuit. On a regular schedule, every 10 to 15 min, the air passages and the waste nitrogen vent passages are switched so that the air flows through clean passages and begins to build up a deposit of moisture

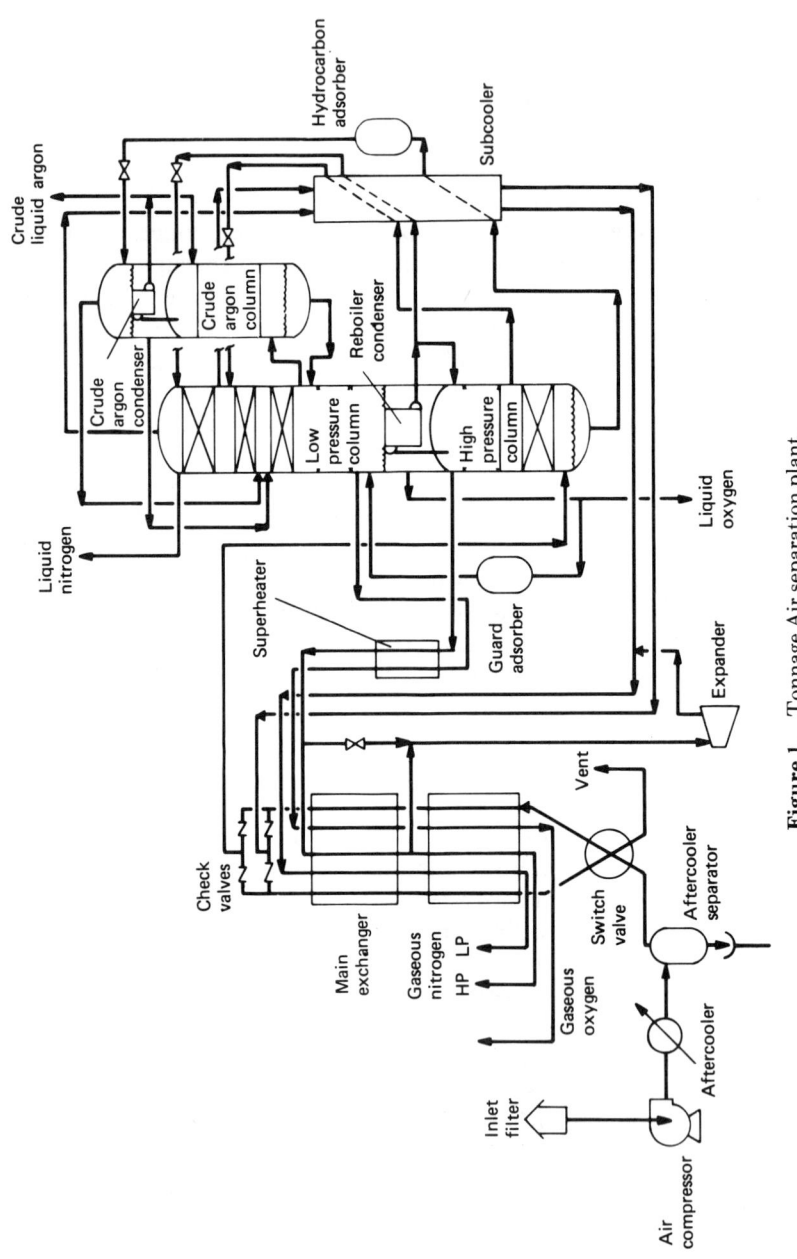

Figure 1. Tonnage Air separation plant.

and CO_2. At the same time the returning waste nitrogen revaporizes the deposited water and CO_2, thereby cleaning the passageways in preparation for the next switch. Since most of the products of the air separation plant are gases at low pressure, only a relatively small quantity of refrigeration is required to overcome heat leak to the equipment and to compensate for heat loss caused by the temperature difference between the incoming air and the outlet products.

The double distillation column of an air separation plant efficiently separates oxygen and nitrogen, resulting in high component recovery from the air. Since the products of air separation are near atmospheric pressure, the low-pressure column and reboiler operate only slightly above this pressure. The pressure of the high-pressure column is set at a level sufficient to condense pure nitrogen in the column overhead condenser by heat transfer with the boiling oxygen in the sump of the low pressure column. A small fraction of the air can be obtained as liquid products by increasing the refrigerating capacity of the expander and withdrawing the liquids from the appropriate column locations. Efficient production of liquid products in tonnage quantities requires the addition of an efficient refrigerant system which most commonly uses nitrogen in an expander cycle (see Oxygen; Nitrogen).

Liquefied Natural Gas. The two primary reasons for liquefying natural gas are for "peakshaving" and for "baseload supply." In the United States, liquefaction of pipeline natural gas during the low demand seasons is referred to as peakshaving. It allows for the production and pipeline transport of a higher yearly quantity by storing natural gas in liquid form at the use points. The liquid is vaporized and injected into the pipeline distribution system during peak demand periods. The advantage of peakshaving is that higher yearly production rates can be achieved without expanding the pipeline and compression facilities required to transport the natural gas from production points to more distant use locations. The liquefaction of natural gas at its source in foreign countries for shipment in tankers is referred to as baseload liquefied natural gas (LNG) since the supply rates are large enough to constitute a steady base supply rate.

The refrigerant cycles used for peakshaving plants are: (*1*) cascade (J-T expansion); (*2*) expansion engine; and (*3*) multicomponent refrigerant. Cascade (stepwise) refrigeration (Fig. 2) uses three liquid vaporization systems. Each has a different working fluid, typically propane, ethylene, and methane. The propane system condenses the refrigerant in the ethylene system, which in turn condenses the refrigerant in the methane system. In addition, each system is internally staged by vaporizing the refrigerant in three fractions that are returned to the compressor as separate streams at progressively lower pressure levels. This improves operating efficiency by reducing the pressure ratio of compression for the fractions which return at higher pressures and thus reduce compression power requirements.

The conventional expansion engine cycle provides refrigeration by circulating nitrogen, which is typically compressed to 4.1 MPa (ca 600 psi) and precooled in a main heat exchanger against the returning expanded streams. Part of the precooled compressed nitrogen is then drawn off from the main exchanger and directed to a centrifugal expansion engine. There the pressure and temperature are reduced. The expanded nitrogen portion flows back from the extreme cold end of the main exchanger to give up its refrigeration to the high pressure cooling nitrogen stream. The remaining unexpanded high-pressure nitrogen continues to flow to the cold end of the main exchanger and then to a J-T valve where its pressure is reduced to near atmospheric.

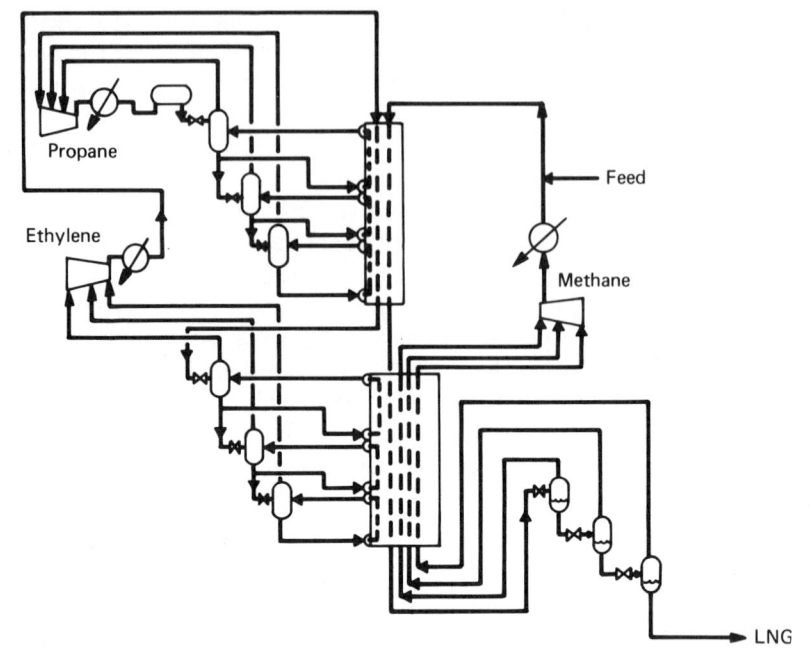

Figure 2. Cascade process for liquefied natural gas.

Natural gas enters the warm end of the main exchanger through parallel passages to be cooled and liquefied to LNG against the returning nitrogen refrigerant.

One variation of the expansion engine cycle uses a relatively large quantity of high pressure pipeline gas, normally reduced for distribution. Instead of reducing the bulk gas through a reducing station, this flow-by natural gas is precooled in a cryogenic cycle and expanded in an expansion engine to refrigerate and liquefy a small quantity of the same gas, which is then sent to LNG storage. Using this flow-by natural gas to supply the refrigeration for liquefaction substantially reduces the capital investment and energy consumption for the facilities.

The most recent development for natural gas liquefaction mixes the refrigerants that are used for the three separate cascade refrigeration circuits, thereby supplying refrigeration through a single circulating refrigerant stream. Nitrogen is added as a fourth component to lower further the refrigerating temperature, thus reducing or eliminating the vapors flashed from the LNG following pressure reduction into LNG storage. This cycle substantially simplifies the equipment arrangement.

One of the more common mixed refrigerant systems consists of a compressor that raises the pressure of the mixed refrigerant to a nominal 2.75 MPa (ca 400 psi), a condenser that partially condenses the mixed refrigerant against cooling water, and a main heat exchanger that precools and liquefies the natural gas and precools, liquefies, and subcools the mixed refrigerant. Following the expansion of the high-pressure mixed refrigerant through J-T valves to a nominal 0.28 MPa (ca 40 psi), the mixed refrigerant is warmed in the main exchanger in counterflow with the natural gas and high-pressure mixed refrigerant to provide refrigeration for the cycle. The mixed refrigerant then returns to the compressor to complete the cycle. One variation of the mixed refrigerant cycle utilizes a precooling propane system (Fig. 3) to precool

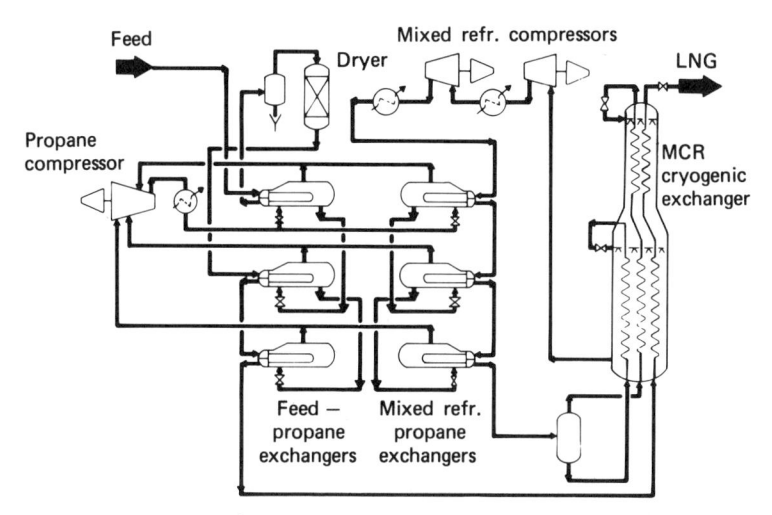

Figure 3. Propane precooled mixed refrigerant process for natural gas liquefaction.

the natural gas and mixed refrigerant. This cycle has the low power characteristics of the cascade cycle, but a simpler equipment arrangement. It is particularly suitable for baseload LNG plants (see Gas, natural; Hydrocarbons, C_1–C_6).

Hydrogen Purification. Perhaps the simplest industrial application of cryogenics is hydrogen purification, which is important in the production of benzene (qv) by the hydrodealkylation of toluene (qv) or hydrotreating pyrolysis gasoline (5–7) (see Hydrogen). Methane, which builds up in the recycle reaction, is the primary inert impurity in the hydrogen. A purge stream is purified to recover the hydrogen from a slipstream of the benzene production recycle loop and to reject the methane, which is used for fuel. This prevents the buildup of methane which would significantly reduce the reaction rate. This cryogenic purification process upgrades the hydrogen purity of the purge stream to 70–90% at a nominal pressure of 4.1 MPa (ca 600 psi) (Fig. 4). The purge feed gas is cooled and partially condensed in a main cryogenic heat exchanger against the returning purified hydrogen and methane streams (J-T cycle). Liquid and vapor phases leaving the cold end of the cryogenic exchanger are separated in a phase separator. The hydrogen vapor flowing back through the cryogenic exchanger is rewarmed to near ambient temperature and returned to the reaction loop. The liquid is reduced in pressure through a Joule-Thomson valve and returned in parallel with the hydrogen through the main cryogenic exchanger to provide the net refrigeration required for the purification process. This is one of the simplest cryogenic processes in that no mechanical machinery is required for compression or expansion, and the process is self-sufficient with regard to refrigeration. An aluminum plate-and-fin heat exchanger is commonly used (see Heat exchange technology).

Ammonia Purge Gas Recovery. The synthesis of ammonia (qv) by combining nitrogen and hydrogen in a catalytic synthesis recycle loop requires a continual vent purge to prevent accumulation of inert components that enter the system as a part of the hydrogen production process. Natural gas and air, both of which contain the inert argon component, are used for the production of the hydrogen. Because full conversion of natural gas is not possible, methane as well as argon builds up in the synthesis loop. The purge gas normally contains approximately 4% argon, 7% methane,

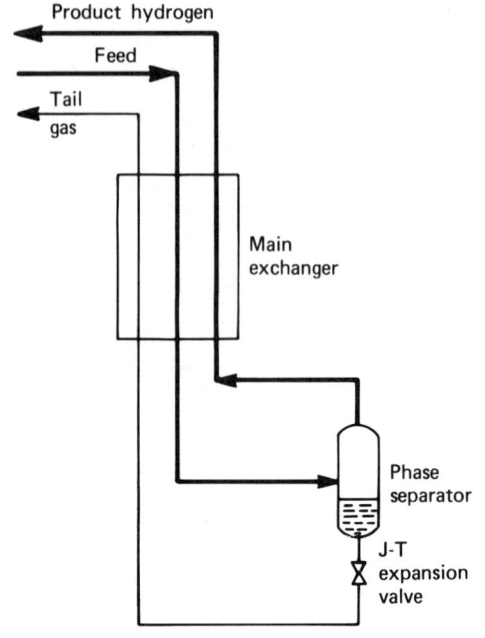

Figure 4. Hydrogen purification cycle.

and 2% ammonia. Increased energy costs have provided greater economic incentive for complete recovery of the hydrogen, methane, and ammonia contained in the purge stream. First, the ammonia is adsorbed at a pressure of 17.2–20.7 MPa (2500–3000 psi), typically in a molecular sieve adsorber bed (see Adsorptive separation; Molecular sieves). The purge stream is then cooled and partially condensed using a recycle refrigeration system (8). The two phases are separated at approximately −187°C, yielding a vapor rich in hydrogen and a liquid containing nitrogen, argon, and methane components. The liquid is reduced in pressure and flashed to a lower temperature. It is then warmed in parallel with the hydrogen in heat exchange with the incoming purge feed. Hydrogen is recycled to the ammonia synthesis loop, and the stream that contains the methane can be used for fuel in a natural gas preheater or an auxiliary steam superheater.

Helium Recovery. In the early 1960s, the U.S. Government instituted a helium conservation program to recover helium from the natural gas being produced from the Kansas–Oklahoma–Texas panhandle fields. The several recovery plants built typically processed 12×10^6 m^3/d of STP natural gas containing slightly more than 0.4% helium (9). Since helium has a much lower boiling point than natural gas and, consequently, a higher relative volatility, distillation was not required for the separation process. The cryogenic process used staged flash steps incorporated into two flash columns (Fig. 5). The feed gas from the natural gas pipeline was almost completely condensed at a pressure of approximately 3.6 MPa (525 psi) in the main cryogenic heat exchangers. These very large exchangers, fabricated of 9.5 mm OD coil wound aluminum tubing, efficiently processed the very large quantities of natural gas feed. The first main flash column provides for the initial enrichment of the helium from the natural gas feed by reducing the pressure and separating the flashed vapor

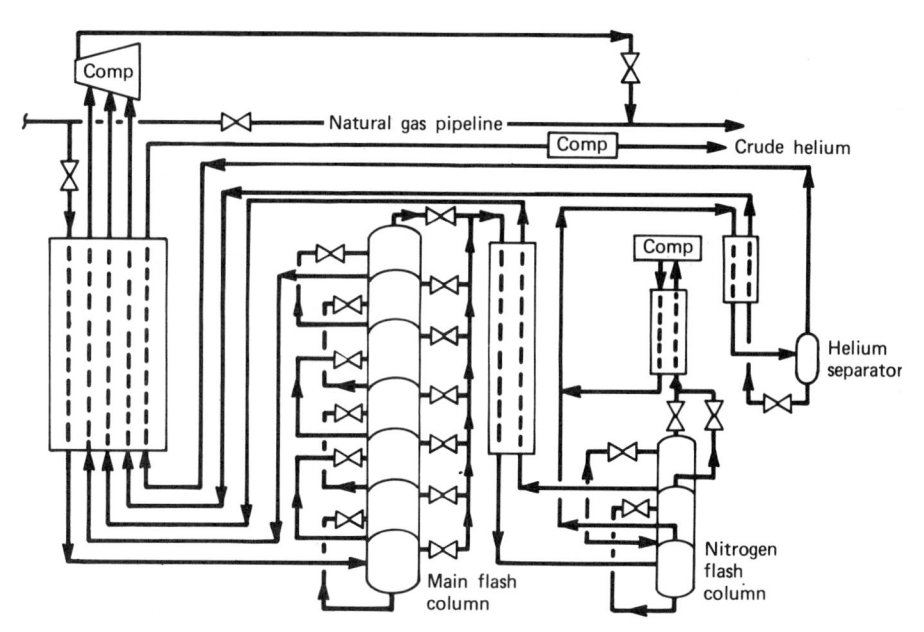

Figure 5. Helium recovery process.

phases in several steps. The several enriched vapor streams are combined, further cooled and partially condensed, then flashed again in the second-stage flash column. The vapors from the first flash stage are further cooled, condensed, and the two phases separated. The vapor phase is rewarmed to become the crude helium product, which is compressed and sent to underground storage for future use. The flash vapors from the subsequent stages of the second column are collected and recycled through a compressor into the crude helium from the first stage. The liquid streams from the main flash column, second flash column, and the final separator are rewarmed against the cooling streams, thus providing the refrigeration for the process. These streams, after being rewarmed to near ambient temperature, are recompressed to the original natural gas pressure and reinjected into the natural gas pipeline (see Helium-group gases).

Ethane and Ethylene Recovery. Cryogenic cycles are commonly used to recover ethane or ethylene (qv) in conjunction with hydrogen from refinery off-gases. These cycles generally incorporate a distillation column and, depending upon the required conditions of the products, may incorporate a centrifugal expander (10). Distillation provides for the separation of the C_1–C_2 components. Hydrogen is flashed and separated from the liquid at cryogenic temperatures in a manner like that of the hydrogen purification cycle. The hydrogen product, C_{2+} product, and the methane reject streams return and are warmed against the incoming feed. The methane reject stream is commonly used for fuel at the plant site (see Hydrocarbons, C_1–C_6).

Nuclear Off-Gas Systems. A recent application of cryogenics is in minimizing the release to the environment of gaseous radionuclides from power reactors and nuclear fuel reprocessing plants (see Nuclear reactors). These radionuclides (and stable isotopes of the same species) are concentrated from the main off-gas stream and held long enough to permit substantial decay prior to their ultimate release. One of the first operating cryogenic systems has been used since 1959 at the Idaho Chemical Processing Facility of DOE to concentrate krypton, and in particular, 10.74-year ^{85}Kr (11–12). About 1×10^{14} disintegrations per second (several thousand curies) of ^{85}Kr are separated annually for tracer gas use. The off-gas, essentially air from fuel dissolution, is treated to remove nitrous oxide and other impurities, then compressed and cooled in switching packed regenerators. Primary (continuous) separation is made in a sieve tray column, with periodic transfer of the column bottoms to the packed secondary (batch) column. Primary column reflux is replaced by direct addition of liquid nitrogen to the top tray. The secondary column is operated to remove, successively, N_2, O_2, Kr, and Xe. The O_2 and Xe cuts containing Kr are collected for subsequent recycling. Product gases are warmed and compressed into shielded product shipping cylinders.

Cryogenic distillation appears to be the most feasible method of hydrogen isotope management where protium (H) removal and D_2 and DT concentration is required (13) as in fusion test facilities and reactors. The distillation system would use three or four columns with equilibrium at ambient temperature to permit necessary separations of H_2, HD, HT, D_2, DT, and T_2 (see Deuterium and tritium).

Helium Refrigeration and Liquefaction. Because of the growth in helium demand in the last 10 to 15 years and the development of effective multilayer insulation, transportation economics favor the movement of helium as liquid or cold supercritical gas rather than as a pressurized gas at ambient temperature. Present day helium liquefiers have a capacity of fifty to several hundred liters per hour and consume energy at the rate of 1000–2000 kW (14). A liquefier recently built for the National Accelerator Laboratory maintains an array of superconducting magnets at 4.5 K. Other helium refrigerators of 300 W to several kilowatts have been built (see Superconducting materials).

Designs are being considered for superconducting accelerators, generators, transmission lines, and electric storage devices. Several of these operate below the 2.1 K Lambda point for ^{4}He, making use of the unique properties of superfluid liquid ^{2}He. The number of possible combinations of thermodynamic cycles and equipment arrangements available for helium liquefaction and refrigeration cycles is large. In general, however, each requires compression at ambient temperature with heat rejection to air or cooling water, recuperative heat exchangers or regenerators, expansion devices such as reciprocating expanders, turboexpanders, and Joule-Thomson valves, and possibly liquid nitrogen for precooling, high efficiency thermal insulation, and vacuum to insulate the equipment. A typical cycle for helium liquefaction (Fig. 6) might consist of a 900-kW helium recycle compressor in combination with liquid nitrogen (LIN) refrigeration, two turboexpanders, and what is referred to as a wet reciprocating expander. The expander, a recent addition to the helium liquefaction cycle, replaces the cold J-T valve and provides additional refrigeration in the critical region. The expander reduces the pressure of the helium as a dense critical fluid, and this results in an increased production of 10 to 25% when incorporated in liquefaction plants.

Hydrogen Liquefaction. From a thermodynamic standpoint, the liquefaction of hydrogen and helium are comparable in that both utilize liquid nitrogen to provide refrigeration at the warmer temperature levels. Expansion engines supply refrigeration at the colder temperature levels. Both hydrogen and helium boil below $-240°$C. Consequently, the temperature ranges over which the components are cooled are comparable and attaining them requires large amounts of power per unit of production.

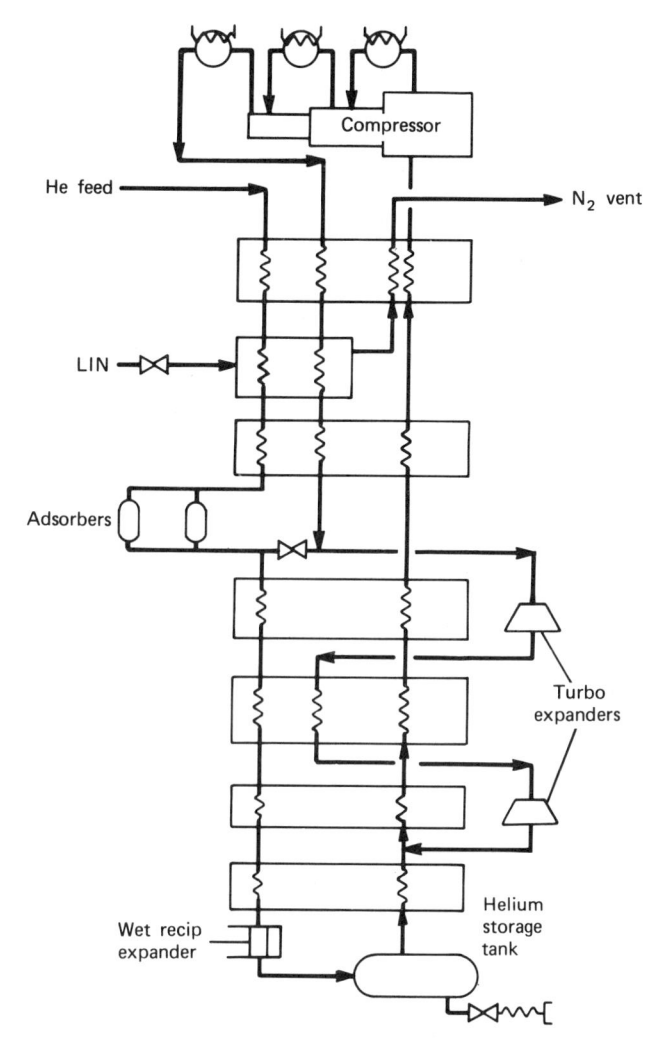

Figure 6. Helium liquefaction

The power requirement is strongly related to heat transfer efficiency, and these cycles are designed with very close temperature approaches at the lower levels, typically 1°C compared to 6°C at ambient temperature levels. Although this requires larger heat exchangers, it benefits cycle efficiency and product cost.

The two basic cycles for tonnage liquid hydrogen production are: (1) a low pressure cycle that incorporates centrifugal expanders; and (2) a high pressure cycle that utilizes reciprocating expanders with higher expansion ratios (15). Both cycles use liquid nitrogen and cold gaseous nitrogen refrigeration, but the low-pressure cycle (Fig. 7) utilizes a portion of its liquid nitrogen under vacuum to lower the refrigerating temperature of the LIN. This is not required of the high-pressure cycle since the expansion ratio and temperature range covered by the expander is larger.

An important feature of liquid hydrogen cycles is the catalytic conversion of the hydrogen molecule from the ortho to the para state. The equilibrium para hydrogen

238 CRYOGENICS

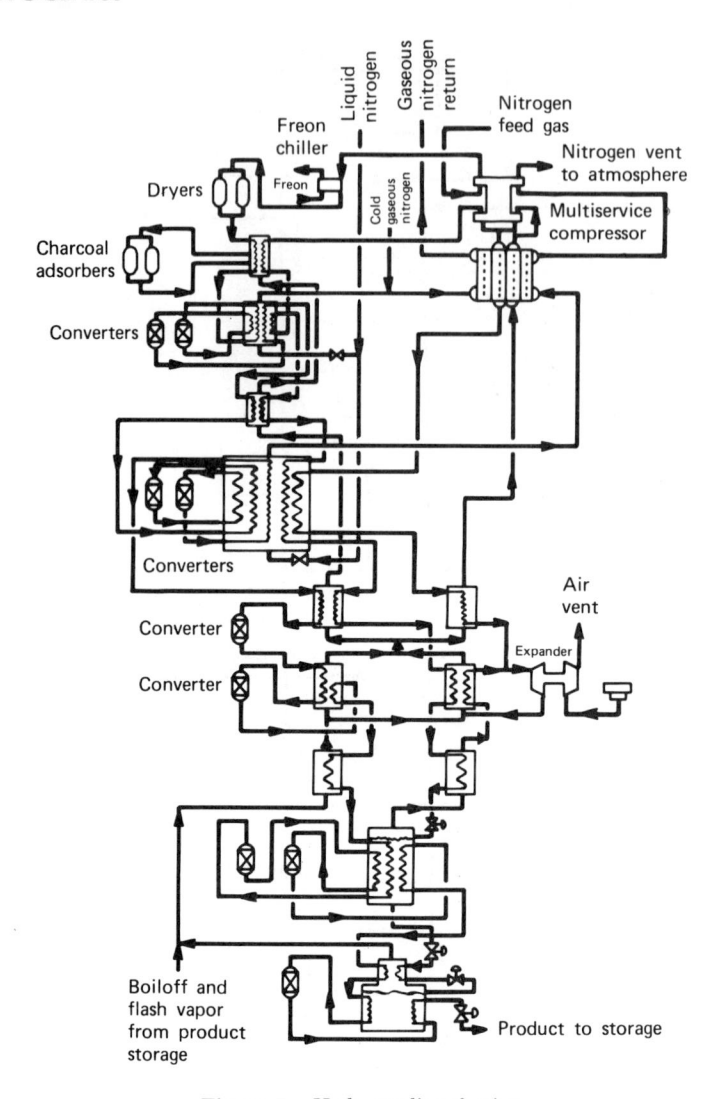

Figure 7. Hydrogen liquefaction.

content at ambient temperature is 25% whereas at liquid hydrogen temperature the equilibrium concentration is near 100%. This exothermic reaction, if allowed to take place in storage, would cause significant liquid product losses. The catalytic conversion steps in the liquefaction process are normally staged to maximize the heat of conversion at the highest temperature levels for the cycle in order to minimize the overall cycle power requirements.

Small Cryogenic Refrigerators. Since the mid 1950s a major application for small cryogenic refrigerators has been in cooling infrared detectors (see Infrared technology). Sensors that are sensitive to thermal radiation from objects at normal ambient temperatures must be cooled below 100 K to eliminate their own thermal "noise." The sensors are typically mounted in evacuated and silvered Dewar flasks, which have a cooling requirement of less than 1 watt. This application more than any other initiated the development of small cryogenic refrigerators that has occurred during the past 25 years.

The Phillips Company design used a small regenerator heat exchanger and piston expander operating on the Stirling cycle. In this cycle the compressor piston is coupled on the same shaft as the expander piston, but operates 90° out of phase. Stirling cycle machines have the highest efficiencies of the small refrigerators, but they must operate at speeds in excess of 15 Hz (cps) in order to exchange kinetic (flywheel) and potential (compressed fluid) energy during a cycle. Gifford and McMahan developed a small cooler in 1958 that separated the compressor from the expander by means of a valve mechanism, thus allowing the expander to run much more slowly than the compressor. This approach based upon a general concept first described by Solvay in 1885, has the advantage of incorporating a conductive heat transfer means to the refrigeration heat load directly from the small expander head. It also provides for isolating compressor vibration and features high reliability since the expander can operate at a much lower speed than the compressor. The self-cleaning characteristics of a regenerator also make it possible to use an air conditioning type of oil-lubricated compressor.

Because of their relatively low cost and high reliability, units that operate on variations of the Solvay cycle are the most widely used small closed-cycle cryogenic refrigerators in research laboratories. Minimum temperature of about 10 K is set by limitation of thermal storage capacity of the regenerator materials. The most popular units produce about 2 W of refrigeration at 20 K and draw about 1.5 kW of power. Applications include spectroscopy, x-ray diffraction, detector cooling, condensing hydrogen for accelerator targets, cooling parametric amplifiers for satellite communication and radio astronomy, and cooling laser diodes for atmospheric analysis. Cryopumping has recently become a major industrial application because its high speed, ultrahigh vacuum capability, and clean vacuum characteristics make it ideal for the production of microelectronic components.

Infrared detectors are also commonly cooled by small J-T heat exchanger units operating from nitrogen or argon stored in small bottles at 19.3–39.3 MPa (2800–5700 psi). Units of this kind, which weigh a few kilograms and can cool down in a few seconds, are widely used to cool heat-seeking ir detectors in guided missiles. A nitrogen–hydrogen system was used in ir studies on the Mariner Mars fly-by in 1969 (see Analytical methods; Planetary exploration).

Cryobiology

The low temperature of liquid nitrogen has found many important applications in the biological sciences (16–17). Certain biological materials, ranging from single cells to more highly organized structures, can be cooled to extremely low temperatures without loss of life. In 1949, British researchers Polge, Smith, and Parkes, at the National Institute for Medical Research in London, discovered that glycerol protected sperm and red blood cells during freezing and thawing. The first commercial application of nitrogen for the preservation of living cells was in the storage and distribution of semen for dairy cattle breeding. Economics and advantages in the distribution of frozen semen are such that most artificial breeding of dairy cattle in the U.S. has been converted to this technique. Liquid nitrogen is also used in the cryogenic preservation of whole blood. The hope for painless hemorrhage-free surgery is closer to realization with recent developments in cryosurgery (18). The ice bond established between tissue and cryogenic instrument, eg, permits the safer removal of cataracts. The disruptive effect of the ice formation on cell structure can also be used for the destruction and removal of tumors and skin cancer.

Equipment

Heat Exchangers. The two most common types of heat exchangers in cryogenic applications are the spirally wound, tube-in-shell exchanger and the plate-fin exchanger. Typically these exchangers are designed to handle three or more streams in each unit. Since process efficiency is highly dependent on obtaining close temperature approaches between the warming and cooling streams, the physical properties of the fluids and the pressure drop correlations of the exchanger types must be known in order to determine accurately the complete duty requirements. These cryogenic heat exchangers are susceptible to the effects of flow maldistribution (19) and require special design features to ensure adequate distribution of each stream in its allocated passages.

The spirally wound exchanger consists of a mandrel which essentially is a pipe onto which tubes are spirally wound. As the mandrel is rotated in a lathe, it pulls a group of tubes to form a layer. Spacers are placed on each tube layer before the next layer is wound. This process is repeated until the desired tube bundle diameter is obtained. The ends of the tubes are expanded into tube sheets located beyond the ends of the bundle. The bundle is finally enclosed by a shell having the appropriate inlet and outlet nozzles to conduct the shellside streams.

The plate-fin exchanger is fabricated with stacked aluminum sheets, every other sheet of which is corrugated. The corrugations form the fins between the flat sheets. The stack is then brazed in a closely controlled salt bath to form a complete unit (see Heat-exchange technology).

Insulation. The insulation designs for cryogenic processing equipment, transfer piping, and storage are many and varied and cannot easily be categorized. The design should economically reduce the heat leak through the hardware to the point that its effect on the overall refrigeration (or equivalent power) requirement becomes minimal. The general types of insulation used are unevacuated bulk insulation, rigid foam, vacuum, vacuum-powder, and multilayer.

Enclosing the process equipment in a steel frame and sheet metal box filled with perlite or rock wool and purging the box with dry air or nitrogen is a common method of insulating the equipment for a H_2 purification, hydrocarbon recovery, air separation plant, or nitrogen refrigerator since all the equipment operates above the temperature at which the box purge would condense. Small refrigerators for which this scheme would be too bulky, commonly utilize evacuated enclosures. Solid urethane insulation on each individual piece of equipment and associated piping has been found economical for baseload LNG plants because of the large equipment size.

The most common design for transporting cryogenic liquids consists of an inner vessel enclosed by an outer vessel with the annulus containing powder insulation under a vacuum (vacuum-powder). For oxygen, nitrogen, and argon, the insulation may be perlite. For liquid hydrogen and helium, multilayer insulation is commonly used. For on-site storage of $>10^3$ m^3 (several hundred thousand gallons) or more requiring field erection, vacuum-powder insulation has been used for liquid hydrogen service and unevacuated powder insulation has been used for LNG.

For transfer of cryogenic fluid from producing equipment to storage to transport and to use point, rigid foam is used for large baseload LNG plants and receiving terminals where the flow rates are high and the cooldown vaporization losses are a major consideration; vacuum transfer lines are commonly used for liquid oxygen, nitrogen, argon, and hydrogen; and multilayer evacuated transfer lines are most often used for liquid hydrogen and helium service where heat leak must be reduced to a minimum (see Insulation, thermal).

Materials of Construction. The primary criteria in materials selection for cryogenic applications are: (1) ductility; (2) thermal conductivity; (3) coefficient of expansion; and (4) combustibility.

The loss of ductility or impact strength at reduced temperatures is usually the primary consideration for metals selection. Carbon steel, preferred because of its lower cost, loses most of its ductility below $-40°C$ and is subject to fracture from impact or vibration. The materials selection for pressure vessels, piping, and heat-exchanger surface, therefore, is generally restricted to aluminum, copper, 18-8 stainless steel, and 9% nickel steel below $-40°C$ service.

The thermal conductivity of a construction material is important to minimize heat leak from supporting structures at ambient temperature to the cryogens. For example, stainless steel, with its low thermal conductivity, is a suitable metal for the transitions between the cold inner and warm outer concentric pipes of a vacuum jacketed transfer line. Plastics have also been used as load-bearing thermal barriers because of their substantially lower thermal conductivities and high compressive strengths.

The relative expansion or contraction of the construction material needs to be determined and analyzed to avoid excessive stresses and resultant failures. For example, convoluted steel bellows have been used to compensate for the contraction of the inner pipe of a vacuum jacketed transfer line while the outer pipe remains at ambient temperature.

The combustibility of the construction materials is important, particularly in oxygen service. Plastic spacers, for example, even though isolated, might present a safety hazard from a line leak caused by excessive stressing, vibration, or external impact.

Safety

The potential for the uncontrolled release of cryogenic liquids must be carefully considered in the design of all facilities that process and handle them. The design must provide for containment of the liquids in the immediate storage area by means of dikes. Procedures to protect personnel from cryogenic burns and asphyxia and to protect the adjoining equipment from embrittlement failure must be established and followed. Trace quantities of impurities in the plant feed may concentrate during the processing and transport of the cryogen. When these impurities can result in the combination of an oxidant with a flammable cryogen (eg, solid oxygen in liquid hydrogen) or a combustible with an oxidant (eg, acetylene in liquid oxygen) special precautions must be taken to eliminate them.

BIBLIOGRAPHY

"Cryogenics" in *ECT* 1st ed., Suppl. 2, pp. 272–281, by H. R. Morrison, Linde Company, Division of Union Carbide Corporation; "Cryogenics" in *ECT* 2nd ed., Vol. 6, pp. 471–481, by H. R. Morrison, Linde Company, Division of Union Carbide Corporation.

1. K. D. Timmerhaus and co-workers, eds., *Advances in Cryogenic Engineering,* Vol. 1 (1955)–Vol. 24 (1978), Plenum Press, New York.
2. C. A. Bailey, *Advanced Cryogenics*, Plenum Press, New York, 1971.
3. "Industrial Gases," *Current Industrial Reports,* U.S. Dept. of Commerce (April, 1978).
4. D. J. Hersh and J. M. Abrardo, *Cryogenics* 383 (July, 1977).
5. S. Dunlap, R. Banks, *Hydrocarbon Processing* 147 (July 1977).
6. L. M. Lehman and E. H. Bausch, *CEP* 44 (Jan., 1978).
7. A. Haslom, *Hydrocarbon Process.* 101 (March, 1972).
8. A. Haslam, P. Brook, and H. Isalski, *Hydrocarbon Process.* 103 (Jan. 1976).
9. L. S. Gaumer, *Chem. Eng. Prog.* **63**(5), (1967).
10. *Oil Gas J.* 60 (July 18, 1977).
11. C. L. Bendixson and F. G. Offutt, "Rare Gas Recovery Facility at the Idaho Chemical Processing Plant," *USAEC Report IN-1221,* April, 1969.
12. C. L. Bendixson and F. O. German, "1974 Operation of the I.C.P.P. Rare Gas Recovery Facility," *US-ERDA Report ICP-1057,* March, 1975.
13. R. H. Sherman, J. R. Barlit, and R. A. Briesmeister, *Cryogenics* **16,** 611 (1976).
14. R. H. Kropschop, B. W. Birmingham, and D. B. Mann, eds., "Technology of Liquid Helium," *Nat. Bur. Stand. U.S. Monogr. No. 111,* Washington (1968).
15. C. L. Newton, *Chem. Process Eng.* 51 (Dec., 1967).
16. A. U. Smith, *Current Trends In Cryobiology,* Plenum Press, New York, 1970.
17. L. K. Lozina-Lozinskii, *Studies In Cryobiology,* John Wiley & Sons, Inc., New York, 1972.
18. R. W. Rand, A. P. Rinfret, and H. von Leden, *Cryosurgery,* Charles C Thomas, Springfield, Illinois, 1968.
19. R. F. Weimer and D. G. Hartzog, "Effects of Maldistribution on the Performance of Multi-Stream, Multi-Passage Heat Exchangers," *Thirteenth National Heat Transfer Conference,* AIChE-ASME, Denver, Colorado, Aug. 6–9, 1972.

C. L. NEWTON
Air Products and Chemicals, Inc.

CRYSTALLIZATION

Crystallization is one of the oldest of the chemical engineering unit operations and a major processing technique in the chemical industry. Vast quantities of crystalline substances are manufactured commercially. For example, sodium chloride, sodium and ammonium sulfates, and sucrose all have worldwide production rates in excess of 100 million metric tons per year. A high proportion of the products of the pharmaceutical and organic fine chemicals industries are crystalline. Many organic liquids are now purified on a large scale by crystallization, as an alternative to distillation, for the separation of azeotropes and close-boiling mixtures. As shown in Table 1, enthalpies of crystallization are generally much lower than enthalpies of vaporization and crystallization operations are usually carried out much nearer the ambient temperature than are distillation processes.

The freezing of water is an important industrial technique in the food industry for the freeze-concentration of fruit juices and other beverages. Freezing is also a potentially successful method for the desalination of sea water because the energy requirement for the crystallization of ice is about one seventh that for the vaporization of water. However, cooling energy is generally more costly than heating energy and the ice-brine separation can be expensive (see Water, supply and desalination).

Crystallization is acknowledged to be a very complex operation and some of the reasons for this complexity are fairly obvious. For example, the growth of crystals in a crystallizer involves the simultaneous processes of heat and mass transfer in a multiphase, multicomponent system. These conditions alone present complications enough, but the crystallization process is also strongly dependent on fluid and particle mechanics in a system where the size and size distribution of the particulate solids, neither property being capable of unique definition, can vary with time. Furthermore, the solution in which the solids are suspended is thermodynamically unstable, frequently fluctuating between so-called metastable and labile states, and sometimes entering the unsaturated condition, eg, in the vicinity of heat exchanger surfaces. Traces of impurity, sometimes a few parts per million, can profoundly affect the nucleation and crystal growth kinetics. In view of these complexities it is perhaps un-

Table 1. Crystallization and Distillation Energy Requirements

Substance	Melting point, °C	Enthalpy of crystallization, kJ/kg[a]	Boiling point, °C	Enthalpy of vaporization, kJ/kg[a]
o-cresol	31	115	191	410
m-cresol	12	117	203	423
p-cresol	35	110	202	435
o-xylene	−25	128	141	347
m-xylene	−48	109	139	343
p-xylene	13	161	138	340
o-nitrotoluene	−4.1	120	222	344
m-nitrotoluene	15.5	109	233	364
p-nitrotoluene	51.9	113	238	366
water	0	334	100	2260

[a] To convert kJ to kcal, divide by 4.184.

244 CRYSTALLIZATION

derstandable why crystallization has been slow to submit to simple analytical procedures.

Nevertheless, crystallizers can be designed and operated successfully despite the uncertainties surrounding the fundamental principles which lie in four main areas: solubility and phase relationships, hydrodynamics of crystal suspensions, nucleation characteristics, and crystal growth rates.

Saturation and Supersaturation

Solubility and Crystal Yield. Solubility may be expressed in many units but the most convenient, and the least liable to misinterpretation, is mass of solute per mass of solvent. For hydrated salts dissolved in water the solute concentration is best expressed in terms of the anhydrous species because no difficulties will then arise if several hydrates exist over the operating range of temperature. The conversion of solubility data from one set of units to another may be made through the equations listed in Table 2.

An estimate of the crystal yield for a simple cooling or evaporating crystallization may be made from a knowledge of the solubility characteristics of the solution. The general equation may be written:

$$Y = \frac{WR[c_o - c_f(1 - V)]}{1 - c_f(R - 1)} \tag{1}$$

where c_o = initial solution concentration, kg anhydrous salt/kg water; c_f = final solution concentration, kg anhydrous salt/kg water; W = initial mass of water, kg; V = water lost by evaporation, kg/kg of original water present; R = ratio of molecular weights of hydrated and anhydrous salts; and Y = crystal yield, kg.

In practice the actual yield may differ slightly from that calculated from equation 1. For example, if the crystals are washed with fresh solvent on the filter, losses may occur owing to dissolution. On the other hand, if mother liquor is retained by the crystals an extra quantity of crystalline material will be deposited on drying. It should be remembered that solubility data published in the literature usually refer to pure solvents and solutes. Since pure systems are rarely encountered industrially, it is generally advisable to check solubilities for the actual working liquors.

Before equation 1 can be applied to the case of vacuum (adiabatic cooling) crystallization, the quantity V must be estimated, eg, using equation 2:

$$V = \frac{qR(c_o - c_f) + C(t_o - t_f)(1 + c_o)[1 - c_f(R - 1)]}{\lambda[1 - c_f(R - 1)] - qRc_f} \tag{2}$$

where λ = latent heat of evaporation of solvent, J/kg; q = heat of crystallization of product, J/kg; t_o = initial temperature of solution, °C; t_f = final temperature of solution, °C; C = heat capacity of solution, J/(kg·K); and c_o and c_f have the same meaning as in equation 1.

Supersaturation. It is usually quite easy to prepare a supersaturated solution, ie, one containing more dissolved solute than that required for equilibrium saturation. For example, a clean solution allowed to cool slowly in a dustfree atmosphere can usually be supercooled many degrees below its normal crystallization point. Supersaturation plays a vital role in the crystallization process, and Ostwald (1897) and Miers (1906) suggested that two types of supersaturation could be recognized, namely, the

Table 2. Solution Concentration Conversion Factors[a]

Concentration	Equivalent expressions						
c_1		$\dfrac{c_2}{1-c_2}$	$\dfrac{c_3}{R-c_3}$	$\dfrac{c_4}{R+(R-1)c_4}$	$\dfrac{c_5}{\rho-c_5}$	$\dfrac{c_6}{\rho R-c_6}$	$\dfrac{M_A c_7}{\rho - M_A c_7}$
c_2	$\dfrac{c_1}{1+c_1}$		$\dfrac{c_3}{R}$	$\dfrac{c_4}{R(1+c_4)}$	$\dfrac{c_5}{\rho}$	$\dfrac{c_6}{\rho R}$	$\dfrac{M_A c_7}{\rho}$
c_3	$\dfrac{Rc_1}{1+c_1}$	Rc_2		$\dfrac{c_4}{1+c_4}$	$\dfrac{Rc_5}{\rho}$	$\dfrac{c_6}{\rho}$	$\dfrac{M_H c_7}{\rho}$
c_4	$\dfrac{Rc_1}{1-(R-1)c_1}$	$\dfrac{Rc_2}{1-Rc_2}$	$\dfrac{c_3}{1-c_3}$		$\dfrac{Rc_5}{\rho - Rc_5}$	$\dfrac{c_6}{\rho - c_6}$	$\dfrac{M_H c_7}{\rho - M_H c_7}$
c_5	$\dfrac{\rho c_1}{1+c_1}$	ρc_2	$\dfrac{\rho c_3}{R}$	$\dfrac{\rho c_4}{R(1+c_4)}$		$\dfrac{c_6}{R}$	$M_A c_7$
c_6	$\dfrac{\rho R c_1}{1+c_1}$	$\rho R c_2$	ρc_3	$\dfrac{\rho c_4}{1+c_4}$	Rc_5		$M_H c_7$
c_7	$\dfrac{\rho c_1}{M_A(1+c_1)}$	$\dfrac{\rho c_2}{M_A}$	$\dfrac{\rho c_3}{M_H}$	$\dfrac{\rho c_4}{M_H(1+c_4)}$	$\dfrac{c_5}{M_A}$	$\dfrac{c_6}{M_H}$	

[a] c_1 = kg of anhydrous substance/kg of water.
c_2 = g of anhydrous substance/kg of solution.
c_3 = kg of hydrate/kg of solution.
c_4 = kg of hydrate/kg of free water.
c_5 = kg of anhydrous substance/m³ of solution.
c_6 = kg of hydrate/m³ of solution.
c_7 = kmol of anhydrous substance/m³ of solution.
c_8 = kmol of hydrate/m³ of solution.
$c_7 = c_8$.
M_A = molecular weight of anhydrous substance.
M_H = molecular weight of hydrate.
$R = M_H/M_A$.
ρ = density of supersaturated solution, kg/m³.

metastable and labile states, respectively. Although this subdivision has long been recognized as an oversimplification, the concept is still very useful.

The two states of supersaturation can be represented on a temperature–concentration diagram (Fig. 1). The solid line is the normal solubility curve for the solute in the solvent. The broken line is the supersolubility curve, representing temperatures and concentrations at which spontaneous nucleation is likely to occur, but it is not so well defined as the solubility curve. Its position in the diagram depends on the intensity of agitation, the presence of trace impurities, and so on.

The diagram can be divided into three parts: (*1*) the stable (unsaturated) zone where crystallization is impossible; (*2*) the metastable (supersaturated) zone, between the two curves, where spontaneous nucleation is improbable (although a crystal located in a metastable solution would grow); and (*3*) the unstable or labile (supersaturated) zone where spontaneous nucleation is probable but not inevitable.

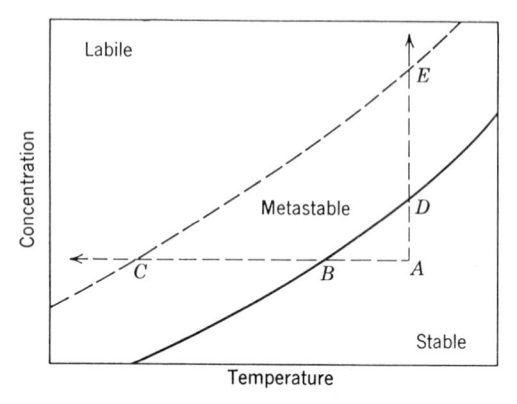

Figure 1. The solubility–supersolubility diagram.

If a solution represented by point A in Figure 1 is cooled without loss of solvent (line ABC), spontaneous nucleation cannot occur until conditions represented by C are reached. Although the tendency to nucleate increases once the labile zone is penetrated, some solutions become so viscous as to prevent nucleation and set to a glass. Supersaturation can also be achieved by removing some of the solvent from the solution by evaporation. Line ADE represents such an operation carried out at constant temperature. Penetration into the labile zone rarely occurs because the solution near the evaporating surface is more supersaturated than is the bulk solution. Crystals generated at the surface fall into the solution and induce nucleation, often before bulk conditions represented by E are reached. In industrial practice some combination of cooling and evaporation is most often used.

Expressions of Supersaturation. Because the supersaturation of a system may be expressed in a number of different ways, considerable confusion can be caused if the basic units of solution concentration are not clearly defined. Among the most common expressions of supersaturation are the concentration driving force Δc, the supersaturation ratio S, and relative supersaturation σ:

$$\Delta c = c - c^* \tag{3}$$

$$S = \frac{c}{c^*} \tag{4}$$

$$\sigma = \frac{\Delta c}{c^*} = S - 1 \tag{5}$$

where c is the solution concentration and c^* is the equilibrium value at the given temperature.

Two problems arise, namely how to express the supersaturation and which units to use for the solution concentration. For mass balances almost any units are acceptable, but in practice complications can often be avoided by making a careful selection. Much depends on the computational aids available and on whether the solute forms hydrates. Of the above three expressions for supersaturation, only Δc is dimensional, unless the concentrations are stated in mole fractions, and the magnitudes of the expressions depend on the units used to express concentrations.

The quantity that changes most in the example shown in Table 3 is Δc, but sig-

Table 3. Examples of the Expression of Supersaturation of an Aqueous Solution of Sucrose (mol wt = 342) at 20°C[a]

Solution composition	c	c^*	Δc	S	σ
kg/kg of water	2.45	2.04	0.41	1.20	0.20
kg/kg of solution	0.710	0.671	0.039	1.06	0.06
kg/m³ of solution (= g/L)	966	893	73	1.08	0.08
kmol/m³ of solution (= mol/L)	2.82	2.61	0.21	1.08	0.08
mole fraction of sucrose	0.114	0.097	0.017	1.18	0.18

[a] Equilibrium saturation $c^* = 2.04$ kg/kg of water; solution density = 1330 kg/m³; concentration of the supersaturated solution $c = 2.45$ kg/kg of water; density = 1360 kg/m³.

nificant changes occur in all expressions of supersaturation, even the dimensionless ones. It is essential, therefore, to quote the concentration units used when expressing the supersaturation of a system and also to record the temperature because the equilibrium saturation concentration is temperature-dependent.

Measurement of Supersaturation. If the concentration of a solution can be measured at a given temperature, and the corresponding equilibrium saturation concentration is known, then it is a simple matter to calculate the supersaturation (eqs. 3–5). However, there are many methods of measuring concentration, and hence of supersaturation, but not all are readily applicable to industrial crystallization practice.

Solution concentration may be determined directly by analysis, or indirectly by measuring some property of the system that is a sensitive function of concentration. Properties frequently chosen for this purpose include density, viscosity, refractive index, and electrical conductivity; these can often be measured with high precision, especially if the actual measurement is made under carefully controlled conditions in the laboratory. However, for the operation of a crystallizer under laboratory or pilot plant conditions the demand is usually for an *in situ* method, preferably capable of continuous operation, and in these circumstances problems may arise from the temperature dependence of the property being measured. In general, density and refractive index are the least temperature-sensitive properties and consequently these are the most commonly used.

For industrial crystallization, where temperature and feedstock conditions cannot be controlled with precision, very crude methods of supersaturation measurement may have to be employed. The most common method consists of a mass balance coupled with feedstock, exit liquor, and crystal production rates taken over a suitable period, eg, several hours, to reduce fluctuations.

Crystallization Phenomena

The starting point for any crystallization operation is the state of supersaturation that may be achieved by cooling, partial evaporation of the solvent, addition of a precipitating agent, as the result of a chemical reaction, etc. However, supersaturation alone is not sufficient to cause crystals to grow. Before growth can commence there must exist in the system a number of seed crystals which may be formed spontaneously, induced artificially or added deliberately.

Nucleation Kinetics. The classical theories of homogeneous nucleation (1–5) are of no help in determining the limits of metastability to be expected under the sort of conditions likely to be encountered in industrial practice. According to these theories the rate of homogeneous nucleation J may be expressed as a function of the supersaturation S by the following relationship:

$$J = A \cdot \exp[-K(\log S)^{-2}] \quad (6)$$

which predicts an explosive increase in the nucleation rate above a certain level of supersaturation. However, the constants A and K are usually inapplicable to industrial conditions where the primary nucleation is predominantly heterogeneous rather than homogeneous: it is virtually impossible to prepare a perfectly clean solution even in the laboratory. Therefore, so far as industrial crystallization is concerned, relationships such as equation 6 are of little use and all that can be justified are simple empirical relationships such as:

$$B = K_N (\Delta c)^b \quad (7)$$

The nucleation rate constant K_N and the order of the nucleation process b depend on the physical properties and hydrodynamics of the system. Values of b frequently lie in the range 2–5. B is the overall nucleation (or birth) rate of crystals.

It is not *primary* nucleation that has the greatest influence on an industrial crystallizer, the greatest hazard comes from *secondary* nucleation, which can be defined as the generation of nuclei by the crystals already present in suspension (Fig. 2). In such cases, only empirical relationships such as equation 7 can be justified, but this sort of information can be very useful in the design of crystallizers. Metastable zone widths and other nucleation data can be measured in the laboratory with a simple apparatus (2,6). A list of maximum allowable supercoolings, above which uncontrolled nucleation occurs, measured in the presence of a few freely suspended crystals, for some common aqueous salt solutions, is given in Table 4. It should be noted, however, that the working value of the supercooling in an actual crystallizer will generally be less than 50% of these values. The relation between supercooling $\Delta \theta$ and supersaturation Δc is:

$$\Delta c = (dc^*/d\theta) \Delta \theta \quad (8)$$

where c^* is the equilibrium saturation concentration.

Secondary nucleation in industrial crystallizers arises predominantly from crystal–crystal or crystal–equipment (especially the agitator) contact. Nuclei up to

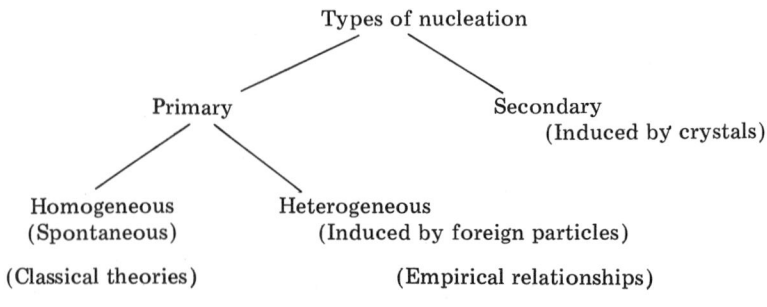

Figure 2. Types of nucleation.

Table 4. Maximum Allowable[a] Supercooling, $\Delta\theta_{max}$, for Some Common Aqueous Salt Solutions at 25°C[b]

Substance	°C	Substance	°C	Substance	°C	Substance	°C
NH$_4$ alum	3.0	MgSO$_4$.7H$_2$O	1.0	NaI	1.0	KBr	1.1
NH$_4$Cl	0.7	NiSO$_4$.7H$_2$O	4.0	NaHPO$_4$.12H$_2$O	0.4	KCl	1.1
NH$_4$NO$_3$	0.6	NaBr.2H$_2$O	0.9	NaNO$_3$	0.9	KI	0.6
(NH$_4$)$_2$SO$_4$	1.8	Na$_2$CO$_3$.10H$_2$O	0.6	NaNO$_2$	0.9	KH$_2$PO$_4$	9.0
NH$_4$H$_2$PO$_4$	2.5	Na$_2$CrO$_4$.10H$_2$O	1.6	Na$_2$SO$_4$.10H$_2$O	0.3	KNO$_3$	0.4
CuSO$_4$.5H$_2$O	1.4	NaCl	4.0	Na$_2$S$_2$O$_3$.5H$_2$O	1.0	KNO$_2$	0.8
FeSO$_4$.7H$_2$O	0.5	Na$_2$B$_4$O$_7$.10H$_2$O	4.0	K alum	4.0	K$_2$SO$_4$	6.0

[a] The working value for normal crystallizer operation may be 50% of these values, or lower. The relation between $\Delta\theta_{max}$ and Δc_{max} is given by equation 8.
[b] Measured in the presence of crystals under conditions of slow cooling and moderate agitation (2).

20 μm can be formed and these appear to have variable growth rates, the smaller crystals growing very slowly or not at all. No general theory of secondary nucleation has yet been developed, but several comprehensive reviews of this subject have recently been made (7–12).

Crystal Growth Kinetics. As in the case of nucleation, the classical theories of crystal growth (1–4,13–15) have not led to convenient working relationships for industrial crystallization. For crystallizer design and assessment, rates of crystallization are most conveniently expressed in terms of the supersaturation by empirical relationships. Two aspects of crystal growth are of interest to the chemical engineer. First, overall mass deposition rates, which can be measured in the laboratory by standard methods in fluidized beds or agitated vessels (2), are needed for the design of industrial crystallizers. Secondly, growth rates of individual crystal faces under different environmental conditions are helpful for specification of the operating conditions.

A useful pictorial representation, although admittedly a gross oversimplification, of the crystallization process is given by the diffusion–integration theory in which two separate steps are postulated. The first is a diffusional process whereby solute is transported from the bulk fluid phase through the solution boundary layer close to the crystal surface. This is followed by the integration of adsorbed solute ions or molecules at the crystal surface into the crystal lattice (see Fig. 3). These two steps, which proceed under the influence of different driving forces, can be represented by equations 9a and 9b.

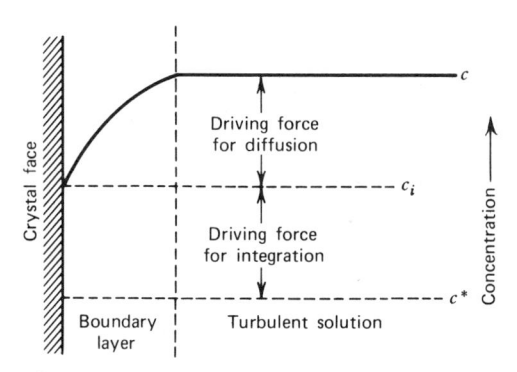

Figure 3. Concentration driving forces for crystal growth from solution.

$$\frac{dm}{dt} = k_d A(c - c_i) \tag{9a}$$

$$= k_r A(c_i - c^*)^r \tag{9b}$$

where m = mass deposited in time t; A = crystal surface area; c, c_i, and c^* are solute concentrations in the bulk solution at the interface and at equilibrium saturation; and k_d and k_r are diffusion and reaction (ie, integration) mass transfer coefficients.

Equations 9a and 9b are not easy to apply in practice because they involve the interfacial concentration which is difficult, if not impossible, to measure. It is usually more convenient to eliminate c_i by considering the over-all concentration driving force $\Delta c = c - c^*$, which is quite easily measured. A general equation for crystallization based on this over-all driving force can be written as:

$$\frac{dm}{dt} = K_G A(\Delta c)^g \tag{10}$$

where K_G is an over-all crystal growth coefficient. The exponents r and g in equations 9a and 10 are usually referred to as the order of the integration and over-all crystal growth processes, respectively. However, the use of this term order should not be confused with its more conventional use in chemical kinetics, where it always refers to the power to which a concentration should be raised to give a factor proportional to the rate of an elementary reaction. In crystallization work the exponent applied to a concentration difference has no fundamental significance and cannot give any indication of the number of elementary species involved in the growth process (2). If $g = 1$ and $r = 1$, the interfacial concentration c_i may be eliminated from equations 9a and 9b to give:

$$K_G = \frac{k_d k_r}{(k_d + k_r)} \tag{11}$$

For cases of extremely rapid integration, ie, large k_r, $K_G \approx k_d$, and the crystallization process is controlled by the diffusional operation. Similarly, if the value of k_d is large, ie, if the diffusional resistance is low, $K_G \approx k_r$, and the process is controlled by the integration step.

The diffusional step (eq. 9a) is generally linearly dependent on the concentration driving force, but the integration process (eq. 9b) is rarely first-order. Many inorganic salts crystallizing from aqueous solution give values of the over-all growth rate order g in the range 1–2.

There is no simple or generally accepted method of expressing the rate of growth of a crystal since it has a complex dependence on temperature, supersaturation, size, habit, system turbulence, etc. However, for carefully defined conditions, crystal growth rates may be expressed as an over-all mass deposition rate R_G (kg/m^2·s), an over-all linear growth rate $G(=dL/dt)$ (m/s), or as a mean linear velocity $\bar{v}(=dr/dt)$ (m/s). L is some characteristic size of the crystal, eg, the equivalent sieve aperture size, and r is the radius corresponding to the equivalent sphere, ie, $L = 2r$.

The relationships between these quantities are:

$$R_G = \frac{1}{A}\left(\frac{dm}{dt}\right) = \frac{3\alpha\rho G}{\beta} = \frac{6\alpha\rho\bar{v}}{\beta} \tag{12}$$

The volume and surface shape factors α and β, respectively, are defined by $m = \alpha \rho L^3$ and $A = \beta L^2$, where m and A are the particle mass and surface area, respectively. For spheres and cubes, $6\alpha/\beta = 1$. For octahedra, $6\alpha/\beta = 0.816$. Some typical values of the mean linear growth velocity $\bar{v}(= \frac{1}{2}G)$ are given in Table 5.

Habit Modification. Under different environmental conditions, different crystal faces grow at different rates. In general, the high index faces grow faster than the low, and changes in the environment (temperature, supersaturation, pH, impurities, etc) can have a profound effect on the individual face growth rates. Changes in the face growth rates give rise to habit (shape) changes in the crystals.

The solvent frequently influences the crystal habit. For example, naphthalene crystallizes in the form of needles from cyclohexane and as thin plates from methanol, and pentaerythritol forms tetragonal bipyramids from water and tetragonal plates from acetone. The possibility of habit change is thus an important factor to be borne in mind when a solvent for crystallization is being chosen. Supersaturation can also influence crystal habit. The individual crystal face growth kinetics usually depend to a different extent on supersaturation, so by raising or lowering the supersaturation it is sometimes possible to effect considerable control over the crystal habit. It may be, of course, that the desired habit can only be grown at a high supersaturation, above the metastable limit, and in such cases a nucleation inhibitor may have to be added to allow growth to proceed as planned.

One of the most common causes of habit modification is the presence of impurities in the crystallizing solution (2,14,16). These may already be present, eg, in the crystallization of beet sugar where the presence of raffinose induces characteristic flat crystals, or they may be added deliberately, eg, traces of borax which change the habit of $MgSO_4 \cdot 7H_2O$ from needles to prisms. Surfactants and polyelectrolytes are commonly used for habit modification purposes and in some cases a few parts per million of an impurity can cause remarkable changes in the crystal habit. Table 6 lists some examples of industrial interest.

In nearly every industrial crystallization some form of habit modification is necessary to control the type of crystal produced. This may be done, eg, by controlling the supersaturation, temperature or pH, or by changing the solvent, or by deliberately adding some impurity which acts as a habit modifier. If the habit is influenced by more than one variable, eg, both supersaturation and impurities, the combined effect may conveniently be indicated on a morphogram as shown in Figure 4 for the crystallization of sodium chloride in the presence of potassium ferrocyanide (17).

There is as yet no generally accepted theory to explain the mode of action of habit modifying impurities, but different impurities can act in different ways. Some substances are undoubtedly adsorbed on certain crystal faces and block active growth sites. Others may change the solution properties (structural or otherwise) or the equilibrium saturation concentration, or they may alter the characteristics of the adsorbed layer that exists at the crystal–solution interface, and thus influence the integration of growth units into the crystal lattice.

Ionic impurities, such as Cr^{3+} and Fe^{3+}, frequently exert a considerable influence on the nucleation and growth behavior of simple inorganic salts in aqueous solution, and it would appear that their powerful effect is to some extent associated with hydration characteristics. Aquo ions, eg, $Cr(H_2O)_6^{3+}$, would be attracted to certain crystallographic planes, although they need not actually be adsorbed since their mere presence would exert a dilution effect at the interface, retard diffusion, hinder the

Table 5. Some Mean Overall Crystal Growth Rates Expressed as a Linear Velocity[a]

Crystallizing substance	°C	S	$\bar{v}$, m/s
$(NH_4)_2SO_4 \cdot Al_2(SO_4)_3 \cdot 24H_2O$	15	1.03	1.1×10^{-8}*
	30	1.03	1.3×10^{-8} *
	30	1.09	1.0×10^{-7} *
	40	1.08	1.2×10^{-7} *
NH_4NO_3	40	1.05	8.5×10^{-7}
$(NH_4)_2SO_4$	30	1.05	2.5×10^{-7}
	60	1.05	4.0×10^{-7}
$NH_4H_2PO_4$	20	1.06	6.5×10^{-8}
	30	1.02	3.0×10^{-8}
	30	1.05	1.1×10^{-7}
	40	1.02	7.0×10^{-8}
$MgSO_4 \cdot 7H_2O$	20	1.02	4.5×10^{-8} *
	30	1.01	8.0×10^{-8} *
	30	1.02	1.5×10^{-7} *
$NiSO_4 \cdot (NH_4)_2SO_4 \cdot 6H_2O$	25	1.03	5.2×10^{-9}
	25	1.09	2.6×10^{-8}
$K_2SO_4 \cdot Al_2(SO_4)_3 \cdot 24H_2O$	15	1.04	1.4×10^{-8} *
	30	1.04	2.8×10^{-8} *
	30	1.09	1.4×10^{-8} *
	40	1.03	5.6×10^{-8} *
KCl	40	1.01	6.0×10^{-7}
KNO_3	20	1.05	4.5×10^{-8}
	40	1.05	1.5×10^{-7}
K_2SO_4	20	1.09	2.8×10^{-8} *
	20	1.18	1.4×10^{-7} *
	30	1.07	4.2×10^{-8} *
	50	1.06	7.0×10^{-8} *
	50	1.12	3.2×10^{-7} *
KH_2PO_4	30	1.07	3.0×10^{-8}
	30	1.21	2.9×10^{-7}
	40	1.06	5.0×10^{-8}
	40	1.18	4.8×10^{-7}
NaCl	50	1.002	2.5×10^{-8}
	50	1.003	6.5×10^{-8}
	70	1.002	9.0×10^{-8}
	70	1.003	1.5×10^{-7}
$Na_2S_2O_3 \cdot 5H_2O$	30	1.02	1.1×10^{-7}
	30	1.08	5.0×10^{-7}
citric acid monohydrate	25	1.05	3.0×10^{-8}
sucrose	30	1.13	1.1×10^{-8} *
	30	1.27	2.1×10^{-8} *
	70	1.09	9.5×10^{-8}
	70	1.15	1.5×10^{-7}

[a] The supersaturation is expressed by $S = c/c^*$ with c and c^* as kg of crystallizing substance per kg of free water. The significance of the mean linear growth velocity, $\bar{v}$ is explained by the relationships in equation 12, and the values recorded here refer to crystals in the approximate size range 0.5–1 mm growing in the presence of other crystals (2). An asterisk denotes that the growth rate is probably size-dependent.

Table 6. Some Habit Modifications of Industrial Interest

Substance	Normal habit	Habit modifier	Changed habit
NH$_4$ alum	octahedra	borax	cubes
NH$_4$Cl	dendrites	Cd^{2+}, Ni^{2+}	cubes
NH$_4$H$_2$PO$_4$	needles	Al^{3+}, Fe^{3+}, Cr^{3+}	tapered prisms
NH$_4$NO$_3$	short crystals	Acid Magenta	needles
(NH$_4$)$_2$SO$_4$	prisms	Fe^{3+}	needles
		H$_2$SO$_4$	needles
H$_3$BO$_3$	needles	gelatin, casein	flakes
CaSO$_4$.2H$_2$O	needles	sodium citrate	prisms
CuSO$_4$.5H$_2$O	small crystals	gelatin	large crystals
MgSO$_4$.7H$_2$O	needles	borax	prisms
K alum	octahedra	borax	cubes
KBr	cubes	phenol	octahedra
KCN	cubes	Fe^{3+}	dendrites
KCl	cubes	Fe(CN)$_6^{4+}$	dendrites
K$_2$SO$_4$	rhombs	Fe^{3+}	irregular crystals
NaBr	cubes	Fe(CN)$_6^{4+}$	dendrites
NaCN	cubes	Fe^{3+}	dendrites
NaCl	cubes	Fe(CN)$_6^{4+}$	dendrites
		nitrilotriacetamide	dendrites
		formamide	octahedra
		Pb^{2+}, Cd^{2+}	large crystals
		poly(vinyl alcohol)	needles
		Na$_6$P$_4$O$_{13}$	octahedra
Na$_2$B$_4$O$_7$.10H$_2$O	needles	carboxymethyl cellulose	flakes

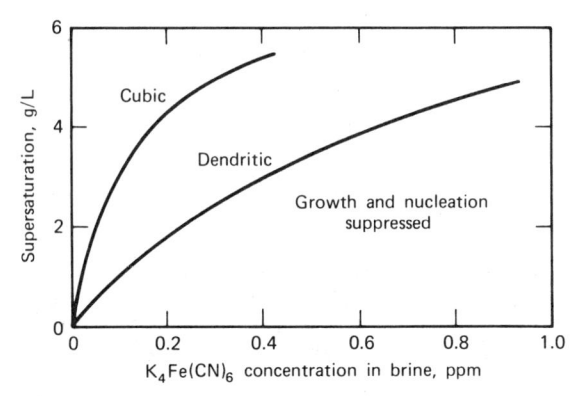

Figure 4. Morphogram for the crystallization of sodium chloride from aqueous solution in the presence of K$_4$Fe(CN)$_6$ at 42°C (17).

aggregation of growth units in the adsorbed layer and thus retard the growth rate. However, if adsorption did take place, the aquo ions would lose some or all of their hydration molecules and the resulting counterflux of water away from the interfacial regions would further retard the crystal growth rate. The ability of an ionic impurity to form complex species in solution is another important factor to consider.

An example of the effect of an ionic impurity on the crystallization of an inorganic

254   CRYSTALLIZATION

salt in a continuously operated mixed-suspension crystallizer (18) is shown in Figure 5. The crystals of ammonium sulfate grown from pure solution exhibit the characteristic habit (Fig. 5a). Figures 5b and c show the crystalline products obtained from solutions containing 10 and 20 ppm Cr^{3+} (added as $CrCl_3.6H_2O$), respectively. The initial effect of the impurity is to encourage the development of 110 or 111 faces, resulting in a pyramidal habit, but later there appears to be a breakdown of normal growth leading to the formation of crystals of grotesque shape. The other pronounced

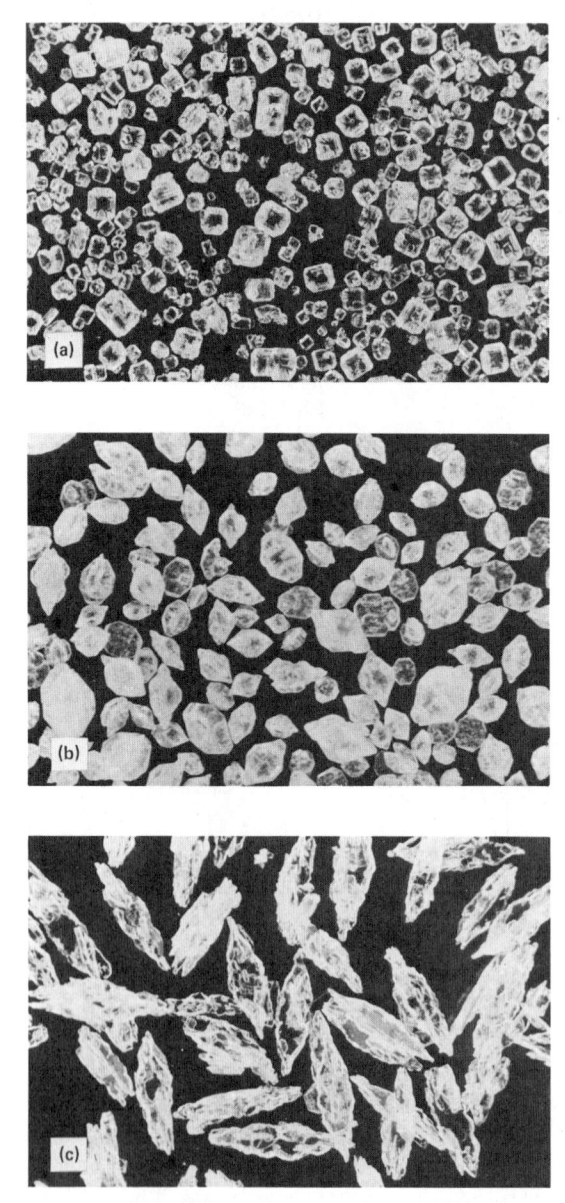

Figure 5. Ammonium sulfate crystallized (**a**) from pure solution (mean size ca 350 μm), (**b**) in the presence of 10 ppm Cr^{3+} (mean size ca 500 μm), and (**c**) in the presence of 20 ppm Cr^{3+} (length ca 5 mm).

effect of Cr³⁺ is the suppression of nucleation which results in the increase of crystal mean size from about 350 μm (Fig. 5a) to a crystal length of about 5 mm (Fig. 5c).

Inclusions in Crystals. Crystals generally contain inclusions (pockets of solid, liquid, or gaseous impurities). The term occlusion is more generally applied to impure fluid adhering to the surfaces of crystals, or trapped between agglomerated crystals, eg, after filtration. A number of terms are used to describe inclusions, some of which are self-explanatory, such as bubbles, fjords (parallel channels), veils (thin sheets of small inclusions), clouds (random clusters of small inclusions), negative crystals (faceted inclusions), and so on. Most frequently inclusions appear in random array but sometimes they show a remarkable regularity.

Fluid inclusions may often be observed with the aid of a simple magnifying glass, although a more detailed picture is revealed under a low-power microscope. A simple technique for this purpose consists of immersing the crystal in an inert liquid of similar refractive index or, alternatively, in its saturated solution. If a solution is used as the medium, the inclusion can be identified under the microscope by raising the temperature slightly to dissolve the crystal. If the inclusion is a liquid, concentration streamlines will be seen as the two fluids meet; if it is a vapor, a bubble will be released.

Inclusions are a frequent source of trouble in industrial crystallization. Crystals grown from aqueous solution can contain up to 0.5 wt % of included liquid, which can significantly affect the purity of the product. Inclusions can cause caking of stored crystals by the seepage of liquid if the crystals become broken. From the industrial crystallization point of view, the main interest lies in finding methods to prevent the formation of inclusions. The crystallizing system must be kept clean to avoid dirt, rust, and other debris being built into the crystals. The formation of air or vapor inclusions can sometimes be minimized by avoiding vigorous agitation or boiling in an industrial crystallizer, although it is interesting to note that inclusions are sometimes formed more readily in large single crystals grown under laboratory conditions with natural convection than with forced convection. The application of ultrasonic irradiations to the system may help to prevent bubbles or particles adhering to a growing crystal face. In general, the faster a crystal grows, the more easily are inclusions formed, so the growth rate should be restricted to an acceptable level. In effect, this means avoiding high supersaturations.

Several comprehensive reviews of the subject of inclusions have been made (19–21) and a worldwide coverage of research on inclusions, although mainly of geological interest, is provided by the annual COFFI reviews (22).

Ripening. The phenomenon of ripening, frequently referred to as Ostwald ripening, can occur in suspensions of very small crystals. Those smaller than some critical size dissolve, despite the fact that the system may apparently be supersaturated, while the larger crystals grow. Precipitates kept in contact with their mother liquor are therefore prone to the ripening process, ie, the particle size distribution changes with time, and the mean particle size increases.

One reason for ripening is the difference in solubility between crystals of different sizes, as expressed by the Gibbs-Thomson (or Ostwald-Freundlich) equation, which may be written in the form:

$$\ln\left[\frac{c(r)}{c^*}\right] = \frac{2\,\gamma v}{\nu\,kTr} \tag{13}$$

where $c(r)$ is the solubility of a particle of radius r; c^* is the normal equilibrium solubility; γ is the surface energy of the solid in contact with its solution; v the molecular volume of the solid; k the Boltzmann constant; T the absolute temperature; and ν the number of ions in the solute, if it is an electrolyte. For nonelectrolytes, $\nu = 1$.

For inorganic salts in aqueous solution, significant deviations from the normal equilibrium solubility occur only for particle sizes < 1 μm. However, a form of ripening can occur over a period of time in both static and stirred suspensions containing crystals of 10 μm or even larger. The most probable explanation of this behavior is that small local temperature differences result in local supersaturation fluctuations, some regions even becoming temporarily unsaturated. Dissolution and crystal growth can then continue virtually side by side.

The deliberate use of temperature cycling has been used successfully as a technique for increasing the mean crystal size and hence reducing the crystal size distribution in industrial batch crystallizers (23–24). On the other hand, temperature cycling can cause considerable trouble in the storage of paints and pigments, pharmaceutical pastes, and similar preparations owing to the development of oversize particles.

Industrial Crystallization Processes

   Growth of Single Crystals.   There is an ever increasing demand for single crystals of a large number of substances for use in the electronics industry where their dielectric, piezoelectric, semiconductor, and other properties are exploited (see Semiconductors; Zone refining). Single crystals are also in demand for masers, lasers, and gem stones (see Gems, synthetic). A large number of publications (1,14–15,25–31) have been devoted to this specialized branch of crystallization practice which is outside the scope of this article.

   Fractional Solidification.   In recent years a considerable number of melt crystallization techniques have been developed. In the Newton-Chambers process (32) for the purification of benzene from a coal-tar benzole fraction, an impure feedstock is mixed with refrigerated brine. The slurry is centrifuged to yield benzene crystals (freezing point 5.4°C) and a mixture of brine and mother liquor. After settling, the brine is returned for refrigeration and the mother liquor is reprocessed for motor fuel. The success of the method depends on the efficiency of removal of impure mother liquor adhering to the benzene crystals. There are several possible methods of operation.

In the thaw–melt method (Fig. 6a) the benzene crystals are washed in the centrifuge with brine at a temperature above 6°C. This partially melts some of the benzene crystals, and the adhering mother liquor is washed away. The thaw liquor can be recycled. Multistage operation is also possible (Fig. 6b). The first crop of crystals is taken as the product and the second, from the mother liquor, is melted for recycle. The purity of the crystals from the second stage should not be less than that of the original feedstock.

The Proabd Refiner (32) is essentially a batch cooling process in which a static liquid feedstock is progressively crystallized on extensive cooling surfaces, eg, fin-tube heat exchangers located inside the crystallization tank. As solidification proceeds, the remaining liquid becomes progressively more impure. In some cases, crystallization may be continued until virtually the entire charge has solidified. The crystallized mass is then slowly melted by heating the circulating heat transfer fluid. The impure fraction melts and drains, and as melting proceeds, the melt run-off becomes progressively

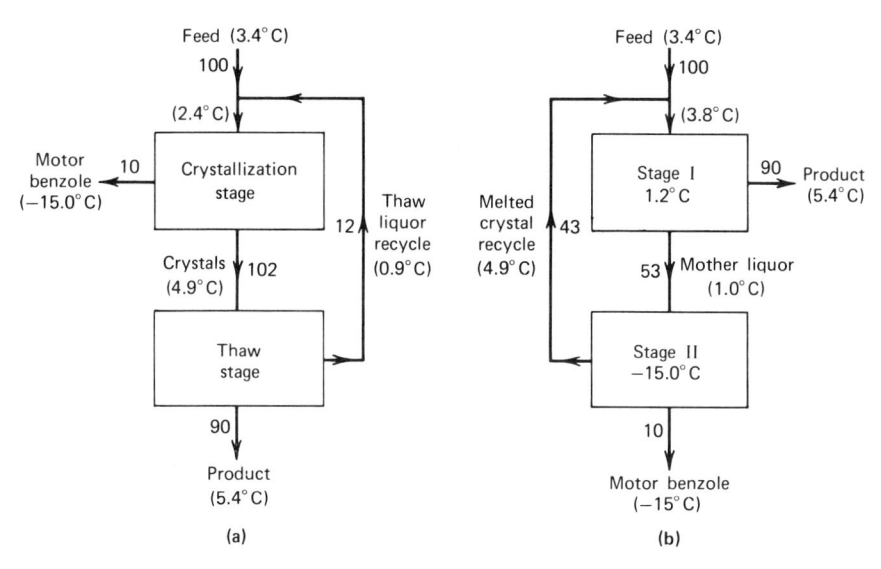

Figure 6. The Newton-Chambers process: (**a**) thaw-melt, (**b**) two-stage. The temperatures in parentheses indicate the freezing points of the various streams. The numbers indicate typical mass flow rates for a mass feed rate of 100 (32).

richer in the desired component. Any number of fractions can be taken during the melting stage. A typical flow diagram based on a scheme for the purification of naphthalene is shown in Figure 7. The circulating fluid is water and heating is facilitated by steam injection.

The MWB process (33), which acts effectively as a multistage countercurrent scheme, is illustrated for a 4-stage operation in Figure 8**a**. The cycle starts at stage 1

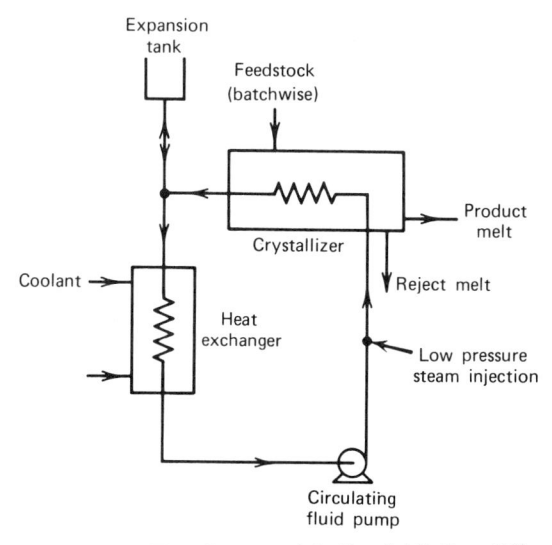

Figure 7. Flow diagram of the Proabd Refiner (32).

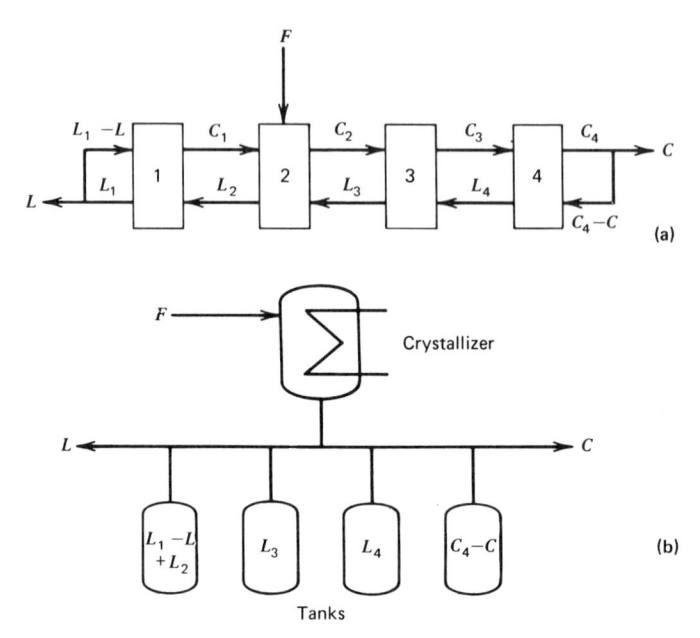

Figure 8. The MWB process: (**a**) multistage scheme, (**b**) practical lay-out (33).

which is fed with melt L_2 and recycle liquor $L_1 - L$, where L is the reject impure liquor stream. A quantity of crystals C_1 is deposited in stage 1. In stage 2 the melted crystals C_1 are contacted with melt L_3 and fresh feedstock F. Crystallization yields crystals C_2 and melt L_3. Stages 3 and 4 follow similar patterns and, in this example, the final high-purity stream C_4, after being remelted, is split into product C and recycle $C_4 - C$. Only one crystallizing vessel is needed (Fig. 8**b**). The crystals are not transported; they remain inside the crystallizer, deposited on the internal heat exchanger surfaces, until they are melted by the appropriate warm incoming liquor stream. The control system linking the storage tanks and crystallizer consists of a program timer, actuating valves, pumps, and cooling loop. The process has commercial applications in the purification of a wide range of organic substances, including benzoic acid and caprolactam.

A process for fractional crystallization in a column crystallizer by the countercurrent contact between crystals and their melt was first patented by Arnold in 1951. The principles of column crystallization are shown in Figure 9.

In the end-fed pulse unit (Fig. 9**a**) the slurry feedstock enters at the top of the column and the crystals fall countercurrently to a pulsed upflow of melt. There is a heat and mass interchange between the solid and liquid phases and pure crystals migrate to the lower zone where they are remelted to provide a high purity liquid for the upflow stream.

The center-fed crystallizer (Fig. 9**b**), developed by Schildkneckt in 1961, utilizes a spiral conveyor to transport the solids through the purification zone countercurrently to the melt. The mode of action is similar to that in a center-fed distillation column. Laboratory crystallizers of this type are frequently operated batchwise under total reflux, but reflux may be provided at both ends (Fig. 9**c**) when the unit functions as a complete fractionator.

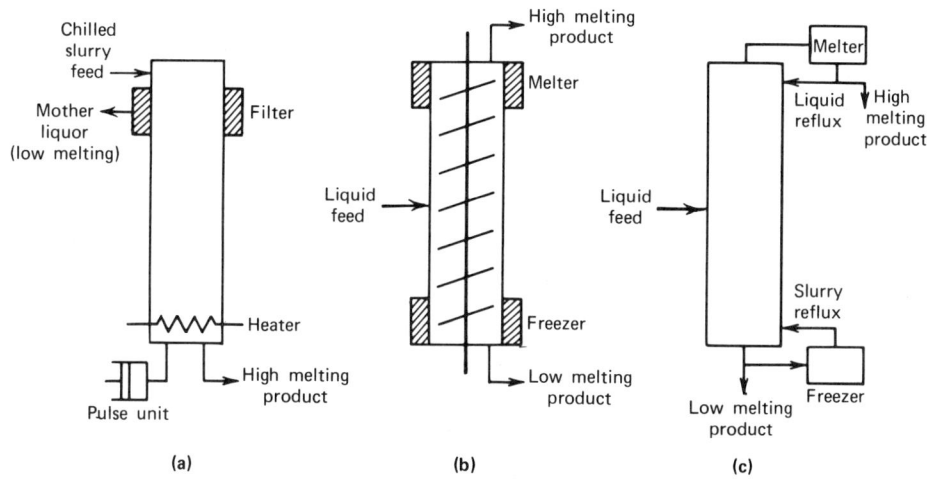

Figure 9. Column crystallizers: (**a**) end-fed pulse column (Arnold type), (**b**) center-fed column with spiral conveyor (Schildknecht type), (**c**) center-fed column with reflux at both ends.

Several comprehensive reviews of the theory and practice of column crystallizers have been made (34–35).

One of the best examples of the successful commercial application of countercurrent column fractional crystallization is the Phillips process (36). The essential features of the crystallization zone are shown in Figure 10a. Chilled slurry feed, from a scraped-surface chiller, enters at the top of the column. Crystals are forced downwards by means of a piston and impure liquor is removed through a wall filter. Wash

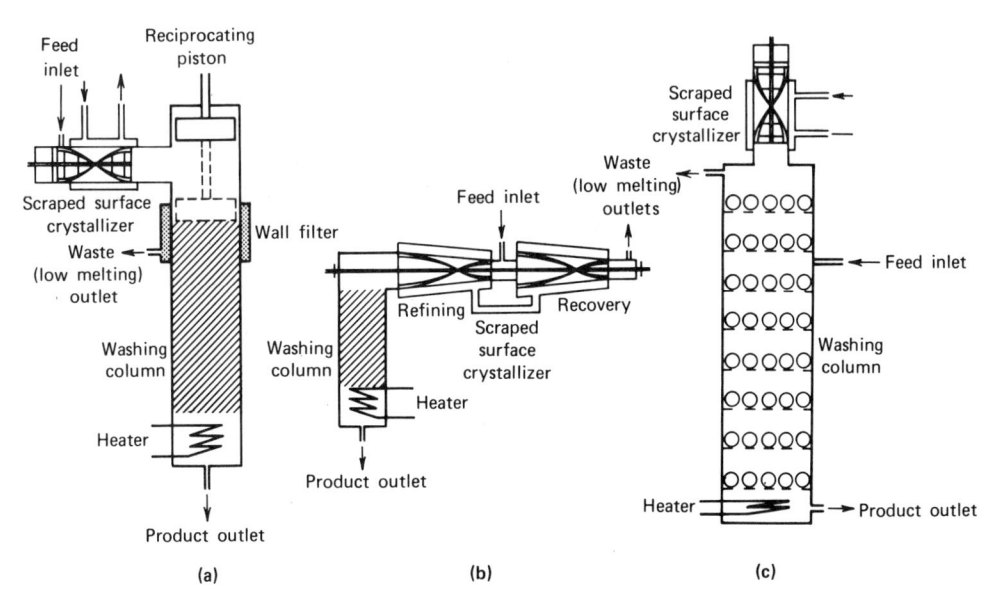

Figure 10. Some column crystallizers used industrially. (**a**) Phillips process, (**b**) Brodie purifier, (**c**) TNO process.

liquor, produced by melting pure crystals at the bottom of the column, is transported upwards countercurrently to the crystals. The wash liquor may be pulsed upwards. The Phillips process is used for the large-scale production of p-xylene. It has also been applied to the freeze-concentration of beer and fruit juices.

The Brodie Purifier (37) is another successful melt crystallizer. It consists basically of a horizontal center-fed column with the special feature (Fig. 10**b**) that the column cross-section reduces in the direction of diminishing flow, ie, towards the cold end, to maintain a reasonably constant axial flow velocity and prevent back-mixing. In practice this is achieved by installing a sequence of scraped-surface chillers, each of progressively smaller diameter. The Brodie Purifier has been applied to the large-scale production of high-purity p-dichlorobenzene.

In the recently developed TNO process (38), separation is effected by countercurrent washing coupled with repeated recrystallization facilitated by grinding the crystals during their transport through a vertical column. The small crystalline fragments melt more easily and impurities trapped as inclusions or as solid solutions are returned to the melt. Grinding is achieved by balls rolling over sieve plates in the vibrated column (Fig. 10**c**). Good agitation and interphase transport are promoted but undesirable top-to-bottom mixing is prevented. Pilot-scale trials with benzene–thiophene separations have been successful and industrial operation looks promising.

Seeding and Controlled Cooling. Excessive production of nuclei during a crystallization operation should be avoided. The solution should be maintained in the metastable condition. It should not be allowed to become labile (Fig. 1). However, it may be necessary to introduce seed crystals into the solution if insufficient numbers of nuclei are generated. The mass of seeds M_s of size L_s that can be added to a crystallizer, assuming that crystallization occurs only on the added seeds, depends on the required crystal yield Y (see eq. 1) and the product crystal size L_p:

$$M_s = Y\left(\frac{L_s^3}{L_p^3 - L_s^3}\right) \tag{14}$$

The effect of cooling an unseeded solution rapidly is shown in Figure 11**a**. The solution cools at constant concentration until the labile zone is penetrated and spontaneous nucleation occurs. The temperature rises slightly, owing to the release of heat of crystallization, but continued cooling reduces it again. Eventually the temperature and concentration fall as indicated. No control is possible over the nucleation or growth processes. The slow cooling of a seeded solution is shown in Figure 11**b**. The temper-

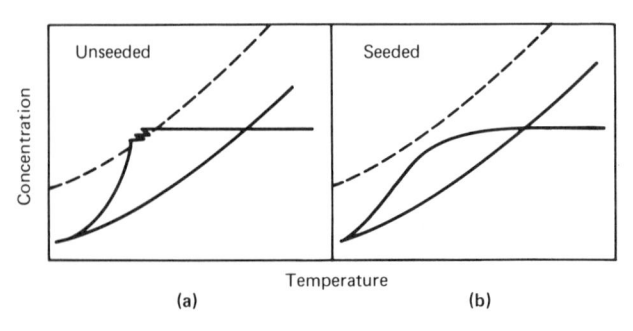

Figure 11. The effect of seeding on a crystallization process.

ature is controlled so that the system remains metastable throughout the operation; spontaneous nucleation does not occur and growth takes place at a controlled rate on the added seeds. This method of operation is used in many large-scale batch crystallizers.

Control can be exercised over the product crystal size from a batch crystallizer by controlling the cooling rate. For example, natural cooling (Fig. 12a) gives a supersaturation peak which induces heavy nucleation. However, by following a cooling path that maintains the supersaturation at a constant low level (Fig. 12b), nucleation can be controlled within acceptable limits. The calculation of optimum cooling curves for different operating conditions present a complex problem (39) but for industrial purposes the following simplified relationship is generally adequate:

$$\theta_t = \theta_o - (\theta_o - \theta_f) \left(\frac{t}{\tau}\right)^3 \qquad (15)$$

where θ_o, θ_f, and θ_t are the temperatures at the beginning, end, and at any time t during the process. τ is the overall batch time.

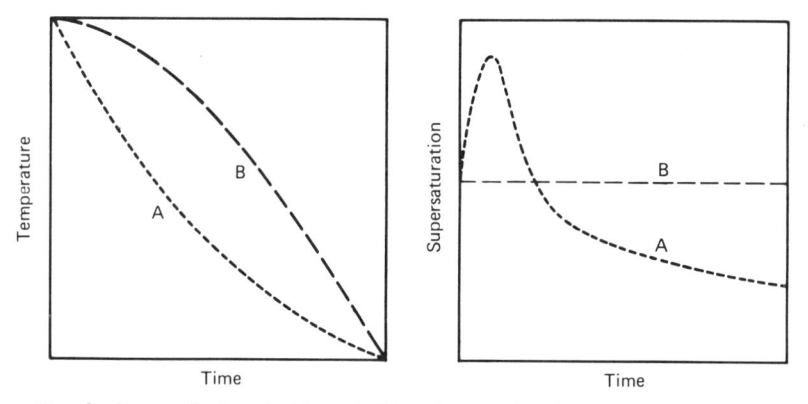

Figure 12. Cooling modes for a batch crystallizer: A, natural cooling; B, controlled cooling (constant supersaturation).

Salting-out Crystallization. A solution can be supersaturated by the addition of a substance, preferably a liquid, which reduces the solubility of the solute in the solvent. The added component should be miscible in all proportions with the solvent and the solute should be relatively insoluble in it. This process, known by such names as salting-out, watering-out, precipitation, etc, is widely employed for the crystallization of organic substances from water-miscible organic solvents, eg, from alcoholic solution by the controlled addition of water. The precipitation of inorganic salts from aqueous solution by an alcohol has also been applied commercially. For example, iron-free alums have been prepared in this way and so have coarse grained anhydrous precipitates of normally hydrated salts (2). Various salts can be crystallized selectively from natural brines by alcoholic precipitation (40).

Salting-out processes offer several advantages. For instance, highly concentrated feedstock solutions can be prepared in a suitable solvent and a high recovery of solute can be effected by the choice of a suitable additive. All this can be done at around room temperature, which is highly desirable if heat labile substances are being processed. Frequently the mixed-solvent mother liquor has a high retention capacity for the impurities and very pure crystals result. The disadvantage of the process is the need for a solvent recovery unit to separate the mixed solvents.

Reaction Crystallization. The precipitation of a solid phase by the chemical interaction of gases or liquids is a common method of preparing many industrial chemicals. Precipitation occurs when the fluid phase becomes supersaturated with respect to the solute. An uncontrolled precipitation process can be transformed into a controlled crystallization operation if the degree of supersaturation can be maintained below the level at which spontaneous nucleation occurs, ie, operating in the metastable rather than in the labile state (Fig. 1).

Reaction crystallization is widely employed in industry, especially when valuable waste gases are produced. For example, sodium bicarbonate can be prepared from flue gases containing 10–20% of carbon dioxide by countercurrent contact with brine in packed towers. Ammonia can be recovered from coke-oven gases by interaction with sulfuric acid to give crystalline ammonium sulfate. Many conventional crystallizers can readily be adapted for use as reaction crystallizers. The reactants are usually fed into a zone where intimate mixing takes place quickly. Any heat of reaction may be utilized for the partial evaporation of the solution (41).

Spray Crystallization. Strictly speaking, the term spray crystallization is a misnomer. The process is not a true crystallization; it has more in common with spray drying. Solids are simply deposited from a highly concentrated solution or a melt by spraying droplets into a large chamber where they fall countercurrently to an upflowing stream of hot air. In some cases, particularly with molten feedstocks, the product consists of near-spherical particles called prills which have good flow properties, a high crushing strength and good storage characteristics. Hygroscopic prills can be coated with an inert anticaking agent. Fertilizer chemicals, particularly ammonium nitrate and urea, are manufactured on a large scale by this method (42).

Substances with an inverted solubility, eg, Na_2SO_4, $FeSO_4$, etc, which cause trouble in conventional evaporative crystallizers owing to scale formation on heat transfer surfaces, are often manufactured by spray crystallization. In these cases the product usually consists of 1–2 mm granules of agglomerated crystalline powder.

Emulsion Crystallization. Organic substances may be purified by fractional crystallization from the melt or from organic solvents, but these operations frequently present difficulties in large-scale production. Apart from expensive solvent losses and the potential fire and explosion hazards, yields are often low on account of the high solubility of the crystals. The method known as emulsion crystallization is generally free of these shortcomings (2). Briefly, crystallization is carried out by cooling from an aqueous emulsion. Impurities remain in the emulsion from which they may be recovered by further cooling. The organic substance should be (1) practically insoluble in water and (2) able to melt and solidify in a heterogenous aqueous medium, and remain stable.

The organic melt is emulsified in water with the aid of a suitable nonionic surfactant and stabilized by a protective colloid. The system is crystallized by cooling, and the crystals separated from the emulsion and washed with water. The operation may be repeated, if required, although considerably fewer steps are usually needed compared with conventional fractional crystallization. The high efficiency of emulsion crystallization is apparently owing to the fact that crystal agglomeration does not occur to any great extent and the impure emulsion is readily washed away with water.

Extractive Crystallization. A binary mixture that forms a eutectic cannot be separated into its pure components by conventional fractional crystallization. However, if a third component, which alters the solid–liquid phase relationships, is added to the system it is sometimes possible to effect the separation by a stepwise crystallization process. Many hydrocarbon isomers and other close-boiling mixtures, such as those encountered in the petroleum industry, are amenable to this treatment.

The use of extractive crystallization has been suggested for the separation of m- and p-cresol, a system in which two eutectics and an equimolecular addition compound are formed. Acetic acid is a suitable extraction solvent for this purpose. The technique of extractive crystallization has also been proposed for the separation of m- and p-xylene using n-heptane as solvent (43–44).

Adductive Crystallization. Liquid mixtures, especially of chemical isomers, that are not amenable to conventional crystallization can often be separated by adding another component to the system that causes a solid phase to be deposited. This phase may be a true molecular compound or a complex, such as a clathrate or an adduct. These complexes are not chemical compounds but strongly bound physical mixtures.

Molecular compound formation can be utilized, eg, in the separation of the eutectic system of m- and p-cresol. If benzidine is added to the mixture a solid compound p-cresol–benzidine can be crystallized. Similarly, m- and p-xylene mixtures can be separated by adding carbon tetrachloride; the compound p-xylene–carbon tetrachloride is deposited.

In a clathrate, molecules of one substance (the guest) are trapped in the open crystal lattice of the other (the host). Clathrates do not have a fixed composition but there is a maximum host:guest ratio for any given system. The guest material can be recovered by melting or dissolving the clathrate in a suitable solvent. It is suggested that clathrates are formed in a thermodynamically unstable crystal lattice which is stabilized by the presence of the guest molecules. Monoammine nickel cyanide has been used as a clathrating agent in the production of thiophene-free benzene (see Clathration).

In an adduct, molecules of one substance are held by molecular attractive forces in holes or channels in the crystals of the other substance. Urea and thiourea have the property of forming adducts, especially with hydrocarbons. Adducts usually have a definite composition but they are not normally expressed in simple integers, eg, the n-octane–urea adduct has a 1:6.7 mole ratio of paraffin:urea.

The main large-scale use of adductive crystallization so far has been in the petroleum industry, eg, for the manufacture of low pour-point oils by urea dewaxing, but many other attractive possibilities exist (2,43–44).

Industrial Crystallization from Solution

Industrial crystallizers (2,6,41,45–46) may be classified into a number of general categories. Terms such as batch or continuous, agitated or nonagitated, controlled or uncontrolled, classifying or nonclassifying, circulating liquor or circulating magma, etc, are useful for this purpose. Classification of crystallizers according to the method by which supersaturation is achieved is still probably the most widely used method; thus we have cooling, evaporating, vacuum, reaction, etc, crystallizers.

Many of the crystallizer classes are self-explanatory, but some require definition.

For example, the term controlled refers to supersaturation control. The term classifying refers to the production of a selected product size by classification in a fluidized bed of crystals. In a circulating-liquor crystallizer the crystals remain in the crystallization zone, only the clear mother liquor is circulated. In the circulating-magma crystallizer the crystals and the mother liquor are circulated together. Crystallizers operating continuously with fully mixed suspensions, discharging a magma (crystal slurry) of average composition, are generally referred to as mixed-suspension, mixed-product removal (MSMPR) crystallizers. The abbreviation MSCPR is sometimes used for a mixed-suspension, classified-product removal unit. Any given crystallizer may well belong to several of the above types.

Cooling Crystallizers. *Unstirred Tanks.* These are the simplest types of cooling crystallizer in use. A hot concentrated solution is charged to the open vessel where it is allowed to cool, often over several days, by natural convection. The batch may be given an occasional stir to prevent the formation of hard crystalline lumps on the bottom of the crystallizer. No seeding is required. Sometimes thin rods or strips of metal are hung in the solution on which crystals grow, thus preventing some of the product falling to the bottom of the vessel. The magma can be pumped out or the mother liquor may be drained off, leaving the crystals to be discharged by hand.

Because of the slow cooling and lack of fluid movement in an unstirred crystallizer, many interlocked crystals are usually obtained and these can retain considerable amounts of mother liquor, even after washing on a filter. The dried crystals, therefore, are generally impure. No control over the product size is possible; the crystals may range from fine dust to large agglomerates, but experience will indicate the size of vessel and cooling time necessary to produce the desired type of crystalline mass.

Handling labor costs are generally high, but for small batches the method is economic because of the low capital outlay and negligible process and maintenance costs. The main disadvantage is that the equipment is generally bulky and occupies valuable floor space.

Agitated Vessels. When a stirrer is employed in an open tank crystallizer, smaller and more uniform crystals are formed and the batch time is reduced. A somewhat purer product results owing to the lower retention of mother liquor by the crystals and the more efficient washing that can be effected.

The vessel may be equipped with a water jacket or cooling coils. Jackets are generally preferred because coils tend to become encrusted with a hard crystalline deposit and cease to function efficiently. The inner cold surfaces of the crystallizer should be as smooth and flat as possible to minimize encrustation. Polished stainless steel is a good material of construction for this purpose.

The operating cost for an agitated-cooler is higher than that for a simple tank crystallizer, but it is small in comparison with the advantages gained by the quicker throughput. Labor costs for handling the product may still be rather high.

Tank crystallizers, stirred or otherwise, vary in design from shallow pans to large cylindrical tanks, according to the needs of the particular process. A large modern agitated cooling crystallizer is shown in diagrammatic form in Figure 13a. This vessel has an upper conical section which reduces the upward velocity of liquor and prevents the crystalline product being swept out with the spent-liquor out-flow. The magma (crystal slurry) is circulated in the growth zone of the crystallizer by an agitator located in the lower region of a draft tube. An internal cooling device may be provided if required. Good mixing within the crystallizer and high rates of heat transfer between

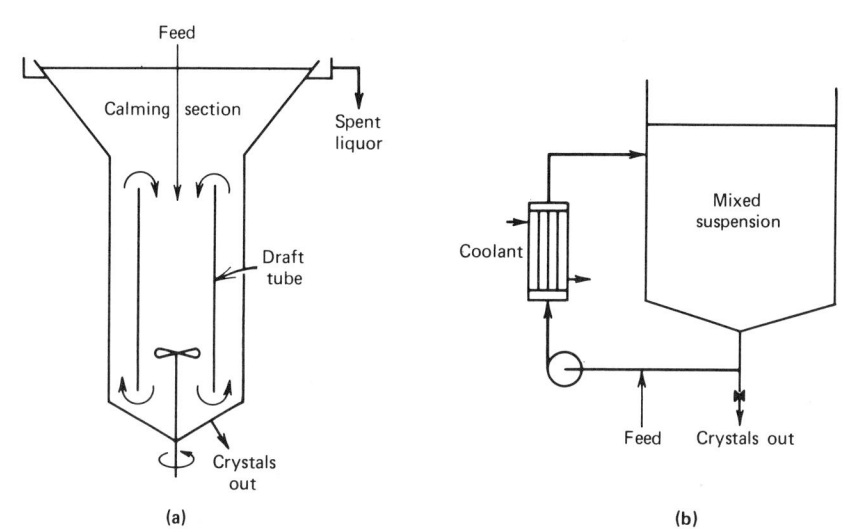

Figure 13. Agitated tank crystallizers: (**a**) internal circulation with a draft tube; (**b**) external circulation through a heat exchanger.

the liquor and coolant can be achieved by the use of circulation through an external heat exchanger (Fig. 13**b**). Because of the high liquor velocity in the tubes, low temperature differences suffice for cooling, and encrustation may be considerably reduced.

Scraped-Surface Crystallizers. The Swenson-Walker scraped-surface crystallizer, developed over 50 years ago, still finds industrial application in the processing of inorganic salts that have a high temperature solubility coefficient with water. It consists of a shallow semicylindrical trough, about 600 mm wide and 3–12 m long, fitted with a water-cooled jacket. A slow speed (5–10 rev/min) helical scraper keeps the cooling surfaces clean and enhances the growth rates of the crystals by gently moving them concurrently through the solution which flows down the sloping trough. Several units can be connected in series. The production capacity of this type of unit is determined by the heat transfer rate which, for economic reasons, should exceed 60 kJ/s (14.3 kcal/s) (46). Heat transfer coefficients of 50–150 W/(m²·K) are usual.

Higher heat transfer coefficients, and hence production capacities, are obtainable with the double-pipe, scraped-surface units such as the Votator and the Armstrong crystallizers. Spring-loaded internal agitators scrape the heat transfer surfaces and the turbulent flow inside the tube maintains heat transfer coefficients in the range 50–700 W/(m²·K) for crystallization operations. The units vary in size from about 75 to 600 mm dia and 0.3–3 m long and may be connected in series. Double-pipe, scraped-surface crystallizers are mainly employed for the processing of fats, waxes, and other organic melts, although several applications to inorganic salts from solution, eg, sodium sulfate from viscose spin-bath liquors, have been reported (47) (see Rayon).

Direct Contact Cooling. Cooling crystallizers operated with conventional heat exchangers, eg, coils, jackets, shell and tube exchangers, etc, often suffer from encrustation on the heat transfer surfaces, which severely reduces the crystallizer performance. One method for avoiding the use of a conventional heat exchanger is to employ vacuum cooling (see below) but encrustation on the inner walls of the crystallizer can still be substantial. Another technique is to use direct contact cooling (DCC) where supersaturation is achieved by contacting the process liquor with a cold heat transfer medium.

Some of the advantages of DCC over the more conventional indirect contacting methods include better heat transfer, smaller coolant requirement and the elimination of heat exchanger encrustation. Problems associated with DCC crystallization arise from the possibility of product contamination and the difficulty of separating and recovering the coolant.

The coolant may be solid, liquid, or gaseous and heat may be exchanged by the transfer of sensible and/or latent heat. The coolant may or may not boil during the operation and it may be miscible or immiscible with the process liquor. Thus several types may be envisaged: (1) immiscible, boiling—solid or liquid coolant (transfer of latent heat of sublimation or vaporization is the main source of heat removal); (2) immiscible, nonboiling—solid, liquid, or gaseous coolant (mainly sensible heat transfer); (3) miscible, boiling—liquid coolant (mainly latent heat transfer); (4) miscible, nonboiling—liquid coolant (mainly sensible heat transfer).

Several successful DCC crystallization processes have been operated in recent years, eg, benzene purification (32), oil dewaxing (48), separation of close-boiling hydrocarbons, notably the production of p-xylene (49) and the desalination of seawater (50–51) (see Water, supply and desalination). In addition there have been a few applications to the production of inorganic salts from aqueous solution (52–53).

A continuous crystallizer that utilizes immiscible, nonboiling DCC crystallization for the production of calcium nitrate tetrahydrate, the Cerny process (2,41) is shown in Figure 14. Aqueous feedstock enters at the top of the crystallizer at 25°C and cools as it flows countercurrently to an upflow of immiscible coolant (eg, petroleum at −15°C) introduced as droplets into a draft tube. The low density coolant collects in the upper layers, but the high density aqueous solution circulates up the draft tube and down the annulus, keeping the small crystals in suspension. Crystals >0.4 mm settle to the lower regions and are discharged in the magma at about −5°C. A slow speed agitator prevents consolidation of the crystals in the magma outlet. The coolant is passed to a cyclone to remove traces of aqueous solution, and recycled through the cooler.

Classifying Crystallizers. The first continuous classifying crystallizer to be developed was the Oslo (1919), also called the Jeremiassen or Krystal crystallizer. A concentrated solution, continuously cycled through the crystallizer, is supersaturated in one part of the apparatus and then conveyed to another part where the supersaturation is released into a mass of growing crystals.

The operation of the Oslo cooling crystallizer can be described with reference to Figure 15. Hot concentrated feed solution enters the vessel located directly above the inlet to the circulation pipe. Recycle saturated solution and feedstock are circulated by a pump through a heat exchanger where supersaturation is created. The supersaturated solution flows through a pipe and emerges from an outlet into a mass of growing crystals. The rate of liquor circulation is such that the crystals are maintained in a fluidized state, and partial classification occurs. Crystals that have grown to the required size sink towards the bottom of the vessel and are discharged from an outlet pipe. Excess fine crystals near the top of the crystallizer are removed in a fines-trap, and the clear liquor is reintroduced into the system.

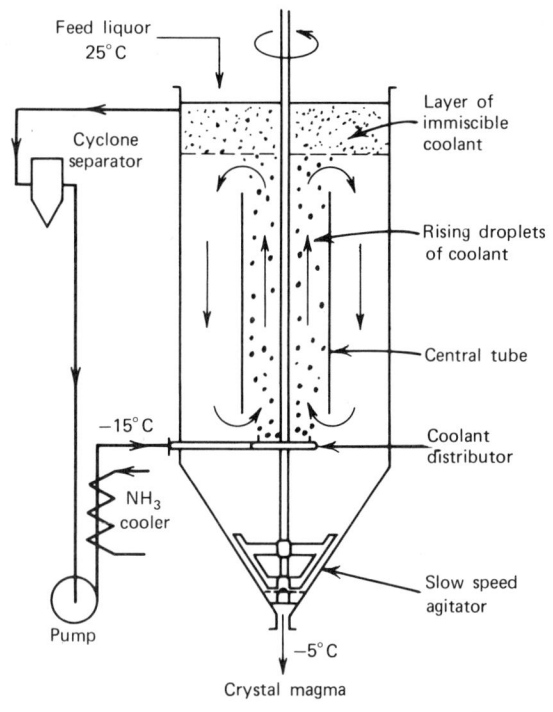

Figure 14. The Cerny direct-contact cooling crystallizer (41).

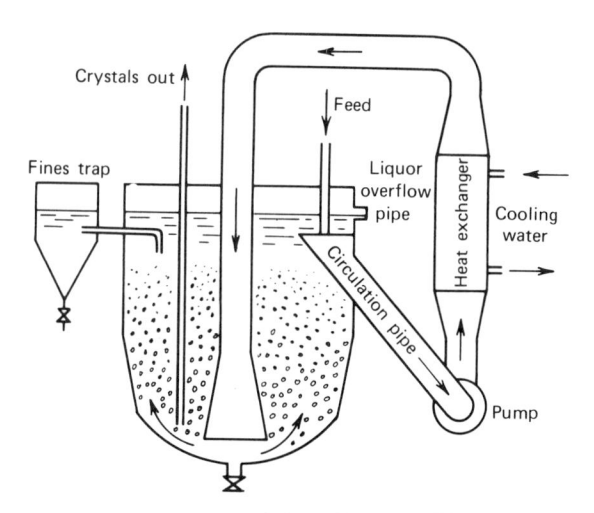

Figure 15. An Oslo cooling crystallizer.

The original practice of circulating clear liquor and maintaining a partially classified bed of crystals within the crystallizer is not essential. In fact, it is quite common nowadays to operate Oslo crystallizers with mixed suspension, ie, to circulate the magma (54). A wide variety of other types of classifying crystallizer are now available incorporating elutriating legs and other devices (2).

Evaporating and Vacuum Crystallizers. When the solubility of a solute in a solvent is not appreciably decreased by a reduction in temperature, supersaturation of the solution can be achieved by removal of some of the solvent. Evaporation techniques for the crystallization of salts have been used for centuries and the simplest method, the utilization of solar heat, is still a commercial proposition in many parts of the world (55–56). Common salt is produced from brine in enclosed calandria evaporators, appropriately called salting evaporators, and evaporating crystallizers of the calandria type, often in multieffect series, are used in sugar refining (see Chemicals from brine; Solar energy; Sugar).

The use of reduced pressure in an evaporator to aid the removal of solvent, to minimize the heat consumption or to decrease the operating temperature of the solution is common practice. However, such units are best described as reduced-pressure evaporating crystallizers. The true vacuum crystallizer operates on a slightly different principle—supersaturation is achieved by the simultaneous evaporation and adiabatic cooling of the feedstock.

The large number of different types of continuously operated crystallizers may be grouped into three basic types as shown in Figure 16. The heat exchanger in each case may be either a heater or a cooler, ie, the crystallizers may be of the evaporative or cooling types. Alternatively, the units may be operated adiabatically, ie, as vacuum crystallizers.

The high rate of recirculation through the external heat exchanger in the forced circulation crystallizer (Fig. 16a) improves heat transfer and minimizes encrustation. In the draft-tube agitated unit (Fig. 16b) a high rate of internal circulation ensures efficient mixing. These crystallizers may be fitted with internal baffles to facilitate fines removal and an elutriating leg to effect some degree of product classification (as in the Swenson DTB crystallizer shown in Fig. 17).

In the Oslo-type crystallizer (Fig. 16c) a fluidized bed is maintained by the upflow of supersaturated liquor in the annular region surrounding the central downcomer. These units were originally designed to act as classifying crystallizers, but they are now more generally operated in a mixed-suspension mode which improves productivity (54) because higher circulation rates and magma densities can be employed. Further, the active growth volume of the crystallizer is increased because the magma circulates through the vaporizer and downcomer.

The position of the feed liquor entry point in a vacuum crystallizer must be selected with considerable care. Feedstock should enter below the boiling surface of the

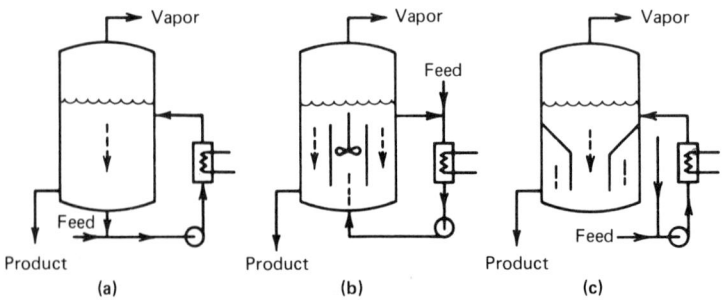

Figure 16. Three basic types of continuous crystallizer: (**a**) forced circulation, (**b**) draft-tube agitated, and (**c**) fluidized bed (Oslo).

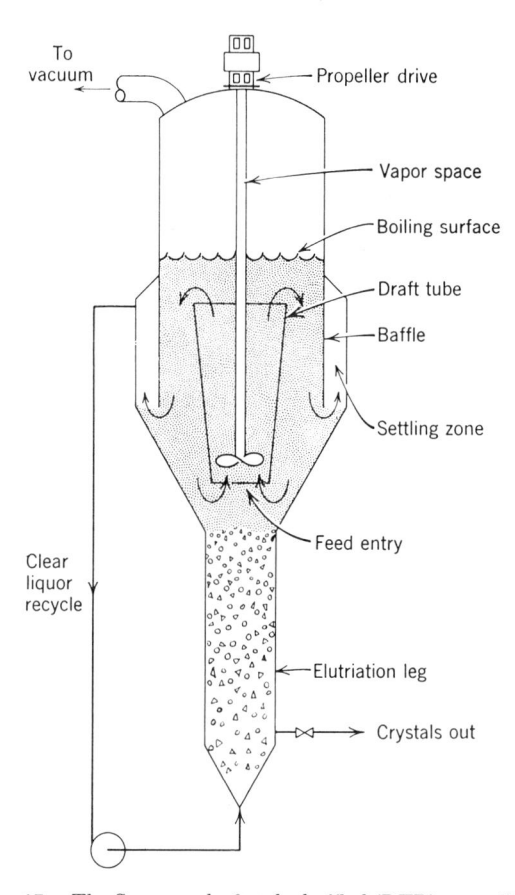

Figure 17. The Swenson draft-tube baffled (DTB) crystallizer.

magma to avoid flashing, which causes excessive nucleation. On the other hand, if it enters at more than 300 mm below the surface, it may not boil at all because its boiling point is increased by the static head. In such circumstances, it is not unusual for the feed liquor to proceed towards the product discharge point without releasing its potential supersaturation. Vacuum crystallizers must, therefore, be well-agitated if they are to operate efficiently; this is another reason why crystal size classification is rarely attempted in these units.

Design and Operation of Crystallizers

Crystallizer Selection. The temperature–solubility relationship between the solute and solvent is of prime importance in the selection of a crystallizer. For solutions that yield appreciable quantities of crystals on cooling, the choice of equipment will generally lie between a simple cooling or a vacuum cooling crystallizer. For solutions that change little in composition with a reduction in temperature, an evaporating crystallizer would normally be used, although the method of salting-out could be employed in certain cases.

The desired shape, size, and size distribution of the crystalline product will also exert an influence on the type of crystallizer selected. For the production of reasonably large uniform crystals the trend is towards the controlled suspension crystallizers, fitted with suitable fines traps, which permit the discharge of a partially classified product. The subsequent washing and drying operations are carried out much more easily; the screening of the crystal product may not be necessary.

Cost and space requirements are also important factors to be considered. Unfortunately, few comparative performances or up-to-date cost data are available for the many units commonly used in practice. However, if the cost of a small unit is known, a rough estimate of the cost of a larger one may be made by the six tenths rule, ie, cost $\propto$ (capacity)$^{0.6}$, for capacities in the range 0.5–50 t/h.

Continuous crystallizers are generally more economical in operating and labor costs than the batch units, especially for large production rates. The batch units are usually cheaper in initial capital cost. One of the main advantages of a continuous unit is that the amount of mother liquor needing reworking is usually small, sometimes less than 5% of the feedstock handled. In batch units, as much as 50% of the mother liquor may require reworking.

Although some of the simpler cooling crystallizers are relatively inexpensive, the initial cost of a mechanical unit can be fairly high, but no costly vacuum-producing or condensing equipment is required. Heavy crystal slurries can be handled in cooling units not requiring liquor circulation. Unfortunately, cooling surfaces can become coated with a hard crust of crystals, thus reducing cooling efficiency. Vacuum crystallizers have no cooling surfaces and do not suffer this disadvantage, but they cannot be used when the liquor has a high boiling point elevation. Vacuum and evaporating crystallizers generally require a considerable height.

Once a particular class of crystallizer is decided upon, the choice of a specific unit depends upon such factors as the initial and operating costs, the space availability, the type and size of crystals required, the physical characteristics of the feed liquor and crystal slurry, the need for corrosion resistance, and so on. The production rate and supply of feed liquor to the crystallizer will generally be the deciding factors in the choice between a batch and continuous unit. For example, production rates >2 t/d or liquor feed rates >20 m^3/h are often best handled on a continuous basis, but no general rules can be specified. Sugar, eg, may be produced batchwise at rates around 20 t/d per crystallizer. Particular attention must be paid, in the design of a crystallizer, to the liquor-mixing zones in the circuit. The liquor circulation loop generally includes many regions where flow streams of different temperature and composition mix, often under adiabatic conditions. These are all potential danger points where high supersaturations may be created temporarily, causing heavy nucleation which can lead to encrustation, bad performance, operating instability, and so on. It is essential that the composition and enthalpy of mixed streams are always such that only one equilibrium phase can exist under the local temperature and pressure conditions—the so-called mixing criterion (57).

Information for Design. It is useful to know what sort of information a designer generally needs in order to specify an industrial crystallizer. A detailed description of the product should be given, including its full chemical name and formula, specifying if a hydrate is required or not. A realistic purity specification should be laid down. The production rate should be given, both as t/yr and kg/h. Very generous allowances should be made for maintenance and other shutdown periods.

A solubility curve of the product in the liquor should be drawn. Care should be taken in doing this because solubility data reported in the literature usually refer to pure solutes and solvents, and these may be quite inapplicable. If impurities are known to be present in the working liquors, the relevant solubility data should be measured.

The shape and size of the crystalline product should be specified, but it is important to remember that the more rigid is the size specification the more difficult the crystallizer design becomes. An actual sample of the desired crystals, if available, is usually very helpful. Single-size specifications, such as "about 450 μm," should not be made as they have no useful meaning, but upper and lower size limits may be quoted. For example, a specification such as "90% to be retained between 600 and 300 μm (ca 30 and 50 mesh) sieves" would be quite acceptable. Alternatively, the desired median size (MS) and coefficient of variation (CV) may be designated. MS is the aperture size of the sieve that retains 50% of the product and CV is defined:

$$\text{CV} = \frac{100(a_{84} - a_{16})}{2 \, (\text{MS})} \qquad (16)$$

where a_{84} and a_{16} are the aperture sizes of the sieves that retain 84 and 16%, respectively. Values of a_{84}, a_{16}, and MS ($= a_{50}$) are readily determined from a plot of the sieve analysis (2). Thus a specification such as "MS 400 μm, CV 40%" gives a clear picture of the required product size distribution.

Scale-Up and Operating Problems. The design of a large-scale crystallizer from data obtained on a small-scale unit is never a simple matter. After installation, trial and error procedures are generally necessary before the correct operating conditions are determined for the production of a given product from the crystallizer. The scale-up crystallization equipment is more difficult than that for any of the other unit operations of chemical engineering.

One of the basic problems of crystallizer design is the choice of method for measuring design data. The mode of operation, type of apparatus, and scale of operation have all to be considered. Basically there are three choices: to measure the data in the laboratory, in a pilot plant, or on a full-scale plant. The latter is not uncommon, because many crystallizers are designed on the basis of past experience, but generally the choice lies between laboratory and pilot units. In terms of approximate physical dimensions:

	Laboratory	Pilot	Industrial
volume, L	1	100	10,000
diameter, mm	100	500	2,500

So despite the large volumetric differences (100:1), the diameter differences (5:1) are really quite small. Therefore, since diameter is an important parameter, laboratory-scale crystallization operations are not vastly different from pilot-plant work. In any case, the perils of pilot-plant operation are well known: they are expensive to design, construct, and operate, and they are frequently unstable in operation owing to the ease of pipe-line blockage and the need for carefully controlled low flow rates. Laboratory-scale operations can be increased, sometimes with advantage, to about 5 L without much trouble, and the above diameter ratios then become even more favorable.

It is now generally accepted that much useful information can be obtained in the laboratory so long as the fundamental differences between the laboratory and industrial plant scales of operation are fully appreciated. The two hydrodynamic conditions are quite different. For example, liquor paths and turn-over times are much shorter in the smaller vessels. The suspension circuit may only be a few centimeters in a beaker, whereas it may be 10 or 20 m in a full-scale plant. If the solution desupersaturates rapidly, very high circulation rates would be needed in the large crystallizer to utilize fully the working volume. In a laboratory beaker there is generally no problem.

The exact scale-up of crystallizers is not possible because it would be necessary to preserve similar flow characteristics of both liquid and solid phases and identical temperatures and supersaturations in all equivalent regions. The scale-up of simple one-phase agitated vessels has long been recognized as a difficult problem: if the conditions in two agitated vessels are to be similar in all respects, geometrical (shape), kinematic (velocity), and dynamic (force) similarity have to be maintained. For example, both the Reynolds and Froude numbers:

$$Re = \frac{\rho n d^2}{\eta} \text{ and } Fr = \frac{n^2 d}{g} \tag{17}$$

should be kept constant (d = agitator diameter, n = agitator speed (rev/min); ρ = density; η = viscosity; g = gravitational acceleration) for exact scale-up. However, if the agitator diameter is increased by a factor of 4, the agitator speed must be decreased by a factor of 2 if Fr is to be kept constant ($n \propto d^{-1/2}$) but by a factor of 16 if Re is to remain the same ($n \propto d^2$). It is impossible, therefore, to satisfy both criteria, and some compromise must be made. For agitated crystallizers, more importance is attached to Re (which influences fluid friction, heat and mass transfer) than to Fr, so scale-up is generally based on the Reynolds number.

The power input P to an agitator is related to Re and Fr by:

$$\frac{P}{\rho n^3 d^5} = f(Re, Fr) \tag{18}$$

At high values of Re (>10,000) the dimensionless group $P/\rho n^3 d^5$ (the power number) is roughly constant so, under these conditions, if the agitator speed is doubled the power input would have to be increased eightfold. If the agitator diameter is doubled, its speed remaining constant, the power input would have to be increased by a factor of 32.

The scale-up of crystallizers is not an easy task. Caution should always be exercised in the use of data from small experimental units. Over the years, many empirical rules for crystallizer scale-up have been postulated. One recent analysis based on considerable practical experience (58) suggests that as crystal-agitator contacts are probably the prime source of nuclei in an industrial crystallizer (see Nucleation Kinetics), scale-up should be done by maintaining the value R^2/T constant, keeping the magma density and residence times constant. R = agitator tip speed and T = turn-over time, ie, crystallizer volume/volumetric liquor circulation rate.

The processes of crystal growth and nucleation are powerfully influenced by the system hydrodynamics (59), thus the method of agitation in a crystallizer is very important. However, there is very little reliable design information available for agitated vessels for solid–liquid systems. Different types of agitators produce different flow patterns. For example, a centrally mounted propeller (Fig. 18a) induces tangential flow and a well-established pattern of circulation. At high speeds a vortex is generally

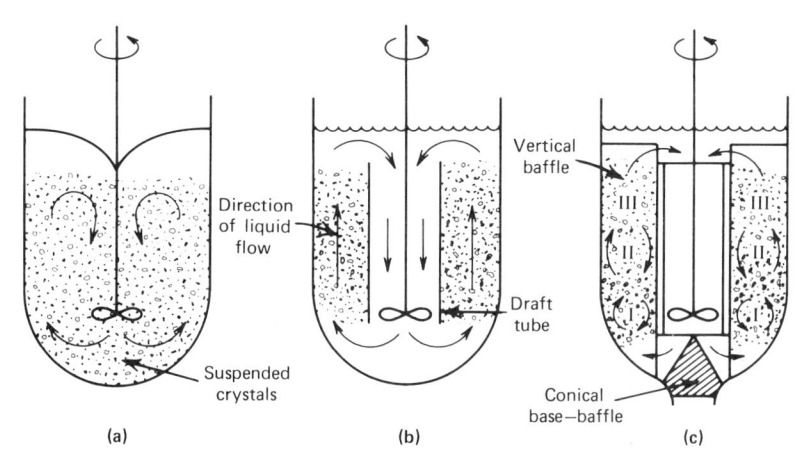

Figure 18. Simplified flow patterns in agitated vessels: (*a*) central propeller, (*b*) draft-tube, and (*c*) draft-tube with wall baffles and conical deflector.

produced, but the introduction of wall baffles creates a chaotic turbulent motion. Draft tube agitation (Fig. 18**b**) gives a very different type of circulation. The propeller acts as a pump and the liquid may be forced to flow either up or down the draft tube. Tangential flow occurs in the annular zone unless wall baffles are installed.

It is very difficult to ensure efficient mixing in suspensions of solid particles. One partially successful arrangement is shown in Figure 18**c**. Four wall baffles extend down to the base of the vessel to create four equal annular zones. A conical deflector eliminates the dead space at the bottom of the vessel underneath the draft tube which operates with a down-flow. However, despite these precautions the contents of the vessel are rarely perfectly mixed and frequently three distinct flow regimes can be identified. In region I crystals circulate vigorously and stay mainly in this zone which is similar to that in a turbine agitated vessel. Region II is less vigorous than region I. Crystals tend to migrate up and down. This zone is rendered unstable by voids which are propagated periodically (a behavior often seen in liquid fluidized beds) due to the pumping action of the impeller. Region III is a low-voidage, gently fluidized zone in which there is a slow circulation of crystals. The overall system voidage can affect considerably the extents of the three zones, but it is very difficult to quantify these effects (60).

Modes of Operation. Many industrial crystallizers are of the mixed-suspension type in which crystals of all sizes are dispersed reasonably uniformly throughout the working zone. Such crystallizers may be divided, accordingly to the type of product discharge, into mixed suspension, mixed product removal (MSMPR) and mixed suspension, classified product removal (MSCPR) units. Both types can be operated so that either all the mother liquor leaving the crystallizer leaves together with the product crystals or a part of the mother liquor is allowed to overflow separately.

Four possibilities are shown diagrammatically in Figure 19. In the simple MSMPR crystallizer (Fig. 19**a**) the crystal and liquor residence times are identical. However, by allowing some of the liquor to overflow from the crystallizer (Fig. 19**b**) the crystal residence time can be increased and rendered independent of the liquor residence time. Alternatively, the liquor overflow can be recycled back into the crystallizer through

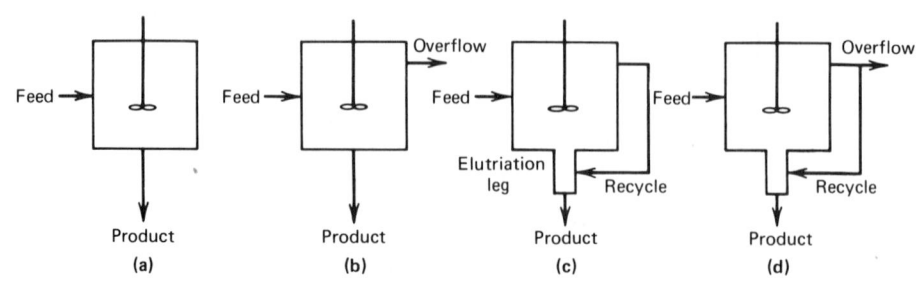

Figure 19. Basic types of continuously operated mixed suspension crystallizers: (**a**) MSMPR, (**b**) MSMPR with liquor overflow, (**c**) MSCPR, and (**d**) MSCPR with liquor overflow and recycle.

an elutriating leg, thus imparting some classifying action on the product crystals (Fig. 19**c**). The MSCPR crystallizer may also be operated with some liquor overflow (Fig. 19**d**). These different modes of action can be seen in the crystallizers illustrated above.

One reason for allowing liquor to overflow from a continuously operated crystallizer is to increase the residence time of crystals in the growth zone. Another is to permit excess fine crystals to be removed from the system: most crystallizers suffer from excessive nucleation, and to produce reasonably large crystals these excess fines must be removed. If the liquor overflow is to be recirculated, it may be passed through a fines trap (eg, a steam-heated dissolver) for this purpose.

The product crystal size distribution can often be altered significantly by using a number of MSMPR crystallizers in series where the entire magma flows from stage to stage. In particular, the CV (eq. 16) may be reduced compared with that from a single stage MSMPR unit, although the mean crystal size is often reduced. The theory of in-series or cascade operation is reasonably well developed (6,61–63) but the idealized models are often difficult to achieve in practice.

Apart from increasing the overall residence time and allowing crystal growth to occur at elevated temperatures in the early stages, which could result in an increase in the product crystal size in some cases, the main practical advantages of cascade operation are generally associated with the problems of heat transfer. For example, owing to the stepwise reduction of the overall temperature drop along a cooling cascade, the load on the cooling circuit is eased.

Population Balance. In order to achieve a complete description of the crystal size distribution in a continuously operated crystallizer, it is necessary to quantify the nucleation and growth processes and to apply the conservation laws of mass, energy, and crystal population. The exploitation of the concept of the population balance, in which all particles in a system must be accounted for, has led to major advances in the analysis of crystallization processes (61).

Application of the population balance is best demonstrated with reference to the simple case of a continuously operated MSMPR crystallizer (see Fig. 19**a**) assuming steady state operation; no crystals in the feed stream; all crystals of the same shape, characterized by a chosen linear dimension L; no break-down of crystals by attrition; and crystal growth rate independent of crystal size. The fundamental relationship between crystal size L and population density n (number of crystals per unit size per unit volume of system) is:

$$n = n^o \exp(-L/G\tau) \qquad (19)$$

where n^o is the population density of nuclei (zero-sized crystals). Equation 19 characterizes the crystal size distribution. Empirical expresssions of the rates of nucleation B and growth, $G\ (= dL/dt)$ written in terms of supersaturation as:

$$B = k_1 \Delta c^b \tag{20}$$

and

$$G = k_2 \Delta c^g \tag{21}$$

respectively, may be combined to give:

$$B = k_3 G^i \tag{22}$$

where

$$i = b/g \tag{23}$$

Furthermore, since the nucleation rate:

$$B = \left.\frac{dN}{dt}\right|_{L=0} = \left.\frac{dN}{dL}\right|_{L=0} \left(\frac{dL}{dt}\right) \tag{24}$$

the relationship between the two processes may be expressed as:

$$B = n^o G \tag{25}$$

or

$$n^o = k_4 G^{i-1} \tag{26}$$

Evaluation of the nucleation and growth processes may be made from experimental measurement of the crystal size distribution in a crystallizer operating under steady-state conditions.

For instance, equation 19 indicates that a plot of log n versus L should give a straight line of slope $-1/G\tau$ and intercept at $L = 0$ equal to n^o (see Fig. 20a). If the residence time τ is known then the crystal growth rate G can be calculated. Similarly, a plot of log n^o versus log G should give a straight line of slope $i - 1$ (eq. 26 and Fig. 20b). Thus the kinetic order of nucleation b may be evaluated if the order of the growth process g is known (eq. 23).

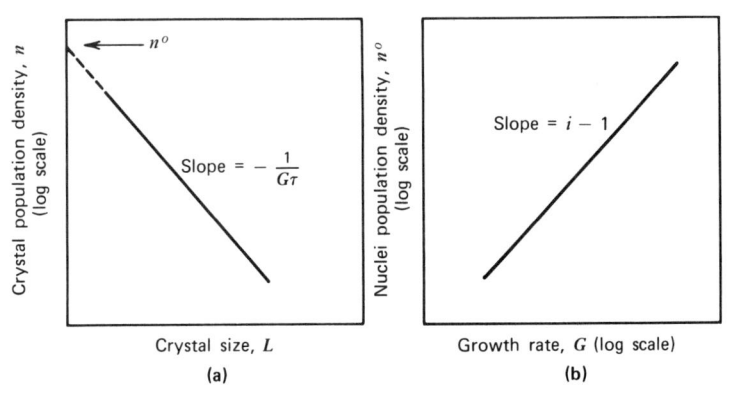

Figure 20. Population plots characterizing (a) the crystal size distribution and (b) the nucleation and growth kinetics for a continuous MSMPR crystallizer.

Particle size distributions may be described in various ways. Equation 19, eg, represents a distribution on a number basis, and this has obvious utility for population balances. On the other hand, a distribution on a mass basis is often required, eg, for mass balance purposes.

The number of crystals N, up to size L, is given by:

$$N = \int_0^L n\, dL$$
$$= n^o G\tau[1 - \exp(-L/G\tau)] \tag{27}$$

which for $L \to \infty$ gives the total number of crystals in the system:

$$N_T = n^o G\tau \tag{28}$$

The mass of crystals M up to size L is given by

$$M = \alpha\rho \int_0^L nL^3\, dL \tag{29}$$

where α is a volume shape factor, defined by $\alpha = $ volume/L^3, and ρ is the crystal density. For $L \to \infty$ equation 29 gives the total mass of crystals in the system:

$$M_T = 6\alpha\rho n^o(G\tau)^4 \tag{30}$$

This quantity, the magma density (mass of crystals per unit volume of suspension) M_T, is controlled by the feedstock and operating conditions of the crystallizer. As the population density of nuclei n^o is related to the growth and nucleation rates (eq. 25), equation 30 may be rewritten in an alternative form:

$$M_T = 6\alpha\rho B G^3 \tau^4 \tag{31}$$

The mass of crystals dM in a given size range dL is:

$$dM = \alpha\rho n L^3\, dL \tag{32}$$

so the mass fraction in the particular size range is dM/M_T. Therefore, from equations 30 and 31, the mass distribution is given by:

$$\frac{dM(L)}{dL} = \frac{dM}{M_T} \cdot \frac{1}{L} = \frac{nL^3}{6\, n^o\, (G\tau)^4} \tag{33}$$

which, from the population density relationship (eq. 19), becomes:

$$\frac{dM(L)}{dL} = \frac{L^3 \exp(-L/G\tau)}{6\, (G\tau)^4} \tag{34}$$

The peak of this mass distribution (the dominant size L_D of the crystal size distribution) is found by maximizing equation 34, which gives:

$$L_D = 3\, G\tau \tag{35}$$

As nucleation in a crystallizer depends on crystal contacts as well as on the level of supersaturation, the empirical nucleation equation 22 can be expanded to include the magma density:

$$B = k_5 M_T^j G^i \tag{36}$$

and combining equations 31, 35, and 36 for the simple, but quite realistic, case of $j = 1$:

$$G = \left[\frac{27}{2\,\alpha\rho k_5 L_D{}^4}\right] 1/i - 1 \tag{37}$$

Combination of equations 35 and 37 yields the interesting relationship:

$$L_D \propto \tau^{(i-1)/(i+3)} \tag{38}$$

which enables the effect of changes in residence time to be evaluated. For example, if $i = 3$, a typical value for many inorganic salt systems, a doubling of the residence time would only increase the dominant product crystal size by 26%. However, to double the residence time it would be necessary either to double the crystallizer volume or halve the volumetric feed rate, and hence halve the production rate. So it is clear that residence time manipulation may not be very effective for controlling the product crystal size, contrary to popular belief.

Although discussion has been confined in this section to the MSMPR configuration, the concepts of the population balance can be applied to crystallizers having other flow patterns and relevant design relationships have been developed (61,64–65).

Fines Removal and Product Classification. Most crystallizers suffer from excessive nucleation and to produce reasonably large crystals it is generally necessary to remove the excess fines. In terms of mass, the small crystals are insignificant; the cumulative mass of crystals in an MSMPR crystallizer varies as the fourth power of their size. Thus, eg, crystals up to half the product mean size constitute only $(\tfrac{1}{2})^4$, one sixteenth, of the total mass in suspension. However, in terms of numbers, the small crystals exert a dominant effect in a crystallizer. The number of crystals smaller than one tenth of the product mean size may exceed that of all the larger crystals present.

For crystallizers operating with full or partial classification, the fine crystals tend to circulate with the liquor, leaving the larger crystals in the crystallization zone. It is possible, therefore, to remove the excess fines by circulating the liquor through an internal or external fines trap (66).

For efficient operation the separation of fines should be at the smallest feasible size, but at a large enough rate to materially decay the population density in the size range of removal (67). Figure 21 shows the effect of an internal fines removal system.

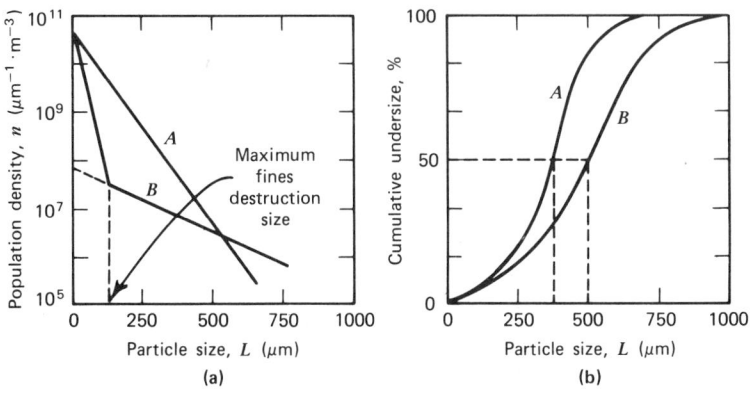

Figure 21. The effect of fines removal on (**a**) the population density and (**b**) the crystal size distribution. A, no fines destruction; B, fines destruction (67).

The small-sized crystals undergo a very rapid decay up to the maximum size of particles in the fines trap. Thereafter, the decay proceeds at the rate appropriate to mixed product removal operation. It is possible to extrapolate the population density decay line (Fig. 21**a**) back to zero to obtain an effective nuclei density, and this is much lower than the actual value produced by the crystallizer. The mean product size (Fig. 21**b**) is increased by fines destruction. Analyses of the influence of fines trapping on the performance and dynamic response of continuous crystallizers have been presented by several authors (68).

Product classification is frequently effected in crystallizer practice with an elutriating device, hydrocyclone, or wet screen located in a recycle leg of the crystallizer. However, it is not generally appreciated that although this practive may narrow the size range, it leads to the production of smaller crystals (67). In other words, it decreases both the median size (MS) and coefficient of variation (CV) of the product crystals.

This behavior is clearly seen in Figure 22. The cases of mixed and classified product removal are compared on the population density plot (Fig. 22**a**). With classified product operation the accelerated removal that occurs for sizes in excess of the classification size L_c can be seen. The net result, however, is the production of a smaller product size (Fig. 22**b**).

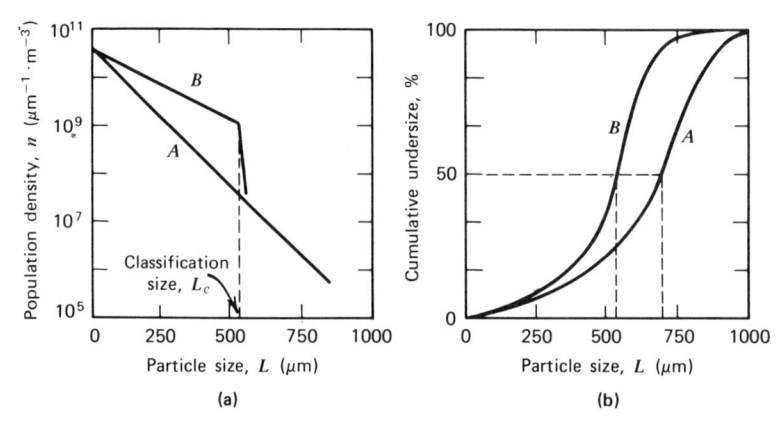

Figure 22. The effect of classified product removal on (**a**) the population density and (**b**) the crystal size distribution. A, mixed product removal; B, classified product removal (67).

Unstable Operation. A continuously operated crystallizer will often exhibit cyclic changes in production rate and product size even though the operating conditions may appear to be steady. It used to be thought that a continuous crystallizer is self-stabilizing. The reasoning was that an increase in supersaturation leads to a higher nucleation rate, but as the total available crystal surface area increases, owing to the growth of the nuclei, the supersaturation decreases again and the nucleation rate is reduced. Similarly, when product crystals are withdrawn, the total crystal surface area decreases and this in turn causes an increase of the supersaturation and the establishment of a high nucleation rate, and so on. However, these effects are subject to a considerable time-lag because newly formed crystals have no appreciable surface area for a long time. Therefore, before the stabilizing action can occur, large numbers of crystals can be formed which will later reduce the supersaturation below its steady-state value.

The consequent slow nucleation will then lead to a decrease of the total crystal surface area below its steady-state value and this in turn will cause an increase of the supersaturation above its steady-state value, and so on. The result of this sequence of events is the occurrence of periodic changes in crystallizer output and the crystallizer is said to cycle.

Periodic behavior is most pronounced just after the start-up of a continuous crystallizer and in most cases the fluctuations are subsequently damped, resulting in near-steady-state operating conditions. However, in certain cases damping is ineffective and cycling may be prolonged or even permanently sustained.

When discussing unstable operation, a clear distinction should be drawn between transients caused by operational changes, eg, in feed rate or condition, product discharge, fines destruction, etc, and instability which arises from the inherent characteristics of the crystallizing system.

It may take 10 or more residence times, eg, 5–20 h in an industrial crystallizer, after an operational change to eliminate transients, and thus to operate under steady-state conditions demands very careful control to avoid system upsets. However, the capacity of a crystallizer to sustain oscillations of crystal size, production rate, etc, depends greatly on the relative order of the nucleation and growth rate, ie, on i ($= b/g$) in equation 22.

Stability tends to increase with increasing crystal growth rate and slurry density, and with decreasing nucleation order (b in eq. 20), product size, and the relative amount of crystals withdrawn from the crystallizer. Continuous seeding can also have a stabilizing effect. Instability is likely when the crystallizer is operated at either very low or very high supersaturations. Fines removal and product classification, and especially both together, tend to make a crystallizer unstable in operation.

A considerable number of attempts have been made in recent years to analyze the problems of crystallizer instability (7,61,69–77).

Encrustation. At some stage during their operating cycle, most crystallizers develop crystalline deposits on their internal surfaces. Such terms as scaling, fouling, salting, incrustation, and encrustation are commonly used in this context. Scaling generally refers to the hard thin layers such as those deposited on heat exchanger tubes, whereas salting is often used for the massive build-up of crystalline material on crystallizer walls. However, for convenience, only one term, encrustation, is used here to cover all types.

Encrustation reduces heat transfer and/or evaporation rates and thus reduces production. Large lumps can break away and cause mechanical damage within the crystallizer, or in associated equipment, and small pieces can contaminate the product.

Encrustation on cooling surfaces can be minimized by reducing the temperature difference across the wall of the heat exchanger. Cooling surfaces should always be as smooth and flat as possible, not only to minimize encrustation but also to facilitate its removal. Crystals should never be chipped from the wall of a crystallizer, for tiny scratches on the surface can become undesirable seed centers. Melting or dissolution is the only safe method of crystal scale removal. The use of ultrasonics (qv) to prevent encrustation seems promising in certain situations (78). Cold spots on vessel and pipe walls are potential sites where scale can form and the lagging of such points can avoid many operating problems (see Insulation, thermal).

In vacuum and evaporative crystallizers, encrustation frequently occurs in the vapor space: very thick layers of solute can build up on the walls, particularly near the liquid–vapor interface. Such a build-up increases the vapor release velocity and intensifies mother liquor entrainment. Problems also occur if large pieces of solid material break away from the wall—they can stop agitators or circulating pumps and block vessel openings and liquor lines.

Encrustation in vapor spaces can often be reduced by limiting the level of superheating, and hence the supersaturation generated in the solution. A small quantity of wash water or dilute liquor is frequently used to keep the walls of the vapor space clean. This may be done on either a continuous or intermittent basis. It is also common practice to stop production at periodic intervals for a complete washout. The time intervals vary widely from a few hours to several months.

The rate at which scale forms and subsequently grows depends not only on the nature of the surface and the supersaturation produced, but also on the nucleation and growth kinetics of the crystallizing system. Since the kinetics can be greatly influenced by trace impurities and added habit modifiers, these substances can have important effects on encrustation in a crystallizer. Any change in feed liquor composition or the addition of habit modifiers may, therefore, alter the scaling characteristics of the system for better or worse. It is important to be aware that such changes are likely and, if possible, make initial trials on the pilot scale.

Washing of Crystals. The actual crystals produced in a crystallizer may themselves be essentially pure, yet after they have been separated from their mother liquor and dried, the resultant crystalline mass may be relatively impure. Even if no inclusions are formed, the removal of mother liquor is frequently inadequate. Crystals retain a small quantity of mother liquor on their surfaces by adsorption and a larger amount within the voids of the particulate mass owing to capillary attraction. If the crystals are irregular, the amount of mother liquor retention within the crevices may be considerable; crystal clusters and agglomerates are notorious in this respect.

Industrial crystallization processes from the liquid phase must be followed by an efficient liquid–solid separation. Centrifugal filtration can often reduce the mother liquor content of granular crystalline masses to about 5–10% although irregular small crystals may retain more than 50%. It is extremely important, therefore, not only to use the most efficient type of filter but to produce regular crystals in the crystallizer.

After filtration, the product is usually given a wash to reduce still further the amount of impure mother liquor retained. Cakes of crystalline substances should not be too thick if a washing operation is required; otherwise the wash liquor becomes saturated long before it passes through the mass and the mother liquor impurities are not removed effectively. If the crystals are very soluble in the mother solvent, another liquid, in which the substance is relatively insoluble, may be used for washing purposes. The wash liquid should be miscible with the mother solvent. For example, water could be used for washing substances crystallized from methanol; whereas methanol could be used for washing substances crystallized from benzene. Unfortunately, this two-solvent method of working usually means that a solvent recovery unit is required. Alternatively, a wash liquor consisting of a cold saturated solution of the pure substance in the pure mother solvent could be used if the crystals were appreciably soluble in the solvent. The contaminated wash liquor could then be recycled back to the crystallizer or reused in some other way.

When simple washing is inadequate, two mother liquor removal stages may be necessary: the wet crystals are removed from the filter, redispersed in wash liquor and then filtered again. There may be an appreciable loss of yield after such a washing process but this would be less than the loss after a complete recrystallization.

When crystallization has been carried out by the reaction method, the crystalline product may be relatively insoluble in the working solvent. On the other hand, the mother liquor may contain large quantities of another soluble material, resulting from the chemical reaction, and simple filtration and washing may be quite inadequate for its complete removal, especially if the crystalline particles are very small or irregular. For example, when barium sulfate is precipitated by mixing solutions of barium chloride and sodium sulfate, it is difficult to remove the resulting sodium chloride contamination by conventional filtration and washing. In such cases, a sequence of dispersion-washing and decantation steps may be very effective.

The wash liquor requirements for decantation washing can be deduced as follows. For simplicity it is assumed that the soluble impurity is in solution and solution concentrations are constant throughout the dispersion vessel.

Batch Operation. If Y_o and Y_n denote the impurity concentrations in the crystalline material (kg of impurity/kg of product) initially and after washing stage n, respectively, and F is the fraction of liquid removed at each decantation, then from a mass balance:

$$Y_n = Y_o(1 - F)^n \tag{39}$$

which may be rewritten in the form:

$$\ln(Y_n/Y_o) = n \ln(1 - F) \tag{40}$$

Continuous Operation. For washing on a continuous basis, where fresh wash liquid enters the vessel continuously, and liquor is continually withdrawn through a filter screen, a mass balance over the unit gives:

$$V dY = -Y dW \tag{41}$$

which may be written as:

$$\ln(Y_n/Y_o) = -W/V \tag{42}$$

where Y_o and Y_n are the initial and final impurity concentrations. V and W are the volumes of the liquor in the vessel and wash water, respectively. Combination of equations 40 and 42 gives:

$$n \ln(1 - F) = -\frac{W}{V} \tag{43}$$

or, rearranging and dividing by F:

$$\frac{W}{nFV} = \frac{-\ln(1 - F)}{F} \tag{44}$$

W and nFV represent the wash liquor requirements for continuous and batch operations, respectively. Equation 44 is useful for comparing the two methods (2).

Caking of Crystals. One of the most troublesome properties of crystalline materials is their tendency to bind together, or cake, on storage. The causes of caking may vary for different materials. The crystal size, shape, moisture content, the pressure under which the product is stored, temperature variations during storage and the storage time can all contribute to the compacting of the crystal mass into a solid lump.

Caking is generally caused by the crystal surfaces becoming damp; the solution which is formed evaporates later and unites adjacent crystals with a cement of recrystallized solid. Crystal surfaces can become damp in a number of ways; the product may contain traces of residual solvent left behind owing to inefficient drying, or the moisture may come from external sources.

If the partial pressure of water in the atmosphere is greater than the vapor pressure that would be exerted by a saturated aqueous solution of the crystals at that temperature, water will be absorbed at the crystal surfaces and cause partial dissolution. If, later, the atmospheric moisture content is reduced below the vapor pressure of the saturated salt solution, the surface moisture will evaporate and the crystal mass will cake.

If the crystals are stored under pressure, eg, when packed in bags stacked on top of one another, the crystals are forced into close contact and may even be crushed. Furthermore, the local pressure at points of contact can be extremely high and if the solubility of the salt in water increases with pressure, traces of supersaturated solution may be formed. This solution will flow into the voids and crystallize.

The presence of impurities in the product may be the cause of trouble. For example, traces of calcium chloride in sodium chloride would cause the crystals to become damp at very low humidities. A simple test for determining the atmospheric conditions under which crystals will absorb moisture consists of placing samples of the crystals in desiccators containing solutions of various strengths of sulfuric acid. The equivalent relative humidities of the atmospheres in the desiccators can be calculated from the known vapor pressures of the solutions. Atmospheres of constant relative humidity can also be obtained by using saturated solutions of various salts such as those listed in Table 7.

One obvious method to reduce the possibility of caking is to pack the crystals in a dry atmosphere, store them in an airtight container, and prevent any pressure being applied during storage. These desirable conditions, however, cannot always be obtained. Caking can be minimized by reducing the number of contacts between the crystals, and this can be done by endeavoring to produce crystals of uniform size and shape.

Table 7. Relative Humidities of the Atmospheres Above Saturated Solutions of Various Pure Salts at 15°C

Substance	Stable phase	Relative humidity, %
lead nitrate	$Pb(NO_3)_2$	98
sodium sulfate	$Na_2SO_4.10H_2O$	93
sodium carbonate	$Na_2CO_3.10H_2O$	90
barium chloride	$BaCl_2.2H_2O$	88
potassium bromide	KBr	84
ammonium sulfate	$(NH_4)_2SO_4$	81
sodium chloride	NaCl	78
sodium chlorate	$NaClO_3$	75
sodium nitrite	$NaNO_2$	66
sodium bromide	$NaBr.2H_2O$	58
sodium dichromate	$Na_2Cr_2O_7.2H_2O$	52
potassium carbonate	$K_2CO_3.2H_2O$	43
calcium chloride	$CaCl_2.6H_2O$	32
lithium chloride	$LiCl.H_2O$	15

It is generally advantageous to produce crystals as large as possible to minimize caking. The larger the crystals the smaller will be the exposed surface area per unit mass. However, the actual size is not as important as the size distribution and shape (habit) of the crystals. The minimum caking tendency is shown by a mass of monosized spherical particles. Crystalline products with wide size distributions are more prone to caking because the voids between the larger crystals are partially filled with smaller crystals, thus increasing the number of points of contact. Any departure from the spherical shape increases the tendency to caking. Needles and plates, which give areas as well as points of contact between crystals, are generally undesirable.

The use of anticaking agents is a popular method for controlling storage properties and retaining the freeflow characteristics of crystalline masses (2,79). Table salt, eg, is usually coated with a very fine dust of magnesium carbonate. Icing sugar is coated with tricalcium phosphate or corn flour. Other anticaking agents that find use for industrial purposes include aluminum powder, chalk, calcium sulfate, kaolin, diatomaceous earth, silica, magnesium aluminum silicate, zinc oxide and miscellaneous synthetic resins. These finely divided substances must have a good covering power so that small quantities will give the required protection. Hydrocarbon oils are sometimes used to coat hygroscopic materials, eg, calcium nitrate, and this method also reduces the dust hazard.

In recent years, trace additives have been used to control caking either by treating the crystals with a solution containing small quantities of a habit modifying substance at the washing stage or by maintaining a critical level of additive in the crystallizing vessel. When the crystals cake, the cement of crystalline material that exists in a modified habit is very weak and the crystal mass easily breaks down on handling. This technique has proved very effective with a wide range of inorganic salts (2,80–81).

Nomenclature

A	= surface area of particle
B	= rate of nucleation
c	= concentration
CV	= coefficient of variation
d	= agitator diameter
F	= fraction of liquid removed at decantation
Fr	= Fronde number
G	= rate of growth
g	= gravitational acceleration
J	= rate of homogeneous nucleation
k	= Boltzmann constant; mass transfer coefficient
K_N	= nucleation rate constant
L	= size of crystals
M	= mass of crystals
m	= particle of mass
N	= number of crystals
n	= population density; agitator speed
q	= heat of crystallization
R	= ratio of molecular weights
Re	= Reynolds number
R_G	= mass deposition rate
S	= supersaturation
V	= water lost by evaporation; volume of liquor in vessel

W	= volume of liquor in wash water
Y	= crystal yield
α	= volume shape factor; aperature size
β	= surface shape factor
γ	= surface energy
η	= viscosity
θ	= temperature
λ	= latent heat of evaporation of solvent
ν	= number of ions
ρ	= crystal density
τ	= batch time
υ	= molecular volume

BIBLIOGRAPHY

"Crystallization" in *ECT* 1st ed., Vol. 4, pp. 619–636, by A. Ralph Thompson, University of Pennsylvania; "Crystallization" in *ECT* 2nd ed., Vol. 6, pp. 482–515, by J. W. Mullin, University of London.

1. P. Hartman, ed., *Crystal Growth,* Elsevier Scientific Publishing Co., Inc., New York, 1973.
2. J. W. Mullin, *Crystallization,* 2nd ed., Butterworths, London, Eng., 1972.
3. A. E. Nielsen, *Kinetics of Precipitation,* Pergamon, Oxford, Eng., 1964.
4. R. F. Strickland-Constable, *Kinetics and Mechanism of Crystallization,* Academic Press, Inc., New York, 1968.
5. A. C. Zettlemoyer, ed., *Nucleation,* Marcel Dekker, Inc., New York, 1969.
6. J. Nyvlt, *Industrial Crystallization from Solutions* (transl. from Czech.), Butterworths, London, Eng., 1971.
7. *AIChE Symp. Ser.* **68**(121), (1972).
8. J. Estrin in W. R. Wilcox, ed., *Chemical Vapor Transport, Secondary Nucleation and Mass Transfer in Crystal Growth,* Marcel Dekker, Inc., New York, 1976.
9. G. D. Botsaris in J. W. Mullin, ed., *Industrial Crystallization,* Plenum Press, New York, 1976, p. 3.
10. R. F. Strickland-Constable in ref. 9, p. 33.
11. N. A. Clontz and W. L. McCabe, *Chem. Eng. Prog. Symp.* **67**(110), 6 (1971).
12. J. Garside, G. Rusli, and M. A. Larson, *paper presented at AIChE meeting at Atlanta, Georgia, Mar. 1978.*
13. J. C. Brice, *Growth of Crystals from the Melt,* 1965, *Growth of Crystals from Liquids,* 1973, North-Holland, Amsterdam, The Netherlands.
14. H. E. Buckley, *Crystal Growth,* John Wiley & Sons, Inc., New York, 1951.
15. B. Pamplin, ed., *Crystal Growth,* Pergamon, Oxford, Eng., 1975.
16. E. V. Khamskii, *Crystallization from Solutions* (transl. from Russian), Consultants Bureau, New York, 1969.
17. E. G. Cooke, *Krist. Tech.* **1,** 119 (1966).
18. M. A. Larson and J. W. Mullin, *J. Crystal Growth* **20,** 183 (1973).
19. G. Deicha, *Lacunes des Cristeaux et leurs Inclusions Fluids,* Masson, Paris, Fr., 1955.
20. H. E. C. Powers, *Sugar Technol. Rev.* **1,** 85 (1970).
21. W. R. Wilcox and V. H. S. Kuo, *J. Crystal Growth* **19,** 221 (1973).
22. E. Roedder, ed., *COFFI Proceedings,* U.S. Geological Survey, U.S. Government Printing Office, Washington, D. C., 1968 ff.
23. J. E. Carless and A. A. Foster, *J. Pharm. Pharmacol.* **18,** 697, 815 (1966).
24. J. Scrivanek and co-workers in ref. 9, p. 173.
25. B. Chalmers, *Principles of Solidification,* John Wiley & Sons, Inc., New York, 1964.
26. D. Elwell and H. J. Scheel, *Crystal Growth from High Temperature Solutions,* Academic Press, Inc., New York, 1975.
27. J. J. Gilman, ed., *The Art and Science of Growing Crystals,* John Wiley & Sons, Inc., New York, 1963.
28. H. K. Hanisch, *Crystal Growth in Gels,* Pennsylvania State University Press, University Park, Pa., 1970.
29. R. A. Laudise, *The Growth of Single Crystals,* Prentice-Hall, Englewood Cliffs, N. J., 1970.
30. W. G. Pfann, *Zone Melting,* 2nd ed., John Wiley & Sons, Inc., New York, 1966.

31. M. Zief and W. R. Wilcox, eds., *Fractional Solidification,* Marcel Dekker, Inc., New York, 1967.
32. J. G. D. Molinari in ref. 31, Chapters 13 and 14.
33. *Metallwerk A.G.,* Buchs, Switz., brochure.
34. J. E. Powers and co-workers in ref. 31, Chapt. 11; *AIChE J.* **16,** 648, 1055 (1970).
35. M. R. Player, *Ind. Eng. Chem. Proc. Des. Dev.* **8,** 210 (1969).
36. D. L. McKay in ref. 31, Chapt. 16.
37. J. A. Brodie, Union Carbide (Australia) Ltd., personal communication; *Chem. Eng.* **85,** 73 (Feb. 13, 1978).
38. G. J. Arkenbout, *Chemtech* **6,** 596 (Sept. 1976).
39. J. W. Mullin, J. Nyvlt, and A. G. Jones, *Chem. Eng. Sci.* **26,** 369 (1971); **29,** 105, 1075 (1974).
40. J. A. Fernandez-Lozano, *Ind. Eng. Chem. Proc. Des. Dev.* **15,** 445 (1976).
41. A. W. Bamforth, *Industrial Crystallization,* Leonard Hill, London, Eng., 1965.
42. A. G. Roberts and K. D. Shah, *Chem. Eng. (London)* 748 (Dec. 1975).
43. R. A. Findlay and J. A. Weedman in K. A. Kobe and J. J. McKetta, eds., *Advances in Petroleum Chemistry and Refining,* Vol. 1, Interscience Publishers, New York, 1958.
44. J. P. Tare and M. R. Chivate, *AIChE Symp. Ser.* **72**(153), 95 (1976).
45. G. Matz, *Kristallization,* 2nd ed., Springer, Berlin, 1969.
46. R. C. Bennett in R. H. Perry and C. H. Chilton, eds., *Chemical Engineers' Handbook,* 5th ed., McGraw Hill, New York, 1973, Section 19.
47. A. J. Armstrong, *Chem. Process Eng.* **51**(11), 59 (1970).
48. D. A. Gudelis, J. F. Eagen, and J. B. Bushnell, *Hydrocarbon Process.,* 141 (Sept. 1973); *Oil Gas J.* **80** (Oct. 1975).
49. R. S. Atkins, *Hydrocarbon Process.* 127 (Nov. 1970).
50. J. McDermott, *Desalination by Freeze Concentration,* Noyes Data Corp., Park Ridge, N. J., 1971.
51. A. J. Barduhn, *Chem. Eng. Progr.* **71**(11), 80 (1975).
52. K. Schmock, *Chemische Technik (Leipzig)* **18**(1), 661 (1966).
53. I. P. Usyukin and co-workers, *Soviet Chem. Ind.* **4,** 267 (1973).
54. W. C. Saeman, *AIChE J.* **2,** 107 (1956).
55. S. Nadel, *Eng. Min. J.* **166**(10), 84 (1965).
56. C. W. Bonython, *2nd Symposium on Salt, Cleveland, Ohio,* 152 (1966).
57. A. G. Toussaint and A. J. M. Donders, *Chem. Eng. Sci.* **29,** 237 (1974).
58. R. C. Bennett, H. Fiedelman, and A. D. Randolph, *Chem. Eng. Progr.* **69**(7), 86 (1973).
59. E. P. K. Ottens and E. J. deJong, *Krist. Techn.* **9,** 873 (1974).
60. A. G. Jones and J. W. Mullin, *Trans. Inst. Chem. Eng. (London)* **51,** 302 (1973).
61. A. D. Randolph and M. A. Larson, *Theory of Particulate Processes,* Academic Press, Inc., New York, 1971.
62. J. N. Robinson and J. E. Roberts, *Can. J. Chem. Eng.* **35,** 105 (1957).
63. J.-S. Wey and J. P. Terwilliger, *Ind. Eng. Chem. Proc. Des. Dev.* **15,** 467 (1976).
64. M. A. Larson and J. Garside, *Chem. Eng. (London),* 318 (June 1973).
65. M. A. Larson, *Chem. Eng.,* 19 (Feb. 13, 1978).
66. W. C. Saeman, *Ind. Eng. Chem.* **8,** 612 (1961).
67. A. D. Randolph, *Chem. Eng.,* 80 (May 1970).
68. *Chem. Eng. Prog. Symp.* **67**(110), 108 (1971).
69. *Chem. Eng. Prog. Symp.* **65**(95), (1969).
70. Ref. 68, all papers.
71. *AIChE Symp. Ser.* **72**(153), (1976).
72. J. W. Mullin, ed., *Industrial Crystallization,* Plenum Press, New York, 1976.
73. J. Nyvlt and J. W. Mullin, *Chem. Eng. Sci* **25,** 131 (1970).
74. S. J. Lei, R. Shinnar, and S. Katz, *AIChE J.* **17,** 1459 (1971).
75. A. D. Randolph, G. L. Beer, and J. P. Keener, *AIChE J.* **19,** 1140 (1973).
76. Y. Song and J. M. Douglas, *AIChE J.* **21,** 924 (1976).
77. A. D. Randolph, J. R. Beckman, and Z. I. Kraljevich, *AIChE J.* **23,** 500 (1977).
78. A. G. Duncan and C. D. West, *Trans. Inst. Chem. Eng. (London)* **50,** 109 (1972).
79. R. R. Irani, C. F. Callis, and T. Liu, *Ind. Eng. Chem.* **51,** 1285 (1959).
80. L. Phoenix, *Br. Chem. Eng.* **11,** 34 (1966).
81. S. Sarig, A. Glasner, and J. A. Epstein, *J. Crystal Growth* **28,** 295, 300 (1975).

J. W. MULLIN
University College London

CRYSTALS. See X-ray technique.

CUMENE

Cumene [98-82-8] (1-methylethylbenzene, 2-phenylpropane, isopropylbenzene), C_9H_{12}, is normally a liquid, substituted aromatic compound in the benzene (qv), toluene (qv), ethylbenzene series (see BTX processing; Xylenes and ethylbenzene). It is the major intermediate chemical used in the worldwide production of phenol (qv), acetone and α-methylstyrene, all of which are components in plastic resins (see Styrene plastics). High purity cumene is normally manufactured from propylene and benzene although it is a minor constituent of most gasolines. Estimated United States production for 1977 was two million metric tons (1).

Properties

Some physical and chemical properties of cumene are listed in Tables 1 and 2, respectively (2–4). The transport and other thermodynamic properties of cumene and other compounds have been listed and presented in correlated form by Yaws (5).

Manufacture

Cumene as a pure chemical is manufactured exclusively from propylene and benzene utilizing an acidic catalyst (see Alkylation; Friedel-Crafts reactions). Many different catalyst systems have been proposed, including both crystalline and noncrystalline silicas, boron fluoride, aluminum chloride, and phosphoric acid, but current manufacturing plants almost exclusively employ solid phosphoric acid catalyst (6–9). The most common manufacturing process (8) is illustrated in Figure 1 (see also Alkylphenols).

The process utilizes an excess of benzene in the reactor charge to minimize side reactions that detract from product purity and yield. The propylene feed must be essentially free of ethylene and butylene but can contain substantial quantities of propane, such as that occurring in a refinery stream from fluid catalytic cracking operations. A small portion of the propylene charge oligomerizes to form materials boiling with the cumene product. These materials, which can interfere with the oxidation of the product, are avoided by the proper selection of processing conditions. Most of the energy required for the cumene synthesis can be recovered at a lower temperature level and used in the oxidation step (8).

Economic Aspects

Since benzene constitutes two thirds of the cumene product mass, and yields approach stoichiometric, the price of cumene follows closely the selling price of benzene, a more widely traded commodity. In some instances, cumene manufacturers have related the sales price of cumene directly to that of benzene. Cumene of high purity

Table 1. Physical Properties of Cumene

Property	Value
freezing point, °C	−96.03
boiling point, °C	152.39
density, g/cm³	
0°C	0.8786
20°C	0.8619
40°C	0.8450
refractive index, n_D^{20}	1.49145
thermal conductivity at 25°C, W/(m·K)	0.124
viscosity, mPa·s (= cP)	
0°C	1.076
20°C	0.791
40°C	0.612
surface tension at 20°C, mN/m (= dyn/cm)	28.2
vapor pressure[a]	

kPa[b]	t, °C
1	33.22
20	99.08
40	119.79
80	143.50
200	180.68

[a] Calculated from the equation: $t = \dfrac{B}{(A - \ln P)} - C$; where: t = temp, °C; P = vapor pressure, kPa[b]; $A = 13.957$; $B = 3363.6$; $C = 207.78$.
[b] To convert kPa to mm Hg, multiply by 7.5.

can be easily produced from either refining or steam cracker-derived propylene (qv). Both petroleum refineries and chemical companies have historically operated cumene manufacturing facilities; however, the recent trend has been toward chemical company installations where the economics of heat integration with a phenol unit can be obtained (8).

In a modern operation of reasonable scale, the cost of production can be approximated as indicated in Table 3. The feedstocks constitute over 80% of the cost of manufacture, with depreciation and energy requirements playing lesser roles.

Specifications and Analytical Methods

Most cumene manufactured for chemical purposes is produced to specifications such as those listed in Table 4. The purity specification is on an aromatic basis and does not include the 200–700 ppm of olefinic materials normally present that are indicated by the bromine index. ANSI/ASTM-D-3760 is a new (1979) ASTM gc procedure for the analysis of cumene.

Table 2. Thermodynamic Properties of Liquid Cumene

Property	Value
relative molar mass	120.2
critical temperature, °C	351.4
critical pressure, kPa (atm)	3220 (31.8)
critical density, g/cm^3	0.280
heat of vaporization at bp, J/g[a]	312
heat of vaporization at 25°C, J/g[a]	367
heat of formation at 25°C, J/mol[a]	−41,300
free energy of vapor at 25°C, J/mol[a]	137,000
heat of combustion at constant pressure and 25°C, J/g[a]	
gross	43,400
net	41,200
heat capacity at 25°C, J/(mol·K)[a]	197
heat capacity ideal vapor at 25°C, J/mol[a]	153

[a] To convert J to cal, divide by 4.184.

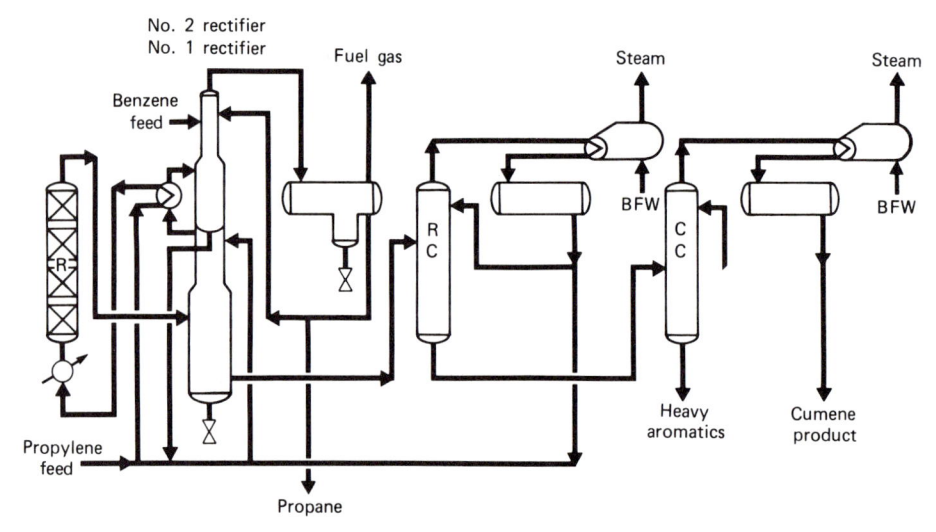

Figure 1. UOP catalytic condensation process for cumene synthesis. R = reactor; RC = recycle column; CC = cumene column; BFW = boiler feed water.

Health and Safety Factors (Toxicology)

The time-weighted average allowable exposure level for the United States occupational safety and health standard lists the permissible airborne limit for cumene as 245 mg/m^3 (10–11). Activated carbon-containing cartridge respirators have been shown to be effective for the removal of cumene from respiratory air (12). Liquid cumene is a primary skin irritant that is slowly adsorbed on contact through intact skin; however, experiments with animals indicate that cumene has a potent narcotic action and is entirely depressant (13–14). The slow elimination of cumene suggests that long exposures may result in cumulative effects.

Table 3. Cost of Cumene Production, Per Metric Ton Cumene

Factor	Consumption, t	Typical cost, $	Manufacturing cost, %
benzene charge	0.67	147.4	62
propylene charge at $160/t at 95% purity	0.39	62.7	26
by-product (credit at $120/t)	0.06	(6.6)	(3)
utilities			
heat and electricity	3.8 MJ[a]	7.2	2
steam credit	1.2	(2.4)	(1)
catalyst and chemicals		0.3	
labor, insurance, maintenance, etc		8.5	4
interest and depreciation at 30% of capital		23.0	10
Total		249.1	

[a] To convert MJ/t cumene to Btu/lb, divide by 2.324.

Table 4. Frequently Used Specifications For Cumene[a]

Property	Test	Value
specific gravity, 15.5/15.5	ASTM D 891	0.864 min/0.867 max
acid wash color	ASTM D 848	2 max
bromine index	ASTM D 1492	100 max
color	ASTM D 1209	15 max
sulfur compounds[b], ppm		2 max
product assay		
cumene, wt %	gas chromatograph	99.9 min
butylbenzenes, ppm	gas chromatograph	500 max
n-propylbenzene, ppm	gas chromatograph	500 max
ethylbenzene, ppm	gas chromatograph	500 max

[a] No standard specifications have been established for this material.
[b] Normally measured via a nickel reduction procedure.

Uses

Over 90% of manufactured high purity cumene is used in the production of phenol (qv) and acetone. Since the alternative route to acetone utilizes propylene, the cumene route is essentially equivalent to the production of phenol from benzene while obtaining high yields of acetone from propylene. Alternative routes to phenol via chlorination of benzene and oxidation of toluene have been almost completely displaced by the cumene oxidation system.

The route from cumene to phenol and acetone is based on the observation that in a basic medium cumene can be oxidized to stable cumene hydroperoxide. Upon acidification the peroxide is cleaved cleanly to phenol and acetone:

$$C_6H_5CH(CH_3)_2 + O_2 \xrightarrow{\text{base}} C_6H_5C(CH_3)_2OOH$$

$$C_6H_5C(CH_3)_2OOH \xrightarrow{\text{acid}} C_6H_5OH + (CH_3)_2CO$$

After removal of unreacted cumene, acetone, and reaction by-products such as α-methylstyrene, the crude phenol product is suitable for plywood resins, bakelite resins, etc. Further purification is required for a pharmaceutical-grade product. The α-methylstyrene is normally hydrogenated back to cumene for recycle to the reactor.

Overall yields for both phenol and acetone are over 90% of stoichiometry which is the primary reason for the commercial popularity of this route to phenol (13).

The second major chemical use for cumene is the production of α-methylstyrene. Although some α-methylstyrene is produced as a by-product from phenol operations (see above), its major source is the catalytic dehydrogenation of cumene, employing steam in the same general fashion as in the production of styrene from ethylbenzene. α-Methylstyrene (1-methylethylenebenzene) is used as a copolymer in specialty resin systems. Though exact figures are not available, the current use level is estimated at 30,000 t/yr for the United States.

Other minor uses include the use of cumene hydroperoxide as a chain initiator in polymer chemistry (see Initiators) and cumene as a component in aviation gasolines where it is used to improve the octane rating for internal combustion engines (see Gasoline).

BIBLIOGRAPHY

"Cumene" in *ECT* 2nd ed., Vol. 6, pp. 543–546, by D. J. Ward, Universal Oil Products Company.

1. *Chem. Week,* 11, 12 (July 20, 1977).
2. R. W. Gallant, *Hydrocarbon Process.* **48**(11), 263 (1969).
3. *Selected Values of Properties of Hydrocarbons, Research Project 44,* American Petroleum Institute, New York, 1953.
4. T. P. Thinh and co-workers, *Hydrocarbon Process.* **50**(1), 98 (1971).
5. C. L. Yaws, *Chem. Eng.* (*N.Y.*) **82**(20), 73 (1975).
6. Y. Mita and N. Kametake, *Hydrocarbon Process.* **47**(10), 122 (1968).
7. J. C. Butler and W. P. Webb, *Chem. Eng. Data Serv.* **2**(1), 42 (1957).
8. P. R. Pujado, J. R. Salazar, and C. V. Berger, *Hydrocarbon Process.* **55**(3), 91 (1976).
9. E. F. Harper, D. Y. Ko, and H. K. Lese, "Alkylation of Benzene with Propylene over a Crystalline Alumina Silicate," *ACS Symposium on Recent Advances in Alkylation Chemistry, New Orleans, Louisiana, Mar. 20–25, 1977.*
10. N. I. Sax, *Dangerous Properties of Industrial Materials,* Van Nostrand Reinhold Company, New York, 1975, pp. 17, 582.
11. U.S. Dept. of Labor, "Toxic Substances," *Fed. Reg.* **40**(196), 77262 (Oct. 8, 1975).
12. U.S. Pat. 3,385,906 (May 28, 1968), S. Kaufman (to Union Carbide Corporation).
13. L. D. Wilson, *Health Hazards from Aromatic Hydrocarbons, UOP Booklet No. 268,* UOP Process Division, Des Plaines, Ill., 1962.
14. *Registry of Toxic Effects of Chemical Substances,* U.S. Dept. of Health, Education, and Welfare, Washington, D.C., 1975.

Dennis J. Ward
UOP Inc.

CUMIN OIL. See Oils, essential; Flavors and spices.

CUPFERRON. See Zinc.

CUPRITE, Cu₂O. See Copper.

CURARE AND CURARE-LIKE DRUGS. See Alkaloids.

CYANAMIDES

It has been suggested that under primordial conditions, cyanamide could have acted as the original peptide-forming and phosphorylating reagent at the beginning of life on earth (1). Calcium cyanamide [156-62-7] was first produced commercially around 1900 as a fertilizer. Later numerous derivatives were developed, the most important of which are dicyandiamide (dimer) and melamine (trimer). The reactions of cyanamide and the products obtained are well established. The different names of each and their structural formulas are given in Table 1.

Table 1. Cyanamide, Its Dimer Dicyanamide, and Trimer Melamine

Compound	CAS Registry No.	Formula
cyanamide, carbamic acid nitrile, carbamodiimide	[420-04-2]	$H_2NC{\equiv}N$
dicyandiamide, cyanoguanidine	[461-58-5]	$(H_2N)_2C{=}N{-}C{\equiv}N$
melamine, cyanurotriamide, cyanuramide, 2,4,6-triamino-1,3,5-triazine	[108-78-1]	2,4,6-triamino-1,3,5-triazine

In North America, calcium cyanamide is no longer used as fertilizer. It is, however, used in agriculture in defoliants, fungicides, and weed killers. Industrial uses as a chemical intermediate have increased in the last decade.

CYANAMIDE

Properties

Cyanamide crystallizes from a variety of solvents as somewhat unstable, colorless, orthorhombic, deliquescent crystals (2). Dimerization is prevented by traces of acidic stabilizers such as monosodium phosphate and by storage at low temperature.

The ir spectrum of cyanamide exhibits an intense, poorly resolved doublet at 2225–2260 cm^{-1}. The uv spectrum shows only weak absorption below 230 nm, with no maximum observable above 208 nm. Studies of the infrared and Raman spectra (3–4) support the N-cyanoamine structure, $NH_2{-}C{\equiv}N$. The properties of cyanamide are listed in Table 2.

Cyanamide is a weak acid with a very high solubility in water. It is completely soluble at 43°C, and has a minimum solubility (eutectic) at −15°C. It is highly soluble in polar organic solvents, such as the lower alcohols, esters, and ketones, and less soluble in nonpolar solvents (5).

Table 2. Properties of Cyanamide[a]

Property	Value	Refs.
molecular weight	42.04	
mp, °C	46	7
bp, °C		
101 kPa[b]	dec	
2.47 kPa[b]	140	
75 Pa[b]	85–87	
density[c], g/mL	1.282	5
refractive index at 48°C	1.4418	
vapor pressure, up to 120°C	$\log \text{kPa}^b = \dfrac{-3.58 \times 10^3}{K} + 8.91$	5,8
dissociation constants,		
K_a	5.4×10^{-11}	9
K_b	2.5×10^{-12}	10
specific heat at 0–39°C, J/(g·K)[d]	2.288	11
heat of formation at 25°C, kJ/mol[d]	58.77	12
heat of solution[e] at 15°C, kJ/mol[d]	−15.05	13
heat of combustion at 25°C, kJ/mol[d]	−737.9	12
heat of neutralization[f], kJ/mol[d]	15.48	14

[a] Refs. 5–6.
[b] To convert kPa to mm Hg, or Pa to μm Hg, multiply by 7.5.
[c] Calculated.
[d] To convert J to cal, divide by 4.184.
[e] In 1000 parts H_2O.
[f] Excess NaOH solution.

Reactions

Reactions of cyanamide are either additions to the nitrile group or substitutions at the amino group. Both are involved in the dimerization to dicyandiamide.

$$H_2NC\equiv N \rightleftharpoons H\bar{N}C\equiv N + H^+$$

$$H_2NC\equiv N + H\bar{N}C\equiv N \longrightarrow H_2N\overset{\overset{NH}{\|}}{C}-\bar{N}C\equiv N$$

$$H_2\overset{\overset{NH}{\|}}{N}CNC\equiv N + H_2\bar{N}C\equiv N \longrightarrow H\bar{N}C\equiv N + \underset{H_2N}{\overset{H_2N}{>}}C=N-C\equiv N$$

Dimerization involves addition of the cyanamide anion to the nitrile group of an undissociated molecule to give the anion of cyanoguanidine, or dicyandiamide. This reaction takes place most readily at pH 8–10 where the reactants are present in favorable proportion. The product is a weaker acid than cyanamide which is protonated at once with generation of a new cyanamide anion.

Most of the reactions occurring at the amino group require the anionic species, sometimes in equivalent amount and occasionally as provided by base catalysis. Therefore, conditions should be such as to avoid base-catalyzed dimerization.

Similarly, nucleophilic reagents are suitable for addition reactions only if they are not so strongly basic as to produce the cyanamide anion in large amounts. In such cases, dicyandiamide is produced or a cyanamide salt is obtained. N,N-Disubstituted cyanamides do not ionize, of course, and react easily with strongly basic nucleophiles.

With nitriles in general, addition reactions take place readily in the presence of acids, because of the enhanced electrophilic character of the central carbon atom in the conjugate acid:

$$NH_2-\overset{\delta+}{C}\equiv\overset{\delta-}{N} \xrightarrow{H^+} NH_2-\overset{+}{C}=NH$$

Substitutions. The cyanamide anion is strongly nucleophilic and reacts with most alkylating or acylating reagents; addition to a variety of unsaturated systems occurs readily. In some cases, a cyanamide salt is used; in others, base catalysis suffices.

Alkylation with a variety of common alkyl halides or sulfates gives stable dialkylcyanamides. However, the intermediate monoalkylated compounds usually cannot be isolated and cyclic trimers or cotrimers with cyanamide are obtained (15). The reaction can be carried out efficiently in water or alcohol. Allyl chloride is an especially useful reagent, producing diallylcyanamide [538-08-9] (5).

$$H_2NCN + NaOH \rightarrow NaHNCN + H_2O$$

$$NaHNCN + 2\ CH_2=CHCH_2Cl + NaOH \rightarrow (CH_2=CHCH_2)_2NCN + 2\ NaCl + H_2O$$

The amino group in cyanamide is very reactive toward formaldehyde. The reaction produces first an unstable hydroxymethyl derivative that resinifies more or less rapidly, passing through a water soluble stage which permits various applications. This product can be dried and thermoset. The reactions leading to resinification include condensation to give methylene and methyl ether bridges, hydration of the cyano group if acidic conditions are used, and polymerization of cyanamide residues to give dicyandiamide, melamine, or possibly isomelamine structures (see Amino resins).

Cyanogen chloride and NaOH give the potentially useful sodium dicyanamide [1934-75-4] salt (16).

$$NH_2CN + ClCN + 2\ NaOH \rightarrow NaN(CN)_2 + NaCl + 2\ H_2O$$

Other substitution reactions have been described with ketones, epoxides, anhydrides, acyl halides, amides, and imidates, among others (5).

Additions. The addition reactions of ammonia and amines to the cyanamide nitrile group have been thoroughly studied (17). For optimum conditions, the reaction should be carried out in an aqueous medium at about 140°C. Gradual addition of the cyanamide to the amine salt minimizes dimerization.

A great variety of amino compounds have been used (5). The two examples given below have practical applications (guanidine nitrate is used in the production of nitroguanidine, dodecylguanidine acetate is a fungicide).

$$H_2NC\equiv N + NH_4NO_3 \rightarrow H_2N\overset{\overset{NH}{\|}}{C}-NH_2\cdot HNO_3$$

$$H_2NC\equiv N + C_{12}H_{25}NH_2\cdot CH_3COOH \rightarrow H_2N\overset{\overset{NH}{\|}}{C}-NHC_{12}H_{25}\cdot CH_3COOH$$

In many cases it is advantageous to use anhydrous cyanamide in alcoholic solvents.

Uses of such guanidine derivatives depend upon their strongly basic character and the cationic character of the salts.

Hydrogen chloride gives a dihydrochloride. Chloroformamidine hydrochloride is a convenient anhydrous form of cyanamide which is easily handled and stored (18).

$$H_2NCN + 2\,HCl \longrightarrow \left[Cl-C\underset{NH_2}{\overset{NH_2}{\diagup}} \right]^+ Cl^-$$

Addition of alcohols and phenols in the presence of anhydrous hydrogen chloride gives O-substituted pseudourea salts (19). The reaction is sluggish except with the lower alcohols, and long reaction time and temperatures up to 100°C are required to obtain good yields.

$$H_2N-CN + ROH + HCl \xrightarrow{25-50\,°C} \left[H_2N-C\underset{OR}{\overset{NH_2}{\diagup}} \right]^+ Cl^-$$

Manufacture

From Limestone and Coal. The basic process for the manufacture of cyanamide comprises four stages.

(1) Lime is made from high-grade limestone (see Lime and limestone)

$$CaCO_3 \rightarrow CaO + CO_2$$

(2) Calcium carbide is manufactured from lime and coal or coke (see Carbides).

$$CaO + 3\,C \rightarrow CaC_2 + CO$$

(3) Calcium cyanamide is produced by passing gaseous nitrogen through a bed of calcium carbide, which is heated to 1000–1100°C in order to start the reaction. The heat source is then removed and the reaction continues because of its strong exothermic character.

$$CaC_2 + N_2 \xrightarrow{1000-1100\,°C} CaCN_2 + C$$

The batch oven consists of a cylindrical refractory-lined steel shell with a steel cover lined with insulating material. A vent in the cover permits the escape of off-gases. The charge is either added directly to the oven or inserted in a perforated basket. The ovens are heated by carbon electrodes. Batch ovens range in size from 1–8 metric tons capacity. A cycle time of 4–5 days is required for 4-t ovens using a single starting electrode; with multiple electrodes, shorter times are possible. The reaction is almost completed in three days; the remaining time is required for cooling to avoid surface oxidation when the pig is withdrawn from the oven. A detailed account of the oven operation is given in ref. 20.

Continuous kilns were first used in Germany at Knapsack (21) and later at Trostberg (22). In the Knapsack process, carbide (0.75–2 mm in size) is fed to a rotary kiln, 3 m in diameter and 12 m long with 1% slope; 1–2% calcium chloride is added to promote the reaction. The kiln produces 12–13 t fixed nitrogen per day. The product is granular and can be sold without further processing.

The Süddeutsche Kalkstickstoffwerke process at Trostberg uses powdered carbide along with recycle product and calcium fluoride in a rotary kiln at 1000–1100°C. The capacity of a unit is 25 t fixed nitrogen per day. The product passes to a rotary cooler and is granular (22).

For the manufacture of calcium cyanamide, crude carbide (ca 3.36 × 1.68 mm or 6 × 12 mesh) can be used, whereas for cyanamide and dicyandiamide, a 74 µm (200 mesh) anhydrous carbide is used.

For agricultural uses, a fully hydrated calcium cyanamide powder containing 2% oil is used for dusting applications; for general fertilizer use, the hydrated material is granulated with water in a rotary drum and dried giving a 2.38 × 0.32 mm (8 × 48 mesh) product.

(4) In the final step cyanamide is manufactured from calcium cyanamide by continuous carbonation in an aqueous medium (see Fig. 1) (23).

$$CaNCN + H_2O + CO_2 \rightarrow H_2NCN + CaCO_3$$

A carbonated slurry of cyanamide solution, solid calcium carbonate, and graphite is cooled to remove the heat of reaction. Part of the slurry is recycled to facilitate temperature control whereas the remainder is filtered yielding cyanamide solution and a cake of calcium carbonate and graphite. The filtered solution is also recycled in order to control the solids content. The final concentration of cyanamide is normally maintained at 25%.

The calcium cyanamide feed is well mixed with the recycled slurry and filtrate in a feed vessel. The calcium cyanamide is added at a rate to maintain a pH of 6.0–6.5 in the cooling tank. The carbonation step can be conducted in a turbine absorber with a residence time of 1–2 min. After the carbonation step, the slurry is held at 30–40°C to complete the formation of calcium carbonate, after which the slurry is cooled and filtered. All equipment for the process is preferably of stainless steel. The resulting solution is used directly for conversion to dicyandiamide.

For production of commercial 50% solution and for recovery of crystalline cyanamide, this process is modified to improve purity and concentration. Calcium and iron may be removed by ion-exchange treatment. The commercial 50% solution of the American Cyanamid Company is stabilized at pH 4.5–5.0 with 2% monosodium phosphate and contains less than 1.5% dicyandiamide and 0.2% urea. Such solutions are expected to show less than 1% change in cyanamide content per month of storage below 10°C. It is advisable, however, to adjust the pH periodically during extended storage. Organic esters may be used instead for improved stability (24).

Of the four steps described above, only the first two consume large amounts of energy, especially the calcium carbide step. With increasing energy shortages and cost increases, attention should be focused upon alternatives for making calcium cyanamide.

Other Processes. (1) Reaction of lime with hydrogen cyanide (25):

$$CaO + 2 HCN \xrightarrow{750-850°C} CaCN_2 + CO + H_2$$

(2) Reaction of limestone with ammonia (26):

$$CaCO_3 + 2 NH_3 \xrightarrow{700-800°C} CaCN_2 + 3 H_2O$$

296 CYANAMIDES

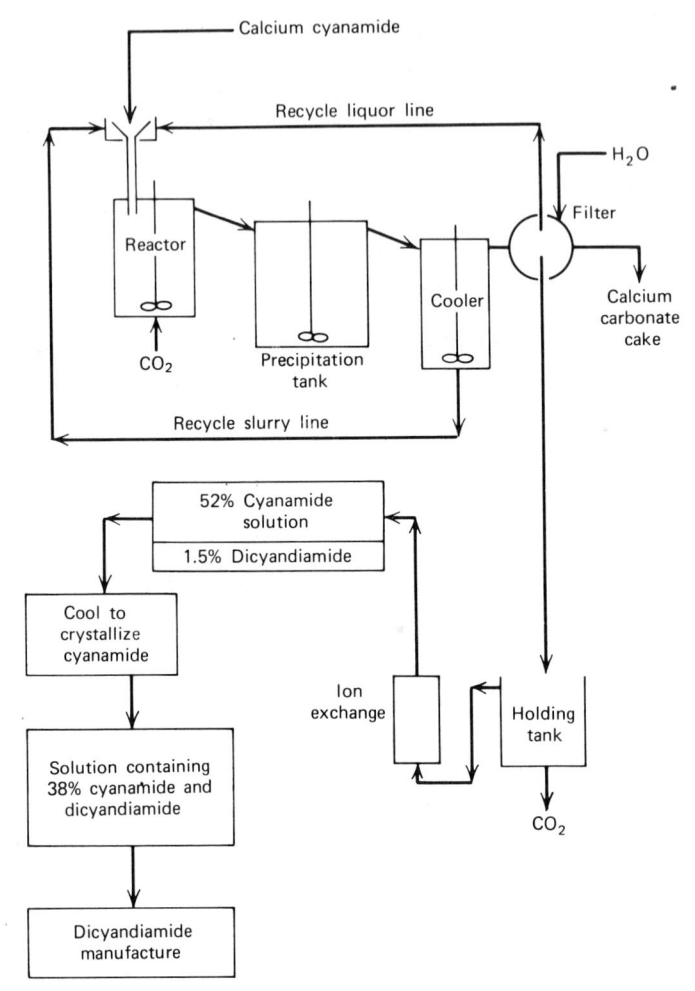

Figure 1. Manufacture of cyanamide from calcium cyanamide.

(3) Reaction of lime and urea to form calcium cyanate (3,27) which is then converted to calcium cyanurate: the latter gives calcium cyanamide at a higher temperature:

$$CaO + 2\ H_2NCONH_2 \xrightarrow{200°C} Ca(OCN)_2 + H_2O + 2\ NH_3$$

$$3\ Ca(OCN)_2 \xrightarrow{400°C} Ca_3(OCN)_6$$

$$Ca_3(OCN)_6 \xrightarrow{700°C} 3\ CaCN_2 + 3\ CO_2$$

The product obtained from these processes is white calcium cyanamide whereas the product obtained from limestone and coal contains carbonaceous (graphite) impurities. So far, none of these processes has been commercially exploited.

Economic Aspects

A peak in calcium cyanamide production was probably reached in 1962 when the world production for fertilizer use was of the order of 1,000,000 metric tons of calcium cyanamide per year, and for industrial use approximately 300,000 t (excluding the U.S.S.R.). The largest producers are in Japan, the Federal Republic of Germany, and Canada.

Altogether approximately forty plants are capable of producing calcium cyanamide. In North America the only producer is Cyanamid Canada, Inc. in Niagara Falls, Ontario, serving the Canadian and United States markets. This company, a subsidiary of American Cyanamid Company, also produces dicyandiamide, melamine, cyanamide, and a number of their derivatives. Cyanamide is occasionally marketed in North America in crystalline form, but mostly as 50% solution at about $2.50/kg (1977 price). In Europe cyanamide is available as 50% solution and in crystalline form from the Süddeutsche Kalkstickstoff-Werke A.G., Trostberg, FRG.

In 1977 the total production of cyanamide products was about half that of 1962.

Specifications and Analysis

Cyanamide is sold as anhydrous, aqueous 50%, and calcium cyanamide. Anhydrous and aqueous 50% cyanamide contain a buffering additive, usually 2% NaH_2PO_4, to stabilize the pH and prevent formation of dicyandiamide and urea. Calcium cyanamide is stable under dry conditions. Table 3 gives typical analyses of the three commercial forms.

Table 3. Typical Analysis of Commercial Cyanamide [a]

Assay	%
Anhydrous, Cyanamide-100	
cyanamide	96
dicyandiamide	1.0
urea	0.5
water	0.5
NaH_2PO_4 (stabilizer)	2.0
Aqueous 50%, Cyanamide-50	
cyanamide	50
dicyandiamide	1.5
urea	0.2
NaH_2PO_4 (stabilizer)	2
water	46.3
Calcium cyanamide	
calcium cyanamide	65
CaO	12
carbon	12.5
CaF_2	0.5
CaS	1.0
calcium carbide	0.5
other nitrogen impurities	1.5
other mineral impurities	7

[a] Ref. 5.

Cyanamide is precipitated by excess ammoniacal silver nitrate as disilver cyanamide [3384-87-0] which is dissolved in acid and titrated with thiocyanate solution (28).

Traces of cyanamide can be determined colorimetrically by complex formation with 1,2-naphthaquinone-4-sodium sulfonate (5).

Handling and Storage

Cyanamide solution dimerizes to dicyandiamide and urea with the evolution of heat and a gradual increase in alkalinity accelerating the reaction. Storage above 30°C without pH stabilizer leads to excessive dimerization and can result in violent exothermic polymerization. Figure 2 illustrates this effect at 40°C.

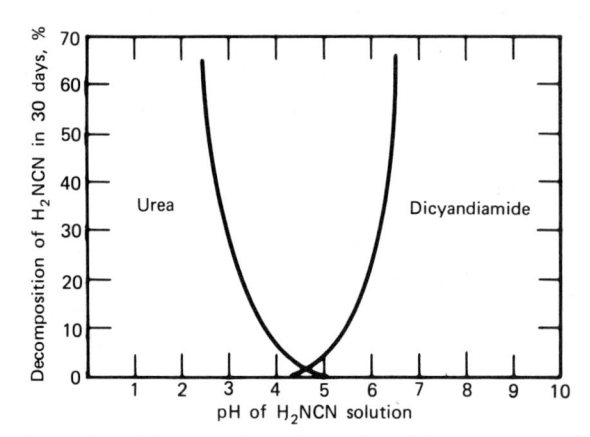

Figure 2. Formation of dicyanamide and urea from 25% aqueous solution at 40°C.

Cyanamide should be stored under refrigeration, and the pH tested periodically. Stabilized cyanamide can be kept at ambient temperature for a few weeks.

In general, cyanamide should be added to a reaction mixture at such a rate that it is used up as it is added. Otherwise a high concentration of cyanamide results which could react violently.

Health and Safety Factors (Toxicology)

Manufacture of cyanamide and calcium cyanamide does not present any serious health hazard. Ingestion of alcoholic beverages by workmen within several hours of leaving work sometimes results in a vasomotor reaction known as cyanamide flush. Cyanamide interferes with the oxidation of alcohol and accumulation of acetaldehyde probably accounts for this temporary phenomenon. Although extremely unpleasant, it has not been known to result in serious illness or to have any permanent effect.

Commercial grades of calcium cyanamide contain lime, and are moderate skin irritants where contact is repeated or prolonged.

Contact or ingestion of cyanamide must be avoided, and precautions taken to prevent inhalation of dust or spray mist. Experiments with rats gave the following results:

(1) Cyanamide-100 toxicity ranges from a single oral dose LD_{50} of 280 mg/kg to

a single dermal dose LD$_{50}$ of 590 (420–820) mg/kg. The compound is, therefore, considered to be moderately toxic both by ingestion in single doses and by single skin applications. An aqueous paste of the product is corrosive to rabbit skin. Small quantities of the dry product produced severe irritation when introduced into the conjunctival sac of the rabbit eye.

(2) Cyanamide-50 is considered to be slightly toxic by ingestion in single dose and moderately toxic by single skin applications (see Table 3). The degree of irritation to rabbit skin and eye produced by this product is only slightly less than that observed with Cyanamide-100.

Uses

In Europe cyanamide and calcium cyanamide are used as fertilizers (qv), weed killers, and defoliants. In North America these applications have been practically discontinued. However, calcium cyanamide is used as a weed killer for asparagus and onions. Heavy applications approximately one month before planting are used to control soil-borne plant disease and weed seeds (see Herbicides).

A direct application of calcium cyanamide is particularly useful to fertilize acid soils where lime values are needed. It also can be used in mixed fertilizers although the high alkalinity reduces the soluble phosphate content if an excess is used.

Calcium cyanamide is used in the treatment of alcoholics. A small pill, taken once a day, subjects the user to cyanamide flush if any alcohol is taken. The result is unpleasant and discourages drinking (29).

Industrial uses make up most of the market for cyanamide. Calcium cyanamide is used directly for steel nitridation (30) and to some extent for desulfurization (31) (see Steel). Cyanamide is used to produce cationic starch (32), and calcium cyanide. Cyanamide is, of course, the raw material for dicyandiamide and melamine (see below). New uses include as intermediates for pesticides.

DICYANDIAMIDE

Properties

Dicyandiamide (see Table 1) is the dimer of cyanamide and crystallizes in colorless monoclinic prisms. It is amphoteric, and generally soluble in polar solvents and insoluble in nonpolar solvents. Its properties are listed in Table 4.

Table 4. Properties of Dicyandiamide[a]

Property	Value
molecular weight	84.08
mp, °C	208
bp, °C	dec
acid dissociation at 25°C, K_a	6 × 10^{-15}
specific heat at 25°C, J/(g·K)[b]	1.41
heat of formation at 25°C, kJ/mol[b]	24.9
heat of combustion at 25°C, kJ/mol[b]	−1382
heat of solution at 15°C, kJ/mol[b]	−24.1

[a] Refs. 33–35.
[b] To convert J to cal, divide by 4.184.

Reactions

The reactions of dicyandiamide resemble those of cyanamide. However, cyclizations take place easily and the nitrile group is less reactive.

Under pressure and in the presence of ammonia, dicyandiamide cyclizes to melamine. Considerable tonnages of melamine have been made in this manner. Today, however, melamine is produced chiefly by the urea process (36):

$$3\,(NH_2)_2C{=}N{-}CN \xrightarrow{NH_3} 2\;\text{melamine}$$

Excellent yields of guanamines (37) are obtained when dicyandiamide is heated with alkyl or aryl nitriles in the presence of small amounts of alkali.

$$(NH_2)_2C{=}N{-}CN + RCN \xrightarrow[108-200\,°C]{alkali} \text{2,4-diamino-6-R-triazine}$$

Heating with aromatic amines in water in the presence of an equivalent amount of mineral acid gives high yields of aryl biguanide salts (38).

$$(NH_2)_2C{=}N{-}CN + ArNH_2 + HCl \xrightarrow{H_2O} H_2N{-}\underset{\|}{C}(NH){-}NH{-}\underset{\|}{C}(NH){-}NHAr \cdot HCl$$

Reaction with ammonium salts gives biguanide salts which react further with the ammonium salt forming guanidine salts. Guanidine nitrate is manufactured by this route (39).

$$(NH_2)_2C{=}N{-}CN + 2\,NH_3 \cdot HNO_3 \longrightarrow 2\,H_2N{-}\underset{\|}{C}(NH){-}NH_2 \cdot HNO_3$$

Hydrolysis (40) of dicyandiamide occurs easily at elevated temperatures in the presence of an equivalent of mineral acid to yield guanylurea salts. This reaction is quantitative and can be used for the determination of dicyandiamide (41).

$$(NH_2)_2C{=}N{-}CN + HX + H_2O \xrightarrow{90-100\,°C} H_2N{-}\underset{\|}{C}(NH){-}NH{-}\underset{\|}{C}(O){-}NH_2 \cdot HX$$

Dicyandiamide may be treated with formaldehyde (42) to produce resinous compositions of varying properties (see Amino resins), under either acid or alkaline conditions. The reaction can be controlled to give mainly monomethyloldicyandiamide, a very water-soluble compound:

$$(NH_2)_2C{=}N{-}CN + HCHO \longrightarrow HOCH_2{-}NH{-}\underset{\|}{C}(NH){-}NH{-}CN$$

Manufacture

Dicyandiamide is manufactured by dimerization of cyanamide in aqueous solution (3). The 25% cyanamide solution produced as described above is adjusted to pH 8–9, and held at approximately 80°C for two hours to give complete conversion. The hot liquor is filtered and transferred to a vacuum crystallizer where it is cooled. The crystals of dicyandiamide are separated in continuous centrifuges and passed to rotary driers. The finished material is stored in bulk or in bags. The mother liquor is partly recycled to the cyanamide extraction system and partly purged to keep the thiourea content low.

Specifications, Analysis, and Toxicity

Dicyandiamide is identified qualitatively by paper chromatography, and quantitatively by ultraviolet spectrometry of the chromatogram (3). More commonly, total nitrogen analysis is used as a purity control or the dicyandiamide is converted by hydrolysis to guanylurea, which is determined gravimetrically as the nickel salt (43). Methods based on the precipitation of silver dicyandiamide picrate are sometimes used (44). Dicyandiamide can also be titrated with tetrabutylammonium hydroxide in pyridine solution (3). Table 5 gives a typical analysis of a commercial sample.

Table 5. Typical Analysis of Commercial Dicyandiamide

Assay	Value
dicyandiamide, %	99.3
water, %	0.01
melamine, %	0.7
thiourea, ppm	200
heavy metals, ppm	10

Dicyandiamide is essentially nontoxic. It may, however, cause dermatitis.

Uses

Dicyandiamide is used as a raw material for the manufacture of several chemicals, such as guanamines, biguanide and guanidine salts, and various resins. Melamine has extensive applications in the resin and plastic industry: guanamines are used as copolymers in many resin compositions, and guanidine phosphate is employed as a fire retardant in applications where water solubility is not a drawback.

Dicyandiamide, in conjunction with phosphoric acid, is used as a flame retardant for cellulosic materials (45). Cotton fabrics have also been treated effectively in this manner. Use as a fire retardant for wood, particularly shingles, has found commercial application (46).

Dicyandiamide reduces the viscosity of certain colloidal solutions. This property is of commercial significance in the manufacture of glues and adhesives, in the coating and sizing of paper and textiles, and in the conditioning of phosphate drilling muds (see Petroleum). This action may prove useful in other applications where control of viscosity is important (47).

Dicyandiamide fluidifies various adhesives and glues. In the presence of dicyandiamide, the initial viscosity of the adhesive formulations is lowered and maintained at a lower range for longer periods of time, thus preventing premature gelling and extending the useful life of the adhesive (see Glue; Adhesives).

Dicyandiamide, although completely stable at lower temperatures, has found use in surface coating formulations as a flame retardant. When exposed to a flame, the coating composition intumesces or bubbles up, thereby insulating the substrate. In this manner, the substrate retains its physical properties and strength. For example, one effective intumescent formulation includes 10 parts of dicyandiamide, 22 parts of pentaerythritol, 56 parts of monoammonium phosphate, 12 parts of titanium dioxide, 50 parts of latex binder (50% solids), and 33 parts of water. Such a formulation gives good foam volume at low coating weights, and reasonable stability of the foam structure under flaming (48) (see Foams).

MELAMINE

Although melamine was prepared for the first time in 1834 by Liebig, almost a century passed before its properties and reactivity were thoroughly investigated. In the late 1930s a commercial process was developed. Since then, uses for melamine reaction products have been developed rapidly in a variety of fields, including plastics, surface coatings, bonding agents, paper and textile finishes, tanning agents, pharmaceuticals, and petroleum and rubber chemicals.

Properties

The outstanding characteristic of melamine, usually a white crystalline material, is its insolubility in most organic solvents. This property is also evident in melamine resins after they are cured. On the other hand, melamine is appreciably soluble in water, its solubility increasing with increased temperature. The properties of melamine are listed in Table 6.

The chemistry of melamine is reviewed in refs. 49 and 50. Melamine, although moderately basic, is better considered as the triamide of cyanuric acid than as an aromatic amine (see Cyanuric and isocyanuric acids). Its reactivity is poor in nearly all reactions considered typical for amines. In part, this may be due to its low solubility (see Amino resins).

Table 6. Properties of Melamine[a]

Property	Value
molecular weight	126.13
mp, °C	350
density, g/mL	1.573
specific heat at 25°C, J/(g·K)[b]	1.23
heat of combustion at 25°C, kJ/mol[b]	−1964
heat of sublimation, kJ/mol[b]	−121

[a] Ref. 49.
[b] To convert J to cal, divide by 4.184.

Manufacture

Dicyandiamide is converted into melamine by heating. Simple pyrolysis above the melting point leads to an exothermic reaction; however, deammoniation occurs, forming products containing two or three triazine rings as well as melamine. After it was discovered in 1940 that deammoniation can be counteracted by conducting the reaction under ammonia pressure, various methods were developed to control the exothermic reaction on an industrial scale.

By now the dehydration condensation of urea has displaced the dicyandiamide process (see Urea). Although the latter is still used occasionally, the urea process predominates in North America. A flow sheet is shown in Figure 3 (36).

A urea melt is supplied to a one-stage reactor containing a fluid-bed catalyst. The

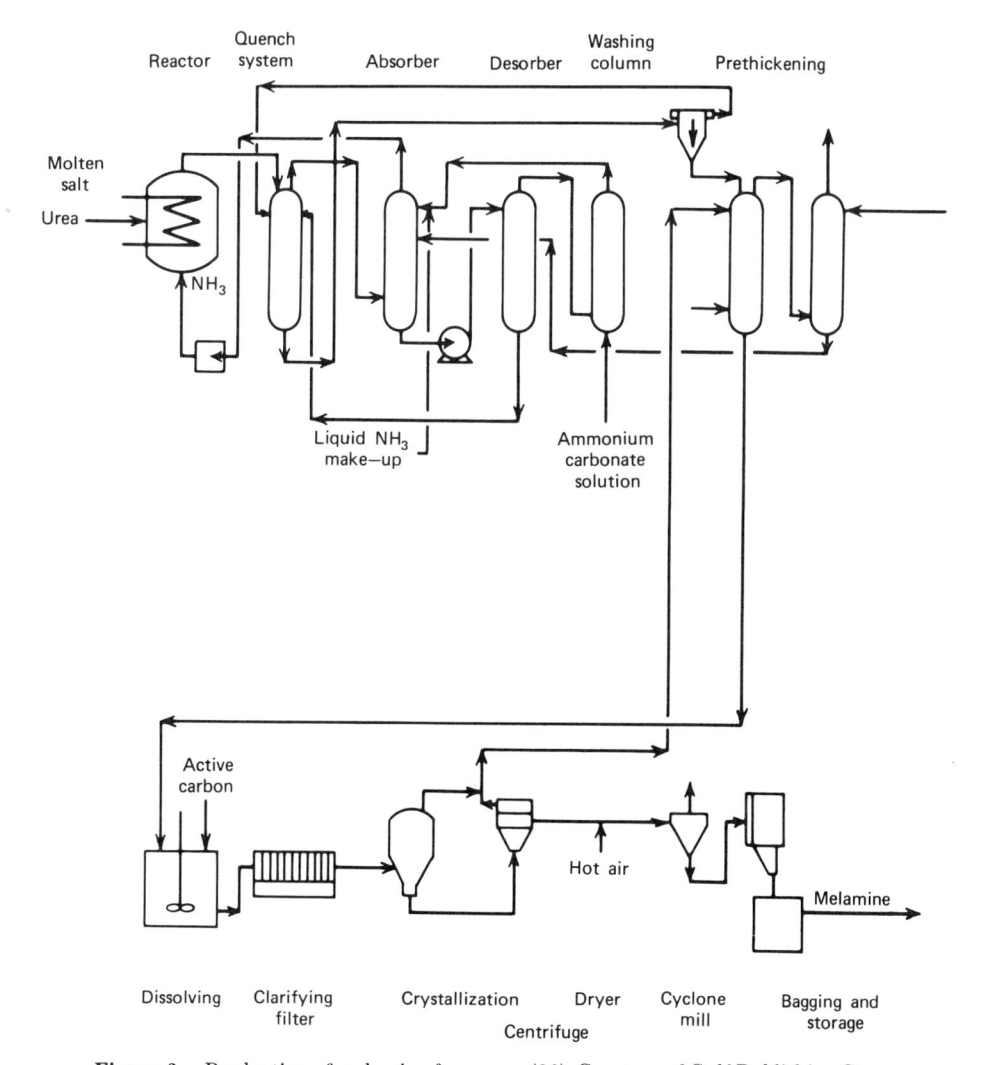

Figure 3. Production of melamine from urea (36). Courtesy of Gulf Publishing Co.

reactor is heated internally by circulating molten salt. Upon entering the reactor, the urea is converted to melamine by the hot catalyst.

$$6\ NH_2CNH_2 \longrightarrow \text{[melamine]} + 6\ NH_3 + 3\ CO_2$$

The hot ammonia used as carrier gas serves simultaneously as fluidizing agent and inhibits deammonization.

The reactor effluent is rapidly quenched with aqueous mother liquor in specially designed equipment operating at pressures essentially equal to the reactor pressure. This operation yields an off-gas consisting of ammonia and carbon dioxide vapor and a crystalline melamine slurry saturated with ammonia and carbon dioxide. The slurry is concentrated in a cyclone mill. The mother liquor overflow is returned to the quenching system. The concentrated slurry is redissolved in the mother liquor of the crystallization system, and the dissolved ammonia is stripped simultaneously.

This ammonia is recycled to the reactor via a compressor and a heater. Liquid ammonia is used as reflux on the top of the absorber. The net amount of carbon dioxide formed in the reactor is removed as bottom product from the absorber in the form of a weak ammonium carbamate solution, which is concentrated in a desorber-washing column system. The bottom product of this washing column is a concentrated ammonium carbamate solution which is reprocessed in a urea plant. The top product, pure ammonia, is liquefied and used as reflux together with liquid make-up ammonia. The desorber bottom product, practically pure water, is used in the quench system in addition to the recycled mother liquor.

In order to upgrade the melamine, the solution obtained after prethickening and stripping is treated with activated carbon passed over a clarifying filter and fed to a two-stage vacuum crystallization system from which the pure melamine is recovered in a continuous centrifuge. Stainless steel is used as construction material for nearly all parts of the equipment exposed to product streams.

Analysis and Toxicity

Melamine is determined gravimetrically by precipitation of the insoluble oxalate from a hot aqueous solution (49).

Extensive toxicity investigations have been performed with melamine in experimental animals. These suggest that the compound may have a low order of biological activity. The acute oral LD_{50} for mice was found to be 4.55 g/kg, approximately the same for rats.

Chronic feeding tests have been carried out on rats over a two-year period at a dietary level of 1000 ppm and on dogs for one year at a level of 30,000 ppm. Throughout the study, the general health of the test animals did not seem significantly different from that of the controls. Nevertheless, after 60–90 days the dogs showed melamine crystalluria, which persisted throughout the remainder of the observation. However, gross and microscopic examination of the tissues revealed no abnormality attributable to the feeding of melamine.

Melamine in a skin test on rabbits produced neither local irritation nor systemic toxicity. As a 10% solution in methyl cellulose, it caused no irritation in the eyes of rabbits. Human subjects were given patch tests with melamine. No evidence of either primary irritation or sensitization was found.

Such results suggest that melamine crystal may be handled in ordinary industrial use without special hygienic precautions.

Uses

Most of the melamine produced is used in the form of melamine–formaldehyde resins (see Amino resins and plastics). Other applications (49) include the use of melamine pyrophosphate [15541-60-3] in fire retardant textile finishes, chlorinated melamine as a bactericide, and melamine as a tarnish inhibitor in detergent compositions.

BIBLIOGRAPHY

"Cyanamides" in *ECT* 1st ed., Vol. 4, pp. 663–675, by V. Alexanderson, American Cyanamid Company, and L. M. Sherman, Chemical Construction Corp.; "Cyanamides" in *ECT* 2nd ed., Vol. 6, pp. 553–573, by J. R. McAdam, Cyanamid of Canada Ltd., and F. C. Schaefer, American Cyanamid Company.

1. B. E. Turner and co-workers, *Astrophys. J.* **201**, 3 (1975).
2. D. Costa and C. Bolis-Cannella, *Ann. Chim. (Rome)* **43**, (1953).
3. American Cyanamid Company, unpublished results.
4. W. H. Fletcher, *J. Chem. Phys.* **39**, 2478 (1963).
5. *Cyanamide*, technical bulletin, American Cyanamid Company, Wayne, N. J., 1966.
6. *SKW-Produktstudie Cyanamid*, Süddeutsche Kalkstickstoff-Werke A.G., Trostberg, FRG, 1976, 85 pp.
7. K. J. Odo, *J. Chem. Soc. (Japan) Pure Chem. Sect.* **74**, (1953).
8. F. Baum, *Biochem. Z.* **26**, 325 (1910).
9. N. Kameyama, *Trans. Am. Electrochem. Soc.* **40**, 131 (1921); S. Kawai, *Sci. Pap. Inst. Phys. Chem. Res.* **16**, 24, 306 (1931).
10. G. Grude and G. Motz, *Z. Phys. Chem.* **118**, 145 (1925); U.S. Pat. 2,443,504 (June 15, 1948), I. Hechenbleikner (to American Cyanamid Company).
11. E. Mulder and J. A. Roorda-Smit, *Ber.* **7**, 1634 (1874).
12. L. A. Pinck and H. C. Hetherington, *Ind. Eng. Chem.* **18**, 629 (1926).
13. M. Kurabayashi and K. Yanagiya, *Rep. Govt. Chem. Eng. Res. Inst. (Tokyo)* **48**, 133 (1953).
14. .H. Pincas, *Ind. Chim.* **21**, 413 (1934).
15. S. A. Miller and B. Bann, *J. Appl. Chem. (London)* **6**, 89 (1956).
16. *Methods of Analysis*, Assoc. of Official Agricultural Chemists, Washington, D.C., 1955; Can. Pat. 956,081 (Oct. 15, 1974), F. C. Schaefer (to American Cyanamid Company).
17. R. H. McKee, *Am. Chem. J.* **36**, 208 (1906); U.S. Pat. 2,476,452 (July 19, 1949), J. H. Paden and A. F. MacLean (to American Cyanamid Company).
18. U.S. Pat. 2,727,922 (Dec. 20, 1955), H. Z. Lecher and C. L. Kosloski (to American Cyanamid Company).
19. S. Basterfield and co-workers, *Can. J. Res.* **1**, 261 (1929); F. Kurzer and A. Lawson, *Org. Syn.* **34**, 67 (1954).
20. M. L. Kasten and W. G. McBurney, *Ind. Eng. Chem.* **43**, 1020 (1951).
21. F. Ritten and J. Krause in K. Winnaker and L. Kuchler, eds. *Chemische Technologie*, Vol. 2, 2nd ed., Carl Hanser Verlag, Munich, FRG, 1959, p. 269.
22. U.S. Pat. 2,838,379 (Sept. 15, 1958), F. Kaess and co-workers (to Süddeutsche Kalkstickstoff Werke).
23. U.S. Pat. 3,300,281 (Jan. 24, 1967), D. R. May (to American Cyanamid Company).
24. U.S. Pat. 3,295,926 (Jan. 3, 1967), R. H. Janes, G. J. Novak, and R. C. Rawlings (to American Cyanamid Company).

25. *Cyanamide by Cyanamid,* American Cyanamid Company, Wayne, N. J., 1961; O. I. Polyanshikov and co-workers, *Vopr. Khim. Khim. Tekhnol.* **39,** 136 (1975).
26. A. A. Pimenova and co-workers, *Tr. Tashk. Politekh. Inst.* **107,** 49 (1973).
27. V. G. Golov and co-workers, *Zh. Khim.,* Abstract No. 23V1s (1975); *Chem. Abstr.* **84,** 159006n (1975).
28. L. A. Pinck, *Ind. Eng. Chem.* **17,** 459 (1925).
29. K. Arikawa and K. Inanaga, *Folia Psychiat. Neurol. Jpn.* **27**(1), 9 (1973).
30. Ger. Pats. 1,771,827 (May 24, 1973), P. Birk and K. Wohlgemuth (to Goerig & Co.); 2,136,450 (Feb. 1, 1973), K. Deutzmann (to W. Banck and O. Vorbach).
31. Ger. Pats. 2,252,795 (May 22, 1974), W. Meichsner (to Süddeutsche Kalkstickstoff-Werke A.G.); 2,252,796 (May 22, 1973), A. Freissmuth (to Süddeutsche Kalkstickstoff-Werke A.G.).
32. Ger. Pat. 2,031,720 (Jan. 5, 1972), H. Prietzel (to Süddeutsche Kalkstickstoff-Werke A.G.).
33. N. Kameyama, *J. Chem. Ind. (Japan)* **24,** 1263 (1921).
34. *AERO® Dicyandiamide,* American Cyanamid Company, Wayne, N. J., 1964.
35. *SKW-Produktstudie Dicyandiamid,* Süddeutsche Kalkstickstoff-Werke A.G., Trostberg, FRG, 1973, 60 pp.
36. T. C. Ponder, *Hydrocarbon Process.* **48**(11), 200 (1969).
37. P. Ostrogovich, *Atti Accad. Lincei* **20**(1), 249 (1911); U.S. Pat. 2,302,162 (Nov. 17, 1942), W. Zerwick and W. Brunner (vested in the Alien Property Custodian).
38. F. H. S. Curd and F. L. Rose, *Chem. Ind.* 75 (1946); Brit. Pat. 581,346 (1946), F. H. S. Curd and F. L. Rose.
39. J. H. Paden, K. C. Martin, and R. C. Swain, *Ind. Eng. Chem.* **39,** 952 (1947).
40. J. Haag, *Ann. Chem.* **122,** 22 (1862).
41. Brit. Pat. 434,961 (Sept. 12, 1935), (to E. I. du Pont de Nemours & Co., Inc.).
42. C. Hasegawa, *J. Soc. Chem. Ind. (Japan)* **45,** 416 (1942).
43. C. D. Garby, *Ind. Eng. Chem.* **17,** 266 (1925).
44. R. N. Harger, *Ind. Eng. Chem.* **12,** 1107 (1920); E. Johnson, *Ind. Eng. Chem.* **13,** 533 (1921).
45. J. W. Lyons, *The Chemistry and Use of Fire Retardants,* John Wiley & Sons, Inc., New York, 1970, p. 136.
46. U.S. Pat. 3,159,503 (Dec. 1, 1964), I. S. Goldstein and W. A. Dreher (to Koppers Co., Inc.).
47. U.S. Pats. 2,280,995 (Apr. 28, 1942), R. B. Booth (to American Cyanamid Company); 2,581,111 (Jan. 1, 1952), C. G. Landes and J. Studeny (to American Cyanamid Company).
48. U.S. Pat. 2,523,626 (Sept. 26, 1950), G. Jones, W. Juda, and S. Soll (to Albi Manufacturing Company, Inc.).
49. *AERO® Melamine,* American Cyanamid Company, Wayne, N. J., 1968.
50. B. Bann and S. A. Miller, *Chem. Res.* **58,** 131 (1958).

DONALD R. MAY
Cyanamid Canada Inc.

CYANIDES

Hydrogen cyanide, 307
Alkali metal cyanides, 320
Alkaline earth metal cyanides, 331

HYDROGEN CYANIDE

Hydrogen cyanide [74-90-8] (hydrocyanic acid, prussic acid, formonitrile), HCN, is a colorless, poisonous, low viscosity liquid with an odor characteristic of bitter almonds. It was prepared in dilute solution by Scheele in 1782 and as the pure compound by Gay-Lussac in 1815. The compound may have been known and used as a poison since early times. It is an acute but not accumulative or chronic poison. It is theorized that HCN played the key role in the origin of plant and animal life on the earth via amino acids (1–2). Its structure is that of a linear, triply-bonded molecule, HC≡N. It is likely that HCN is formed, usually in trace quantities, whenever hydrocarbons are burned in air (3) or when photosynthesis takes place (4).

In early times HCN was manufactured from beet sugar residues and from coke oven gas. These outmoded methods have been supplanted by large-scale modern plants utilizing catalytic synthesis from ammonia and hydrocarbons.

Recently, hydrogen cyanide has become a basic chemical building block for such chemical products as sodium cyanide, potassium cyanide, methyl methacrylate, methionine, triazines, iron cyanides, adiponitrile, and chelates. Its use for fumigation is now minor, whereas thirty years ago fumigation was a major application.

Physical Properties

The physical properties of hydrogen cyanide are listed in Table 1.

Chemical Properties

Hydrogen cyanide is a very weak acid; its ionization constant has the same magnitude as that of the natural amino acids. The reactions of the cyanide ion are discussed under Alkali metal cyanides.

Hydrogen cyanide, as the nitrile of formic acid, undergoes many of the typical nitrile reactions (see Nitriles). For example, it can be hydrolyzed to formic acid by aqueous sulfuric acid (5); it can be hydrogenated to methylamine (6); and it can be converted to phenylformamidine with aniline and hydrogen chloride (7).

Hydrogen cyanide can be oxidized by air at 300–650°C over silver (8) or gold (9) catalysts to give yields of up to 64% cyanic acid, HOCN, and 26% cyanogen, $(CN)_2$. Reaction with chlorine in the liquid phase gives cyanogen chloride (10–11) which is the basic route to triazines of which melamine is an important derivative (see Cyanamides; Urea). Bromine reacts similarly, but the reaction with iodine is incomplete. Chlorination in the vapor phase can be controlled to give cyanogen (12), cyanogen chloride (12), or cyanuric chloride (13) (see Cyanuric and isocyanuric acids).

Hydrogen cyanide normally adds across carbon–carbon double bonds only if they are situated adjacent to a powerful electron-withdrawing group, eg, carbonyl or cyano

Table 1. Physical Properties of Hydrogen Cyanide

Property	Value
mol wt	27.03
mp, °C	−13.24
triple point, °C	−13.32
bp, °C	25.70
d_4^t, liquid, g/mL	
0°C	0.7150
10°C	0.7017
20°C	0.6884
specific gravity, aqueous solution, d_{18}^{18}	
10.04% HCN	0.9838
20.29% HCN	0.9578
60.23% HCN	0.829
vapor pressure, kPa^a	
−29.5°C	6.697
0°C	35.24
27.2°C	107.6
vapor density, at 31°C (air = 1)	0.947
surface tension at 20°C, mN/m (= dyn/cm)	19.68
viscosity at 20.2°C, mPa·s (= cP)	0.2014
specific heat, J/mol^b	
−33.1°C, liquid	58.36
16.9°C, liquid	70.88
27°C, gas	36.03
heat of fusion at −14°C, kJ/mol^b	7.1×10^3
heat of formation, kJ/mol^b	
gas	−128.6
liquid at 18°C, 100 kPa^a	−10.1
heat of combustion, kJ/mol^b	667
critical temperature, °C	183.5
critical density, g/mL	0.195
critical pressure, MPa^c	5.4
dielectric constant	
0°C	158.1
20°C	114.9
dipole moment, gas, C·m^d, at 3–15°C	7.0×10^{-30}
dissociation constant, K_{18}^d	1.3×10^{-9}
conductivity, S/cm	3.3×10^{-6}
heat of vaporization, kJ/mol^b	25.2
heat of polymerization, kJ/mol^b	42.7
entropy, gas at 27°C, 100 kPa^a, J/(mol·°C)^b	202.0
flash point, closed cup, °C	−17.8
explosive limits at 100 kPa^a and 20°C	6–41 vol % in air
autoignition temperature, °C	538
light sensitivity	not sensitive to light
n_D^{10}	1.2675
enthalpy, kJ/mol^b	140

^{a} To convert kPa to mm Hg, multiply by 7.5.
^{b} To convert J to cal, divide by 4.184.
^{c} To convert MPa to atm, divide by 0.101.
^{d} To convert C·m to debeye, divide by 3.336×10^{-30}.

groups. In these cases a Michael addition proceeds readily under basic catalysis, as with acrylonitrile to yield succinonitrile in high yield (14). Formation of acrylonitrile by addition across the acetylenic bond can be accomplished under catalytic conditions. This is the obsolete process for acrylonitrile manufacture (see Acetylene-derived chemicals; Acrylonitrile).

Hydrogen cyanide adds across the carbonyl group of aldehydes and ketones and opens the oxirane ring of epoxides, both under mildly basic conditions (see Cyanohydrins). Several cyanohydrins are commercially important. Lactic acid is made by the hydrolysis of lactonitrile which is formed from the reaction of acetaldehyde and HCN. Acetone cyanohydrin is an intermediate in the manufacture of methyl methacrylate (see Methacrylic acid and derivatives). Hydrogen cyanide reacts with phenols in the presence of hydrogen chloride and aluminum chloride to give aromatic aldehydes (the Gattermann synthesis). Hydrogen cyanide is a starting chemical for alanine, phenylalanine, valine, and α-aminobutyric acid.

HCN reacts with formaldehyde and aniline to form N-phenylglycinonitrile, and with formaldehyde alone to form glycolic nitrile (15). HCN reacts with NaOH, KOH, and Ca(OH)$_2$ to form the corresponding cyanides. Amines can be derived from olefins and HCN via the Ritter reaction (16). Nitrilotriacetonitrile is produced by the reaction of a salt of ammonia and a nonoxidizing acid with formaldehyde and hydrogen cyanide (17). Cyanogen can be formed by oxidative cleavage of HCN over a Ag catalyst (18). Melamine can be prepared from HCN and NH$_3$ by way of cyanamide (19). Iminodiacetonitrile can be prepared from hexamethylenetetramine, HCHO, and HCN (20). Oxamide is formed from HCN and H$_2$O$_2$ (21). Malononitrile can be produced from ClCN, CH$_3$CN, and HCN at 700°C (22). Diiminosuccinonitrile is formed from HCN, Cl$_2$, and trimethylamine (23).

Cyanohydrins (qv) are formed by the reaction of glucose and similar compounds with HCN.

The corresponding aminonitrile from MIBK (methyl isobutyl ketone) can be formed with ammonia and HCN.

HCN reacts violently with sulfuric acid in strong concentrations. The HCN:H$_2$SO$_4$ complex decomposes to give SO$_2$, CO$_2$, and derivatives of ammonia. Dimethylformamide can be produced from the reaction of HCN and methanol. Adenine can be prepared from HCN in liquid ammonia. Thioformamide can be produced from HCN and H$_2$S.

Under certain conditions hydrogen cyanide can polymerize to black solid compounds, eg, the homopolymer [26746-21-4] (1) and the tetramer [27027-02-2] (2). There is usually an incubation period prior to rapid onset of polymer formation. Temperature has an inverse logarithmic effect on the incubation time. Acid stabilizers such as sulfuric and phosphoric acids avoid polymerization. The presence of water reduces the incubation period.

Although HCN is a weak acid and normally not considered corrosive, it has a corrosive effect under two special conditions: (1) water solutions of HCN cause trans-

310 CYANIDES (HCN)

crystalline stress-cracking of carbon steels under stress even at room temperature and in dilute solution; (2) water solutions of HCN containing sulfuric acid as a stabilizer severely corrode steel above 40°C and stainless steels above 80°C (see Corrosion).

Manufacture and Processing

Hydrogen cyanide had been manufactured from sodium cyanide and mineral acid, and from dehydration of formamide. At present neither of these methods is economically important, owing primarily to high raw material costs.

Two synthesis processes account for most of the hydrogen cyanide produced. The dominant commercial process for direct production of hydrogen cyanide is based on the classical technology developed by Andrussow (24–33). This is the reaction of ammonia, methane (natural gas), and air over platinum metals as catalysts. The second process involves the reaction of ammonia with methane and is called the BMA process (30,34–36); it was developed by Degussa, Federal Republic of Germany. HCN is also obtained as a by-product in the manufacture of acrylonitrile by the ammoxidation of propylene (Sohio technology) (see Acrylonitrile). Limited quantities of HCN are also recovered from coke oven gases.

Hydrogen cyanide is produced when hydrogen, nitrogen, and carbon-containing compounds are brought together at high temperatures with or without a catalyst. This fact plus the waning supply and escalating prices for methane have resulted in extensive research for alternative raw materials and systems (see Gas, natural). The Shawinigan process, utilizing a unique reactor system, was operated for several years by Shawinigan Chemicals Ltd. (now Gulf Oil Canada Ltd.) (37–38). The process utilized the Fluohmic furnace, consisting of a fluidized bed of coke at 1350–1650°C, and the passage of electrical current between electrodes immersed in the bed. Feed gas was ammonia and a hydrocarbon, preferably propane. High yield and high concentration of HCN in the off-gas was claimed. A process developed by the U.S. Bureau of Mines synthesizes HCN from coal and ammonia but has not yet been commercialized (39).

Examples of the variety of systems that yield hydrogen cyanide but which have not yet shown economic significance are: at 1100–1300°C acetonitrile and ammonia produce a gas containing HCN 47.3, NH_3 1.1, N 0.8, CH_4 0.4, CH_3CN 0.04, and H 49.9 mol % (37,40); formamide is decomposed at low pressure and elevated temperature to produce HCN (41); HCN is produced from the reduction of nitric oxide over precious metal and perovskite catalysts at 400–800°C (42) (see Exhaust control); methanol and ammonia react in the absence of O_2 and in the presence of catalysts at 600–950°C to produce HCN (38).

The stoichiometry of the Andrussow process may be represented by the following:

$$CH_4 + NH_3 + 1.5\ O_2 \rightarrow HCN + 3\ H_2O \quad 481.9\ kJ/mol\ (115.2\ kcal/mol)$$

Usually, the overall reaction is carried out adiabatically. The catalyst temperatures are ca 1100°C. Precious metal catalysts (90% Pt/10% Rh in gauze form) are normally used in the commercial processes (31). The converters are similar to the ammonia reactors used in the production of nitric acid although the latter operate at somewhat lower temperatures. Optimum operating range is above the upper flammability limit, but caution must be exercised to avoid entering the explosive range.

The actual reactions taking place are more complex than the basic reaction shown above. Most of the endothermic heat requirement for HCN formation is supplied by combustion of methane. The endotherm is 251 kJ/mol (60 kcal/mol). Controlling the decomposition of ammonia and HCN is a critical aspect of the synthesis process. After reaction, the mixture is quickly quenched to <400°C in a steam-generating waste-heat boiler to minimize HCN decomposition. Conversion, yields, and productivity of the synthesis unit are influenced by the extent of preheat, purity of the feeds, reactor geometry, feed-gas ratios, contact time, catalyst and catalyst purity, quench time, and materials of construction. The process consists of two systems, only one of which recovers and recycles ammonia to the converters. Figure 1 is a flow diagram of the process in which ammonia is recycled. The process in which ammonia is not recovered simply substitutes a sulfuric acid scrubber for the ammonia recovery–recycle system. The resulting ammonium sulfate can pose a significant pollution problem. Figure 2 is a flow diagram of the nonrecovery process.

In one patent, a filtered, heated (above the dew point) mixture of air, methane, and ammonia in a volume ratio of 5:1:1 was passed over a 90% platinum–10% rhodium

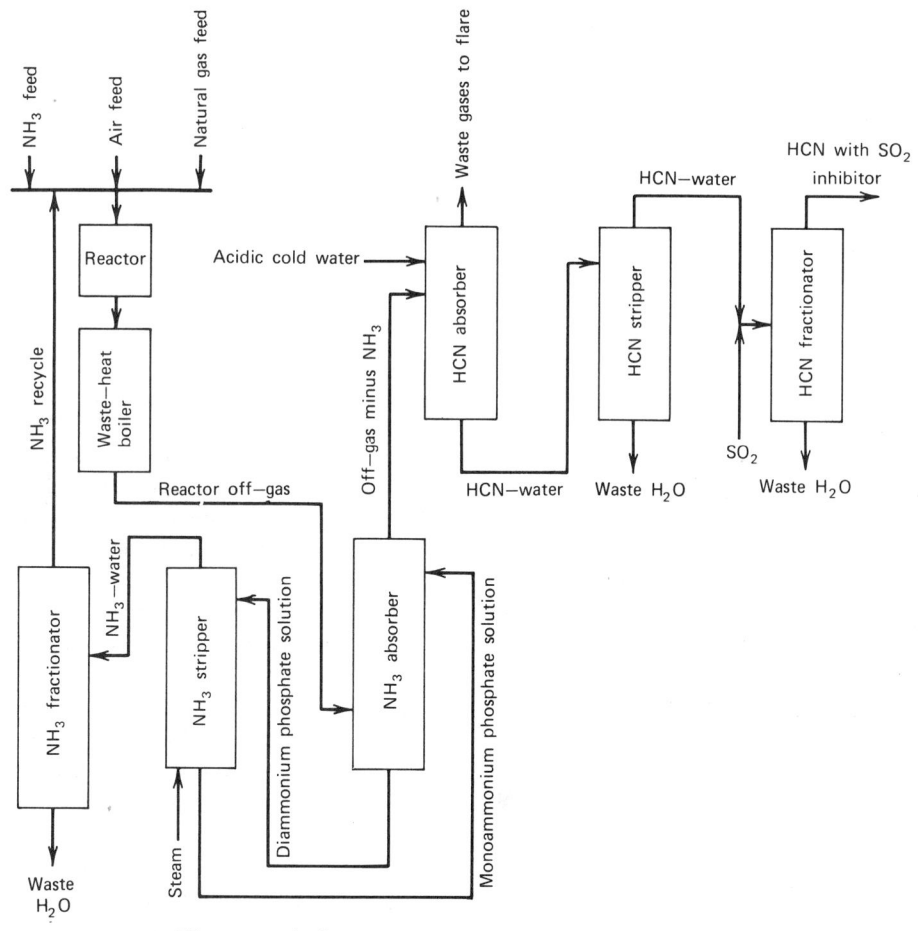

Figure 1. Andrussow process plant with ammonia recycle.

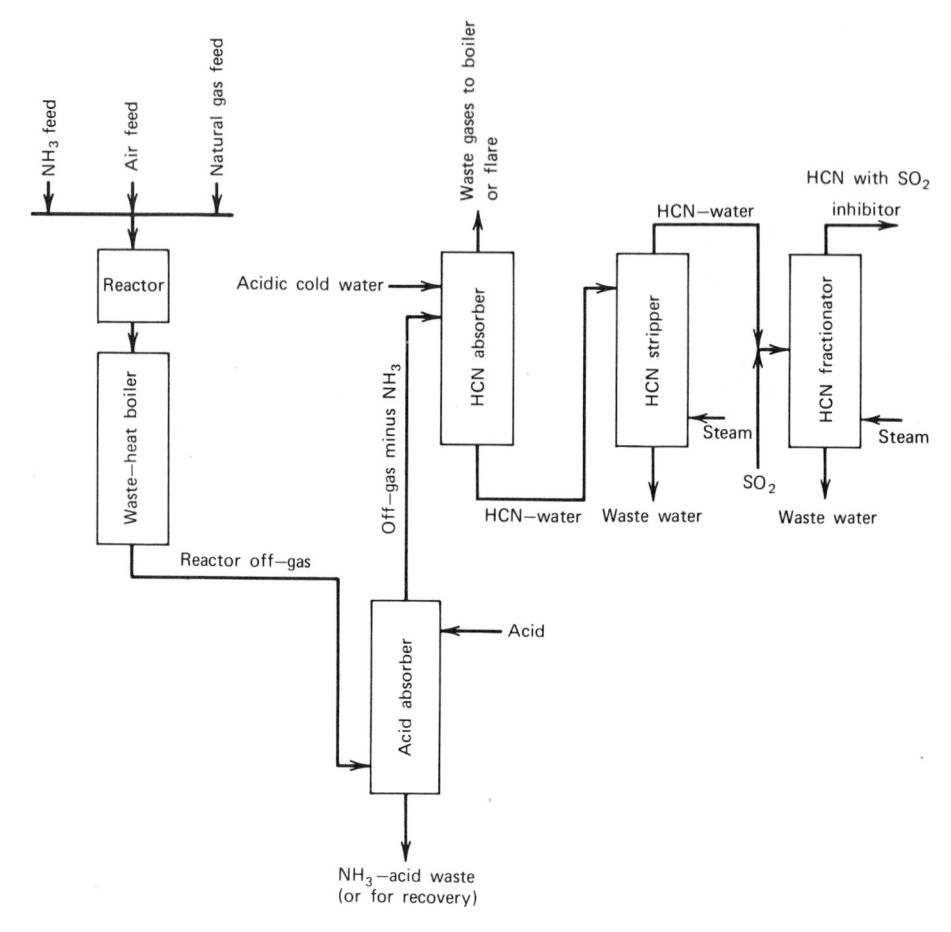

Figure 2. Andrussow process without ammonia recycle.

gauze catalyst at 200 kPa (2 atm) (32). The ammonia was absorbed from the off-gas in a phosphate solution which was subsequently stripped and refined to 90% ammonia–10% water and recycled to the converter. The yield of hydrogen cyanide from ammonia was ca 80%. On the basis of these data, the off-gas mol % composition can be calculated: nitrogen 46.5%; water 15%; hydrogen 22%; hydrogen cyanide 8%; carbon monoxide 5%; carbon dioxide 0.5%; methane 0.5%; and ammonia 2.5%.

Modern HCN plants are energy-sensitive and recovery of heat is very important. Use of a waste-heat boiler directly below the catalyst bed accomplishes a twofold purpose: (1) quenching of the gases to <400°C to minimize decomposition, and (2) the generation of steam to recover energy. Approximately 5 kg of steam is generated per kilogram of hydrogen cyanide. HCN can be recovered and refined after ammonia has been removed from the gas stream. In processes where ammonia is not recovered and recycled, a sulfuric acid scrubber is used to remove ammonia as ammonium sulfate, which can be recovered as dry crystalline ammonium sulfate. An alternative approach recycles ammonia gas to the converter. A commercial system involves passing the product gases from the converters through a solution of monoammonium phosphate at 60–80°C where the ammonia reacts to form diammonium phosphate. The diammonium phosphate solution is stripped with steam and the overheads are fractionated under pressure to yield ammonia gas for recycle to the converter (43–44). The off-gas from the monoammonium phosphate absorber passes to a cold water absorber to remove hydrogen cyanide as a dilute solution in water. The dilute solution is stripped and fractionated by conventional means to 99.5% purity. The absorber and downstream equipment are acidified with a trace of sulfuric acid to avoid polymerization. The product HCN is stabilized with sulfur dioxide which co-condenses in the product condenser. Sulfuric acid is added for supplemental stabilization. To avoid corrosion problems, all recovery equipment should be constructed of austenitic stainless steels. Trace quantities of the following impurities are present in the product: cyanogen, acrylonitrile, acetonitrile, and propionitrile. Maximum energy conservation is obtained by using many process to process and reheat exchangers. Additional energy is recovered by utilizing the large volume, low heat content off-gas from the cold water absorber for fuel in a steam boiler. Low concentrations of HCN and ammonium sulfate constitute the major waste materials in the process effluent.

As with most chemical processes today, environmental considerations are paramount. Air pollution is a minimal problem since all waste gases are flared (burned to CO_2, nitrogen, and H_2O). Strict limits for liquid effluents require treatment to neutralize or remove active, free cyanide. The most common pretreatment or treatment method is alkaline chlorination which converts cyanide to the essentially nontoxic cyanate. The volume of water in the effluent can be reduced about eightfold by recycling the HCN stripper bottoms back to the absorber via refrigerated exchangers. Although this reduces the volume of waste to be treated, a larger amount of energy and investment are required for refrigeration.

The Andrussow process is by far the dominant one in use today. This process inherently has an off-gas stream that is dilute in hydrogen cyanide compared with the Degussa or BMA process. Recovery equipment must be large to accommodate the gas volume. Preheating Andrussow feed gases is a method of raising HCN concentration by supplying part of the required reaction heat in the feed (33). Advantages of the Andrussow process are: low synthesis investment, low maintenance costs, and high natural gas yields.

In the Degussa process, methane (natural gas) or mixed hydrocarbons and ammonia react in the absence of air (45). The endothermic heat is supplied from an externally fired furnace. Yield based on ammonia is 85%. The off-gas from the reactors contains more than 20 mol % HCN, more than 70 mol % H_2, 2–3 mol % NH_3, 1–2 mol % CH_4, and about 1 mol % N_2. Each reactor consists of 8 furnaces, each with 13 tubes of sintered alumina lined with 70% platinum. The furnaces are fired to 1200–1300°C. Each furnace produces 45 t HCN/mo. Furnace design is critical and operating conditions are rigid. The reaction tubes are fragile and require special design features. This type of plant is especially suited for sites where a use for hydrogen exists.

A major source of HCN is the by-product from the Sohio-type acrylonitrile process. Approximately 15 kg of HCN is produced per 100 kg acrylonitrile (qv).

Recovery of hydrogen cyanide from coke-oven gases has been dormant for the last 10–20 yr. Recently, however, new methods involving environmental control of off-gas pollutants are leading the way for a modest return to the recovery of cyanide from coke-oven gases (46) (see Coal, carbonization).

Economic Aspects

Annual capacity of HCN in 1977 was 313,400 metric tons (47) broken down as in Table 2.

Actual United States production for 1955–1977 has been estimated as shown in Table 3.

Table 2. United States Producers of Hydrogen Cyanide and Their Capacities

Producer	Capacity[a], 10^3 t/yr
American Cyanamid, Fortier, La.[b]	14.5
Dow, Freeport, Tex.[c]	2.3
DuPont, Beaumont, Tex.[b,c]	20.4
DuPont, Memphis, Tenn.[b,c]	83.9
Monsanto, Alvin, Tex.[b]	30.0
Monsanto, Texas City, Tex.[b,c]	62.6
Rohm and Haas, Houston, Tex.[c]	81.6
Vistron, Lima, Ohio[b]	18.1
Total	313.4

[a] Excludes DuPont's HCN capacity for adiponitrile at Victoria, Tex.; Orange, Tex.; and LaPlace, La. Monsanto's acrylonitrile expansion increased HCN capacity by 28,500 metric tons in the first quarter of 1977; production is for captive use only. Ciba-Geigy began an HCN facility at St. Gabriel, La. in mid-1978 with capacity of 38,000 annual metric tons.
[b] By-product.
[c] Primary.

Table 3. United States Estimated Production of Hydrogen Cyanide for 1955–1977

Year	Production, 1000 t/yr (million lb/yr)	
1955	58	(128)
1958	84	(186)
1961	117	(259)
1963	163	(360)
1969	204	(450)
1970	222	(490)
1973	236	(520)
1975	249	(550)
1977	259	(570)

Output of hydrogen cyanide in the United States is expected to rise to 295,000 t by 1980.

Worldwide annual production and capacity of hydrogen cyanide are estimated to be 500,000 and 590,000 metric tons, respectively.

Specifications and Analysis

Hydrogen cyanide is classified by the DOT as a Class A Flammable Poison and is subject to rigid packaging, labeling, and shipping regulations. Nominal capacities of authorized steel cylinders are 4.5, 22.7, and 45.4 kg of hydrocyanic acid. Purchases may be made in tank car sizes of 13.6, 24.0, and 45.4 metric tons at 73¢/kg. Specifications are: 99.5% hydrogen cyanide (minimum), 0.5% water (max), 0.06–0.10% acidity, and the color not darker than APHA 20. A combination of H_2SO_4 (or H_3PO_4) and SO_2 acts as a stabilizer to prevent polymerization; H_2SO_4 stabilizes the liquid phase and SO_2 stabilizes the vapor phase.

Assay of hydrogen cyanide can be done by specific gravity or silver nitrate titration similar to that to be described under the section Alkali metal cyanides. Sulfur dioxide in hydrogen cyanide can be determined by ir analysis, or by reaction of excess standard iodine solution and titration with standard sodium thiosulfate or by measurement of total acidity by evaporation to dryness and remeasurement of acidity. Total acidity, which includes SO_2 plus H_2SO_4 or H_3PO_4, can be obtained by titration with 0.1 N NaOH using methyl red–methylene blue mixed indicator. Water can be determined by Karl Fischer titration (see Analytical methods).

Health and Safety Factors (Toxicology)

The cyanides are true noncumulative protoplasmic poisons, ie, they can be detoxified readily. They combine with those enzymes at the blood tissue interfaces that regulate oxygen transfer to the cellular tissues. Unless the cyanide is removed, death results through asphyxia. The warning signs of cyanide poisoning include: dizziness, numbness, headache, rapid pulse, nausea, reddened skin, and blood-shot eyes. More prolonged exposure can cause vomiting and labored breathing followed by unconsciousness, cessation of breathing, rapid, weak heart beat, and death. Severe exposure (by inhalation) can cause immediate unconsciousness; this rapid knockdown power without an irritating odor makes hydrogen cyanide more dangerous than materials of comparable toxicity, eg, H_2S. Hydrogen cyanide can enter the body by inhalation, oral absorption, or skin absorption (see below).

Inhalation. The threshold limit is 10 ppm. This is the maximum safe limit for an 8-h period. OSHA is currently considering a reduction to 5 ppm. Exposure to 20 ppm of HCN in air causes slight symptoms after several hours; 50 ppm causes disturbances within an hour; 100 ppm is dangerous for exposures of 30–60 min; and 300 ppm can be rapidly fatal unless prompt, effective first aid is administered. The body has a mechanism for continuous removal of small amounts of hydrogen cyanide by converting it to thiocyanate which is removed in the urine.

Skin Absorption. Normal skin absorbs HCN slowly. However, 2% HCN in air may cause poisoning in 3 min, 1% is dangerous in 10 min, and 0.05% may produce symptoms after 30 min even though a gas mask or air mask is worn. Cuts and abrasions absorb HCN rapidly and 50 mg of HCN absorbed through the skin can be fatal. Mucous membranes also absorb HCN rapidly.

Oral Absorption. Absorption in the stomach, unless prompt first aid or medical treatment is given, is rapidly fatal; 1 mg HCN/kg body wt can be fatal. Immediate and repeated administration of emetics and regurgitation (if the victim is conscious), followed or accompanied by the first aid and medical treatments described below should be carried out. If the victim is unconscious, stomach lavage should be performed by a physician or trained personnel.

First Aid and Medical Treatment. In case of an accident, action should be fast and efficient. With the protection of a gas mask remove or drag the victim to fresh air. Remove contaminated clothing and rinse contaminated body areas. Keep victim warm. If the victim is breathing, break an amyl nitrite pearl and hold it under the victim's nose for 15 s. Repeat every 15 s for five times if necessary. Amyl nitrite is a powerful cardiac stimulant and should not be used more than necessary. If the patient is not breathing, apply artificial respiration; this can best be done using a resuscitator containing oxygen if available. Mouth to mouth resuscitation is the next best method followed by the Holger-Nielsen arm-lift method.

Notify a physician immediately. A suggested procedure for physicians or nurses is intravenous administration of 0.3 g (10 mL of a 3% solution) of sodium nitrite at the rate of 2.5–5 mL/min followed by 12.5 g (50 mL of a 25% solution) of sodium thiosulfate at the same rate. Watch the patient for 24–48 h. If symptoms reappear, repeat the injections in half the original amounts. These solutions should be kept readily available. In some cases first aid personnel have been trained to use the intravenous medication subject to government regulations.

Detection. Most people can detect hydrogen cyanide by odor or sensation at the 5 ppm concentration in air, but a very few people cannot smell it even at toxic levels. Anyone planning to work with cyanide should be checked by a sniff-test employing a known safe concentration. This test should be given periodically. Several chemical detection and warning methods can be employed. A test paper for HCN can be made by moistening a piece of filter paper with a mixture of equal parts of a solution of 2.86 g cupric acetate in 1 L of water, and a solution of 675 mL of saturated benzidine acetate mixed with 525 mL water. This paper will turn blue on exposure to 20 ppm hydrogen cyanide. The mixed solution deteriorates in a few days; therefore, it should be mixed as needed. Alternatives to the above detection method are: (1) paper treated with N,N'-diethyl-p-phenylenediamine sulfate and $Cu(O_2CCH_3)_2$ or (2) paper treated with p-nitrobenzaldehyde and K_2CO_3. On exposure to 10 ppm HCN for 5 s, the first method produces a pink stain and the second method produces a reddish purple stain. The first method is more sensitive but is limited by a short shelf life of reagents and interferences by Cl_2, NH_3, and NO_2; the second method does not exhibit these limitations (48). A device called Toxguard, manufactured by Mine Safety Appliance Co., can detect 0–50 ppm HCN in air. The device can be used to indicate, alarm, or interlock.

Disposal. Small quantities of hydrogen cyanide can be burned in a hood in an open metal vessel. Large-scale burning in outdoor pans can be performed, but special safety precautions must be employed. A cyanide solution can be decontaminated by making the solution strongly basic (pH 12) with caustic and pouring it into ferrous sulfate solution. The resulting ferrocyanide is relatively nontoxic. Cyanide solution can be converted to less toxic cyanate by treatment with chlorine, sodium–calcium hypochlorite, or ozone at pH 9–11. A maximum of 10% hypochlorite should be employed. The final solution should be checked for absence of free cyanide. The hypochlorite or Cl_2 + NaOH method is by far the most widely used commercially (49). However, other methods are (49): ozonation to cyanate; hydrogen peroxide oxidation to cyanate; electrolysis to CO_2, NH_3, and cyanate; hydrolysis or saponification at elevated temperatures to NH_3 and salts of formic acid; air stripping at low pH to the atmosphere at higher elevations; oxidation and activated carbon treatment to cyanate; chlorination and activated carbon treatment; permanganate oxidation to cyanate;

biological decomposition to CO_2 and N_2; chlorite and chlorine dioxide oxidation to cyanate; chromium treatment to give oxamides; nitrite or nitrate treatment to give carbonates and nitrogen; removal by ion exchange and recovery; lime–sulfur reaction to give sulfate, carbonate, and free sulfur; and gamma-ray treatment to give nontoxic constituents.

Environmental and Occupational Exposure. The toxicity of cyanide that occurs in the aquatic environment or natural waters is dependent upon the so-called free forms of cyanide, ie, as HCN and CN^-. These forms, rather than complexed forms such as iron cyanides, determine the lethal toxicity to fish. Complexed cyanides may revert to free cyanide under uv radiation but the rate is too slow to be a significant toxicity factor. Much work has been done to establish stream and effluent limits for cyanide to avoid harmful effects on aquatic life. The many tests indicate that a free cyanide stream concentration of 0.04 mg/L is acceptable (50).

NIOSH recommends that employee exposure to hydrogen cyanide be controlled as follows: maximum employee exposure to HCN 5 mg/m^3 (4.7 ppm); annual medical examinations and medical records; medical treatment and first aid kits available; labeling, posting and use instructions; protective equipment clothing; and use, training, control practices.

General Safety Aspects. Laboratory work with hydrogen cyanide should be carried out only in a well-ventilated fume hood. Special safety equipment such as air masks, face masks, plastic aprons, and rubber gloves should be used. A chemical-proof suit should be used for emergency. Where hydrogen cyanide is handled inside a building, suitable ventilation must be provided.

The most important rule when working with HCN is: *Never work alone.* This applies especially to sampling, and opening lines and equipment. The extra person must be in view ca 9–10 m away at all times, must be equipped to make a rescue, and must be trained in first aid for HCN exposure.

Hydrogen cyanide undergoes an exothermic polymerization under basic conditions. This polymerization can become explosively violent, especially if confined. Presence of water and heat contribute to the onset of this polymerization. Stored hydrogen cyanide should contain less than 1% water, should be kept cool, and inhibited with 0.06–0.5% sulfuric or phosphoric acid. Manufacturers recommend a maximum of 90 d storage for inhibited hydrogen cyanide in metal containers (cylinders). Storage containers should not be used as heated vaporizers; remove as a liquid and feed to a suitably designed vaporizer as needed. The presence of any compound or surface that depletes the acid stabilizer can cause polymerization, which is first indicated by a yellow–brown color followed by development of heat. Prompt action to be taken includes reacidification and cooling.

Explosively violent hydrolysis can occur if an excess of a strong acid (H_2SO_4) is added to confined HCN. This can occur with >10% H_2SO_4 in HCN.

From a practical-use standpoint there are four rather unique properties of HCN that have safety implications: (*1*) small leaks will form milk-white icicles (even in summer); (*2*) HCN has an unusually steep density change with temperature; (*3*) CO_2 is very soluble in HCN, which creates violent foaming and spouting when warmed; and (*4*) this violent action also occurs when liquid HCN is depressurized from the superheated state.

Because of its low boiling point, hydrogen cyanide can be a fire and explosion hazard.

Uses

Estimates of United States production of HCN in various uses are: methyl methacrylate, 60%; cyanuric chloride, 15%; sodium cyanide, 10%; nitrilotriacetic acid (NTA), ethylenediaminetetraacetic acid (EDTA), and other chelates, 10%; others, 5% (47). There has been a sizeable shift in distribution of uses in the last 10 years. A decade ago about half of the HCN was consumed in the production of acrylonitrile (qv). Acrylonitrile plants now use Sohio's process or similar technology and produce HCN as a by-product rather than consume it. Methyl methacrylate is produced by the conventional acetone cyanohydrin route in which HCN reacts with acetone in the presence of an alkaline catalyst (51) (see Methacrylic acid and derivatives). Herbicides (qv) based on cyanuric chloride have shown the fastest growth in the past 10–15 yr; triazines produced from HCN and Cl_2 are used as starting materials. A new process (Uhde GmbH, FRG) for manufacture of cyanogen uses hydrogen cyanide and oxygen (52).

Another use that has shown considerable growth is methionine, an animal food supplement (see Pet and other livestock feeds). Another fairly large use of HCN is for the chelating agents (qv) such as EDTA. Modest quantities of HCN go to a large number of relatively small uses including: fumigation, ferrocyanides (see Iron compounds), acrylates, lactic acid, pharmaceuticals, and specialty chemicals.

BIBLIOGRAPHY

"Hydrogen Cyanide" under "Cyanides and Cyanogen Compounds" in *ECT* 1st ed., Vol. 4, pp. 689–697, by W. Jane Miller, E. I. du Pont de Nemours & Co., Inc.; "Hydrogen Cyanide" under "Cyanides" in *ECT* 2nd ed., Vol. 6, pp. 574–585, by P. D. Montgomery, Monsanto Company.

1. C. Ponnamperuma, *Sci. J.* **1**(3), 39045 (1965).
2. J. P. Ferris **203**, 1135 (1979); C. N. Matthews, *ibid.*, 1136.
3. D. M. L. Griffiths and H. A. Standing, *Adv. Chem. Ser.* **55**, 666 (1966).
4. A. Blumenthal and G. Kiss, *Mitt. Geb. Lebensmittelunters. Hyg.* **61**, 394 (1970).
5. A. W. Cobb and J. H. Walton, *J. Chem. Phys.* **41**, 351 (1957).
6. S. Barrett and A. F. Titley, *J. Chem. Soc.* **115**, 902 (1919).
7. A. Bernthsen, *Ann* **192**, 1 (1878).
8. U.S. Pat. 2,712,493 (July 5, 1955), W. Moje (to E. I. du Pont de Nemours & Co., Inc.).
9. Ger. Pat. 1,056,101 (Apr. 30, 1959), H. Zima (to Rohm & Haas G.m.b.H.).
10. U.S. Pat. 2,672,398 (Mar. 16, 1954), H. Huemer, H. Schulz, and W. Pohl (to Deutsche Gold und Silber-Scheideanstalt).
11. R. Oleinek, *Rev. Chim.* **16**, 19 (1965).
12. U.S. Pat. 2,399,361 (Apr. 30, 1946), B. S. Lacy and W. S. Hinegardner (to E. I. du Pont de Nemours & Co., Inc.).
13. U.S. Pat. 2,762,798 (Sept. 11, 1956), N. L. Hardwicke and G. W. Walpert (to Monsanto Co.).
14. U.S. Pat. 2,434,606 (Jan. 13, 1948), E. L. Carpenter (to American Cyanamid Co.).
15. G. Schlesinger and S. L. Miller, *J. Am. Chem. Soc.* **95**, 3729 (1973).
16. L. I. Krimen and J. Donald, *Org. React.* **17**, 213 (1969).
17. Can. Pat. 902,099 (Apr. 26, 1971), H. Neumaier and co-workers (to Knapsack Aktiengesellschaft).
18. Ger. Pat. 2,205,983 (Aug. 16, 1973), W. Gruber and co-workers (to Rohm G.m.b.H.).
19. U.S. Pat. 3,177,215 (Apr. 6, 1965), R. W. Foreman and co-workers (to Standard Oil Co.).
20. U.S. Pat. 3,988,360 (Oct. 26, 1976), R. Gaudette and J. Philbrook (to W. R. Grace & Co.).
21. R. C. Sheridan and E. H. Brown, *J. Org. Chem.* **30**(2), 668 (1965).
22. Jpn. Pat. 74 04,207 (Jan. 31, 1974), Y. Hamamoto (to Kyowa Gas Chemical Industry Co., Ltd.).
23. U.S. Pat. 3,666,787 (May 30, 1972), O. Webster (to E. I. du Pont de Nemours & Co., Inc.).
24. U.S. Pat. 1,934,839 (Nov. 14, 1933), L. Andrussow (to I. G. Farbenindustrie, A.G.).

25. L. Andrussow, *Angew. Chem.* **48,** 593 (1935).
26. L. Andrussow, *Bull. Soc. Chim. Fr.* **18,** 45 (1951).
27. L. Andrussow, *Chem. Ing. Tech.* **27,** 469 (1955).
28. L. Andrussow, *Genie Chim.* **86,** 39 (1961).
29. C. T. Kautter and W. Leitberger, *Chem. Ing. Tech.* **25,** 699 (1953).
30. P. W. Sherwood, *Pet. Eng.* **31,** C-22, C-51 (1959).
31. U.S. Pat. 3,215,495 (Nov. 2, 1965), W. R. Jenks and A. W. Andresen (to E. I. du Pont de Nemours & Co., Inc.).
32. U.S. Pat. 3,360,335 (Dec. 26, 1967), W. R. Jenks (to E. I. du Pont de Nemours & Co., Inc.).
33. U.S. Pat. 3,104,095 (Sept. 24, 1963), W. R. Jenks and R. M. Shepherd (to E. I. du Pont de Nemours & Co., Inc.).
34. R. Rodiger, *Chem. Tech. (Berlin)* **10,** 135 (1958).
35. F. Endter, *Chem. Ing. Tech.* **30,** 305 (1958).
36. *Chem. Week* **83**(4), 70 (1958).
37. Ger. Pat. 2,014,523 (Apr. 17, 1970), (to Degussa).
38. Ger. Pat. 1,143,497 (Feb. 14, 1963), G. Schulze and G. Weiss (to Badische Anilin- und Soda-Fabrik A.G.).
39. G. E. Johnson and co-workers, *U.S. Bur. Mines Rep. RI6994,* (Aug. 1967).
40. Ital. Pat. 845,992 (July 2, 1969), (to Montecatini Edison).
41. Ger. Pat. 1,961,484 (June 16, 1971), K. Sennewald and co-workers (to Knapsack A.G.).
42. R. J. H. Voorhoeve and co-workers, *J. Catal.* **45,** 297 (1976).
43. U.S. Pat. 2,797,148 (June 25, 1957), H. C. Carlson (to E. I. du Pont de Nemours & Co., Inc.).
44. U.S. Pat. 3,718,731 (Feb. 27, 1973), W. T. Hess and H. C. Carlson (to E. I. du Pont de Nemours & Co., Inc.).
45. F. Endter, *DECHEMA Monogr.* **33,** 28 (1959).
46. Ger. Pat. 2,260,248 (June 12, 1974), H. Karwat (to Linde A.G.).
47. *Chem. Mark. Rep.,* (Apr. 1, 1977).
48. R. Hill, J. Holt, and B. Miller, *Ann. Occup. Hyg.* **14,** 289 (1971).
49. H. Roos and M. Schmidt, *Chem. Ing. Tech.* **49,** 369 (1977).
50. D. Lind, L. Smith, and S. Broderius, *J. Water Pollut. Control Fed.,* 262 (Feb. 1977).
51. Jpn. Pat. 73 23,426 (July 13, 1973), T. Ikeda, M. Kezuka, and T. Yoshida (to Nitto Chemical Industry Co., Ltd. and Mitsubishi Rayon Co., Ltd.).
52. *Chem. Week* 45 (Apr. 11, 1979).

WILLIAM R. JENKS
E. I. du Pont de Nemours & Co., Inc.

ALKALI METAL CYANIDES

SODIUM CYANIDE

Sodium cyanide [143-33-9], NaCN, is a white cubic crystalline solid commonly called white cyanide. It was first prepared in 1834 by F. and E. Rodgers who heated prussian blue and sodium carbonate together and extracted sodium cyanide from the cooled mixture using alcohol. It remained a laboratory curiosity until 1887, when J. S. MacArthur and the Forrest brothers patented a process for the extraction of gold and silver from their ores by means of a dilute solution of cyanide (see Extractive metallurgy). A mixture of sodium and potassium cyanides, produced by Erlenmeyer's improvement of the Rodgers' process, was marketed in 1890.

The Beilby process started in 1891 and by 1899 accounted for half of the total European production of cyanide. In this process, a fused mixture of sodium and potassium carbonates reacts with ammonia in the presence of carbon to give a product very similar in composition to that obtained from the Erlenmeyer process. In 1900, the Castner process, whereby sodium, ammonia, and charcoal react to give a high grade (98%) sodium cyanide, superseded the Beilby process. Sodium cyanide became an article of commerce and soon replaced potassium cyanide in all except special uses. The Castner process has now been replaced by the so-called neutralization or wet process in which liquid HCN and sodium hydroxide solution react and water is evaporated. The resulting crystals are briquetted or made into granular form. During the 1950s, essentially all sodium cyanide was used in electroplating and case hardening (see Metal surface treatments). Today, case hardening is a minor use and the major applications of sodium cyanide are: electroplating; gold and silver extraction; iron blues; and synthesis of a large number of chemicals.

Physical Properties

The physical properties of sodium cyanide are listed in Table 1.

The solid phase in contact with a saturated aqueous solution at temperatures above 34.7°C is the anhydrous salt; below 34.7°C, the solid phase is the dihydrate [25178-25-0]. The solubility of the dihydrate in g NaCN/100 g satd soln is 16.01 at −15°C; 32.8 at 10°C; 34.2 at 15°C; and 45 at 34.7°C. The solubility of the anhydrous salt is less dependent on temperature.

Sodium cyanide is very soluble in liquid ammonia. At temperatures below −31°C the pentaammoniate [69331-34-6], NaCN·5NH$_3$, separates in large flat crystals. At 15°C, 100 g anhydrous methanol dissolves 6.44 g anhydrous NaCN; at 67.4°C it dissolves 4.10 g. The hemihydrate, NaCN·0.5H$_2$O [69331-35-7], has been obtained by recrystallization of sodium cyanide from cold 85% alcohol. The system NaCN–NaOH–H$_2$O has been studied (1–2). Sodium cyanide is slightly soluble in formamide, ethanol, methanol, SO$_2$, furfural, and dimethylformamide.

At room temperature (20°C) sodium cyanide is a crystalline ionic compound with a structure similar to that of sodium chloride or bromide, ie, an fcc crystal lattice.

Table 1. Physical Properties of Sodium Cyanide

Property	Value
mp, °C	
100%	563.7 (±1)
98%	560
bp, °C (extrapolated)	1500
density, g/cm^3	
cubic	1.60
orthorhombic	1.62–1.624
molten, at 700°C	1.22 (approx)
vapor pressure, kPa[a], at temp	
800°C	0.1013
900°C	0.4452
1000°C	1.652
1100°C	4.799
1200°C	11.9
1300°C	27.2
1360°C	41.8
heat capacity[b], 25–72°C, J/g[c]	1.38
heat of fusion, J/g[c]	314
heat of vaporization, J/g[c]	3190
heat of formation, $\Delta H_f°$, NaCN(c), J/mol[c]	-89.9×10^3
heat of solution[d], ΔH_{soln}, J/mol[c]	+1510
hydrolysis constant, K_h, 25°C	2.51×10^{-5}
viscosity, 26 wt % NaCN–H$_2$O, 30°C, mPa·s (= cP)	4

[a] To convert kPa to mm Hg, multiply by 7.5.
[b] The heat capacity of sodium cyanide has been measured between 100 and 345 K (1).
[c] To convert J to cal, divide by 4.184.
[d] In 200 mol H$_2$O.

Sodium chloride and sodium cyanide are isomorphous and form an uninterrupted series of mixed crystals. The ferrocyanide ion has a marked effect on the habit of sodium cyanide crystallized from aqueous solution (3). Sodium cyanide and sodium carbonate form a eutectic at approximately 53 wt % sodium carbonate and 465°C. The specific conductivity of molten 98% sodium cyanide is 1.17 S/cm (4).

Chemical Properties

When heated in a dry CO$_2$ atmosphere, sodium cyanide fuses without much decomposition. A brown-black color appears when water vapor and CO$_2$ are present at temperatures of 100°C below the fusion point. This color is presumably produced from the HCN polymer. Thermal dissociation of sodium cyanide has been studied in an atmosphere of helium at 600–1050°C and in an atmosphere of nitrogen at 1050–1255°C. It has been shown that the vapor phase over melt contains decomposition products (5). In the presence of a trace of iron or nickel oxide, rapid oxidation occurs when cyanide is heated in air, first to cyanate and then to carbonate:

$$2 \text{ NaCN} + \text{O}_2 \rightarrow 2 \text{ NaCNO}$$

$$2 \text{ NaCNO} + 1\tfrac{1}{2} \text{ O}_2 \rightarrow \text{Na}_2\text{CO}_3 + \text{N}_2 + \text{CO}_2$$

Case hardening of steels using a sodium cyanide molten bath depends on these reactions where the active carbon and nitrogen are absorbed into the steel surface; hence the names carburizing and nitriding (see Metal surface treatments; Steel). The ease with which it is oxidized makes sodium cyanide a good reducing agent and the oxides of several metals, such as lead, tin, manganese, or copper, are readily reduced.

No reaction takes place below 500°C when sodium cyanide and sodium hydroxide are heated in the absence of water and oxygen. Above 500°C sodium carbonate, sodium cyanamide, sodium oxide, and hydrogen are produced. In the presence of small amounts of water at 500°C decomposition occurs with the formation of ammonia and sodium formate, and the latter is converted into sodium carbonate and hydrogen by the caustic soda. In the presence of excess oxygen, sodium carbonate, nitrogen, and water are produced (6).

Molten sodium cyanide reacts with strong oxidizing agents such as nitrates and chlorates with explosive violence. In aqueous solution, sodium cyanide is oxidized to sodium cyanate by oxidizing agents such as potassium permanganate or hypochlorous acid. The reaction with chlorine in alkaline solution is a basis of a successful method of neutralizing industrial cyanide waste liquors (7):

$$NaCN + 2\ NaOH + Cl_2 \rightarrow NaCNO + 2\ NaCl + H_2O$$

$$2\ NaCNO + 4\ NaOH + Cl_2 \rightarrow 6\ NaCl + 2\ CO_2 + N_2 + 2\ H_2O$$

The pH value is usually maintained above 9 to avoid formation of nitrogen trichloride. At lower pH values, aqueous solutions react with chlorine to form cyanogen chloride (5).

Sodium cyanide when fused with sulfur or a polysulfide is converted into sodium thiocyanate; this compound is also formed when a solution of sodium cyanide is boiled with sulfur or a polysulfide:

$$NaCN + S \rightarrow NaCNS$$

$$4\ NaCN + Na_2S_5 \rightarrow 4\ NaCNS + Na_2S$$

A solution of sodium cyanide shaken with freshly precipitated ferrous hydroxide is converted to a ferrocyanide:

$$6\ NaCN + Fe(OH)_2 \rightarrow 2\ NaOH + Na_4Fe(CN)_6$$

If the solution is acidified and a little ferric sulfate added, ferric ferrocyanide is produced. This salt has a characteristic deep-blue color, and the reaction may be used to test for cyanide.

Aqueous solutions of sodium cyanide are slightly hydrolyzed at room temperature (8) according to the reversible reaction:

$$NaCN + H_2O \rightleftharpoons NaOH + HCN$$

At $\geq 50°C$, irreversible hydrolysis to formate and ammonia becomes important, as shown below. If the heat of reaction is not removed, the increased temperature accelerates the decomposition and can create high pressure in a closed vessel.

$$NaCN + 2\ H_2O \rightarrow NH_3 + HCOONa$$

Hydrogen cyanide is a weak acid and can readily be displaced from a solution of sodium cyanide by weak mineral acids or by reaction with carbon dioxide, eg, from the atmosphere; however, the latter takes place at a slow rate.

In the presence of oxygen, aqueous sodium cyanide dissolves most metals in the finely divided state, with the exception of lead and platinum. This is the basis of the MacArthur process for the extraction of gold and silver from their ores which, in the case of gold, may be represented as follows:

$$4\ NaCN + 2\ Au + \tfrac{1}{2}\ O_2 + H_2O \rightarrow 2\ NaAu(CN)_2 + 2\ NaOH$$

The gold is then recovered from solution by precipitation with zinc dust. Sodium cyanide is used extensively in organic syntheses, especially in the preparation of nitriles (qv).

Manufacture

Alkali cyanides became commercially significant when J. R. and H. Elkington discovered the electroplating process for gold and silver in ca 1840. In 1899, the MacArthur-Forrest cyanide process, used for the extraction of gold and silver from low grade ores, was tested successfully on a commercial scale in New Zealand and South Africa. The MacArthur process was rapidly adopted in goldfields and resulted in a phenomenal growth in the cyanide manufacturing industry. World production of alkali cyanide rose from 5900 t/yr in 1899 to 21,000 t/yr by 1915.

The total world capacity is estimated to be in excess of 136,000 metric tons. Annual usage in 1977 was ca 113,000 metric tons.

Early demands for cyanide were met by the Rodgers process, in which potassium ferrocyanide was fused with potassium carbonate or sodium carbonate. The Erlenmeyer process, using the reaction shown below, became commercial in 1876.

$$K_4Fe(CN)_6 + 2\ Na \rightarrow 4\ KCN + 2\ NaCN + Fe$$

Later in a process developed by G. T. Beilby (9), alkali carbonates were used in a reaction with carbon and ammonia:

$$K_2CO_3 + 4\ C + 2\ NH_3 \rightarrow 2\ KCN + 3\ CO + 3\ H_2$$

Present-Day Processes. Today, almost all sodium cyanide is manufactured by the neutralization or so-called wet processes in which hydrogen cyanide reacts with sodium hydroxide solution. This process has replaced the historic Castner process which was developed in the 1800s and operated throughout the world until the late 1960s. Lower yields and higher costs were the major reasons for obsolescence of the Castner process:

Castner process: $2\ Na + 2\ NH_3 + C \rightarrow 2\ NaCN + 3\ H_2$ charcoal

Wet process: $HCN + NaOH \rightarrow NaCN + H_2O$

Most plants use essentially purified anhydrous liquid HCN (sometimes vapors) to react with NaOH, generally fed as a 50% solution to the reactor. These raw materials must be sufficiently pure to yield a high quality, 99% NaCN. Sodium hydroxide with low iron content is necessary to avoid interference with crystallization in the subsequent processing steps (3). Purified hydrogen cyanide from the Andrussow (see Hydrogen cyanide) and by-product HCN from acrylonitrile (qv) processes are used in most commercial NaCN processes. Direct absorption of HCN in the product gases from Andrussow reactors into NaOH solution is practiced to some extent; however, the assay of the dry product is lower, 96–97%, in this case owing to higher concentrations of sodium carbonate and sodium formate.

Most modern high-tonnage plants use the reaction of high purity HCN and aqueous sodium hydroxide in a unit system that embodies evaporation of water and crystallization (10). Control of the system is critical to avoid HCN polymer formation, which would produce an off-white product; to minimize sodium formate production, which would reduce the product purity; and to maximize the average crystal size.

A German process produces a high NaCN assay (99%) by absorbing the gases from the HCN reactor directly in sodium hydroxide solution (11). A BMA reactor (Degussa type) is used wherein ammonia and natural gas or hydrocarbons react in the absence of air in ceramic catalyst tubes enclosed in a furnace chamber. The resulting sodium cyanide solution is heated in a crystallizer to remove water, and thus form NaCN crystals.

The formation of larger crystals facilitates dewatering in the filtration step. In most modern processes, the moist salt from the filter is passed through a mixing conveyor to break up lumps. In some processes, air heated to 450°C is passed through the cake on the filter and through the mixing conveyor (12). Drying is completed in a hot-air conveyor–dryer (see Drying). The above types of adiabatic hot-air drying avoid overheating NaCN when moisture is present, thereby minimizing the creation of sodium formate in the dried product. Inherently, sodium cyanide is a small crystal (50 μm dia), that gives a very dusty solid of low bulk density that must be compacted or fused into larger particles for safer handling. Melting the dry, dusty material and casting it into molds is rarely done because of the high energy requirement. Most modern processes employ mechanical compacting devices that produce either briquettes, granular products, or both (13). Briquettes or tablets are formed in approximate 15-g and 30-g sizes. The granular size is ca 4 mm (see Pelleting and briquetting).

In a typical process the finely divided dry crystals are compacted under heat and pressure in a roll press into a translucent, noncrystalline sheet or ribbon with a density of 1.550–1.590. Under these conditions and with pressure the fine particles are converted to a continuous semiplastic sheet or ribbon. The ribbon is fed to a granulator containing an internal screen to control the maximum size of particles. The granulated ribbon is classified in a hot pneumatic classifier to remove the undersized particles and give a product that is essentially dust-free. Alternatively, the classified granular NaCN is fed to a double-roll briquetting machine with matching pockets in the rolls to give desired product size, weight, and quality (13). The briquettes are passed to a rotary screen where the "fins" or thin layers of material attached to the periphery of the briquette centerline, are removed and reprocessed. The finished briquettes pass into large storage bins from which they are packaged into drums and other shipping containers.

The residues from beet manufacture (*Schlempe* in German) can be a source of cyanides (14). They contain nitrogen as betaine, $(CH_3)_3N^+CH_2COO^-$, which upon heating to above 1000°C, is largely converted to hydrogen cyanide which in turn can be converted to NaCN. However, use of this process has declined significantly.

Plants for the production of sodium cyanide from hydrogen cyanide obtained from the Andrussow process or from by-product HCN in acrylonitrile synthesis are operating in the United States, Italy, Japan, and England. In the Federal Republic of Germany, Degussa converts HCN directly from the reactor to NaCN (11). The neutralization process is not energy intensive; added heat evaporates water formed in the reaction and water entering the system with the raw material, 50% NaOH. The significant waste effluent contains 10–100 ppm NaCN and must be treated before disposal.

A slight excess of NaOH must be maintained at all stages of processing to avoid color formation owing to generation of black or brown HCN polymer. The product as shipped must also contain a slight excess of NaOH so that clear, colorless NaCN solutions will result. Most sodium cyanide is sold in dry form to minimize the higher freight costs associated with shipping water. Formerly, an appreciable tonnage was sold as 30% aqueous solution.

Grades and Shipping

Sodium cyanide is sold in the following forms: granular or powder, pillow-shaped briquettes of 15 g and 30 g, tablets of 30 g, and 30% aq soln. List price in the *Chemical Marketing Reporter* is 88¢/kg (1978).

Typical analysis for the neutralization or wet process product is given in Table 2.

Table 2. Typical Analysis of NaCN

Constituent	Value
sodium cyanide, NaCN, %	98–99
sodium carbonate, Na_2CO_3, %	0.3–1.3
sodium formate, HCOONa, %	0.4
sodium hydroxide, NaOH, %	0.2
sodium chloride, NaCl, %	<0.05
sodium sulfide, Na_2S ppm	1
water, H_2O, %	<0.05
iron, as Fe, ppm	10–20

Sodium cyanide is packed in mild-steel or fiber drums and in 1.4-t flo-bins. Dry sodium cyanide is also shipped in wet-flo tank cars of up to 32 metric tons net. Water is circulated through the car to dissolve the dry NaCN. This type of car shipment reduces freight cost and improves safety compared to 30% aqueous solution shipment. Safety regulations are imposed by the various shipping lines and by the countries in which cyanide is transported.

Analysis

The cyanide content of a sodium cyanide sample may be determined by titration of an aqueous solution containing 10 g sample/L with 0.1 N silver nitrate solution with potassium iodide as the indicator. Use of the indicator p-dimethylaminobenzalrhodamine in the above procedure gives a more sensitive end point and can therefore be used for greater accuracy for very low concentrations of cyanide. For still lower concentrations of cyanide (less than 0.2 ppm), the selective or specific ion electrode can be employed; this has the added advantage of functioning in colored or dirty solutions (see Ion-specific electrodes).

The sodium carbonate content may be determined on the same sample after a slight excess of silver nitrate has been added. An excess of barium chloride solution is added and, after the barium carbonate has settled, it is filtered, washed, and decomposed by boiling with an excess of standard hydrochloric acid. The excess of acid is then titrated with standard sodium hydroxide solution, using methyl red as indicator, and the sodium carbonate content is calculated.

Sulfide content is determined by titration with standard lead nitrate solution (1 g/L). The titration is continued until a drop of the test solution on a filter paper ceases to produce a stain with a drop of lead nitrate solution.

Other ions (eg, ferrate and chloride formate) are determined by first removing the cyanide ion at ca pH 3.5 (methyl orange end point). Iron is titrated with thioglycolic acid and the optical density of the resulting pink solution is measured at 538 nm. Formate is oxidized in titration with mercuric chloride. The mercurous chloride produced is determined gravimetrically. Chloride ion is determined by a titration with 0.1 N silver nitrate. The end point is determined electrometrically.

Health and Safety Factors (Toxicology)

Environmental aspects and toxicity of cyanides are discussed under Hydrogen cyanide. Handling, storage, and use of the alkali metal cyanides must be carried out by trained persons under strict supervision. Most serious injuries and fatalities have been caused by inadvertently mixing these cyanides with acids, thereby causing liberation of HCN. A prime safety rule is to avoid storage near, or mixing with acids. The present TLV for eight-hour exposure is 10 mg/m^3 for cyanide salts; there is a proposal by NIOSH to reduce this to 5 mg/m^3. All necessary precautions should be taken to prevent cyanide salts from contacting liquid or airborne acids. Cyanide salts must be protected from large concentrations of carbon dioxide to avoid HCN liberation. Carbon dioxide fire extinguishers should not be used. Cyanide salts as solids or solutions must be stored in tightly closed containers which must be protected from corrosion or damage. They should be stored so there is no contact with nitrate–nitrite mixtures or peroxides. Storage areas must be adequately ventilated and proper container labeling is mandatory. Rubber gloves should be worn when handling dry salts. In addition, the following protective items should be used with solutions or dusty salts: protective sleeves, aprons, shoes, boots or overshoes made of rubber, chemical safety goggles, full face shield, and filter-type respirator (where dust is present). Cyanide spills should be flushed to a contained area where treatment to destroy the cyanide can be carried out. Avoid flushing to drains or ditches that may contain acids. In the event that cyanide salts or solutions contact the eyes, they should be flushed for 15 min with a copious, gentle flow of water followed by immediate medical attention. Eating, smoking, and chewing of tobacco or gum should be forbidden in areas where cyanide salts are handled. Carrying food, gum, and tobacco should also be forbidden. Employees should be required to wash carefully after working with cyanide salts and before eating, smoking, or chewing. Meals should be eaten in a room separate from areas where cyanides are stored or used (15).

Uses

Electroplating has been the largest single market for sodium cyanide, especially for zinc (16), copper, brass, and cadmium. Gold, gold alloys, and silver are also plated onto base metals from sodium cyanide solutions. Some decline in cyanide electroplating has occurred in recent years owing to tighter restrictions on cyanide discharge and conservation of plating and rinse solutions (see Electroplating).

Use for heat-treating is now small compared to 15 years ago; it has been supplanted to a large extent by furnaces using special atmospheric conditions. Heat-treatment salts containing sodium cyanide are used for small metal parts when selective case hardening is required (see Metal surface treatments).

Sodium cyanide has three specific uses for extraction and recovery of minerals and metals from ores (17): cyanidation recovery of gold and silver; froth flotation (qv) beneficiation of sulfide ores; and refinement of mineral concentrates. Recovery of gold by cyanidation is the largest single mining use for NaCN and has been growing in recent years owing to the sharp rise in gold prices. Gold in the presence of oxygen and weak NaCN solution forms the water soluble cyanide complex salt $NaAu(CN)_2$ [15280-09-8]. Metallic silver frequently occurring with the gold is recovered at the same time.

Chemical uses for sodium cyanide may constitute as much as 25% of the total, and are in five general categories: dyes, including optical brighteners (see Brighteners, optical); agricultural chemicals; pharmaceuticals; chelating or sequestering agents; and specialties. NaCN is used in the preparation of nitriles (qv), carbylamines (isonitriles), cyano fatty acids, and heavy metal cyanides (see also Cyanocarbons). NaCN reacts with dextrose hydrate in aqueous solution to produce glucoheptonate (18). Phenylglycine, the intermediate in the manufacture of indigo, is manufactured using NaCN. Substantial quantities of NaCN are used in sodium ferrocyanide for the manufacture of prussian blue. Sodium cyanide solution treated with $FeCl_2$ in the presence of alkali metal hydroxide with careful adjustment of pH produces yellow $Na_4Fe(CN)_6$ (19). NaCN is used to produce sodium nitrilotriacetate (SNTA) which is a popular and growing chelating agent, as well as a phosphate replacement in certain areas of the world (see Chelating agents). Cyanuric chloride is produced in moderate quantities from NaCN (see Cyanuric and isocyanuric acids).

Miscellaneous uses for NaCN include heat-treating, metal stripping, and compounds used for clearing smut. Treatment of wood chips with NaCN and $CaCl_2$ reportedly increases the kraft cooking yield of pulp (qv) (20).

POTASSIUM CYANIDE

Potassium cyanide [151-50-8], KCN, is a white crystalline, deliquescent solid.

It was initially used as a flux and later for electroplating; electroplating (qv) is the single greatest use today. The demand for potassium cyanide was met by the ferrocyanide process until the latter part of the 19th century, when the extraordinary demands of the gold mining industry for alkali cyanide resulted in the development of direct synthesis processes. When cheaper sodium cyanide became available, potassium cyanide was displaced in many uses. The total world production in 1976 was estimated to be ca 7300 t.

Commercial potassium cyanide made by the neutralization or wet process rather than the obsolete Beilby process contains 99% KCN; the principal impurities are potassium carbonate, formate, and hydroxide. To prepare 99.5+% KCN, high quality HCN and KOH must be used.

Physical Properties

The physical properties of potassium cyanide are given in Table 3. Unlike sodium cyanide, potassium cyanide does not form a dihydrate.

Table 3. Physical Properties of Potassium Cyanide

Property	Value
mp, °C	
100%	634.5
96.05%	622
density, g/cm^3	
cubic at 20°C	1.553
cubic at 25°C	1.56
orthorhombic at −60°C	1.62
specific heat, 25–72°C, J/g[a]	1.01
heat of fusion, J/mol[a]	14.7 × 10^3
heat of formation, ΔH_f°, J/mol[a]	−113 × 10^3
heat of soln, ΔH_{soln}, J/mol[a]	+11.7 × 10^3
hydrolysis constant, 25°C	2.54 × 10^{-5}
sol in water at 25°C, g/100 g H$_2$O	71.6
resistivity, Ω·cm	
0.25 normal soln	70
0.5 normal soln	15
1.0 normal soln	10
2.0 normal soln	5

[a] To convert J to cal, divide by 4.184.

The solubility of potassium cyanide in nonaqueous solvents is as follows: in anhydrous liquid ammonia, 4.55 g KCN/100 g NH$_3$ at −33.9°C; 4.91 g/100 g methanol at 19.5°C; 0.57 g/100 g ethanol at 19.5°C; 146 g/L solution in formamide at 25°C; 41 g/100 g hydroxylamine at 17.5°C; 24.24 g/100 g glycerol of sp gr 1.2561 at 15.5°C; 0.73 g/L solution in phosphorus oxychloride at 20°C; 0.017 g/100 g liquid sulfur dioxide at 0°C; 0.22 g/100 g dimethylformamide at 25°C.

At room temperature potassium cyanide has fcc crystal structure (21). Average KCN crystals inherently have 3 to 4 times the mass or size of NaCN crystals.

Chemical Properties

Potassium cyanide is readily oxidized to potassium cyanate by heating in the presence of oxygen or easily reduced oxides such as those of lead or tin, or manganese dioxide, and in aqueous solution by reaction with hypochlorites or hydrogen peroxide.

Dry KCN in sealed containers is stable for many years. An aqueous solution of potassium cyanide is very slowly converted to ammonia and potassium formate; the decomposition rate accelerates with increasing temperature. However, at comparable temperatures the rate of conversion is far lower than that for sodium cyanide; only about 25% as great.

Many reactions can be carried out between potassium cyanide and organic compounds with the alkalinity of the KCN acting as a catalyst; these reactions are analogous to reactions of sodium cyanide. The reactions of potassium cyanide with sulfur and sulfur compounds are also analogous to those of sodium cyanide. Potassium cyanide is reduced to potassium metal and carbon by heating it out of contact with air in the presence of powdered magnesium. Magnesium is converted to the nitride:

$$2\,KCN + 3\,Mg \rightarrow 2\,K + 2\,C + Mg_3N_2$$

Beryllium, calcium, boron, and aluminum act in a similar manner. Malonic acid is made from monochloroacetic acid by reaction with potassium cyanide followed by hydrolysis. The acid and the intermediate cyanoacetic acid are used for the synthesis of polymethine dyes, synthetic caffeine, and for the manufacture of diethyl malonate, which is used in the synthesis of barbiturates. Most metals dissolve in aqueous potassium cyanide solutions in the presence of oxygen to form complex cyanides (see Coordination compounds).

Manufacture

Potassium cyanide was made by the Beilby process prior to the introduction of the neutralization or wet process. In the Beilby process, cyanide is made according to the following overall reaction:

$$K_2CO_3 + 4\,C + 2\,NH_3 \rightarrow 2\,KCN + 3\,CO + 3\,H_2$$

In this dry process, ammonia gas passes into a molten mixture of potassium carbonate and charcoal. Although purity of the product is high, this process became obsolete owing to higher costs than the neutralization process.

The Neutralization or Wet Process. At the present time, potassium cyanide is manufactured by the reaction of an aqueous solution of potassium hydroxide with hydrogen cyanide:

$$KOH + HCN \rightarrow KCN + H_2O$$

The cyanide, which crystallizes in the anhydrous state from aqueous solution, is recovered by evaporation under reduced pressure, filtration, and drying. Since the crystal size is significantly larger than sodium cyanide it can be sold in powder form without excessive dusting. However, it tends to cake in the shipping container and it often compacted and granulated to larger sizes.

Uses

Potassium cyanide is primarily used for fine silver plating but is also used for dyes and speciality products (see Electroplating). Electrolytic refining of platinum is carried out in fused potassium cyanide baths, in which a separation from silver is effected. Potassium cyanide is also a component of the electrolyte for the analytical separation of gold, silver, and copper from platinum. It is used with sodium cyanide for nitriding steel and also in mixtures for metal coloring by chemical or electrolytic processes.

Potassium cyanide like sodium cyanide, is shipped in steel or fiber drums. Potassium cyanide costs more than sodium cyanide primarily because of the higher price of potassium hydroxide. Current price is $1.76/kg (1978).

LITHIUM, RUBIDIUM, AND CESIUM CYANIDES

Lithium cyanide [2408-36-8], rubidium cyanide [19073-56-4], and cesium cyanide [21159-32-0] are white or colorless salts, isomorphous with potassium cyanide. In their physical and chemical properties these cyanides closely resemble sodium and potassium cyanide. They have as yet no industrial uses.

All of these cyanides may be prepared by passing hydrogen cyanide into an aqueous solution of the hydroxide, or by precipitating a solution of barium cyanide with lithium, rubidium, or cesium sulfate. A product with fewer contaminants may be obtained by the reaction of the base in absolute alcohol or dry ether with anhydrous hydrogen cyanide (22). In another method of preparation, a suspension of rubidium, cesium, or lithium metals in anhydrous benzene is treated with anhydrous hydrogen cyanide and the benzene subsequently removed by evaporation under reduced pressure.

These cyanides are all very soluble in water. The cyanide ion is weakly held so that water solutions have a much stronger odor of HCN above them than sodium and potassium cyanide solution.

Lithium cyanide melts at 160°C. In the fused state the specific gravity at 18°C is 1.075. It is highly hygroscopic. Rubidium cyanide is not hygroscopic and is insoluble in alcohol or ether. Cesium cyanide is highly hygroscopic.

Lithium cyanide decomposes to cyanamide and carbon below ca 600°C. This decomposition is similar to the alkaline earth cyanides (23). Iron accelerates decomposition and, when heated with 10% iron at 500°C for 15 h, lithium cyanide is completely converted to lithium cyanamide.

AMMONIUM CYANIDE

Ammonium cyanide [*12211-52-8*] NH_4CN, a colorless crystalline solid, is relatively unstable, and decomposes into ammonia and hydrogen cyanide at 36°C.

Ammonium cyanide reacts with ketones to yield aminonitriles. Reaction of ammonium cyanide with glyoxal produces glycine. Because of its unstable nature, ammonium cyanide is not shipped or sold commercially. Unless it is kept cool and dry, decomposition releases vapors and forms black HCN polymer.

Ammonium cyanide may be prepared in solution by passing hydrogen cyanide into aqueous ammonia at low temperatures. It may also be prepared from barium cyanide and ammonium sulfate, or calcium cyanide with ammonium carbonate. It may be prepared in the dry state by gently heating a mixture of potassium cyanide or ferrocyanide and ammonium chloride, and condensing the vapor in a cooled receiver.

Ammonium cyanide is very soluble in water or alcohol. The vapor above solid NH_4CN contains free NH_3 and HCN, a very toxic mixture.

BIBLIOGRAPHY

"Alkali Metal Cyanides" under "Cyanides and Cyanogen Compounds" in *ECT* 1st ed., Vol. 4, pp. 698–708, by Dorothy E. Nye, E. I. du Pont de Nemours & Co., Inc.; "Cyanides (Alkali Metal)" in *ECT* 2nd ed., Vol. 6, pp. 585–601, by R. B. Mooney and J. P. Quin, Imperial Chemical Industries Ltd.

1. E. D. Oliver and S. E. J. Johnson, *J. Am. Chem. Soc.* **76**, 4721 (1954).
2. V. A. Kireev and L. I. Vogranskaya, *J. Gen. Chem. U.S.S.R.* **5**, 963 (1935).
3. U.S. Pat. 2,949,341 (Aug. 16, 1960), C. P. Green (to E. I. du Pont de Nemours & Co., Inc.).
4. E. Rishkevich, *Z. Electrochem.* **39**, 531 (1933).
5. E. W. Guernsey and M. S. Sherman, *J. Am. Chem. Soc.* **48**, 695 (1926).
6. R. Holtze, *Z. Anorg. Allgem. Chem.* **214**, 65 (1933).
7. H. Roos and M. Schmidt, *Chem. Ing. Tech.* **49**, 369 (1977).
8. B. Ricca and G. D'Amore, *Gazz. Chim. Ital.* **79**, 308 (1949).
9. Brit. Pat. 4820 (Mar. 18, 1891), G. T. Beilby (to ICI).
10. U.S. Pat. 2,993,754 (July 25, 1961), W. R. Jenks and J. S. Linder (to E. I. du Pont de Nemours & Co., Inc.).

11. U.S. Pat. 3,619,132 (Nov. 9, 1971), H. J. Mann and co-workers (to Degussa).
12. U.S. Pat. 2,944,344 (July 12, 1960), C. P. Green and W. R. Jenks (to E. I. du Pont de Nemours & Co., Inc.).
13. U.S. Pat. 3,615,176 (Oct. 26, 1971), W. R. Jenks and O. W. Shannon (to E. I. du Pont de Nemours & Co., Inc.).
14. R. W. E. B. Harman and F. P. Worley, *Trans. Faraday Soc.* **20,** 502 (1925).
15. *National Institute of Occupational Safety and Health Recommendations; Occupational S and H. Reporter,* U.S. Bureau of National Affairs, Inc., 1976.
16. H. Geduld, *Met. Finish.* **74**(1), 45 (1976).
17. J. E. Laschinger and co-workers, *Natl. Inst. Metall. Repub. S. Afr. Rep.* **154,** 7 (1976).
18. U.S. Pat. 3,679,659, (July 25, 1972), C. Zak (to Belzak Corp.).
19. U.S. Pat. 3,695,833 (Oct. 3, 1972), O. Wiedeman and co-workers (to American Cyanamid Co.).
20. U.S. Pat. 3,883,391 (May 13, 1975), R. B. Phillips (to International Paper Co.).
21. R. M. Bozorth, *J. Am. Chem. Soc.* **44,** 317 (1922).
22. J. Meyer, *Z. Anorg. Allgem. Chem.* **115,** 203 (1920).
23. A. Perret and J. Riethmann, *Helv. Chim. Acta* **26,** 740 (1943).

WILLIAM R. JENKS
E. I. du Pont de Nemours & Co., Inc.

ALKALINE EARTH METAL CYANIDES

CALCIUM CYANIDE

Crude calcium cyanide [592-01-8], about 48–50 eq % NaCN, is the only commercially important alkaline earth metal cyanide currently produced, although output tonnage has been greatly reduced in recent years. This product, commonly called black cyanide, is marketed in flake form, as a powder, or as cast blocks under the trademarks Aero and Cyanogas of the American Cyanamid Company.

Following the discovery of the Cyanamid process in Germany (1), attempts were made to convert calcium cyanamide into calcium cyanide (see Cyanamides). Only 17.5 eq % NaCN was attainable up to 1914. Because of a shortage of cyanide in the United States in 1916, the American Cyanamid Company developed a batch process that gave higher yields and which evolved into a continuous electric-furnace process in 1919. Subsequent improvements increased the strength to 48–50 eq % NaCN. Calcium cyanide solutions of ca 15% concentration have been prepared for use in gold extraction in South Africa from hydrogen cyanide and lime.

Physical and Chemical Properties

Because of decomposition, the melting point of calcium cyanide can only be estimated by extrapolation to be 640°C (2).

Calcium cyanide as the diammoniate [69365-88-4], Ca(CN)$_2$·2NH$_3$, is formed in liquid ammonia by reaction of calcium hydroxide or nitrate with ammonium cyanide. Deammoniation under heat and high vacuum yields calcium cyanide, a white powder, which is readily hydrolyzed to hydrogen cyanide. The normal humidity content of air is sufficient to cause evolution of hydrogen cyanide. The diammoniate is more stable than the cyanide containing no ammonia. Calcium cyanide can be made by passing ammonia and hydrogen cyanide over calcium hydroxide. A compound having even higher cyanide content is obtained by subjecting calcium carbide (see Carbides) to the action of liquid hydrogen cyanide in the presence of 0.5–5% of water; it has also been formed in a nonaqueous medium such as dimethylformamide. These processes are distinct from those for the fused cyanides in which nitrification of calcium carbide yields calcium cyanamide which in turn produces calcium cyanide by reaction with carbon.

Aqueous solutions of calcium cyanide prepared even at low temperature turn yellow or brown owing to the formation of HCN polymer. Calcium cyanide hydrolyzes readily.

$$Ca(CN)_2 + 2\ H_2O \leftrightarrows 2\ HCN\ \text{black polymer} + Ca(OH)_2$$

The presence of brown polymer in solid form is sometimes noted even in dry calcium cyanide that has been stored for long periods. Calcium cyanide is decomposed by carbon dioxide, acids, and acidic salts liberating hydrogen cyanide.

Ferrocyanides are produced by reaction with ferrous salts, as shown below:

$$3\ Ca(CN)_2 + FeSO_4 \rightarrow Ca_2[Fe(CN)_6] + CaSO_4$$

With sulfur in aqueous medium, calcium cyanide forms calcium thiocyanate, as follows:

$$Ca(CN)_2 + 2\ S \rightarrow Ca(SCN)_2$$

Manufacture

Calcium cyanide is made commercially by heating crude calcium cyanamide (which contains carbon) in an electric furnace above 1000°C in the presence of sodium chloride, as follows:

$$CaNCN + C \xrightarrow[NaCl]{\Delta} Ca(CN)_2$$

The resulting melt is cooled rapidly to prevent reversion to calcium cyanamide. The product is marketed in the form of flakes, dark gray because of the presence of carbon. Because the rate of hydrogen cyanide evolution is relatively high, it is readily adaptable to fumigation. Specific gravity of the product is 1.8–1.9.

The price of black cyanide is generally lower than sodium cyanide; it is manufactured in Canada and South Africa.

Analysis

The major diluents are chlorides and calcium oxide with smaller quantities of silica, iron and aluminum oxides, and calcium cyanamide, carbide, and sulfide. The cyanide content is determined by titration with silver nitrate solution with potassium iodide indicator after removal of sulfides with sodium carbonate–lead acetate solution.

Safety Precautions

Precautions similar to those used for sodium cyanide should be used for black cyanide; contact with acidic compounds will generate hydrogen cyanide gas. Some additional precaution must be taken to minimize contact with water or even the humidity of the atmosphere since decomposition and hydrogen cyanide evolution will occur. Black cyanide is shipped in steel drums and is designated as a Class B poison. It is extremely toxic to humans, animals, and fish. Ingestion, contact with the skin, inhalation of the dust or HCN evolved should be avoided. Black cyanide should be stored in tight containers under moisture-free conditions; the small amount of contained carbide can evolve acetylene. Adequate ventilation and protective equipment similar to that for sodium cyanide should be provided in handling the solid cyanide and in the preparation of aqueous solutions.

Uses

The extraction or cyanidation of precious metal ores was the first and is still the largest use for black cyanide (3). The leaching action of the cyanide is due to the formation of soluble cyanide complexes by the following typical equation, which is shown for silver but also applies to gold:

$$4\,Ag + 4\,Ca(CN)_2 + O_2 + 2\,H_2O \rightarrow 2\,Ca[Ag(CN)_2]_2 + 2\,Ca(OH)_2$$

The flake cyanide is added to suspensions of finely ground ore under agitation and in the presence of air. The dissolved gold (qv) and silver (qv) are precipitated by the addition of zinc dust and the crude precipitate is refined to produce bullion.

Black cyanide is used in the froth flotation (qv) of minerals as a depressant or inhibitor of the flotation of certain minerals; it allows higher concentrates of the desired metal. Typical applications of this type are the depression of zinc in the flotation of lead–zinc ores, the depression of zinc and iron in the treatment of complex lead–zinc ores, and the removal of iron from copper concentrates.

The use of black cyanide as a fumigant and rodenticide utilizes the action when atmospheric humidity liberates hydrogen cyanide gas. It can only be used effectively in confined spaces where HCN builds up to lethal concentrations for the particular application (see Poisons, economic).

Black cyanide is also used in limited quantities in the production of prussiates or ferrocyanides (see Iron compounds).

OTHER ALKALINE EARTH CYANIDES

Magnesium cyanide [22400-99-3], Mg(CN)$_2$, may be formed by reaction of finely divided magnesium metal with ammonium cyanide [12211-52-8] in liquid ammonia. Removal of the ammonia yields a white solid, the diammoniate [69309-45-1]; deammoniation by heating under vacuum gives the cyanide, which is similar in properties to calcium cyanide. Strontium cyanide [60448-24-0] and barium cyanide [60448-23-9] are somewhat more stable than calcium and magnesium cyanides and their preparation is reported by reaction of strontium or barium hydroxide with hydrocyanic acid in water followed by low temperature evaporation under vacuum. The cyanides of magnesium, barium, and strontium have not been marketed.

BIBLIOGRAPHY

"Alkaline Earth Metal Cyanides" under "Cyanides and Cyanogen Compounds" in *ECT* 1st ed., Vol. 4, pp. 708–712, by H. W. Boynton, American Cyanamid Company; "Alkaline Earth Metal Cyanides" under "Cyanides" in *ECT* 2nd ed., Vol. 6, pp. 601–704, by R. B. Booth, American Cyanamid Company.

1. U.S. Pat. 776,314 (Nov. 29, 1904), A. Frank.
2. G. Peterson and H. H. Franck, *Z. Anorg. Allgem. Chem.* **237**, 1 (1938).
3. J. V. N. Dorr and F. L. Bosqui, *Cyanidation and Concentration of Gold and Silver Ores*, McGraw-Hill Book Co., Inc., New York, 1950.

General References

W. S. Landis, *Can. Chem. J.* **4**, 130 (1920).
W. S. Landis, *Chem. Met. Eng.* **22**, 265 (1920).
Ger. Pat. 505,208 (Aug. 15, 1930), I. G. Farbenindustrie.
G. H. Buchanan, *Trans. Electrochem. Soc.* **60**, 93 (1931).
N. Hedley and H. Tabachnick, "Chemistry of Cyanidation," *Mineral Dressing Notes, No. 23,* American Cyanamid Company, Wayne, N.J., 1958.
H. E. Williams, *Cyanogen Compounds,* Edward Arnold and Co., London, Eng., 1948.

WILLIAM R. JENKS
E. I. du Pont de Nemours & Co., Inc.

CYANINE DYES

The cyanine dyes, which are among the oldest synthetic dyes known, comprise a large group of dyes with a wide variety of colors. Absorption spectra of cyanines range in position from the ultraviolet to the infrared region and, as a group, cover a wider span of the spectrum than those of any other dye class. The first dye was discovered in 1856 (1). The name cyanine (from the Greek *kyanos*) was attributed to its beautiful blue color; however, the dye was extremely fugitive to light and of no practical use at that time for ordinary (fabric) dyeing purposes (see Color). The great usefulness of cyanines was discovered later in photography, and they include the most powerful photographic sensitizing dyes known (2–3) (see Color photography). There are several important reasons for the cyanines' prominence as sensitizers: (*1*) high light absorption per molecule coupled in many cases with a single narrow absorption band in the visible or infrared spectral region which gives efficient and very color-selective absorption of light; (*2*) the tendency to form dye aggregates that have even narrower, more color-selective absorptions than the monomeric dyes themselves; and (*3*) high chemical and photochemical reactivity which leads to efficient participation in photographic sensitization processes (see Dyes, sensitizing). In addition, certain cyanines marketed as Povan [*3546-41-6*] (4) and Dithiazanine [*7187-55-5*] (5) have proved useful as anthelmintics (see Chemotherapeutics, anthelmintics), as well as Indocyanine Green [*3599-32-4*] (6) which is used as an infrared-absorbing tracer for blood dilution studies in medical diagnoses (see Infrared technology).

The cyanine dye literature has been reviewed by Brooker (2,7), Hamer (8), Ficken (9), Kiprianov (10), Heseltine (11), and Sturmer (11–12). The detailed review by Hamer (8) is the best source for the heterocyclic chemistry of cyanine and related dyes. Several new syntheses and many new dye structures were discussed in later reviews (9,12), combined with compilations of physical and photophysical data (12).

The Discovery Period, 1850–1910. Particularly knowledgeable accounts of these events were written by Brooker (13) and Hamer (8). In 1856 Williams characterized quinoline obtained by the distillation of alkaloids with caustic alkali (1). After quaternization with alkyl iodides, these samples of quinoline gave reddish-blue dyes by treatment with silver oxide. Although the dyes attracted little commercial attention despite their superior color, Brooker noted that "the dye from the amyl iodide salt was, however, manufactured in small quantities by the Paris firm of Menier" (13).

Spectral sensitization was discovered in 1873 by Vogel during a study of the spectral sensitivity of silver halide dry collodion plates (14). Certain plates with dyes added for halation protection gave spectral sensitivity maxima in the green. This spectral sensitivity was removed by washing the plate prior to exposure and regained by addition of other dyes, including Williams' original cyanine. By 1910 spectral sensitivity extended throughout the visible region with isocyanine dyes (from 2-methylquinoline salts), erythrosin, and hundreds of other dyes (3).

Structures of the Early Dyes, 1915–1925. Most spectral sensitizers were made in Germany prior to 1915; thus, during World War I, Pope and Mills at Cambridge were asked to supply photographic sensitizers for Britain. The Cambridge group prepared and established the structures of a great many dyes shown in Figure 1 (8).

Most of the cyanine dyes had monomethine (CH) links between two heterocyclic nuclei. The important discovery in this work was that pinacyanol contains two quin-

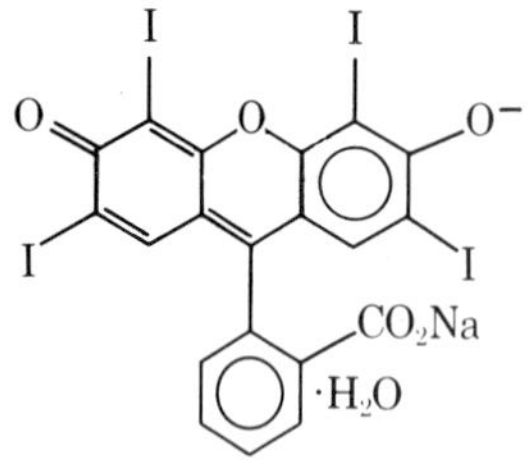

Erythrosin hydrate
[*49746-10-3*]

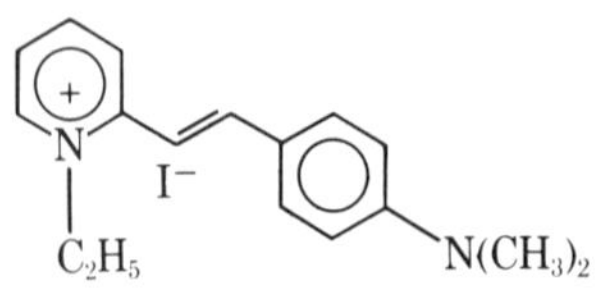

Pinaflavol
[*3785-01-1*]

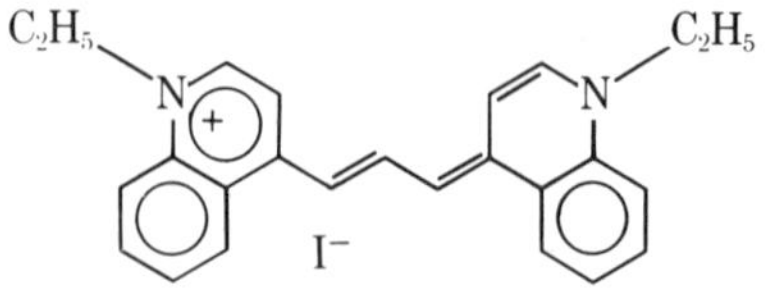

Kryptocyanine
[*4727-50-8*]

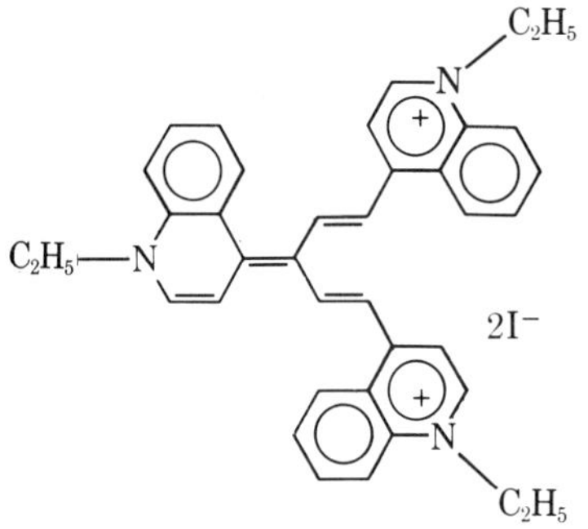

Neocyanine
[*4846-34-8*]

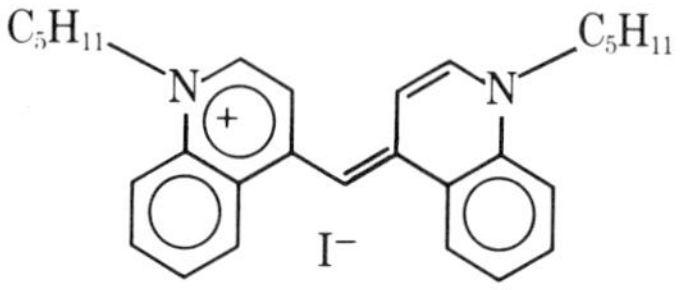

Williams' cyanine (1856)
[*862-57-7*]

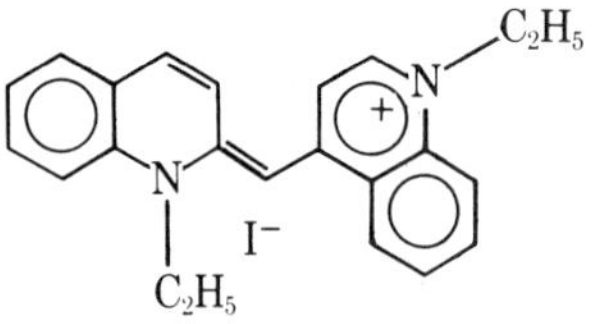

Isocyanine (1902)
[*634-21-9*]

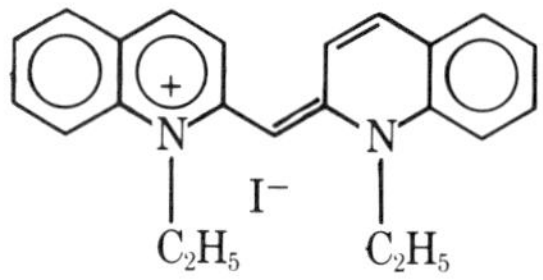

Pseudoisocyanine (1912)
[*977-96-8*]

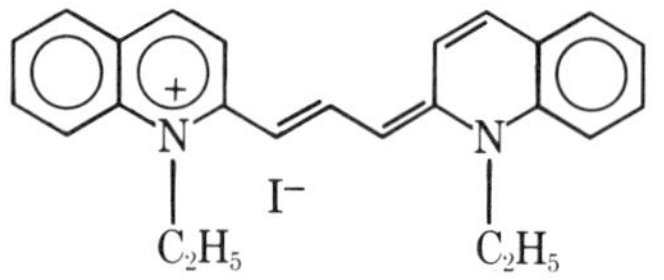

Pinacyanol (1904)
[*605-91-4*]

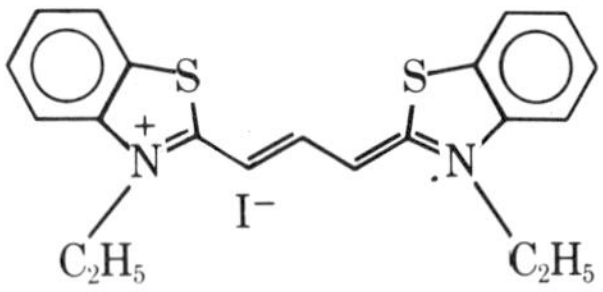

Thiacarbocyanine (1887)
[*905-97-5*]

Figure 1. Structures of early cyanine dyes. Generally, counterions (usually anions) for dyes are not specified in the text.

olines linked not by one but by three methine carbons (CH⋯CH⋯CH). The structure of the benzothiazole dye, thought to be a dye with a monomethine link since 1887, was also found to have three methine carbon atoms. Since the monomethine dyes related to these trimethine dyes absorbed at much shorter wavelengths, one of the first important relations between dye color and constitution was established (see Polymethine dyes).

Synthetic Methods and Theories about Color, After 1925. Following Pope and Mills' structural assignments, many generalized synthetic procedures were discovered that provided new intermediates, new classes of dyes, and a variety of dyes less symmetrical than those in Figure 1. Definitive theories on color–structure relations were proposed. The color–structure relation for symmetrical dyes were essentially defined by the early structural work on cyanines, but as noted below, unsymmetrical dyes were important in understanding color and constitution. Historical accounts of the early experimental work on unsymmetrical cyanine dyes are contained in references 2, 7, and 11. Extensive experimental and theoretical data for ground- and excited-state cyanine dyes are now available (12).

Properties

Color and Constitution. The color and constitution of cyanine dyes may be understood through detailed consideration of their component parts, ie, chromophoric systems, terminal groups, and solvent sensitivity of the dyes. Resonance theories have been developed to accommodate all the major trends in a very successful manner. For an experienced dye chemist, these are still useful in the design of dyes with a specified color, band shape, or solvent sensitivity. More recently, the cataloguing of oxidation and reduction potentials has led to new predictive methods for sensitizers and the computer-assisted design of new molecules.

Chromophoric Systems. The primary types of chromophores for cyanine dyes are the amidinium-ion system (**A**), the carbonyl-ion system (**B**), and the dipolar amidic system (**C**). Several examples of dyes with chromophores (**A**) and (**C**) are shown in Figure 2. For each system two extreme resonance structures are shown in (**A**), (**B**), and (**C**) where the formal charges are located at the ends of the chromophore. Intermediate resonance structures, with the charges closer to the center of the chromophore or with additional dipoles, are less important in the resonance picture of dyes. However, structural changes that favor intermediate forms have significant effects on the color of symmetrical dyes containing (**A**). For the amidic dyes (**C**), structural features stabilizing both neutral and dipolar extreme resonance forms in an equivalent manner gave dyes absorbing at longer wavelengths.

The important characteristics that influence the absorption wavelengths for these dyes are the length of the conjugated chain and the nature of the terminal group. Many of the early cyanine dyes comprised a chain with an odd number of methine carbon atoms (C—H) and two heterocycles, eg, quinoline or benzothiazole (Fig. 1). Historically, the terms simple cyanine, carbocyanine, dicarbocyanine, and such were used to designate both the specific dyes derived from quinoline as well as generic dye structures (from other heterocycles) with one, three, five, etc, methine carbon atoms (see structure (**1**)). In the dyes from quinoline, which can also be designated quinocyanines, the ring position attached to the methine chain and the N-substituent are usually specified. For example, 1,1′-diethyl-2,2′-cyanine (or 1,1′-diethyl-2,2′-quinocyanine) is dye (**1**), X = —CH=CH—, and n = 0. Pinacyanol (Fig. 1) can be named as 1,1′-diethyl-2,2′-carbocyanine iodide, or as 1,1′-diethyl-2,2′-quinocarbocyanine iodide. Chemical Abstracts names pinacyanol as a substituted quinolinium iodide, ie, quinolinium, 1-ethyl-2-[3-(1-ethyl-2(1H)-quinolylidene)-1-propenyl]-iodide.

Dyes derived from the primary chromophores (**B**) and (**C**) are designated oxonols and merocyanines, although the term neutrocyanine has also been used for (**C**). For

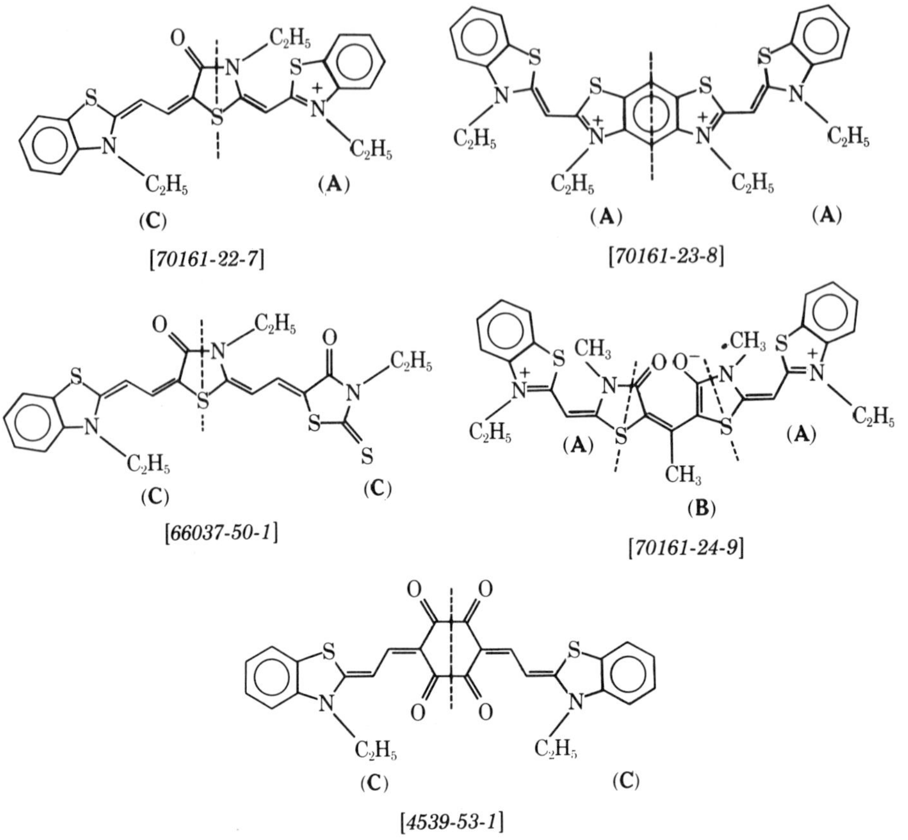

338 CYANINE DYES

Figure 2. Chromophoric systems. Note that (**A**) = amidinium–ion system (a cyanine), (**B**) = carbonyl–ion system (an oxonol), and (**C**) = dipolar amidic system (a merocyanine).

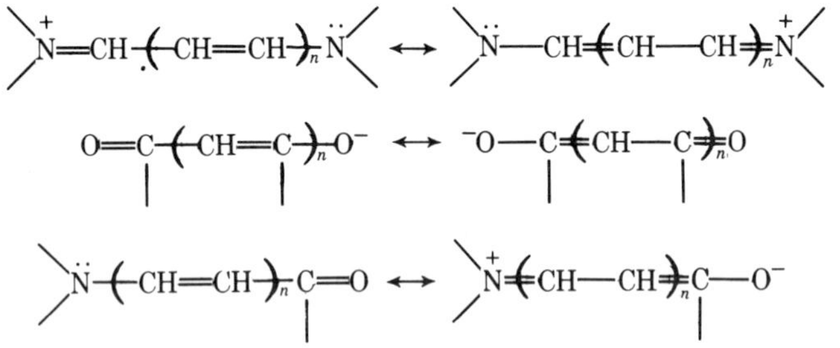

$$\text{\textbackslash N}^+{=}CH{-}(CH{=}CH)_n{-}N{:} \longleftrightarrow N{-}CH{=}(CH{-}CH)_n{=}N^+ \quad \textbf{(A)}$$

$$O{=}C{-}(CH{=}C)_n{-}O^- \longleftrightarrow {}^-O{-}C{=}(CH{-}C)_n{=}O \quad \textbf{(B)}$$

$$N{-}(CH{=}CH)_n{-}C{=}O \longleftrightarrow N^+{=}(CH{-}CH)_n{=}C{-}O^- \quad \textbf{(C)}$$

certain cyanines and merocyanines, the simple cyanine (or simple merocyanine) designation refers to a dye with zero methine carbon atoms (ie, the shortest possible linkage that retains the chromophore between the terminal groups); the term merocarbocyanine designates dyes having chromophore (**C**) with two methine carbon atoms.

Dyes that differ only by the number of vinyl groups +CH=CH+ in the methine chain are termed a vinylogous series. Absorption maxima for vinylogous series of dyes like (2) to (6) in Figure 3 shift to longer wavelengths as the methine chain length increases. With the usual exception of the first member of each series, the shift ap-

(1)

X = —CH=CH—

n = 0, a simple cyanine [977-96-8]
n = 1, a carbocyanine [605-91-4]
n = 2, a dicarbocyanine [14187-31-6]
n = 3, a tricarbocyanine [17695-32-8]

proximates 100 nm per vinyl group in most symmetric chromophores like (2) (amidinium–ion system, **A**) and (3) (carboxyl-ion system, **B**), and these are termed nonconverging series. Less symmetric chromophores including those of the dipolar amidic systems ((4)–(6) and **C**) show markedly reduced shifts as each vinyl unit is added (converging series). In the dyes the degree of asymmetry and the absorption shifts are both related to the structures of the heterocyclic terminal groups. Infrared absorbing dyes are primarily derived from symmetric structures such as those in Figure 3.

Terminal Groups. The cyanines (**A**), oxonols (**B**), and merocyanines (**C**) were originally considered as polymethine dyes with two heterocyclic terminal groups. Hemicyanines were defined as dyes with one heterocyclic and one noncyclic terminal group. Dye bases such as (5), Figure 3, were the nonalkylated analogues of cyanine dyes like (2), Figure 3. Currently, dyes without heterocyclic terminal groups are designated as either cyanines or polymethine dyes. In fact, almost any atom or group of atoms can function as a terminal group for dyes if the nitrogens and oxygens in the primary chromophores (**A**), (**B**), and (**C**) are replaced by electronically equivalent atoms. Dyes from novel terminal groups are quite numerous (18). However, the fundamental concepts and perhaps the largest class of useful cyanines are derived from dyes with heterocyclic terminal groups. The heterocycles are of two major types: (*1*) basic or electron-donating, and (*2*) acidic or electron-accepting. Typical examples of terminal groups are shown in Table 1.

Basic Heterocycles. In addition to the superior benzothiazole dyes, eg, (2), other heterocyclic thiazoles as well as related oxazoles, pyrroles, and imidazoles were subsequently used for cyanines. When two different terminal groups were incorporated, certain unsymmetrical dyes absorbed at unexpectedly short wavelengths, whereas the absorption of others more closely approximated the mean wavelength for the related symmetrical dyes. These observations resulted in the concept of deviation (19–21), which related the absorption characteristics of unsymmetrical dyes to the electron-donating abilities (basicities) of the various heterocycles. Brooker (2) selected the *p*-dimethylaminophenyl group as a weakly basic terminal group and suggested that the importance of the two extreme amidinium resonance structures would depend on the basicity of the other terminal group. For the symmetrical dye (7) and in the example below, the two resonance structures are equivalent, but in the unsymmetrical "styryl" dye (8) the structure with the charged heterocycle is favored particularly for

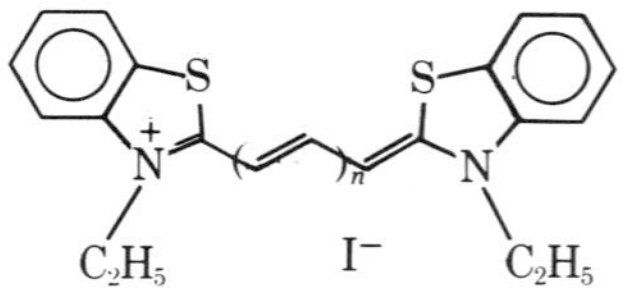

(2)

n = 0 [2197-01-5]
n = 1 [905-97-5]
n = 2 [514-73-8]
n = 3 [3071-70-3]
n = 4 [17094-08-5]
n = 5 [15979-18-7]

n =	0	1	2	3
λ_{max}	423	557	650	758
$\epsilon \times 10^{-4}$	(8.5)	(15)	(23)	(25)

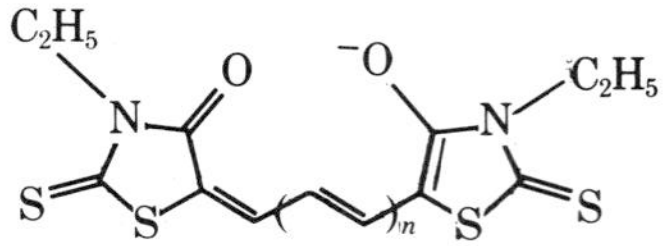

(3)

n = 0 [21812-35-1]
n = 1 [50962-54-4]
n = 2 [50962-55-5]

n =	0	1	2	3
λ_{max}	542	613	714	—
$\epsilon \times 10^{-4}$	(8.2)	(11)	(14)	

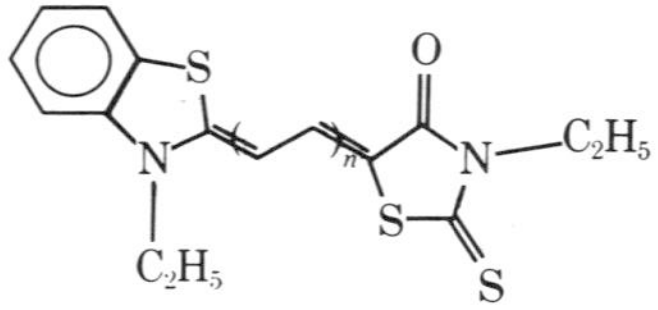

(4)

n = 0 [3377-05-7]
n = 1 [3568-36-3]
n = 2 [20185-07-3]
n = 3 [36652-36-5]

n =	0	1	2	3
λ_{max}	432	528	605	635
$\epsilon \times 10^{-4}$	(6.0)	(9.3)	(8.5)	(6.9)

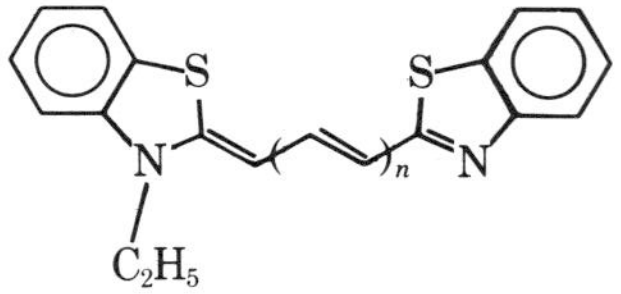

(5)

n = 0 [66037-47-6]
n = 1 [41504-99-8]
n = 2 [66037-46-5]
n = 3 [66037-54-5]

n =	0	1	2	3
λ_{max}	396	458	490	510
$\epsilon \times 10^{-4}$	(5.9)	(5.7)	(6.4)	

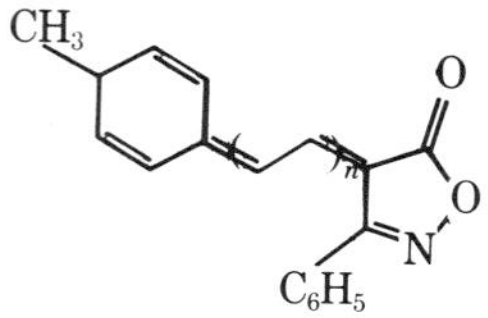

(6)

n = 0 [66037-53-4]
n = 1 [66037-52-3]
n = 2 [66037-51-2]
n = 3 [15736-88-6]

n =	0	1	2	3	Pyridine–water
λ_{max}	380	470	530	553	1:1
λ_{max}	400	515	610	710	1:0

Highly polar merocyanine

Figure 3. Absorption of vinylogous dyes. Absorption wavelengths and extinction coefficients. Spectral data in methanol or ethanol for dye series 2, 3, 5, and pyridine for 4, listed as λ_{max} (nm) and $\epsilon \times 10^{-4}$ L/(mol·cm). References for dyes 2 and 5 (15); dye 3 (16); and dyes 3, 4, and 6 (17).

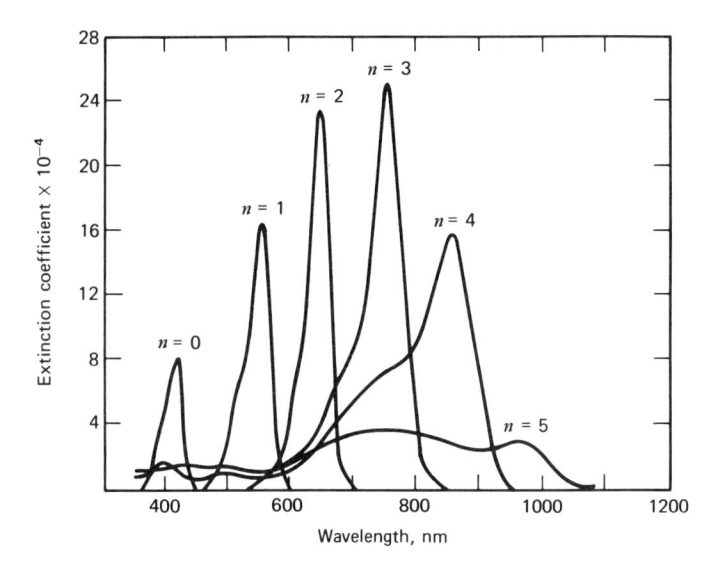

Figure 3. (*continued*)

good electron donors (highly basic heterocycles). The resulting bond alternation induces a polyene character to the dye chromophore, and the absorption is shifted accordingly to a shorter wavelength.

A quantitative expression of these observations is shown in equation 1. The λ_{obs} is the observed absorption

$$\text{deviation} = \Delta\lambda = \lambda_I - \lambda_{obs} \quad (1)$$

maximum for the unsymmetrical carbocyanine and λ_I is the arithmetic mean (isoenergetic wavelength) for the absorption maxima of the related symmetrical dyes. Many deviations are tabulated in reference 12 along with the structures of the basic heterocycles. The higher the deviation, the more readily the nucleus accommodates a positive charge (19–20).

Acidic Heterocycles. A similar classification is made for the acidic electron-accepting terminal groups used in dipolar (merocyanine) chromophores. The unsymmetrical dyes again incorporated the *p*-dimethylaminophenyl group, connected to the acidic group (Table 1) by one or three methine carbon atoms as in the merocyanine (**9**). For the unsymmetrical dye, the nonpolar resonance form was expected to be the dominant one, except when highly electron-accepting terminal groups were present. Nonpolar dyes exhibited the characteristics of polyenes, absorbing at much shorter wavelengths than expected from the arithmetic mean absorption for the two symmetrical dyes. Because of this, the least acidic (electron-accepting) groups showed the highest deviations. The least acidic heterocycles are typified by rhodanines and hydantoins (Table 1); the more acidic are isoxazolones and pyrazolidinediones. Acidic terminal groups with intermediate deviations include indandiones and malononitrile.

Solvent Sensitivity. There is a large influence of solvents on absorption spectra for dyes, particularly for the merocyanines, eg, (**C**) (7,12,21). Brooker and co-workers (21) constructed merocyanines with widely different terminal groups, and a few serve

Table 1. Examples of Nuclei Occurring in Important Cyanine Dyes

Basic nuclei for cyanines and merocyanines

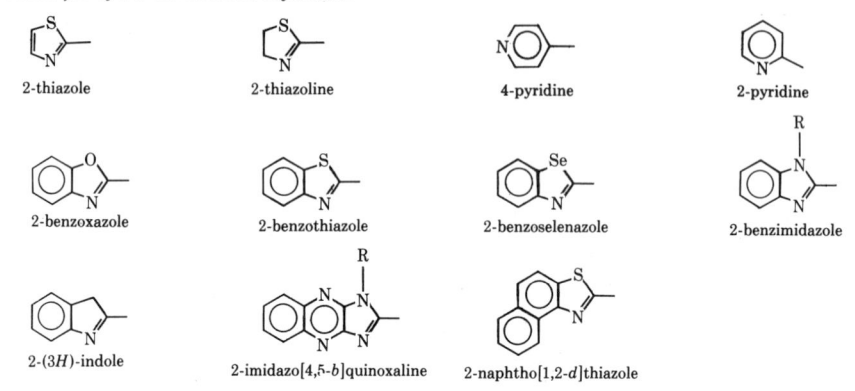

2-thiazole, 2-thiazoline, 4-pyridine, 2-pyridine

2-benzoxazole, 2-benzothiazole, 2-benzoselenazole, 2-benzimidazole

2-(3H)-indole, 2-imidazo[4,5-b]quinoxaline, 2-naphtho[1,2-d]thiazole

Acidic nuclei for merocyanines and oxonols

rhodanine, 2-thiohydantoin, malononitrile, indan-1,3-dione

isoxazolin-5-one, pyrazolidin-3,5-dione, 2-thiobarbituric acid

---

$(CH_3)_2\overset{+}{N}=\!\!\!\!=\!\!\!\!\!-CH-\!\!\!\!-\!\!\!\!-N(CH_3)_2 \longleftrightarrow (CH_3)_2N-\!\!\!\!-\!\!\!\!-CH=\!\!\!\!=\!\!\!\!\!\overset{+}{N}(CH_3)_2$

(7) a symmetrical dye
[14844-71-4], Michler's hydrol

(8) an unsymmetrical carbocyanine (a *styryl* dye)
[41017-17-8]

to illustrate the general pattern of solvent effects on the absorption maxima and peak intensities: (1) weakly polar dyes like (4), Figure 3, or (10) (χ_R) were red-shifted (longer wavelength) and showed increased extinction coefficients (ϵ_{max}) as solvent polarity increased; and (2) highly polar dyes like (6) or (11) (χ_B) were blue-shifted and showed decreased ϵ_{max} values for increasingly polar solvents.

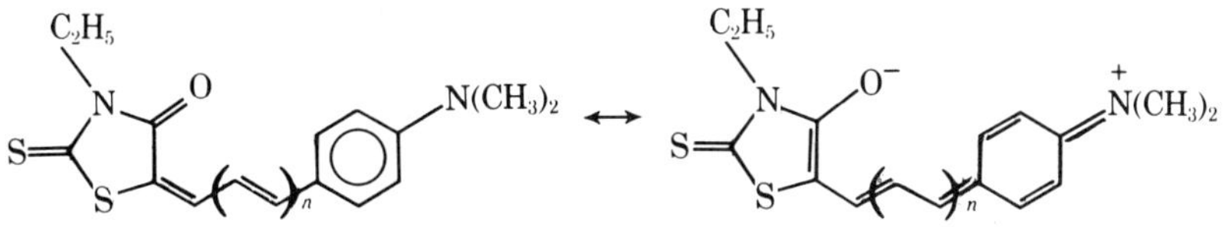

(9) an unsymmetrical dye (a benzylidene, n = 0)
n = 0 [23517-90-0]
n = 1 [42906-02-5]
n = 2 [66037-49-8]
n = 3 [66037-48-7]

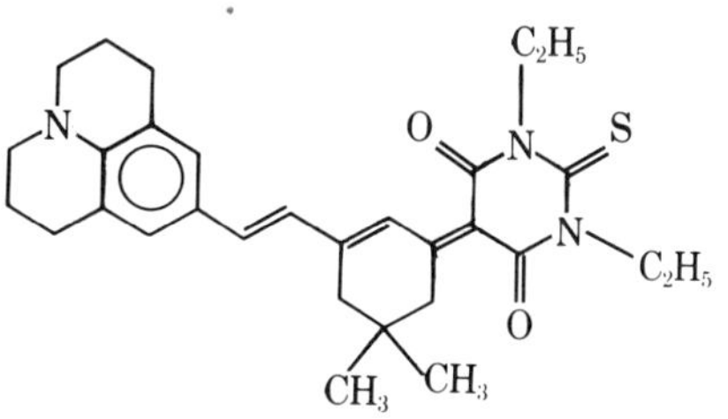

(10) χ_R [2913-22-6]

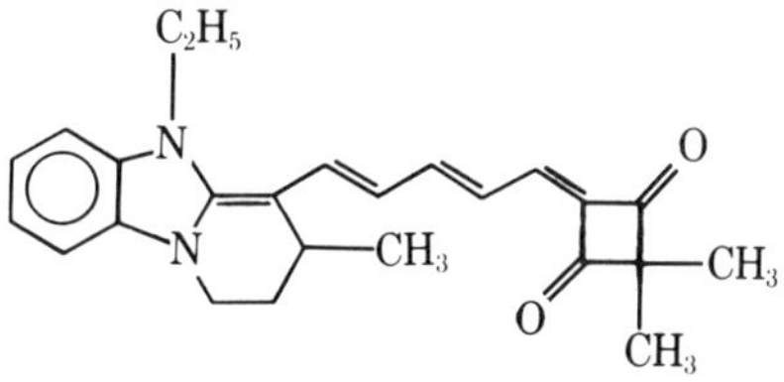

(11) χ_B [3210-95-5]

Two other concepts are important for an understanding of solvent sensitivity. First, in contrast to dyes such as (10) and (11), moderately polar merocyanines do not always show a continuous shift in λ_{max} and ϵ_{max} as solvent polarity is changed; instead, the maximum shift to long wavelength and high ϵ_{max} occurs in mixtures of solvents, eg, pyridine–water. They are isoenergetic points in which both the polar and nonpolar resonance forms are equally stabilized (21). Shorter λ_{max} and lower ϵ_{max} values are observed in either pure solvent. For dyes of (12) (X = S or C(CH$_3$)$_2$), the less basic

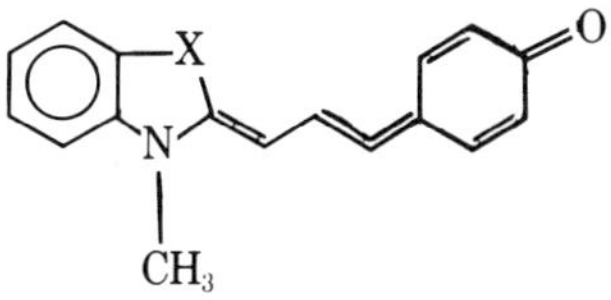

(12)
isoenergetic points

X = S [5894-19-9] 92.5 vol % pyridine–water
X = C(CH$_3$)$_2$ [10017-85-3] 50 vol % pyridine–water

heterocycle ($X = C(CH_3)_2$) requires less of the nonpolar solvent pyridine to reach the isoenergetic point. Secondly, the conformation and thus the absorption of charged dyes may change as a function of solvent, eg, (6) Figure 3, is a solvent-sensitive, infrared dye that shows more long-wavelength absorption in pyridine than in methanol because of increased all-trans conformations in pyridine where the charged terminal groups are separated by the maximum distance. A second example includes the allopolar (partly polar) cyanine dyes investigated extensively by Brooker (7). These dyes exist in two distinct conformations: one with complete charge separation (holopolar, H) and the other with at least one nonionized resonance form (meropolar, M). The major absorption shifts that occur between polar and nonpolar solvents are assigned to these isomers; the holopolar form predominates in polar solvents.

Steric Effects. In addition to solvent and degree of symmetry in cyanine dyes, increased steric hindrance can alter absorption wavelengths, decrease extinction coefficients, decrease photographic sensitization, and alter the patterns of intermolecular association (crystallization and aggregation). The early work on steric effects concentrated on substituted anilines and pyrrole dyes (22), but the prime documentation of steric influences was based on a large series of dyes from quinolines (23) (Fig. 4). For a variety of substituent patterns in symmetrical dyes, a linear relation between increased absorption wavelength and lowered extinction was also shown to correlate with increased steric hindrance. Highly unsymmetrical styryl dyes containing quinolines generally showed decreased extinction coefficients as steric hindrance increased, but the absorption maxima were shifted to shorter wavelengths.

Additional Structure–Property Relationships. Color–structure relationships are important in the design of cyanine dyes for specific wavelengths. In the normal visible spectrum (400–700 nm) compounds with a wide variety of structures absorb at wavelengths required for spectral sensitization, whereas in the infrared spectrum most dyes are electronically symmetrical. In 1945, the importance of the electrochemical properties of sensitizers was recognized (24). However, the significance was not fully appreciated until recently when electrochemical potentials were available for large series of dyes (25–27). The detailed aspects of spectral sensitization are described in references 28–30. Effective sensitizers must have both the correct absorption wavelengths and suitable electrochemical potentials, as well as certain desirable physical properties. The relationship between structure and electrochemical potentials is an important new aspect of the structure–property study of cyanine dyes as spectral sensitizers.

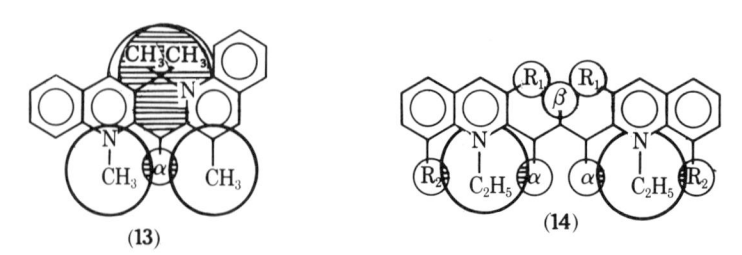

Figure 4. Two conformations of substituted quinoline dyes.

Photophysical Properties

Physical Characterization of Ground-State Dyes. Crystal structure analyses of cyanine and related dyes are reviewed in reference 31. Most typical sensitizers are nearly planar, with angles of less than 15° between planes defined by heterocyclic rings. Distinct solvent of crystallization is present in most of the cationic dyes. X-ray crystal analyses also provide intermolecular data (31). Because of photographic use of J-aggregates of cyanine and carbocyanine dyes, the cation–cation arrangements of most interest have been those for 1,1'-diethyl-2,2'-quinocyanine chloride [2402-42-8], 5,5',6,6'-tetrachloro-1,1',3,3'-tetraethylbenzimidazolocarbocyanine iodide [3520-43-2], and 5,5'-dichloro-3,3',9-triethylthiacarbocyanine bromide [18426-56-7].

The electrochemical properties of dyes have been related to their photographic effects (24). Although many correlations are established using electrochemical data for several vinylogous series of cyanine and related dyes (25–27), a review suggests that completely reversible electrochemical potentials are rare (26). Some apparent deviations are caused by specific substituents like nitro (1) and increased chain length for vinylogous series of dyes (2). Other estimates of redox properties of dyes result from ionization-potential and electron-affinity data. The most direct determinations of these properties are reported in references 32–34.

Protonation equilibria have been studied for a large number of cyanines and carbocyanines, as well as longer chain analogues, vinylogous dye bases, and diquinolyl methanes (35–37). Protonation equilibria for carbocyanines and longer-chain dyes are less affected by steric interactions, and there are correlations between pK_a and other data (30). Brooker's deviation for ten heterocycles was linearly related to the pK_a for the symmetrical carbocyanines (Fig. 5). Increased chain length increases the pK_a.

The Electronic Structure of Sensitizers. Large conjugated molecules such as the cyanines and related dyes are described by general quantum theory (38). The filled orbitals of the ground states contain sigma (σ), pi (π), and lone-pair (n) electrons. Antibonding orbitals are typically pi* (π^*) and sigma* (σ^*). The long-wavelength transitions for cyanine dyes are generally high-extinction transitions involving the π-electrons and show long-axis polarizations in cyanine dyes, as determined by dichroic absorption of polarized light (stretched polymer films with dye) and by fluorescence polarization. Shorter-wavelength transitions are polarized both parallel and perpendicular to the long axis (38).

Most cyanines show prominent vibrational shoulders at shorter wavelength associated with the long-wavelength electronic transition (Fig. 3). These vibrational shoulders included one or two vibrational quanta ($0 \to 1'$, $0 \to 2'$), in addition to the absorption energy at λ_{max} ($0 \to 0'$). Intensity ratios for the $0 \to 0'/0 \to 1'$ bands increased as the chromophoric chain was lengthened from $n = 0$ to $n = 3$ in the thiacyanine series (**2**), Figure 3. Two distinct vibrational-frequency progressions are identified in the low-temperature absorption spectra of azacyanine-, and several cyanine- and carbocyanine-dyes. Dyes with very long methine chains show short-wavelength absorptions that are not vibrational bands (Fig. 3).

Excited-state properties of the cyanine and related dyes are related to their efficiencies as photographic sensitizers and as laser dyes (see Laser dyes). Most cyanine dyes exhibit small Stokes shifts for fluorescence maxima (39). Typical carbocyanines (structure (**1**) with $n = 1$) showed 14- to 16-nm shifts in methanol solution with low

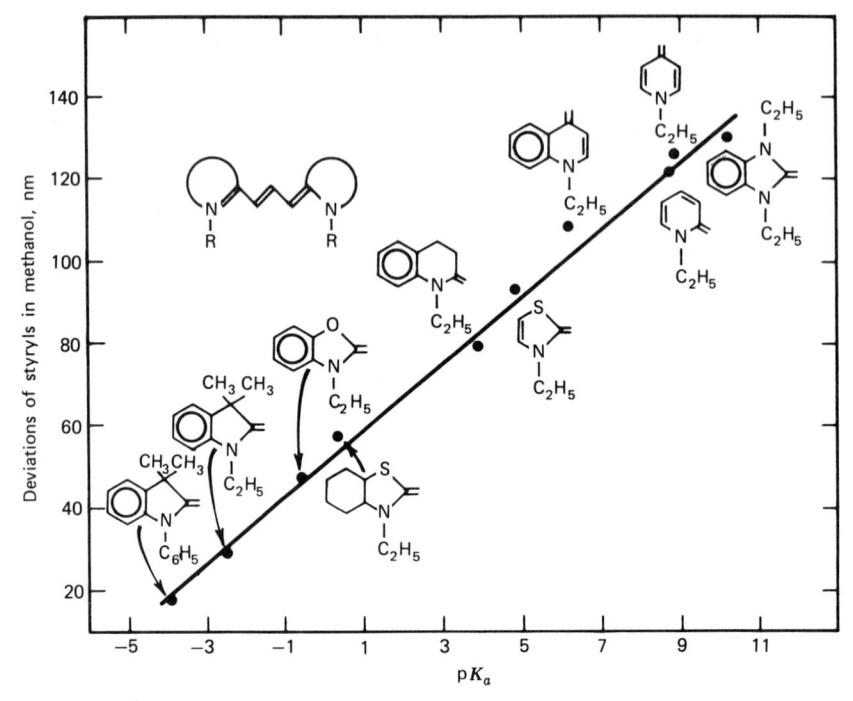

Figure 5. Deviations for heterocyclic terminal groups (based on styryl dye absorptions in methanol and equation 1) and pK_a for carbocyanine dyes from the same heterocycles.

quantum efficiencies for fluorescence (Φ_{Fl}) of less than 0.05. The dicarbocyanine analogues also showed small Stokes shifts but higher quantum yields (Φ_{Fl} = 0.3–0.5).

Data on fluorescence (39–40), phosphorescence (41), excited-state lifetimes, (39), transient absorption spectra (42–44), and dye lasers (45–46) are tabulated in reference 12. The main nonfluorescent process in cyanine dyes is the radiationless deactivation $S_1 \rightarrow S_0$ (39–40). Maximum singlet-triplet interconversion (Φ_{ST}) in methanol for carbocyanines is about 3% (max $\Phi_{ST} \leq 0.03$), and the sum [$\Phi_{Fl} + \Phi_{ST}$] is less than 0.10. In contrast to this, hydrocarbons and other classes of dyes exhibit significant triplet yields (Φ_{ST} = 0.2–0.9), and the sum [$\Phi_{Fl} + \Phi_{ST}$] is close to 1.0.

Aggregation, General Characteristics. Interactions among dye molecules produce large spectral shifts and distinct changes in band shape (47–50). In aqueous solution at room temperature, these changes are most evident as the concentration of the dye increases or as dye is concentrated by adsorption to crystal surfaces or oppositely-charged polymer sites. Relative to the absorptions for monomeric dye molecules (M-band, Fig. 6), progressive shifts of absorption maxima to shorter wavelength are designated as H-bands (hypsochromic) although the term metachromic has been used as well. Shifts to longer wavelength are less progressive. The new absorption is designated as a J-band (bathochromic). Reviews of H- and J-aggregation are included in descriptions of metachromasia, general aggregate phenomena (48), dye structure and detailed aggregation tendencies (12), the relation of x-ray crystal structure to possible aggregate structure (31), and the characteristics of aggregates absorbed to solid substrates (47).

The degree of self-association for aggregated dyes is determined spectroscopically

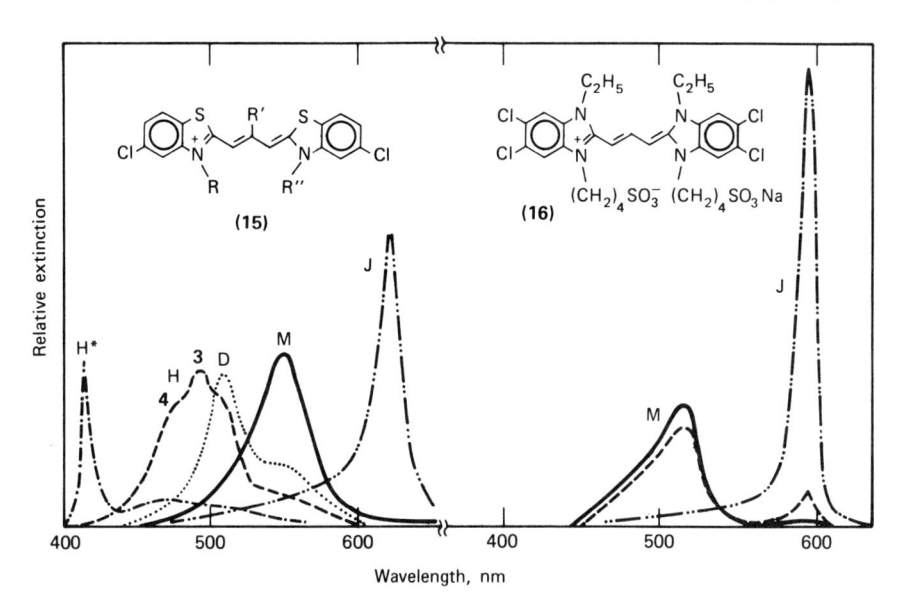

Figure 6. Relative absorption of aggregates. For (**15a**) R = C$_2$H$_4$CO$_2^-$, R' = CH$_3$, R" = C$_2$H$_4$CO$_2$H [*18281-98-6*]; (**15b**) R = (CH$_2$)$_4$SO$_3^-$, R' = C$_2$H$_5$, R" = (CH$_2$)$_4$SO$_3$Na [*42905-44-2*]; and (**16**) [*18462-64-1*].

for several H-aggregates (dimers, trimers, and tetramers) and for J-aggregates (48). Under ideal circumstances aggregation in solution involves two-component equilibria (Fig. 6): monomeric dye (curve M) with either dimer (curve D) or J-aggregate (curve J). Larger H-aggregates, formed at increasing concentration, shift the absorption peaks progressively to shorter wavelengths (peaks H-*3,4*). Most H-bands have been considered to have larger half-band width than the monomer until recently, when an extremely sharp H-band designated H* was noted for two thiacarbocyanines. On the other hand, most J-bands, which are favored also by increased concentrations of certain dyes, exhibit unusually sharp absorptions. Herz (48) estimates a minimum of four monomer units per absorbing unit for the J-aggregate although physically larger J- and H-aggregates are demonstrated to exist by solution centrifugation of J-aggregates, electron microscopy of H-aggregate threads, and J-band absorption spectra in crystalline dyes. Dye aggregates were shown to be dichroic by the absorption of plane-polarized light by aggregates oriented on silver halide crystals or as thin dye crystals and circular dichroism in the presence of optically-active substrates.

Substituent Effects. Ring substituents for the carbocyanine chromophore (**17**) have been studied extensively. Up to 40 groups in the 5,5'- and 6,6'-positions are tabulated for the benzothiazole and benzimidazole dyes (9). For the thiacarbocyanines, most substituents provided absorption maxima at longer wavelengths. Halogens, alkoxy, and alkyl groups in either the 5- or 6-position provided similar shifts. Electron-withdrawing groups and conjugated substituents in the 6,6'-positions show greater bathochromic effects than in the 5,5'-positions. Certain unsaturated substituents provided polymerizable dyes. Hammett $\sigma\rho$ relations for many of these ring substituents correlate with data other than absorption maxima. Both pK_a and polarographic reduction potential were correlated with σ values for several dye series (X = N—R, O,

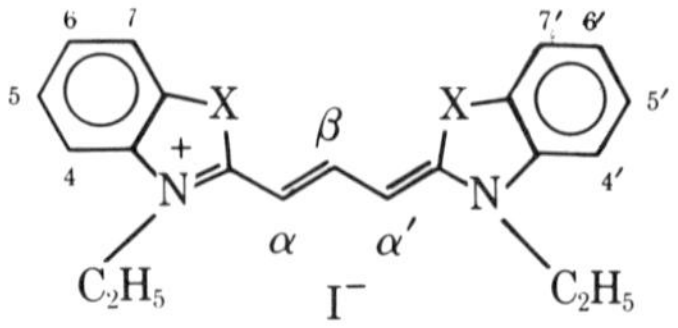

(17)

X = —CH=CH— [605-91-4]
X = S [905-97-5]

S, and C(CH$_3$)$_2$) (27,37). Additional benzenes or heterocycles have been fused to the essential chromophores of thiazolocarbocyanine, imidazolocarbocyanine, and other dyes. These new rings are similar to substituents, since fusing benzene rings to either the 4,5- and 4′,5′-positions or the 5,6- and 5,′6′-positions in (17) gives a similar color change. The resulting electrochemical properties (and, therefore, chemical reactivity) depend strongly on the position of the benzene ring (51). The many variations in heterocyclic structure obtained in this way have been reviewed (8–9,12), and some examples are included in Table 1.

Substituents on the methine chain are most important historically (Hamer (8) devotes several chapters to chain-substituted cyanines). These substituents cause two types of changes in the properties of the dyes. First, the absorption maxima shift depending on the inductive (electronic) effect of the substituent and its position in the methine chain. Azacyanines absorb at significantly shorter wavelength than the monomethine analogues, and related simple cyanines with phosphorus and arsenic atoms in the chain also absorb at shorter wavelengths. In azacarbocyanines the α-aza dyes absorb at shorter wavelength than the carbon analogues. (Similar effects were found for carbocyanines with α-cyano, α-formyl, and α-nitro groups, whereas groups like α-alkyl, α-alkoxy, α-alkylthio, and α-fluoro generally shift the absorption to longer wavelengths.) On the other hand, β-aza and β-cyano (nitro, C$_3$F$_7$, etc) carbocyanines (18) absorb at longer wavelengths. Secondly, substituents that change the steric properties of the dye can affect both the color and the aggregation. The "compact" dyes aggregate readily whereas loose or crowded dyes, or both, do not (1,23). As judged by the type of dyes in Figure 4, (13) would be very crowded, whereas (14) would be compact. The spacing of dyes in aggregates is, to some degree, determined by substituents that project out of the plane of an otherwise planar chromophore. For alkyl groups, an ethyl group on the center methine carbon of (17) leads to improved J-aggregation, whereas a methyl group reinforces H-aggregation.

Several types of nitrogen substituents have been incorporated into typical dye structures (12). The most useful is substituted alkyl, but other types of substituents

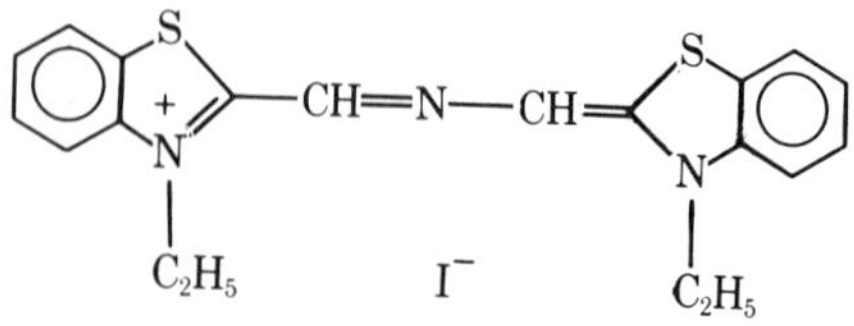

(18)

[22268-68-4]

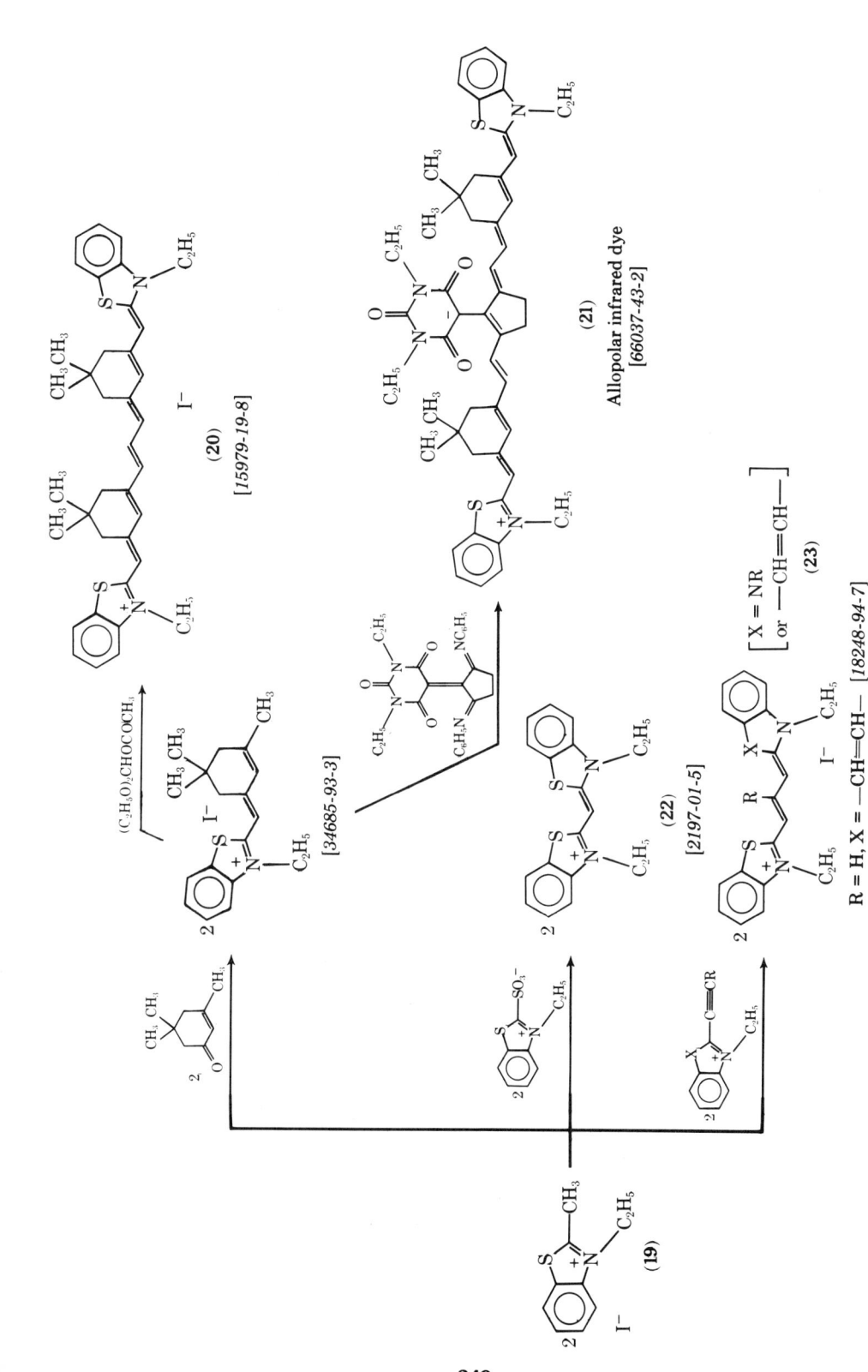

Figure 7. Nonoxidative cyanine syntheses. Note that OTs = tosylate = *p*-toluenesulfonate.

(25) Hemicyanine iodide [16384-23-9]

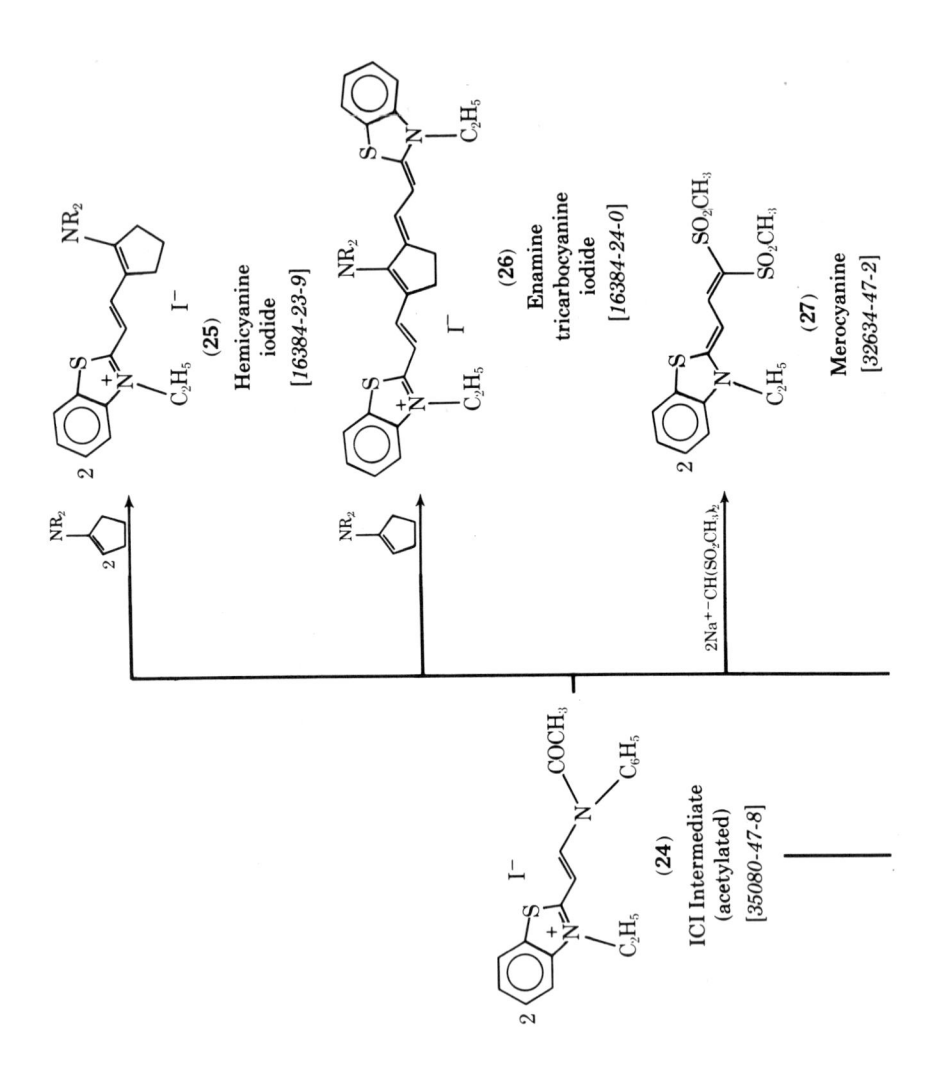

(26) Enamine tricarbocyanine iodide [16384-24-0]

(27) Merocyanine [32634-47-2]

(24) ICI Intermediate (acetylated) [35080-47-8]

350

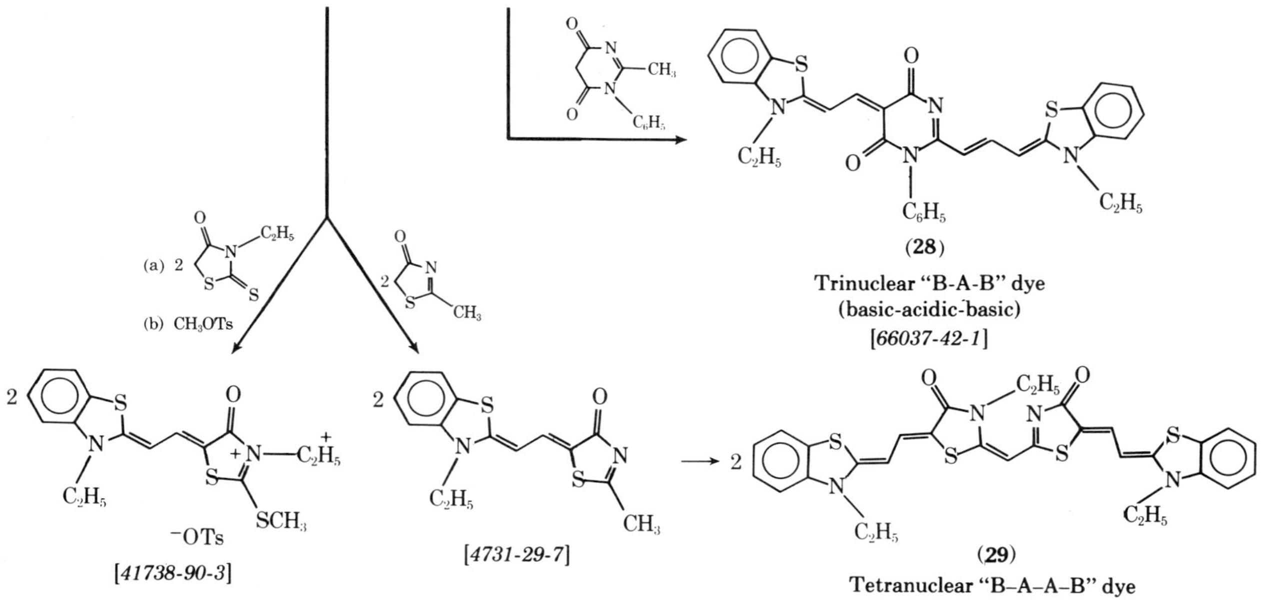

Figure 7. (*continued*)

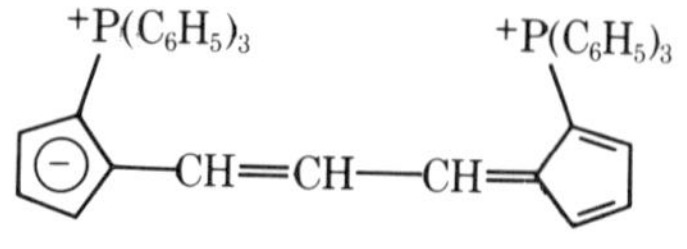

352 CYANINE DYES

(30)
[47879-81-2]

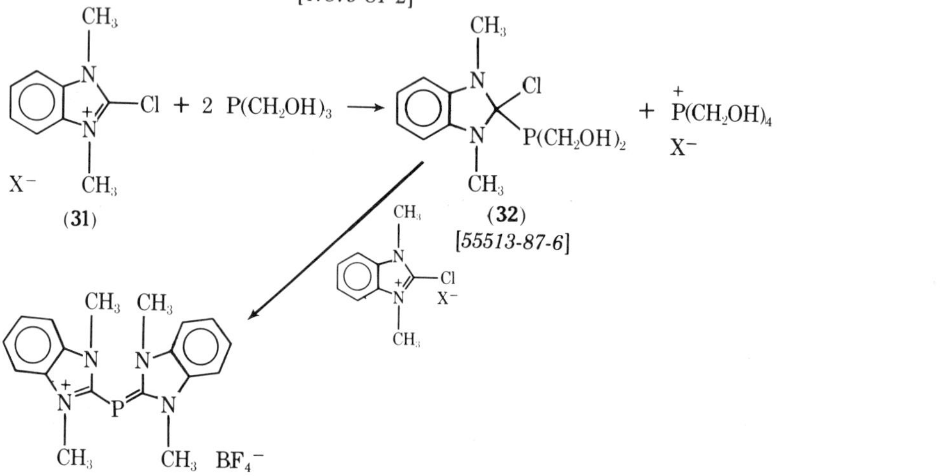

(31)   →   (32)
       [55513-87-6]

(33)
[55620-79-6]

Figure 7. (*continued*)

have included N-aryl groups, heterocyclic substituents, and complexes of dye bases with metal ions (iridium, platinum, zinc, copper, and nickel) (36,52–53).

Nitrogen substituents derived from alkyl groups are of primary importance for sensitizing dyes. Aside from the simple alkyl groups, there are three functional types of nitrogen substituents that have been used extensively: (*1*) The acidic N-substituents provided desirable solubility and adsorption characteristics for many practical cyanine and merocyanine sensitizers. The patents in this area have been reviewed (54). (*2*) Long N-alkyl chains, like octadecyl, gave surface-active dyes for monolayer studies (55). (*3*) Heteroatom substituents directly bonded to nitrogen (N—O$^-$, N—NR$_2$, N—OR) provided photochemically reactive dyes (56).

Synthesis of Cyanines and Related Dyes

As noted above, quite general synthetic methods were developed after 1920 and extended to many new systems. Dye-forming reactions may be classified as oxidative or nonoxidative. The oxidative syntheses (57) are primarily of historical interest, whereas nonoxidative syntheses are the most versatile and employ varied combinations of nucleophilic and electrophilic reagents. Selected examples to illustrate nonoxidative syntheses are shown in Figure 7, and more specific information may be obtained from other reviews. The review by Hamer lists references for the synthesis of dyes prepared before 1959 (8), and Ficken's review provides supplemental references to more recent compounds (9). Specific references to the many nucleophilic and electrophilic reagents used to synthesize cyanine and related dyes are tabulated in ref. 12.

Nonoxidative Cyanine Syntheses. Dye syntheses are characterized as "the condensation type, two intermediates reacting under suitable conditions with elimination of some simple molecule" (7). Most nonoxidative combinations of nucleophilic and electrophilic reagents are described in this way, including those necessary for syntheses of complex dyes. In contrast to oxidative dye syntheses in which a methylene base and the related quaternary salt can combine to form a dye, nonoxidative syntheses often combine reagents of quite different structure. For example, two molecules of methylene base are often combined with an ortho ester, $HC(OR)_3$. The latter reagent contributes only one methine CH to the chromophore of the dye, all the other atoms being derived from the two methylene base molecules.

Typical synthetic reactions are illustrated in Figure 7. Nucleophilic methylene bases derived from the reactions of quaternary salts, such as 3-ethyl-2-methylbenzothiazolium iodide (Fig. 7, (**19**)), and a variety of electrophilic reagents provide dyes in which the nucleophilic reagent (**19**) serves as the terminal heterocyclic group for the chromophore. The quaternary salts are also converted with ortho esters or anilides to generally useful electrophilic reagents, eg, intermediate (**24**). Reaction with various nucleophilic reagents provides several types of dyes. Those with simple chromophores include: (*1*) hemicyanines (**25**) in which one of the terminal nitrogens is nonheterocyclic; (*2*) enamine tricarbocyanines (**26**) useful as laser dyes; and (*3*) merocyanines (**27**). More complex polynuclear dyes from reagents with more than one reactive site include the trinuclear "B-A-B" dye (**28**) containing basic-acidic-basic heterocycles, and the tetranuclear "B-A-A-B" dye (**29**) containing basic-acidic-acidic-basic heterocycles. Structural variations of the reagents used in these reactions have been a primary source of progress in dye synthesis.

Extensive investigations of phosphorus reagents led to phosphocyanines (terminal R_3P group, dye (**30**)) and phosphacyanines (chromophoric methine replaced by phosphorus) (58). Simple phosphacyanines (**33**) were formed from (**31**) through a tetrahedral intermediate (**32**). Arsenic and sulfur analogues of the phosphocyanines were also prepared (59).

Acetylenic reagents for cyanine dye synthesis include the well-known acetylenic quaternary salts as general electrophilic reagents for the preparation of carbocyanine dyes. A number of tautomeric pairs of acetylenic dyes have been prepared and their tautomeric equilibria determined (60).

Ring-Closure Reactions. Some dyes are prepared by ring-closure reactions at or near the dye-forming step of a synthetic sequence. The constitution of thiacyanine was originally established by the reaction of diethyl malonate and *o*-aminothiophenol (8).

Quaternary salts like (**34**) (with reactive *N*-alkyl groups) are formed readily from acrolein (R' = H) and heterocyclic hydrobromide salts (Fig. 8). Cyclodehydration of these adducts by brief heating in dimethylacetamide provide dihydropyridinium salts, and dehydrogenation of the dihydropyridinium salts give the related pyridinium salts (61). Direct formation of dyes from the uncyclized adduct (**34**) lead to the thiacarbocyanine (**35**). The reactive *N*-alkyl groups in this dye are ring-closed to give either the partly rigid thiacarbocyanine (**36**) or the completely rigid and highly fluorescent (**37**).

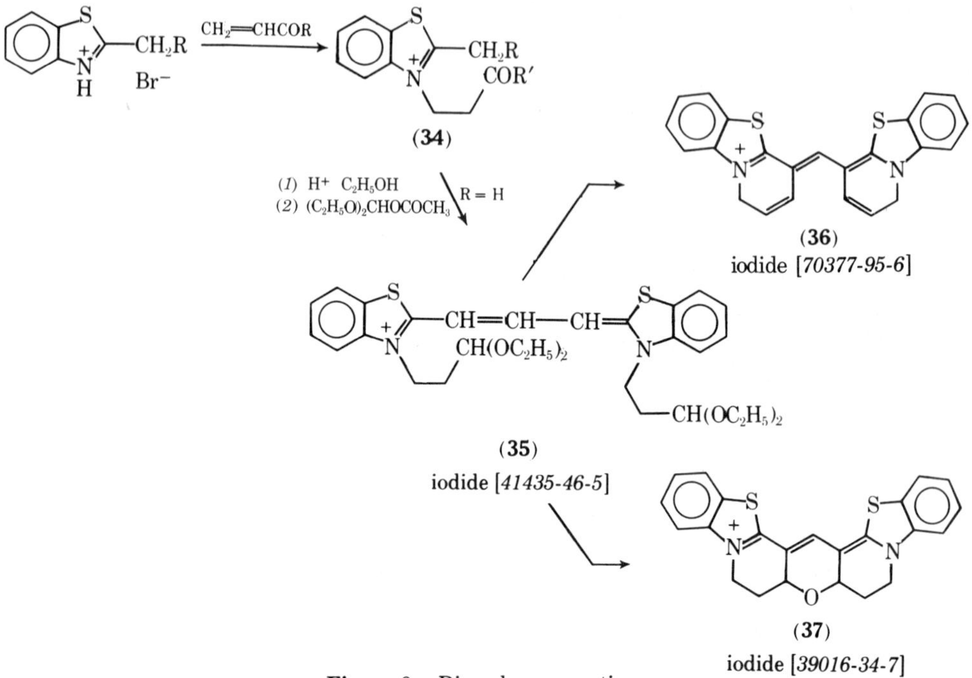

Figure 8. Ring-closure reactions.

Reactivity of Cyanine Dyes

Photoreactions. Photooxidation and photoreduction of dyes in solution have been studied extensively for noncyanine dye classes, such as the xanthene, thiazine, and acridine dyes. The early work of Oster and co-workers (62) and some more recent results have been reviewed (63). Photoreactions of cyanines and related dyes have been studied less, perhaps because of the low singlet-triplet intersystem crossing yields and the high propensity of cyanines to undergo rapid $S_1 \rightarrow S_0$ radiationless deactivation (39). Low quantum yield photofading reactions of cyanine dyes are observed under both oxidative conditions (oxygen, reducible metal ions) and reductive conditions (ascorbic acid, gelatin). Cyanine dyes interact with oxidizing or reducing agents in solution to produce changes in photoconductivity (64).

Both direct and sensitized photooxidations of cyanine dyes have been investigated (65). The naphthothiacyanine (38) was photodegraded more readily as the H-aggregate. Rate constants (k_{pr}) for sensitized dye bleaching are generally parallel to electrochemical oxidation potentials (ie, thiatricarbocyanine [3071-70-3] reacts more readily than thiacarbocyanine). The products from these reactions included those shown in the reaction of (39) which were consistent with 1,2-addition of singlet oxygen followed by cleavage of the C—C bond between the first methine carbon and the heterocyclic ring.

Photoreactive substituents on quaternary nitrogen atoms have been characterized (56). Both N—OR and N—NR$_2$ for the N—X part of (40) provide photobleachable and thermally-bleachable dyes. The photolytic reaction for the N-methoxy dyes lead primarily to dye base (41), chain-substituted analogues of (41), and other heterocyclic products.

Chemical Reactions. Many cyanines and related dyes are formed by addition–elimination reactions in which a simple terminal group is replaced by more complex ones (see, for example, the reactions of (**24**), Fig. 7). Aside from such syntheses and the relatively reversible protonation of vinylogous cyanines (**37**), many dyes can also react with amines, sulfite, and other reagents to produce new products with distinctly altered absorption spectra. Chemical degradation of oxidized and reduced dyes may contribute to irreversible electrochemical potentials (26).

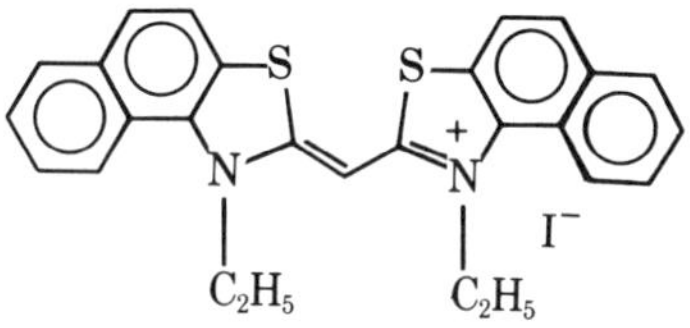

photooxidation
monomer, $\Phi \leq 10^{-6}$
H-aggregate, $\Phi \approx 10^{-2}$

(**38**) [18474-32-3]

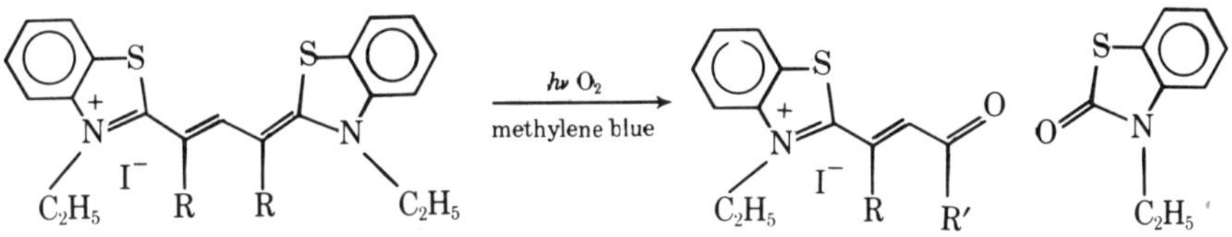

(**39**) R = CH$_3$ [62222-86-0]

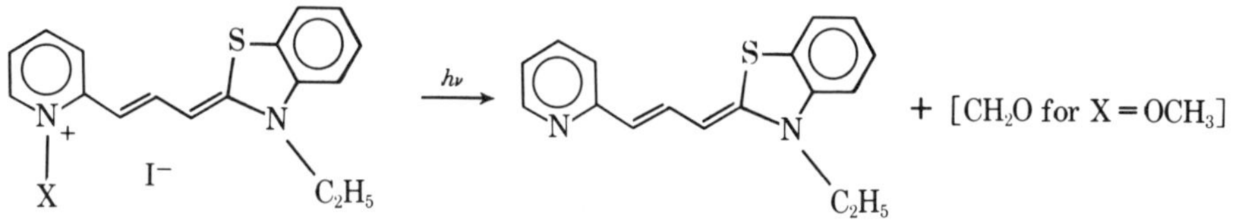

$\xrightarrow{h\nu\, O_2}$ methylene blue

(**40**) X = OCH$_3$
[37829-75-7]

$\xrightarrow{h\nu}$

+ [CH$_2$O for X = OCH$_3$]
+ other products

(**41**) [61109-40-8]

The irreversible reactions of vinylogous cyanine dyes in basic media include removal of the substituent on the heterocyclic nitrogen by amines or strong base (a general preparative method for the related dye bases (**5**)) (66). Substituted N-alkyl groups with acidic β-protons (eg, 2-carbethoxyethyl) allow dealkylation under milder conditions through reverse Michael reactions. Other carbocyanines from pyrrole or pyrrocoline are sensitive to small amounts of base at ambient temperature. This high reactivity is desirable for decolorizing sensitizers and filter dyes during photographic processing, but stable solutions of these dyes usually require dissolution of the dyes in slightly acidified solvents (see Dyes, sensitizing).

Benzoxazole dyes exhibit irreversible degradations that involve opening of the oxazole (67). Oxacarbocyanines (eg, **42**) react most readily with aqueous acid, whereas benzoxazole merocarbocyanines (**43**) react with sulfite or hydroxide ion to produce ring-opened products such as (**44**). The rate of reaction with sulfite is ca 100 times faster than the rate of reaction with hydroxide, and decreasing the Brooker basicity of the ketomethylene portion of the merocarbocyanine increases the rate of reaction.

The reactions of halogens with cyanine dyes proceed readily at silver halide surfaces (28) and in solution (68). The equilibria and halogenation products for simple cyanines from quinoline and benzothiazole have been reviewed (68). Initial halogenation is analogous to protonation and decolorizes the simple cyanines (45). Proton loss from (46) produces chain-halogenated dyes like (47) that can absorb at significantly longer wavelength than the unsubstituted cyanines. Reactions of simple cyanines and carbocyanines with excess halogen were also studied.

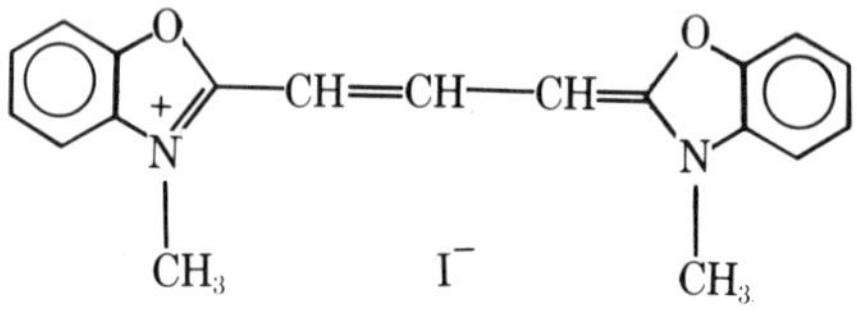

(42) [48198-86-3]

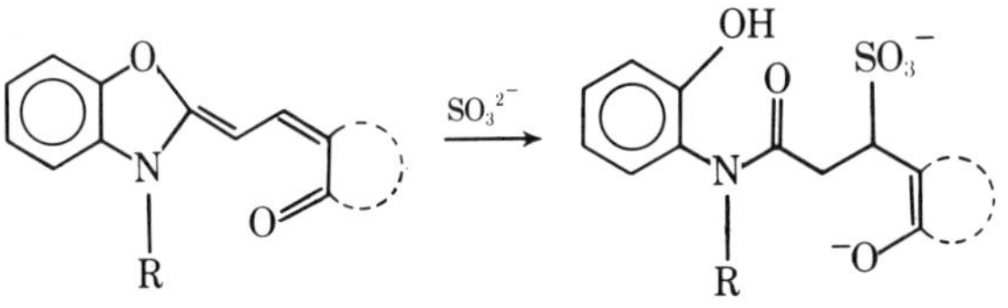

(43)   (44)

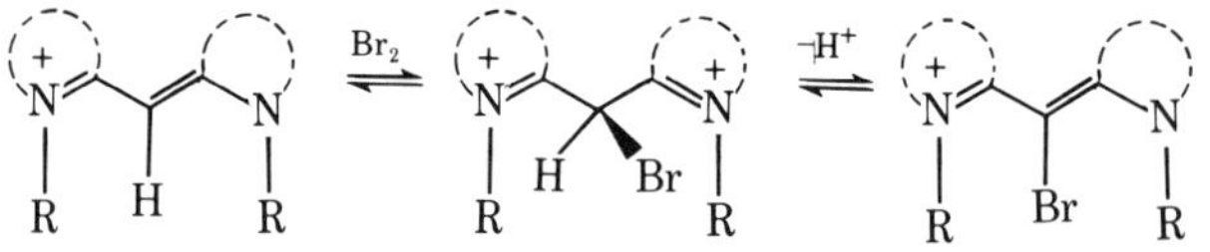

(45)   (46)   (47)

Uses and Suppliers

Cyanine dyes are used primarily for specialty purposes: photographic sensitizers and desensitizers, laser dyes, and certain medicinal applications. Because of this, their manufacture is limited to significantly smaller quantities than for fabric dyes or other widely used coloring agents. However, the photographic, laser, and medicinal uses place high demands on the degree of purity required, and the reproducibility of synthetic methods and purification steps is very important. Suppliers of cyanine dyes include manufacturers of other specialty organic and photographic chemicals: Aldrich Chemical Company (Milwaukee, Wisc.), Eastman Organic Chemicals (Rochester, N.Y.), Japanese Institute for Photosensitizing Dyes (Okayama, Japan), and Pfaltz and Bauer (Stamford, Conn.) are some examples. More importantly, these firms provide sources of generally useful reagents which, in two or three synthetic steps (8), lead to many of the commonly used cyanine dyes.

BIBLIOGRAPHY

"Cyanine Dyes" in *ECT* 1st ed., Vol. 4, pp. 742–754, by G. H. Keyes and E. J. Van Lare, Eastman Kodak Company; "Cyanine Dyes" in *ECT* 2nd ed., Vol. 6, pp. 605–624, by E. J. Van Lare, Eastman Kodak Company.

1. C. H. Greville Williams, *Trans. R. Soc. Edinburgh* **21,** 377 (1856).
2. L. G. S. Brooker in C. E. K. Mees, ed., *The Theory of the Photographic Process,* 1st ed., Macmillan, New York, 1942, p. 987.
3. W. West, *Photogr. Sci. Eng.* **18,** 35 (1974).
4. U.S. Pat. 2,925,417 (Feb. 16, 1960), E. F. Elslager and D. F. Worth (to Parke, Davis and Co.); U.S. Pat. 2,515,912 (July 18, 1950), E. Van Lare and L. G. S. Brooker (to Eastman Kodak Co.).
5. U.S. Pat. 2,893,914 (July 7, 1959), M. C. McCowen and P. F. Wiley (to Eli Lilly and Co.).
6. U.S. Pat. 2,895,955 (July 21, 1959), D. W. Heseltine and L. G. S. Brooker (to Eastman Kodak Co.).
7. L. G. S. Brooker in C. E. K. Mees and T. H. James, eds., *The Theory of the Photographic Process,* 3rd ed., Macmillan, New York, 1966, p. 198.
8. F. M. Hamer in A. Weissberger, ed., "The Cyanine Dyes and Related Compounds," *The Chemistry of Heterocyclic Compounds,* Vol. 18, Interscience Publishers Inc., a division of John Wiley & Sons, Inc., New York, 1964.
9. G. E. Ficken in K. Venkataraman, ed., *The Chemistry of Synthetic Dyes,* Vol. 4, Academic Press, Inc., New York, 1971, p. 211.
10. A. I. Kiprianov, *Usp. Khim.* **29,** 1336 (1960); **35,** 361 (1966); **40,** 594 (1971).
11. D. M. Sturmer and D. W. Heseltine in T. H. James, ed., *The Theory of the Photographic Process,* 4th ed., Macmillan, New York, 1977, p. 194.
12. D. M. Sturmer in E. C. Taylor and A. Weissberger, eds., *The Chemistry of Heterocyclic Compounds,* Vol. 30, Wiley-Interscience, New York, 1977, p. 441.
13. L. G. S. Brooker and P. W. Vittum, *Photogr. Sci. Eng.* **5,** 71 (1957).
14. H. W. Vogel, *Phil. Photogr.* **11,** 25 (1874); republished in *Photogr. Sci. Eng.* **18,** 33 (1974).
15. L. G. S. Brooker and co-workers, *J. Am. Chem. Soc.* **62,** 1116 (1940).
16. M. V. Deichmeister, I. I. Levkoev, and E. B. Lifshits, *Zh. Obshch. Khim.* **23,** 1529 (1953).
17. L. G. S. Brooker and co-workers, *J. Am. Chem. Soc.* **73,** 5332 (1951).
18. W. Jung, ed., *Optische Anregung Organische Systeme (Internationale Farbensymposium, Elman, 1964),* Verlag Chem., Weinheim, 1966.
19. L. G. S. Brooker and co-workers, *J. Am. Chem. Soc.* **67,** 1875 (1945).
20. J. R. Platt, *J. Chem. Phys.* **25,** 80 (1956).
21. L. G. S. Brooker and co-workers, *J. Am. Chem. Soc.* **87,** 2443 (1965); previous papers in this series.
22. L. G. S. Brooker and co-workers, *Chem. Rev.* **41,** 325 (1947).
23. L. G. S. Brooker and co-workers, *J. Photogr. Sci.* **1,** 173 (1953).
24. S. E. Sheppard, R. H. Lambert, and R. D. Walker, *J. Phys. Chem.* **50,** 210 (1946).
25. A. Stanienda, *Naturwissenschaften* **47,** 353 (1960).
26. R. F. Large in R. J. Cox, ed., *Photographic Sensitivity (Cambridge, 1972),* Academic Press, Inc., New York, 1973, p. 241.
27. P. Beretta and A. Jaboli, *Photogr. Sci. Eng.* **18,** 197 (1974).
28. W. West and P. B. Gilman, Jr. in T. H. James, ed., *The Theory of the Photographic Process,* 4th ed., Macmillan, New York, 1977, p. 251.
29. R. C. Nelson, *J. Photogr. Sci.,* **24,** 13 (1976).
30. B. H. Carroll, *Photogr. Sci. Eng.* **21,** 151 (1977).
31. D. L. Smith, *Photogr. Sci. Eng.* **18,** 309 (1974).
32. F. I. Vilesov, *Dokl. Akad. Nauk SSSR* **132,** 632 (1960).
33. J. W. Trusty and R. C. Nelson, *Photogr. Sci. Eng.* **16,** 421 (1972).
34. P. Yianoulis and R. C. Nelson, *Photogr. Sci. Eng.* **18,** 94 (1974).
35. G. Scheibe, *Chimia* **15,** 10 (1961).
36. G. Scheibe and E. Daltrozzo in A. R. Katrizky and A. J. Boulton, eds., *Advances in Heterocyclic Chemistry,* Vol. 7, Academic Press, Inc., New York, 1966, p. 153.
37. A. H. Herz, *Photogr. Sci. Eng.* **18,** 207 (1974).
38. S. F. Mason in K. Venkataraman, ed., *The Chemistry of Synthetic Dyes,* Vol. 3, Academic Press, Inc., New York, 1970, p. 169.
39. D. F. O'Brien, T. M. Kelly, and L. F. Costa, *Photogr. Sci. Eng.* **18,** 76 (1974).
40. A. V. Buettner, *J. Phys. Chem.* **68,** 3253 (1964).
41. W. West in H. Sauvenier, ed., *Scientific Photography (Liège, 1959),* Pergamon Press, London, 1962, p. 557.
42. S. H. Ehrlich, *Photogr. Sci. Eng.* **18,** 179 (1974); **20,** 5 (1976).
43. P. J. McCartin, *J. Chem. Phys.* **48,** 2980 (1965).
44. F. Doerr, J. Kotschy, and H. Kausen, *Ber. Bunsenges. Phys. Chem.* **69,** 11 (1965).

45. B. B. Snavely in J. B. Birks, ed., *Org. Mol. Photophys.*, Vol. 1, John Wiley & Sons, Inc., London, 1973, p. 239.
46. F. P. Schaefer, *Top. Curr. Chem.* **61,** 1 (1976).
47. A. H. Herz in T. H. James, ed., *The Theory of the Photographic Process*, 4th ed., Macmillan, New York, 1977, p. 235.
48. A. H. Herz, *Photogr. Sci. Eng.* **18,** 323 (1974).
49. J. F. Padday, *J. Phys. Chem.* **72,** 1259 (1968).
50. F. Dietz, *J. Signalaufzeichnungsmaterialien* **1,** 157, 237, 381 (1973).
51. D. M. Sturmer, W. S. Gaugh, and B. J. Bruschi, *Photgr. Sci. Eng.* **18,** 56 (1974).
52. J. W. Faller, A. Mueller, and J. P. Phillips, *J. Org. Chem.,* **29,** 3450 (1964).
53. T. Winkler and C. Mayer, *Helv. Chim. Acta* **55,** 2351 (1972).
54. E. J. Poppe, *Z. Wiss. Photogr. Photophys. Photochem.* **63,** 149 (1969).
55. J. Sondermann, *Justus Liebigs Ann. Chem.* **749,** 183 (1971).
56. J. D. Mee, D. W. Heseltine, and E. C. Taylor, *J. Am. Chem. Soc.* **92,** 5814 (1970).
57. J. Metzger and co-workers, *Bull. Soc. Chim. Fr.,* 3156 (1969); references therein.
58. K. Dimroth, *Fortschr. Chem. Forsch.* **38,** 5 (1973).
59. H. Depoorter, M. J. Libeer, and G. Van Mierlo, *Bull. Soc. Chim. Belg.* **77,** 521 (1968).
60. J. D. Mee, *J. Am. Chem. Soc.* **96,** 4712 (1974); see also *J. Org. Chem.* **42,** 1035, 1041 (1977).
61. L. L. Lincoln and co-workers, Eastman Kodak Co., personal communication; *Chem. Commun.,* 648 (1974).
62. G. Oster, *Trans. Faraday Soc.* **47,** 660 (1951).
63. H. Meier in ref. 9, p. 389.
64. H. Gerischer, *Photochem. Photobiol.* **16,** 243 (1973).
65. G. W. Byers, S. Gross, and P. M. Henrichs, *Photochem. Photobiol.* **23,** 37 (1976).
66. Ref. 8, p. 362.
67. F. J. Sauter, Eastman Kodak Co., personal communication.
68. A. van Beek and co-workers, *Rec. Trav. Chim.* **94,** 31 (1975).

DAVID M. STURMER
Eastman Kodak Company

CYANOCARBONS

Cyanocarbons are compounds having such a large number of cyano groups that the chemical reactions of the class are essentially new in kind and not shared by analogous compounds free of such groups. The unique reactivity of cyanocarbons was first recognized with the synthesis of tetracyanoethylene [670-54-2] (TCNE) (1–2). Previously, dicyanoacetylene [1071-98-3] (3) and dicyanodiacetylene [16419-78-6] (4) were the only known "per" cyanocarbons, and cyanoform [454-50-2] (5), CH(CN)$_3$, was the best known compound whose properties were exceptional as the consequence of a large number of cyano groups.

The cyano group is a powerful electron-withdrawing group. It is sufficiently small to present no great steric problem. Tetracyanoethylene must be regarded as a highly electron-deficient olefin that should be a strongly electrophilic reagent. This character is reflected in the ease with which TCNE is attacked by electron-rich olefins, eg, Diels-Alder additions and cyclobutane formation, and other nucleophiles, eg, methanol and dimethylaniline. Moreover, the affinity of TCNE for electrons is so great that the stable anion radical, TCNE$^{\cdot-}$, is formed by treatment with many mild reducing agents, such as I$^-$. For similar reasons, a compound with a hydrogen atom on a saturated carbon atom bearing cyanocarbon groups is highly ionized and is a very strong acid, eg, cyanoform and 1,1,2,3,3-pentacyanopropene [45078-17-9].

The most important members of the cyano class are the alkenes: tetracyanoethylene, hexacyanobutadiene, and tetracyanoquinodimethan; the alkanes: tetracyanomethane and hexacyanoethane; dicyanoacetylene; hexacyanobenzene; tetracyanoquinone; cyanocarbon acids; oxacyanocarbons; thiacyanocarbons; and azacyanocarbons (see below). Tetracyanoethylene is described first because its chemical versatility makes it a rich source of other polycyano compounds. Moreover, an understanding of its chemistry is helpful in understanding the chemistry of other cyanocarbons.

Tetracyanoethylene

Tetracyanoethylene, ethenetetracarbonitrile (TCNE) (2) has a high positive heat of formation (Table 1), but is stable to shock. It has good thermal stability and can thus be sublimed unchanged through a tube at 600°C. Tetracyanoethylene resists oxidation, but once ignited in oxygen it burns with an intensely hot flame that may be above 4000 K and is, at any rate, hotter than the oxygen–acetvlene flame.

Table 1. Physical Properties of Tetracyanoethylene[a]

color	colorless
mp, °C	200–202
bp, °C	223
density[b], g/mL	1.318
$\lambda_{max}^{CH_2Cl_2}$, nm(ϵ)	267 (13,600)
	277 (12,050)
heat of combustion, kJ/mol (kcal/mol)	−3022 (−722)
heat of formation, kJ/mol (kcal/mol)	628 (150)

[a] Ref. 6.
[b] Ref. 7.

360 CYANOCARBONS

Tetracyanoethylene forms intensely colored complexes with olefins or aromatic hydrocarbons, eg, benzene solutions are yellow, xylene solutions orange, and mesitylene solutions red. The colors arise from complexes of a Lewis acid–base type, with partial transfer of a π-electron from the aromatic hydrocarbon to TCNE (8). TCNE is conveniently prepared in the laboratory by debromination of dibromomalononitrile [1885-23-0] (1) with copper powder (9). The debromination can also be done by pyrolysis at ca 500°C (10).

$$CH_2(CN)_2 + Br_2 \rightarrow Br_2C(CN)_2 \rightarrow (NC)_2C=C(CN)_2 \quad (1)$$
$$[109\text{-}77\text{-}3] \qquad\qquad (1) \qquad\qquad TCNE$$

With substances that give up an electron more readily than aromatic hydrocarbons, such as potassium, nickel carbonyl, cyanide ion, or iodide ion, complete transfer of an electron occurs and the TCNE anion radical is formed (11). Potassium iodide is a particularly useful reagent for this purpose, and merely dissolving potassium iodide in an acetonitrile solution of TCNE causes the potassium salt of the anion radical to precipitate as bronze-colored crystals.

$$2\,KI + 2\,(NC)_2C=C(CN)_2 \rightarrow 2\,K^+\,[(NC)_2C=C(CN)_2]^{\cdot -} + I_2 \quad (2)$$

The anion radical is quite stable in the solid state. It is paramagnetic, and its intense electron paramagnetic resonance (epr) spectrum has nine principal lines with the intensity ratios expected for four equivalent ^{14}N nuclei (12) and may be used as an internal reference in epr work (see Analytical Methods).

Tetracyanoethylene undergoes two principal types of reactions, addition to the double bond and replacement of a cyano group.

$$TCNE + H_2 \xrightarrow{Pd} (NC)_2CHCH(CN)_2 \quad (3)$$
$$[14778\text{-}29\text{-}1]$$

$$TCNE + \underset{CH_3}{\overset{CH_3}{>}}C=C\underset{CH_2}{\overset{CH_2}{<}} \longrightarrow \text{(cyclohexene with } H_3C, H_3C, (CN)_2, (CN)_2) \quad (4)$$
$$[69155\text{-}29\text{-}9]$$

$$TCNE + CH_3OCH=CH_2 \longrightarrow \text{(cyclobutane with } CH_3O, (CN)_2, (CN)_2) \quad (5)$$
$$[69155\text{-}30\text{-}2]$$

$$TCNE + CH_3COCH_3 \xrightarrow{BF_3} CH_3\overset{O}{\overset{\|}{C}}CH_2-\underset{CN}{\overset{CN}{C}}-\underset{CN}{\overset{CN}{CH}} \quad (6)$$
$$[69155\text{-}31\text{-}3]$$

$$TCNE + 2\,CH_3-\underset{CH_3}{\overset{CN}{C}} \longrightarrow CH_3-\underset{CH_3}{\overset{CN}{C}}-\underset{CN}{\overset{CN}{C}}-\underset{CN}{\overset{CN}{C}}-\underset{CN}{\overset{CH_3}{C}}-CH_3 \quad (7)$$
$$[69178\text{-}29\text{-}6]$$

Among the reagents that add readily to TCNE are hydrogen (eq. 3) (13), 1,3-dienes (eq. 4) (14), electron-rich alkenes (eq. 5) (15), and ketones (eq. 6) (13). Nucleophilic radicals also add readily to TCNE (eq. 7) (13).

Of these reactions, the Diels-Alder addition to 1,3-dienes is particularly interesting because of the exceptional ease with which it takes place. In a study of the relative rates of addition of 20 dienophiles to cyclopentadiene, TCNE was at the head of the list, eg, it added 7700 times as rapidly as maleic anhydride (14). Reaction with most 1,3-dienes takes place rapidly and in high yield at room temperature. TCNE has often been used to characterize 1,3-dienes, including those that are unstable and difficult to isolate (16).

Although a C—CN bond is normally strong, one or two cyano groups in TCNE can be replaced easily, about as easily as the one in an acyl cyanide. The replacing group can be hydroxyl, alkoxyl, amino, or a nucleophilic aryl group. Thus, hydrolysis of TCNE under neutral or mildly acidic conditions leads to tricyanoethenol [27062-39-1] (2), a strong acid isolated only in the form of salts (17).

$$(NC)_2C=C(CN)_2 + H_2O \longrightarrow (NC)_2C=C(CN)(O^-H^+) + HCN \quad (8)$$
$$(2)$$

Heating TCNE with an alcohol in the presence of a mild base such as urea causes replacement of either one (18) or two (19) cyano groups by alkoxyl.

Aromatic compounds that are sufficiently nucleophilic to condense with benzenediazonium chloride and form azo compounds generally condense with TCNE, eg, the reaction of N,N-dimethylaniline proceeds stepwise (20–21).

p-Tricyanovinylarylamines such as (3) [6673-15-0] are highly colored substances that give bright orange, red, or blue dyeings on poly(ethylene terephthalate) and other hydrophobic fibers. Among other aromatic compounds that have been tricyanovinylated are phenanthrene (22), o-alkylphenols (23), pyrrole (22), indoles (22,24), 2-methylfuran (25), azulenes (25–26), diazocyclopentadiene (27), and a variety of phenylhydrazones (25).

A few kg of TCNE are sold by chemical suppliers each year in 10- and 50-g lots.

$$(NC)_2C=C(CN)_2 \xrightarrow[-HCN]{C_2H_5OH} (NC)(C_2H_5O)C=C(CN)_2 \xrightarrow[-HCN]{C_2H_5OH} (C_2H_5O)_2C=C(CN)_2 \quad (9)$$
$$[69155\text{-}32\text{-}4] \qquad [17618\text{-}65\text{-}4]$$

$$(CH_3)_2NC_6H_5 \xrightarrow{TCNE} (CH_3)_2N^+=C_6H_4(H)(C(CN)_2\bar{C}(CN)_2) \xrightarrow[-CN]{-H^+} (CH_3)_2N-C_6H_4-C(CN)=C(CN)_2 \quad (10)$$

$$(3)$$

Hexacyanobutadiene

Hexacyanobutadiene [5104-27-8] (4), 1,3-butadiene-1,1,2,3,4,4-hexacarbonitrile, is prepared in good yield by a two-step process from the disodium salt of tetracyanoethane (28). It is like TCNE in forming colored π-complexes and an anion radical.

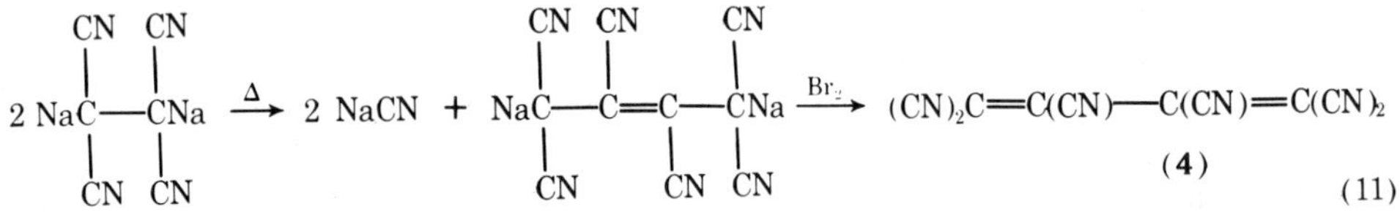

$$2\ \underset{\underset{CN}{|}}{\overset{\overset{CN}{|}}{NaC}}-\underset{\underset{CN}{|}}{\overset{\overset{CN}{|}}{CNa}} \xrightarrow{\Delta} 2\ NaCN + \underset{\underset{CN}{|}}{\overset{\overset{CN}{|}}{NaC}}-\overset{\overset{CN}{|}}{C}=C-\underset{\underset{CN}{|}}{\overset{\overset{CN}{|}}{CNa}} \xrightarrow{Br_2} (CN)_2C=C(CN)-C(CN)=C(CN)_2 \qquad (11)$$

$$(4)$$

Tetracyanoquinodimethan

Tetracyanoquinodimethan [1518-16-7] (5), 2,2'-(2,5-cyclohexadiene-1,4-diyli-

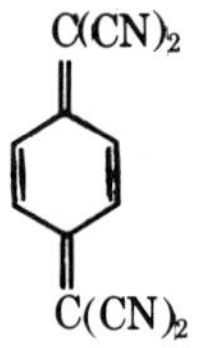

(5)

dene)bispropanedinitrile (TCNQ), is prepared by condensing 1,4-cyclohexanedione with malononitrile to give 1,4-bis(dicyanomethylene)cyclohexane [1518-15-6], which is oxidized with bromine (29). It resembles tetracyanoethylene in that it adds reagents such as hydrogen (29), sulfurous acid (29), and tetrahydrofuran (30) to the ends of the conjugated system of carbon atoms; suffers displacement of one or two cyano groups by nucleophilic reagents such as amines (31) or sodiomalononitrile (32); forms π-complexes with aromatic compounds (33); and takes an electron from iodide ion, copper, or tertiary amines to form an anion radical (33–34). The anion radical has been isolated as salts of the formula M^+ $(TCNQ)_n^{-}$ where M^+ is a metal or ammonium cation, and n = 1, 1.5, or 2. Some of these salts have unusual electrical properties; their conductivities are in the remarkably high range 0.01–100 S/cm and differ by several orders of magnitude along the three major crystal axes (33,35).

The most remarkable substance of this kind is tetrathiafulvalenium tetracyanoquinodimethanide [40210-84-2] (6) (TTF–TCNQ). It is the most highly conducting organic compound known (10^4 S/cm at 70 K) (36–37) (see Semiconductors, organic). Moreover, the conduction is metallic-like in that it increases as temperature decreases. Superconduction properties for highly purified crystals of TTF–TCNQ at 60 K are reported (38–39) (see Superconducting materials).

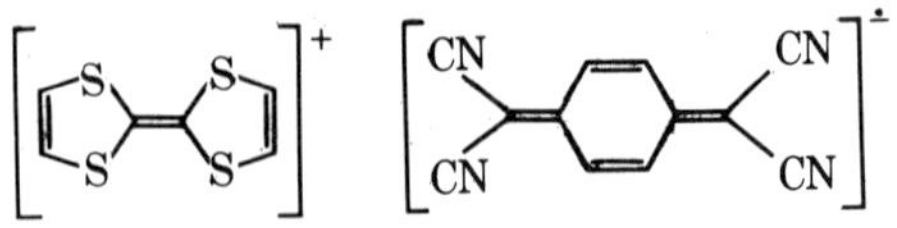

(6)

Tetracyanomethane

Tetracyanomethane [24331-09-7], methanetetracarbonitrile, is prepared by heating silver tricyanomethanide [36603-81-3] in liquid cyanogen chloride (40). It is a very strong cyanating agent; eg, it converts chloride ion to cyanogen chloride [506-77-4] (7).

$$AgC(CN)_3 + ClCN \rightarrow C(CN)_4 + AgCl \qquad (12)$$

$$C(CN)_4 + Cl^- \rightarrow ClCN + {}^-C(CN)_3 \qquad (13)$$
$$(7)$$

Hexacyanoethane

Hexacyanoethane [4383-67-9] (8), ethanehexacarbonitrile, is quite unstable and readily decomposes to TCNE and cyanogen [460-19-5] (41). It is prepared follows:

$$TCNE \xrightarrow{NaCN} NaC(CN)_2-C(CN)_2 \xrightarrow{+\,(7)} (CN)_3CC(CN)_3 + NaCl \qquad (14)$$
$$(8)$$

Dicyanoacetylene

Dicyanoacetylene, 2-butynedinitrile, is obtained from dimethyl acetylenedicarboxylate by ammonolysis to the diamide, which is dehydrated with phosphorus pentoxide (42). It burns in oxygen to give a flame with a temperature of 5260 K, the hottest flame temperature known (43). Alcohols and amines add readily to its acetylenic bond (44). It is a powerful dienophile in the Diels-Alder reaction; it adds to many dienes at room temperature, and at 180°C actually adds 1,4- to benzene to give (9) [18341-68-9] (45).

Hexacyanobenzene

Hexacyanobenzene [1217-44-3] (10), benzenehexacarbonitrile, is prepared from 2,4,6-trifluorobenzene-1,3,5-tricarbonitrile [3638-97-9] by substitution with calcium cyanide (46–47). It forms colored π-complexes with aromatic hydrocarbons.

$$\bigcirc + NCC \equiv CCN \longrightarrow \text{(9)} \qquad (15)$$

364 CYANOCARBONS

$$2 \; \underset{(10)}{\text{C}_6\text{F}_3(\text{CN})_3} + 3 \; \text{Ca(CN)}_2 \rightarrow \text{C}_6(\text{CN})_6 + 3 \; \text{CaF}_2 \quad (16)$$

Tetracyanobenzoquinone

Tetracyanobenzoquinone [4032-03-5], 3,6-dioxo-1,4-cyclohexadiene-1,2,4,5-tetracarbonitrile, is a remarkably strong oxidizing agent for a quinone; it abstracts hydrogen from tetralin or ethanol even at room temperature (48). It is a stronger π-acid then TCNE because it forms more deeply colored π-complexes with aromatic hydrocarbons.

Cyanocarbon Acids

These acids (49) are organic molecules that contain a plurality of cyano groups and are readily ionized to hydrogen ions and resonance-stabilized anions. Typical cyanocarbon acids are cyanoform, methanetricarbonitrile (5); 1,1,3,3-tetracyanopropene [32019-26-4], 1-propene-1,1,3,3-tetracarbonitrile (50); 1,1,2,3,3-pentacyanopropene [45078-17-9], 1-propene-1,1,2,3,3-pentacarbonitrile (49); 1,1,2,6,7,7-hexacyano-1,3,5-heptatriene [69239-39-0] (51); 2-dicyanomethylene-1,1,3,3-tetracyanopropane [32019-27-5] (49); and 1,3-cyclopentadiene-1,2,3,4,5-pentacarbonitrile [69239-40-3] (52–53). Many of these acids rival mineral acids in strength (54) and are usually isolable only as salts with metal or ammonium ions. The remarkable strength of these acids results from resonance stabilization in the anions that is not possible in the protonated forms.

Some of the cyanocarbon anions are highly colored, eg, salts of 1,1,2,6,7,7-hexacyano-1,3,5-heptatriene are deep blue, $\log_{10} \epsilon = 4.22$ at 635 nm (51).

Oxacyanocarbons

Tetracyanoethylene oxide [3189-43-3] (11), oxiranetetracarbonitrile, is the most

$$\underset{(11)}{\text{(NC)}_2\text{C}-\text{O}-\text{C(CN)}_2} + \text{CH}_2\!\!=\!\!\text{CH}_2 \rightarrow \underset{[3041\text{-}31\text{-}4]}{\text{(NC)}_2\text{C}-\text{O}-\text{C(CN)}_2\text{-CH}_2\text{-CH}_2} \quad (17)$$

$$(11) + \text{C}_6\text{H}_6 \rightarrow \underset{[3041\text{-}36\text{-}9]}{\text{product}} \quad (18)$$

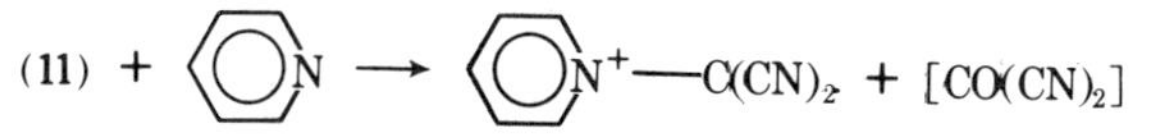

$$(11) + \underset{}{\bigcirc}\!N \longrightarrow \underset{}{\bigcirc}\!N^+\!\!-\!\!C(CN)_2 + [CO(CN)_2] \qquad (19)$$
$$[27032\text{-}01\text{-}5]$$

$$(11) + (CH_3)_2S \longrightarrow (CH_3)_2\overset{+}{S}\!\!-\!\!\overset{-}{C}(CN)_2 + CO(CN)_2 \qquad (20)$$
$$[5362\text{-}78\text{-}7]$$

notable member of the class of oxacyanocarbons (55). It is made by treating TCNE with hydrogen peroxide in acetonitrile. In a reaction unprecedented for olefin oxides, it adds to olefins (eq. 17), acetylenes, and aromatic hydrocarbons (eq. 18) via cleavage of the ring C—C bond.

With pyridine, reaction takes place at the nitrogen rather than at a double bond, and an ylid is formed (eq. 19) (55–56). Sulfides react similarly to give sulfilidenes and carbonyl cyanide (eq. 20) (57); this is the most convenient synthesis of carbonyl cyanide [1115-12-4].

Tetracyanofuran [17989-87-6] (12), 2,3,4,5-furantetracarbonitrile, has been obtained by dehydration of the corresponding tetracarboxamide (58). The α-cyano groups are more reactive, preferentially adding water, hydroxylamine, or phenylhydrazine.

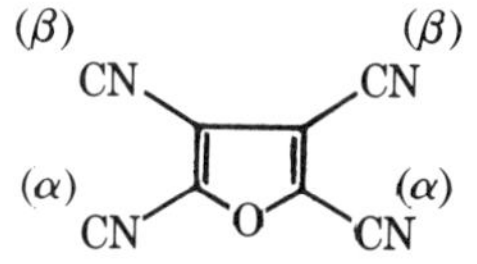

(12)

Thiacyanocarbons

For the most part, thiacyanocarbons are derived from "Bähr's Salt" [18820-77-4] (13), prepared from carbon disulfide and sodium cyanide (59).

Oxidation of Bähr's Salt with iodine or thionyl chloride gives tetracyano-1,4-dithiin [2448-55-7] (14), 1,4-dithiin-2,3,5,6-tetracarbonitrile (60–61). The dithiin loses sulfur at 210°C to give tetracyanothiophene [4506-96-1] (15), 2,3,4,5-thiophenetetracarbonitrile, and adds sulfur under the influence of sodium ethoxide to give the

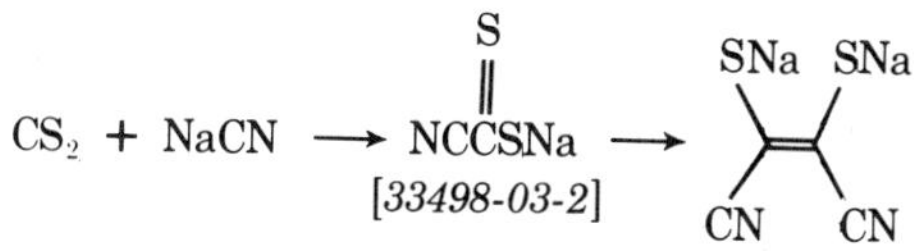

$$CS_2 + NaCN \longrightarrow \underset{[33498\text{-}03\text{-}2]}{NC\overset{S}{\overset{\|}{C}}SNa} \longrightarrow \underset{CN\quad CN}{\overset{SNa\ SNa}{\underset{}{\bigvee}}} \qquad (21)$$
(13)

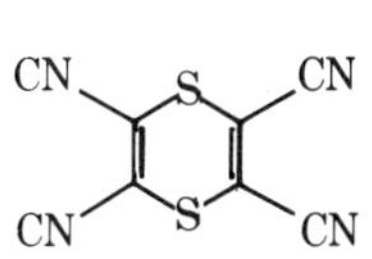

(14) 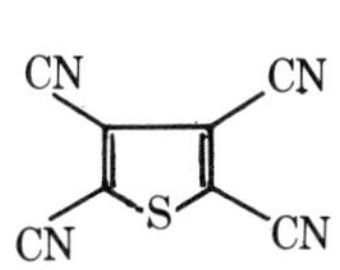 (15) 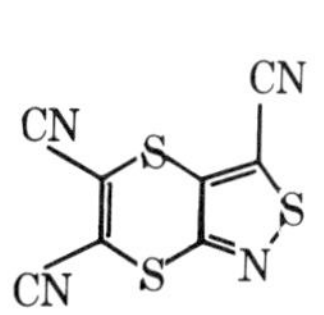 (16)

366   CYANOCARBONS

isothiazole [4656-27-3] (16) (55). The nitrile groups of these three heterocyclic thiacyanocarbons behave normally, and are convertible to amide, imide, or carboxylic acid groups.

Dicyano-1,2-dithiete [53562-16-6] (17) is thought to be an intermediate when Bähr's Salt is oxidized (62–63). If the oxidation is carried out in the presence of vinyl ethers, dihydrodithiins (18) can be obtained in yields up to 60%.

$$(13) \xrightarrow{SOCl_2} \left[\text{(17)} \right] \xrightarrow{ROCH=CH_2} \text{(18)} \qquad (22)$$

Azacyanocarbons

Hydrogen cyanide tetramer [1187-42-4] (19), 2,3-diamino-2-butenedinitrile, (Z-), an azacyanocarbon, is produced by Nippon Soda in pilot-plant quantities for development as a chemical intermediate (64–65). On oxidation it forms 2,3-diiminobutanedinitrile [28321-79-7] (20) (66). These two, in turn, combine to give pyrazinetetracarbonitrile [33420-37-0] (21) (67).

$$4 \text{ HCN} \xrightarrow{base} (19) \xrightarrow{[O]} (20) \qquad (23)$$

$$(19) + (20) \xrightarrow{H^+} (21) \qquad (24)$$

Treatment of (19) with nitrous acid gives 1H-1,2,3-triazole-4,5-dicarbonitrile [53817-16-6] (22) (68). The action of thionyl chloride on (19) gives 1,2,5-thiadiazole-3,4-dicarbonitrile [23347-22-0] (23) (69).

Reaction of (19) with cyanogen chloride gives 2-amino-1H-imidazole-4,5-dicarbonitrile [40953-34-2] (24) (70). This amino compound reacts with nitrous acid to form 2-diazo-1H-imidazole-4,5-dicarbonitrile [51285-29-1] (25) (70).

Thermolysis of (25) generates an extremely electrophilic carbene (78).

5-Diazo-1,3-cyclopentadiene-1,2,3,4-tetracarbonitrile [54747-37-4] (26) (52) and 2-diazopropanedinitrile [1618-08-2] (27) (71) also possess reactive diazo groups.

Other azacyanocarbons are: 1H-imidazole-2,4,5-tricarbonitrile [69239-41-4] (28)

$$(22) \xleftarrow{HNO_2} (19) \xrightarrow{SOCl_2} (23) \qquad (25)$$

(72); 1H-pyrazole-3,4,5-tricarbonitrile [69239-42-5] (**29**) (73); 1H-pyrrole-2,3,4,5-tetracarbonitrile [5231-17-4] (**30**) (74); 2,3,4,5,6-pyridinepentacarbonitrile [69239-43-6] (**31**) (75); and 1,3,5-triazine-2,4,6-tricarbonitrile [7615-57-8] (**32**) (76).

Health and Safety Factors (Toxicology)

Unless specifically tested, all cyanocarbons should be considered as toxic as sodium cyanide or hydrogen cyanide (see Cyanides). They should be used only in a fume hood, and rubber gloves should be worn.

BIBLIOGRAPHY

"Cyanocarbons" in *ECT* 2nd ed., Vol. 6, pp. 625–633, by B. C. McKusick and T. L. Cairns, E. I. du Pont de Nemours & Co., Inc.

1. T. L. Cairns and co-workers, *J. Am. Chem. Soc.* **79**, 2340 (1957); *J. Am. Chem. Soc.* **80**, 2775 (1958).

2. U.S. Pat. 3,166,584 (Jan. 19, 1965), T. L. Cairns and E. A. Graef (to E. I. du Pont de Nemours & Co.).
3. C. Moureu and J. Bongrand, *Compt. Rend. Acad. Sci.* **170,** 1025 (1920).
4. F. J. Brockman, *Can. J. Chem.* **33,** 507 (1955).
5. E. Cox and A. Fontaine, *Bull. Soc. Chim. Fr.,* 948 (1954); S. Trofimenko, *J. Org. Chem.* **28,** 2755 (1963).
6. C. E. Looney and J. R. Downing, *J. Am. Chem. Soc.* **80,** 2840 (1958).
7. D. A. Bekoe and K. N. Trueblood, *Z. Kristallogr.* **113,** 1 (1960).
8. R. E. Merrifield and W. D. Phillips, *J. Am. Chem. Soc.* **80,** 2778 (1958); G. D. Farnum, E. R. Atkinson, and W. C. Lothrop, *J. Org. Chem.* **26,** 3204 (1961); M. J. S. Dewar and H. Rogers, *J. Am. Chem. Soc.* **84,** 395 (1962).
9. R. A. Carboni, *Org. Synth.* **39,** 64 (1959).
10. U.S. Pat. 3,118,929 (Jan. 21, 1964), E. L. Martin (to E. I. du Pont de Nemours & Co., Inc.).
11. O. W. Webster, W. Mahler, and R. E. Benson, *J. Am. Chem. Soc.* **84,** 3678 (1962).
12. W. D. Phillips, J. C. Rowell, and S. I. Weissman, *J. Chem. Phys.* **33,** 626 (1960); P. H. Rieger and co-workers, *J. Am. Chem. Soc.* **85,** 683 (1963).
13. W. J. Middleton and co-workers, *J. Am. Chem. Soc.* **80,** 2783 (1958).
14. J. Sauer, *Angew. Chem.* **73,** 545 (1961).
15. A. T. Blomquist and Y. C. Meinwald, *J. Am. Chem. Soc.* **81,** 667 (1959); J. K. Williams, D. W. Wiley, and B. C. McKusick, *J. Am. Chem. Soc.* **84,** 2210 (1962); C. A. Stewart, *J. Org. Chem.* **28,** 3320 (1963).
16. M. Ozolins and G. H. Schenk, *Anal. Chem.* **33,** 1035 (1961); G. N. Schrauzer and S. Eichler, *Angew. Chem. Int. Ed. Engl.* **1,** 454 (1962); H. Prinzbach, *Angew. Chem.* **73,** 169 (1961); W. E. Doering and D. W. Wiley, *Tetrahedron* **11,** 183 (1960); K. Hafner and J. Schneider, *Justus Liebigs Ann. Chem.* **624,** 37 (1959); R. Criegie, *Angew. Chem. Int. Ed. Engl.* **1,** 519 (1962). These references contain examples of the use of TCNE to determine or characterize dienes.
17. W. J. Middleton and co-workers, *J. Am. Chem. Soc.* **80,** 2795 (1958).
18. C. L. Dickinson, D. W. Wiley, and B. C. McKusick, *J. Am. Chem. Soc.* **82,** 6132 (1960).
19. W. J. Middleton and V. A. Engelhardt, *J. Am. Chem. Soc.* **80,** 2788 (1958).
20. B. C. McKusick and co-workers, *J. Am. Chem. Soc.* **80,** 2806 (1958).
21. Z. Rappoport, *J. Chem. Soc.* 4498 (1963); P. G. Farrell and J. Newton, *Tetrahedron Lett.,* 189 (1964).
22. G. N. Sausen, V. A. Engelhardt, and W. J. Middleton, *J. Am. Chem. Soc.* **80,** 2815 (1958).
23. B. Smith and U. Persmark, *Acta Chem. Scand.* **17,** 651 (1963).
24. W. E. Noland, W. C. Kuryla, and R. F. Lange, *J. Am. Chem. Soc.* **81,** 6010 (1959).
25. J. R. Roland and B. C. McKusick, *J. Am. Chem. Soc.* **83,** 1652 (1961).
26. K. Hafner and K. Moritz, *Justus Liebigs Ann. Chem.* **650,** 92 (1961).
27. D. J. Cram and R. D. Partos, *J. Am. Chem. Soc.* **85,** 1273 (1963).
28. O. W. Webster, *J. Am. Chem. Soc.* **84,** 3370 (1962).
29. D. S. Acker and W. R. Hertler, *J. Am. Chem. Soc.* **84,** 3370 (1962).
30. J. Diekmann and C. J. Pedersen, *J. Org. Chem.* **28,** 2879 (1963).
31. W. R. Hertler and co-workers, *J. Am. Chem. Soc.* **84,** 3387 (1962).
32. J. K. Williams, *J. Am. Chem. Soc.* **84,** 3478 (1962).
33. L. R. Melby and co-workers, *J. Am. Chem. Soc.* **84,** 3374 (1962).
34. P. H. H. Fischer and C. A. McDowell, *J. Am. Chem. Soc.* **85,** 2694 (1963).
35. W. J. Siemons, P. E. Bierstedt, and R. G. Kepler, *J. Chem. Phys.* **39,** 3523 (1963).
36. J. Ferraris and co-workers, *J. Am. Chem. Soc.* **95,** 948 (1973).
37. L. B. Coleman and co-workers, *Solid State Commun.* **12,** 1125 (1973).
38. M. J. Cohen and co-workers, *Phys. Rev. Sect. B* **13,** 5111 (1976).
39. G. A. Thomas and co-workers, *Phys. Rev. Sect. B* **13,** 5105 (1976).
40. E. Mayer, *Monatsh. Chem.* **100,** 462 (1969).
41. S. Trofimenko and B. C. McKusick, *J. Am. Chem. Soc.* **84,** 3677 (1962).
42. A. T. Blomquist and E. C. Winslow, *J. Org. Chem.* **10,** 149 (1945).
43. A. D. Kirshenbaum and A. V. Grosse, *J. Am. Chem. Soc.* **75,** 499 (1953).
44. A. W. Johnson, *The Chemistry of the Acetylenic Compounds,* Vol. II, Longmans, Green & Co., New York, 1950, pp. 258–260.
45. E. Ciganek, *Tetrahedron Lett.,* 3321 (1967).
46. K. Wallenfels and K. Friedrich, *Tetrahedron Lett.,* 1223 (1963).

47. K. Friedrich, *Chem. Ber.* **103,** 3951 (1970).
48. K. Wallenfels and G. Bachmann, *Angew. Chem.* **73,** 142 (1961); K. Wallenfels and co-workers, *Tetrahedron* **21,** 2239 (1965); U.S. Pat. 3,114,756 (Dec. 17, 1963), (to Farbwerke Hoechst).
49. W. J. Middleton and co-workers, *J. Am. Chem. Soc.* **80,** 2795 (1958).
50. Y. Urushibara, *Bull. Chem. Soc. Jpn.* **2,** 278 (1927).
51. J. K. Williams, D. W. Wiley, and B. C. McKusick, *J. Am. Chem. Soc.* **84,** 2216 (1962).
52. O. W. Webster, *J. Am. Chem. Soc.* **88,** 4055 (1966).
53. *Ibid.,* 3046 (1966).
54. R. H. Boyd, *J. Phys. Chem.* **67,** 737 (1963).
55. W. J. Linn, O. W. Webster, and R. E. Benson, *J. Am. Chem. Soc.* **85,** 2032 (1963).
56. A. Rieche and P. Dietrich, *Chem. Ber.* **96,** 3044 (1963).
57. U.S. Pat. 3,115,517 (Dec. 24, 1963), W. J. Linn (to E. I. du Pont de Nemours & Co., Inc.).
58. C. Weis, *J. Org. Chem.* **27,** 3514 (1962).
59. G. Bähr and G. Schleitzer, *Chem. Ber.* **88,** 1771 (1955); **90,** 438 (1957); G. Bähr, *Angew. Chem.* **68,** 525 (1956).
60. G. Bähr, *Angew. Chem.* **70,** 606 (1958).
61. H. E. Simmons and co-workers, *J. Am. Chem. Soc.* **84,** 4746 (1962).
62. H. E. Simmons, D. C. Blomstrom, and R. D. Vest, *J. Am. Chem. Soc.* **84,** 4756 (1962).
63. *Ibid.,* 4773, 4782 (1962).
64. H. Bredereck, G. Schmötzer, and E. Oehler, *Justus Liebigs Ann. Chem.* **600,** 81 (1956).
65. U.S. Pat. 3,701,797 (Oct. 31, 1972), T. Okada and N. Asai (to Sagami Chemical Research Center).
66. O. W. Webster and co-workers, *J. Org. Chem.* **37,** 4133 (1972).
67. R. W. Begland and co-workers, *J. Org. Chem.* **39,** 1235 (1974).
68. H. Bredereck and G. Schmötzer, *Justus Liebigs Ann. Chem.* **600,** 95 (1956).
69. U.S. Pats. 2,990,408 and 2,990,409 (June 27, 1961), M. Carmack, D. Shew, and L. M. Weinstock.
70. W. A. Sheppard and O. W. Webster, *J. Am. Chem. Soc.* **95,** 2695 (1973).
71. E. Ciganek, *J. Am. Chem. Soc.* **88,** 1979 (1966); **89,** 1454 (1967).
72. U.S. Pat. 3,793,339 (Feb. 19, 1974), O. W. Webster (to E. I. du Pont de Nemours & Co., Inc.).
73. C. D. Weis, *J. Org. Chem.* **27,** 3695 (1962).
74. U.S. Pat. 3,221,024 (Nov. 30, 1965), H. E. Simmons (to E. I. du Pont de Nemours & Co., Inc.).
75. P. Neumann, Dissertation, University of Freiburg, 1967.
76. E. Ott, *Chem. Ber.* **52B,** 656 (1919).

General References

E. Ciganek, W. J. Linn, and O. W. Webster, in Z. Rapport, ed., *The Chemistry of the Cyano Group,* Interscience Publishers, London, 1970, pp. 423–638. An extensive review of cyanocarbon chemistry.
T. L. Cairns and B. C. McKusick, *Angew. Chem.* **73,** 520 (1961).
F. Freeman, *Chem. Revs.* **9,** 591 (1969). Review of malonitrile chemistry.
V. Migrdichian, *The Chemistry of Organic Cyanogen Compounds,* Reinhold, New York, 1947.
K. Wallenfels and co-workers, *Angew. Chem. Int. Ed. Engl.* **15,** 261 (1976).

O. W. WEBSTER
E. I. du Pont de Nemours & Co., Inc.

CYANOETHYLATION

Cyanoethylation, the reaction in which a compound possessing an active hydrogen adds across the double bond of acrylonitrile (qv), offers the synthetic chemist a convenient route for incorporating the propanenitrile moiety into a molecule. The characteristic feature of compounds undergoing this reaction is their possession of a labile hydrogen atom that, once removed, produces a nucleophilic group which then attacks the most positive position in acrylonitrile. Examples of chemical types that undergo cyanoethylation include hydroxyl compounds (water, alcohols, phenols, and oximes), thiols (hydrogen sulfide, aliphatic mercaptans, and thiophenols), nitrogen compounds (ammonia, amines, amides, imides, hydrazine, and various heterocyclic compounds), some carbon compounds (malononitrile, cyanoacetic esters, benzyl cyanide, and cyclopentadiene), hydrogen chloride, hydrogen bromide, sodium bisulfite, as well as arsines, boranes, germanes, phosphines, silanes, and stannanes. Because of its versatility, this reaction has assumed great importance in synthesis and new examples are constantly being reported.

The reaction is most frequently performed with a basic catalyst, but acids and metallic compounds have also been used. Yields are usually high. The strongly exothermic character of the reaction necessitates the use of solvents and other means to keep the operation under control. The reaction is reversible with high temperatures and strong base catalysts.

Mechanism

Because of the strong electron-withdrawing character of the nitrile group, the β-carbon atom of acrylonitrile is rendered relatively positive whereas the nitrile portion of the molecule is relatively negative. This effect is illustrated in structure (1).

$$\overset{\delta+}{CH_2}=CHC\overset{\delta-}{\equiv}N$$
(1)

This molecule, therefore, is susceptible to attack at the β position by a nucleophilic reagent. These nucleophiles may be anions, ie, alkoxide ions, or even neutral molecules possessing nucleophilic character such as amines.

The generally accepted mechanism is summarized in equations 1–3 in which an alcohol is used as a reactant with acrylonitrile.

$$ROH + base \rightleftharpoons RO^- + (base-H)^+ \qquad (1)$$

$$RO^- + CH_2=CHC\equiv N \rightleftharpoons \begin{cases} [ROCH_2CHC\equiv N] \\ \updownarrow \\ [ROCH_2CH=C=\bar{N}] \end{cases} \qquad (2)$$
(2)

$$(2) + ROH \rightleftharpoons ROCH_2CH_2C\equiv N + RO^- \qquad (3)$$

The catalyst is regenerated in equation 3 thus requiring only a small quantity to achieve a good yield of the cyanoethylated product. Also, this is a reversible reaction sequence. This reaction mechanism has been confirmed by kinetic investigations (1–2). The reaction rate is proportional to the concentrations of alkoxide ion and of acrylonitrile and increases with increasing basic strength of the alkoxide ion. Thus the rate of reaction decreases: $(CH_3)_2CHO^- > CH_3(CH_2)_2CH_2O^- > CH_3CH_2CH_2O^- > CH_3CH_2O^- > CH_3O^-$. Similarly, the reaction rate increases with the decrease in the dielectric constant in a mixed solvent system containing methanol and dioxane. The necessity of using a catalyst with sufficient basicity to remove the labile proton from the compound undergoing cyanoethylation is implied in the reaction mechanism. Studies show that the rate-determining step is equation 2, whereas equation 3 is very rapid. These results agree with the known behavior of anions. The intermediate anion (2) is capable of adding additional molecules of acrylonitrile as shown in equation 4.

$$(2) + CH=CHCN \rightleftharpoons ROCH_2CH\bar{C}NCH_2\bar{C}HCN \qquad (4)$$

This side reaction indicates a potential pitfall in the use of an excess of acrylonitrile. This product is obtained in the presence of a large excess of acrylonitrile, but essentially no polymer is formed under ordinary conditions.

The mechanism outlined above is believed to apply to all base-catalyzed cyanoethylations, such as those of alcohols, mercaptans, phosphines, and activated carbon compounds. In the latter case, carbanions of the types (3), (4), and (5) are formed with strong bases and then react in the normal way with acrylonitrile. This variation, shown in equation 5, is a specific case of the Michael addition of nucleophilic reagents to α,β-unsaturated compounds and illustrates the formal analogy of cyanoethylation to this well-known reaction.

$$H_5C_2OOC\bar{C}HCOOC_2H_5 \qquad CH_3-\overset{\overset{O}{\|}}{\underset{\underset{CH_3}{|}}{\bar{C}}}-C-H \qquad CH_3\bar{C}HNO_2$$
$$(3) \qquad\qquad (4) \qquad\qquad (5)$$

$$CH_2(CO_2C_2H_5)_2 + C_6H_5CH=CHCO_2C_2H_5 \xrightarrow{\text{NaOR}} \underset{\underset{CH(CO_2C_2H_5)_2}{|}}{C_6H_5CHCH_2CO_2C_2H_5} \qquad (5)$$

Strong nucleophiles such as amines require no catalyst. The reaction mechanism is believed to be as shown in equations 6 and 7.

$$RNH_2 + CH_2=CHCN \rightleftharpoons R\overset{+}{N}H_2CH_2\bar{C}HC\equiv N \leftrightarrow R\overset{+}{N}H_2CH_2CH=C=\bar{N} \qquad (6)$$
$$(6)$$
$$(6) \rightleftharpoons RNHCH_2CH_2CN \qquad (7)$$

Catalysts. The usual catalyst for cyanoethylation is a strong base, although acids are effective under certain circumstances. Useful catalysts include the hydroxides, oxides, alkoxides, cyanides, hydrides, and amides of the alkali metals, or the alkali metals themselves. Quaternary ammonium hydroxides, such as benzyltrimethylammonium hydroxide, are strongly basic and are advantageously used; these compounds are quite soluble in organic solvents. In general, 1–5% of catalyst, based on the weight of acrylonitrile, is satisfactory.

Because acrylonitrile may also be polymerized by the action of strong bases in a reaction that is vigorous and, at times, violent, it is recommended that contact between large quantities of acrylonitrile and substances likely to initiate polymerization be avoided.

Solvents. Cyanoethylations are frequently highly exothermic and require precautionary measures to control the reaction. This may be achieved in part by the use of inert solvents, such as benzene, toluene, pyridine, acetonitrile, or dioxane. Below 50°C, *tert*-butyl alcohol is a particularly good solvent because it dissolves appreciable amounts of potassium hydroxide. Cyanoethylation of *tert*-butyl alcohol at higher temperatures limits its utility.

An alternative is the use of the compound to be cyanoethylated as the solvent. This is feasible because the cyanoethylated product is typically a high boiling liquid or solid, thus permitting easy separation from the starting material.

Reaction Chemistry. *Reversibility.* Because cyanoethylation is reasonably vigorous under moderate conditions, the reaction is frequently carried out at room temperature. Another reason for avoiding elevated temperatures is that the reaction tends to reverse under these conditions. As a result, the yields are lower. Reaction reversibility can be readily demonstrated. For example, 3,3'-oxydipropanenitrile yields acrylonitrile and water upon being heated to about 200°C in the presence of an alkaline substance (3).

$$O(CH_2CH_2CN)_2 \rightleftharpoons 2\ CH_2\!\!=\!\!CHCN + H_2O \tag{8}$$

Similarly, dicyanoethylated amino acids form the monocyanoethylated substance upon being heated with a base (4). Trans-cyanoethylation may also take place (5–7). There is evidence, however, that at least some reactions of cyanoethylated compounds involve a substitution mechanism rather than elimination of acrylonitrile followed by cyanoethylation of the substrate (5).

In addition to avoiding high temperatures, care must be exercised during work-up and subsequent reaction sequences. For example, attempts to hydrolyze poly(vinyl cyanoethyl ether) or cyanoethyl cellulose with aqueous sodium hydroxide causes cleavage primarily to starting materials as seen in equation 9, rather than hydrolysis (8–10).

Although high concentrations of amine catalyst usually increase the yield, high concentrations of alkaline catalysts have a harmful effect.

$$ROCH_2CH_2CN \xrightarrow{NaOH} \underset{\text{major products}}{ROH + CH_2\!\!=\!\!CHCN} + \underset{\text{minor products}}{ROCH_2CH_2COONa} \tag{9}$$

On Carbon. For cyanoethylation on carbon, at least one hydrogen atom adjacent or alpha to one or more strong electron-withdrawing groups is required. If three hydrogen atoms are available, a mixture of mono-, di-, and tricyanoethylated products may result. Some examples are given in Table 1.

Several comments should be made relative to cyanoethylation activated by the carbonyl group. In the case of aldehydes, the aldol condensation is a competing reaction and, therefore, acetaldehyde may not be conveniently cyanoethylated. Methylene groups are more reactive than methyl groups. Thus methyl ethyl ketone undergoes cyanoethylation on the CH_2 group rather than on the CH_3 adjacent to the carbonyl.

In order to overcome this tendency toward multiple substitution, the so-called enamine procedure for cyanoethylating carbonyl compounds was introduced (13). In this procedure (eq. 10) a ketone, eg, cyclohexanone (**7**), reacts with a secondary

Table 1. Cyanoethylation on Carbon[a]

Compound	Name	Formula	Yield, %
isobutyraldehyde	4-methylpentanenitrile-4-carboxaldehyde	$NC(CH_2)_2C(CH_3)_2CHO$	85
acetone	4-acetyl-4-(2-cyanoethyl)-heptanedinitrile	$CH_3COC(CH_2CH_2CN)_3$	77
diethyl malonate	diethyl bis(cyanoethyl)-malonate	$(C_2H_5OOC)_2C(CH_2CH_2CN)_2$	83
1-nitropropane	4-nitrohexanenitrile	$CH_3CH_2CH(NO_2)CH_2CH_2CN$	80
2-methylquinoline	2-(3-cyanopropyl)quinoline[b]		46
fluorene	9,9-dicyanoethylfluorene		74
2-naphthol	1-cyanoethyl-2-naphthol		93

[a] Ref. 11, except where otherwise stated.
[b] Ref. 12.

amine, pyrrolidine (8), to form the enamine 1-pyrrolidinocyclohexene (9), which then undergoes cyanoethylation. The enamine acts as its own catalyst and neither aldol condensation, acrylonitrile polymerization, nor multiple substitution is a problem. Yield from the enamine to the cyanoethylated ketone is about 80%. Another difference in this procedure is demonstrated (eq. 11) with methylcyclohexanone (10) where substitution tends to occur at the least substituted position using the enamine and at the more substituted position using base catalysis.

A variation (14) of the enamine procedure, in which the ketone is premixed with catalytic amounts of primary amine and acetic acid, gives high yields of product. Although cyanoethylation normally takes place at the position α to the carbonyl, there is at least one report of cyanoethylation in the γ position (15). Another procedure for the monocyanoethylation of ketones has been patented (16). In this process, a pressure vessel containing the ketone, acrylonitrile, solvent (benzene), and an acid–salt catalyst is sealed and heated.

With hydrocarbons or similar material, cyanoethylation can be achieved where there is an acidic or easily removed proton caused by activation of multiple double bonds or aromatic systems such as cyclopentadiene (11), fulvene (11), and ferrocene (17). Other examples of cyanoethylation activated by the ketone group have been reported in the literature (14,18). Examples in which reaction occurs on a carbon site are provided by hydrogen cyanide (11), malononitrile (19), certain imine structures (11,20), benzyl phenyl sulfone (11), chloroform (21), and poly(acrylonitrile) (22).

On Oxygen. In the presence of a strongly basic catalyst, hydroxy compounds react readily with acrylonitrile to produce 3-alkoxypropionitriles. The reaction proceeds rapidly with primary alcohols, somewhat more slowly with secondary alcohols, and requires drastic conditions with tertiary alcohols.

$$CH_2=CHCN + ROH \rightleftharpoons ROCH_2CH_2CN \qquad (12)$$

Simple alcohols give fully cyanoethylated products in high yield; however, substances with a large number of OH groups, such as cellulose (qv) or poly(vinyl alcohol), lead to intermediate levels of substitution (see Vinyl polymers). With alcohol cyanoethylation, ion-exchange resins are frequently employed as catalysts (11). Phenols require temperatures of 90–150°C for cyanoethylation (11). Strong electronegative groups tend to hinder reaction. For example, a 70% yield was obtained in the cyanoethylation of phenol but only 10% yield in the case of *p*-chlorophenol (11). Higher yields with aromatic phenols containing electron-withdrawing substituents have been achieved recently (23–24). A special case is 2-naphthol, which may be cyanoethylated at the oxygen or at the carbon at the 1 position (25). Sulfinic acids cyanoethylate to give high yields of sulfonylpropanenitriles, but sulfonic and carboxylic acids reportedly do not react. Both formaldehyde and H_2O undergo biscyanoethylation (11). Oximes and hydroperoxides require strong bases for cyanoethylation. Examples of cyanoethylation on oxygen are given in Table 2.

On Nitrogen. As might be anticipated from their strong nucleophilic character, nitrogen compounds readily undergo cyanoethylation. However, since many nitrogen compounds have several active sites, selectivity of reaction to achieve the degree of substitution desired has presented problems. In some cases, reaction proceeds without any catalyst or again basic, acid, or metal catalysts may be required. Ammonia and simple aliphatic amines react without catalyst whereas *tert*-carbinamines require a catalyst (11,26).

As a rule, acyclic secondary amines react more slowly than primary amines. The size and branching of the substituent alkyl groups affect the rate of reaction and yield. Thus *n*-dipropylamine gives an 88% yield of cyanoethylation product, whereas diisopropylamine gives only a 12% yield (27). Cyclic secondary amines such as piperidine and morpholine are somewhat more reactive than acyclic secondary amines such as diethylamine (see Amines, cyclic). Use of a strong base has been recommended as the catalyst for cyanoethylation of certain secondary amines (28).

Table 2. Cyanoethylation on Oxygen[a]

Compound	Product Name	Product Formula	Yield, %
ethanol	3-ethoxypropanenitrile	$CH_3CH_2OCH_2CH_2CN$	95
trimethylsilylmethanol	3-(trimethylsilyl)methoxy-propanenitrile	$(CH_3)_3SiCH_2OCH_2CH_2CN$	79
methyl ethyl ketoxime	2-butanone, O-(2-cyanoethyl)oxime	$\begin{array}{c}CH_3\\ \diagdown\\ C=NOCH_2CH_2CN\\ \diagup\\ C_2H_5\end{array}$	70
tert-butyl hydroperoxide	tert-butyl 2-cyanoethyl peroxide	$(CH_3)_3COOCH_2CH_2CN$	
benzenesulfinic acid	3-phenylsulfonylpropane-nitrile	$C_6H_5SO_2CH_2CH_2CN$	97

[a] Ref. 11.

Aromatic amines react in the presence of acetic acid, sometimes with added copper(I) chloride. Copper(II) acetate favors monocyanoethylation and is effective even with sterically hindered ortho-substituted amines (29). The reaction of aromatic amines also proceeds in high yields and selectivity toward monocyanoethylation when carried out in the presence of a diethylamine salt (5,30).

Ion-exchange resins have been employed as catalysts for aromatic amines (31). Of interest also is a patent reporting cyanoethylation of aniline using aluminum chloride as a catalyst (32). Catalysts such as zinc chloride, zinc bromide, boron trifluoride, and boron trifluoride etherate promote the cyanoethylation of secondary aromatic amines (33). Strong basic catalysts have been used in the case of many heterocyclic amines and amino acids. The influence of nucleophilicity on the course of cyanoethylation has been studied with a number of uracil derivatives (34).

Ammonia reacts with up to three moles of acrylonitrile (11,35). The disubstituted product is readily obtained; however, a large excess of ammonia is needed to obtain monosubstitution (11,36).

Cyanoethylation with amines is reversible and care must be exercised in the distillation of products. Because the reaction is an equilibrium, yields are higher when an excess of one of the reactants is used (35).

Amides, imides, and lactams are readily cyanoethylated (11). Unsubstituted amides may react with one or two moles of acrylonitrile; N-alkyl amides react normally with one mole. N-methylformamide is an exception and does not react well in the presence of a basic catalyst. Higher homologues add easily to acrylonitrile. A number of aromatic and aliphatic sulfonamides can be cyanoethylated.

The hydrogen atoms of hydrazine are active in this reaction. In the case of alkyl substituted hydrazines (CH_3, $C_6H_5CH_2CH_2$, benzyl, and cyclohexyl), cyanoethylation occurs on the substituted nitrogen atom (37–38). This contrasts with the results obtained with phenylhydrazine (38). The amino hydrogen atom of hydroxylamine is more reactive than the hydroxyl hydrogen, and cyanoethylation yields 3-(hydroxylamine)-propanenitrile, $HONHCH_2CH_2CN$. Bis-substituted nitrogen derivatives were obtained when the amino group was attached to an alkyl group. The introduction of alkyl groups at the carbon α to the amino group progressively reduced reactivity so that only mono derivatives are produced (39). Examples of cyanoethylation on nitrogen are given in Table 3.

Table 3. Cyanoethylation on Nitrogen[a]

Compound	Product Name	Product Formula	Yield, %
ammonia	3-aminopropanenitrile	H$_2$NCH$_2$CH$_2$CN	1.7
	3,3'-iminodipropanenitrile	HN(CH$_2$CH$_2$CN)$_2$	88.5
	3,3',3''-nitrilotripropanenitrile	N(CH$_2$CH$_2$CN)$_3$	6.0
acetamide	N-(2-cyanoethyl)glycine	HOOCCH$_2$NHCH$_2$CH$_2$CN	59–70
p-toluenesulfonamide[b]	N,N-bis(2-cyanoethyl)-p-toluenesulfonamide	CH$_3$—C$_6$H$_4$—SO$_2$N(CH$_2$CH$_2$CN)$_2$	88
hydrazoic acid	3-azidopropanenitrile	N$_3$CH$_2$CH$_2$CN	17
hydrazine hydrate	3,3'-hydrazonopropanenitrile	H$_2$NN(CH$_2$CH$_2$CN)$_2$	80

[a] Ref. 11.
[b] Ref. 5, 40.

On Sulfur. Sulfhydryl compounds react like other compounds containing active hydrogen. The labile nature of the proton attached to sulfur and the strong nucleophilic character of mercaptans strongly favors the cyanoethylation of aliphatic mercaptans. Representative examples of cyanoethylation on sulfur are given in Table 4. In addition to the expected thiols, hydrogen sulfide, bisulfites, and thioacids undergo cyanoethylation. The strong reactivity of thiols can be seen by the fact that 2-aminobenzenethiol reacts only at the sulfhydryl group; 2-aminoethanethiol hydrochloride reacts solely at the thiol group, but under basic conditions both the thiol and amino groups are reactive (21). Similarly, cysteine also reacts preferentially at the –SH group (22). An exception to preferential reaction at the sulfur moiety is the case of 2-thiazoline-2-thiol which is cyanoethylated exclusively on the nitrogen atom (41). As with the corresponding aniline analogues, the cyanoethylation of benzenethiols provides a ready route to cyclization (42).

Table 4. Cyanoethylation on Sulfur[a]

Compound	Product Name	Product Formula	Yield, %
hydrogen sulfide	3,3'-thiodipropanenitrile	S(CH$_2$CH$_2$CN)$_2$	86–93
sodium bisulfite	sodium 2-cyanoethane sulfonate	NaO$_3$SCH$_2$CH$_2$CN	96
thioacetic acid	thioacetic acid, 2-cyanoethyl S-ester	CH$_3$COSCH$_2$CH$_2$CN	88
2-aminobenzenethiol	3-(2-aminophenylthio)propanenitrile	C$_6$H$_4$(NH$_2$)(SCH$_2$CH$_2$CN)	

[a] Ref. 11.

Miscellaneous. Compounds with phosphorus–hydrogen bonds, such as phosphines, phosphine oxides, phosphonates, and phosphinates, readily undergo cyanoethylation (11,43–45). Some of these compounds require no catalyst, but the presence of a strong base is usually preferred. White phosphorus undergoes addition to acrylonitrile under the influence of aqueous solutions of strong bases to yield tris(2-cyanoethyl)phosphine oxide (45). Stannanes, germanes, and arylarsines may be cyanoethylated in the absence of catalyst, but alkylarsines require a strong base.

The cyanoethylation of silanes has been investigated (see Table 5). Beta-cyanoethylation takes place exclusively when trimethylsilyldiethylamine, $(CH_3)_3SiN(C_2H_5)_2$, is the catalyst (47). Some examples of miscellaneous cyanoethylations are given in Table 5.

Cellulose

The cyanoethylation of cellulosic material has been the subject of continuing research. Although first mentioned in a German patent in 1938, this development was pioneered in the 1940s by MacGregor in England and by Houtz and Stallings in the United States. About 1953 a major effort was initiated to commercialize the cyanoethylation of cotton (qv). Particularly active in this work were the American Cyanamid Company, the Institute of Textile Technology, the Monsanto Company, and the Southern Regional Research Center of the USDA. Although these efforts were technically successful, there are no known producers of important quantities of cyanoethylated cotton today. Product considerations were related to heat resistance and rot resistance and the change in dye affinity caused by cyanoethylation. Present research on cyanoethylation is academic in character.

Some work has also been carried out with cotton linters and various wood pulps.

Table 5. Miscellaneous Cyanoethylations[a]

Compound	Product Name	Formula	Yield, %
phosphine[b]	2-cyanoethylphosphine	$H_2PCH_2CH_2CN$	6
	bis(2-cyanoethyl)phosphine	$HP(CH_2CH_2CN)_2$	56
	tris(2-cyanoethyl)phosphine	$P(CH_2CH_2CN)_3$	28
dimethyl phosphonate	dimethyl(2-cyanoethyl)phosphonate	$(CH_3O)_2P(O)CH_2CH_2CN$	86
di-*n*-octylphosphine oxide	3-(dioctylphosphoroso)propanenitrile	$(C_8H_{17})_2P(O)CH_2CH_2CN$	39
triphenylstannane	3-(triphenylstannyl)propanenitrile	$(C_6H_5)_3SnCH_2CH_2CN$	94
phenylarsine	3,3′-(phenylarsylene)dipropanenitrile	$C_6H_5As(CH_2CH_2CN)_2$	77
trichlorosilane[c]	2-(trichlorosilyl)propanenitrile	$Cl_3SiCH(CH_3)CN$	31
	3-(trichlorosilyl)propanenitrile	$Cl_3SiCH_2CH_2CN$	74
hydrochloric acid	3-chloropropanenitrile	$ClCH_2CH_2CN$	97

[a] Ref. 11, unless otherwise stated.
[b] Ref. 44.
[c] Ref. 46–47.

Regenerated cellulose is highly reactive and offers advantages in the preparation of products of high degree of substitution. The vegetable fibers, ie, flax, jute, manila, and sisal, react with greater difficulty (see Fibers, vegetable). Wood may be treated by the two-step batch process but lack of penetration has presented some problems (48).

Cyanoethylated cellulose derivatives (qv) include methyl cellulose (11), ethyl cellulose (11), hydroxyethyl cellulose (49–50), and sodium carboxymethyl cellulose (51).

Processes. Cellulose is cyanoethylated by three major processes. All are heterogeneous systems in which the cellulose is immersed in the liquid reagents. The choice depends largely on the characteristics desired of the product and on the equipment available. An important economic consideration in selecting a given process is its efficiency in utilizing acrylonitrile. Since most practical processes employ aqueous sodium hydroxide as the catalyst, significant amounts of acrylonitrile are consumed in side reactions to form primarily 3,3'-oxydipropanenitrile.

The degree of cyanoethylation is usually expressed as the percent nitrogen of the cellulose or by the degree of substitution (DS). Fully cyanoethylated cellulose has 13.2% N, which corresponds to a DS of 3.0. In general, cyanoethylated cotton has ca 3.5% N (DS = 0.5) and highly cyanoethylated cellulose ca 12% N (DS = 2.5).

Two-Step Batch Process. The two-step process (52–54) consists of first impregnating cellulose in a large volume of dilute sodium hydroxide, removing the excess alkaline solution and then immersing the alkali-pretreated cellulose in a large volume of acrylonitrile. Circulation of liquids through the cellulose is maintained throughout the reaction. The reaction is stopped by neutralization of the alkali. Unused acrylonitrile is then recovered by distillation (55). The cellulose thus obtained has a nitrogen content of 3.5%. The most notable disadvantages are slow reaction, high waste of acrylonitrile to by-products, and the need for a pretreatment (56) with liquid ammonia, ammonium hydroxide, or other swelling agents such as dimethyl sulfoxide (57–60). A two-step process utilizing simple block-and-tackle equipment has been described for use by nontechnical personnel, such as fishermen wanting to treat nets (61).

A modification of the two-step procedure controls the position of the cyanoethylated group on the fabric, minimizes use of reagents, and reduces side products (62). Both the caustic and acrylonitrile are applied by means of a kiss-roll of the type shown in Figure 1 (see Coating processes). Wet pickups are reduced to ca 15–20% with considerably less excess acrylonitrile present at the end of the reaction. Similarly, the cyanoethyl groups (nitrogen content 2–4%) are concentrated at the fabric surface so that the positive effects of cyanoethylation are maintained while minimizing the detrimental effects on fabric properties.

Attempts to eliminate water in cyanoethylation have been generally unsuccessful.

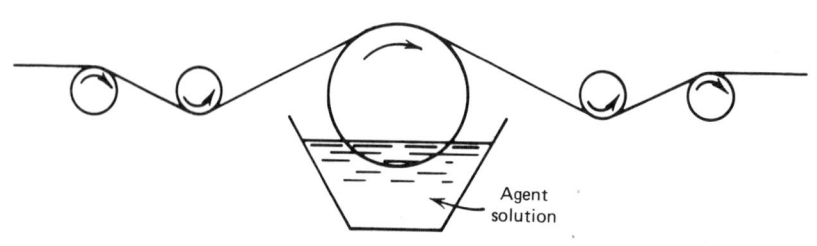

Figure 1. Kiss-roll application with a one-roll pad.

A possible problem occurring under anhydrous conditions is polyacrylonitrile formation or even some graft polymerization (63) (see Copolymers).

One-Step Batch Process. To avoid a pretreatment, cellulose may be treated (56,64) with only small amounts of dilute aqueous alkali together with the required amount of acrylonitrile. However, a dual-temperature procedure is needed. The acrylonitrile–aqueous alkali mixture is first allowed to impregnate the cellulose uniformly; then the temperature is raised to initiate reaction as in the two-step process. In this process, by-product formation is reduced (56) and yield is improved (65–66).

The one-step process is particularly well suited for the production of highly cyanoethylated cellulose, ie, products with sufficiently high substitution to render them soluble in organic solvents (67–68).

A variation of the one-step procedure used to cyanoethylate paper pulp (69) and cotton (70) consists of cyanoethylating the cellulosic material in essentially aqueous medium, ie, small amounts of acrylonitrile are used with large amounts of aqueous alkali. A further development of the aqueous procedure consists of carrying out the reaction at very high solids contents (71–73).

Continuous Process. A major advance was the continuous process for cyanoethylating cotton developed by Monsanto (74). It consisted of passing alkali-treated cotton into a reaction chamber where hot vapors of acrylonitrile are allowed to contact the fabric for short periods. Figure 2 shows a schematic diagram of this process. An important feature is the maintenance of low levels of water in the reaction chamber by azeotropic removal of excess water with acrylonitrile. Procedures for continuous treatment of yarn have also been described (75–76).

Cyanoethylated Products. The degree of substitution is usually considered to be the single most important characteristic affecting the properties of cellulose. Cyanoethylated products may be classified into those partially cyanoethylated, which still retain the morphology of cellulose (up to ca 6% N), and those with high levels of substitution, which are soluble in polar organic solvents (10–13% N) (77). Most work has been carried out with cyanoethylated cotton, particularly that in the 3.5% N range.

Cotton. Partially cyanoethylated cotton with a nitrogen content of $3.8 \pm 0.5\%$ is similar in appearance and processing characteristics to ordinary cotton, but differs primarily and significantly in other properties. The treated cotton is permanently rot- and mildew-proof. It may be buried in soil for extended periods of time and retain full strength. Cyanoethylated cotton or manila may also be immersed in water without rotting (78).

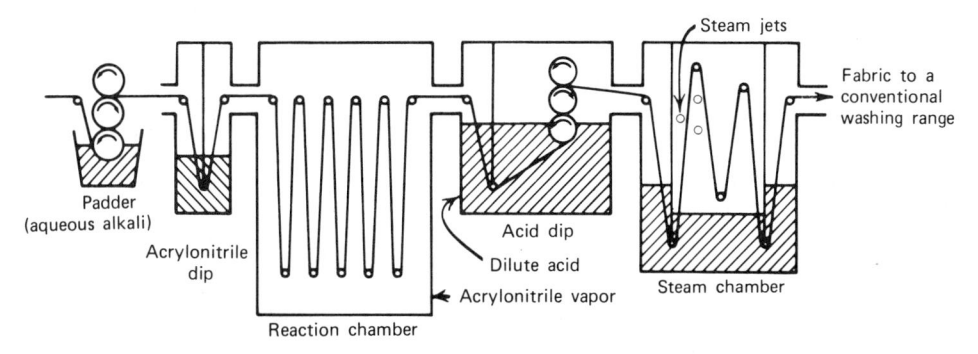

Figure 2. Continuous cyanoethylation of cotton fabric.

The degree of microbiological resistance increases with increasing nitrogen content and becomes essentially complete at about 2.8–3.5% N. However, the presence of relatively small amounts of carboxyethyl groups or nonuniformity of treatment may be detrimental (79).

Cyanoethylated cotton has increased resistance to degradation by either wet or dry heat (80). In general, the tensile strength of cyanoethylated fibers and yarns is increased to about the same extent as the increase in weight owing to reaction. For fabrics, decreases sometimes occur in tear strength. The crystallinity of cotton is retained essentially intact in partially cyanoethylated cotton (DS of 0.5) (54,70,81). At a DS of about 1.1, the crystalline structure gives way to an amorphous state and this process appears to be complete at a DS of about 2.0. This DS corresponds to the point where the cyanoethylated cellulose becomes soluble in organic solvents.

Alkaline treatment results not only in hydrolysis but also in decyanoethylation. Cyanoethyl cellulose is converted to carboxyethyl cellulose through the amide intermediate by a mild treatment with hydrogen peroxide. Acid treatment then produces the carboxylic derivative (82–83). The amide structure can also be cross-linked by treatment with formaldehyde (83) and the nitrile group can be reduced to the amine (84–85).

Graft polymerization may also be carried out on cyanoethylated cotton (86). Treatment with strong oxidizing agents (eg, chromic acid) causes the formation of water-soluble fibers (87–88). The unsubstituted hydroxyl groups of cyanoethyl cellulose may also react, eg, to form acetates (89).

Paper. Cyanoethylation to nitrogen contents of about 2.3% increases the dielectric constant of paper (69,71). Very low degrees of cyanoethylation (0.5–1.8% N) also improve the fold (90) and tear (91) strengths of handsheets made from cyanoethylated linters.

More highly cyanoethylated paper (8.7 to >10.0%) has good electrical characteristics and is useful as a condenser paper (92–93). Other advantages are improved heat (94) and rot resistance. Cyanoethylated paper may be sized with rosin and aluminum salts (95) and may be pressed into laminates that possess good resistance to voltage breakdown and ageing (96).

Highly Cyanoethylated Cellulose. Highly cyanoethylated cellulose (12–13% N) is available as a fluffy, amorphous, and thermoplastic solid. It is readily soluble in certain organic solvents (52) and concentrated aqueous solutions of zinc chloride and sodium thiocyanate. The outstanding characteristic of highly cyanoethylated cellulose is its unusually high dielectric constant coupled with a low dissipation factor. The dielectric constant is considerably higher than that of the common plastic materials. The material may also be molded. Highly cyanoethylated cellulose may be plasticized, eg, with dicyanoethyl phthalate (97).

Wood. Cyanoethylation preserves wood (98) and renders it resistant to rot (48) and to attack by termites (see Wood).

Other Macromolecular Substances

Starch and Gums. The basic procedure for cyanoethylation of starch (qv) (99–100) consists in dispersing the starch in dilute sodium hydroxide and stirring the mixture in the presence of acrylonitrile until the desired DS is obtained.

Ungelatinized starch is cyanoethylated by treating suspensions of starch granules

(101), preferably in the presence of a salt (102). Cyanoethyl starch containing up to about 6% N is water soluble (99–100,103). With increasing degrees of substitution, the product becomes soluble in organic solvents (99–100). Various galactomannan and glucomannan gums have been cyanoethylated by procedures similar to those used for starch (104) (see Gums).

Cyanoethylated starch is highly resistant to microbiological degradation. Unlike untreated starch, it does not form a complex with iodine.

Poly(vinyl Alcohol). Poly(vinyl alcohol) may be cyanoethylated by procedures similar to those employed for starch (85,99–100). Highly cyanoethylated poly(vinyl alcohol) is a yellow, tacky solid soluble in acrylonitrile, acetonitrile, dimethylformamide, and other highly polar organic liquids (67,105–106). When free from ionic contaminants, it has electrical properties similar to those of highly cyanoethylated cellulose (see Vinyl polymers).

Proteins. Proteins (qv) that form homogeneous dispersions in dilute aqueous alkali are readily cyanoethylated. Among these are casein, gluten, zein, soybean protein, and gelatin (107–109). A property imparted by cyanoethylation is high resistance to putrefaction. Cyanoethylated soybean protein may be used as a substrate for graft copolymerization with acrylonitrile and the product may be spun into textile fibers (110).

Wool. Cyanoethylation of wool (qv) may be carried out by first impregnating with dilute sodium hydroxide, and then immersing the pretreated wool in acrylonitrile. Provided that the concentration of base is very low (<0.4%), no significant amount of degradation occurs (111). Alternatively, the reaction may be carried out with vapors of acrylonitrile (112). The treated wool shows improved dyeability (111–112), as well as decreased shrinkage (112). Solutions of keratin may be stabilized for physicochemical studies by treatment with mercaptoethanol followed by cyanoethylation (113).

Miscellaneous Materials. Alkaline salts of lignin (qv) react with acrylonitrile (114). Polystyrene, polyacrylonitrile, and styrene–acrylonitrile copolymers have been reported to be cyanoethylated under anhydrous conditions (22). Polyamides have also been reported to react (115–116).

Uses

Because of the versatility of the reaction, cyanoethylated products have hundreds of uses and only a few will be mentioned. Cyanoethylation is important in organic synthesis, eg, in the preparation of amino acids, chemotherapeutic agents, dyes, blowing agents, and numerous intermediates (11,117).

Cyanoethylated cotton is particularly suited for use in fishing nets, ironing-board covers, sandbags, filter fabrics, enameling duck, and tobacco shade cloth (78) (see Textiles). Alkali-soluble derivatives of cyanoethylated cotton can be employed as backing fabric in the Schiffli process for making lace (118). Cyanoethylated manila is useful in marine ropes (78). The improved heat resistance and electrical properties of cyanoethylated paper have resulted in its use as insulation in certain transformers (94,119).

Highly cyanoethylated cellulose is the preferred material for embedding phosphors in electroluminescent devices (97) (see Luminescent materials). It may be plasticized with other cyanoethylated materials (49,67,97). Films may be used in capacitors (67) and as insulation (120).

382 CYANOETHYLATION

Cyanoethylated starch is an excellent detergent additive to prevent the redeposition of soil in laundering (103). It may also be used as a flocculating agent (qv) for ores (121), as a drilling-fluid additive (122) (see Petroleum) and as an emulsion stabilizer in paints (123). Cyanoethyl sucrose has been suggested as a dielectric fluid (124). Cyanoethylated proteins are useful as emulsion stabilizers in coating compositions (109) and as pigment binders in paper coatings (107). Cyanoethylated casein adhesives are more water resistant than unmodified casein. Cyanoethylated polyamides are curing agents for epoxy resins (qv) (116).

Cyanoethylated silanes may be converted to silicones with high polarity, low surface tension, and good hydrolytic stability (125) (see Silicon compounds).

Although it is a grafting reaction rather than a cyanoethylation, highly absorbent polymers have been made by grafting acrylonitrile to starch and following this with a hydrolysis reaction (126–128). These materials may gain wide utility (129).

As a facile, high yielding reaction, cyanoethylation provides a ready tool for the production of new chemicals and new materials. The extent to which these reactions and products gain utility in the marketplace depends upon relative costs and advantages. Because of this, new uses are frequently arising as older applications are being phased out.

BIBLIOGRAPHY

"Cyanoethylation" in *ECT* 1st ed., Suppl. 1, pp. 177–190, by Glen E. Journeay, Monsanto Chemical Company; "Cyanoethylation" in *ECT* 2nd ed., Vol. 6, pp. 634–668, by Norbert M. Bikales, Gaylord Associates, Inc.

1. Y. Ogata and M. Okano, *J. Am. Chem. Soc.* **78,** 5426 (1956).
2. B. A. Feit and A. Zilkha, *J. Org. Chem.* **28,** 406 (1963).
3. U.S. Pat. 2,832,798 (Apr. 29, 1958), L. Rapoport (to American Cyanamid Co.).
4. P. Buekus and N. Ragoutiene, *Zh. Obshch. Khim.* **34,** 593 (1964); *Chem. Abstr.* **60,** 13311 (1964).
5. J. Cymerman-Craig and co-workers, *J. Chem. Soc.,* 3628 (1955).
6. U.S. Pat. 2,842,541 (July 8, 1958), G. E. Journeay (to Monsanto Chemical Co.).
7. U.S. Pat. 2,853,510 (Sept. 23, 1958), A. E. Montagna and E. C. Stout (to Union Carbide Corp.).
8. J. F. Wright and L. M. Minsk, *J. Am. Chem. Soc.* **75,** 98 (1953).
9. G. C. Daul, R. M. Reinhardt, and J. D. Reid, *Text. Res. J.* **25,** 246 (1955).
10. L. W. Mazzeno and co-workers, *Text. Res. J.* **26,** 597 (1956).
11. *The Chemistry of Acrylonitrile,* 2nd ed., American Cyanamid Co., New York, 1959.
12. E. Lathwood and H. Suschitzky, *J. Chem. Soc.,* 2477 (1964).
13. G. Stork and co-workers, *J. Am. Chem. Soc.* **85,** 207 (1963).
14. Ger. Pat. 1,002,342 (Feb. 14, 1957), U.S. Pat. 2,850,519 (Sept. 2, 1958), H. Krimm (to Farbenfabriken Bayer, A.G.).
15. C. R. Engel and J. Lessard, *J. Am. Chem. Soc.* **85,** 638 (1963).
16. U.S. Pat. 3,150,142 (Sept. 22, 1964), C. J. Eby (to Monsanto Co.).
17. A. N. Nesmeyanov, *Synth. Inorg. Met. Org. Chem.* **1,** 279 (1971).
18. M. H. Elnagdi and M. Ohta, *Bull. Chem. Soc. Jpn.* **46,** 3818 (1973).
19. J. A. Adamcik and E. J. Miklasiewicz, *J. Org. Chem.* **28,** 336 (1963).
20. A. V. Gomez and J. Barluenga, *Tetrahedron Lett.,* (30), 2819 (1973).
21. R. J. Gaul, W. J. Fremuth, and M. N. O'Connor, *J. Org. Chem.* **25,** 725 (1960).
22. M. Freidman, J. F. Cavins, and J. S. Wall, *Abstr. 148th. Meeting Am. Chem. Soc., Sept. 1964,* p. 418.
23. U.S. Pat. 3,732,279 (May 8, 1973), Y. Suzuki (to Sumitomo Chemical Co., Ltd.).
24. J. Hirao and co-workers, *Mem. Kyushu Inst. Technol. Eng.* (3), 25 (1973).
25. J. Lichtenberger, J. Core, and R. Geyer, *Bull. Soc. Chim. Fr.,* 997 (1962).
26. K. M. Taylor and co-workers, *J. Am. Chem. Soc.* **81,** 5333 (1959).
27. F. C. Whitmore and co-workers, *J. Am. Chem. Soc.* **66,** 725 (1944).

28. J. Corse, J. T. Bryant, and H. A. Schonle, *J. Am. Chem. Soc.* **68,** 1905, 1911 (1946).
29. S. A. Heininger, *J. Org. Chem.* **22,** 1213 (1957).
30. J. Cymerman-Craig and M. Moyle, *Org. Synth.* **36,** 6 (1956).
31. A. Gauvreau and A. Lattes, *C. R. Acad. Sci. (Paris) Ser. C* **266,** 1162 (1968).
32. East Ger. Pat. 98,918 (July 12, 1973), E. Keller.
33. West Ger. Pat. 1,947,933 (Apr. 2, 1970), R. H Meen (to Eastman Kodak Co.).
34. A. Novacek and M. Lissnerova, *Coll. Czech. Chem. Commun.* **33,** 604 (1968).
35. F. C. Whitmore and co-workers, *J. Am. Chem. Soc.* **66,** 725 (1944).
36. E. M. Smolin and L. C. Beegle, *Ind. Eng. Chem.* **50,** 1115 (1958).
37. H. Dorn, A. Zubek, and G. Hilgetag, *Chem. Ber.* **98,** 3377 (1965).
38. H. Dorn and K. Walter, *Z. Chem.* **7**(4), 151 (1967).
39. J. A. Bell and C. Kenworthy, *Synthesis* (12), 650 (1971).
40. A. F. Bekhli, *Dokl. Akad. Nauk SSSR* **113,** 588 (1957).
41. R. J. Gaul, W. J. Fremuth, and M. N. O'Connor, *J. Org. Chem.* **26,** 5106 (1961).
42. S. M. Thakkar and J. R. Merchant, *Curr. Sci.* **45**(5), 178 (1976).
43. M. M. Rauhut and co-workers, *J. Am. Chem. Soc.* **81,** 1103 (1959).
44. M. M. Rauhut and co-workers, *J. Org. Chem.* **26,** 5138 (1961).
45. M. Grayson, *Chem. Eng. News* **40,** 90 (Dec. 3, 1962).
46. V. M. Udovin and A. D. Petrov, *Usp. Khim.* **31,** 793 (1962).
47. R. A. Pike and R. L. Schank, *J. Org. Chem.* **27,** 2190 (1962).
48. I. S. Goldstein and co-workers, *Ind. Eng. Chem.* **51,** 1313 (1959).
49. Belg. Pats. 619,830, 619,831 (Jan. 7, 1962), W. O. Fugate and E. C. McClenachan (to American Cyanamid Co.).
50. Jpn. Pat. 73 26,881 (Apr. 9, 1973), N. Kawamoto (to Fuji Chemical Co., Ltd.).
51. R. M. Reinhardt and co-workers, *Text. Res. J.* **29,** 802 (1959).
52. U.S. Pat. 2,375,847 (May 15, 1945), R. C. Houtz (to E. I. du Pont de Nemours & Co., Inc.).
53. U.S. Pat. 2,473,308 (June 14, 1949), J. W. Stallings (to Rohm and Haas Co.).
54. J. Compton and co-workers, *Text. Inds.* **117,** 138A (1953); U.S. Pat. 2,786,258 (Mar. 26, 1957), J. Compton and C. P. Jones (to Institute of Textile Technology).
55. U.S. Pat. 2,731,401 (Jan. 17, 1956), H. F. Karnes and A. H. Gruber (to American Cyanamid Co.).
56. A. H. Gruber and N. M. Bikales, *Text. Res. J.* **26,** 67 (1956).
57. H. Schleicher, B. Lukanoff, and B. Philipp, *Faserforsch. Textiltech.* **25**(5), 179 (1974).
58. H. Schleicher, C. Daniels, and B. Philipp, *J. Polym. Sci. Polym. Symp.*, **47,** 251 (1974).
59. A. Koura, H. El Saied, and M. Fadl, *Faserforsch. Textiltech.* **26**(2), 61 (1975).
60. Jpn. Pat. 65 22,397 (Oct. 4, 1965), S. Enomoto and T. Sano (to Kokoku Rayon and Pulp Co.).
61. U.S. Pats. 3,026,168 (Mar. 20, 1962), 3,046,775 (July 31, 1962), V. E. Wellman and N. M. Bikales (to American Cyanamid Co.).
62. J. Compton, W. H. Martin, and D. M. Gagarine, *Text. Res. J.* **40,** 813 (1970).
63. B. A. Feit and co-workers, *J. Appl. Polym. Sci.* **8,** 1869 (1964).
64. Can. Pat. 586,257 (Nov. 3, 1959), N. M. Bikales, A. H. Gruber, and E. L. Carpenter (to American Cyanamid Co.).
65. U.S. Pat. 2,907,625 (Oct. 6, 1959), N. M. Bikales, A. H. Gruber, and L. Rapoport (to American Cyanamid Co.).
66. J. Compton and co-workers, *Text. Res. J.* **25,** 58 (1955).
67. U.S. Pat. 3,067,141 (Dec. 4, 1962), N. M. Bikales and W. O. Fugate (to American Cyanamid Co.).
68. Can. Pat. 628,194 (Sept. 26, 1961), D. H. Hogle and T. W. Dakin (to Westinghouse Electric Corp.).
69. U.S. Pat. 2,535,690 (Dec. 26, 1950), H. F. Miller and R. G. Flowers (to General Electric Co.).
70. N. M. Bikales and L. Rapoport, *Text. Res. J.* **28,** 737 (1958); *Text. Rec.* **78** (936), 71 (1961).
71. J. L. Morton and N. M. Bikales, *Tappi* **42**(10), 855 (1959).
72. Can. Pat. 622,868 (June 27, 1961), E. D. McKay (to Southern Chemical Cotton Co., Inc.) and N. M. Bikales (to American Cyanamid Co.).
73. U.S. Pat. 2,904,386 (Sept. 15, 1959), D. M. Gagarine (to Deering Milliken Research Corp.).
74. U.S. Pat. 2,860,946 (Nov. 18, 1958), S. F. Belt (to Monsanto Chemical Co.).
75. U.S. Pat. 2,786,735 (Mar. 26, 1957), J. Compton and C. P. Jones (to Institute of Textile Technology).
76. H. J. Janssen and co-workers, *Ind. Eng. Chem.* **50,** 76 (1958).
77. N. M. Bikales, *Macromol. Synth.* **5,** 35 (1974).
78. J. Compton, *Text. Res. J.* **27,** 222 (1957).

79. F. Fordemwalt and R. E. Kourtz, *Text. Res. J.* **25,** 84 (1955).
80. *Cyanoethylation of Cotton,* a joint report by Institute of Textile Technology, American Cyanamid Co. and Monsanto Chemical Co., Sept. 1956.
81. C. M. Conrad and co-workers, *Text. Res. J.* **30,** 339 (1960); *J. Appl. Polym. Sci.* **5,** 163 (1963).
82. U.S. Pat. 2,820,691 (Jan. 21, 1958), J. R. Stephens and L. Rapoport (to American Cyanamid Co.).
83. M. Negishi and N. Aida, *Text. Res. J.* **29,** 982 (1959).
84. U.S. Pat. 3,122,526 (Feb. 25, 1964), W. H. Schuller and L. C. Beegle (to American Cyanamid Co.).
85. L. Alexandru, M. Opris, and A. Ciocanel, *J. Polym. Sci.* **59,** 129 (1962).
86. U.S. Pat. 3,157,460 (Nov. 17, 1964), J. C. Arthur, Jr., and R. J. Demint (to U.S. Department of Agriculture).
87. U.S. Pat. 2,724,632 (Nov. 22, 1955), H. Weisberg (to Londat Aetz Fabric Co.).
88. R. M. Reinhardt, J. D. Reid, and T. W. Fenner, *Ind. Eng. Chem.* **50,** 83 (1958).
89. U.S. Pat. 2,825,623 (Mar. 4, 1958), J. R. Stephens and L. Rapoport (to American Cyanamid Co.).
90. J. J. Spadaro and co-workers, *Tappi* **41,** 674 (1958).
91. J. A. Harpham, A. R. Reid, and H. W. Turner, *Tappi* **41,** 629 (1958).
92. Brit. Pat. 1,060,983 (Mar. 8, 1967), (to General Electric Company).
93. L. C. Flowers and D. Berg, *J. Electrochem. Soc.* **111,** 1239 (1964).
94. *Pap. Trade J.* **143**(42), 44 (1959); *Chem. Week* **85,** 61 (Aug. 15, 1959).
95. U.S. Pat. 2,794,736 (June 4, 1957), O. P. Cohen and J. F. Heaps (to Monsanto Chemical Co.).
96. U.S. Pat. 2,994,634 (Aug. 1, 1961), J. E. Jayne (to Kimberly-Clark Corp.).
97. U.S. Pats. 2,901,652 (Aug. 25, 1959), 2,918,594 (Dec. 22, 1959), E. G. Fridrich (to General Electric Co.).
98. G. Fuse, A. Endo, and K. Nishimoto, *Mokuzai Kenkyu* **27,** 15 (1962).
99. J. H. MacGregor and C. Pugh, *Proc. Int. Congr. Pure Appl. Chem. 11th, London, Eng., 1947* **V,** 123 (1953).
100. J. H. MacGregor, *J. Soc. Dyers Colour.* **67,** 66 (1951); J. H. MacGregor and C. Pugh, *J. Soc. Dyers Colour.* **67,** 74 (1951).
101. Can. Pat. 563,046 (Sept. 9, 1958), T. E. Sample (to Monsanto Chemical Co.).
102. U.S. Pat. 2,965,632 (Dec. 20, 1960), E. F. Paschall and W. H. Minkema (to Corn Products Co.).
103. Can. Pat. 581,696 (Aug. 18, 1959), E. A. Vitalis and F. L. Andrew (to American Cyanamid Co.).
104. U.S. Pat. 2,461,502 (Feb. 8, 1949), O. A. Moe (to General Mills, Inc.).
105. C. W. Lewis and D. H. Hogle, *J. Polym. Sci.* **21,** 411 (1956).
106. U.S. Pat. 2,341,553 (Feb. 15, 1944), R. C. Houtz (to E. I. du Pont de Nemours & Co., Inc.).
107. U.S. Pat. 2,562,534 (July 31, 1951), J. R. Coffman (to General Mills, Inc.).
108. U.S. Pat. 2,594,239 (Apr. 29, 1952), J. C. Cowan, C. D. Evans, and L. L. McKinney (to United States of America).
109. U.S. Pat. 2,775,565 (Dec. 25, 1956), L. L. McKinney, J. C. Cowan, and C. D. Evans (to United States of America).
110. Jpn. Pat. 64 90 (Jan. 11, 1964), S. Morimoto, A. Yamamoto, and K. Hamada (to Toyo Spinning Co.).
111. N. M. Bikales, J. J. Black, and L. Rapoport, *Text. Res. J.* **27,** 80 (1957); U.S. Pat. 2,890,925 (June 16, 1959), N. M. Bikales (to American Cyanamid Co.).
112. M. Oku and H. Ishibashi, *J. Text. Inst.* **51,** T637 (1960).
113. Y. Tomimatsu, J. J. Bartulovich, and W. H. Ward, *Text. Res. J.* **29,** 593 (1959).
114. U.S. Pat. 2,816,100 (Dec. 10, 1957), H. M. Walker (to Monsanto Chemical Co.).
115. Fr. Pat. 1,230,797 (Sept. 20, 1960), I. Minter and F. E. Gherghel (to Ministerul Industrei Petrolului si Chimiei).
116. U.S. Pat. 3,091,595 (May 28, 1963), T. F. Mika (to Shell Oil Co.).
117. U.S. Pats. 2,766,227, 2,766,229 (Oct. 9, 1956), W. B. Hardy and F. H. Adams (to American Cyanamid Co.).
118. R. W. Reinhardt, T. W. Fenner, and J. D. Reid, *Am. Dyestuff Rep.* **50**(19), 67 (Sept. 18, 1961).
119. U. Usmanov and co-workers, *Dokl. Akad. Nauk Uz. SSSR* **20**(3), 25 (1963).
120. Brit. Pat. 825,166 (Dec. 9, 1959), (to Westinghouse Electric Corp.).
121. U.S. Pat. 3,001,933 (Sept. 26, 1961), T. P. Malinowski (to Monsanto Chemical Co.).
122. U.S. Pat. 3,032,498 (May 1, 1962), H. M. Walker (to Monsanto Chemical Co.).
123. U.S. Pat. 2,881,143 (Apr. 7, 1959), A. R. Wilson (to Monsanto Chemical Co.).
124. *Cyanoethyl Sucrose, Technical Data Sheet X-132,* Eastman Chemical Products, Inc., Kingsport, Tenn., 1962.

125. *Chem. Eng. News* **40,** 35 (Aug. 27, 1962).
126. M. O. Weaver and co-workers, *J. Appl. Polym. Sci.* **15,** 3015 (1971).
127. L. A. Gugliemelli and co-workers, *J. Polym. Sci.* **12**(11), 2683 (1974).
128. U.S. Pat. 3,809,664 (May 7, 1974), G. F. Fanta and R. C. Burr (to U.S. Department of Agriculture).
129. G. F. Fanta, M. O. Weaver, and W. M. Doane, *Chem. Technol.,* **4**(11) 675 (Nov. 1974).

General References

C. H. Bamford and G. C. Eastmond, "Acrylonitrile Polymers" in N. M. Bikales, ed., *Encyclopedia of Polymer Science and Technology,* Vol. 1, Interscience Publishers, a division of John Wiley & Sons, Inc., New York, 1964.
H. A. Bruson, "Cyanoethylation" in R. Adams, ed., *Organic Reactions,* Vol. 5, John Wiley & Sons, Inc., New York, 1949.
The Chemistry of Acrylonitrile, 2nd ed., American Cyanamid Co., New York, 1959.

ROBERT J. HARPER, JR.
United States Department of Agriculture

CYANOHYDRINS

A cyanohydrin is an organic compound that contains both a cyanide and a hydroxy group on an aliphatic section of the molecule. The cyanohydrin category is usually comprised of α-hydroxy nitriles which are the products of addition of hydrogen cyanide to the carbonyl group of aldehydes and ketones. The IUPAC name for cyanohydrins is based on the α-hydroxy nitrile name. Common names of cyanohydrins are derived from the aldehyde or ketone from which they are formed (Table 1).

Table 1. Cyanohydrin Nomenclature

Name	CAS Registry No.	Formula	Synonyms
hydroxyacetonitrile	[107-16-4]	HOCH$_2$CN	formaldehyde cyanohydrin, glycolonitrile, hydroxyacetonitrile
2-hydroxypropanenitrile	[78-97-7]	HOCHCN|CH$_3$	acetaldehyde cyanohydrin, lactonitrile
2-hydroxy-2-methylpropanenitrile	[75-86-5]	CH$_3$C(OH)CN|CH$_3$	acetone cyanohydrin, α-hydroxybutyronitrile, 2-methyllactonitrile
α-hydroxybenzeneacetonitrile	[532-28-5]	C$_6$H$_5$CH(OH)CN	benzaldehyde cyanohydrin, mandelonitrile
3-hydroxypropanenitrile (a β-hydroxy nitrile)	[109-78-4]	HOCH$_2$CH$_2$CN	ethylene cyanohydrin, hydracrylonitrile

CYANOHYDRINS

The outstanding chemical property of cyanohydrins is their ready conversion to α-hydroxy acids and their derivatives, especially α-amino and unsaturated acids. All commercially important cyanohydrins occur as chemical intermediates. For this reason, data on production and prices of cyanohydrins are not usually published. The industrial significance of some cyanohydrins is waning as more direct and efficient routes to the desired products are developed. Today, acetone cyanohydrin is the world's most prominent industrial cyanohydrin as it is the only current route of methyl methacrylate manufacture.

Physical Properties

Cyanohydrins are usually colorless to straw yellow liquids with an objectionable odor akin to that of hydrogen cyanide. The lower molecular weight cyanohydrins can be distilled under reduced pressure provided the cyanohydrin is kept at a slightly acidic pH. Table 2 lists physical properties of some common cyanohydrins.

Chemical Properties

Cyanohydrins can react either at the nitrile group or at the hydroxy group. Hydrolysis of the nitrile group proceeds through the amide to the corresponding carboxylic

Table 2. Physical Properties of Some Cyanohydrins

	CAS Registry No.	Mol wt	Mp, °C	Bp, °C	Sp gr	η_D^{20}	Flash point, °C
formaldehyde cyanohydrin, (glycolonitrile)		57.05	<−72	119 at 3.2 kPa^a	1.104	1.4117	
acetaldehyde cyanohydrin, (lactonitrile)		71.08	−40	182–184, dec	0.988	1.405	77
acetone cyanohydrin, (methyllactonitrile)		85.11	−19	82 at 3 1 kPa^a	0.932	1.3992	63
cyclohexanone cyanohydrin,	[931-97-5]	121.17	29	109–113 at 1.2 kPa^a	1.032	1.4576	60
benzaldehyde cyanohydrin,		133.15	−10	170, dec	1.117	1.5315	
ethylene cyanohydrin, (hydracrylonitrile)		71.08	−46.2	228	1.059	1.4256	>112
propylene cyanohydrin, HOCH(CH$_3$)CH$_2$CN (2-methylhydracrylonitrile)	[2567-01-3]	85.11		116–118 at 2.7 kPa^a 207		1.4280	

^{a} To convert kPa to mm Hg, multiply by 7.5.

acid. Owing to the instability of cyanohydrins at high pH, this hydrolysis must be carried out with an acid catalyst. In cases where amide hydrolysis is slower than nitrile hydrolysis, the amide may be isolated.

Thus, acid hydrolysis of acetophenone cyanohydrin (R = C_6H_5, R' = CH_3) yields the corresponding amide which can be isolated. Further hydrolysis, usually with sodium hydroxide, gives atrolactic acid in a 30% overall yield (1).

The hydroxy analogue of methionine (2-hydroxy-4-methylthiobutyric acid) is manufactured by acid hydrolysis of 3-methylthiopropionaldehyde cyanohydrin [3268-49-3], which is produced by the reaction of methyl mercaptan with acrolein.

$$CH_3S\text{-}CH_2CH_2\text{-}CH(OH)\text{-}CN + H_2O \xrightarrow{H^+} CH_3S\text{-}CH_2CH_2\text{-}CH(OH)\text{-}COOH$$

The mixture of D and L optical forms of this hydroxy analogue of methionine is converted to the calcium salt which is used in animal feed supplements. Cyanohydrins react with ammonium carbonate to form hydantoins (2), which yield amino acids upon hydrolysis. Commercial DL-methionine is produced by hydrolysis of the hydantoin of 3-methylthiopropionaldehyde.

$$CH_3S\text{-}CH_2CH_2\text{-}CHO + HCN \xrightarrow{(NH_4)_2CO_3} \text{hydantoin} + H_2O \xrightarrow{OH^-} CH_3S\text{-}CH_2CH_2\text{-}CH(NH_2)\text{-}COOH$$

Reaction of cyanohydrins with absolute ethanol in the presence of HCl yields the ethyl esters of α-hydroxy acids (3). N-Substituted amides can be synthesized by heating a cyanohydrin and an amine in water. Thus formaldehyde cyanohydrin and β-hydroxyethylamine lead to N-(β-hydroxyethyl)hydroxyacetamide (4).

$$HOCH_2CN + HOCH_2CH_2NH_2 \xrightarrow{H_2O} HOCH_2CNHCH_2CH_2OH + NH_3$$

Catalytic hydrogenation of the nitrile function of cyanohydrins can give amines. As in the case of ordinary nitriles, catalytic reduction of cyanohydrins can yield a mixture of primary, secondary, and tertiary amines unless special steps are taken to minimize formation of the secondary and tertiary amines. Addition of acid or acetic anhydride to the reaction medium minimizes formation of secondary or tertiary amines through formation of the amine salt or acetamide derivative of the primary amine.

The hydroxy group of cyanohydrins is subject to displacement with other electronegative groups. Cyanohydrins react with ammonia to yield amino nitriles. This is a step in the Strecker synthesis of amino acids. A one step synthesis of α-amino acids involves treatment of cyanohydrins with ammonia and ammonium carbonate under pressure. Thus acetone cyanohydrin, when heated at 160°C with ammonia and ammonium carbonate for 6 h gives an 86% yield of α-aminoisobutyric acid (5). Primary and secondary amines can also be used to displace the hydroxyl group. Similarly, hydrazine reacts with two molecules of cyanohydrin to give the disubstituted hydrazine.

Cyanohydrins react with hydrogen halides or PCl_5 to give α-halo nitriles which can be further hydrolyzed to the α-halo carboxy acids. The α-hydroxy group of cyanohydrins can be esterified with an acid or acid chloride. Dehydration of cyanohydrins with phosphorus pentoxide gives >80% yields of alkylacrylonitriles (6).

High yields of optically active cyanohydrins have been prepared from hydrogen cyanide and carbonyl compounds using an enzyme as catalyst. Reduction of these optically active cyanohydrins with lithium aluminum hydride in ether affords the corresponding substituted, optically active ethanolamine (see Alkanolamines) (7). Addition of hydrogen cyanide to an aldose to form a cyanohydrin is the first step in the Kiliani-Fischer method for increasing the carbon chain of aldoses by one unit. Cyanohydrins react with Grignard reagents (see Grignard reaction) to give α-hydroxy ketones.

$$\underset{R'}{\overset{OH}{\underset{|}{R\overset{|}{C}CN}}} \xrightarrow{R''MgBr} \underset{R'}{\overset{OH}{\underset{|}{R\overset{|}{C}\overset{O}{\overset{\parallel}{C}}R''}}}$$

Acetone cyanohydrin and methyl isobutyl ketone cyanohydrin, dissolved in an organic solvent such as diethyl ether or methyl isobutyl ketone, undergo cyanide exchange with aqueous cyanide ion to yield a significant cyanide carbon isotope separation. The two-phase system yields cyanohydrin enriched in carbon-13 and aqueous cyanide depleted in carbon-13. The equilibrium

$$\underset{R'}{\overset{OH}{\underset{|}{R\overset{|}{C}CN_{(org)}}}} + {}^{13}CN^-_{(aq)} \rightleftharpoons \underset{R'}{\overset{OH}{\underset{|}{R\overset{|}{C}{}^{13}CN_{(org)}}}} + CN^-_{(aq)}$$

is obtained in seconds. Some nitrogen isotope separation is also observed (8).

Preparation

Cyanohydrins can be formed by (1) the acid- or base-catalyzed reaction of hydrogen cyanide with an aldehyde or ketone:

$$\overset{O}{\underset{}{\overset{\parallel}{RCR'}}} + HCN \xrightarrow{acid\ or\ base} \underset{R'}{\overset{OH}{\underset{|}{R\overset{|}{C}CN}}}$$

(2) the displacement of bisulfite ion by cyanide ion on the bisulfite addition compounds of aldehydes and ketones:

$$\underset{R'}{\overset{OH}{\underset{|}{R\overset{|}{C}SO_3^-}}} + CN^- \longrightarrow \underset{R'}{\overset{OH}{\underset{|}{R\overset{|}{C}CN}}} + SO_3^{2-}$$

or (3) the exchange of cyanide ion between a ketone cyanohydrin and an aldehyde to give the more stable (usually) aldehyde cyanohydrin:

$$\underset{R'}{\overset{OH}{\underset{|}{R\overset{|}{C}CN}}} + R''CHO \rightleftharpoons \overset{O}{\underset{}{\overset{\parallel}{RCR'}}} + \underset{H}{\overset{OH}{\underset{|}{R''\overset{|}{C}CN}}}$$

Direct combination of hydrogen cyanide and a carbonyl compound is the commercial, and most common route to cyanohydrins.

The base-catalyzed addition of hydrogen cyanide to carbonyl compounds was one of the first mechanistic studies in organic chemistry (9). Addition of the cyanide ion to the carbonyl group is the rate controlling step. The reaction is slightly subject to general acid catalysis (10–11), an observation common to many other nucleophilic additions to carbonyl groups. Cyanohydrin formation probably proceeds by attack of cyanide ion on a carbonyl group that is already hydrogen-bonded to a proton donor, ie, hydrogen cyanide (12).

$$\overset{-}{NC} + \underset{R'}{\overset{R}{\underset{|}{\overset{|}{C}}}}=O + HCN \rightleftarrows NC-\underset{R'}{\overset{R}{\underset{|}{\overset{|}{C}}}}-OH + \overset{-}{CN}$$

The reaction is exothermic and reversible. Stabilization of the product mixture with acid is necessary before the cyanohydrin can be isolated, usually by distillation at reduced pressure.

All aliphatic aldehydes and most ketones react to form cyanohydrins. Ketones are usually less reactive than aldehydes. This behavior has been attributed to a combination of electron-donating effects and increased steric hindrance of the second alkyl group in the ketones. The magnitude of the equilibrium constants for the addition of hydrogen cyanide to a carbonyl group is a measure of the stability of the cyanohydrin relative to the carbonyl compound plus hydrogen cyanide. This formation equilibrium constant (13–14) is expressed in units of M^{-1} (L/mol):

$$K = \frac{[\text{cyanohydrin}]}{[\text{HCN}][\text{carbonyl compound}]}$$

The large values of the equilibrium constants for the aldehydes listed in Table 3 indicate the ease of forming aldehyde cyanohydrins relative to the majority of the ketone cyanohydrins. The most sterically hindered ketones (diisopropyl ketone and isopropyl t-butyl ketone) do not form appreciable amounts of cyanohydrins. Aromatic aldehydes form cyanohydrins, but an excess of aromatic aldehyde in basic cyanide solution can lead to further reaction to give benzoin condensation products. Alkyl aryl ketones can form cyanohydrins to a limited extent although the equilibrium usually falls far on the side of the reactants, ketone plus hydrogen cyanide. Diaryl ketones do not form cyanohydrins at all. The large cyanohydrin-formation constants of some cyclohexanones listed in Table 3 are attributed to a reduction of steric strain of the six-membered ring in the cyanohydrin relative to that of the ketone (13). The stability of cyclohexanone cyanohydrin is such that treatment of the cyanohydrin with potassium hydroxide does not give HCN and ketone but yields instead the potassium salt of the cyanohydrin.

Production of cyanohydrins is accomplished through the base-catalyzed combination of hydrogen cyanide and the carbonyl compound in a solvent, usually the cyanohydrin itself (15). The reaction is carried out at high dilution of the feeds, at 10–15°C and a pH of 6.5–7.5. The product is continuously removed from the reaction zone, cooled to push the equilibrium toward cyanohydrin formation, and then stabilized with mineral acid. Purification is usually effected by distillation.

Table 3. Equilibrium Constants for Formation of Cyanohydrins from Hydrogen Cyanide Plus Carbonyl Compounds[a]

Carbonyl compound	K, M^{-1}	Reference
acetaldehyde	7100	13
propionaldehyde	476	14
n-butyraldehyde	1042	14
isobutyraldehyde	1042	14
acetone	28	13
methyl ethyl ketone	33	13
methyl isopropyl ketone	52	13
methyl t-butyl ketone	29	13
isopropyl t-butyl ketone	0.12	13
methyl cyclohexyl ketone	56	13
cyclobutanone	108	13
cyclopentanone	34	13
cyclohexanone	1700	13
2-methylcyclohexanone	930	13
2-t-butylcyclohexanone	7.6	13
4-t-butylcyclohexanone	1700	13
cycloheptanone	7.7	14
benzaldehyde	200	13
p-chlorobenzaldehyde	204	14
p-nitrobenzaldehyde	55	14
p-methoxybenzaldehyde	32	14
m-chlorobenzaldehyde	400	14
m-nitrobenzaldehyde	370	14
m-methoxybenzaldehyde	233	14
o-chlorobenzaldehyde	1000	14
o-nitrobenzaldehyde	1429	14
o-methoxybenzaldehyde	385	14
acetophenone	0.8	14
ethyl phenyl ketone	1.7	14
n-propyl phenyl ketone	1.1	14
isopropyl phenyl ketone	4.0	14
isobutyl phenyl ketone	0.6	14
t-butyl phenyl ketone	11.1	14

[a] Equilibrium constants measured at 20–25°C, except isopropyl t-butyl ketone which was measured at 35°C.

Shipping, Storage, and Handling

Acetone cyanohydrin and lactonitrile are usually manufactured and used at the same site without shipping. When shipped, steel drums, carboys, tank cars, and barges are used (16).

Cyanohydrins should be stabilized with acid to pH 3–4 to prevent decomposition to hydrogen cyanide and carbonyl compound (17). Formaldehyde cyanohydrin should be considered an explosion and fire risk, particularly if impure. In general, cyanohydrins are combustible liquids and many decompose upon heating. They should be stored in a cool dry place, preferably outside and separated from other storage. Containers should be protected against physical damage (16,18).

Health and Safety

Cyanohydrins are highly toxic by inhalation or ingestion, and moderately toxic through skin absorption (19). All α-hydroxy nitriles are potential sources of hydrogen cyanide or cyanides and must be handled with considerable caution. Contact with the skin and inhalation should be rigorously avoided. Special protective clothing should be worn and any exposure should be avoided (16,18). The area should be adequately ventilated. Immediate medical attention is essential in case of cyanohydrin poisoning.

Uses

Cyanohydrins are used primarily as intermediates in the production of other chemicals. Manufacture of methyl methacrylate, used in molding powders and clear sheet (eg, Plexiglas), from acetone cyanohydrin is the most economically important cyanohydrin process (see Methacrylic acid). Cyanohydrins are also used as solvents in applications including fiber-spinning and metals refining. Cyanohydrins and their derivatives reportedly act as anti-knock agents in fuel oil and motor fuels and serve as electrolytes in electrolytic capacitors. Lactonitrile and acetone cyanohydrin are claimed to be effective as avicides.

Specific Compounds

Formaldehyde Cyanohydrin. Formaldehyde cyanohydrin (Toxic Substances List (19), MC 75250), also known as glycolonitrile, is a colorless liquid with a cyanide odor. It is soluble in water, alcohol, and diethyl ether. Equimolar amounts of 37% formaldehyde and aqueous hydrogen cyanide mixed with a sodium hydroxide catalyst at 2°C for one hour give formaldehyde cyanohydrin in 79.5% yield (20).

Although usually handled as an aqueous solution, formaldehyde cyanohydrin can be isolated in the anhydrous form by ether extraction, followed by drying and vacuum distillation (21). Pure formaldehyde cyanohydrin tends to be unstable, especially at high pH. Small amounts of phosphoric acid or monochloroacetic acid are usually added as a stabilizer. Monochloroacetic acid is especially suited to this purpose because it co-distills with formaldehyde cyanohydrin (22). Properly purified formaldehyde cyanohydrin is reported to have excellent stability (23).

Direct reaction of formaldehyde cyanohydrin and ethylenediamine in the presence of a sulfuric acid catalyst gives ethylenediaminetetraacetonitrile, hydrolysis of which leads to ethylenediaminetetraacetic acid (EDTA), a widely used sequestering agent (24).

$$HOCH_2CN + H_2NCH_2CH_2NH_2 \rightarrow (NCCH_2)_2NCH_2CH_2N(CH_2CN)_2$$

$$\xrightarrow{H_2O} (HO_2CCH_2)_2NCH_2CH_2N(CH_2CO_2H)_2$$

Nitrilotriacetonitrile, $N(CH_2CN)_3$, a precursor to nitrilotriacetic acid NTA, $N(CH_2CO_2H)_3$, can be prepared from the reaction of formaldehyde cyanohydrin with ammonia (24). Commercial formaldehyde cyanohydrin is available as a 70% aqueous solution stabilized by phosphoric acid. The March 1979 price was 88¢/kg.

392 CYANOHYDRINS

Acetaldehyde Cyanohydrin. Acetaldehyde cyanohydrin (Toxic Substances List (19), OD 82250), commonly known as lactonitrile, is soluble in water and alcohol, but insoluble in diethyl ether and carbon disulfide. Lactonitrile is used chiefly to manufacture lactic acid and its derivatives, primarily ethyl lactate. Lactonitrile is manufactured from equimolar amounts of acetaldehyde and hydrogen cyanide containing 1.5% of 20% NaOH at −10 to 20°C. The product is stabilized with sulfuric acid (25). Sulfuric acid hydrolyzes the nitrile to give a mixture of lactic acid and ammonium bisulfate. This mixture can be purified by adding methanol to form

$$CH_3\overset{OH}{\underset{|}{C}}HCN + 2\ H_2O \xrightarrow{H_2SO_4} CH_3\overset{OH}{\underset{|}{C}}HCOOH + NH_4HSO_4 \xrightarrow{CH_3OH} CH_3\overset{OH}{\underset{|}{C}}COOCH_3$$

methyl lactate which is separated from the ammonium bisulfate. The methyl lactate is distilled, then hydrolyzed back to the aqueous acid. Removal of most of the water yields 90% lactic acid (26).

Acetone Cyanohydrin. Acetone cyanohydrin (Toxic Substances List (19), OD 92750), also known as 2-methyllactonitrile and α-hydroxyisobutyronitrile, is very soluble in water, diethyl ether, and alcohol but only slightly soluble in carbon disulfide or petroleum ether. Acetone cyanohydrin is the most important commercial cyanohydrin. As of 1978 acetone cyanohydrin offered the only commercial route to methacrylic acid and its derivatives, principally methyl methacrylate (see Methacrylic acid and derivatives).

Acetone cyanohydrin is used as a raw material for insecticide manufacture and also to produce ethyl α-hydroxyisobutyrate, a pharmaceutical intermediate. It has been used as a complexing agent for metals refining and separation. Acetone cyanohydrin complexes can be used to separate Ni^{2+}, Cu^{2+}, Hg^{2+}, Zn^{2+}, Cd^{2+}, or Fe^{2+} from Mg^{2+}, Ba^{2+}, Ca^{2+}, Na^+, or K^+ on strongly basic ion-exchange resins (27). Acetone cyanohydrin is also used as a reagent in the formation of aldehyde cyanohydrins from aldehydes and in combination with a KCN–crown ether complex (see Chelating agents), it acts as an effective, stereoselective hydrocyanating reagent (28).

Reaction of hydrazine with acetone cyanohydrin gives the disubstituted

$$CH_3\overset{OH}{\underset{\underset{CH_3}{|}}{\overset{|}{C}}}CN + H_2NNH_2 \longrightarrow CH_3\overset{CN}{\underset{\underset{CH_3}{|}}{\overset{|}{C}}}-NHNH-\overset{CN}{\underset{\underset{CH_3}{|}}{\overset{|}{C}}}CH_3 \xrightarrow[H_2O]{Cl_2} CH_3\overset{CN}{\underset{\underset{CH_3}{|}}{\overset{|}{C}}}-N=N-\overset{CN}{\underset{\underset{CH_3}{|}}{\overset{|}{C}}}CH_3$$

$$\ A\phantom{CH_3 \xrightarrow[H_2O]{Cl_2} CH_3C-N=N-}\ AIBN$$

hydrazine A, which upon oxidation with chlorine water gives 2,2′-azobisisobutyronitrile (AIBN), a stable, colorless, crystalline material at room temperature. When heated to 120°C, AIBN decomposes to form nitrogen and two 2-cyanoisopropyl radicals. The ease with which AIBN forms radicals, and the fact that the rate of formation does not vary much in various solvents has resulted in wide use of AIBN as a free radical initiator (see Initiators). AIBN is used commercially as a catalyst for vinyl polymerization.

Acetone cyanohydrin is manufactured by the direct reaction of hydrogen cyanide with acetone catalyzed by base, generally in a continuous process (Fig. 1). Acetone and hydrogen cyanide are fed continuously with a small amount of sodium hydroxide catalyst to the stirred acetone cyanohydrin generator. Since the reaction is exothermic, acetone cyanohydrin is also used as the reaction solvent to aid in dissipating the heat of reaction.

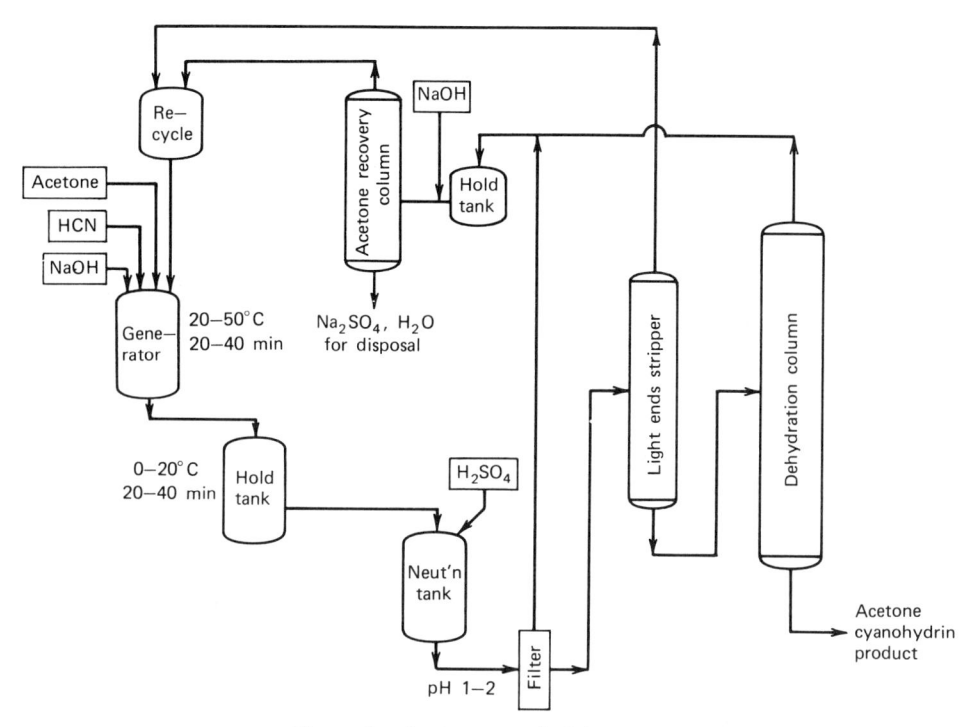

Figure 1. Acetone cyanohydrin process.

$$(CH_3)_2CO + HCN \xrightarrow{\text{base}} (CH_3)_2C(OH)CN$$

To push the equilibrium toward cyanohydrin formation, the reaction mixture is chilled first in the generator, then in the hold tank. The crude acetone cyanohydrin is neutralized with sulfuric acid to a pH of 1–2 (17). The acidification causes precipitation of the sodium catalyst as sodium sulfate salts which are then filtered. The crude acetone cyanohydrin is fed to a light-ends stripping column where the hydrogen cyanide, acetone, and some of the water are removed overhead and recycled to the acetone cyanohydrin generator. The concentrated acetone cyanohydrin bottoms are then sent to a dehydration column where the remainder of the water is removed under vacuum. The anhydrous acetone cyanohydrin is removed as the bottoms product and used directly in methacrylate manufacture.

Single-pass conversions of acetone cyanohydrin are 90–95% depending on the residence times and temperatures in the generator and hold tank. Overall yields of product from acetone and hydrogen cyanide can be ≥95%. There are no significant by-products of the reaction other than the sodium salts produced by neutralization of the catalyst. However, conversion of acetone cyanohydrin to methyl methacrylate does produce a large amount of ammonium bisulfate by-product which lacks ready marketability and is usually converted to sulfuric acid for reuse in the conversion of acetone cyanohydrin to methacrylates. The nitrogen values of the ammonium bisulfate are lost through the recycle process. The necessity for the ammonium bisulfate recycle plants has created increased interest in routes to methacrylates that do not co-produce ammonium bisulfate (29).

394 CYANOHYDRINS

The hydrogen cyanide used in the process is sometimes produced on-site through catalytic conversion of an ammonia–methane–air mixture directly to hydrogen cyanide (the Andrussow process). However, by-product hydrogen cyanide from acrylonitrile (qv) manufacture is increasingly being used in acetone cyanohydrin production. Although sodium hydroxide was used as the catalyst in the above example, any of a number of other basic catalysts would suffice. Other catalysts include KOH, K_2CO_3 (17), ion-exchange resins (30), and acetic acid–sodium acetate buffers (31).

The acetone cyanohydrin is used for methacrylate manufacture. Sulfuric acid is added in greater than an equimolar amount, and by thermal cracking, the acetone cyanohydrin is converted to methacrylamide sulfate. The methacrylamide sulfate

$$CH_3\underset{CH_3}{\overset{OH}{C}}CN + H_2SO_4 \xrightarrow{heat} CH_2=C\underset{CH_3}{\overset{CNH_2 \cdot H_2SO_4}{\|}}$$

is esterified with methanol to give methyl methacrylate and ammonium bisulfate as a by-product. The basic process for manufacture of acetone cyanohydrin was

$$CH_2=C\underset{CH_3}{\overset{CNH_2 \cdot H_2SO_4}{\|}} + CH_3OH \longrightarrow CH_2=C\underset{CH_3}{\overset{COCH_3}{\|}} + NH_4HSO_4$$

developed in the 1930s by Imperial Chemical Industries and has been improved over the years by the producing companies.

Benzaldehyde Cyanohydrin. Benzaldehyde cyanohydrin (Toxic Substances List (19), OO 84000), also known as mandelonitrile, is a yellow oily liquid, insoluble in water, but soluble in alcohol and diethyl ether. Mandelonitrile is a component of the glycoside amygdalin, a precursor of laetrile found in the leaves and seeds of most *Prunus* species (plum, peach, apricot, etc).

amygdalin [29883-15-6] $\xrightarrow[(2)[O]]{(1) H_2O, H^+}$ Laetrile [1332-94-1] $\xrightarrow{H_2O, H^+}$ mandelonitrile [532-28-5]

Mandelonitrile was the first cyanohydrin to be synthesized (in 1832). It is commercially prepared from benzaldehyde and hydrogen cyanide.

Ethylene Cyanohydrin. Ethylene cyanohydrin (Toxic Substances List (19), MU 52500), also known as hydracrylonitrile and glycocyanohydrin, is a straw-colored liquid miscible with water, acetone, methyl ethyl ketone, and ethanol, and insoluble in benzene, carbon disulfide, and carbon tetrachloride. Ethylene cyanohydrin differs from the other cyanohydrins discussed here in that it is a β-cyanohydrin. It is formed by the reaction of ethylene oxide with hydrogen cyanide. Like the formation of the α-cyanohydrins, this reaction is

$$\text{(ethylene oxide)} + \text{HCN} \xrightarrow{\text{NaCN}} \text{HOCH}_2\text{CH}_2\text{CN}$$

catalyzed by bases or cyanide ion, but unlike the α-cyanohydrin case this reaction is not reversible, and under certain conditions it can proceed with violence. Ethylene cyanohydrin can also be prepared by the reaction of ethylene chlorohydrin and alkali cyanides (32) (see Chlorohydrins).

The first U.S. plant for acrylonitrile manufacture used an ethylene cyanohydrin feedstock. This was the primary route for acrylonitrile manufacture until the acetylene-based process began to replace it in 1953 (33) (see Acrylonitrile). Maximum use of ethylene cyanohydrin to produce acrylonitrile occurred in 1963. Acrylonitrile has not been produced by this route since 1970.

The first commercial process for manufacture of acrylic acid (qv) and acrylates involved hydrolysis of ethylene cyanohydrin in aqueous sulfuric acid.

$$\text{HOCH}_2\text{CH}_2\text{CN} \xrightarrow{\text{H}_2\text{SO}_4} \text{CH}_2{=}\text{CHCO}_2\text{H} + \text{NH}_4\text{HSO}_4$$

This route is no longer commercially significant.

BIBLIOGRAPHY

"Cyanohydrins" in *ECT* 1st ed., Vol. 4, pp. 755–759, by R. H. F. Manske, Dominican Rubber Co. Ltd. and H. S. Davis, American Cyanamid Co.; "Cyanohydrins" in *ECT* 2nd ed., Vol. 6, pp. 668–675, by Arnold P. Lurie, Eastman Kodak Company.

1. E. L. Eliel and J. P. Freeman in N. Rabjohn, ed., *Organic Syntheses,* Coll. Vol. IV, John Wiley & Sons, Inc., New York, 1963, p. 58.
2. E. C. Wagner and M. Baizer in E. C. Horning, ed., *Organic Syntheses,* Coll. Vol. III, John Wiley & Sons, Inc., New York, 1955, p. 323.
3. V. P. Belikov and co-workers, *Izv. Akad. Nauk SSSR Ser. Khim.* **8,** 1862 (1967).
4. U.S. Pat. 3,190,916 (June 22, 1965), N. B. Rainer (to Coastal Interchemical Co.).
5. Neth. Appl. 6,607,754 (Dec. 5, 1966), (to Rohm and Haas G.m.b.H.).
6. S. I. Mekhtiev and R. G. Mamedov, *Azerb. Khim. Zh.* **2,** 110 (1971); *Chem. Abstr.* **76,** 58924g (1972).
7. W. Becker, H. Freund, and E. Pfeil, *Angew. Chem. Int. Ed. Engl.* 4(12), 1079 (1965).
8. L. L. Brown and J. S. Drury, *J. Inorg. Nucl. Chem.* **35,** 2897 (1973); U.S. Pat. 3,607,010 (Sept. 21, 1971), L. L. Brown (to U.S. Atomic Energy Commission).
9. A. Lapworth, *J. Chem. Soc.,* 995 (1903); A. Lapworth and R. H. F. Manske, *J. Chem. Soc.,* 2533 (1928); 1976 (1930).
10. W. J. Svirbely and J. F. Roth, *J. Am. Chem. Soc.* **75,** 3106 (1953).
11. J. Hine, *Physical Organic Chemistry,* McGraw-Hill Book Co., Inc., New York, 1956, p. 251.
12. H. H. Hustedt and E. Pfeil, *Justus Liebigs Ann. Chem.* **640,** 15 (1961).
13. J. Hine, *Structural Effects on Equilibria in Organic Chemistry,* John Wiley & Sons, Inc., New York, 1975, p. 259, and references cited therein.

14. V. Migrdichian, *The Chemistry of Organic Cyanogen Compounds,* Reinhold Publishing Co., New York, 1947, Chapt. 9.
15. Ger. Pat. 1,087,122 (Aug. 18, 1960), K. Sennewald and co-workers (to Knapsack-Grieshein Akt. Ges.).
16. *Fire Protection Guide on Hazardous Materials,* 6th ed., National Fire Protection Association, 1975.
17. U.S. Pat. 2,537,814 (Jan. 9, 1957), H. S. Davis (to American Cyanamid Co.).
18. N. I. Sax, ed., *Dangerous Properties of Industrial Materials,* 4th Ed., Van Nostrand Reinhold Co., New York, 1975; *Toxic and Hazardous Industrial Chemicals Safety Manual,* The International Technical Information Institute, Tokyo, 1976.
19. H. E. Christensen, ed., *Registry of Toxic Effects of Chemical Substances,* U.S. Dept. of Health, Education and Welfare, Rockville, Md., 1976.
20. U.S. Pat. 2,890,238 (June 9, 1959), A. R. Sexton (to Dow Chemical).
21. R. Gaudry in E. C. Horning, ed., *Organic Syntheses,* Coll. Vol. III, John Wiley & Sons, Inc., New York, 1955, p. 436.
22. U.S. Pat. 2,623,896 (Dec. 30, 1952), H. Beier (to Rohm and Haas, G.m.b.H.).
23. U.S. Pat. 3,057,903 (Oct. 9, 1962), J. W. Nemec and C. H. McKeever (to Rohm and Haas Co.).
24. U.S. Pat. 2,855,428 (Oct. 7, 1958), J. J. Singer and M. Weisberg (to Hampshire Chemical Co.).
25. Jpn. Pat. 68 29,576 (Dec. 18, 1968), S. Yamaguchi and co-workers (to Mitsubishi Chem. Ind. Co., Ltd.).
26. *Hydrocarbon Process.,* 156 (Nov. 1975); Jpn. Pat. 65 2333 (Feb. 6, 1965), T. Kodama and co-workers (to Toyama Chem. Ind. Co. Ltd.).
27. L. Legradi, *J. Chromatogr.* **102,** 319 (1974).
28. C. L. Liotta, A. M. Dabdoub, and L. H. Zalkow, *Tetrahedron Lett.* **13,** 1117 (1977).
29. For an example of an alternative route see: Y. Oda and co-workers, *Hydrocarbon Process.,* 115 (Oct. 1975).
30. M. Borreal and J. Modiano, *Chim. Ind. Paris* **78,** 632 (1957).
31. U.S. Pat. 2,748,154 (May 29, 1956), G. E. Journeay (to Monsanto Chemical Co.).
32. E. C. Kendall and B. McKenzie in H. Gilman, ed., *Organic Syntheses,* Coll. Vol. I, John Wiley & Sons, Inc., New York, 1932, p. 256.
33. M. Sittig, *Acrylonitrile,* Noyes Development Corp., Park Ridge, N.J., 1965, p. 56.

MICHAEL S. CHOLOD
Rohm and Haas Company

CYANURIC AND ISOCYANURIC ACIDS

Although cyanuric acid [108-80-5] has been known for two hundred years, it did not achieve commercial importance until the mid 1950s and even then mainly via its derivatives. N-chlorination of cyanuric acid produces products that have gained wide acceptance as dry bleaches and sanitizers for use in dishwasher formulations, institutional and industrial cleansers and in swimming pool disinfection. The triallyl- and tris(2-hydroxyethyl) derivatives are employed as additives to impart special properties to polyester resins. Tris(2,3-epoxypropyl) isocyanurate is used in weather-resistant powder coatings (qv).

Nomenclature is based on two of the several possible equilibrating tautomeric species, ie, the trioxo and the trihydroxy forms. The trihydroxy form is variously designated cyanuric acid, s-triazine-2,4,6-triol or 2,4,6-trihydroxy-s-triazine. The trioxo structure, or s-triazine-2,4,6($1H,3H,5H$)-trione is the basis for the isocyanuric acid nomenclature. The triazine-based system is preferred in the scientific literature and is used by government regulatory agencies; however, the less cumbersome (iso)cyanurate designations persist in commerce.

The literature is replete with reports, sometimes conflicting, on which equilibrium form predominates in the crystalline state and which in solution; current evidence suggests that the trioxo species prevails in both. A review of the complex reasoning for this conclusion based upon infrared, Raman and ultraviolet spectroscopic, and x-ray crystallographic data can be found in refs. 1–3. Through common usage, both forms are collectively called cyanuric acid (CA). In alkaline solution, the anion formed is that of the hydroxy tautomer.

isocyanuric acid   cyanuric acid (CA)

Physical Properties

(Iso)cyanuric acid is a white, odorless, crystalline solid that does not melt up to 330°C; at higher temperatures isocyanic acid (HNCO) has been entrapped as a decomposition product and indeed has been used for further reaction (4). (Iso)cyanuric acid can be envisioned as a trimer of isocyanic acid, HNCO. (Iso)cyanuric acid is isolable in dihydrate form from aqueous solution as colorless monoclinic prisms; these effloresce in dry air to give the anhydrous compound (5).

(Iso)cyanuric acid is titratable as a weak acid, either potentiometrically or with phenolphthalein indicator using 0.1 N sodium hydroxide; there is general agreement that at about 25°C the first dissociation constant lies near 1×10^{-7} (6). The pH of a saturated aqueous solution at room temperature is approximately 4.5. Smolin and Rapoport in their review on (iso)cyanuric acid (7) list numerous other physical and thermal constants including: heat of combustion, 918.4 kJ/mol (219.5 kcal/mol);

specific heat, 0.327; heat of formation, −690.8 kJ/mol (−165.1 kcal/mol); heat of vaporization, 160.2 kJ/mol (38.3 kcal/mol); and heat of neutralization (NaOH), 1st hydrogen, 28.2 kJ/mol (6.74 kcal/mol), 2nd hydrogen, 17.2 kJ/mol (4.12 kcal/mol). Spectroscopic data are available in refs. 1–3 (and refs. therein). Proton magnetic resonance spectroscopy is of limited usefulness because of CA's symmetry, exchangeable protons and generally low solubility. Nuclear quadrupole resonance measurements of ^{14}N in (iso)cyanuric acid have been reported (8).

Solubilization of (iso)cyanuric acid is a frequently encountered problem in developing applications for the material, eg, water solubility at 25°C is only 0.2% (wt/wt), and at 90°C 2.6% (wt/wt). It is less than 0.1% soluble (wt/wt) at room temperature in most common organic solvents such as acetone, benzene, diethyl ether, ethanol, hexane, and isopropyl alcohol. See Table 1 for those solvents in which CA has significant solubility.

Solubility is frequently achieved through salt formation, eg, appreciable quantities of CA dissolve in aqueous bases such as solutions of sodium- or potassium hydroxide. Chemical reaction to form a new compound will also solubilize (iso)cyanuric acid as in the case of dissolution of CA in aqueous formaldehyde or in ethylene oxide at elevated temperature (see below under Chemical Properties).

Chemical Properties

The chemistry of (iso)cyanuric acid is best interpreted in terms of its cyclic triimide structure. Virtually all of the reactions discussed below find their counterparts in straightforward imide chemistry, eg, salt formation, hydrolysis, N-halogenation, and alkylation. See ref. 9 for an excellent review of cyclic imide chemistry.

Many mono-, numerous di-, and some tri-salts of cyanuric acid have been reported in the literature. Inorganic cations include: ammonium, barium, cadmium, calcium, cobalt, copper, mixed copper–sodium, lead, magnesium, manganese, mercury, nickel, potassium, silver, mixed silver–lead, mixed silver–potassium, sodium, and zinc. Organic salts reported are those of: caffeine, cinchonine, guanidine, quinine, quinoline, strychnine, and trimethylamine. Descriptions of these and other (iso)cyanurate salts and literature references appear in the Smolin and Rapoport review (7). None of these salts has achieved commercial significance.

The unusually stable isocyanurate ring is slowly hydrolyzed by hot aqueous alkaline solution. It is much less susceptible to acidic hydrolysis; in fact, CA can be heated under pressure in sulfuric acid solution at 200°C with minimal decomposition. This stability toward acidic hydrolysis is utilized in acidic digestion of the aminotriazine impurities in crude (iso)cyanuric acid to yield purified CA (see below).

Table 1. Solubility of (Iso)Cyanuric Acid at 25°C

Solvent	Solubility (wt/wt) %
N,N-dimethylformamide	7.2
dimethyl sulfoxide	17.4
N,N-dimethylacetamide	3.0
N-methyl-2-pyrrolidinone	6.3
pyridine	2.2
sulfuric acid (96%)	14.1

The chemical reactivity of (iso)cyanuric acid is frequently limited by its low solubility in common organic solvents. As in general imide chemistry, reaction of (iso)cyanuric acid normally occurs by replacement of hydrogens at nitrogen; this, of course, results in derivatives of the isocyanurate series. Reaction at oxygen produces cyanurate compounds:

isocyanurates   cyanurates

Mixed derivatives are possible, as in the case of the sodium salt of dichloroisocyanuric acid, sodium dichloroisocyanurate or sodium 1,3-dichloro-s-triazine-2,4,6(1H,3H,5H)-trione [2893-78-9], where the chlorines are on nitrogen and sodium is at the oxygen anion.

The N-chloro derivatives of isocyanuric acid (chlorinated s-triazinetriones) are the most important commercial products derived from (iso)cyanuric acid. Trichloroisocyanuric acid (TCCA) or 1,3,5-trichloro-s-triazine-2,4,6(1H,3H,5H)-trione [87-90-1]:

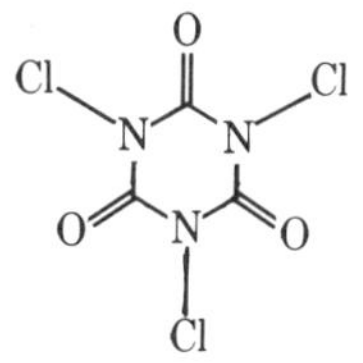

TCCA

is obtained in 90% yield by chlorination of a 3:1 mole ratio mixture of NaOH to (iso)cyanuric acid in aqueous solution (10).

$$3\ NaOH + CA + 3\ Cl_2 \rightarrow TCCA + 3\ NaCl + 3\ H_2O$$

Chlorine monoxide can also be used (11).

$$2\ CA + 3\ Cl_2O \rightarrow 2\ TCCA + 3\ H_2O$$

Dichloroisocyanuric acid (DCCA) or 1,3-dichloro-s-triazine-2,4,6(1H,3H,5H)-trione [2782-57-2] is the product obtained also in 90% yield when a 2:1 mole ratio of sodium hydroxide to CA is used during chlorination (10).

$$2\ NaOH + CA + 2\ Cl_2 \rightarrow DCCA + 2\ NaCl + 2\ H_2O$$

DCCA behaves like a normal acid and salts are readily prepared by neutralization with one equivalent of the appropriate base in aqueous medium (12). Only the sodium [2893-78-9] and potassium [2244-21-5] salts of dichloroisocyanuric acid or sodium [2893-78-9] and potassium [2244-21-5] 1,3-dichloro-s-triazine-2,4,6(1H,3H,5H)-trione have achieved commercial significance, although numerous other metal salts appear in the literature. Sodium dichloroisocyanurate (NaDCC) forms a dihydrate (NaDCC.2H$_2$O) or sodium 1,3-dichloro-s-triazine-2,4,6(1H,3H,5H)-trione dihydrate [51580-86-0] (13) and both the anhydrous salt and the dihydrate are items of commerce.

NaDCC

A complex composed of four molecules of potassium dichloroisocyanurate (KDCC) and one molecule of TCCA [30622-37-8] is produced commercially. It can be prepared, by the reaction of TCCA in acetone with aqueous KDCC (14).

Some analogous N-bromoisocyanurates have been synthesized; these include tribromoisocyanuric acid [17497-85-7], dibromoisocyanuric acid [15114-43-9], and its potassium [15114-46-2] and sodium salts [15114-34-8] (15).

Although numerous mono-, di-, and trisubstituted organic derivatives of cyanuric and isocyanuric acids appear in the literature, many are not accessible via (iso)cyanuric acid. Cyanuric chloride [108-77-0], 2,4,6-trichloro-s-triazine, is generally employed as the intermediate to most of the cyanurates. Trisubstituted isocyanurates can also be produced by trimerization of either aliphatic or aromatic isocyanates with appropriate catalysts (16) (see Isocyanates, organic). Direct synthesis of isocyanic acid and its trimerization has now been demonstrated as well (see under Manufacture).

Alkylation of (iso)cyanuric acid generally produces trisubstituted isocyanurates even when a conscious attempt is made to produce mono- or disubstituted derivatives. There are exceptions, as in the production of mono-2-aminoethyl isocyanurate [18503-66-7] in nearly quantitative yield by reaction of CA and aziridine in dimethylformamide (17).

Table 2 contains some specific examples of isocyanurate derivatives synthesized directly from (iso)cyanuric acid. As mentioned previously, virtually all of the derivatives are produced by reactions that are analogous to straightforward cyclic imide chemistry; some of these are discussed below. Most of the chemistry is readily interpreted in terms of the isocyanurate nitrogen (frequently as the anion) engaging in nucleophilic attack at a positively polarized carbon of the second reactant.

CA and ethylene oxide react nearly quantitatively at 100°C to form tris(2-hydroxyethyl) isocyanurate (THEIC). Substitution of propylene oxide yields the hydroxypropyl analogue (18–19). Alternatively, THEIC can be prepared by reaction of CA and 2-chloroethanol in aqueous caustic (20). THEIC obviously may react further to form esters, ethers, urethanes, phosphites, etc (30).

Epichlorohydrin engages in the typical CA–epoxide reaction in alkaline dioxane to form tris(3-chloro-2-hydroxypropyl) isocyanurate (see Chlorohydrins). This material can be converted into the commercial product, tris(2,3-epoxypropyl) isocyanurate (triglycidyl isocyanurate) by dehydrohalogenation with alkali (21).

Table 2. Trisubstituted Isocyanuric Acid Derivatives

Compound	CAS Registry Number	Substituent group R	Refs.
tris(2-hydroxyethyl) isocyanurate	[839-90-7]	—CH$_2$CH$_2$OH	18–20
tris(3-chloro-2-hydroxypropyl) isocyanurate	[7423-53-2]	—CH$_2$CHOHCH$_2$Cl	21
tris(2,3-epoxypropyl) isocyanurate	[2451-62-9]	—CH$_2$CH—CH$_2$ \ O /	21
triallyl isocyanurate	[1025-15-6]	—CH$_2$CH=CH$_2$	19, 22
tribenzyl isocyanurate	[606-03-1]	—CH$_2$C$_6$H$_5$	22
trimethallyl isocyanurate	[6291-95-8]	—CH$_2$C(CH$_3$)=CH$_2$	22
tris(2-ketopropyl) isocyanurate	[61050-97-3]	—CH$_2$COCH$_3$	23
tris(carbethoxymethyl) isocyanurate	[69455-18-1]	—CH$_2$CO$_2$C$_2$H$_5$	19, 23
tris(carboxymethyl) isocyanurate	[1968-52-1]	—CH$_2$CO$_2$H	23–24
tris(carbamoylmethyl) isocyanurate	[1843-48-7]	—CH$_2$CONH$_2$	24
tri-n-hexyl isocyanurate	[36761-61-2]	—(CH$_2$)$_5$CH$_3$	25
trimethyl isocyanurate	[827-16-7]	—CH$_3$	7, 23, 26–27
tris(hydroxymethyl) isocyanurate	[10471-40-6]	—CH$_2$OH	28
tris(2-cyanoethyl) isocyanurate	[2904-28-1]	—CH$_2$CH$_2$CN	19
tris(2-carbethoxyethyl) isocyanurate	[2904-39-4]	—CH$_2$CH$_2$CO$_2$C$_2$H$_5$	19
tris(triphenylstannyl) isocyanurate	[752-74-9]	—Sn(C$_6$H$_5$)$_3$	29

In a reaction analogous to the nucleophilic attack of the CA anion on 2-chloroethanol, the commercial product triallyl isocyanurate (TAIC) is formed in 82% yield by action of cyanuric acid upon allyl chloride at 130°C in o-dichlorobenzene with triethylamine as proton acceptor. Similarly, by following the same general procedure but with substitution of the appropriate organo halide, the following trisubstituted isocyanurates can be synthesized: tribenzyl isocyanurate (92% yield); trimethallyl isocyanurate (89% yield) (19,22); tris(2-ketopropyl) isocyanurate (78% yield); and tris(carbethoxymethyl) isocyanurate (94% yield). The last compound upon acidic hydrolysis has been reported to yield tris(carboxymethyl) isocyanurate although ring rupture occurs under alkaline conditions (19,23). A mixed isocyanurate having both hydroxyethyl and allyl substituents is obtainable by the simultaneous action of 2-chloroethanol and allyl chloride upon an alkaline solution of CA (31).

Other compounds prepared from reactions of CA and organic halo compounds at elevated temperatures for 16–18 h include the production of tris(carbamoylmethyl) isocyanurate from 2-chloroacetamide in 80% yield (24) and tri-n-hexyl isocyanurate from n-hexyl chloride in 71% yield (25).

Trimethyl isocyanurate can be synthesized by the reaction of (iso)cyanuric acid with dimethyl sulfate in alkaline medium in 60% yield (23) or with diazomethane (26); and by thermal rearrangement of trimethyl cyanurate [877-89-4] during several hours of reflux, in essentially quantitative yield. Transformation of alkyl cyanurates to the corresponding isocyanurates is frequently observed (7,27).

(Iso)cyanuric acid readily dissolves in aqueous formaldehyde with formation of tris(hydroxymethyl) isocyanurate (THMIC); this is isolated by evaporation of the water. THMIC in turn reacts with: acetic anhydride to yield tris(acetoxymethyl) isocyanurate [54635-07-3]; either thionyl chloride or phosphorus pentachloride to give tris(chloromethyl) isocyanurate [63579-00-0]; and with phenyl isocyanate in pyridine to yield tris(N-phenylcarbamoxymethyl) isocyanurate [21253-39-4], the last in 87% yield (28).

(Iso)cyanuric acid will add across the activated double bond of acrylonitrile in alkaline N,N-dimethylformamide solution at 130°C. Both the tris(2-cyanoethyl) and bis(2-cyanoethyl) isocyanurates [2904-27-0] can be isolated; the trisubstituted derivative is obtained by precipitation, whereas the disubstituted compound remains in solution. Each readily undergoes acidic hydrolysis to form the corresponding tris- and bis(2-carboxyethyl) isocyanurate [2904-40-7]s. Tris(2-carboxyethyl) isocyanurate [2904-41-8] is easily converted to the ethyl ester, tris(2-carbethoxyethyl) isocyanurate (19). Interestingly, hydrogenation of the tris(cyanoethyl) derivative in the presence of ammonia produces the bis- [69455-19-2] and mono-3-aminopropyl [69455-20-5] isocyanurates (19).

An organometallic compound, tris(triphenylstannyl)isocyanurate, is obtained by heating CA and triphenyltin hydroxide for one hour at 160–180°C under reduced pressure (29).

Ketene in acetone reacts with (iso)cyanuric acid under alkaline catalysis to give a triacetyl derivative in high yield; the product is reported to be triacetyl cyanurate [13483-16-4] rather than triacetyl isocyanurate (32). However, the literature does describe the preparation of the N,N',N"-triacetyl isocyanurate [35843-57-3] by the reaction of N,N',N"-tris(tributylstannyl)isocyanurate [752-58-9] and acetyl chloride (33).

(Iso)cyanuric acid and ammonia react under pressure and catalysis at 350–400°C to produce melamine [108-78-1] (see under Manufacture). At this temperature, (iso)-cyanuric acid decomposes and isocyanic acid may be an intermediate (34).

$$CA + 3\ NH_3 \rightarrow C_3N_3(NH_2)_3 + 3\ H_2O$$

Another example of reaction at carbon is the conversion of CA into cyanuric chloride [108-77-0], $C_3N_3Cl_3$, by phosphorus pentachloride (35).

Cyanuric chloride as an imidoyl chloride engages in chemical reactions analogous to those of an acid chloride and it is unlike the various active chlorine N-chloroisocyanurates. Cyanuric chloride is a convenient intermediate to alkyl or aryl cyanurates by reaction with alcohols or phenols, or to substituted melamines by reaction with

$$CA\ +\ 3\ PCl_5 \longrightarrow \underset{\text{(cyanuric chloride)}}{\text{C}_3\text{N}_3\text{Cl}_3}\ +\ 3\ POCl_3\ +\ 3\ HCl$$

amines; alkaline conditions are employed in both cases and yields are generally high. For a review of cyanuric chloride chemistry, see ref. 36.

$$\text{(Cl}_3\text{-triazine)} + 3\,ROH + 3\,NaOH \rightarrow \text{(OR}_3\text{-triazine)} + 3\,NaCl + 3\,H_2O$$

$$\text{(Cl}_3\text{-triazine)} + 3\,RNH_2 + 3\,NaOH \rightarrow \text{(NHR}_3\text{-triazine)} + 3\,NaCl + 3\,H_2O$$

Cyanuric chloride reacts with sodium sulfide to form trithio(iso)cyanuric acid [638-16-4] (7).

Triazine ring destruction is still another mode of chemical reaction available for (iso)cyanuric acid. Boron nitride is produced when a mixture of CA and boric acid is heated above 200°C (37). Heating (iso)cyanuric acid and high boiling fatty acids at 250°C for less than three hours gives high yields of the corresponding nitriles (38). Metal cyanates are obtained by treating CA with alkali metal carbonates at 350–550°C in an atmosphere of carbon dioxide (39); as has been discussed, isocyanic acid is a thermal decomposition product of (iso)cyanuric acid (4).

At elevated temperatures (ca 200°C) CA and alkylene oxides react in inert solvent to give N-hydroxyalkyloxazolidones in approximately 70% yield (40).

Carbohydrazide is produced in 71% yield by refluxing a mixture of (iso)cyanuric acid and hydrazine hydrate for 17 h (41).

Manufacture

Commercial production of (iso)cyanuric acid involves pyrolysis of urea. Although a number of processing options are available, historically and, in fact currently, pyrolysis of solid urea in a kiln is virtually the exclusive method of large-volume manufacture. The solid urea is heated at 200–300°C for several hours and (iso)cyanuric acid is produced according to the following equation.

$$3\,H_2NCONH_2 \xrightarrow{200-300°C} CA + 3\,NH_3$$

The equation is deceptively simple and does not, in fact, reveal the entire situation since it ignores intermediates and impurities. During the pyrolysis urea initially forms a free-flowing melt which subsequently thickens and finally solidifies. Components of the pyrolyzate include urea, biuret ($H_2NCONHCONH_2$) and presumably triuret ($H_2NCONHCONHCONH_2$), all intermediates to (iso)cyanuric acid. Also present are ammelide [645-93-2], some ammeline [645-92-1] and minor amounts of melamine as aminotriazine impurities. Concentrations of these vary according to the stage of pyrolysis. Since the reactivity of triuret precludes its isolation under kiln conditions, it would react as rapidly as it forms, triuret is presumed to be an intermediate. At the later stages of pyrolysis, heat transfer becomes poor and care must be exercised to prevent the mass from being heated above 300°C since yields are lowered by decomposition of the product.

Another problem occurring during pyrolysis is adhesion of the reaction products to the walls of the kiln. The crude cyanuric acid so produced generally contains 20–30% of impurities, consisting of mainly ammelide and ammeline and minor amounts of melamine, biuret, urea, and triuret. Purified CA (>98% purity) is produced by digesting the entire body of crude product in 15–20% sulfuric acid; this hydrolyzes the acyclic impurities to carbon dioxide and ammonia and the aminotriazines to (iso)cyanuric acid.

$$\text{ammelide} + \text{ammeline} + \text{melamine} \xrightarrow[H_2O]{H^+} CA + NH_3$$

Further processing entails filtering or centrifuging the slurry, washing the residual sulfuric acid from the solids, and drying the product in any conventional dryer at temperatures up to 200°C. The cyanuric acid is then ready for sale; however, the majority of the CA produced is converted to N-chloro- or N-alkylated isocyanurates.

The ammonia by-product from digestion is trapped as ammonium bisulfate; this can represent a significant water pollution burden and recent patents concerning the manufacture of CA seek alternatives to aminotriazine formation and subsequent hydrolysis. It should be noted that gaseous ammonia by-product from the pyrolysis of urea can be recovered or burned and thus does not represent an exceptional pollution problem. Other pollution constraints associated with isocyanurate manufacture appear in the production of the N-chlorinated derivatives. Biocidal active chlorine in waste streams must be destroyed by hydrogen peroxide or other suitable reagents, and triazine compounds must be reduced in the effluent since they represent a nitrogen pollution load. This is generally achieved by dechlorination and acidification to yield (iso)cyanuric acid which, because of its low solubility, can be removed by filtration and recycled.

Some procedures designed to overcome one or more of the various problems associated with the kiln pyrolysis of solid urea are: (1) urea pyrolysis over molten tin or alloys (42); (2) recycling 60–90% of the crude CA into the urea feed prior to pyrolysis (43); (3) pyrolysis of solid urea under vacuum (44); (4) reaction of urea in high boiling solvents such as sulfolane or N-cyclohexyl-2-pyrrolidinone, usually under vacuum or with a low boiling solvent added to strip ammonia from the reactor (45); (5) prior formation of urea cyanurate [69455-21-6] and subsequent pyrolysis (46); (6) pyrolysis of urea cyanurate in a fluidized bed (47); (7) pretreatment of urea at 165°C for approximately 2 h to yield a liquid mixture of primarily urea, biuret, CA and triuret with final pyrolysis on a rotating, heated drum at 240–270°C for 10 min (48); and (8) pyrolysis of urea salts, eg, urea hydrochloride, urea sulfate (49).

The use of vacuum or acid salts in several of the above processes is designed to remove or affix ammonia during the pyrolysis since it has been shown that aminotriazine impurity formation requires the presence of ammonia (44–45). The solvent process under vacuum and in combination with wiped-film evaporation–purification promises significant processing advantages. The reactor produces an aminotriazine-free mixture of (iso)cyanuric acid, urea, and biuret; urea and biuret are volatile under vacuum wiped-film still conditions and codistill with the solvent for recycling.

In addition to pyrolysis of urea, cyanuric acid may be produced in the laboratory by: (1) hydrolysis of cyanuric chloride, or (2) acid digestion of aminotriazines, eg, melamine (50). Cyanuric chloride can be prepared by trimerization of cyanogen chloride, ClCN, which is the product of reaction between chlorine and hydrogen cyanide (36). Neither of these methods appear commercially practical primarily because of the expense of the starting materials. However, the conversion of by-product or waste cyanuric chloride or aminotriazines to CA on a break-even or near break-even basis may be economic as a pollution control measure. Additional laboratory preparations appear in ref. 7. Several options for the purification of (iso)cyanuric acid appear in the patent literature; these include dissolution in aqueous ammonia, in aqueous formaldehyde or in hot N,N-dimethylformamide followed by filtration to remove most of the impurities. The CA is then recoverable by acidifying the ammonium hydroxide solution (51), or cooling the N,N-dimethylformamide solution with further precipitation of CA by addition of carbon tetrachloride (52). Sodium hydroxide addition will precipitate monosodium cyanurate from the formaldehyde solution (53). Direct synthesis of isocyanic acid, HNCO, from H_2, CO, and NO in 60–75% yields over palladium or iridium catalysts at 280–450°C has also been disclosed (54). Trimerization gives (iso)cyanuric acid and ammonium cyanate, NH_4OCN, is a by-product.

Economic Aspects

FMC Corporation and the Monsanto Company are the two largest bulk suppliers of (iso)cyanuric acid and the N-chloroisocyanurates in the United States. These two manufacturers have a combined CA production capability estimated at 38,000 metric tons per year, 90% of which is converted into the chlorinated derivatives. Also, in 1977 the Olin Corporation was test marketing chloroisocyanurates and announced construction of a commercial plant (55). Both tris(2-hydroxyethyl) isocyanurate and triallyl isocyanurate have been available in commercial quantities from Allied Chemical Corporation; since the mid 1970s Toyomenka (America) Inc. has supplied THEIC.

Non-United States producers of CA include: Chlor-Chem Ltd. (England), APC (France), Delsa (Spain), Shikoku Kasei Kogyo Co. and Nissan Chemical Industries Ltd. (Japan). The world capacity for chlorinated isocyanurates in 1977 was estimated to be 50,000 metric tons per year (55).

During the five year period 1972–1977, the United States detergent market for isocyanurates remained relatively constant at about 9000 t/yr, whereas the United States swimming pool market for these products grew rapidly to 20,500 t in 1977. The 1977 isocyanurate swimming pool sales represented 23% of the total chlorine demand for residential pool treatment in that year and 5% of the commercial pool chlorine demand (55). Future United States growth over the decade 1978–1988 is expected to be 4–5%/yr in detergents and over 10%/yr in pools. More detailed market information is available in ref. 55. The 1979 bulk quantity price for both (iso)cyanuric acid and its chlorinated derivatives was approximately $2.29–2.62/kg.

Specifications and Standards

(Iso)cyanuric acid is generally sold in a coarse granulation, minimum of ca 85% 0.14–2 mm (−10 to 100 mesh). A typical analysis of commercial CA can be found in Table 3.

Table 3. Analysis of (Iso)Cyanuric Acid

appearance	white crystalline solid
(iso)cyanuric acid	98.5% min
ammelide	1% max
moisture	0.6% max
pH, ca 1% slurry	2.8 min

Analytical and Test Methods

Titration of (iso)cyanuric acid as a monobasic acid to the end point, ca pH 8.5, is of limited value as an analytical method since weakly acidic materials such as acetic acid interfere, as do usual impurities such as ammelide. (Iso)cyanuric acid,in solution can be analyzed gravimetrically by precipitation of the very insoluble melamine–(iso)cyanuric acid complex (56). Precipitation via melamine also serves as a spot test for CA.

Thin layer chromatography is valuable in determining the presence of (iso)cyanuric acid in a complex mixture including intermediates during manufacture, urea and biuret and aminotriazine impurities, ammelide, ammeline, and melamine (57).

Health and Safety (Toxicology, Biodegradability)

Extensive toxicity testing has been conducted on (iso)cyanuric acid and the N-chloroisocyanurates; the results from earlier testing are reviewed in ref. 58. Reported LD_{50} values in g/kg for rats are: (1) (iso)cyanuric acid, >5.0; (2) sodium dichloroisocyanurate, 1.67; (3) potassium dichloroisocyanurate, 1.22; and (4) trichloroisocyanuric acid, 0.75 (58). Unpublished data from more recent investigations are available from the United States manufacturers. The studies show that the compounds are safe for use in the bleaching, sanitizing, and swimming pool disinfection applications when handled and used as directed. (Iso)cyanuric acid is not troublesome to handle since it is stable and relatively inert. Chlorinated isocyanurates are stable when dry, uncontaminated, and kept away from fire or a source of high heat. They are, however, active chlorine compounds and thus are strong oxidizing agents; care must be exercised to prevent hazardous contamination. Most uses of N-chloroisocyanurates are regulated by the EPA under the Federal Insecticide, Fungicide and Rodenticide Act.

(Iso)cyanuric acid is ultimately the end product of use of chlorinated isocyanurates in bleaching, sanitizing, and disinfection applications. Since the N-chloro derivatives are biocidal, biodegradation studies have centered on the residual (iso)cyanuric acid. Three studies that demonstrate the biodegradability of (iso)cyanuric acid appear in the literature (59).

Uses

Most of the cyanuric acid produced commercially is chlorinated to produce sodium dichloroisocyanurate (NaDCC), sodium dichloroisocyanurate dihydrate (NaDCC.2H$_2$O), potassium dichloroisocyanurate (KDCC), trichloroisocyanuric acid (TCCA), and the mixed complex 4 KDCC:1 TCCA. These have become standard ingredients in formulations for scouring powders, household bleaches, institutional and industrial cleansers, automatic dishwasher compounds, general sanitizers, and most importantly, in swimming pool disinfection. The choice of the N-chloroisocyanurate for any particular application depends on the desired combination of solubility, available chlorine content, and pH.

The salts of dichloroisocyanuric acid have a far greater water solubility than TCCA. Trichloroisocyanuric acid, however, is highest in available chlorine content and is far slower to dissolve than the salts. This slow rate of TCCA dissolution is used to advantage in pool sanitizing where it is tabletted and then dispensed into the water by means of slow-erosion feeders, either floating on the surface or in-line with the filtration system. The salts are usually broadcast into the swimming pool. See Table 4 for some properties of the N-chloroisocyanurates.

The increased cost of the various N-chloroisocyanurates over inexpensive chlorine bleaches such as sodium- or calcium hypochlorites is offset by significant advantages in storage stability, ease of handling and dispensing, solubility, and formulation stability with adjunct ingredients (see Bleaching agents).

Trichloroisocyanuric acid and the salts of dichloroisocyanuric acid retain their active chlorine values over long-term storage. In addition, both trichloroisocyanuric acid and sodium dichloroisocyanurate dihydrate are resistant to sustained thermal decomposition. Upon being subjected to an intense heat source, TCCA volatilizes; NaDCC.2H$_2$O when stored in proper containers has a greatly retarded rate of thermal decomposition in bulk as compared to the anhydrous dichloroisocyanurate salts and inorganic hypochlorites. The reduced rate of thermal decomposition of the hydrate is primarily owing to hydrate water evaporation. This increased thermal stability is useful in meeting warehousing and shipping requirements.

The N-chlorinated isocyanurates can be used in the bleaching of cotton, synthetics, and their blends; they do, however, attack proteinaceous fibers, such as silk or wool, presumably via active chlorine reaction on the peptide (amide) linkage. However, the chlorinated isocyanurates can be used as shrink-proofing agents in wool finishing (60), (see Textiles; Wool). The same action of chlorine upon proteins contributes to the effectiveness of chloroisocyanurates in automatic dishwashers.

(Iso)cyanuric acid itself finds wide use as a stabilizer against sunlight (ultraviolet) destruction of available chlorine in swimming pools; chlorine stabilization is achieved at CA levels of 30 ppm and above (61). Since (iso)cyanuric acid is the by-product of N-chloroisocyanurate usage, available chlorine stabilization is most efficiently effected by pool sanitation treatments combining an initial CA charge with continued use of the various N-chloroisocyanurates. Thus an effective CA level is maintained through filter backwashing and swimmer splashout. Since pure, unformulated agent will be stored at home by pool owners, sodium dichloroisocyanurate dihydrate offers distinct safety advantages. Parameters for using (iso)cyanuric acid and the chlorinated iso-

Table 4. Properties of N-Chlorinated Derivatives of Isocyanuric Acid

Property	NaDCC	NaDCC.2H$_2$O	KDCC	TCCA
available chlorine (theory), %	64.5	55.4[a]	60.1	91.5
pH of 1% solution	5.8–7.0	5.8–7.0	5.9–6.7	2.0–3.7
solubility in H$_2$O at 25°C, g/100 mL solution	22.7	>22.7[a]	9.0	1.2

[a] Actual values vary according to hydrate water content.

cyanurates in swimming pool disinfection can be found in refs. 58 and 62 (see Water, treatment of swimming pools).

Tris(2-hydroxyethyl) isocyanurate (THEIC) is used as an additive in the production of high performance polyester magnet-wire enamels and in electrical varnishes. THEIC increases adhesion and imparts improved resistance to heat and weather in a variety of resin systems. In addition, the use of THEIC produces a significantly improved combination of mechanical, chemical, and electrical properties. The enamels and varnishes are used in the manufacture of electric motors, television tubes, and transformer coils (30). THEIC is supplied by Toyomenka (America) Inc. In the mid 1970s, THEIC from Allied Chemical Corp. was selling for $2.75/kg.

Triallyl isocyanurate (TAIC) is a polyfunctional olefin monomer that can be homopolymerized to clear, hard, heat-resistant polymers having excellent electrical properties. It can also be copolymerized with unsaturated polyester resins (see Polyesters, unsaturated) or other unsaturated monomers including diallyl phthalate, methyl methacrylate, and styrene to form a variety of materials with special properties. Because of its polyfunctionality, it is an effective cross-linking agent in polyethylene, poly(vinyl chloride) and ethylene–propylene systems (63). In the mid 1970s, TAIC was sold for $10.36/kg (see Allyl monomers and polymers).

Tris(2,3-epoxypropyl) isocyanurate (ie, triglycidyl isocyanurate) is manufactured in Switzerland by the Ciba-Geigy Corporation and is used for weather-resistant powder coatings (qv) (64).

BIBLIOGRAPHY

"Triazinetriol" in *ECT* 2nd ed., Vol. 20, pp. 662–671, by Raymond N. Mesiah, FMC Corporation.

1. J. Elguero and co-workers in A. R. Katritzky and A. J. Boulton, eds., *The Tautomerism of Heterocycles*, Suppl. 1, Academic Press, Inc., New York, 1976, pp. 138–139.
2. M. Cignitti and L. Paoloni, *Spectrochim. Acta* **20**, 211 (1964).
3. A. R. Katritzky and J. M. Lagowski in A. R. Katritzky, ed., *Advances in Heterocyclic Chemistry*, Vol. 1, Academic Press, Inc., New York, 1963, p. 387.
4. F. W. Hoover, H. B. Stevenson, and H. S. Rothrock, *J. Org. Chem.* **28**, 1825 (1963); F. Zobrist and H. Schinz, *Helv. Chim. Acta* **35**, 2380 (1952).
5. E. Billows, *Z. Kristollogr. Mineral.* **46**, 481 (1908).
6. A. P. Brady, K. M. Sancier, and G. Sirine, *J. Am. Chem. Soc.* **85**, 3101 (1963); J. P. De Busscher and co-workers, *Chim. Anal. (Paris)* **54**, 69 (1972); J. Gardiner, *Water Res.* **7**, 823 (1973).
7. E. M. Smolin and L. Rapoport in A. Weissberger, ed., *The Chemistry of Heterocyclic Compounds*, Vol. 13, Interscience Publishers, a division of John Wiley & Sons, Inc., New York, 1967, pp. 17–146.
8. R. H. Widman, *J. Chem. Phys.* **43**, 2922 (1965).
9. M. K. Hargreaves, J. G. Pritchard, and H. R. Dave, *Chem. Rev.* **70**, 439 (1970).
10. U.S. Pat. 2,969,360 (Jan. 24, 1961), R. H. Westfall (to FMC Corporation).
11. U.S. Pat. 3,993,649 (Nov. 23, 1976), D. L. Sawhill and H. W. Schiessl (to Olin Corporation).
12. U.S. Pats. 3,299,060 (Jan. 17, 1967), S. J. Kovalsky and R. A. Olson (to FMC Corporation); 3,035,054 (May 15, 1962), W. F. Symes and N. S. Hadzekyriakides (to Monsanto Co.).
13. U.S. Pats. 3,803,144 (Apr. 9, 1974), 3,818,004 (June 18, 1974), S. Berkowitz (to FMC Corporation).
14. U.S. Pat. 3,272,813 (Sept. 13, 1966), W. F. Symes (to Monsanto Co.).
15. K. Morita, *Bull Chem. Soc. Jpn.* **31**, 347 (1958); W. Gottardi, *Monatsh Chem.* **98**, 507, 1613 (1967).
16. J. H. Saunders and R. J. Slocombe, *Chem. Rev.* **43**, 203 (1948); R. G. Arnold, J. A. Nelson, and J. J. Verbanc, *Chem. Rev.* **57**, 47 (1957).
17. N. Milstein, *J. Chem. Eng. Data* **13**, 275 (1968).
18. R. W. Cummins, *J. Org. Chem.* **28**, 85 (1963).
19. T. C. Frazier, E. D. Little, and B. E. Lloyd, *J. Org. Chem.* **25**, 1944 (1960).
20. A. A. Sayigh and H. Ulrich, *J. Chem. Soc.*, 3148 (1961).

21. U.S. Pat. 2,809,942 (Oct. 15, 1957), H. G. Cooke, Jr., (to Devoe and Raynolds Co., Inc.).
22. U.S. Pat. 3,075,979 (Jan. 29, 1963), J. J. Tazuma and R. Miller (to FMC Corporation).
23. *Cyanuric Acid,* technical data bulletin, FMC Corporation, New York (currently Philadelphia, Pa.), 1965.
24. Brit. Pat. 988,631 (Apr. 7, 1965), (to Spencer Chemical Company).
25. U.S. Pat. 3,249,607 (May 3, 1966), B. Taub and J. B. Hino (to Allied Chemical Corp.).
26. K. H. Slotta and R. Tschesche, *Ber.* **60B,** 301 (1927).
27. L. Paoloni and M. L. Tosato, *Ann. Chim. (Rome)* **54,** 897 (1964).
28. Z. N. Pazenko and L. I. Chovnik, *Ukr. Khim. Zh.* **30,** 195 (1964).
29. U.S. Pat. 3,326,906 (June 20, 1967), W. A. Stamm (to Stauffer Chemical Co.).
30. *Tris(2-hydroxyethyl) Isocyanurate,* technical data bulletin, Allied Chemical Corp., Morristown, N.J., 1960.
31. Brit. Pat. 961,624 (June 24, 1964), (to Spencer Chemical Company).
32. U.S. Pat. 3,318,888 (May 9, 1967), J. H. Blumbergs and D. G. MacKellar (to FMC Corporation).
33. S. Freireich, D. Gertner, and A. Zilkha, *J. Organometal. Chem.* **35,** 303 (1972).
34. U.S. Pat. 3,112,312 (Nov. 26, 1963), P. L. Veltman and E. Fisher (to W. R. Grace and Co.).
35. Z. Yoshida and R. Oda, *J. Chem. Soc. (Jpn.)* **56,** 92 (1953); *Chem. Abstr.* **49,** 4679d (1955).
36. H.-F. Piepenbrink in E. Müller, ed., *Methoden der Organischen Chemie (Houben-Weyl),* Sauerstoffverbindungen III, Vol. 8, Georg Thieme Verlag, Stuttgart, FRG, 1952, p. 226 ff.
37. Brit. Pat. 951,280 (Mar. 4, 1964), T. E. O'Connor (to E. I. du Pont de Nemours & Co., Inc.).
38. U.S. Pat. 2,444,828 (July 6, 1948), W. Kaplan (to Sun Chemical Corp.).
39. Brit. Pat. 710,143 (June 9, 1954), B. Bann and S. A. Miller (to British Oxygen Co.).
40. U.S. Pat. 3,194,810 (July 13, 1965), R. L. Formaini and E. D. Little (to Allied Chemical Corp.).
41. U.S. Pat. 3,258,485 (June 28, 1966), C. S. Argyle (to Whiffen and Sons).
42. U.S. Pat. 3,275,631 (Sept. 27, 1966), H. Yanagizawa (to Shikoku Kasei Kogyo Co.).
43. U.S. Pat. 2,943,088 (June 28, 1960), R. H. Westfall.
44. N. P. Shishkin and A. I. Finkel'shtein, *J. Appl. Chem. USSR* **42,** 446 (1969); *Zh. Prikl. Khim.* **42,** 474 (1969).
45. U.S. Pat. 3,563,987 (Feb. 16, 1971), S. Berkowitz (to FMC Corporation); U.S. Pat. 3,954,751 (May 4, 1976), H. Fuchs and co-workers (to BASF Aktiengesellschaft); Ger. Offen 2,630,811 (Jan. 27, 1977), A. P. H. Schouteten and M. J. A. M. Den Otter (to Stamicarbon B.V.).
46. U.S. Pats. 3,318,887 (May 9, 1967), W. P. Moore and C. B. R. Fitz-William, Jr., (to Allied Chem. Corp.); 3,154,545 (Oct. 27, 1964), W. F. Symes and co-workers (to Monsanto Co.).
47. U.S. Pat. 3,394,136 (July 23, 1968), W. P. Moore and D. E. Elliott (to Allied Chem. Corp.).
48. U.S. Pat. 3,093,641 (June 11, 1963), R. L. Formaini (to Allied Chem. Corp.).
49. F. J. Schiltknecht, *Promotionsarb. (Zurich),* **3403** (1963); *Chem. Abstr.* **63,** 5647g (1965); N. I. Malkina, *Zh. Prikl. Khim.* **34,** 1630 (1961).
50. U.S. Pat. 2,768,167 (Oct. 23, 1956), W. F. Marzluff and L. H. Sutherland (to American Cyanamid Co.)
51. U.S. Pat. 3,172,886 (Mar. 9, 1965), I. Christoffel and D. P. Schutz (to Allied Chem. Corp.).
52. U.S. Pat. 2,905,671 (Sept. 22, 1959), J. D. Christian and E. W. Lard (to W. R. Grace & Co.).
53. U.S. Pat. 2,712,002 (June 28, 1955), L. A. Lundberg (to American Cyanamid Co.).
54. R. J. H. Voorhoeve and L. E. Trimble, *Science* **202,** 525 (1978).
55. *Chem. Age,* 2 (July 1, 1977); *Chem. Mark. Rep.,* 7 (May 30, 1977); *Chem. Week,* 34 (Mar. 2, 1977).
56. L. Nebbia and co-workers, *Chim. Ind. (Milan)* **39,** 81 (1957).
57. J. Mlochowski and Z. Skrowaczewska, *Chem. Anal. (Warsaw)* **15,** 871 (1970); *Chem. Abstr.* **74,** 13646t (1971).
58. E. Canelli, *Am. J. Public Health* **64,** 155 (1974).
59. J. Saldick, *Appl. Microbiol.* **28,** 1004 (1975); D. C. Wolf and J. P. Martin, *J. Environ. Qual.* **4,** 134 (1975); H. L. Jensen and A. S. Abdel-Ghaffar, *Arch. Mikrobiol.* **67,** 1 (1969).
60. Ger. Offen. 2,326,463 (Dec. 19, 1974), H. Bille and T. Siemenc (to BASF, A.G.); Jpn. Kokai 74, 110,988 (Oct. 22, 1974), T. Hanyu and co-workers (to Toyobo Co., Ltd.).
61. U.S. Pat 2,988,471 (June 13, 1961), R. J. Fuchs and I. A. Lichtman (to FMC Corporation).
62. *Pool Serivce Technician's Handbook,* National Swimming Pool Institute, Washington, D.C., 1970; M. A. Gabrielsen, ed., *Swimming Pools, A Guide to their Planning Design and Operation,* Hoffman Publications, Inc., Fort Lauderdale, Fla., 1972.
63. J. K. Gillham, "Allyl Polymers" in N. Bikales, ed., *Encyclopedia of Polymer Science and Technology,*

CYCLIZATION. See Petroleum.

CYCLOHEXANE. See Hydrocarbons, C_1–C_6.

CYCLOHEXANOL AND CYCLOHEXANONE

CYCLOHEXANOL

Cyclohexanol [108-93-0], $CH_2(CH_2)_4CHOH$, is a colorless, viscous liquid with a camphoraceous odor. It is used chiefly as a chemical intermediate, a stabilizer, and a homogenizer for various soap and detergent emulsions, and as a solvent for lacquers and varnishes.

Cyclohexanol was first prepared by the treatment of 4-iodocyclohexanol with zinc dust in glacial acetic acid, and later by the catalytic hydrogenation of phenol at elevated temperatures and pressures.

Properties

Important physical properties of cyclohexanol are shown in Table 1.

Cyclohexanol shows most of the typical reactions of secondary alcohols. It reacts with organic acids to form esters, and with halogen acids to form the corresponding halides. Dehydrating agents convert cyclohexanol into cyclohexene $CH_2(CH_2)_3CH=CH$. Catalytic dehydrogenation or mild oxidation of cyclohexanol yields cyclohexanone. Strong oxidizing agents such as nitric acid or alkaline potassium permanganate convert cyclohexanol into adipic acid (qv), $HOOC(CH_2)_4COOH$.

Manufacture

Cyclohexanol is prepared commercially by the catalytic air oxidation of cyclohexane or catalytic hydrogenation of phenol. The oxidation of cyclohexane to a mixture of cyclohexanol and cyclohexanone, known as KA-oil (ketone–alcohol, cyclohexa-

Table 1. Properties of Cyclohexanol and Cyclohexanone

Property	Cyclohexanol	Cyclohexanone
mp, °C	25.15	−47
bp, °C	161.1	156.7
d_4^{20}, g/mL	0.9493	0.9478
n_D^{25}	1.4648	
n_D^{20}		1.4507
vapor pressure, kPa[a], 25°C	0.15	
20°C		0.53
sp heat, 15–18°C, J/g[b]	1.75	1.81
viscosity, 25°C, mPa·s(= cP)	4.6	2.2
flash point, open cup, °C	67.2	54
soly in water, g/100 g, 10°C	4.2	15
30°C	4.3	5
soly of water in compound, g/100 g, 20°C	12.6	9.5
miscibility	miscible in all proportions with most organic solvents including those customarily used in lacquers	miscible with methanol, ethanol, acetone, benzene, n-hexane, nitrobenzene, diethyl ether, naphtha, xylene, ethylene, glycol, isoamyl acetate, diethylamine, and most organic solvents
dissolves	many oils, waxes, gums and resins	cellulose nitrate, acetate, and ethers, vinyl resins, raw rubber, waxes, fats, shellac, basic dyes, oils, latex, bitumen, kaure, elemi, and many other organic compounds

[a] To convert kPa to mm Hg, multiply by 7.5.
[b] To convert J to cal, divide by 4.184.

none–cyclohexanol crude mixture), is used for most production (1). The earlier technology that used an oxidation catalyst such as cobalt naphthenate at 180–250°C at low conversions (2) has been improved. Cyclohexanol can be obtained through a boric acid-catalyzed cyclohexane oxidation at 140–180°C with up to 10% conversion (3). Unreacted cyclohexane is recycled and the product mixture is separated by vacuum distillation. The hydrogenation of phenol to a mixture of cyclohexanol and cyclohexanone is usually carried out at elevated temperatures and pressure in either the liquid (4) or in the vapor phase (5) catalyzed by nickel.

Specifications and Containers

Commercial products are offered in two grades, ie, technical and high grade. Typical specifications are listed below (6).

	Technical	High grade
mp, °C, minimum	−10	18
distillation range	10–19% within 1.5°C of 161.7°C	100% 159–163, 5–95% within 1.5°C of 161.7°C
$d_{15.5}^{25}$, g/mL	0.9375–0.9405	0.9425–0.9455
phenol content	none to trace	none
n_D^{25}	1.4575–1.4605	1.463–1.465

Some cyclohexanol is shipped with 2.25% of methanol as antifreeze.

Cyclohexanol is shipped in 208-L (55-gal) drums, in tank cars, and tank trucks. DOT regulations classify cyclohexanol as a combustible solvent. Drums containing less than 416 L (110 gal) do not require hazardous material labeling. Larger quantities must be labeled "Combustible Liquid" (7). The U.S. International Trade Commission price in 1971 was 31¢/kg; the 1977 price was ca 75¢/kg.

Analytical Methods

Cyclohexanol is best determined by gas–liquid chromatography using a DC-710 or carbowax 20M-on-chromosorb column. Impurities such as cyclohexane, benzene, cyclohexanone, and phenol do not interfere. Other analytical procedures include acylation with acetyl chloride and pyridine at elevated temperatures. Following hydrolysis, the excess acid is titrated with sodium hydroxide solution (8). Aldehydes and easily hydrolyzed esters interfere with this method. For samples containing high concentrations of acids and easily hydrolyzed esters but no aldehydes or ketones, an alternative procedure is used (9). This procedure involves treatment at elevated temperatures of the sample containing cyclohexanol with acetic acid containing boron trifluoride catalyst. The water formed in the esterification reaction is then determined by the Karl Fischer method. For analysis based on micro procedures, acetic anhydride is the preferred acetylating agent (10).

Cyclohexanol can be determined colorimetrically by reaction with p-hydroxybenzaldehyde in sulfuric acid (11). This method can be used in the presence of cyclohexanone and cyclohexane. Cyclohexanol and cyclohexanone both show a maximum absorbency at 535 nm but at 625 nm the absorption by cyclohexanone is negligible, whereas cyclohexanol shows appreciable absorption.

Health and Safety Factors

Cyclohexanol is slightly to moderately toxic (12–13). The toxic dose with respect to inhalation is 75 ppm (14). The time-weighted average OSHA standard for air is 50 ppm (15). The aquatic toxicity rating at 96 h static or continuous flow is 100–110 ppm (14). Contact with the skin may cause dermatitis in sensitive persons (12). It is also irritating to the mucous membranes. Prolonged inhalation or ingestion of small amounts can cause nausea, gastrointestinal disturbances, slight nervous symptoms, and trembling. The hazards of industrial exposure to vapors of cyclohexanol at room temperature, however, are believed to be limited by the low vapor pressure and the low rate of evaporation of this compound.

Precautions that should be observed as a matter of course in using cyclohexanol include adequate and proper ventilation, avoidance of prolonged breathing of vapor or contact of the liquid with the skin, avoidance of internal consumption, and protection of the eyes against splashing liquids.

Uses

The production of adipic acid (qv) through oxidation of cyclohexanol is the most important use; fully 90% of adipic acid produced in the United States is used in the manufacture of nylon 66, a polymer of adipic acid and hexamethylenediamine. Adipic acid may also be used to produce the hexamethylenediamine. The next important usage of cyclohexanol, pure or admixed with cyclohexanone as KA-oil, is in the production of ε-caprolactam, NH(CH$_2$)$_5$CO, which is used in the manufacture of nylon 6 polymer (see Polyamides).

The amount of refined cyclohexanol used in other applications, according to U.S. International Trade Commission figures, was ca 4300 metric tons in 1971. One major use has been in the manufacture of esters for use as plasticizers (qv), ie, cyclohexyl and dicyclohexyl phthalates. In the finishes industry, cyclohexanol is used as a solvent for lacquers, shellacs, and varnishes. Its low volatility helps to improve secondary flow and to prevent blushing. Also, it improves the miscibility of cellulose nitrate and resin solutions and helps to maintain homogeneity during drying of lacquers. Cyclohexanol is used as a stabilizer and homogenizer for soaps and synthetic detergent emulsions. It is used also by the textile industry as a dye solvent and kier-boiling assistant (see Dye carriers).

Derivatives

Commercial methylcyclohexanol, CH$_3$C$_6$H$_{10}$OH, is a slightly viscous, straw-colored liquid which is a mixture of 2-methylcyclohexanol (bp$_{99.3\ kPa}$ (745 mm Hg) 164.5–165.5°C), 3-methylcyclohexanol (bp$_{102.5\ kPa}$ (769 mm Hg) 173.7–174°C) and 4-methylcyclohexanol (bp$_{99.3\ kPa}$ 172.5–173°C). The commercial product, made by the catalytic hydrogenation of mixed cresols, has a boiling range of 155–180°C, d$_{15.5}^{15.5}$ 0.924 ± 0.003 g/mL, and n_D^{20}, 1.461. It is used as a solvent in lacquers, as an ingredient in soap-based spot removers, as a blending agent for special textile soaps and detergents, and in the manufacture of lubricating-oil additives.

CYCLOHEXANONE

Cyclohexanone [108-94-1], CH$_2$(CH$_2$)$_4$CO, is a colorless, mobile liquid with an odor suggestive of peppermint and acetone. Cyclohexanone is used chiefly as a chemical intermediate and as a solvent for resins, lacquers, dyes, and insecticides.

Cyclohexanone was first prepared by the dry distillation of calcium pimelate, OOC(CH$_2$)$_5$COOCa, and later by Bouveault by the catalytic dehydrogenation of cyclohexanol.

Properties

Physical properties of cyclohexanone are listed in Table 1.

Cyclohexanone shows most of the typical reactions of aliphatic ketones. It reacts with hydroxylamine, phenylhydrazine, semicarbazide, Grignard reagents, hydrogen cyanide, sodium bisulfite, etc, to form the usual addition products, and it undergoes the various condensation reactions that are typical of ketones having α-methylene groups. Reduction converts cyclohexanone to cyclohexanol or cyclohexane, and oxidation with nitric acid converts cyclohexanone almost quantitatively to adipic acid.

Manufacture

Cyclohexanone may be produced by the catalytic hydrogenation of phenol, by the catalytic air-oxidation of cyclohexane, by the catalytic dehydrogenation of cyclohexanol, or by the oxidation of cyclohexanol. The hydrogenation of phenol to cyclohexanone is the most efficient route. This process can be carried out either in the liquid phase, catalyzed by palladium on carbon (15), or in the vapor phase, catalyzed by palladium on alumina (16). Of the two processes, the liquid phase is more selective and requires much less catalyst inventory. Allied Chemical Corporation operates an inherently safe liquid-phase phenol hydrogenation process at temperatures below the atmospheric boiling point of the reaction liquids, permitting over 99% selectivity at 90% conversion (17).

Vapor-phase oxidation of cyclohexane is commercially feasible, but the preferred route is liquid-phase cyclohexane oxidation (2). In the latter process conversions are usually below 10%. The product is a mixture of cyclohexanone and cyclohexanol which may be separated by distillation (2) or converted to cyclohexanone by dehydrogenation over either zinc oxide (5) or nickel on magnesium oxide (18). Mild oxidation of cyclohexanol to cyclohexanone can be carried out with (19) or without (20) catalyst.

Specifications and Standards

Commercial cyclohexanone is offered in various grades. The specifications of two typical grades are listed in Table 2 (6).

Cyclohexanone is shipped in 208-L (55-gal) drums, in tank cars, and tank trucks. DOT regulations classify cyclohexanone as a combustible liquid. Drums containing less than 416 L (110 gal) do not require hazardous material labeling. Larger quantities must be labeled "Combustible Liquid" (7). The U.S. International Trade Commission price in 1974 was 60¢/kg; the 1975 price was 77¢/kg (21).

Table 2. Specifications of Two Typical Cyclohexanones

Property	Commercial grade	High purity
appearance	colorless liquid	colorless liquid
$d_{15.5}^{15.5}$, g/mL		0.950–0.951
d_{20}^{20}, g/mL	0.940–0.950	
distillation range, 95% at 101.3 kPa[a]	152–157°C	152–157°C
ash, % max	0.01	0.01
acidity, % max	none	0.03
alkalinity to phenolphthalein	none	none
ketone content, % min	89	99.5
soly in water at 20°C, %		5
water content, % max	0.2	0.2
flash point (Abel closed cup), °C		42
(Cleveland open cup), °C	46	
evaporation rate (ether = 1)	38	40

[a] To convert kPa to mm Hg, multiply by 7.5.

Analytical Methods

Cyclohexanone is most readily determined by gas–liquid chromatography over DC-710 or carbowax 20M on chromosorb. Impurities such as cyclohexane, benzene, cyclohexanol, and phenol do not interfere. In the absence of other carbonyl compounds cyclohexanone may be determined by treatment with hydroxylamine hydrochloride, which forms the oxime, as follows:

$$\text{C}_6\text{H}_{10}\text{=O} + \text{NH}_2\text{OH·HCl} \longrightarrow \text{C}_6\text{H}_{10}\text{=NOH} + \text{H}_2\text{O} + \text{HCl}$$

With materials containing low concentrations of acids, the liberated hydrogen chloride is titrated with sodium hydroxide, either electrometrically (22) or with an indicator (23).

Health and Safety Factors

The toxic dose limitation for inhalation is 75 ppm (14). The time-weighted average OSHA standard for air is 50 ppm (14). The aquatic toxicity rating at 96 h with static or continuous flow is 100–110 ppm (14). Primary irritation and defatting of the skin can result from substantial or prolonged contact with cyclohexanone. Inhalation exposure in excess of the recommended threshold limit value of 50 ppm (vol) (24) is uncomfortably irritating and not readily tolerated for prolonged periods.

The precautions usually observed when handling volatile solvents should be observed as a matter of course with cyclohexanone. These include adequate and proper ventilation, avoidance of prolonged breathing of vapor or contact of the liquid with the skin, avoidance of internal consumption, and protection of the eyes against splashing liquids.

Uses

The most important use of cyclohexanone is as a chemical intermediate in nylon manufacture; 97% of all cyclohexanone output is used either to make caprolactam for nylon 6, or adipic acid for nylon 66. In the caprolactam process cyclohexanone is converted to cyclohexanone oxime (mp, 89–90°C), which is rearranged with sulfuric acid to ϵ-caprolactam (mp, 65°C). The overall efficiency is >97%. In the production of adipic acid, cyclohexanone is oxidized with nitric acid in the presence of catalysts. Cyclohexanone is also used as a solvent (qv) and thinner for lacquers, especially those containing nitrocellulose or vinyl chloride polymer and copolymers and as a general solvent for synthetic resins and polymers. Cyclohexanone is an excellent solvent for insecticides, and many other similar materials. Cyclohexanone is used as a building block in the synthesis of many organic compounds, such as pharmaceuticals, insecticides and herbicides. Cyclohexanone is used in the manufacture of magnetic and video tapes (see Magnetic tape). Cyclohexanone is not restricted under California Rule 66 and has, therefore, found use as a substitute for isophorone as a solvent for resins and polymers (25) (see Air pollution). Cyclohexanone sales were a small fraction of total production and amounted to ca 13,000 t in 1975 (21).

BIBLIOGRAPHY

"Cyclohexanol and Cyclohexanone" in *ECT* 1st ed., Vol. 4, pp. 769–773, by J. D. Young, E.I. du Pont de Nemours & Co., Inc.; "Cyclohexanol and Cyclohexanone" in *ECT* 2nd ed., Vol. 6, pp. 683–688, by R. D. Kralovec and H. B. Louderback, E.I. du Pont de Nemours & Co., Inc.

1. *Chem. Week,* **93**(20), 74 (1963).
2. U.S. Pats. 2,223,493 and 2,223,494 (Dec. 3, 1940), D. J. Doder (to E.I. du Pont de Nemours & Co., Inc.).
3. "Cyclohexanol and Cyclohexanone: Salient Statistics," *Chemical Economics Handbook,* SRI International, Menlo Park, Ca., June 1976.
4. Brit. Pat. 22,523 (Oct. 6, 1913), A. Brochet.
5. U.S. Pat. 3,998,884 (Dec. 21, 1976), C. A. Gibson (to Union Carbide).
6. H. C. Reis and R. G. Muller, *Caprolactam, Part II,* Report No. 7, Process Economics Program, SRI International, Menlo Park, Ca., Nov. 1965, pp. 441–442.
7. R. M. Graziano, *Tariff #31,* Hazardous Material Regulations of the Department of Transportation, Association of American Railroads, Bureau of Explosives, effective date: March 31, 1977.
8. D. M. Smith and W. M. D. Bryant, *J. Am. Chem. Soc.* **57,** 61 (1935).
9. W. M. D. Bryant, J. Mitchell, Jr., and D. M. Smith, *J. Am. Chem. Soc.* **62,** 1 (1940).
10. N. A. Cheronics and T. S. Ma, *Organic Functional Group Analysis by Micro and Semimicro Methods,* Interscience Publishers, Inc., a division of John Wiley & Sons, Inc., New York, 1964, pp. 493–496.
11. S. D. Nogare and J. Mitchell, Jr., *Anal. Chem.* **25,** 1376 (1953).
12. K. B. Lehmann and F. Flury, *Toxicology and Hygiene of Industrial Solvents,* The Williams & Wilkins Co., Baltimore, 1943, p. 213.
13. J. F. Treon, W. E. Crutchfield, Jr., and K. V. Kitzmiller, *J. Ind. Hyg. Toxicol.* **25,** 199, 323 (1943).
14. *Registry of Toxic Effects of Chemical Substances,* 1976 ed., U.S. Department of Health, Education, and Welfare, 1976.
15. U.S. Pat. 3,076,810 (Feb. 5, 1963), R. J. Duggan and co-workers (to Allied Chemical Corp.).
16. U.S. Pat. 3,305,586 (Feb. 21, 1967), P. Phielix (to Stamicarbon).
17. U.S. Pat. 4,092,360 (May 30, 1978), J. F. Van Peppen and W. B. Fisher (to Allied Chemical Corp.).
18. U.S.S.R. Pat. 232,231 (July 25, 1977) (to Scientific-Research Physical-Chemical Institute).
19. U.S. Pat. 3,974,221 (Aug. 10, 1976), R. J. Duggan (to Allied Chemical Corp.).
20. U.S. Pat. 2,285,914 (June 9, 1942), O. Drossbach (to E.I. du Pont de Nemours & Co., Inc.).
21. *Chemical Economics Handbook: Manual of Current Indicators,* SRI International, Menlo Park, Ca., Dec. 1976, p. 239.
22. J. Mitchell, Jr., D. M. Smith, and W. M. D. Bryant, *J. Am. Chem. Soc.* **63,** 573 (1941).
23. W. M. D. Bryant and D. M. Smith, *J. Am. Chem. Soc.,* **57,** 57 (1935).
24. *American Conference of Governmental Industrial Hygientists,* 1977 ed.
25. W. L. Faith, D. B. Keyes, and R. L. Clark, *Industrial Chemicals,* John Wiley & Sons, Inc., New York, 1975.

WILLIAM B. FISHER
JAN F. VAN PEPPEN
Allied Chemical Corp.

CYCLOPENTADIENE AND DICYCLOPENTADIENE

Two well-established chemical building blocks are cyclopentadiene (CP), C_5H_6, and its more stable and available form, the dimer dicyclopentadiene (DCP), $C_{10}H_{12}$. Although both are used, the monomer (CP) obtained via cracking of the dimer (DCP) has extensive use. Worldwide studies are related to this molecule owing to its versatility and reactive conjugated diolefin system. Four review articles dealing with the chemistry of cyclopentadiene have been published (1–4). Another article deals with dicyclopentadiene, specifically in Europe (5). The discovery in 1951 of stable metal derivatives has given additional impetus to the study of the chemistry of cyclopentadiene (see Organometallics).

Physical Properties

Dicyclopentadiene (DCP) exists in two stereoisomeric forms, the endo and exo isomers. The commercial product is predominantly the endo isomer. DCP, 4,7-methano-3a,4,7,7a-tetrahydroindene, is the form in which cyclopentadiene is sold commercially.

CP   DCP

The physical properties of CP and DCP are given in Table 1. In addition to more detailed data relating to boiling point, density, index of refraction, etc, the strong absorption bands in the ir and uv spectra as well as the Raman spectra are discussed in reference 2. The nmr spectrum of DCP is described in reference 6.

Chemical Reactions

Cyclopentadiene contains two conjugated double bonds and an active methylene group and can thus undergo a diene addition reaction with almost any unsaturated compound, eg, ethylene and its derivatives, and acetylene. The number of derivatives prepared from cyclopentadiene is extensive; therefore only the reactions and derivatives considered most important are discussed.

Polymerization. Cyclopentadiene (CP) polymerizes spontaneously at ordinary temperature to dicyclopentadiene (DCP). At temperatures above 100°C, cyclopentadiene can be made to polymerize noncatalytically to tri-, tetra-, and higher polymers. The polymerization involves a series of consecutive Diels-Alder reactions. For example, the trimer (1) is formed by the monomer adding to the dimer.

CP + DCP $\xrightarrow{\Delta}$

(1)   pentacyclo[6.5.1.1^{3,6}.O^{2,7}.O^{9,13}]trideca-4,10-diene

418 CYCLOPENTADIENE AND DICYCLOPENTADIENE

Table 1. Physical Properties of Cyclopentadiene and Dicyclopentadiene

Physical properties	Cyclopentadiene [542-92-7]	Dicyclopentadiene[a] [77-73-6]
bp, 101.3 kPa[b], °C	41.5	170[c]
mp, °C	−85	33.6
physical form	colorless liquid	colorless crystals
odor	sweet terpenic	camphoraceous
d_4^{20}, g/mL	0.8024	
d_4^{35}, g/mL		0.9770
n_D^{20}	1.4429	
n_D^{35}		1.5061
heat of combustion, kJ/mol[d]	700	1378.4
heat of vaporization, kJ/mol[d]	7.0	9.2
heat of cracking, kJ/mol[d]		24.6
heat of fusion, kJ/mol[d]		0.5
specific heat, kJ/(kg·K)[d]		0.411
spontaneous ignition temperature, °C		
in oxygen		510
in air	640	680
dielectric constant at 40°C	2.43	

[a] Endo isomer.
[b] To convert kPa to mm Hg, multiply by 7.5.
[c] Depolymerizes at boiling point to form two molecules of cyclopentadiene.
[d] To convert J to cal, divide by 4.184.

The yields of polymers at varying contact times and different temperatures are shown in Table 2.

The thermal polymers are crystalline compounds with little odor. The trimer melts at 60–68°C. Eight stereoisomers are possible for the trimer (1), but only the endo and exo forms have been isolated thus far. The trimer is easily hydrogenated, first to the dihydro (at the 4,5-positions), and then to the tetrahydro derivative. Alder and Stein prepared these compounds using palladium as the catalyst. Tetrahydrotricyclopentadiene has also been prepared with hydrogen over Adams' platinum at room temperature and 345 kPa (50 psi). The endo isomers of the hydrogenated compounds are waxlike solids. Dihydro-*endo*-tricyclopentadiene melts at 37°C and boils at 145–146°C at 1.7 kPa (13 mm Hg) and 271°C at 102 kPa (766 mm Hg). Tetrahydro-*endo*-tricy-

Table 2. Yield of Polymers at Various Contact Times and Temperatures

	Yield, wt %		
Polymer	At 150–160°C, 14 h	At 170–180°C, 22 h	At 200°C, 90 h
tricyclopentadiene	40	50	25
tetracyclopentadiene	10	30	45
pentacyclopentadiene	2	5	10
unchanged	50	10	5

clopentadiene melts at 49–53°C, and boils at 151°C at 2 kPa (15 mm Hg) and ca 285°C at 102 kPa (766 mm Hg).

In addition to thermal polymerization, it is possible to polymerize cyclopentadiene with inorganic halides as catalyst. With trichloroacetic acid as catalyst, it is possible to obtain deeply colored, blue polymers that conduct electricity in nonpolar solvents such as benzene in the presence of acid. The conductivity and color are caused by blocks of conjugated double bonds present in the polymers (7–8).

Diene Addition (Diels-Alder). Owing to the two conjugated double bonds, cyclopentadiene can undergo a diene addition reaction. The reaction involves the addition of an ethylenic group contained in the first reactant, called the dienophile, across the 1,4-position of cyclopentadiene. The Diels-Alder-addition products are usually bicyclo[2.2.1]heptene derivatives. The reaction is carried out by simply bringing the two reactants together in the presence or absence of a solvent at temperatures ranging from essentially room temperature in some cases to ca 200°C. The reaction is highly exothermic (71–75 kJ/mol or 17–18 kcal/mol) and in many cases the yield is essentially quantitative. Although cyclopentadiene monomer is usually used, it is possible to start with dicyclopentadiene provided the reaction is carried out at temperatures at which the dicyclopentadiene dissociates into the monomer, which then adds to the dienophile. With maleic anhydride as the dienophile, the reaction leads to (2).

The diene addition is stereochemically specific (9). Two stereoisomers are possible, the endo (3) and the exo (4) forms. This is illustrated schematically below for the ad-

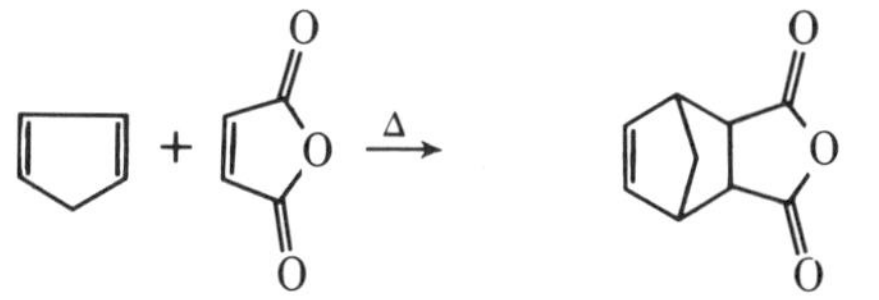

(2) bicyclo[2,2,1]hept-5-ene-2,3-dicarboxylic acid anhydride

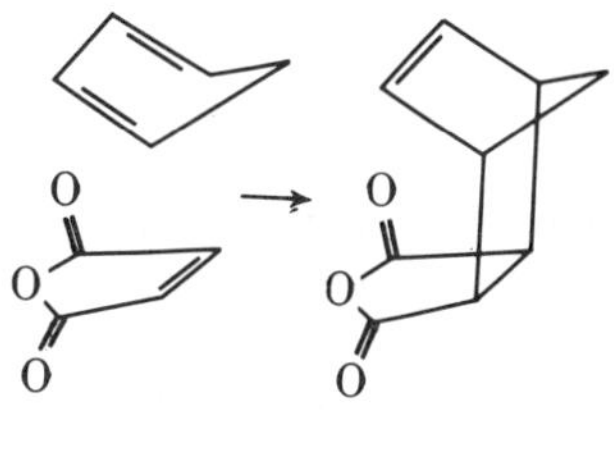

(3) endo

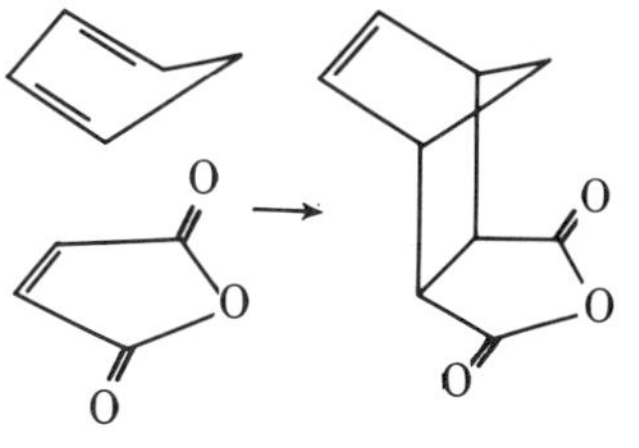

(4) exo

dition of maleic anhydride to cyclopentadiene, which gives the endo adduct exclusively (10).

As a result of the diene addition reaction, cyclopentadiene can be converted into innumerable derivatives, some of which are tabulated in Table 3.

The products obtained from the diene addition reaction are extremely versatile chemicals suitable as intermediates for the production of plasticizers, pharmaceuticals, pesticides, resins, paint driers, perfumes, and many other products. Methylenenorbornene, 5-methylenebicyclo[2.2.1]hept-2-ene, a diene addition product of cyclopentadiene and allene, can be used as a constituent for the manufacture of a terpolymer elastomer with ethylene and propylene, which gives high-quality vulcanizates possessing high tensile strength, high resilience, and excellent electrical properties (8). Heat aging and ozone resistance are also outstanding. In the case of methylenenorbornene, the double bond of the bicycloheptene ring system is involved in the polymerization (see Elastomers, synthetic–Polypentanamers).

Condensation Involving the Methylene Group. The methylene group in cyclopentadiene is extremely reactive and can undergo condensation-type reactions. A large number of derivatives have been prepared this way.

Aldehydes and ketones condense with cyclopentadiene in the presence of alkaline condensing agents to produce colored fulvene derivatives. Such ketones and aldehydes as acetone, methyl ethyl ketone, benzaldehyde, acetaldehyde, acetophenone, cyclohexanone, and cyclopentanone have been condensed with cyclopentadiene. A typical condensation with a ketone is depicted as follows:

$$\underset{R'}{\overset{R}{>}}=O \; + \; \underset{CP}{\bigcirc} \; \longrightarrow \; \underset{R'}{\overset{R}{>}}\!\!=\!\!\bigcirc \; + \; H_2O$$

fulvene, R = R' = H
dimethyl fulvene, R = R' = CH$_3$

The color of a fulvene deepens when R and R' change from hydrogen to methyl, and deepens further with increasing conjugation with R, eg, R = vinyl. With aromatic substituents the fulvenes are deep red. The aldehyde condensation products are also strongly colored but resinify so easily that it is difficult to isolate them in the pure form.

In addition to the condensations carried out with aldehydes and ketones, benzenediazonium chloride reacts with cyclopentadiene in the presence of potassium hydroxide or sodium acetate to form cyclopentadienyl azobenzene. Isonitrosocyclopentadiene is produced when ethyl nitrite is added drop by drop to a solution of cyclopentadiene in sodium ethylate and absolute ethanol. Dimethylfulvene prepared from acetone and cyclopentadiene has been copolymerized with isobutylene to a rubber. N,N-Dimethyl-α-aminofulvene-3-carboxaldehyde (5) and N,N-dimethyl-α-aminofulvene-3,4-dicarboxaldehyde (6) can be prepared from phosphorus oxychloride, dimethylformamide, and CP (11).

CP + DMF + POCl$_3$ $\longrightarrow$ (structure 5 with (CH$_3$)$_2$N and CHO) + (structure 6 with (CH$_3$)$_2$N and two CHO groups)

(5)   (6)

Table 3. Diels-Alder Adducts from Cyclopentadiene

Dienophile	Structure of adduct
Dibasic acids and derivatives	
chloromaleic anhydride	R = Cl
maleic anhydride	H
Monobasic acids	
crotonic acid	R = H R' = CH$_3$
methacrylic acid	CH$_3$ H
Aldehydes	
acrolein	R = H
crotonaldehyde	CH$_3$
Ketones	
propenyl methyl ketone (3-penten-2-one)	R = CH$_3$ R' = CH$_3$
vinyl methyl ketone (3-buten-2-one)	H CH$_3$
Ketene	
Vinyl compounds	
ethylene	R = H
styrene	C$_6$H$_5$
vinyl acetate	OOCCH$_3$
allene	=CH$_2$; no H on C attached to =CH$_2$
Acetylenes	
acetylene	R = H
acetylenedicarbonitrile	CN
A quinone	
p-benzoquinone	
A nitroso compound	
nitrosobenzene	

Hydrogenation. Dicyclopentadiene has been progressively hydrogenated through the dihydro to the tetrahydro derivative (12–13). The double bond in the bicycloheptene ring is the more reactive, and the heat of hydrogenation is 139 kJ/mol (33.2 kcal/mol). The heat of hydrogenation for the pentagonal ring is 110 kJ/mol (26.2 kcal/mol) (14).

The higher polymers can also be hydrogenated. Alder and co-workers (15) dissolved tricyclopentadiene in methanol, added colloidal platinum, and agitated in an atmosphere of hydrogen. Data on the physical properties of the hydrogenated polymers of cyclopentadiene are in references 14 and 16.

Halogenation. Halogen and halogen acids add readily to the unsaturated carbon linkages of the cyclopentadiene molecule. By such additions a series of halogenated derivatives range, in the case of the chloride, from 3-chlorocyclopentene to tetrachlorocyclopentane. Of all the possible chloro derivatives of cyclopentadiene, only the hexachlorocyclopentadiene has reached commercial status; it can be prepared by a liquid-phase chlorination of cyclopentadiene below 50°C (17). Tetrachlorocyclopentane is produced first, by addition, and is then converted to octachlorocyclopentane in 97.5% yield by catalytic chlorination over arsenious oxide or phosphorus pentachloride at 175–250°C. Octachlorocyclopentane is then dehydrochlorinated thermally to hexachlorocyclopentadiene in yields of 85–97% (see Chlorocarbons).

Oxidation. Cyclopentadiene reacts spontaneously with oxygen to form brown, gummy products that usually contain substantial amounts of peroxides. Catalytic vapor-phase oxidation of cyclopentadiene at 400–525°C over vanadium oxide leads to the production of maleic anhydride. Dihydroxycyclopentenes and tetrahydroxycyclopentane can be prepared by the oxidation of cyclopentadiene with hydrogen peroxide (18–21).

Possible uses for these polyhydroxy compounds include the preparation of alkyd-type resins with polybasic acids, the formation of ester plasticizers, and the preparation of surface-active agents.

Source and Production

No commercial process for the sole production of cyclopentadiene exists. It is obtained as a by-product from thermal operations. In the carbonization of coal, the tar, light oil, and coke-oven gas contain cyclopentadiene and dicyclopentadiene in very low concentrations (see Coal, carbonization). Thermal cracking of hydrocarbons, particularly gas oil and naphtha, in the presence of steam represents a major source of cyclopentadiene and also methylcyclopentadienes consisting of a mixture of 1-methyl- and 2-methylcyclopentadiene with the 2-isomer predominating. Another relatively recent source of cyclopentadiene is the ethylene (qv) production resulting from the thermal cracking of ethane, propane, and other raw materials. The yield of cyclopentadiene is usually less than 2 wt %.

Thermal-cracking operations also produce other hydrocarbons including benzene, naphthalene, isoprene, piperylene, ethylene, propylene, etc. Cyclopentadiene is recovered from the other hydrocarbons by distilling the total cracked product to recover a distillate consisting of C_5-hydrocarbons and lighter components. The distillate is heated to a temperature of ca 100°C to convert monomeric cyclopentadiene to dicyclopentadiene. Depending on the temperature and concentration, the heat-soaking operation is 5–24 h. The dicyclopentadiene, which boils higher than the unreacted hydrocarbons of the distillate, is recovered as distillation bottoms.

Regardless of its source, only the dimer is available since the monomer spontaneously reacts to form the dimer. The dimer normally occurs as a relatively low percentage component of a stream and is concentrated by a series of distillations. These distillations produce crude streams of ≥50 wt % DCP concentration which are available commercially. High purity DCP is obtained by cracking the DCP in the crude stream, separating the low-boiling monomer CP by distillation, and allowing the concentrated CP to dimerize under controlled conditions. The process of preparing CP from DCP is illustrated in reference 22. This method of cracking DCP to CP can be used with either crude or pure DCP streams and includes detailed temperature data on the process described.

The total potential supply of readily available cyclopentadiene from various sources has been estimated at >4500 metric tons per year. About 910 t/yr can be recovered from the forerunnings of coke-oven benzene, and the rest from hydrocarbon-cracking operations. Based on the total operations, more cyclopentadiene is produced than indicated above, but because of the relatively small quantities available at a single location, much of the cyclopentadiene cannot be recovered profitably.

The purity of a dicyclopentadiene stream is usually expressed in terms of available monomer. The analysis is determined by gas chromatography. The sample is charged to the gc equipment under conditions that cause essentially complete reduction of the dimers to monomers.

In the United States and Europe, DCP streams of 70–95 wt % purity are available (5). This is particularly true with large scale steam-cracking operations in Europe. Estimates of dicyclopentadiene production for 1980 in Western Europe are 350–700 t, and U.S. production of refined or higher purity dicyclopentadiene (95 wt %) is estimated to be 16,000–25,000 t. The U.S. capacity for all purity levels of DCP in 1977 was 78,000 t (23).

The production of DCP is a relatively low energy process since only one distillation is required to separate the cracked monomers, ie, the crude DCP stream from the higher boiling dimers. However, the consumer of cyclopentadiene monomer must go through the same cracking operation to obtain the monomer. From the values in Table 1, the energy required to crack a kilogram of DCP to CP is ca 2800 kJ (ca 2650 Btu). Although this is not very large, it illustrates the process problem facing the user relative to other diolefins that do not require a separate operation to produce the monomer.

Handling and Storage

Cyclopentadiene monomer spontaneously dimerizes at room temperature. The approximate dimerization rates of liquid cyclopentadiene at temperatures normally encountered in handling cyclopentadiene monomer are as follows:

Temp, °C	Approx dimerization rate, mol %/h	Temp, °C	Approx dimerization rate, mol %/h
−20	0.05	25	3.5
0	0.5	30	6
10	1	35	9
15	1.5	40	15
20	2.5		

The dimerization is highly exothermic (ca 75 kJ/mol or 18 kcal/mol) and, as the above rates indicate, increases rapidly with increase in temperature. Therefore, if the temperature around a closed container is not low enough to dissipate the heat of reaction through the available surface, the temperature rises, the dimerization reaction accelerates, and the pressure can increase rapidly enough to cause a rupture of the container. Because of this spontaneous dimerization and the fact that dicyclopentadiene is readily reconverted to the monomer, the dimer forms a safe and convenient way of handling cyclopentadiene commercially. Generally, dicyclopentadiene is moved in heated barges, or heated tank cars or trucks.

Dicyclopentadiene can be easily cracked to the monomer. The dissociation is a monomolecular reaction; in the pure liquid state the rate is expressed by the equation $k = 6 \times 10^{12} \exp(-34,000/(RT \cdot s))$ The cracking can be accomplished by distilling the dicyclopentadiene under atmospheric pressure. As dicyclopentadiene boils at 170°C, it cracks at a rate of ca 36%/h; by maintaining the top temperature of the fractionating column at 41–42°C, a distillate consisting of pure cyclopentadiene monomer is obtained. Another procedure involves adding dicyclopentadiene to some kind of hot liquid, preferably a high-boiling oil, at 250–260°C. The resulting vapors are fractionated to remove refluxing, uncracked dimer, and entrained liquid. The most practical method, from a commercial standpoint, consists of cracking the dimer to the monomer in the vapor phase with relatively short contact times at 350–400°C. Virtually complete cracking of the dimer to the monomer occurs under these conditions.

The distillation and/or cracking of dicyclopentadiene should be carried out in an inert atmosphere to prevent the formation of peroxides. Dicyclopentadiene should never be distilled to dryness since there is danger of explosion if peroxides are present. Also, when cooling a piece of equipment that contains the hot bottoms from the distillation or cracking of dicyclopentadiene, care should be taken to exclude air. Contacting the hot bottoms with air could lead to rapid peroxide formation with a resulting disruptive polymerization.

If pure monomer is to be used in a reaction, it must be used immediately or stored at ca −20°C to prevent dimerization to any appreciable extent. Chemical inhibition does not prevent dimerization; low temperature is preferred. If the monomer has to be stored over any length of time, it must be protected against oxygen to repress peroxidation and polymer formation. Cyclopentadiene monomer reacts spontaneously with oxygen of the air to form brown, gummy peroxide-containing products.

Toxicology

Toxicological studies conducted on dicyclopentadiene indicate that it is a moderately toxic material and, to some extent, an irritant and a narcotic. By oral administration in the rat, the LD_{50} is 0.82 g/kg of body wt, and by skin absorption in the rabbit, the LD_{50} is 6.72 mL/kg. An atmospheric concentration of 2000 ppm causes death in rats exposed for a period of 4 h.

Studies with rats indicate that dicyclopentadiene is not a myelotoxic chemical; ie, it has no deleterious effects on the blood and blood-forming organs. Toxicologically, it appears to be similar in action to the terpenes rather than to benzene.

The 1978 TLV for DCP was established at 5 ppm by the ACGIH. This is for a daily 8-h exposure on a time-weighted average.

Uses of Dicyclopentadiene

Owing to the unusual reactivity of the DCP molecule, there are a number of wide and varying end use areas. The primary uses are: (1) DCP-based hydrocarbon type resins; (2) elastomers via a third monomer: ethylidenenorbornene; (3) polychlorinated pesticides; and (4) polyhalogenated flame retardants.

Hydrocarbon Resins. There has been a growing interest in the use of dicyclopentadiene in both crude and purified form as a monomer in hydrocarbon resin production (see Hydrocarbon resins). These resins, produced in both the United States and Europe, are polymerized from concentrated DCP streams or from so-called resin oils which are obtained as by-products from steamcracking naphtha or gas oil. The DCP-containing stream may be polymerized with Friedel-Crafts (qv) (24) catalysts either alone or in admixture with reactive aromatics (the resin oil usually contains unsaturated aromatics, eg, styrene, vinyl toluene, indenes, etc), or with aliphatic olefins and diolefins (25) or by thermal polymerization (26).

Producers of hydrocarbon resins buy either high or low purity cyclic streams to make a specific type of resin characterized by softening point, color, unsaturation, etc. In many cases these variables can be controlled or modified as a function of the wt % DCP present. These products are generally marketed as solids in the form of flakes or pellets in bags, drums, and bulk quantities. The major end use areas are adhesives, rubber tackification, surface coatings, and resin replacement.

Elastomers. Ethylene–propylene terpolymer (diene monomer) elastomers (EPDM) use a variety of third monomers during polymerization (see Elastomers). Ethylidenenorbornene is one of these important monomers and requires dicyclopentadiene as a precursor. The amount of third monomer used in any polymerization varies, but it is usually present at less than 10 wt % of the finished polymer. Ethylidenenorbornene is synthesized in a two step preparation, ie, a Diels-Alder reaction of cyclopentadiene (via cracking of DCP) with butadiene to yield the bicyclic, 5-vinylbicyclo[2.2.1]-hept-2-ene (7) where the external double bond is then catalytically isomerized toward the ring yielding 5-ethylidenebicyclo[2.2.1]hept-2-ene (ENB) (8).

Polychlorinated Pesticides. A once substantial but diminishing use for DCP is in the preparation of chlorinated derivatives for further use or synthesis into pesticide compounds (see Insect control technology). Soil permanence and solubility of the

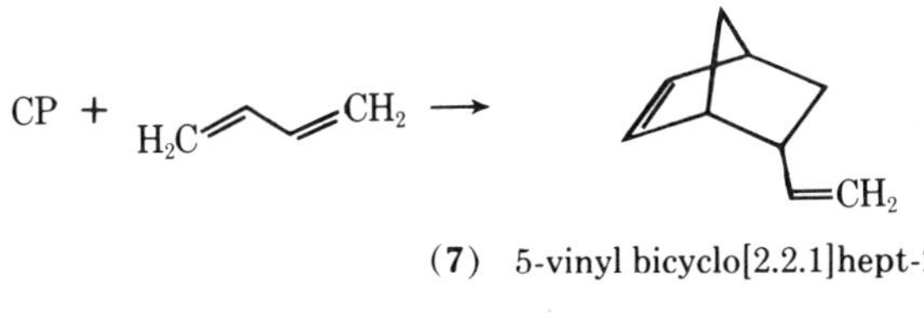

CP + H₂C=CH-CH=CH₂ →

(7) 5-vinyl bicyclo[2.2.1]hept-2-ene

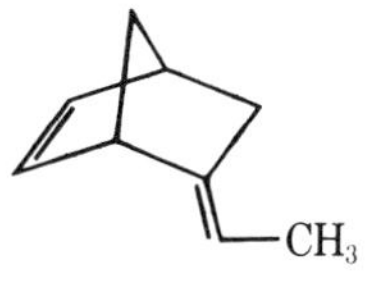

(8) ethylidenenorbornene, ENB

CYCLOPENTADIENE AND DICYCLOPENTADIENE

products in human fatty tissues have considerably restricted the use of this type of compound. The more prominent chlorinated pesticides were aldrin, dieldrin, chlordane, and heptachlor, all of which use hexachlorocyclopentadiene (9) as a starting material (see Chlorocarbons).

Aldrin (10) and dieldrin (11) are no longer used; in January, 1978, the producer of chlordane (12) and heptachlor (13) and the EPA agreed on phase-out of the latter two. In an *agreement in principle,* production of chlordane and heptachlor would be halted immediately (1977) for certain agricultural uses and phased out completely over the next few years for other uses. Since 1974, the EPA has sought to ban every use except for eradication and control of subterranean termites.

Flame Retardants. Although the use of DCP has been restricted in the pesticide area, an increasing area of interest is in flame and fire retardant chemicals (see Flame retardants). The starting material is the fully chlorinated DCP cracked to monomeric hexachlorocyclopentadiene, which is then converted via a Diels-Alder reaction with maleic anhydride to a reactive bicyclo anhydride (14). The bicyclic anhydride can then be esterified with polyols yielding a polyester with a residual double bond available if necessary for further reaction.

$$(14) + HOCH_2(CH_2)_n CH_2OH \rightarrow \text{polyester}$$

Fuels. Owing to their high density, both DCP and CP produce high heat levels when burned. This property would make them excellent high energy fuels except for their reactive characteristics and, therefore, poor storage stability. The tetrahydrogenated products of these reactive diolefins have been claimed to be high energy fuels for racing cars, missiles, etc (27–28) (see Fuels).

Derivatives

Dicyclopentadienedicarboxylic Acid. Sodium cyclopentadienide (15) can be carboxylated to give dicyclopentadienedicarboxylic acid (16) in 85–95% yields. Spectroscopic studies indicate conjugation of carboxyl groups with ring double bonds.

The melting point of the acid varies with the temperature of carboxylation. This is probably caused by a variation in the ratio of stereoisomers, endo and exo, formed at the various temperatures.

Temp of carboxylation, °C	mp, °C
−20--−30	200–203 (dec)
20–30	205–208 (dec)

The dicyclopentadienedicarboxylic acid can be used in making alkyd resins (qv) that have excellent air-drying and baked-film properties. It can also be converted into many other products characteristic of dibasic acids.

Similar to DCP, dicyclopentadienedicarboxylic acid splits into the monomeric form in the presence of a dienophile to give an adduct of the monomeric acid. The reaction takes place with stoichiometric amounts of dienophile and the dibasic acid at 130–190°C. The yield of adduct is usually 75–95%. A large number of polyfunctional compounds are easily prepared in this manner. The reaction with maleic anhydride gives a tribasic acid (17).

Other Derivatives. Like CP, DCP can be polymerized readily to form resinous products. Copolymerization of DCP with other unsaturated substances has received even wider attention, and several useful applications have been developed. With drying oils, copolymerization leads to the production of resinous products that give improved drying and result in coatings of greater resistance to weathering. Dicyclopentadiene also reacts with alkyd resins, with styrene, and with sulfur to form copolymer resins having useful applications (5,29).

Since World War II, a number of useful reactions of dicyclopentadiene (and higher polymers) have been described in the patent literature. These involve addition of the reacting chemical at the double bond of the bicycloheptene ring portion of the dicyclopentadiene molecule to give such products as secondary alcohols, ethers, esters, and halides. When *endo*-dicyclopentadiene is used as the starting material, the resulting products have the exo configuration. For example, in the presence of dilute sulfuric acid, water adds to dicyclopentadiene to form a secondary alcohol (18).

$$DCP + H_2O \xrightarrow[95-100°C]{(20-50\%)\ H_2SO_4} HO-\text{(18)}$$

In an analogous manner, dicyclopentadiene reacts with alcohols and phenols to form ether derivatives; with halogen acids, with thiocyanic acid, and with various carboxylic acids to form esters. With formic acid, a 96% yield of dicyclopentadiene formate has been reported. The products formed are described as useful in varied applications, such as plasticizers, solvents, hydraulic fluids, intermediates for synthetic waxes, resins, drugs, wetting agents, and insecticides.

In recent years an interesting rearrangement (30) has been found to take place when tetrahydrodicyclopentadiene (19) is heated in the presence of aluminum trichloride (30). The rearrangement is as follows (19) and (20):

(19) → (20)

Adamantane (20), tricyclo[3.3.1.1^{3,7}]decane, is a base for drugs that control German measles and influenza (31–32) (see Chemotherapeutics, antiviral).

BIBLIOGRAPHY

"Cyclopentadiene and Dicyclopentadiene" in *ECT* 1st ed., Suppl. 2, pp. 282–296, by H. K. Wiese, Esso Research and Engineering Company; "Cyclopentadiene and Dicyclopentadiene" in *ECT* 2nd ed., Vol. 6, pp. 688–704, by H. K. Wiese, Esso Research and Engineering Company.

1. A. F. Plame and J. M. Terent'eva, *Usp. Khim* **20**, 560 (1951).
2. J. H. Wells and P. J. Wilson, *Chem. Rev.* **34**, 1 (1944).
3. M. Moulin, *Bull. Assoc. Fr. Tech. Pet.* **135**, 563 (1959).
4. A. S. Onishchenko, *Diene Synthesis*, Old Bourne Press, London, 1964, pp. 274–320.
5. W. Meyer, *Hydrocarbon Process.*, (Sept. 1976).
6. K. C. Ramey and D. C. Lini, *J. Magn. Res.* **3**, 94 (1970).

7. P. V. French, L. Roubinek, and A. Wassermann, *Proc. Chem. Soc.* 248 (1960).
8. *Chem. Eng. News* **38,** 46 (Dec. 19, 1960).
9. J. C. Martin and R. K. Hill, *Chem. Rev.* **61,** 537 (1961).
10. D. R. Eckroth, *J. Org. Chem.* **41,** 394 (1976).
11. K. Hafner and K. H. Vöpel, *Angew. Chem.* **71,** 672 (1959).
12. K. Alder and G. Stein, *Ann.* **504,** 205 (1933).
13. J. Pirsch, *Ber.* **67B,** 1115 (1934).
14. K. Alder and G. Stein, *Ber.* **67B,** 613 (1934).
15. K. Alder and co-workers, *Ann.* **496,** 204 (1932).
16. G. Becker and W. A. Roth, *Ber.* **67B,** 627 (1934).
17. E. T. McBee and H. E. Ungnade, *Chem. Rev.* **58,** 249 (1958).
18. N. A. Milas and L. S. Maloney, *J. Am. Chem. Soc.* **62,** 1841 (1940).
19. R. Criegee, *Ann. Chem.* **481,** 263 (1940).
20. P. Seguin, *Compt. Rend.* **216,** 667 (1943).
21. G. O. Schenk and D. E. Dunlap, *Angew. Chem.* **68,** 248 (1956).
22. U.S. Pat. 2,831,904 (Apr. 22, 1958), R. W. F. Krept (to Shell Dev. Co.).
23. *Chem. Mark. Rep.,* (Mar. 26, 1977).
24. U.S. Pat. 3,023,200 (Feb. 27, 1962), M. E. Epstein and N. C. Gangemi (to Pennsylvania Ind. Chem. Corp.).
25. Ger. Offen. 2,014,424 (Dec. 23, 1970), H. L. Bullard, R. A. Osborn, and D. R. St. Lyr, (to Goodyear Tire & Rubber Co.).
26. U.S. Pat. 3,084,147 (Apr. 2, 1963), L. P. Wilks (to Velsicol Chem. Corp.).
27. U.S. Pat. 3,381,046 (Apr. 30, 1968), C. A. Cohen and C. W. Muessig (to Esso Research & Engineering Co.).
28. U.S. Pat. 3,002,829 (Oct. 3, 1961), J. J. Kolfenbach and H. K. Wiese (to Esso Research & Engineering Co.).
29. P. L. Smith, C. W. McGary, Jr., and L. R. Comstock, *Society of the Plastics Industry, Inc., 22nd Annual Meeting,* Washington, D.C.
30. P. von R. Schleyer, *J. Am. Chem. Soc.* **79,** 3292 (1957).
31. *Chem. Week* **94,** 119 (Apr. 25, 1964); A. W. Galbraith and co-workers, *Lancet ii,* 113 (1971).
32. R. C. Fort, Jr. and P. von R. Schleyer, *Chem. Rev.* **64,** 277 (1964).

M. FEFER
A. B. SMALL
Exxon Chemical Company

D

DAIRY PRODUCTS. See Milk products.

DAMMAR, DAMAR. See Resins, natural.

DATA INTERPRETATION AND CORRELATION. See Engineering and chemical data correlation; Programmable pocket calculators.

DDT (ClC$_6$H$_4$)$_2$CHCl$_3$. See Insect control technology.

DECAHYDRONAPHTHALENE, DECALIN. See Naphthalene.

DECARBOXYLASES. See Enzymes.

DEFLOCCULATING AGENTS. See Clays; Dispersants; Flocculating agents.

DEFOAMERS

A defoamer is a material that, on addition to a foaming liquid in low concentration, controls the foam problem. During the last twenty years the use of formulated chemical defoamers in many aqueous industrial processes has grown in importance. Instead of being used as a convenience or for simple removal of unsightly surface foam, defoamers have become a necessary process aid in many industries. Some industries that rely heavily on the efficient and economical use of defoamers are the pulp and paper industry, the paint and latex industry, coating processes of all types, the fertilizer industry, the textile industry, fermentation-based processes, the metal working industry, adhesive manufacture, polymer manufacture, the sugar beet industry, and various chemical processes. Defoamers are used as process aids or to increase the quality of the finished product. As process aids, defoamers improve filtration, dewatering, washing, and drainage of many types of suspensions, mixtures, or slurries. They are also needed to increase holding capacity of various kinds of vessels, improve efficiency of distillation or evaporation equipment, and to improve lubrication efficiency in both aqueous and nonaqueous applications.

Finished products such as paints, coatings, or adhesives are improved in quality through proper use of defoamers because the thin films or coatings that are laid down are free of uneven spotting or pin holes caused by foam. The quality and cleanliness as well as the speed of pulp and paper production is improved by the use of formulated defoamers. Recently, defoamers have found application in energy and water conservation by making possible more efficient use of mechanical equipment. In certain cases, drastic reductions in water requirements have been documented. Many municipal and industrial treatment facilities now rely upon defoamers to rid the aeration basins and lagoons of unsightly and troublesome foam. This type of foam can cause electrical failures if left unchecked. These are just some of the varied applications for defoamers.

The Market

Defoamer application is a large and very specialized market. Many chemical companies currently provide proprietary products, technical service, and sales departments specifically for the marketing of defoamers to industry. For example, there are more than fifty listed suppliers of defoamers for the paper industry (1). The market for defoamers of all types is probably greater than 136,000 metric tons per year in the United States, at least equal to that in the rest of the world.

Use Requirements of a Defoamer

To be marketable, a proprietary defoamer must meet five basic requirements in addition to any special criteria for a particular industry or process: (1) cost efficiency, (2) ease of handling, (3) specificity of action, (4) absence of any adverse effect on the final product, and (5) environmental safety.

Cost-Effectiveness. To meet most industry requirements, a defoamer must be effective and economical in the specific process for which it was intended. Defoamer economics are usually measured in cost per unit of product produced, such as cost per liter of paint or coating, cost per ton of pulp or paper, cost per million liters of plant effluent, or cost per batch of chemical produced. In this way different defoamers can be compared fairly no matter what the relative cost per kilogram of defoamer. One industrial application engineer may find that controlled use of a relatively expensive defoamer results in a lower cost per unit of product than use of an inexpensive defoamer. Another may determine that his process is so variable or uncontrollable that even though the expensive defoamer performs more efficiently there is significant waste and over-use, making the inexpensive product more attractive. Defoamer suppliers have technical sales personnel who are well versed in the applications being considered and in the use of their products, and who can suggest and help in implementing the most efficient application methods. They are usually backed by extensive technical service groups and regional laboratories whose main mission is to solve specific defoamer application problems.

Many types of equipment are available to control and minimize defoamer usage.

One example is a level controller which activates the defoamer feed pump as the foam reaches a certain predetermined level. This equipment may be based upon the conductance or capacitance principle.

Generally, equipment designed to control or minimize defoamer consumption has met with success. However, mechanical equipment, such as sonic devices, designed to control foam has not met with wide industry acceptance. Defoaming systems vary in concentration of foaming ingredients, pH, temperature, contaminants, etc. A sonic device is thought to be unable to cope with these changes in the nature of the system (see Ultrasonics).

One of the few mechanical devices that is used is a simple high pressure water spray or shower which simply dissipates surface foam. The addition of low levels of chemical defoamer increases the effectiveness of this type of spray device.

Convenience and Ease of Handling. Formulated defoamers are currently supplied as liquids, emulsions, pastes, bricks, flakes, powders, etc. By far the bulk of all chemical defoamers are pumpable fluids which are the most convenient to handle. These pumpable fluids may be 100% active, or they may be emulsions.

In many instances defoamers are diluted into aqueous emulsions prior to being fed into the foaming process. Care must be observed when defoamers are diluted because factors such as inadequate mixing, hard water, water temperature, bacterial contamination, and length of emulsion storage time may have pronounced deleterious effects on the defoamer.

In general, if a pumpable liquid product can be used, and if such a product can be fed at 100% activity directly into the foamy system, then these handling problems are reduced or minimized. When a chemical defoamer is fed directly into a system at 100% activity, it should be added at a point of good agitation just prior to the foam problem. The mixing ensures good distribution of the defoamer throughout the aqueous system.

Specificity of Action. In the manufacture of pulp and paper the most economical and effective application of defoamers is to use a product tailor-made for the particular requirement of each phase of the operation (see Paper; Pulp). In the pulp mill brown stock washing area, temperatures of over 82°C and soap and high alkalinity conditions are present. In most mills insoluble defoamers are necessary for this severe condition. In the screen room area of the pulp mill, conditions are not as severe (ie, temperatures are usually below 53°C, soap solids and alkalinity are less). In the screen room, emulsifiable defoamers are the most effective; the insoluble products used in the brown stock area simply do not spread well enough to become distributed in the foamy pulp slurry. The bleach plant offers another defoaming challenge. Here, depending upon the specific bleaching stage and sequence of pulp bleaching, the pH can vary from 1.5 to 11.0 and the temperature from 21°C to 65°C. Specific defoamers designed for wide range effectiveness are used in the bleaching of pulps.

On the paper machine, there are no naturally occurring soaps and the pH usually ranges from 4.2 to 7.0. The defoamer is needed to help drainage and formation of the sheet on high speed continuous paper-machine wire. There are blends of esters and mineral oils that are extremely effective in this use. Paper coating also presents a unique problem and there are specific products designed for this application. They must be highly effective yet have no adverse effect on the finished coating.

Mill effluents require products that are long lasting and will carry through kilometers of piping and will be effective in aerated treatment ponds. Both oil-based and water-based defoamers are effective in this area.

Thus a paper mill may easily use six or more defoamers for routine coated paper manufacture. The same holds true for other defoamer applications. For example, a paint manufacturer may need different defoamers for the various phases of paint production such as in the pigment grind, letdown, can filling, and mill effluent. Manufacturers of textiles, sugar from sugar beets, and fertilizers require different defoamers for the various phases of production.

A manufacturer must weigh the convenience of using one defoamer throughout the operation versus the economy of using several tailor-made products. In most cases (as the paper industry has found), it was determined that the use of multiple defoamers is preferable because of cost savings.

Effects of the Defoamer on the Finished Product. Another basic requirement is obvious but of utmost importance: a defoamer must not have an adverse effect on the finished product. Some deleterious effects that can be caused by improper use of chemical defoamers are loss of sizing or water repelling in paper, the appearance of fisheyes (uneven spotting in paint or coatings), and the increase of BOD (Biological Oxygen Demand) in plant effluents. There are many ways of offsetting these potential adverse effects which are caused chiefly by insolubility or incompatability problems. Changing the feeding method or points of addition often eliminate them; so can changing the defoamer solubility through minor formula alterations such as the reduction or increase of emulsifier levels.

Environmental Effects. The newest basic requirement that must be met by a useful and serviceable chemical defoamer is that it must be ecologically safe. Components used in defoamers must pass not only specific industry standards but also state and federal regulations. There are now state laws prohibiting the use of certain chemicals in any industrial plant effluent. There are also federal regulatory lists of chemical components that can be used in defoamer formulations for the various industries. For example, the *Code of Federal Regulations of 1978* lists chemicals that can be used as defoaming agents for foods, defoamers for pulp and paper production, and for coatings (2). These regulations are becoming more stringent each day as more is learned about the characteristics and shortcomings of the chemicals currently in use. In general, however, formulated defoamers are rather inert chemical mixtures, because by their very nature, they must reside for a reasonable amount of time in the foaming system and, therefore, cannot be chemically reactive. Personnel safety goes hand in hand with ecological safety, and most suppliers of chemical defoamers have on-going programs of improving the safety characteristics of their products. In addition, new OSHA regulations prohibit the use of flammable or toxic material as defoamers for any industrial process. Future regulations of defoamers will become more restrictive as they will with all chemical additives. Present suppliers of these products will undoubtedly be required to have available general and perhaps more specific formulation data for government agency review. The Toxic Substance Control Act of 1976 (TSCA) is the first step in this direction. This requirement, although necessary, may create a problem for the defoamer supplier in that the proprietary nature of the product may no longer exist.

Recent History of Defoamers in Industry

Defoamers are complex blends of many different organic compounds. To understand the need for these different blends and compositions, it is useful to review the recent history of defoamers in various industrial applications.

Pulp and Paper. The paper industry has always had some of the most critical and troublesome foam problems, so this industry has led the way in the use of defoamers. Only three decades ago many pulp mills were using large amounts of kerosene or fuel oil to reduce foam in the pulp mill or on the paper machine. The use of these hydrocarbons caused not only ecological problems but cleanliness problems throughout the mill. It was discovered that blends of esters, waxes, sulfated tallows, and saturated soaps were far more effective than the hydrocarbon oils for defoaming many pulp and paper mill systems (3). These mixtures were the first paste-type defoamers which are still in use today. They worked well in all systems such as pulp screening, bleaching, and paper formation but not in the Kraft black-liquor application with its rigorous conditions of alkalinity, high concentrations of soaps, and extreme operating temperatures. The so-called black-liquor defoamers were commercialized specifically for this application. The very first products were simply mixtures of silicones (dimethylpolysiloxane) and kerosene, but later, more sophisticated products based on the use of hydrophobic (silicone treated) silica were tried with excellent results (4) (see Silicon compounds). The use of these defoamers not only enabled more Kraft pulp mills to run at capacity, but sometimes also to produce pulp at speeds greater than designed capacity. These defoamers met with wide commercial success and hydrophobic silica products soon found application in other industries such as paint, latex, and polymer manufacture.

Sugar Beet Industry. In the production of sugar from sugar beets, foaming was a serious problem, especially in the diffusion area (see Sugar). Before regulatory agencies were made aware of the problem, many producers used spent motor oils as defoamers in this process. Today this practice has ceased, and only defoamers formulated from specifically approved raw materials can be used for sugar production (5).

Paint. The paint industry began to use defoamers as soon as water-based paints gained widespread use. The basic problem here was one of a need to make the process more efficient and expedient (and, therefore, more profitable) without adversely affecting the quality of the finished paint. It was found immediately that the silicones (100% dimethylpolysiloxane products) were excellent defoamers for latex-based paints of almost all types, but no matter how rigorously they were mixed into the paint, small spots called fisheyes always appeared as the paint was applied by brush or roller. The chemical defoamer suppliers attacked this problem customer-by-customer and even paint-by-paint, and within a few short years there was a family of effective products that the paint manufacturer could choose from (6). These defoamers, which are compatible with most latex paint systems, are mostly complex mixtures of various surfactants, hydrocarbon oils, polymers, and proprietary silicone derivatives. The paint, latex, and industrial coating industries remain a growing market for the chemical defoamer supplier to service (see Latex technology; Paint; Coatings).

Following is a summary of a few other industries and the types of products used as defoamers.

Textiles. Although silicone antifoam agents are effective in the jet dyeing process, they were found to inhibit the fire retardancy of the fabric. Today, the trend is definitely toward proprietary formulated products based on fatty amides, mineral oils, and emulsifiers. Modified silicone products, such as mixtures of silicone glycol copolymers, silicone oils, and silicas, are also commercially successful (7) (see Textiles).

Fermentation Processes. Fermentation processes such as the production of penicillin and yeast require defoamers to control evolution of gases produced during the reaction. Fermentation that is allowed to proceed freely without control is extremely inefficient. In the past, fatty oils such as tallow or lard oil were used as combination fermentation defoamer and nutrient; however, today much more effective proprietary products are widely used. Modern fermentation defoamers are based on blends of fats, polyols, fatty esters, and fatty alcohols (8) (see Antibiotics; Fermentation; Yeast).

Metal Working. The metal working industry has always been plagued with foaming problems with cutting oils and coolants which are sprayed onto the tool–workpiece interface to help provide cooling, controlled lubrication, corrosion protection, and increased tool life. For years this large industry simply lived with this situation. Coolant manufacturers are currently incorporating formulated defoamers into their products to prevent foaming. These products may be either formulated silicones or dispersions of fatty amides in mineral oils. They have been found to be very effective, especially in large central systems.

Fertilizer Industry. Finally, the fertilizer industry has for years used tall oil fatty acids as defoamers for the digestion of phosphate rock into sulfuric and phosphoric acid. This process is extremely foamy and obviously very acidic conditions are encountered. Recently, formulated products have proven much more satisfactory. Many of these products continue to contain tall oil fatty acids, but they are formulated with emulsifiers and sulfated products which greatly increase their effectiveness. Certain surfactants are also very effective in this application.

Other industries that use formulated defoamers are oil well cementing, adhesive and caulking compound manufacture, cleaning compounds of all types, dispersants for cooling towers, and primary and secondary waste treatment facilities of all types.

Defoamer Suppliers

Many suppliers compete in the defoamer market for the pulp and paper, paint, latex, and coating, textile, fertilizer, fermentation, waste treatment and chemical industries. Some are relatively small companies; others are simply small divisions or departments within large chemical corporations. Owing to the very competitive nature of the defoamer market, the profitability associated with other specialty chemicals is sometimes lacking. In some cases, the high volume associated with the larger defoamer markets overrides the depressed profit.

Functional Classifications of Defoamer Components

The components in most defoamers can be divided into five general functional classifications. Many defoamers contain more than five components and some contain only two or three, so obviously there is some overlapping of the function of each component. Most defoamers contain: a primary antifoam agent, a secondary antifoam agent, a carrier or vehicle, an emulsifier or spreading agent, and a coupler or stabilizing agent.

Primary Antifoam Agent. The primary antifoam agent, the main active ingredient in the defoamer, frequently consists of highly insoluble particulate materials such as hydrophobic (silicone-treated) silica, fatty amides, hydrocarbon waxes, fatty acids, and fatty esters. In most cases, the primary antifoam agent is insoluble in the carrier as well as in the foaming media. Some chemical defoamers function by the inverted cloud-point mechanism. These products are soluble at low temperature and precipitate when the solution is warmed to a certain temperature. When precipitated these defoamer–surfactants function as defoamers; when dissolved they may even act as foam producers.

These primary components are the backbone of the defoamer formulation and are usually the most costly ingredient in the product. The mechanisms of function are discussed below.

Secondary Antifoam Agents. Secondary antifoam agents have a much less obvious effect on the formulation. They function by providing a synergistic effect with the primary antifoam agent. The secondary antifoam agent is the defoamer formulator's tool in that these components usually modify the surface effect of the primary antifoam agent by modifying the spreading, solubility, or crystallinity of that material with regard to the carrier and the foaming media. Some secondary antifoam agents are fatty alcohols, fatty esters, silicones, and certain oil insoluble polymers.

Carriers. Carriers, or vehicles, comprise the bulk of the defoamer formulation. They are usually hydrocarbon oils or water, although other products such as fatty alcohols, solvents, or fatty esters are sometimes used. Ideally the carrier should not only serve to introduce the main and secondary antifoam agent into the system, but should also contribute to product performance. Certain carriers have been found to be optimal under specific pH, temperature, and application situations, and these are preferred. Today many defoamer suppliers are considering water as a carrier for emulsion-type defoamers owing to the ever increasing cost and somewhat unreliable supply of hydrocarbon raw materials.

Emulsifiers. Even though a defoamer is insoluble, it must spread or emulsify rapidly into the system. Since application of a chemical defoamer actually pits the rate of foam collapse (owing to the activity of the product) against the rate of foam formation, it is imperative that the defoamer contact as much of the surface as possible. Thus the defoamer must be made to disperse quickly.

Emulsifiers or spreading agents function by introducing the main components (primary antifoam agent, secondary antifoam agent, and carrier) into the system. These products dictate the speed of foam decay by spreading the product quickly throughout the foaming media. Spreading effectiveness is one of the most important measurable characteristics of a defoamer no matter what the application. Examples of emulsifying or spreading agents are esters, ethoxylated products, sorbitan esters, silicones, and alcohol sulfates (see Emulsions).

Some chemical defoamers, such as those used to defoam Kraft black liquor, do not make use of spreading or emulsifying agents because there are sufficient soaps and surfactants in the liquor to provide more than enough emulsifying agent. This is also true in many latex paint systems, as latex paints contain large amounts of surfactants and dispersants.

Coupling or Stabilizing Agents. Coupling, or stabilizing agents are additives that contribute to defoamer stability or shelf life. Frequently these components are cosolvents for the primary or secondary antifoam agents and the carrier. In a water-based defoamer the stabilizing agent may be simply a preservative added to prevent bacterial spoilage in the drum or other shipping container. Examples of coupling agents are red oil (oleic acid), hexylene glycol, fatty alcohols, naphthalene sulfonates, butyl alcohol, and formaldehyde.

A defoamer must be reasonably inert in the foaming system, because it must exhibit excellent longevity to be economically justified. A defoamer must not be capable of reacting with the finished product or the aqueous system for which it was designed. Almost all defoamers are considered unreactive in the particular process to which they are applied.

Clearly one component, such as a fatty ester, may fulfill many of these fundamental, functional classifications such as primary and secondary antifoam agents, and even carrier. On the other hand, some defoamers may have two or more primary antifoam agents, several secondary antifoam agents, one carrier, and several stabilizing agents. This is especially true of the newer water-based defoamers that are quickly becoming the fastest growing type of defoamer used in the pulp and paper industry. These water borne products and many other types of defoamers are somewhat unstable by their very nature as emulsions (oil-in-water or water-in-oil) or dispersions of solids in liquids, both thermodynamically unstable states. Coupling or stabilizing agents are one method of achieving product stability. However, another concept cannot be overlooked (9). Many dispersion-type formulated defoamers achieve greater stability by mechanical means such as high pressure homogenization, ball or sand milling, high shear mixing, and the like (10). In addition, optimization of process techniques, such as heating, cooling, and mixing, invariably adds to product stability, especially of emulsion-type defoamers. For example, when long-chain fatty alcohols or paraffinic waxes are used in emulsion-type defoamers, the rate of cooling of the emulsion affects product stability. Therefore, optimum process parameters are always sought. Process variables, such as cooling rate and homogenization techniques, also affect other factors such as product viscosity and performance, by determining both emulsion particle size and particle crystallinity (11).

Theory of Defoamer Action

There are a number of mechanisms whereby defoamers inhibit the formation of foam. Some mechanistic ideas are presented here that are novel and have not gained general acceptance because the theory of action of defoamers is a dynamic field where ideas are in a state of ferment.

This discussion focuses on foam control in aqueous systems although the principles are the same for nonaqueous systems.

Defoamers, in general, are insoluble in the foaming liquid and are more surface-active than the foaming liquid. The insoluble defoamer functions largely by spreading on the surface of the foam or entering the foam, as described later. Because the film formed by the spread of defoamer on the surface of a foaming liquid does not support foam, the foam situation is alleviated. It is well known that a low defoamer viscosity tends to promote defoaming action. This relates to the spreading velocity of a defoamer and to its dispersal throughout the foaming system.

The extraordinary stability of the delicate foam lamella is described by E, the Gibbs elasticity:

$$E = \frac{2\,dv}{d\ln A}$$

where v is the surface tension, $\ln A$ is the logarithm of the area of the film. The essential reason for the stability of foam, as indicated by the definition of E, is the ability of the lamella to respond to stress by local adjustment of the surface tension. A stressed area is an area of locally increased surface tension which draws at adjacent areas where the surface tension is normal. The underlying liquid is drawn as well so that the equilibrium film is healed. This is known as the Marongoni effect. A pure liquid cannot have adjacent surface areas differing in surface tension, and consequently, stable foams are unknown in one component systems.

The Young-Laplace equation indicates that the pressure inside a bubble is greater than the external atmospheric pressure:

$$\Delta P = 2 v \left[\frac{1}{R_1} + \frac{1}{R_2} \right]$$

where ΔP is the pressure difference; v is the surface tension; R_1 and R_2 are radii of curvature. Contrary to intuition, this equation implies that the pressure inside a small bubble is greater than that inside a larger bubble. Small bubble foams are more stable than large bubble foams because there is a greater relative amount of liquid available to counter the liquid loss by drainage that occurs as a result of the influence of gravity on the liquid in bubble walls. The action of a defoamer with time may often be observed in terms of an increase in average bubble size as well as a decrease in foam volume.

Spreading Action of Defoamers. The spreading of a liquid film on a film of foam is a highly effective mechanism of foam destruction. For spreading to occur the two liquids must be immiscible. The shock generated by the act of spreading is the cause of foam destruction (12). Schulman and Teorell (13) demonstrated that a film of oleic acid spreads over water at a velocity of 13 m/s (30 miles per hour). This effect may readily be observed by adding a drop of oleic acid (or other immiscible surface-active liquid) to a water surface dusted with talc. The talc is rapidly pushed away from the surface-active drop and a clean circular area of liquid is generated.

The thermodynamic requirement for the spreading of a duplex film (at least two molecules in depth) over a liquid surface is:

$$S = v_F - v_{DF} - v_D$$

where S, the spreading coefficient, is the negative of the free energy change. S should have a positive value for spreading to occur. v_F is the surface tension of the foaming liquid, v_D is the surface tension of the defoamer, and v_{DF} is the interfacial tension between foamer–defoamer phases.

When the defoamer phase becomes saturated with the foaming phase, the surface tension of the defoamer and the interfacial tension are affected, decreasing the value of the spreading coefficient. The foaming phase probably never becomes saturated with the defoamer phase since the volume of the former is so much greater than that of the latter. The ability of a defoamer to spread is diminished by solubilization of the defoamer in the foaming liquid.

When the spreading coefficient has become zero or negative, spreading is not necessarily halted. The spreading coefficient refers to spreading of a bulk or duplex film on a substrate. The spreading of a monomolecular film is also effective in the destruction of foam. Where a duplex film has not been removed by solubilization, the residual state is that of a monolayer with excess liquid collected as a lens (14). Spreading as a monomolecular film may also occur from the surface of solid particles utilized as defoamers in certain cases (15).

It would seem that spreading is of value not only as an important defoaming mechanism in its own right but also as a means of distribution of a defoamer formulation throughout a foaming system. Frequently, the spreading agent added to a hydrocarbon oil base is a water-in-oil emulsification agent as well, so that dispersion of a defoamer formulation is promoted by absorption of the aqueous phase into the minor amount of nonaqueous defoamer phase. Despite the tendency to think of surfactants as having a single function, there is no reason why an oil-soluble surfactant cannot function simultaneously as a spreading agent, emulsifying agent, dispersant, and even as a viscosity modifier.

Spreading involves the action of a low surface-tension liquid moving over the surface of a relatively high surface-tension aqueous foaming system. Low boiling materials such as acetone and diethyl ether are characterized by low surface tension values, and consequently, should make good spreading defoamers. In point of fact, they do but are impractical because of their ready loss by volatilization. The breath of diabetic patients contains acetone which explains the ability of such people to destroy the foam on a glass of beer.

The number of spreading agents is legion—propoxylates, ethoxylates, fatty alcohols, fatty acids, fats and oils, silicones, etc. Any material that decreases the surface tension of the defoamer and decreases the interfacial tension (foamer–defoamer) is a candidate. Such materials may be used neat if liquid, or diluted with a vehicle such as hydrocarbon oil, a defoamer in its own right. The diluted formulation is more common for economic reasons.

Entering Action of Defoamers. Robinson and Woods (16) indicated that entry of a defoamer droplet into a foam surface is associated with film collapse. The thermodynamic requirements for entry are as follows:

$$E = v_F + v_{DF} - v_D$$

where E is the entry coefficient, the decrease in free energy on entry.

For some years the entry coefficient was regarded as an invalid concept which reduced in mathematical terms to a trivial concept—a positive value of the entry coefficient indicates that the foaming agent does not spread on the surface of the defoamer. The spreading coefficient of the foamer on defoamer surface is given by:

$$S_{DF} = v_D - v_{DF} - v_F$$

Since S_{DF} is negative:

$$-S_{DF} = v_F + v_{DF} - v_D$$

and the term on the right is the entry coefficient of the defoamer on a foam lamella. Hence, a positive value of the entry coefficient does have the trivial implication that the foaming agent will not spread on the defoamer surface. However, the fact that the entry coefficient has a trivial implication in one respect is not inconsistent with its importance in another respect, namely, that it serves as a measure of the ability of a defoamer droplet to enter the foaming surface. There is no doubt today that entry is a valid concept since many cases can be cited in which the spreading mechanism and other mechanisms are inoperative.

The question arises as to what happens after entry of the defoamer droplet into the foam surface; bubble rupture does not instantly occur. According to Ross (17) the film is thinned by gravitational drainage until the defoamer lens bridges the surface. At this point the foam becomes highly vulnerable to rupture.

Recent work by Roberts and co-workers (18) based on cinematic microphotography indicates that emulsified drops of defoamer locate in the body of the foam lamella and cause the foam to rupture. Photographs demonstrate that the defoamer droplet remains stationary and intact after rupture while the foam rearranges itself.

Since the defoamer droplet was mobile in the foam lamella prior to rupture, the authors conclude that radially symmetrical forces were acting on the defoamer at the instant of foam rupture. Speculation as to mechanism takes two lines of approach. (*1*) If it is assumed that diffusion of the defoamer droplet in the foam lamella is relatively slow, an idea suggested by Ross may be applicable. As previously indicated, a positive entry coefficient necessarily implies that spreading of the foaming phase on the defoamer phase is thermodynamically barred. The withdrawal of the foaming phase from the surface of the defoamer droplet results in rupture. (*2*) If it is assumed that the rate of diffusion of the emulsified droplet is substantial while the droplet resides in the foam lamella, the surface tension at the air–lamella surface will be locally decreased in the vicinity of the droplet. This would result in stretching of the area of locally altered surface tension. If the Gibbs elasticity value of the film is exceeded, film rupture occurs.

It is well known that it becomes more difficult to defoam a system as the concentration of foam surfactant increases. Recently, this matter has been studied for the case of silicone defoamers. It was found that the loss of efficiency with increasing concentration of foam surfactant is caused by ionic surfactant adsorption (19). This results in a zeta potential (potential between the outer layer of an electric double layer and bulk solution) on the surfaces of defoamer droplets and foam bubbles which inhibits approach of droplet and bubble. Similar behavior might apply with other defoamer formulations, eg, hydrophobic silica. The distinction between the surfactant adsorption discussed in the cited work and the term solubilization may be purely a semantic one.

The solubilization capacity of foaming liquids may be diminished by pretreatment with a relatively large quantity of hydrocarbon oil so that later addition of a hydrocarbon-vehicle defoamer becomes quite effective. However, economically this is not a useful approach. The detailed relationship of the solubilization of a defoamer to the composition of the foamer is an area that remains to be explored. Clarification of this area must await clarification of solubilization phenomena in general, a field still in its infancy.

Hydrophobic Silica. Hydrophobic silica is finely divided silica coated with chemisorbed silica. Usually the silica is dispersed in hydrocarbon oil. Like silicone defoamers, hydrophobic silica particles present a low energy silicone surface to the foaming environment but they have the economic advantage of a low-cost silica interior. From an economic as well as a scientific viewpoint, the development of hydrophobic silica as a defoamer component by Boylan (20–21) represents one of the most significant recent developments in defoamer technology (see Silica). A major use for defoamers is in wood pulping. Many current pulping defoamers, as well as defoamers utilized for other purposes, consist in part of hydrophobic silica. An interesting consumer use of this composition is as an antifoam component in over-the-counter antacid preparations (see Gastrointestinal agents).

Prior to the development of hydrophobic silica formulations, relatively expensive silicone formulations were in use, eg, Dow Corning Antifoam A. Another defoamer used at the time was kerosene applied in large concentrations. The problem with kerosene and other hydrocarbon oils is that they are solubilized rapidly by the surfactants of foaming solutions. This behavior is to be contrasted with the behavior of hydrophobic silica which, by itself, in the absence of hydrocarbon oil has no defoaming ability. The combination of hydrophobic silica and hydrocarbon oil results in remarkable improvement in defoaming properties. Lichtman and co-workers have demonstrated that there is a critical hydrophobicity above which silica will act as a defoamer (22). Silica increases in hydrophobicity as the silica surface is increasingly covered with chemisorbed silicone. Defoaming performance continues to improve up to the point at which the silica surface is saturated with silicone. The critical degree of hydrophobicity is such that hydrocarbon oil preferentially wets the silica in the presence of foaming liquid, ie, the contact angle of hydrocarbon oil on silica in the presence of foaming liquid is less than 90°. It is significant that the conditions necessary for silica to act as a defoamer component are the same conditions required for silica to act as a water-in-oil emulsifying agent (23). Further, particle-stabilized emulsions are unusually stable compared to emulsions prepared with chemical emulsifiers. It is suggested that the function of hydrophobic silica is to form a tough coating about the emulsified oil droplet, thereby slowing solubilization of hydrocarbon oil, the active defoamer component.

Powdered Teflon, a fluorinated ethylene polymer more hydrophobic than silicone-treated silica, might be expected to perform impressively as a defoamer component (see Fluorine compounds, organic). However, with the usual hydrocarbon oil formulations, Teflon works poorly. The problem is that Teflon is so hydrophobic that there is little attraction to relatively polar hydrocarbon oil, and consequently, particles aggregate. If a low surface tension-vehicle is used, eg, a siliconized or fluorinated oil, powdered Teflon particles do not aggregate and the formulation performs well as a defoamer, although economically it is impractical.

Amide Particles. Amides, eg, lauramide (mp 102°C), in particulate form have been used in the past as defoamers for boiler water. The defoaming mechanism involved is quite different from that for hydrophobic silica as evidenced by the fact that direct addition of amide particles is effective in control of boiler water foam. Reference 24 discusses the nature of the monolayers formed as amide spreads directly from the solid particle surface onto water. The unformulated amides are not very effective at controlling severe foam problems.

Recently, comparatively high melting amide defoamers have been pioneered by the Diamond Shamrock Corporation, eg, distearamide of ethylenediamine (mp 140°C), for use as components of alkaline pulping defoamers and other applications (10). The problem with the high melting amides is that they are not well dispersed in hydrocarbon oil, nor in most other economic vehicles. The Diamond Shamrock patents show how amides can be dispersed in oil. The amide is first heated to achieve dissolution in oil and is subsequently chilled rapidly. The resultant product, after homogenization, consists of fine particles infused with oil and dispersed in the oil phase.

The mechanism of defoaming of the high melting amide particles is believed similar to that of hydrophobic silica, namely, deceleration of the rate of solubilization of hydrocarbon oil by the foaming medium is evidenced by the fact that amide particles in the absence of hydrocarbon oil are as ineffective as hydrophobic silica particles in the control of foam.

It is striking that contact angle measurements on an amide surface indicate that hydrocarbon oil wets the amide in the presence of foaming liquid to the same extent as that observed in the case of hydrophobic silica. The amide surface cannot be as hydrophobic as the silica surface; hence, this behavior is ascribed to a compensation of factors involved in the Young equation which governs the contact angle behavior of liquid drops on surfaces:

$$v_{SW} = v_{SO} + v_{OW} \cos \theta$$

where v_{SW} is solid–water interfacial tension, v_{SO} is solid–oil interfacial tension, θ is the contact angle measured in the oil phase. The compensatory factors must involve the terms v_{SW} and v_{SO}.

Liquid Crystals and Foam Stability. The viscosity of a film surface has a direct bearing on the stability of foam (25). Friberg and co-workers (26) have pioneered an approach related to this phenomenon which correlates and systematizes information. Three-component phase rule diagrams have been established for systems such as water–surfactant–medium chain length alcohol. A convincing explanation is offered for the fact that the alcohol component may serve as a foam stabilizer or a foam breaker: when the alcohol is added in low concentration, the liquid crystalline phase formed enhances film stability. On addition of alcohol in large concentration, a phase region is formed which is associated with film instability. This kind of approach offers promise for further understanding of defoaming mechanisms and a route to new advances in practical defoamer technology (see Liquid crystals).

Product Selection

In determining the best and most economical defoamer for a specific application, trial-and-error methods are still most often used. The experience and skill of defoamer supplier personnel is put to this test every day, but rarely does a supplier make a product recommendation without gathering copious background information. In most cases, at least cursory laboratory-scale evaluation of several different products is conducted before a recommendation is made. There are many laboratory methods for testing the relative merits of one defoamer against another or against a control foaming system with no defoamer (27).

Obtaining good correlation between laboratory and actual plant conditions has always been a problem. The laboratory test usually consists of obtaining a representative sample of the foaming medium and subjecting it to conditions similar to actual process conditions. Examples of these conditions are agitation, recirculatory pumping, turbulence, injection of air, shaking, addition of chemicals creating gases, and vacuum.

One of the most common devices used to evaluate defoamers in the laboratory is a simple recirculating system consisting of a graduated glass cylinder, a small pump and a piping system that recirculates foamy liquor through the system (Fig. 1). In this way the foaming tendency of the medium with and without defoamer can be observed by the operator. Various defoamers are added to this system at equivalent cost levels, and their relative defoaming tendencies are recorded. Measurements taken usually include the time required to collapse the foam already formed, the degree of deaeration obtained, and the length of hold down, or longevity, of the defoamer. A graph that illustrates the relative effectiveness of each defoamer tested can be drawn and one can choose the most efficient product. In cases of nearly equal product performance, it has been found that an experienced observer can usually detect the best product through consideration of size of bubbles and formation of the foam. With the use of this type of laboratory device as many as thirty defoamers can easily be tested in a few hours, and the correlation with actual industrial application is excellent. Many defoamer suppliers have either technical service personnel or sales personnel who are trained to conduct this type of laboratory testing. Much of this work is done at a potential customer's plant site to obtain fresh foaming liquors.

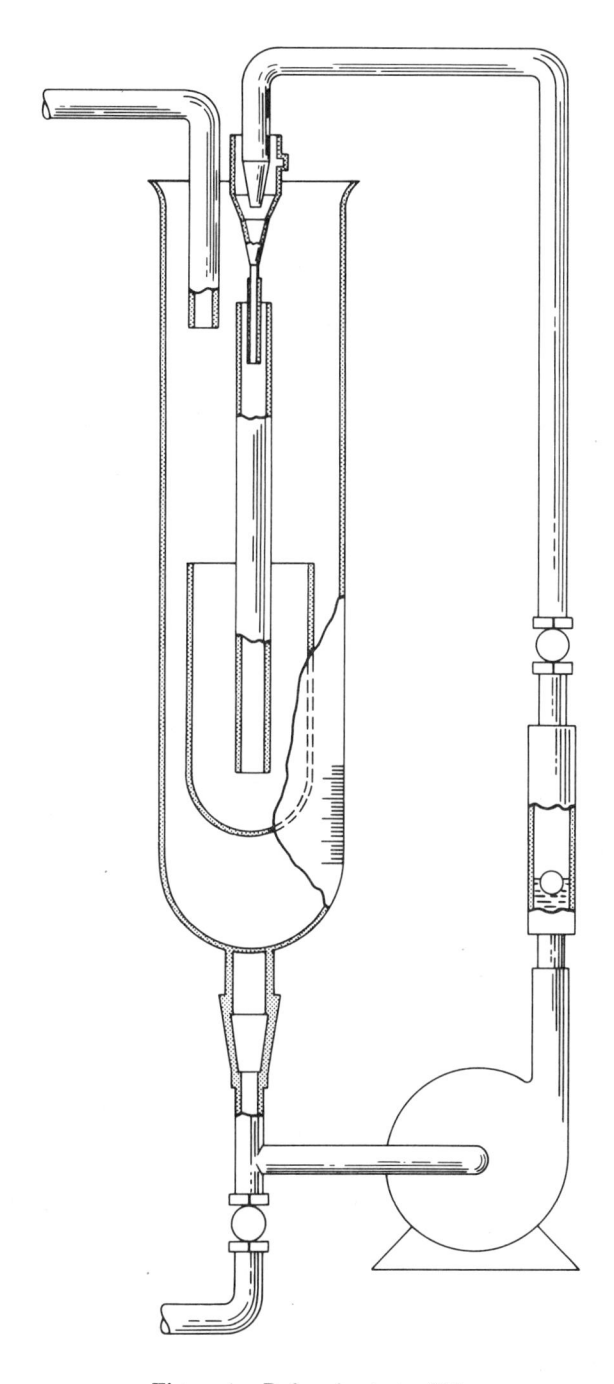

Figure 1. Defoaming tester (27).

444 DEFOAMERS

The chemical composition of the foaming system along with the mechanical conditions to which it is subjected are, of course, the most important data in determining an effective defoamer. Two other very important criteria are the pH and temperature of the foaming system. With only chemical and mechanical data along with pH and temperature, many experienced workers in the field can recommend a chemical defoamer that will perform well enough to control foam in an emergency situation. Over the years certain defoamer formulations have been found to defoam especially well when used on specific applications or under specific conditions of temperature or pH. The reasons for this phenomenon have not been clearly understood in all cases. Much detailed history on success or failure of defoamers in specific cases has been compiled by the defoamer suppliers. This information helps provide meaningful guidance in the selection of the most economical defoamer for a system.

Defoamers in Current Use

Many different types of defoamers are currently in use in a multitude of application areas. Some of the most common of these are described below.

Blends of Fats and Oils. Blended products have been on the market for years for use in pulp and paper, sugar beet, fertilizer, and effluent applications (28). Typical formulations may consist of solutions of saturated fatty acids and fatty alcohols in a mineral oil carrier or blends of polyethylene glycol esters in the same type of carrier. The carrier is not always necessary, and a simple blend of esters may be used. Inexpensive tall oil fatty acid (see Carboxylic acids) which has been sulfated or treated with a nonionic emulsifier is used as a defoamer for the digestion of phosphate rock. Blended products may be emulsifiable or may contain no emulsifiers, depending upon the composition and mechanical turbulence present in the foaming medium. The fatty blends are still widely used in many industries. The disadvantage of this class of defoamer is the titer effect that occurs when dissolved fats or fatty acids come out of solution at low temperatures. The product thus loses effectiveness owing to lack of homogeneity. Such products should be warmed gently to room temperature before use, and instructions stating this are usually provided or stenciled on drums or containers. These blends of fatty materials are slowly being replaced by other types. This is especially true in the pulp and paper industry.

Paste-Type Defoamers. Paste defoamers, consisting of immobile oil-in-water emulsions of fatty acid soaps, esters of saturated fats, paraffin waxes, mineral oils, emulsifiers, and the like, were the first large volume defoamers (29). These products still are used in the pulp and paper industry, adhesives, coating, paint, and the chemical industries. These rather stiff pastes are usually converted to dilute emulsions before introduction to the foamy system. Owing to this handling inconvenience, these products, although very effective, were replaced by the relatively easy to handle liquid blends mentioned earlier. The increasing costs of liquid carriers such as hydrocarbon oils have reversed this trend, however, and has made paste-type products more attractive, especially in the pulp and paper industry. Over the years automatic pumping systems for paste-type defoamers have been perfected. Many such systems have been in existence for over a decade with good results. The dilution of the paste defoamers has the disadvantage of being affected by temperature, formation of hard water soaps, creaming-out owing to lack of agitation, and the formation of insoluble deposits in the feed system. Although these products are currently still selling well, they are expected soon to be replaced by newer defoamers that are easier to handle and equally effective.

Dispersion-Type Defoamers. Many patents have been issued for dispersion-type defoamers (10,30–32). Basically these products are dispersions of finely divided particulate matter in various insoluble vehicles such as mineral oils, kerosene, fatty alcohols, silicone oils, and the like. The dispersed particle, usually a crystalline structure, has a high surface area that has been treated or modified by one or more of a variety of polymeric compounds. Typical dispersed particles are the silicas, talc, clays, fatty amides, heavy metal soaps, and high melting polymerics (33). The particles are treated by silicones or other chemicals, such as various polymers, to render them hydrophobic or to modify crystallinity and surface tension properties (34). Various techniques for the dispersion of the particulates have been found to produce uniform, stable products. Homogenization and high shear mixing are perhaps the most common.

Dispersions have the advantage of being easy to handle owing to relatively low viscosities and low pour points. They are free of the titer problem of blended fats. In addition, no solvents are necessary as cloud point depressants. These solvents are frequently used in the blended products. Thus dispersion-type defoamers have significantly higher (and safer) flash points. These dispersions have replaced both blended fatty defoamers and paste-type products in many applications in the pulp and paper industry. The textile and paint and latex industries were hesitant to utilize defoamers at all until this type was commercialized.

There are also large and growing markets for dispersion defoamers in water and waste treatment facilities, coating applications, and in the chemical industry. Currently these defoamers seem to be the most effective and versatile, and they are being constantly evaluated in new applications.

Water-Based Defoamers. The changing availability of mineral oils and the desire to economize and conserve supplies has led to the marketing of new, highly effective water-based chemical defoamers. These products are quite different from paste defoamers in that they are fluid and, therefore, easy to handle. They may be fed directly, siphoned into a water source, or prediluted in an emulsion vessel. Water-based defoamers have proven to be equal to the dispersion products in ease of handling and the pumpability of some of them rivals that of the blended fats at proper temperature. The water-based defoamers are usually considered combination dispersion and oil-in-water emulsion products. Some components are fatty alcohols, fatty acid soaps, fatty amides, emulsifiers, ethoxylates, mineral oils, and esters.

Several recent patents have been granted for this defoamer type (11,35). The markets now just beginning to be served are pulp and paper, paint, waste treatment, chemical, and metal working. There probably is no limit to the application potential for water-based defoamers, and indeed products are now being formulated and tested for one of the most difficult of all applications—Kraft pulp black liquor. Production of stable fluid water-based defoamers requires good manufacturing techniques and know-how. The production of a single batch of defoamer may entail blending, emulsification, emulsion inversion, and homogenization.

Other Types of Defoamers

Silicone Defoamers. The silicones as a class are highly effective defoamers and can be used in many applications. Only their relatively high cost prohibits more widespread use. These products are dispersions of finely divided particles of silica dispersed in polydimethylsiloxane or similar silicones. They might also be oil-in-water emulsions of these products at 10–30% activity. Silicone defoamers are used in many specialty applications such as antibiotics, winemaking, distillations, peeling of vegetables (potatoes), metal working fluids, and as additives to surfactants. Some silicone products of specific molecular weight are generally recognized as safe by the FDA as direct and indirect food additives; others are permissible under other specific regulations. Most defoamers suppliers have FDA information included on product data sheets or will supply such information on request.

Solid Defoamers. Another minor class of defoamer is the solid (brick or flake type), which has proven to be especially effective for use on paper machines to promote better sheet formation by removal of entrained air (36). These defoamers are usually made into dilute (less than 0.50%) solutions and fed into the application area. Solid defoamers are composed of waxes, esters, fatty alcohols, soaps, and the like.

Surfactants. Certain surfactants, such as ethylene oxide–propylene oxide block copolymers, are used as defoamers. These products are used in starch solution tanks, evaporation processes, and certain distillation procedures. They must be used in relatively high temperature operations because they may become insoluble in water at elevated temperatures by the cloud point mechanism function (37) (see Surfactants).

Powdered Defoamers. Powdered defoamers are used for incorporation into dry products such as starch, cements, and caulking compounds (38–39). These are usually high surface area, highly absorbent inorganic fillers that have been surface-treated with liquid proprietary defoamers. They function by defoaming the slurry or dispersion of cement as it is mixed with water and are becoming increasingly important, especially for oil well drilling cements (see Fillers).

Simple Compounds. Some simple compounds such as fatty oils (triglycerides) and polyethylene glycol esters are used as defoamers in the enzyme and yeast making industries, as well as in vegetable peeling and bleaching. These products generally have FDA approval for use as direct additives for food and, therefore, are considered safe for those industries.

Defoamers for Nonaqueous Systems. There has been comparatively little study in the area of defoamers for nonaqueous systems because of their limited economic importance. Foam problems in this category include degassing of jet fuel on ascension of aircraft, automotive antifreeze, fractional distillation in chemical plants, etc.

If the surface tension of the foaming system is not too low, eg, 40 mN/m (= dyn/cm), formulations composed of conventional defoamer components suffice. For instance, mixtures of waxes and hydrocarbon oils are used to control foam in the stripping of monomers and low molecular weight polymers from butadiene–styrene latex (40).

The foaming of low surface tension liquids such as hydrocarbon fuels poses a more difficult problem. The defoamer formulation is restricted to the limited number of materials with surface tensions low enough to be useful. Silicones provide an answer, although a comparatively costly one. In cases where silicones cannot be dispersed in the foaming medium, copolymers of silicone and ethylene oxide and/or propylene oxide have proved effective. Fluorinated compounds provide another solution to the problem but one that is even more costly than the use of silicones.

Future Developments

The future of industrial markets for defoamers will be dictated by new regulatory and ecological restrictions, raw material costs, and availability considerations. The defoamer supplier will simply face the problems that all other industries will face in the next few decades. One lead that has already proven of interest is the use of water-based products. Emulsions or dispersions of highly effective formulated products in a water carrier have been found to be functionally equal on a cost basis to the more expensive oil-based blended and dispersion-type defoamers. The water-based products are easier to handle than the paste-type defoamers. They can be equally as convenient as the dispersion types if formulated and processed correctly. And they have both ecological and safety advantages over the oil-based defoamers.

BIBLIOGRAPHY

1. H. T. Kerner, *Foam Control Agents,* Noyes Data Corp., Park Ridge, N.J., 1976.
2. *Code of Federal Regulations,* U.S. Government Printing Office, Washington, D.C., 1978.
3. U.S. Pat. 2,715,614 (Aug. 16, 1955), C. E. Snook (to Nopco Chemical Company).
4. U.S. Pat. 3,408,306 (Oct. 29, 1968), F. J. Boylan (to Hercules Incorporated).
5. U.S. Pat. 3,990,905 (Nov. 9, 1976), R. J. Wachala and R. E. Svetic (to Nalco Chemical Co.).
6. U.S. Pat. 3,673,105 (June 27, 1972), J. A. Curtis and F. E. Woodward (to Diamond Shamrock Co.).
7. U.S. Pat. 3,746,653 (July 17, 1973), J. W. Churchfield (to Dow Corning Corp.).
8. Jpn. Pat. 74 27751 (July 20, 1974), H. Akabayashi and co-workers (to Nissan Kagaku Kogyo K.K.).
9. U.S. Pat. 4,021,365 (May 3, 1977), I. A. Lichtman and J. V. Sinka (to Diamond Shamrock Co.).
10. U.S. Pat. 3,677,963 (July 18, 1972), I. A. Lichtman and A. M. Rosengart (to Diamond Shamrock Co.).
11. U.S. Pat. 4,009,119 (Feb. 22, 1977), F. Poschmann and W. Bergold (to Badische Anilin- und Soda-Fabrik Aktiengesellschaft).
12. S. Ross, *The Inhibitors of Foaming, Engineering and Science Series 63,* Rensselaer, New York, 1950.
13. J. H. Schulman and T. Teorell, *Trans. Faraday Soc.* **54,** 1337 (1938).
14. A. W. Adamson, *Physical Chemistry of Surfaces,* 3rd ed., John Wiley & Sons, Inc., New York, 1976.
15. G. L. Gaines, *Insoluble Monolayers at Liquid–Gas Interfaces,* Interscience Publishers, New York, 1966.
16. J. Robinson and W. Woods, *A General Method of Selecting Foam Inhibitors, Technical Note No. 1025,* National Advisory Committee for Aeronautics, Washington, D.C., 1946.
17. S. Ross, private communication, 1971, Diamond Shamrock Chemical Co., Morristown, N. J.
18. K. Roberts, C. Axberg, and R. Osterlund, *J. Colloid Interface Sci.* **62**(2), (Nov. 1977).
19. R. Kulkami, E. Goddard, and B. Kanner, *J. Colloid Interface Sci.* **59**(3), (May, 1977).
20. U.S. Pat. 3,076,768 (Feb. 5, 1963), F. J. Boylan (to Hercules Powder Company).
21. U.S. Pat. 3,408,306 (Oct. 29, 1968), F. J. Boylan (to Hercules Powder Company).
22. I. Lichtman, J. Sinka, and D. Evans, *Assoc. Mex. Technol. Ind. Cellul. Pap.,* **XV(1)** 26 (1975).
23. J. Schulman and J. Leja, *Trans. Faraday Soc.* **50,** 598 (1954).
24. A. L. Jacoby, *J. Phys. Colloid Chem.* **52,** 689 (1948).
25. S. Ross and J. Butler, *J. Phys. Chem.* **60,** 1255 (1956).
26. H. Saito and S. Friberg, *Pramana* **Suppl. No. 1,** 537 (1955).
27. U.S. Pat. 3,107,519 (Apr. 27, 1960), E. P. McGinn.
28. U.S. Pat. 2,868,734 (Jan. 13, 1959), A. DeCastro, J. Groll, and C. Lighthipe (to Nopco Chemical Company).
29. U.S. Pat. 2,715,614 (Aug. 16, 1955), C. E. Snook (to Nopco Chemical Company).
30. U.S. Pat. 3,923,683 (Dec. 2, 1975), R. J. Michalsik and R. W. Youngs (to Nalco Chemical Co.).
31. U.S. Pat. 4,021,355 (May 3, 1977), I. A. Lichtman and J. V. Sinka (to Diamond Shamrock Co.).
32. U.S. Pat. 3,652,453 (Mar. 28, 1972), T. F. MacDonnell (to Diamond Shamrock Co.).
33. U.S. Pat. 2,843,551 (July 15, 1968), F. J. Leonard, A. DeCastro, and T. F. Groll (to Nopco Chemical Company).

448 DEFOAMERS

34. U.S. Pat. 3,951,883 (Apr. 20, 1976), (to Diamond Shamrock Co.).
35. U.S. Pat. 4,032,473 (June 28, 1977), R. K. Berg and D. S. Salley (to Associated Chemists, Inc.).
36. U.S. Pat. 2,797,198 (June 25, 1957), F. L. Chappell (to Hercules Powder Company).
37. U.S. Pat. 2,834,748 (May 13, 1958), D. L. Baily and F. M. O'Conner (to Union Carbide Corporation).
38. U.S. Pat. 3,834,919 (Sept. 10, 1974), E. T. Parker (to BASF Wyandotte Corporation).
39. Can. Pat. 867,321 (Mar. 30, 1971), I. A. Lichtman and M. D. Kellert (to Diamond Shamrock Co.).
40. U.S. Pat. 2,379,268 (June 26, 1945), J. C. Zimmer (to Standard Oil Development Co.).

IRWIN LICHTMAN
TED GAMMON
Diamond Shamrock Chemical Co.

DEFORMATION RECORDING MEDIA

Deformation recording media are deformable materials used for recording information in the form of ripples or grooves. The deformations, produced by selectively depressing the surface of the recording material, represent relief images that can be projected by optical techniques to give black-and-white or colored images. The deformations may be temporary or permanent. In applications such as real-time video recording and display, the deformations are temporary and disappear spontaneously after an appropriate time period representing a single frame. In recording media designed for information storage, the images are permanent but, if desired, can be erased under the proper conditions.

Deformation recording media have several unique characteristics. Images are developed by heat without need for fixing, and processing requires no chemicals; development is dry and fast. The images can be erased and the media can, in principle, be reused. Fluid media are spontaneously developed (and erased) at ambient temperature without deliberate heat input. Even photosensitive media do not require storage or handling in the dark because they are electrically sensitized to light just before or during exposure. Photographic speed is adequate for document copying but generally not sufficiently fast for snapshot or movie applications. The major shortcoming of deformation images is that they are not easily seen by the unaided eye. In most instances special optical systems must be used to project the image.

Deformation of electrically charged insulating resins was observed (1) as early as 1897, and was investigated further in the 1950s (2). When a pattern of electric charges is placed on the surface of a dielectric film supported on a conductive substrate, a charge pattern of opposite polarity is induced at the interface of the dielectric and conductor (see Fig. 1) (see Electrophotography). The attraction of opposite charges gives rise to an electrostatic force which is opposed by the viscosity and surface tension of the dielectric. When the electrostatic force exceeds these opposing forces, as in the case of an appropriate thin fluid layer (or a solid softened by heat), the surface of the fluid deforms in a pattern corresponding to the charge pattern. When the charge is

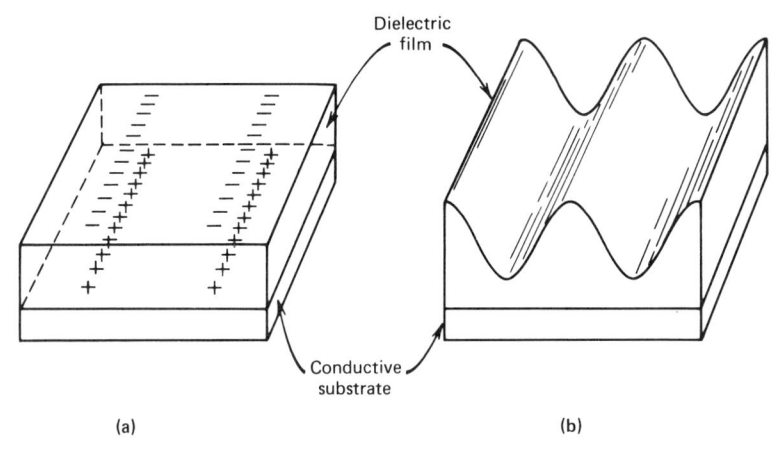

Figure 1. (a) Charge pattern on dielectric film. (b) Deformation image corresponding to charge pattern.

gradually lost, the deformations disappear because of the restoring force of surface tension. If the viscosity is increased after the deformations have been produced, eg, by cooling a previously softened solid, the deformations are frozen-in permanently. However, even in this case the deformations can be removed by heating the layer to reduce its viscosity and allow the charges to decay.

Optical Readout

Generally, deformation images cannot be effectively projected by ordinary projectors designed for images of variable optical density because most deformation images exhibit no useful variation in optical density. The depth of deformations is usually very small. Typical values are 0.1–1 μm. Consequently, the optical density of a groove is practically the same as that of an adjacent undeformed region. However, the grooves cause a retardation of transmitted (or reflected) light and change the phase of light. This change can be detected with a schlieren optical system. The overall effect of this system is to convert an invisible change in phase to a visible change in amplitude.

A schlieren projection system (3) contains the usual components of a projector such as light source, condenser, and projection lens. In addition, the projector (Fig. 2) contains a schlieren lens and two plates A and B with apertures that work in a cooperative fashion. When there are no grooves in the light path, light transmitted by plate A is stopped by plate B and the screen is therefore dark. An array of grooves represents a grating that diffracts light into a number of diffraction orders. Plate B intercepts the undeviated zero-order beam and passes the higher-order beams which produce a bright spot on the screen. The optical system acts as a light valve and performs a spatial filtering operation. Light modulation by a single groove can be readily visualized by geometric optics as shown in the insert of Figure 2.

Projectors can be designed for dark field projection (as in Fig. 2) or bright field projection (ie, the screen is dark or bright, respectively, in areas corresponding to undeformed areas on the recording medium or zero-diffraction order). Projectors can also be designed with reflection optics for readout of images of high reflectivity. The reflectivity is provided by a reflective coating located either underneath or on top of the deformable film.

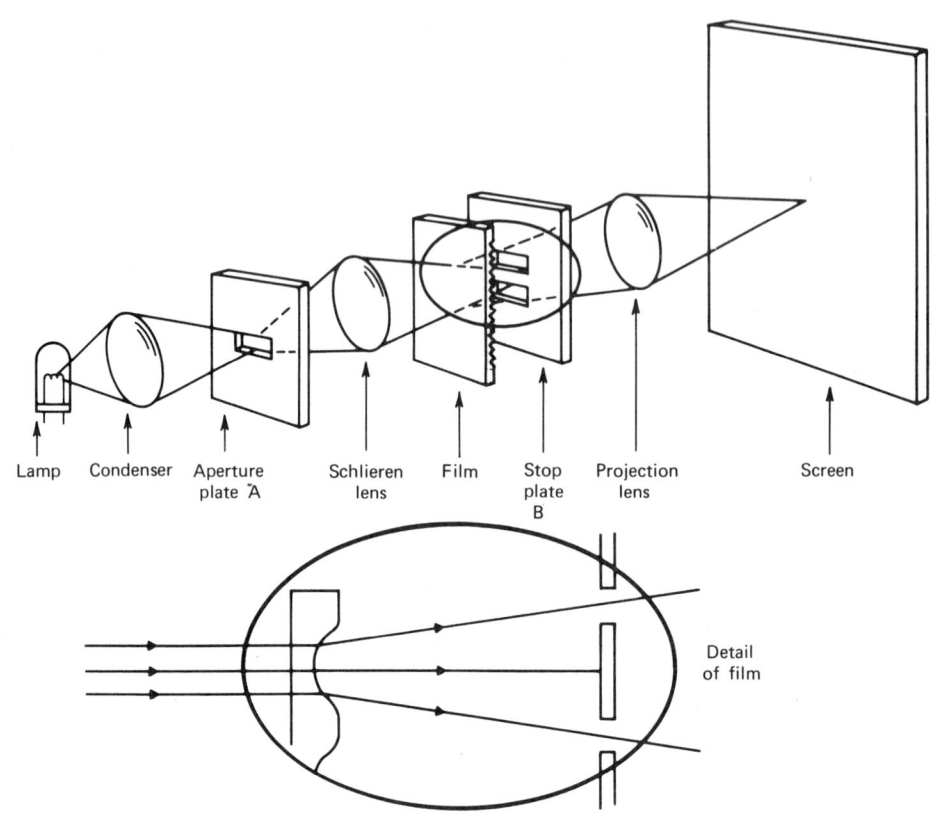

Figure 2. Schematic of schlieren projector and light path through a groove.

Recording Media

Recording media can be classified according to the type of stimulus used for producing the electrostatic image, electron beam or light, and according to the physical state of the deformable component. This article covers electron-beam-sensitive fluids and solids and photosensitive solids (see Photoreactive polymers; Polymers, conducting).

Practically all recording media consist of at least three components, ie, deformable layer, electrically conductive layer, and substrate. Many requirements for the substrate and conductive layer are common for all deformation media. Extremely high optical quality is particularly essential because the schlieren method detects variations in refractive index. Thus, haze or cloudiness, air bubbles, and scratches, which would be of little consequence with ordinary projection optics, show up prominently as bright spots on a dark field.

Glass is the substrate par excellence but can be used only in certain applications. The substrate must meet several requirements and plastic tape is more versatile. High tensile strength and dimensional stability for temperatures up to 100–130°C are needed to prevent distortion at the development temperature. The tape must be flexible to permit travel over rollers during use and resistant to solvents used in coating the deformable layer.

Base tape made from polyester of photographic quality, such as Cronar or Mylar, has given good results (4). The backside of the tape should have antistatic properties to prevent buildup of electrostatic charge and consequent sticking of the tape when it is rolled up. This polyester is practically transparent to visible light and is inert to most solvents. For adequate strength the thickness should be about 0.05 to 0.1 mm.

The conductive layer serves as an electrical ground plane. It must be transparent (for transmissive readout) and have good adhesion to the base support but need not have low resistivity. A resistivity of up to about 10 kΩ per square is adequate. When used as a resistance heating element, the resistance must be sufficiently low and uniform.

Transparent conductive layers can be made from nichrome alloys deposited by vacuum evaporation (see Film deposition techniques). The transmission in the visible region is about 50% for 5-nm thick layers. Other suitable conductors are gold, copper(I) iodide containing small amounts of oxygen (5), copper sulfide (6), tin oxide and indium oxide.

Electron-Beam-Sensitive Fluid Media. These materials form the basis of light valve television systems capable of producing monochrome and color images projected on large screens (7–8). Whereas the operations of the two systems vary in detail, both systems employ a scanning electron beam to deposit a raster of electron charges on the surface of a fluid layer. The electron beam is modulated by electric signals corresponding to the images to be projected. The electrostatic forces associated with the deposited charges deform the surface of the fluid—called eidophor (image bearer)—and the resulting deformation images are projected with schlieren optics. In one system, the fluid layer and electron gun providing the raster are in a continuously pumped evacuated enclosure and full-color images are obtained by using separate deformable layers and optics for each primary color. The other system employs a sealed tube housing the gun and fluid (9–10), and full-color images are obtained from a single deformable layer and a single set of optics by writing diffraction gratings for each primary color.

The fluids must have appropriate dielectric and physical properties for use in television (11). The electrical resistivity must be sufficiently high (about 10^{11} Ω-cm at the operating temperature) to permit the buildup of a proper potential for deformation and to prevent loss of resolution of the raster by lateral dissipation of charge. After the fluid has deformed, the charge has to decay so that self-erasure due to the restoring forces of surface tension can take place within a TV time frame ($1/30$ s). The charge is lost to the conductive coating underneath the fluid layer. This coating, deposited on a glass substrate, can be transparent or reflective.

The viscosity of the fluid determines the rates of deformation and erasure. The viscosity plays a role in erasure not only for mechanical reasons but also in its effect on electrical characteristics because the mobility of charge carriers varies inversely with viscosity. A viscosity of about 10^4 mPa·s(=cP) at the operating temperature is optimal. The vapor pressure of the fluid should be low to minimize contamination of the electron gun. Bombardment by electrons can cause cross-linking or bond scission. The former raises the molecular weight and viscosity, whereas the latter results in volatile fragments, gas formation, decrease of viscosity, and loss of fluid. Therefore, the fluid should have high resistance to electron radiation. It must also be stable photochemically so as not to be degraded by the radiation of the projector light source.

From the standpoint of radiation resistance, the preferred materials are polysiloxanes (see Silicon compounds, silicones) and products made from benzyl chloride and aromatic compounds such as naphthalene and toluene. Poly(methylphenylsiloxanes) have better radiation resistance than poly(dimethylsiloxanes) and have lower melting points than the diphenyl analogues. Typical Friedel-Crafts reactions of benzyl chloride with benzene, toluene, naphthalene, or biphenyl give complex products from which suitable fractions have been isolated by distillation (12–14). Similar types of products have been synthesized from benzyl alcohol instead of benzyl chloride (15).

High molecular weight additives such as polystyrene (mol wt of ca 20,000), poly(methylphenylsiloxane), and poly(2,6-dimethyl-1,4-phenylene ether) have been used in small percentages to improve the operation of the base fluids (16). When the thickness of the fluid layer is below a certain critical value, which ranges between 5–15 μm for different fluids, there is a tendency for the formation of a hydrodynamic instability in the fluid that results in "optical noise" and degrades the projected image. Polymer additives impart non-Newtonian (viscoelastic) behavior to the fluid and raise its critical thickness to permit operation with thin layers which is desirable from the standpoint of dynamic response characteristics.

Electron-Beam-Sensitive Solid Media. Information recording with an electron beam on thermoplastic films was first described (17) as thermoplastic recording. Recording and readout are similar to the eidophor, except that the images are developed by heating and fixed by cooling. The transparent recording medium is a thin film of a dielectric thermoplastic on a conductive tape substrate.

The image is recorded by a scanning electron beam in a vacuum (see Fig. 3). The beam deposits a charge pattern on the thermoplastic, and the electrostatic pattern is developed to a deformation image by heating the thermoplastic to its softening temperature. Electrostatic forces deform the surface of the thermoplastic in proportion to the deposited charge, and the deformations remain embossed in the surface when the thermoplastic is cooled below the softening temperature. Deformation time constant is in the millisecond range. The image is read out by projection with a schlieren system. Color images can be projected by recording the patterns as diffraction gratings. To erase the image, the thermoplastic is heated above the softening temperature to dissipate any remaining charge and to permit surface tension forces to restore uniformity. The medium can then be reused.

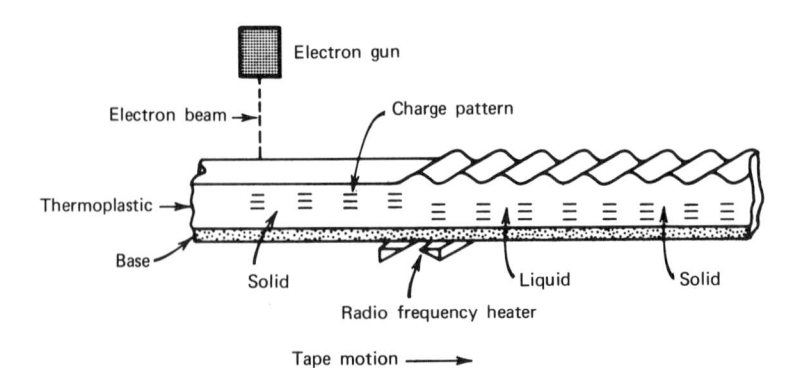

Figure 3. Thermoplastic recording (17).

The thermoplastic film must possess several specific properties (4,18). It must be a good insulator with a resistivity of at least 10^{11} Ω·cm in the fluid state at the development temperature. If the charge is lost, the appropriate groove depth cannot be developed. The film should be colorless and have high light transmission. The glass transition temperature and softening temperature or melting point should be well above room temperature to prevent "cold flow" at the storage temperature which would result in gradual erasure. A high softening temperature also prevents sticking of the thermoplastic to the backside of the tape when the tape is wound up after image development. The softening temperature should not be so high that it impairs the dimensional stability and strength of the base tape. The rate of deformation is inversely proportional to viscosity. A viscosity of 10^5–10^7 mPa·s (=cP) has proven adequate. The vapor pressure of the thermoplastic should not exceed 1.3 mPa (0.01 μm Hg) in the liquid state.

Thermoplastic polymers and polymer mixtures have been developed specifically for thermoplastic recording. For high thermal and radiation stability, the polymers should have a high proportion of aromatic groups. However, such polymers tend to have high softening temperature and lack flexibility. Therefore, it is necessary to plasticize the polymers externally or introduce aliphatic groups for internal plasticization. The polymers should have polar groups to promote adhesion to the substrate. A satisfactory material (19) is a blend of a poly(diphenylsiloxane) with a molecular weight of about 1300 and a commercially available poly(2,6-dimethyl-1,4-phenylene ether) marketed under the trade name PPO (see Polyethers). The viscosity and softening temperature of the blend can be controlled by varying the proportions of these polymers. Other thermoplastic materials include polystyrene plasticized with o-terphenyl (18) and copolymers (20–21) of styrene with methyl, butyl, hexyl, and octadecyl methacrylate. In practice, thermoplastic film thickness is in the range of 5–15 μm, and the films are coated on the tape from solutions in appropriate solvents.

The sensitivity of thermoplastic film is about 10^{-7}–10^{-8} C/cm^2, and resolving power is in excess of 500 lp (line pairs)/mm (22). No data have been published on the shelf life of unrecorded film, but it may be presumed to be indefinite for certain thermoplastics such as polysiloxanes. Reusability depends on many factors, such as stability of the substrate and thermoplastic, and the extent to which dust and contaminants can be controlled. Styrene-octadecyl methacrylate copolymer (18), for example, showed no evidence of degradation after 50,000 write-and-erase cycles. Although no quantitative studies on archival properties appear to have been made, it has been noted that recorded film kept at room temperature for over ten years has not undergone self-erasure.

Photosensitive Solid Media. Photosensitive solid media (23), designed for photographic or reprographic applications, are basically photoconductors (see Electrophotography). The general recording technique consists in sensitizing the film in the dark by an electric charge, exposing it to an optical image, and then heating it to develop the image. Sensitization can be effected with a corona charging device comprising one or more wires on which a high potential (typically 6–8 kV) is impressed to ionize the air. Ions and electrons are attracted to the film. The corona device and film may be stationary or one may move with respect to the other. Charging may also be simultaneous with exposure. Since the media are not light sensitive until charged, they need not be kept in the dark before use. The heat is supplied by a heater or by passing a current throughout the conductive layer. Heating time can be as fast as 10 ms. Images can be erased by heating the film above the development temperature or holding the film for some time at the development temperature. Since the photosensitivity is not destroyed by image erasure, the media can be reused.

The mechanism of image formation may be envisioned as follows: when a thermoplastic film is charged to a given potential and heated to reduce its viscosity, the film tends to deform spontaneously into ripples even without exposure to light. Deformation takes place because repulsion of like charges makes a charged surface unstable, and the electrical energy stored by the charged system can be decreased by spreading the charge over the larger surface area produced by deformation. The ripples have a certain periodicity, characterized by a wavelength that is an inverse function of film thickness. Because deformation rate and depth are proportional to the potential, deformation imagewise takes place when the potential is modulated by exposure to light. Thus a spontaneous process is channeled into useful deformation, whereas random ripples constitute noise. The deepest grooves are produced for spatial frequencies coinciding with the wavelength of the spontaneous deformations. In recording a scene that contains a distribution of spatial frequencies, eg, from 0.1 to 500 cycles per millimeter, only those frequencies falling within the bandwidth capability of the film are reproduced. In other words, the medium cannot render scenes of continuous tone. To reproduce continuous tone and grey scale, it is necessary to introduce a carrier as is the practice in half-tone processes. The carrier may be a grating with a periodicity corresponding to that of the predominant wavelength response of the film. No grating is required to reproduce line copy having the proper frequency.

When the surface potential exceeds a certain value, the film tends to deform into worm-like light-scattering patterns called frost (24) because of their frosted appearance (see Fig. 4). The quasi-random pattern of ripples in frost has a wavelength on the order of film thickness. Frost may serve as a natural carrier for continuous tone rendition (25). Frost images scatter light at large angles and do not require schlieren projection for readout. An ordinary projector with small aperture, eg, f/16 or smaller, gives satisfactory results.

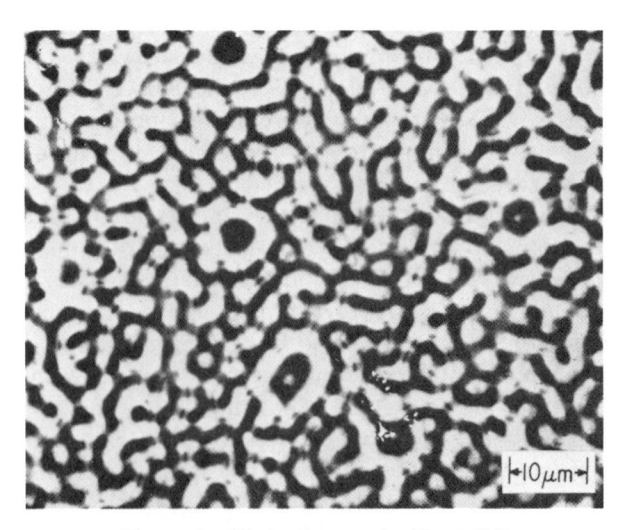

Figure 4. Photomicrograph of frost (24).

Single-Layer Configuration. In single-layer configurations the layer is photoconductive and thermoplastic. The medium may be designed for transmissive or reflective readout. Development of single-layer media was pioneered by the General Electric Company (photoplastic film, photoplastic recording) (26) and RCA (27–28). Another type of single-layer medium, which is not based on photoconductivity and does not require electrical charging for sensitization, has also been disclosed as a photocharge process (29).

To record an image (see Fig. 5), the film is charged in the dark to a uniform potential by a corona charging device. It is then exposed to an optical image in a camera. Exposure creates charge carriers that discharge the surface potential in proportion to exposure and yield a latent electrostatic image. A deformation image results when the film is heated to reduce its viscosity. The image is fixed by allowing the film to cool to increase its viscosity. Since optical density is not a meaningful concept for deformable media, the criterion of modulation efficiency (ME), ie, the ratio of deviated to total light transmitted by the film has been suggested for sensitometry (30).

Single layers have three types of compositions:

(*1*) Dispersions of insoluble photoconductor particles in a thermoplastic polymer.
(*2*) Solid solutions of an organic photoconductor in a thermoplastic polymer.
(*3*) Inherently photoconductive thermoplastic polymers.

Insoluble photoconductors must be of small particle size to be dispersed in a deformable polymer matrix and must be dispersible without agglomeration. Films have been prepared by dispersing blue photoconductive copper phthalocyanine at a concentration of 5% in plasticized polystyrene (31). Copper phthalocyanine (average particle size of 0.1 µm) is first suspended in a solution containing the polymer, and the suspension is coated over the conductive substrate. Light transmission of the film

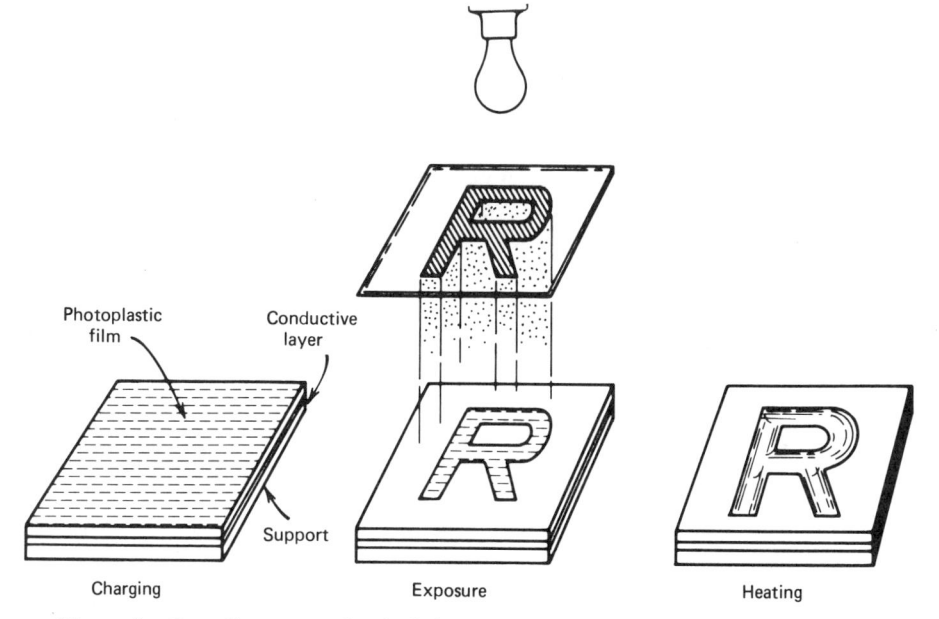

Figure 5. Recording process for single-layer configuration (photoplastic recording) (38).

depends on photoconductor concentration and film thickness. This film is sensitive over the entire visible region of the spectrum with a maximum at 600–700 nm and a minimum at 480 nm. The photoconductivity ratio at these wavelengths is about 10. The absolute sensitivity at a given wavelength depends on the potential to which the film is charged, the film thickness and spatial frequency to be recorded, and upon the thermal development condition. Images have been recorded with an exposure of 1–20 μJ/cm^2 using white light. Resolving power exceeds 400 lp/mm, and the response at a given spatial frequency depends on film thickness as mentioned earlier.

Dispersions of photoconductive copper-doped cadmium sulfide have also been investigated (32). The specially prepared particles (<0.2 μm diameter) were used at a concentration of 1–2% in plasticized polystyrene. Spectral response of the light yellow film is in the range of 380–500 nm and decreases beyond 500 nm. Images were produced with an exposure of 2–20 μJ/cm^2. Resolving power in 25-μm thick films was 50–100 lp/mm.

Films consisting of a solid solution of an organic photoconductor in a polymer matrix are homogenous and grainless. Their spectral response, corresponding to that of the photoconductor, can be modified by the addition of dyes. The following photoconductors have been reported: charge transfer complex of pyrene and tetracyanoethylene (33) beta-carotene (26), carbazole (26), phenazine (26), and leucobases of dyes such as malachite green (27). The latter tend to oxidize to the corresponding dyes, and trace amounts of these dyes extend the spectral sensitivity to longer wavelengths.

Inherently photoconductive polymers with appropriate deformability are more difficult to prepare than dispersions or solutions. One approach has been to make charge transfer complexes from thermoplastic polymers, eg, by dissolving tetracyanoethylene in polystyrene or in a naphthalene–formaldehyde polymer (34) or from 2,4,5,7-tetranitro-9-fluorenone and a carbazole–formaldehyde condensation polymer (35). Reactions of aromatic amines with epoxy resins (36) or polyaldehydes (37), such as acrolein–styrene copolymer, have given photoconductive thermoplastic products. Internal plasticization of poly(vinyl carbazole) by copolymerization with aliphatic monomers, such as methacrylates, and subsequent formation of charge transfer complexes should also provide deformable media.

Readout by transmitted light has certain disadvantages. For maximum projection brightness, the absorption of the film should be low and this limits the sensitivity of the medium. The projected image is colored because of the color of the photoconductor. Most important, however, imperfections in the substrate may contribute to noise in the projected image. These disadvantages are eliminated in media designed for reflective readout. Reflective deformable media have been made by overcoating the single-layer film with a thin metal layer of specular reflectivity (38). Images are recorded by corona charging, exposure through the transparent substrate and thermal development, and read out by front surface reflection in a schlieren system. Alternatively, the film may be exposed through the partially transmissive reflective layer. Since light is not transmitted through the film during readout, color and absorption of the film and substrate imperfections cannot degrade projected image quality.

The reflective coating must be smooth for specular reflectivity. It must not offer excessive resistance to deformation as this would impair photographic sensitivity, and it should be resistant to oxidation. Indium is a suitable metal. Its reflectivity is relatively uniform in the range of 400–700 nm. Thin layers of up to a few tens of nanometers have an island structure (island size less than 1 μm) and therefore have high lateral resistivity. Films thinner than about 30 nm exhibit a steep dependence of reflectivity on thickness. At a wavelength of 500 nm, the reflectivity is 40% for 10 nm thickness, 58% for 20 nm, and 68% for 30 nm. Layers with thickness below 30 nm deform readily with the underlying photosensitive thermoplastic. However, these films require a slightly higher heat input for thermal development and erasure. Overcoated films have a lower voltage threshold for frost formation. Frost can be avoided by charging the film to a lower potential and this has a corresponding effect on photosensitivity. Resolving power is a function of thermoplastic film thickness. In 7-μm thick film, 400 lp/mm have been resolved. Tin is also a suitable metal, whereas metals such as aluminum, zinc, and copper tend to craze on thermal development (39).

A recording technique that does not require electrical charging, the photocharge process (29), employs a single photosensitive layer supported on a substrate which need not be conductive. Before an image is recorded, the film is heated in the dark to dissipate any latent images that may have been produced during preparation or handling in light. The film is then exposed to the image to be reproduced without need for earlier or subsequent application of an electric field. The latent image is developed to a deformation image by heating the film to the development temperature. The deformations are located at the boundary of illuminated and dark areas, and the image is read out by schlieren projection. The image may be erased by additional heat input, and the film can be reused.

The photosensitive layers are prepared by dissolving photosensitive compounds at a concentration of 1–5% in an inert thermoplastic polymer, such as polystyrene. The photosensitive compounds include: polyhalogenated hydrocarbons, such as iodoform, methylene iodide, tetraiodoethylene, chloranil; aniline derivatives, such as m-nitroaniline, 2-nitrodiphenylamine, and azo compounds, such as azobenzene, p-phenylazoaniline or N,N-dimethyl-p-phenylazoaniline. Mixtures of these classes of compounds have also been used. Films are cast from solution to a thickness of 5–25 μm. For a given polymer, the development temperature depends on the concentration of the photosensitive component as the latter acts as a plasticizer. Development temperature is typically 80°C. The photosensitive material need not be photoconductive as photoconductivity does not play a role. The mechanism of the process has not been elucidated, but photochemical reactions occur during exposure as evidenced by bleaching. This process is apparently reversible to an extent because the original density is reestablished upon heating for erasure. It is conceivable that charged species are produced during exposure and flow temperature is reduced owing to the formation of photogenerated molecules (see Photoreactive polymers).

As with the preceding media, continuous tone can be achieved by the use of a carrier. The sensitivity depends on the photosensitive compound, residual solvent content, film thickness, and light intensity, ie, the film does not obey reciprocity. Images have been recorded with exposure of 0.1–1 mJ/cm^2 (0.024–0.240 mcal/cm^2). Resolving power of 360 lp/mm has been demonstrated.

Double-Layer Configuration. The double-layer configuration comprises adjacent photoconductor and thermoplastic layers. This technique, pioneered by Xerox Corporation, has been called thermoplastic xerography or frost process (24).

To record an image on the double-layer medium (see Fig. 6), the film is corona-charged to a uniform surface charge density (24). The voltage divides itself between photoconductor and thermoplastic according to their respective capacitances. Sub-

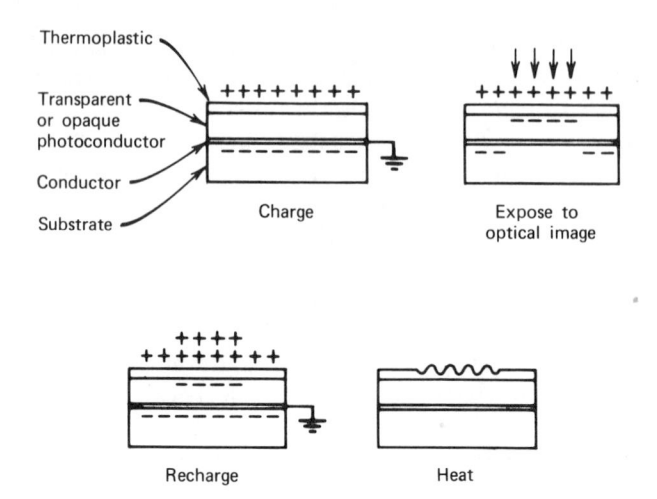

Figure 6. Recording process for double-layer configuration (frost process) (24).

sequent exposure to an optical image decreases the resistivity of the photoconductor and lowers the voltage across it while the surface charge density remains unchanged. The medium is then recharged to a uniform potential to increase the force in the exposed areas, and the image is developed by heat. Corona charging and exposing can also be concurrent. Frost images are obtained when the charge density is high, and these images can be displayed in a conventional projector with a small aperture. The projected image is a negative of the original. The image can be erased by additional heating.

The photoconductor is a thin layer (typically 1.5–4 μm thick) of poly(vinyl carbazole) complexed with 2,4,7-trinitro-9-fluorenone (40–41) or sensitized with Brilliant Green (42). Both materials respond in the visible region of the spectrum, and Brilliant Green conveys peak sensitivity at 640 nm. The layers are suitable for transmissive readout. The photoconductivity of the poly(vinyl carbazole) complex has been extensively characterized (43). The thermoplastic layer is kept very thin, on the order of 0.5–2 μm. It is made of Staybelite Ester 10 (41), polystyrene (44), and some specially prepared polymers such as a terpolymer of styrene, octyl methacrylate, and decyl methacrylate (44) in molar ratio of 85:9:6. The Staybelite ester tends to oxidize in air, particularly in the presence of light.

As with single-layer media, the sensitivity depends on many factors. Under certain conditions, images have been recorded with minimum exposure of 1–10 μJ/cm^2 (0.24–2.4 μcal/cm^2).

Since the thermoplastic and photoconductor are separate elements, the thickness of the thermoplastic can be made very small for maximum resolving power. Owing to their high resolving power and other features, such as on-site development and phase-modulated images, these media have been of considerable interest for hologram recording (41, 44–45) when used in a frost-free mode (see Holography).

Separate Photoconductor–Thermoplastic Configuration. In this technique an electrostatic image produced on a photoconductor is transferred to a separate thermoplastic film, and the latter is developed by heat. The photoconductor does not become part of the record.

The recording medium consists of a nonphotosensitive thermoplastic film facing a photoconductor plate. The gap between these two structures may be air (46) or a fluid of controlled conductivity (47). To record an image (see Fig. 7), the photoconductor is exposed to an optical image while a potential is simultaneously applied across the conductive layers of the photoconductor and thermoplastic. The thermoplastic is separated from the photoconductor and subsequently heated to produce deformations. When the gap is air, reproducibility is difficult to control and the resolution is limited to about 40 lp/mm. More reproducible results and resolution in excess of 100 lp/mm have been obtained with fluids of controlled conductivity in the gap. These results were obtained under conditions yielding frost images using thermoplastic films of Staybelite Ester-10 and Velsicol X-37. Fluids used for charge transfer were silicones and fluorocarbons doped with dibutyltin dilaurate or bis(tributyltin) oxide to control the conductivity. The photoconductor, a selenium–tellurium alloy (48) sandwiched between selenium and a transparent substrate, had panchromatic response. Recordings can also be made under conditions that do not yield frost. The thermoplastic films described under Electron-Beam-Sensitive Solid Media are also suitable.

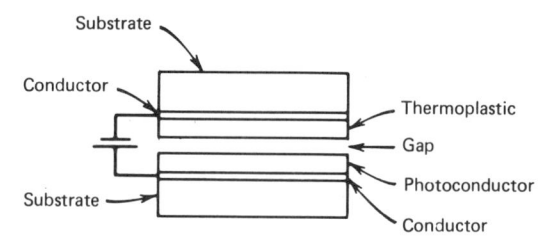

Figure 7. Recording process for separate photoconductor–thermoplastic configurations (charge transfer) (46).

Photosensitive Elastomeric Media. These media are similar to the double-layer configuration, except that an elastomer layer is substituted for the thermoplastic (49) (see Elastomers, synthetic). The device, called ruticon (*rutis* = wrinkle, *icon* = image), is made of a transparent conductive substrate, thin photoconductive layer, thin elastomer layer, and a deformable electrode. The latter may be a conductive liquid such as mercury or gallium–indium alloy (α-ruticon), a conductive gas such as argon operated in the gas discharge mode (β-ruticon), or a thin metal mirror (γ-ruticon) (50). The latter type has been used as a TV projection display (51). To obtain continuous tone images, a grating made of opaque and clear lines (eg, 40 lp/mm) is included between the transparent conductor and photoconductor. To record an image (see Fig. 8), a bias of 200–400 V (dc) is applied between the conductive layer and metal mirror, and an optical image is focused on the photoconductor. The charge pattern created near the photoconductor–elastomer interface gives rise to an electric field in the

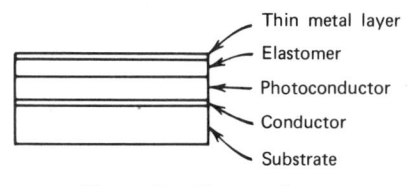

Figure 8. The γ-ruticon.

elastomer which leads to imagewise deformation of the elastomer. For a photoconductor of amorphous selenium, the exposure is less than 1.5 μJ/cm^2. The image is read out from the free metal surface by reflective schlieren optics. The image persists for a few minutes to a few hours but can be erased rapidly by exposure with a flash tube. In this device the thickness of the photoconductor and elastomer was 3–7 μm and 5 μm, respectively. Poly(vinyl carbazole) sensitized with a green dye has also been used as photoconductor. The elastomer is a siloxane-type polymer.

BIBLIOGRAPHY

1. J. W. Swan, *Proc. R. Soc. London* **62,** 38 (1897).
2. A. M. Thomas, *Brit. J. Appl. Phys.* **2,** 98 (1951).
3. A. G. Dewey, *IEEE Trans. Electron Devices* **24,** 918 (1977).
4. N. Kirk, *J. Soc. Motion Pict. Telev. Eng.* **74,** 666 (1965).
5. C. S. Herrick and A. D. Tevebaugh, *J. Electrochem. Soc.* **110,** 119 (1963).
6. U. S. Pat. 3,203,836 (Aug. 31, 1965), J. Gaynor, J. F. Burgess, and B. C. Wagner (to General Electric Co.).
7. E. Baumann, *J. Soc. Motion Pict. Telev. Eng.* **60,** 344 (1953).
8. W. E. Glenn, *J. Opt. Soc. Am.* **48,** 841 (1958).
9. W. E. Good, *IEEE Trans. Broadcast Telev. Rec.* **15,** 21 (1969).
10. T. T. True, *Proc. 1973 IEEE Intl. Convention New York*, paper 26/1.
11. W. E. Glenn, *J. Soc. Motion Pict. Telev. Eng.* **79,** 788 (1970).
12. U. S. Pat. 3,288,927 (Nov. 29, 1966), R. E. Plump (to General Electric Co.).
13. U. S. Pat. 3,317,664 (May 2, 1967), E. F. Perlowski, Jr. (to General Electric Co.).
14. U. S. Pat. 3,317,665 (May 2, 1967), E. F. Perlowski, Jr. (to General Electric Co.).
15. U. S. Pat. 3,761,616 (Sept. 25, 1973), C. E. Timberlake (to General Electric Co.).
16. U. S. Pat. 3,541,992 (Nov. 24, 1970), C. S. Herrick and F. E. Holub (to General Electric Co.).
17. W. E. Glenn, *J. Appl. Phys.* **30,** 1870 (1959).
18. H. R. Anderson, Jr., E. A. Bartkus, and J. A. Reynolds, *IBM J. Res. Develop.*, 140 (1971).
19. U. S. Pat. 3,063,872 (Nov. 13, 1962), E. M. Boldebuck (to General Electric Co.).
20. U. S. Pat. 3,118,786 (Jan. 21, 1964), A. Katchman and B. T. McKenzie, Jr. (to General Electric Co.).
21. U. S. Pat. 3,413,146 (Nov. 26, 1968), H. R. Anderson, Jr. and P. Levine (to International Business Machines Corp.).
22. W. E. Glenn and J. E. Wolfe, *Intl. Sci. Technol.*, 28 (June 1962).
23. J. Gaynor, *IEEE Trans. Electron Devices* **19,** 512 (1972).
24. R. W. Gundlach and C. J. Claus, *Photogr. Sci. Eng.* **7,** 14 (1963).
25. J. C. Urbach, *Photogr. Sci. Eng.* **10,** 287 (1966).
26. J. Gaynor and S. Aftergut, *Photogr. Sci. Eng.* **7,** 209 (1963).
27. N. E. Wolff, *RCA Rev.* **25,** 200 (1964).
28. E. C. Giaimo, *RCA Rev.* **25,** 692 (1964).
29. J. Gaynor and G. J. Sewell, *Photogr. Sci. Eng.* **11,** 204 (1967).
30. S. Aftergut, J. Gaynor, and B. C. Wagner, *Photogr. Sci. Eng.* **9,** 30 (1965).
31. J. J. Bartfai, V. Ozarow, and J. Gaynor, *Photogr. Sci. Eng.* **10,** 60 (1966).
32. S. Aftergut, J. J. Bartfai, and B. C. Wagner, *Appl. Opt. Suppl.* **3,** 161 (1969).
33. R. F. Kopczewski and H. S. Cole, *Appl. Opt., Suppl. 3,* 156 (1969).
34. Fr. Pat. 1,361,838 (May 22, 1964), S. Aftergut (to Compagnie Francaise Thomson-Houston).
35. Jpn. Kokai 76,126,148 (Nov. 4, 1976), A. Hirai (to Matsushita Electric Industrial Co., Ltd.).
36. U. S. Pat. 3,373,140 (Mar. 12, 1968), S. Aftergut (to General Electric Co.).
37. U. S. Pat. 3,373,141 (Mar. 12, 1968), S. Aftergut (to General Electric Co.).
38. S. Aftergut, R. F. Kopczewski, and J. F. Burgess, *Photogr. Sci. Eng.* **15,** 495 (1971).
39. J. F. Burgess, S. Aftergut, and R. F. Kopczewski, *Thin Solid Films* **13,** 385 (1972).
40. L. H. Lin and H. L. Beauchamp, *Appl. Opt.* **9,** 2088 (1970).
41. T. C. Lee, *Appl. Opt.* **13,** 888 (1974).
42. R. F. Bergen, *Photogr. Sci. Eng.* **17,** 473 (1973).
43. R. M. Schaffert, *IBM J. Res. Dev.* **15,** 75 (1971).
44. W. S. Colburn and E. N. Tompkins, *Appl. Opt.* **13,** 2934 (1974).

45. T. L. Credelle and F. W. Spong, *RCA Rev.* **33,** 206 (1972).
46. N. F. D'Antonio, *Appl. Opt., Suppl. 3,* 142 (1969).
47. J. T. Bickmore and C. J. Claus, *Photogr. Sci. Eng.* **9,** 283 (1965).
48. J. H. Neyhart, *Photogr. Sci. Eng.* **10,** 126 (1966).
49. N. K. Sheridon, *IEEE Trans. Electron Devices* **19,** 1003 (1972).
50. A. I. Lakatos, *J. Appl. Phys.* **45,** 4857 (1974).
51. A. I. Lakatos and R. F. Bergen, *IEEE Trans. Electron Devices* **24,** 930 (1977).

SIEGFRIED AFTERGUT
General Electric Company

DEGREASING AGENTS.   See Metal surface treatment.

DEHUMIDIFICATION.   See Air conditioning.

DELRIN.   See Acetal resins.

DENSITY AND SPECIFIC GRAVITY.   See Analytical methods.

DENTAL MATERIALS

> Ceramic and ceramic-organic materials,   463
> Colloids,   475
> Elastomers,   477
> Metals and alloys,   480
> Polymers,   501
> Dental waxes,   508
> Abrasives,   514
> Adhesives,   515
> Biological and clinical evaluators,   517

The official definition of dentistry as approved by the Federation Dentaire Internationale (FDI) is: "Dentistry—The science and art of preventing, diagnosing and treating diseases and malformations of and injuries to the teeth, mouth and jaws, and of replacing lost teeth and associated structures." Dental therapy involves the replacement of hard and soft oral tissues that are lost through disease with inert materials (metallic, ceramic, and organic) or with composites employing combinations of these three broad classes. The operative restorations and prostheses are made of amalgam, chromium-based alloys, precious metal alloys, special cements, synthetic polymers, and porcelain, all of which must withstand the rigors of the oral environment (see also Prosthetic and biomedical devices). The accessory materials needed in the fabrication procedures include synthetic polymers, gums, and waxes, both synthetic and natural, hydrocolloids, gypsums, and refractories.

These materials are used by ca 106,000 practicing dentists (1975) and >8,500 commercial dental laboratories employing ca 37,500 technicians. The limited market is small because the public does not buy dental materials directly except for toothpastes, mouth washes, tooth brushes, denture aids, etc (see Dentifrices).

The total value of dental supplies and equipment manufactured in the United States in 1976 is estimated at ca $520,000,000 (1). In the 1972 Census of Manufacturers (2) the value of these products was $409,200,000. This reflects the relatively small size of the dental industry which had 429 establishments in 1972, 107 of which employed more than 20 persons.

In 1975 there were 1,275 dental research projects sponsored by government and private organizations in the United States for $61,000,000. Although dentistry achieved professional status 150 years ago, it developed no criteria in the form of standards for its various unique restorative materials until 1926. In that year Dr. Wilmer Souder and co-workers, at the National Bureau of Standards formulated the first dental specification for amalgam. In 1928, a cooperative program began with the American Dental Association (ADA) and the Federal Government. The cooperative dental research program continues at the NBS but in 1965, the specification program moved to the headquarters of the ADA in Chicago, Ill. The responsibility for the development and promulgation of ADA specifications for dental materials and devices is delegated to the association's Council on Dental Materials and Devices.

The ADA has adopted or is formulating the following specifications. The date in parentheses after the title of the specification is the effective date of the last revision or reaffirmation or expected completion date as noted: (1) Alloy for dental amalgam (1970); (2) Casting investment for dental gold alloy (1962); (3) Dental impression compound (1975); (4) Dental inlay casting wax (1976); (5) Dental casting gold alloy (1976); (6) Dental mercury (1975); (7) Dental wrought gold wire alloy (1975); (8) Dental zinc phosphate cement (1970); (9) Dental silicate cement (1975); (10) Denture rubber (obsolete); (11) Dental agar impression material (1976); (12) Denture base polymer (1976); (13) Denture self-curing repair resin (1975); (14) Dental chromium–cobalt casting alloy (1975); (15) Acrylic resin teeth (1975); (16) Dental impression paste–zinc oxide–eugenol type (1962); (17) Denture base temporary relining resin (1975); (18) Dental alginate impression material (1976); (19) Nonaqueous elastomeric impression material (1977); (20) Dental duplicating material (1968); (21) Dental zinc silico–phosphate cement (1976); (22) Dental radiographic film (1971); (23) Dental excavating burs (1975); (24) Dental base plate wax (1976); (25) Dental gypsum products (1973); (26) Dental x-ray equipment (1975); (27) Direct filling resins (1977); (28) Endodontic files and reamers (1976); (29) General specification for hand instruments (1976); (30) Zinc oxide–eugenol restorative materials (1977); (31) Exposure time designation for timers of dental x-ray machines (1975); (32) Orthodontic wire not containing precious metals (1977); (33) Standard terminology and definitions (1976), with supplement (1976); (34) Asperating syringes (1977); (35) Handpieces (not completed); (36) Diamond rotary cutting instruments (1980); (37) Prophylactic materials (1979); (38) Porcelain–metal systems; (39) Fissure sealants (1980); (40) Dental implants; (41) Standard practices for toxicity testing (1978); (42) Phosphate-bonded investments; (43) Mechanical amalgamators; (44) Dental electrosurgical equipment (1978); (45) Porcelain teeth; (46) Chairs; (47) Units; (48) Ultraviolet activator and disclosing lights (1980); (49) Analgesia–anesthesia equipment; (50) Casting machines; (51) Gas furnaces; (52) Dental porcelain (1978); (53) Crown and bridge plastics (1980); and (54) Needles (1980).

The formulation of new specifications and the revision or reaffirmation of any of the foregoing are the responsibility of American National Standards Committee MD156 of ANSI for Dental Materials and Devices. This committee coordinates its work closely with Technical Committee 106—Dentistry of the ISO. A major responsibility of the FDA is the enforcement of the Federal Food, Drug, and Cosmetic Act of 1938 and its various amendments (eg, May 1976). Dental materials and devices are included and premarketing clearance requirements apply for establishing their safety and effectiveness.

There is a close liaison between the FDA and the ADA specification and certification programs and the FDA is beginning to adopt the ADA specifications.

The ADA maintains a List of Certified Dental Materials and Devices based upon: (1) the certification of the maker to the Association that the item complies with the requirements of the ADA specification; and (2) the testing for compliance with the specification of the item procured on the market in the laboratories of the Association. The ADA also maintains a List of Classified Dental Materials and Devices which have proved "acceptable" or "provisionally acceptable" to the Association's satisfaction based upon data submitted by the applicant and data available in the literature.

Ceramic and Ceramic–Organic Materials

Dental Cements. Dental cements are used in approximately 50% of all dental restorations as cementing (luting) agents for fixed restorations and orthodontic appliances, as a base under permanent restorations to insulate against thermal and chemical shock, for pulp capping, root-canal sealers, and temporary fillings.

Cements are classified into seven types based on composition along with their principal uses, as shown in Table 1. Several important physical properties of cements such as hardness, strength in compression and tension, brittleness, and excessive disintegration in oral fluids, are generally inferior to the ideal; however, they have many favorable qualities for their unique uses.

Extensive research for the improvment of current cements and the development of new ones has yielded products of several distinct characteristics. Most important new developments are the o-ethoxybenzoic acid [134-11-2] (EBA) modifications of the zinc oxide–eugenol cements and the polycarboxylate cements.

Zinc Phosphate Cements. Properties. Zinc phosphate cements must be protected from water loss or gain until the setting is essentially complete. They are not adhesive. Retention is done by mechanical locking in the irregularities of surfaces being cemented. The setting times in the mouth are 3–11 min, typically 7–8 min. Dimensional change upon setting may be less than 0.1% shrinkage, if the mass is protected from water loss, compressive strength at 24 h is 88–143 MPa [(12.8–20.8) × 10^3 psi]. The solubility of the set cement in distilled water will be from 0.1–0.15% for specimens exposed for 7 d.

Composition. The powder consists of calcined zinc and magnesium oxides in ratio of ca 9:1. The liquids are phosphoric acid solutions buffered by aluminum salts and in some instances by both aluminum and zinc salts; the water content is 33 ± 5% (3). The compounds formed by reaction of the powder and liquid are noncrystalline phosphates of zinc [7779-90-0], magnesium [7757-86-0], and aluminum [7784-30-7] (4). The set cement consists of unreacted cores of powder particles bound together with structureless phosphates of zinc, magnesium, and aluminum. The proportion

Table 1. Classification of Dental Cements

Class	Cement type	CAS Registry No.	Composition (principal ingredients) Powder	Composition (principal ingredients) Liquid	Composition (principal ingredients) Set cement	Use Primary	Use Secondary
I	zinc oxide–eugenol	[1314-22-3]	zinc oxide (uncalcined), zinc acetate rosin plastic	eugenol with mineral and/or vegetable oils as diluents	unreacted zinc oxide and eugenol plus zinc eugenolate	cavity lining, base, temporary filling	temporary cement
	zinc oxide–eugenol–EBA	[97-53-0]	zinc oxide (uncalcined) quartz or alumina, rosin	eugenol-(EBA-ortho-ethoxybenzoic acid)	(not known)		
II	zinc phosphate regular	[7779-90-0]	calcined zinc oxide and magnesium oxide; sometimes copper, silver or salts	orthophosphoric acid plus aluminum phosphate and/or zinc phosphate	unreacted powder in a noncrystalline phosphate matrix		
	water settable		zinc oxide, tertiary zinc phosphate, zinc monophosphate; others have monocalcium phosphate and monomanganese phosphate	water	unreacted zinc oxide and zinc phosphates and sometimes manganese and calcium phosphates	cement cavity lining, base	temporary filling

III	copper phosphate	[10103-48-7]	red	Class II regular with some cuprous oxide	unreacted powder with zinc, magnesium, and copper phosphates	Class II—regular
			black	cupric oxide, cobalt oxide, with or without Class II regular	Class II regular	Class II—regular
IV	silicate[a]			complex glass aluminosilicates containing magnesium, fluorine, calcium, sodium, and phosphorus	Class II regular with less acid and more water	unreacted glass particles sheathed by silicate gel and colloidal phosphates	semipermanent filling	temporary facing
V	zinc silicophosphate (Stein cements)			mechanical mixture of Class II regular and Class IV	Class IV	probably combination of Class II regular and Class IV	semipermanent filling	semitranslucent
				Class IV plus zinc and magnesium oxide	Class II or IV			
VI	resin			acrylic polymers mineral fillers	acrylic monomers	acrylic polymers and inert mineral fillers	cement (translucent or opaque)	base
VII	polycarboxylate			zinc oxide and magnesium oxide	polycarboxylic acid	unreacted zinc oxide polycarboxylate	adhesive cement, base, root canal fillings	adhesive base, temporary cement

[a] Water settable type is described but is inferior to the standard type. Apparently the powder comprises active acid phosphates, the usual silicate clinker and magnesium dihydrogen phosphate dihydrate.

of unused powder particles and phosphate matrix in the set cement varies with the amount of powder incorporated into a given amount of liquid to produce the optimum consistency for the specific use. The set cement with minimum amount of matrix and maximum amount of particles has the best physical properties for use in the mouth (see also Phosphoric acids and phosphates).

Silicate Cements. Silicate cements were an important class of restorative materials for the anterior teeth but have been almost completely replaced by composite restorative materials.

Properties. The properly proportioned, mixed, and set silicate cement has many of the properties of porcelain. It is hard, translucent, and can be pigmented to closely duplicate natural tooth structure. The crushing strength of dental silicate cements is 160–187 MPa [(23.2–27.2) × 10^3 psi].

The set silicate cement is an irreversible gel that must be protected from dehydration or the translucency is lost and excessive shrinkage occurs. The average life of a silicate cement restoration is from three to five years because of solubility.

Composition. The powders of these cements are glasses of alumino-silicates containing magnesium, fluorine, calcium, sodium, and phosphorus (5). Most of the powders contain large quantities of fluorides, some as much as 50 wt %. This high fluoride content is associated with the apparent anticariogenic properties of silicate restorations, as there is less incidence of caries on proximal contact of adjacent teeth than with silver or copper amalgam restorations (6).

Liquids are aqueous solutions of phosphoric acid buffered by alumina or zinc salts, or both (5). Their water content is ca 40 ± 5%. The liquids have ca 2 pH.

Zinc Silicate Cements. Zinc silicate [13597-65-4] cements are the luting subgroup of the silicate class of cements. They are primarily used for cementation of orthodontic appliances and translucent porcelain crowns.

Properties. Compressive strengths are comparable to those of silicate cements. The zinc silicate cements are less esthetic than the silicate cements.

Composition. Zinc silicate cement powders are a combination of zinc phosphate and silicate cement powders. The liquid is similar to that used for the silicate cement.

Resin Cements. Resin cements, based primarily on poly(methyl methacrylate) [9011-14-7], have been available since 1952 (7). They have not been generally accepted because of their rapid setting and difficulty in removing excess hardened cement from the interproximal spaces and from beneath free gingival margins (see Methacrylic polymers).

Properties. They have excellent esthetic qualities and are insoluble in mouth fluids. Compressive strength is low, and they have no adhesion to the tooth when wetted. Retention is dependent on mechanical locking when the cement flows into irregularities on the surfaces of the substances being cemented.

Composition. The powder consists largely of methyl methacrylate polymer, to which various fillers (qv) may be added. These include calcium carbonate [471-34-1], silica [7631-86-9], barium carbonate [513-77-9], and calcium tungstate [7790-75-2]. An organic peroxide capable of free-radical formation is also present (see Initiators). The most commonly used organic peroxide in dental application is benzoyl peroxide [94-36-0].

The liquid is basically methyl methacrylate with a suitable inhibitor to ensure adequate shelf life. *N,N*-Dimethyl-*p*-toluidine [99-97-8] is probably the most common promoter; p-toluenesulfonic acid [104-15-4], tri-n-butyl amine [102-82-9], tri-n-hexylamine [102-86-3], and tri-n-octylamine [1116-76-3] have also been used.

Zinc Oxide–Eugenol Cements. Zinc oxide–eugenol cements have many uses in dentistry. Powdered zinc oxide [1314-13-2] and liquid eugenol are mixed together with a spatula to form a paste. The mixture is slow-setting and is used in temporary fillings. More sophisticated compositions result in complex mixtures that serve a variety of dental needs. Some of these compositions are of a cementing nature, but many are useful in other applications, eg, impression pastes and surgical packs.

The more extensive uses of the zinc oxide–eugenol cements include root-canal fillings, linings for deep-seated cavities under restorations, the cementing of temporary or semipermanent acrylic crown forms or splinted crowns on prepared tooth stumps, sealing-in treatments, and temporary fillings.

Properties. Suitably formulated, the zinc oxide–eugenol system can be designed with useful viscosities from thin fluid mixes to heavy puttylike masses. The setting times can be varied from almost instantaneous to several hours. Crushing strengths, in 24 h, of up to 34 MPa (4.9×10^3 psi) have been reported. The degree of hardness or plasticity of the final set mass can vary from a hard friable mass to a hard strong mass, or to a soft gummy mass according to the conditions required. The cements show excellent dimensional stability upon aging. Shrinkages as low as 0.2% after 48 h have been measured. Long-term disintegration (20–22 wk) measurements on conventional zinc oxide–eugenol materials indicate a continual leaching of the eugenol. This loss of eugenol leads to a matrix breakdown and a marked reduction in mechanical properties.

Composition. The cementing applications employ a variety of products compounded either as powder-and-liquid two-part formulations, or as two-part paste-type products. The zinc oxide portion of the formula may contain modifying agents to improve the physical properties of the set paste, or to act as accelerating agents to control the setting rate of the mix. These materials may include zinc stearate [557-05-1], zinc acetate [557-34-6], rosin [8050-09-7], polymerized rosin, hydrogenated rosin, calcium phosphate [7758-23-8], magnesium oxide [1309-48-4], quartz [14808-60-7], aluminum oxide [1344-28-1] (8), medicinal additives, and many other types of additives.

The eugenol portion of the formula may consist of eugenol alone or it may be compounded into a viscous pastelike fluid. The addition of rosin, polymerized rosin, hydrogenated rosin, EBA (9), or other resins, of plasticizing oils, ie, olive oil, and of setting-time control agents, medicinal additives, and other modifiers serve to extend the usefulness of the basic system.

Polycarboxylate Cements. The most recently introduced class of cements (10) are made by mixing powdered zinc oxide and an aqueous solution of poly(acrylic acid) [9003-01-4] (see Acrylic acid). Biological evaluations of these cements indicate that they are safe to use without an intermediate protection between the dentin and the cement (11). When freshly mixed, the carboxyl bonds evidently attach to the calcium of the hydroxyapatite in the enamel and give some adhesion. The adhesion to dentin is reduced since there is less calcium available in the dentin. This cement also has adhesion to stainless steel but not to dental porcelain, resins, or gold alloys (12–14).

Properties. The compressive strength of 7-day-old specimens stored in distilled water at 37°C is in the range of 54–124 MPa [(7.8–18) × 10^3 psi]. The solubility and disintegration in distilled water for 7 d at 37°C, 0.04–0.08 wt %, is not reflected in clinical performance. It expands on setting 50–420 µm/cm (13,15).

Composition. The powder contains zinc oxide and magnesia (13), and the liquid contains the polycarboxylic acid. The set cement consists of unreacted zinc oxide with a matrix of polycarboxylate (10).

Glass-Ionomer Cement. This variation of the polycarboxylate cement was developed for restoration of anterior teeth, erosion cavities, general cementation, cavity liner, and pit and fissure sealant (16). The cement is based on the reaction between alumino-silicate glass powders and aqueous solutions of polymers and copolymers of acrylic acid.

Properties. The glass-ionomer cement has some translucency in contrast to that of the opaque polycarboxylate cement. The setting time is approximately 4 minutes. Compressive strength in 24 h is 154–175 MPa [(22.3–25.4) × 10^3 psi], and tensile strength is 10.1–12.9 MPa [(1.5–1.9) × 10^3 psi]. The solubility is ca 0.44%.

Composition. The powder is an alumino-silicate glass fused with a fluoride flux. The liquid is an aqueous solution of acrylic acid and itaconic acid copolymer [*25948-33-8*] containing methanol as an inhibitor and tartaric acid [*87-69-4*] (5%) to improve working and hardening characteristics (17).

Gypsum Products. Gypsum is widely distributed naturally as calcium sulfate dihydrate [*10101-41-4*] ($CaSO_4 \cdot 2H_2O$). When partially calcined, the hemihydrate ($CaSO_4 \cdot \frac{1}{2}H_2O$) is formed (see Calcium compounds). Gypsum has been used since 1756 for making dental casts, and in another form for dental impressions since 1844 (18).

Plaster is the rehydrated calcined gypsum. The American Dental Association classifies four types according to the physical properties. These are: Type I—plaster, impression; Type II—plaster, model; Type III—dental stone; and Type IV—dental stone, high strength.

Although plaster has been a very successful and serviceable material, it is seriously lacking in hardness, edge strength, chip resistance, abrasion resistance, and strength to fulfill many needs of dentistry. Some of these requirements have been partially filled by the development of the Type III and Type IV plasters.

These different types are the result of various calcining methods. To form plaster, gypsum is ground and subjected to temperatures of 110–120°C in open kettles to drive off part of the water of crystallization. The crystals thus formed are large and porous and are the Type I and Type II plasters. These crystals require a 2:1 powder-gaging water ratio for proper consistency. Type I, impression plaster, and Type II, model plaster differ in additives that control working and setting times. When gypsum is calcined under steam pressure in an autoclave, rod- or prism-shaped crystals are formed. These crystals form the Type III plaster commonly called dental stone and require 28–35 mL of water for 100 g of powder. The Type IV dental stone, high strength, is formed by calcining gypsum in a 30% solution of calcium chloride. The crystals resulting from this process are slightly larger and denser than the Type II crystals and require even less gaging water (20–22 mL/100 g of powder).

Although gypsum products develop a linear expansion during hardening, the true volume of the final dihydrate is ca 7% less (19–20) than the total volume of the hemihydrate plus that of the water required for the chemical conversion of the hemihydrate to dihydrate. This apparent paradox is easily explained: in order to produce a pourable consistency, an excess of water is always used. As the gypsum crystals grow in the form of needle-like projections from each center of crystallization, these growing spiny shapes floating in an excess of water push themselves apart from each other. Voids form between the growing crystals which ultimately interlock to make the mass rigid and hard. Setting expansion is a result of this apparent increase in volume owing to the formation of voids.

As a result of the linear expansion, the reduced volume of the dihydrate and the evaporation of excess water, the percentage of void spaces in plaster is ca 45%, in stone 15%, and in improved stone 10%. Thus, the additional amount of water required for plaster contributes to the volume but not to the strength of the hardened material (21).

Each 100 g of calcined gypsum theoretically requires only 18.6 mL of water to complete the chemical reaction from the hemihydrate to the dihydrate. Any amount of water greater than 18.6 mL/100 g of powder is excess and reduces the strength of the hardened plaster. Water in excess of the amount required for chemical conversion is needed to make a workable consistency. Evaporation of the excess water occurs which leaves voids. When a mixture of the hemihydrate and water hardens, a linear expansion takes place. This expansion may amount to as much as 0.5% for plaster. Dental stones also expand on setting, but the amount is significantly less than that permitted in plaster, 0.2% for Type III and 0.08–0.13% for Type IV.

Impression Plasters. Impression plasters are prepared by mixing with water. Types I and II plasters are weaker than dental stone (Types III and IV) because of their particle morphology and void content. Two factors contribute to the weakness of plaster compared to that of dental stone: (*1*) the porosity of the particles makes it necessary to use more water for a mix, and (*2*) the irregular shapes of the particles prevent them from fitting together tightly. Thus, for equally pourable consistencies, less gypsum per unit volume will be present in plaster than in dental stone, and the plaster will be considerably weaker.

Properties. Impression plasters are formulated to produce a thin, fluid slurry when mixed with the proper amount of water. A satisfactory impression plaster should have the following characteristics: a setting time of 4 ± 1.5 min; fineness, 98% should pass a number 100 sieve (ca 0.15 mm), and 90% pass a number 200 sieve (ca 0.07 mm); setting expansion at 2 h should be <0.15%; the compressive strength at one hour should be 5.9 ± 2 MPa (850 ± 290 psi); and testing consistency as determined by the diameter of the slump in the consistency test should be 90 ± 3 mm.

Composition. Impression plasters are manufactured from the finest finishing plasters that are selected for color and purity. Setting time accelerators, setting expansion control agents, fillers, flavors, colors, or other special modifying agents may be added, eg, starch to cause disintegration of the plaster impression when it is boiled.

Uses. Impression plasters are used to obtain an impression (or negative) of the hard and soft tissues of the mouth. The plaster slurry is placed in a tray, inserted into the mouth and pressed in place against the area in question, and held still until it hardens.

Model Plasters. *Properties.* Model plaster should have a setting time of approximately 10 min, the fineness of the powder should be such that 98% passes a number 100 sieve (ca 0.15 mm) and 90% passes a number 200 sieve (ca 0.07 mm), setting expansion should be less than 0.30%, and compressive strength at the end of one hour should have a minimum of 8.8 MPa (1275 psi), and the consistency should form a disk during the slump test of 30 ± 2 mm dia.

Composition. Model plasters are manufactured from select finishing plasters with special emphasis on a clean, white color. Setting-time control agents, setting-expansion control additives, fillers, and pigments may be added.

Uses. Model plaster, after it is mixed with water, is poured into an impression (negative) to produce a cast (positive) of oral structures. This plaster cast is produced chiefly for study or record purposes, ie, successive casts allow an orthodontist to follow and demonstrate the results of corrective treatment. Additional uses for model plaster include the making of study casts, denture repair casts, mounting interarch registration assemblies on articulators, flasking dentures during processing, and for a variety of other applications where strength and abrasion resistance are not of prime importance.

Dental Stones. Dental stones, produced from gypsum, are sold as dry powders in sealed containers. They are prepared for use by mixing with water, in proportions recommended by the manufacturer.

Properties. Dental stone is generally used at a water–powder ratio of about 30 parts water to 100 parts of stone. The mix is not easily poured, but can flow readily under mechanical vibration. The physical property requirements are: setting time 10 ± 3 min; fineness of powder, 98% should pass a number 100 sieve (ca 0.15 mm) and 90% pass a number 200 sieve (ca 0.07 mm); linear setting expansion at 2 h should be <0.20% and the compressive strength at 1 h >20.6 MPa (3000 psi); and the consistency such that the slump test disk is 30 ± 2 mm dia.

Composition. Setting-time control agents, setting-expansion control agents, fillers, colors, or other modifying additions may be added. Calcium sulfate dihydrate and sodium chloride [7647-14-5] are effective accelerators. Potassium sulfate [7778-80-5] and sodium potassium tartrate [304-59-6] accelerate the setting rate and decrease the setting expansion. Borax [1303-96-4], potassium carbonate [584-08-7], sodium carbonate [497-19-8], and sodium citrate [68-04-2] are all effective retarders and reduce the setting expansion.

Uses. These Type III dental stones are used for casts requiring higher compressive strength and abrasion resistance than for the casts that are formed with the Type II plaster. These dental casts are used for the processing of denture base materials.

Type IV—Dental Stone, High Strength. This dental stone is often referred to as "improved dental stone." The crystals are of a cubic or rectangular shape and show a reduction in cracking and porosity. These changes in crystals reduce the amount of water required to wet the powder and produce a workable consistency (20–22 parts water per 100 parts powder).

Properties. Mixes of improved dental stone (Type IV) using 22 parts of water to 100 parts of powder produce a mass that is not fluid and pourable but can be easily vibrated into place. The physical properties of the improved dental stone are: setting time 10 ± 3 min, fineness of powder such that 98% passes a 100 sieve (ca 0.15 mm) and 90% passes a 200 sieve (ca 0.07 mm); the setting expansion at 2 h is limited to a max of 0.10%, the compressive strength at 1 h should be at least 34.3 MPa (5000 psi); and the disk formed in the slump test for consistency is a 30 ± 2 mm dia.

Uses. Type IV dental stones are used to make casts for a single tooth, for crown or inlay work, and for a complete dental arch.

Dental Investments. Dental investments are refractory compositions suitable for forming a mold into which molten metal may be cast. The cavity is formed in the mold by burning out an expendable wax pattern, which has been imbedded in the mold. Dental casting is a specialized form of the lost wax casting process.

Properties. Some of the requirements for casting investment for dental gold alloy are listed in Table 2 (22). At the temperature tested, the compressive strengths specified for inlay or partial denture types are adequate to prevent fracture during handling and in casting. The strength may vary at casting temperature. The strengths of gypsum bonded investments at high temperatures vary with the additive used to reduce shrinkage during heating (23). The type of binder also influences the effect of the temperature on strength. Table 2 presents some of the physical and mechanical property limits defined by ADA Specification No. 2 for Casting Investment for Dental Gold Alloy (22).

One of the most important functions of an investment is to provide sufficient expansion to compensate for the shrinkage of the wax pattern (up to 1%) and the shrinkage of the cast alloys (1.2–2.6%). The higher the solidus temperature of the alloy the greater the casting shrinkage.

The compensating expansion of dental investment may be separated into two types, setting and thermal. Setting expansion includes: (*1*) normal expansion owing to the hardening of the investment, and (*2*) expansion caused by an excess of water, in addition to the original gaging water in contact with the setting investment (24). Water may be added after the investment has been formed or as excess in the asbestos liner placed between the investment and casting ring. Thermal expansion is a result of phase changes (25).

Composition. The various investments consist generally of a powdered refractory, a binder, and modifiers. The refractory base is usually some crystalline form of silica (SiO_2); mainly quartz [*14808-60-7*] and cristobalite are used. Investments containing cristobalite expand at lower temperatures and exhibit greater expansions than those containing only quartz. Quartz is found in great abundance in nature, but cristobalite is prepared commercially by calcining selected quartz at 1500°C. Cristobalite can also be made by heating silicic acid at 1000°C for several hours.

The usual binder in investments for casting gold alloys is the calcined gypsum, dental stone.

Some of the investments used for casting cobalt–chromium–nickel alloys, that solidify at a temperature of about 1300°C, also have a gypsum binder. Casting of these lower fusing alloys is made possible by the presence of additives that decompose into CO_2 in the investment; this keeps the mold flushed with this gas. Gypsum decomposes rapidly, especially in the presence of carbon, at temperatures required for casting many of the cobalt–chromium alloys (26). Accordingly, investments for use with Type II cobalt–chromium alloys which have solidifying temperatures above 1300°C must have a binder other than gypsum. Organic and inorganic silicates (27–28) as well as some phosphates (29–30) are used. These types of investments are recommended also for use with high-fusing gold and other alloys to which porcelain is fused. They are also recommended for use with alloys offered as substitutes for the gold alloys. Trapped air between the investment and wax pattern may be a problem, even with the use of

Table 2. Some Requirements of ADA Specification No. 2 for Casting Investment for Dental Gold Alloy[a]

Classification	Time of setting at 23.0 ± 2.0°C min, min	Time of setting at 23.0 ± 2.0°C max, min	Compressive strength min, MPa[b]	Linear expansion Setting In air min, %	Linear expansion Setting In air max, %	Linear expansion Setting In water min, %	Linear expansion Setting In water max, %	Temp, °C	Thermal Expansion min, %	Thermal Expansion max, %	Combined setting and thermal expansion min, %	Combined setting and thermal expansion max, %
Type I inlay, thermal	5	25	2.45	0.0	0.5			700	1.0	2.0	1.3	2.0
Type II inlay, hygroscopic	5	25	2.45			1.2	2.2	500	0.0	0.6	1.3	2.7
Type III partial denture, thermal	5	25	4.90	0.0	0.4			700	1.0	1.5	1.2	1.9

[a] Ref. 22.
[b] To convert MPa to psi, multiply by 145.

a wetting agent and vacuum-investing (31). Although the phosphate investments are preferred for the high-melting gold alloys, they are suitable for standard gold alloys.

A number of modifiers are employed to prevent most of the shrinkage of gypsum when it is heated above 300°C. Among these are boric acid (32) and sodium and other chlorides (33). Graphite (34), colloidal copper, and other substances that produce a reducing atmosphere in the mold have been used to protect imbedded metal pieces, such as backings, during the burnout process (35). A review of the development and composition of investments is given in reference 35.

Dental Porcelain. Dental porcelain may be classified into three types according to use. One type is used for artificial teeth, another for jacket crowns and inlays, and the third type, more properly designated as an enamel, is used as a veneer over cast metal (see also Enamels). Composition and technique are similar for all three types.

Dental porcelain, a fine ceramic powder, pigmented to produce the color and translucency of human teeth, is mixed with water to form a paste that is formed into the desired shape and fused to form a ceramic body which is relatively strong, insoluble in oral fluids, and has excellent esthetic qualities.

The excellent service and tissue compatibility of porcelain inspired considerable work directed toward use of porcelain and porcelain-processing techniques, not only for artificial teeth but for inlays, crowns, bridges, and complete denture bases. One undesirable characteristic, brittleness, limits the use of porcelain in many restorative procedures and often requires adequate metallic reinforcement and proper design of the restoration.

Dental porcelain may also be classified according to the temperature at which it matures, to produce both its physical and esthetic qualities. These are also divided into three classes: (1) high-temperature maturing 1290–1370°C, (2) medium-temperature maturing 1090–1260°C, and (3) low-temperature maturing 870–1070°C.

Artificial teeth are generally constructed using the high-temperature maturing porcelain. The material is a mixture of fine particles of feldspar, quartz, and kaolin [1332-58-7]. The feldspar melts first to provide a glassy phase and serves as a matrix for the quartz which is held in suspension in the fired body. Quartz is used as a strengthener; although it reacts with feldspar to provide a bond, it acts chiefly as a coring agent or filler.

During the manufacture, when feldspar is melted at approximately 1250–1500°C, the alkalis (Na_2O and K_2O) unite with the alumina and silica to form sodium aluminum silicate [1344-00-9] and potassium aluminum silicate [1327-44-2]. A glassy phase is formed with a free crystalline silica phase. In general, the lower the soda to potash ratio, the lower the fusion temperature. However, the potash form provides greater viscosity of the molten glass and less slump or pyroplastic flow of the porcelain during firing. This pyroplastic flow of a dental porcelain should be low in order to prevent the rounding of margins, the loss of tooth form, and the obliteration of surface marking so important to a lifelike appearance.

The low and medium temperature maturing porcelains are ground from blocks of matured porcelain. The raw ingredients are mixed and fused. The fused mass is then quenched in water. The glass is stressed so that considerable cracking and fracturing occur (fritting); the product is a frit. With such a brittle structure, it can be readily ground to a fine powder of almost colloidal dimensions. During subsequent firing little or no pyrochemical reaction occurs. The particles are merely fused together, but the temperature must be controlled to minimize the pyroplastic flow.

Alkalis are introduced as either carbonates or the naturally-occurring minerals (feldspar or nepheline syenite [63919-78-8], or both). In the latter case, some silica and alumina are introduced. Boron can be added as borax or boric acid. When lime is present, it can be added as calcium carbonate that reverts to calcium oxide during fritting (see Calcium compounds). Other additions are made using the oxide itself.

These porcelains are translucent. Opaqueing agents such as the oxides of zirconium [1314-23-4], tin [1332-29-2], or titanium [13463-67-7] are used to mask the color of dentin, or more frequently the color of the metal to which it is fused.

A ceramic veneer or glaze may be added to a porcelain restoration after it has been fired. This veneer consists of a layer of transparent glass with a maturing temperature lower than that of the body porcelain. The result is a glossy or semiglossy, nonporous surface. The coefficient of thermal expansion of this glaze should be equal to that of the porcelain body to which it is applied. If the glaze has a higher coefficient of thermal expansion than the body, it cools under radial tension which may cause a crazing of the surface; the greater the stressed condition, the finer the crazing network.

Porcelains are completely nonductile after vitrification. Dislocations and slip cannot occur. When it breaks, a brittle fracture occurs. However, porcelain may be strengthened by adding a filler that can resist or inhibit the propagation of the crack. A core structure is formed with glass as the matrix and the stronger filler as the core. An efficient coring material is recrystallized alumina [1344-28-1] (Al_2O_3) (36).

The alumina particles are much stronger with a higher modulus of elasticity than quartz and are more effective in the interruption of crack propagation. As with quartz, a two-phase system is present. Also like quartz, the alumina is slightly attacked by the flux, and a primary bond is formed. However, no change in the alumina itself occurs during firing. Glass–alumina composites have been termed aluminous porcelains.

Dental porcelains are pigmented by oxides in the frit to provide desired colors. Generally, these powders are highly pigmented with brilliant hues of the desired color; the colors vary from brilliant red, yellow, or brown to pure white. The powders are blended with the unpigmented powdered frit to provide the proper hue and shade.

Porcelain-to-Metal Bonded Systems. When porcelain or glass are used in porcelain-to-metal (metal–ceramic) systems they are called ceramic enamels (see also Ceramics). The technique is as follows: a thin metal casting is made to fit the prepared tooth and the ceramic is then fused as a veneer on the metal casting so that little or no metal is visible. A layer of opaque porcelain is fused against the casting, and then the tooth contour is built by fusing an overlay of translucent enamel. The final enamel-veneered structure is then cemented on the prepared tooth.

The alloys used for the construction of metal–ceramic restorations have a number of stringent requirements. Both the metal and ceramic must have coefficients of thermal expansion closely matched to prevent undesirable tensile stresses at the interface. The metal should have a high proportional limit and particularly a high modulus of elasticity; the magnitudes, together with the bulk of the alloy, determine the ability of the casting to resist deformation. Adhesion resulting from van der Waals forces is among the several explanations of the bonding mechanism (37–38). Additional explanations include the presence of tin oxide on the surface of the metal that combines with the porcelain to produce a bond, and the possibility of a bond being produced by compressive forces resulting from small differences in thermal expansion of the two phases (39).

Compositions of some veneering porcelains show high silica and alumina contents (40).

Chemical bonding is the most probable mechanism, and the bond probably occurs through the diffusion of oxides from the metal to the porcelain. Roughening of gold castings does not appear to be necessary, since it does not increase the bond strength of porcelain to the alloy (38,41–42). Manufacturers generally recommend degassing the metal by heating the alloy to ca 1035°C before veneering. Handling the metal with the fingers before fusing produces a volatile contaminant.

Colloids

Agar-Based Impression Materials (Reversible Hydrocolloids). Dentistry needed an elastic impression material that would take accurate, one-piece impressions of undercut areas. The agar-based impression materials are thermally reversible, aqueous gels (43–44). They become viscous fluids in boiling water and set to an elastic gel when cooled below 35°C. Although the popularity of agar-based impression materials has diminished with the subsequent introduction of competing types of elastic impression materials such as alginate-based, polysulfide, silicone, and polyether impression materials, agar [*9002-18-0*] materials are still used in substantial quantities.

Impressions of inlay and crown preparations, and all gingival areas, are best obtained by filling the preparation or gingival area with impression material injected from a hypodermic syringe. This eliminates trapping air in the corners and recesses and gives a more faithful reproduction without nodules or other imperfections.

The agar-based impression materials are used extensively for duplicating casts. Frequently, it is desired to retain the original model for reference and do the actual work on a duplicate cast. Partial-denture fabrication requires that the original stone cast be duplicated in an investment. For duplicating, the agar-based impression material is usually diluted with water, boiled, cooled to the desired temperature, and carefully poured over the model to be duplicated.

Properties. Requirements of an elastic impression material include that it must: (*1*) be easily and quickly prepared; (*2*) set quickly to an elastic mass in the mouth; (*3*) not be harmful or cause discomfort to the oral tissues; (*4*) flow to all areas without the need of excessive force; (*5*) copy detail accurately; (*6*) possess sufficient strength, toughness, and elasticity to resist permanent deformation when removed from the mouth; (*7*) not adversely affect the set properties of the cast material; (*8*) be capable of being manipulated within a temperature range tolerated by the oral tissues; and (*9*) have sufficient viscosity to remain in a tray.

The natural characteristics that make agar good for impression compositions include unusually good elastic behavior and hysteresis between its liquefying and gelation temperatures. Agar sols liquefy at 70–100°C. On cooling, they form a solid gel at 30–50°C. The delay in gelling brings the temperature within a range tolerated in the mouth. Another useful characteristic is the dimensional stability of the material during gelation, cooling, and storage as long as the water content remains constant.

Agar-based impression materials must have a compressive strength of at least 0.2 MPa (28 psi). They should have a strain in compression of 4–20% in stresses of 9.8–98 kPa (1.4–14 psi) per specification test method. They should not have a permanent deformation exceeding 3% after 12% strain is applied for 30 s.

Composition. The basic ingredient is agar, a sulfuric acid ester of a galactan complex.

About 6–12% agar is generally used with a 75–85% water content. Fillers may include zinc oxide and clays. Small percentages of boron compounds increase the viscosity, strength, toughness, and resiliency of the composition. Borax, calcium metaborate [13701-64-9], and organic borate compounds have been used; waxes or fatty acids plus emulsifying agents may be present; plaster-accelerating agents are frequently added and may include potassium sulfate, magnesium sulfate [7487-88-9], and zinc sulfate [7733-02-0]. Such additions can be kept to a minimum if the cold-water-soluble salts and organic contaminants are washed from the agar before compounding. Agents for pH control and fibrous reinforcements have been included in formulations. Agar impression materials must be formulated to remain stable in storage. They should be free of salts or additives that crystallize and salt out, degrade the complex agar molecule, or induce syneresis in the agar gel. Any syneresis will destroy the dimensional accuracy of the composition.

Alginate-Based Impression Materials (Irreversible Hydrocolloids). The alginate impression materials are supplied as a dry powder. When the correct proportions of the powder and water are mixed, a viscous but slightly fluid mass is formed. The gel formation is based upon the conversion of soluble salts, usually potassium or sodium salts of alginic acid [9005-32-7], to insoluble salts of alginic acid, usually calcium alginate [7440-70-2].

The improvements in available alginate products, their convenience of use, and the cost advantage have contributed to increased growth and popularity of the alginates.

The technique of preparing an alginate impression material requires only the mixing of measured amounts of the powder and water for 30–60 s. The paste is placed in a suitably prepared tray and seated over the area of which an impression is desired. Within 1–4 min the material sets to a strong, tough elastic gel. The impression should be unseated by a quick, deft movement to minimize tearing or distortion of the impression. The precautions required to stabilize the water content of the alginate impression materials are the same as those described for the agar impression materials. The casts should be poured immediately. Most modern alginates do not require treatment in a fixing solution before pouring the cast.

Properties. The alginate impression materials are chemically reactive mixtures. All factors that influence reaction rates are, therefore, important in the use of these materials, ie, correct proportioning; temperature of the water, powder, and mixing equipment; and spatulation rate and duration.

Alginate impression materials must have a compressive strength of at least 0.34 MPa (50 psi) 8 min after the start of the mix; at least 3.5 min of this time interval should be in storage at $37 \pm 1°C$. They should have a strain in compression of 4–20% between stresses of 9.8–98 kPa (1.4–14 psi) per specification method. They should not have a permanent deformation exceeding 3% after a 12% strain is applied for 30 s.

Composition. Alginate impression materials are usually the potassium or sodium salts of alginic acid. These commercially available materials have various degrees of polymerization, controlled pH ranges, and particle sizes. Alginic acid is derived from specific varieties of kelp, a marine plant. It is a high molecular-weight linear polymer of anhydro-β-d-mannuronic acid [1986-14-7].

The cold-water-soluble sodium or potassium alginates must be precipitated as insoluble alginate salts to be useful impression materials. The precipitating compounds may be any heavy-metal ion whose rate of availability for precipitation can be controlled. Calcium sulfate [7778-18-9] (45), lead silicate [11120-22-2] (46), and chromic sulfate [10101-53-8] (47) have been used. Calcium ions are used most frequently. Retardation of the precipitation reaction, to give useful working time, is generally accomplished by the introduction of a phosphate compound; disodium phosphate [7558-79-4], trisodium phosphate [7601-54-9], alkali metal polyphosphates, tetrasodium pyrophosphate [7722-88-5], and sodium hexametaphosphate [10124-56-8] are typical retarders. It is desirable to have the precipitation of the insoluble alginate completed as rapidly as possible, after an adequate working time. The addition of accelerating materials promotes a fast "snap over." Aluminum, sodium, potassium, zinc, manganese, and magnesium fluosilicates are helpful. The inclusion of suitable filler gives bulk to the dry powder, and blends and controls the viscosity and body of the mixed material. Calcium carbonate, magnesium oxide, zinc oxide, diatomaceous earth [7631-86-9], or other silica types may be used.

Elastomers

Polysulfide Impression Materials. The very extensive need for elastic impression materials by the dental profession has spurred a continuous search for better systems. The agar- and alginate-based products fulfill a basic need, but owing to their dimensional instability with any loss or gain of water, improved systems have been needed. In 1953 the first nonaqueous, elastic dental impression material, Static, was introduced. This was closely followed by Permlastic in 1954. Both are based on the room-temperature conversion of a liquid polymer, a polyfunctional mercaptan (polysulfide), to a strong, tough, dimensionally accurate elastomer. The liquid polymers are identified as polysulfides under the trade name Thiokol (see Polymers containing sulfur). Many brands of polysulfide impression materials are now available to dentistry. The conversion of the liquid polymer to an elastic solid has been achieved in most products by lead peroxide [1309-60-0]. Great improvements in strength, toughness, and dimensional stability of the set elastomers over the aqueous elastic impression materials make the new compositions popular.

The materials are available in three basic grades, ie, light-bodied or syringe, regular, and heavy-bodied types. The products are supplied as a two-part paste system, usually packaged in collapsible tubes. The polysulfide-containing base and the setting-agent paste (usually lead peroxide) are mixed together in approximately equal amounts to give a homogeneous, streak-free mass, which is transferred to a tray of adhesive-coated metal or of custom-made acrylic resin. The filled tray is then placed over the area of interest and held motionless until the "rubber" has set, usually within 5–10 min.

Properties. The polysulfide impression materials can be formulated with a wide range of physical and chemical characteristics. Generally, some of the differences are in the base (polysulfide portion), and some are in the "catalyst" (misnomer for PbO_2) portion of the formulation. Further changes may be obtained by varying the proportion of the base to the "catalyst" in the final mix. Three basic objectives can be satisfied by these mechanisms: (1) viscosity control from thin fluid mixes to heavy thixotropic mixes; (2) setting-time control; and (3) control of the set-rubber hardness from a Shore A Durometer scale of 20 to 60. Variations in strength, toughness, and elasticity can

478   DENTAL MATERIALS

also be achieved. The shrinkage, upon conversion from the fluid state to the elastic solid state, has been reported in the range of 0.03% in 24 h for one composition. Generally the max shrinkage in 24 h is 0.15% for most products. The strain in compression, 9.8–98 kPa (1.4–14 psi), varies from 7 to 15%. The permanent deformation, after 12% strain is applied for 30 s, varies from 4 to 6.5%. The better polysulfide compositions, although they undergo change with age, if properly packaged remain workable and useful for many years.

The liquid polysulfide polymers, as manufactured, are slightly acidic. In this condition they are very stable. In an alkaline oxidizing medium they polymerize rapidly. Heat, moisture, and an alkaline environment accelerate the reaction. Stearic acid [57-11-4], oleic acid [112-80-1], lead stearate [7428-48-0], and aluminum stearate [637-12-7] are active retarders.

The polysulfide rubbers are capable of being electroplated by a special technique. Either copper or silver can be used, but the alkaline silver cyanide plating system gives more reliable results (see Electroplating).

Composition.   The base material (polysulfide) contains 50–80% of the polyfunctional mercaptan which is a clear, amber, syrupy liquid polymer with a viscosity at 25°C of 35 Pa·s (350 P), an average mol wt of 4000, a pH range of 6–8, and a mild, characteristic mercaptan odor. Fillers are added to extend, reinforce, harden, and color the base. They may include silica, calcium sulfate, zinc oxide, zinc sulfide [1314-98-3], alumina, titanium dioxide [13463-67-7], and calcium carbonate. The high shear strength of the liquid polymer makes the compositions difficult to mix. The addition of limited amounts of diluents improves the mix without reducing the set-rubber characteristics unduly. Diluents are dibutyl phthalate [84-74-2], tricresyl phosphate [1330-78-5], and tributyl citrate [77-94-1].

A number of modifying agents are desirable. Sulfur, in fractional percentages, is a powerful activator. Stearic acid and oleic acid may be used as retarding agents. Additions of acid must be carefully selected because strong acid attacks the methylenedioxy linkages and degrades the polymer. Buffering agents for pH control and perfumes for odor-masking are also frequently used.

The second paste of the two-paste product (the "catalyst") is a curing agent. A wide variety of materials convert the liquid polysulfide polymers to elastomeric products. Alkalies, sulfur, metallic oxides, metallic peroxides, organic peroxides, and many metal–organic salts, ie, paint driers, are all potential curing agents.

Only two types of systems have found application in dentistry. Lead peroxide is the curing agent most frequently used for the polysulfide polymers that serve as dental impression materials. Lead peroxide converts the liquid polymer to an elastic solid within a time short enough for oral applications.

Many curing systems bring about a liquid-to-solid conversion of the polysulfide polymers. Most curing agents produce an initial solid mass characterized by a high degree of plasticity and poor elasticity. The development of elastic properties is so slow that the materials are not suitable for the accurate reproduction of undercut areas found in oral structures.

The curing paste is usually dark brown owing to the high concentration (20–78%) of lead dioxide. A nonlead peroxide curing system offers curing pastes in yellow and red that produce lighter shades of set-rubber colors. A suitable vehicle may be selected from the materials listed as diluents for the base. Suitable fillers may also constitute part of the curing paste formula. Buffering agents, thixotropic additives, and other modifiers may be added.

Silicone Impression Materials. Odor, color, and stickiness of the polysulfide rubbers deterred universal acceptance. With the development of the room-temperature-vulcanizing (RTV) silicone rubbers, another acceptable elastomeric system was made available to dentistry (see Silicon compounds, silicones; Elastomers, synthetic). The early offerings of silicone impression materials had several serious problems: (*1*) The silicone base had limited shelf-life; (*2*) the catalyst systems were air-sensitive and subject to deterioration; and (*3*) the set elastomers evolved hydrogen and produced a spongy surface on casts. The materials presently used have a shelf-life of two years or longer and the setting reaction is completely free from gas evolution.

Properties. The silicone impression materials have not produced the variety of products that are found with the polysulfide rubbers. Suppliers have chosen to market only compounded products and not the base gums. The dental-material manufacturers have added little to the silicone compounds, except perhaps a coloring agent. Some product individuality has been evident in the catalyst systems. Here again, the differences from product to product are less than for the Thiokol-based products. The silicone bases have a mild, pleasant odor. They are generally white, although some manufacturers add a pink pigment. The materials are nontacky and can be wiped away from instruments, hands, etc, at any stage of the mix or set.

The silicones offer a selection of setting rates and mix viscosities. Regardless of base viscosity, the set rubbers are about of the same hardness, a Shore A Durometer value of about 55. The linear shrinkage upon curing exceeds that of the mercaptan rubbers, and is in the range of 0.2% in 24 h. The strain in compression, 9.8–98 kPa (1.4–14 psi) is 8.8–34%. The permanent deformation, after 12% strain is applied for 30 s, is 1–4.5%.

Both the silicone base and the catalyst compositions are subject to deterioration when exposed to the atmosphere. The bases tend to polymerize in the tubes and the catalysts lose their activity, and in some cases crystallize.

Composition. The silicone bases are siloxanes with terminal hydroxy groups. They are cured by means of a condensation reaction initiated by suitable catalysts. The reaction can be represented as follows:

$$\underset{\text{fluid polymer}}{HO-\underset{\underset{R}{|}}{\overset{\overset{R}{|}}{Si}}-OH + HO-\underset{\underset{R}{|}}{\overset{\overset{R}{|}}{Si}}-OH} \xrightarrow{\text{catalyst}} \underset{\text{elastic solid}}{HO-\underset{\underset{R}{|}}{\overset{\overset{R}{|}}{Si}}-O-\underset{\underset{R}{|}}{\overset{\overset{R}{|}}{Si}}-OH + H_2O}$$

In the presence of an organic silicate, certain heavy-metal-soaps trigger the reaction. The useful metal soaps include stannous octanoate [*1912-83-0*], zinc octanoate [*557-09-5*], dibutyltin dilaurate [*77-58-7*], and dibutyltin diacetate [*1067-33-0*]. The reactivity of the different soaps varies considerably. Stannous octanoate effects a cure in 0.5–2 min. Zinc octanoate may require 24–96 h; the dibutyltin dilaurate, 10–20 min. Heat and moisture accelerate the curing rate, but to a lesser degree than in the case of the polysulfide rubbers.

A silicone product has been recently introduced which is polymerized by an addition reaction. This is supplied in a two-paste system. One of the tubes of paste contains a low molecular weight silicone with terminal vinyl groups, reinforcing filler and a chloroplatinic acid catalyst. The second tube contains a paste having a low molecular weight silicone with terminal silane hydrogens, and reinforcing filler. The addition reaction occurs between the hydrogen and the vinyl groups, and no by-product is formed.

The silicone impression materials are very compatible with gypsum products. The casts have excellent hard surfaces. The silicone-based impression materials can be electroplated with either copper or silver. However, the acidic copper sulfate bath gives more acceptable results.

Uses. The technique of taking a silicone rubber impression is similar to that employed for the polysulfide–rubber impressions. The materials are available in three basic types, ie, syringe, regular, and heavy-bodied. The products are supplied as two-part products. The silicone bases are viscous fluids, usually packaged in collapsible metal tubes, or in jars for the heavy-bodied. The catalysts are either liquid or paste products. The liquids are packaged in plastic bottles or collapsible metal tubes. The impression procedures are the same as those given for the polysulfide impression materials.

Polyether Impression Materials. Another type of elastomeric impression material, a polyether-base polymer, was introduced in 1964 (48–49) (see Polyethers). This material is cured by the reaction between aziridine rings

$$\underset{H}{\underset{|}{N}}\diagdown\underset{R}{\overset{R}{\diagup}}C-CH(CH_3)-$$

which are at one end of branched polyether molecules. The main chain is probably an ethylene oxide-tetrahydrofuran copolymer [27637-03-2]. Cross-linking and thus setting is brought about by an aromatic sulfonate ester of the type:

$$C_6H_5-SO_3R$$

This produces cross-linking by cationic polymerization via the imine end groups. These polyethers are supplied as two pastes. Just as with the rubber and silicone light-body materials, one mix is used in the syringe, and the remainder in the tray. Setting time for these materials is relatively short, ie, 2.5–3 min.

The polyether material has water sorption values of about 14% at equilibrium in water (50). The material should not be in contact with water during setting and storage (51).

Metals and Alloys

The chemical and physical properties of some metals and alloys, such as hardness, strength, stiffness, toughness, resistance to corrosion, and biocompatibility have provided materials capable of withstanding the most severe demands of restorative dentistry, namely the harsh corrosive environment of the mouth and high stresses on small cross-sectional areas. Metallurgical and dental science have given dentistry many excellent alloys for restorative dentistry, prosthetic dentistry, orthodontics, and dental techniques (see also Prosthetic and biomedical materials). The following classes of metals and alloys serve these needs: amalgams; base metal alloys (cobalt–chromium casting alloys for partial dentures, chromium-containing casting alloys for crowns and bridges); gold and gold alloys (foil, crystal, Au–Ca alloy and combination, casting and wrought alloys); platinum and platinum alloys; solders (qv) (gold solders, silver solders); electrodeposited metals (copper, silver).

Amalgams. Since the earliest recorded use of amalgam as a dental restorative material in 1528 (52), it has achieved the status of one of dentistry's main therapeutic agents as it is used in 70–80% of all restorations of single teeth. It is also a striking example of the many unique dental materials that have been developed over the centuries. The modern amalgams are metallurgical triumphs. They remain plastic at room and mouth temperatures for several minutes after preparation so they can be compacted into the prepared cavity and then harden quickly with very small dimensional change. Their strength and corrosion resistance need improvement, but they are remarkable considering the dental requirements.

Properties. Dental amalgam is a novel alloy. Within a few seconds of the start of trituration, the alloy and mercury [7439-97-6] must amalgamate to a smooth plastic mass. Within 3–5 min it should set to a carvable mass and remain so for 15 min. Within 2 h it must develop sufficient strength, hardness, and toughness to resist mild biting and chewing forces, and should not tarnish or corrode much in the mouth nor react to produce toxic or soluble salts; it must maintain its color. Most present-day amalgams can be produced to meet a considerable degree of these formidable requirements; however, owing to the peculiarities of the alloy system, rigid control of all aspects of processing is necessary. A satisfactory restoration should have a mercury content of less than 50%. Higher mercury contents in the final restoration result in low-strength amalgam with high creep, excessive setting expansion and poor corrosion resistance. Properly proportioned, mixed and processed amalgams will have compressive strengths in the range of 210–345 MPa [(30–50) $\times 10^3$ psi] in 24 h. Tensile strengths are 48–55 MPa (7000–8000 psi). Static creep of 8×4 mm cylinders containing 48% residual mercury, after aging for one week at 20°C and then subjected to a static load of 36 MPa (5200 psi) for 4 h, shortened from ca 0.9 to 8.0%. When the residual mercury content is increased to 53% the values are 1.52 times as great (53). The linear setting expansion from 5 min after the end of trituration to 24 h is considered satisfactory if it falls within 0 ± 20 μm/cm. All attempts to relate the dimensional change on setting, and many mechanical properties, to clinical behavior have not been very successful except for the findings of Mahler and co-workers who showed that creep is seemingly effective in predicting marginal fracture of amalgam restorations (54). This correlation has been demonstrated by many other investigators. However, it was not until 1977 that a plausible theory was presented to show a linear relationship between the amount of extrusion of amalgam fillings placed in a steel block and the percentage of creep for the three amalgams that have had long-time rigorous clinical comparisons (55–56). This theory suggested that volume changes caused by phase changes were restricted in proportion to the creep of the amalgams (55). Therefore, it was postulated that the amount of extrusion of amalgam restorations from the cavity, which is correlated with marginal integrity (57), is dependent on the relative resistance of the amalgams to deformations caused by phase changes (56). Not until adequate clinical data are available on other amalgams and further data on the expansion of such confined amalgam specimens can this creep–marginal integrity relationship theory be confirmed.

Composition. The compositions of powdered alloys used in preparing dental amalgams are usually silver, 66.7–75.5%; tin, 25.3–27.0%; copper 0.0–6.0%; and zinc, 0–1.9%. These are commonly referred to as conventional alloys and the amalgams made from them as conventional amalgams. This composition range was in use for almost a century until two Canadian metallurgists designed an alloy consisting of a mechanical mixture of two powders; one with the customary flakes of the aforementioned composition range and the other, spherical shapes of the silver–copper eutectic (Ag 72%, Cu 28%) (58). The ratio of the powders was approximately 2:1, respectively, giving an approximate composition of silver, 70%; tin, 18%; copper, 11%; and zinc, 1%. Amalgams made with this powder have high compressive strengths and low creep and have better clinical performance than conventional amalgams (56). They are sometimes called dispersant amalgams. Other powdered alloys having a uniform composition appeared on the market with about the same overall composition but with high compressive strength of 462 MPa (67,000 psi), high tensile strength of 55 MPa (8000 psi) and low creep (0.2%) compared to the amalgams made with the mixture of powdered alloys (59). Other alloys appeared soon with even higher copper content. The range of composition of some of these alloys was silver, 42.2–70.3%; tin, 17.7–30.0%; copper, 12.0–27.8%; and zinc, 0–0.3%. In addition, one of the powdered alloys contained 3.4% of indium. Many clinical performance tests are in progress on amalgams made from such alloys; they should continue for a minimum of three years. These amalgams have also been called non-γ_2 amalgams. The γ_2 refers to $Sn_{7-8}Hg$ (Sn_7Hg [*11092-11-8*]).

The complete role of each of the metals in dental amalgams is not fully understood. It is generally conceded that when the powdered conventional alloy is triturated with mercury the gamma and beta phases of the silver–tin system are attacked by the mercury and that principally Ag_2Hg_3 (γ_1) [*12249-78-4*] and $Sn_{7-8}Hg$ (γ_2) are formed. These phases form a matrix that binds the remaining unreacted part of the original powder particles together. Usually the less the matrix, the better the values for pertinent physical properties.

One of the serious defects of the conventional amalgams is the corrosion of the $Sn_{7-8}Hg$ phase, which is absent or, if it forms, mostly disappears in amalgams with the increased copper content; hence the term as non-γ_2 amalgams. The following structures are present in the two conventional amalgams used in the clinical testing (54,56,60):

    Ag–Sn phase (γ) Ag_3Sn [*12041-38-2*]
    Ag–Hg phase (γ_1) $Ag_{22}SnHg_{27}$ [*69484-15-7*]
    Sn–Hg phase (γ_2) Sn_8Hg [*11092-12-9*]
    Cu–Sn phase (ϵ) Cu_3Sn [*12019-61-3*]

In the dispersant amalgam the following additional phases were detected: (*1*) eutectic of the Ag–Cu system and reaction ring of γ_1 + Cu_6Sn_5 [*12019-69-1*] around the particles. If 10% gold was added to conventional alloys at the expense of the silver content, non-γ_2 amalgams were formed (61). (*2*) The use of elements such as tellurium and manganese as modifying agents to improve the corrosion resistance and other pertinent properties are the subject of some current research.

The phases present in hardened amalgam and their proportions are controlled by many factors. The composition of the alloy, the size, shape, and size distribution of the particles, the thermal history of the cast ingot and the comminuted alloy, and the surface treatment of the particles are some of the factors for which the manufacturer is responsible. The proper cavity preparation and mixing, and the compacting and finishing techniques of the dentist can make the difference between satisfactory and unsatisfactory restorations with the best of alloys. A minimal amount of residual mercury and of porosity are needed to obtain the most serviceable restorations (62).

Uses. Amalgam restorations are prepared by the dentist from a powdered alloy by mixing it with mercury to form a plastic moldable mass that is packed or condensed into the prepared cavity. The cavity is designed to provide mechanical retention, maximum marginal mass, support to absorb the functional stresses transmitted through the restoration, and maximum protection to the remaining tooth structure. The restoration reestablishes the normal tooth anatomical form and function. The success enjoyed by amalgam restorations is amazing, considering the abuses to which many restorations are subjected, both in preparation and use. Amalgams have shown some sensitivity to manipulation variables. It is therefore essential to follow the instructions for proportioning, preparation, condensation, and finishing provided with each individual alloy. The general procedure conforms to the following outline:

The alloy and mercury are proportioned by weight. Current alloy improvements, better trituration equipment, and improved condensation methods have all tended to permit a reduction of the mercury content in the mix and in the restoration.

The alloy and mercury should be triturated together as directed in mechanical amalgamators with controlling timers. The practice of mulling in the hand following trituration is obsolete and hazardous.

There should be very little excess mercury in the mix.

The amalgam must be condensed properly in the prepared, clean, dry cavity. In general, the maximum effective load tolerated by the patient should be used on a condenser of optimum size, to remove as much mercury from the mass as possible.

The amalgam must be carved to the contour of the lost tooth tissue it is replacing.

Final polishing should be delayed for at least 24 h after placing the restoration. Amalgam restorations should not be subjected to biting or chewing forces for at least 2 h, and preferably 6 h, after insertion to avoid fracture of the restoration.

A smooth, well-polished amalgam restoration will retain its color and appearance much better than a poorly finished one. Roughness promotes uncleanliness, discoloration, and corrosion.

Above all the patient must be taught proper oral hygiene.

Comparisons. Amalgams made with spherical particles may predominate over those made with flake-shaped particles because desirable plasticity is obtained with lower mercury content, satisfactory compaction is achieved with lower packing pressures, and there is less influence of manipulative variables upon values for appropriate physical properties.

Generally, the new amalgams with the higher copper content have better values for the properties of interest than the so-called conventional amalgams. The high copper content (non-γ_2) amalgams are more corrosion resistant, have less creep, higher compressive strength and lower tensile strength, better marginal integrity in restorations, less discoloration, and a much higher liquidus temperature than the conventional amalgams.

Manufacturing. The general procedure of manufacturing processes to produce alloys for dental amalgams is as follows: (*1*) metals of the required purity are melted together, in their proper proportions, under nonoxidizing conditions; (*2*) atomization is used to produce spherical particles or ingots are cast and cooled at a rate that produces some Widmanstätten structure, a geometrical pattern caused by the formation of a new phase along certain crystallographic planes of the parent solid solution; splat cooling is also being used experimentally; (*3*) the ingot may be given a homogenization heat treatment at 400°C; (*4*) the ingots are reduced to filings by machining on a lathe or a cutter; (*5*) iron particles are removed magnetically; (*6*) the filings may be ground to reduce their size; (*7*) undesirable particle sizes are screened out to the desired gap-graded sizes; (*8*) the particles are given a heat treatment at 60–100°C for 1–6 h to pre-age the metal, remove the cold-work effects, and reduce the rate of amalgamation and setting; and (*9*) the particles may be acid-cleaned to remove any oxide film and other contaminants, then washed and dried.

Copper Amalgams and Admixtures of Copper and Conventional Amalgam. Copper amalgam, a binary amalgam of copper (27–38%) and mercury (62–72%), is now seldom used in the United States as a restorative material. It corrodes and wastes away in the mouth and discolors the teeth, but it has bacteriostatic qualities. Copper amalgam can be made by treating precipitated washed copper with a dilute solution of mercuric nitrate. This causes a precipitation of mercury on the copper powder which is then easily amalgamated by triturating with at least two times as much mercury as powder. After the amalgam hardens it can be replasticized by heating and compacted into a cavity where it will harden. Some dentists use a mixture of copper amalgam and the conventional amalgam as a restorative material. Such a mixture seems to produce non-γ_2 amalgam that has many of the attributes of the high copper amalgams (63), and the serviceability and appearance of the restorations (see Copper alloys).

Gallium-Based Alloys. Gallium-based alloys have not been useful in dentistry because of their toxicity (64) (see Gallium).

Base-Metal Alloys. Base-metal casting alloys, although inferior to gold-based casting alloys in some dental applications, are superior in others. For inlay, crown, and bridge castings the gold-based alloys are preferred because the wax pattern can be reproduced more accurately in the casting. Gold alloys are lower melting and easier to use; they can also be fabricated conveniently. The noble metal alloys seem to be superior in the porcelain–metal bonding combinations. However, base-metal alloys are continuing to make inroads in dental procedures at the expense of the noble metal alloys. Several base-metal alloys containing sufficient chromium to make them passive have essentially displaced gold-based alloys in casting skeletons for partial dentures since their introduction several decades ago (65–67). Practically all partial dentures are fabricated in commercial dental laboratories from Co–Cr, Cr–Ni, or Ni–Cr–Co alloys. Some inlays, crowns, or bridges are made of nonprecious alloys. Wrought stainless steels of the 18-8 type are used infrequently as denture bases but have largely supplanted gold alloys in orthodontic appliances for preventing and correcting malocclusion and associated dental and facial disharmonies. Chromium-containing alloys are the only base metal alloys that can be used with dental techniques and have almost no tarnish or corrosion in the mouth. Titanium-based alloys have a potential use because means have been found to fabricate them using dental methods that require every inlay, crown, bridge, or denture to be tailor-made.

Cast Chromium-Containing Alloys for Removable Partial Dentures. The use of these alloys is attributed to their light weight, low cost, stiffness, and general passiveness (see Chromium and chromium alloys). The high temperatures necessary to get proper fluidity for casting precludes their use in gypsum-bonded investments; silicate or phosphate-bonded investments had to be developed. The special materials and techniques that have evolved to fabricate restorations from the cobalt–chromium alloys have essentially restricted the use of these alloys to commercial dental laboratories. Some progress has been made in the development of alloys with lower melting ranges that can be cast in the gypsum-bonded investments using the customary techniques. The ADA Specification No. 14 for dental chromium–cobalt casting alloy names two types: high-fusing (minimum liquidus temperature 1300°C), and low-fusing (maximum liquidus temperature 1300°C). There appears to be only one brand of the low-fusing type.

Properties. It is generally conceded that the casting shrinkage of the cobalt–chromium alloys is greater than that of the gold alloys. The linear value is approximately 2.3% (68) depending on the diameter of the cast rod. The casting shrinkage of gold alloys is much lower (1.25–1.56%). Values are affected by the size and shape of the castings, spruing, composition, and other variables. There is no such thing as a fixed linear casting shrinkage of a metal or any of its alloys. A value of 1.2 ± 0.2% might be considered a reasonable range for the linear casting shrinkage of dental gold alloys. The lighter weight of the base metal alloys, 8–9 g/cm^3, is an advantage over the higher-density gold alloys, 17–18 g/cm^3, in certain types of bulky restorations. The cobalt–chromium alloys have Knoop hardness of 310–415 (Table 3). They show yield strengths of 414–711 MPa [(60–103) × 10^3 psi] and tensile strengths of 672–1200 MPa [(97.5–174) × 10^3 psi], with elongations of 1–16% (Table 3). These characteristics compare very favorably with those of the partial denture gold alloys. However, the modulus of elasticity (stiffness) of the cobalt–chromium alloys is 1.5–2 times that of the gold alloys, ie, in the range of 216 ± 7 GPa [(31.3 ± 1) × 10^6 psi] compared with 103 GPa (15 × 10^6 psi) for the gold alloys. The physical characteristics given are typical for metals melted and cast under controlled and optimum conditions. A suitable technique must be properly carried out if acceptable castings are to be obtained.

Table 3. Properties of Base Metal Alloys for Partial Dentures

Alloy	Tensile strength, MPa[a]	Yield strength, MPa[a]	Elongation, %	Modulus of elasticity, MPa[a]	HK
Vitallium	871	711	1.5	218,000	415
Platinore	802	664	1.7	226,000	411
Niranium N/N	836	686	1.5	218,000	391
Dentillium P-D	840	702	9.0	202,000	335
Crutanium	920–1200		10.0–15.6		
A	745	441	3.4		388
C	758	690	5		388
E	672	414	10		310

[a] To convert MPa to psi, multiply by 145.

Compositions. The compositions of the dental cobalt–chromium casting alloys have been established, within limits, by specification. The stipulation is that the alloy shall contain no less than 85 wt % chromium, cobalt, and nickel combined. This broad classification has apparently defined the range of compositions that has the necessary corrosion and tarnish resistance to withstand the oral environment. Most alloys exceed this minimum requirement. No known corrosion test has so far shown acceptable correlation with the clinical performance of these alloys, hence the composition requirement serves as a practical guide. These complex alloys may contain 4–9 major constituents plus several minor ingredients. The major metals may include chromium, cobalt, nickel, molybdenum, and beryllium; the more important minor constituents may be carbon, iron, silicon, manganese, and nitrogen (Table 4). These trace elements, however, may exert a controlling influence on the alloy and must be carefully controlled. Carbon may effectively strengthen the alloys; in excess, it induces brittleness owing to excessive carbide formation. Small amounts of nitrogen produce a brittle alloy.

Chromium-Containing Casting Alloys for Crowns and Bridges. Since gold is so expensive there has been an intensive development of alloys based mostly on nickel–chromium with as many as eight modifying elements. These nickel–chromium alloys have not been fully successful with the advocated techniques. They cannot be cast as easily or as accurately as gold-based alloys, nor can they be fabricated as easily by soldering. A modification of the manufacturers' techniques using hand- instead of mechanical-spatulation of the investments, a dry asbestos liner in the casting ring, and hygroscopic expansion can produce oversize crown castings (0.00–0.45%) (73). Castings made by the manufacturers were consistently undersize (−0.04 to −0.18%) (73).

Properties. The available alloys show a wide range of physical properties (Table 5). No data are available on the casting shrinkage, but in alloys with the same approximate nickel and chromium contents, pattern-makers usually allow roughly 3% linear casting shrinkage. Tensile strength ranges are 640–1140 MPa [(93–165) × 10^3 psi], yield strengths 290–780 MPa [(42–113) × 10^3 psi], and elongations 0.8–23% (Table 5). Since the melting temperature range is 1220–1345°C it is necessary to heat the investment molds at 800–935°C. The castings should not be pickled in acid because of their high nickel content, they should be cleaned with a sandblast. Since the alloys

Table 4. Composition of Base-Metal Alloys for Partial Dentures

Alloy	Co	Cr	Ni	Mo	Fe	Mn	Si	C	W	Pt	Cu	Be
Vitallium[a]	62.5	30.0		5.0	1.0	0.5	0.5	0.5				
Platinore[a]	60.7	26.7	2.7	5.8	2.6	0.5	0.6	0.3	0.3	0.1		
Niranium N/N[a]	64.7	27.5		5.4	0.4	0.5	1.0	0.4				
Dentillium P-D[a]	6.0	24.0	4.0	2.5	63.3							
Crutanium[b]	bal	5–15	5–15	0.3	1.0	4–10	Ti, Si, Mn	C				
A[c]	43.5	21.6	20.1	7	0.25	3.0	0.35	0.05			3.5	10.9
C[c]		13	68	4.5	2.5			Ti 1	Al 6	Nb 2		
E[c]	52	26.1	14.2	4.0	1.2	0.7	0.58	0.22				

[a] Ref. 69.
[b] Refs. 70 and 71.
[c] Ref. 72.

Table 5. Properties of Some Chromium-Containing Alloys for Crowns and Bridges

Alloy	Tensile strength, MPa[a]	Yield strength, MPa[a]	Modulus of elasticity, MPa[a]	Elongation, %	HV
Nobil Ceran	1,142	732	165,000	23.2	326
Microbond 2000	1,139	780	193,000	11.6	348
Wiron S	706	620	181,000	5.9	315
Gemini II	1,103	695	176,000	11.3	340
Neydium	641	476	214,000	0.8	290
Ceramalloy	752	290	186,000	1.2	330
Ticon	150–800			5–6	290–300
Microbond NP	400–460			2–7	222
Jelspan	777	583	184,000	3	310
Ultratek	919	544	207,000	16	270

[a] To convert MPa to psi, multiply by 145.

are so hard, they are difficult to finish and to abrade for clinical adjustment in the mouth.

Compositions. The major metals include nickel, chromium, with lesser amounts of molybdenum, aluminum, manganese, beryllium, cobalt, silicon, palladium, and carbon (Table 6).

Table 6. Composition of Some Chromium-Containing Alloys for Crowns and Bridges

Alloy	Ni %	Cr %	Fe %	Al %	No %	Si %	Be %	Cu %	Mn %	Co %	Sn %	Other %
Nobil Ceran[a]	80.75	12.58	0.34	3.42	1.53	0.29	0.37	0.15	0.13			
Microbond 2000[a]	79.67	13.24	0.11	3.87	1.52	0.30	0.65		0.12			
Qualimet[a]	78.51	19.47	0.43	0.21		1.10						
Wiron S[a]	68.96	16.54	0.37	4.15	5.10	0.83			3.05	0.42		
Gemini II[a]	80.86	11.93	0.20	2.95	1.85	0.18	1.55	0.13	0.14			
Howmet III[a]	68.75	19.57	0.38		4.22	2.72		1.54	1.24		1.25	
Dentillium[a]	5.65	26.28	59.20		2.12	0.44			0.49	5.65		
Neydium[b]	79	11	0.05	1.8	3.6	1.1						Nb 3.2
Ceramalloy[b]	70	20	0.2		5.6	3.96			0.22	0.02		Ti C 0.09
Ticon[c]	70.0	16.0	0.8	3.0	4.1	0.5	0.5		3.9	0.9		C 0.03
Microbond NP[c]	76.0	13.8		3.0	5.0	1.1			0.1			0.05
Jelspan[d,e]	39.0	18.62	4.0	0.005	1.73	1.0	1.20	0.005	0.1	10.93	0.02	Ti Pd 24.41
Ultratec[d,f]	80.48	11.39	2.0	2.22	1.99	0.5	1.6	0.005	0.02	0.5	0.02	Ti

[a] Ref. 74. Elements checked but not found: Zn, Cd, In, Bi, Sb, As, Tl, Ga, Ge, Pb, V, W, Nb, and Ta.
[b] Ref. 75.
[c] Ref. 76.
[d] Ref. 77.
[e] Ag 0.001%, Au 0.02%.
[f] Ag 0.005%, Au 0.005%.

There is no meaningful correlation between any laboratory corrosion tests and clinical performance (78) but *in vitro* corrosion rates were high on several high nickel dental prosthetic alloys (79), together with the high prevalence of allergy to nickel caused by plated jewelry, gives cause for caution. In any event the dentist should inquire if the patient has a history of allergy to nickel before restorations of nickel-containing alloys are used and should inform the dental laboratory that is manufacturing the appliance.

Some of the base-metal crown and bridge alloys are used with porcelain veneers (Table 6). Special porcelains are needed in order to have compatible coefficients of thermal expansion. Alloys with a high modulus of elasticity and a high proportional limit are required because they have to be used in small cross-sections and because, if the bridge flexes under masticatory loading, the porcelain will fracture or shear off. The metal–ceramic bond is thought to have both chemical and mechanical components, that is, oxides on the surface that combine with the enamel are interlocked in the comparatively rough surface of the alloy. The alloys also have to be high-fusing to accommodate the high-fusing temperature of the porcelain and must have little tendency to hot shortness.

The composition limits are within the ranges given for the cast chromium-containing alloys for crowns and bridges (Table 6). Many of these alloys have at least six metals.

Stainless Steels. Most of the materials and compositions discussed in this review have been developed expressly for dental applications (see Steel). The category of stainless steels finds dentistry attempting to utilize commercially available materials, intended for less specialized applications. Stainless steels of various types are being used in dentistry for orthodontic appliances, space maintainers, crown forms, and a variety of instruments including root-canal files and reamers for endodontic treatment. Only orthodontic appliances are discussed below.

Of the broad classifications of stainless steel—ferritic, martensitic, and austenitic—the austenitic group is most widely used for oral appliances. Only the low-carbon austenitic steels are resistant to the tarnish and corrosive action of the oral fluids. Orthodontics is concerned with the fabrication of corrective appliances to move the teeth. Such fabrication is achieved by the bending, shaping, and soldering of wire, sheet, or strips to construct the necessary appliance.

Properties. The detailed properties of 18-8 stainless steel can be obtained by reference to appropriate handbooks. The characteristics of 18-8 orthodontic wire and those of gold-alloy wires used for the same purpose in dentistry are somewhat similar. The 18-8 alloys have higher melting ranges, higher modulus, and generally more elongation than the wrought gold alloy wires. The important difference in the two alloy systems, from the working viewpoint, is the mode of obtaining the range of properties from the soft state to the hard state and how this relates to the practical use of the alloys. The gold alloys can be bent, shaped, and soldered with relative ease. The final desired values for the pertinent properties can then be adjusted by suitable heat treatment. The 18-8 stainless steels, on the other hand, do not respond to heat treatment, but must depend upon work-hardening to achieve the maximum physical properties. This presents a constant challenge when components of the appliance must be soldered together after contouring. Careless use of the torch, and improper solder or flux may result in a ruined appliance through overheating and loss of the desired chemical and work-hardened characteristics.

Composition. The alloys most often used for orthodontic appliances, for reasons of tarnish and corrosion resistance to the oral fluids, are selected from the low-carbon austenitic stainless steels. These 18-8 alloys contain ca 17–19% chromium, 7–9% nickel, and 0.08–0.2% carbon. Silicon is limited to 0.75% max and manganese, to 0.60% max.

Other wrought nickel–chromium alloys (80% Ni–20% Cr), wrought-cobalt–nickel alloys (40% Co, 20% Cr, 15% Ni, 7% Mo, 2% Mn, 0.15% C, 0.04% Be, and balance Fe), as well as wrought nickel–titanium alloys (55% Ni, 1.5% Co, and balance Ti) are used in orthodontics instead of 18-8 stainless steel.

Electrodeposited Metals. Electrodeposited, electroplated, or electroformed metals are of interest to the dental technician for producing an accurate metal-clad die or cast, from a compound, polysulfide rubber, or silicone rubber impression (see Electroplating). Dies and casts with an improved working surface having strength, abrasion resistance, chip resistance, and faithful accuracy have been a basic need of dentistry. Two metals, copper and silver, are used. The acidic copper sulfate plating bath has been more compatible with the compound and silicone rubber impressions than with others. The alkaline silver cyanide plating solutions give more consistent results and less trouble with the polysulfide (eg, Thiokol) rubber impressions. The impression surfaces are rendered conductive by a coating of powdered graphite, powdered metallic copper, or powdered metallic silver. Careful control of plating bath composition, cleanliness, careful application of the conductive layer to the impression surface, current densities, and other factors are essential to achieving consistent and successful results. Usually 12–18 h are allowed to produce a satisfactory metal deposit of 25–30 μm thickness. The metal shell obtained must be filled or reinforced with some solid base to prevent distortion or collapse. Artificial stone or self-curing resins may be used for this purpose, before the metal shell is removed from the impression.

Properties. The metal plating baths, as mentioned, are acidic copper sulfate and alkaline silver cyanide. Acid contamination in the alkaline silver cyanide bath will release extremely poisonous hydrogen cyanide gas. For this reason, the two plating setups should be isolated from each other. Both plating baths should be well ventilated and covered when not in use, to reduce evaporation losses and contamination.

Composition. Copper-plating bath compositions of various types have been used. A typical bath formulation consists of:

copper sulfate, crystals	200 g
sulfuric acid, conc	30 mL
phenolsulfonic acid	2 mL
distilled water	1000 mL

A pure copper anode may be used; a copper anode containing a trace of phosphorus reduces sludge accumulation in the plating bath.

Silver-plating bath compositions are somewhat variable; a typical composition contains:

silver cyanide	36 g
potassium cyanide	60 g
potassium carbonate	45 g
distilled water	1000 mL

A pure silver anode is required.

An excellent review on metals and alloys in dentistry is reference 80.

Gold and Gold Alloys. Gold foil, crystal, an Au–Ca alloy, and combinations in the cohesive and noncohesive states are used in dentistry as direct-filling materials. Gold alloys are used for cast restorations and prosthetic devices. Wrought gold alloys are used for wire clasps and fabricated orthodontic appliances (see Gold and gold compounds).

Gold and gold alloys serve the needs of dentistry better than any other metals or alloy systems. Gold alloys have a broad range of working characteristics and physical properties, coupled with excellent resistance to tarnish and corrosion in the mouth.

Pure or alloyed gold (Au–Ca) is used as a direct filling material in restorative dentistry as: (1) foil (ca 0.635 μm thick), also available as gold–platinum laminated foil containing up to 40% platinum; and (2) powdered pure gold crystals. Gold, in either of these forms welds to itself at room temperature when condensed by force. Certain other metals have this characteristic, but owing to the practical difficulty of achieving and maintaining a sufficient degree of surface purity, adequate welds are difficult to obtain. Gold is sufficiently inert for the required surface purity to be retained for cold welding. A gold restoration [foil or other type, see (1) and (2) above] of pure or alloyed gold, or both, is relatively weak, soft, and malleable. The cohesive characteristic of the gold foil is partially lost with time as the metal is exposed to the atmosphere. The cohesive characteristic is restored, immediately before use, by heating in a clean flame to rid the surface of contaminating substances, including sorbed gases. The gold restoration is produced by malleting or condensing the clean metal into the clean, dry cavity preparation. Hardness of the final restoration exceeds the hardness of cast pure gold, owing to the cold-working. The hardness approaches that of 22 kt gold.

Gold Casting and Wrought Alloys. Gold alloys that are useful in dentistry may contain gold, silver, platinum, palladium, iridium, indium, copper, nickel, tin, iron, and zinc. Occasionally, other metals are found in minor amounts. The effect of each of the constituents is really empirical, but the following observations have been made:

Gold [7440-57-5] is the principal constituent of gold-colored alloys. It contributes gold color; increases the specific gravity; raises the melting point, if that of the alloy is below that of gold; increases ductility, malleability, corrosion and stain resistance; and produces heat-treatable compositions with copper, platinum, and zinc. It has a useful range of 25–100 wt %.

Copper [7440-50-8] produces a reddish color; reduces the melting point of the alloy; and produces heat-treatable compositions with gold, platinum, and palladium resulting in increased hardness, strength, and generally improved physical properties. The tarnish resistance of the alloy is usually decreased. The gold–copper system is the fundamental system of many dental gold alloys. Copper has a useful range of 0–20 wt %.

Silver [7440-22-4] alters the color of binary gold–silver alloys to green–gold and finally to a silver-colored alloy, as the percentage of silver increases. In gold–silver alloys containing copper and platinum, silver improves the color of the alloy. Silver also improves ductility and exerts an influence upon the rate and intensity of response to heat treatment. Excessive silver content may cause alloys to absorb a large volume of oxygen while molten. Upon solidification of the alloy this absorbed gas may be released and cause "spitting" or gas inclusions in the metal. This problem is seldom evident within the range of 0–20 wt % silver.

Platinum [7440-06-4] (qv) detracts from the gold color, producing an undesirable grayish-red color; increased platinum produces a platinum-colored alloy. It increases the strength, proportional limit, and solidification temperatures; reduces grain size; produces a heat-treatable alloy with gold; absorbs large volumes of oxygen when molten which may cause gas inclusions upon solidification of the alloy and expulsion of the gas. It has a useful range of 0–18 wt %.

Palladium [7440-05-3] produces a very great color change (5–6 wt % may produce a white alloy); raises the melting point rapidly; increases hardness, tensile strength, and proportional limit of gold–copper–silver alloys; lowers tarnish resistance slightly; absorbs large volumes of oxygen and hydrogen when molten which may cause inclusions in the alloy upon solidification and expulsion of the gases.

Zinc [7440-66-6] (qv) whitens gold alloys, but not rapidly; lowers the melting point rapidly; increases the hardness of gold alloys; embrittles gold alloys when present in high percentages; increases strength and elastic limit slightly; reduces surface tension; and protects other metals from oxidation.

Iridium [7439-88-5] and rhodium [7440-16-6] individually increase corrosion resistance, hardness, and strength of platinum alloys and can be used to reduce grain size (81).

Indium [7440-74-6] is sometimes used as a scavenger and promotes uniform grain size and casting fluidity.

Iron [7439-89-6] (qv) is used to produce age-hardening in Au–Pd–Pt alloys at temperatures involved in ceramo–metallic techniques (82).

Tin [7440-31-5] (qv) is used up to 4.6 wt % to aid in porcelain fused to metal processes and to decrease the melting point of gold solders.

Lead (qv) and mercury (qv) are contaminants.

Gold Alloys, Cast Types. Four types of alloys are recognized for cast dental restorations in American Dental Association Specification No. 5 for dental casting gold alloys. They are Type I (soft), Type II (medium), Type III (hard), and Type IV (extra hard, partial denture).

These alloys provide the desired material for specific uses.

Type I, soft alloys, 20–22 kt golds, are used for inlays of the simpler nonstress-bearing types. The alloys have HVs (hardness values) of 50–90, and can be burnished. Type I alloys are not heat-treatable. They are composed essentially of gold–silver–copper with minor modifying additions, eg, zinc.

Type II, medium–hard alloys, are harder, stronger, and have lower elongation than the Type I alloys. They are used for moderate stress application eg, three-quarter crowns; abutments; pontics, full crowns; and saddles. The alloys have HVs of 40–120 and are difficult to burnish. The Type II alloys can usually be heat-treated.

Type III, hard alloys, are the hardest, strongest, and least ductile of the inlay casting golds. They are indicated for restorations required to resist large forces such as three-quarter crowns, abutments, pontics, supports for appliances, and precision-fitting inlays. The alloys have HVs of 120–150 and cannot be burnished. Heat treatment improves all their physical properties, except ductility which is greatly decreased.

Type IV, extra-hard (partial denture) alloys, are indicated where high strength, hardness, and stiffness are required. They are harder, stronger, and less ductile than the Type III alloys. These partial-denture alloys are used for cast removable partial dentures, precision-cast fixed bridges, certain three-quarter crowns, saddles, bars, arches, and clasps. These alloys have HVs of 150 min; values as high as 300 have been reported.

Properties. The desirable characteristics that a casting gold alloy should possess are: (*1*) The alloy should melt in an air–gas flame. (*2*) It should melt and cast without much change in composition. (*3*). It should be very fluid when melted and have low surface tension. (*4*) The molten alloy should not absorb gases or oxidize excessively. (*5*) The total shrinkage, from the temperature at the beginning of a solidification to room temperature, should be small. (*6*) The alloy should have a pleasing, acceptable color. (*7*) The physical properties—hardness, tensile strength, elastic limit, and fatigue resistance—should be adequate for the intended use. (*8*) The physical properties should respond to heat treatment. (*9*) The alloy should resist all corrosion or tarnish in the mouth (9). The alloy should be passive (10).

Some physical properties of gold cast alloys are given in Table 7.

Table 7. Approximate Range of Values of Mechanical Properties and Liquidus Temperatures of Dental Cast Gold Alloys[a]

No. and type	Heat treatment[b]	HV[c]	Tensile strength, MPa[d]	Proportional limit, MPa[d]	Elongation, %	Liquidus, °C
Yellow gold alloys							
1 Type I soft	Q	65–90	207–310	55–103	20–35	950–1,050
2 Type II medium	Q	100–110	310–379	138–173	20–35	930–970
3 Type III hard	Q	115–135	331–393	159–207	20–25	
3 Type III hard	A	135–185	414–565	200–400	6–20	950–1,000
4 Type IV extra hard	Q	150–180	414–517	241–324	44–25	
	A	230–255	690–827	414–634	1–6	870–985
White gold alloys							
5 Type II hard	Q	125–135	345–393	165–207	9–18	
5 Type III hard	A	140–190	414–517	193–310	2–12	1,030–1,070
6 Type IV	Q	150–210	448–517	270–310	9–15	1,025–1,050
6 extra hard	A	245–280	758–827	517–572	1–3	
7 Type IV	Q	180–200	565–600	345–379	9–12	
7 extra hard	A	240–300	793–896	448–690	2–3	930–1,000
modulus of elasticity 76,000–83,000 MPa							

[a] Data obtained by methods used in ref. 83.
[b] Q = Quenched after holding for 10 min at 700°C; A = aged, uniformly cooled from 440 to 240°C in 30 min after quenching.
[c] Approximate conversion from "Baby Brinell" numbers to Vickers numbers by data from reference 84. Table adapted from ref. 85.
[d] To convert MPa to psi, multiply by 145.

Composition. The lack of a satisfactory corrosion test that correlates with clinical experience has resulted in establishing certain composition limitations that have given satisfactory clinical service. The current ADA Specification sets the following limitations in composition:

Alloy	Gold and metals of the platinum group, wt % (*min*)
Type I soft	83
Type II medium	78
Type III hard	78
Type IV extra hard (partial denture)	75

It was shown that for the three alloys studied there was a correlation between an *in vitro* tarnish test using a 5% aqueous solution of sodium sulfide and clinical observations, and that the percentage of noble metals was not in itself a guarantee that the alloy would not tarnish in the mouth (86). Thus, an 18 kt gold alloy (75% Au, 12.5% Ag, and 12.5% Cu) tarnished badly, whereas an experimental alloy, 39–46% Au, 30–33% Ag, 20–22% copper, and 4–6% Pd, did not.

Table 8 gives the range of compositions of the dental gold casting alloys. In addition to the elements listed in Table 8, In, Ni, Sn, Ir, and Co have been used as modifying metals.

There are also a number of casting gold alloys on which porcelain veneers are fused. Published compositions show high gold, platinum, and palladium contents (79.6–98.8%) as shown in Table 9. There are other alloys on the market with lower Au–Pt–Pd contents of ca 71–87 wt % and still others with contents as low as 46 wt %.

The average coefficient of thermal expansion of the first six alloys in Table 9 is $(14–15) \times 10^{-6}/°C$ from room temperature to ca 1000°C, and two opaque porcelains that are used with them have coefficients of 6.45 and $7.88 \times 10^{-6}/°C$ from room temperature to 820°C (87). The HVs are 109–193 and the tensile strengths are 464–509

Table 8. Composition Limits, wt % of Cast Dental Gold Alloys[a]

No. and type	Au	Ag	Cu	Pd	Pt	Zn
	Yellow golds					
Type I soft	79–92.5	3–12	2–4.5	0.5 max	0.5 max	0.5 max
Type II medium	75–78	12–14.5	7–10	1–4	1	0.5
Type III hard	62–78	8–26	8–11	2–4	3	1
Type IV extra hard (partial denture)	60–71.5	4.5–20	11–16	5 max	8.5 max	1–2
	White golds					
Type III hard	65–70	7–12	6–10	10–12	4	1–2
Type IV extra hard (partial denture)[b]	60–65	10–15	9–12	6–10	4.8	1–2
Type IV extra hard (partial denture)[c]	23–30	25–30	20–25	15–20	3–7	0.5–1.5

[a] Table adapted from ref. 85.
[b] Yellowish white.
[c] Very white.

Table 9. Composition of Gold-Based Alloys Used in Ceramo–Metallic Prosthesis

Trade name	Au %	Ag %	Pt %	Pd %	In %	Ir %	Zn %	Fe %	Sn %
Amator[a]	82.6	2.4	12.4	0.8			1.8		
Amator 2[a]	82.0	2.5	11.9	1.8	1.8				
Ceramco[a]	87.7	1.0	6.1	4.6	0.6				
Degudent[a]	84.8	1.3	7.9	4.6	1.25	.15			
Herador[a]	83.2		15.6		0.9	0.3			
V4[a]	81.5	2.7	11.7	2.3	1.8	3.			
Ceramco[b]	87.5	0.9	4.2	6.7				0.3	0.4
Ceramco "O"[b]	87.1	1.3	4.6	6.5				0.3	0.2
Cameo[c]	51.5	12.1		29.5	6.8				
Vivostar[c]	54.2	15.7		25.4					4.6

[a] Ref. 87.
[b] Ref. 88.
[c] Ref. 89.

MPa [(67–74) × 10^3 psi]. For the last four alloys in Table 9 the HVs are 102–216 and the tensile strengths are 358–662 MPa [(52–96) × 10^3 psi], depending upon thermal history.

Gold Alloys, Wrought Type. Two types of wrought gold alloys are currently recognized by the ADA Specification No. 7 for the fabrication of orthodontic and prosthetic dental appliances. They are (*1*) Type I, high-precious-metal alloys; (2) Type II, low-precious-metal alloys (gold color).

The wrought alloys are formed into plate, bands, wire of various sizes, cross sections, and shapes for clasps and orthodontic applications, bars, and tooth backings. The specifications and most of the available data cover only wires. It is anticipated that alloys of similar composition and treatment would have similar properties, regardless of form. Most dental wrought gold alloys respond to heat treatment.

Type I. High-precious-metal alloys that contain a minimum of 75 wt % gold and metals of the platinum group. Type I alloys are white.

Type II. Low-precious-metal alloys that contain a minimum of 65 wt % gold and metals of the platinum group. Type II alloys are gold-color.

Properties. The desirable properties of the wrought gold alloys should conform to those given for the cast gold alloys. In addition, the fusion temperatures must be known and stated to permit the fabricator to select a proper solder. Typical physical properties of wrought gold wire alloys are given in Table 10.

Composition. The lack of a satisfactory corrosion test that correlates with clinical experience has resulted in establishing certain composition limits, which have given satisfactory clinical service. These are broadly given in the wire-type definitions. Table 11 gives the composition ranges for several wires.

The alloys shown in Table 11 can be manufactured within a broad range of composition. Within each numbered alloy composition range, it is frequently possible to formulate either a Type I or Type II alloy.

In regard to gold alloy color, the white alloys are not used as extensively as the gold-colored alloys. This is largely a matter of preference. The physical properties of the white gold alloy wires are superior, but appearance and the desire to distinguish a gold restoration from a base metal (chrome–cobalt) restoration are also considered.

Platinum and Platinum Alloys. *Properties.* Platinum, a bluish-white metal, is soft, tough, ductile, and malleable. It has a melting point of 1773°C. The coefficient of thermal expansion is $9 \times 10^{-6}/°C$.

Platinum has excellent resistance to oxidation at elevated temperatures and to strong acids. It must be protected from low-fusing elements or the oxides of low-fusing elements at high temperatures, under reducing conditions. Easily reduced metals at high temperatures may form low-fusing alloys with platinum.

Carbon or silicon may produce brittleness and loss of useful properties. Caustic alkalies and many alkaline earth salts or hydroxides may attack platinum at elevated temperatures.

Uses. Platinum has many uses in dentistry. Pure platinum foil serves as the matrix in the construction of fused-porcelain restorations. Platinum foil may be laminated with gold foil for cold-welded foil restorations. Platinum wire has found use as retention posts and pins in crown and bridge restorations. Heating elements and thermocouples in high-fusing porcelain furnaces are usually made of platinum or its alloys (see Platinum-group metals).

Table 10. Some Physical Properties of High-Strength Noble Metal Wires Used in Dentistry[a]

Alloy[b]	Tensile strength[a] Soft[c], MPa[e]	Tensile strength[a] Hardened[d], MPa[e]	Proportional limit Soft, MPa[e]	Proportional limit Hardened, MPa[e]	Elongation[a] (/200 mm gage) Soft, %	Elongation[a] (/200 mm gage) Hardened, %	Approx HV Soft	Approx HV Hardened	Fusion temp (wire method), °C	Dens, g/cm³	Electrical resistivity[f], μΩ·cm
1	862–1241	see below[g]	552–1034	see below[g]	14–15	see below[g]	219–264	see below[g]	1499–1532	16.9–17.6	42–43
2	758–896	1103–1282	496–703	896–1041	12–22	5–10	169–206	259–304	1004–1099	15.0–18.5	28–33
3	965–1034	1103–1172	758–827	896–965	8–10	7–9	229–239	269–289	1066–1121	15.5–15.8	
4	620–793	827–1138	379–552	586–965	14–26	2–8	285–314	259–314	943–1016	14.5–15.6	
5	565–827	896–1138	365–503	710–958	14–20	1–3	154–219	249–309	899–932	14.1–15.2	20–22
6	579–690	662–1082	358–400	483–855	20–28	1–2	157–189	239–299	877–899	13.7–14.0	
7	662–1020	1034–1324	414–793	758–1103	9–20	1–8	169–294	199–289	941–1079	11.5–15.6	
8	690–758	896–1172	434–600	938–876	16–24	8–15	169–219	254–289	1043–1077		

[a] Tension tests on wires 1 mm in dia. Most elongation data on 200 mm gage lengths. Table adapted from ref. 85.
[b] See Table 11 for chemical composition of these alloys.
[c] Quenched from 705 to 870°C depending on type of alloy.
[d] Cooled slowly and uniformly from 450 to 250°C in 30 min. This is a severe hardening treatment used for test purposes in determining the behavior of wire under adverse conditions. Manufacturers recommend hardening treatments for specific uses.
[e] To convert MPa to psi, multiply by 145.
[f] Very few data are available on the electrical resistivity of dental alloys, since resistance is not an important factor in their use.
[g] Not appreciably affected by heat treatment.

Table 11. Composition Limits of High-Strength Noble Metal Wires Used in Dentistry[a]

Alloy	Au	Pt	Pd	Ag	Cu	Ni	Zn	Color
1	25–30	40–50	25–30					platinum
2	54–60	14–18	1–8	7–11	11–14	1 max	2 max	platinum
3	45–50	8–12	20–25	5–8	7–12		1 max	platinum
4	62–64	7–13	6 max	9–16	7–14	2 max	1 max	light gold
5	64–70	2–7	5 max	9–15	12–18	2 max	1 max	gold
6	56–63	5 max	5 max	14–25	11–18	3 max	1 max	gold
7	10–23	25 max	20–37	6–30	14–21	2 max	2 max	platinum
8		1 max	42–44	38–41	16–17	1 max		platinum

Composition[b], wt %

[a] Table from ref. 85.
[b] Fractional percentages of iridium, indium, and rhodium are omitted.

Platinum, as an alloying element, is used in many dental casting golds (Tables 8 and 9) and wrought gold alloy compositions (Table 11) to improve hardness and elastic qualities. Platinum in combination with palladium and iridium has limited use for dental pins and wires.

Palladium and Palladium Alloys. Palladium is not used in the pure state in dentistry. However, it is a useful component of many gold casting alloys, as shown in Tables 8 and 9 (see Platinum-group metals).

Alloys based on Ag–Pd have been used for a number of years and are available from most gold alloy manufacturers (90). Their palladium content is 22–50 wt %, and their silver content from 35–66 wt %. Minor amounts of Zn, In, or Sn are often present to increase the fluidity. Both In and Sn form intermetallic compounds with both Pd and Ag and, therefore, some of the commercial alloys are susceptible to age hardening (91). These alloys are difficult to fabricate and require meticulous processing. There appears to be no satisfactory clinical data published that would warrant their routine use. Cost would be the primary criterion for their use.

Solders. In the fabrication of certain types of prosthetic and orthodontic appliances, it is frequently desirable to join parts made from wrought gold alloys to cast gold structures, or to join several wrought gold parts into a single appliance, or to join two or more castings into a single appliance. In the joining of wrought parts, welding may be utilized. Generally, in the joining of castings or wrought parts to castings, gold solders are used. Certain orthodontic appliances constructed of stainless steel may require the use of silver solders (see Solders).

Properties. Phillips (91) has given the following requisites for a satisfactory solder:

(1) It must not corrode or tarnish in the mouth fluids, which implies that it must not only be sufficiently noble in composition, but also that its composition should be such that its solution potential approximates that of the metal upon which it is used.

(2) Its fusion temperature must be lower than that of the alloy upon which it is employed, in order that the latter may not be fused during the soldering; it should be at least 100°C below that of the parts to be soldered. In the case of wires, it is necessary that its melting temperature be less than the recrystallization temperature of the metal.

(3) Its color should match that of the metal employed.

(4) It should flow promptly and smoothly over the surfaces of the parts to be joined. This property depends upon the surface tension, viscosity, and adhesive properties of the molten solder.

(5) Its physical properties should be at least as good as those of the metal, in order for the joint not to be a source of weakness.

Compositions. Dental gold solders are composed essentially of gold, silver and copper, and tin and zinc. Tables 12 and 13 give some mechanical properties, melting range, and the composition of a number of dental gold solders (92).

Silver Solders. Sometimes silver solders are used in the fabrication of orthodontic or other dental appliances. Table 14 compares the compositions and fusion temperatures of a number of silver solders.

Metal–Organic Products. For many years dentistry has utilized a number of products based on the interaction of an organic substance and a metal oxide or salt. The most exploited system has been the zinc oxide–eugenol system. As early as 1894 this system was in use (94). Many modifications and improvements have been suggested, such as the use of rosin (95), hydrogenated rosin (96), polystyrene (97), and accelerating systems (98–99). It was not until 1955 that a critical study of the basic reaction showed that zinc oxide will form chelate compounds with many materials

Table 12. Mechanical Properties and Melting Range of Dental Gold Solders[a]

Solder	Heat[b]	Proportional limit, MPa[c]	Tensile strength, MPa[c]	Modulus of elasticity, MPa[c]	Elongation[d], %	HV[e]	Melting range, °C
S I	S	186	293	75,845	14	130	745–785
	H	379	434	82,740	1	204	
S II	S	203	307	75,845	12	122	750–805
	H	534	576	82,740		212	
S III	S	207	303	82,740	9	130	765–800
	H	530	634	89,635		218	
S IV	S	165	248	75,845	7	122	755–835
	H	424	483	82,740		199	
S V		141	534	75,845	18	97	845–870

[a] Adapted from ref. 92.
[b] S = Quenched in water from 700°C; H = Q + slow cooling from 450°C; S V was not appreciably affected by heat treatment.
[c] To convert MPa to psi, multiply by 145.
[d] 75 mm gage length. Where value is missing it was less than 1%.
[e] HV = hardness value.

Table 13. Composition of Dental Gold Solders, wt %[a]

Alloy designation	Au	Ag	Cu	Zn	Sn
S I	65.4	15.4	12.4	3.9	3.1
S II	66.1	12.4	16.4	3.4	2.0
S III	65.0	16.3	13.1	3.9	1.7
S IV	72.9	12.1	10.0	3.0	2.0
S V	80.9	8.1	6.8	2.1	2.0

[a] Ref. 92.

498 DENTAL MATERIALS

Table 14. Composition and Fusion Temperatures of Silver Solders[a]

Grade no.	Composition, wt %				Mp, °C	Flow pt, °C	Color
	Ag	Cu	Zn	Cd			
1	10	52	38	0.5 max	820	870	yellow
2	20	45	35	0.5 max	777	815	yellow
3	20	45	30	5.0 max	777	815	yellow
4	45	30	25		677	743	nearly white
5	50	34	16		693	774	nearly white
6	65	20	15		693	718	white
7	70	20	10		724	754	white
8	80	16	4		738	793	white

[a] Adapted from ref. 93.

that have a structure similar to eugenol (100). It appears that it is necessary for the compound to have a methoxy group ortho to a phenolic hydroxyl group.

The chelation of zinc oxide with eugenol (see Ethers) can be represented by the following equation:

$$2\,\text{eugenol} + \text{ZnO} \longrightarrow \text{zinc eugenolate} + \text{H}_2\text{O}$$

zinc eugenolate [32601-56-2]

Zinc Oxide–Eugenol Impression Pastes. These two-paste systems are usually supplied in collapsible tubes, one tube contains the white zinc oxide and the other contains the brown eugenol. Approximately equal amounts of the two pastes are mixed to form a smooth, homogeneous streak-free mass. The impression paste is spread thinly over the surface of a custom-made tray. The coated tray is positioned in the mouth and held motionless until the paste mix hardens, usually within 2–5 min. The fluid consistency of the impression paste allows the wash to faithfully copy the finest detail of mucosal surface contours. The various brands of impression paste have differing characteristics of mix consistency, setting rate, and set-paste hardness. Hard-setting and soft-setting pastes are recognized in the ADA specifications.

Properties. The zinc oxide–eugenol impression pastes produce a rigid impression with good accuracy and good reproduction of surface detail. Among the more favorable properties of these materials are that: they have sufficient resistance so that borders can be built up if the tray is slightly deficient in any area; and they set to a cement-like hardness and the resulting impression can be taken in and out of the mouth repeatedly, giving an opportunity to test for stability and tissue adaptation. In addition they register detail well, and are quite stable dimensionally.

Compositions. The compositions of zinc oxide–eugenol impression pastes are similar to those of the zinc oxide–eugenol cements (101). Variations in specific characteristics are achieved by the proportions of the ingredients (102). Properties vary in commercial products (103). The modifications of the zinc oxide–eugenol system intended for bite-registration pastes may include agents to increase the body or thixotropic character of the unset mix to improve the handling and utility for this specialized use.

Uses. Zinc oxide–eugenol impression pastes are used primarily as corrective washes over compound impressions, as veneer impressions, as temporary liners or stabilizers in base-plates and dentures, and as bite-registration pastes for recording occlusal relationships in inlay, crown, and bridge techniques.

Specifications. The test methods for zinc oxide impression pastes are outlined in ADA Specification No. 16 and Federal Specification U-I-500.

Surgical Pastes. Zinc oxide–eugenol pastes having essentially the same formulation as impression pastes are used as dressing for gingival protection and healing following periodontal surgery. These pastes have delayed setting times, are less brittle and weaker after hardening than the impression pastes.

In some mouths, tissues are sensitive to eugenol. Also in some patients, after wearing a surgical pack for 2–3 wk, a chronic gastric disturbance may occur. A material similar to zinc oxide–eugenol is formed by a reaction between zinc oxide and any carboxylic acid. *o*-Ethoxybenzoic acid is valuable in this use. Bacteriocides and other medicines can be added without interfering with the reaction.

There is no specification for the surgical paste.

Calcium Chelates (Salicylates). The first commercially successful dental products based on a calcium chelate system were the dental cements Dycal (The L. D. Caulk Co.) and Hydrex (Kerr Manufacturing Co.). The calcium salicylate [824-35-1] system offers certain advantages over the more widely used zinc oxide–eugenol system. These products can be completely bland, they do not produce burning or tissue irritation, and they do not retard the polymerization reaction of the free-radical-initiated acrylic-curing systems. The principal deficiency of the system is the relatively high solubility of the set products that are currently available.

Properties. The physical characteristics of the calcium salicylate system compare favorably with those of the zinc oxide–eugenol products. The setting times are subject to modification over a wide range. However, they are somewhat more critical to control than those of the zinc–oxide–eugenol system owing to the greater solubility of calcium hydroxide as compared with that of zinc oxide. Crushing strengths in excess of 14 MPa (2000 psi) have been obtained in 24 h. The set product may be formulated to have a hard, brittle, resinous fracture or may be modified to a tough, strong, semiplastic mass. The dimensional stability within a 48-h period varies somewhat with the basic system. Shrinkages of only 0.2% in 48 h have been obtained, but 0.4% may be reached in some compositions. Solubilities, when tested according to ADA Specification No. 8 for zinc phosphate cements, range from <1 to 14% for the formulations based on polyhydric esters of salicylic acid. The calcium salicylate cements are not retarders for the free-radical-initiated polymerization of methyl methacrylate compositions. They permit the direct placement of acrylic fillings.

Composition. The formulation of this class of materials is based upon the formation of a low-solubility chelate between calcium hydroxide and a salicylate. Dycal utilizes the reaction product of a polyhydric compound and salicylic acid [69-72-7]. Other salicylic acid esters can be similarly used. Vehicles used to carry the calcium hydroxide, extenders, and fillers may include mineral oil, N-ethyl-p-toluenesulfonamide [80-39-7], and polymeric fluids. The filler additions may include titanium dioxide, zinc oxide, silica, calcium sulfate, and barium sulfate. Zinc oxide and barium sulfate [7727-43-7] are useful as x-ray opacifying agents to ensure a density greater than that of normal tooth structure. Resins, rosin, limed rosins, and modified rosins may serve as modifiers of the physical characteristics in both the unset and set states.

Uses. Calcium hydroxide [1305-62-0] had been recognized as a desirable lining for deep-seated cavity preparations involving pulp exposures, or near exposures. Calcium hydroxide stimulates the formation of secondary dentin that forms a natural pulp-protecting wall. Water slurries or mucilage suspensions of calcium hydroxide provided the therapeutic benefits of the material but were without mechanical strength. These cements are hard-setting calcium hydroxide-containing compositions that compare favorably with the better qualities of zinc oxide–eugenol cements currently available. They provide both the therapeutic values of calcium hydroxide and the structurally strong bases needed for permanent restorations of metal, porcelain, plastic, or silicate cement.

The calcium chelate cements are limited to the use of a cavity liner. They may be placed directly over an exposed tooth pulp, to protect the pulp and stimulate the growth of secondary dentin, or used as a therapeutic insulating base under permanent restorations. The high alkalinity and high solubility of these materials prohibits their use in close proximity to soft tissues or in contact with oral fluids.

The chelated calcium cementing materials are supplied as two-part paste products. In use, equal parts of the two pastes are thoroughly mixed together to give a fluid mass that can be applied without pressure over an exposed tooth pulp or in a deep-seated cavity. Under the influence of the oral temperature and humidity, the fluid mass sets to a hard, strong, therapeutic protective seal.

Cavity Varnishes and Liners. Various types of cavity-lining agents are used to cover the walls and floor of prepared cavities. There are two types of these materials: (*1*) a cavity varnish, which is principally a natural gum such as copal, rosin, or a synthetic resin, dissolved in an organic solvent, such as acetone, chloroform, or ether; and (*2*) a cavity liner, which is a liquid in which calcium hydroxide or zinc oxide, or both are suspended in solutions of natural or synthetic resins.

These varnish or lining materials differ from the cements used in placing bases or for luting in that they are formulated to provide a fluid substance that can be readily painted on the surface of the prepared cavity. The solvent evaporates rapidly leaving a relatively thin film protecting the underlying tooth structure.

Cavity Varnish. Cavity varnishes are used primarily for protection of pulpal tissue from chemical irritation. This irritation may come directly from the restorative material or in the form of microleakage of oral fluids at the interface of the restoration and the tooth material. Recent investigations have revealed that varnishes applied in several layers between dentin and phosphoric acid cements significantly reduce the diffusion of acid from the cements into the dentin. Therefore, several layers of varnish are recommended in deep cavities and should be applied to the dentin to protect the pulp when the restoration or the cement base to be used contains an irritating acid. The base that contains either calcium hydroxide or zinc oxide–eugenol should be applied directly over the dentin and subsequently covered by the varnish (104). Varnishes aid in the reduction of postoperative sensitivity when applied to dentinal surfaces under newly-placed restorations. This reduction in irritation is attributed to the sealing of the margin and preventing microleakage of oral fluids into the cavity between the filling and the tooth. Although the tooth is less sensitive to thermal changes, the thickness of the varnish is inadequate to insulate against thermal conduction. Therefore, the reduced sensitivity results from reduced irritation or inflamation in the pulp, and thus, the pulp is less sensitive to temperature changes (104).

The conventional cavity varnishes should not be used under direct resin restorations. The solvent in the varnish may react with or soften the resin. Likewise, the varnish prevents proper wetting of the resin to the prepared cavity. Only cavity-lining agents supplied or recommended by the manufacturer for a particular resin restorative material should be used (104).

When a cavity varnish is used in conjunction with silicate restorations or with zinc silicophosphate cementation materials, the varnish should be carefully removed from the cavo-surface margins. Enamel adjacent to the silicate cements has been observed to possess an unusual resistance to caries; this has been associated with the fluoride content of the silicate cement powder. The presence of cavity varnish at the margins adjacent to the enamel inhibits the penetration of the fluoride ions from the silicate cement, reducing the effectiveness of the fluoride approximately 50% (104).

Cavity Liners. Cavity liners are similar to the varnishes in that they are solutions of synthetic or natural resins in a volatile organic solvent; they differ in that they contain suspensions of calcium hydroxide or zinc oxide, or both, or calcium hydroxide in methyl cellulose (104). These liners can be applied as a relatively thin layer to the dentinal surface of the cavity preparation and serve as a barrier against the irritation of the overlying zinc phosphate, silicate cements, and composite restorative materials. Since cavity liners are fluids, they can be painted or flowed over the dentinal surfaces, and their organic solvents quickly evaporate to leave a film residue that protects the underlying pulp (105). When the cavity liner is used, none of the liner should remain on the enamel margins. Calcium hydroxide is soluble in saliva, and as it dissolves, the residual porous film allows increased marginal leakage (106).

Specifications. There are no specification or acceptance programs for cavity varnishes and cavity liners. Their use and contents are discussed in reference 107.

Polymers

Synthetic Polymers in Dentistry. The use of synthetic polymers in dentistry involves combinations of unique properties. There are overall requirements that have general applications such as minimal dimensional changes during and subsequent to polymerization, lifelike appearance, color stability, passivity, biocompatibility, and suitability for dental techniques. The polymers that best satisfy the foregoing are the acrylic resins. Polysulfides, polyethers, and silicones are discussed above under impression materials. Many other polymers have been used but are unable to compete with the acrylic resins (see Acrylic ester polymers).

An estimate of the consumption of plastics in dentistry is shown in Table 15 (108).

Table 15. Estimate of U.S. Consumption of Plastics in Dentistry in 1974[a]

Appliance	Units made per year (million)	Weight of resin used per unit, g	Total resin used, kg
complete denture	7.1	25	180,000
partial denture	5.0	6	30,000
acrylic teeth	125[b]	0.5	55,000
denture relined or rebased	2.4	4	10,000
fillings	35[c]	0.15	5,000
impression materials (polysulfide, polyether, silicones)			100,000
miscellaneous			100,000
		Total	480,000

[a] Ref. 109, excluding maxillofacial prostheses and mouth protectors.
[b] 45% of all teeth used.
[c] 75% of all anterior fillings.

Miscellaneous uses include mouth protectors, tissue conditioners, splints for periodontic and orthodontic treatment, impression trays, and patterns for metal castings.

Polymerization is induced by heat, chemical means (see Initiators) or electromagnetic irradiation (see Radiation curing). For the polymerization of denture bases, a dough is formed by blending a powder (polymer), and liquid (monomer). This dough is heated and compressed in a gypsum mold. The molecular weight of the resultant denture-base acrylic is as high as 600,000 and such a polymer will have a higher molecular weight if cross-linked. Heat is the most common medium for polymerizing. Benzoyl peroxide is the usual initiator. It decomposes at relatively low temperatures (50–100°C) generating free radicals which are potent initiators of the polymerization (see Polymers).

The presence of impurities impedes the reaction, and some chemicals inhibit the polymerization. Butylated hydroxytoluene [128-37-0] (BHT) is the inhibitor commonly used in the monomer (0.006% or less) to block polymerization until curing is desired. Other inhibitors are used to slow the reaction, reduce the exothermic temperature, and provide an extended working time for the manipulation and packing of the dough, especially in chemically activated polymerization. There are several chemicals, eg, tertiary amines, that activate the peroxide. In addition, a benzoin methyl ether is activated by ultraviolet radiation, thus activating the peroxide initiator.

An unwanted inhibitor of the polymerization reaction is air. Any exposed surfaces will be incompletely polymerized and also have excessive surface porosity as a result of evaporation of unreacted monomer. For denture bases the acrylic doughs are molded into the final dental form by compression and injection molding or by a pouring technique.

Denture Resins, Compression Molding. Compression molding is the most widely used molding technique for acrylic resin dentures. A split plaster mold is produced by investing a wax pattern, complete with teeth and mounted on its own cast, in a suitable split metal flask. The assembly is heated in boiling water until the wax pattern is softened. The split mold is opened, the wax is removed, and the resultant cavity flushed out with boiling water containing detergent to remove traces of wax. The surface of the cavity is then painted with a suitable film, usually an aqueous solution of an alginate, to aid in separating the cured acrylic molding from the plaster mold. The acrylic dough is packed into the cavity. A sheet of cellophane is placed over the acrylic dough and the flask halves are closed slowly to force the dough into the mold. The cellophane separator is removed from the halves and the mold is closed until metal-to-metal contact is obtained. The flask is then clamped and transferred to a water bath, or other heat source of controlled temperature. Time–temperature rates of 71°C for ≥6 h, 76.7°C for ≥3 h, or 82°C for ≤2 h are typical. The flasks should be allowed to cool to room temperature in air. Once cool, the molding should be deflasked and separated from the investing material with care to avoid cracking the teeth or the denture base. Trimming and polishing completes the process.

Denture Resins, Injection Molding. Injection molding of denture bases uses a special flask to provide an injection chamber and sprues to connect the injection chamber with the mold cavity (109).

Pour Molding. Modified chemically-cured acrylics, the so-called fluid resins, have gained increased popularity. Polymerization is accomplished in flexible agar or alginate molds or in a soft gypsum investing medium rather than in hard gypsum molds.

One disadvantage of the fluid resin technique is that comparatively large shrinkage of the resin tends to pull the posterior teeth from the resilient mold toward the cast and out of the occlusal pattern to which they were set. Thus, premature contact in the anterior teeth, which is difficult to relieve without disfigurement of the teeth, is often encountered. The fluid resin technique shares a weakness with the other chemically cured resins, that is, inadequate bonding of the resin to plastic teeth. This bonding can be improved by treating the acrylic teeth for 4 min with a 50:50 mixture of dichloromethane [75-09-2] and methyl methacrylate [80-62-6] cold-curing monomer (110) (see Methacrylic polymers). The teeth must remain damp from this soak while packing or pouring the acrylic resin.

Properties. About 3 vol of polymer to 1 vol of monomer makes a workable dough. The lowest liquid content that produces a workable dough is needed to minimize polymerization shrinkage. Methyl methacrylate shrinks ca 21 vol % during polymerization. The polymer–monomer dough technique reduces this shrinkage to 6–7 vol %. Clinically, normal processing results in a linear shrinkage of about 0.5% across the posterior part of a denture (109). The acrylic denture base absorbs water in use, which causes a partially compensating linear expansion of 0.1–0.2%. The net linear shrinkage in the acrylic denture base, of 0.3–0.4%, is not clinically significant since the tissue on which the denture rests adjusts to such changes (109).

The polymerization of methyl methacrylate is an exothermic reaction. The peak temperature within the polymerizing mass depends upon several factors, such as (*1*) polymerization accelerators; (*2*) polymerization inhibitors; (*3*) cross-section of the mass; and (*4*) time-temperature cycle for heating.

Specifications. ADA Specification No. 12 and Federal Specification U-R-179 for Denture Base Polymers cover the requirements of the test methods for both heat-cured and self-curing acrylic compositions intended for this use.

Radiopaque Denture Materials. The detection of dentures or fragments of dentures that have been swallowed, inhaled, or imbedded in the tissues has been difficult because the materials are radiolucent. Radiopaque materials, that satisfy current ADA specifications for denture base polymers and are not deficient in strength or color stability, are reported in reference 111 (see Radiopaques).

Repair Resin. Denture repairs are usually made with cold-cured resins. The repaired strength of the denture is approximately 50% of that of the original heat-cured denture (112). Cold-cured polymers are, however, widely used because there is less dimensional change (113), and the simple repair process is quickly done; consequently, the patients are not deprived of their dentures for long periods.

Specification. ADA Specification No. 13 covers the tests and test limits for repair resins.

Resilient Liners. Natural rubber, vinyl, acrylic, and silicone liners have been used as resilient liners for dentures. A recent innovation in soft liners has been the use of hydroxy-substituted acrylic polymers (114). These materials may be processed along with the hard denture base and can be finished and polished when dry by typical methods. When placed in water the hardness decreases. The corresponding water sorption is 20% and swelling occurs which may cause unwanted changes in the contour of the tissue-bearing surface.

At present, none of the soft liners has a life expectancy comparable to the denture base. The permanent use of soft liners is unwarranted. When the patient cannot tolerate a rigid denture surface it may be desirable to employ a soft lining material with the expectation that it will have to be replaced within a relatively short time. The Council on Dental Materials and Devices of the American Dental Association recommends that soft liners be considered only as temporary expedients (115).

Mouth Protectors. During the past two decades pilot programs demonstrated that mouth protectors were so effective that the National Alliance Football Rules Committee made both face guards and mouth protectors mandatory. This action was influenced by a 1962 report by a joint committee on mouth protectors of the American Association for Health, Physical Education, and Recreation and the American Dental Association (116). Materials that have been used as mouth protectors include rubber latex, vinyl acetate–ethylene copolymers [*24937-78-8*], plasticized acrylic polymers, polyurethane, and silicone rubber. The latex materials are painted on models followed by drying. The thermoplastic polymers are molded on a gypsum model using vacuum or pressure molding, or both.

Restorative Resins. Resins were introduced as tooth restorative material in the early 1940s. These materials can be classified as (*1*) unfilled tooth-restorative resins, (*2*) composite or filled restorative resins, and (*3*) pit and fissure sealants.

Unfilled Tooth, Restorative Resins. Chemically-cured resins are used in this direct filling esthetic material for conservative restorations. Their physical properties limit the use of unfilled resins to restoring for esthetics those surfaces of the teeth not subject to load bearing. The color stability of the unfilled resins is unsatisfactory because of the straw-colored reaction products of amine initiators used to decompose the benzoyl peroxide. Some products use a sulfinic acid activator giving marked improvements in color stability. Ultraviolet absorbers (qv) are also added to most present-day tooth restorative resins to lessen discoloration. Weaknesses of these unfilled resins include a high volume shrinkage during curing (6–7 vol %), a coefficient of thermal expansion (100 ppm/°C) which is excessively different from that of tooth structure (10 ppm/°C), and low resistance to abrasion.

Composite or Filled Tooth Restorative Resins. The addition of silane-treated inorganic fillers to unfilled resins has improved the properties of the resin restorative materials. The addition reaction product of bisphenol A [*80-05-7*] and glycidyl methacrylate [*106-91-2*] is really a compromise between epoxy and methacrylate resins

(117). This BIS–GMA resin, which polymerizes through the methacrylate groups, is not an epoxy resin molecule (118) (see Epoxy resins). Mineral fillers coated with a silane coupling agent were incorporated into this BIS–GMA monomer which was diluted with other methacrylate monomers to make it less viscous and which bonded the powdered inorganic fillers chemically to the resin matrix (117). Almost all brands of composites use the BIS–GMA polymer.

Another composite was developed in England (119). This composite contains a resin-coated, silane-bonded, alumino–silicate filler as the powder and methyl methacrylate with methacrylic acid as the liquid. In 1977 and 1978, two new composite restorative materials were introduced. Both are based on urethane dimethacrylate [69766-88-7] polymerizable at mouth temperature; one with an amine–peroxide system, the other with electromagnetic irradiation.

The composites are dispensed in powder–liquid, paste–paste, and paste–liquid containers, the ingredients of which are hand-mixed usually before application to the cavity. Ingredients of typical composites are listed in Table 16 (108).

The curing shrinkage values for composites have been reduced by one fourth that of the unfilled resins. The combination of the high molecular weight (BIS–GMA) and the filler bonded to the resin reduced the total shrinkage to 1.5–3 vol % (120). In addition to reducing the shrinkage during hardening, the coefficients of thermal expansion have also been reduced for the composite restorative materials (120). These coefficient-of-thermal-expansion values are approximately 3–4 times that of teeth as compared to 8–10 times that of teeth for the unfilled resins.

The composite resins also have improved physical and mechanical properties such as the compressive strength, diametral tensile strength, and modulus of elasticity (120).

Specification. ADA Specification No. 27 is for Direct Filling Resins.

Pit and Fissure Sealants. The BIS–GMA resin portion of the composite restorative material has been further diluted with methyl methacrylate monomer or other low viscosity monomers and used to seal poorly developed but not decayed pits and fis-

Table 16. Ingredients of Typical Composite Restoratives

Component	Approx percentage in typical composite	Typical ingredients	CAS Registry No.
monomer	21	BIS–GMA or other aromatic dimethacrylate	[1565-94-2]
comonomer (diluent)	9	tetra(tri-or di-)ethylene glycol dimethacrylate	[4353-28-0], [2351-42-0], [2358-84-1]
inhibitor	0.06	butylated hydroxytoluene, hydroquinone methyl ether	[128-37-0], [150-76-5]
accelerator	0.15	N,N-dimethyl-p-toluidine, N,N-bis(2-hydroxyethyl)-p-toluidine	[99-978], [3077-12-1]
reinforcing fillers	68.5	treated SiO_2, Al_2O_3, or lithium aluminium silicate (1–40 μm)	
coupling agent	0.4	[3-(methacryloxy)propyl]trimethoxysilane	[2530-85-0]
initiator	0.2	benzoyl peroxide	[94-36-0]
ultraviolet stabilizer	0.5	Uvistat 247	[3550-43-4]
dye or pigment	?		

sures. When incompletely closed or exceptionally deep they tend to decay soon after the teeth erupt. By properly acid-etching the enamel surface within and surrounding the pits and fissures, good adhesion is attained. The low viscosity, cold-curing, modified BIS–GMA flows into these porosities created by the etching, and fills and effectively seals the pits and fissures (121). This sealing has proved to be effective clinically in reducing the rate of decay on these vulnerable surfaces (122–124).

Specification. Pit and fissure resins are not covered by an ADA specification. These products are currently covered by the ADA Acceptance Program.

Dental Impression Compounds. Dental impression compounds are thermoplastic compositions designed for taking oral impressions. The materials are softened by dry heat or in warm water at 50–60°C, depending upon the type of compound, until they are thoroughly plastic and free of lumps. The material is adapted to a previously prepared metal tray. The tray is placed in the oral cavity where an impression is desired. The assorted compounds are designed with variations in softening temperatures, plastic ranges, plasticity, and body. The higher-softening compounds are used as "tray" compounds to support the lower-softening "impression compounds" and the still lower-softening corrective compounds, without plastic distortion. Each successive step of the procedure uses a material of lower-softening range to minimize any change in the completed work. Impression compound is supplied in the form of cakes, sticks, cones, and cylinders, as a convenience for various needs.

Properties. Close control of raw materials and manufacturing processes, as well as constant laboratory testing of the finished product, are necessary to ensure uniform thermoplastic characteristics and working properties.

The thermal expansion or contraction of impression compounds has an important bearing on the dimensional accuracy of a compound impression. The rather high expansion on warming, or shrinkage on cooling, has required that considerable skill be employed to minimize this effect. For an all-compound impression, the final registration of detail must be done without warming of the mass of the compound, but only the immediate surface (a procedure made possible by the low heat conductivity of the compounds). The difficulty of mastering this technique has resulted in the use of a wash technique. A wash of plaster, a zinc oxide–eugenol impression paste, or a rubber-base impression material is applied over the compound surface to procure the final detail.

Characteristics. Ideal impression compounds (*1*) are free of poisonous or irritating ingredients; (*2*) should harden at, or slightly above, mouth temperature; (*3*) are plastic at a temperature that is sufficiently low so as not to injure the mouth tissues, and not unduly uncomfortable for the patient; (*4*) should harden uniformly without warpage or distortion of any sort; (*5*) are sufficiently cohesive with a consistency that will enable them to reproduce all details of crevices and other small markings, and retain such detail after solidification (without being adhesive in nature); (*6*) are not permanently deformed or fractured while being removed from the mouth, yet reproduce all unyielding undercuts, dovetails, etc, in the withdrawn impression; (*7*) exhibit a smooth, glossy surface after flaming; (*8*) do not flake when trimmed at room temperature; and (*9*) maintain their shape and dimensions indefinitely at normal room temperature and humidity under reasonable care (125).

The plasticity characteristics of impression compounds are defined by a flow curve

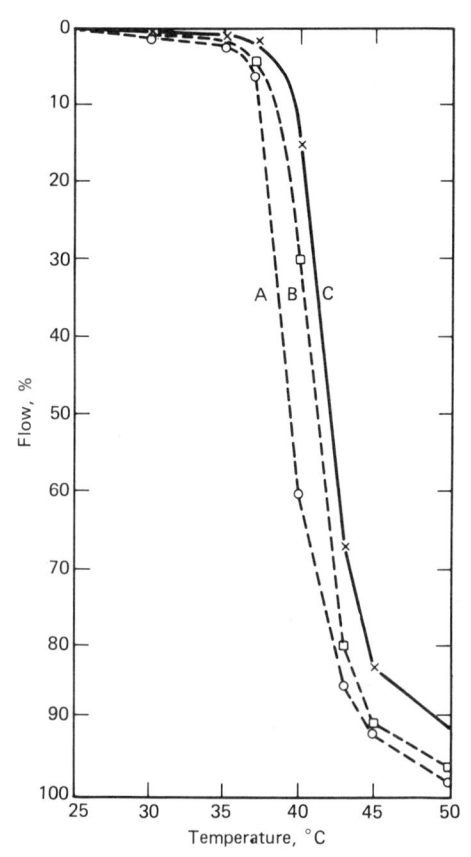

Figure 1. ADA flow curves of dental compounds. A = Corrective compound; B = impression compound; C = tray compound.

(Fig. 1). The thermal expansion of an impression compound is illustrated by Figure 2.

Composition. The gums, resins, and waxes of dental impression compounds are essentially natural products. The objective of many manufacturers has been to improve formulas and to take advantage of cleaner, cheaper, more uniform, and more available synthetic products. Frequently, such substitutions are most difficult to achieve without sacrificing some of the more subtle qualities that characterize a particular product. Essentially, the impression compound formulas include materials from the following classifications: (1) thermoplastic resins, which may include the coumarone–indene resins, shellac; (2) plasticizers, which may be selected from stearic acid, carnauba wax, Japan wax (see Waxes), palm oil, or synthetics; (3) fillers, for body, which may include talc and calcium carbonate; and (4) pigments for color, which are selected from the iron oxides, bone black, and titanium dioxide (125–126).

Specifications. ADA Specification No. 3 and Federal Specification U-I-490 for dental impression compounds cover the requirements and test procedures for this class of materials.

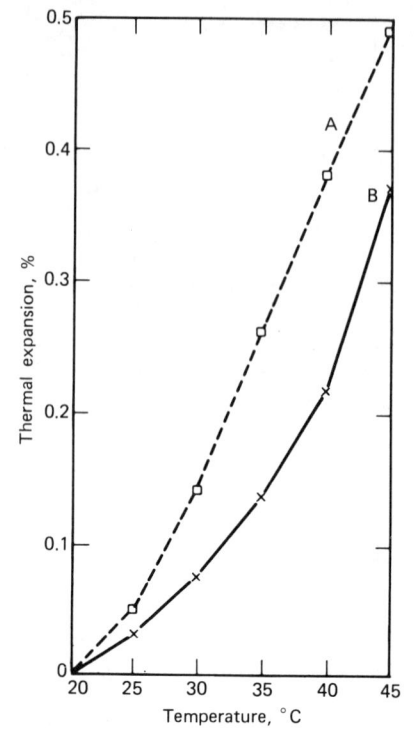

Figure 2. Thermal expansion of dental compounds. A = Impression and corrective compounds; B = tray compound.

Dental Waxes

Waxes (qv) and wax compositions have been important in the dental profession for many years (127). The diversification of their usefulness is illustrated by the following classification:
Pattern waxes:
    Inlay casting waxes:
        Type A (hard) direct technique when an extra hard wax is preferred by the dentist in making patterns in the mouth.
        Type B (medium) direct technique as used by most dentists in making patterns in the mouth.
        Type C (soft) indirect technique for making patterns outside of the mouth.
    Base-plate waxes:
        Type I (soft) for building contours and veneers.
        Type II (medium) for pattern production in the mouth in temperate weather.
        Type III (hard) for pattern production in the mouth in hot weather.
    Sheet and shape waxes:
Impression waxes:
    Impression wax.
    Bite-registration waxes.
    Disclosing waxes.

Processing waxes:
  Boxing wax.
  Sticky wax.
  Utility wax.
Study waxes (carving wax).

In the discussion of dental waxes, two characteristics are repeatedly mentioned, flow and thermal expansion. Flow testing defines plasticity characteristics vs temperature. Most dental waxes find application within the oral cavity during some phase of their use. Impression waxes must be plastic and moldable at mouth temperatures, and chill to a stiff nonplastic mass upon cooling within a few degrees below mouth temperature.

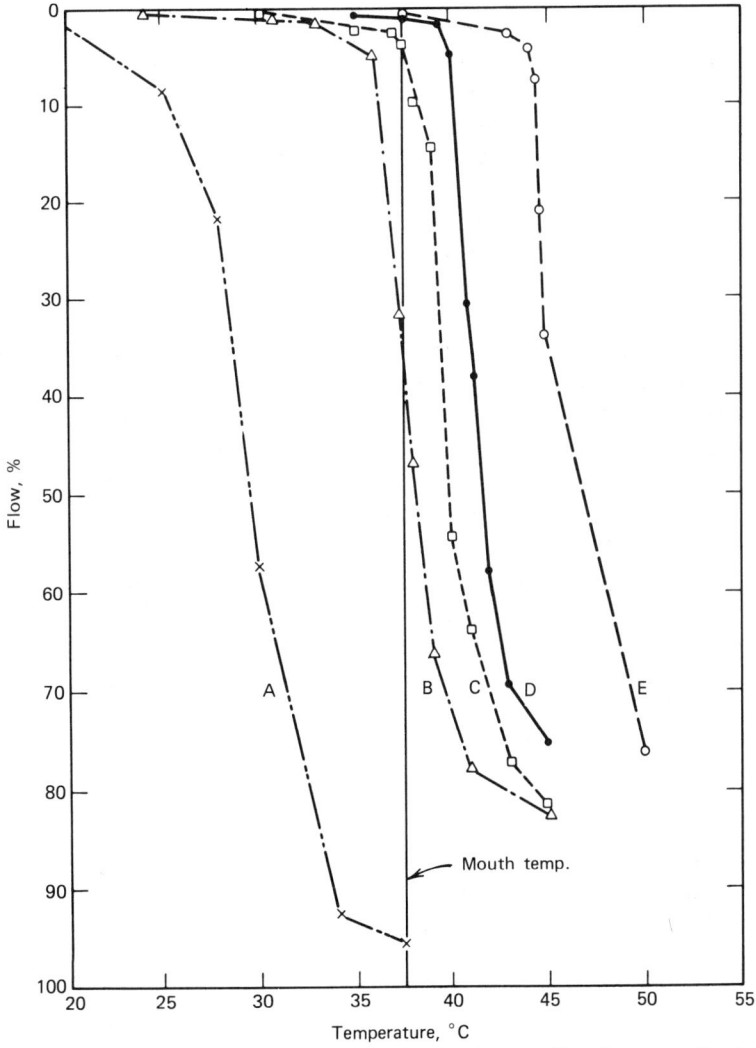

Figure 3. ADA flow curves of dental waxes. A = Impression wax; B = sheet wax; C = indirect-inlay wax; D = direct-inlay wax; E = base-plate wax.

Flow. Wax cylinders 6.0 mm thick by 10.0 mm in diameter are prepared by a controlled technique (ADA Specification No. 4, para. 4.3.1.1). Typical flow curves are shown in Figure 3.

Flow characteristics at 37°C of a ternary wax mixture of spermaceti, paraffin, and ceresin are shown in Figure 4 (128). Such a mixture with the composition in preparations of 60 wt % paraffin 124, 20 wt % spermaceti, and 20 wt % ceresin is sometimes referred to as the Iowa impression wax (129). It has a melting point of 48.5°C, a flow of 78.6% at 30°C, 95.4% at 37°C, 96.9% at 40°C, and 97.6% at 45°C. The dotted area shows where the waxes are not soluble in each other in the solid state.

Pattern Waxes. The pattern waxes are used to construct the prototype or pattern from which a finished dental restoration is produced.

Inlay Waxes. These are used to produce the wax patterns for the lost wax casting process, in the production of cast gold inlays, crowns, and bridges. Some inlay wax is also used to produce patterns for acrylic restorations.

Composition. The exact formulations are considered trade secrets and little has been published on the subject. McCrorie suggests a binary mixture of 65–75 wt % paraffin wax (60–63°C) and a microcrystalline wax with a melting-point >60°C (130). This produces a mixture with a solid–solid transition point at about 37°C with little plastic deformation (1–3%) at 37°C and a desirable plasticity at 45°C (73–77%) (130).

Inlay waxes are generally dark blue, green, black, or purple; however, inlay waxes for acrylic work are uncolored and are a natural ivory. The waxes are generally produced in stick form, about 75 mm long and 6.5 mm across; the cross section may be round or hexagonal.

Base-Plate Waxes. Base-plate waxes are used to substitute for, or be used in conjunction with, a base plate to form a pattern for the production of complete or partial dentures and certain orthodontic appliances to be molded of acrylic resin, modified vinyl resins, or other denture-base polymers.

Base-plate waxes are formulated for specific uses or working conditions into Types I, II, and III. Consequently, the flow requirements differ as shown below:

Flow Requirements for Base-Plate Waxes
(ADA Specification No. 24, para. 3.2.8)

Temp, °C	Type I min, %	Type I max, %	Type II min, %	Type II max, %	Type III min, %	Type III max, %
23.0		1.0		0.6		0.2
37.0	45.0	85.0		2.5		1.2
45.0			50.0	90.0	5.0	50.0

The linear thermal expansion of base-plate waxes should not exceed 0.8% at 25.0–40.0°C. It is desirable to invest any waxed-up case as soon after waxing is completed as possible. This minimizes changes in articulation, owing to tooth shift and changes in palatal thickness owing to lifting of the palatal section by wax shrinkage caused by variations in room temperature or by the release of stress.

Composition. Base-plate wax compositions are generally regarded as trade secrets. A substantial percentage of paraffin is usually present, probably 50–80 wt %. Beeswax, carnauba wax, ceresin, microcrystalline waxes, Acrawax C (Glyco Products

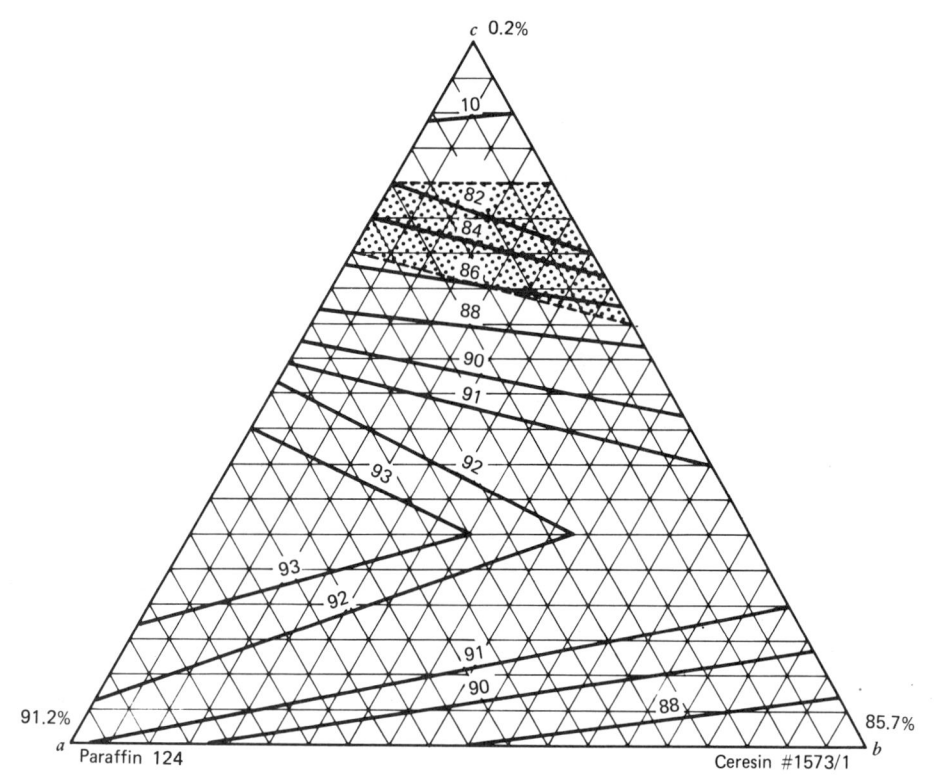

Figure 4. Flow diagram of ternary wax mixtures of Stevenson spermaceti, Cornelius paraffin 124, and Ross 1573/1 ceresin at 37°C.

Co., Inc.), mastic gum, rosin [8050-09-7], and synthetic resins may make up the balance of the formulation. Base-plate waxes are generally sold in sheet form about 1.3 mm thick, 75 mm wide, and 140 mm long.

Sheet and Shape Waxes. These are used to produce the patterns from which complete or partial dentures are cast of gold or base metal alloys. They are used to fabricate the restoration prototype directly upon a refractory investment cast.

Properties. The flow of sheet and shape wax is much higher than that of inlay wax and base-plate wax, reflecting increased pliability, ductility, and plasticity. At 35°C these waxes should have a flow of 10% max, and at 38°C a flow of 60% min. At 23 ± 1°C they should not fracture when bent double.

Composition. The compositions of sheet and shape waxes are also trade secrets. However, they are blends of various proportions of paraffin, microcrystalline waxes, carnauba wax, ceresin, beeswax, gum dammar, mastic gum, and possibly other resins. Sheet waxes are marketed in square sheets approximately 80 × 90 mm. Various thicknesses are available from 32 gage (0.5 mm) to 14 gage (1.63 mm).

Shape wax is similar in composition and properties to the nontacky sheet waxes. It is processed into preformed shapes or definite shape and gage to facilitate the "waxing-up" of partial-denture patterns. The shapes and sizes conform to those most needed for lingual bars, palatal bars, clasps, saddle construction, retainers, etc. The available sizes and shapes include rounds from 10 gage (2.0 mm) to 20 gage (0.84 mm), half-rounds from 6 gage (3.36 mm) to 14 gage (1.63 mm), and half-pear shapes of 6 gage (3.36 mm). The pieces are usually cut approximately 10 cm long.

Impression Waxes. Impression waxes include those waxes used to obtain a negative cast of the mouth structure (impression waxes), waxes used to establish tooth articulation (bite-registration waxes), and waxes used to detect tooth interference and high spots or improper fit of denture bases (disclosing waxes).

Impression Waxes. Impression waxes proper are generally considered to be soft, low-melting, plastic waxes used as the last "wash" on an impression to obtain fine detail of the oral structures. However, complete wax-impression techniques are available that involve the use of hard, high-melting, rigid base waxes, with successively softer wax additions to build a complete impression. The last low-melting wax gives the final detail without thermally destroying the higher-softening waxes of the base impression.

Properties. Flow properties of a set of impression waxes are:

Impression wax	Flow, %	Temp, °C
grade #1	5	37.5
grade #2	79	37.5
grade #3	28	30.0
grade #4	75	30.0

Composition. The exact compositions of impression waxes are trade secrets. The materials that have been identified in the compositions are paraffin, ceresin, vegetable waxes, rosin, mastic gum, and spermaceti. The composition of the Iowa impression wax was given previously.

Bite-Registration Waxes. These are used to establish the occlusion or horizontal relationship of the lower jaw to the upper jaw when there are opposing teeth present. Bite waxes may have a flow of 84% at 30°C.

Composition. Bite waxes are generally compounded from high-flow, low-melting paraffins, microcrystalline waxes, and resins.

Disclosing Waxes. Disclosing waxes or pressure indicating pastes are used in fitting a dental appliance to establish the location and extent of high spots or pressure areas, as on impressions and complete dentures. They are very soft, salve-like compositions that are painted onto the tissue side of an impression or denture. When the wax-coated denture is seated in the mouth, the soft paste is forced out of the areas showing hard contact between the denture and the mucous membrane. These areas on the denture can be easily marked and material removed from the denture to relieve the premature or hard contact.

Composition. Disclosing waxes include soft paraffins, petrolatums, coconut oil, zinc oxide, titanium dioxide, and suitable dyes.

Processing Waxes. The extensive amount of handwork and craftmanship necessary in the fabrication of most dental restorations and appliances has created a need for several types of wax compositions.

Boxing Waxes. Boxing wax, as the name implies, was developed to box impressions, ie, to construct a retaining ring of wax around an impression to confine the plaster or stone when the cast is poured. These waxes may have melting points of 65.5–71.1°C, and a flow of 60–80% at 37.5°C.

Compositions. Boxing waxes are generally formulated from various microcrystalline waxes to give the desired tackiness and pliability. Various resinous additions may also be made to increase strength and tackiness.

Sticky Waxes. Sticky waxes are used as thermoplastic cements. The broken pieces of a plaster impression are reassembled and held in position with sticky wax. Broken denture bases may be held in proper alignment for repair. Orthodontic appliances may be assembled with a sticky wax prior to investing and soldering. Plaster splints may be sealed to stone models in the production of porcelain or resin facings. Thus, sticky wax is useful in almost any operation where it is desired to position and hold several small pieces in a temporary relationship.

Composition. Sticky waxes are generally composed of resins and wax. A high resin content gives viscosity to the melt, a long plastic range, and a brittle fracture when cooled. No modern formulas are available but the older recipes usually had rosin, beeswax, and gum dammar (see Gums) as the essential constituents.

Utility Waxes. Utility waxes, as the name suggests, are waxes of many uses. They are generally useful for sealing, filling, contouring, building up areas, positioning, and many other functional purposes.

Composition. Utility wax compositions are considered trade secrets; they are probably microcrystalline waxes or blends of beeswax, petrolatum, and rosin.

Study Waxes. Waxes for carving are useful in the study and modeling of tooth forms and the teaching of anatomical detail.

Composition. Carving wax compositions include paraffin, ceresin, ozokerite, carnauba wax, montan waxes, and Acrawax C. Fillers and pigments may be added.

Wax Manufacture. Wax compositions for dental usage are usually compounded in simple melting and blending equipment. The melting is done with the least destruction of the wax properties. Direct-fired equipment can be used but jacketed melting kettles reduce the chance of localized overheating, ie, control of temperatures and heat distribution is desired. Some compositions may require a carbon dioxide or other protective cover to minimize oxidation. Foaming or swelling of the batch in the kettle may occur on melting from thermal expansion of the waxes and resins, raw materials that contain water, volatilization, or low-boiling constituents.

Agitation of the wax melt facilitates heat transfer, reduces localized overheating, and increases the melting rates. Overagitation should be avoided to minimize oxidation of the materials. The melting and processing of waxes should be done as quickly as possible and at the minimum temperature compatible with the process to avoid the loss of volatile constituents, degradation of the materials, and oxidation changes. A cover of inert gas may be used to protect susceptible compositions from excessive oxidation.

Wax compositions are processed into rods, sheets, cakes, and special forms by a variety of processes. Casting, roll forming, extrusion (by mechanical or hydraulic-piston-type extruders, or by continuous screw extrusion) are operations in practice. Stamping, roll cutting, or molding are also employed.

Specifications. The applicable specifications for each type of wax are:

Dental inlay casting wax	ADA Specification No. 4
Base plate wax	ADA Specification No. 24
Impression wax	none
Bite registration wax	none
Disclosing wax	none

514   DENTAL MATERIALS

Boxing wax	Federal Specification U-W-138
Sticky wax	Federal Specification U-W-149
Utility wax	Federal Specification U-W-156
Sheet and shape wax	Federal Specification U-W-140

Abrasives

Dental abrasives range in fineness from those that do not damage tooth structure to those which are coarse and hard enough to cut tooth enamel or the hardest of the restorative materials (see Abrasives).

Characteristics. Abrasive particles should be irregular and jagged so they always present a sharp edge, and should be harder than the material abraded.

Another property of an abrasive is its impact strength; if the particle shatters on impact it is ineffective. The other extreme is also not desirable, because if it never fractures, the edge becomes dull. The most effective abrasives continuously present a sharp edge.

In addition, a desirable characteristic is the ability of the abrasive to resist wear and solubility.

Uses. Dental abrasives can be classified according to their use or according to their degree of abrasivity (see Dentifrices). The use classification, adopted for the ADA Specification No. 37 for powdered dental abrasive materials, is based on removal of stain from natural teeth or on restorations of all types. Several abrasives are used in dentistry in a variety of grit sizes and shapes.

Aluminum Oxide. Emery [57407-26-8] is a natural oxide of aluminum with various impurities. One of these impurities, iron oxide, also acts as an abrasive. Pure aluminum oxide is made from bauxite [1318-16-7] and has partially replaced emery.

Silicates. Garnet [12178-41-5] is any one of many siliceous combinations with aluminum, cobalt, magnesium, iron or manganese, or both. Garnet for dental use is usually coated on paper with a glue binder and formed in disks and strips. Pumice is of volcanic origin. Kieselguhr is the siliceous remains of minute aquatic plants known as diatoms. Its common name is diatomaceous earth, and in a coarser form is used as a filler in many materials. The finer particles are excellent mild abrasive and polishing agents (see Diatomite).

Tripoli. Tripoli is a mild abrasive and polishing agent from porous rocks near Tripoli.

Rouge. Rouge is a fine red powder of ferric oxide [1309-37-1] (Fe_2O_3). It is usually used in the cake form, but is also impregnated in paper or cloth known as crocus cloth.

Tin Oxide. Tin oxide is a pure white powder. Mixed with water, glycerol, or alcohol into a paste, it is used for polishing teeth and metallic restorations.

Chalk. Chalk is a calcium carbonate prepared by precipitation. It is used in many polishing compounds including dentifrices.

Sand. Sand and other forms of quartz are used as a powder in sandblasting. If a gentler abrasive material is wanted, powdered walnut shells are often used.

Carbides and Diamond. Small particles of carbides and diamond [7782-40-3] are imbedded in a binder and are used to cut tooth structure. Diamond chips are the hardest and most effective abrasive for tooth enamel (see Hardness).

Zirconium Silicate. Zirconium silicate [10101-52-7] is a polishing agent with the unique characteristic of gradually reducing its abrasivity during use. When first applied the degree of coarseness permits it to cut rapidly.

Polishing. The term polishing usually denotes the gradual reduction in the size of surface irregularities or scratches by using successively finer abrasives. The final abrasive should leave a mirror-like high luster.

It is essential to select the appropriate abrasive for specific purposes. For example, levigated alumina will impart a high polish to metal without removing much metal, but if used on tooth an excessive amount of tooth tissue will be removed without obtaining the desired polish.

Adhesives

The fundamentals of adhesion have a role in many aspects of dentistry, especially the adhesion of restorative materials to the tooth tissues, enamel, and dentin (see Adhesives).

Adhesion is the force with which the molecules of one substance adhere or are attracted to the molecules of another substance. The retention of restorative materials to tooth structure, in addition to holding restorations to the teeth, seals the interface between the tooth and restorative material. Prevention of this microleakage is important since it reduces pulpal irritation and a potential for recurrent decay.

The retention of restorative materials is dependent on either mechanical locking in cavities prepared with undercuts, or microstructural roughness etched in the enamel (created by acid treatment) into which tags of the resin project.

Two chemical bonding systems have promise. One is the chelating effect of carboxyl groups in poly(acrylic acid) to the calcium in the tooth (13,131–132). This type of adhesive material includes polycarboxylate cements and glass-ionomer cements. Another bonding system is the use of coupling agents.

Tooth enamel and dentin are not homogeneous masses. The enamel has an extremely high inorganic content, and the dentin is comparatively low in inorganic material, but higher in organic structures. A material that could potentially adhere to one would be difficult to adhere to the other.

As a tooth is prepared by either cutting or acid etching, irregularities are developed in the surface. As the tooth and restoration are subjected to function and temperature-change forces, stresses are concentrated in the irregularities. These stress concentrations tend to develop a zipper-like effect by crack propagation starting at the interface, defects that weaken or break the bond.

Surface treatments to enhance adhesion have included agents to cleanse the surface or selectively remove some of the organic or inorganic portions of the dentin and enamel. Other techniques have modified tooth surfaces by chemical means to improve adhesion. The effect of many different compounds used in the surface treatment of bovine enamel on the adhesion of acrylic restoratives is shown in Table 17 (133). These compounds, with the exception of the sealant, etch the enamel with a variety of effects. The most desirable effect is a combination of increasing the surface area and reducing the surface energy to permit the resin to wet the surface and form the microscopic tags or projections into the surface irregularities for mechanical retention.

A detailed description of various aspects of coupling agents, surface grafting, and

Table 17. Effect of Surface Pretreatment on Adhesion of Acrylic Restoratives to Bovine Enamel[a]

Pretreatment[b]	Bond strength[c], MPa[d]
none	0
50% phosphoric acid	4.8
20% citric acid	4.7
5% lactic acid	3.9
20% lactic acid	6.0
42.5% lactic acid	5.7
5% dihydroxymaleic acid	4.2
5% dihydroxytartaric acid	6.4
5% 1,3,5-pentanetricarboxylic acid	5.0
5% cis-1,2,3,4-cyclopentanetetracarboxylic dianhydride	5.9
5% 1,2,4-benzenetricarboxylic acid	3.5
5% benzenepentacarboxylic acid	5.5
5% benzenehexacarboxylic acid	0
5% tetrahydrofuran-2,3,4,5-tetracarboxylic acid	6.2
pit and fissure sealant	1.2

[a] Ref. 133.
[b] Surface treated twice for 10 s at 3 min intervals.
[c] Mean value of at least five specimens stored in water for 24 h at 37°C.
[d] To convert MPa to psi, multiply by 145.

individual chemicals for use in the adhesion of restorative materials to tooth structures is given in references 133–134.

Coupling agents are multifunctional molecules that promote adhesion of the solid substrate to the restorative. These adhesion promoters function by adsorbing onto the surface of the solid to facilitate interaction with the restorative. The portion of the molecule that is not adsorbed presents a surface that can be more easily wetted by the restorative. An adhesion promoter is ideally adsorbed as a monomolecular layer that prevents contamination of the solid by organic substances. It should be chemically inert, have a low surface energy, and form a hydrophobic barrier film to avoid the accumulation of water that could prevent efficient wetting by an uncured restorative (134).

There are several coupling agents used in dentistry. One of them, glycerophosphoric acid dimethacrylate [69727-44-2], is a component of a cavity seal liquid used with Sevriton, a direct filling resin (135–137). The compound, when incorporated into methylmethacrylate autopolymerizing resin, adheres to some degree to tooth surfaces. Another coupling is N-(2-hydroxy-3-methacryloxypropyl)-N-phenylglycine [4896-81-5] (NPG–GMA) (138). It produces a significant increase in adhesive strength over a short-term water exposure of dentin or enamel joined to methacrylate resins or restoratives. The adhesion developed by this coupling agent was enhanced after pretreatment of the tooth surface with dilute acid, base, or a 10% solution of EDTA (138).

Another technique employing a coupling agent pretreated the enamel surface with a phosphoric acid solution and a silanizing agent (139). Tributylborane [122-56-5] is used to initiate the polymerization of methyl methacrylate which adheres to this pretreated enamel surface. However, this development has not been successful in practice.

Another system is the surface grafting of monomers to collagen of soft and hard tissues (140). After polymerization of this monomer, the proteinaceous surface is bonded covalently to the polymer side chain, and this offers an attractive technique for improving surface characteristics of skin, bone, or dentin. The reaction presumably involves the hydroxy groups in the side chains of hydroxyproline, hydroxylysine, serine, and threonine residues of collagen. Monomers that react with these soft tissues include acrylic acid, methacrylic acid and many vinyl monomers containing a variety of functional groups using ceric ions as initiators (140).

An additional adhesive system is the zinc polycarboxylate cement system that offers a good adhesion under moist conditions (131). This cement system contains 90–95% zinc oxide and some magnesium oxide powder which is mixed with poly(acrylic acid) or an acrylic acid–itaconic acid copolymer. The setting mechanism results from the neutralization of the polycarboxylic acid with metal oxides, as well as the stiffening of the polymeric network by ionic interactions between carboxylate groups with zinc ions and by the reinforcing action of excess metal oxide (141). Poly(acrylic acid) interacts with the Ca^{2+} of tooth enamel to form ionized carboxyl groups (142). The adhesion of the carboxylate cement arises from two effects: (1) etching of the enamel by the poly(acrylic acid) which provides an easily wettable, porous surface, amenable to mechanical interlocking with hardened cement, and (2) formation of ionic bonds between calcium and carboxylate ions at the enamel–resin interface. These polyacrylate cements adhere to enamel surfaces and to a lesser extent to dentin surfaces because of the lower calcium content in the dentin.

This cement also adheres well to many metals, including stainless steel, silver, and amalgam, but does not bond to porcelain or dental restorative resins.

Another group of compounds proposed for adhesion to restorative materials and tooth structure are the alkyl cyanoacrylates. Thin films of alkyl 2-cyanoacrylates are readily polymerized anionically by hydroxyl or amino groups that are present in dentin (see Acrylic ester polymers). Monomers such as ethyl- [7085-85-0] or 2,2,2-trifluoro-1-methylethyl-2-cyanoacrylate [23023-91-8] bond to dentin and to acrylic resins, and the adhesive strength increases on prolonged water exposure of the joint (143). These materials also adhere to etched enamel.

Biological and Clinical Evaluations

Toxicity Tests. In addition to specifications for dental materials, whereby such materials are characterized by chemical and physical tests, it is necessary to have clinical and biological evaluations of any new material, or of any material that has been in use for a long time if substantial changes are made in its formulation.

The American Dental Association, through its Council on Dental Materials and Devices, has adopted *ad interim* Recommended Standard Practices for Biological and Clinical Evaluations of Dental Materials. These *ad interim* recommendations adopted in 1971 were developed by a Subcommittee of American National Standards Committee MD156 of the American National Standards Institute. The Council serves as the Administrative Sponsor of MD156. The recommended standard practices are given in the Guide to Dental Materials and Devices, published by the American Dental Association biennially. The recommended standard practices for toxicity tests applies to all dental materials except drugs, medicines or dental instruments. Individual kinds of specific materials are classified under five types depending upon their use. The test procedures include those for implantation, acute systemic toxicity, mucous membrane irritation, irritation of the tooth pulp, and for pulp capping.

DENTAL MATERIALS

Clinical Tests. Uniform guidelines for the clinical evaluation for dental materials were also adopted by the American Dental Association in 1971 via its Council on Dental Materials and Devices. These, too, are printed in the Guide to Dental Materials and Devices.

Status reports on the safety and efficiency of various materials and devices are published from time to time in the *Journal of the American Dental Association.* They are usually reprinted in the biennially revised Guide to Dental Materials and Devices.

The FDA has the responsibility of enforcing the Federal Food, Drug, and Cosmetic Act as amended and the Radiation Control for Health and Safety Act. Dental materials and devices come under the amended Act so any new material or substantially modified old materials have to have premarket clearance from the FDA prior to its introduction in interstate commerce.

BIBLIOGRAPHY

"Dental Materials" in *ECT* 1st ed., Vol. 4, pp. 894–928, by F. H. Freeman, Kerr Manufacturing Co.; "Dental Materials" in *ECT* 2nd ed., Vol. 6, pp. 777–847, by F. H. Freeman, Kerr Manufacturing Co.

1. D. A. Ellis, personal communication, American Dental Trade Association, 1976.
2. U.S. Bureau of the Census, Census of Manufacturers, *Industry Sales, Medical Instruments; Ophthalmic Goods; Photographic Equipment; Clocks, Watches, and Watchcases, MC72(2)-38B,* Superintendent of Documents, U.S. Government Printing Office, Washington, D.C., 1972.
3. G. C. Paffenbarger, W. T. Sweeney, and A. A. Isaacs, *J. Am. Dent. Assoc.* **20,** 1960 (1933).
4. G. E. Servais and L. Cartz, *J. Dent. Res.* **50,** 613 (1971).
5. G. C. Paffenbarger, I. C. Schoonover, and W. Souder, *J. Am. Dent. Assoc.* **25,** 32 (1938).
6. V. Lind, G. Wennerholm, and S. Nystrom, *Acta Odontol. Scand.* **22,** 333 (1964).
7. P. J. Schouboe, G. C. Paffenbarger, and W. T. Sweeney, *J. Am. Dent. Assoc.* **52,** 584 (1956).
8. G. M. Brauer, R. McLaughlin, and E. Huget, *J. Dent. Res.* **47,** 622 (1968).
9. S. Civjan and G. M. Brauer, *J. Dent. Res.* **43,** 281 (1964).
10. D. C. Smith, *Br. Dent. J.* **124,** 381 (1968).
11. E. L. Truelove, D. F. Mitchell, and R. W. Phillips, *J. Dent. Res.* **50,** 166 (1966).
12. R. W. Phillips, M. L. Swartz, and B. Rhodes, *J. Am. Dent. Assoc.* **81,** 1353 (1970).
13. D. C. Smith, *J. Can. Dent. Assoc.* **37,** 22 (1971).
14. E. Mizrahi and D. C. Smith, *Br. Dent. J.* **127,** 371 (1969).
15. K. V. Mortimer and T. C. Tranter in ref. 14, p. 365.
16. A. D. Wilson and B. E. Kent, *Br. Dent. J.* **132,** 133 (1972).
17. S. Crisp and co-workers, *J. Dent.* **3,** 125 (1975).
18. C. L. Hadden, *Chem. Ind.* **21,** 190 (1944).
19. D. B. Mahler and K. Asgarzadeh, *J. Dent. Res.* **32,** 354 (1953).
20. K. D. Jorgensen, *Odontol. Tidskr.* **61,** 305 (1953).
21. C. W. Fairhurst, *J. Dent. Res.* **39,** 812 (1960).
22. *Guide to Dental Materials and Devices,* 8th ed., American Dental Assoc., Chicago, Ill., 1976–1978, p. 112.
23. R. Earnshaw, *Aust. Dent. J.* **14,** 264 (1969).
24. D. B. Mahler and A. B. Ady in ref. 22, p. 578.
25. R. H. Volland and G. C. Paffenbarger, *J. Am. Dent. Assoc.* **19,** 185 (1932).
26. W. J. O'Brien and J. P. Nielsen, *J. Dent. Res.* **38,** 541 (1959).
27. U.S. Pat. 1,909,008 (May 16, 1933), C. H. Prange (to Austenal Laboratories, Inc.).
28. C. P. Mabie, *J. Dent. Res.* **52,** 758 (1973).
29. U.S. Pat. 2,222,781 (Nov. 26, 1940), T. E. Moore (to Ransom and Randolph Co.).
30. C. P. Mabie in ref. 29, p. 96.
31. R. J. Schnell, G. Mumford, and R. W. Phillips, *J. Prosthet. Dent.* **13,** 324 (1963).
32. U.S. Pat. 1,708,436 (Apr. 9, 1929), L. J. Weinstein.
33. U.S. Pat. 1,924,874 (Aug. 29, 1933), T. E. Moore (to Ransom and Randolph Co.).

34. N. O. Taylor, G. C. Paffenbarger, and W. T. Sweeney, *J. Am. Dent. Assoc.* **17,** 2266 (1930).
35. J. S. Shell, *Dent. Surv.* **36,** 616 (1960); 770 (1960).
36. J. W. McLean and T. H. Hughes, *Br. Dent. J.* **119,** 251 (1965).
37. W. J. O'Brien and G. Ryge, *J. Prosthet. Dent.* **15,** 1094 (1965).
38. F. J. Knap and G. Ryge, *J. Dent. Res.* **45,** 1047 (1966).
39. R. C. Vickery and L. A. Badinelli in ref. 8, p. 683.
40. J. N. Nally and J. M. Meyer, *Schweiz. Mschr. Zahnheilk* **80,** 250 (1970).
41. J. S. Shell and J. P. Nielsen, *J. Dent. Res.* **41,** 1424 (1962).
42. M. Lavine and F. Custer, *Int. Assoc. Dent. Res. Prog.*, Abst. No. M26, Pittsburgh, Pa. (1963).
43. Brit. Pat. 252,112 (May 14, 1925), A. Poller (to DeTrey Bros. Ltd.).
44. U.S. Pat. 1,672,776 (June 5, 1928), A. Poller (to DeTrey Bros. Ltd.).
45. U.S. Pat. 2,249,694 (July 5, 1939), S. W. Wilding (to the Amalgamated Dental Co.).
46. U.S. Pat. 2,434,005 (Jan. 6, 1948), S. E. Noyes.
47. U.S. Pat. 2,454,709 (Nov. 23, 1948), E. J. Molnar (to Montclair Research Corporation).
48. Brit. Pat. 1,044,753 (Oct. 5, 1966), (to ESPE Fabrik Pharmazeutischer Praeparate G.m.b.H.).
49. U.S. Pat. 3,453,242 (July 1, 1969), W. Schmitt and R. Purrman (to ESPE Fabrik Pharmazeutischer Praeparate G.m.b.H.).
50. M. A. Braden, *J. Dent. Res.* **51,** 889 (1972).
51. J. H. Hembree and L. J. Munez, *J. Am. Dent. Assoc.* **89,** 1134 (1974).
52. P. Reithe, *Tsch. Zahnaertzl. Z.* **21,** 301 (1966).
53. D. B. Mahler and J. van Eysden, *J. Dent. Res.* **48,** 501 (1969).
54. D. B. Mahler and co-workers, *J. Dent. Res.* **49,** 1452 (1970).
55. G. C. Paffenbarger, N. W. Rupp, and P. R. Patel, to be published in *J. Am. Dent. Assoc.*, 1979.
56. D. B. Mahler, L. G. Terkla, and J. van Eysden in ref. 29, p. 823.
57. D. B. Mahler and J. van Eysden, *J. Dent. Res.* **53,** Special Issue, Abst. No. 26 (1974).
58. D. B. K. Innes and W. V. Youdelis, *J. Can. Dent. Assoc.* **29,** 587 (1963).
59. K. Asgar, *J. Dent. Res.* (Special Issue) **53,** Abst. No. 23 (1974).
60. D. B. Mahler, J. D. Adey, and J. van Eysden, *J. Dent. Res.* **54,** 218 (1975).
61. L. B. Johnson, *J. Biomed. Mater. Res.* **4,** 269 (1970).
62. R. Nadal, R. W. Phillips, and M. L. Swartz, *J. Am. Dent. Assoc.* **63,** 488 (1961).
63. A. Archarya and co-workers in ref. 29, p. 187.
64. R. M. Waterstrat, *J. Am. Dent. Assoc.* **78,** 536 (1969).
65. U.S. Pat. 1,909,008 (May 16, 1933), C. H. Prange (to Austenal Laboratories).
66. U.S. Pat. 1,956,278 (Apr. 24, 1934), R. W. Erdle and C. H. Prange (to Austenal Laboratories).
67. U.S. Pat. 1,958,446 (May 15, 1934), C. H. Prange (to Austenal Laboratories).
68. R. Earnshaw, *Aust. Dent. J.* **3,** 159 (1958).
69. H. F. Morris and K. Asgar, *J. Pros. Dent.* **33,** 36 (1975).
70. J. Jahtzen, *Quintessence* (1), 1 (1970).
71. J. F. Bates and A. G. Knapton, *Int. Metall. Rev.* **22,** 39 (1977).
72. R. W. Phillips, *Skinner's Science of Dental Materials*, 7th ed., W. B. Saunders Co., Philadelphia, 1973, p. 593.
73. G. T. Eden and co-workers, *J. Dent. Res.* (Special Issue) **56,** Abst. No. 648 (1977).
74. J. P. Moffa in T. M. Valega, Sr., ed., *Physical and Mechanical Properties of Gold and Base Metal Alloys, Alternatives to Gold Alloys*, DHEW Publication (NIH) No. 77-1227, conference proceedings, Jan. 24–26, 1977.
75. E. F. Huget, J. M. Vilca, and R. M. Wall, *J. Dent. Res.* (Special Issue B) **56,** Abstract 642 (1977).
76. E. Huget, *Metals and Alloys in Dentistry, International Metals Reviews,* International Symposium for Dental Materials, International Association for Dental Research, London, March 1977, as reported in ref. 71 by J. F. Bates and A. G. Knapton.
77. J. P. Moffa and co-workers, *J. Pros. Dent.* **30,** 424 (1973).
78. J. P. Moffa and W. A. Jenkins in ref. 52, p. 652.
79. N. K. Sarkar and E. H. Greener, *American Society for Metals,* Materials Engineering Conference, Chicago, Oct. 1973.
80. J. F. Bates and A. G. Knapton, *Int. Metall. Rev.* **22,** 39 (March 1977).
81. J. P. Nielsen and J. T. Tuccillo in ref. 39, p. 964.
82. D. L. Smith and co-workers in ref. 55, p. 283.
83. R. C. Brumfield, *J. Am. Dent. Assoc.* **49,** 17 (1954).
84. J. A. Barton, J. D. Eick and G. Dickson in ref. 29, p. 163.

85. A. S. M. *Metals Handbook,* 8th ed., American Society for Metals, 1961, p. 1189.
86. A. B. Burse and co-workers, *J. Biomed. Mater. Res.* **6,** 267 (1972).
87. J. N. Nally, *Int. Dent. J.* **18,** 309 (1968).
88. S. Civjan, E. F. Huget, and J. E. Marsden, *J. Am. Dent. Assoc.* **85,** 1309 (1972).
89. S. Civjan and co-workers, *J. Am. Dent. Assoc.* **90,** 659 (1975).
90. E. F. Huget and S. Civjan in ref. 52, p. 383.
91. Ref. 72, p. 565.
92. R. L. Coleman, *Bur. Stand. U.S. J. Res.* **1,** 867 (1928).
93. L. Addicks, *Silver in Industry,* Reinhold Publishing Corp., New York, 1940, p. 202.
94. S. B. Lukie, *Dent. Items Interest* **20,** 490 (1898).
95. G. V. Black, *Operative Dentistry,* 7th ed., Vol. 3, Medico-Dental Publishing Co., Chicago, 1936, p. 233.
96. D. A. Wallace and H. L. Hansen, *J. Am. Dent. Assoc.* **26,** 1536 (1939).
97. J. J. Messing, *Br. Dent. J.* **110,** 95 (1961).
98. W. Harvey and W. J. Pitch, *Br. Dent. J.* **80,** 1 (1946).
99. E. J. Molnar and E. W. Skinner, *J. Am. Dent. Assoc.* **29,** 744 (1942).
100. H. I. Copeland and co-workers, *J. Dent. Res.* **34,** 740 (1955).
101. A. R. Ross, *J. Am. Dent. Assoc.* **21,** 2029 (1934).
102. D. F. Vieira, *J. Prosthet. Dent.* **9,** 70 (1959).
103. M. T. Tyas and H. J. Wilson, *Br. Dent. J.* **129,** 461 (1970).
104. Ref. 72, pp. 222, 519–522.
105. H. R. Stanley, R. E. Going, and H. H. Chauncey, *J. Am. Dent. Assoc.* **91,** 817 (1975).
106. R. W. Phillips, M. L. Swartz, and R. D. Norman, *Materials for the Practicing Dentist,* C. V. Mosby Co., St. Louis, 1969.
107. *Accepted Dental Therapeutics,* 37th ed., The American Dental Association, Chicago, Ill.
108. G. M. Brauer in R. G. Craig, ed., *Polymers in Dentistry, Dental Material Reviews,* University of Michigan School of Dentistry, Ann Arbor, Michigan, 1977, p. 49.
109. J. B. Woelfel, G. C. Paffenbarger, and W. T. Sweeney, *J. Am. Dent. Assoc.* **61,** 413 (1960).
110. N. W. Rupp, G. C. Paffenbarger, and R. L. Bowen, *J. Am. Dent. Assoc.* **83,** 601 (1971).
111. H. H. Chandler, R. L. Bowen, and G. C. Paffenbarger, *J. Biomed. Mater. Res.* **5,** 335 (1971).
112. J. W. Stanford, C. L. Burns, and G. C. Paffenbarger, *J. Am. Dent. Assoc.* **61,** 307 (1955).
113. D. H. Anthony, M. S. Thesis, University of Michigan School of Dentistry, 1961.
114. W. J. O'Brien, J. Herman, and J. H. Shepherd, *J. Biomed. Mater. Res.* **6,** 15 (1972).
115. Council on Dental Research, *J. Am. Dent. Assoc.* **67,** 559 (1963).
116. *Report of the Joint Committee on Mouth Protectors of the American Association for Health, Physical Education, and Recreation and the American Dental Association,* Chicago, Ill., 1962.
117. U.S. Pat. 3,066,112 (Nov. 27, 1962), R. L. Bowen (to U.S. Department of Commerce).
118. U.S. Pat. 3,179,623 (Apr. 20, 1965), R. L. Bowen (to U.S. Department of Commerce).
119. J. W. McLean and I. G. Short, *Br. Dent. J.* **127,** 9 (1969).
120. R. L. Bowen, *J. Am. Dent. Assoc.* **66,** 57 (1963).
121. E. I. Cueto and M. G. Buonocore, *J. Am. Dent. Assoc.* **75,** 121 (1967).
122. M. G. Buonocore, *J. Am. Dent. Assoc.* **82,** 1090 (1971).
123. W. P. Rock, *Br. Dent. J.* **134,** 193 (1973).
124. R. J. McCune and J. F. Cvar, presentation before *Int. Assoc. Dent. Res. Program and Abstracts of Papers,* Abst. No. 745, Chicago, Ill., 1971.
125. Ref. 72, p. 84.
126. F. A. Peyton and R. G. Craig, *Restorative Dental Materials,* 4th ed., C. V. Mosby Co., St. Louis, 1971, p. 171.
127. R. G. Craig, W. J. O'Brien, and J. M. Powers, *Dental Materials: Properties and Manipulation,* 5th ed., C. V. Mosby Co., St. Louis, Missouri, 1975.
128. M. Ohashi and G. C. Paffenbarger, *J. Nihon Univ. Sch. Dent.* **11,** 109 (1969).
129. L. C. Dirksen in ref. 97, p. 273.
130. J. W. McCrorie in ref. 17, p. 189.
131. D. C. Smith, *Br. Dent. J.* **125,** 381 (1968).
132. D. C. Smith, *Dent. Clin. North Am.* **15,** 3 (1971).
133. G. M. Brauer and D. J. Termini, *XXIII International Congress of Pure and Applied Chemistry,* Vol. 1, Boston, Mass. 1971, p. 601.

134. G. M. Brauer in J. A. von Fraunhofer, ed., *Scientific Aspects of Dental Materials,* Butterworth, London, 1975.
135. M. G. Buonocore, W. Wileman, and F. Brudevold, *J. Dent. Res.* **35,** 846 (1956).
136. I. R. H. Kramer and J. W. McLean, *Br. Dent. J.* **93,** 150 (1952).
137. I. R. H. Kramer and K. W. Lee in ref. 22, p. 1003.
138. R. L. Bowen, *J. Dent. Res.* **44,** 895, 903, 906 (1969).
139. E. Masuhara and co-workers, *Shika Zairyo Kenkyusho Hokoku* **2,** 782 (1966).
140. G. M. Brauer and D. J. Termini, *J. Appl. Polym. Sci.* **17,** 2557 (1973).
141. A. Jurecik and C. W. Fairhurst, presentation, *Int. Assoc. Dent. Res.,* 50th General Meeting, Las Vegas, Nevada, Abstracts of Papers, 1972, Abst. No. 379.
142. D. R. Beech, *Arch. Oral Biol.* **17,** 907 (1972).
143. D. R. Beech in ref. 51, p. 1438.

General References

R. W. Phillips, *Skinner's Science of Dental Materials,* 7th Ed., W. B. Saunders Co., Philadelphia, Pa., 1973.
R. G. Craig, W. J. O'Brien, and J. M. Powers, *Dental Materials: Properties and Manipulation,* 5th ed., C. V. Mosby Co., St. Louis, 1975.
Guide to Dental Materials and Devices, American Dental Association, Chicago, Ill., (latest edition 1976–1978).
A. H. Warth, *The Chemistry and Technology of Waxes,* 2nd ed., Reinhold Publishing Corporation, New York, 1956.

Journals

Index of Dental Literature, American Dental Association, Chicago, Ill., yearly publication.
Journal of the American Dental Association, American Dental Association, Chicago, Ill.
Journal of Dental Research, International Association for Dental Research, Washington, D.C.

Retrieval Services

The Science Information Exchange, Smithsonian Institution, Washington D.C., provides a national registry of research in progress or retrieval services for queries are available.
Reports of Research Contractors, National Institute of Dental Research, Westwood Building, Bethesda, Md., the Research Contracts Office will furnish a list of Clearinghouse Numbers of reports available from the National Technical Information Service, Springfield, Va.
The Federal Supply Catalog (FSC 6520), Index of Specifications and Standards, Superintendent of Documents, U.S. Gov't. Printing Office, Washington, D.C.

G. C. PAFFENBARGER
N. W. RUPP
American Dental Association Health Foundation, Research Unit
National Bureau of Standards

DENTIFRICES

A dentifrice is a substance or preparation used with a toothbrush to aid mechanical cleaning of the accessible surfaces of the teeth (1). A typical formulation for a dentifrice paste contains abrasives, flavoring mixture, humectants (glycerol, sorbitol), thickening agents (carboxymethyl cellulose, carrageenan), foaming agents, and water (2). Commercial dentifrices are available as pastes, powders, and gels.

Mouthwashes are pleasant-tasting solutions used for rinsing and spraying the mouth (see Mouthwashes). Topical oral medication for the prevention of dental caries became a subject of scientific investigation about 1940. This interest was further developed by evidence of the effectiveness of topical fluorides (3) followed by the announcement of very considerable reductions in caries rates among small groups using 45% urea solutions and 0.1% benzalkonium chloride solutions as dentifrices (4).

On the basis of clinical studies involving the use of dentifrices containing sodium monofluorophosphate, Na_2PFO_3, as the active ingredient, two dentifrices, Colgate with MFP and Macleans Fluoride, have been accepted by the Council on Dental Therapeutics of the American Dental Association (ADA) as effective in helping to prevent caries. Sodium monofluorophosphate is a relatively stable, inorganic salt which should be distinguished from the sodium fluoride–phosphoric acid mixture used for topical applications. Sodium monofluorophosphate occurs as a white to slightly gray odorless powder which is freely soluble in water. When incorporated in dentifrice formulations at a level of approximately 0.76%, it has been shown to be of benefit in 17–38% reduction of dental caries (5).

These findings are in the same general range as those reported for dentifrices already accepted by the ADA which contain 0.4% stannous fluoride (SnF_2) (Crest and Aim).

Variations in the effectiveness of caries-preventive dentifrices also depends on the thoroughness with which the dentifrice is used and whether the studies were or were not conducted in naturally fluoridated areas.

Dentifrices, for which therapeutic value is claimed, are covered by the Federal Food, Drug and Cosmetic Act. The FDA is responsible for enforcement of this act and other relevant statutes (see Regulatory agencies). An amendment in 1962 requires adequate controls to assure the safety, identity, strength, quality, and purity of the product.

The FDA has recently formed an OTC (over-the-counter) Panel on Dentifrices and Dental Care Agents. This panel is preparing a monograph to evaluate the safety and efficacy of various oral products (6). At present, the FDA and the Council on Dental Therapeutics of the American Dental Association do not evaluate dentifrices or mouthwashes that do not claim therapeutic value.

Dentifrices that claim therapeutic value are listed in *Accepted Dental Therapeutics* if they meet standards of acceptance with respect to usefulness, composition, advertising, and labeling. Products that are obsolete, markedly inferior, useless, or dangerous to the health of the user are declared unacceptable.

The Council on Dental Therapeutics of the American Dental Association classifies products as "accepted," "provisionally accepted," or "unacceptable." This method of dentifrice evaluation assists in determining the safety and usefulness of any dentifrice by considering all ingredients in a dentifrice separately and collectively. Accepted products include those for which there is adequate evidence for safety and effectiveness.

"Provisionally accepted" products are those for which there is reasonable evidence of usefulness and safety, but which lack sufficient evidence of dental usefulness to justify being "accepted." "Unaccepted" products include those for which there is no substantial evidence of usefulness, or a question of safety exists (7).

A large amount of clinical research is continually undertaken in search of a dentifrice that will significantly reduce the occurrence of dental caries and/or periodontal disease in humans.

The flavoring agents used in commercial dentifrices are diverse and complex (see Flavors and spices). The sweetener is usually saccharin (see Sweeteners). The detergents are sodium lauryl sulfate, sodium N-lauroyl sarcosinate, and sodium cocomonoglyceride sulfonate. In addition, dentifrices contain humectants such as glycerol, propylene glycol, sorbitol solution, and water, and thickeners such as gum tragacanth, sodium alginate, and cellulose derivatives. The thickening agents stabilize dentifrice formulations and prevent separation of the liquid and solid phases (8).

Abrasion

The abrasives used in dentifrices are generally relatively insoluble inorganic salts. They include several phosphate salts such as dicalcium phosphate, insoluble sodium metaphosphate, calcium pyrophosphate, calcium carbonate, magnesium carbonate, hydrated aluminum oxide, as well as silicates and dehydrated silica gels.

From a mechanical standpoint, the most important ingredient in a dentifrice is the abrasive agent. The abrasive requirement for individuals varies, eg, heavy smokers require more abrasion to remove the stain. Ideally, the abrasion should be strong enough to remove the stain without harming the tooth. Highly abrasive agents such as pumice and levigated alumina remove excessive amounts of tooth substance and should not be used. There are very little sound data on the effectiveness of various dentifrices. Individual abrasives may be classified according to the amount of tooth structure removed, but combinations of abrasives have unique characteristics; some have additive effects and others cause no increase in abrasion. The type of detergent and other constituents may dissipate or agglomerate the abrasives to give markedly different degrees of abrasion. The most desirable toothpaste, from the mechanical aspect, is one that gives a high luster to the tooth.

For those individuals with exposed root-portions of their teeth it is especially important to use milder abrasives. Just a brush and water can be used for 10–12 d. If stain accumulates in that time, sodium bicarbonate can be used as a dentifrice. If this fails to remove the stain, gradually more abrasive dentifrices should be tried. Table 1 lists the qualitative abrasive ranking of several dentifrices. Current, authoritative data are needed though they would be valid only as long as the manufacturers did not change the formulas of their products (10).

The procedure for routine comparisons of abrasives in dentifrices was considered useful and reasonably convenient by the American Dentifrice Abrasion Committee (11). It is, however, a means of comparing dentifrices fairly simply and quickly in the laboratory and is not intended to predict the average abrasiveness of dentifrices used in the mouth. The degree to which the results may be translated into presumed wear of teeth among dentifrice users is a matter of conjecture and experiment (12).

Table 1. Qualitative Abrasive Ranking of Dentifrices[a]

Low	Medium	High
Amm-i-dent	Close-up	Iodent #2
Listerine	Colgate	Walgreens Smokers
Pepsodent	Crest	
T-Lak	Gleem II	
	Macleans	
	Pearl Drops	
	Ultra-Brite	

[a] Ref. 9.

Dental hypersensitivity is usually associated with gingival recession, erosions, and abrasions. The surgical treatment of periodontal pockets often results in hypersensitivity.

A clinical study of a new desensitizing dentifrice compared the desensitizing effect of a flavored 2% dibasic sodium citrate Pluronic F-127 gel and an F-127 preparation to a commercially available stannous fluoride (0.4%) glycerol gel and a 10% strontium chloride dentifrice. The number of sensitive surfaces of the participants using the pluronic gels was significantly lower than those observed in the control group. Patients using the stannous fluoride or the strontium chloride dentifrices did not exhibit a significant improvement over the control group (13).

Special dentifrices, based on the dentifrices' action of softening, loosening, and removing film, molds, stains, and tarnish on both acrylics and metals, are sometimes used for cleaning dentures. Care must be taken, however, in the unsupervised use of these products by the public as some materials present in commercial denture cleansers may be unsuitable for general use in the mouth.

Table 2. U. S. Sales of Oral Hygiene Products in 1974

Oral hygiene product	Millions of dollars
Dentifrices	
tooth paste	419.80
tooth powders	4.09
Total	*423.89*
Denture products	
denture cleansers	38.68
denture adhesives	46.09
denture brushes	6.71
Total	*91.48*
Other oral hygiene products	
toothbrushes, nonelectric	98.05
toothbrushes, electric	15.38
oral lavages	25.67
dental floss	3.64
mouthwashes and gargles	252.75
breath fresheners	34.47
Total	*429.96*
Total for all oral hygiene products	*945.33*

MOUTHWASHES

A mouthwash is a solution for rinsing the teeth and mouth (1). Generally, a mouthwash serves as an adjunct in cleaning the mouth and as an aid in removing loose debris after brushing.

Although mouthwashes are considered cosmetic, early reports suggest their therapeutic value (14). More recent studies involved a double-blind clinical trial in which both the *in vivo* effects and the plaque-inhibiting ability of chlorhexidine and picloxydine mouthwashes on oral flora were compared (15).

The Council on Dental Therapeutics of the American Dental Association does not recognize any substantial contribution to oral health in the unsupervised use of medicated mouthwashes by the general public (11).

Economic Aspects

Table 2 shows the amount spent in the U.S. in 1974 on oral hygiene products.

BIBLIOGRAPHY

"Dentifrices" in *ECT* 1st ed., Vol. 4, pp. 928–935, by Erwin DiCyan, DiCyan & Brown; "Dentifrices" in *ECT* 2nd ed., Vol. 6, pp. 848–852, by Anthony F. Posteraro, College of Dentistry, New York University.

1. A. Osol, ed., *Blakiston's Gould Medical Dictionary*, 3rd ed., McGraw-Hill, New York, 1972.
2. *J. Am. Dent. Assoc.* **81,** 1177 (1970).
3. J. W. Knutson and W. D. Armstrong, *Public Health Rep.* **58,** 1701 (1943).
4. K. W. Lu and D. R. Porter, *J. Dent. Res.* **41,** 1277 (1962).
5. *Accepted Dental Therapeutics,* 37th ed., American Dental Association, Chicago, Ill., 1977, pp. 302–304.
6. *Food and Drug Administration Public Advisory Committees: Authority, Structure, Functions, Members,* April 1976, pp. 64–65.
7. Ref. 5, p. xvi.
8. Ref. 5, pp. 310–311.
9. J. J. Hefferren, *Pharmacy Times,* 50 (July, 1974).
10. G. C. Paffenbarger and N. W. Rupp, private communication, April 27, 1978.
11. *Accepted Dental Therapeutics,* 37th ed., American Dental Association, Chicago, Ill., 1977, pp. 312–313.
12. J. J. Hefferren, *J. Dent. Res.* **55,** 563 (1976).
13. D. D. Zinner, L. F. Duany, and H. J. Lutz, *J. Am. Dent. Assoc.* **95,** 982 (1977).
14. E. J. Alderman and V. L. Scallon, *Chronicle Omaha D.S.* **28,** 284 (1965).
15. G. M. Newcomb, G. M. McKellar, and B. D. Rawal, *J. Periodontol.* **48,** 282 (1977).

General References

F. D. Stevens, I. L. Shannon, and H. J. Williams, "Effects of Fluoride-Containing Commercial Dentifrices on Solubility of Intact Root Surfaces," *Dent. Hyg.* **50,** 160 (April, 1976).
M. S. Putt and co-workers, "Physical Characteristics of a New Cleaning and Polishing Agent in a Prophylaxis Paste," *J. Dent. Res.* **54,** 527 (1975).
J. R. Heath and H. J. Wilson, "Abrasion of Restorative Materials by Toothpaste," *J. Oral Rehabilitation* **3,** 121 (1976).
J. J. Hefferren, "A Laboratory Method for Assessment of Dentifrice Abrasivity," *J. Dent. Res.,* **55**(4), 563 (1976).
J. J. Hefferren, "Interfaces of Laboratory and Clinical Assessment of Dentifrice Abrasivity," *J. Soc. Cosmet. Chem.* **24,** 815 (1973).

D. B. Harte and R. S. Manly, "Four Variables Affecting Magnitude of Dentifrice Abrasiveness," *J. Dent. Res.* **55**(3), (1976).
Consumer Reports **37**(4), 251 (1972).
Accepted Dental Therapeutics, 37th ed., American Dental Association, Chicago, Ill., 1977.

ANTHONY F. POSTERARO
College of Dentistry
New York University

DESIGN OF EXPERIMENTS

Obtaining valid results from a test program calls for commitment to sound statistical design. In fact, proper experimental design is often more important than sophisticated statistical analysis. Results of a well planned experiment are often evident from simple graphical analyses. However, the world's best statistical analysis cannot rescue a poorly planned experimental program.

The main reason for designing an experiment statistically is to obtain unambiguous results at a minimum cost. The need to learn about interactions among variables and to measure experimental error are some of the added reasons (see below).

Many chemists and engineers think of experimental design solely in terms of standard plans for assigning treatments to experimental units, such as the Latin square, factorial, fractional factorial, and central composite (or Box) designs. These designs are described in various books, such as those summarized in the General References, and catalogued in various reports and articles. Important as such formal plans are, the final selection of test points represents only the proverbial tip of the iceberg, the culmination of a careful planning process.

Statistically planned experiments are characterized by (1) the proper consideration of extraneous variables; (2) the fact that primary variables are changed together, rather than one at a time, in order to obtain information about the magnitude and nature of the interactions of interest and to gain improved precision in the final estimates; and (3) built-in procedures for measuring the various sources of random variation and for obtaining a valid measure of experimental error.

A well planned experiment is often tailor-made to meet specific objectives and to satisfy practical constraints. The final plan may or may not involve a standard textbook design. If possible, a statistician knowledgeable in the design of experiments should be called in early, made a full-fledged team member and be fully apprised of the objectives of the test program and of the practical considerations and constraints. He or she may contribute significantly merely by asking probing questions. After the problem and the constraints have been clearly defined, the statistician can evolve an experimental layout to minimize the required testing effort to obtain the desired information. This may involve one of the formal statistical designs. However, designing an experiment is often an iterative process, requiring reworking as new information and preliminary data become available. With a full understanding of the problem, the statistician is in an improved position to respond rapidly if last-minute changes are required and to provide meaningful analyses of the subsequent experimental results.

Much of this article previously appeared in refs. 1 and 2. In addition, the books listed and described in the General References provide information concerning the technical and practical aspects of experimental design. Ref. 3 describes a computer program for evaluating the precision of an experiment before running it. Ref. 4 describes a case study that illustrates many of the considerations discussed here and introduces a few others. It provides good follow-up reading to this article. Finally, refs. 5 and 6 provide two other surveys dealing with the design of experiments, including more detailed discussions of specific designs and their analysis.

Purpose and Scope

Designing an experiment is like designing a product. Every product serves a purpose; so should every experiment. This purpose must be clearly defined at the outset. It may, eg, be to optimize a process, to estimate the probability that a component operates properly under a given stress for a specified number of years, to evaluate the relative effects on product performance of different sources of variability, or to determine whether a new process is superior to an existing one.

In addition to defining the purpose of a program, one must decide upon its scope. An experiment is generally a vehicle for drawing inferences about the real world. Since it is highly risky to draw inferences about situations beyond the scope of the experiment, care must be exercised to make this scope sufficiently broad. This must especially be kept in mind if laboratory findings are to apply to the manufacturing line. Thus if the results are material dependent, the material for fabricating experimental units must be representative of what one might expect to encounter in production. If the test program were limited to a single heat of steel or a single batch of raw material, the conclusions might be applicable only to that heat or batch, irrespective of the sample size. Similarly, in deciding whether or not temperature should be included as an experimental variable to compare different preparations, it must be decided whether the possible result that one preparation outperforms another at a constant temperature would also apply for other temperatures of practical interest. If this is not expected to be the case, one would generally wish to include temperature as an experimental variable.

Variables

An important part of planning an experimental program is the identification of the variables that affect the response and deciding what to do about them. The decision as to how to deal with each of the candidate variables can be made jointly by the experimenter and the statistician. However, identifying the variables is the experimenter's responsibility. Controllable or independent variables in a statistical experiment can be dealt with in four different ways. The assignment of a particular variable to a category often involves a trade-off among information, cost, and time.

Primary Variables. The most obvious variables are those whose effects upon performance are to be evaluated directly; these are the variables that, most likely, created the need for the investigation in the first place. Such variables may be quantitative, such as concentration of catalyst, temperature, or pressure, or they may be qualitative, such as method of preparation, catalyst type, or batch of material (see below).

Quantitative controllable variables are frequently related to the performance variable by some assumed statistical relationship or model. The minimum number of conditions or levels per variable is determined by the form of the assumed model. For example, if a straight-line relationship can be assumed, two levels (or conditions) may be sufficient; for a quadratic relationship a minimum of three levels is required.

Qualitative variables can be broken down into two categories. The first consists of those variables whose specific effects are to be compared directly; eg, comparison of the effect on performance of two proposed preparation methods or of three specific types of catalysts. The required number of conditions for such variables is generally evident. Such variables are sometimes referred to as fixed effects or Type I variables.

The second type of qualitative variables are those whose individual contributions to performance variability are to be evaluated. The specific conditions of such variables are randomly determined. Material batch would be such a variable if one is not interested in the behavior of specific batches *per se,* but instead needs information concerning the magnitude of variation in performance caused by differences between batches. In this case, batches would be selected randomly from a large population of batches. For variables of this type, it is generally desirable to have a reasonably large sample of conditions (eg, five or more) so as to obtain an adequate degree of precision in estimating performance variability attributable to the variable. Such variables are sometimes referred to as random effects or Type II variables.

When there are two or more variables, they might interact with one another, ie, the effect of one variable upon the response depends on the condition of the other variable. Figure 1 shows a situation where two noninteracting variables, preparation type and temperature, independently affect time to rupture, ie, the effect of temperature on time to rupture is the same for both preparation types. In contrast, Figure 2 shows two examples of interactions between preparation and temperature. In Figure 2**a**, an increase in temperature is beneficial for preparation A but does not make any difference for preparation B. In Figure 2**b**, an increase in temperature raises time to rupture for preparation A but decreases it for preparation B.

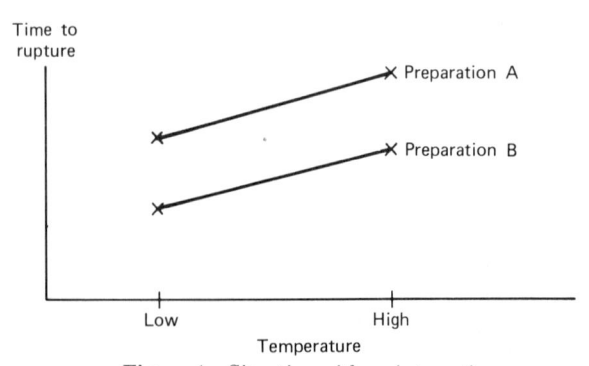

Figure 1. Situation with no interaction.

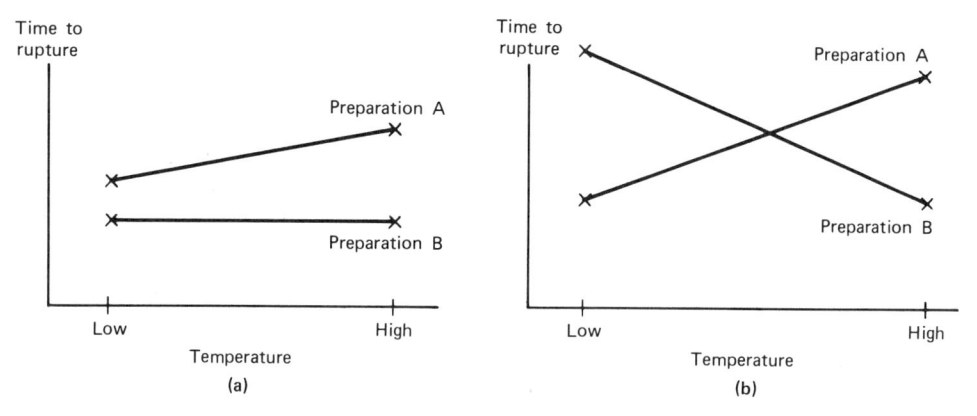

Figure 2. Situations with interactions.

An important purpose of a designed experiment is to obtain information about interactions among the primary variables. This is accomplished by varying factors simultaneously rather than one at a time. Thus in Figure 2, each of the two preparations would be run at both low and high temperatures using, eg, a full factorial or a fractional factorial experiment (see below).

Background Variables and Blocking. In addition to the primary controllable variables there are those variables, though not of primary interest, that cannot, and perhaps should not, be held constant. Such variables arise, for example, when an experiment is to be run over a number of days, or when different machines or operators or both are to be used. It is of crucial importance that such background variables are not varied in complete conjunction with the primary variables. For example, if preparation A were run only on day 1 and preparation B only on day 2, it is impossible to determine how much of any observed difference in performance between the two preparations is due to normal day-to-day process variation (such mixing of effects is called confounding). If the background variables do not interact with the primary variables, they may be introduced into the experiment in the form of experimental blocks. An experimental block represents a relatively homogenous set of conditions within which different conditions of the primary variables are compared.

A well-known example of blocking arises in the comparison of wear for different types of automobile tires. Tire wear may vary from one automobile to the next, irrespective of the tire type, because of differences among automobiles, variability among drivers, and so forth. Assume, for example, that for the comparison of four tire types (A, B, C, and D), four automobiles are available. A poor procedure would be to use the same type of tire on each of the four wheels of an automobile and vary the tire type between automobiles, as in the following tabulation:

	Automobile		
1	*2*	*3*	*4*
A	B	C	D
A	B	C	D
A	B	C	D
A	B	C	D

Such an assignment would be undesirable because the differences among tires cannot be separated from the differences among automobiles in the subsequent analysis. Separation of these effects can be obtained by treating automobiles as experimental blocks and randomly assigning tires of each of the four types to each automobile as follows:

Automobile

1	2	3	4
A	A	A	A
B	B	B	B
C	C	C	C
D	D	D	D

The above arrangement is known as a randomized block design.

The symmetry of the preceding example is not always found in practice. For example, there may be six tire types under comparison and fifteen available automobiles. Tires are then assigned to automobiles to obtain the most precise comparison among tire types, using a so-called incomplete block design. Similar concepts apply if there are two or more primary variables, rather than tire type alone.

A main reason for running an experiment in blocks is to ensure that the effect of a background variable does not contaminate evaluation of the primary variables. However, blocking also permits the effect of the blocked variables to be removed from the experimental error, thus providing more precise evaluations of the primary variables. Finally, in many situations, the effect of the blocking variables on performance can also be readily evaluated—a further advantage of blocking.

In some situations, there may be more than one background variable whose possible contaminating effect is removed by blocking. Thus in the automobile tire comparison, the differences between wheel positions may be of concern in addition to differences between automobiles. In this case, wheel position might be introduced into the experiment as a second blocking variable. If there are four tire types to be compared, this might, for example be done by randomly assigning the tires of each of the four types according to the following plan, known as a Latin square design:

Wheel position	*Automobile*			
	1	2	3	4
1	A	D	C	B
2	B	A	D	C
3	C	B	A	D
4	D	C	B	A

In this plan, the effects of both automobile and wheel position are removed by blocking. It should, however, be kept in mind that for the Latin square design, as for other blocking plans, it is generally assumed that the blocking variables do not interact with the main variable to be evaluated.

Uncontrolled Variables and Randomization. A number of further variables, such as ambient conditions, can be identified but not controlled, or are only hazily identified or not identified at all but affect the results of the experiment. To ensure that such uncontrolled variables do not bias the results, randomization is introduced in various ways into the experiment to the extent that so doing is practical.

Randomization means that the sequence of preparing experimental units, assigning treatments, running tests, taking measurements, and so forth, is randomly determined, based, for example, on numbers selected from a random number table. The total effect of the uncontrolled variables is thus lumped together into experimental error as unaccounted variability. The more influential the effect of such uncontrolled variables, the larger the resulting experimental error, and the more imprecise the evaluations of the effects of the primary variables. (Sometimes, when the uncontrolled variables can be measured, their effect can be removed from experimental error statistically; see below.)

Background variables could also be introduced into the experiment by randomization, rather than by blocking techniques. Thus in the previous example, the four tires of each type could have been assigned to automobiles and wheel positions completely randomly, instead of treating automobiles and wheel positions as experimental blocks. This could have resulted in an assignment such as the following:

| | \multicolumn{4}{c}{Automobile} |
Wheel position	1	2	3	4
1	B	C	B	D
2	A	D	D	A
3	C	C	A	D
4	B	B	C	A

Both blocking and randomization generally ensure that the background variables do not contaminate the evaluation of the primary variables. Randomization sometimes offers the advantage of greater simplicity compared to blocking. However, under blocking the effect of a background variable is removed from the experimental error, whereas under randomization it usually is not. Thus one might aim to remove the effects of the one or two most important background variables by blocking, while counteracting the possible contaminating effect of others by randomization.

Variables Held Constant. Finally, some variables should be held constant in the experiment. Holding a variable constant limits the size and complexity of the experiment but, as previously noted, can also limit the scope of the resulting inferences. The variables to be held constant in the experiment must be identified and the mechanisms for keeping them constant defined. The experimental technique should be clearly specified at the outset of the experiment and closely followed.

Experimental Environment and Constraints

The operational conditions under which the experiment is to be conducted and the manner in which each of the factors is varied must be clearly spelled out.

Variables are not created equal; some can be varied more easily than others. For example, a change in pressure may be implemented by a simple dial adjustment; on the other hand, stabilization requirements might make it difficult to change the temperature. In such situations, completely randomizing the sequence of testing is impractical. However, basing the experimental plan on convenience alone can lead to ambiguous and unanalyzable results. For example, if the first half of the experiment were run at one temperature and the second half at another temperature, there is no way of knowing whether the observed difference in the results is due to the difference in temperature or to some other factors that varied during the course of the experiment, such as raw material, ambient conditions, or operator technique. Thus the final experimental plan must be a compromise between cost and information. The experiment must be practical to run, yet still yield statistically valid results (7–8).

Practical considerations enter into the experimental plan in various other ways. In many programs, variables are introduced at different operational levels. For example, in evaluating the effect on tensile strength of alloy composition, oven temperature, and varnish coat, it may be convenient to make a number of master alloys with each composition, split the alloys into separate parts to be subjected to different heat treatments, and then cut the treated samples into subsamples to which different coatings are applied. Tensile strength measurements are then obtained on all coated subsamples.

Situations such as the preceding arise frequently in practice and are referred to as split-plot experiments. (The terminology is due to the agricultural origins of experimental design, eg, a farmer needed to compare different fertilizer types on a plot of land with varying normal fertility.) A characteristic of split-plot plans is that more precise information is obtained on the low level variables (varnish coats in the preceding example) than on the high level variables (alloy composition). The split-plot nature of the experimental environment, if present, is important information both in the planning and in the analysis of the experiment.

Prior Knowledge. Prior knowledge is often available concerning the expected outcome at certain experimental conditions. For example, some combinations of conditions might be known to yield poor results or might not be attainable with the equipment available, or worse yet, could result in blowing up the plant. Furthermore, all proposed conditions in the experiment need to make sense. (We recently heard of an experiment with a condition that required a resistance of 50 Ω but called for a circuit without resistors.) Clearly unreasonable conditions must be omitted from the experimental design, irrespective of whether they happen to fall in with a standard statistical pattern. Thus the experiment must be adjusted to accommodate reality and not the reverse.

The Performance Variable. A clear statement is required of the performance characteristics or dependent variables to be evaluated. Even a well designed experiment fails if the response cannot be measured properly. Frequently, there may be a number of response variables; for example, tensile strength, yield strength, % elongation, etc. It is important that standard procedures for measuring each variable be established and documented. Sometimes the performance variable is on a semiquantitative scale; eg, material appearance may be in one of five categories, such as outstanding, superior, good, fair, and poor. In this case, it is particularly important that standards be developed initially, and especially so if judgments are to be made at different times and perhaps by different observers.

Different Types of Repeat Information. The various ways of obtaining repeat results in the experiment need to be specified. Different information about repeatability is obtained by (*1*) taking replicate measurements on the same experimental unit, (*2*) cutting a sample in half at the end of the experiment and obtaining a reading on each half, and (*3*) taking readings on two samples prepared independently of one another (eg, on different runs) at the same aimed-at conditions. Greater homogeneity among replicate measurements on the same sample than among measurements on different samples would be expected. The latter reflect the random unexplained variation of repeat runs conducted under identical conditions. A skillfully planned experiment imparts information about each component of variability, if such information is not initially available, and uses a mixture of replication and repeat runs so as to yield the most precise information for the available testing budget. The way in which such information is obtained must also be known to perform a valid analysis of the results.

Preliminary Estimates of Repeatability. Initial estimates of overall repeatability should be obtained before embarking on any major test program. Such information may be available from previous testing; if it is not, valid preliminary runs should be conducted at different times under supposedly identical conditions. If these runs result in large variability in performance, the important variables that affect the results have not been identified, and further research may be needed before the experiment can commence.

Consistent Data-Recording Procedures. Clear procedures for recording all pertinent data from the experiment must be developed and documented, and unambiguous data recording forms must be established. These should include provisions not only for recording the values of the measured responses and the desired experimental conditions, but also the conditions that actually resulted if these differ from those planned. It is generally preferable to use the values of the actual conditions in the statistical analysis of the experimental results. For example, if a test was supposed to have been conducted at 150°C but was run at 148.3°C, the actual temperature would be used in the analysis. In experimentation with industrial processes, process equilibrium should be reached before the responses are measured. This is particularly important when complex chemical reactions are involved.

The values of any other variables that might affect the responses should also be recorded, if possible. For example, although it may not be possible to control the ambient humidity, its value should be measured if it might affect the results. In addition, variations in the factors to be held constant, special happenings, and other unplanned events should be recorded. The values of such covariates can be factored into the statistical analysis, thereby reducing the unexplained variability or experimental error. If the covariates do indeed have an effect, this leads to more precise evaluations of the primary variables. Alternatively, such covariates may be related to the unexplained or residual variation that remains after the analysis, using graphical or other techniques.

Running the Experiment in Stages

Contrary to popular belief, a statistically planned experiment does not require all testing to be conducted at one time. Instead, the program can be conducted in stages of, eg, 8–20 runs; this permits changes to be made in later tests based on early results and allows preliminary findings to be reported. A recent experiment to improve the properties of a plastic material involved such variables as mold temperature, cylinder temperature, pressure, ram speed, and material aging. The experiment was conducted in three stages. After the first stage, the overall or main effects of each of the variables were evaluated; after the second stage, interactions between pairs of variables were analyzed, and after the third stage nonlinear effects were assessed. Each stage involved about a month of elapsed time, and management was apprised of progress each month. If unexpected results had been obtained at an early stage, eg, poor results at one of the selected ram speeds, the later stages of the experiment could have been changed.

Whether or not to conduct a particular experiment in stages depends upon the program objectives and the specific experimental situation; a stage-wise approach is recommended when units are made in groups or one at a time and a rapid feedback of results is possible. Running the experiment in stages is also attractive in searching for an optimum response, because it might permit a move closer to the optimum from stage to stage. On the other hand, a single-stage experiment may be desirable if there are large start-up costs at each stage or if there is a long waiting time between fabricating the units and measuring their performance. This is the case in many agricultural experiments and also when the measured variable is product life.

If the experiment is conducted in stages, precautions must be taken to ensure that possible differences between the stages do not invalidate the results. Appropriate procedures to compare the stages must be included, both in the test plan and in the statistical analysis. For example, some standard test conditions, known as controls, may be included in each stage of the experiment.

Other Considerations

Many other questions must be considered in planning the experiment. In particular:

(1) What is the most meaningful way to express the controllable or independent variables? For example, should current density and time be taken as the experimental variables, or are time and the product of current density and time the real variables affecting response? Judicious selection of the independent variables often reduces or eliminates interactions between variables, thereby leading to a simpler experiment and analysis.

(2) What is a proper experimental range for the selected quantitative controllable variables? Assuming a linear relationship between these variables and performance, the wider the range of conditions or settings, the better are the chances of detecting the effects of the variable. (However, the wider the range, the less reasonable is the assumption of a linear relationship.) On the other hand, one generally would not want to conduct experiments appreciably beyond the range of physically or practically useful conditions. The selection of the range of the variables depends in part upon the ultimate purpose of the experiment: is it to learn about performance over a broad region or to search for an optimum condition? A wider range of experimentation would be more appropriate in the first case than in the second.

(3) What is a reasonable statistical model, or equation, to approximate the relationship between the independent variables and each response variable? Can the relationship be approximated by an equation involving linear terms for the quantitative independent variables and two-factor interaction terms only or is a more complex model, involving quadratic and perhaps even multifactor interaction terms, necessary? As indicated, a more sophisticated statistical model may be required to describe adequately relationships over a relatively large experimental range than over a limited range. A linear relationship may thus be appropriate over a narrow range, but not over a wide one. The more complex the assumed model, the more runs are usually required.

(4) What is the desired degree of precision of the final results and conclusions? The greater the desired precision, the larger is the required number of experimental runs.

(5) Are there any previous benchmarks of performance? If so, it might be judicious to include the benchmark conditions in the experiment to check the results.

(6) What statistical techniques are required for the analysis of the resulting data, and can these tools be rapidly brought to bear after the experiment has been conducted?

Formal Experimental Plans

After the preceding considerations have been attended to, a formal statistical test plan may be evolved. This might involve one of the standard plans developed by statisticians. Such plans are described in various texts (see General References) and are considered only briefly here.

Blocking Designs. Such designs use blocking techniques to remove the effect of extraneous variables from experimental error (see above). Well-known blocking designs include randomized and balanced incomplete block designs to remove the effects of a single extraneous variable, Latin square and Youden square designs to remove the effects of two extraneous variables, and Greco-Latin square and hyper-Latin square plans to remove the effects of three or more extraneous variables.

Complete Factorial and Fractional Factorial Designs. Such designs apply for two or more primary independent variables. Factors are varied simultaneously, rather than one at a time, so as to obtain information about interactions among variables and to obtain a maximum degree of precision in the resulting estimates. In complete factorial plans, all combinations of conditions of the independent variables are run. For example, a $3 \times 3 \times 2 \times 2$ factorial design requires running all 36 combinations of two variables with three conditions each and two variables with two conditions each. A fractional factorial design is often used when there is a large number of combinations of possible test points, arising from many variables or conditions per variable, and it is not possible or practical to run all combinations. Instead, a specially selected fraction is run. For example, a $2^{(6-1)}$ fractional factorial plan is one where there are six variables each at two conditions, resulting in a total of 64 possible combinations, but only a specially selected one half, or 32, of these combinations are actually run. Ref. 9 provides a comprehensive catalogue of fractional factorial designs.

An additional advantage of full factorial and fractional factorial designs is that by providing a comprehensive scanning of the experimental region they can often identify without any further analyses one or two test conditions that are better than any others seen previously.

Response Surface Designs. These are special multivariable designs for quantitative independent variables (temperature, pressure, etc). The relationship between the independent variables and the performance variable is fitted to develop prediction equations using the technique of least squares regression analysis. Popular response surface designs include the orthogonal central composite (or Box) designs and the rotatable designs.

Frequently, combinations of the preceding plans are encountered, such as a factorial experiment conducted in blocks or a central composite design using a fractional factorial base. There are also plans to accommodate special situations; eg, so-called mixture experiments apply for the situation where the experimental variables consist of the percentages of the ingredients that make up a material and must therefore add up to 100% (10).

BIBLIOGRAPHY

1. G. J. Hahn, *Chem. Technol.* **5,** 496 (Aug. 1975); **5,** 561 (Sept. 1975).
2. G. J. Hahn, *J. Quality Technol.* **9**(1), 13 (Jan. 1977).
3. G. J. Hahn, W. Q. Meeker, Jr., and P. I. Feder, *J. Quality Technol* **8**(3), 140 (July 1976).
4. F. X. Mueller and D. M. Olsson, *J. Paint Technol.* **43,** 54 (May 1971).
5. M. G. Natrella in J. M. Juran, ed., *Quality Control Handbook,* 3rd ed., McGraw-Hill Book Company, New York, 1974, Chapt. 27.
6. F. G. Tingey in I. M. Kolthoff and P. J. Elving, eds., *Treatise on Analytical Chemistry,* Pt. 1, Wiley-Interscience, New York, 1972, Chapt. 106.
7. G. J. Hahn, *Chem. Technol.* **7,** 630 (Oct. 1977).
8. G. J. Hahn, *Chem. Technol.* **8,** 164 (Mar. 1978).
9. G. J. Hahn and S. S. Shapiro *A Catalog and Computer Program for the Design and Analysis of Orthogonal Symmetric and Asymmetric Fractional Factorial Experiments, General Electric Company TIS Report 66C165,* General Electric Company, Schenectady, N.Y., May 1966.
10. R. D. Snee, *J. Quality Technol.* **3**(4), 159 (Oct. 1971).

General References

The following are books that deal mainly with the application of experimental design to scientific, industrial, and general situations. Thus, with some exceptions, books directed principally at educational, psychological, or related applications and those that deal mostly with the theory of experimental design or the analysis of experimental data are excluded. Most, but not all, of the books presume that the reader has had at least one or two introductory statistics courses. Items in quotation marks are taken directly from the book, generally from its preface.

V. L. Anderson and R. A. McLean, *Design of Experiments—A Realistic Approach,* Marcel Dekker, Inc., New York, 1974; an extensive exposition of experimental design is provided at a relatively elementary level. Most of the standard material is included, as well as detailed discussions of such subjects as nested and split-plot experiments. Special emphasis is given to restrictions on randomization.

G. E. P. Box, W. G. Hunter, and J. S. Hunter, *Statistics for Experimenters,* John Wiley & Sons, Inc., New York, 1978; this recent book by three eminent practitioners is an introduction to the philosophy of experimentation and the part that statistics plays in experimentation. Subtitled, *An Introduction to Design, Data Analysis, and Model Building,* it provides a highly readable and practically motivated introduction to basic concepts and methods of experimental design, without assuming a background in statistics. This book "is written for those who collect data and try to make sense of it, and it is an introduction to those ideas and techniques" that the authors "have found especially useful. Statistical theory is introduced as it becomes necessary. Readers are assumed to have no previous knowledge of the subject, the mathematics needed is elementary." The book relies heavily on numerous practical examples and case studies and provides detail about many elementary and a few more advanced designs. It is, however, an introductory treatment and more advanced situations are not discussed. The principle that "frequently conclusions are easily drawn from a well-designed experiment, even when rather elementary methods of analysis are employed" is strictly adhered to; heavy emphasis is placed on graphical tools for analyzing experimental results. In contrast, the more computationally involved analysis of variance tools receives relatively little attention. The four parts of the book deal with: comparing two treatments, comparing more than two treatments, measuring the effects of variables, and building models and using them.

V. Chew, ed., *Experimental Designs in Industry,* John Wiley & Sons, Inc., New York, 1958; this is a compilation of a series of papers given at a symposium on industrial applications of experimental design. It is written with the realization that the "basic designs originally constructed for agricultural experiments ... (and treated in most of the standard texts on experimental design) ... may not take into account conditions peculiar to industry." Included are papers by well-known practitioners on basic experimental designs, complete factorials, fractional factorials, and confounding; simple and multiple regression analysis; experimental designs for exploring response surfaces; experiences with incomplete block designs; experiences with fractional factorials; application of fractional factorials in a food research laboratory; experiences with response surface designs; and experiences and needs for design in ordnance experimentation.

W. G. Cochran and G. M. Cox, *Experimental Designs,* 2nd ed., John Wiley & Sons, Inc., New York, 1957; this is one of the earliest, best known, and most detailed books in the field. The classical experimental designs are described in a relatively simple manner. The treatment is oriented toward agricultural and biological applications. Extensive catalogues of designs are included, making the book a useful reference guide.

D. R. Cox, *Planning of Experiments,* John Wiley & Sons, Inc., New York, 1958; a simple survey of the principles of experimental design and some of the most useful schemes is provided. It tries "as far as possible, to avoid statistical and mathematical technicalities and to concentrate on a treatment that will be intuitively acceptable to the experimental worker, for whom the book is primarily intended." Emphasis is on basic concepts, rather than on calculations or technical details. Thus chapters are devoted to such topics as "Some Key Assumptions," "Randomization," and "Choice of Units, Treatments, and Observations."

C. Daniel, *Applications of Statistics to Industrial Experimentation,* John Wiley & Sons, Inc., New York, 1976; this book is based upon the personal experiences of the author, an eminent practitioner of industrial applications of experimental design. It provides extensive discussions and new concepts, especially in the areas of factorial and fractional factorial designs. "The book should be of use to experimenters who have some knowledge of elementary statistics and to statisticians who want simple explanations, detailed examples, and a documentation of the variety of outcomes that may be encountered." Some of the unusual features are chapters on "Sequences of Fractional Replicates" and "Trend-Robust Plans," and sections entitled "What is the Answer? (What is the Question?)," and "Conclusions and Apologies."

O. L. Davies and co-workers, *Design and Analysis of Industrial Experiments,* 2nd ed., Hafner Publishing Company, New York, 1956; a sequel to the authors' basic text *Statistical Methods in Research and Production,* this volume is highly readable, despite its size. It is directed specifically at industrial situations and chemical applications with three chapters devoted to factorial experiments and one to fractional factorial plans. A lengthy chapter discusses the determination of optimum conditions and response surface designs, which are associated with the name of G. Box, one of the seven co-authors. Some theoretical material is presented in appendices.

W. T. Federer, *Experimental Design: Theory and Application,* Macmillan Company, New York, 1955; "classical" plans of experimentation with heavy emphasis on blocking and factorial designs are described. Consideration is also given to covariance analysis, missing data, transformations, and to multiple range and multiple F tests. Orientation is toward biological and agricultural applications.

D. J. Finney, *Experimental Design and Its Statistical Basis,* The University of Chicago Press, Chicago, Ill., 1955; this small introductory volume is intended to be "in a form that will be intelligible to students and research workers in most fields of biology.... Even a reader who lacks both mathematical ability and acquaintance with standard methods of statistical analysis ought to be able to understand the relevance of these principles to his work." A chapter on biological assay is included.

R. A. Fisher, *Design of Experiments,* 8th ed., Hafner Publishing Company, New York, 1965; an early book on the subject and a classic in the field, this work includes detailed and readable discussions of experimental principles and confounding in experimentation. The treatment is directed at agricultural situations and many of the plans common in modern industrial situations are, thus, not considered in detail.

C. R. Hicks, *Fundamental Concepts in the Design of Experiments,* 2nd ed., Holt, Rinehart and Winston, Inc., New York, 1973; "it is the primary purpose of this book to present the fundamental concepts in the design of experiments using simple numerical problems, many from actual research work.... The book is written for anyone engaged in experimental work who has a good background in statistical inference. It will be most profitable reading to those with a background in statistical methods including analysis of variance." This work provides an intermediate level coverage of most of the basic experimental designs.

N. L. Johnson and F. C. Leone, *Statistics and Experimental Design in Engineering and the Physical Sciences,* 2nd ed, John Wiley & Sons, Inc., New York, 1977, although this is the second volume of a two volume book, it stands reasonably well on its own and provides an intermediate level treatment of various experimental plans. The calculational aspects of the analysis of variance associated with different designs are emphasized and some relatively complex actual situations are illustrated.

O. Kempthorne, *Design and Analysis of Experiments,* John Wiley & Sons, Inc., New York, 1952; this book is directed principally at readers with some background and interest in statistical theory. The early chapters deal with the general linear hypothesis and related subjects using matrix notation. Factorial experiments are discussed in detail. Chapters are devoted to fractional replication and split-plot experiments. Examples are generally taken from biology or agriculture.

C. Lipson and N. J. Sheth, *Statistical Design and Analysis of Engineering Experiments,* McGraw-Hill Book Company, New York, 1973; "this book is written in a relatively simple style so that a reader with a moderate knowledge of mathematics may follow the subject matter. No prior knowledge of statistics is necessary." Appreciably more space is devoted to the discussion of statistical analysis than to the planning of experiments. Treatments of such relatively nonstandard subjects (for an introductory text) as accelerated experiments, fatigue experiments, and renewal analysis are included.

W. Mendenhall, *Introduction to Linear Models and the Design and Analysis of Experiments,* Wadsworth Publishing Company, Belmont, Calif., 1968; a readable review and introduction to basic concepts and the most popular experimental designs is provided without going into extensive detail. In contrast to most other books on the design of experiments, the emphasis in the development of many of the stated models and analysis methods is principally on a regression rather than an analysis of variance viewpoint, thus providing a more modern outlook.

J. L. Myers, *Fundamentals of Experimental Design,* 2nd ed., Allyn and Bacon, Inc., Boston, Mass., 1972; this book is intended to "provide a reasonably sound foundation in experimental design and analysis .... The reader should be familiar with material usually covered in a one semester introductory statistics course." Although the treatment is quite general, many of the examples are taken from psychology.

R. H. Myers, *Response Surface Methodology,* Allyn and Bacon, Inc., Boston, Mass., 1971; this is a specialized book devoted to a detailed exposition of one type of experimental design, namely response surface plans. Such designs are appropriate when independent variables on a quantitative scale, such as temperature, pressure, etc, are to be related to the dependent variable using regression analysis. (This is in contrast to the situation where some or all of the independent variables are on a qualitative scale, such as operators, machines, batches of material etc.) It is frequently desired to determine the combination of the independent variables which leads to the optimum response. "The primary purpose of *Response Surface Methodology* is to aid the statistician and other users of statistics in applying response surface procedures to appropriate problems in manỳ technical fields .... It is assumed that the reader has some background in matrix algebra, elementary experimental design, and the method of least squares ...."

D. C. Montgomery, *Design and Analysis of Experiments,* John Wiley & Sons, Inc., New York, 1976; this "introductory textbook dealing with the statistical design and analysis of experiments ... is intended for readers who have completed a first course in statistical methods." It provides a basic treatment of standard experimental plans and techniques for analyzing the resulting data.

K. C. Peng, *The Design and Analysis of Scientific Experiments,* Addison-Wesley Publishing Company, Reading, Mass., 1967; this book is subtitled *An Introduction with Some Emphasis on Computations.* "It is written primarily for statisticians, computer programmers, and persons engaged in experimental work who have some background in mathematics and statistics. The mathematical background should include calculus and elementary matrix theory. The statistical background should be equivalent to a one-year course in statistics."

Elementary Statistics

O. L. Davies and P. L. Goldsmith, *Statistical Methods in Research and Production,* rev. ed., Longman, London, Eng., 1976.
I. Miller and J. E. Freund, *Probability and Statistics for Engineers,* 2nd ed., Prentice-Hall, Inc., Englewood Cliffs, N.J., 1977.
M. B. Natrella, *Experimental Statistics, National Bureau of Standards Handbook 91,* U.S. Government Printing Office, Washington, D.C., 1963.

Gerald J. Hahn
General Electric Company

DETERGENCY. See Surfactants and detersive systems.

DETONATING AGENTS. See Explosives.

DEUTERIUM AND TRITIUM

Deuterium, 539
Tritium, 554

The element hydrogen has three known isotopes. These hydrogen species are identical in atomic number, ie, they have identical extranuclear electronic configurations ($1s^1$), but they differ in nuclear mass. Over 99.98% of the hydrogen in nature has a nucleus consisting of a single proton and, therefore, has mass 1 (symbol ^1H). Two heavier isotopes of hydrogen, present in small amount in nature, of mass 2 and 3 are also known; these have nuclei consisting of one proton and one neutron (deuterium) or two neutrons (tritium). The three isotopes of hydrogen resemble each other closely in chemical and physical properties, but because the ratios of their masses are the largest for any set of isotopes in the periodic table, differences in chemical and physical properties exist that are larger than those encountered in any other set of isotopes. Whereas ordinary hydrogen and deuterium are stable isotopes, tritium is unstable and its nucleus undergoes radioactive decay (see Radioactive elements; Radioisotopes). The two heavy isotopes of hydrogen are always present to a small extent in any compound or substance containing the light isotope of hydrogen. Because both isotopically pure deuterium and tritium can now be manufactured on a large industrial scale, a large variety of isotopically pure deuterium and tritium compounds is readily available. Heavy water, D_2O, is the most important compound of deuterium and the only form of deuterium produced and used on a large scale.

DEUTERIUM

Deuterium [7782-39-0] (symbol ^2H or D) occurs in nature in all hydrogen-containing compounds to the extent of about 0.0145 atom %. Small but real differences in the deuterium content of water from various sources (rain, snow, glaciers, freshwater, seawater from different oceans) can readily be detected, and variations in the natural abundance of deuterium resulting from evaporation, precipitation, and molecular exchange make it possible to draw far-reaching conclusions about the genesis and geological history of natural waters.

Deuterium was first isolated in relatively pure form by Urey and co-workers at Columbia University in 1931 (1), and nearly pure D_2O was prepared by G. N. Lewis shortly thereafter by electrolysis (2). Numerous studies on the comparative properties of ^1H and D were initiated, and applications of deuterium as a tracer for the path of hydrogen in biological systems were developed and became widely used. In physical organic chemistry, the differences in rates of reaction between corresponding ^1H and D compounds became an important tool for the elucidation of organic reaction mechanisms. The significance of deuterium isotope effects in a biological context attracted attention very early (3–4). The discovery in 1959 that it was possible, contrary to earlier conclusions stated in the literature, to grow fully deuterated organisms (5–7) opened new areas of isotope chemistry and biology for exploration.

The recognition in 1940 that deuterium as heavy water has nuclear properties that make it a highly desirable moderator and coolant for nuclear reactors (qv) (8–9) fueled by uranium (qv) of natural isotopic composition stimulated the development of industrial processes for the manufacture of heavy water. Between 1940 and 1945 four heavy water production plants were operated by the United States Government to produce small amounts of heavy water for nuclear research. One of these plants was located in Canada at Trail, British Columbia, and three were at the U.S. Army Ordinance Works operated by the DuPont Company at Morgantown, West Virginia; Childersburg, Alabama; and Dana, Indiana. The plant at Trail used chemical exchange between hydrogen gas and steam for the initial isotope separation followed by electrolysis for final concentration. The three plants in the United States used vacuum distillation of water for the initial separation followed by electrolysis. A total of 32 metric tons of D_2O was produced by these four plants before they were shut down in 1945. Details of these plants and their operations have been given by Murphy (10).

In 1950 construction of a truly large-scale facility was initiated at the Savannah River site near Aiken, South Carolina, to produce heavy water for the nuclear reactors operated there. The newly developed dual temperature exchange of deuterium between hydrogen sulfide and water (the GS process) was selected (11–12) for isotopic enrichment of deuterium. This plant was designed to produce 450 metric tons of heavy water per year, and a plant of similar capacity was constructed at Dana, Indiana. At present, the plant at Savannah River is the only heavy water plant in operation in the United States, producing about 160 t of heavy water annually. The Canadian Government, because of its commitment to heavy water-moderated natural uranium reactors, has plants planned, under construction, or operating with a total capacity of about 1450 t D_2O/yr (13). France, Norway, Switzerland, and India also have small production plants. The continued large-scale production of heavy water is closely coupled to the large-scale use of heavy water in nuclear reactor technology. As the economics of such reactors is very complex and not subject to normal economic considerations, prediction of future trends is difficult. Unless new large-scale uses for deuterium emerge, however, there would appear to be no compelling reasons for the expansion of deuterium production facilities.

Physical Properties

Although the chemical and physical properties of all isotopes of an element are qualitatively the same, there are quantitative differences among them. The physical and chemical differences between the hydrogen isotopes are relatively much greater than those among the isotopes of all other elements because of their large relative differences in mass (H:D:T = 1:2:3).

As in the case of hydrogen and tritium, deuterium exhibits nuclear spin isomerism (14). However, the spin of the deuteron [12597-73-8] is 1 instead of $1/2$ as in the case of hydrogen and tritium. As a consequence, and in contrast to hydrogen, the ortho form of deuterium is more stable than the para form at low temperatures, and at normal temperatures the ratio of ortho- to para-deuterium is 2:1 in contrast to the 3:1 ratio for hydrogen.

The following list of physical and thermodynamic properties of elemental hydrogen and deuterium and of their respective oxides illustrates the effect of isotopic mass differences.

Properties of Light and Heavy Hydrogen. Vapor pressures from the triple point to the critical point for hydrogen, deuterium, tritium and their diatomic combinations are listed in Table 1 (15). These data are derived by an analysis of published experi-

Table 1. Vapor Pressures, Triple Points, and Critical Points of Hydrogen Isotopes[a,b]

Factor	e-H_2	n-H_2	HD	HT	e-D_2	n-D_2	DT	n-T_2
triple point, K	13.81	13.96	16.60	17.62	18.69	18.73	21.71	20.62
vapor pressure, kPa[c]	7.04	7.20	12.37	14.59	17.13	17.14	19.43	21.60
20.0	93.22	90.31	51.02	38.50	29.66	29.41	21.98	
22.5	185.1	179.1	112.3	89.49	71.82	71.17	56.46	44.80
25.0	327.65	316.63	214.42	177.66	147.19	145.79	120.82	100.03
27.5	534.07	515.13	369.98	315.36	268.02	265.28	227.11	194.10
30.0	819.56	788.49	592.74	515.65	447.55	442.56	388.42	339.88
32.5	1203.4	1153.8	898.50	792.91	700.58	691.91	619.04	548.47
35.0			1307.4	1164.0	1044.7	1030.0	935.27	843.06
critical point, K	32.99	33.24	35.91	37.13	38.26	38.35	39.42	40.44
vapor pressure, kPa[c]	1293.9	1298.0	1484.41	1570.53	1649.59	1664.79	1773.18	1850.24

[a] Adapted from Ref. 15.
[b] The prefixes e- and n- refer to equilibrium and normal states. For T, data are available only for the normal state. The equilibrium state for these substances is the low temperature ortho–para composition existing at 20.39 K, the normal boiling point of normal hydrogen. The normal state is the ortho–para composition, which is essentially constant above 200 K.
[c] To convert kPa to mm Hg, multiply by 7.5.

mental data (16–21). Data are presented for the equilibrium and normal states. The equilibrium state for these substances is the low temperature ortho–para composition existing at 20.39 K, the normal boiling point of normal hydrogen. The normal state is the high temperature ortho–para composition, which remains essentially constant, above 200 K.

Brickwedde and co-workers (22) have tabulated thermodynamic data on H_2, HD, and D_2, including values for entropy, enthalpy, free energy, and specific heat. Extensive PVT data are also presented in ref. 22 as are data on the equilibrium–temperature behavior of the ortho and para forms of H_2 and D_2. Some physical properties of liquid H_2 and D_2 at 20.4 K are presented in Table 2.

Properties of Light and Heavy Water. Selected physical and thermodynamic properties of light and heavy water are listed in Table 3 (23).

Table 2. Some Properties of Liquid H_2 and D_2 at 20.4 K

Property	Equilibrium H_2 (0.21% o-H_2)	Equilibrium D_2 (97.8% o-D_2)
density, g/L	70	169
viscosity, mPa·s(= cP)	1.4×10^{-2}	4.0×10^{-2}
surface tension, mN/m (= dyn/cm)	2.1_7	3.7_2
thermal conductivity, W/(cm·K)	11.6	12.6_4
dielectric constant	1.22_6	1.27_5
Property	Equilibrium n-H_2 (75% o-H_2)	Equilibrium n-D_2 (67% o-D_2)
molar volume at 20 K, mL	28.3	23.5
heat of vaporization, J/mol[a]	904	1226
heat of fusion, J/mol[a]	117	197

[a] To convert J to cal, divided by 4.184.

Table 3. Physical Properties of Light and Heavy Water[a]

Property	H_2O	Ref.[a]	D_2O	Ref.[b]
molecular weight, ^{12}C scale	18.015		20.028	
melting point, T_m, °C	0.00		3.81	
triple point, T_{tr}, °C	0.01		3.82	
temp of max. density, °C	3.98		11.23	
normal boiling point, T_b, °C	100.00		101.42	
critical constants		25		26
temperature, °C	374.1		371.1	
pressure, MPa[c]	22.12		21.88	
volume, cm^3/mol	55.3		55.0	
density at 25°, g/cm^3	0.99701		1.1044	
ΔH fusion at T_m, kJ/mol[d]	6.008		6.339	22
ΔS fusion at T_m, J/K[d]	22.0		22.6$_7$	
ΔH_{vap} at 3.82°C, kJ/mol[d]	44.77		46.48	
ΔE_{vap} at 3.82°C, kJ/mol[d]	42.47	e	44.18	e
ΔS_{vap} at 3.82°C, J/K[d]	161.62		167.8	
ΔH_{subl} at T_{tr}, kJ/mol[d]	50.92		52.84	
ΔE_{subl} at T_{tr}, kJ/mol[d]	48.66	e	50.54	
ΔS_{subl} at T_{tr}, J/K[d]	186.4		190.7	
$\Delta H°_{vap}$ at 25°C, kJ/mol[d]	44.02	27	45.39	
$\Delta G°_{vap}$ at 25°C, kJ/mol[d]	8.62	27	8.95	25
$\Delta S°_{vap}$ at 25°C, J/K[d]	118.8	e	122.25	e
$\Delta E°_{vap}$ at 25°C, kJ/mol[d]	41.55	e	42.93	e
$(\Delta H°_f)_{298}$ (liquid), kJ/mol[d]	−285.9	28	−294.6	27
$(\Delta G°_{298})$ (liquid), kJ/mol[d]	−237.2	28	−243.5	27
$S°_{298}$ (liquid), J/K[d]	70.08	28	76.11	27
$(\Delta H°_f)_{298}$ (gas), kJ/mol[d]	−241.8	28	−249.2	27
$(\Delta G°_f)_{298}$ (gas), kJ/mol[d]	−228.6	28	−234.6	27
$S°_{298}$ (gas), J/K[d]	188.8	28	198.3	29
$H°_{298} − H°_0$ (gas), kJ/mol[d]	9.908	28	9.954	27
C_p (liquid at 25°C), J/(mol·K)[c]	75.27		84.35	
C_v (liquid at 25°C), J/(mol·K)[c]	74.48		83.7	
vapor pressure (liquid at 25°C), kPa[f]	3.166		2.734	
molar volumes, cm^3/mol				
solid at T_m	19.65		18.679	
liquid at T_m	18.018		18.118	
liquid at T_{max} density	18.016		18.110	
liquid at 25°C	18.069		18.134	
coefficients of thermal expansion, °C^{-1}				
solid at T_m	1.39×10^{-4}		1.39×10^{-4}	
liquid at T_m	-5.9×10^{-5}		-3.2×10^{-5}	
liquid at 25°C	26.2×10^{-5}		21.8×10^{-5}	
compressibility at 20°C, Pa^{-1} [g]	4.45	25	4.59	25
crystallographic parameters at 0°C, nm		30		30
a	0.45228		0.45258	
c	0.73673		0.73689	
c/a	0.1629		0.1628	
length of the hydrogen bond, nm	0.2765		0.2760	
dipole moment, D				
benzene soln at 25°C	1.76		1.78	
vapor at 100–200°C	1.84		1.84	
dielectric const at 25°C	78.304		77.937	
refractive index, n_D^{20}	1.3330		1.3283	
polarizability of vapor near 100°C, cm^3/mol	58.5		61.7	
viscosity at 25°C, mPa·s(= cP)	0.8903	31	1.107	
surface tension at 25°C, mN/m(= dyn/cm)	71.97		71.93	

Table 3 (*continued*)

Property	H$_2$O	Ref.[a]	D$_2$O	Ref.[b]
moments of inertia, 10^{40} g-cm^2				
I_e^A	1.0224		1.833	
I_e^B	1.9180		3.841	
I_e^C	2.9404		5.674	
vibrational fundamentals, cm^{-1}		32		32
ν_1	3657.05		2671.69	
ν_2	1594.59		1178.33	
ν_3	3755.79		2788.03	
ionization const at 25°C	1.01×10^{-14}	33	1.11×10^{-15}	33
heat of ionization at 25°C, kJ/mol[d]	56.27	33	60.33	33

[a] Ref. 23.
[b] Unless otherwise indicated, data are from Nemethy and Sheraga (24).
[c] To convert MPa to atm, multiply by 10.1.
[d] To convert J to cal, divide by 4.187.
[e] Calculated from data given elsewhere in this table.
[f] To convert kPa to mm Hg, multiply by 7.5.
[g] To convert Pa to bar, multiply by 10^{-5}.

Structure. Arnett and McKelvey (23) concluded that both light and heavy water at room temperature are highly structured (extensively hydrogen bonded) and that D$_2$O is the more structured. This is reflected in a higher temperature of maximum density, a higher boiling point, and a generally lower solubility of inorganic salts in heavy water (34). At room temperature, the viscosity of D$_2$O is about 25% higher than that of H$_2$O (35), again arguing for more structure in D$_2$O. Nonpolar solutes elicit greater structure-making properties from D$_2$O than for H$_2$O, and structure-breaking salts disrupt more structure in D$_2$O because there is more structure to break. Structure breaks down faster in D$_2$O than H$_2$O with increase in temperature, so that D$_2$O may be more structured or less structured than H$_2$O, depending on the temperature.

Ionic Equilibria. The ion product constant of D$_2$O is 1.11×10^{-15} at 25°C compared to a value of 1.01×10^{-14} for H$_2$O, a difference of an order of magnitude (33,36–37). Covington and co-workers (38) have established the relationship pD = pH + 0.41 (molar scale; 0.45 molal scale) for pD in the range 2–9 as measured by a glass electrode standardized in H$_2$O. For many phenomena strongly dependent on hydrogen ion activity, as is the case in many biological contexts, the difference between pH and pD may have a large effect on the interpretation of experiments.

Deuterium as Neutron Moderator. Deuterium has very desirable properties as a moderator for neutrons. A good moderator is a material in which there is, on the average, a considerable decrease in neutron energy per collision (the logarithmic energy decrement ξ is large) one in which the probability of a scattering interaction with neutrons (the macroscopic scattering cross section Σ_s) is large, and finally a material whose absorption cross section for neutrons Σ_a is low. These factors all combine in a term called the moderating ratio, the ratio of the slowing-down power of the material to its neutron-absorption characteristics. As illustrated in Table 4 heavy water has a much greater moderating ratio than any of the other materials commonly used as moderators (39).

Table 4. Properties of Neutron Moderators

Moderator	Slowing-down power $\xi \times \Sigma_s$, cm^{-1}	Macroscopic absorption cross section Σ_a, cm^{-1}	Moderating ratio, $\xi \times \Sigma_s/\Sigma_a$
water	1.28	2.2×10^{-2}	58
heavy water	0.18	8.5×10^{-6}	21,000
helium	10^{-5}	2.2×10^{-7}	45
beryllium	0.16	1.2×10^{-3}	130
graphite	0.065	3.3×10^{-4}	200

Kinetic Isotope Effects. The principal difference in chemical behavior between H and D derives from the generally greater stability of chemical bonds formed by D. Thus a C—D bond is less reactive than a comparable C—H bond, a greater amount of energy is required to activate a C—D bond for reaction, and in general reactions involving the rupture of a C—D bond proceed considerably more slowly than for a comparable C—H bond. The most important factor contributing to the difference in bond energy and the kinetic isotope effect is the lower (5.021–5.275 kJ/mol) (1.2–1.5 kcal/mol) zero-point vibrational energy for D bonds. A comprehensive quantum and statistical mechanical theory of kinetic isotope effects has been developed (40). Kinetic isotope effects are particularly important in the elucidation of organic reaction mechanisms, and are the subject of a large and growing literature (41).

Biological Effects of Deuterium. Replacement of more than one third of the hydrogen by deuterium in the body fluids of mammals or two thirds of the hydrogen in higher green plants has catastrophic consequences for the organisms. At lower deuterium levels in higher plants, growth is markedly slowed. In mice and rats low levels of deuterium result in sterility, and at higher concentrations neuromuscular disturbances, fine muscle tumors, and a tendency to convulsions can be noted (42–43). Thomson found impairment of kidney function, anemia, disturbed carbohydrate metabolism, central nervous system disturbances, and altered adrenal function in deuterated mice (44–45). Hemoglobin and red blood cell count, serum glucose, and cholesterol all decrease in deuterated dogs (43).

Extensive replacement of H by D in living organisms is, however, not invariably fatal to living organisms. Numerous green and blue–green algae have been grown in which >99.5% of the hydrogen has been replaced by D. These fully deuterated organisms can be used to start a food chain to provide fully deuterated nutrients for organisms that have more demanding nutritional requirements. Numerous varieties of bacteria, molds, fungi, and even a protozoan have been successfully grown in fully deuterated form. These organisms of unnatural isotopic composition and the deuterated compounds that can be extracted from them have found uses in many areas of biological research (46).

Production of Heavy Water

Because of the low natural abundance of deuterium, very large amounts of starting material, which is water, must be processed to produce relatively small amounts of highly enriched deuterium. No water or other hydrogen compound has been found either in nature or as a by-product of an industrial operation that is significantly enriched in deuterium. The cost of subsequent enrichment to 99% is negligible compared to the costs incurred in the initial enrichment from natural abundance to 1%. For small-scale preparations, a highly efficient but very expensive process such as electrolysis can be used. For large-scale use, however, a high enrichment factor per stage is of only secondary importance to the overall costs of operation both in power and in capital investment. The isotope separation methods that have attracted the greatest interest include chemical exchange between water and hydrogen sulfide, hydrogen and water, and hydrogen and ammonia; distillation of water or hydrogen; and electrolysis of water in combination with other procedures (47).

Chemical Exchange Processes. Isotope exchange reactions (26) between hydrogen gas and water or ammonia, and between water and hydrogen sulfide provide the basis for the most efficient large-scale methods known for the concentration of deuterium. Equilibrium constants for these reactions are shown in Table 5. These equilibria are temperature dependent. The efficiency of chemical exchange processes can be increased by taking advantage of the difference in the equilibrium constants as a function of temperature in the form of a dual-temperature exchange process.

In the dual-temperature H_2O/H_2S process (50–51), exchange of deuterium between $H_2O(l)$ and $H_2S(g)$ is carried out at pressures of ca 2 MPa (20 atm). At elevated temperatures deuterium tends to displace hydrogen (^1H) in the hydrogen sulfide and thus concentrates in the gas. At lower temperatures the driving force is reversed and the deuterium concentrates in H_2S in contact with water on the liquid phase.

The deuterium exchange reactions in the H_2S/H_2O process (the GS process) occur in the liquid phase without the necessity for a catalyst. The dual-temperature feature of the process is illustrated in Figure 1a. Dual-temperature operation avoids the necessity for an expensive chemical reflux operation that is essential in a single-temperature process (11,52) (Fig. 1b).

As shown in Figure 1a, the basic element of the H_2S/H_2O process is a pair of gas–liquid contacting towers each containing a number of sieve or bubble-cap plates (see Distillation). The cold tower operates at a temperature of 30°C and the hot tower at 120–140°C. Water entering this system flows downward through the cold tower and then through the hot tower countercurrent to a stream of hydrogen sulfide gas at 1896 kPa (275 psig). The water is progressively enriched in deuterium as it passes through the cold tower and progressively depleted as it passes through the hot tower, eventually leaving the hot tower at a concentration below that at which it entered the system. The HDO and HDS that build up within the process are withdrawn from the base of the cold tower and top of the hot tower, respectively, by withdrawing a fraction of the water and gas flow. These enriched fractions are fed to a succeeding stage for further

Table 5. Chemical Exchange Reactions

Reaction	Equilibrium constant	Ref.
$H_2O(l) + HDS(g) \rightleftharpoons HDO(l) + H_2S(g)$	$K_{30} = 2.18$ $K_{130} = 1.83$	48
$NH_3(l) + HD(g) \underset{}{\overset{catalyst}{\rightleftharpoons}} NH_2D(l) + H_2(g)$	$K^*_{-50} = 6.60$ $K^*_0 = 4.42$	49
$H_2O(g) + HD(g) \underset{}{\overset{catalyst}{\rightleftharpoons}} HDO(g) + H_2(g)$	$K_{25} = 3.62$ $K_{125} = 2.43$	48

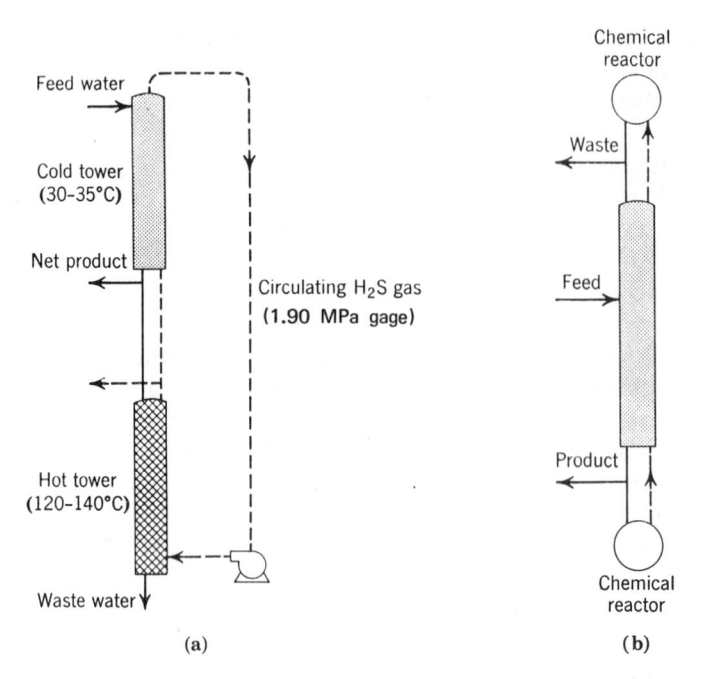

Figure 1. Simplified flow diagrams for H_2S/H_2O heavy-water processes. To convert MPa to atm, divide by 0.100. Courtesy of George Newnes, Ltd., London.

concentration. The hydrogen sulfide gas, which acts as a transport medium for the deuterium, circulates in a closed loop within the several stages of the process. About 20% of the deuterium in natural water can be economically extracted in this manner.

In producing tonnage quantities of heavy water, large equipment must be used to perform the initial separation because of the low feed concentration and the high throughput. To produce one metric ton of D_2O the plant must process 41,000 t of water, and must cycle 135,000 t of hydrogen sulfide. By appropriate process staging, the amount of materials handled and the size of equipment can be reduced almost in proportion to the increase in concentration of heavy water through the plant. This staging may be done with any of the possible heavy-water processes, or the process may be changed from stage to stage. In a properly staged plant most of the capital investment is incurred in the first hundredfold concentration gain, ie, from about 0.015 to 1.5 mol% D_2O.

In the heavy-water plants constructed at Savannah River and at Dana, these considerations led to designs in which the relatively economical GS process was used to concentrate the deuterium content of natural water to about 15 mol%. Vacuum distillation of water was selected (because there is little likelihood of product loss) for the additional concentration of the GS product from 15 to 90% D_2O, and an electrolytic process was used to produce the final reactor-grade concentrate of 99.75% D_2O.

In addition to the large contacting towers, a large amount of heat-recovery equipment was required to improve thermal efficiency. As illustrated in the process flow diagram for one of the Savannah River production units (Fig. 2), there were three principal heat-recovery circuits. These circuits are (1) the humidifier, comprising the

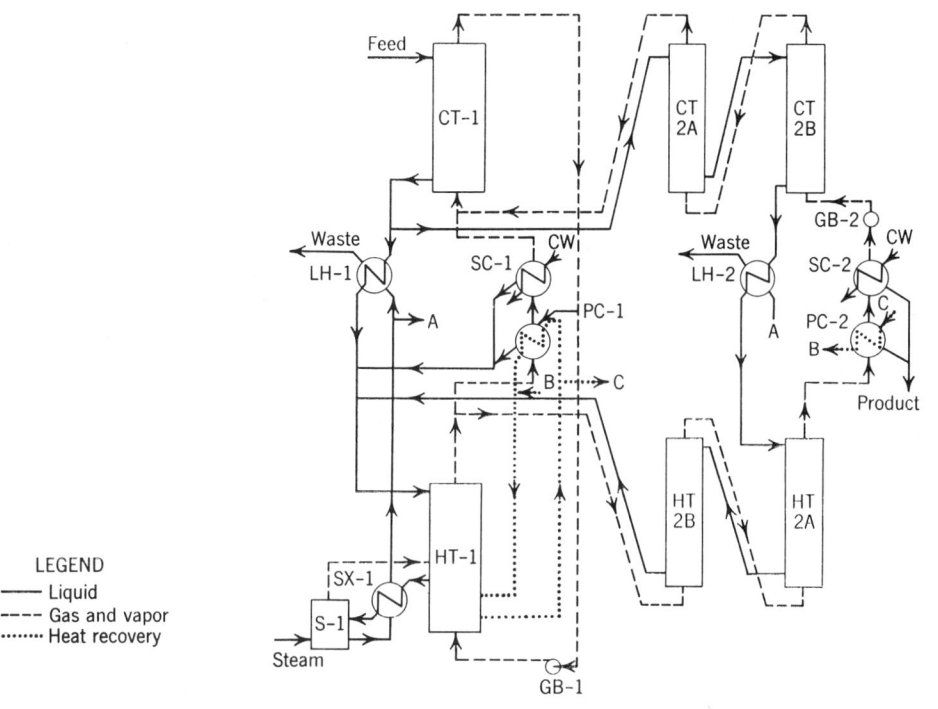

Figure 2. Diagram of the Savannah River GS process. CT, cold tower; HT, hot tower; LH, liquor heater; PC, primary condenser; SC, secondary condenser; SX, stripper exchanger; GB, gas blower; S, stripper; suffix 1 denotes first stage, and suffix 2 denotes second stage. Courtesy of *Chemical Engineering Progress*.

bottom ten plates of the hot tower and the primary gas coolers (PCs); (2) the liquid heaters (LHs); and (3) the waste–stripper feed preheaters (SXs). The most important of these is the humidifier circuit in which about half of the heat contained in the gas stream leaving the top of the hot tower is exchanged with a recirculating water stream that in turn heats and humidifies the gas entering the base of the hot tower. To make up for heat losses, steam is added to the hot tower after first being used in the stripper to remove the dissolved H_2S from the waste.

A dual-temperature system requires twice the number of separating stages (hot and cold columns) required for a single temperature system with the same effective separation factor. However, the requirement for a refluxing step is eliminated. The heat energy needed to maintain the temperature difference between the hot and cold columns can be minimized by heat exchange at appropriate points to preheat the gas and liquid streams entering the hot column and to precool the gas entering the cold column (53).

All of the major installations currently producing highly enriched deuterium in quantities greater than 20 metric tons D_2O per year use the H_2S/H_2O dual-temperature exchange system.

The fundamental parameters for the ammonia–hydrogen exchange (Table 5) are much more favorable than the corresponding factors in the H_2S/H_2O system, but the exchange reaction must be catalyzed to achieve a useful rate of exchange. The discovery (54) that the amide ion NH_2^- (produced by addition of alkali metal to liquid ammonia) is an efficient catalyst for the NH$_3$/H$_2$ exchange has stimulated intensive interest in this system (55–56). A dual-temperature system operating with a hot column at 70°C (single stage separation factor of 2.9) and a cold column at −40°C (single stage separation factor of 5.9) would have an effective separation factor of 2.0, which would permit extraction of 50% of the deuterium from the ammonia feed. Catalysis of the exchange by potassium amide is sufficiently effective even at −40°C to attain equilibrium in reasonably sized exchange columns. A single-temperature plant has been operated in France to produce about 20 t D$_2$O/yr (57). The major limitation on the use of this process has been the availability of sufficient quantities of ammonia for plant feed. Even a plant producing 1000 t ammonia/d would provide sufficient feed only to permit production of 60–70 t D$_2$O/yr. A concept using an enrichment stripping system with a regeneration column has been developed (48–49) in which water is the deuterium feed via a hydrogen–water exchange step. This would, in principle, allow H$_2$/NH$_3$ chemical exchange plants of unlimited production capacity.

A variant of the H$_2$/NH$_3$ chemical exchange process uses alkyl amines in place of ammonia. Hydrogen exchange catalyzed by NH$_2^-$ is generally faster with alkyl amines than ammonia and a dual-temperature flow sheet for a H$_2$/CH$_3$NH$_2$ process has been developed (58).

Chemical exchange between hydrogen and steam (catalyzed by nickel–chromia, platinum, or supported nickel catalysts) has served as a pre-enrichment step in an electrolytic separation plant (10,59). If the H$_2$(g)/H$_2$O(l) exchange could be operated as a dual-temperature process, it very likely would displace the H$_2$S/H$_2$O process. However, suitable catalysts for liquid-phase use have not been reported.

Distillation. Vacuum distillation (qv) of water was the first method used for the large-scale extraction of deuterium (10,47). Water contains the three molecular species H$_2$O, HDO, and D$_2$O (Fig. 3). From the equilibrium constant in the liquid phase it is evident that the distribution of ^{1}H and D is not statistical. The differences in vapor pressure between H$_2$O and D$_2$O are significant, and a fractionation factor (see footnote to Table 6) of 1.05 can be obtained at about 50°C. As the fractionation factor decreases with increasing temperature, vacuum distillation is used which can be carried out at low temperatures. However, low pressure distillation requires larger towers to handle a given mass flow rate of vapor at lower pressure. In practice, tower top pressures be-

Table 6. Distillation Process Requirements for the Production of Deuterium[a]

Requirement	Water distillation	Hydrogen distillation
deuterium content of feed, %	0.0149	0.0149
separation factor, α	1.05[b]	1.52[c]
temperature, K	323	23
pressure	ca 13.3 kPa (ca 100 mm Hg)	ca 0.16 MPa (ca 1.6 atm)
min no. stages	308	41
min reflux ratio	141,000	19,600
recovery, %	ca 5	90
mol feed/mol product	133,940	7,442
ratio of operating costs	11	1

[a] Ref. 50.
[b] $\alpha = \sqrt{p_{H_2O}/p_{D_2O}}$.
[c] $\alpha = \sqrt{p_{H_2}/p_{D_2}}$.

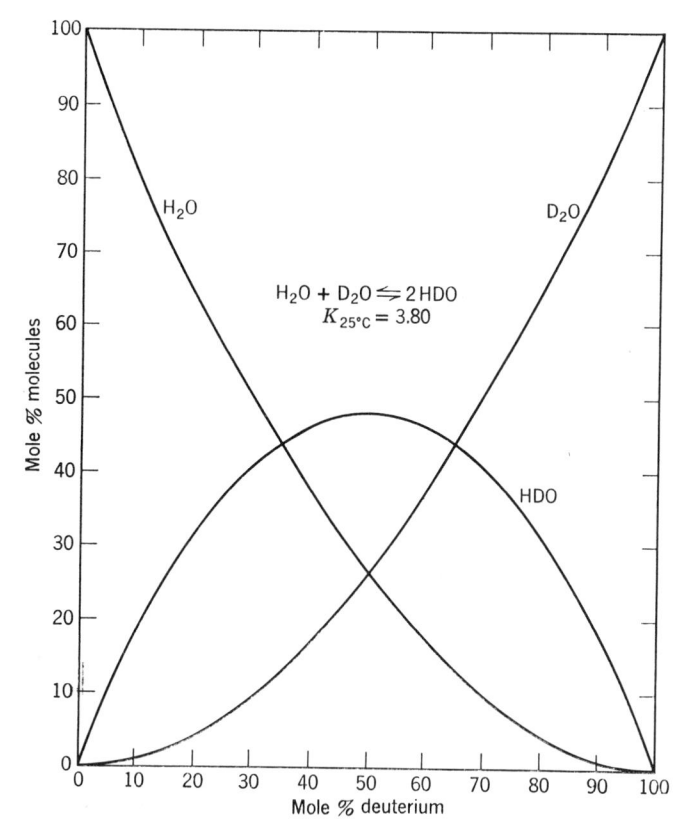

Figure 3. Equilibrium concentrations of H₂O, HDO, and D₂O vs overall concentration of deuterium in water.

tween 6.7–16.7 kPa (50–125 mm Hg) were used, corresponding to temperatures between 30–50°C and separation factors between 1.06–1.05. Maximum deuterium recovery in a water distillation plant was considerably less than 5%, and the deuterium was concentrated from an initial value of 0.0143 atom % to 87–91% atom %. Further concentration to 99.8% was achieved by electrolysis. Operating parameters for a water distillation plant for deuterium recovery are shown in Table 6. Distillation of water is now only used for final enrichment of D_2O (60).

Distillation of liquid hydrogen as a method for separating deuterium received early consideration (10,47) because of the excellent fractionation factor that can be attained, and the relatively modest power requirements. The cryogenic temperatures, and the requirement that the large hydrogen feed required be extremely pure (traces of air, carbon monoxide, etc, are solids at liquid hydrogen temperature) have been deterrents in the past to the use of this process (see Cryogenics).

Another problem in the liquid-hydrogen distillation process is the presence in ordinary hydrogen or deuterium gas of nuclear spin isomers. Ordinary hydrogen gas consists of three parts of ortho-H_2 (nuclear spins parallel) and one part para-H_2 (nuclear spins antiparallel) (see Hydrogen). Spontaneous transitions between the ortho and para states are strictly prohibited by quantum mechanical considerations. Equilibrium is attained only slowly unless catalyzed by a paramagnetic species (eg, oxygen absorbed on charcoal). Normal hydrogen fed to a cryogenic liquid distillation plant releases considerable heat in the conversion from the ortho to the para form, and care must be taken to eliminate materials of construction or impurities that can accelerate the ortho–para conversion.

Current developments in liquid hydrogen technology, and the possibility that hydrogen may achieve large-scale use as a fuel, brightens the prospects for liquid hydrogen distillation (see Hydrogen energy). A comparison of the operating characteristics of a water distillation plant with a liquid hydrogen distillation plant (Table 6) and other studies (61–62) indicate that hydrogen distillation may well become important as a sizeable source of deuterium if the production of hydrogen continues to expand.

Electrolysis. For reasons not fully understood (63), the isotope separation factor commonly observed in the electrolysis of water is between 7 and 8. Because of the high separation factor and the ease with which it can be operated on the small scale, electrolysis has been the method of choice for the further enrichment of moderately enriched H_2O–D_2O mixtures. Its usefulness for the production of heavy water from natural water is limited by the large amounts of water that must be handled, the relatively high unit costs of electrolysis, and the low recovery.

Economics. D_2O produced by the H_2S/H_2O route is offered for sale by the United States government at \$220/kg in bulk quantities. As there is no free market in D_2O, the fixed price may not have a simple relationship to the cost of production.

Methods of Analysis

The principal methods for determination of the deuterium content of hydrogen and water are based upon measurements of density, or on measurements of mass or infrared spectra. Other methods are based on proton magnetic resonance techniques (64–65), ^{19}F nuclear magnetic resonance (66), interferometry (67), osmometry (68), nuclear reaction (69), combustion (70), and falling drop methods (71).

Density. Measurement of the density of water by pycnometry is the classical method used by Kirshenbaum (35) in most of his original analytical work on establishing deuterium concentrations in heavy water. Very precise measurements can be made by this method, provided the sample is prepared free of suspended or dissolved impurities and the concentration of oxygen-18 in water is about 0.2 mol %. However, in nearly all heavy water manufactured since 1950 in the United States, the concentration of oxygen-18 is about 0.4 mol % as the result of its concentration in the water-distillation portion of the production facilities. With correction for oxygen-18, the sensitivity of the densimetric method is ±0.2% D_2O but the accuracy is not better than 0.03% D_2O because of an uncertainty of 30 ppm in the density of pure heavy water.

Mass Spectrometry. Mass spectrometric methods are used for both hydrogen gas and water. They are capable of determining directly the concentration of different isotopic species in samples of H_2O, HDO, and D_2O, eg, by measuring the relative abundance of ions of mass 18 and 19 or mass 19 and 20. The principal limitation of the method is the contamination of the sample-introduction system with water vapor. This contamination causes significant memory effects on succeeding samples. To obtain accurate analyses it is necessary to use a comparative standard with an isotopic concentration within 7% of that of the unknown. The precision of this method is 0.005–0.1 mol% over the range 0.05–15% and is 0.02 mol% in the range 99–100% D_2O.

Infrared Spectrophotometry. The isotope effect on the vibrational spectrum of D_2O makes infrared spectrophotometry the method of choice for deuterium analysis. It is as rapid as mass spectrometry, does not suffer from memory effects, and requires only ordinary laboratory equipment. Measurement at either the O–H fundamental vibration at 2.94 μm (O–H) or 3.82 μm (O–D) can be used. This method is equally applicable to low concentrations of D_2O in H_2O, or the reverse (72–73). Absorption in the near infrared can also be used (74) and this procedure is particularly useful, as it can be carried out in a conventional spectrophotometer (see Analytical methods; Infrared technology).

Uses

The only large-scale use of deuterium in industry is as a moderator (in the form of D_2O) for nuclear reactors (see Nuclear reactors). Because of its favorable slowing-down properties and its small capture cross section for neutrons, deuterium moderation permits the use of uranium containing the natural abundance of uranium-235, thus avoiding an isotope enrichment step in the preparation of reactor fuel. Heavy water-moderated thermal neutron reactors fueled with uranium-233 and surrounded with a natural thorium blanket offer the prospect of successful fuel breeding, ie, production of greater amounts of ^{233}U (by neutron capture in thorium) than are consumed by nuclear fission in the operation of the reactor. The advantages of heavy water-moderated reactors are difficult to assess. The economics of D_2O-moderated reactors is determined by the costs of isolating deuterium relative to the cost of enriching uranium, the inventory charges for the large amount of expensive D_2O required, and a number of other factors that could be strongly affected by new technological developments. For instance, if fast breeder reactors using plutonium-239 as fuel become important in energy production, the role of D_2O in nuclear technology would likely be adversely affected.

The nuclear properties of deuterium may become important in energy production by nuclear fusion:

$$^2_1D + ^2_1D \rightarrow ^3_1T + ^1_1H + 4.0 \text{ MeV}$$

$$^3_1T + ^2_1D \rightarrow ^4_2He + ^1_0n + 17.6 \text{ MeV}$$

$$^3_1T + ^1_1H \rightarrow ^4_2He + 19.6 \text{ MeV}$$

The tritium–deuteron reaction that yields helium-4 and a neutron is comparable in energy production to the tritium–proton reaction that forms helium-4. In these nuclear reactions, deuterium and/or tritium undergo nuclear fusion with the evolution of an enormous amount of energy (see Tritium; Fusion energy). It has been estimated (75) that the entire energy requirements of the United States in the year 2020 could be supplied by the D–D fusion reaction with 1000 metric tons of D_2 (5000 t D_2O) per year. The applications of deuterium to problems in biology and chemistry at the present time are small, but new large-scale uses of deuterium may evolve from such research.

BIBLIOGRAPHY

"Deuterium" under "Deuterium and Tritium" in *ECT* 2nd ed., Vol. 6, pp. 895–910, by J. F. Proctor, E. I. du Pont de Nemours & Co., Inc.

1. H. C. Urey, *Science* **78**, 566 (1933).

2. G. N. Lewis and R. T. MacDonald, *J. Chem. Phys.* **1**, 341 (1933).
3. G. N. Lewis, *Science* **79**, 151 (1934).
4. H. J. Morowitz and L. M. Brown, *U.S. Nat. Bur. Stand. Rep.* **2179**, (1953).
5. H. L. Crespi, S. M. Archer, and J. J. Katz, *Nature* **184**, 729 (1959).
6. W. Chorney and co-workers, *Biochim. Biophys. Acta* **37**, 280 (1960).
7. D. Kritchevsky, ed., *Ann. N.Y. Acad. Sci.* **84**, 573 (1960).
8. W. H. Zinn, and C. A. Trilling, *Proc. Int. Conf. Peaceful Uses At. Energy, 3rd Geneva* **28**, 209 (1964).
9. W. B. Lewis and T. G. Church in ref. 8, p. 1.
10. G. M. Murphy, ed., *Production of Heavy Water*, McGraw-Hill Book Co., Inc., New York, 1955, p. 4.
11. W. P. Bebbington and V. R. Thayer, *Chem. Eng. Progr.* **55**(9), 70 (1959).
12. *Technica Ed Economia Della Produzione Di Acqua Pesante*, Comitato Nazionale Energia Nucleare Symposium, Turin, Italy, Sept. 30, 1970.
13. *U.S. Atomic Energy Commission Document 1174-71*, U.S. Government Printing Office, Washington, D.C., 1971.
14. A. Farkas, *Orthohydrogen, Parahydrogen, and Heavy Hydrogen*, Cambridge University Press, Cambridge, Eng., 1935.
15. H. M. Mittelhauser and G. Thodos, *Cryogenics* **4**, 368 (1963).
16. A. S. Friedman, D. White, and H. L. Johnston, *J. Chem. Phys.* **19**, 126 (1951).
17. A. S. Friedman, D. White, and H. L. Johnston, *J. Am. Chem. Soc.* **73**, 1310 (1951).
18. D. White, A. S. Friedman, and H. L. Johnston, *J. Am. Chem. Soc.* **72**, 3565 (1950).
19. E. R. Grilly, *J. Am. Chem. Soc.* **78**, 843 (1951).
20. H. J. Hoge and R. D. Arnold, *J. Res. Natl. Bur. Stand.* **47**, 71 (1951).
21. H. J. Hoge and J. W. Lassiter, *J. Res. Natl. Bur. Stds.* **47**, 75 (1951).
22. H. W. Wooley, R. B. Scott, and F. G. Brickwedde, *J. Res. Natl. Bur. Stand.* **41**, 379 (1948).
23. E. C. Arnett and D. R. McKelvey in J. F. Coetzee and C. D. Ritchie, eds., *Solute–Solvent Interactions*, Marcel Dekker, Inc., New York, 1969, pp. 343–398.
24. G. Nemethy and H. A. Scheraga, *J. Chem. Phys.* **41**, 680 (1964).
25. E. Whalley, *Proc. Conf. Thermodynamics Transport Properties of Fluids, 10–12 July 1957*, Institute of Mechanical Eng., London, Eng., 1958, p. 15.
26. F. T. Barr and W. P. Drews, *Chem. Eng. Progr.* **56**, 49 (1960).
27. F. D. Rossini, J. W. Knowlton, and H. L. Johnston, *J. Res. Natl. Bur. Stand.* **24**, 369 (1940).
28. F. D. Rossini, *J. Res. Natl. Bur. Stand.* **22**, 407 (1939).
29. E. A. Long and J. D. Kemp, *J. Am. Chem. Soc.* **58**, 1829 (1936).
30. H. D. Megaw, *Nature* **134**, 900 (1934).
31. J. R. Coe and T. B. Godfrey, *J. Appl. Phys.* **15**, 625 (1944).
32. A. S. Friedman and L. Haar, *J. Chem. Phys.* **22**, 2051 (1954).
33. A. K. Covington, R. A. Robinson, and R. G. Bates, *J. Phys. Chem.* **70**, 3820 (1966).
34. E. C. Noonan, *J. Am. Chem. Soc.* **70**, 2915 (1948).
35. I. Kirschenbaum, *Physical Properties and Analysis of Heavy Water*, McGraw-Hill Book Co., Inc., New York, 1951, p. 54.
36. A. K. Covington, R. A. Robinson, and R. G. Bates, *J. Chem. Ed.* **44**, 635 (1967).
37. H. L. Clever, *J. Chem. Ed.* **45**, 231 (1968).
38. A. K. Covington and co-workers, *Anal. Chem.* **40**, 700 (1968).
39. S. Glasstone and A. Sesonske, *Nuclear Reactor Engineering*, D. Van Nostrand Co., Inc., Princeton, N.J., 1963, p. 134.
40. J. Bigeleisen in P. A. Rock, ed., *Isotopes and Chemical Principles*, American Chemical Society Symposium Series No. 11, 1975, pp. 1–43.
41. C. J. Collins and N. S. Bowman, eds., *Isotope Effects in Chemical Reactions*, American Chemical Society Monograph 167, Van Nostrand Reinhold Co., New York, 1971.
42. J. J. Katz and co-workers, *Am. J. Physiol.* **203**, 357 (1961).
43. D. M. Czajka and co-workers, *Am. J. Physiol.* **201**, 357 (1961).
44. J. F. Thomson, *Ann. N.Y. Acad. Sci.* **84**, 736 (1960).
45. J. F. Thomson, *Biological Effects of Deuterium*, Pergamon Press Ltd., Oxford, Eng., 1963.
46. J. J. Katz and H. L. Crespi in ref. 41, Chapt. 5, pp. 286–363.
47. M. Benedict in *Progress in Nuclear Energy, Series IV, Technology and Engineering*, Pergamon Press, Ltd., Oxford, Eng., 1956, pp. 3–56.

48. U. Schindewolf and G. Lang in ref. 12, pp. 77–88.
49. S. Walter and E. Nitschke in ref. 12, pp. 175–193.
50. J. Spevack in ref. 12, pp. 25–43.
51. U.S. Pats. 2,787,526 (Apr. 2, 1957), 2,895,803 (Sept. 29, 1950), 3,142,540 (Sept. 29, 1950), 4,008,046 (Mar. 21, 1971), J. Spevack (to U.S. Atomic Energy Commission).
52. H. London, ed., *Separation of Isotopes,* George Newnes, Ltd., London, Eng., 1961, p. 7.
53. W. Spindel in P. A. Rick, ed., *Isotopes and Chemical Principles,* American Chemical Society Symposium Series No. 11, 1975, pp. 77–100.
54. Y. Claeys, J. C. Dayton, and W. K. Wilmarth, *J. Chem. Phys.* **18,** 759 (1950).
55. M. Perlman, J. Bigeleisen, and N. Elliot, *J. Chem. Phys.* **21,** 70 (1953).
56. J. Bigeleisen in J. Kistenmaker, J. Bigeleisen, and A. O. C. Nier, eds., *Proceedings of the International Symposium on Isotope Separation,* North Holland Publishing Co., Amsterdam, The Netherlands, 1958, p. 121.
57. E. Roth and M. Rostain in ref. 12, pp. 69–74.
58. A. R. Bancroft and H. K. Ral in ref. 12, pp. 49–60.
59. Ref. 47, pp. 32–52.
60. P. Baertschi and W. Kuhn in ref. 47, pp. 57–61.
61. H. Gutowski in ref. 12, pp. 93–100.
62. H. Gutowski, *Kerntechnik* **12**(5/6), (1970).
63. B. E. Conway, *Proc. Royal Soc. London* **A247,** 400 (1958).
64. D. E. Leyden and C. N. Reilley, *Anal. Chem.* **37,** 1333 (1965).
65. W. Johnson and R. A. Keller, *Anal. Lett.* **2,** 99 (1969).
66. C. Deverell and K. Schaumberg, *Anal. Chem.* **39,** 1879 (1967).
67. J. Mercea, *Chem. Ing. Tech.* **41,** 508 (1969).
68. E. Lazzarini, *Nature* **204,** 875 (1964).
69. S. Amiel and M. Peisach, *Anal. Chem.* **34,** 1305 (1962).
70. R. N. Jones and M. A. MacKenzie, *Talanta* **7,** 124 (1960).
71. M. F. Clarke, *Anal. Biochem.* **31,** 81 (1969).
72. W. H. Stevens and W. Thurston, *Atomic Energy of Canada, Ltd., Report No. 295,* Chalk River, Ontario, Canada, 1954.
73. M. D. Turner and co-workers, *Microchem. J.* **8,** 395 (1964).
74. H. L. Crespi and J. J. Katz, *Anal. Biochem.* **2,** 274 (1961).
75. J. Powell and co-workers, *Brookhaven National Laboratory Report BNL-18430,* Upton, N.Y., 1972.

JOSEPH J. KATZ
Argonne National Laboratory

TRITIUM

Tritium [*10028-17-8*] is the name given to the hydrogen isotope of mass 3 (symbol ^3_1H or more commonly T). Its isotopic mass on the ^{12}C scale (^{12}C = 12.000000) is 3.0160497 (1). The molecular form is T_2, analogous to that for the other hydrogen isotopes. The tritium nucleus is energetically unstable and decays radioactively by the emission of a low-energy β particle. The half-life is relatively short (12.26 yr), and therefore tritium occurs in nature only in equilibrium with amounts produced by cosmic rays or man-made nuclear devices (see Radioactive elements; Radioisotopes).

Tritium was first prepared in the Cavendish Laboratory by Rutherford, Oliphant, and Harteck in 1934 (2–3) by the bombardment of deuterophosphoric acid with fast deuterons. The D-D nuclear reaction produced tritium ($^2_1D + {}^2_1D \rightarrow {}^3_1T + {}^1_1H$ + energy) but also produced some ^3He by a second reaction ($^2_1D + {}^2_1D \rightarrow {}^3_2He + {}^1_0n$ + energy). It was not immediately known which of the two mass-3 isotopes was radioactive. In 1939 Alvarez and Cornog (4) at Berkeley established that ^3He occurred in nature and was stable, and later proved that tritium was radioactive.

Physical Properties

Tritium is the subject of various reviews (5–7) and a recent book (8) provides a comprehensive survey of the preparation, properties, and uses of tritium compounds.

Selected physical properties for molecular tritium (T_2) are given in Table 1.

Calculated vapor pressure relationships of T_2, HT, and DT have been reported (9) and are listed in Table 1 of the accompanying article on Deuterium. An equation for the vapor pressure of solid tritium has been given (10) as:

$$\log p \text{ (kPa)} = 5.6023 - 88.002/T \quad (\log p_{\text{mm Hg}} = 6.4773 - 88.002/T)$$

Relative volatilities for the isotopic system deuterium–deuterium tritide–tritium have been found (12) to be 5–6% below the values predicted for ideal mixtures.

The three-phase region of D_2–DT–T_2 has been studied (13). All components appear miscible in both liquid and solid phases from 17 to 22 K. For a 50–50 mol % mixture of liquid D–T at ca 19.7 K, the gas phase will contain ca 42% T and the solid phase 52%.

The T–T bond energy has been estimated at 4.5881 eV (14).

The entropy of T_2 at 298.16 K is 164.82 kJ/mol (39.394 kcal/mol) and the specific heat is 29.20 J/(mol·K) (6.978 cal/(mol·°C)) (15).

Ortho-Para Tritium. As in the case of molecular hydrogen, molecular tritium exhibits nuclear spin isomerism. The spin of the tritium nucleus is $\frac{1}{2}$, the same as that for the hydrogen nucleus, and therefore H_2 and T_2 obey the same nuclear isomeric statistics (16). Below 5 K, molecular tritium is 100% para at equilibrium. At high temperatures (100°C) the equilibrium concentration is 25% para and 75% ortho. The kinetic parameters of conversion for T_2 at low temperatures are faster than rates at corresponding temperatures for H_2. In the solid phase the conversion of molecular

Table 1. Physical Properties of Tritium[a]

Property	Value
melting point, at 21.6 kPa[b], K	20.62
boiling point, 101 kPa (=1 atm), K	25.04
critical temperature, K	40.44[c]
critical pressure, MPa[d]	1.850
critical volume, cm³/mol	57.1 (calc.)
heat of sublimation, J/mol[e]	1640
heat of vaporization, J/mol[e]	1390
entropy of vaporization, J/(mol·K)[e]	54.0
molar density of liquid, mol/L	45.35[f,g]
	42.65[f,h]
	39.66[f,i]

[a] Unless otherwise indicated, values are from ref. 10.
[b] At the triple point (162 mm Hg).
[c] From ref. 9.
[d] To convert MPa to atm, divide by 0.101.
[e] To convert J to cal, divide by 4.184.
[f] From ref. 11.
[g] At the triple point.
[h] At 25 K.
[i] At 29 K.

tritium to a state of ortho-para equilibrium is 210 times as fast as that for molecular hydrogen.

The experimental and theoretical aspects of the radiation and self-induced conversion kinetics and equilibria between the ortho and para forms of hydrogen, deuterium, and tritium have been correlated (17). In general, the radiation-induced transitions are faster than the self-induced transitions.

Properties of T_2O. Some important physical properties of T_2O are listed in Table 2. Tritium oxide can be prepared by catalytic oxidation of T_2 or by reduction of copper oxide with tritium gas. T_2O even of low isotopic abundance (2–19% T) undergoes radiation decomposition to form HT and O_2. Decomposition continues when the water is frozen even at 77 K. Pure tritiated water irradiates itself at the rate of 10 MGy/d (10^9 rad/d). A stationary concentration of tritium peroxide (T_2O_2) is always present

Table 2. Physical Properties of T_2O

Property	Value	Reference
molecular weight (^{12}C scale)	22.032	
triple point, °C	4.49	18
temperature of maximum density, °C	13.4	
boiling point, °C	101.51	18
density at 25°C, g/cm³	1.2138	19
$\Delta H°_{vap}$ at 25°C, kJ/mol[a]	ca 45.81	18
liquid vapor pressure at 25°C, kPa[b]	2.64	
vibrational fundamentals, cm^{-1}	1017, 2438	20
ionization constant at 25°C	ca 6×10^{-16}	21

[a] To convert J to cal, divide by 4.184.
[b] To convert kPa to mm Hg, multiply by 7.5.

(8). All of these factors must be taken into account in evaluating the physical constants of a particular sample of T_2O.

Nuclear Properties

Radioactivity. Tritium decays by β emission, $^3_1T \rightarrow {}^3_2He + \beta^-$. A summary of the radioactive properties of T, adapted from Evans (22), is given in Table 3.

Nuclear Fusion Reactions. Tritium reacts with deuterium or protons (at sufficiently high temperatures) to undergo nuclear fusion:

$$^3_1T + {}^2_1D \rightarrow {}^4_2He + {}^1_0n + 17.6 \text{ MeV}$$

$$^3_1T + {}^1_1H \rightarrow {}^4_2He + 19.6 \text{ MeV}$$

The first of these nuclear fusion reactions produces a neutron that can be used to form a new atom of tritium (see Production of Tritium), as well as evolving a very large amount of energy in the form of extremely hot helium. Nuclear fusion with tritium can be initiated and sustained at the lowest temperature (at least in principle) of any nuclear fusion reaction known. Tritium thus becomes the key element for thermonuclear energy sources (see below) (see Fusion energy).

Nuclear Magnetic Resonance. All three hydrogen isotopes have nuclear spins I $\neq$ 0 and consequently can all be used in nmr spectroscopy (Table 4) (see Analytical methods). Tritium is an even more favorable nucleus for nmr than is ^1H, which is by far the most widely used nucleus in nmr spectroscopy. The radioactivity of T and the ensuing handling problems are a deterrent to its widespread use for nmr, but even so considerable progress has been made in the applications of tritium nmr (24–25).

Table 3. Radioactive Properties of Tritium

Property	Value
half-life, yr	12.26[a]
decay constant, s^{-1}	1.765 × 10^{-9}
maximum β energy, keV	18.6[a]
mean β energy, keV	5.7
maximum specific activity, GBq/mmol[b]	1078.9
molar activity, TBq/mol[b]	2157

[a] From ref. 23.
[b] To convert Bq to Ci, divide by 3.7 × 10^{10}.

Table 4. Nuclear Magnetic Resonance Properties of Hydrogen Isotopes

Isotope	Nuclear spin	Resonance frequency,[a] MHz	Relative sensitivity[b]	Magnetic moment, 10^{-4} J/T[c]
H	1/2	100.56	1.000	25.8995
D	1	15.360	9.64 × 10^{-3}	7.9513
T	1/2	104.68	1.21	27.625

[a] At a field of 2.35 T (23.5 kilogauss).
[b] At constant field.
[c] To convert J/T to μ_B (nuclear Bohr magnetons), divide by 9.274 × 10^{-24}.

Chemical Properties

Most of the chemical properties of tritium are common with those of the other hydrogen isotopes. However, notable deviations in chemical behavior result from isotope effects and from enhanced reaction kinetics induced by the β emission in tritium systems. Isotope exchange between tritium and other hydrogen isotopes is also an interesting manifestation of the special chemical properties of tritium.

Isotope Effects. Any difference in the chemical or physical properties of two substances that differ only in isotopic composition constitutes an isotope effect. Isotope effects are usually largest when the isotope is directly involved in the rate-determining step of a reaction. The greater the mass of an atom, the lower its zero-point bond energy, and the greater the activation energy required to cleave a chemical bond. In general, therefore, reactions involving rupture of a —C—T bond may proceed at a markedly slower rate than cleavage of a corresponding —C—H bond. Kinetic isotope effects arising from bond cleavage are termed primary isotope effects. Secondary isotope effects occur as a result of the presence of the isotope in nearby molecular sites, bonds involving the isotope being neither broken or formed in the reaction. Secondary isotope effects are generally smaller than are primary effects, which may easily be 10 to 100 times greater. Solvent isotope effects include many important primary effects, eg, solvolyses and acid-base reactions, as well as secondary effects such as solvation. Although primary and secondary kinetic isotope effects are the most extensively studied, numerous other isotope effects have been observed with tritium. Thus, an H/T separation factor of about 14 occurs in the electrolysis of HOT (26–27). Other tritium isotope effects of significant magnitude have been observed in ion exchange (28), and gas chromatography (29–30). Many other examples have been described (8).

Enhanced Reaction Kinetics. For reactions involving tritium, the reaction rates are frequently larger than expected because of the ionizing effects of the tritium β-decay. For example, the uncatalyzed reaction $2 T_2 + O_2 \rightarrow 2 T_2O$ can be observed under conditions (25°C) for which the analogous reaction of H_2 or D_2 would be too slow for detection (31).

Isotopic Exchange Reactions. Exchange reactions between the isotopes of hydrogen are well known and well substantiated. The equilibrium constants for exchange between the various hydrogen molecular species have been documented (18). Kinetics of the radiation-induced exchange reactions of hydrogen, deuterium, and tritium have been critically and authoritatively reviewed (32). The reaction $T_2 + H_2 \rightarrow 2\ HT$ equilibrates at room temperature even without a catalyst (31).

In 1957, Wilzbach (33) demonstrated that tritium could be introduced into organic compounds by merely exposing them to tritium gas. Since that time hundreds of compounds, of types as simple as methane and as complex as insulin, have been labeled with tritium by this basic method of isotope exchange. Much work has been done to optimize conditions for the exchange technique through control of operating variables such as temperature, pressure, and addition of noble gases to facilitate energy transfer (34). Exchange of the hydrogen of organic compounds with tritium gas has been facilitated by activating the gas by ultraviolet light, γ- and x-irradiation, and microwave discharge. Although high chemical yields are obtained by the Wilzbach technique, highly tritiated impurities of structures similar to but not identical with that of the starting material are formed by direct irradiation damage of the target material and by decomposition of the tritiated products by self-irradiation. Separation of radioactive impurities produced in Wilzbach labeling may prove difficult, as is the task of proving that the radioactive T introduced into the substrate compound is actually in the compound of interest.

Catalytic exchange in solution between an organic compound and tritium gas or T_2O is a general procedure for introducing tritium with high specificity. Much higher molar specific activities (greater than 1.85 GBq/mmol (50 mCi/mmol)) can be attained than with the Wilzbach method. Both homogeneous and heterogeneous catalysts, as well as acid-base catalysis can be used.

Natural Production and Occurrence

Tritium arises in nature by the action of primary cosmic rays (high-energy protons) or cosmic-ray neutrons on a number of elements. Natural tritium was first detected in the atmosphere by Faltings and Harteck (35) and was later shown by Libby and co-workers (36) to be present in rainwater. Because of its relatively short radioactive half-life, naturally produced tritium does not accumulate indefinitely and the amount of tritium found in nature is very small. The unit for measurement of natural tritium, the TU (Tritium Unit) signifies a ratio of 1 atom of tritium per 10^{18} atoms of hydrogen. Libby (37) has compiled a very excellent world-wide survey of tritium levels in natural waters. Values range from less than one TU for water in certain deep wells and at extreme sea depths to several hundred TU in samples of rainwater taken during periods of active thermonuclear weapons testing. The level of tritium in atmospheric hydrogen increased from 3800 TU in 1948–1949 to 490,000 TU in 1959 (38). In 1958 the world's inventory of natural tritium (38) was estimated to be 20 kg in rainwater and 200 g in atmospheric hydrogen.

The principal source of natural tritium is the nuclear reactions induced by cosmic radiation in the upper atmosphere, where fast neutrons, protons, and deuterons collide with components of the stratosphere to produce tritium:

$$^{14}_{7}N + ^{1}_{0}n \rightarrow ^{3}_{1}T + ^{12}_{6}C$$

$$^{14}_{7}N + ^{1}_{1}H \rightarrow ^{3}_{1}T + \text{fragments}$$

$$^{2}_{1}D + ^{2}_{1}D \rightarrow ^{3}_{1}T + ^{1}_{1}H$$

The most important of these reactions by far is the $^{14}N(n, ^{12}C)^3T$ reaction (39). The energetic tritons so produced are incorporated into water molecules by exchange or oxidation, and the tritium reaches the earth's surface as rainwater.

Tritium has also been observed in meteorites and recovered material from man-made satellites. The tritium activity in meteorites can be reasonably well explained by the interaction of cosmic-ray particles and meteoritic material. The tritium contents of recovered satellite materials have not in general agreed with predictions based on cosmic-ray exposure. For observations higher than predicted (Discoverer XVII and satellites), a theory of exposure to incident tritium flux in solar flares has been proposed. For observations lower than predicted (Sputnik 4), the suggested explanation is a diffusive loss of tritium during heating on reentry.

Production in Nuclear Reactors

Nuclear Reactions for Production. The primary reaction for the production of tritium is:

$$^6_3\text{Li} + ^1_0\text{n} \rightarrow ^3_1\text{T} + ^4_2\text{He}$$

The capture cross section of ^{6}Li for this reaction with thermal neutrons is 930×10^{-28} m^2 (930 b) (40). A second, more favorable reaction is given by the following:

$$^3_2\text{He} + ^1_0\text{n} \rightarrow ^3_1\text{T} + ^1_1\text{H}$$

for which the capture cross section of ^{3}He to thermal neutrons is 5200×10^{-28} m^2 (5200 b). The limited availability of ^{3}He (0.00013–0.00017% natural abundance) (41) restricts the practical importance of this mode of production. Tritium is also produced by the action of high energy protons (such as primary cosmic rays) on a number of elements, and by the reaction of cosmic ray neutrons with ^{14}N (see above). Reaction cross sections are small, generally a few to a few hundred m$^2 \times 10^{-31}$ (millibarns).

Production in Target Elements. Tritium is produced on a large scale by neutron irradiation of ^{6}Li. The principal U.S. site of production is the Savannah River plant near Aiken, South Carolina. Tritium is produced there in large heavy-water moderated, uranium-fueled reactors. The tritium may be produced either as a primary product by placing target elements of Li–Al alloy in the reactor, or as a secondary product by using Li–Al elements as an absorber for control of the neutron flux.

Production in Heavy-Water Moderator. A small quantity of tritium is produced through neutron capture by deuterium in the heavy water used as moderator in the reactors. The thermal neutron capture cross section for deuterium is extremely small (about 6×10^{-32} m^2), and consequently the tritium produced in heavy-water moderated reactors is generally significant only as a potential health hazard. However, in a high-flux reactor such as that at the Institute Max von Laue-Paul Langerin (Grenoble, France), the heavy water moderator is a useful source of tritium (39).

Production in Fission of Heavy Elements. Tritium is produced as a minor product of nuclear fission (42). The yield of tritium is one to two atoms in 10,000 fissions of natural uranium, enriched uranium, or a mixture of transuranium nuclides (see Uranium; Actinides).

Recovery and Purification

Production-Scale Processing. The tritium produced by neutron irradiation of ^{6}Li must be recovered and purified after target elements are discharged from nuclear reactors. The targets contain tritium and ^{4}He as direct products of the nuclear reaction, a small amount of ^{3}He from decay of the tritium, and a small amount of other hydrogen isotopes present as surface or metal contaminants.

In the recovery process the gaseous constituents of the target are evolved, and the hydrogen isotopes separated from other components of the gas mixture. A number of methods are available for separating the tritium from the hydrogen and deuterium that can be applied to a process mixture or to naturally occurring sources. Because of the military importance of tritium, details of the large-scale production of this isotope have not been published. A report, however, is available that describes the large-scale production of tritium and its uses in France (43).

Isotopic Concentration. A number of techniques have been reported in the open literature for concentrating tritium from naturally occurring sources. Kaufman and Libby (44) concentrated tritium by electrolysis of tritiated water. Separation factors (H/T) of 6.6 to 29 were observed, with tritium being concentrated in the undecomposed water.

Low temperature (20–25 K) distillation has been widely used to separate hydrogen and deuterium and has also been successfully applied to the separation of tritium from the other hydrogen isotopes (45–46). At Los Alamos Scientific Laboratory, a system of four interlinked cryogenic fractionation columns has been designed for the separation of an approximately equal mixture of deuterium and tritium containing a small amount of ordinary hydrogen into a tritium-free stream of HD for waste disposal, and streams of high-purity D_2, DT, and T_2 (47) (see Cryogenics).

Concentration by gas chromatography has also been demonstrated. Elution chromatography has been used on an activated alumina column to resolve the molecular species H_2, HT, and T_2, thereby indicating a technique for separation or concentration of tritium (48). This method was extended (49) to include deuterium components. The technique was first demonstrated in 1964 with macro quantities of all six hydrogen molecular species (50).

Analysis and Detection

Tritium is readily detectable because of its radioactivity. Under certain conditions concentrations as low as 370 μBq/mL (10^{-8} μCi/mL) can be detected. Most detection devices and many analytical techniques exploit the ionizing effect of the tritium β-decay as a principle of operation (51–52). Some of the most widely used instruments for analysis of tritium are discussed below (see Analytical methods).

Ionization Chamber. The ionization chamber is a simple, sensitive, and sturdy device filled with gas and containing two electrodes between which a potential difference is maintained. When gas containing tritium is admitted to the chamber, the radiation ionizes the gas, the ions are drawn to the electrodes, and a flow of current results that is proportional to the tritium concentration. Because of the ability to detect very small currents, this technique is very sensitive and can be adapted to either static or flow systems. Ionization chambers are used as process-stream monitors, leak detectors, stack monitors, breathing-air monitors, detectors for surface contamination, and detectors instrumented with vibrating-reed electrometers. With the latter instrument, concentrations of tritiated water vapor in air of 370 μBq/mL (10^{-8} μCi/mL) (STP) can be measured.

Mass Spectrometer. The mass spectrometer is the principal analytical tool of direct process control for the estimation of tritium. Gas samples are taken from several process points and analyzed rapidly and continually to ensure proper operation of the system. Mass spectrometry is particularly useful in the detection of diatomic hydrogen species such as HD, HT, and DT.

Thermal-Conductivity Analyzer. The thermal-conductivity analyzer operates on the principle that the loss of heat from a hot wire by gaseous conduction to a surface at a lower temperature varies with the thermal conductivity of the gas, and is virtually independent of pressure between 1.3 kPa (10 mm Hg) and 101 kPa (1 atm). This technique is frequently used in continuous monitors for tritium in binary gas mixtures for immediate detection of process change.

Calorimeter. The β-decay energy of tritium is very precisely known (8). The thermal energy generated by the decay can thus be used with a specially designed calorimeter to measure the quantity of tritium in a system of known heat capacity (see Calorimetry).

Liquid Scintillation Counter. The rapid and sensitive technique of liquid scintillation counting is applied for the determination of tritium in liquid systems. The tritiated sample is dissolved in a solvent that contains an organic scintillator. The emitted β particles excite the organic molecules which, in returning to their normal energy levels, emit light pulses that are detected by a photomultiplier tube, amplified, and electronically counted. Liquid scintillation counting is by far the most widely used technique in tritium tracer studies and has superseded most other analytical techniques for general use (53).

Health Physics Aspects

Hazards. Since tritium decays with emission of low-energy radiation ($E_{av} = 5.7$ keV), it does not constitute an external radiation hazard. However, tritium presents a serious hazard through ingestion and subsequent exposure of vital body tissue to internal radiation. The body will assimilate tritiated water and distribute it throughout body fluids with remarkable speed and efficiency. When exposed to tritiated water vapor via inhalation, people will absorb 98–99% of the activity inspired through the respiratory system (54). Uniform distribution throughout body fluids occurs within 90 min. Also, when exposed to such an atmosphere, tritium entering the body through the total skin area will approximately equal that entering the lungs. Elemental tritium (T_2 or HT) is much less readily assimilated. Approximately 0.004% of such activity inspired is absorbed, apparently after preliminary oxidation in the lungs. Negligible amounts of elemental tritium are absorbed through the skin.

The body excretes tritium with a biological half-life of 8–14 d (10.5 d average) (55), which can be reduced significantly with forced fluid intake. For humans, the estimated maximum permissible total body burden is 37 MBq (1 mCi). The median lethal dose (LD$_{50}$) of tritium assimilated by the body is estimated to be 370 GBq (10 Ci). Higher doses can be tolerated with forced fluid intake to reduce the biological half-life.

Monitoring and Control. A detailed description of methods in use at the Savannah River Plant for air, surface, personnel, and environmental monitoring for tritium has been published (56). A widely used instrument for air monitoring is a type of ionization chamber called a Kanné chamber. Surface contamination is normally detected by means of smears, which are simply disks of filter paper wiped over the suspected surface and counted in a windowless proportional-flow counter. Uptake of tritium by personnel is most effectively monitored by urinalyses normally made by liquid scintillation counting on a routine or special basis. Environmental monitoring includes surveillance for tritium content of samples of air, rainwater, river water, and milk.

The radiological hazard of tritium to operating personnel and the general population is controlled by limiting the rates of exposure and release of material. Maximum permissible concentrations (MPC) of radionuclides were specified in 1959 by the International Commission on Radiological Protection (57). For purposes of control all tritium is assumed to be tritiated water, the most readily assimilated form. The MPC of tritium in breathing air (continuous exposure for 40 h/wk) is specified as 185 kBq/mL (5 µCi/mL) and the MPC for tritium in drinking water is set at 3.7 GBq/mL (0.1 Ci/mL) (57). The maximum permitted body burden is 37 MBq (one millicurie). Whenever bioassay indicates this value has been expected, the individual is withdrawn from further work with tritium until the level of tritium is reduced.

Personnel are protected in working with tritium primarily by containment of all active material. Containment devices such as process lines and storage media are normally placed in well-ventilated secondary enclosures (hoods or process rooms). The ventilating air is monitored and released through tall stacks; environmental tritium is limited to safe levels by atmospheric dilution of the stack effluent. Tritium can be efficiently removed from air streams by catalytic oxidation followed by water adsorption on a microporous solid absorbent (58) (see Absorption). Personnel who must work in areas in which tritium contamination exceeds permitted levels are safeguarded by protective clothing, such as ventilated plastic suits. Detailed descriptions of laboratories suitable for manipulation of tritium can be found in reference 8.

Uses

Thermonuclear Energy Sources. As discussed in the section on Nuclear Properties (see above), tritium is a reactant in nuclear fusion reactions. The large energy release of these reactions is the basis for the use of tritium as a source of thermonuclear energy in weapons and in controlled fusion reactions. Fusion power reactors are currently under development, although such controlled thermonuclear fusion may not be realized in practice for some years to come. Tritium is a desirable constituent for a nuclear fusion device because the D–T fusion reaction will proceed at a temperature of approximately 50×10^6 °C, whereas the D–D fusion reaction requires a temperature of about 400×10^6 °C for a self-sustaining chain reaction (52) (see Fusion energy).

BIBLIOGRAPHY

"Deuterium and Tritium" in *ECT* 2nd ed., Vol. 6, pp. 895–910, by J. F. Proctor, E. I. du Pont de Nemours & Co., Inc.

1. A. H. Wapstra, *Natl. Bur. Stds. (US) Spec. Publ.* **343**, 151 (1971).
2. M. L. E. Oliphant, P. Harteck, and E. Rutherford, *Proc. Roy. Soc.* **A144**, 692 (1934).
3. T. W. Bonner, *Phys. Rev.* **53**, 711 (1938).
4. L. W. Alvarez and R. Cornog, *Phys. Rev.* **56**, 613 (1939).
5. R. Viallard, "Tritium," in P. Pascal, ed., *Nouveau Traite de Chimie Mineral* tome 1, Masson et Cie, Paris, 1956, p. 911.
6. E. L. Meutterties, *Transition Metal Hydrides,* Marcel Dekker, Inc., New York, 1971, Chapt. 1, pp 1–7.
7. K. M. MacKay and M. F. A. Dove, in *Comprehensive Inorganic Chemistry,* Pergamon Press, Ltd., Oxford, 1973, Chapt. 3, pp. 77–116.
8. E. A. Evans, *Tritium and Its Compounds,* 2nd ed., John Wiley & Sons, Inc., New York, 1974.
9. H. M. Mittelhauser and G. Thodos, *Cryogenics* **4**, 368 (1964).
10. E. R. Grilly, *J. Am. Chem. Soc.* **73**, 843 (1951).
11. *Ibid.,* 5307 (1951).
12. R. H. Sherman, J. R. Bartlit, and R. A. Briesmeister, *Cryogenics* **16**, 611 (1976).
13. P. C. Souers and co-workers, *Three-phase Region of D_2-DT-T_2, Lawrence Livermore Laboratory Report* UCRL-79036, 1977, 24 pp.
14. F. D. Rossini and co-workers, *Natl. Bur. Stds. (US) Circular 500,* U.S. Government Printing Office, Washington, D.C., 1950.

15. R. E. Elson and co-workers, *University of California Livermore Report UCRL-4519*, U.S. Atomic Energy Commission, 1956.
16. E. W. Albers, P. Harteck, and R. R. Reeves, *J. Am. Chem. Soc.* **86,** 204 (1964).
17. J. W. Pyper and C. K. Briggs, *The Ortho-Para Forms of Hydrogen, Deuterium and Tritium: Radiation and Self-induced Conversion Kinetics and Equilibrium*, Lawrence Livermore Laboratory Report UCRL-52278, 1977, 20 pp.
18. W. M. Jones, *J. Chem. Phys.* **48,** 207 (1968).
19. M. Coldblatt, *J. Phys. Chem.* **68,** 147 (1964).
20. P. A. Staats, H. W. Morgan, and J. H. Goldstein, *J. Chem. Phys.* **24,** 916 (1956).
21. M. Goldblatt and W. M. Jones, *J. Chem. Phys.* **51,** 1881 (1969).
22. Ref. 8, p. 9.
23. C. M. Lederer, J. M. Hollander, and I. Perlman, *Table of Isotopes*, 6th ed., John Wiley & Sons, Inc., New York, 1967, p. 3.
24. P. Diehl, in T. Axenrod and G. Webb, eds. *Nuclear Resonance Spectroscopy of Nuclei Other Than Protons*, John Wiley & Sons, Inc., New York, 1974, pp. 275–285.
25. J. P. Bloxsidge and co-workers, *J. Chem. Res. (Part S)*, 42 (1977).
26. M. L. Eidenoff, *J. Am. Chem. Soc.* **69,** 977 (1947).
27. *Tritium in the Physical and Biological Sciences*, Vol. I, International Atomic Energy Agency, Vienna, 1962, p. 162.
28. H. Gottschling and E. Freese, *Nature* **196,** 829 (1962).
29. K. E. Wilzbach and P. Riesz, *Science* **126,** 748 (1957).
30. P. D. Klein, *Adv. Chromatog.* **3,** 3 (1966).
31. L. M. Dorfman and B. A. Hemmer, *Phys. Rev.* **94,** 754 (1954).
32. J. W. Pyper and C. K. Briggs, *Kinetics of the Radiation-induced Exchange Reactions of Hydrogen, Deuterium, and Tritium*, Lawrence Livermore Laboratory Report UCRL-52380, 1978, 12 pp.
33. K. Wilzbach, *J. Am. Chem. Soc.* **79,** 1013 (1957).
34. M. Wenzel and P. E. Schulze, *Tritium Markierung, Preparation, Measurement and Uses of Wilzbach Labelled Compounds*, Berlin, Walter de Gruzter, 1962.
35. V. Fallings and P. Harteck, *Z. Naturforsch.* **5a,** 438 (1950).
36. A. V. Grosse and co-workers, *Science* **113,** 1 (1951).
37. Ref. 25, pp. 5–32.
38. Ref. 27, pp. 56–67.
39. P. Pautrot and J. P. Arnauld, *Trans. Am. Nucl. Soc.* **20,** 202 (1975).
40. G. Friedlander and J. W. Kennedy, *Nuclear and Radiochemistry*, John Wiley & Sons, Inc., New York, 1955, p. 404.
41. *Ibid.*, p. 415.
42. E. L. Albenesius, *Phys. Rev. Lett.* **3,** 274 (1959).
43. *Le Tritium*, Commissariat a l'Energie Atomique, Bull. Inform. Scientifiques et Techniques, No. 178, Feb. 1973.
44. S. Kaufman and W. F. Libby, *Phys. Rev.* **93,** 1337 (1954).
45. T. M. Flynn and co-workers, *Proceedings of the 1957 Cryogenic Engineering Conference*, U.S. National Bureau of Standards, Boulder, Col., 1958, p. 58.
46. M. Damiani, R. Getraud, and A. Senn, *Sulzer Tech. Rev.* **4,** 41 (1972).
47. J. R. Bartlet, W. H. Denton, and R. H. Sherman, "Hydrogen Isotope Distillation for the Tritium Systems Test Assembly," *American Nuclear Society Conference on the Technology of Controlled Nuclear Fusion, May 9–11, 1978*, Sante Fe, N.M.
48. Ref. 27, pp. 121–133.
49. J. King, *J. Phys. Chem.* **67,** 1397 (1963).
50. D. L. West and A. L. Marston, *J. Am. Chem. Soc.* **86,** 4731 (1964).
51. E. L. Albenesius and L. H. Meyer, *DP-771*, Savannah River Plant, E. I. du Pont de Nemours & Co., Inc., Aiken, South Carolina, 1962.
52. D. B. Hoisington, *Nucleonics Fundamentals*, McGraw-Hill Book Co., Inc., New York, 1959, pp. 316–319.
53. K. D. Neame and C. A. Homewood, *Liquid Scintillation Counting*, John Wiley & Sons, Inc., New York, 1974.
54. E. A. Pinson and W. H. Langham, *J. Appl. Physiol.* **10,** 108 (1957).
55. H. L. Butler and R. W. Van Wyck, *DP-329*, Savannah River Plant, E. I. du Pont de Nemours & Co., Inc., Aiken, South Carolina, 1962.

56. W. C. Reinig and E. L. Albenesius, *Am. Ind. Hyg. Assoc. J.* **24,** 276 (1963).
57. International Commission on Radiological Protection Publication 2, *Report of Committee II on Permissible Dose for Internal Radiation,* Pergamon Press, Ltd., Oxford 1959.
58. A. E. Sherwood, *Tritium Removal from Air Streams by Catalytic Oxidation and Water Adsorption,* Lawrence Livermore Laboratory Report UCRL-78173, 1976, 16 pp.

JOSEPH J. KATZ
Argonne National Laboratory

DEXTROSE AND STARCH SYRUPS. See Syrups.

DIALYSIS

Dialysis is a membrane transport process in which solute molecules are exchanged between two liquids. The process proceeds in response to differences in chemical potentials between the liquids. Dialysis requires that the membrane separating the liquids permit diffusional exchange between at least some of the molecular species present while effectively preventing any convective exchange between, or commingling of, the solutions. Dialysis is used in industry, medicine, and the laboratory as a separation and dilution process (see Separation systems synthesis). In practice, dialysis is almost always accompanied by other membrane processes that result in both further solute transport (that may augment or decrease that effected by dialysis) and solvent transport (see Membrane technology). Dialysis may be defined as any process in which the principal effect is transfer of solute molecules from one liquid to another through a membrane when the important differences in state of the liquids are the chemical potentials of their solutes.

Originally and classically, dialysis has been used to separate molecules with large differences in size, typically crystalloids (low molecular weight substances), from colloids. Such were the separations reported in 1861 by Graham, who is generally conceded to be the discoverer of dialysis (1). In recent years smaller differences in molecular size have been found adequate to achieve significant separations, and in some cases size has been an inadequate parameter to represent the separations. The simple concepts of the membrane as a rigid sieve and molecules as hard objects of precise dimension have given way to a complex of ideas about membrane–solution interactions involving ionic, covalent, and hydrogen bonding, chemisorption, physical adsorption, electrostatic effects, and the concept of the membrane as a flexible matrix whose permeable pathways change dimensions randomly over time. New dialysis membranes have contributed to development of this more comprehensive view and of a broader range of industrial applications for dialysis. Many of the new membrane materials have membrane–molecule spacings which, during the synthesis of the polymer and the forming of the membrane, can be adjusted to a desired tightness or can be made to accommodate molecular substituents that exert forces selectively on different solute molecules. These membranes are also capable of withstanding many chemical environments that were entirely destructive to classical simple cellulose materials.

Dialysis has not achieved the breadth and frequency of use as an industrial process that the development of new membranes in the 1950s seemed to be predicting. Rather these membranes have been more thoroughly exploited in other membrane processes, notably ultrafiltration (qv), reverse osmosis (qv), and electrodialysis (qv). The latter three processes can often be operated at higher fluxes (rates of transport per unit area of membrane) than dialysis since they utilize extrinsic driving forces (pressure for the first two, electrical potential for the third) to establish the flux. In contrast, the driving force for dialysis depends only upon solution concentrations. Because capital costs, which increase nearly linearly with membrane area, and the amortized cost of the membrane itself, are relatively high, overall processing costs have often been found to be lower with high flux processes despite their higher energy costs per unit of product. Dialysis has persisted as a competitive process where the intrinsic driving force is large (eg, feed concentrations are high) or where extrinsic driving forces are not effective. The latter may occur because of concentration polarization phenomena, or because the extrinsic forces are potentially damaging to the fluid being processed, or in special situations where energy costs predominate over other costs.

Dialysis as a contemporary industrial process is a passive separation process with a low operating cost because it uses no external thermal or chemical energy sources. It requires a high capital investment, essentially because of the large membrane areas required per volume of fluid to be processed per unit of time, and accidentally because of the slow rate of development of adequate, compact, commercial processing equipment. Dialysis is most attractive in separating dissolved materials present in high concentration or when processing sensitive systems. The future of dialysis as an important industrial separation process is uncertain, depending upon several factors: (*1*) changing economics—the cost of energy, the investment climate, and the expected process lifetime which determines how plant investment is to be amortized; (*2*) improvement in equipment design—increasing the amount of membrane area effective for transfer that can be fitted into a unit volume of process equipment whose overall performance characteristics are acceptable in other respects, and the introduction of versatile dialysis modules that permit *a priori* process design and reduced scale-up risks; (*3*) development of new membranes—membranes that offer special transport characteristics best realized or only realizable in a dialysis regimen; (*4*) innovations in process design—notably the use of staging concepts with reflux, discussed below, to enhance overall selectivity; and (*5*) development of certain schemes involving removal of products from large-scale cultures of suspended cells.

Theory

Theories of dialysis can be considered to be of three kinds: (*1*) theories of intramembrane transport that seek to (*a*) interrelate different fluxes, concentration gradients across the membrane, and concentration levels within it, and (*b*) identify how the fluxes are affected both by the chemical composition and physical–chemical state of the membrane, as well as by the nature of the permeating molecules; (*2*) boundary layer transport theories designed to elucidate the relationship between fluxes and the fluid–mechanical conditions adjacent to the membrane, often with particular emphasis on the prediction of accumulation of impermeable constituents to an extent that fouling of the membrane or at least a sharp reduction in flux occurs; (*3*) theories that show how fluxes and concentrations vary along the fluid pathway and how certain recycle and reflux schemes may be used to alter the molecular selectivity intrinsic to the membrane. Each of these categories of theory is considered here although their application to dialysis is a special case of membrane technology (qv) as a whole.

Intramembrane Transport. Most theories of intramembrane transport treat each permeating solute independently of other solutes although all allow for solute-membrane interaction and for solute–solvent interaction. This approach is approximately equivalent to a dilute solution theory and is often justified by what is observed, but it is misleading to the novice because the essential purpose of a dialysis operation is the separation of different solute molecules from each other. Thus, although the behavior of each kind of solute can often be predicted without reference to the others, a dialysis operation can be rationalized only by predicting the behavior of at least two solutes and comparing these behaviors as they occur throughout the apparatus.

The basic formulation that is now commonly used to compare specific models of intramembrane transport is irreversible thermodynamics, especially in the form of frictional coefficients (2–3). Each group of molecules in a region that belong to the same species is characterized by an average velocity (see Fig. 1). The relative velocities between pairs of species in the same region, when multiplied by a frictional coefficient, gives rise to a frictional force that characterizes the interaction of the two molecular groups. The total frictional force acting on any one species is the sum of forces between it and the others that it encounters. Thus, in ideal, single-solute solutions when there is negligible solvent flux:

$$J_s = \frac{K_s RT}{\Delta X (f_{sw} + f_{sm})} \cdot (C_s^o - C_s^{\Delta x})$$

Here J_s is the solute flux; the membrane is supposed to have one of its faces at $x = o$ and the other at $x = \Delta x$; K_s is the coefficient of distribution of solute between solution and membrane; f_{sw} and f_{sm} are the frictional coefficients, respectively, between solute and solvent and solute and membrane; and C_s^o and $C_s^{\Delta x}$ are the concentrations of solute in solution at the two faces of the membrane. Formulations for multicomponent solutions are available but have not been much used. The single-solute formulation is applicable to ionized species as long as there is no current flow. The quantity $K_s/\Delta X (f_{sw} + f_{sm})$ has been termed the permeability coefficient ω by Katchalsky and Curran (4) and the quantity ωRT is the membrane permeability K_m used by most authors in treatments of dialyzer design.

In practical dialysis situations the solvent flux J_w is not insignificant. In fact the dilution of feed by counterdiffusing solvent is a principal, usually undesired, concomitant of dialysis. In terms of the frictional coefficients, the molal ratio of solvent flux to solute flux when there is no pressure difference across the membrane is approximately:

$$-\sigma \left[\frac{\psi_w \overline{V}_w \overline{C}_s}{K_s} \right] \cdot \left[\frac{f_{sw} + f_{sm}}{f_{wm}} \right]$$

The fluxes are opposed, hence the negative sign, and proportional to the reflection coefficient σ. (The reflection coefficient is defined as the fraction of solute molecules

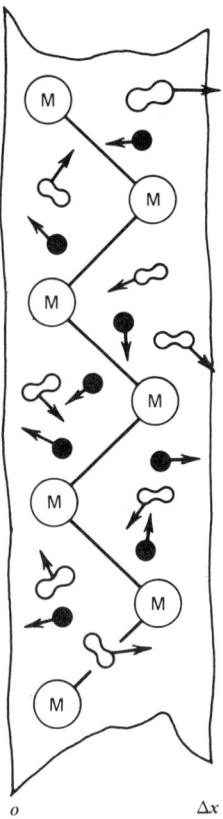

Figure 1. Intramembrane transport vector average velocities. M, membrane; o and x, the two faces of the outer membrane. Δx = the membrane width.

reflected from a membrane; it has a value of unity for large impermeable molecules and approaches zero for small molecules (5).) The first bracket is the ratio of solute to solvent molecules in the membrane expressed in terms of solvent volume fraction ψ_w and specific volume $\overline{V}_w$; and solute mean concentration $\overline{C}_s$ and distribution coefficient K_s. The second bracket is a ratio of solute to solvent frictional coefficient.

It is possible both to reduce solvent flow and to augment the rate of dialysis by establishing a higher pressure on the side of the membrane bathed by the solute-rich solution; this approach is only practical when hollow-fiber membranes are used because membranes in this configuration are uniquely capable of withstanding such pressures over large areas without complex supports (see Hollow-fiber membranes). Solvent flow can also be reduced by introduction of a second, impermeable solute opposite the solution that is rich in the solute to be transferred; this approach is seldom practical.

For many years the pore model (6) has been used to relate dialysis rates to properties of the membrane and the size of the solute molecule (see Fig. 2). According to this model:

$$K_m = \frac{D A_p}{\tau \Delta x} f\left(\frac{d_s}{d_p}\right)$$

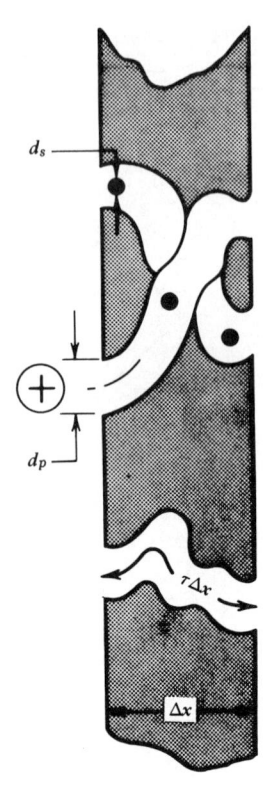

Figure 2. The pore model. d_s = solute diameter; d_p = pore diameter.

where the membrane permeability is considered proportional to the free-liquid diffusivity D of the solute and the fraction of the membrane surface occupied by pores A_p (usually taken to equal to void volume ν of the membrane). Membrane thickness Δx is multiplied by a tortuosity factor τ to give a corrected distance that the average molecule must travel to cross the membrane. The function f is a drag factor dependent on the ratio of solute-diameter to pore diameter d_s/d_p and is usually taken to approach unity for small values of the ratio, to approach zero as the ratio approaches unity and to be zero for ratios of unity and larger. Methods for determining A_p or ν, τ, and d_p have been described by several authors (7). Slightly different versions of the function f have been proposed and approximate conformity of its predictions with experimental results has been shown for a special membrane whose pores had a known, sharp size distribution (8). Practical membranes have a range of pore sizes, which must be taken into account in accurate work. The pore theory is essentially mechanical and accounts neither for chemical interactions between the solute and the membrane nor for distribution coefficients between the solution and the membrane that are different from unity. Because the principal basis of selectivity in all practical dialysis membranes is molecular size, these deficiencies in pore theory do not nullify its usefulness. Only ions of different charge dialyzing through charged membranes exhibit selectivities that are not predominantly size related, and dialysis is not a preferred process for separating such species. Craig (9) has shown that the effective size for dialysis of many natural molecules can vary significantly as the physical–chemical environment of the solution in which they are dissolved is varied. He has also demonstrated striking changes in membrane permeability for such molecules as the membrane is stretched; the results are different according to whether the stretch is uniaxial or biaxial (10).

Dialysis membranes are not sharply selective and that fact has been a principal deterrent to wider use of the process. If the high selectivity found in some natural membranes could be duplicated in artificial systems at workable flux levels, significant new applications could be anticipated. When natural membranes exhibit such selectivity they appear to do so because they contain a mobile carrier molecule that can associate selectively with a specific solute on one side of the membrane and dissociate on the other. This process may be either random, in which case the net flux is proportional to the concentration difference, or as is determined by a more extensive reaction system in which case chemical energy may be used to transport the species against an electrochemical potential gradient (see Filtration). Not only have such features not yet been incorporated in practical artificial systems, but if they were, a dramatic decrease in thickness of the artificial membranes would probably be needed since natural membranes involving carriers are only about a thousandth as thick as artificial membranes. Such a reduction in thickness might only be practical if the membranes were arranged as so-called artificial cells similar to those described by Chang (11).

Juxtamembrane Transport. Useful dialysis always involves the movement of molecules from the bulk of one fluid to the bulk of another. The theory of intramembrane transport considered above assumes that regardless of the rate at which molecules enter the membrane on one side and leave it on the other, their concentration in the immediately adjacent fluid layers is somehow maintained. Maintaining these concentrations when there is a finite flux through the membrane requires that there be a tandem process in each fluid, matching the membrane flux with a supply process to deliver molecules from the fluid bulk on one side, and a removal process to take molecules from the membrane into the fluid bulk on the other. These necessary components of useful dialysis are termed juxtamembrane transport; they may be so slow that they rather than intramembrane transport may determine the overall rate of the process. Juxtamembrane transport is driven by juxtamembrane concentration differences, illustrated along with the transmembrane concentration difference in Figure 3.

Transport in fluids may occur by convection (movement of a particular molecular species as part of a moving stream) and by diffusion (movement of a particular molecular species separately from other species in the same region as a summed result of random, thermal movements of individual molecules). In practice effective juxtamembrane transport is by convective diffusion, involving both of these fundamental processes, with convection occurring principally parallel to the membrane surface and diffusion being responsible for transverse transport, especially near the membrane surface. Convection is either forced, as by pumping or gravity feed, or free, as induced by density gradients arising from dilution of the feed and concentration of the opposite solution. Free convection induced by mass transfer is closely analogous to thermal free convection, induced by heat transfer, whose effects are commonly seen in waves of warmed air rising from heated surfaces (see Mass transfer).

As transport through a membrane can be characterized by a dialysis coefficient K_m, so transport from the bulk of one fluid to the membrane surface and from the opposite surface to a second fluid can be characterized by two fluid–film coefficients of mass transfer, $k_{f,1}$ and $k_{f,2}$:

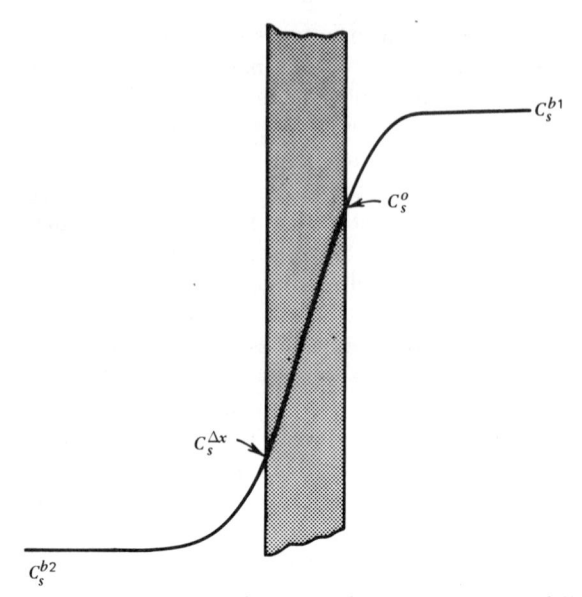

Figure 3. Juxtamembrane transport and transmembrane concentration difference. C_s, solute concentration; $b1$ and $b2$, concentrations of bulk fluids.

$$J_s = k_{f,1}(C_s^{b1} - C_s^o) = k_{f,2}(C_s^{\Delta x} - C_s^{b2})$$

where the superscripts $b1$ and $b2$ refer to the concentrations in the bulk of fluids 1 and 2. Usually the surface concentrations are not known and the definition of K_m is used with the foregoing equations to obtain the transfer rate or flux in terms of bulk concentrations only:

$$J_s = \frac{1}{1/k_{f1} + 1/k_{f2} + 1/K_m}(C^{b1} - C^{b2}) = K(C^{b1} - C^{b2})$$

which also defines an overall transport coefficient K (12). This coefficient will be predominantly determined by the smallest of its three component coefficients.

Forced convection may be either turbulent or laminar, depending on the geometry of the fluid compartment and the dynamics of the flow as summarized by the Reynolds number (see Rheological measurements). Turbulence can be enhanced by placing obstructions in the fluid path to cause fluid elements to exchange position (mix) in the plane perpendicular to fluid movement; the obstructions are called turbulence promoters and they have been incorporated in a variety of dialyzer designs (13). At lower Reynolds numbers, laminar forced convection predominates. Very little information has been gathered about juxtamembrane transport in equipment used for industrial dialysis, but information obtained for heat transfer in similar geometries can be easily translated into a good estimate of each fluid's mass transfer coefficient using well established analogies such as the Chilton-Colburn j-factor (14) (see Absorption). These analogies are somewhat limited by the fact that in dialysis the convective flux through the duct wall (membrane) is not zero as it is in the case of most heat transfer surfaces. The juxtamembrane transport for laminar flow in hollow-fiber dialyzers has been worked out in considerable detail (15). More general design equations have been worked out specifically for the design of hemodialyzers but they are more broadly useful (16).

Free convection has been recognized as an important determinant of juxtamembrane transport in cell or plate-and-frame types of dialyzers such as that shown in Figure 4 but little use has been made of available theory and the obvious analogy with thermal convection in optimizing the design and operation of commercial dialyzers. Barrera (18) has analyzed the role of free convection in facilitating juxtamembrane transport in open, vertical dialysis cells for several cases including cells bounded by membrane on one or both sides and with equal and unequal flow rates on opposite sides of the membrane. As in forced convection, the observable, overall transport coefficient K can be resolved into components. For the case of a cell bounded on both faces by membrane with equal, opposite flows on each side of the membrane:

$$K^{-1} = [1/K_m + WH/Q + 0.75\ T/D(1/a_1 + 1/a_2)]$$

where W, and H, and T are, respectively, the width (along the membrane surface) the height, and the thickness (equal to the intermembrane spacing) of the cell. The quantities Q and D are the flow rate and free-liquid diffusion coefficients and a_1 and a_2 are related to the Rayleigh numbers of the fluids on each side of the membrane.

$$A = Ra^{1/4}/2$$

$$Ra = \rho \beta g A T^4/\mu D$$

ρ being the mean fluid density, β the volumetric mass expansion coefficient, g the acceleration of gravity, A the vertical gradient of concentration, and μ the liquid viscosity. The expression for K^{-1} is additive and the components separately identifiable: the first, R_m^{-1} is the membrane resistance; the second, WH/Q a component attributable to forced convection, and the last, the free convection component. However, unlike the situation in forced convection, here the terms cannot be independently evaluated. The membrane resistance affects the free convection and there is no unique liquid resistance related only to liquid conditions. The equations are however useful for design (18), both for this case and the more general case of unequal flow.

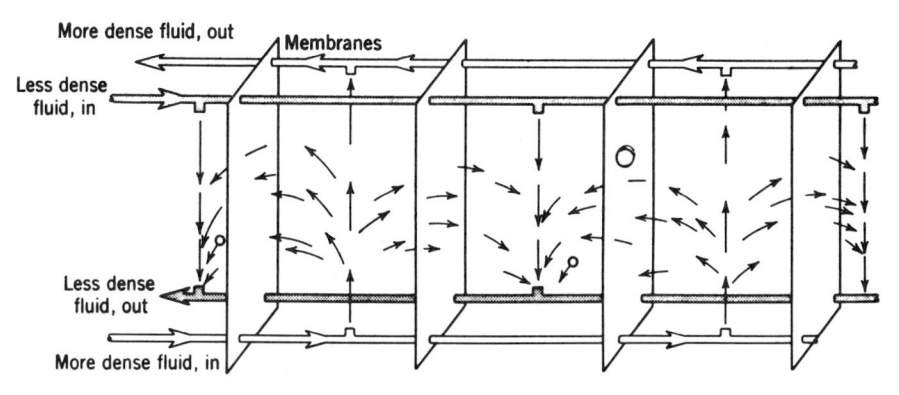

Figure 4. Flow paths in a filter-press dialyzer (17).

Staging, Recycling, and Refluxing. The traditional utilization of dialysis as an industrial process has emulated the process of heat transfer far more than it has other mass transfer processes with which it might more directly compete for a given molecular separation. This is true not only in terms of the internal configuration of its fluid paths which, as noted above, are similar to those of heat exchangers, but also in the basic design of essentially all dialysis systems, none of which use the staging so common in other mass transfer operations to enhance the separation between solutes whose transfer coefficients are only moderately different. As noted above, dialysis is used primarily as a separation process rather than a dilution or concentration process and the flux of one solute must not only be absolutely large enough to keep the membrane area requirement within practical limits but it must also be large relative to that of another solute from which it is to be separated. Thus, in a dialyzer in which two streams of equal flow rate Q are contacted countercurrently, ignoring any solvent flux, the fraction E of solute in one stream that dialyzes into the other is:

$$E = \theta/(1 + \theta)$$

where

$$\theta = KS/Q$$

S being the membrane area and Q the flow rate (19). If 90% extraction of the more permeable and only 10% extraction of the less permeable solute is desired, the two permeabilities must differ by a factor of more than 80. Such calculations show clearly why dialysis has been confined to a process for separating colloids from crystalloids. Noda and Gryte (19) demonstrate how simple cascading cannot improve separation of two substances of given relative permeability and they proceed to analyze theoretically units in which a dialyzer and concentrator are combined. (The nature of the concentrator—ultrafiltrator, evaporator, freezing device—is immaterial to the theory.) They then show that very large separations, even between solutes with similar permeabilities, can be obtained with multiple units of this kind, recognizing however that the practicality of such systems is limited by the high cost of the concentrator relative to the dialyzer. A system is then described in which multiple dialyzers can be supported by a single concentrator (see Fig. 5) which handles amounts of solvent that are large relative to the feed flow rate, finally suggesting a hybrid system containing many dialysis stages and multiple but fewer concentrators.

Applications

For many years the archetypal industrial application of dialysis has been the recovery of caustic from hemicellulose (20) (see Pulp). This application illustrates well two of the three prerequisites that render dialysis an attractive process: the existence of a large concentration difference of the substance to be diffused between two process streams, and a large molecular weight (and thus permeability) difference between the molecules to be separated. The third prerequisite, avoidance of chemical and mechanical stress to the process fluids, is not present here but has been of dominant importance in some more recent applications.

In the 1970s somewhat standardized equipment and membranes became available for the processing of industrial fluids according to several basic membrane processes (21). For dialysis such equipment had long been available. These competitive processes

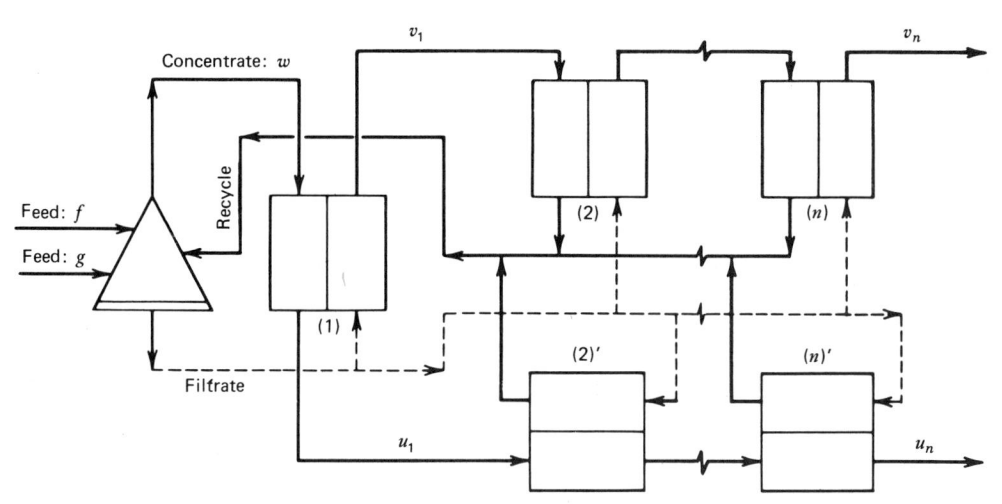

Figure 5. A system in which multiple dialyzers can be supported by a single concentrator (19).

are reverse osmosis (qv) for solvent purification, ultrafiltration (qv) for separation and concentration of solutes, electrodialysis (qv) for the separation of charged species from solvent and other solutes or from each other according to charge and mobility, and electroosmosis for concentrating solutions by removal of water associated with ions (see Filtration; Membrane technology; Water). To some extent ion exchange (qv) has also become a stronger competitor of dialysis. All of these processes allow for the provision of a driving force or source of energy for separation in addition to the chemical potential contained in the feed solution. However, each of these active processes carries greater potential for the development of undesired precipitation or chemical change at some points in the system consequent to the development of high solute concentrations or changes in solution composition such as pH shifts.

These alternative processes offer two other advantages over dialysis: each of them utilizes tight membranes that allow for sharper separation of molecules, and each intrinsically provides a somewhat sharper separation with a given membrane by overcoming an overlapping of separations that occurs with a simple diffusional process such as dialysis. This overlapping occurs because all molecules that permeate through a medium in response to a concentration gradient have a maximum velocity proportional to their molecular diffusivity, those undergoing specific interaction with the membrane travelling slower. Thus even in the absence of specific membrane effects, a retardation of movement occurs that is dependent only on molecular size and consequently is not sharp because it roughly corresponds to the cube root of molecular weight (if the Stokes-Einstein equation is followed). When the molecule moves by convection or forced diffusion, the membrane interaction is not affected in this way and the separation is intrinsically sharper.

Dialysis is under active consideration in conjunction with development of biomass projects (22) (see Fuels from biomass; Foods non-conventional). In these projects, single-celled organisms are cultured, suspended within large liquid masses, either to augment natural photosynthesis or to convert its products into more useful forms, eg, cellulose into sugars, organic acids and aldehydes, and alcohol. The organisms are often inhibited in activity by their products which must be continuously removed with minimum trauma to the suspended organism. Dialysis provides the gentleness wanted, but the separation is not very discriminating between desired products and intermediates. However, the dialysate is cell-free and can be treated by other processes with appropriate recycling of intermediates (23). Dialysis has also been proposed for various waste recovery operations, notably in the plating and metal finishing industries (24). It is unlikely that these processes will be competitive except where highly concentrated wastes of intermediate value are involved (see Recycling).

By far the largest contemporary use of dialysis is not for processing of industrial streams but in hemodialysis, the treatment of the blood of persons with end-stage renal disease in which the kidneys are no longer capable of removing the products of metabolism from the blood and excreting them. The application of dialysis is essential here because of two requirements: that the processing be gentle and that the basic separation is between quite large and quite small molecules. The former are typified by proteins and the latter by urea, the principal form in which catabolic nitrogen is eliminated. The absence of large concentration differences is apparently outweighed by the benefits just cited and is offset by using large areas of membranes to compensate for the small fluxes. Typically, areas of ca 1 m^2 are used to process a relatively small flow of ca 200 mL/min. Both the clinical and technical investigators of dialysis have sought, over many years, to establish a clear division of middle molecules into those that should be retained and those that should be removed, but the concentrations of suspected essential as well as toxic molecules in this group are so low and their metabolic effects so indirect that the criteria on which a closer separation might be based have not yet been found.

Meeting gross excretory needs of patients with end-stage renal disease by dialysis is possible in a practical sense only if concomitant ultrafiltration is employed and carefully controlled, and if the dialyzing solution with which the blood is in contact contains all the crystalloids (such as inorganic salts and glucose) that must be retained and are not constrained by available membranes. Both problems have been solved. With the membranes presently used, ultrafiltration is easily effected using convenient pressure differences (ca 13 kPa or 100 mm Hg). Commercial concentrates for blending dialyzing solution are widely available but the inconvenience and cost of the large volume of solution required has spawned many efforts at *in situ* processing of dialyzing solution, notably by enzymatic degradation of urea, ion exchange of salts, and nonspecific sorption of other toxins. So far these efforts have been, at best, marginally acceptable because of their cost.

The principal innovator and proponent of hemodialysis has been Kolff, head of the Artificial Organs Division at the University of Utah, who first treated patients by dialysis in wartime Holland (1944) and invented several improved artificial kidneys. Kolff developed, following an earlier design of Inouye and Clark, a twin-coil dialyzer (Fig. 6) which became the progenitor of a large family of spiral-flow cartridges that are still widely used for hemodialysis. Other designs have used multiple parallel paths of rectangular cross-section, where the opposed wide faces of the rectangles are the membrane. An advanced modern version of this type of device, developed by Hoeltzenbein, is shown in Figure 7.

Throughout the history of the development of the artificial kidney, which has been more intense by far than any other development process involving dialysis, membranes other than cellulose have been sought; none has been found to offer suf-

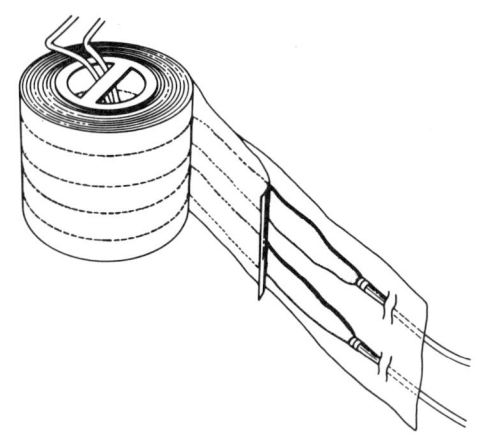

Figure 6. Twin-coil artificial kidney. Courtesy of Baxter Travenol Laboratories, Inc.

ficient advantage to justify its greater cost, although one, a polyacrylonitrile developed by Rhone-Poulenc in France, is in clinical use. Among cellulose membranes there are significant variations, both in relative permeabilities among solutes and in the balance between the desirable properties of high absolute permeabilities, limited hydraulic permeability (ultrafiltration flux), and adequate mechanical strength. Cellulose membranes prepared by the copper–ammonium process are generally preferred (see Cellulose). The subject of hemodialysis has a voluminous literature, an excellent, recent introduction has been provided (25); see also ref. 16.

Probably the most important technical innovation in dialysis technology in support of hemodialysis is the development of hollow-fiber dialysis membranes and devices. As with sheet membranes, most hollow fibers are cellulosic although other materials (26) have also been prepared and evaluated. Fibers have diameters in the neighborhood of 200 μm with wall thicknesses of 20–30 μm. The current designs have several thousand fibers, each 10–30 cm long in order to compete with dialyzers using sheet membranes of ca 1 m^2 in area operated at pressures of ca 13 kPa (100 mm Hg) for propelling blood through the unit. These fibers are deployed so that a typical dialyzer looks like a miniature shell-and-tube heat exchange replete with headers, tube sheets, baffles, and shell (see Heat exchange technology). Most of the theory and design variations pertinent to shell-and-tube heat exchangers have been explored and the possibility of emulating various compact heat exchanger designs has been extensively considered. A contemporary hollow-fiber hemodialyzer is illustrated in Figure 8.

Hollow-fiber artificial kidneys are mass-produced, more than 10^6 units per year. Their designs as well as procedures for analysis of their performance and their optimization have been considered. Some direct use has been made of these devices as prototype separators for industrial processes but none are known to be in regular use on a large-scale. It is likely that efficient use in this context would require redesign. However, it is also likely that these units are used in small-scale processing of drugs, flavors, and other sensitive materials of high unit value. The impact of the development of hollow fibers for dialysis has, however, been blunted by the concomitant development of fibers for ultrafiltration and reverse osmosis. If dialysis is to occupy a significant place commercially, development of membranes of great specificity or staging processes that achieve specificity through appropriate concentration and recycling are required.

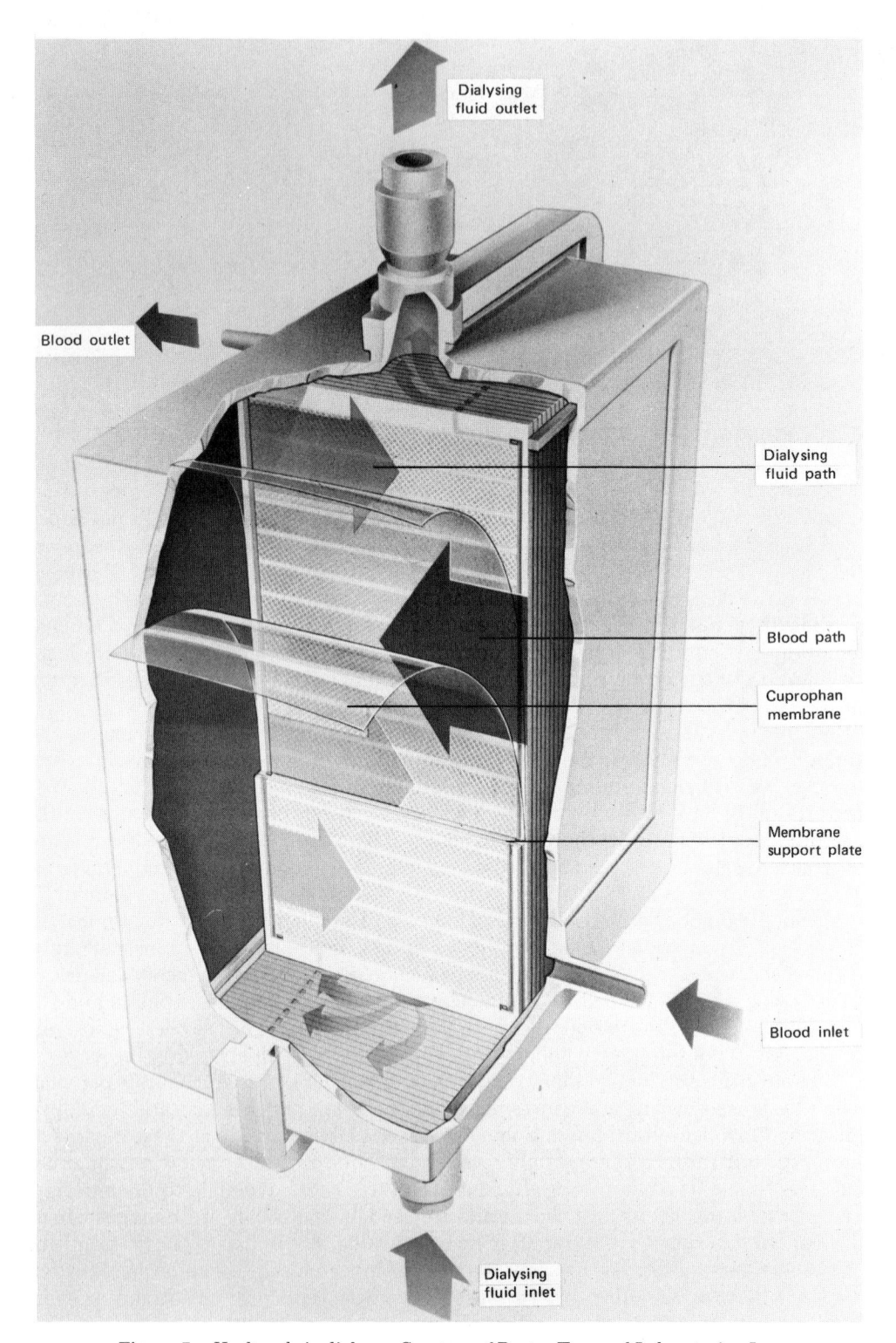

Figure 7. Hoeltzenbein dialyzer. Courtesy of Baxter Travenol Laboratories, Inc.

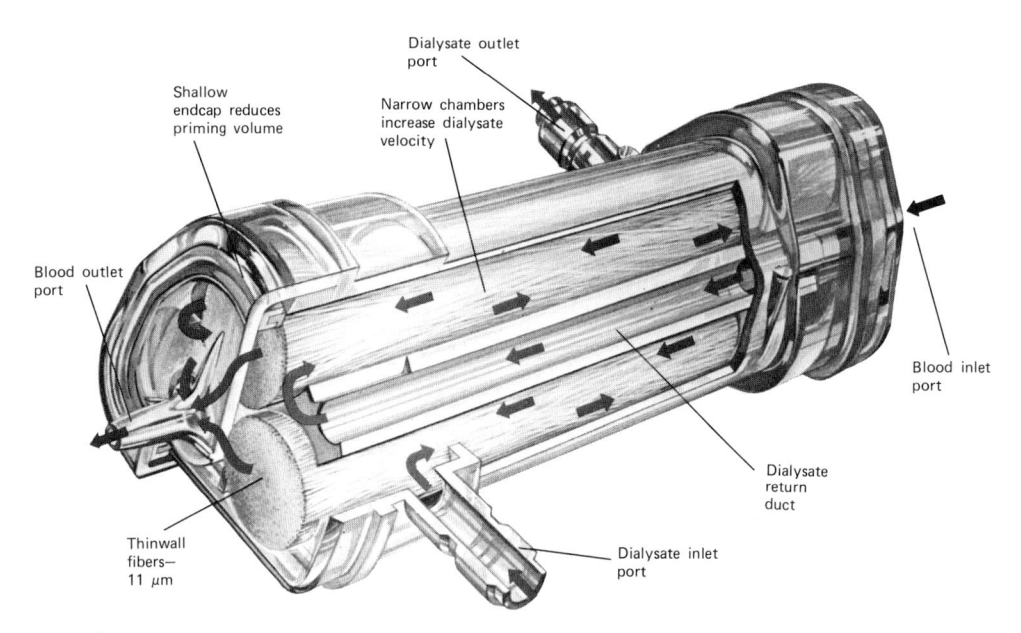

Figure 8. Hollow-fiber hemodialyzer. Courtesy of Extracorporeal Medical Specialties, Inc.

The increased use of enzymes (qv) in chemical and biochemical processing has created a new role for membranes, especially but not exclusively hollow fibers, as extensive surfaces in which to fix the enzymes and as diffusion barriers to protect the enzymes from unwanted, sometimes destructive reactants (27). The use of dialysis in such applications will be determined by the particulars of each situation.

Dialysis has been very frequently mentioned as a purification step in the preparation of small quantities of large molecules, or the separation of biochemical products from large molecules on a small scale. Craig (10) developed an ingenious thin-film dialyzer for laboratory use (see Fig. 9) that is still favored by some researchers and occasional mention is made of the use of small hemodialyzers for laboratory separations. Because of the lack of specificity, slowness of the process, frequent dilution of product, and increasing availability of laboratory ultrafiltrators with a specifiable cutoff range (often stated in effective pore size or molecular weight), the use of the time-honored cellulose sausage-casing dialysis bag is disappearing.

In analytical chemistry it is frequently necessary to separate interfering colloids before quantifying the amount of a crystalloid that is present or *vice versa*. Although some analytical procedures accordingly call for batch dialysis of samples, the principal use of dialysis in clinical chemistry is in conjuction with automated, continuous analytical machines (see Biomedical instrumentation). From the viewpoint of dialysis theory no innovations are involved: the membranes are usually cellulosic sheets over which a succession of samples, each separated from its neighbors by a gas bubble, is passed along a spiral pathway. Each sample, so separated, is gently mixed as it flows through the system and its concentration change over time mimics what it would experience if it were treated in a batch device.

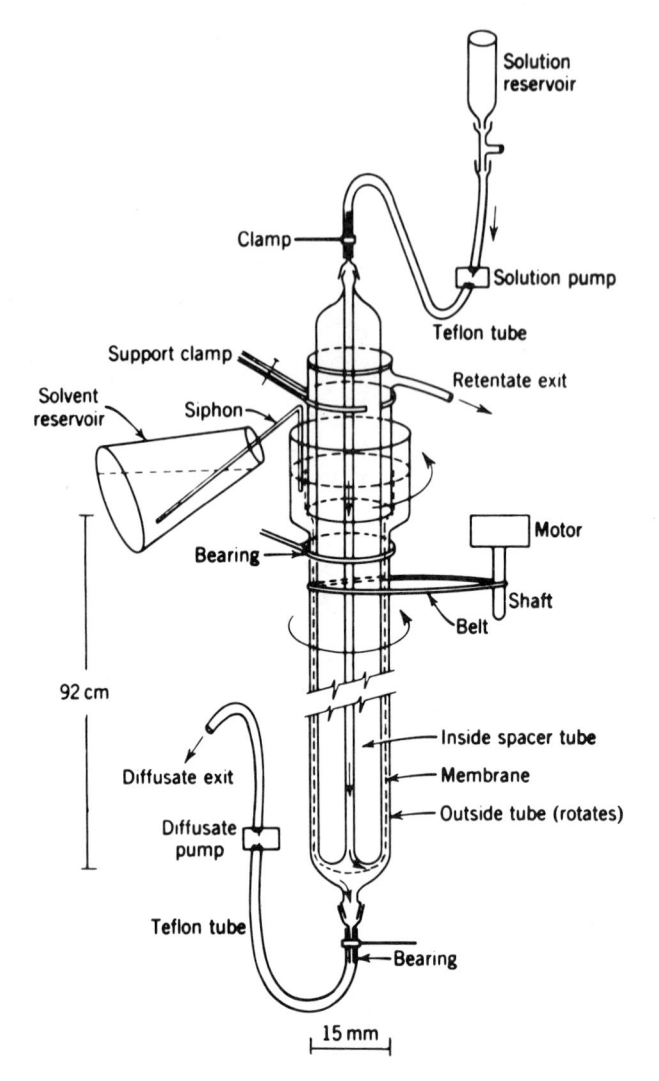

Figure 9. Thin-film dialyzer.

BIBLIOGRAPHY

"Dialysis" under "Dialysis and Electrodialysis" in *ECT* 1st ed., Vol. 5, pp. 1–20, by F. K. Daniel, Chemical Consultant; "Dialysis" in *ECT* 2nd ed., Vol. 7, pp. 1–21, by E. F. Leonard, Columbia University.

1. T. Graham, *Phil. Trans. Roy. Soc. London* **144,** 177 (1854).
2. K. S. Spiegler, *Trans. Farad. Soc.* **54,** 1409 (1958); S. R. Partington, *A History of Chemistry,* Macmillan, New York, 1964.
3. O. Kedem and A. Katchalsky, *Biochim. Biophys. Acta* **27,** 229 (1958).
4. A. Katchalsky and P. F. Curran *Non-Equilibrium Thermodynamics in Biophysics,* Harvard University Press, Cambridge, Mass., 1967, p. 124.
5. A. J. Staverman, *Rec. Trav. Chem.* **70,** 344 (1951).
6. J. A. Lane and J. W. Riggle, *Chem. Eng. Prog. Symp. Ser.* **55,** 127 (1959); S. B. Tuwiner, *Diffusion and Membrane Technology,* American Chemical Society Monograph No. 156, Reinhold, New York, 1962; J. D. Ferry, *Chem. Rev.* **18,** 414 (1936); J. D. Ferry, *J. Gen. Physiol.* **20,** 95 (1936); E. M. Renkin, *J. Gen. Physiol.* **38,** 225 (1954).

7. V. W. Sidel and A. K. Soloman, *J. Gen. Physiol.* **41,** (1957); D. A. Goldstein and A. K. Solomon, *J. Gen. Physiol.* **44,** (1960); J. R. Pappenheimer, E. M. Renkin, and L. M. Borrero, *Am. J. Physiol.* **167,** 13 (1951); J. R. Pappenheimer, *Physiol. Rev.* **33,** 387 (1953).
8. R. E. Beck and J. S. Schultz, *Science* **170,** 1302 (1970).
9. L. C. Craig, *Science* **144,** 1093 (1964).
10. L. C. Craig in N. M. Bikales, ed., *Encyclopedia of Polymer Science and Technology,* Vol. 4, John Wiley & Sons, Inc., New York, 1966, p. 824.
11. T. M. S. Chang, *Artificial Cells,* Charles C Thomas, Publisher, Springfield, Ill., 1966.
12. E. F. Leonard and L. W. Bluemle, Jr., *Trans. Am. Inst. Artif. Int. Organs* **6,** 33 (1960); E. F. Leonard and L. W. Bluemle, Jr., *Trans. N.Y. Acad. Sci.* **21,** 585 (1959).
13. B. H. Vroman, *Ind. Eng. Chem.* **54,** 21 (1962); R. Dvorin, *Met. Finish.* **57**(4), 52, 62 (1959); J. F. Muldoon and E. F. Leonard, *Trans. Am. Inst. Chem. Eng.* (1960).
14. R. B. Bird, W. E. Stewart, and E. N. Lightfoot, *Transport Phenomena,* John Wiley & Sons, Inc., New York, 1960, p. 647.
15. I. Noda, *Development of Dialysis Systems and Theoretical and Experimental Characterization of Counter-Current Hollow Fiber Dialyzers,* thesis, Columbia University, New York, 1978; J. E. Sigdell, *A Mathematical Theory for the Capillary Artificial Kidney* (in Engl.), Hippokrates Verlag, Stuttgart, FRG, 1974.
16. E. Klein and co-workers, *Evaluation of Hemodialyzers and Dialysis Membranes,* DHEW Pub. No. (*NIH*) 77-1294, U.S. Government Printing Office, Washington, D.C., 1977.
17. *Chem. Eng.* **66**(9), 118 (1959).
18. A. Barrera, *Free Convection Effects in Dialysis Cells,* thesis, Columbia University, New York, 1970.
19. I. Noda and C. C. Gryte, *AIChE J.* **25**(1), 113 (1979).
20. H. B. Volrath, *Chem. Met. Eng.* **43,** 303 (1936); D. J. Eynon, *J. Soc. Chem. Ind.* **52,** 173T (1936).
21. M. Bier, ed., *Membrane Processes in Industry and Biomedicine,* Plenum Press, New York, 1971; R. E. Lacey and S. Loeb, eds., *Industrial Processing with Membranes,* Wiley-Intersicence, New York, 1972.
22. H. P. Gregor and J. D. Gregor, *Sci. Am.* **239,** 112 (1978).
23. M. R. Friedman and E. L. Gaden, Jr., *Biotechnol. Bioeng.* **12,** 961 (1970); P. Landwall and T. Holme, *J. Appl. Chem. Biotechnol.* **27,** 165 (1977); G. A. Coulman, R. W. Stieber, and P. Gerhardt, *Appl. Environ. Microbiol.* **34,** 725 (1977); R. W. Stieber, G. A. Coulman, and P. Gerhardt, *Appl. Environ. Microbiol.* **34,** 733 (1977).
24. S. B. Tuwiner, *Diffusion and Membrane Technology,* American Chemical Society Monograph No. 156, Reinhold, New York, 1962; R. Dvorin, *Met. Finish.* **57**(4), 52, 62 (1959).
25. E. Klein, *Clin. Nephrol.* (*Tokyo*) (*Rinsho Shinkeigaku*) **9,** 131 (1978).
26. E. Klein and co-workers, *J. Membr. Sci.* **1,** 371 (1976).
27. W. R. Vieth and K. Venkatasubramanian, *Chem. Tech.* **4,** 434 (1974); L. R. Waterland, A. S. Michaels, and C. R. Robertson, *AIChE J.* **20**(1), 50 (1974).

EDWARD F. LEONARD
Columbia University

DIAMINES AND HIGHER AMINES, ALIPHATIC

The ethyleneamine family of products consists of ethylenediamine (EDA), diethylenetriamine (DETA), triethylenetetramine (TETA), tetraethylenepentamine (TEPA), pentaethylenehexamine (PEHA), aminoethylpiperazine (AEP), and diethylenediamine (DEDA), more commonly known as piperazine. In this family of compounds, EDA, DETA, AEP, and DEDA are available in a pure form (>98% purity), whereas the higher molecular weight members, TETA, TEPA, and PEHA, are mixtures of linear, branched, and cyclic isomers (1).

The propyleneamine family of products consists of 1,2-propanediamine (1,2-PDA), 1,3-propanediamine (1,3-PDA), iminobispropylamine (IBPA), and dimethylaminopropylamine (DMAPA) (see also Amines, diarylamines; Amines, cyclic).

Table 1. Properties and Manufacturers of Commercial Diamines and Higher Amines

Commercial name	CAS Registry No.	Abbreviation	Molecular formula
ethylenediamine	[107-15-3]	EDA	$H_2N(CH_2)_2HN_2$
diethylenetriamine	[111-40-0]	DETA	$H_2N(CH_2)_2NH(CH_2)_2NH_2$
triethylenetetramine	[112-24-3]	TETA	$H_2N(CH_2CH_2NH)_2CH_2CH_2NH_2$
tetraethylenepentamine	[112-57-2]	TEPA	$H_2N(CH_2CH_2NH)_3CH_2CH_2NH_2$
pentaethylenehexamine	[4067-16-7]	PEHA	$H_2N(CH_2CH_2NH)_4CH_2CH_2NH_2$
aminoethylpiperazine	[140-31-8]	AEP	$NH_2CH_2CH_2N(CH_2CH_2)_2NH$
piperazine	[110-85-0]	DEDA	$HN(CH_2CH_2)_2NH$
propylenediamine	[78-90-0]	1,2-PDA	$H_2NCH(CH_3)CH_2NH_2$
1,3-diaminopropane	[109-76-2]	1,3-PDA	$H_2NCH_2CH_2CH_2NH_2$
iminobispropylamine	[56-18-8]	IBPA	$H_2N(CH_2)_3NH(CH_2)_3NH_2$
dimethylaminopropylamine	[109-55-7]	DMAPA	$(CH_3)_2NCH_2CH_2CH_2NH_2$
menthanediamine	[80-52-4]	MDA	(structure: cyclohexane with CH$_3$, H$_2$N, C(CH$_3$)$_2$, NH$_2$ substituents)
triethylenediamine	[280-57-9]	TEDA	$N(CH_2CH_2)_3N$
N,N,N',N'-tetramethylethylenediamine	[110-18-9]	TMEDA	$(CH_3)_2NCH_2CH_2N(CH_3)_2$
N,N,N',N'-tetramethyl-1,3-butanediamine	[97-84-7]	TMBDA	$(CH_3)_2NCH(CH_3)CH_2CH_2N(CH_3)_2$
hexamethylenediamine	[124-09-4]	HMDA	$H_2N(CH_2)_6NH_2$

[a] 101.3 kPa = 1 atm.
[b] Key to manufacturers: AP = Air Products Co.; B = BASF-Wyandotte Corp.; D = The Dow Chemical Co.; & Haas Co.; UC = Union Carbide Corp.
[c] Tag closed cup.
[d] Pensky-Martin closed cup.
[e] At 0.67 kPa, 10–60% distills in this range.
[f] Tag open cup.
[g] At 93.3°C.
[h] At 1.3 kPa.
[i] At 25°C.
[j] Heat of sublimation, below 78°C.
[k] For manufacture of HMDA in preparation of nylon-6,6, see Polyamides.

Physical Properties

Physical properties of the commercially available polyamines are tabulated in Table 1. These products are usually slightly viscous liquids with a strong ammoniacal odor. Most are completely miscible in water. Exceptions are PEHA which forms a gel when mixed with water at ca 50:50 wt proportions, and TEDA which has a solubility of 33 g/100 g H$_2$O. They are also completely miscible with alcohol, acetone, benzene, and ethyl ether, but only slightly soluble in heptane.

Molecular weight	Freezing point, °C	Boiling point at 101.3 kPa, °C	Heat of vaporization at 101.3 kPa[a], kJ/mol	n_D^{20}	Viscosity at 20°C, mPa·s (= cP)	Flash point, °C	Manufacturers[b]
60.1	10.8	117.0	40.4	1.4565	1.6	40[c]	D, UC, J
103.2	−35	206.7	51.3	1.4859	7.1	98[d]	D, UC
146.2	−39	277.4	54.4	1.4986	26.7	118[d]	D, UC
189.3	<−40	340.3	63.3	1.5067	96.2	154[d]	D, UC
232.4	−30	180–280[e]			300–400	166[d]	D, UC
129.2	−17.6	222.0	51.1	1.5003	14.1	93[f]	D, UCC, J
86.1	109.6	148.5				85[f]	D, UCC, J
74.1	−37.2	120.9	38.2	1.4455	1.6	43[f]	B
74.1	−12	139.7	46.4[g]	1.4555	2.0	50[f]	J, B
131.2	−15.1	238.3	76.2[g]	1.4791	9.6	121[d]	J
102.2	−100	134.9	35.6	1.4350	1.1	45[f]	J
170.3	−45	107–126[h]		1.479	17.5[i]	102[f]	R & H
112.2	158	174	61.9[j]			>50	AP, J
116.2	−55.1	119–122		1.4160		17[f]	R & H
114.3	<−100	165.1	30.0	1.4311	1.0	46[f]	UC
116.2	40.9	204.5	51.0			85[f]	DP[k]

DP = E. I. du Pont de Nemours & Co., Inc.; J = Jefferson Chemical Div., Texaco Corp.; R & H = Rohm

Ethylenediamine forms the following binary azeotropes:

Secondary component	bp, °C	EDA, wt %
water	119	81.6
ethylene glycol monomethyl ether	130	31–32
n-butanol	125	35.7
isobutanol	120	50
toluene	104	30.8

Chemical Properties

The aliphatic polyamines are, of course, strong bases. The pK_b values for some of the lower members, listed below, indicate that they are stronger bases than ammonia and only slightly weaker than alkyl amines:

Amine	pK_1	pK_2	Ref.	Amine	pK_1	pK_2	Ref.
EDA	3.83	6.56	2	TETA	3.84	4.47	4
1,2-PDA	4.03	6.90	2	NH_3	4.76		5
1,3-PDA	3.70	5.71	3	$C_2H_5NH_2$	3.25		5
DETA	3.89	4.60	4				

Reactions. The aliphatic polyamines undergo reactions typical of compounds containing primary and secondary amines. The more important reactions with a variety of reagents are summarized below. Ethyleneamines are most often used in the examples cited; however, the other polyfunctional amines exhibit similar chemical behavior.

Inorganic Acids. The aliphatic polyamines react readily with the common inorganic acids to form the corresponding salts. These products are water soluble, crystalline materials that react with strong bases to regenerate the original amine.

At elevated temperatures, the reaction with nitric acid takes a somewhat different course; two molecules of water are split from the initial reaction product, yielding the explosive compound ethylenedinitramine [505-71-5] (6).

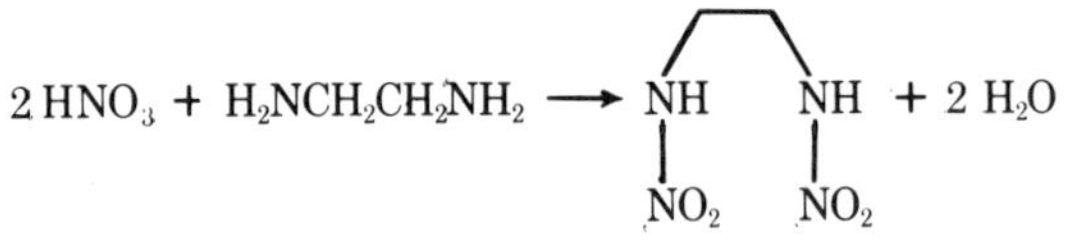

$$2\ HNO_3 + H_2NCH_2CH_2NH_2 \longrightarrow \underset{\underset{NO_2}{|}}{NH}\frown\underset{\underset{NO_2}{|}}{NH} + 2\ H_2O$$

Alkylene Oxides and Glycols. The aliphatic polyamines react with alkylene oxides, such as ethylene oxide or propylene oxide. The mole ratio of the reactants and the reaction conditions control the products obtained. With a large excess of oxide and a caustic soda catalyst, all four primary-amine hydrogen atoms are replaced, oxyethylene chains are formed, and a product having the structure shown below results (7–8).

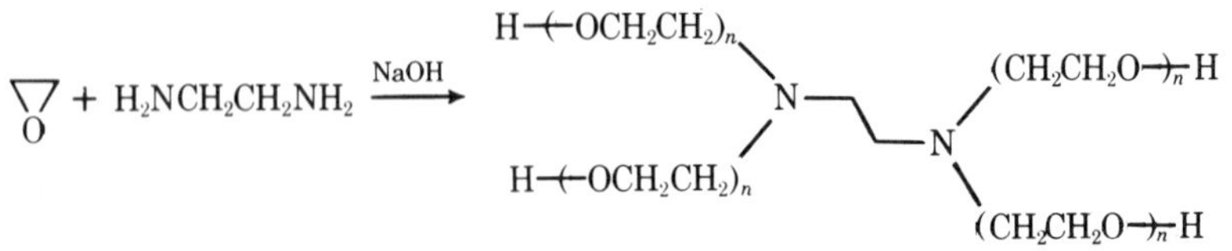

$$\underset{O}{\triangledown} + H_2NCH_2CH_2NH_2 \xrightarrow{NaOH} \begin{array}{c} H\text{-}(OCH_2CH_2)_n \\ \diagdown \\ N\text{---}\diagup\diagdown\text{---}N \\ \diagup \\ H\text{-}(OCH_2CH_2)_n \end{array} \begin{array}{c} (CH_2CH_2O)_n\text{-}H \\ \\ (CH_2CH_2O)_n\text{-}H \end{array}$$

With less oxide and no catalyst, mono-, di-, tri-, and tetraoxyalkylated products are obtained in which one oxide unit, $HOCH_2CH_2$—, is substituted for each amine hydrogen replaced. For example, a 1:1 amine:oxide ratio gives a 45% yield of the monooxyalkylated product (9) (see Polyethers).

The gas-phase reaction of EDA and 1,2-propylene glycol over an alumina catalyst gives 2-methylpiperazine [109-07-9] (10).

$$H_2NCH_2CH_2NH_2 + HOCH_2CHOH\underset{CH_3}{|} \longrightarrow HN\underset{CH_3}{\bigcirc}NH + 2H_2O$$

Aldehydes. Ethyleneamines react exothermically with aliphatic aldehydes to form a variety of adducts. The adducts formed depend upon the stoichiometry and reaction conditions employed. Reaction of the aldehyde with one of the primary amine groups results in the formation of a Schiff base that undergoes cyclization to a 2-substituted imidazolidine (11).

$$H_2NCH_2CH_2NH_2 + RCHO \longrightarrow RCH{=}NCH_2CH_2NH_2 \longrightarrow R{-}\overset{H}{\underset{H}{\underset{N}{\overset{N}{\bigg\langle}}}}\bigg]$$

Higher ethyleneamines yield 1,2-disubstituted imidazolidines.

$$HN(CH_2CH_2NH_2)_2 + RCHO \longrightarrow R{-}\overset{H}{\underset{\underset{CH_2CH_2NH_2}{|}}{\underset{N}{\overset{N}{\bigg\langle}}}}\bigg] + H_2O$$

Analogous reactions with 1,3-PDA yield 2-alkylhexahydropyrimidines (12). When an excess of aldehyde is used, a complex series of reactions ensues, leading to highly cross-linked resins (13–14).

Ethyleneamines and formaldehyde can react to give complex quinuclidine-like structures (15).

Organic Acids and Acid Derivatives. EDA and its homologues react when heated with an excess of a monobasic carboxylic acid to yield mono- and disubstituted amides (16).

$$RCOOH + H_2NCH_2CH_2NH_2 \longrightarrow RCONHCH_2CH_2NH_2 + H_2O$$

$$2\,RCOOH + H_2NCH_2CH_2NH_2 \longrightarrow R\underset{H}{\overset{O}{\overset{\|}{N}}}C\frown N\underset{H}{\overset{O}{\overset{\|}{C}}}R + 2H_2O$$

The same products are obtained with acyl halides, acid anhydrides, or esters (17).

The monoamides undergo further condensation to yield the cyclic 2-substituted imidazoline derivatives. Monoacetylethylenediamine, heated with calcium oxide, yields 2-methyl-2-imidazoline (18).

$$\underset{NH_2}{\overset{H}{\underset{}{\overset{N}{\bigg\langle}}}}\overset{}{\underset{O}{\overset{\|}{C}}}{-}CH_3 \xrightarrow{CaO} \underset{N}{\overset{H}{\underset{}{\overset{N}{\bigg\langle}}}}{-}CH_3$$

Similar results are obtained by distillation at reduced pressure in the absence of catalyst (19). One-step procedures, ie, without isolation of the intermediate monoamides, have been used to prepare 2-pentyl-2-imadazoline (20) and 2-heptadecyl-2-imidazoline (21) from EDA and caproic or stearic acids, respectively. The formation of 2-alkyl-2-imidazolines in high boiling organic solvents such as ethylene glycol in the presence of acidic catalysts has also been reported (22).

The higher ethyleneamines, such as DETA, react with carboxylic acids (23) or their lower alkyl esters (24) to form 1,2-disubstituted imidazolines.

$$HN(CH_2CH_2NH_2)_2 + RCOOH \longrightarrow R-\underset{\underset{CH_2CH_2NH_2}{|}}{\overset{N}{\underset{N}{\bigvee}}} + 2\ H_2O$$

The diamines of ethylenediamine condense with formaldehyde to form 1,3-diacyl imidazolidines (25).

$$R-\overset{O}{\overset{\|}{C}}-\underset{H}{\overset{H}{N}}\diagdown\diagup\underset{-N}{\overset{H}{|}}-\overset{O}{\overset{\|}{C}}-R + HCHO \longrightarrow R-\overset{O}{\overset{\|}{C}}-N\diagdown\diagup N-\overset{O}{\overset{\|}{C}}-R + H_2O$$

Dibasic acids and their derivatives, as well as certain dimerized and trimerized vegetable oils (qv), polymerize with ethyleneamines, forming polyamide resins (26) (see Polyamides).

Acid–amine reaction products may be further condensed with a wide variety of compounds to yield additional derivatives. Compounds that may be used in carrying out such condensations include formaldehyde, dibasic acids, dimethyl sulfate, and epoxy resins (qv) (27–29).

Other Carboxylic Acid Derivatives. EDA reacts with urea (qv) to give 2-imidazolidinone (ethyleneurea) (30–31).

$$NH_2-\overset{O}{\overset{\|}{C}}-NH_2 + H_2NCH_2CH_2NH_2 \longrightarrow HN\diagdown\diagup NH + 2\ NH_3$$

Ethyleneurea undergoes further condensation with formaldehyde to form 1,3-dimethylolimidazolidone (32).

$$HN\diagdown\diagup NH + 2\ CH_2O \longrightarrow HOCH_2N\diagdown\diagup NCH_2OH$$

With DETA and the higher ethyleneamine homologues, a number of urea reaction products are possible (33).

Organic Halides. Alkyl halides and those aryl halides that are activated by NO_2 or similar groups in the ortho or para positions can be condensed with ethylenediamine to give mono- and disubstituted derivatives, depending upon the ratios used (34).

$$RX + H_2NCH_2CH_2NH_2 \rightarrow RNHCH_2CH_2NH_2 \cdot HX$$
$$2\ RX + H_2NCH_2CH_2NH_2 \rightarrow RNHCH_2CH_2NHR \cdot 2HX$$

With aliphatic dihalides, polymeric, cross-linked, water soluble cationic products are obtained (35–36).

$$2n\ XRX\ +\ n\ HN(CH_2CH_2NH_2)_2 \longrightarrow \displaystyle -\!\!\!\left[RNHCH_2CH_2\underset{\underset{HX}{|}}{\overset{\overset{R}{|}}{N}}CH_2CH_2NH\right]\!\!\!-_n$$
$$\overset{}{HX}\overset{}{HX}\overset{}{HX}$$

Treating EDA with 2,2'-dichloroethyl ether in the presence of Na_2CO_3 followed by extraction with an aromatic hydrocarbon gives 1,2-dimorpholinoethane [1723-94-0] (37).

$$H_2NCH_2CH_2NH_2\ +\ O(CH_2CH_2Cl)_2\ \longrightarrow\ \text{[1,2-dimorpholinoethane]}\ +\ 4\ HCl$$

α,α,α-Trichlorotoluene reacts with EDA to give 2-phenyl-2-imidazoline (38).

$$\text{Ph}-CCl_3\ +\ H_2NCH_2CH_2NH_2\ \longrightarrow\ \text{[2-phenyl-2-imidazoline]}\ +\ 3\ HCl$$

With chloroacetic acid, EDA gives ethylenediaminetetraacetic acid

$$4\ ClCH_2COOH\ +\ H_2NCH_2CH_2NH_2\ \longrightarrow\ \text{[EDTA structure]}\ +\ 4\ HCl$$

Cyanides and Nitriles. Under suitable alkaline conditions EDA reacts with sodium cyanide and formaldehyde to give the tetrasodium salt of ethylenediaminetetraacetic acid (39). The reaction is carried out at elevated temperatures and under a partial vacuum to remove by-product ammonia. The tetrasodium salt is easily converted to the tri-, di-, and monosodium salts, and to the free acid, by treatment with sulfuric or hydrochloric acid. These products find wide use in a variety of industries as chelating agents (see Chelating agents).

EDA and hydrogen cyanide react in diethyl ether at 100°C to give 2-imidazoline (40).

$$HCN\ +\ H_2NCH_2CH_2NH_2\ \longrightarrow\ \text{[2-imidazoline]}\ +\ NH_3$$

2-Alkyl-2-imidazolines are formed by the reaction of ethylenediamine with aliphatic nitriles in the presence of sulfur (41) or polysulfides (42) as catalysts.

$$RCN\ +\ H_2NCH_2CH_2NH_2\ \longrightarrow\ R\!-\!\text{[2-alkyl-2-imidazoline]}\ +\ NH_3$$

Ethyleneamines add readily to acrylonitrile (qv) in the presence of a catalyst, forming cyanoethyl derivatives (43–44).

$$\text{H}_2\text{NCH}_2\text{CH}_2\text{NH}_2 + 2\ \text{CH}_2\!\!=\!\!\text{CHCN} \longrightarrow \underset{\substack{|\\ \text{H}}}{\overset{\substack{\text{H}\\|}}{\text{N}}}\!\!\diagup\!\!\diagdown\!\!\text{CN}\ \ \ \text{N}\!\!\diagdown\!\!\text{CN}$$

The reaction is exothermic, and hence requires cooling to prevent polymerization of the acrylonitrile (see also Cyanoethylation).

Upon prolonged heating at 100–140°C, DETA and higher homologues combine with cyanamide or dicyandiamide producing water-soluble condensation products (45) (see Cyanamides).

Sulfur and Sulfur Compounds. Sulfur reacts exothermically with ethyleneamines to form a mixture of products of uncertain composition. The product mixture can be solubilized by heating at 80–100°C for a short period or by aging for several days (46–47). In the presence of a base catalyst, such as sodium hydroxide, EDA readily adds two moles of carbon disulfide to yield ethylenebisdithiocarbamate (48).

$$\text{H}_2\text{NCH}_2\text{CH}_2\text{NH}_2 + 2\ \text{CS}_2 \xrightarrow{\text{NaOH}} \begin{array}{c}\text{H}\ \ \ \ \text{S}\\ |\ \ \ \ \|\\ \text{N}\!\!-\!\!\text{C}\!\!-\!\!\text{SNa}\\ |\\ \text{N}\!\!-\!\!\text{C}\!\!-\!\!\text{SNa}\\ |\ \ \ \ \|\\ \text{H}\ \ \ \ \text{S}\end{array}$$

This product can also be prepared from zinc dithiocarbamate and EDA (49).

With a 1:1 mole ratios of reactants, cyclization occurs and 2-mercapto-2-imidazoline is formed.

$$\text{H}_2\text{NCH}_2\text{CH}_2\text{NH}_2 + \text{CS}_2 \longrightarrow \begin{array}{c}\text{N}\\ \diagup\ \ \diagdown\\ |\ \ \ \ \ \ \rangle\!\!-\!\!\text{SH}\\ \diagdown\ \ \diagup\\ \text{N}\\ |\\ \text{H}\end{array}$$

Sulfonamides can be prepared from ethyleneamines; eg, by heating EDA with a saturated aliphatic sulfonyl chloride (50) or with mono-, di-, or tetrasubstituted benzenesulfonyl chloride, or with p-toluenesulfonyl chloride (51).

Cyclization. At elevated temperatures and pressures, in the presence of various fixed-bed catalysts, EDA and the higher ethyleneamines are transformed into piperazine, substituted piperazines, and similar cyclic compounds (see Amines, cyclic). Over a kaolin–alumina catalyst at 280–400°C, eg, DETA is converted into a mixture of piperazine, pyrazine and triethylenediamine (52–53).

$$\text{HN}(\text{CH}_2\text{CH}_2\text{NH}_2)_2 \longrightarrow \text{HN}\!\!\bigcirc\!\!\text{NH} + \text{N}\!\!\bigcirc\!\!\text{N} + \text{N}(\text{CH}_2\text{CH}_2)_3\text{N}$$

Higher temperatures appear to favor pyrazine formation, and lower temperatures favor formation of piperazine and substituted piperazines (54). Other catalysts that have been used in these cyclization reactions include mixed transition metal oxides (55–56) and Al–Ti–Ni alloys (57). AEP may also be obtained in good yields (56).

Manufacture

Ethyleneamine Process. Two processes are currently used to prepare ethyleneamines on a commercial scale: (1) the reaction of aqueous ammonia with 1,2-dichloroethane (EDC), and (2) the catalytic amination of monoethanolamine (MEA). The EDC–ammonia process is the older method and represents the major commercial route. This process yields the entire family of ethyleneamines: EDA, DEDA, DETA, TETA, TEPA, PEHA, and AEP (58–59). These polyamines are produced as their hydrochloride salts, and must be neutralized, typically with aqueous caustic soda (59), to obtain the free amines. The by-product salt produced in the neutralization step is separated and the individual products are isolated by fractional distillation (60).

The product distribution obtained in the EDC–NH_3 reaction is influenced by the ratio of EDC:NH_3 employed, the reaction time and the reaction temperature (61). The EDC:NH_3 ratio is most often used to control the product distribution. A typical distribution obtained at a 15:1 mole ratio of NH_3:EDC, 100°C and 4.82 MPa (47.6 atm) is EDA 52%, DETA 19%, TETA 12%, TEPA plus PEHA 11%, and residue 6% (62). At NH_3:EDC ratios in excess of 30:1 the EDA percentage can be increased to 87% (63). At ratios of NH_3:EDC as low as 2.6:1 yields of the higher polyethylenepolyamines can be maximized (64). Another method of controlling product distribution is to recycle one or more of the product polyamines to the primary reactor (65). Recycling of either EDA (62) or DETA (66) increases the percentage of higher polyethylenepolyamines in the product mix. The condensation of EDA in the presence of hydrogen over Cu–Ni–Co catalysts has been used to prepare DETA in good yields (67). The thermodynamic and kinetic aspects of the EDC–NH_3 reaction (68) and various reactor configurations (69) have been studied.

The catalytic reductive amination of MEA (70–71) is more selective than the EDC process, producing primarily EDA. Small amounts of DEDA, DETA, AEP and other amines are coproduced, however. In this process, MEA, excess ammonia, and, optionally, hydrogen are passed over a fixed-bed catalyst near 200°C and 13.9–27.7 MPa (2000–4000 psig) (72–73). The catalysts may be Raney nickel or various combinations of Ni with Mg, Co, Cu, Cr, and other transition metals (73–74). In addition to MEA, ethylene oxide (75–78), ethylene glycol (77,79) or mixtures of MEA and diethanolamine (72) may be used as feedstocks in the reductive amination reaction (see Amines by reduction). In the commercial production of piperazine, via the reductive amination of MEA, EDA (80), and DETA (81) are obtained as coproducts. Piperazine and EDA may also be prepared from the reductive amination of sucrose (82).

The reaction of formaldehyde, hydrogen cyanide, ammonia, and hydrogen over Co or Ni catalysts at 110–120°C gives EDA in good yields (83). This process may also be conducted by using glycolonitrile ($HOCH_2CN$), previously prepared by the condensation of hydrogen cyanide and formaldehyde, as the feedstock (84–85). Some quantities of DETA and TETA may also be isolated.

Propyleneamine Process. The reductive amination of propylene oxide (78), 1,2-propylene glycol (77), or monoisopropanolamine (86) are useful procedures for the preparation of 1,2-PDA. Similarly, 1,3-PDA may be prepared from 1,3-propanediol (77).

Acrolein (qv) and ammonia may be combined and the resulting aminonitrile can be catalytically reduced to provide a mixture of 1,3-PDA, IBPA, and higher molecular weight polypropylenepolyamines (87). Acrylonitrile may be used as the raw material in this reaction sequence to yield a similar mixture of products (88–89). The oligomerization of 1,3-PDA over supported Cu–Co–Ni catalysts in the presence of hydrogen gives a mixture of polypropyleneamines (90–91). 1,2-PDA and 1,3-PDA can be produced from aqueous ammonia and the appropriate dichloropropanes (92). These processes are similar to the production of EDA from EDC in that a mixture of polypropyleneamines is produced and the product mix is dependent upon the mole ratios of ammonia to dichloropropane used.

Economic Aspects

Annual production capacities for the ethyleneamine family of products in 1979, broken down according to geographical areas and major producers in each area, are shown in Table 2. Of the estimated 181,000 t/yr of total capacity, about 85,000 t/yr represents EDA capacity, and the remaining 96,000 t/yr represents capacity for DETA and higher homologues. In Europe, the capacities are somewhat skewed toward EDA since some of the European capacity is based on the reductive amination of MEA, which gives an abnormally high ratio of EDA:higher homologues relative to the EDC route. A further breakdown of the capacity for ethyleneamine higher homologues into capacities for specific products is not feasible since most manufacturers have the capability for varying production mix according to market demands.

Announced expansions are expected to increase total United States ethyleneamines capacity to approximately 136,000 t/yr by 1983. In Europe, expansions and new plant construction will increase total capacity to ca 85,000 t/yr by 1983. Of the latter 54,000 t/yr capacity increase, 22,000 t/yr will be in EDA and 32,000 t/yr will be in higher homologues. Thus by 1983 world capacity for ethyleneamines is expected to reach 235,000 t/yr, with 107,000 t/yr as EDA and 128,000 t/yr as DETA and higher homologues.

Prices of some of the major commercial polyamines in effect in 1979, in the United States, are given in Table 3. European pricing tends to be 10–15% higher than United States prices on the higher amines, and somewhat lower than United States prices on EDA.

Table 2. Ethyleneamines Production Capacities, 1979

Area, company	Capacity, 10^3 t/yr
United States	91
Dow	28
Union Carbide	63
Western Europe	76
BASF[a]	25
Bayer	5
Berolkemi[a]	10
Delamine	15
Dow	14
Montedison	7
Japan	14
Kento Denka	2
Seitetsu Kagaku	2
Shunan Petrochemical	10
Approximate free world capacity[b]	181

[a] By reductive amination of MEA, others use EDC route.
[b] Expected to reach 235,000 metric tons by 1983.

Table 3. Prices of Commercially Available Polyamines

Product	Price, $/kg[a]
ethylenediamine	1.84
diethylenetriamine	2.34
triethylenetetramine	2.27
tetraethylenepentamine	2.69
ethyleneamine highers[b]	2.01
propylenediamine	3.84
1,3-propanediamine	2.07
iminobispropylamine	6.61
hexamethylenediamine	1.76
dimethylaminopropylamine	1.98
N,N,N',N'-tetramethyl-1,3-butanediamine	7.72
N,N,N',N'-tetramethylethylenediamine	8.11
menthanediamine	8.02

[a] Bulk or truckload drum prices, Apr. 1979.
[b] Product is 50–60% PEHA and remainder mostly higher polyethylenepolyamines.

Specifications and Test Methods

Specifications for a number of the commercially available aliphatic polyamines are given in Table 4. For more detailed specifications on these products and for specifications of other polyamines listed in Table 1, the individual manufacturers should be consulted.

The assay of the lower ethyleneamines, especially EDA and DETA, is frequently accomplished by gas chromatography (93). The gas chromatographic procedures should be applicable to other aliphatic polyamines as well. Owing to the high polarity of these compounds, severe tailing is often encountered in their assay by gas chro-

Table 4. Specifications of Some Aliphatic Polyamines

Commercial name	Assay, min %	Amine value, mg KOH/g	Specific gravity	Color, APHA, max
ethylenediamine	99.0		0.839–0.906[a]	15
diethylenetriamine	98.5		0.947–0.951[a]	30
triethylenetetramine		1410–1480	0.973–0.981[a]	50
tetraethylenepentamine		1290–1375	0.989–0.995[a]	
pentaethylenehexamine[b]		1100–1300	1.00–1.03[a]	
propylenediamine	95.0		0.855–0.885[c]	15
1,3-diaminopropane	99.0		0.886–0.890[c]	20
iminobispropylamine	98.0		0.926–0.936[c]	30
dimethylaminopropylamine	98.5		0.815–0.820[c]	15
hexamethylenediamine	99.9[d]		0.889–0.900[a]	10

[a] At 25°C/25°C.
[b] Commercial product contains 50–60% PEHA, remainder mostly highers.
[c] At 20°C/20°C.
[d] Anhydrous basis, commercial material contains 9–16% water.

matography when conventional supports are used (94). To counteract this, the use of silane-treated (95) or alkali hydroxide-treated (96) solid supports has been recommended. Fluorocarbon polymers may also be used as the solid support to reduce tailing (93). Poly(oxyethylene)glycols are frequently used as liquid phases.

Gas chromatography can also be used for the analysis of the higher ethyleneamines. Uncoated porous column packings based on poly(2,6-diphenyl-p-phenylene oxide) give good resolution of individual isomers from these products (97).

Owing to the complexity of the higher commercial ethyleneamines, it is more common to assay these materials by titration, either in aqueous or nonaqueous media (93). These procedures lead to an amine value for the product being analyzed, ie, a measure of the total basic nitrogen content of the product. Titrimetric procedures are also available to distinguish between primary, secondary, and tertiary amine content (93,98).

Thin layer chromatography has been used for the separation of polyethylenepolyamines from mixtures of these materials (99). In one such method the silica gel layer is impregnated with transition metal ions to improve resolution of the components (100). Liquid chromatography on cellulose columns has also been used (101).

Infrared spectroscopy can be used to estimate the quantity of piperazine ring structures in the higher polyethylenepolyamines (102).

The determination of EDA, DETA, and TETA in air has been accomplished by absorbing the amines in dilute aqueous HCl, converting the absorbed amines to colored reaction products with 1-chloro-2,4-dinitrobenzene, and reading the optical density at 430 nm (103). Sensitivity is reported as 10 μg of amine per sample.

Health and Safety Factors (Toxicology)

The aliphatic polyamines do not present any unusual flammability problem at ordinary temperatures. With the exception of TMEDA, all have flash points of at least 40°C (see Table 1).

The vapors of these products are painful and irritating to the eyes, nose, throat, and respiratory system. The liquids severely damage the eye and may cause serious burns upon contact with the skin. Both vapors and liquid are capable of producing hypersensitivity in some people, resulting in contact dermatitis or an asthmatic respiratory response, or both. When swallowed, the concentrated liquid materials may produce considerable local injury. They are not, however, particularly toxic systemically, and small amounts accidentally ingested should not cause systemic injury. The LD$_{50}$ values for white rats are shown in Table 5.

Strict precautions should be observed to prevent direct contact with the polyamines. Where there is any possibility of eye contamination, face shields or chemical workers' goggles should be worn. Eye showers and facilities for complete washing should be available in working areas. Rubber gloves and other suitable protective clothing should be worn as necessary to prevent skin contact. Breathing of vapors should be avoided.

If the eyes become contaminated with any of these materials, they should be flushed immediately and thoroughly with flowing water for at least 15 min, and medical aid should be obtained as soon as possible. In the event of skin contact, contaminated shoes and clothing should be removed as quickly as possible and the affected area flushed with copious amounts of flowing water. In cases of especially severe contam-

Table 5. Single-Dose Oral Toxicity, For White Rats, of Various Alkylenepolyamines

Product	LD_{50}, g/kg body wt
ethylenediamine	1.2
diethylenetriamine	1.4
triethylenetetramine	2.5
tetraethylenepentamine	2.1
pentaethylenehexamine	1.6
1,3-propanediamine	0.5
iminobispropylamine	1.4
dimethylaminopropylamine	1.9

ination, showering may be started while clothing is being removed. Medical attention should be obtained if skin injury or irritation is apparent or if pain persists. Contaminated clothing should be washed before reuse, and contaminated shoes discarded. If any of the polyamines are swallowed, copious amounts of water or milk should be given promptly and a physician called immediately. Vomiting should be induced only at the physician's instruction.

The ACGIH has adopted TLVs of 10 ppm (25 mg/m^3) and 1 ppm (4 mg/m^3) for ethylenediamine and diethylenetriamine, respectively. Thus for the normal 8-h workday, the time-weighted average concentrations of EDA and DETA in the air of the workplace should not be allowed to exceed these levels.

The information presented above on the physiological effects of the aliphatic polyamines is, necessarily, of a very general nature. Before handling one of these products, it is important to contact the manufacturer for complete toxicological information and safe handling recommendations.

Applications

The aliphatic polyamines are versatile chemical intermediates with a broad spectrum of industrial applications. The major applications of the ethyleneamine family of products are listed in Table 6 along with estimated consumption figures in the various application areas for 1975.

The technology of the major applications is discussed below.

Fungicides. The ethylenebisdithiocarbamates were introduced in the early 1940s and currently represent the most important class of general purpose, broad spectrum fungicides (qv) used for the control of mildew, scabs, rusts and blights on fruits, vegetables, cotton and, to some extent, on grains.

These fungicides are prepared in a two-step process. First, EDA and carbon disulfide react in the presence of sodium or ammonium hydroxide to form the water soluble disodium (104) or diammonium (105) salt of ethylenebisdithiocarbamate. The water-soluble salt is then converted to the poorly soluble Zn and/or Mn salts by reaction with inorganic salts of these metals (106–109). The fungicides are commonly known as zineb (Zn salt) and maneb (Mn salt).

Another class of fungicides based on EDA consists of the substituted imidazolines (110). The most important compound in this class is 2-heptadecyl-2-imidazoline (111)

Table 6. United States and World Application Profiles For Ethyleneamines, 1975, 10^3 t/yr

Application	United States EDA	United States DETA	United States Higher homologues	World EDA	World DETA	World Higher homologues
fungicides	4.5			23.1		
chelating agents	6.0	0.4		11.4	1.2	
wet strength resins		6.8	1.4		7.7	1.8
epoxy curing agents	0.1	1.3	2.6	0.2	2.2	4.4
polyamide resins	0.8[a]	0.4[a]		2.6	1.4	
surfactant–softeners	2.3[b]	2.8[b]		3.2	3.9	
corrosion inhibitors		1.8	0.4		2.0	0.9
gas and lube additives	0.3	1.4	8.4	0.5	1.6	12.5
asphalt emulsifiers		0.4	2.3		0.7	2.7
other	5.0	3.8	2.2	10.2	6.6	4.3
Total	19.0	19.1	17.3	51.2	27.3	26.6

[a] Exclusive of epoxy curing application.
[b] Includes ore flotation agents.

which may be prepared by the condensation of EDA and stearic acid. This compound is usually employed as the acetate salt and is especially effective against apple scab and cherry leaf spot.

The 2:1 EDA–copper sulfate complex has been proposed for the control of aquatic fungi and algae (112).

Chelating Agents. A major industrial application of EDA is the manufacture of ethylenediaminetetraacetic acid (EDTA), a well established chelating agent. All of the industrially used methods of manufacture involve the addition of formaldehyde and hydrogen cyanide, or an alkali metal cyanide to an aqueous solution of EDA. The tetrasodium salt of EDTA (Na_4EDTA) may be prepared directly by using excess sodium hydroxide in the reaction (113–117) or the intermediate ethylenediaminetetracetonitrile may be isolated and hydrolyzed to Na_4EDTA in a separate step (118–119).

DETA or aminoethylethanolamine, prepared from EDA and ethylene oxide, can be substituted for EDA in these procedures to prepare two other important industrial chelating agents: pentasodium diethylenetriaminepentaacetic acid and trisodium N-hydroxyethylethylenediaminetriacetic acid (see Alkanolamines).

Wet-Strength Resins. Various thermosetting resins are used to impart high wet strength to certain paper products, especially those designed for towelling, tissue, and packaging applications. The polyalkylenepolyamines are often used as components of these wet strength additives.

Cationic urea–formaldehyde resins can be prepared by the reaction of polyethylenepolyamines with urea–formaldehyde condensation products (120), by the coreaction of polyethylenepolyamines with formaldehyde and urea (121) or by condensing the polyamine with urea and then adding formaldehyde (122). These resins have greater substantivity to cellulose (qv) and, consequently, are more efficient wet strength additives than unmodified urea–formaldehyde resins. Melamine–formaldehyde resins have also been modified with polyalkylenepolyamines for use as wet strength additives (123–124) (see Amino resins).

The epichlorohydrin-modified polyamide resins constitute another class of compounds widely used as wet-strength additives for paper (125–129). These cationic, thermosetting resins are highly substantive to the paper. Additionally, they may be used under neutral to slightly alkaline conditions in contrast to the urea–formaldehyde or melamine–formaldehyde types which require acidic conditions for best results. They are prepared by the reaction of a polyethylenepolyamine (typically DETA or TETA) with a dicarboxylic acid, such as adipic acid (qv), to form a low molecular weight polyamide which is further modified by reaction with epichlorohydrin (see Chlorohydrins).

Anionic polyamide resins, prepared by the condensation of polyethylenepolyamines with polymeric fatty acids followed by neutralization with an organic base, are readily dispersed in aqueous media and may be used to improve the wet strength of paper and impart waterproofing properties as well (130).

Epoxy Curing Agents. Compounds with primary and secondary amine functionality are used as reactive hardeners in epoxy resin (qv) formulations employed for protective coatings, electrical embedments, and adhesives (131). All of the polyethylenepolyamines, from EDA through PEHA, have been used for such applications but DETA and TETA are the most widely used members of this series (132–134).

The utility of the polyethylenepolyamines as curing agents or hardeners for epoxy resin compositions is greatly extended through the use of numerous derivatives. One such class of derivatives consists of the hydroxy-substituted amines prepared by the reaction of a polyethylenepolyamine (typically DETA or TETA) with ethylene oxide or propylene oxide (135). Cyanoethylated polyethylenepolyamines represent another useful class of derivatives. For example, the reaction of DETA with acrylonitrile in equimolar amounts gives N-cyanoethyldiethylenetriamine which, in epoxy formulations, gives reduced cure rates and a longer effective pot-life (133). Polymers obtained by heating the cyanoethylated polyamines in the presence of catalytic amounts of sulfur are also effective hardeners (136).

Substituted imidazolines prepared by the condensation of polyethylenepolyamines with carboxylic acids or carboxylic acid derivatives have also been employed as epoxy curing agents (137–142). These curing agents are proposed for use in aqueous emulsion systems (138–139) and to provide coatings with improved hydrolytic stability (141).

Liquid, reactive polyamide resins may be prepared from the condensation of the polyethylenepolyamines with di- and trifunctional fatty carboxylic acids obtained from the thermal polymerization of vegetable oil fatty acids (143–144). Various polyethylenepolyamines may be used in the preparation of the polyamides, which may be characterized as low molecular weight, highly branched, highly viscous polymers with high amine functionality (145). The main advantage of the liquid polyamides in epoxy resin formulations is in the substantial improvements in the flexibility (145) of the cured composite, leading to higher impact resistance and flexural strength.

Derivatives with lower viscosities can be prepared by the reaction of fatty polyamides with alpha–beta unsaturated esters (146). These derivatives are somewhat less reactive, leading to lower reaction isotherms and longer pot lives. Viscosity reduction can also be effected by blending the polyamides with the higher polyethylenepolyamines (147). Other polyamides useful in epoxy curing include the reaction products of a polyethylenepolyamine with an aliphatic, hydroxyl-containing carboxylic acid (148), with tall oil (qv) fatty acids (149), and with keto-acids (150).

Other adducts useful as epoxy curing agents may be prepared by the reaction of polyethylenepolyamines with aromatic monoisocyanates (151), the lower aliphatic aldehydes (152), epoxidized fatty nitriles (153), and with propylene sulfide (154).

Polyamide Resins. In addition to their use as epoxy curing agents, discussed above, the fatty polyamide resins, prepared by the reaction of aliphatic polyamines with dimer acids, are useful in adhesives (qv), coatings (qv), and inks (qv).

In the preparation of hot melt adhesives, a difunctional amine, typically EDA (155), is used. Polyamines of higher functionality such as DETA or TETA may also be included in the reaction but usually only in minor amounts. The softening point of these polyamide resins may be lowered by replacing about one third of the EDA with 1,2- or 1,3-PDA (156). Blends of the EDA-based polyamide resins and DETA-based polyamides give adhesives with improved toughness (157). The DETA-based polyamide resins may also be used in pressure-sensitive adhesives (158). Solvent-type adhesives may be prepared by dissolving the fatty polyamide resins and an acrylic acid–ethylene copolymer in an alcohol–hydrocarbon mixture (159). Suspensoids useful as wet-stick adhesives are prepared by dispersing the polyamide resins in hot, acidified water (160).

Other carboxylic acids that may be used in the preparation of these polyamides for hot melt adhesive applications include polymerized tall oil fatty acids (162–163), adipic acid (162), sebacic acid, and azelaic acid (161,164). Piperazine and substituted piperazines may be used in combination with EDA to provide the amine functionality (161,164).

In the coatings area, fatty polyamide resins are used in spray coatings for hot-rolled steel panels (165), to modify alkyd resins (qv) in the manufacture of thixotropic paint vehicles (166), and as plasticizers in nitrocellulose lacquers (167). Polyamides prepared from DETA and tall oil fatty acids are useful as corrosion inhibitors in alkyd paints (168). Castor oil–TEPA adducts, neutralized with phosphoric acid, are used as emulsifiers in nitrocellulose lacquers (169).

The condensation of polymerized fatty acids with a mixture of aliphatic diamines (typically EDA and PDA) leads to a class of fatty polyamide resins widely used as binders for flexographic printing inks (170). These resins are characterized by high solubility in alcoholic solvents and good adhesion to paper, polyolefins, polyesters, and cellophane. Minor amounts of monocarboxylic acids, such as 2-ethylhexanoic acid (170) or acetic acid (171), are frequently incorporated as chain terminators.

Surfactants and Softeners. Diamidoamines, prepared by the condensation of one mole of DETA with two moles of a fatty acid (typically stearic acid), provide the starting point for a number of compounds used as commercial fabric softeners. Dehydration–cyclization of the diamidoamine to the substituted imidazoline structure followed by quaternization with dimethyl sulfate gives 2-heptadecyl-1-methyl-1-(stearylaminoethyl)-2-imidazolinium methosulfate, a popular fabric softener for household formulations (172–174). Other alkylating agents such as octadecyl bromide or benzyl chloride may be used to quaternize the substituted imidazoline (175). The diamidoamine may be neutralized with glycolic acid (176), condensed with epichlorohydrin (177) or treated with propylene oxide and quaternized to prepare useful softeners (178).

Bisimidazoline compounds, prepared by the reaction of TETA or TEPA with fatty acids (179) or the parent glycerides (180), may be quaternized with dimethyl sulfate to yield cationic fabric softeners or with sodium chloroacetate (181) to give amphoteric softeners. The reaction of 1,3-PDA with an alkyl isocyanate (182) or epoxide (183) is used to prepare softeners compatible with anionic detergents.

The condensation of EDA with fatty acid derivatives gives 2-alkylimidazolines (184). Salts of the 2-alkylimadazolines exhibit good foaming action (185) and their condensation with two moles of sodium chloroacetate leads to surface-active agents with amphoteric character (186). Substituted imidazolines prepared by the reaction of DETA with coconut oil fatty acids, when neutralized with lactic acid, are high foaming surfactants useful in shampoo preparations (187). These substituted imidazolines may also be quaternized with sodium chloroacetate (188), condensed with ethylene oxide, and then quaternized with chloroacetamide (189) or oxidized to imidazoline oxides with hydrogen peroxide (190) to give surfactants useful in shampoos.

Polyols prepared by the reaction of EDA with ethylene and propylene oxides (191) and bisamides prepared by the condensation of one mole of EDA with two moles of stearic (192) or hydrogenated tallow acids (193) are effective antifoam agents (see Defoamers). Tetraacetylethylenediamine is useful as an activator for perborate bleaching agents (qv) in powdered detergent compositions (194) (see Surfactants).

Corrosion Inhibitors. Amidoamines and substituted imidazolines, prepared by the reaction of DETA with fatty acids, are used as corrosion inhibitors (qv) in petroleum production operations (195–196). Also used in this application are cyclic amidine polymers from EDA and vinyl nitriles or cyanohydrins in the presence of thiourea (197), and polymers prepared by the reaction of polyethylenepolyamines with acrylate esters (198). The substituted imidazolines may be further derivatized by reaction with benzyl halides, propane sultone and thiourea (199) to give acid corrosion inhibitors for mild steel. Nitrobenzoic acid salts of polyethylenepolyamines function as metal corrosion inhibitors in coating formulations (200). EDA acts as a corrosion inhibitor for aluminum alloys (201) in the presence of mineral acids and provides corrosion protection in distillation columns (202).

Lubricating Oil and Fuel Additives. The polyisobutenylsuccinimides are a widely used class of ashless, dispersant–detergent additives for high quality motor oils. These compounds are prepared by the reaction of a polyisobutenylsuccinic anhydride, wherein the polyisobutylene radical contains at least 50 aliphatic carbon atoms, with a polyalkylenepolyamine (203–205). Other hydrophobes such as polybutene (206), alpha-olefin polymers (207), and poly(propylene glycol) (208) may be substituted for the polyisobutylene radical. The polyamines that are useful include the entire family of ethyleneamines (209), a number of polypropyleneamines (209), DMAPA (210), cyanoethylated TEPA (211), and urea–TEPA condensation products (212). Similar compounds with dispersant–detergent properties are prepared by alkylating the polyamines with halogenated polyolefins (213) (see also Dispersants).

Lubricating oil additives with both dispersant and viscosity-index improvement properties may be prepared by the reaction of TEPA or PEHA with chlorinated ethylene–propylene–nonconjugated diene terpolymers (214), oxidized ethylene–propylene copolymers (215) or carboxylated polyisoprene (216) and by the mastication of ethylene–propylene copolymers in the presence of polyethylenepolyamines (217). Antirust additives for lube oils are obtained by condensing polyalkylenepolyamines with fatty monocarboxylic acids and then with an alkenylsuccinic anhydride (218–219).

The reaction of EDA with chlorinated polyisobutylene leads to compounds useful as detergent additives for hydrocarbon fuels (220–221). Neutralization of these compounds with monocarboxylic acids is reported to improve their rust inhibiting properties (222). Derivatives of DETA via reaction with a fatty acid (223), an octadecyl halide (224), or chlorinated polypropylene glycol (225) also act as dispersant additives in fuels. Metal salts of alkyl phenol–EDA adducts are useful as antioxidants in fuels (226).

Asphalt Emulsifiers. The reaction of polyalkylenepolyamines with fatty acids or fatty acid derivatives is used to prepare surface-active agents useful in the preparation of cationic asphalt emulsions (227–229). These derivatives, which are complex mixtures of amine salts, amides, and substituted imidazolines (230), also improve the adhesion between the bituminous materials and rock aggregates (231–234). Similar derivatives useful in these applications may be prepared by the reaction of polyalkylenepolyamines with chlorinated paraffin oil (235), octadecyl bromide (236), ozonized tall oil fatty acids (237), or epoxidized asphalt (238) (see Asphalt).

Other Applications. 1,3-Dimethylolethyleneurea (239) and 1,3-dimethylolpropyleneurea (240) are useful in the treatment of cotton (qv) fabrics to impart durable press and crease resistance properties (241) (see Textiles). These compounds are prepared by the reaction of EDA or 1,3-PDA with urea followed by the addition of formaldehyde to the resulting cyclic ureas. Durable antistatic properties may be imparted to synthetic fibers by treatment with polymers prepared by the reaction of DETA with stearic acid, urea, and formaldehyde (242) or DMAPA with urea and dimethyl sulfate (243) (see Antistatic agents).

The reaction products obtained by condensing 1 mole of DETA with 1–3 moles of tall oil fatty acids (244) or naphthenic acid (245) are used in the separation of silica from phosphate ores by frothing (see Flotation). Aqueous solutions of EDA or DETA may be used to leach lead from lead-bearing ores (246) and wastes (247).

The higher molecular weight polyethylenepolyamines are used as coagulation aids for synthetic rubber latexes (248). Synthetic elastomers (qv) with a high degree of resistance to heat and lubricating oils are prepared by the vulcanization of alkyl acrylate copolymers with TETA (249). Vulcanization of compounded rubber stocks containing small amounts of EDA gives cured composites with superior properties (250). 2-Mercapto-2-imidazoline, a well-known vulcanization accelerator (251), is prepared by the reaction of EDA with carbon disulfide.

BIBLIOGRAPHY

"Ethylene Amines" in *ECT* 1st ed., Vol. 5, pp. 898–905, by John Conway, Carbide and Carbon Chemicals Division, Union Carbide and Carbon Corporation; "Diamines and Higher Amines, Aliphatic" in *ECT* 2nd ed., Vol. 7, pp. 22–39, by Andrew W. Hart, The Dow Chemical Company.

1. L. Bergstedt and G. Widmark, *Acta Chem. Scand.* **24**, 2713 (1970).
2. R. N. Keller and L. J. Edwards, *J. Am. Chem. Soc.* **74**, 2931 (1952).
3. A. Gero, *J. Am. Chem. Soc.* **76**, 5159 (1954).
4. H. B. Jonassen and co-workers, *J. Am. Chem. Soc.* **72**, 2430 (1950).
5. H. H. Willard and N. H. Furman, *Elementary Quantitative Analysis*, D. Van Nostrand, Inc., New York, 1940, p. 474.
6. U.S. Pat. 2,011,578 (Aug. 20, 1935), G. C. Hale.
7. U.S. Pat. 1,737,458 (Nov. 26, 1929), M. Hartmann and J. Kaji (to Society of Chemical Industry in Basle).
8. U.S. Pat. 2,701,239 (Feb. 1, 1955), J. W. Ryznar (to National Aluminate Corp.).
9. L. J. Kitchen and C. B. Pollard, *J. Org. Chem.* **8**, 342 (1943).
10. J. Okada and K. Hayakawa, *Yakugaku Zasshi* **96**, 783 (1976).
11. J. Hine and K. W. Narducy, *J. Am. Chem. Soc.* **95**, 3362 (1973).
12. M. Tsuchiya and co-workers, *Yakugaku Zasshi* **96**, 1005 (1976).

13. U.S. Pat. 2,226,534 (Dec. 31, 1940), J. G. Lichty (to Wingfoot Corp.).
14. U.S. Pat. 2,643,977 (June 30, 1953), W. B. Hughes (to Cities Service Research and Development Co.).
15. H. Krassig, *Makromol. Chem.* **17,** 77 (1955).
16. U.S. Pat. 2,387,201 (Oct. 16, 1945), N. Weiner (to Bonneville, Ltd.).
17. U.S. Pat. 1,926,015 (Sept. 5, 1933), K. W. Rosenmund.
18. U.S. Pat. 2,392,326 (Jan. 8, 1946), L. P. Kyrides (to Monsanto Chemical Co.).
19. M. Kita and Y. Yamada, *Yukagaku* **24**(2), 112 (1975).
20. U.S. Pat. 2,508,415 (May 23, 1950), H. L. Morril (to Monsanto Chemical Co.).
21. U.S. Pat. 2,992,230 (July 11, 1961), G. A. Lescisin (to Union Carbide Corp.).
22. V. B. Piskov, V. P. Kasperovich, and L. M. Yakovleva, *Khim. Geterotsikl. Soedin.* (8), 1112 (1976); *Chem. Abstr.* **86,** 5372h (1977).
23. U.S. Pat. 2,355,837 (Aug. 15, 1944), A. L. Wilson (to Carbide and Carbon Chemicals Corp.).
24. T. Tkaczynski, A. Wojciechowska, and A. Urbanska, *Acta Pol. Pharm.* **32,** 663 (1975); *Chem. Abstr.* **86,** 72515v (1977).
25. U.S. Pat. 3,917,638 (Nov. 4, 1975), R. R. Mod, F. C. Magne, G. Sumrell, A. F. Novak, and J. M. Solar (to the United States of America as represented by the Secretary of Agriculture).
26. U.S. Pat. 2,371,104 (Mar. 6, 1945), R. H. Keinle and co-workers (to American Cyanamid Co.).
27. U.S. Pat. 2,570,895 (Oct. 9, 1951), M. W. Wilson (to B. F. Goodrich Co.).
28. U.S. Pat. 2,583,771 (Jan. 29, 1952), L. O. Gunderson (to Dearborn Chemical Co.).
29. U.S. Pat. 2,609,381 (Sept. 2, 1952), H. B. Goldstein and S. T. Charles (to Sun Oil Corp.).
30. U.S. Pat. 2,145,242 (Jan. 31, 1939), H. W. Arnold (to E. I. du Pont de Nemours & Co., Inc.).
31. C. E. Schweitzer, *J. Org. Chem.* **15,** 471 (1950).
32. Brit. Pat. 732,059 (June 15, 1955), (to Farbenfabriken Bayer A.G.).
33. U.S. Pat. 2,554,475 (May 22, 1951), T. J. Suen and J. H. Daniel, Jr., (to American Cyanamid Co.).
34. F. Linsker and R. L. Evans, *J. Am. Chem. Soc.* **67,** 1581 (1945).
35. U.S. Pat. 2,683,134 (July 6, 1954), J. B. Davidsen and E. J. Romatowski (to Allied Chemical & Dye Corp.).
36. U.S. Pat. 2,595,935 (May 6, 1952), J. H. Daniel, Jr., and C. G. Landes (to American Cyanamid Co.).
37. U.S.S.R. Pat. 517,592 (June 15, 1976), E. I. Krasulina, F. P. Grigor'ev, and M. I. Aleksandrova; *Chem. Abstr.* **85,** 123938z (1976).
38. T. Suzuki and K. Mitsuhashi, *Seikei Daigaku Kogakubu Kogaku Hokoku* **22,** 1579 (1976); *Chem. Abstr.* **86,** 139916x (1977).
39. E. V. Anderson and J. A. Gaunt, *Ind. Eng. Chem.* **52,** 190 (1960).
40. W. Jentzsch and M. Seefelder, *Chem. Ber.* **98,** 1342 (1965).
41. N. Sawa, *Nippon Kagaku Zasshi* **89,** 780 (1968).
42. Ger. Offen. 2,512,513 (Oct. 7, 1976), A. Frank and T. Dockner (to Badische Anilin- und Soda-Fabrik A.G.).
43. H. A. Bruson in R. Adams, ed., *Organic Reactions*, Vol. 5, John Wiley & Sons, Inc., New York, 1949, pp. 79–87.
44. Ger. Offen. 2,446,489 (Apr. 15, 1976), H. Graefje and co-workers (to Badische Anilin- und Soda-Fabrik A.G.).
45. U.S. Pat. 2,622,075 (Dec. 16, 1952), H. M. Hemmi and P. Trefzer (to Sandoz, Ltd.).
46. A. L. Wilson, *Ind. Eng. Chem.* **27,** 867 (1935).
47. U.S. Pat. 2,616,875 (Nov. 4, 1952), J. W. Adams and J. A. Reynolds (to United States Rubber Co.).
48. U.S. Pat. 2,326,643 (Aug. 10, 1943), W. F. Hester (to Rohm & Haas Co.).
49. U.S. Pat. 2,600,245 (June 10, 1952), S. L. Hoffenstead (to B. F. Goodrich Co.).
50. U.S. Pat. 2,361,188 (Oct. 24, 1944), A. L. Fox (to E. I. du Pont de Nemours & Co., Inc.).
51. L. O. Gunderson and W. L. Denman, *Ind. Eng. Chem.* **40,** 1363 (1948).
52. A. Andersons, S. Jurels, and M. V. Shimanskaya, *Khim. Geterotsikl. Soedin.* (2), 346 (1967); *Chem. Abstr.* **67,** 100107s (1967).
53. U.S. Pat. 2,937,176 (May 17, 1960), E. C. Herrick (to Houdry Process Corp.).
54. A. Anders, S. Jurels, and M. V. Shimanskaya, *Latv. PSR Zinat. Acad. Vestis Kim. Ser.* (1), 47 (1971); *Chem. Abstr.* **75,** 98037p (1971).
55. U.S.S.R. Pat. 271,525 (May 26, 1970), S. Hillers and co-workers; *Chem. Abstr.* **73,** 109263w (1970).
56. S. A. Jurels, U. A. Pinka, and M. Shimanskaya, *Laty. PSR Zinat. Akad. Vestis Kim. Ser.* (6), 691 (1974); *Chem. Abstr.* **82,** 155164y (1975).

57. M. V. Shimanskaya and co-workers, *Geterogennyi Katal. Reakts. Poluch. Prevrashch. Geterosikl. Soedin,* 185 (1971); *Chem. Abstr.* **76,** 99615h (1972).
58. U.S. Pat. 1,832,534 (Nov. 17, 1932), G. O. Curme, Jr., and F. W. Lommen (to Carbide and Carbon Chemicals Corp.).
59. Ger. Offen. 2,427,440 (Jan. 2, 1975), C. S. Steele (to Jefferson Chemical Co., Inc.).
60. P. J. Garner and C. P. Nunes, *Rev. Part. Quim* **15,** 158 (1973).
61. S. Ropuszynski, E. Mularczyk, and J. Kuchar, *Przem. Chem.* **47,** 542 (1968); *Chem. Abstr.* **70,** 46733q (1969).
62. Fr. Pat. 1,555,162 (Jan. 24, 1969), M. Lichtenwalter and T. H. Cour (to Jefferson Chemical Co., Inc.).
63. Brit. Pat. 1,147,984 (Apr. 10, 1969), J. G. Blears and P. Simpson (to Simon-Carves Ltd.).
64. U.S. Pat. 3,751,474 (Aug. 7, 1973), K. G. Phillips and M. J. Geerts (to Nalco Chemical Company).
65. Jpn. Pat. 69 12,723 (June 9, 1969), S. Wakiyama and Y. Kashida (to Toyo Soda Manufg. Co., Ltd.).
66. U.S. Pat. 2,769,841 (Nov. 6, 1956), S. W. Dylewski, H. G. Dulude, and G. W. Warren (to The Dow Chemical Co.).
67. Ger. Offen. 2,439,275 (Mar. 4, 1976), H. Graefje, W. Mesch S. Winderl, and H. Hoffman (to Badische Anilin- und Soda-Fabrik A.G.).
68. E. Dutaki, *Rev. Chim. (Bucharest)* **17,** 536 (1966); *Chem. Abstr.* **66,** 94719w (1967).
69. Z. Leszczynski and co-workers, *Przem. Chem.* **46,** 210 (1967); *Chem. Abstr.* **67,** 32273r (1967).
70. Ger. Pat. 1,170,960 (May 27, 1964), G. F. MacKenzie (to The Dow Chemical Co.).
71. U.S. Pat. 2,861,995 (Nov. 25, 1958), G. F. MacKenzie (to The Dow Chemical Co.).
72. Fr. Pat. 1,347,648 (Dec. 27, 1963), S. Winderl, E. Haarer, H. Corr, and P. Hornberger (to Badische Anilin- und Soda-Fabrik A.G.).
73. U.S. Pat. 3,151,115 (Sept. 29, 1964), P. H. Moss and N. B. Godfrey (to Jefferson Chemical Co., Inc.).
74. Ger. Pat. 1,100,645 (July 7, 1958), R. Lichtenberger and F. Weiss.
75. Fr. Pat. 1,179,771 (May 28, 1959), G. D. Gillies.
76. East Ger. Pat. 14,480 (Mar. 17, 1958), H. Schade; *Chem. Abstr.* **53,** 14129h (1959).
77. U.S. Pat. 3,270,059 (Aug. 30, 1966), S. Winderl and co-workers (to Badische Anilin- und Soda-Fabrik).
78. U.S. Pat. 3,597,483 (Aug. 3, 1971), E. Haarer, H. Corr, and S. Winderl (to Badische Anilin- und Soda-Fabrik).
79. U.S. Pat. 3,137,730 (June 16, 1964), C. B. Fitz-William (to Allied Chemical Corp.).
80. Brit. Pat. 833,589 (Apr. 27, 1960), (to Jefferson Chemical Co., Inc.).
81. U.S. Pat. 3,331,756 (July 18, 1967), V. S. Currier and J. G. Milligan (to Jefferson Chemical Co., Inc.).
82. U.S. Pat. 2,978,451 (Apr. 4, 1961), H. B. Hass and P. S. Skell (to Research Corp.).
83. Ger. Pat. 1,154,121 (Sept. 12, 1963), J. W. Nemec, C. H. McKeever, and E. L. Wolffe (to Rohm & Haas Co.).
84. Brit. Pat. 798,075 (July 16, 1958), C. Scandura (to Badische Anilin- und Soda-Fabrik A.G.).
85. U.S. Pat. 3,067,255 (Dec. 4, 1962), H. Scholz and P. Gunthert (to Badische Anilin- und Soda-Fabrik A.G.).
86. U.S. Pat. 2,519,560 (Aug. 22, 1950), G. W. Fowler (to Union Carbide & Carbon Corp.).
87. Brit. Pat. 658,422 (Oct. 10, 1951), H. D. Finch and E. A. Peterson (to N. V. de Bataafsche Petroleum Maatschappij).
88. U.S. Pat. 3,260,752 (July 12, 1966), A. F. Miller, M. Salehar, and J. Williams (to Standard Oil Co., Ohio).
89. A. P. Terent'ev, K. I. Chursina, and A. N. Kost, *Zhur. Obshchei Khim.* **20,** 1073 (1950).
90. Ger. Offen. 2,540,871 (Mar. 24, 1977), H. Graefje and co-workers (to Badische Anilin- und Soda-Fabrik A.G.).
91. Ger. Offen, 2,605,212 (Aug. 25, 1977), W. Mesch, H. Hoffmann, and D. Voges (to Badische Anilin- und Soda-Fabrik A.G.).
92. U.S.S.R. Pat. 509,578 (Apr. 5, 1976), R. N. Zagidullin and co-workers (to Sterlitamak Chemical Plant); *Chem. Abstr.* **85,** 5187f (1976).
93. J. M. Weber in F. D. Snell and L. S. Ettre, eds., *Encyclopedia of Industrial Chemical Analysis,* Vol. 11, John Wiley & Sons, Inc., 1971, pp. 421–428.
94. E. D. Smith and R. D. Radford, *Anal. Chem.* **33,** 1160 (1961).
95. J. Bohemen and co-workers, *J. Chem. Soc.,* 2444 (1960).

96. J. Tornquist, *Acta Chem. Scand.* **19,** 777 (1965).
97. L. H. Ponder, *J. Chromatog.* **97,** 77 (1974).
98. E. A. Emelin, G. S. Kitukiuna, and V. V. Zharkov, *Zh. Anal. Khim.* **28,** 1599 (1973); *Chem. Abstr.* **80,** 60438x (1974).
99. I. S. Dukhovanaya and N. F. Kazarinova, *Zh. Anal. Khim.* **31,** 2018 (1976); *Chem. Abstr.* **86,** 17533z (1977).
100. H. Nascu, T. Hodisan, and C. Liteanu, *Stud. Univ. Babes-Bolyai Ser. Chem.* **20,** 63 (1975); *Chem. Abstr.* **84,** 98916k (1976).
101. Rom. Pat. 58,125 (Dec. 28, 1974), S. Gocan, T. Hodisan, H. Nascu, and C. Liteanu (to Institutul "Chimigaz"); *Chem. Abstr.* **85,** 33930m (1976).
102. H. L. Spell, *Anal. Chem.* **41,** 902 (1969).
103. A. Krynska, *Pr. Cent. Inst. Ochr. Pr.* **19**(60), 71 (1969); *Chem. Abstr.* **71,** 6350b (1969).
104. U.S. Pat. 2,504,404 (Apr. 18, 1950), A. L. Flenner (to E. I. du Pont de Nemours & Co., Inc.).
105. U.S. Pat. 2,844,623 (July 22, 1968), E. A. Fike (to Roberts Chemicals, Inc.).
106. U.S. Pat. 2,317,765 (Apr. 1943), W. F. Hester (to Rohm & Haas Co.).
107. Belg. Pat. 617,407 (Nov. 9, 1962), C. B. Lyon, J. W. Nemec, and V. H. Unger (to Rohm & Haas Co.).
108. Belg. Pat. 617,408 (Nov. 9, 1962), J. M. Nemec, V. H. Unger, and C. B. Lyon (to Rohm & Haas Co.).
109. U.S. Pat. 3,379,610 (Apr. 23, 1968), C. B. Lyon, J. W. Nemec, and V. H. Unger (to Rohm & Haas Co.).
110. R. H. Wellman and S. E. A. McCallan, *Contrib. Boyce Thompson Inst.* **14,** 151 (1946).
111. H. S. Cunningham and E. G. Sharvelle, *Phytopathology* **30,** 4 (1940).
112. Ger. Pat. 2,506,431 (Sept. 18, 1975), M. D. Meyers and G. A. Stoner (to Sandox G.m.b.H.).
113. U.S. Pat. 2,387,735 (Oct. 29, 1945), F. C. Bersworth (to Martin Dennis Co.).
114. U.S. Pat. 2,407,645 (Sept. 17, 1946), F. C. Bersworth (to Martin Dennis Co.).
115. U.S. Pat. 2,461,519 (Feb. 15, 1949), F. C. Bersworth.
116. U.S. Pat. 2,845,457 (July 29, 1958), H. Kroll and M. Dexter (to Geigy Chemical Corp.).
117. U.S. Pat. 2,860,164 (Nov. 11, 1958), H. Kroll and F. P. Butler (to Geigy Chemical Corp.).
118. U.S. Pat. 2,205,995 (June 25, 1940), H. Ulrich and E. Ploetz (to I. G. Farbenindustrie).
119. U.S. Pat. 2,855,428 (Oct. 7, 1958), J. J. Singer (to Hampshire Chemical Corp.).
120. U.S. Pat. 2,554,475 (May 22, 1951), T. Suen and J. H. Daniel, Jr., (to American Cyanamid Co.).
121. U.S. Pat. 2,683,134 (July 6, 1954), J. B. Davidson and E. J. Romatowski (to Allied Chemical and Dye Corp.).
122. U.S. Pat. 2,742,450 (Apr. 17, 1956), R. S. Yost and R. W. Auten (to Rohm & Haas Co.).
123. U.S. Pat. 2,769,799 (Nov. 6, 1956), T. Suen and Y. Jen (to American Cyanamid Co.).
124. U.S. Pat. 2,769,800 (Nov. 6, 1956), T. Suen and Y. Jen (to American Cyanamid Co.).
125. U.S. Pat. 3,227,671 (Jan. 4, 1966), G. I. Keim (to Hercules Powder Co.).
126. U.S. Pat. 3,565,754 (Feb. 23, 1971), N. W. Dachs and G. M. Wagner (to Hooker Chemical Corp.).
127. U.S. Pat. 3,058,873 (Oct. 16, 1962), G. I. Keim and A. C. Schmalz (to Hercules Powder Co.).
128. U.S. Pat. 3,609,126 (Sept. 28, 1971), H. Asao, F. Yoshida, K. Tomihara, M. Akimoto, and G. Kubota (to Toho Kagaku Kogyo Kabushiki Kaisha).
129. U.S. Pat. 3,442,754 (May 6, 1969), H. H. Espy (to Hercules, Inc.).
130. U.S. Pat. 2,926,117 (Feb. 23, 1960), H. Wittcoff (to General Mills Inc.).
131. I. Skeist and G. R. Somerville, *Epoxy Resins,* Reinhold Publishing Co., New York, 1958, pp. 21, 167, 185, 233.
132. *Ibid.,* p. 35.
133. F. J. Allen and W. M. Hunter, *J. Appl. Chem.* **7,** 86 (1957).
134. T. Izumo, *Shikizai Kyokaishi* **49,** 538 (1976).
135. F. Pitt and M. N. Paul, *Mod. Plast.* **34,** 124 (1957).
136. Jpn. Kokai 73 90,000 (Nov. 24, 1973), S. Kiriyama and H. Ikeda (to Showa Highpolymer Co., Ltd.).
137. Jpn. Kokai 75 117,900 (Sept. 16, 1975), A. Kotone, T. Hori, and M. Hoda (to Sakai Chemical Industry Co., Ltd.).
138. Jpn. Kokai 77 26,598 (Feb. 28, 1977), M. Nagakura and Y. Higaki (to Nisshin Oil Mills, Ltd.).
139. Jpn. Kokai 75 98,960 (Aug. 6, 1975), H. Suzuki, T. Inoue, and T. Hatano (to Asahi Denka Kogyo K. K.).
140. Jpn. Kokai 77 00,898 (Jan. 6, 1977), H. Suzuki and co-workers (to Asahi Denka Kogyo K. K.).
141. Ger. Offen. 2,405,111 (Aug. 14, 1975), G. Johannes and co-workers (to Hoechst A.G.).

142. Ger. Offen. 2,326,668 (Dec. 12, 1974), F. Schuelde, J. Obendorf, and V. Kulisch (to Veba-Chemie A.G.).
143. U.S. Pat. 2,450,940 (Oct. 12, 1948), J. C. Cowan, L. B. Falkenburg, and A. Lewis (to U.S. Department of Agriculture).
144. R. Anderson and D. H. Wheeler, *J. Am. Chem. Soc.* **70,** 760 (1948).
145. U.S. Pat. 2,705,223 (Mar. 29, 1955), M. Renfrew and H. Wittcoff (to General Mills, Inc.).
146. U.S. Pat. 2,999,825 (Sept. 12, 1961), D. E. Peermon (to General Mills, Inc.).
147. U.S. Pat. 2,881,194 (Apr. 7, 1959), D. Peerman and D. Floyd (to General Mills, Inc.).
148. Jpn. Kokai 75 33,693 (Nov. 1, 1975), M. Nakajima and A. Yanaguchi (to Nippon Carbide Industries Co., Inc.).
149. Jpn. Kokai 75 123,200 (Mar. 27, 1975), T. Hoki, K. Toyomoto, and H. Komoto (to Asahi Chemical Industry Co., Ltd.).
150. Jpn. Kokai 75 151,998 (Dec. 6, 1975), Y. Nakamura, S. Aoyama, and T. Suzuki (to Toto Chemical Industry Co., Ltd.).
151. U.S. Pat. 3,407,175 (Oct. 22, 1968), W. E. Presley and T. J. Hairston (to The Dow Chemical Co.).
152. U.S. Pat. 3,026,285 (Mar. 20, 1962), F. N. Hirosawa and J. Delmonte.
153. U.S. Pat. 3,356,647 (Dec. 5, 1967), W. M. Bunde (to Ashland Oil & Refining).
154. U.S. Pat. 3,548,002 (Dec. 15, 1970), L. Levine (to The Dow Chemical Co.).
155. U.S. Pat. 3,595,816 (July 27, 1971), F. O. Barrett (to Emery Industries, Inc.).
156. U.S. Pat. 3,408,317 (Oct. 29, 1968), L. R. Vertnik (to General Mills, Inc.).
157. U.S. Pat. 2,839,219 (June 17, 1958), J. H. Groves and G. G. Wilson (to General Mills, Inc.).
158. U.S. Pat. 3,462,284 (Aug. 19, 1969), L. R. Vertnik (to General Mills, Inc.).
159. Jpn. Kokai 76 47,927 (Apr. 24, 1976), H. Takahashi, H. Suenaga, S. Ishikawa, and Y. Michii (to Nippon Paint Co., Ltd.).
160. U.S. Pat. 2,811,459 (Oct. 29, 1957), H. Wittcoff and W. A. Jordan (to General Mills, Inc.).
161. Ger. Offen. 2,361,486 (June 12, 1975), W. Imoehl and M. Drawert (to Schering A.G.).
162. Fr. Addn. 94,402 (Aug. 14, 1969), (to Schering A.G.).
163. Ger. Pat. 1,745,447 (Feb. 12, 1976), M. Drawert, C. Burba, and E. Griebsch (to Schering A.G.).
164. Ger. Offen. 2,148,264 (Apr. 5, 1973), M. Drawert and E. Griebsch (to Schering A.G.).
165. U.S. Pat. 2,550,682 (May 1, 1951), L. B. Falkenburg, A. J. Lewis, and J. C. Cowan (to the United States of America, as represented by the Secretary of Agriculture).
166. U.S. Pat. 2,663,649 (Dec. 22, 1953), W. B. Winkler (to T. F. Washburn Co.).
167. U.S. Pat. 2,379,413 (July 3, 1945), T. F. Bradley (to American Cyanamid Co.).
168. Pol. Pat. 60,441 (Aug. 5, 1970), L. Chromy, J. Polaczy, and E. Smieszek.
169. Pol. Pat. 80,751 (Feb. 20, 1976), S. Ropuszynski and H. Matyschok (to Politechnika Wroclawska).
170. U.S. Pat. 3,412,115 (Nov. 19, 1968), D. E. Floyd and D. W. Glaser (to General Mills, Inc.).
171. Neth. Appl. 6,512,985 (Apr. 18, 1966), (to Schering A.G.).
172. Ger. Offen. 2,165,947 (July 20, 1972), T. V. Kandathil (to S. C. Johnson and Son, Inc.).
173. U.S. Pat. 3,954,634 (May 4, 1976), J. A. Monson, W. L. Stewart, and H. F. Gruhn (to S. C. Johnson and Son, Inc.
174. Ger. Offen. 2,520,150 (Jan. 2, 1976) P. Goullet (to Azote et Produits Chimiques S. A.).
175. U.S. Pat. 2,874,074 (Feb. 17, 1959), C. E. Johnson (to National Aluminate Corp.).
176. Fr. Pat. 1,487,570 (July 7, 1967), J. E. Clark and J. A. Bungener (to Standard Chemical Products Inc.).
177. Jpn. Kokai 77 70,194 (Jun. 10, 1977), Y. Minegishi and H. Arai (to Kao Soap Co., Ltd.).
178. U.S. Pat. 3,933,871 (Jan. 20, 1976), L. J. Armstrong (to Armstrong Chemical Co., Inc.).
179. U.S. Pat. 3,887,476 (June 3, 1975), R. B. McConnel (to Ashland Oil, Inc.).
180. U.S. Pat. 3,855,235 (Dec. 17, 1974), R. B. McConnell (to Ashland Oil, Inc.).
181. U.S. Pat. 3,898,244 (Aug. 5, 1975), R. B. McConnel (to Ashland Oil, Inc.).
182. U.S. Pat. 3,965,015 (June 22, 1976), R. A. Bauman (to Colgate-Palmolive Co.).
183. U.S. Pat. 4,049,557 (Sept. 20, 1977), H. E. Wixon (to Colgate-Palmolive Co.).
184. U.S. Pat. 2,215,864 (Sept. 24, 1941), E. Waldmann and A. Chwala (to General Aniline & Film Corp.).
185. Brit. Pat. 479,491 (Feb. 7, 1938), E. Waldmann and A. Chwala.
186. U.S. Pat. 2,781,349 (Feb. 12, 1957), H. S. Mannheimer.
187. U.S. Pat. 3,003,969 (Oct. 10, 1961), O. Albrecht (to Ciba Ltd.).
188. Jpn. Kokai 75 137,917 (Nov. 1, 1975), Y. Nakamura and co-workers (to Toho Chemical Industry Co., Ltd.).

189. U.S. Pat. 2,794,808 (June 4, 1957), O. Albrecht and E. Matter (to Ciba Ltd.).
190. U.S. Pat. 3,951,878 (Apr. 20, 1976), R. L. Wakeman, Z. J. Dudzinski, and A. Lada (to Millmaster Onyx Corp.).
191. Ger. Offen. 1,944,569 (Mar. 11, 1971), H. J. Schuestler and R. Scharf (to Henkel and Cie., G.m.b.H.).
192. U.S. Pat. 3,951,853 (Apr. 20, 1976), D. W. Suwala (to Diamond Shamrock Corp.).
193. Ger. Offen. 2,031,827 (Jan. 14, 1971), I. A. Lichtman and F. E. Woodward (to Diamond Shamrock Corp.).
194. Ger. Offen, 2,616,350 (Oct. 28, 1976), R. S. Heslam (to Unilever N. V.).
195. Brit. Pat. 1,177,134 (Jan. 7, 1970), C. O. Bundrant, C. R. Hainebach, and F. H. Mays (to Champion Chemicals, Inc.).
196. U.S. Pat. 3,758,493 (Sept. 11, 1973), J. Maddox, Jr. (to Texaco, Inc.).
197. U.S. Pat. 3,514,251 (May 26, 1970), R. R. Annand, D. Redmore, and B. M. Rushton (to Petrolite Corp.).
198. U.S. Pat. 3,514,250 (May 26, 1970), B. M. Rushton (to Petrolite Corp.).
199. Jpn. Kokai 75 51,046 (May 7, 1975), Y. Higaki and T. Kataoka (to Nisshin Oil Mills, Ltd.).
200. L. I. Voronchikhina and co-workers, *Fazovye Ravnovesiya* **2**, 148 (1975); *Chem. Abstr.* **86**, 91859b (1977).
201. J. D. Talati and J. M. Pandya, *Anti-Corros. Methods Mater.* **21**(12), 7 (1974).
202. U.S. Pat. 3,819,328 (June 25, 1974), T. S. Go (to Petrolite Corp.).
203. U.S. Pat. 3,623,985 (Nov. 30, 1971), Y. G. Hendrickson (to Chevron Research Co.).
204. U.S. Pat. 3,451,931 (June 24, 1969), D. J. Kahn and M. L. Robbins (to Esso Research and Engineering Co.).
205. U.S. Pat. 3,390,082 (June 25, 1968), W. M. LeSuer and G. R. Norman (to Lubrizol Corp.).
206. U.S. Pat. 3,361,673 (Jan. 2, 1968), F. A. Stuart, R. G. Anderson, and A. Y. Drummond (to Chevron Research Co.).
207. U.S. Pat. 3,235,503 (Feb. 15, 1966), L. DeVries (to Chevron Research Co.).
208. U.S. Pat. 4,039,461 (Aug. 2, 1977), T. L. Hankins and T. I. Wang (to Atlantic Richfield Co.).
209. U.S. Pat. 3,272,746 (Sept. 13, 1966), W. M. LeSuer and G. R. Norman (to Lubrizol Corp.).
210. U.S. Pat. 3,018,250 (Jan. 23, 1962), R. G. Anderson, F. A. Stuart, and A. Y. Drummond (to California Research Corp.).
211. U.S. Pat. 3,366,569 (Jan. 30, 1968), G. R. Norman and W. M. LeSuer (to Lubrizol Corp.).
212. Ger. Pat. 1,810,852 (Dec. 11, 1969), E. J. Piasek, R. E. Karll, and R. J. F. Lee.
213. U.S. Pat. 3,996,285 (Dec. 7, 1976), G. S. Culbertson (to Standard Oil Co., Indiana).
214. Ger. Offen. 2,429,819 (Jan. 16, 1975), W. R. Song, J. B. Gardiner, and L. J. Engel (to Esso Research and Engineering Co.).
215. U.S. Pat. 3,864,268 (Feb. 4, 1975), G. S. Culbertson and R. E. Karll (to Standard Oil Co., Indiana).
216. U.S. Pat. 3,903,003 (Sept. 2, 1975), Z. L. Murphy and R. T. Schlobohm (to Shell Oil Co.).
217. Ger. Offen. 2,606,823 (Sept. 23, 1976), L. J. Engel and J. B. Gardiner (to Exxon Research and Engineering Co.).
218. U.S. Pat. 2,568,876 (Sept. 25, 1961), R. V. White and P. S. Landis (to Socony-Vacuum Oil Co., Inc.).
219. U.S. Pat. 2,794,782 (June 4, 1957), E. P. Cunningham and D. W. Dinsmore (to Monsanto Chemical Co.).
220. U.S. Pat. 3,438,757 (Apr. 15, 1969), L. R. Honnen and R. G. Anderson (to Chevron Research Co.).
221. U.S. Pat. 3,960,515 (June 1, 1976), L. R. Honnen (to Chevron Research Co.).
222. U.S. Pat. 3,996,024 (Dec. 7, 1976), M. D. Coon (to Chevron Research Co.).
223. U.S. Pat. 2,922,708 (Jan. 26, 1960), E. G. Lindstrom and M. R. Barusch (to California Research Corp.).
224. U.S. Pat. 3,231,348 (Jan. 25, 1966), E. G. Lindstrom and W. L. Richardson (to Chevron Research Co.).
225. Ger. Offen. 2,401,930 (July 24, 1975), D. Wagnitz (to B. P. Benzin and Petroleum A.G.).
226. U.S. Pat. 3,493,508 (Feb. 3, 1970), H. J. Andress, Jr., (to Mobil Oil Corp.).
227. U.S. Pat. 3,097,292 (July 2, 1963), E. W. Mertens (to California Research Corp.).
228. U.S. Pat. 3,230,104 (Jan. 18, 1966), C. W. Falkenberg, R. A. Paley, and J. J. Patti (to Components Corp. of America).
229. Brit. Pat. 1,174,577 (Dec. 17, 1969), N. H. Greatorex and K. N. Shaw (to Swan, Thomas and Co., Ltd.).

230. O. K. Dobozy, *Egypt K. Chem.* **16,** 419 (1973).
231. U.S. Pat. 2,426,220 (Aug. 25, 1947), J. M. Johnson (to Nostrip, Inc.).
232. U.S. Pat. 2,812,339 (Nov. 5, 1957), M. L. Kalinowski and L. T. Crews (to Standard Oil Co., Indiana).
233. O. Dobozyk, *Tenside* **7**(2), 83 (1970).
234. U.S. Pat. 2,766,132 (Oct. 9, 1956), C. M. Blair, Jr., W. Groves, and K. L. Lissant (to Petrolite Corp.).
235. U.S. Pat. 2,721,807 (Oct. 25, 1955), J. L. Rendall and D. R. Husted (to Minnesota Mining and Manufacturing Co.).
236. U.S. Pat. 2,623,831 (Dec. 30, 1952), L. A. Mikeska (to Standard Oil Development Co.).
237. U.S. Pat. 3,249,451 (May 3, 1966), E. D. Evans and C. H. Hopkins (to Skelly Oil Co.).
238. Ger. Pat. 2,513,843 (Oct. 9, 1975), M. Fujita, and S. Okada.
239. U.S. Pat. 2,373,136 (Apr. 10, 1945), F. W. Hoover and G. T. Vasla (to E. I. du Pont de Nemours & Co., Inc.).
240. T. F. Cooke and W. F. Baitinger in N. M. Bikales, ed., *Encyclopedia of Polymer Science and Technology*, Vol. 13, John Wiley & Sons, Inc., New York, 1970, p. 733.
241. U.S. Pat. 2,661,312 (Dec. 1, 1953), G. M. Richardson (to E. I. du Pont de Nemours & Co., Inc.).
242. Ger. Pat. 2,402,258 (Aug. 14, 1975), J. Ibrahim, G. Pusch, and H. Singer (to Chemische Fabrik Pfersee G.m.b.H.).
243. Fr. Pat. 1,589,742 (May 15, 1970) (to Badische Anilin- und Soda-Fabrik A.G.).
244. U.S. Pat. 2,857,331 (Oct. 21, 1958), C. A. Hollingsworth, K. F. Schilling, and J. L. Wester (to Smith-Douglass Co., Inc.).
245. U.S. Pat. 2,322,201 (June 15, 1943), D. W. Jayne, Jr., J. M. Day, and S. E. Erickson (to American Cyanamid Co.).
246. U.S. Pat. 2,940,964 (Aug. 30, 1960), F. A. Forward, H. Veltman, and A. I. Vizsolyi (to Sherritt Gordon Mines Ltd.).
247. N. Lyakov and T. Nikolov, *Metalurgiya* (*Sofia*) **32**(3), 23 (1977); *Chem. Abstr.* **87,** 71416x (1977).
248. U.S. Pat. 3,751,474 (Aug. 7, 1973), K. G. Phillips and M. J. Geerts (to Nalco Chemical Co.).
249. U.S. Pat. 2,732,361 (Jan. 24, 1956), E. M. Filachione C. F. Woodward, and J. E. Hansen (to the United States of America as represented by the Secretary of Agriculture).
250. U.S. Pat. 2,692,871 (Oct. 25, 1954), A. Pechukas (to Columbia-Southern Chemical Corporation).
251. Jpn. Kokai 75 04,032 (Feb. 13, 1975), T. Saito, H. Sakurai, and K. Ishii (to Nippon Zeon Co., Ltd.).

R. D. Spitz
Dow Chemical U.S.A.

DIAMOND. See Carbon.

DIARYLAMINES. See Amines—Aromatic amines—Diarylamines.

DIASPORE, β-ALUMINA MONOHYDRATE, Al$_2$O$_3$.H$_2$O. See Aluminum monohydrate.

DIASTASES. See Enzymes.

DIATOMITE

Diatomite (diatomaceous earth, kieselguhr) is a sedimentary rock of marine or lacustrine deposition. It consists mainly of accumulated shells or frustules of hydrous silica secreted by diatoms, which are microscopic, one-celled, flowerless plants of the class *Bacillarieae*. Some are mobile, others stationary. From 60,000,000 years ago to as late as perhaps 100,000 years ago, the diatom plants thrived in the waters of the earth. They still exist, although in only a fraction of their prehistoric population.

Diatoms are single-celled photosynthetic plants consisting of two shells that fit together in the same manner as the two halves of a pill box (1). Reproduction is by division and at such a rate that it is estimated that one diatom could produce 10^{10} descendants in thirty days under the most favorable conditions. The prehistoric diatom plants extracted silica from the water and used this substance to form an encasing shell or exoskeleton. It is believed that during a period of many thousands of years intense volcanic activity caused both fresh and salt waters to develop a high silica content on which the diatoms flourished. At the end of a very brief life, the diatom settles to the bottom of the body of water and the organic matter decomposes, leaving the siliceous skeleton. These fossil skeletons (specifically frustules) are in the shape of the original diatom plant in designs as varied and intricate as snow flakes (2). Over 10,000 varieties have been classified (1). A few of these are shown in Figures 1 and 2.

Origin of Deposits

Diatoms still inhabit fresh, brackish, or sea waters, as undoubtedly was the case millions of years ago. Salinity changes produce different types of diatoms which may appear at different levels of the same deposit as changes in the salinity of the water occurred. Today's commercial deposits of diatomite are accumulations of the fossil skeletons in beds varying in thickness to as much as 900 meters (2) in some locations. Marine deposits must have been formed on the bottoms of bays, lagoons, or other bodies of quiet water, undisturbed by strong currents. Ocean currents, however, could have had considerable effect in carrying the diatoms from some distance and concentrating them in the quiet waters.

The main deposits of fresh water diatomite were laid down in large lakes. Possibly earthquakes, landslides, or lava flows formed dams in rivers to create lakes where the diatoms thrived. Several tens of square kilometers in Nevada, west of Tonopah, are covered with diatomite as are other large areas in that state.

The principal marine deposits were formed during the Tertiary era and more particularly the upper Miocene period. Major deposits of fresh water origin are more recent, dating from 1,000,000 years to as late as 100,000 years ago. During many thousands of years, owing both to earth upheavals and other causes, either the waters receded or the land was uplifted and the present day dry land deposits were formed. Most commercial deposits are at or comparatively near the surface.

Location of Deposits. Deposits of diatomite are known to exist in every continent, in nearly every country, in widely diversified and even unexpected environments. Over half of the states of the United States reportedly contain diatomaceous deposits. In some cases, deposits of marine and fresh water origin occur almost side by side, as do deposits of widely varying ages (2). Most of the deposits are not large enough or suf-

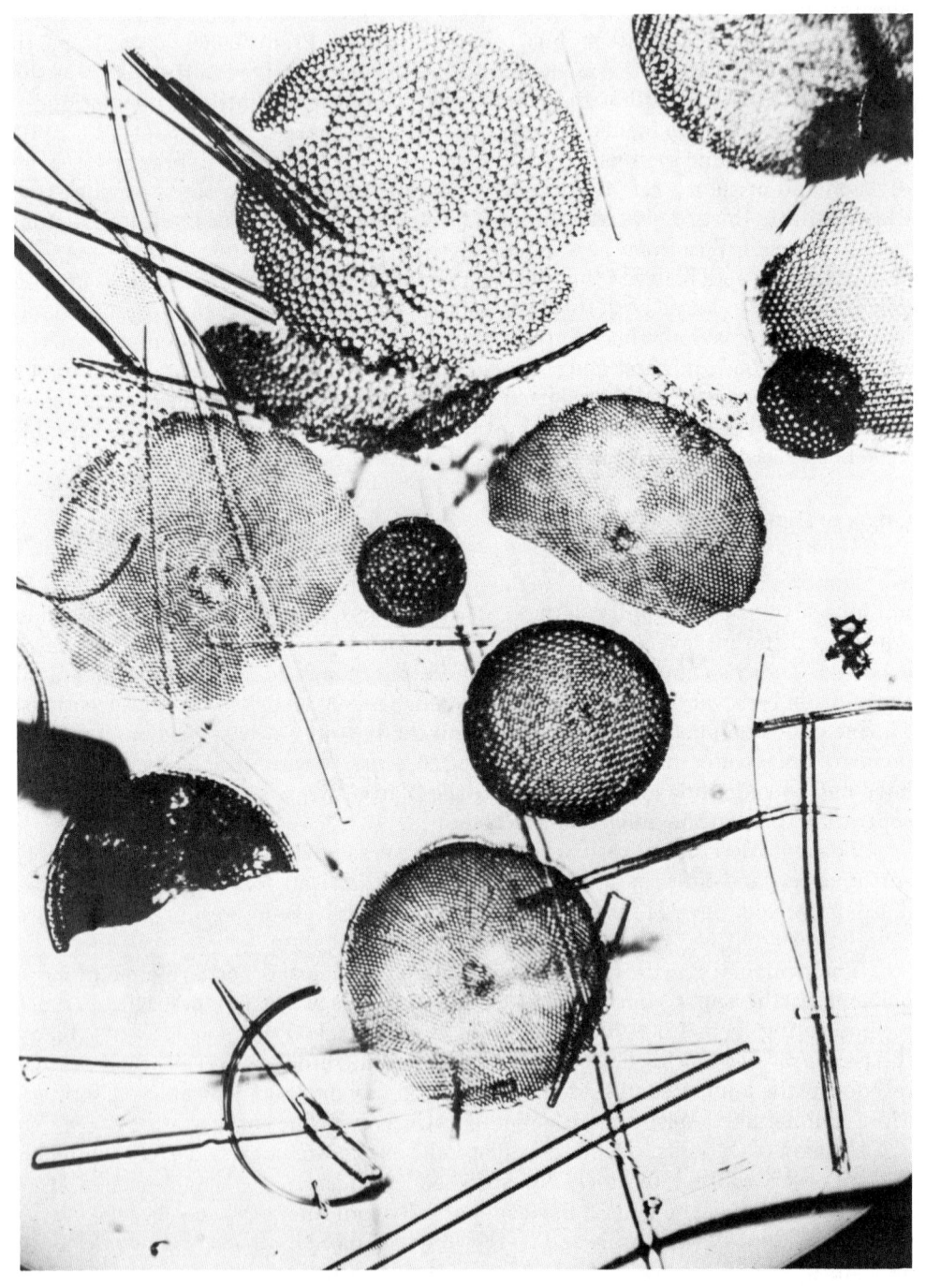

Figure 1. A typical field of diatoms showing various species, particularly *Coscinodiscus* (800×).

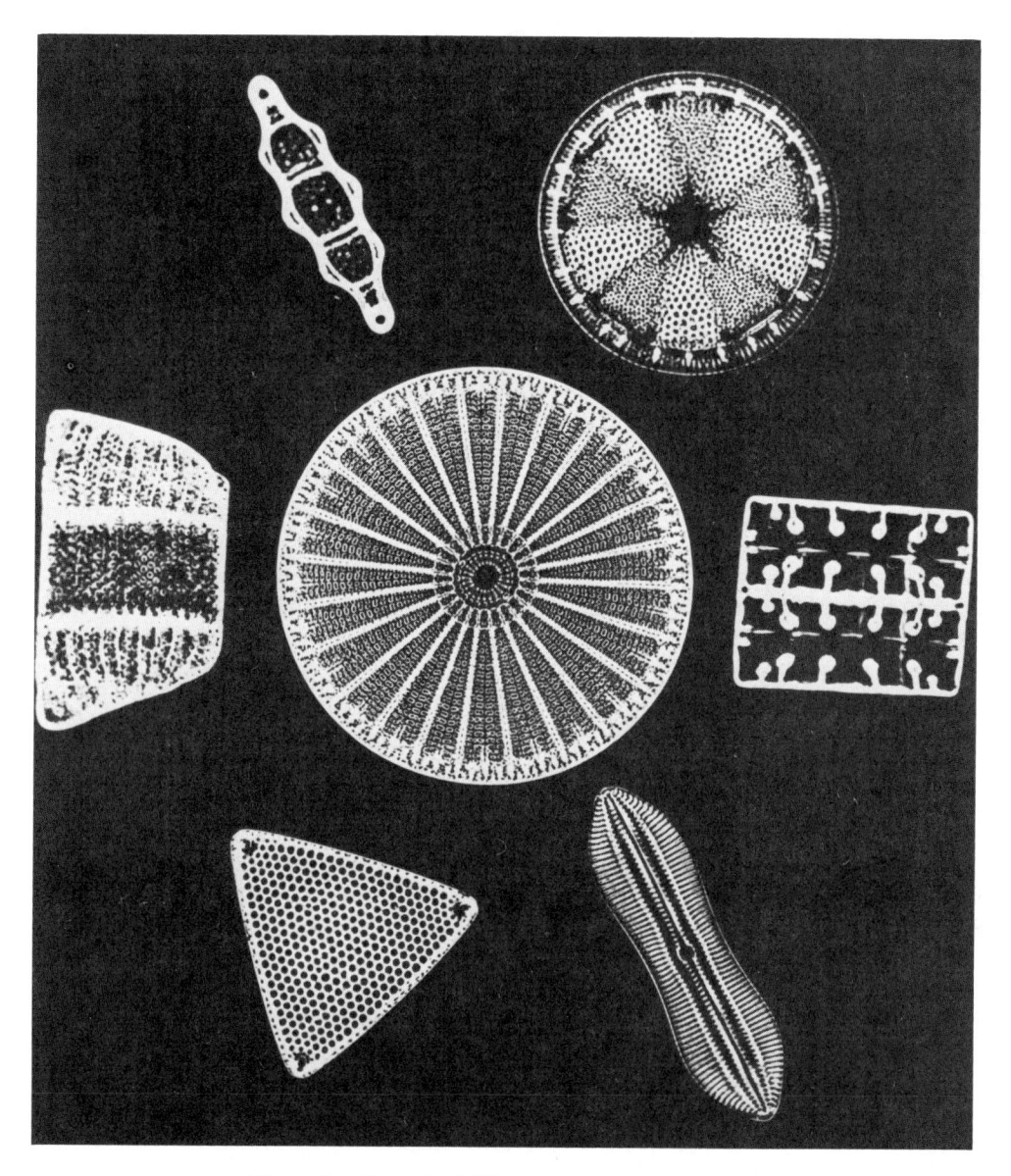

Figure 2. Example of different diatom shapes (500×).

ficiently pure to have commercial value; production figures show the location of the deposits that meet commercial standards in both respects.

The states in the United States having the most extensive commercial deposits are California, Nevada, Oregon, and Washington. The U.S. Bureau of Mines also reports the commercial operation of diatomite deposits in Arizona, Connecticut, Florida, Maryland, New Hampshire, New Mexico, New York, Virginia, and Idaho.

California contains the largest formation of diatomite in the United States; the Monterey formation extends from Point Arena in Mendocino County in the north to San Onofre in the south. The most extensive deposit in the area is near Lompoc, Santa

Barbara County, and it is of marine origin. Other important deposits (not now mined) are located in Monterey, Fresno, Shasta, Inyo, Kern, Orange, San Bernardino, San Joaquin, Sonoma, and San Luis Obispo counties. A large deposit in Los Angeles county was operated from 1930 to 1958. It had reached the point where the quantity and quality of crude diatomite available were overbalanced by the value of the land for real-estate development.

An extensive deposit in Oregon near Terrebonne was operated intensively from 1936 to 1961. Current Oregon production is from two deposits near Christmas Valley. Three different companies operate at least four deposits in Nevada. Of the several comparatively large deposits in Washington, only one, near Quincy, is being worked on a significant commercial scale (3).

Large fresh-water deposits are found in Canada, in British Columbia. Small deposits are located in Nova Scotia, New Brunswick, Quebec, and Ontario. Although deposits exist in other continents, the most important are in Europe, Africa, Japan, and the Union of Soviet Socialist Republics. Judging from reported output of finished diatomaceous earth products, the leading producers in Europe are France, the Federal Republic of Germany, Italy, and Denmark. Important deposits are located in Algeria and Kenya in Africa, with smaller operations in Mozambique, Rhodesia, and the Republic of South Africa. The Union of Soviet Socialist Republics is said to have very large deposits in the Caucasus Mountains.

Physical and Chemical Properties

Chemically, diatomite consists primarily of silicon dioxide, and is essentially inert. It is attacked by strong alkalies and by hydrofluoric acid, but it is practically unaffected by other acids. Because of the intricate structure of the diatom skeletons that form diatomite, the silicon dioxide has a very different physical structure from other forms in which it occurs (see Silicon compounds). The chemically combined water content varies 2–10%. Impurities are other aquatic fossils (sponge residues, *Radiolaria, Silico-flagellatae*), sand clay, volcanic ash, calcium carbonate, magnesium carbonate, soluble salts, and organic matter. The types and amounts of impurities are highly variable and depend upon the conditions of sedimentation at the time of diatom deposition. Variations exist among deposits as well as among parts of the same deposit. Typical chemical analyses of diatomite from different deposits are given in Table 1.

Table 1. Typical Spectrographic Analysis of Various Diatomites (Dry Basis)

Constituent, %	Lompoc, Calif.	Basalt, Nev.	Sparks, Nev.
SiO_2	88.90	83.13	87.81
Al_2O_3	3.00	4.60	4.51
CaO	0.53	2.50	1.15
MgO	0.56	0.64	0.17
Fe_2O_3	1,69	2.00	1.49
Na_2O	1.44	1.60	0.77
V_2O_5	0.11	0.05	0.77
TiO_2	0.14	0.18	0.77
ignition loss	3.60	5.30	4.10

The color of pure diatomite is white, or near white, but impurities such as carbonaceous matter, clay, iron oxide, volcanic ash, etc, may darken it. The refractive index ranges from 1.41 to 1.48, almost that of opaline silica. Diatomite is isotropic.

The apparent density of powdered diatomite varies from 112 to 320 kg/m^3 and may reach 960 kg/m^3 for impure lump material. The true specific gravity of diatomite is 2.1–2.2, the same as for opaline silica, or opal (1).

Bed moisture may be over 65%, but in arid regions it often varies 15–25% (4).

The thermal conductivity is low but increases with the increase in percentage of impurities and the weight per unit volume. The fusion point depends on the purity, but averages about 1590°C for pure material, slightly less than the pure silica. The addition of certain chemical agents reduces the fusion point.

Diatomite has only weak adsorption powers, but shows excellent absorption (qv). Acids, liquid fertilizers, alcohol, water, and other fluids are absorbed by diatomite.

Mining and Processing

Diatomite deposits are usually discovered by observation of outcroppings and the value of the deposit is determined by geological prospecting and exploration. Usually samples are taken from the surface outcrops by digging or trenching; underground samples are secured from test holes, core-drill holes, or tunnels. Samples are examined microscopically, physically, and chemically to determine the suitability of the diatomite for various uses. Accessibility of the deposit to transportation is important. Estimates of tonnages available, based on the dry weight of the material per unit volume, are necessary to determine the potential profitability of operation as well as the life of the deposit.

Mining. Mining diatomite is relatively simple, especially in the larger deposits. In the past, some underground mining was practiced utilizing tunneling and room-and-pillar techniques. Today this method is rarely employed except in small-scale operations and in some nations where low-cost labor is available.

The most common method of mining diatomite is by quarrying or open-pit operation (5). The first step in opening a quarry area is removal of the overburden. This is highly mechanized: power shovels lift the crude material, which is then hauled to the processing plant. The diatomite is sometimes stockpiled, for air drying or for storage as blending crude, or for both reasons.

Processing. Manufacture of finished products from the crude diatomite has greatly advanced since the first serious research and development work in 1912. The principal reason for the large growth of the industry in the United States is the high degree of technological efficiency attained by the major producers. Every step from the selection of the crude material to the testing of the finished product before shipment is carefully monitored. The industry devotes much of its energy to the development of improved products and processes, including specially designed production equipment. The principal producers make every effort to supply custom-made materials for specialized applications. Figure 3 is a flow sheet showing the processing of diatomite at a plant at Lompoc, California.

Three general types are produced, with a range of grades in each. The term grade, as used here, refers not to the quality of the product but designates one of the series of products made for specific uses. Regardless of the ultimate use, the processing of each type is essentially the same. The crude diatomite, which may contain up to 60%

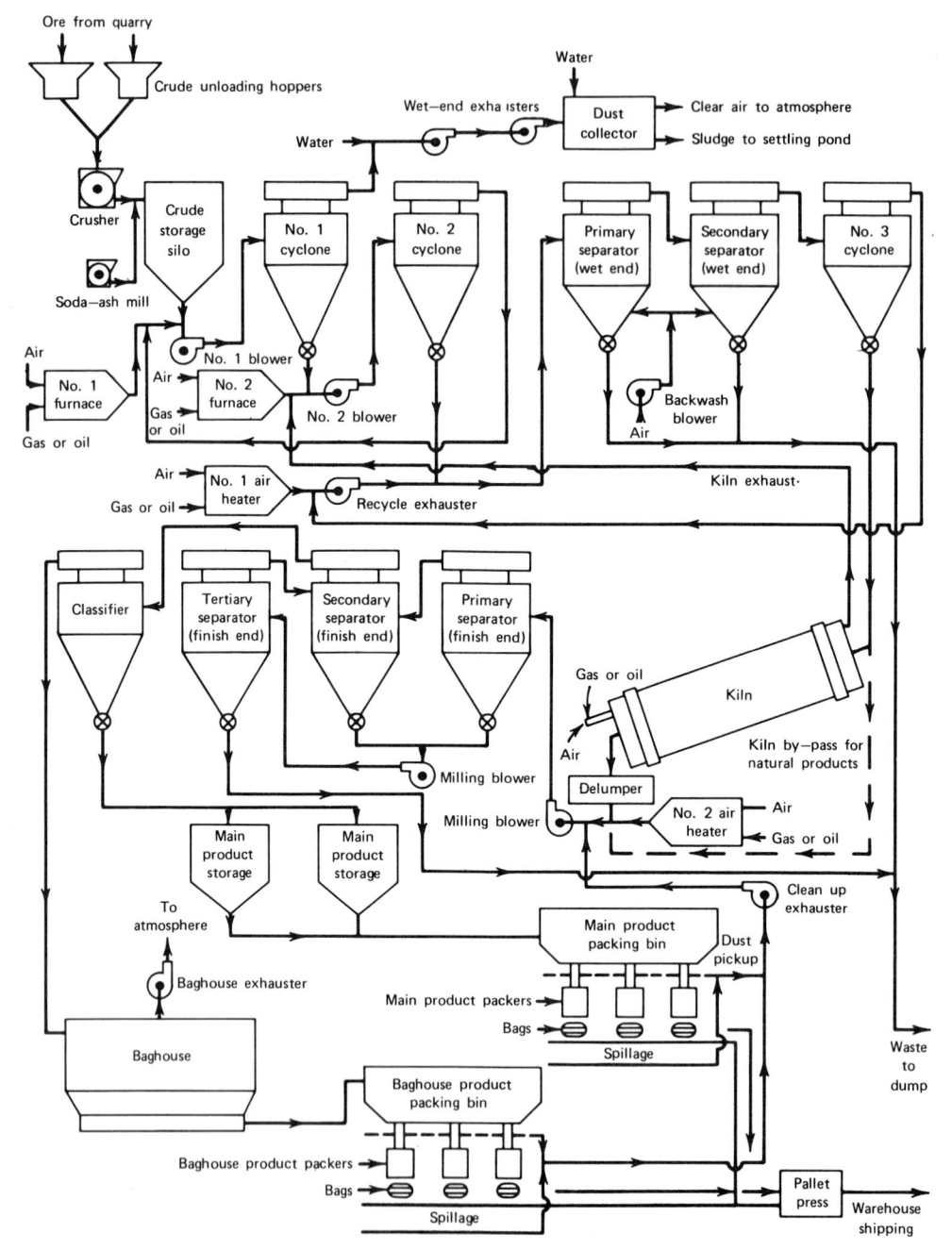

Figure 3. Flow sheet for the processing of diatomite (4).

moisture, is first broken up in hammer mills. This material is fed to dryers operating at relatively low temperatures, where virtually all of the moisture is removed. All manufacturing processes described with the exception of the calcination step, take place while the material is pneumatically conveyed. Coarse and gritty nondiatomaceous earth material is removed in separators and preliminary particle-size separation is made in cyclones. The resultant material is termed a natural product, and is the only type made by some of the smaller producers (6).

The second type is produced from natural diatomite, processed as above and also subjected to high temperature calcination in a rotary kiln at about 980°C. The calcined material is then again milled and classified to remove coarse agglomerates as well as extreme fines.

The third type of product is the so-called white or flux-calcinated diatomite material. This is obtained by calcination of the natural product in the presence of a flux, generally soda ash, although sodium chloride can also be employed. Such processing has the effect of reducing the surface area of the particles, changing the color from the natural buff cast to a true white, and rendering various impurities insoluble. Some of the diatomite is converted on calcination to cristobalite. The Threshold Limit Value (TLV) in mg/m^3 for calcined products is in the range of 0.094–0.24 and for flux-calcined diatomite is 0.074 (7).

In all processing of diatomite, the selection of the proper crude ore, the milling, calcination, and classification are of extreme importance. Each grade of material is manufactured to rigid specifications. The finished product must pass certain performance, chemical, and physical tests to ensure compliance with these specifications (see under Uses).

Economic Aspects

Owing to the light weight of diatomite, freight and trucking rates (on a weight basis) are high. The finished products are packed and shipped in laminated kraft-paper bags, usually containing 22.5 kg, for United States exports, or the product is shipped in bulk. Bagged products are shipped by truck or rail box car, normal box car loading being 27–36 metric tons. It may also be palletized and wrapped with a polyethylene shrink wrap. Bulk shipments are made in bulk pressure-differential trucks of 45 and 74 m^3 (1600 and 2600 ft^3). These carry 12–15 t and 16–21 t, respectively. Distribution by bulk truck is confined to the economical trucking distance from the producing plant.

For longer hauls, pressure-differential rail hopper cars are used. These hold between 36–45 t per car. In-plant storage is in conventional silos usually with a 60° cone bottom. The silo may be pressurized for discharge, but this is exceptional. Normally, diatomite is moved from the silo to an adjacent small pressure vessel for transfer to user sites. Or the material may be aerated and pumped using a modified diaphram pump. Bulk handling has the advantage of improved environmental conditions as well as lower handling cost in comparison to bagged material.

There are so many different products, and variations of these among manufacturers, that it is impossible to cite an average price for diatomite in the United States. The processed powders and aggregates range in price from $20.00–225.00/t for what might be called standard products in carload quantities. Materials for specialized applications, which require special processing, range up to $900.00/t in carload quantities. All these prices are FOB the diatomite plant.

610 DIATOMITE

Producers. Principal companies mining diatomite and processing it into finished products in the United States are Grefco, Inc., Lompoc, California, and Mina, Nevada; Johns-Manville Products Corporation, Lompoc; Eagle-Picher Industries, Inc., Sparks and Lovelock, Nevada; and Witco Chemical, Quincy, Washington. The remaining producers listed in *Minerals Yearbook* are Excel-Minerals Company, Taft, California; Airox Earth Resources, Inc., near Santa Maria, California; N L Industries, Inc., near Wallace, Kansas; Cyprus Mines, near Fernley, Nevada; and A. M. Matlock and American Fossil, Inc., both near Christmas Valley, Oregon.

Annual United States Production. The development of the diatomite industry in the United States is well illustrated by production figures for diatomite products as given in the yearbooks and special bulletins of the U.S. Bureau of Mines. In twenty-four years the tonnage has almost doubled, as shown in Table 2.

World Production. After the United States, the Union of Soviet Socialist Republics is the largest producer of diatomite products with an estimated 1974 production of 399,432 t. The ten highest producers among other countries, with annual production, are shown in Table 3.

Uses

It is reported that in 532 AD the Roman Emperor Justinian used diatomite bricks for lightweight construction of the dome when building the Church of St. Sophia in Constantinople (Istanbul). The names bergmehl, fossil flour, farine fossile, and mountain flour apparently originated when early poverty-driven people extended their

Table 2. Annual Production of Diatomite in the United States[a]

Years	Production, t	Year	Production, t
1954–1956	334,457[b]	1971	485,962
1957–1959	408,310[b]	1972	522,974
1960–1962	437,748[b]	1973	552,765
1963–1965	526,775[b]	1974	603,054
1966–1968	569,456[b]	1975	520,169
1969	543,302	1976	563,744
1970	542,962	1977	580,656

[a] Refs. 3, 8.
[b] Average annual production.

Table 3. Annual World Production of Diatomite[a]

Country	Production, t	Country	Production, t
U.S.S.R.	399,432	Iceland	22,513
France	208,794	Spain	20,879
Italy	59,007	Mexico	19,972
FRG	44,482	Denmark	19,972
Costa Rica	35,404	Argentina	19,972

[a] Estimated tonnage for all countries from Vol. 1 of *Minerals Yearbook*, U.S. Dept. of Interior, Bureau of Mines, U.S. Government Printing Office, Washington, D.C., 1974 (3); see also ref. 8.

meager supply of meal and flour by dilution with diatomaceous earth. Tripolite is a name given to diatomite formerly mined in Tripoli, North Africa. Kieselguhr is the name given to the diatomite first mined in Hanover, Germany, in 1836 or earlier, and is still used as a general name for all diatomite products in Europe. An incorrect name which persisted for many years was infusorial earth, incorrect because Infusoria comprises a group of the animal kingdom (1). Nobel developed the first important industrial use of diatomite as an absorbent for liquid nitroglycerin in the making of dynamite late in the nineteenth century (see Explosives and propellants).

The first substantial commercial shipment of diatomite in the United States was made in 1893, and consisted of material from a one-man quarry operation in the vast deposit near Lompoc, California. It went to San Francisco to be used for pipe insulation. Small-scale operation of parts of the Lompoc deposit continued from 1893 for about fifteen years until it was acquired by the Kieselguhr Company of America. This name was later changed to The Celite Company (9). The first significant research and development work on the production of diatomite materials for industrial use was started in 1912 (10). In the ca 70 years since, the industry has grown immensely and diatomite products are used in almost every country of the world.

Types of Products. Several hundred diatomite products are available, many of them processed for specific purposes. In general, these can be grouped according to use as follows: filteraids, fillers or extenders, thermal insulation, absorbents, catalyst carriers, insecticide carriers and diluents, fertilizer conditioners, and miscellaneous.

In physical form, powders make up by far the greatest proportion of diatomite products. Mean particle diameters range from 20 to 0.75 μm. Aggregates are available for special uses and range from 1.27-cm particles to fine powder. The other most common form is molded insulating brick. This does not take into account a number of types and shapes of thermal insulation that are composed of other ingredients mixed with diatomite.

Uses of Products. There are three principal ways in which finished diatomite products are utilized in manufacturing plants. One type of diatomite product, a filteraid, is used as an expendable processing aid. Another type, filler, becomes a component and remains as part of the manufactured product. A third type, insulation, is installed as a structural member in walls, bases, and other parts of heated equipment to prevent heat loss.

The value of diatomite for most uses depends on its unique physical structure, which is responsible for its high porosity, low density, and great surface area. The particles are quite irregular in shape. Regardless of seeming fragility, each fossil particle, because of its silica composition, is a rigid individual shape. These physical characteristics, in addition to chemical inertness, are of considerable importance to the uses of the material. Table 4 gives some properties of the products of typical diatomite products (11).

According to 1973 figures from the U.S. Bureau of Mines, the use of diatomite products is divided as follows: filtration, 61%; thermal insulation, 4%; miscellaneous, including fillers, 35%.

Filtration. The first use of diatomite as a positive filteraid was by Slater (9) for the filtration of sewage sludge (see Filtration; Water, sewage). The filteraid use developed, however, from the application in cane sugar refining (12). This is still one of the principal uses, including filtration of various liquors in refining cane, beet, and

Table 4. Properties of Typical Diatomite Products

Property	Filteraids Natural	Filteraids Calcined	Filteraids White	Fillers Natural	Fillers Calcined	Fillers White
relative filtration rates, based on natural with flow rate of unit value	1	1–3	3–25			
wet cake density, kg/m^3	2.4–3.5	2.4–3.7	2.6–3.4	2.4–4.5	2.4–4.8	2.4–6.4
(lb/ft^3)	(15–22)	(15–23)	(16–21)	(15–28)	(15–30)	(15–40)
sedimentation particle size distribution, %						
+40 μm	2–4	5–12	5–24	0–4	5–12	0–4
20–40 μm	8–12	5–12	7–34	0–10	5–12	1–7
10–20 μm	12–16	10–15	20–30	0–14	10–15	10–25
6–10 μm	12–18	15–20	8–33	1–32	15–20	15–35
2–6 μm	35–40	15–45	4–30	33–40	15–45	30–50
−2 μm	10–20	8–12	1–3	14–66	8–12	3–20
moisture, max, %	6.0	0.5	0.5	6.0	0.5	0.5
specific gravity	2.00	2.25	2.33	2.00	2.25	2.33
pH	6.0–8.0	6.0–8.0	8.0–10.0	6.0–8.0	6.0–8.0	8.0–10.0
refractive index	1.46	1.46	1.46	1.46	1.46	1.46
Gardner-Coleman oil absorption				150–170	130–160	95–160
ASTM rubout oil absorption				130–160	130–160	130–160
retained on 44 μm (325 mesh) screen, %	0–12	0–12	12–35	0–12	0–12	0–35
surface area by nitrogen adsorption, m^2/g	12–40	2–5	1–3	12–40	2–5	1–3

corn sugar and in clarification of syrups, molasses, etc (see Sugar). From this application, the material and technique was applied to a wide range of separation problems involving beer (qv), various chemicals, water, solvents, antibiotics, oils and fat, phosphoric acids, and many others. In recent years, the necessity of producing clean plant effluents has spurred the application of diatomite filtration for solids removal from waste streams from manufacturing plants (see Adsorptive separation).

Fillers. In a broad sense, diatomite mineral fillers (qv) are used primarily (1) where bulk is needed with minimum weight increase, (2) as an extender where economy of more expensive ingredients is a factor, or (3) where the structure of the particle is important. In other applications, diatomite can add strength, toughness, and resistance to abrasion, and in still others may act as a mild abrasive and polishing agent (see Abrasives).

The paint and paper industries are typical of those employing diatomite extensively as fillers and extender pigments. Also diatomite is extensively used as a polyethylene antiblock (see Abherents).

Insulation. Diatomite makes an efficient thermal insulator because of its high resistance to heat (fusion point about 1593°C) and its high porosity. Materials in the form of powders, aggregates, and bricks are most commonly used. At one time solid bricks were sawed directly from strata in a deposit, dried in a kiln, and then milled to size. In this natural form the diatomite would withstand direct service temperatures up to 870°C without undue shrinkage. This type is no longer available, although at one time it was very widely used as a brick course in walls, bases, and tops of heated equipment. Diatomite insulating bricks for all temperatures are now formed by adding a binder, molding the mixture to sizes and shapes as desired, and then firing in a kiln. Diatomite powders and aggregates are often installed loose over tops or in hollow wall-spaces of furnaces, kilns, ovens, etc. Calcined aggregates are supplied for mixture with water and portland cement to make insulating concrete for casting bases, doors, baffles, etc, for various types of heated equipment (see Insulation, thermal).

Calcium Silicate Insulation. This type of thermal insulation is produced by a number of manufacturers utilizing the reaction of lime and natural diatomite. It is usually precast in molds for pipe covering and other special shapes, in blocks for covering large areas, and in curved slabs for insulating tanks and other equipment. A fast reaction rate is essential for efficient production of calcium silicate insulation. Since the reaction time depends largely on a high surface area of diatomite used, special grades have been developed for this purpose. Most of this material is manufactured from a fresh-water diatomite which is characterized by its high surface area. The diatomite and lime are mixed with water in a slurry together with a catalyst and heated to ca 79°C. This heated slurry is poured into molds where an initial set takes place rapidly so that the desired shapes can be removed from the molds in a minimum of time. The material is then autoclaved and dried. This calcium silicate insulation is characterized by high temperature resistance, which permits its use at temperatures up to 677°C, whereas the well-known 85% magnesia insulation cannot be employed at temperatures exceeding 316°C.

Other Uses. Miscellaneous uses of diatomite are many, some being highly specialized and extensive. As a pozzolanic admixture in concrete mixes, it improves the workability of the mix, permitting easier chuting and placement in intricate forms (see Cement). Special grades are used as carriers for catalysts in petroleum refining, hydrogenation of oils, and in manufacture of certain acids. Diatomite powders are used as carriers for insecticides, and as a fluffing agent for heavier dusts (see Insect control technology). Certain diatomite powders, with nothing else added, act as a natural insecticide and are used to protect seeds and stored grain. This use is not extensive since outdoor application is very sensitive to low humidity and with stored grains, diatomite changes the density and grain value. The fertilizer industry uses large quantities of diatomite as an anticaking agent or conditioner, particularly for prilled ammonium nitrate. The diatomite greatly reduces absorption of moisture by the fertilizer, thus preventing caking and balling in the bag and making spreading easier (see Fertilizers).

BIBLIOGRAPHY

"Diatomite in *ECT,* 1st ed., Vol. 5, pp. 33–37, by H. Mulryan, Sierra Talc & Clay Company; "Diatomite" in *ECT* 2nd ed., Vol. 7, pp. 53–63, by E. L. Neu, Great Lakes Carbon Corporation.

1. R. Calvert, *Diatomaceous Earth,* American Chemical Society Monograph Series, Chemical Catalog Company, J. J. Little & Ives Co., New York, 1930.
2. W. W. Wornardt, Jr., *Miocene and Pliocene Marine Diatoms From California, Occasional Papers of the California Academy of Sciences, No. 63,* The California Academy of Sciences, Los Angeles, Calif., 1967.
3. *Minerals Yearbook,* Vol. 1, U.S. Department of Interior, Bureau of Mines, U.S. Government Printing Office, Washington, D.C., 1974, pp. 539–541.
4. W. G. Hull and co-workers, *Ind. Eng. Chem.* **45,** 256 (Feb. 1953).
5. C. V. O. Hughes, Jr., *Diatomaceous Earth Mining Engineering,* American Institute of Mining, Metal, and Petroleum Engineers, Littleton, Colorado, March 1953.

6. P. W. Leppla, *Diatomite,* Mineral Information Service, Department of Natural Resources, State of California, **6**(11), 1–5 (1953).
7. *Fed. Reg.* **36**(123), 242 (1971).
8. G. Coombs, *Diatomite Mining Engineering,* American Institute of Mining, Metal, and Petroleum Engineers, Littleton, Colorado, March 1977.
9. A. B. Cummins, *The Development of Diatomite Filter Aid Filtration, Filtration and Separation,* Uplands Press Ltd., Craydon, Eng., Jan–Feb 1973.
10. P. A. Boeck, *Chem. Met. Eng.* **12,** 109 (1914).
11. E. Brannigan, *Product Characteristics,* Internal Publication, Dicalite Division-Grefco, Inc., Los Angeles, Calif., 1977.
12. H. S. Thatcher, *Sugar Filtration Improved Methods Filtration,* Kieselguhr Co. of America, Lompoc, Calif., 1915.

E. L. NEU
A. F. ALCIATORE
Grefco, Inc.

DICARBOXYLIC ACIDS

This article describes the saturated, linear aliphatic series of dicarboxylic acids (diacids), the first member of which is oxalic acid. The general formula for the group is $HOOC(CH_2)_n COOH$ for the best known diacids in which $n = 0$–11. Higher members of the series are known but are not as readily available and have not achieved the industrial importance or degree of evaluation of many of the lower members. Some members of this class are reviewed in detail elsewhere in the Encyclopedia (see Oxalic acid; Malonic acid; Succinic acid; Adipic acid; see also Acids, carboxylic). Aromatic dibasic acids and unsaturated aliphatic dicarboxylic acids are not discussed here (see Phthalic acids; Maleic anhydride, maleic acid and fumaric acid; Dimer acids). The melting points of diacids for which n is 0–19 are given in Table 1. Additional properties for acids with $n = 0$–10 are listed in Tables 1 and 2; see also refs. 1 and 15 and the following text.

Nomenclature

The aliphatic dicarboxylic acids, $HOOC(CH_2)_n COOH$, are best known by trivial names for $n = 0$–8. IUPAC names, which are derived from the name of the parent hydrocarbon plus the suffix dioic, are given for all of the acids. Indexes to *Chemical Abstracts* in recent years use the IUPAC names and enter trivial names only as cross references. However, the older literature uses the trivial names as well as a third system based on a hydrocarbon name for the $(CH_2)_n$ carbon segment and the suffix, dicarboxylic acid. Thus the acid

$$\overset{1}{HOOC}-\underset{\alpha}{CH_2}\underset{\beta}{CH_2}\underset{\gamma}{CH_2}\underset{\delta}{CH_2}\underset{\delta'}{CH_2}\underset{\gamma'}{CH_2}\underset{\beta'}{CH_2}\underset{\alpha'}{CH_2}-\overset{10}{COOH}$$

Table 1. Melting Points and Boiling Points of Linear Dicarboxylic Acids, HOOC(CH$_2$)$_n$COOH

Total number of carbon atoms	IUPAC name	CAS Registry No.	Common name	mp, °C[a]	bp, °C[a,b]
2	ethanedioic	[144-62-7]	oxalic	187 (dec)	
3	propanedioic	[141-82-2]	malonic	134–136 (dec)	
4	butanedioic	[110-15-6]	succinic	187.6–187.9	
5	pentanedioic	[110-94-1]	glutaric	98–99	200$_{2.7\ kPa}$[c]
6	hexanedioic	[124-04-9]	adipic	153.0–153.1	265$_{13.3\ kPa}$, 216.5$_{2.0\ kPa}$[d]
7	heptanedioic	[111-16-0]	pimelic	105.7–105.8	272$_{13.3\ kPa}$, 223$_{2.0\ kPa}$
8	octanedioic	[505-48-6]	suberic	143.0–143.3	279$_{13.3\ kPa}$, 230$_{2.0\ kPa}$
9	nonanedioic	[123-99-9]	azelaic	107–108	286.5$_{13.3\ kPa}$, 237$_{2.0\ kPa}$[d]
10	decanedioic	[110-20-6]	sebacic	134.0–134.4	294.5$_{13.3\ kPa}$, 243.5$_{2.0\ kPa}$[d]
11	undecanedioic	[1852-04-6]		110.5–112[e]	
12	dodecanedioic	[693-23-2]		128.7–129.0	254$_{2.0\ kPa}$[f]
13	tridecanedioic	[505-52-2]	brassylic	114	
14	tetradecanedioic	[821-38-5]		126.5[g]	
15	pentadecanedioic	[1460-18-0]		114.7[g]	
16	hexadecanedioic	[505-54-4]	thapsic	125[g]	
17	heptadecanedioic	[2424-90-0]		117–118[h]	
18	octadecanedioic	[871-70-5]		124.6–124.8[i]	
19	nonadecanedioic	[6250-70-0]		118–119.5	
20	eicosanedioic	[2424-92-2]		124–125[j]	
21	heneicosanedioic	[505-55-5]	japanic	118–120[g]	

[a] Data from ref. 1 except as noted.
[b] To convert kPa to mm Hg, multiply by 7.5.
[c] Ref. 2.
[d] Ref. 3.
[e] Ref. 4.
[f] Ref. 5.
[g] Ref. 6.
[h] Ref. 7.
[i] Ref. 8.
[j] Ref. 9.

with a total of ten carbon atoms is sebacic acid or decanedioic acid and may also be referred to as 1,8-octanedicarboxylic acid. To identify the location of substituents, numbers are used as shown for the IUPAC system. In the older dicarboxylic acid names the numbering excludes the carboxyl groups. Greek letters have also been used for position designation with α and α' being adjacent to the carboxyl groups as shown.

Physical Properties

All of the diacids are colorless, crystalline solids. As the aliphatic chain becomes longer, the trend of melting points is downward and the solids tend to become more waxy. Many of the dibasic acids exhibit more than one crystalline form.

There is an alteration of some physical properties of neighboring members of the series. Properties showing the oscillating effect are melting point, decarboxylation temperature, solubility, and index of refraction. Some properties not showing an alternating effect are boiling point, molar heat of combustion, density, and dielectric constant.

Table 2. Decarboxylation Temperatures and Molar Heats of Combustion of Dicarboxylic Acids

Dicarboxylic acid	Decarboxylation temp[a], °C	Molar heat of combustion[b], MJ/mol[c]
oxalic	166–180	0.246
malonic	140–160	0.864
succinic	290–310	1.492
glutaric	280–290	2.151
adipic	300–320	2.800
pimelic	290–310	3.464
suberic	340–360	4.115
azelaic	320–340	4.778
sebacic	350–370	5.429
dodecanedioic		6.740

[a] Refs. 10–12.
[b] Refs. 13–14.
[c] To convert J to cal, divide by 4.184.

As an example of the alternating effect, glutaric acid with 5 carbon atoms melts much lower than its neighbors: succinic with 4 carbon atoms and adipic with 6 carbon atoms (Table 1). This alternation persists throughout the series, the odd-numbered acids always melting lower than their neighbors, but the effect diminishes as the number of carbon atoms increases (Fig. 1). Empirical equations (6) and graphical means (16) have been applied to relate the melting points and attempt to define a convergence point for the plot.

Such alternating melting point effects are known for other related series of compounds (3) and are quite pronounced in certain groups of high polymers derived from aliphatic diacids, eg, polyamides (qv), polyurethanes, and polyesters (qv) (17–19) (see Urethane polymers).

The alternating relationship does not extend to the boiling points of the diacids, which rise with increasing molecular weight as shown in Table 1. Boiling points for

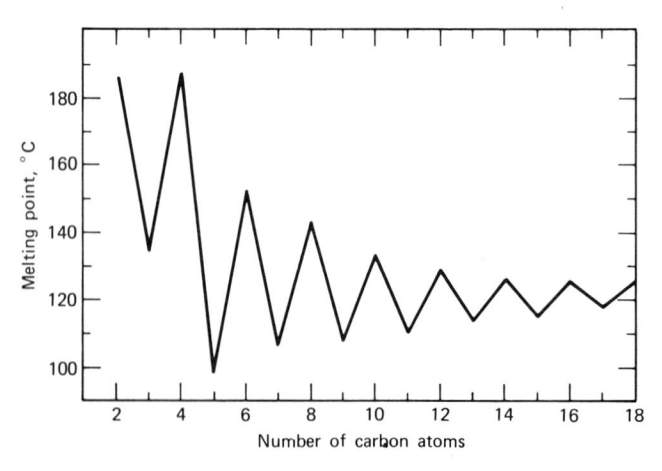

Figure 1. Melting points of linear saturated aliphatic dicarboxylic acids.

lower members of the series are difficult to measure because of sublimation, dehydration, or decomposition.

The alternating effect between adjacent odd and even structured acids has been demonstrated in the ease of decarboxylation of the acids as shown in Table 2. The acids with even numbers of carbon atoms are more resistant to decomposition than the odd-numbered acids preceding them. The activation energy of this reaction for the higher members of this series is fairly high, on the order of 251 kJ/mol (60 kcal/mol) for adipic acid whereas malonic acid has an activation energy of decarboxylation of ca 126 kJ/mol (30 kcal/mol). The order of decomposition temperatures of the diacids obtained by DTA (20) differs from the data in Table 2; ie, sebacic (330°C) > azelaic = pimelic (320°C) > suberic = adipic = glutaric (290°C) > succinic (255°C) > oxalic (200°C) > malonic (185°C). The differences no doubt depend in part on such factors as sample size and manner of heating. Thermal decompositions also are quite sensitive to catalytic effects of impurities, environment, and products of the decomposition itself. Therefore, in reaction systems such as polymer preparation the decarboxylation temperatures may be appreciably lower.

The molar heats of combustion of the diacids do not show the alternating effect (Table 2); like the boiling point they rise steadily with increasing molecular weight.

A property of aliphatic diacids showing the alternating effect is water solubility (Fig. 2). In this case, malonic and glutaric acids are extremely water soluble whereas the even-membered diacids are only moderately soluble and this solubility diminishes rapidly for all acids as the hydrocarbon segment is lengthened (28). There are equations for the relationship (29). In addition, the alternating effect has been observed in acetone and ethyl acetate solvents (30).

All of the aliphatic diacids give acidic reactions in water. Oxalic and malonic acids are strong acids and higher homologues are progressively weaker. Ionization constants are assembled in Table 3. No alternation is apparent, and the values are essentially constant above adipic acid (21–22). A change in values at dodecanedioic acid is owing to a change of the medium to ethanol–water. It is noteworthy that the two carboxyl groups are not equivalent even in the tridecanedioic acid. Ionization constants in the form of pK_1 and pK_2 are given for oxalic through sebacic acid in ref. 2.

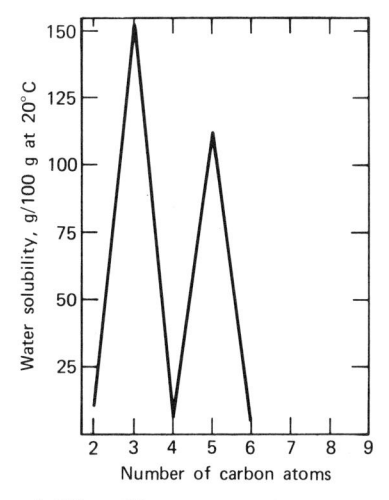

Figure 2. Water solubility of linear saturated aliphatic dicarboxylic acids.

Table 3. Ionization Constants of Dicarboxylic Acids

Name of diacid	Medium	Temp, °C	K_1	K_2	Ref.
oxalic	water	25	5.36×10^{-2}	5.42×10^{-5}	21
malonic	water	25	1.42×10^{-3}	2.01×10^{-6}	22
succinic	water	25	6.21×10^{-5}	2.31×10^{-6}	23
glutaric	water	25	4.58×10^{-5}	3.89×10^{-6}	24
adipic	water	25	3.85×10^{-5}	3.89×10^{-6}	25
pimelic	water	18	3.19×10^{-5}	3.74×10^{-6}	25
suberic	water	18	3.05×10^{-5}	3.85×10^{-6}	25–27
azelaic	water	18	2.88×10^{-5}	3.86×10^{-6}	25, 27
sebacic	water	25	3.1×10^{-5}	3.6×10^{-6}	27
dodecanedioic	ethanol–water 40:60	25	2.0×10^{-6}	2.5×10^{-7}	27
tridecanedioic	ethanol–water 40:60	25	1.6×10^{-6}	2.9×10^{-7}	27

Chemical Properties

The chemical reactions of the dicarboxylic acids depend mainly on the terminal carboxyls with their reactive hydroxyl groups. Heat-induced chemical decomposition of the acids results either in decarboxylation to a monocarboxylic acid or in anhydride formation with water elimination. Oxalic acid gives carbon dioxide and formic acid, which may undergo further decomposition.

$$\text{HOOCCOOH} \rightarrow CO_2 + \text{HCOOH} \rightarrow H_2O + CO$$
oxalic acid

The dibasic acids undergo all the reactions of the monocarboxylic acids (see Acids, carboxylic) with some interaction peculiar to the presence of two carboxyl groups in proximity. This is demonstrated by the facile elimination of carbon dioxide from malonic acid where the carboxyls are both attached to the same carbon.

$$\text{HOOCCH}_2\text{COOH} \rightarrow CO_2 + CH_3\text{COOH}$$
malonic acid

Under some conditions, carbon suboxide (C_3O_2) may also result.

Succinic and glutaric acids, on heating or on removal of water by acetic anhydride or acetyl chloride, produce cyclic anhydrides.

succinic acid → succinic anhydride + H_2O

When adipic acid is heated, it gives a high molecular weight (polymeric) anhydride, but careful vacuum distillation of the polymer results in the unstable cyclic anhydride. Polyanhydrides are obtained from all of the higher diacids (C_6 and above) upon heating with acetic anhydride (31). Cyclization together with decarboxylation is also possible. This occurs when adipic acid is heated in the presence of some metal oxides such as barium or thorium oxide as shown (Dieckmann reaction).

$$\text{HOOC(CH}_2)_4\text{COOH} \xrightarrow{\text{BaO}} \text{cyclopentanone} {=} \text{O} + \text{H}_2\text{O} + \text{CO}_2$$

Each carboxyl group in the diacids is subject to the formation of many normal acid derivatives; this capability is the basis for much of the utility of the higher members, particularly in polymers, plasticizers, and lubricants. In recent years diacid dichlorides have been much used in low temperature syntheses of condensation polymers, primarily polyamides (12,18–19,32–33). Diacid dichlorides are readily formed by reaction with thionyl chloride (18).

Esters with Monoalcohols. Esterification (qv) of the acids with monoalcohols in the presence of acid catalysts has led to a series of commercially important diesters (see also Esters, organic). These materials are used as plasticizers, lubricants, and hydraulic fluids. Several well-known diesters are bis(3-methylbutyl) azelate [*10340-99-5*], bis(2-ethylhexyl) adipate [*103-23-1*], and bis(2-ethylhexyl) sebacate [*122-62-3*]. Some properties of diesters are tabulated in refs. 15 and 34.

Higher alkyl esters are normally prepared by transesterification of the lower alkyl diesters.

$$2 \text{ ROH} + \text{CH}_3\text{OOC(CH}_2)_2\text{COOCH}_3 \rightarrow \text{ROOC(CH}_2)_2\text{COOR} + 2 \text{ CH}_3\text{OH}$$

Polyesters with Aliphatic and Aromatic Diols. Wholly aliphatic polyesters have been known since the work of Carothers in the late 1920s (35). A polyester is made by the reaction of an aliphatic diol with the diacid in the presence of mildly acidic or basic catalysts to accelerate the reaction and minimize dehydration of the diol to olefin (36) (see Polyesters). The wholly aliphatic polyesters are low melting (many below 100°C). Although they are film- and fiber-forming, they do not crystallize well and are hydrolytically sensitive and, therefore, have limited use by themselves. However, such polyesters of low molecular weight with hydroxyl end groups are currently used to some extent as the soft segments in polyurethane foams and coatings (37–38). They have been displaced from this use to a considerable extent by polyether–polyols which have greater hydrolytic stability and are less expensive. The aliphatic polyesters are also used as polymeric plasticizers (qv).

Higher melting polyesters can be prepared from aliphatic dicarboxylic acids and aromatic diols. The ester from hydroquinone (1,4-dihydroxybenzene) and adipyl chloride melts at ca 190°C (39).

[*31800-02-9*]

Polyesters in this category are most easily synthesized by low temperature procedures with alkali and diacid chlorides (18,32,40). They have been examined extensively on a laboratory scale but have not become commercially important because of the superior performance and lower cost of several other types of condensation polymers (eg, polycarbonates) (41).

Network Polyesters. Aliphatic dibasic acids such as adipic, azelaic, and sebacic are used to a limited extent in alkyd resins (qv) (42) and contribute greater flexibility in unsaturated polyester resins (see Polyesters, unsaturated) (43–44). One important use is in glass fiber-reinforced paneling and other shaped structures (see Laminated materials, glass).

Polyamides. The principal use of adipic, sebacic, and dodecanedioic acids is in production of polyamides. Aliphatic polyamides (nylon) are made by condensation reactions between aliphatic diamines and aliphatic diacids or their derivatives (see Polyamides) (12,45). The best known polyamide, nylon-6,6, is based on hexamethylenediamine and adipic acid (35). It was introduced commercially by the Dupont Co. in 1938 as a textile fiber. Today the polymer is used on a large scale in textile fibers (see Fibers, synthetic), tire cord (qv) (46–48), bristles and moldings. Polyamides and copolyamides for molding uses are derived from adipic, azelaic, sebacic, and dodecanedioic acids (49–51). In 1968 DuPont Co. introduced a new, easy-care, silklike polyamide fiber under the trademark Qiana (46). It is reported to be derived from dodecanedioic acid and bis(4-aminocyclohexyl)methane (47,52). Today aliphatic polyamides are made and sold by many companies around the world. Many hundreds of polyamides have been and are being prepared and evaluated in a search for better products to surpass incumbent materials or to suit a special use (see also Aramid fibers).

Manufacture

A one-source reference to the laboratory preparation of diacids from oxalic to brassylic is ref. 14. Generally, the acids from succinic to dodecanedioic are prepared by controlled oxidation or other reactions on available cyclic compounds. Some typical examples, but not necessarily the preferred procedures, of preparation from cyclic compounds are listed in Table 4.

Many diacids are synthesized from linear compounds containing a carbon chain similar to that required in the final product. Some typical examples are given in Table 4. The hydrolysis of a dinitrile is a general procedure. Dinitriles can be prepared from α,ω-dihaloalkanes with alkali cyanides. Electrolytic coupling of monomethyl esters of the lower dicarboxylic acids is a route to the esters of long chain diacids. Carbon dioxide is a by-product. Methyl hydrogen adipate [627-91-8] has been coupled to dimethyl sebacate [106-79-6] in 75–85% yield (53). Some diacids can be obtained in good yield by oxidative cleavage of unsaturated fatty acids from naturally occurring oils. Suberic, azelaic, sebacic, and brassylic acids are obtained in this way. A relatively recent approach to diacid synthesis, which could have important potential, is the conversion of n-alkanes to diacids by oxidative fermentation (54–57) (see Fermentation).

Among the aliphatic dicarboxylic acids, adipic acid has achieved the greatest production and use, nylon fiber and plastics being the main end-uses. Nearly 5.9×10^6 metric tons of the acid was produced in the United States in 1976 (58). This high production is accompanied by a low selling price (Table 5). Azelaic, sebacic, and dodecanedioic acids are the other principal, commercially-used, longer-chain diacids at present. Their higher price reflects in part the lower scale of production. These di-

Table 4. Some Intermediates for Dicarboxylic Acid Production

Acid produced	Cyclic intermediates Raw material	Synthetic route	Linear intermediates Raw material	Synthetic route
oxalic			ethylene; propylene	HNO_3–H_2SO_4 oxidation
succinic	maleic anhydride	hydrogenation	maleic acid	reduction
	tetrahydrofuran	HNO_3 oxidation	butanediol; paraffin waxes	oxidation
glutaric	cyclopentanol cyclopentanone	oxidation	glutaronitrile	hydrolysis
adipic	cyclohexane	oxidation	sebacic acid	HNO_3 oxidation
pimelic	2-cyanocyclohexanone	concentrated alkali	pimelonitrile 1,7-heptanediol	hydrolysis N_2O_4 oxidation
suberic	cyclooctane	air and HNO_3 oxidation	1,6-hexanediol ricinoleic acid	carbonylation HNO_3 oxidation
azelaic			oleic acid	ozonolysis–oxidation
sebacic			ricinoleic acid methyl hydrogen adipate	alkali fusion electrolysis
dodecanedioic	cyclododecene	oxidation		

Table 5. Dicarboxylic Acid Prices, $/kg [a]

| | 1962 | | 1979 | |
Acid	High	Low	High	Low
adipic	0.71	0.64	0.91	0.89
pimelic	110[b]		132[b,c]	
suberic	3.31[d]		65[b,c]	
azelaic	0.82	0.79	1.85[e]	1.67[e]
sebacic	1.44	1.44	3.48	3.48
dodecanedioic[f]	3.31[d]		58[b,c]	

[a] Ref. 59.
[b] Research chemical listing for small amounts.
[c] Not in United States market listings as an industrial chemical.
[d] Offered for developmental use.
[e] Technical grade.
[f] Dodecanedioic acid became available from the DuPont Co. in 1978–1979 as high polymer grade in bulk quantities at $2.65/kg.

acids, as well as most of the other linear aliphatic diacids up to C_{16}, are available from chemical supply houses at research chemical prices.

Toxicology

The toxicity of the diacids varies markedly throughout the series (1). The variation appears to be related in part to acidity (Table 3), spacing of the carboxyl groups and solubility (Fig. 2). Oxalic acid is a hazardous chemical because it is a strong acid and a precipitant of blood calcium. Prolonged skin exposure can cause dermatitis and contact with the dust or vapor severely irritates the eyes and respiratory tract. Severe exposure or ingestion causes such symptoms as vomiting, coughing, pain, and even death. Malonic acid likewise is a hazardous chemical because it is a strong acid and forms insoluble calcium salts. The toxic effects are somewhat less severe than for oxalic acid. Succinic acid has reduced acidity and is reported to have a low order of toxicity in animals. Adipic acid has low toxicity, is only slightly irritating to the kidneys, and is less toxic than tartaric or citric acid upon intravenous administration. Both succinic and adipic acids have been used as acidulants in foods. Glutaric acid, on the other hand, is appreciably toxic. Sodium glutarate is severely nephrotoxic to rabbits in subcutaneous treatment (60). Suberic and azelaic acids are mildly nephropathic as the sodium salts administered subcutaneously to rabbits (61). Sebacic acid is essentially nontoxic in humans (62) and dodecanedioic acid also has low toxicity (63). Although the higher dibasic acids (C_8 and higher) are reported to have low internal toxicity, it is suggested that inhalation of dusts during handling of the powdered acids may have an irritating effect in the respiratory system. Note that water solubility decreases with molecular weight (see Fig. 2). The original literature and various manuals on hazardous chemicals should be consulted for specific information (64–66).

Higher Members of the Dicarboxylic Acid Series

The preceding discussion and tables provide information on the preparation, properties, and uses of the higher homologues of the aliphatic dicarboxylic acid series. The following is further discussion of the synthesis and uses of the better known diacids from C_7 to C_{21}. Additional information can be found in ref. 1 and *Chemical Abstracts*.

Pimelic Acid. Pimelic acid, $HOOC(CH_2)_5COOH$, formula wt 160.17; water solubility 4.97 g/100 g soln at 22°C; easily soluble in alcohol, diethyl ether, and hot benzene; two crystalline forms, both monoclinic. The name pimelic is derived from the Greek word for fat, *pimele*. The first synthesis of the pure acid was by nitric acid oxidation of cycloheptanone. Many other synthetic routes, in addition to the usual conversion of diols and dinitriles, are: reduction and cleavage of salicylic acid with sodium in amyl alcohol (67); in 86% yield from furfural by way of furylacrylic acid and diethyl 4-ketopimelate (68); alkali cleavage of 2-cyanocyclohexanone in 85% yield (69).

Pimelic acid has been incorporated in various polyamides (12,18–19,70) and polyesters (18,40,71), usually as members of a homologous series. Most of the common esters have been prepared, but not much interest in their use is evident.

Suberic Acid. Suberic acid, $HOOC(CH_2)_6COOH$, formula wt 174.19; water solubility 0.16 g/100 g at 20°C; soluble in alcohols and other organic solvents. It is dimorphic and crystallizes in long needles and rough plates. The acid was first obtained by the oxidation of cork (qv) and its name is from the Latin word for cork, *suber*.

Suberic acid can be synthesized in many ways. Early routes were by oxidation of castor oil (qv) or ricinoleic acid (72), hydrolysis of the dinitrile, and electrochemical coupling of monomethyl glutarate. Several routes from cyclooctane and cyclooctene are available. These intermediates are obtained by hydrogenation of 1,5-cyclooctadiene (from cyclic oligomerization of butadiene; see related synthesis of dodecanedioic acid) or cyclooctatetraene [from acetylene (73)]. Cyclooctane can be converted by air oxidation to a cyclooctanol–cyclooctanone mixture which is further oxidized with nitric acid (1). Cyclooctene is cleaved by various oxidizing agents to suberic acid (74). Ozonolysis/oxidation procedures have been used by several groups (75–76).

As with other diacids, suberic acid has been used to prepare polyesters and polyamides for studying structure–property relationships (12,40,77). A hydrolytically stable, high melting polyamide based on suberic acid and 1,4-cyclohexanebis(methylamine) was developed by the Eastman Kodak Co. for a molding plastic (78–79). The acid has also been used in combination with other diacids and bis(4-aminocyclohexyl)methane to yield polyamides which give clear molded parts and films (80). It appears not to be economically competitive with related diacids in most uses.

Azelaic Acid. Azelaic acid, $HOOC(CH_2)_7COOH$, formula wt 188.22; solubility in water 0.2 g/100 g at 15°C, 1.65 g/100 g at 55°C; ether solubility 2.7 g/100 g at 15°C; crystallizes in platelets from acetone and needles from the melt. The name azelaic acid is derived from the French *azotique* (nitric) and elaidic acid ($C_{18}H_{34}O_2$, a stereoisomer of oleic acid). It is found in many natural products containing long-chain fatty acids and related compounds. Commercial production is by ozonolysis of oleic acid combined with oxidation of the product with air or oxygen to yield the diacid and pelargonic acid (81). The reaction may be promoted by various agents and is conveniently carried out in a solvent medium, such as acetic or pelargonic acids. Other preparative procedures are nitric acid oxidation of oleic acid (some C_{10}, C_8, C_7, and related diacids are formed), dichromate oxidation of oleic acid, potassium permanganate oxidation of potassium ricinoleate, and alkali cleavage of oleic acid derivatives such as 9,10-dihydroxystearic acid.

The acid is used in the form of esters in hydraulic fluids (qv) and lubricants (see Lubrication). Metal salts, such as aluminum and lithium, are used in greases. Among the many diesters described are cyclohexyl [*18803-77-5*], n-hexyl [*109-31-9*], isooctyl [*26544-17-2*], and 2-ethylhexyl [*103-24-2*] (15,34). These esters and others are used as plasticizers (qv) for poly(vinyl chloride), polystyrene, other plastics, elastomers, and propellants (see Elastomers, synthetic). The acid is used to modify alkyd resins (qv) and in nylon molding resins. Many other polyamides have been evaluated. Fibers from the polyamide derived from bis(4-aminocyclohexyl)methane are silklike and yield a fabric with wash-and wear characteristics (82).

Sebacic Acid. Sebacic acid, $HOOC(CH_2)_8COOH$, formula wt 202.24, crystallizes as leaflets; water solubility 0.1 g/100 g at 15°C; soluble in alcohols and ether. The name appears to be derived from the Latin *sebaceus* (made from tallow). The major route to sebacic acid is the high temperature, alkali treatment of ricinoleic acid (12-hydroxy-9-octadecenoic acid) (83–85), which is found in castor oil (qv) in amounts up to 90% of the fatty acids present. The principal by-product of the synthesis is 2-octanol. The process has been the subject of many variations and refinements. Other syntheses are the electrolytic coupling of adipic mono esters, nitric acid oxidation of cyclodecane and cyclodecanol, and ozonolysis of undecylenic acid or its ethyl ester.

A product, given the trade name isosebacic acid, was produced on a development scale in the late 1950s, which was a mixture of a 5–10% sebacic acid, 12–18% 2,5-diethyladipic acid [*13087-17-7*], and 72–80% 2-ethylsuberic acid [*3971-33-3*]. This diacid mixture was obtained by carbonation and hydrogenation of an isomeric mixture of disodio octadienes from butadiene (1,86). The material is not in the 1978 trade lists.

Sebacic acid and its derivatives are used in the plastics industry for plasticizers, alkyd resins, polyester resins, polyurethanes, and polyamide molding resins. The diesters are used as lubricants, thinners for castor oil lubricants, and as stable plasticizers for poly(vinyl chloride) and other film-forming polymers. Well known esters are methyl [*106-79-6*], isopropyl [*7491-02-3*], butyl [*109-43-3*], octyl [*2432-87-3*], isooctyl [*27214-90-0*], 2-ethylhexyl [*122-62-3*], nonyl [*4121-16-8*], and benzyl [*140-24-9*] (15,34). Approximately 770 t of sebacic esters were produced in the United States in

1976. The average selling price was ca $2.60/kg (58). The low toxicity of the sebacate esters makes them especially useful in food packaging films (see Packaging materials) and plastics. Polyesters from polyglycols and sebacic acid yield nonmigrating plasticizers. Nylon molding resin-6,10 from hexamethylenediamine has high toughness and lower moisture absorption than nylon-6 or nylon-6,6 as well as better dimensional stability and electrical properties (49–51).

Dodecanedioic Acid. Dodecanedioic acid, $HOOC(CH_2)_{10}COOH$, formula wt 230.30, crystallizes in the monoclinic system; water solubility 0.003 g/100 g at 23°C, 0.37 g/100 g at 100°C. The synthesis of dodecanedioic acid has been reviewed (1,87). The key intermediates for economical synthesis are cyclododecane and its precursors and derivatives, which can be prepared from butadiene. Figure 3 presents an outline of two routes to the diacid. Many variables are involved in attaining the best yields.

Dodecanedioic acid has become industrially important since DuPont's introduction of Qiana nylon fibers in 1968 and nylon molding resins (49) in 1970 (52). The diacid is available through research chemical supply houses and in bulk from the DuPont Co. It is also produced on an industrial scale by Chemische Werke Hüls, FRG (87–88).

Tridecanedioic Acid. Tridecanedioic acid, $HOOC(CH_2)_{11}COOH$, formula wt 244.32; water solubility 0.004 g/100 g at 24°C; readily soluble in alcohol, ether and

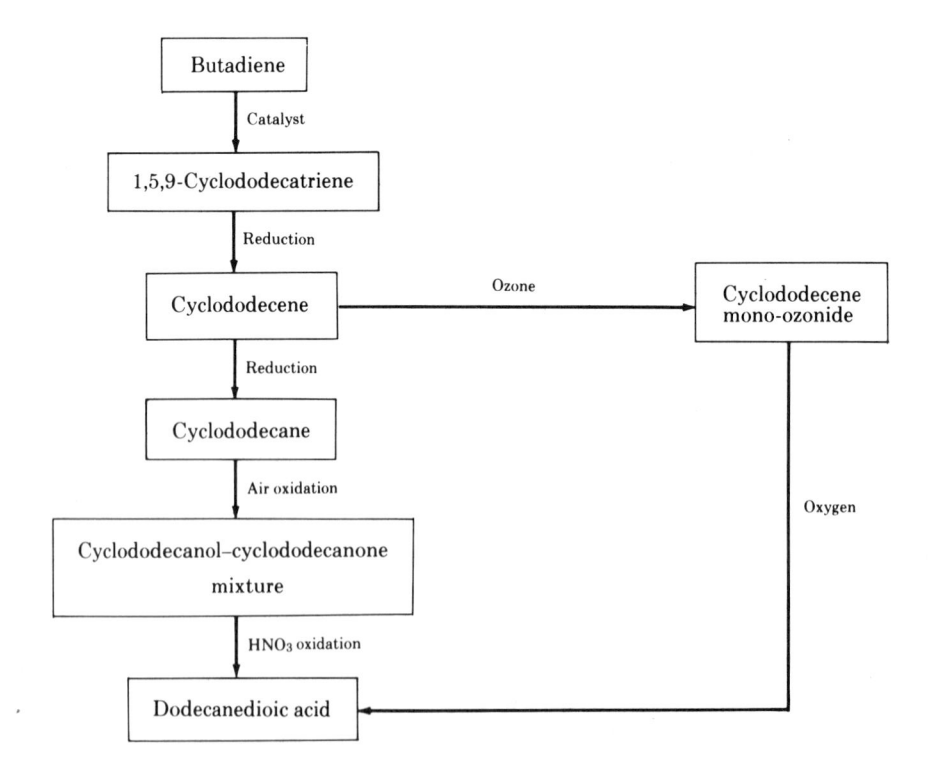

Figure 3. Preparation of dodecanedioic acid.

chloroform. The principal source is erucic acid, $CH_3(CH_2)_7CH=CH(CH_2)_{11}COOH$, which occurs in rapeseed oil and to the extent of 55–60% in the seed oil of *Crambe abyssinica* (1,89). The latter is an industrial crop being promoted by the USDA and first planted on a large scale in 1965 (89) (see Vegetable oils). Tridecanedioic acid has been prepared with 99% purity and yields of 82–92% by ozonolysis–oxidation of erucic acid in acetic acid on a small pilot scale (89). The by-product pelargonic acid was easily separated.

This diacid has been used to prepare various polyamides and polyesters, among which are 6,13 (90) and 13,13 (90–91). The C_{13} diamine was prepared from the diacid. Fibers were made from the 13,13 polyamide.

Others. Many higher members of the aliphatic diacid series are known (Table 1). Most have been described in terms of experimental testing and uses, but at present have limited availability. Many branched-chain dicarboxylic acids are known as chemical entities or as mixtures. The "isosebacic acid" system, mentioned earlier under sebacic acid, is an example of such a mixture. Another mixture containing the C_{19} dicarboxylic acids is prepared in a different way. This product contains little, if any, of the linear nonadecanedioic acid (Table 1) and has a structure dependent upon the method of preparation. The C_{19} dicarboxylic acids are obtained from oleic acid and/or methyl oleate by three routes (92): the Koch reaction (CO in sulfuric acid) (93), the Reppe reaction, and the Oxo reaction (hydroformylation) (see Oxo process). Extensive isomerization of the double bond usually occurs in all processes, but can be avoided by using hydroformylation and rhodium–triphenylphosphine catalyst to yield, as a final product, a mixture of 9- and 10-carboxystearic acids (2-nonyldecanedioic acid [4165-02-0] and 2-octylundecanedioic acid [4124-87-2]) (94). The two isomers have been separated and used in polyamide preparation (95). The C_{19} diacids have potential uses (1,92) in polyamides as adhesives (96), polyester components in urethanes for foams, and for lubricants and plasticizers. There were no known commercial producers for C_{19} diacids in 1976 although the Union Camp Corp. and BASF had offered developmental products earlier.

A diacid, which has been called C_{21} dicarboxylic acid, is produced by Westvaco Corp. by condensing acrylic acid with the diunsaturated (linoleic) portion of tall oil fatty acids using a Diels-Alder reaction (97). The product consists primarily of two isomers, 5- and 6-carboxy-4-hexyl-2-cyclohexene-1-octanoic acids [42763-47-3] and [42763-46-2], respectively (see Dimer acids).

Potential uses are in polymers for application as textile lubricants, automotive lubricants, in plasticizers, and as a general purpose emulsifier.

BIBLIOGRAPHY

"Acids, Dicarboxylic" in *ECT* 1st ed., Vol. 1, pp. 152–157, by C. J. Knuth and P. F. Bruins, Polytechnic Institute of Brooklyn, and R. R. Umbdenstock, Chas. Pfizer and Co., Inc.; "Acids, Dicarboxylic" in *ECT* 2nd ed., Vol. 1, pp. 240–254, by W. M. Muir, Harris Research Laboratories, Inc.

1. E. H. Pryde and J. C. Cowan, "Aliphatic Dibasic Acids" in J. K. Stille and T. W. Campbell, eds., *Condensation Monomers,* Wiley-Interscience, New York, 1972, pp. 1–153 (a comprehensive review with 1380 refs.).
2. J. A. Dean, ed., *Lange's Handbook of Chemistry,* 11th ed., McGraw-Hill, New York, 1973.
3. F. Krafft and H. Noerdlinger, *Chem. Ber.* **22,** 816 (1889).
4. L. J. Durham, D. J. McLeod, and J. Cason, *Org. Syn. Coll. Vol. IV,* 510 (1963).
5. H. Noerdlinger, *Chem. Ber.* **23,** 2356 (1890).
6. J. G. Erickson, *J. Am. Chem. Soc.,* **71,** 307 (1949).
7. C. Hell and C. Jordanoff, *Chem. Ber.,* **24,** 991 (1891).
8. L. Ruzicka, Pl. A. Plattner, and W. Widmer, *Helv. Chim. Acta.* **25,** 1086 (1942).
9. *Ibid.,* p. 604.
10. V. V. Korshak and S. V. Rogozhin, *Izv. Akad. Nauk SSSR Otd. Khim. Nauk,* 531 (1952).
11. V. V. Korshak and S. V. Rogozhin, *Dokl. Akad. Nauk SSSR* **76,** 539 (1951).
12. W. Sweeny and J. Zimmerman, "Polyamides" in N. M. Bikales, ed., *Encyclopedia of Polymer Science and Technology,* Vol. 10, Interscience Publishers, a division of John Wiley & Sons, Inc., New York, 1969, pp. 483–507.
13. R. C. Wilhoit and D. Shiao, *J. Chem. Eng. Data* **9,** 595 (1964).
14. P. E. Verkade, H. Hartman, and J. Coops, *Rec. Trav. Chim. Pays-Bas,* **45,** 380 (1926).
15. A. M. Schiller, "Aliphatic Dicarboxylic Acids and Derivatives" in ref. 12, Vol. 1, 1964, pp. 109–122.
16. D. E. F. Armstead, *Sch. Sci. Rev.,* **55,** 416 (1973).
17. R. Hill, ed., *Fibres from Synthetic Polymers,* Elsevier, New York, 1953, pp. 135, 164, 311.
18. P. W. Morgan, *Condensation Polymers by Interfacial and Solution Methods,* Wiley-Interscience, New York, 1965, pp. 244, 382.
19. P. W. Morgan and S. L. Kwolek, *Macromolecules* **8,** 104 (1975).
20. S. Gal, T. Meisel, and L. Erdey, *J. Therm. Anal.,* **1,** 159 (1969).
21. L. S. Darken, *J. Am. Chem. Soc.* **63,** 1007 (1941); G. D. Pinching and R. G. Bates, *J. Res. Natl. Bur. Stand.* **40,** 405 (1948).
22. S. N. Das and D. L. G. Ives, *Proc. Chem. Soc.,* 373 (1961); W. J. Hamer, J. O. Burton, and S. F. Acree, *J. Res. Natl. Bur. Stand.* **24,** 269 (1940).
23. G. D. Pinching and R. G. Bates, *J. Res. Natl. Bur. Stand.* **45,** 322, 444 (1950).
24. I. Jones and F. G. Soper, *J. Chem. Soc.,* 133 (1936).
25. B. Adell, *Z. Phys. Chem. (Leipzig)* **185,** 161 (1939).
26. G. Kortüm, W. Vogel, and K. Andrussow, *Pure Appl. Chem.* **1,** 190 (1960).
27. G. Bonhomme, *Bull. Soc. Chim. Fr.,* 60 (1968); *Chem. Abstr.* **68,** 117603 (1968); *ibid.* **82,** 72408 (1975).
28. P. E. Verkade, H. Hartman, and J. Coops, *Rec. Trav. Chim. Pays-Bas* **49,** 578 (1930).
29. C. Saracco and E. S. Marchetti, *Ann. Chem. (Rome)* **48,** 1357 (1958).
30. F. L. Breusch and E. Ulusoy, *Fette Seifen Austrichm* **66,** 739 (1964).
31. N. Yoda, "Polyanhydrides" in ref. 12, Vol. 10, 1969, pp. 631.
32. F. Millich and C. E. Carraher, Jr., eds., *Interfacial Syntheses,* Vols. I and II, Marcel Dekker, Inc., New York, 1977.
33. P. W. Morgan, *Macromolecules* **10,** 1381 (1977).
34. *Modern Plastics Encyclopedia, 1977–1978,* McGraw-Hill, New York, 1978, pp. 690–697; J. R. Darby and J. K. Sears, "Plasticizers" in ref. 15, Vol. 10, 1969, pp. 228–306.
35. H. Mark and G. Whitby, eds., *Collected Papers of Wallace Hume Carothers on High Polymeric Substances,* Interscience Publishers, New York, 1940.
36. R. W. Lenz, *Organic Chemistry of High Polymers,* Wiley-Interscience, New York, 1967, p. 91.
37. J. K. Backus, "Polyurethanes" in C. E. Schildknecht and I. Skeist, eds., *Polymerization Processes,* Wiley-Interscience, New York, 1977, pp. 642–680.
38. J. H. Saunders and K. C. Frisch, *Polyurethanes: Chemistry and Technology,* Vol. I, Wiley-Interscience, New York, 1962.
39. W. M. Eareckson, *J. Polym. Sci.* **40,** 399 (1959).
40. P. W. Morgan, *J. Polym. Sci.* **A2,** 437 (1964); *Macromolecules* **3,** 536 (1970).
41. H. Schnell, *Chemistry and Physics of Polycarbonates,* Wiley-Interscience, New York, 1964.
42. R. G. Mraz and R. P. Silver, "Alkyd Resins" in ref. 12, Vol. 1, 1964, pp. 663–734.
43. H. V. Boenig, "Unsaturated Polyesters" in ref. 12, Vol. 11, 1969, p. 149.
44. E. E. Parker and J. R. Peffer, "Unsaturated Polyester Resins" in ref. 37, pp. 68–87.

45. D. B. Jacobs and J. Zimmerman, "Preparation of 6,6-Nylon and Related Polyamides" in ref. 37, pp. 424–467.
46. E. E. Magat and R. E. Morrison, *Chem. Tech.* **6,** 702 (1976).
47. R. W. Moncrieff, *Man-made Fibers*, 6th ed., John Wiley & Sons, Inc., New York, 1975.
48. O. E. Snyder and R. J. Richardson, "Polyamide Fibers" in ref. 12, Vol. 10, 1969, pp. 347–460.
49. M. I. Kohan, ed., *Nylon Plastics*, Wiley-Interscience, New York, 1973 (contains data on history, chemical properties, processing, market volumes, prices, and manufacturers).
50. Ref. 34, p. 36.
51. E. C. Schulze, "Polyamide Plastics" in ref. 12, Vol. 10, 1969, pp. 460–482.
52. Ref. 1, pp. 82, 87, 161; ref. 37, p. 729.
53. Ger. Pat. 854,508 (Nov. 4, 1952), H. A. Offe.
54. S. J. Gutcho, *Chemicals by Fermentation*, Noyes Data Corp., Park Ridge, N.J., 1973.
55. E. J. McKenna, *Degradation Syn. Org. Mol. Biosphere, Proc. Conf. 1971*, pp. 73–97; National Academy of Science, Washington, D.C., 1972.
56. U.S. Pat. 3,975,234 (July 3, 1975), D. O. Hitzman (to Phillips Petroleum Co.).
57. U.S. Pat. 3,784,445 (Jan. 8, 1974), R. V. Dahlstrom and J. H. Jaehnig (to E. I. du Pont de Nemours & Co., Inc.).
58. *Synthetic Organic Chemicals: United States Production and Sales, 1976, Publication 833*, United States International Trade Commission, Washington, D.C., 1976.
59. *Oil Paint Drug Rep.* Hi–Lo Chemical Price Issue, (1962); *Chem. Mark. Rep.* **215**(14), (Apr. 2, 1979).
60. W. C. Rose, *J. Pharmacol.* **24,** 147 (1924).
61. W. C. Rose and co-workers, *J. Pharmacol.* **25,** 59 (1925).
62. R. Emmrich, *Dtsch. Arch. Klin. Med.* **187,** 504 (1941).
63. R. Emmrich and I. Emmrich-Glaser, *Hoppe-Seyler's Z. Physiol. Chem.* **266,** 183 (1940).
64. *Registry of Toxic Effects of Chemical Substances*, NIOSH, U.S. Dept. Health, Education, and Welfare, Rockville, Md., 1976.
65. N. I. Sax, *Dangerous Properties of Industrial Materials*, 5th ed., Van Nostrand, New York, 1979.
66. G. D. Clayton and F. E. Clayton, eds., *Patty's Industrial Hygiene and Toxicology*, 3rd ed., Wiley-Interscience, New York, 1978.
67. A. Müller, *Org. Synth. Coll. Vol II*, 535 (1943).
68. P. D. Gardner, L. Rand, and G. R. Haynes, *J. Am. Chem. Soc.* **78,** 3425 (1956).
69. R. E. Meyer, *Helv. Chim. Acta* **16,** 1291 (1933).
70. V. V. Korshak and T. M. Frunze, *Synthetic Hetero-Chain Polyamides*, N. Kaner, trans., Davey, New York, 1964.
71. V. V. Korshak and S. V. Vinogradova, *Polyesters*, B. J. Hazzard, trans., Pergamon Press, Oxford, Eng., 1965.
72. P. E. Verkade, *Rec. Trav. Chim. Pays-Bas* **46,** 137 (1927).
73. W. Reppe and co-workers, *Justus Liebigs Ann. Chem.* **560,** 1 (1948); Ger. Pat. 890,950 (Sept. 24, 1953), W. Reppe and co-workers (to Badische Anilin- & Soda-Fabrik, A.G.).
74. J. E. Franz and W. S. Knowles, *Chem. Ind. (London)*, 250 (1961).
75. M. I. Fremery and E. K. Fields, *J. Org. Chem.* **28,** 2537 (1963); U.S. Pat. 3,284,492 (Nov. 8, 1966), M. I. Fremery and E. K. Fields (to Standard Oil Co.).
76. U.S. Pat. 3,280,183 (Oct. 18, 1966), A. Maggiolo (to Wallace and Tiernan, Inc.).
77. J. S. Ridgeway, *J. Polym. Sci. Polym. Chem. Ed.* **12,** 2005 (1974).
78. A. Bell, J. G. Smith, and C. J. Kibler, *J. Polym. Sci.* **A3,** 19 (1965).
79. M. T. Watson and G. M. Armstrong, *Soc. Plast. Eng. J.* **21,** 475 (1965).
80. U.S. Pats. 3,936,426 (Feb. 3, 1976), R. W. Campbell; 3,817,903 (June 18, 1974), H. W. Hill and R. W. Campbell (to Phillips Petroleum Co.).
81. U.S. Pat. 2,813,113 (Nov. 12, 1957), C. G. Goebel and co-workers (to Emery Industries, Inc.).
82. U.S. Pat. 3,249,591 (May 3, 1966), F. A. Gadecki and S. B. Speck (to E.I. du Pont de Nemours & Co., Inc.).
83. U.S. Pats. 2,217,515, 2,217,516 (Oct. 8, 1940), A. Houpt (to American Cyanamid Co.).
84. U.S. Pat. 2,182,056 (Dec. 5, 1939), H. A. Bruson and L. W. Covert (to Rohm & Haas Co.).
85. M. J. Diamond and co-workers, *J. Am. Oil Chem. Soc.* **44,** 656 (1967); *ibid.* **42,** 882 (1965).
86. C. E. Frank and W. E. Foster, *J. Org. Chem.* **26,** 303 (1961).
87. O. S. Shchetinskaya and co-workers, *Sov. Chem. Ind.*, 502 (1976).
88. *Eur. Chem. News* (629), 20 (1974).

89. K. D. Carlson and co-workers, *Ind. Eng. Chem. Prod. Res. Dev.* **16**(1), 95 (1977).
90. R. B. Perkins and co-workers, *Mod. Plast.* **46**(5), 136 (1969).
91. H. J. Nieschlag and co-workers, *Ind. Eng. Chem. Prod. Res. Dev.* **16**(1), 101 (1977).
92. E. H. Pryde, E. N. Frankel, and J. C. Cowan, *J. Am. Oil Chem. Soc.* **49**, 451 (1972).
93. N. E. Lawson, T. T. Cheng, and F. B. Slezak, *J. Am. Oil Chem. Soc.* **54**, 215 (1977).
94. E. N. Frankel, *J. Am. Oil Chem. Soc.* **48**, 248 (1971).
95. W. L. Kohlhase, W. E. Neff, and E. H. Pryde, *J. Am. Oil Chem. Soc.*, **54**, 521 (1977).
96. W. L. Kohlhase, E. N. Frankel, and E. H. Pryde, *J. Am. Oil Chem. Soc.* **54**, 506 (1977).
97. B. F. Ward and co-workers, *J. Am. Oil Chem. Soc.* **52**, 219 (1974).

PAUL MORGAN
Consultant

DIENE POLYMERS. See Elastomers, synthetic.

DIESEL FUEL. See Gasoline and other motor fuels; Petroleum products.

DIETARY FIBER

Dietary fiber is defined, at present, as the plant-derived food polysaccharides (see Table 1) and lignin [9005-53-2] which are not digested by enzymes of the human gastrointestinal tract. Whereas the major component cellulose (qv) is fibrous in nature, some of the other components are not. Therefore, there is a contradiction in terms in using this definition.

Dietary fiber is a complex mixture composed of cellulose with lesser amounts of other polysaccharides (hemicellulose, pectin [9000-69-5], plant mucilages) and lignin. The precise composition in terms of polysaccharide type and the proportion of each is related to the plant source and also to the stage of maturity. The composition and the physical properties of the dietary fiber are also affected by the food processing (see Lignin; Cellulose).

A resurgence of interest in dietary fiber was stimulated by epidemiological evidence of differences in colonic disease patterns between certain cultures whose diets contain larger quantities of fiber and western people who consume more highly refined diets. Thus, many African countries are relatively free of diverticular disease, ulcerative colitis, hemorrhoids, polyps, and cancer of the colon (1). Although most of the interest has focused on the beneficial role of dietary fiber, there is concern that high fiber diets may cause disturbances in the absorption of nutrients such as minerals and vitamins. The interrelationships between consumption of dietary fiber and health status are confused by a lack of knowledge about the composition of dietary fiber from various plant sources and the physiological roles played by the various fiber components.

Various names have been proposed for the nondigestible part of plant cells. The proliferation of names reflects the uncertainty of which plant cellular polysaccharides

Table 1. Some Polysaccharides of Land Plants

Polysaccharide	Structure
Structural	
cellulose [9004-34-6]	(1→4)-β-D-glucan (1)
hemicelluloses [9034-32-6]	
xylan [9014-63-5]	(1→4)-β-D-xylan (2)
arabinoxylan [9040-27-1]	(1→4)-β-D-xylan with L-arabinose branches
glucomannan [37230-82-3]	(1→4)-β-D-glucomannan (3)
pectic substances [9046-38-2]	(1→4)-α-D-galacturonan
[39280-21-2]	(1→2)-L-rhamno-(1→4)-α-D-galacturonan
associated polysaccharides	
arabinan [9060-75-7]	(1→5)-α-L-arabinan (4) with L-arabinose branches
arabinogalactan [9036-66-2]	(1→4)-β-D-galactan (5) with (1→5)-L-arabinose branches
galactan [9051-94-9]	(1→4)-β-D-galactan (5)
xyloglucan [37294-28-3]	(1→4)-β-D-glucan (1) with (1→6)-α-D-xylose branches
Nonstructural	
fructan [9037-90-5]	(2→1)-β-D-fructan (6) (2→6)-β-D-fructan (7)
galactomannan [11078-30-1]	(1→4)-β-D-mannan (8) with single (1→6)-α-D-galactose branches
β-glucan [55965-23-6]	(1→3;1→4)-β-D-glucan
starch [9005-25-8]	(1→4)-α-D-glucan and (1→4;1→6)-α-D-glucan

to designate and the tendency to consider a complex mixture as a single entity. In general the names suggest either the fibrous nature of the cell wall polysaccharides or the resistance of the polysaccharides to digestive tract enzymes. Food composition tables have been based on crude fiber analyses but it is now recognized that this assay does not give a true picture even of the fibrous cell wall material. Likewise the term unavailable carbohydrate fails to recognize the presence of lignin or the microbial fermentation that occurs in the colon. More recently other names have been proposed such as edible fiber and nonnutritive fiber. Recognizing that not all of the components present in dietary fiber are fibrous, the term plantix has been coined from the Latin *planta* and *matrix* (2). Other names have been proposed also based on specific methods of analysis. In the Van Soest method of analysis (3–4), the terms neutral detergent fiber (NDF) and acid detergent fiber (ADF) refer, respectively, to the hemicellulose-cellulose-lignin complex and to the cellulose-lignin complex of the plant cell wall.

As originally defined dietary fiber referred to cell wall material not hydrolyzed by the human digestive tract enzymes. This definition has recently been enlarged to include other nondigestible polysaccharides of the cell such as certain storage polysaccharides. There is not total agreement on including the water-soluble polysaccharides. Perhaps no definition or terminology will be adequate until the physiological role of the individual dietary fiber components is understood and the analytical methodology is available to assay for the relevant components.

The primary sources of fiber in the diet are vegetables, cereals, and to a lesser extent fruits. Currently, certain processed foods and breads are supplemented in their fiber content by incorporation of purified cellulose or cereal bran. In addition cellulose and its water-soluble derivatives, seaweed polysaccharides (alginates [9005-38-3], carrageenan [9000-07-1]), seed mucilaginous polysaccharides (eg, guar and carob galactomannans), highly complex plant exudate polysaccharides (gum arabic

630   DIETARY FIBER

(1) (2) (3) (4) (5) (6) (7) (8)

[9000-01-5], tragacanth [9000-65-1], etc), and microbially-synthesized polysaccharides such as xanthan gum [11138-66-2] (eg, Merck's Keltrol) are added to foods for a variety of purposes. Since these are also nondigestible, they may contribute to the total effect of dietary fiber.

Composition

Dietary fiber is a mixture of simple and complex polysaccharides and lignin. The lignin (qv), a highly polymerized alkyl aromatic substance, is covalently linked to some of the cell wall polysaccharides decreasing their digestibility by microbial enzymes. Lignification also renders the polysaccharides less soluble and highly lignified plant tissue must be delignified to facilitate extraction of the structural hemicelluloses. The degree of lignification is peculiar to the type of plant and increases with maturity as the lignin infiltrates the primary and secondary cell walls.

Within the structural and nonstructural groups of polysaccharides, there is great diversity of chemically distinct polysaccharides. Table 1 illustrates some of the main types of polysaccharides occurring in land plants. The structures shown here are simplified. Some of the complex and diverse structures that occur naturally are shown in structures (1–8). In addition, for many of the polysaccharides, the structures have not yet been determined.

Physiological Properties

Dietary fiber has a profound effect on the quality and appearance of the feces and on the passage time of digesta through the gastrointestinal tract. The particle size and shape, density, and water-holding capacity influence the flow rates. Since water-holding capacity is determined by the chemical structure, crystallinity, and lignin content of the fiber polysaccharides, food processing conditions of grinding and cooking also indirectly affect the flow rates. Increased levels of fiber in the diet increase the flow rates and frequency of defecation and the fecal weight. However, there are large variations among individuals.

High fiber diets play a role in excretion of bile acids and cholesterol. There is concern that essential minerals and other nutrients may be lost in this manner. The composition and level of dietary fiber may affect the generation and action of tumorigenic substances in the colon. However, evidence is not yet available to prove this hypothesis. When the significance of the specific fiber components to these various roles is better understood, it should be possible to predict, specify, and market the dietary fiber sources that are optimum for good health.

Since the only known human digestive enzymes in the gastrointestinal tract with strong activity on polysaccharides are the amylases ($(1\rightarrow 4)$-α-D-glucanases), it is assumed that all of the cell wall fiber components and most of the nonstructural polysaccharides pass into the colon largely unchanged. However, a recent study presented evidence to the contrary; in six ileostomy patients only 84.5% of the dietary cellulose and 27.5% of the dietary hemicellulose were excreted from the small intestine (5). It is not known whether this apparent digestion of cellulose and hemicellulose derived from bacterial or human enzyme activity. A control group of ten normal subjects excreted 22.4% of the dietary cellulose and 4% of the dietary hemicellulose in the feces. The colon bacteria possess inducible enzymes with activity to degrade and ferment many of the fiber polysaccharides (6). Some of the bacteria possess cellulolytic activity. The volatile fatty acids generated by bacterial action are partially absorbed through the colon wall and are available as a supplementary energy source.

Physicochemical Properties

Several physicochemical properties of dietary fiber contribute to the physiological role of the fiber. Water-holding capacity, ion-exchange capacity, solution viscosity, density, and molecular interactions are characteristics determined by the chemical structure of the component polysaccharides, the crystallinity and surface area.

Water-Holding Capacity. All polysaccharides are hydrophilic and bind variable amounts of water through hydrogen bonds. Fibrous polymers also absorb water by capillary action to saturate their interstitial spaces. The more highly crystalline fiber components are less accessible to water of hydration and have less tendency to swell. Structural features and other factors (including grinding and cooking) which decrease crystallinity increase the hydration capacity and solubility. In general, branched polysaccharides are more soluble than linear since close packing of molecular chains is precluded.

Water-holding capacity is strongly influenced by the pentosan components of dietary fiber and therefore varies with the structure of those hemicelluloses: 4.5 and 5.5 g H_2O/g pentosan for the soluble and insoluble arabinoxylans of Durum wheat and 6.3 and 6.7 g H_2O/g pentosan for the soluble and insoluble arabinoxylans of hard red spring wheat (7). This variation reflects relatively small differences in chemical structure.

Soluble polysaccharides such as agar [9002-18-0] and pectin, which can form three-dimensional networks stabilized by physical or covalent interactions, imbibe large quantities of water to form rigid gels (8). This gelling behavior as well as the high viscosity of the single-unit branched polysaccharides (eg, galactomannans) affects gastrointestinal function.

Ion-Exchange. Acidic polysaccharides containing uronic acids, sulfate or phosphate groups, are effective as cation exchangers in binding metal ions. The type of bound cation may influence the physical properties of the polysaccharide. For example, alginic acid [9005-32-7], which is relatively insoluble as the free acid, is soluble as the sodium salt and forms a gel with calcium ion. The cation-exchange capacity of dietary fiber is primarily dependent on the pectic acid content and certain glucuronic acid-containing hemicelluloses. Ion-exchange capacities have been determined for several dietary fiber sources (9) and the evidence suggests that cooking may alter the exchange capacity.

Molecular Interactions. Polysaccharides associate readily with a variety of other substances including proteins, bile acids and cholesterol, low molecular weight organic molecules, inorganic salts, and ions. Complexes with salts occur even with neutral polysaccharides (10). Although relatively little is known about the nature of the interactions in general, certain interactions are structure-specific. Inclusion complexes are formed between starch amylose and several classes of polar molecules including fatty acids, glycerides, alcohols, esters, ketones, and iodine/iodide. The absorbed molecule occupies the cavity of the amylose helix which has the capacity to expand somewhat in order to accommodate larger molecules. The primary structural requirement is one of spatial accommodation. The starch–lipid complex is important in food systems. Whether similar inclusion complexes can form with any of the dietary fiber components is not known.

Polysaccharides interact with proteins in a nonspecific manner, which is essentially a competition for solvent, with the consequence of lowering the solubility of the protein. There is also the structure-specific interaction exemplified by the lectins. These proteins interact with carbohydrates having a specific structure; the consequence of such interactions between the protein and a polysaccharide is to cause aggregation and precipitation of the insoluble complex. This interaction displays the characteristics of an antigen–antibody reaction since it is characterized by a strict structural requirement for both carbohydrate and protein, competitive inhibition by the monomer sugar, a reversible reaction, and dissolution of the complex at high carbohydrate concentrations. If such interactions occur in the milieu of the gastrointestinal tract, the nutritional availability of the protein could be impaired.

Bile acids and cholesterol are bound by some dietary fibers which probably facilitate their excretion from the body and removal from the metabolic pool. This interaction is more pronounced for some fiber components but the structural requirements are not known.

Analysis

Analytical methods that are suitable for routine assays measure groups of fiber components having similar solubility properties. Quantitative analysis of specific fiber components (other than cellulose and lignin) is time-consuming since it necessarily involves fractionation of complex, chemically similar polysaccharides. The commonly used fiber analysis methods are described here. They are of three general types, ie, crude fiber detergent methods for cell wall fiber, and those methods that attempt to assay soluble and insoluble polysaccharides of the plant cells.

Crude Fiber. The official crude fiber (CF) method of the Association of Official Agricultural Chemists (AOAC) is an empirical procedure widely used in food analysis, and on which food composition tables have been based (11). This method is no longer recommended because of the inherent errors. Strict adherence to the procedure gives replicable data but the results do not correlate closely with any component or meaningful class of components in the foodstuff. It is an approximate measure of the cellulose-lignin complex but frequently the cellulose content is grossly underestimated.

The procedure consists of successive digestions of a fat-free sample with hot 1.25% sulfuric acid (wt/vol) and hot 1.25% sodium hydroxide (wt/vol) under rigidly controlled conditions and in the presence of asbestos as a filter aid. The hot acid and alkali degrade and extract most of the plant components except cellulose and lignin. Errors arise from partial digestion and solubilization of the cellulose itself. The insoluble residue from the acid and alkaline digestions is dried, weighed, and ashed at 600°C. Crude fiber is represented by the loss in weight on ashing.

Detergent Methods. The neutral detergent fiber (NDF) and acid detergent fiber (ADF) methods developed by Van Soest and associates (3–4,12) measure total plant cell wall material (NDF) and the cellulose-lignin complex (ADF). The ADF can be assayed subsequently for each component. The NDF residue is not digested by mammalian secretory or mucosal enzymes, but it is susceptible to extensive microbial enzyme activity in the gut. The method is valid for certain plant materials such as forages for which it was derived but is subject to errors in certain foodstuffs. Thus the easily-solubilized pectins, galactomannans of legume seeds, various plant gums, and seaweed polysaccharides would be excluded from the NDF although they too would be unavailable to the enzymes in the upper gastrointestinal tract.

The detergent methods are simple, reproducible, and amenable to routine assays. The disadvantages are that other nondigestible plant polysaccharides, including cell wall pectins, are not measured and specific hemicelluloses are not distinguished from each other.

Neutral Detergent Fiber. In the NDF method the ground sample is extracted for 1 h under reflux at pH 7.0 ± 0.1 with a buffered detergent solution containing sodium lauryl sulfate and ethylenediaminetetraacetic acid. Prior extraction of lipids from foods high in fat is desirable. Starch in high concentration also interferes and can be removed by amylase digestion prior to extraction with the detergent solution. Certain foodstuffs that contain highly viscous water-soluble gums are also difficult to analyze because of filtration problems.

Acid Detergent Fiber. The ground sample is heated for 1 h under reflux in a solution of 2% cetyltrimethylammonium bromide in 1 N H_2SO_4. The acid causes hydrolysis of glycosidic bonds in the noncellulosic polysaccharides, making them water soluble. The ADF is therefore relatively free of hemicelluloses and contains all the cellulose and lignin together with some inorganic salts.

Cellulose may be extracted from the ADF with 72% H_2SO_4 (wt/wt) at 0–4°C for 24 h leaving an insoluble residue of lignin. In some processed foods insoluble artifacts are generated during heat processing and remain in the residue together with the true lignin.

An alternative procedure for lignin that avoids this problem utilizes oxidation with permanganate. In this method the ADF is treated with an acidic solution of potassium permanganate at room temperature, oxidizing and solubilizing the lignin. The white fibrous residue contains cellulose and inorganic matter, and the loss in weight is equivalent to the lignin. Lignin-like artifacts are not included in this lignin value.

Analysis of Dietary Fiber. *Method 1.* A method is suggested for the analysis of cell wall polysaccharides and lignin by successive extraction of pectin substances, 5% (wt/vol) potassium hydroxide-soluble hemicelluloses, and 24% (wt/vol) potassium hydroxide-soluble hemicelluloses (13). The residue remaining after hydrolysis of the 24% alkali-insoluble residue in 1 N H_2SO_4 is cellulose. Lignin is determined separately on a protein-free sample by dissolving the cellulose with 72% (wt/vol) H_2SO_4 at RT and 3% (wt/vol) H_2SO_4 under reflux. The residue contains lignin and ash and the loss in weight on ignition corresponds to lignin.

Fructans, present in several plants, can be determined approximately by measuring the glucose and fructose released by hydrolysis with 0.1 N H_2SO_4 at 95–100°C for 1 h. Simple sugars and oligosaccharides must be removed by preliminary extraction of the sample with 90% (vol/vol) ethanol.

This method is too time-consuming for routine assays but it does give information on classes of polysaccharides. Analysis for component sugars in these fractions can give further information on the specific polysaccharides.

Method 2. An alternative method for determining total dietary fiber is based on analysis of the component sugars (hexoses, pentoses, uronic acids) in the polysaccharide fractions of the sample (13–14). The sample is first freed from starch by digestion with amylase and the hot water-soluble polysaccharides are extracted from the residue and hydrolyzed. The residue is heated at 95–100°C with 1 N H_2SO_4 for 2.5 h and the supernatant solution is separated by filtration or centrifugation. The residue is treated with 72% (wt/wt) H_2SO_4 at 0–4°C for 24 h and filtered. The residue is lignin. The hydrolyzates of the starch-free water-soluble polysaccharides, the water-insoluble polysaccharides and the 72% H_2SO_4 hydrolyzate are each analyzed for the component sugars.

Although this method gives quantitative data on the component sugars of the dietary fiber, it is difficult, if not impossible, to deduce the amount of the original polysaccharides present.

Method 3. In a simplified procedure for total dietary fiber, a defatted sample is extracted with hot 80% ethanol to remove the simple sugars and oligosaccharides (15). The residue is digested successively with pepsin and then with glucoamylase and pancreatin to remove protein and starch. The water-soluble polysaccharides can be recovered from the soluble digest by precipitation with four volumes of ethanol. The insoluble polysaccharide residue from the enzyme digest and the water-soluble polysaccharide fraction can be analyzed further by any of the methods already described. The advantage of this method is that the total dietary fiber is analyzed by solubility groups, whereas in the other methods water-soluble polysaccharides are lost.

Dietary Fiber Supplements

Several commercial food-grade purified cellulose products are available. These together with cereal bran are used in food products either as a fiber supplement or for other functional uses. In addition natural dietary fiber preparations are available in "natural" food stores. The characteristics of the readily available celluloses and cereal bran are described below.

Preparation and Physical Properties of Commercial Cellulose. High quality commercial cellulose is usually prepared from soft or hard woods by pulping methods which remove most of the associated hemicellulose, lignin, waxes, and other constituents. Depending on the specific process, the wood is cooked at high temperatures in solutions of alkaline sulfide (Kraft process), acidic bisulfite and sulfur dioxide, or sodium hydroxide. Although the main intent is to degrade and remove the lignin and hemicelluloses, the cellulose too may be partially depolymerized. The pulp is further purified by bleaching with chlorine and chlorine dioxide to give a white cellulose (alpha-cellulose) which is a relatively pure $(1\rightarrow 4)$-β-D-glucan (see Cellulose; Pulp).

Purified cellulose is white, odorless, and nonabrasive. Although devoid of inherent flavor, it possesses an undesirable mouth feel which is related to its fibrous nature and insolubility. Cellulose is relatively inert to weak acids and alkalis, is insoluble in most polar and nonpolar solvents, and does not present any special storage problems. Strong alkali causes swelling of the cellulose fibers and certain strong bases such as liquid ammonia will dissolve the fibers. Concentrated H_3PO_4 (85%) also promotes solubility, probably accompanied by partial hydrolysis of the polymer. The cellulose is recovered from this acid solution by precipitation with water, but the product has the physical appearance of a free-flowing powder.

Fibrous Cellulose. A purified fibrous cellulose manufactured from wood pulp is available from Brown Co, Berlin-Gorham Division, New York City, marketed under the trade name Solka-Floc (in Canada, Alpha-Floc). This cellulose is claimed to be 99.5% cellulose on a dry weight basis, virtually lignin-free, free of fat and protein and with low ash content. Food grades are available in fiber lengths averaging 20–25 to 100–140 μm. The product is sold in 13.6–27.2-kg bags. For experimental use, 0.5-kg samples are available.

Microcrystalline Cellulose. Limited hydrolysis of fibrous cellulose pulp gives a product that is more highly crystalline than native cellulose. Mineral acids hydrolyze the glycosidic bonds in the less crystalline regions of the cellulose fibers. The resulting cellulose microcrystals can be mechanically dispersed into water to give a gel-like colloidal suspension. These microcrystals, like the native cellulose, do not melt and therefore can be used in food systems under heat processing conditions that would destroy the emulsion stability of other food systems. One product manufactured by FMC Corporation, Philadelphia, Pa., is marketed under the trade name Avicel. The physical characteristics of this microcrystalline cellulose are quite different from the original cellulose. The free-flowing powder has a particle size of 18–40 μm. Avicel celluloses are also available in which the finely milled microcrystalline cellulose particles are coated with carboxymethyl cellulose.

Cellulose Derivatives. Chemical modification drastically alters the crystallinity and other physical properties of cellulose. Common derivatives include methyl, ethyl, and propyl cellulose, hydroxyethyl and hydroxypropyl cellulose, carboxymethyl cellulose, as well as mixed ethers of cellulose. These derivatives are obtained by alkylation of the cellulose under basic conditions. The ether substituents disrupt the intensive hydrogen bonding in the cellulose and permit increased hydration. At certain degrees of substitution (DS) the insoluble cellulose is transformed into a water-soluble colloid. The nature and size of the substituent, the regularity of substitution and the mol wt of the cellulose determine the DS at which water-solubility occurs. In addition to increased hydration, carboxymethyl cellulose displays increased ion-exchange capacity. These cellulose ethers are extensively used in food products and are nondigestible (see Cellulose derivatives).

Bran. As a fiber supplement, cereal bran is a more representative form of dietary fiber than purified cellulose in that the bran contains cellulose, hemicelluloses, and lignin. The composition and characteristics of bran are dependent on the cereal source, plant variety, and the milling practices. Wheat bran is readily available from any flour mill. Frequently, different wheat varieties are blended to give the desired flour properties. In addition, varying amounts of the wheat endosperm components will be present and the particle size will also depend on which break roll the bran is taken. Wheat bran contains large quantities of arabinoxylans, considerable lignin and some starch in addition to cellulose (see Wheat and other cereal grains).

The American Association of Cereal Chemists, St. Paul, Minn., has made a certified food grade wheat bran available for research purposes. This bran is made from

Table 2. Analysis of AACC Certified Food Grade Wheat Bran R07-3691[a]

Component	%	Component	%
crude fiber	8.91	lignin	3.2
protein	14.3	pectin	3.0
water	10.4	pentosan	22.1
fat	5.22	phytic acid (9)	3.36
ash	5.12	starch	17.4
ADF[b]	11.9	water-holding capacity	
NDF[c]	40.2	of NDF[c]	9.5 g/g

[a] Some of the analytical procedures are not mutually exclusive but measure different combinations of specific chemical entities. Thus the pentosans and crude fiber are also assayed by the NDF and ADF procedures. Therefore the percentages cannot be summed to give the total composition. Vitamins, minerals, sugars, and plant sterols are also present but not included in this table.
[b] ADF is acid detergent fiber.
[c] NDF is neutral detergent fiber.

(9)

phytic acid, R = PO₃H₂ [83-86-3]

inositol, R = H [87-89-8] (see Vitamins)

a blend of soft white wheat (87.3%) and club white wheat (12.7%). Some of the components of this bran are summarized in Table 2 as an illustration of the composition and properties of bran. The bran is available in 1.13-kg quantities at $4.41/kg or in 13.6-kg bags at $3.30/kg.

Uses

Food Use of Cellulose Fiber. Purified cellulose is extensively used in the food industry. Because of its ability to absorb water, oils, and fats it is used to control consistency of various food products and improve texture and appearance. It enhances the stability of food formulations at temperatures encountered during cooking and baking (see Bakery processes).

The excellent binding characteristics of cellulose improve the mechanical strength of snack products such as chips and help prevent crumbling of tender baked products. Its absorptive and binding characteristics make it an excellent carrier for flavors, fragrances, vitamins, coloring materials, and other additives. Cellulose is used as an additive at levels up to 10% of the wheat flour in flour-based products, snack foods, breaded coatings, processed meat products, sauces, salad dressings, and liquid food concentrates (see Food additives).

Cellulose also is used to control the caloric density of foods. With the present concern about the low levels of fiber in the diet, it is used increasingly in baked products to increase fiber levels (16).

Nonfood Use. In addition to its use in human and pet foods (qv), purified fibrous cellulose is used as filter aid in processing of products as widely divergent as lard, gelatin, alcoholic beverages, and pharmaceuticals (see Filtration).

Cellulose is also used as binder in pharmaceuticals and as binder and filler in plastic, rubber and textile printing (see Fillers). For purposes other than foods and pharmaceuticals, less highly purified grades may be used.

BIBLIOGRAPHY

1. H. Trowell, *Nutr. Rev.* **35**, 6 (1977).
2. G. A. Spiller and G. Fassett-Cornellius, *Am. J. Clin. Nutr.* **29**, 934 (1976).
3. P. J. Van Soest, *J. Assoc. Off. Agric. Chem.* **46**, 825, 829 (1963).
4. P. J. Van Soest and R. H. Wine, *J. Assoc. Off. Agric. Chem.* **50**, 50 (1967).
5. W. D. Holloway, C. Tasman-Jones, and S. P. Lee, *Am. J. Clin. Nutr.* **31**, 927 (1978).
6. A. A. Salyers and co-workers, *Appl. Environ. Microbiol.* **33**, 319 (1977); **34**, 529 (1977).
7. S. L. Jelaca and I. Hlynka, *Cereal Chem.* **48**, 211 (1971).

8. D. A. Rees, *Adv. Carbohyd. Chem.* **24,** 267 (1969).
9. A. A. McConnell, M. A. Eastwood, and W. D. Mitchell, *J. Sci. Food Agric.* **25,** 1457 (1974).
10. J. A. Rendleman, Jr., *Adv. Carbohyd. Chem.* **21,** 209 (1966).
11. AOAC, *Official Methods of Analysis,* Association of Official Analytical Chemists, Washington, D.C., 1970, p. 129.
12. P. J. Van Soest and R. W. McQueen, *Proc. Nutr. Soc.* **32,** 123 (1973).
13. D.A.T. Southgate, *Determination of Food Carbohydrates,* Applied Science Publ., Ltd., London, 1976.
14. D. A. T. Southgate, *J. Sci. Food. Agric.* **20,** 331 (1969).
15. T. Schweizer and P. Weurst, private communication, July 1978.
16. *Chem. Eng. News* **57**(13), 43 (March 26, 1979).

General References

J. T. Dwyer and co-workers. *Am. J. Hosp. Pharm.* **35,** 278 (1978).
J. L. Kelsay, *Am. J. Clin. Nutr.* **31,** 142 (1978).
Marabou Symposium: "Food and Fibre," *Nutr. Rev.* **35,** 1 (1977).
W. W. Hawkins, ed., *Dietary Fibre,* Miles Symposium, The Nutrition Society of Canada and Miles Laboratories, Ltd., 1976, pp. 1–56.
G. A. Spiller and R. J. Amen, *Fiber in Human Nutrition,* Plenum Press, New York, 1976, pp. 1–278.
G. A. Spiller, *Topics in Dietary Fiber Research,* Plenum Press, New York, 1978, pp. 1–223.
H. Trowell, *Am. J. Clin. Nutr.* **29,** 417 (1976).

BETTY LEWIS
Cornell University

DIETHANOLAMINE, NH(C$_2$H$_4$OH)$_2$. See Alkanolamines.

DIETHYLENE GLYCOL, HOCH$_2$CH$_2$OCH.CH$_2$OH. See Glycols.

DIFFUSION. See Diffusion separation methods.

DIFFUSION SEPARATION METHODS

Ordinary diffusion involves molecular mixing caused by the random motion of molecules. It is much more pronounced in gases and liquids than in solids. The effects of diffusion in fluids are also greatly affected by convection or turbulence. These phenomena are involved in mass-transfer processes, and therefore in separation processes (see Mass transfer; Separation systems synthesis).

In chemical engineering the term diffusional unit operations normally refers to the separation processes in which mass is transferred from one phase to another often across a fluid interface, and in which diffusion is considered to be the rate-controlling mechanism governing the interphase mass transfer. Thus, the standard unit operations such as distillation (qv), gas absorption, extraction (qv), drying (qv), and the sorption processes, as well as the less conventional separation processes, are usually classified under this heading (see Absorption).

Since the advent of nuclear energy, a number of special processes have been developed for difficult separations such as stable isotope separation (see Nuclear reactors). One of these processes, the gaseous diffusion process, has been used on a very large scale to separate the isotopes of uranium. In the United States during the 1940s and 1950s, it led to the investment of well over $2 billion in process facilities and consumed as much as 10% of the national electrical output.

Owing to the importance of the diffusion phenomena in these special processes, that is, because the separation of the components is based upon different rates of diffusion they are often referred to collectively as diffusion separation methods. Since the more traditional unit operations are considered elsewhere (see Distillation; Extraction), emphasis is given here to the process fundamentals and design considerations connected with these more novel diffusional separation methods, particularly with gaseous, mass and sweep, and thermal diffusion processes, and gas centrifugation or other pressure or gravity diffusion, particularly for the separation of the uranium isotopes.

Industrial Uranium Enrichment

The most important industrial application of the diffusion separation methods has been for the enrichment of the ^{235}U isotope of uranium (qv). The United States was the first country to employ the gaseous diffusion process for the enrichment of ^{235}U, the fissionable natural uranium isotope. Natural uranium consists mostly of ^{238}U and 0.711 wt % ^{235}U. An inconsequential amount of ^{234}U is also present. The original K-25 and K-27 gaseous diffusion plants were built in 1943–1945 in Oak Ridge, Tennessee, as part of the Manhattan Project of World War II. They were designed to enrich the ^{235}U isotope to high assay, allowing the product to be used as the fissionable ingredient in nuclear explosive devices.

A thermal diffusion plant and an electromagnetic process plant using the calutron separator were also constructed at Oak Ridge during World War II. However, they were not cost-competitive and were soon shut down. Other processes, such as the gas centrifuge process, were explored on a small scale but were also abandoned in favor of gaseous diffusion.

It should be noted that such isotope separation methods as the electromagnetic or laser-based processes are not diffusional separations and other sources should be consulted for their descriptions (1).

After World War II, when attention turned to peacetime applications of nuclear energy, nuclear reactor designers found uranium enriched in ^{235}U to be useful also in reducing the sizes of critical assemblies for continuing, controlled fission reactions, thereby reducing the required capital investments for commercial power reactors. In the special case of such United States or foreign reactors as the HTGRs or graphite-moderated, high-temperature gas-cooled reactors, the desired ^{235}U enrichment is near 90%. However, the more frequently employed LWRs, or light-water cooled reactors, require uranium fuel that has been enriched to only about 3% in the ^{235}U isotope.

For reasons to be explained later, the more modest degree of ^{235}U enrichment required for LWR fuel saves only the smaller sections or stages *in series* in a gaseous diffusion (or centrifuge) plant. It actually places greater demands on the high-capacity low-assay sections of the cascade that contain the more critically needed and costly power and separation equipment. Consequently, it became obvious that neither the sizes and power levels, nor the efficiencies and costs of the wartime enrichment plants would be adequate for a practical, peacetime fuel-enrichment plant. Fortunately, Union Carbide and other contractors of the U.S. Atomic Energy Commission (later the Energy Research and Development Administration and the Department of Energy) had embarked on an aggressive development, redesign, and expansion program even before the end of the war. Over the years, this program reduced the unit cost of enriched ^{235}U product from that of the original K-25 product by about a hundred-fold. The quality of the separation membrane or barrier, and improved stage equipment and design were among the key improvements (2–3). Since the 1950s, the program was also extended to develop the gas centrifuge process toward commercial application (4), and to investigate other isotope separation processes (5) for possible future expansions, if needed (see Centrifugal separation).

Multibillion-dollar additions, modifications, and improvements to the gaseous diffusion plants have made it possible not only to keep pace with the emerging (LWR) nuclear industry in the United States and abroad, but also to preproduce (in advance of demand) uranium fuel, thereby postponing the need for additional multibillion-dollar capital commitments for still more separations capacity. The time gained by preproduction has also been used to develop the newer processes.

The capacity of the postwar gaseous diffusion installations at Oak Ridge, Tennessee; Paducah, Kentucky, and Portsmouth, Ohio, is again currently being improved and uprated in power by almost 60% from 17 to 28 million SWU per year (SWU, separative work units, are generally applicable measures of separative capacity and will be discussed later.)

The Ford Administration had planned to follow these "cascade improvement and uprating programs" (CIP/CUP) by new plant additions. However, as in the case of gaseous diffusion, tremendous progress has been achieved in the efficiency and costs of the centrifuges (see below) and President Carter announced instead a new gas centrifuge plant at Portsmouth, Ohio. This plant has been estimated to cost about 4.4 billion dollars (1977 basis) to construct, and is expected to add about 8.8 million SWU per year, needed to supply the fuel demands of additional nuclear reactors in the middle and late 1980s.

The more recently developed gas centrifuge process presents somewhat greater uncertainties, but also the possibility of greater opportunities for further process and equipment improvements. At a time of rising energy costs, the relatively low, direct power consumption of the centrifuge process (only a few percent of that for gaseous diffusion) looks particularly appealing, though it must be weighed against considerably greater capital and (nonenergy) operating costs.

As far as the future developments of the gaseous diffusion process are concerned, it is doubtful, at this point, whether any additional plants will be built in the United States. Assuming that the demands for uranium enrichment continue to rise through the 1980s and 1990s, much depends upon future performance and improvements of the gas centrifuges and on other, newer processes for uranium isotope separation based, for example, on lasers and on various aerodynamic phenomena. Although further basic discoveries in gaseous diffusion seem to be rather remote, new scale-ups in stage equipment and other engineering and equipment design improvements are still possible, providing that prototype equipment is built and tested. Scaled-up stages (plant sections) could conceivably be constructed in the coal-rich, semiarid regions of the West, where the alternative use value of the coal is likely to be quite low for some time to come. Stages could be added according to customer's commitments, and the plant's product and tails streams could be shipped to the other existing plant sites for further processing. Without requiring new research, improved gaseous diffusion plants could, therefore, still be constructed if unexpectedly needed.

Gaseous diffusion equipment is also being scaled up by the multinational Eurodif project for a 10.8 million SWU per year plant at Tricastin in southern France, planned to become operational during the period 1978–1982. Belgium, Italy, and Spain have joined France in this venture. The Europeans are also pursuing the centrifuge route through the Urenco-Centec combine. Pilot or demonstration plants are operating and expansions are underway at Almelo, Holland, and Capenhurst, England. Projected overall capacity is 10 million SWU per year by 1990. Great Britain, the Netherlands, and Germany are the partners in this combine. Japan has also announced tentative plans for a centrifuge plant of 4 million SWU per year to be operational by 1990. Brazil and the Federal Republic of Germany are collaborating on plans for a small plant based on the Becker separation nozzle, and South Africa has also been considering a scale-up of its pilot plant, based on an undisclosed fluid dynamic process. Finally, the Soviet Union has been offering product from its gaseous diffusion capacity (6).

Owing to the worldwide interest in multiple suppliers, foreign enrichment plants would be able to secure contracts even at less favorable prices and conditions than in the United States.

General Process and Design Selection

When dealing with difficult separations, such as isotope separations, involving the separation of molecules with very similar physical and chemical properties, the enrichment that can be obtained in a single equilibrium stage or transfer unit of the process is quite small; hence, an extremely large number of these elementary separating units must be connected to form a separation cascade in order to achieve most desired separations. Consequently, very large separation systems requiring large amounts of energy are needed. Such processes are relatively expensive.

Usually, the designer is faced with a choice among a number of different feasible separation methods to accomplish a given separation task. The final selection is generally based on a comparison of the combined costs of the required separating systems and their operation, including energy requirements. The designer endeavors to select the least expensive method and plant design of those available to him, in order to obtain a product at a desired rate by a desired time. In addition to costs, the availability of energy, critical materials and equipment, and simplicity of operation must be considered. Among the parameters that must be expressed in monetary terms are the product delay owing to time necessary to produce enriched holdup and attain equilibrium and steady production, instrumentation and control, site, services and utilities, transportation, auxiliary systems (feed, product, waste), administration, laboratories, and engineering (see also Economic evaluation).

In the process of minimizing costs associated with process variables, it is necessary to relate equipment performance and costs to the selected independent process variables. For instance, in the design of a typical distillation column, the heating and cooling requirements are determined by the desired interstage flow or throughput, the cross section of the column is determined by the flow together with the allowable vapor velocity, and the height is fixed by the required number of theoretical separation units, the unit efficiencies, and the height of an equivalent theoretical unit. Allowable economical velocities determine cross sections of passages in pipes and pumps, and required liquid heads or compression ratios determine other pump or compressor dimensions, such as the radii and speeds of revolution. Thus, for any selected set of process variables, the size and amount of equipment and its cost can be defined. In this way, it is possible to evolve different designs and select the optimum or minimum cost combination of variables in each process.

The total energy requirement is one of the most important cost considerations. The energy or power required by any separation process is related more or less directly to its thermodynamic classification. There are, broadly speaking, three general types of continuous separation processes:

(*1*) Reversible Processes. Distillation is an example of a theoretically reversible separation process. In fractional distillation, heat is introduced at the bottom stillpot to produce the column upflow in the form of vapor which is then condensed and turned back down as liquid reflux or column downflow. This system is fed at some intermediate point, and product and waste are withdrawn at the ends. Except for losses, through the column wall, etc, the heat energy spent at the botton vaporizer can be recovered at the top condenser, but at a lower temperature. Ideally, the energy input of such a process is dependent only on the properties of feed, product, and waste. Among the diffusion separation methods discussed here, the centrifuge process (pressure diffusion) constitues a theoretically reversible separation process (see Distillation).

(*2*) Partially Reversible Processes. In a partially reversible type of process, exemplified by chemical exchange, the reflux system is generally derived from a chemical process and involves the consumption of chemicals needed to transfer the components from the upflow into the downflow at the top of the cascade, and to accomplish the reverse at the bottom. Therefore, although the separation process itself may be reversible, the entire process is not, if the reflux is not accomplished reversibly.

Insofar as the consumption of chemicals is concerned, it is obvious that the total consumption of reflux-producing chemicals is proportional to the interstage flows, or width of the cascade, but independent of the number of stages in series, or length of the system.

(3) *Irreversible Processes.* Irreversible processes are among the most expensive continuous processes and are used only in special situations, such as when the separation factors of more efficient processes (that is, processes that are theoretically more efficient from an energy point of view) are found to be uneconomically small. Except for pressure diffusion, the diffusion methods discussed here are essentially irreversible processes. Thus, gaseous diffusion, in which gas expands from a region of high pressure to one of low pressure, mass diffusion, in which a vapor flows from a region of high partial pressure to one of low partial pressure, and thermal diffusion, in which heat flows from a high-temperature source to a low-temperature sink, are all irreversible processes. In contrast with reversible and partially reversible processes, the energy demand in an irreversible process is distributed over the whole cascade in direct proportion to the distribution flow.

In gaseous diffusion, the cascade consists of individual stages that are connected in series. In each stage part of the gaseous feed is forced through a diffusion membrane or barrier with holes smaller than the mean free path of the gas. Because of their slightly greater mobility, the lighter components flow preferentially through the barrier. This enriched portion of the feed is transported to a neighboring stage, up the cascade, where the lighter components tend to concentrate. The other portion of the gas that does not pass through the barrier is rejected to a neighboring stage, down the cascade, where the heavier components tend to concentrate. The feed to each stage is thus composed of combined upflow and downflow from neighboring stages.

In mass and sweep diffusion, the lighter components of the gas mixture flow preferentially against a stream of readily condensible vapor. The optional presence of a screen or barrier in these processes serves only an auxiliary (fluid flow) role since the pores are much larger than the mean free path of the gas mixture. Again, the effect of the separation is multiplied by cascading.

In thermal diffusion, some components of the gas mixture flow preferentially toward a source of heat, where the hotter gases collect. Countercurrent-flow columns and cascades multiply this separation effect.

Finally, in pressure diffusion, a pressure gradient is established by gravity in a centrifugal field. The lighter components concentrate nearly to the low-pressure (center) portion of the fluid. Countercurrent flow and cascading extend the separation effect.

Electromigration is an example of a process based on diffusion in combination with externally applied forces, or forced diffusion. This type of process is not discussed here.

These processes are mainly applied for the separation of stable isotopes, where the separation factors of the more reversible methods, like distillation, absorption, or chemical exchange, are so low that the diffusion separation methods become economically more attractive. Although the application of these processes is discussed in the context of isotope separation, the results presented are equally valid for the description of separation processes for any ideal mixture of very similar constituents such as close-cut petroleum fractions, members of a homologous series of organic compounds, isomeric chemical compounds, or biological materials.

Cascade Design

Less conventional diffusional separation operations are characterized by the relatively small separations that can be obtained by the elementary separation mechanism. That is, the changes in fluid composition attained in gaseous diffusion across the barrier, in thermal diffusion between the hot and cold walls, in mass diffusion between the inlet and the condensing surface for the sweep vapor, and in the centrifuge between the axis of the rotor and its periphery, are all quite small. Thus, in order to achieve most desired separations, a large number of separating units must be employed. Cascade is the term given to the aggregation of separating units that have been interconnected so as to be able to produce the desired material. The optimum arrangement of the separating units in a separation cascade generally minimizes the unit cost of product and its design is a problem common to all separation processes.

In a stagewise separation process such as gaseous diffusion, each unit of equipment consists of one separation stage. However, in a continuous separation process such as thermal diffusion, where an internal countercurrent flow is used to multiply the basic separation effect, each separating unit (eg, thermal diffusion column) usually consists of a large number of transfer units or equivalent theoretical plates which are analogous to individual separation stages.

The Separation Stage. A fundamental quantity, α, exists in all stochastic separation processes which is an index of the steady-state separation that can be attained in an element of the process equipment. The numerical value of α is developed for each process under consideration in the subsequent sections. The separation stage or in a continuous separation process the transfer unit or equivalent theoretical plate (see Distillation) may be considered as a device separating a feed stream, or streams, into two product streams (often called heads and tails, or product and waste) such that the concentrations of the components in the two effluent streams are related by the quantity, α. For the case of the separation of a binary mixture this relationship is (eq. 1):

$$\left(\frac{y}{1-y}\right) \bigg/ \left(\frac{x}{1-x}\right) = \alpha \qquad (1)$$

where y is the mol fraction of the desired component in the upflowing (heads) stream from the stage and x the mol fraction of the same component in the downflowing (tails) stream from the stage. The quantity α is usually called the stage separation factor.

For the case of separating a binary mixture, the following conventions are used in assigning nomenclature. The concentrations of the streams are specified by the mole fraction of the desired component. The purpose of the separation process is usually to obtain one component of the mixture in an enriched form; if both components are desired, the choice of the desired component is an arbitrary one. The "upflowing" stream from the separation stage is the one in which the desired component is enriched and by virtue of this convention, α is defined as a quantity whose value is greater than unity. However, for the processes considered here, α exceeds unity by only a very small fraction, and the relationship between the concentrations leaving the stage can be written in the form (eq. 2):

$$y - x = (\alpha - 1)x(1 - x) \qquad (2)$$

without appreciable error.

A separation stage or transfer unit operating on a binary mixture is shown schematically in Figure 1. In a cascade of separating stages, the feed stream can be formed by mixing the downflowing stream from the stage above and the upflowing stream from the stage below. The quantity, θ, ie, the fraction of the combined stage feed that goes into the stage upflow stream is termed the cut of the stage. In cascades ordinarily designed for difficult separations, the stage cut is normally very nearly equal to one-half. In the case of a theoretical plate in a continuous process, the feed consists of two separate streams, one from above and one from below. In cascades for either stagewise or continuous processes the upflow rate L and the downflow rate L' (or $L(1-\theta)/\theta$) are very nearly equal. For continuous process units, the length S is the length of equipment necessary to satisfy the requirement of equation 1, that the streams leaving the unit be related by α; it is usually called the height of a transfer unit (HTU) or the height equivalent to a theoretical plate (HETP). Although the HTU and HETP are defined differently and are not precisely equivalent to each other, the difference between them becomes negligible when the value of the quantity $\alpha - 1$ is small.

The Separative Capacity. As should be evident, the separation stage is characterized not only by the separation factor α but also by its capacity or throughput of which the upflow L is a measure, and in the case of the continuous process also by the length S. It is therefore desirable to define and determine a quantity indicative of the amount of useful separative work that can be done per unit time by a single stage. Such a quantity was derived by Dirac and is called the separative capacity of the stage. It can be developed from the following argument: it is postulated that the separation stage does useful work on the streams it processes, and hence, increases their net value. The value of a stream must be a function of its concentration; let this value function

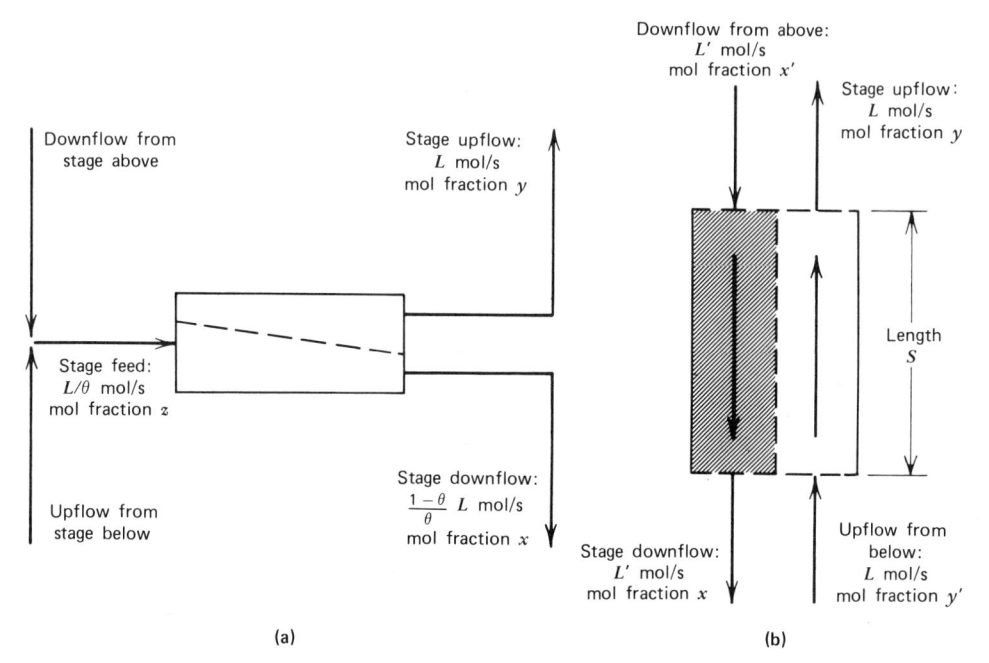

Figure 1. The analogy between the separation stage: (**a**) in a stagewise process and the transfer unit or equivalent theoretical plate; (**b**) in a continuous-separation process.

be designated by $v(x)$. Then the separative capacity of the stage by definition is set equal to the increase in value it creates. The separative capacity of a unit is a very useful concept and permits comparisons to be made between different separation processes.

The Separation Stage. Following the procedure indicated above, the separative capacity of the stage termed δU is set equal to the net increase in value of the streams it processes; thus, δU is given by (eq. 3):

$$\delta U = Lv(y) + \frac{1-\theta}{\theta} Lv(x) - \frac{1}{\theta} Lv(z) \tag{3}$$

The value functions appearing in the above equation may be expanded in Taylor series about x and, since the concentration changes effected by a single stage are relatively small, only the first nonvanishing term is retained. When the value of z is replaced by its material balance equivalent (eq. 4):

$$z = (1-\theta)x + \theta y \tag{4}$$

the separative capacity of the stage is given by (eq. 5):

$$\delta U = \tfrac{1}{2} L(1-\theta)(y-x)^2 v''(x) \tag{5}$$

where $v''(x)$ is the second derivative of the value function. Since the concentrations y and x are related by equation 2 the separative capacity can also be written in the form (eq. 6):

$$\delta U = \tfrac{1}{2} L(1-\theta)(\alpha-1)^2 x^2 (1-x)^2 v''(x) \tag{6}$$

As it is desirable that the separative capacity of the stage be independent of the concentration of the material with which it is operating, the terms in the equation involving the concentration are set equal to a constant, taken for convenience to be unity, and the separative capacity of a single stage operating with a cut of one-half is seen to be (eq. 7):

$$\delta U = \tfrac{1}{4} L(\alpha-1)^2 \tag{7}$$

Thus, the separative capacity of a stage is directly proportional to the stage upflow as well as to the square of the separation effected. This is a basic relationship too frequently overlooked by laboratory research workers who often forecast great commercial advantages for a process that showed an appreciable separation factor in the laboratory without giving any consideration whatever to the throughput of the process.

Equivalent Theoretical Plate. The separative capacity of a theoretical plate in a continuous process can be obtained in the same manner. The following expression is obtained by equating the separative capacity of the unit to the net increase in value of the streams handled (four streams in this case) (eq. 8):

$$\delta U = Lv(y) + L'v(x) - Lv(y') - L'v(x') \tag{8}$$

After expansion in Taylor series about the concentration x and replacing the concentration y' by its material balance equivalent, equation 9 is obtained:

$$\delta U = L(y-x')(x'-x)v''(x) \tag{9}$$

The separative capacity of the equivalent theoretical stage in the continuous process is seen to depend on the concentration difference between the countercurrent streams as well as on the concentration difference between the top and bottom of the stage. The separative capacity is zero when x' is equal to y or x' is equal to x; inspection shows that it attains a maximum value when x' is equal to the arithmetic average of x and y and that this maximum value is (eq. 10):

$$(\delta U)_{\max} = \tfrac{1}{4} L(y-x)^2 v''(x) = \tfrac{1}{4} L(\alpha - 1)^2 \qquad (10)$$

Thus, the maximum value of the separative capacity of a theoretical plate in a continuous process is equal to that of a single separation stage when both units have the same value of the $L(\alpha - 1)^2$ product. It may be noted that when the continuous process is operated so as to yield its maximum separative capacity, the concentrations y' and x' of the streams entering the unit are equal and the similarity between the separation stage and the theoretical plate is accentuated since, for this case, both may be considered to separate a single feed stream into two product streams with concentrations related by α. Perhaps it should be pointed out that the definition of a theoretical plate in the continuous process is essentially arbitrary and not required; however, it is a useful concept, permitting both the stagewise and continuous processes to be treated with the same set of cascade equations.

The Value Function. The value function itself is defined, as has been indicated above, by the second-order differential equation 11:

$$v''(x) = 1/[x^2(1-x)^2] \qquad (11)$$

In the design of cascades, a tabulation of $v(x)$ and of $v'(x)$ is useful. The solution of the above differential equation contains two arbitrary constants. A simple form of this solution results when the constants are evaluated from the boundary conditions $v(0.5) = v'(0.5) = 0$. The expression for the value function is then (eq. 12):

$$v(x) = (2x-1)\ln[x/(1-x)] \qquad (12)$$

and for the derivative of the value function (eq. 13):

$$v'(x) = \frac{2x-1}{x(1-x)} + 2\ln\frac{x}{1-x} \qquad (13)$$

Therefore, $v(1-x) = v(x)$ and $v'(1-x) = -v'(x)$. Selected values are given in Table 1.

Application of the Separative Capacity Concept. In addition to providing a relatively simple means for estimating the production of separation cascades, the separative capacity is useful for solving some basic cascade design problems. For example, the problem of determining the optimum size of the stripping section may be approached as follows:

It is assumed that P, y_P, and x_F for the cascade have been specified, and that the cost of feed and the cost per unit of separative work, the product of separative capacity and time, are known. The basic assumption is that the unit cost of separative work remains essentially constant for small changes in the total plant size. The cost of the operation can then be expressed as the sum of the feed cost and cost of separative work (eq. 14):

$$C_{\text{total}} = (C_F)(F)(\Delta t) + C_{\Delta U}[Pv(y_P) + Wv(x_W) - Fv(x_F)]\Delta t \qquad (14)$$

where C_{total} is the total cost of operation for the period of time Δt, and C_F and $C_{\Delta U}$ are the cost per unit of feed and the cost per unit of separative work, respectively. The

Table 1. The Value Function and Its First Derivative

x	$v(x)$	$-v'(x)$
0.0025	5.959017	410.975
0.0050	5.240372	209.582
0.0072	4.855507	147.735
0.0100	4.503217	108.180
0.015	4.059054	74.0206
0.020	3.736147	56.7632
0.025	3.480384	46.3015
0.030	3.267533	39.2546
0.040	2.923810	30.3144
0.050	2.649995	24.8362
0.070	2.224553	18.3838
0.100	1.757780	13.2833
0.150	1.214221	8.95940
0.200	0.831777	6.52259
0.250	0.549306	4.86389
0.300	0.338919	3.59936
0.350	0.185712	2.55676
0.400	0.081093	1.64426
0.450	0.020067	0.805382
0.500	0.00	0.00

optimum value of x_W is that which minimizes the total cost and can be found by differentiating the total cost with respect to x_W under the restrictions that P, y_P, and x_F remain constant and setting the result equal to zero. The result of this procedure is that the optimum x_W is the solution to equation 15:

$$v(x_F) - v(x_W) - (x_F - x_W)v'(x_W) = C_F/C_{\Delta U} \qquad (15)$$

When the cascade is operated with the optimum x_W, the cost of producing material at any other concentration, y_P, is given by (eq. 16):

$$C_{\text{total}} = C_{\Delta U} P[v(y_P) - v(x_W) - (y_P - x_W)v'(x_W)]\Delta t \qquad (16)$$

obtained by combining equations 14 and 15. An equation of this form can be used to establish the value of material of different concentrations from separation cascades (see below).

Cascade Gradient Equations. An arrangement of separation stages to form a simple cascade is shown in Figure 2. A simple cascade is one that divides a single cascade feed stream into a product stream and a waste stream. Additional side streams, however, could easily be handled. To be consistent with the conventions given for the single stage, the desired component is assumed to be enriched in the product stream at the top of the cascade. The cascade feed is introduced at some intermediate stage between the top and bottom of the cascade. The portion of the cascade that lies above the feed point is termed the enriching section; that which lies below the feed point is termed the stripping section. Though P, y_P, W, and x_W refer to the product and waste streams, as shown in Figure 2, they can be more generally defined, for any section of the plant, as follows: $P = -W$ = net upflow of both components, and $Py_P = -Wx_W$ = net upflow of desired or lighter component. The gradient equations for the cascade are obtained from a combination of the material balance equations, frequently called the operat-

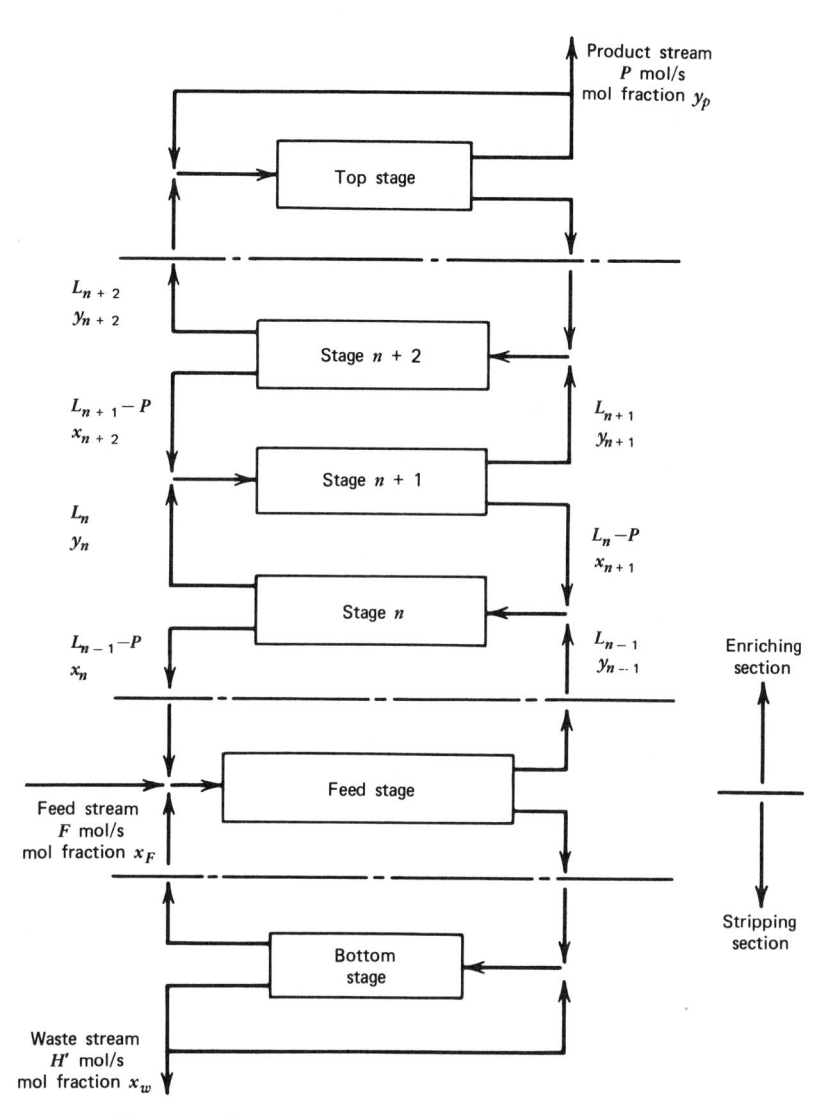

Figure 2. Separation stages arranged to form a simple cascade.

ing-line equations, and the α relationship, usually called the equilibrium-line equation. From a material balance around the top of the cascade down to, but not including, stage n of the enriching section, one obtains the operating-line equation (eq. 17):

$$L_n y_n = (L_n - P)x_{n+1} + Py_P \qquad (17)$$

which can be combined with the equilibrium-line (eq. 2) to give (eq. 18):

$$x_{n+1} - x_n = \frac{L_n}{L_n - P}[(\alpha - 1)x_n(1 - x_n) - (P/L_n)(y_P - x_n)] \qquad (18)$$

For the case under consideration, where the value of $\alpha - 1$ is quite small, it follows that everywhere in the cascade, except possibly at the extreme ends, the stage upflow is many times greater than the product withdrawal rate. Thus $L/(L - P)$ can be set equal to unity. Furthermore when the value of $\alpha - 1$ is small, the stage enrichment $x_{n+1} - x_n$ can be approximated by the differential ratio dx/dn without appreciable error. The gradient equation for the enriching section of a simple cascade therefore takes the form (eq. 19):

$$dx/dn = (\alpha - 1)x(1 - x) - (P/L)(y_P - x) \qquad (19)$$

Similarly, one obtains a gradient equation for the stripping section that has the form (eq. 20):

$$dx/dn = (\alpha - 1)x(1 - x) - (W/L)(x - x_W) \qquad (20)$$

Equations 19 and 20 are the basic equations for cascade design. Although these equations were derived from a consideration of a cascade composed of discrete separation stages, equations of the same form are also obtained for cascade designs based on continuous or differential separation processes. For use in the case of continuous separation processes, however, the term dx/dn, which is the enrichment per stage, is usually replaced by the equivalent terms $S\, dx/dz$, where S is the stage length and dx/dz the enrichment per unit length of process equipment. These equations may then be used to calculate the output from a given cascade configuration, that is, from a cascade for which the variation of $\alpha - 1$ and L is known as a function of the stage number. Thus, by the use of some systematic calculation procedure, an optimum value can be obtained for any parameter.

The terms length and width of a separation cascade are often used. The length refers to the dimensions along which the change in the composition of the mixture proceeds. It might be measured in terms of the height of columns or in terms of the number of stages or separation units connected in series between the terminal points of a cascade. The width refers to the total cross-sectional area available for upflow, or to the upflow itself, at a specific point or value of concentration in a cascade. Thus, the width might be measured in area, in units of upflow, or in terms of the number of stages or separation units connected in parallel. Although the selection of the best combination of lengths and widths for a cascade generally involves a lengthy optimization procedure, it is frequently desirable in preliminary studies to establish only their approximate magnitudes. Lower limits can easily be computed for both dimensions.

Minimum Length or Minimum Number of Stages. It is evident from the gradient equations that the enrichment per stage decreases as the withdrawal rate increases. Thus the minimum number of stages required to span a given concentration difference is obtained when no material is withdrawn from the cascade. This mode of operation ($P = W = F = 0$) is frequently called total reflux operation. Integration of the gradient equation for this case with $\alpha = 1$ taken to be constant gives (eq. 21):

$$N_{\min} = \frac{1}{\alpha - 1} \ln\left(\frac{x_T}{1 - x_T} \Big/ \frac{x_B}{1 - x_B}\right) \qquad (21)$$

where the concentration range to be spanned is from the concentration x_B at the bottom to concentration x_T at the top. As an example of the magnitudes involved, consider for the moment the enrichment of ^{235}U by gaseous diffusion from $x_B = 0.005$ to $x_T = 0.90$. For a value of α equal to 1.0043 the minimum number of diffusion stages required is 1742.

Minimum Width or Minimum Stage Upflow. It also follows directly from the gradient equations that if the withdrawal rates from the cascade are nonzero, it is necessary that the stage upflow from the stage at which the cascade concentration is x must exceed some critical value in order that there be any enrichment at that point in the cascade. This critical value is called the minimum stage upflow and is obtained by setting dx/dn equal to zero in the gradient equation. Thus, for any point in the enriching section the minimum stage upflow is given by (eq. 22):

$$L_{\min} = P(y_P - x)/[(\alpha - 1)x(1 - x)] \tag{22}$$

For the case of enriching ^{235}U to 90 mol % product concentration, the stage upflow at the feed point ($x_F = 0.0072$) must therefore exceed 29,046 times the product withdrawal rate. It can now be seen from a consideration of the minimum stage upflow that the approximation made in deriving the gradient equation, ie, taking the quantity $(1 - P/L)$ equal to unity, introduces negligible error except possibly in the immediate vicinity of the withdrawal points. The condition which arises in a cascade at points where the stage upflow approaches the value $L_{\min}$ is commonly called pinching.

Gradient Equations for a Square Section. A section of a cascade composed of identical stages, that is, a number of stages with the same separation factor and the same stage upflow, is called a square section. For sections of this type the gradient equations are readily integrable. For a section in the enricher of a cascade producing material at rate P and concentration y_P, the solution can be written (eq. 23):

$$N_{\text{sect}} = [(\alpha - 1)(X_1 - X_0)]^{-1} \ln \frac{(X_1 - x_B)(x_T - X_0)}{(X_1 - x_T)(x_B - X_0)} \tag{23}$$

where X_1 and X_0 are the roots of a quadratic equation and are given by (eq. 24):

$$X_1, X_0 = \{\{L(\alpha - 1) + P \pm \{[L(\alpha - 1) + P]^2 - 4L(\alpha - 1)Py_P\}^{1/2}\}\}/[2L(\alpha - 1)] \tag{24}$$

and N_{sect} is the number of stages necessary to span the concentration difference from x_B at the bottom of the section to x_T at the top of the section. Equation 23 is also obtained for a square section in the stripping section, but in the case of the stripper the value of X_1 and X_0 are given by (eq. 25):

$$X_1, X_0 = \{\{L(\alpha - 1) - W \pm \{[L(\alpha - 1) - W]^2 + 4L(\alpha - 1)Wx_W\}^{1/2}\}\}/[2L(\alpha - 1)] \tag{25}$$

Graphical Solution. Some of the preceding concepts can be illustrated graphically by means of a McCabe-Thiele diagram, shown in Figure 3. In such a diagram the equilibrium line, equations 1 or 2, and the operating line, equation 17, are plotted on a set of $x - y$ coordinates. For the case of a square cascade section, the operating line is straight and has the slope $(1 - P/L)$; if the section lies at the product withdrawal end of the cascade, its opening line passes through the point $x = y = y_P$, as shown. The number of stages required to span a given concentration difference, or conversely the concentration difference obtained across a square section with a given number of stages can be illustrated in such a figure.

It is evident from the construction that the closer the operating line lies to the

652   DIFFUSION SEPARATION METHODS

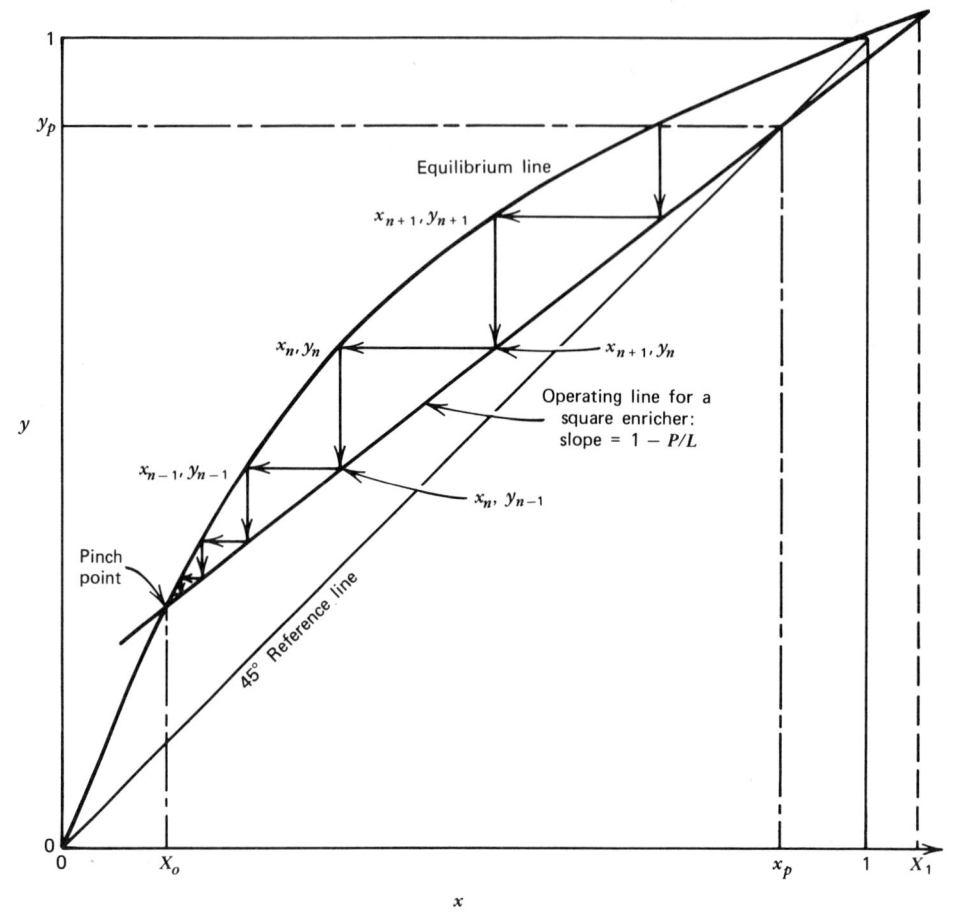

Figure 3. A McCabe-Thiele diagram for a hypothetical square cascade section illustrating pinching.

45-degree reference line, the fewer the number of stages required to span a given concentration difference. The minimum number of stages is therefore required when the operating line coincides with the 45-degree line, that is, when P/L is equal to zero. It is also evident from the construction that in the neighborhood of a point of intersection of the operating and equilibrium lines the enrichment per stage becomes quite small and is equal to zero at the point of intersection itself. At such a point pinching is said to occur, and the origin of the term is made clear by the diagram. In order for a cascade section to span a concentration difference from x_B to y_P, it follows that the operating line may not intersect the equilibrium line before the concentration x_B is reached; the value of L for which the two curves intersect at x_B is the minimum upflow corresponding to the cascade concentration x_B. It is also noteworthy that the values X_0 and X_1 appearing in equations 23 through 25 are the x coordinates of the two points of intersection of the operating line of a square section with the equilibrium line. Although the graphical solution of the cascade gradient equation is simple in principle and exact in theory, it becomes quite cumbersome in practice when processes with separation factors close to unity and hence cascades with thousands of stages are under consideration. For this reason analytic solutions to the gradient equation are usually preferred.

The Ideal Cascade. A cascade of particular interest to design engineers is the so-called ideal cascade; it is a continuously tapered cascade (ie, L is a continuously varying function of x, or n) that has the property of minimizing the sum of the stage upflows of all the stages required to achieve a given separation task. Since, in general, the total volume of the equipment required and the total power requirement of the cascade are directly proportional to the sum of the stage upflows, a consideration of the ideal plant requirements often permits a good economic estimate of the unit cost of product to be made without having to resort to the much more painstaking labor of designing a real (as opposed to ideal) cascade to accomplish the separation job. A simple, intuitive approach to the ideal cascade concept in the case of a cascade composed of discrete stages is given below. Again, the resulting equations are also valid for a cascade based on a continuous or differential separation process.

For the case of a stagewise enrichment process the ideal cascade may be defined as one in which there is no mixing of streams of unequal concentrations. Clearly, the mixing of streams of unequal concentrations in the cascade to form the feed to a separation stage constitutes an inefficiency since it is precisely the reverse of the process taking place in the stage itself. Figure 2 shows that the no-mixing condition at the entrance to stage $n + 1$ requires that $y_n = x_{n+2}$. If the enrichment per stage is essentially constant, x_{n+2} may be written as $x_n + 2(dx/dn)$. The concentration y_n is related to x_n by the α-relationship, (eq. 2). Thus the no-mixing condition leads directly to the gradient equation for the ideal plant (eq. 26):

$$\frac{dx}{dn} = \frac{\alpha - 1}{2} x(1 - x) \qquad (26)$$

The number of stages required to span a given concentration difference in an ideal plant in which all stages have the same separation factor is therefore (eq. 27):

$$N_{\text{ideal}} = \frac{2}{\alpha - 1} \ln \left(\frac{x_T}{1 - x_T} \bigg/ \frac{x_B}{1 - x_B} \right) = 2 N_{\min} \qquad (27)$$

and is twice the minimum number of stages required. The combination of equations 19 and 26 gives the equation for the stage upflow at any point in the enricher of an ideal cascade which is twice the minimum upflow (eq. 28)

$$L_{\text{ideal}} = \frac{2 P(y_P - x)}{(\alpha - 1)x(1 - x)} = 2 L_{\min} \qquad (28)$$

Equations 27 and 28 can be used in conjunction, along with the corresponding equations for the stripping section to produce an ideal plant profile such as is shown in Figure 4 where L_{ideal} is plotted against N_{ideal} for the example of an ideal cascade to produce one mol of uranium per unit time enriched to 90 mol % in ^{235}U from natural feed containing 0.72 mol % ^{235}U, with a waste stream rejected at a concentration of 0.5% ^{235}U. The characteristic lozenge shape of the ideal cascade is evident, no two stages in either the enricher or stripper are the same size. It can be deduced from the above statements that except at the terminals the operating line for an ideal cascade on a McCabe-Thiele diagram is a curved line lying midway between the equilibrium line and the 45-degree reference line.

654   DIFFUSION SEPARATION METHODS

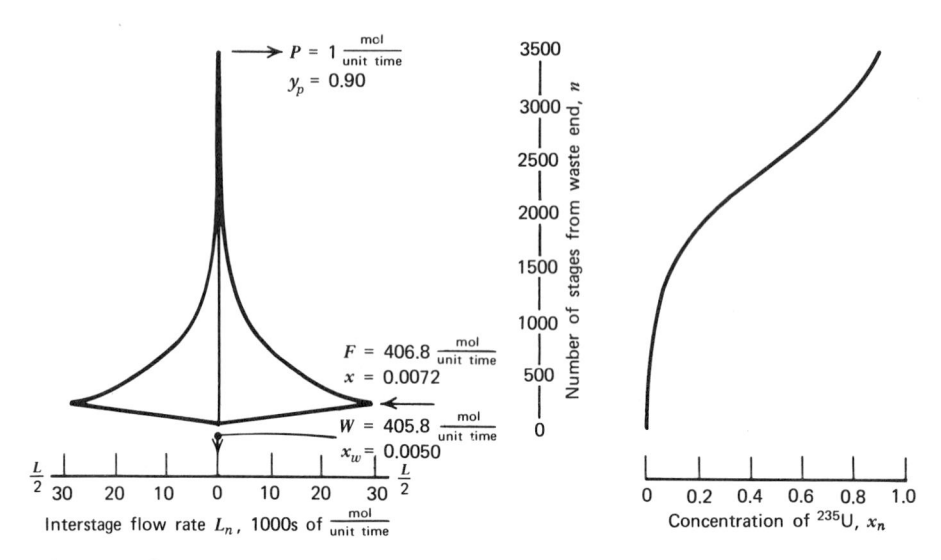

Figure 4. Characteristics of an ideal separation cascade for uranium isotope separation. For this cascade: $\alpha = 1.0043$; $N_T = 3484$ stages: $\Delta U = 153.08$ mol/unit time; $\Sigma L_T = 33.116 \times 10^2$ mol/unit time.

Total Upflow in an Ideal Plant. The sum of the upflows from all of the stages in the ideal plant, or more simply, the total upflow, is the area enclosed by the cascade shown in Figure 4. An analytical expression for this quantity is obtained as follows, with the summation of all the stage upflows in the enriching section expressed as an integral:

$$\sum^{\text{enr}} L_n = \int L \, dn = \int_{x_F}^{y_P} L \frac{dn}{dx} dx \tag{29}$$

For the ideal cascade L is given by equation 28 and dx/dn by equation 26. Making these substitutions, one obtains (eq. 30):

$$\sum^{\text{enr}} L_n = \int_{i_F}^{y_P} \frac{4P}{(\alpha-1)^2} \frac{(y_P - x)}{x^2(1-x)^2} dx \tag{30}$$

However, recalling the definition of the value function, equation 11, and assuming that the value of α is the same for all stages, the integral may be written in the form (eq. 31):

$$\sum^{\text{enr}} L_n = \frac{4P}{(\alpha-1)^2} \int_{x_F}^{y_P} (y_P - x) v''(x) dx \tag{31}$$

which is readily integrated by parts to give (eq. 32):

$$(\text{total upflow})_{\text{enr}} = [4P/(\alpha-1)^2][v(y_P) - v(x_F) - (y_P - x_F)v'(x_F)] \tag{32}$$

The equation for the total flow in the stripping section is obtained in the same manner (eq. 33):

$$(\text{total upflow})_{\text{str}} = [4W/(\alpha-1)^2][v(x_W) - v(x_F) - (x_W - x_F)v'(x_F)] \tag{33}$$

The total flow in the cascade is then given by the sum of equations 32 and 33 which can be simplified with the use of the cascade material balances, (eqs. 34–35):

$$P + W = F \tag{34}$$

and
$$Py_P + Wx_W = Fx_F \tag{35}$$

to give the convenient form (eq. 36):

$$(\text{total upflow})_{\text{cascade}} = [4/(\alpha - 1)^2][Pv(y_P) + Wv(x_W) - Fv(x_F)] \tag{36}$$

For the example considered above, the total cascade upflow is found to be 33×10^6 moles per unit time.

The second term in brackets in the preceding equation is called the separative capacity of the cascade. It is a function only of the rates and concentrations of the separation task being performed and can be calculated quite easily with the aid of Table 1. The great utility of the separative capacity concept lies in the fact that if the separative capacity of a single separation element can be determined—perhaps from equations 7 or 10—then the total number of such identical elements required in an ideal cascade to perform a desired separation job is simply the ratio of the separative capacity of the cascade to that of the element. The concept of an ideal plant is perhaps more useful than one might believe since moderate departures from ideality do not appreciably affect the results presented above. For example, if the upflow in a cascade is everywhere a factor of m times the ideal upflow, the actual total upflow required to perform a separative task will be $m^2/(2m - 1)$ times the ideal cascade total upflow. Thus, if the upflow is 20% greater than ideal at every point in that cascade ($m = 1.2$), the number of separation elements would be only 2.86% greater than that calculated from ideal cascade considerations.

Equations for Large Stage Separation Factors. The preceding results have been obtained with the use of equation 2 and by replacing the finite difference, $x_{n+1} - x_n$, by the differential, dx/dn, both of which are valid only when the quantity $(\alpha - 1)$ is very small compared with unity. In recent years there has been renewed interest, partly because of the development of the gas centrifuge process to commercial status, in the design of cascades composed of stages with large stage separation factors. When the stage separation factor is large, the number of stages required in an ideal cascade in which all stages have the same separation factor is given by (eq. 37):

$$N_{\text{ideal}} = \frac{2}{\ln \alpha} \ln \left(\frac{y_P}{1 - y_P} \bigg/ \frac{x_W}{1 - x_W} \right) - 1 \tag{37}$$

instead of equation (27). When dealing with cascades composed of stages with large separation factors, it is somewhat more convenient to calculate the sum of all the stage feed flows in the cascade rather than the sum of all the stage upflows as was done in the case when $(\alpha - 1)$ is small. When $(\alpha - 1)$ is small with respect to unity, the stage feed flow is essentially just twice the stage upflow rate, and the stage feed flow rate in an ideal cascade (see eq. 28) is:

$$(L/\theta)_{\text{ideal}} = \frac{4 P(y_P - x)}{(\alpha - 1)x(1 - x)} \tag{38}$$

However, when α is large, the corresponding equation for the stage feed rate takes the form (eq. 39):

$$(L/\theta)_{\text{ideal}} = \frac{\sqrt{\alpha} + 1}{\sqrt{\alpha} - 1} \frac{P(y_P - z)}{z(1 - z)} \tag{39}$$

The sum of the stage feed flow rates of all of the stages in an ideal cascade is just twice the total cascade upflow rate when $(\alpha - 1)$ is small with respect to unity, or (eq. 40):

$$\text{(total stage feed)}_{\text{cascade}} = \frac{8}{(\alpha - 1)^2} [Pv(y_P) + Wv(x_W) - F_V(x_F)] \quad (40)$$

but is given by (eq. 41):

$$\text{(total stage feed)}_{\text{cascade}} = \frac{2}{\ln \alpha} \frac{\sqrt{\alpha} + 1}{\sqrt{\alpha} - 1} [Pv(y_P) + WV(x_W) - FV(x_F)] \quad (41)$$

when α is larger. It can be seen that when α is close to unity, equation 41 gives the same result as equation 40. However, if α is equal to 1.1, the total stage feed would be underestimated by 9.2%; if α is equal to 2.0, the total stage feed would be underestimated by 42.4%, if one used equation 40 instead of equation 41. Some of the recent work in this area of cascade theory has dealt with the design of cascades using asymmetric isotope separation stages, for example (7–8) with the analysis of "two-up, one-down cascades," that is, cascades in which the upflow from the nth stage in the cascade bypasses the $(n+1)^{\text{th}}$ stage and is reintroduced at the $(n+2)^{\text{th}}$ stage.

Real Cascades. Although the ideal cascade minimizes the volume of equipment and the energy requirements, it does not generally minimize the cost of the cascade because production economies are realized in the manufacture of the process equipment when a large number of identical units are produced. Thus, a minimum-cost cascade consists of a number of square cascade sections rather than uniformly tapered nonidentical stages. A first approximation to the optimum practical cascade, once the size (length and width) of the individual separating units available is known, is obtained by fitting the ideal plant shape with square sections in some intuitively appealing manner, as illustrated in Figure 5.

This figure shows an ideally tapered enricher that has been replaced by three square cascade sections, a process called squaring-off the cascade (9–12). During the squaring-off process two essential requirements must be kept in mind: The interstage flow in all-square sections must always exceed the local value of $L_{\min}$ at all points in the cascade, and the squared-off cascade must contain a total number of stages which exceeds $N_{\min}$. In order for the squared-off cascade to give a performance closely resembling that of the ideal cascade, the shape of the squared-off cascade should approximate the shape of the ideal cascade.

In the final analysis the problem of determining the optimum practical cascade is rather complex. Equipment performance and costs need to be related to the selected independent process variables. The main process equipment usually consists of a large number of separating units, pumping and heat exchange equipment, control devices, and connecting piping. The whole process is provided with services, and auxiliary systems and feed and withdrawal facilities. It is usually enclosed by a building and surrounded by land of the proper type. Sizes and cost of these important items must be related to the process variables. The cost of most equipment varies according to its size, but is not directly proportional to it. The unit cost of a device is often proportional to its size raised to the 0.6^{th} power. The cost of a specific type of column may be such a function of its surface area, the cost of a pump such a function of its flow capacity, and the cost of a motor such a function of its power rating. Thus, for any selected set of values for the process variables, the number of units of equipment, the

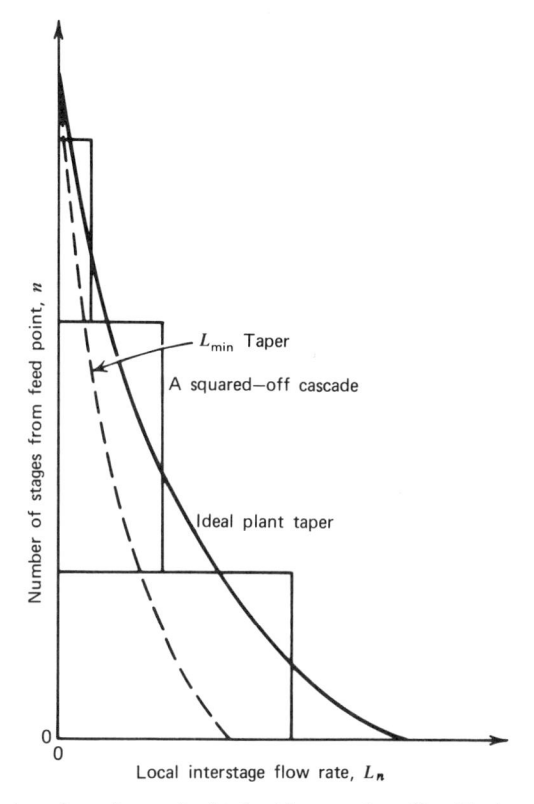

Figure 5. Design of a real cascade obtained by squaring off an ideal enriching section.

required sizes, and associated cost can, in principle, be defined. In this way it is possible to select many different sets of values for the variables and, by a trial and error procedure, solve for the optimum- or minimum-cost design region. The details of procedures used for the optimization of real gaseous diffusion cascades are presented in reference 13, and the problems of optimization of real gas centrifuge cascades are discussed in reference 14.

Time-Dependent Cascade Behavior. The period of time during which a cascade must be operated from start-up until the desired product material can be withdrawn is called the equilibrium time of the cascade. The equilibrium time of cascades utilizing processes with small values of $\alpha - 1$ is a very important quantity since often a cascade may prove to be quite impractical because of an excessively long equilibrium time. An estimate of the equilibrium time of a cascade can be obtained from the ratio of the enriched inventory of desired component at steady state H to the average net upward transport of desired component over the entire transient period from start-up to steady state $\bar{\tau}$. In equation form this definition can be written as (eq. 42):

$$T_{eq} = H/\bar{\tau} = \frac{1}{\bar{\tau}} \int_0^N h_n(x_n - x_F)dn \tag{42}$$

where h_n is the holdup of the nth stage. The average net upward transport for the entire transient period is not usually known; the initial and final values of the net transport, however, are known. At start-up the concentration gradient is flat, since the column is filled with material at feed concentration, and the transport is a maximum. Using this transport in equation 42 gives a lower limit for the equilibrium time, (eq. 43):

$$(T_{eq})_{min} = H/[L(\alpha - 1)x_F(1 - x_F)] \qquad (43)$$

At steady state, with a fully developed gradient, the net transport is $P(y_P - x_F)$, which is a lower limit for net upward transport. Substituting this into equation 42, leads to an expression for an upper limit for the equilibrium time:

$$(T_{eq})_{max} = H/[P(y_P - x_F)] \qquad (44)$$

Equations 43 and 44 thus yield a lower and upper limit, respectively, and used together usually give a satisfactory estimate for the equilibrium time of a cascade.

Examination of equation 42 shows that T_{eq} is directly proportional to the average stage hold-up of process material. Thus, in conjunction with the fact that liquid densities are on the order of a thousand times larger than gas densities at normal conditions, the reason for the widespread use of gas-phase processes in preference to liquid-phase processes in cascades for achieving difficult separations becomes clear.

The unsteady-state behavior of a separation unit is, furthermore, of interest because it can be used for the experimental determination of the separation parameters of the unit. If the holdup of the separating unit is known, the separation factor, $\alpha - 1$, can be obtained from a knowledge of the transient behavior of the unit during start-up. Figure 6 shows the concentration gradient in a square column shortly after start-up. As long as the gradient is flat at the feed point as shown, an equation much like equation 44 for the maximum equilibrium time can be used to relate the enriched holdup to the elapsed time. From a knowledge of the gradient, the enriched holdup H can be computed, and with L known the separation factor $\alpha - 1$ can be computed from (eq. 45):

$$(\alpha - 1) = H/[Lx_F(1 - x_F)\Delta t] \qquad (45)$$

where Δt is the elapsed time since start-up when the column was filled with material at feed concentration.

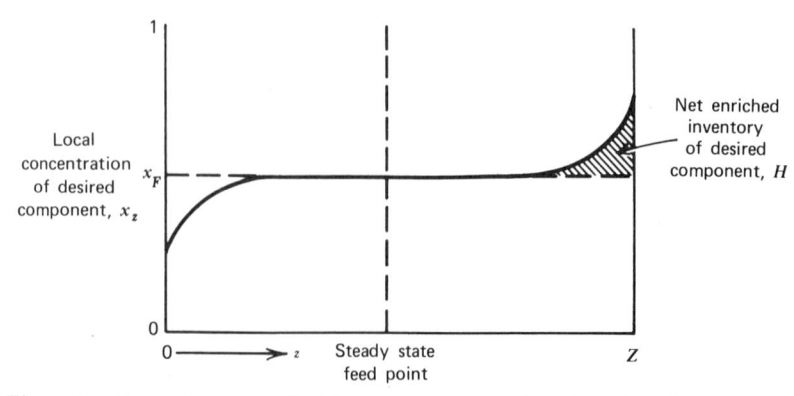

Figure 6. Concentration gradient in a column or cascade a short time after start-up.

The Gaseous Diffusion Process

The gaseous diffusion separation process depends on the separation effect arising from the phenomenon of molecular effusion (that is, the flow of gas through small orifices). When a mixture of two gases is confined in a vessel and is in thermal equilibrium with its surroundings, the molecules of the lighter gas strike the walls of the vessel more frequently, relative to its concentration, than the molecules of the heavier gas. This is caused by the greater average thermal velocity of the lighter molecules. If the walls of the container are porous with holes large enough to permit the escape of individual molecules, but sufficiently small so that bulk flow of the gas as a whole is prevented (that is, with pore diameters approaching mean-free-path dimensions of the gas), then the lighter molecules escape more readily than the heavier ones, and the escaping gas is enriched with respect to the lighter component of the mixture. As derived later, the equation for the separation factor α for this process reflects the relative ease of light versus heavy molecules in escaping through the pores. It turns out that α^*, the ideal separation factor, is the ratio of the two molecular velocities. Since the kinetic energies, $\frac{1}{2} mv^2$, of the two species are the same, α^*, the ratio of the two velocities is equal also to the square root of the inverse ratio of the two molecular weights. This is the basis for equation 49. This separation effect was first discovered experimentally by Graham in 1846 and later explained theoretically with the advent of Maxwell's kinetic theory of gases. Graham showed that mixtures of oxygen and hydrogen and of oxygen and nitrogen could be separated by this method. Following the work of Graham, the effusion process was used by several other investigators. In 1895 Rayleigh and Ramsey used it to separate argon from nitrogen, and in 1920 Aston employed it to slightly enrich the concentration of the neon-22 isotope. A major improvement in diffusion separation technology was the development of a cascade of diffusion stages for isotope separation by Hertz (15) in 1932 by an arrangement similar to that shown in Figure 7. With a 24-stage cascade Hertz was able to obtain appreciable enrichment in the isotopes of neon. Subsequently (16), almost pure deuterium was obtained from a cascade of 48 stages, and the isotopes of nitrogen and of carbon were enriched (17) with a 34-stage cascade. In the latter case, with methane as the process gas, the concentration of ^{13}C isotopes was raised from 1.1 to 16%.

The large plants built for the separation of uranium isotopes following the wartime development of the United States Manhattan Projects are the most outstanding applications of gaseous diffusion. This work culminated in the construction of the gaseous diffusion cascade called the K-25 plant at Oak Ridge, Tennessee. Other gaseous diffusion plants were built at Oak Ridge, and Paducah, Kentucky, and Portsmouth, Ohio, in the United States, and in Capenhurst, England and Pierrelatte, France. Diffusion plants are also reported to be in operation in the U.S.S.R. The United States diffusion plants have been partially described in the Smyth (18) and other reports (19–20), and the British plant in the Jay report (21).

Process Description. The basic unit of a gaseous diffusion cascade is the gaseous diffusion stage. Its main components, shown in Figures 7 and 8, are the converter holding the barrier in tubular form, motors, and compressors moving the gas between stages, a heat exchanger removing the heat of compression introduced by the stage compressors, the interstage piping, and special instruments and controls to maintain the desired pressures and temperatures. Figure 7 is a schematic representation of a section of a cascade; Figure 8 illustrates how the components might be combined in practice.

660 **DIFFUSION SEPARATION METHODS**

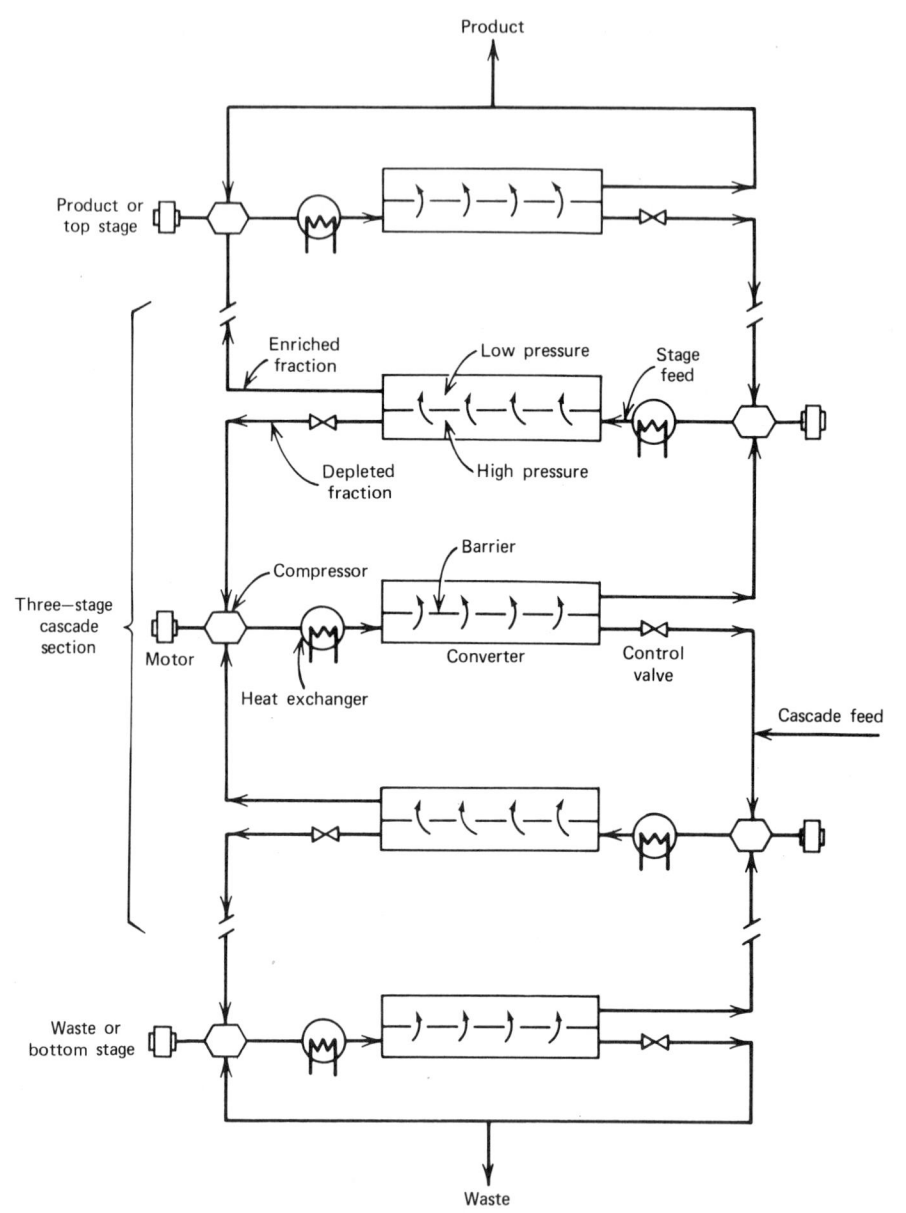

Figure 7. A cascade of gaseous diffusion stages.

As illustrated in Figure 7, the feed stream to a stage consists of the depleted stream from the stage above and the enriched stream from the stage below. This mixture is first compressed and then cooled so that it enters the diffusion chamber at some predetermined optimum temperature and pressure. In the case of uranium isotope separation the process gas is uranium hexafluoride. Within the diffusion chamber the gas flows along a porous membrane or diffusion barrier. Approximately one-half of the gas passes through the barrier into a region of lower pressure. This gas is enriched in the component of lower molecular weight (^{235}U). The enriched fraction, upon leaving

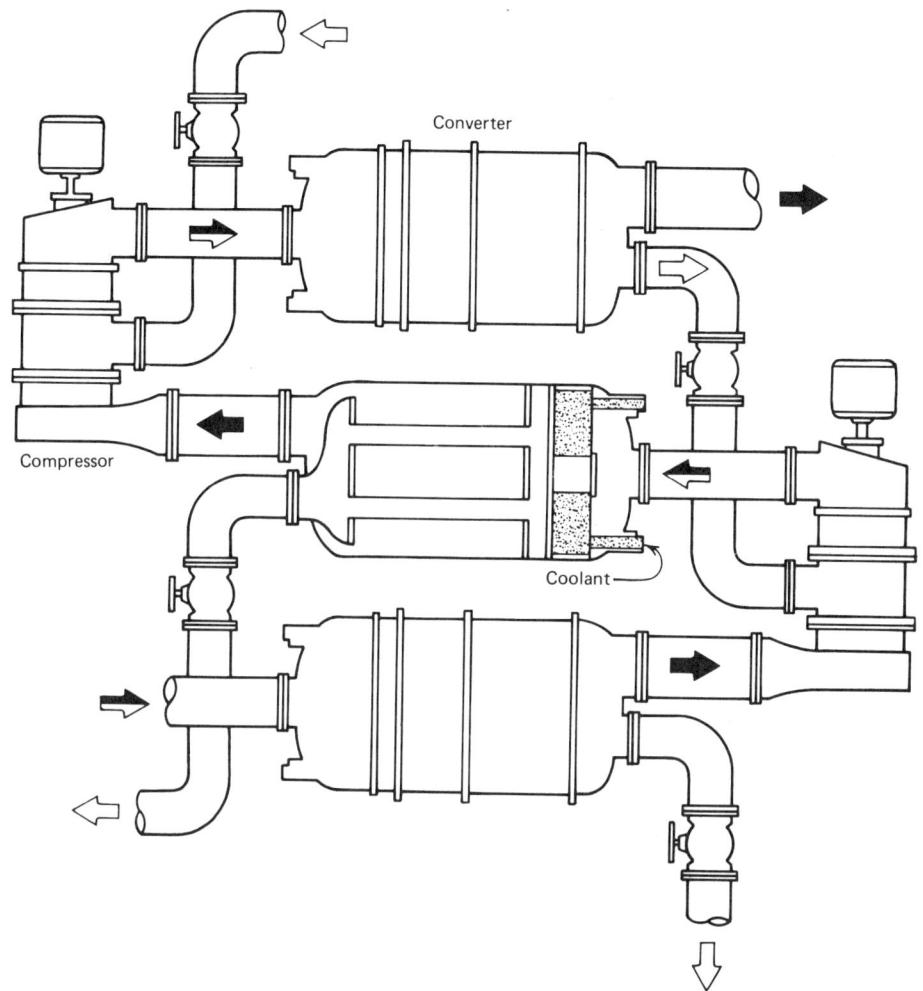

Figure 8. Arrangement of gaseous diffusion stages.

the diffusion chamber, is directed to the stage above where it is recompressed to the barrier high-side pressure. The gas that does not pass through the barrier is depleted with respect to the light component. This depleted fraction, upon leaving the chamber, passes through a control valve, and is directed to the stage below where it too is recompressed to the barrier high-side pressure. However, since it is necessary in this case to compensate only for the frictional losses and the control valve pressure drop, the compression ratio may not need to be as high as that for the enriched fraction. Thus there is some freedom in the design of the stage; that is, two compressors might be used, a larger one for the enriched fraction and a smaller one for the depleted fraction; or the pressure drop across the control valve may be made equal to the pressure drop across the barrier so that both streams can be recompressed by the same compressor (this scheme, although wasteful of power, might make up for it in savings in equipment costs), or some other compromise mode of operation might be used. For operational efficiency a number of gaseous diffusion stages are operated together in units referred to as cells and buildings. Cells and buildings can be removed from operation for routine maintenance and bypassed without disturbing the diffusion cascade. Figure 9 shows a schematic arrangement of a sixteen-stage gaseous diffusion cell.

Successful operation of the gaseous diffusion process requires a special, fine-pored diffusion barrier, mechanically reliable and chemically resistant to corrosive attack by the process gas. For an effective separating barrier, the diameter of the pores must approach the range of the mean free path of the gas molecules, and in order to keep the total barrier area required as small as possible, the number of pores per unit area must be large. Seals are needed on the compressors to prevent both the escape of process gas and the inflow of harmful impurities. The control valves adjust the stage flows and thereby provide for the stability of pressure and gas holdup, and of cascade operation. Some of these problems are discussed in reference 22.

The need for a large number of stages and for the special equipment mentioned above makes gaseous diffusion an expensive process. The three United States gaseous diffusion plants represent a capital expenditure of close to 2.5×10^9 dollars (23).

In addition, according to published figures for 1956 and 1957 (24), these plants consumed over $200,000,000/yr of power; in 1956 this represented more than 10% of the total U.S. electric power consumption. Nevertheless, because of the limited ability of the more efficient separation methods to distinguish between the isotopes of uranium, the gaseous diffusion process is one of the more economical processes yet devised for the separation of these isotopes on a large scale.

Stage Design. A separation cascade employing gaseous diffusion stages can be designed using the principles set forth in the preceding section, once the performance characteristics of the individual stages are known. The important parameters are the stage separation factor and the size of a stage required to handle the desired stage flows. As would be expected, both of these important parameters depend on the characteristics of the barrier.

Barrier Characteristics. As mentioned above, the barrier material must be fine-pored and have many pores per unit area. Preparation and characterization of such a material presents a difficult technological problem. The characteristics of a barrier suitable for the separation of isotopes by gaseous diffusion are discussed in reference 25, including various effective pore sizes and pore size distributions. Experimental techniques used to evaluate barrier characteristics are presented in reference 26. These techniques include adsorption methods, electron microscopy, x-ray analysis, porosity measurements with mercury, permeability measurements with liquids and gases, and measurements of separation effectiveness. Various techniques for barrier preparation result in materials having pore size in the range of 10–30 nm (27). An electrolytic technique leads to a thin sheet material having about 10^{10} pores per square centimeter with radii on the order of 15 nm (28). The separating performance of barrier has been evaluated by means of a 12-stage pilot plant (29).

Barrier Flow. An ideal separation barrier is one that permits flow through it to take place by effusion only. This is the case when the diameter of the pores in the barrier is sufficiently small compared with the mean free path of the gas molecules. If the pores in the barrier are treated as a collection of straight circular capillaries, the rate of effusion through the barrier is governed by Knudsen's law (eq. 46):

$$N = \frac{4}{3} \sqrt{\frac{1}{2\pi MRT}} \frac{\phi d}{l} (p_f - p_b) \tag{46}$$

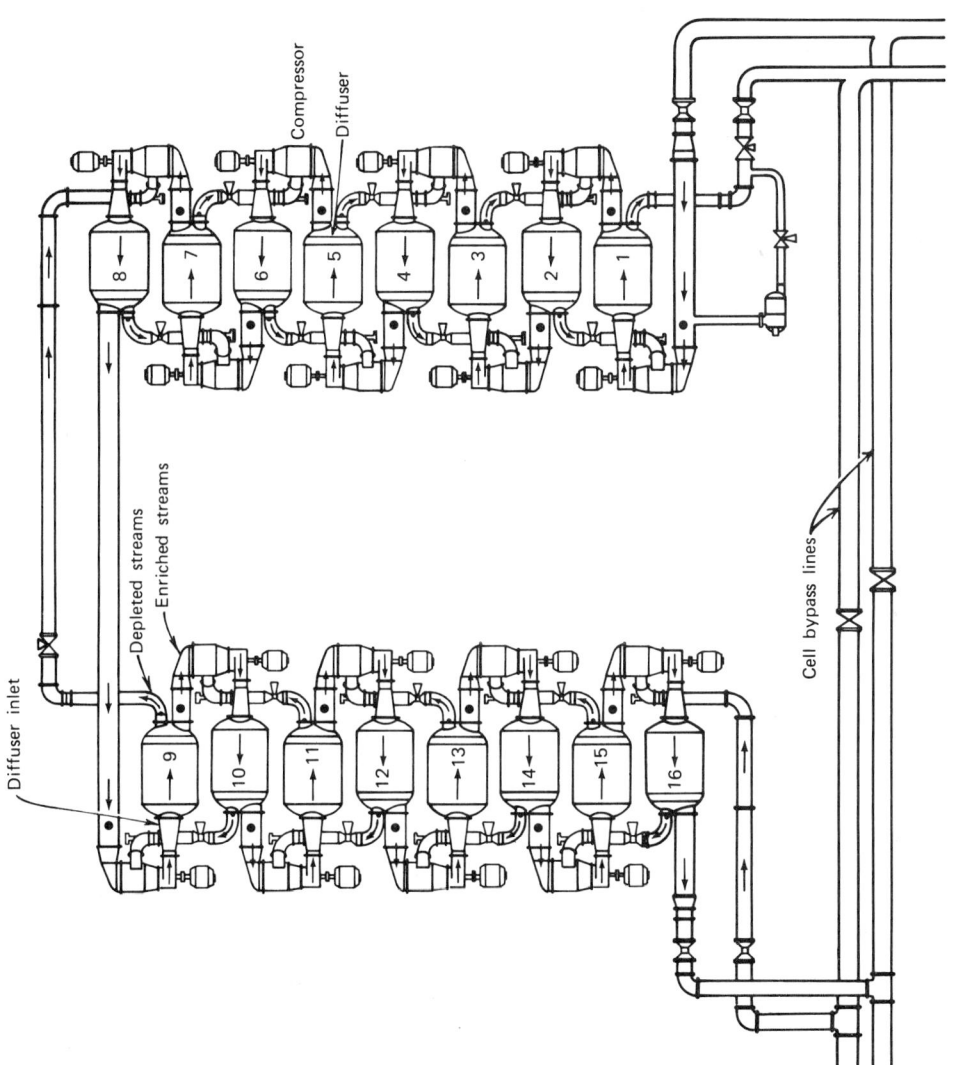

Figure 9. 16-Stage process cell.

where N is the molar flow of gas per unit area through the barrier, M is its molecular weight, R is the gas constant, T is the absolute temperature, ϕ is the fraction of the barrier area open to flow, d is the effective pore diameter, l is the pore length or thickness of the barrier, and p_f and p_b are the high- and low-side pressures of the barrier, respectively. In practice not all of the flow through the barrier is effusive flow. Through those pores whose diameters are of the order of the mean free path or greater, a nonseparative Poiseuille flow occurs. The two types of flow are additive and the total flow can be represented by (30) (eq. 47):

$$N = \frac{a}{\sqrt{M}}(p_f - p_b) + \frac{b}{\mu}(p_f^2 - p_b^2) \qquad (47)$$

where a and b are functions of temperature for a particular barrier and μ is the viscosity of the gas. The first term on the right is the contribution to the total flow of the effusive flow, the second that of the nonseparative Poiseuille flow. Since the pressure dependence of each type of flow is different, the constants a and b can be evaluated from a series of measurements at different pressures (31).

The Fundamental Separation Effect. An ideal-point separation factor can be defined on the basis of the separation obtained when a binary mixture flows through an ideal barrier into a region of zero back pressure. For this case an expression of the form of equation 46 can be written for each component. The flow of light component through the barrier is proportional to $p_f x'/\sqrt{M_A}$, and the flow of heavy component is proportional to $p_f(1 - x')/\sqrt{M_B}$, where x' is the mole fraction of the light component on the high-pressure side of the barrier and M_A and M_B are the molecular weights of the light and heavy components, respectively. The concentration, y', of the effusing gas is therefore (eq. 48):

$$y' = \frac{x'/\sqrt{M_A}}{x'/\sqrt{M_A} + (1 - x')/\sqrt{M_B}} \qquad (48)$$

and from the definition of α (eq. 1), it follows that the ideal-point separation factor is equal to (eq. 49):

$$\alpha^* = \frac{y'/(1-y')}{x'/(1-x')} = \sqrt{M_B/M_A} \qquad (49)$$

which, for example, for the case of uranium isotope separation with UF_6, is equal to 1.00429. It has been pointed out that this is also the expression for the ratio of the velocity of the light molecules to that of the heavy molecules.

The Stage Separation Factor. The stage separation factor, however, in all probability is appreciably different from the ideal-point separation factor because of the existence of four efficiency terms that must be taken into account:

(1) A Barrier Efficiency Factor. In practice, diffusion plant barriers do not behave ideally; that is, a portion of the flow through the barrier is bulk or Poiseuille flow which is of a nonseparative nature. In addition, at finite pressure the Knudsen flow (32) is not separative to the ideal extent, that is, $\sqrt{M_A/M_B}$. Instead, the degree of separation associated with the Knudsen flow is less separative by an amount that depends on the pressure of operation. To a first approximation, the barrier efficiency is equal to the Knudsen flow multiplied by a pressure-dependent term associated with its degree of separation, divided by the total flow.

(2) A Back-Pressure Efficiency Factor. Since a gaseous diffusion stage operates

with a low-side pressure p_b which is not negligible with respect to p_f, there is also some tendency for the lighter component to effuse preferentially back through the barrier. To a first approximation the back-pressure efficiency factor is equal to $(1 - r)$, where r is the pressure ratio p_b/p_f.

(3) *A Mixing Efficiency Factor.* As the gas flows along the high-pressure side of the diffusion barrier, it becomes, as a result of the effusion process, preferentially depleted with respect to the lighter component in the neighborhood immediately adjacent to the barrier. As a result of this, a concentration gradient perpendicular to the barrier is set up on the high-pressure side, and the average concentration x' of the light component in the bulk of the gas flowing past a point is greater than x'', the concentration of the light component at the surface of the barrier at that point. The mixing efficiency factor is equal to the ratio $(y' - x')/(y' - x'')$ as indicated in Figure 10. A value for the point mixing efficiency factor can be calculated from a consideration of diffusion through an effective film representing the resistance to diffusion. It is given by an expression of the form: $\exp(-Nl_f/\rho D)$, where l_f is the thickness of the effective film.

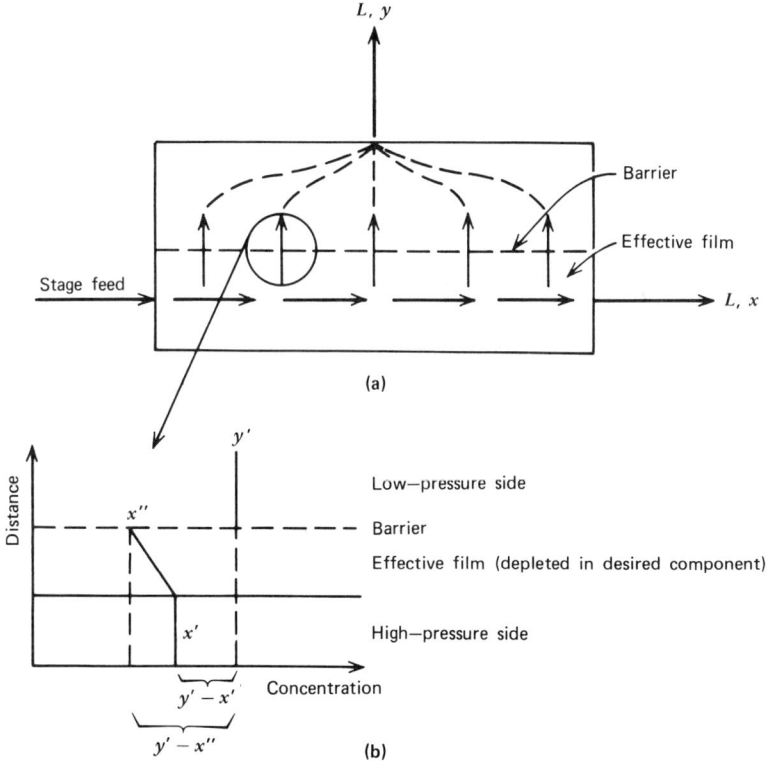

Figure 10. Flow of process gas through a gaseous diffusion stage. (**a**) Gaseous diffusion stage; (**b**) local concentration profiles near the diffusion barrier. y' = point concentration of light component on low-pressure side; x' = average concentration of light component in bulk of gas on high-pressure side of barrier flowing past the specified point; x'' = point concentration of light component at surface of barrier on the high-pressure side; $y' - x''$ = separation which would be obtained across barrier in the absence of the effective film; and $y' - x'$ = actual separation obtained across barrier, taking a film into account.

(4) A Cut-Correction Factor. The stage separation factor has been defined as relating the concentrations in the streams leaving the stage. Since the concentrations, x' and y', on each side of the barrier are changing continuously as the gas flows through the diffusion stage, the relationship between the concentrations of the streams differs from the point relationship. This difference is taken into account with the cut correction factor. If the gas on the high-pressure side flows through the stage with no appreciable mixing taking place in the direction of flow and if the effused fraction is withdrawn from the stage directly upon passing through the barrier, the cut correction can be calculated from material balance considerations. For this case, the exit or stream concentration of the stage upflow is equal to the average concentration of the effused gas, whereas the exit or stream concentration of the downflow is equal to the terminal concentration of the uneffused gas which is, of course, at maximum and not average depletion. Consequently, the stage separation factor relating to exit concentration is greater than the point separation factor, and the cut correction factor exceeds unity. (This phenomenon is analogous to cross-flow in a plate distillation column.)

The stage separation factor can therefore be related to the ideal-point separation factor by an equation of the form

$$(\alpha - 1) = (E_b)(E_p)(E_M)(E_c)(\alpha^* - 1) \qquad (50)$$

where (eqs. 51–53):

$$E_b = \text{barrier efficiency} = \frac{\text{separative flow through barrier}}{\text{total flow through barrier}} \qquad (51)$$

$$E_p = \text{back-pressure efficiency} = 1 - r \qquad (52)$$

$$E_M = \text{mixing efficiency} = e^{-Nl_f/\rho D} \qquad (53)$$

and (eq. 54):

$$E_c = \text{cut correction} = \frac{1}{\theta} \ln \frac{1}{1 - \theta} \qquad (54)$$

For the usual case where approximately one-half of the gas entering the stage passes through the barrier, the value of the cut, θ, is equal to 0.5 and the cut correction takes on the value 1.386.

These efficiency factors are discussed in more detail in references 33 and 34. Actually, the barrier and back-pressure efficiencies are interrelated and cannot be formulated independently, except only as an approximation. A better formation which has been found to fit the experimental results is (eq. 55):

$$(E_b)(E_p) = (1 - r)/[1 + (1 - r)(p_f/p^*)] \qquad (55)$$

where p^* is a constant whose value must be determined experimentally. It may be noted that for an ideal barrier, $E_b = 1$, p^* is equal to ∞ and the back-pressure efficiency is given by equation 52. Figure 10 shows the flow of material in a gaseous diffusion stage in a schematic manner.

Separative Capacity. An expression for the separative capacity of a single gaseous diffusion stage whose upflow rate is L moles per unit time, given in equation 7, can be written in the following terms (eq. 56):

$$\delta U = \tfrac{1}{4} L \left(\frac{1 - r}{1 + (1 - r)(p_f/p^*)} \right)^2 E_M^2 \left(\frac{1}{\theta} \ln \frac{1}{1 - \theta} \right)^2 (\alpha^* - 1)^2 \qquad (56)$$

For a very high quality barrier ($p^* \to \infty$), the separative capacity of a stage with a mixing efficiency of 100% and operating at a cut of one-half would be (eq. 57):

$$\delta U = \frac{1.921}{4} L(1-r)^2(\alpha^* - 1)^2 \tag{57}$$

If the power requirement of the gaseous diffusion process were no greater than the power required to recompress the stage upflow from the pressure on the low-pressure side of the barrier to that on the high-pressure side, then the power requirement of the stage would be $LRT \ln(1/r)$ for the case where the compression is performed isothermally. The power requirement per unit of separative capacity would then be given simply by the ratio (eq. 58):

$$\frac{\text{power requirement}}{\text{unit separative capacity}} = \frac{2.082 \, RT \ln(1/r)}{(1-r)^2(\alpha^* - 1)^2} \tag{58}$$

This quantity is minimized when the stage is operated at a pressure ratio across the barrier corresponding to $r = 0.285$. Furthermore, if power were the only economic consideration, the stage would be operated at this pressure ratio. However, as the value of r is decreased from this optimum, although the cost of power is increased, the number of stages required and hence the capital cost of the plant is decreased. Thus, in practice a compromise between these factors is to be made.

The optimum pressure level for gaseous diffusion operation is also determined by comparison; at some pressure level the decrease in equipment size and volume to be expected from increasing the pressure and density is outweighed by the losses which occur in the barrier efficiency. Nevertheless, since it is well known that the cost of power constitutes a large part of the total cost of operation of gaseous diffusion plants, it can perhaps be assumed that a practical value of r does not differ greatly from the above optimum. Inclusion of this value in the preceding equations yields (eq. 59):

$$\delta U = 0.246 \, L(\alpha^* - 1)^2 \tag{59}$$

and (eq. 60):

$$\frac{\text{power requirement}}{\text{unit separative capacity}} = \frac{5.11 \, RT}{(\alpha^* - 1)^2} \tag{60}$$

The actual power requirement is greater than that given by equation 58 or 60 because of the occurrence of frictional losses in the cascade piping, compressor inefficiencies, and losses in the power distribution system.

Gaseous Diffusion Cascade Design. The results from the analysis of a gaseous diffusion stage can be used with the principles of cascade design to calculate the requirements of real or ideal cascades to perform a given separation job. As an example of the use of the equations, the requirements of an ideal gaseous diffusion cascade for the production of 100 grams per day of methane containing 90% ^{13}C from normal feed material have been computed. The main operating conditions of the cascade are shown in Table 2.

Plant Operation and Costs. In 1972 the USAEC published two reports (35–36) revealing a great deal of information about the operation and the economics of the United States gaseous diffusion plants. The typical mode of operation of the three United States gaseous diffusion plants in 1972 is illustrated in Figure 11. It is clear from the figure that the three plants are not operated independently, but as a single

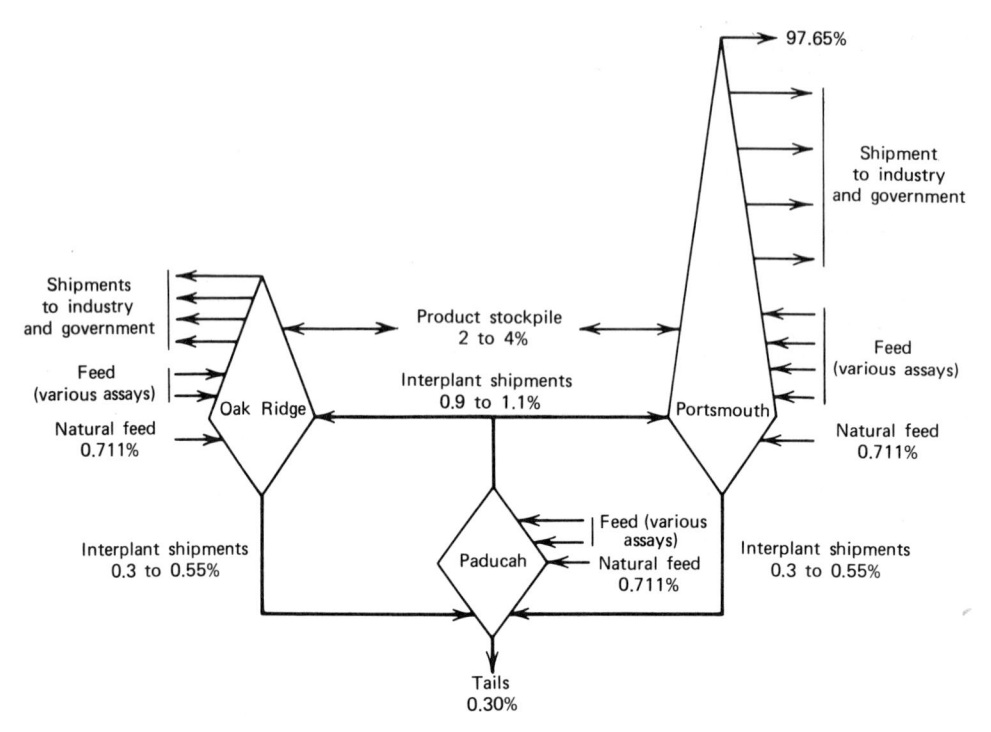

Figure 11. Mode of operation for gaseous diffusion plants (% values are wt % ^{235}U).

Table 2. Requirements of an Ideal Gaseous Diffusion Cascade for the Production of Enriched ^{13}C

Operating conditions	Value	Source of value
product concentration, y_p, mol fraction, ^{13}C	0.90	given[a]
feed concentration, x_F, mol fraction ^{13}C	0.0106	given[a]
waste concentration, x_W, mol fraction ^{13}C	0.00955	given[a]
product rate, P, kg CH$_4$/d	1.0	given[a]
feed rate, F, kg CH$_4$/d	848	eqs. 34–35
waste rate, W, kg CH$_4$/d	847	eqs. 34–35
separative capacity, kg CH$_4$/d	92.9	eq. 36
pressure ratio, r	0.285	assumed for minimum power
$\alpha^* - 1$	0.0308	eq. 49
$\alpha - 1$	0.0305	eqs. 50–54
total number of stages	448	eq. 27
upflow at feed point, kg CH$_4$/d	5560	eq. 28
power required per unit separative capacity, kW·h/kg at 317 K	246	eq. 60
total power requirement, kW	952	

[a] Predetermined by equipment or operation.

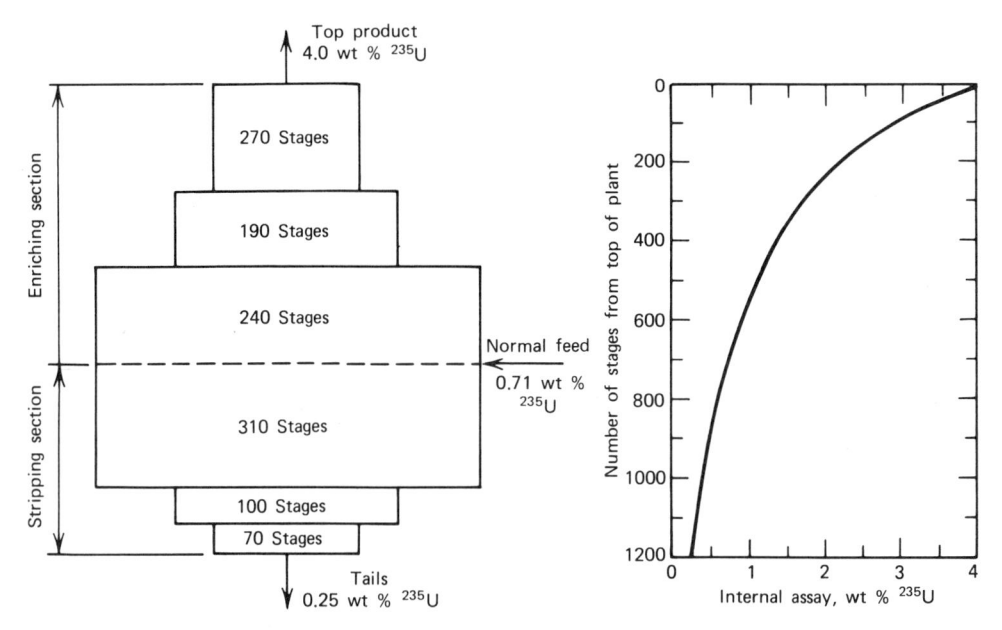

Figure 12. Schematic diagram and internal gradient for an 8.75 million-SWU/yr plant.

gaseous diffusion complex; the concentrations and rates of the interplant shipments are chosen so as to optimize the overall system. Independent operation of the three plants would result in about a 1% loss in separative work.

The three-plant complex, when fully powered to 6100 MW, has a nominal capacity of 17.2 million SWU/yr. Two programs are currently underway that, when completed in 1982, will increase the capacity of these plants to 28 million SWU/yr. The first of these is called the cascade improvement program (CIP). It involves the installation of new and more efficient barriers and the modification of the stage compressors, piping, and control system in order to minimize energy losses. The cost of the total CIP program is estimated at 1.0 billion dollars and will increase the separative capacity of the plants by 5.5 million SWU/yr. The second program is called the cascade uprating program (CUP). It involves uprating of the electrical equipment, heat removal equipment, and compressor shafts so that additional power can be added to the cascades. The completed CUP is estimated at $460,000,000 and will permit the addition of 1300 MW to the cascades, and result in an additional separative capacity of 4.6 million SWU/yr.

Information on the design of new gaseous diffusion plants is presented in reference 37. The shape of a gaseous diffusion cascade, based on 1970 U.S. technology, having 8.75 million SWU/yr separative capacity, designed to produce uranium containing 4% ^{235}U from natural feed with tails at 0.25% ^{235}U is shown in Figure 12. Selected economic data are presented in Tables 3–5. Figure 13 shows the arrangement of the major pieces of process equipment, the diffuser and compressor, in the existing United States diffusion plants.

In 1973, Eurodif, a multinational consortium organized under French leadership, decided to build a large gaseous diffusion plant at Tricastin in France. This plant, to be completed in 1982, is planned for a separative capacity of 10.7 million SWU/yr and

Table 3. Selected Economic Data[a] **for an 8.75 Million SWU/yr New Gaseous Diffusion Plant Using 1970 Technology**

Facility and operating requirements	Value
capital investment, $ millions	1200
specific investment, $/(SWU·yr)	137
power, MW	2430
specific power, kW·h/SWU	2433
operating cost[b]	
$ million/yr	14
manpower requirement	900

[a] Based on June, 1971, dollars.
[b] Excluding power costs.

Table 4. Estimated Plant Capital Cost Breakdowns by Process Stage Components

Plant and equipment	Total capital cost, %
gas diffusers	8.9
gas compressors	12.5
compressor drive motors	7.6
electrical system	13.1
heat removal system	4.4
process buildings and enclosures	7.1
process piping and valves	7.9
instrumentation	2.3
miscellaneous systems[a]	2.9
plant start-up and support	0.9
process support facilities[b]	8.0
subtotal	75.6
engineering at 4.5%	3.4
subtotal	79.0
contingency at 15%	11.8
subtotal	90.8
interest during construction	9.2
Grand total	100.0

[a] Includes process ventilation, fire protection, sanitary water, and sewage.
[b] Includes such facilities as administration building, technical services building, maintenance building, shops, cleaning and decontamination, purge and product building, and site preparation.

is based on French gaseous diffusion technology. Some design and progress information regarding this cascade was published by the French in 1975 (38) and 1976 (39–40). Table 6 presents some of the main features of the Tricastin plant. The engineering design of the French gaseous diffusion stage, although functionally the same as the United States stage, differs appreciably in appearance, and the motor, compressor, and diffuser are arranged vertically and are contained in a single housing in the French plant.

From equation 60 one can obtain a theoretical power requirement of about 900 kW·h/SWU for uranium isotope separation assuming a reasonable operating temperature. A comparison of this number with the specific power requirements of the United States or Eurodif plants indicates that real gaseous diffusion plants have an

Table 5. Estimated Capital Cost Breakdown of Process Stage Components

Stage characteristics	Stage Small	Stage Medium	Stage Large
shaft power, kW	930	1640	3025
number of stages	340	290	550
separative work distribution, %	9	20	71
capital cost distribution, %	20	23	57
Stage components			
stage equipment cost distribution, %			
gas diffuser	6.1	8.0	10.2
gas compressor	10.6	12.0	13.4
compressor drive motor	5.5	7.0	8.5
electrical system	13.8	13.5	12.7
heat removal system	3.4	4.2	4.8
process buildings and enclosures	5.9	6.8	7.7
process piping and valves	10.2	8.6	6.9
instrumentation	3.3	2.5	1.8
miscellaneous systems	3.9	3.1	2.4
plant start-up and support	1.3	1.0	0.7
process support facilities	11.6	8.9	6.5
engineering	3.4	3.4	3.4
contingency	11.8	11.8	11.8
interest during construction	9.2	9.2	9.2
Grand total	*100.0*	*100.0*	*100.0*

Figure 13. View of diffusers and compressor. This photograph shows installed process equipment typical of the largest size used in existing U.S. plants. The large stages of a new 8.75 million SWU/yr plant would use equipment of about this size.

efficiency of about 37%. This represents not only the barrier efficiency, the value of which has not been reported, but also electrical distribution losses, motor and compressor efficiencies, and frictional losses in the process gas flow.

The cost of enriched material from a gaseous diffusion plant depends both on the cost of separative work and of feed material. It can be seen from equation 15 that if the optimum tails concentration from a gaseous diffusion plant is 0.25%, the ratio of the cost of a kilogram of normal uranium to the cost of a kilogram of separative work equal to 0.80 is implied. Since the cost of separative work in new gaseous diffusion plants is expected to be about $100/SWU, equation 16 gives the cost per kilogram of uranium containing 4% ^{235}U as about $1240.

The Sweep and Mass Diffusion Processes

A partial separation of a binary gas mixture can be obtained if the gas mixture is allowed to diffuse through a third auxiliary gas at a constant total pressure provided there exists a difference in the diffusion coefficients of two components of the gas mixture with respect to the auxiliary gas. A porous screen is occasionally used to divide the process region; however, unlike the gaseous diffusion barrier, this screen is not essential to the separation and may be dispensed with entirely. If the screen is included, the process is generally referred to as mass diffusion; if the screen is not used, the process is called sweep diffusion.

In 1922 Hertz (41) recognized that diffusion through an auxiliary gas would yield a separation based on the difference in diffusion coefficients alone. He built and patented mass diffusion stages that were essentially mercury diffusion pumps (42).

Table 6. The Eurodif Plant at Tricastin

Operating conditions	Value
rates, kg UF$_6$/h	
product	600
feed	3025
waste	2425
concentration, % ^{235}U	
product	1.35–3.15
feed	0.72
waste	0.25
stages in enricher,[a] MW	
small, 220 sizes	0.6
medium, 280 sizes	1.6
large, 320 sizes	3.3
stages in stripper,[a] MW	
small, 60 sizes	0.6
medium, 120 sizes	1.6
large, 400 sizes	3.3
total cascade power, MW	3100
total cascade separative capacity,	
million SWU/yr	10.7
specific power requirement, kW·h/SWU	2538

[a] Equipment in three sizes.

Using a cascade of twelve such stages, Hertz (43) reported in 1934 a ten-fold enrichment of ^{22}Ne with respect to ^{20}Ne. Maier (44), who called the process atmolysis, investigated the mass diffusion stage and in an extensive experimental program demonstrated the applicability of the process to the separation of hydrogen, nitrogen, and sulfur dioxide from air using water and carbon tetrachloride vapors as auxiliary gases.

Recently, a detailed theoretical examination of the stagewise mass diffusion process was made at MIT (45) as it might be applied to the large-scale separation of uranium isotopes. It was concluded that the power consumption of the mass diffusion plant would be three times that of a gaseous diffusion plant of the same capacity and the capital cost of a mass diffusion plant would be five times that of a gaseous diffusion plant with its associated power plant.

Although the earlier investigations of the sweep or mass diffusion process were limited to the consideration of the stagewise process, there is an obvious advantage in carrying out the process in a countercurrent flow column since a multiplication of the stage separation effect can be obtained thereby. This means that in principle, at least, any degree of enrichment can be obtained by the simple expedient of making the column longer. A theory was developed (46) for the mass diffusion column process, using the screen to separate the up-flowing stream from the down-flowing stream. Other authors (47) based their theoretical treatment on a column without the screen. Their sweep diffusion column theory was corroborated by an extensive experimental investigation that proved the workability of a column without the screen. Inclusion of the screen enables the two streams to be pumped through the column over a range of rates, and its presence probably suppresses convective remixing of the two streams along their common boundary.

No industrial applications of the sweep or mass diffusion processes are known, but laboratory-scale separations of some gas pairs, including the separation of some light isotopes, have been reported. Data illustrating the size and performance of sweep diffusion columns are summarized in Table 7.

Table 7. Separations Obtained in Sweep Diffusion Columns

Process gas	Sweep vapor	Column length, cm	HETP[a], cm	Sweep vapor flux, kg/(m²·h)	Separation factor, top/bottom	Reference
H_2, natural gas	water	45.7		23.0	4.43	47
H_2, natural gas	water	45.7		24.7	4.48	47
H_2, natural gas	ammonia	45.7		22.1	5.22	47
O_2, N_2 (air)	water	45.7	about 1.2	33.4	1.082	47
neon-20, -22	water	90	2–3	5.6	1.52	48
argon-36, -40	methanol	90	6–7	6.7	1.17	48
H_2 and D_2	methanol	90		7.1	1.39	48
nitrogen-28, -29	methanol	90		8.0	1.04	48
neon-20, -22	xylene	100		43.2	4.25	49
neon-20, -22	xylene	100	about 2	64.8	6.50	49
helium and methane	water	274.3		9.0	35.9	50

[a] HETP = height equivalent to a theoretical plate.

Process Description. The sweep and mass diffusion processes can be carried out either in countercurrent-flow columns containing many equivalent equilibrium stages, or as stagewise processes (46) in which the equilibrium stages are contained in separate and distinct pieces of process equipment. Only the column application is discussed here.

In a column application of the sweep or mass diffusion process, the process gas is confined in a region bounded by a porous wall and a cold wall as shown schematically in Figure 14. The auxiliary gas or sweep vapor is a chemically inert vapor that can easily be condensed on the cold wall at the prevailing pressure. The sweep vapor is introduced to the process region through the porous wall, serving as a vapor distributor, and is made to flow across the process region by its partial pressure gradient to the cold wall where it is condensed. The condensed sweep vapor then forms a continuous film on the cold wall and flows by gravity to a reservoir at the bottom of the column from which it is withdrawn, vaporized in a boiler, and returned to the process region through the porous wall. In the case of sweep diffusion, where no screen is used, the falling liquid film drags the gas adjacent to it toward the bottom of the column, thus inducing a countercurrent circulation of the process gas. It follows that the magnitude of the countercurrent circulation is directly related to the linear surface velocity of the falling film which is in turn a function of the film mass flow rate, viscosity, and density. Thus in sweep diffusion the countercurrent circulation rate L is a dependent variable that is fixed once the operating conditions have been chosen. In the case of mass diffusion,

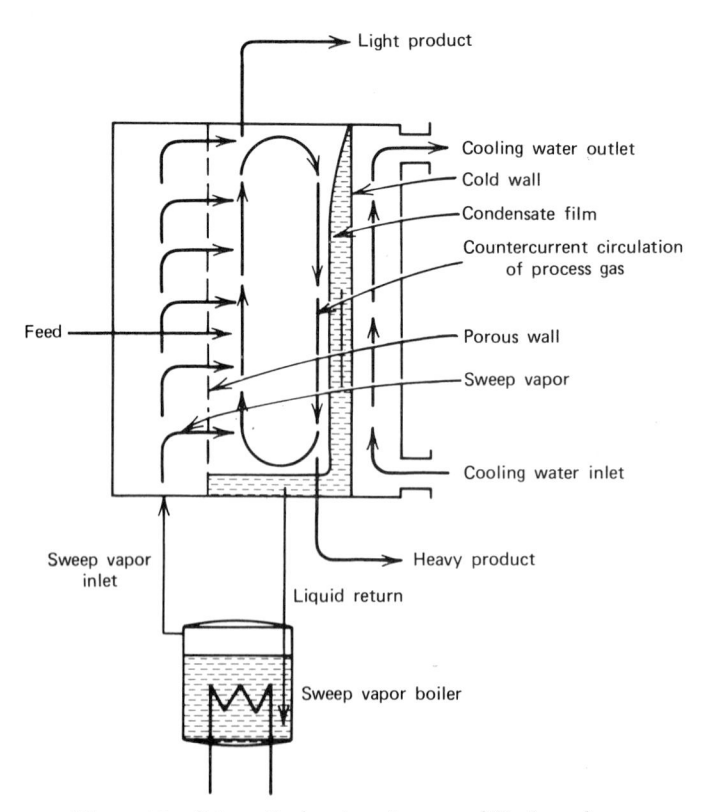

Figure 14. Schematic drawing of a sweep diffusion column.

where a screen dividing the process region is employed, external control of the countercurrent circulation rate can easily be obtained by pumping the two streams. Thus in mass diffusion the countercurrent circulation rate L is an independent variable.

Characteristics and Requirements. Although the process is simple in concept and the process equipment is not complicated or difficult to construct, several problems that have not been fully treated in the literature could reduce the effectiveness of the process. In addition to being chemically inert with respect to both the process gas and the materials of construction of the column itself, the sweep vapor when condensed must not be a solvent for the process gas. Sharp reductions in separative work occur when the process gas is soluble, since this implies removing some fraction of the inventory of the column at all concentrations. Furthermore, since economic considerations dictate that the condensed sweep vapor be reused in large-scale applications, the sweep vapor entering the processing region contains the process gas at an average composition. Thus two mixing losses are sustained as a result of process-gas solubility. The process gas is continuously removed from the column at each point, mixed when the condensed sweep vapor is collected, and evaporated for reuse. When this sweep vapor, containing process gas at an average composition, is fed back to the column it is mixed again with process gas at varying composition in the process region, thereby raising the concentration of desired component in the stripping portion of the column and lowering the concentration of desired component in the enriching portion of the column. If solubility of the process gas is a serious problem, a cascade of mass diffusion stages would be the best choice for the separation job. In the stagewise process, provision can be made, with little additional complication, to isolate the sweep vapor within each stage, thus diminishing the overall loss.

If, in a sweep diffusion column, the molecular weights of the sweep vapor and process gas are not approximately equal, the sharp (exponential) transverse gradient of sweep vapor can lead to large changes in the density of the sweep vapor-process gas mixture across the working space of the column. Since the sweep vapor is relatively abundant at the porous wall and rare at the cold wall, it follows that for a sweep vapor heavier than the process gas, the gas mixture near the porous wall is more dense than that near the cold wall. This effect tends to set up a countercurrent flow pattern, driven by the density gradient, which opposes that driven by the falling film of sweep vapor. The result can be a suppression of reversal of the normally expected flow pattern. This effect was first reported in 1956 (48) when the suppression of the expected axial transport was observed during an attempted separation of neon isotopes with methanol as the sweep vapor. If the process gas is heavier than the sweep vapor, on the other hand, the density gradient effect reinforces the circulation caused by the viscous pumping of the falling film. This density effect does not occur in mass diffusion columns since the streams are pumped at preselected rates by mechanical means.

In a sweep diffusion column the countercurrent circulation of the process gas is induced in part by the falling film of condensate. The surface velocity of the film is a function of the thickness of the film, among other things; at the top of the column, where the liquid film is thin and therefore flowing slowly, the circulation rate is low compared with the bottom of the column where the film is thicker and flowing more rapidly. Therefore, an axial variation in the circulation rate is present which makes the column process difficult to analyze by elementary methods. The axial variation of countercurrent circulation rate can be suppressed (47) somewhat by introducing a stream of condensed sweep vapor on the condenser wall at the top of the column, at a rate several times that of the overall transverse sweep vapor flux. This liquid overflow develops an initial film flow that is only slightly affected by additional condensation of sweep vapor in the column, thus tending to minimize the variation in the surface velocity of the falling film and hence the countercurrent flow rate. The sweep liquid overflow rate cannot be increased indefinitely, however, since the surface of the liquid film develops ripples at low Reynolds numbers while the flow is still in the laminar regime. These ripples, and the waves into which they develop as the film becomes turbulent, distort the laminar flow profile of the countercurrent circulation and can lead to local remixing of the two gas streams that reduces the overall column efficiency. Obviously, in the case of mass diffusion where the countercurrent circulation is caused by pumping, the film flow characteristics are relatively unimportant.

The process, by its very nature, is thermodynamically irreversible. In both sweep and mass diffusion the sweep vapor is vaporized at the operating pressure of the column and subsequently condensed at low partial pressure, thus causing an irreversible flow of heat. It follows as a consequence of this irreversibility that a sweep or mass diffusion plant of large size uses large amounts of power and is, therefore, not able to compete economically with processes such as distillation and gas absorption where the separation process is reversible. It seems likely that the sweep or mass diffusion process is limited in use to the production of enriched isotopes or for performing other difficult separations. Some interest has been shown in sweep or mass diffusion because its energy needs are merely for boiling the sweep vapor. Provided a sweep vapor with low-boiling temperature can be used, low-grade waste heat might be used to drive the process.

Column Design. Development of the phenomenological theory of the sweep diffusion column process for the separation of binary mixtures of gases proceeds in three steps. An expression for the transverse flux of one component of the process gas mixture is developed by first using multicomponent diffusion equations. The second step is the consideration of the overall separation problem in a column with two streams in countercurrent flow and is essentially a column material balance. The third step is to solve the hydrodynamic problem of the column, deriving the velocity profile of the countercurrent flow pattern. The coordinate system and nomenclature shown in Figure 15 are used.

Since a general rigorous solution cannot be presented in a convenient form, the development is restricted by a number of simplifying assumptions. It is assumed first that the process gas is a binary mixture of isotopes (denoted by the subscript 1 for the light component and by subscript 2 for the heavy component) so that their diffusion coefficients in the sweep vapor (denoted by the subscript 0) are nearly equal. If D_0 is the arithmetic mean of the two diffusion coefficients D_{01} and D_{02} of the components of the process gas with respect to the sweep vapor, then the separability γ, a number close to unity for isotope separation, can be defined as follows (eq. 61):

$$\gamma \equiv (D_{01} - D_{02})/D_0 = \frac{M_2 - M_1}{M_2 + M_1} \frac{M_0}{M_0 + (M_1 + M_2)/2} \qquad (61)$$

where M_i is the molecular weight of the ith component. Note that the larger M_0 becomes compared to M_1 and M_2, the larger is γ.

The starting point for deriving expressions for the transverse fluxes is the multicomponent diffusion equation of Maxwell and Stefan relating the concentration and molar fluxes of the components of a mixture. Denoting the molar flux of the jth

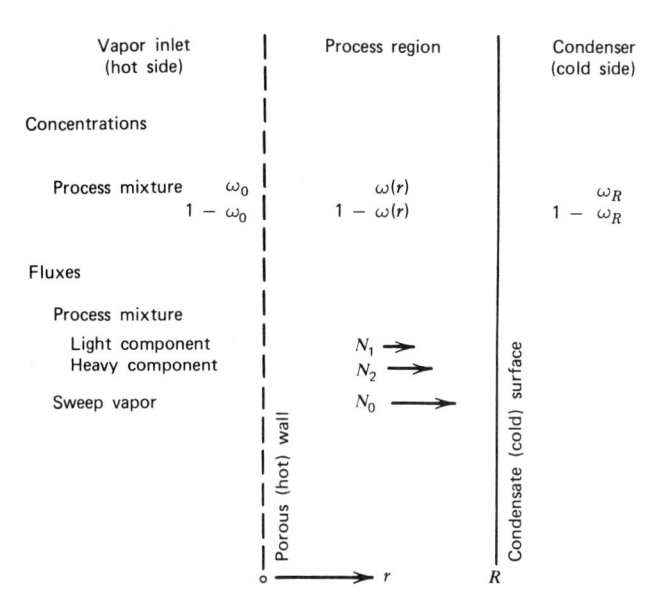

Figure 15. Coordinate system and nomenclature for sweep diffusion column theory.

component by N_j and denoting by ω the concentration of total process gas in the mixture of process gas and sweep vapor, an equation for the flux of sweep vapor N_0 can be derived as follows (eq. 62):

$$N_0 = \frac{cD_0}{R} \ln \frac{\omega_R}{\omega_0} \tag{62}$$

where c is the molar density of the mixture, R is the spacing between the porous wall and the cold wall, and the subscripts 0 and R on ω indicate its evaluation at the porous wall and the cold wall, respectively. The value of ω_R is set by the sweep vapor pressure at the cold wall temperature and the total operating pressure of the columns.

The magnitude of the countercurrent flow depends on the properties of the liquid film of condensate flowing downward on the cold wall. If the film has a surface velocity of magnitude V, and the treatment is limited to the case where ω varies little enough across the column that it can be replaced by a constant value $\bar{\omega}$ in calculating the countercurrent flow profile, then the following column design equations can be derived:

$$\text{interstage flow rate:} \quad L = \frac{4}{27} c\bar{\omega}\, VR \tag{63}$$

$$\text{stage separation factor:} \quad (\alpha - 1) = \frac{9}{16} \gamma \ln \frac{\omega_R}{\omega_0} \tag{64}$$

stage length:

$$S = \frac{9}{140} \frac{V\bar{\omega}R^2}{D_0} \left(\frac{D_0}{D_{12}} + \frac{1-\bar{\omega}}{\bar{\omega}}\right) + \frac{27}{4} \frac{D_0}{V\bar{\omega}} \left(\frac{D_0}{D_{12}} + \frac{1-\bar{\omega}}{\bar{\omega}}\right)^{-1} \tag{65}$$

where the first term of the stage length is the convective contribution and the second term is the contribution caused by back diffusion. The convective term is directly proportional to the liquid film velocity and the back diffusion component is inversely proportional to the liquid film velocity. It follows that there is a value for V which is optimum in the sense of minimizing the stage length. The optimum liquid film velocity, V^*, is given by (eq. 66):

$$V^* = \sqrt{105}\, D_0 \bigg/ \left[R\bar{\omega}\left(\frac{D_0}{D_{12}} + \frac{1-\bar{\omega}}{\bar{\omega}}\right)\right] \qquad (66)$$

Although these results are based on such assumptions as a small transverse variation in ω, they provide some qualitative insight into the operation of a sweep diffusion column and permit first-order estimates of its performance to be made. As shown in equation 19, the gradient equation for a column process can be started as (eq. 67):

$$S\frac{dx}{dz} = (\alpha - 1)x(1 - x) - \frac{P}{L}(y_p - x) \qquad (67)$$

where x denotes the concentration of the light component of the process on a sweep vapor-free basis. The separation parameters given in equations 63 and 65 can be used to design a plant. Plant design requires calculation of the separative capacity of a column which is given in terms of S, L, and $(\alpha - 1)$ by (eq. 68):

$$\delta U = \frac{ZL}{4S}(\alpha - 1)^2 \qquad (68)$$

where Z is the column length and Z/S is the number of stages comprising the column. Equation 68 is a modification of equation 10 which is written for a single stage.

If sweep diffusion is carried out using a sweep vapor in which the process gas is soluble, as mentioned above, there can be rather severe losses in column separative work. An estimate of these losses can be made through the use of Figure 16 which gives a solubility efficiency E defined as the ratio of separative work actually obtainable in a system with solubility effects to the separative work predicted by the equation above in which the solubility effects were ignored. The curves in Figure 16 give this solubility efficiency as a function of a solubility fraction f for columns with various numbers of stages. The dimensionless solubility fraction f is the ratio of process gas crossflow due to solubility to the process gas upflow due to the countercurrent circulation and is defined by (eq. 69):

$$f = N_0 s/L \qquad (69)$$

where N_0 is the molar sweep vapor rate, s is the solubility in mole fraction of process gas in sweep vapor, and L is the net molar upflow of light component. The curves were derived assuming a center-fed column with the average concentration of light material in the dissolved process gas being equal to its concentration in the feed stream. Figure 16 illustrates the sharp reduction in separative work caused by solubility effects, especially for columns with many stages, and should give an approximate estimate of the separative work loss under conditions not greatly different from the assumed model.

As an example of the use of Figure 16 in estimating separative work losses caused by the solubility of process gas in condensed sweep vapor, consider a column one meter in length operated in such a way that the stage length is 2.5 cm, the flux of the sweep vapor is 0.36 kg of gas/(m²·h) and the interstage upflow is 1.20 kg of gas/(h·m of column

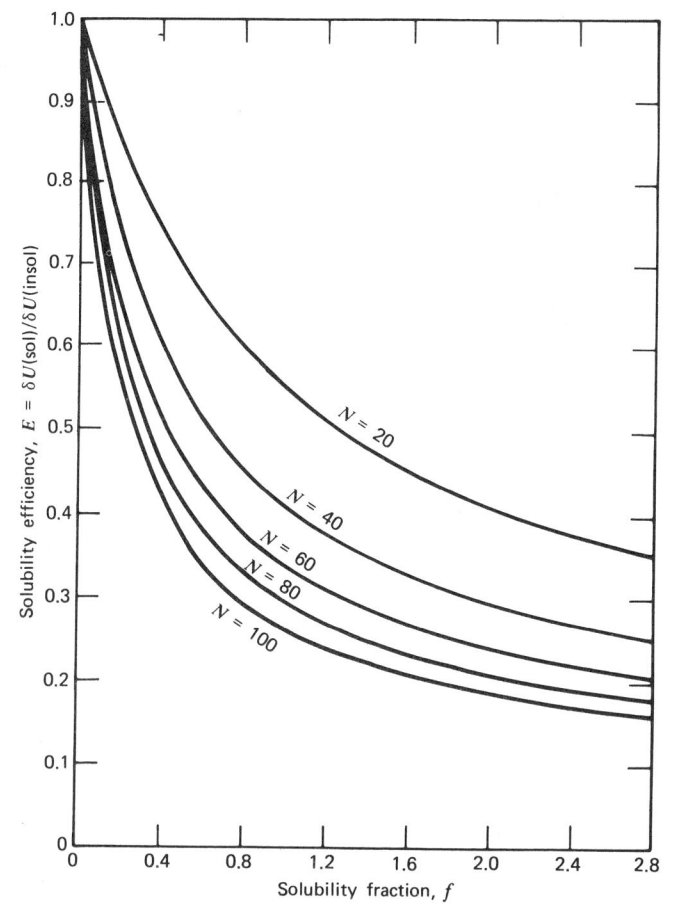

Figure 16. The effect of process-gas solubility on the separative work of a sweep diffusion column.

width). If the process gas is soluble in condensed sweep vapor to the extent of one mol % at the temperature and partial pressure conditions of the cold wall (eq. 70):

$$f = N_0 sZ/L = (0.36)(0.01)(100)/1.20 = 0.30 \qquad (70)$$

Since the column is 100 cm long and the stage length is 2.5 cm, the column must contain 100/2.5 or $N = 40$ stages. From Figure 16 with $f = 0.30$ and $N = 40$ stages, the solubility efficiency $E = 0.667$ is obtained which means that, with solubility effects taken into account, the column produces only 66.7% of the predicted separative work assuming no solubility.

Cascade Design. Equations 63 to 66 may be used to outline the design of a cascade of sweep diffusion columns for the production of 1 kilogram per day of neon enriched to 0.90 mol fraction in ^{22}Ne starting with feed containing 0.09 mol fraction ^{22}Ne and stripping out half of the ^{22}Ne in the feed. Neon, for our purposes here, may be considered to be a binary mixture of ^{20}Ne and ^{22}Ne. Steam is used as the sweep vapor. The sweep diffusion columns are assumed to be flat plate columns 1 meter in height and 0.5 m wide. Reasonable operating conditions include cold wall temperatures that can be reached with cooling tower water, which is on the order of 40°C, and operating near atmospheric pressure so that equipment leakage is not a major concern.

680 DIFFUSION SEPARATION METHODS

Using the sweep diffusion design equations (eqs. 63–66) a reasonable parameter space can be investigated and from this a design basis is chosen. For a sweep diffusion column, the results of such calculations show that, with all other things being equal: (1) Increasing the spacing between the walls R leads to fewer stages, greater upflow, and lower sweep vapor rates; (2) Increasing the cold wall temperature T_R leads to less upflow and less separative capacity but the number of stages and the sweep vapor rates are relatively constant; (3) Increasing the sweep vapor rate N_0 increases separative capacity, decreases upflow and leaves the number of stages about the same; and (4) Increasing the film velocity V increases upflow and generally increases separative capacity. Sweep vapor rate does not change and the number of stages may increase or decrease depending on whether the range of variation includes V^*.

After examining a reasonable parameter space, the design given in Table 8 was selected. The cascade feed stream is introduced halfway up the columns in the next to bottom set. The plant is probably 11 columns in length. The number of columns in parallel at the feed point must be at least half of the ideal number shown in Table 8. Usually 70–80% of the ideal number provides a good cascade design. With regard

Table 8. Cascade Design Calculations for an Ideal Cascade of Sweep Diffusion Columns

Function	Value	Source of value
column operating conditions		
hot wall temperature, T_0, K	373	given[a]
cold wall temperature, T_R, K	337	given[a]
total pressure, kPa (atm)	101.3 (1.0)	given[a]
wall spacing, R, mm	10.0	given[a]
column dimensions		
height, Z, m	1.0	given[a]
width, W, m	0.5	given[a]
area, m²	0.5	
column separation parameters		
separability, γ	0.0220	eq. 61
stage separation factor, α-1	0.0124	eq. 64
interstage flow rate, L, kg Ne/d	2.72	eq. 63
stage length, mm	25.4	eq. 65
sweep vapor flux, N_0, kg H$_2$O/h	5.98	eq. 62
separative capacity, δU, kg Ne/d	4.11×10^{-3}	eq. 68
process gas concentrations		
at hot wall, ω_R	0.184	given[a]
at cold wall, ω_0	0.500	corresponds with T_R
ideal cascade properties		
product rate, P, kg Ne/d	1.0	given[a]
feed rate, F, kg Ne/d	19.0	material balance
separative capacity, ΔU, kg Ne/d	15.75	eq. 36
total number of columns, C_T	3,832	$\Delta U/\delta U$
total number of stages, N_T	424	eq. 27
number of columns in length, C_Z	10.8	$N_T S/Z$
feed point upflow, L_f, kg Ne/d	1,684	eq. 28
number of columns in parallel at feed point, C_f	619	L_f/L
total steam rate, kg H$_2$O/s	22,900	$C_T N_0$
total plant power, kW	15,300	

[a] Predetermined by equipment or operation.

to the discussion of the solubility of process gas in sweep vapor, handbook values of Henry's law constants for the neon–water system can be used to estimate their mutual solubility under process conditions. When this is done, the result is that the mole fraction of neon in water is approximately 4×10^{-5}. Using the N_0 and L values from Table 8, converting them to molar rates, and performing the indicated calculation gives $f = 2.4 \times 10^{-3}$ which Figure 16 shows to have a negligible effect on column separative capacity.

Conclusions. Since the sweep diffusion process involves the vaporization of the sweep vapor at operating pressure and subsequent condensation of the sweep vapor at low partial pressure, the heat flows irreversibly across the process region. It is therefore an irreversible process. It is also a continuous process since many effective separation stages can be attained in a single column. The process depends, for its elementary separation effect, on ordinary molecular diffusion against a stream of sweep vapor and it can therefore be classified as a physical process. The separation effect of the sweep diffusion process depends not only on the mass difference of the components being separated, but also on the molecular weight of the sweep vapor as shown by equations 61 and 64. To a first approximation, the separative work that can be produced by a single column is proportional to $\omega_R Z (\Delta M/M)^2 M_0^2 / (M_0 + M)^2$ where M and M_0 are the molecular weights of the process gas and sweep vapor, respectively. Therefore, it follows that the sweep diffusion process is best suited to the separation of light process gas and is most economical if heavy sweep vapors with low latent heats of vaporization are employed. A further requirement is that the process gas should be nearly insoluble in the condensed sweep vapor and that no chemical reaction takes place between them.

The sweep diffusion process has never been applied on an industrial scale but may have had some application in laboratory-scale separations for some isotopic separations. Since the process is thermodynamically irreversible, it will probably never compete with conventional separation processes such as distillation and extraction where they are applicable. For making difficult separations such as separating isotopes, sweep diffusion has two advantages which might make its use attractive. As in the case of other gas-phase separation methods, centrifugation and gaseous diffusion for example, the process gas holdup is small and therefore the equilibrium time is short, often being only a few hours before the gradient is established. The second advantage is that the equipment required for sweep diffusion is simpler and probably less expensive than gaseous diffusion stages or gas centrifuges.

The Thermal Diffusion Process

Thermal diffusion arises when a mixture is subjected to a temperature gradient leading to partial separation of the components. For example, if a homogeneous binary mixture in a closed vessel is heated at one end and cooled at the other, a relative motion of the two components is induced in the mixture, one component tending to concentrate in the hotter region, the other in the colder region. The thermal diffusion effect was first observed in a liquid solution more than a century ago, but only later has it been considered a practical method for separating the components of gaseous or liquid mixtures. In contrast to the empirical early history of the thermal diffusion effect in liquids, the presence of the effect in gases was first predicted theoretically by the kinetic theory work of Chapman (51) and Enskog (52) before it was confirmed experimentally (53) in 1917.

Process Description. Before 1938, thermal diffusion experiments were conducted only in single-stage cells. In the usual single-stage unit for liquid-phase thermal diffusion, a stationary fluid mixture is subjected to a vertical temperature gradient, the higher temperature being applied at the top of the cell in order to eliminate thermal convection currents. The lighter component of the mixture is usually, but not always, enriched at the top of the cell by thermal diffusion. The usual single-stage apparatus for gas-phase thermal diffusion consists of two bulbs, one maintained at a temperature much higher than the other, connected by a capillary tube so as to inhibit any convective mixing between them. However, the enrichment obtained in a single stage of thermal diffusion is quite small since the temperature difference that can be applied to the system is limited by practical considerations.

Although single-stage thermal diffusion equipment is quite inefficient for effecting the separation of gas mixtures, experiments of this type remain of great theoretical interest. Of the various transport phenomena in gases, thermal diffusion is the most sensitive to the nature of molecular interactions. In fact, kinetic theory predicts that the thermal diffusion effect should vanish completely for molecules that repel each other inversely as the fifth power of the distance between their centers (Maxwellian molecules) and for this reason the thermal diffusion effect was overlooked by Maxwell in his formulation of the kinetic theory of gases. Thus in gaseous thermal diffusion experiments physicists find a valuable tool for the study of intermolecular forces (54–56).

The first multistage thermal diffusion apparatus was devised in 1938 (57). In this apparatus, now commonly called a thermal diffusion column or a thermogravitational column, a confined fluid mixture is subjected to a horizontal temperature gradient. Natural thermal convection currents are thereby established, the flow being upward in the neighborhood of the hot wall and downward near the cold wall. This countercurrent flow multiplies the single-stage separation effect and concentrates at the top of the column that component which diffuses preferentially toward the hot wall and at the bottom of the column that component which diffuses preferentially toward the cold wall. The salient features of the single-stage thermal diffusion cell and of the thermal column are shown in Figure 17. The introduction of the multistage thermal diffusion column greatly stimulated the interest in thermal diffusion, and transformed thermal diffusion from a laboratory phenomenon to a practical method for achieving

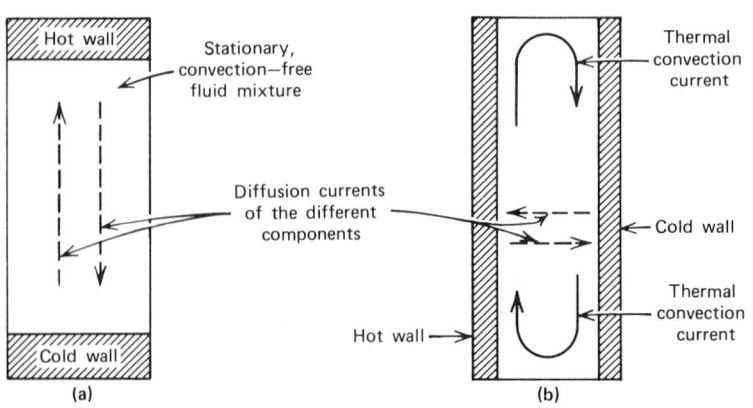

Figure 17. Schematic diagram of conventional thermal diffusion apparatuses. (a) Single-stage cell. (b) Thermogravitational column.

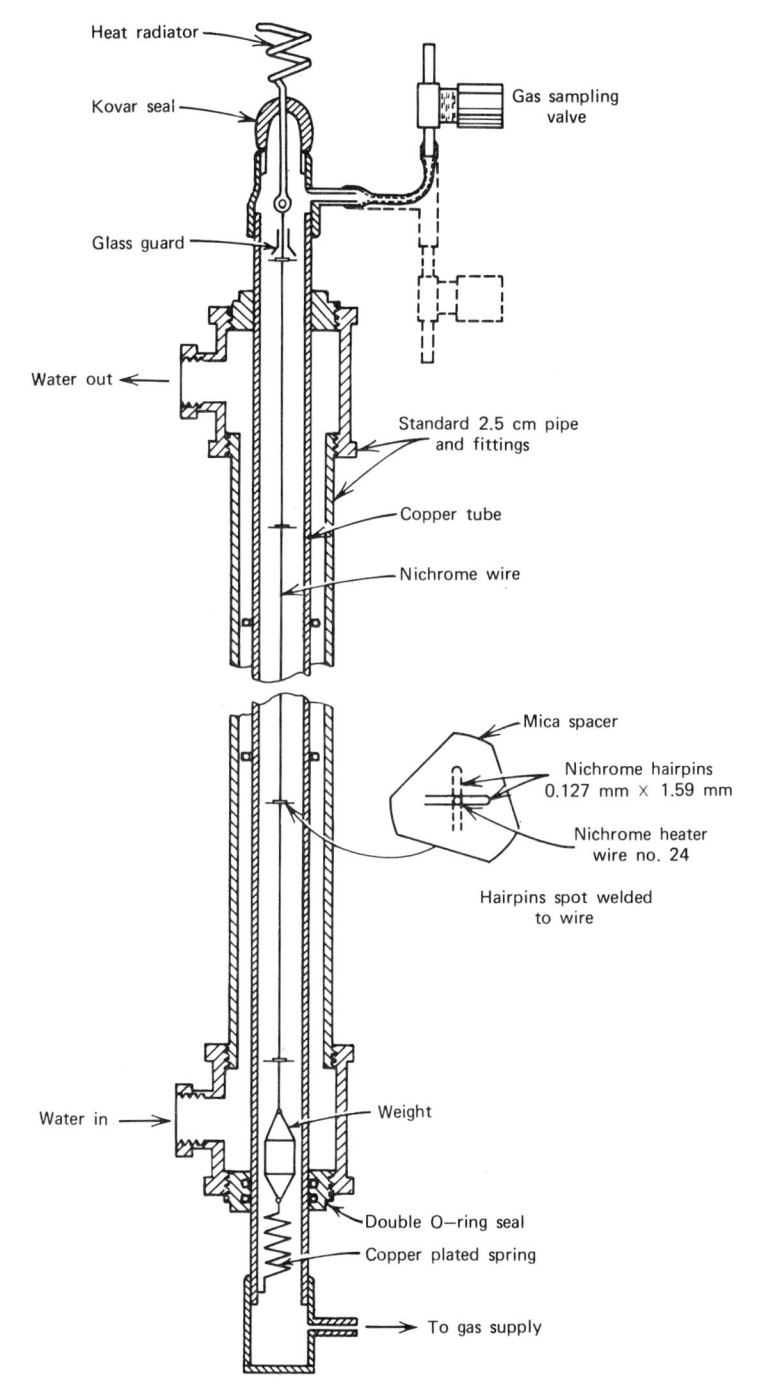

Figure 18. Details of a hot-wire thermal diffusion column.

difficult separations. Since its discovery the thermal diffusion column has been widely used for separating the components of both liquid and gaseous mixtures, including isotopes. Most of the applications of thermal diffusion, however, have been on a relatively small, or laboratory scale. The outstanding exception was the thermal diffusion plant consisting of 2100 columns, each about 15 m long (58), built at Oak Ridge, Tennessee during World War II for the separation of uranium isotopes in the liquid phase. The plant has since been dismantled.

The first thermal diffusion column (57) consisted of a vertical tube about one centimeter or more in diameter and several meters long which, cooled from the outside, served as the cold wall for the process. An electrically heated wire at the axis of the tube served as the hot wall. With the help of an apparatus of this type 36 m long, it became possible to separate the isotopes of chlorine in 1939, producing HCl which contained 99.6% [35]Cl at one end of the column and HCl which contained 99.4% [37]Cl at the other end. Columns of this type, termed hot-wire columns, are still frequently used for thermal diffusion. The details of construction of a hot-wire thermal diffusion column are shown in Figure 18. A modification of the original Clusius-Dickel column consists of a heated cylinder, concentric with the cooled outer wall, in place of the hot wire. This inner cylinder may be an electrically heated calrod, or a tube heated by a circulating fluid. The concentric-tube thermal diffusion column has the advantage of producing more separative work per unit length than the hot-wire column. The spacing between the hot and cold walls of this type of column is usually less than that in the hot-wire column, being a fraction of a centimeter for gas-phase separations and only a fraction of a millimeter for liquid-phase separations. A third type of apparatus, commonly used for liquid-phase separations, is the parallel-plate column in which the liquid mixture is confined in the space between two flat plates, one heated and the other cooled.

These thermal diffusion columns have the disadvantage that once the dimensions of the column and the operating temperatures and pressure have been set, the velocity of the countercurrent flow cannot be controlled independently. Several attempts have been made in recent years to overcome this problem. Thermal diffusion in packed columns, in rotating systems, in columns with moving walls, and with external pumping have all been investigated. Although the use of these devices may be beneficial for some specific applications, the hot-wire and concentric-tube columns, because of their basic simplicity, are the standard equipment for obtaining separations by thermal diffusion.

The thermal diffusion column has been used to achieve a great many separations. A partial list is given in Table 9. It can be seen from the table that thermal diffusion is widely used for the separation of gaseous isotopic mixtures.

The Thermal Diffusion Constant. The degree of separation that can be obtained between the hot and cold walls of a single-stage thermal diffusion cell or between the ends of a thermal diffusion column depends directly upon the value of the thermal diffusion constant for the mixture being processed. For the process in which pressure and forced diffusion effects are absent and only ordinary and thermal diffusion effects need to be considered, the basic equation for diffusion in a binary system can be written (eq. 71):

$$J_A = cD_{AB}\left[\alpha x(1-x)\frac{dT}{Tdr} - \frac{dx}{dr}\right] \tag{71}$$

This is the flux equation for thermal diffusion. The quantity α is termed the thermal diffusion constant for the mixture; it depends upon the nature of the intermolecular forces between dissimilar molecules of the mixture. When α is positive, the lighter component tends to concentrate in the hotter region. Frequently in the literature instead of values of the thermal diffusion constant, values of the Soret coefficient, usually denoted by σ, are reported for liquid mixtures and values of the thermal diffusion ratio, usually denoted by k_T, are reported for gas mixtures. These quantities are formally related to the thermal diffusion constant by (eq. 72):

$$\sigma T = \frac{k_T}{x(1-x)} = \alpha \qquad (72)$$

However, there has been little conformity in the literature regarding the sign convention; it is therefore a wise practice to determine from the experimental results in each case of interest which of the components is enriched in the hot region and thereby determine the proper sign for the thermal diffusion constant.

Estimates for Gases. The thermal diffusion constant for gases can be estimated by means of relationships derived from the kinetic theory of gases. Its value is affected by the relative masses, sizes, and shapes of the process gas molecules and is a function of the concentrations of the components in the mixture. The calculated values for α depend to some extent on the potential model assumed for the molecular interactions. For computing the value of k_T and, hence, of α for isotopic mixtures as well as for mixtures of dissimilar gases, equations can be used (59) based on the Lennard Jones 6:12 potential model. To use these equations it is necessary to know the values of the Lennard Jones potential parameters, ϵ/k and σ; however, the equations for calculating the value of the thermal diffusion constant for binary and multicomponent mixtures of dissimilar gases are rather complex and are not presented here. Generally, the theory based on the Lennard Jones potential does not predict α as accurately as it predicts the other gas transport coefficients. Other intermolecular potential models give a much better representation of the thermal diffusion constant, but tabulations of the parameters needed for calculations for specific gases are often not available. The extended law of corresponding states (60–61) gives reasonably good predictions for the isotopes of the noble gases. Selected comparisons with experimental values are given in Table 10. At this time, the kinetic theory does not provide a completely satisfactory description of the thermal diffusion behavior of mixtures of polyatomic molecules; however, the monatomic theory can be used with caution for estimating the effect. Selected values for such mixtures, based on the Lennard Jones 6:12 potential are also presented in Table 10.

Although for dissimilar gas molecules the thermal diffusion constant α is a function of concentration, this can generally be neglected for similar molecules such as isotopes and therefore the equation for calculating α for gaseous mixtures of isotopes can be written in the following simple form (eq. 73):

$$\alpha = \frac{105}{118} \frac{M_B - M_A}{M_A + M_B} k_T{}^* = \frac{M_B - M_A}{M_A + M_B} \alpha_0 \qquad (73)$$

where M_A and M_B are the molecular weights of the light and heavy components, respectively; $k_T{}^*$ is the ratio of the value of k_T evaluated from the assumed potential model to the value of k_T evaluated under the assumption that the gas is composed of rigid elastic spheres, and α_0 is the reduced thermal diffusion factor. For the Lennard

Table 9. Some Separations Effected by Thermal Diffusion

Mixtures separated	Product	CAS Registry No.	Product conc, mol % of isotope[a]	Phase	Apparatus[b]
^{35}HCl–^{37}HCl	^{35}Cl	[13981-72-1]	99.6	gas	S
	^{37}Cl	[13981-73-2]	99.4		
^{20}Ne–^{22}Ne	^{20}Ne	[13981-34-5]	"pure"	gas	S
	^{22}Ne	[13886-72-1]	"pure"		
Kr isotopes	^{85}Kr	[13983-27-2]	99.5	gas	S
	^{84}Kr	[14993-91-0]	98.2		
$^{235}UF_6$–$^{238}UF_6$	^{235}U	[15117-96-1]		liq	C
O_2 isotopes	^{18}O	[14797-71-8]	99.5	gas	C
	^{17}O	[13968-48-4]	0.5		
$^{235}U_6$–$^{238}UF_6$	^{235}U	[15117-96-1]	0.86	liq	C
N_2 isotopes	$^{15}N_2$	[7727-37-9]	1.1	gas	S
	^{14}N	[17778-88-0]	98.9	gas	S
	^{15}N	[14390-96-6]			
Ne isotopes	^{22}Ne	[13886-72-1]	93.4	gas	S
Xe isotopes	^{136}Xe	[15751-79-8]	44.8	gas	S
N_2 isotopes	$^{15}N_2$	[7727-37-9]	99.8	gas	S
petroleum products					
Li isotopes	^6Li	[14258-72-1]	8.8	liq	S
Ar isotopes	^{36}Ar	[13965-95-2]	"pure"	liq	C
	^{38}Ar	[13994-72-4]	90	gas	S
biological materials					
alloys				liq	S
Xe isotopes	^{136}Xe	[15751-79-8]	99	liq	S
	^{134}Xe	[15751-43-6]	1	gas	C
lubricating oil					
Ne isotopes	^{21}Ne	[13981-35-6]	99.6	liq	S
Ar isotopes	^{36}Ar	[13965-95-2]	99.6	gas	S
	^{38}Ar	[13994-72-4]	99.6		
N_2 isotopes	$^{15}N_2$	[7727-37-9]	12.9	gas	C
hydrocarbon isomers				gas	S
fats and oils				liq	S
^3He–^4He	^3He	[14762-55-1]	10	gas	C
$^{13}CH_4$–$^{12}CH_4$	^{13}C	[14762-74-4]	21.5	gas	C
Kr isotopes	^{85}Kr	[13983-27-2]	97.2	gas	C
	^{78}Kr	[33580-79-9]	4.2		
Xe isotopes	^{124}Xe	[15687-60-2]	0.3	gas	C
KCl–^{42}KCl				gas	S
$CaCl_2$–$^{45}CaCl_2$					
$^{12}CH_4$–$^{13}CH_4$	^{13}C	[14762-74-4]	93.4	gas	C
O_2 isotopes	^{16}O	[17778-80-2]	99.998	gas	C
N_2 isotopes	^{15}N	[14390-96-6]	97.2	gas	C
Ne isotopes	^{20}Ne	[13981-34-5]	99.95	gas	C
	^{22}Ne	[13886-72-1]	99.9	gas	S
$ZrCl_4$–$HFCl_4$					
Kr isotopes	^{85}Kr	[13983-27-2]	1000 fold depletion	gas	C
inertial guidance fluids				liq	S
O_2 isotopes	^{18}O	[14797-71-8]	80.8	gas	C
Si isotopes via SiF_4				gas	S
$H^{79}Br$–$H^{80}Br$	^{79}Br	[14336-94-8]	91.2	gas	C
Kr isotopes	^{85}Kr	[13983-27-2]		gas	C
Li isotopes	^6Li	[14258-72-1]		liq	S

Table 9 (*continued*)

Mixtures separated	Product	CAS Registry No.	Product conc, mol % of isotope[a]	Phase	Apparatus[b]
argon isotopes	^{36}Ar	[13965-95-2]	99.996	gas	C
Xe isotopes	^{131}Xe	[14683-11-5]	64	gas	
	^{131}Xe		31		
enzymes				liq	S
oxygen isotopes	^{17}O	[13968-48-4]	96	gas	C
	^{18}O	[14797-71-8]	99.9	gas	
$C_4H_9^{79}Br$–$C_4H_9^{81}Br$	^{79}Br	[14336-94-8]	72	liq	C
Ar isotopes	^{38}Ar	[13994-72-4]	95	gas	C
	^{40}Ar	[7440-37-1]	99.99		
Xe isotopes	^{124}Xe	[15687-60-2]	64		
	^{126}Xe	[27982-81-6]	13.7	gas	C
	^{128}Xe	[27818-78-6]	20		
	^{129}Xe	[13965-99-6]	80		
Kr isotopes	^{84}Kr	[14993-91-0]	92	gas	C
Ne isotopes	^{21}Ne	[13981-35-6]	90	gas	C
$Y_2(SO_4)_3$–$Dy_2(SO_4)_3$				liq	S
Xe isotopes	^{136}Xe	[15751-79-8]	99	gas	
Kr isotopes	^{85}Kr	[13983-27-2]	99.95	gas	
Kr isotopes	^{78}Kr	[33580-79-9]	99	gas	
	^{83}Kr	[13965-98-5]	70		
Ar isotopes	^{38}Ar	[13994-72-4]	99.9997	gas	
$^{12}CH_4$–$^{13}CH_4$	^{13}C	[14762-74-4]	99.9	gas	
	^{12}C	[7440-44-0]	99.999		
Xe isotopes	^{134}Xe	[15751-43-6]	50	gas	
Kr isotopes	^{82}Kr	[14191-81-2]	90	gas	
$C_6H_5^{35}Cl$–$C_6H_5^{37}Cl$	^{35}Cl	[13981-72-1]	97	liq	
	^{37}Cl	[13981-73-2]	95		
He isotope	^{3}He	[14762-55-1]	99.9999	gas	
$C^{32}S_2$–$C^{34}S_2$	^{34}S	[13965-97-4]	93	liq	
Kr isotopes	^{80}Kr	[26110-67-8]	97	gas	

[a] Where applicable.
[b] S = column; C = cascade.

Jones 6:12 potential model and for the Kestin correlation, the quantity k_T^* can be expressed as a function of the single parameters T^* which is defined as the ratio of the absolute temperature to the potential parameter, ϵ/k. Figure 19, based on values tabulated (62) for the Kestin model, presents the values of α for a broad range of values of T^*. As can be seen from Figure 19, α_0 is small at low reduced temperatures and rises to a maximum value of nearly 0.6 at T^* values of about 30. Table 10 presents a comparison of the values of the thermal diffusion factor calculated from Figure 19 and equation 73, with the experimentally determined values for some selected isotopic gas pairs. The agreement is quite good in several cases; however, in others the theoretical values tend to be somewhat high.

Figure 20 illustrates the variation of α_0 with temperature for isotopic mixtures of the several noble gases. The curves shown are derived from smoothed experimental data (62).

Estimates for Liquids. Although kinetic theory has provided a reasonably accurate basis for predicting the thermal diffusion constant for gases, it provides only a qualitative description of the thermal diffusion mechanism in the case of liquids. Attempts

688 DIFFUSION SEPARATION METHODS

Table 10. Selected Theoretical and Experimental Values of the Thermal Diffusion Constant α for Binary Gas Mixtures

Gas mixture	Conc of lighter component, mol %	$\overline{T}^a$, K	α Exptl	α Theory	Reference[b]
binary[c]					
H_2–He	50.0	330	0.192	0.150	63
H_2–N_2	42	264	0.397	0.303	64
H_2–CO	53	246	0.296	0.308	64
H_2–Ar	47	258	0.255	0.310	64
H_2–CO_2	53	370	0.361	0.385	65
He–Ne	50	330	0.388	0.326	66
He–N_2	53	261	0.430	0.398	63
He–Ar	50	330	0.372	0.392	66
Ne–Ar	50	324	0.183	0.185	66
N_2–Ar	46	252	10.073	0.053	64
Ne–Xe	50	373	0.324	0.325	67
H_2–D_2	50	358	0.154	0.183	68
H_2–HD	50	358	0.0981	0.0866	68
binary isotopic[d]					
^{3}He–^{4}He		450	0.0625	0.0787	69
^{20}Ne–^{22}Ne		450	0.0249	0.0252	70
^{36}Ar–^{40}Ar		450	0.0206	0.0196	70
^{78}Kr–^{86}Kr		450	0.0138	0.0147	70
^{124}Xe–^{136}Xe		450	0.0086	0.0101	71
$^{16}O^{16}O$–$^{16}O^{18}O$		443	0.013	0.012	72
$^{14}N^{14}N$–$^{14}N^{15}N$		408	0.011	0.014	73
$^{12}CH_4$–$^{13}CH_4$		407	0.0073	0.0096	74

[a] $T = [T_H T_C/(T_H - T_C)]\ln(T_H/T_C)$.
[b] For experimental value.
[c] Based on the Lennard-Jones 6:12 potential.
[d] Based on Kestin corresponding states model.

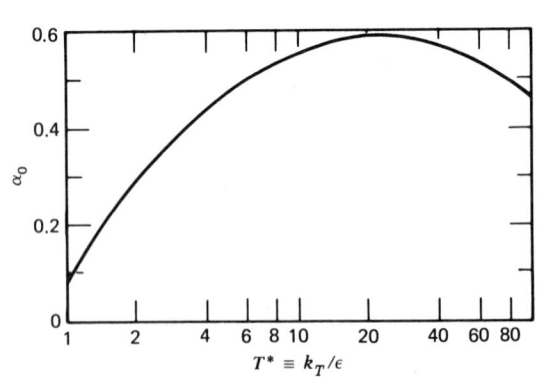

Figure 19. The reduced thermal diffusion factor for the Kestin corresponding states model.

to predict the value of the Soret coefficient, or of α, for liquid mixtures have frequently been in error by orders of magnitude. Consequently, a reasonably accurate estimate of the thermal diffusion effect in liquids can be obtained only by experiment. Reference 75 presents an extensive tabulation containing several hundred experimentally determined values of the Soret coefficient for liquid mixtures. For isotopic mixtures it

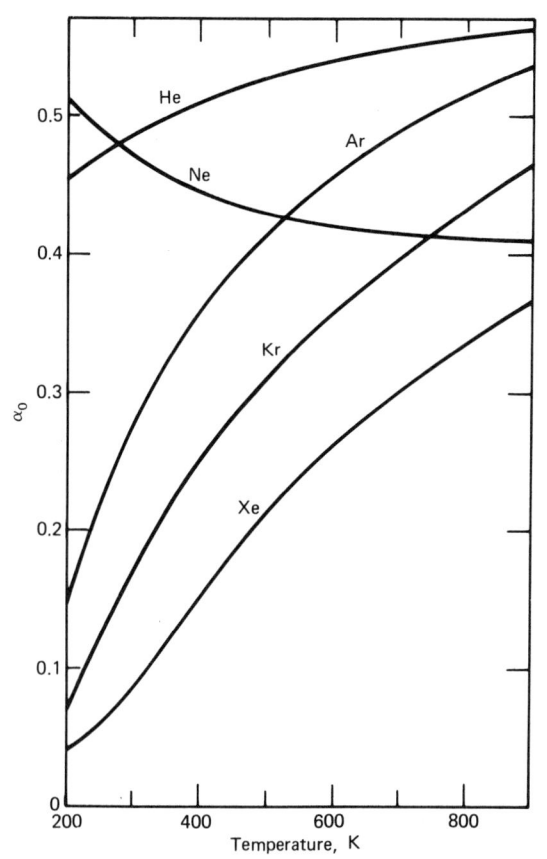

Figure 20. Variation of α_0 with temperature for isotopes of the noble gases. Smoothed experimental data of Taylor and co-workers (62).

is suggested (76) that α_0 is approximately constant at 5.4; however, recent data (77) are on the order of one-half this value.

Experimental Determination. The accuracy of the theory in predicting the value of the thermal diffusion constant for gases is greatest for the noble gases and tends to diminish with the complexity of the molecule. Therefore, it is desirable for many applications of thermal diffusion to determine the thermal diffusion constant experimentally. In the case of gases, the method commonly used involves the measurement of the concentrations attained at steady state in each of the bulbs of a simple two-bulb apparatus. Since the flux must vanish at steady state, a relationship among α, the temperature of the bulbs, and the steady-state concentration is obtained directly from the integration of equation 71 (eq. 74):

$$\alpha \ln(T_H/T_C) = \ln\left[\left(\frac{x}{1-x}\right)_H \bigg/ \left(\frac{x}{1-x}\right)_C\right] \quad (74)$$

where the subscripts C and H refer to the colder and hotter bulbs, respectively. If α is assumed to vary with temperature as $A - B/T$, then the value of α obtained from equation 74 should be associated with the temperature $T = [T_H T_C \ln(T_H/T_C)]/(T_H - T_C)$. The accuracy with which α could be determined was greatly improved by the development of a device in 1955 (78) called a swing separator. It consists of a number of parallel vertical tubes, heated at the top and cooled at the bottom, connected in series by capillary tubing, and having a small piston pump attached to the ends. A multistage effect is obtained by the use of this apparatus thus increasing the accuracy to which α can be determined.

Column Design. The derivation of the gradient equation for predicting the separative performance of a thermal diffusion column, although not complex in principle, involves extensive mathematical manipulation (79–83). Only the principal results are presented below. The procedure for obtaining an analytical expression for the gradient equation of a thermal diffusion column is the same as that used for the other column separation processes (see under The Sweep and Mass Diffusion Processes above and the Gas Centrifuge Process below). The theoretical development may be divided into the solution of the hydrodynamical problem for the convective flow in the column, and the solution to the diffusion problem for the net transport of desired component in the axial direction in the column. After solving the resulting system of equations, the gradient equation can be written in the form (eq. 75):

$$(K_c + K_d)\frac{dx}{dz} = Hx(1-x) - P(y_P - x) \tag{75}$$

for the enriching section of a column, or in the form (eq. 76):

$$(K_c + K_d)\frac{dx}{dz} = Hx(1-x) - W(x - x_W) \tag{76}$$

for the stripping section (84). The quantities H, K_c, and K_d, usually referred to as the column transport coefficients, are the parameters that determine the separative performance of the column. The coefficient H is proportional to α and is thus a measure of the thermal diffusion effect. The coefficients K_c and K_d represent convective remixing effects opposed to the thermal diffusion separating effect. A similarity in the form of these equations with the gradient equations given under Cascade Design is apparent. The two sets of equations differ only in the coefficients that represent the separation parameters. Thus, $K_c + K_d$ is equivalent to the product of the stage length and the interstage flow, and H is equivalent to the product of the stage separation factor and the interstage flow. The column transport coefficients can be computed by means of the following (eqs. 77–78):

$$H = \frac{2\pi g}{6!}\left(\frac{\alpha\rho^2}{\mu}\right)_1 r_1^4 h\left(\frac{T_2}{T_1}, \frac{r_1}{r_2}, T_1^*\right) \tag{77}$$

$$K_c = \frac{2\pi g^2}{9!}\left(\frac{\rho^3}{\mu^2 D}\right)_1 r_1^8 k_c\left(\frac{T_2}{T_1}, \frac{r_1}{r_2}, T_1^*\right) \tag{78}$$

and (eq. 79):

$$K_d = 2\pi(\rho D)_1 r_1^2 k_d\left(\frac{T_2}{T_1}, \frac{r_1}{r_2}, T_1^*\right) \tag{79}$$

The subscripts 1 and 2 refer to the cold and hot walls, respectively. The subscript 1 associated with the gas properties signifies that their values when used with these equations should be taken at the cold-wall temperature. These equations are simplified forms of more complex expressions, containing certain integrals indicated here by the lower case symbols, h, k_c, and k_d, which are generally referred to as shape factors. The shape factors, and hence the transport coefficients, are dependent upon the potential model for the gas as well as the indicated temperature and radius ratios. Shape factors based on the Lennard Jones 6:12 potential model have been evaluated and tabulated for wide ranges of the arguments T_2/T_1, r_1/r_2, and T^* (79,81). The Lennard Jones model, however, generally predicts values of α that are too large. Shape factors based on this model thus lead to values of H that are also too large. One can circumvent this problem by using an experimental value of α or a value derived from some other model as the cold wall value in equation 77.

When using a theoretical value of α based on the Lennard Jones 6:12 potential, it is consistent to use the aforementioned set of shape factors to compute the column transport coefficients. However, if one has a set of experimental α values from which an empirical α-versus-T relationship can be determined, a set of shape factors based on the inverse power repulsion law (80) can be used to incorporate these experimental data into the computation of a semiempirical value of the column transport coefficient, H.

Separation Factor. For the case of total reflux operation ($P = W = 0$) at steady state, integration of the transport equation 75 over the length of a column leads to the following result (eq. 80):

$$\ln q = HZ/(K_c + K_d) \tag{80}$$

where Z is the effective column heater length. The quantity q is referred to as the column separation factor and is defined by (eq. 81):

$$q \equiv \left(\frac{x}{1-x}\right)_T \bigg/ \left(\frac{x}{1-x}\right)_B \tag{81}$$

where T and B refer to the top and bottom of the column, respectively. The column separation factor q can be maximized by the proper choice of operating pressure. An examination of equations 77–79 show that the transport coefficient H is proportional to the square of the column pressure, K_c is proportional to the fourth power of the column pressure, and K_d is independent of the column pressure. It follows that equation 80 is pressure dependent and in fact it can be shown that the column separation factor, q, is a maximum when the column operating pressure is chosen so that K_c is equal to K_d. In batch operation a maximum value of q is usually desired but for continuous column operation another criterion applies.

Separative Capacity. The separative capacity of a thermal diffusion column operating in an ideal cascade is given by (eq. 82):

$$\delta U = \frac{H^2 Z}{4(K_c + K_d)} \tag{82}$$

The column separative capacity increases monotonically with the column operating pressure approaching an upper limit asymptotically. However, equation 82 holds only as long as the countercurrent circulatory flow is in the laminar regime. Turbulence in the column results in the column operating below its theoretical separative capacity. The pressure must be such that K_c is less than $25 K_d$ to ensure laminar internal flow (85). It is generally accepted that if column operating pressure is so chosen that K_c is equal to $10 K_d$, the separative capacity is very nearly a maximum.

To obtain maximum yield from a column or a cascade of columns operated with continuous product, reject, and feed streams, the operating pressure should be such as to yield maximum separative capacity. To obtain maximum enrichment as, for example, in batch operation at total reflux, the operation might be at such a pressure that the column separation factor q is a maximum.

Power Consumption. When choosing a process for medium- or large-scale production, the power consumption of the individual unit is usually of considerable importance; it results from conduction through the process medium and radiation from the hot to the cold wall. The power consumption of a thermal diffusion column of length Z is given by (eq. 83):

$$Q = \frac{2\pi Z \lambda_0 (T_2^{n+1} - T_1^{n+1})}{(n+1)\ln\frac{r_1}{r_2}} + \frac{2\pi r_2 Z \sigma (T_2^4 - T_1^4)}{\frac{1}{\epsilon_2} + \frac{r_2}{r_1}\left(\frac{1}{\epsilon_1} - 1\right)} \qquad (83)$$

where λ_0 and n are defined by the temperature dependency of the thermal conductivity, assumed to be $\lambda = \lambda_0 T^n$, σ is the Stefan-Boltzmann constant, ϵ is the emissivity of the given surface, and Q is the power consumption. The first term in the equation represents heat loss owing to conduction and the second term represents the loss owing to radiation. Both terms represent irreversible heat flows and hence the process is thermodynamically irreversible.

Cascade Design. For many practical applications the desired product rates and concentrations are high enough so that the separative capacity of many columns is required. Among the usual methods of interconnecting thermal diffusion columns are the following: (1) The use of a pump and two tubes connecting the top of one column with the bottom of the one ahead of it to maintain identical concentrations at these two column ends; (2) the use of two tubes, one of them heated, to obtain circulation between the column ends by thermal convection; or (3) the use of a single tube to interconnect the columns and a bellows pump at one end of the cascade to force material periodically from the top of one column to the bottom of the next one, and back again. From the standpoint of a column and its associated interconnecting tubes, the first two methods add significantly more volume to the system than does the third method. This extra volume is undesirable because it increases the process gas holdup and hence the equilibrium time of the system. The third method, which uses only a single interconnecting tube, adds less volume to the columns themselves but is more likely than the first two methods to cause losses in column separative work because of periodic remixing of appreciable amounts of gas at the ends of each column.

As an illustration of the cascade size required to perform a specific separation task by thermal diffusion, computations have been made for the design of a facility for the production of 1 kg of methane per day enriched to 0.90 mol fraction ^{13}C using concentric tube columns, 2.13 m long. The basis for the plant design is an ideal cascade. The results are summarized in Table 11.

Technology. Thermal diffusion, both liquid and gas phases, has received wide attention during the 1950s and 1960s from many theoreticians and experimentalists (75,86–92). Though observation of such a slow process as thermal diffusion in the solid state would appear both impractical and difficult, several papers have been published reporting measurements in solids.

In general the theoreticians have been notably more successful in deriving relationships between the variables affecting thermal diffusion in gases, than they have

Table 11. Characteristics of an Ideal Thermal Diffusion Cascade[a]

Characteristics	Value
column description	
cold wall radius, r_1, mm	8.64
hot wall radius, t_2, mm	5.59
heater length, Z, m	2.13
cold wall temperature, T_1, K	295
hot wall temperature, T_2, K	600
operating pressure ($K_c = 10\,K_d$), P, kPa (atm)	476 (4.7)
column transport coefficients[b]	
H, kg CH$_4$/d	0.0037
K_c, kg CH$_4$ per cm/d	0.283
K_d, kg CH$_4$ per cm/d	0.028
column separative work	
δU, kg CH$_4$/d	0.0026
cascade description	
product rate, P, kg CH$_4$/d	1.0
product concentration, y_P, mole fraction ^{13}C	0.90
separative work, ΔU, kg CH$_4$/d	88.2
total columns required, C_T, columns	33,900
ideal cascade width at feed point, columns	14,500
power requirement, kW	57,000

[a] Cascade to yield 1 kg methane per day enriched to 0.90 mol fraction in ^{13}C.
[b] Shape factors based on Lennard Jones 6:12 potential model.

been in the case of liquids. A considerable number of theoretical studies have been directed at establishing a potential function for molecular interaction in the gaseous state that adequately predicts the magnitude of the thermal diffusion constant and its temperature dependence for mixtures of dissimilar molecules and mixtures of isotopes.

Some of the experimental work with liquids has involved the measurement of Soret coefficients for binary solutions of inorganic salts. Liquids were mostly investigated in regard to the feasibility of fractionating various complex mixtures, including crude petroleum, petroleum distillates, mixtures of higher alcohols, sugars, fats, and oils, chlorinated hydrocarbons, polymers of various kinds, and viruses and antibiotically active fractions from biochemical materials. The petroleum industry has intensively studied thermal diffusion and has obtained a number of related patents chiefly covering column designs. The inherently large energy requirements of the process, however, have precluded its commercial development.

Investigations of isotope separation by thermal diffusion in the liquid state have been relatively few in number. A large-scale uranium isotope separation plant was built ca 1945, but operated only for a very short time since it proved to be uneconomical compared with gaseous diffusion. Cascades have recently been operated to separate chlorine, bromine, and sulfur isotopes. There have been some experimental studies of the separation of isomeric compounds by liquid thermal diffusion. In this type of application, since the molecular masses are identical and the separation effect depends upon differences in molecular shapes, there is no theoretical method for making a priori estimates of α.

The experimental effort in the field of gas-phase thermal diffusion has proceeded in two principal directions. Part of the work has been carried out on the separation of dissimilar gas pairs in order to evaluate and verify proposed models of potential functions describing intermolecular forces. For the bulk of this work common gases such as N_2, O_2, CO, CO_2, H_2, and the noble gases were used. The remainder of the experimental work has been directed toward the production of enriched isotopes. In this connection, the isotopes of oxygen, nitrogen, hydrogen, carbon, chlorine, bromine, boron, sulfur, silicon, and all of the noble gases have been separated using columns of the concentric-tube and hot-wire designs. A number of studies have been made to design optimum cascades to carry out certain specific isotope separations, including a cascade to enrich a dilute middle component such as ^{21}Ne, ^{38}Ar, and ^{17}O. The isotopes of the noble gases, oxygen, and carbon are now being separated in thermal diffusion facilities. When using numbers of columns in a cascade, a bundle-type arrangement of columns results in economy of space and costs.

The use of an auxiliary gas improves the isotopic separation obtained in a thermogravitational column. In batch operations this technique can be used to obtain almost complete quantitative separations of isotopes. Some work has been done on the combination of thermal diffusion and chemical exchange in gas-phase isotopic separations. The utility of this process is limited by the additional requirement of finding suitable chemical exchange reactions with favorable equilibria and reaction kinetics.

Conclusions. The separation effect in the thermal diffusion process depends on the flow of heat through the process-gas mixture from the hot to the cold wall. It is, therefore, an irreversible process and uses a large amount of power to achieve a given separation. In the column application the process is continuous since many equivalent stages of separation may be contained in a single column. Although the overall separation obtainable in a single column is quite large, the column throughput is usually small so that many columns connected in series and parallel are required to form a cascade if appreciable production is desired. The elementary separation mechanism in thermal diffusion depends on the intermolecular forces between the molecules of the process material, and the process can therefore be classified as a physical rather than a chemical one.

The thermal diffusion separation effect is largest for rigid sphere molecules and decreases as the molecules become softer, vanishing entirely for Maxwellian molecules that obey the inverse fifth-power repulsion law. The magnitude of the effect is in general molecular-weight dependent and is larger for light materials than for heavier ones. However, the effect may exist where molecular masses are identical if differences in the shapes or sizes of the molecules exist since the molecular interactions of the components would differ because of these steric effects. Investigations of the effect using various substances under a wide range of conditions serve as a valuable research tool in the study of molecular forces.

Thermal diffusion, though its effect is small, has an area of application in difficult separations such as complex mixtures of close-boiling constituents, mixtures of thermally stable isomeric compounds, and mixtures of isotopes. Since the process is thermodynamically irreversible and the column throughput is small, it is unlikely that thermal diffusion can be economically competitive with other feasible separation methods when large amounts of product are desired. However, since thermal diffusion columns are simple and relatively inexpensive and can run for long periods of time without attention, thermal diffusion can become competitive when small amounts of highly enriched product are desired.

Pressure Diffusion Processes

The development of the kinetic theory of gases led to the conclusion that a partial separation of the components of a gaseous mixture results when the gas is subjected to a pressure gradient. Thus, a column of gas standing in the earth's gravity field should show a separation effect, the lighter components concentrating at the top of the column, the heavier components at the bottom. This is indeed the case, but the effect is too slight to be utilized in a practical separation process, a column about one quarter of a mile in height would be required to give an enrichment, in the case of the isotopes of uranium, equal to that of a single gaseous diffusion stage. Therefore, in order to utilize the pressure diffusion phenomenon, steeper pressure gradients than are normally available are needed. In recent years several devices have been developed for the purpose of producing such pressure gradients, and their separating ability has been investigated. The best known is the gas centrifuge (see also Centrifugal separation). High-speed centrifuges can develop gravitational fields equal to many thousand times that of the earth. Thus relatively large pressure gradients can exist between the axis and periphery of a centrifuge giving rise to appreciable separation effects. By moving streams of gas at the periphery and at the axis countercurrently, the centrifuge can be made equivalent to a multistage separating column.

A second type of apparatus based on the pressure diffusion effect is the separation nozzle developed in the FRG. Pressure gradients in a curved expanding jet produce an isotopic separation similar to that in a centrifuge. The separation effect obtained with a single jet is relatively small, and separation nozzle stages, similar to gaseous diffusion stages, must be used in a cascade to realize most of the desired separations.

A third device that utilizes pressure diffusion is the vortex chamber. Here, as in the centrifuge, angular acceleration effects in a rapidly rotating gas provide the pressure gradient. The vortex chamber may be considered as a centrifuge with a stationary outer wall. The mechanical difficulties of high-speed rotating machinery are avoided at the expense of friction effects between the gas and the stationary wall. The literature concerning the use of such a device for isotope separation is limited. Results of experimentation indicate the effect of some of the process variables and the separation factors in H_2CO_2, H_2HD, and ^{36}Ar–^{40}Ar binary gas mixtures have been measured (93–94). A vortex tube has been used for isotope separation (95), and for the separation of gases in nuclear rocket or ram-jet engines.

With sufficient experimentation and further development, the vortex tube may provide a possible alternative method of separating isotopes, its use depending on the economics of the particular separation. However, it does appear from the data available at present, that a high compression ratio between stages may be an inherent requirement of cascades of vortex tubes. Thus cascades of this type would be characterized by an extremely high power consumption.

The Gas Centrifuge. The first suggestion that centrifugal gravitational fields might be used to effect separation of isotopes was made by Lindemann and Aston in 1919 (96). Attempts at such a separation were unsuccessful until 1934 when J. W. Beams of the University of Virginia developed the convection-free vacuum ultracentrifuge (97–99). Groth and associates (97,100–104) have reported extensively on the construction and operation of high-speed centrifuges and presented experimental data on the separation of the isotopes of argon, xenon, and uranium. Further development of larger and higher-speed centrifuges is planned. Zippe has contributed to centrifuge development projects in the USSR, FRG, and the United States since 1946. A report of his work at the University of Virginia (105) includes details of machine development and separation data on uranium isotopes. The centrifuge developed by Zippe differs appreciably in mechanical details from that developed by Groth.

Theoretical developments since 1940 are discussed in references 105–112.

Because of the increased interest in gas centrifuges and the difficulty of making quantitative experimental measurements of the rotating gas, the literature on the theory of the countercurrent flow has expanded significantly since the early 1960s. At first simplified approximate models of the flow were developed (113–115), followed by more accurate approximations (116–123). More recently, the Japanese have made a sizable contribution to the field (124–128). A range of topics is covered in a collection of papers that were presented at a colloquium in the FRG (129–130) which provide a rather complete coverage of the theory of the countercurrent flow and have a very good list of references. Other surveys can be found in references 4 and 131–133.

The Groth and Zippe Centrifuges. A schematic drawing of the ZG 5 gas centrifuge, in Figure 21 is typical of Groth's machines and gives the essential features needed to illustrate the method of operation (97,103–104). It is suspended and driven from above directly by an electric motor. The rotor spins in a vacuum-tight casing. Gas is introduced through a central tube and removed through scoop tubes at the ends of the rotor shielded by baffles from the main part of the bowl to prevent disturbance of the internal gas flow pattern. The gas is caused to undergo countercurrent axial flow by maintaining a temperature difference between the ends of the rotor. The top end of the bowl is heated by eddy currents in an aluminum ring at the top end cap; the bottom end cap is cooled by a cooling coil. Thermocouples are used to measure the end cap temperatures and the internal pressure at the centrifuge axis is measured by connections to the center tube. Labyrinth seals are used at the ends to maintain a gas seal.

The short bowl centrifuge of Zippe (105) shown in Figure 22 is somewhat simpler than that of Groth. It is supported on a needle bearing at the base and driven by an electric motor, the armature of which is a steel plate rigidly attached to the bottom of the rotor. The stator consists of a flat winding on an iron core positioned so that the poles are separated from the armature by only a small gap (about 6 mm). Power is supplied by an alternator. Zippe also used damping bearings to resist vibrations at both ends of the rotor. The centrifuge is completely closed at the bottom. The other end is connected with the top region of the outer vacuum casing only by a small annular gap around the feed tube. A small amount of gas that leaks from the interior of the bowl at the low pressure near the axis is confined to the region above the top of the rotor by a Holweck-type spiral groove molecular pump surrounding the rotor near the top, and pumped out of this region to maintain the necessary vacuum. The internal gas flow in Zippe's centrifuge experiments was induced by a difference in aerodynamic drag caused by the scoops at the ends of the bowl as modified by the baffle design or lack of baffle at each end. A large variety of baffles and scoops were tested. Some of Zippe's baffles, in contrast to Groth's, have holes near the periphery for gas to flow to and from the scoop chambers as well as the central annular holes, since the scoop and baffle systems are used to control the internal circulation rate of gas as well as to remove the product and waste streams from the centrifuge bowl. The dimensions of Groth's and Zippe's centrifuges are given in Table 12.

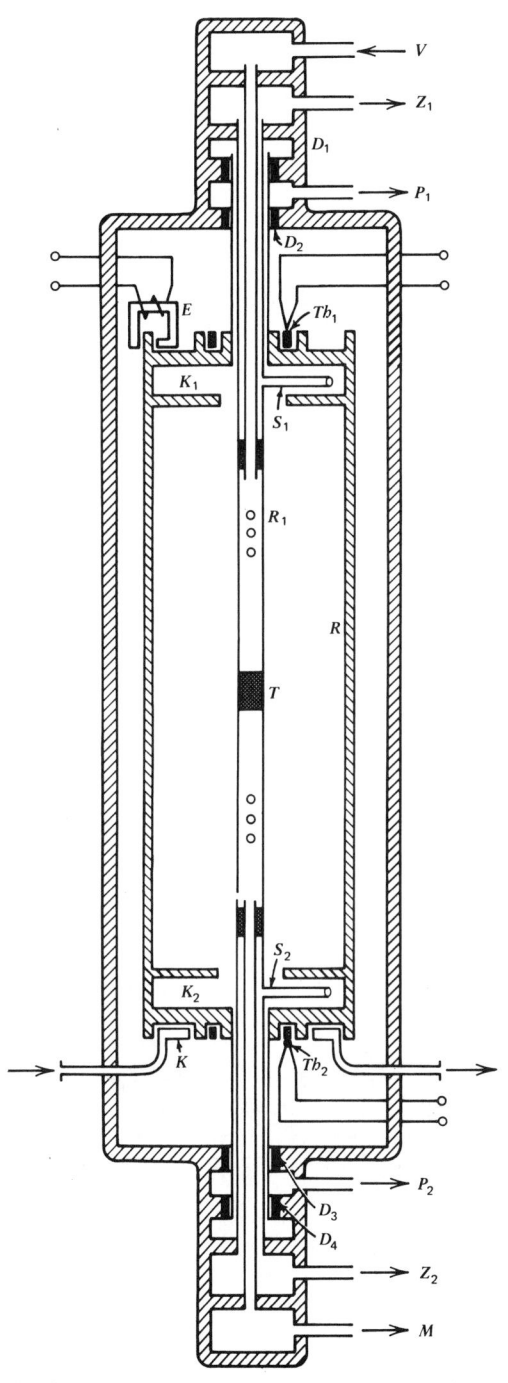

Figure 21. The Groth ZG 5 centrifuge. R, rotor; R_1, stationary shaft; T, Teflon seal; K_1, K_2, chambers for gas scoops; S_1, S_2, scoops; V, gas supply; M, manometer; Z_1, Z_2, tapping points for enriched and depleted gas; P_1, P_2, vacuum chambers; E, electromagnet for eddy current heating; Th_1, Th_2, temperature measuring devices; K, cooling coil; and D_1, D_2, D_3, D_4, labyrinth seals.

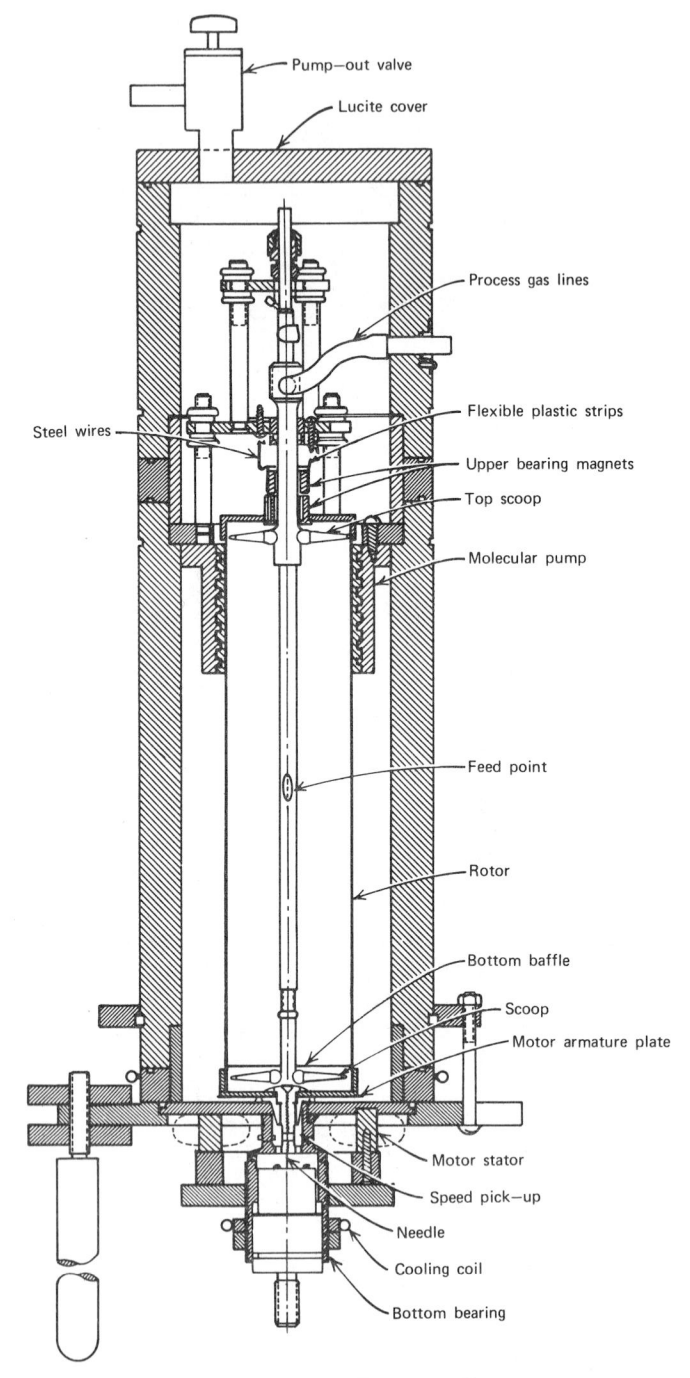

Figure 22. The Zippe centrifuge.

Table 12. Dimensions of Centrifuge Bowls[a]

Bowl	Length, cm	Radius, cm
UZ I	40	6.0
UZ IIIB	63.5	6.7
ZG 3	66.5	9.25
ZG 5	113.0	9.25
ZG 6[b]	240.0	20.0
ZG 7[b]	316.0	22.5
Zippe	30–38	3.81

[a] See refs. 103, 105.
[b] Proposed.

The maximum theoretical separative capacity of a centrifuge is proportional to length and to the fourth power of the peripheral velocity, putting a premium particularly on peripheral speed and to a lesser extent on length. The allowable speed of a cylindrical rotor is limited by the density and tensile strength of the material of construction. Groth estimated the maximum permissible velocity of thin-walled cylinders as 318 m/s for tempered steel, 366 m/s for aluminum alloy, 422 m/s for titanium, and 527 m/s for a glass fiber structure. Since the rotor must be resistant to the process gas, only certain materials may be usable with a given process gas. Both Groth and Zippe have used an aluminum alloy for rotors for use in UF_6.

Mechanical Features. The construction and operation of a precision, high-speed centrifuge in a high-vacuum environment presents some formidable mechanical problems apart from problems associated with the flow and separation phenomena. Although a number of problems have been solved in making single centrifuges operable over relatively short time periods, many other problems must be overcome before centrifuges can be constructed and operated sufficiently economically to present long-term cascade operation at a cost competitive with the other separation processes (97).

A major difficulty with high-speed rotating machinery is the critical-speed phenomena. A long rod, or its equivalent, undergoes resonant vibrations at its fundamental and higher natural frequencies. This can cause large displacements from the axis of rotation unless the rod is properly restrained at high frequencies by damping devices capable of applying sufficient restraining forces. Since essentially perfect balancing is probably not feasible, proper damping for restraining lateral movement must be used at all speeds. Zippe avoided the resonance frequency problem by limiting the ratio of length to diameter of his centrifuge to less than four so that the customary operating speeds (300–350 m/s) are below the first fundamental flexural critical frequency, a so-called subcritical centrifuge. On the other hand Groth's models, the ZG 6 and ZG 7, are long bowl or supercritical centrifuges. These run at rotational speeds above that corresponding to one or more flexural critical values and operate at a speed not too close to any of the critical values.

The major power consumption of a centrifuge at operating speed occurs in friction in the bearing systems and in gas drag on the internal parts, particularly the scoops. A poorly balanced rotor results in high power consumption. Wide variations result from variations in the number, length, diameter of tubing, and tip design of the scoops (105). The long-term maintenance and lubrication of the bearing and support systems are a major problem of centrifuge technology. The cost of power for operation and for maintaining the necessary vacuum is a major economic factor.

The scoop system in Zippe's centrifuge is used to control the internal circulation of the gas in the centrifuge, and in common with Groth's machines, must also extract a sufficient volume of gas at a pressure adequate to pump the gas to the feed point of the next centrifuge in a cascade. If this can be accomplished, no intermachine pumps are required in a cascade. This becomes an increasingly difficult problem at higher speeds since the scoop tips must be close to the centrifuge bowl wall in order to have access to the process gas at a higher pressure. The size and length of the piping in the scoops and the feed insertion tubing is critical because of limitations of their conductance for gas flow at low pressures. Other problems include long-term fatigue and creep of structural materials at high speeds and possibly stress corrosion in some systems.

Separation Data. Groth has published a large amount of data on the separation of xenon, argon, and uranium isotopes (97,100–103). Figure 23 gives the separation factor for argon at a peripheral speed of 298 m/s in the ZG 5 as a function of the throughput rate and temperature difference between the end caps. With UF_6, Groth has reportedly obtained as high as 75% of the maximum theoretical separative work. Zippe (105) obtained as high as 35% of the maximum theoretical separative work with UF_6.

Design Principles. Although the separation of fluid mixtures can be accomplished with several different types of centrifuges, the discussion of the centrifuge separation theory is here confined to the consideration of the countercurrent gas centrifuge. In order to design separation cascades consisting of countercurrent gas centrifuges, it is necessary to know the separative performance of the individual units. Gas centrifuge theory serves fairly well for predicting the performance of a single centrifuge. However, the separation behavior of a particular gas centrifuge depends on the flow pattern of the gas circulating within it which in turn depends on the geometry of any baffles and

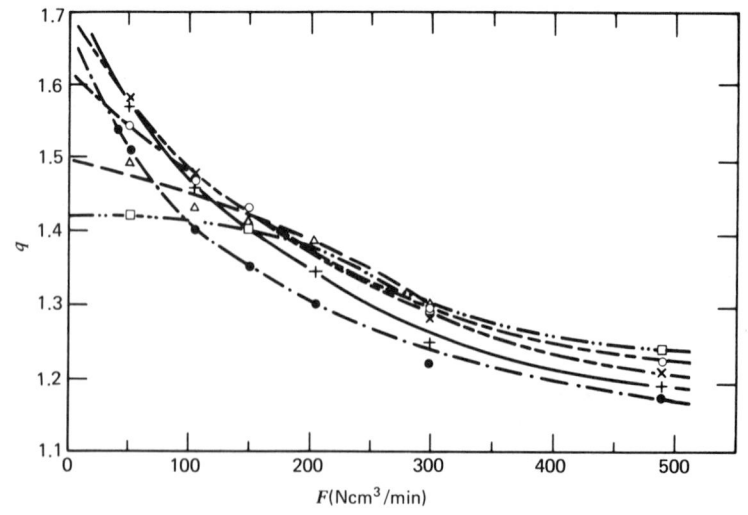

Figure 23. Enrichment q vs throughput F (ZG 5, Argon, 298 m/s). Note that for $-\cdot-\cdot-$ ●, $\Delta T = 20°C$; for —— +, $\Delta T = 25°C$; for $-\cdot\cdot-$ ×, $\Delta T = 30°C$; for $-\cdot-\cdot-$ ○, $\Delta T = 35°C$; for $----$ △, $\Delta T = 40°C$; and for $-\cdot\cdot-\cdot\cdot-$ □, $\Delta T = 45°C$.

scoops within the centrifuge bowl as well as on any temperature gradients in the gas and on the method used to introduce feed to the centrifuge. Since the effects of all these variables are both too numerous to consider, the equations for centrifuge performance are worked out only for a few idealized circulation patterns.

Radial Density and Pressure Gradients. Consider a centrifuge of length Z and of radius r_2 (these are the internal dimensions of the centrifuge bowl) that rotates at a constant angular velocity of ω radians per second. If the centrifuge contains a single pure gas rotating at the same angular velocity as the centrifuge bowl, each element of the gas has a force impressed on it by virtue of its angular acceleration. This force is directed outward in a cylindrical coordinate system and can be expressed as $(\rho \omega^2 r)(r dr d\phi dz)$. At steady state this force must be balanced by a force resulting from the radial pressure gradient established in the centrifuge bowl. The inward force on an element of the gas owing to this pressure gradient is given by $(dp/dr)(r dr d\phi dz)$ Equating these two forces, gives (eq. 84):

$$dp/dr = \rho \omega^2 r \qquad (84)$$

where p is the pressure, N/m^2 or Pa, r is the spatial coordinate in the radial direction, m, ρ is the density of the gas, kg/m^3, and ω is the angular velocity of the centrifuge, rad/s.

The pressure and the density of a gas are related by an equation of state. If the maximum pressure permitted within the centrifuge bowl is not too high, the equation of state for an ideal gas will suffice. The relationship between the pressure and density of an ideal gas is given by the well-known equation 85:

$$p = \rho RT/M \qquad (85)$$

where T is the absolute temperature of the gas, K; M is the molecular weight of the gas, kg/mol; and R is the gas constant, 8.3147 J/(mol·K). Elimination of the density from equations 84 and 85 yields the differential equation for the pressure gradient in the centrifuge (eq. 86),

$$\frac{dp}{dr} = \frac{Mp}{RT} \omega^2 r \qquad (86)$$

which, for the case of an isothermal centrifuge, is readily integrated to yield (eq. 87):

$$p(r) = p(0) e^{(M\omega^2 r^2/2RT)} \qquad (87)$$

Equation 87 gives the pressure at any point within the centrifuge, $p(r)$, as a function of the coordinate r, the pressure at the axis $p(0)$, the angular velocity of the centrifuge, and the temperature and molecular weight of the gas. Should the centrifuge contain not a single pure gas, but a gas mixture, equations of the above forms could be written for each species present. In particular for the case of a binary gas mixture, consisting of species A and B, one would have the equation 88:

$$p_A(r) = p_A(0) e^{(M_A \omega^2 r^2/2RT)} \qquad (88)$$

and 89:

$$p_B(r) = p_B(0) e^{(M_B \omega^2 r^2/2RT)} \qquad (89)$$

The ratio of these two equations gives the radial separation afforded by the gas centrifuge under equilibrium conditions, that is, for no internal gas circulation. An equilibrium separation factor between gas at the axis of the centrifuge and gas at the periphery is therefore given by (eq. 90):

$$\alpha_0 \equiv \frac{x_A(0)}{x_B(0)} \bigg/ \frac{x_A(r_2)}{x_B(r_2)} = e^{[(M_B - M_A)\omega^2 r^2 / 2RT]} \tag{90}$$

It should be noted that the separation factor for the centrifuge process is a function of the difference in the molecular weights of the components being separated rather than, as is the case in the other diffusion separation processes previously discussed, a function of their ratio. The gas centrifuge process would therefore be expected to be relatively more suitable for the separation of heavy molecules. As an example of the equilibrium separation factor of a gas centrifuge, consider the Zippe centrifuge, operating at 60°C with a peripheral velocity ωr_2 of 350 m/s. From equation 90, α_0 is calculated to be 1.0686 for uranium isotopes in the form of UF_6.

Mass Transport. An expression for the diffusive transport of the light component of a binary gas mixture in the radial direction in the gas centrifuge can be obtained directly from the general diffusion equation and an expression for the radial pressure gradient in the centrifuge. For diffusion in a binary system in the absence of temperature gradients and external forces, the general diffusion equation retains only the pressure diffusion and ordinary diffusion effects and takes the form (eq. 91):

$$J_A = -cD_{AB} \left[\frac{M_A x_A}{RT} \left(\frac{\overline{V}_A}{M_A} - \frac{1}{\rho} \right) \frac{dp}{dr} + \frac{dx_A}{dr} \right] \tag{91}$$

Since the total pressure gradient is the sum of the partial pressure gradients, the following substitution can be made in equation 91 (eq. 92):

$$\frac{dp}{dr} = \frac{dp_A}{dr} + \frac{dp_B}{dr} = \frac{(M_A p_A + M_B p_B)}{RT} \omega^2 r \tag{92}$$

and the equation for the radial flux of component A in a mixture of ideal gases is found to be (eq. 93):

$$J_A = cD_{AB} \left[\frac{(M_B - M_A) x_A (1 - x_A)}{RT} \omega^2 r + \frac{dx_A}{dr} \right] \tag{93}$$

The countercurrent centrifuge in which there is convective circulation of the gas in the axial direction will now be considered. Figure 24 is a schematic drawing of a section of a countercurrent gas centrifuge in which an arbitrary axial convective flow pattern is shown. It is assumed that the convective velocity v in the centrifuge can be expressed as a function of r only, and is independent of z. The convective velocity is assumed to be in the z direction only, and the regions at the ends of the centrifuge in which the direction of the flow is changed are neglected.

The net transport of component A in the $+z$ direction in the centrifuge τ_A is equal to the sum of the convective transport and the axial diffusive transport. At the steady state the net transport of component A toward the product withdrawal point must be equal to the rate at which component A is being withdrawn from the top of the centrifuge. Thus, the transport of component A is given by (eq. 94):

$$\tau_A = P x_P = \int_0^{r_2} 2\pi r c v x \, dr - \int_0^{r_2} 2\pi r c D_{AB} \frac{\partial x}{\partial z} \, dr \tag{94}$$

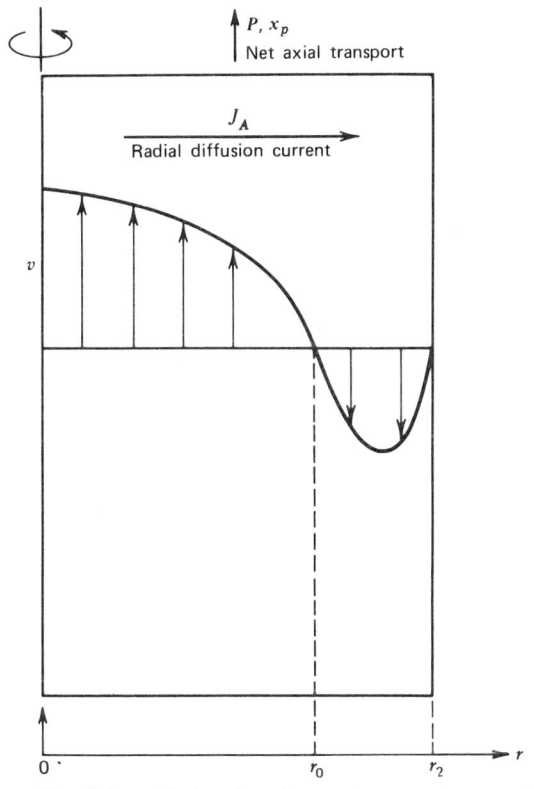

Figure 24. Schematic drawing of a countercurrent gas centrifuge.

where x is used in place of x_A for the mole fraction of component A, and the net transport of both components toward the product withdrawal point is given by (eq. 95):

$$\tau = P = \int_0^{r_2} 2\pi r c v \, dr \qquad (95)$$

where P is the product withdrawal rate, mol/s, and x_P is the concentration of component A in the product material. The integrals appearing in equation 94 are evaluated using the flux equation 93 (134). Several approximations are involved and are most satisfactory for the case of relatively long units in which the axial concentration difference is large compared with the concentration differences in the radial direction and in which the magnitudes of the feed and withdrawal rates are small with respect to the circulation rate of the internal convective flow. The results are not completely satisfactory for application to short-bowl centrifuges with relatively high throughput rates. For the case of the gas centrifuge, application of the method leads to a gradient equation for the enriching section of the centrifuge that can be written in the form (eq. 96):

$$S \, dx/dz = (\alpha - 1)x(1 - x) - (P/L)(x_P - x) \qquad (96)$$

where S is the stage length in the centrifuge and α is the stage separation factor. The quantity L is a measure of the convective circulation rate of the gas in the centrifuge and may be evaluated from the integral (eq. 97):

704   DIFFUSION SEPARATION METHODS

$$L = \int_0^{r_0} 2\pi rcv\,dr \tag{97}$$

where r_0 is the radius at which the convective velocity is equal to zero (see Fig. 24). The stage length S is the sum of two terms (eq. 98):

$$S = \frac{\int_0^{r_2} \frac{dr}{rcD_{AB}} \left[\int_0^r r'cv\,dr'\right]^2}{\int_0^{r_0} rcv\,dr} + \frac{\int_0^{r_2} rcD_{AB}\,dr}{\int_0^{r_0} rcv\,dr} \tag{98}$$

The first term may be considered as the contribution of the internal circulation or convective flow to the stage length, the second term as the contribution of the axial diffusion to the stage length. The stage separation factor is given by

$$\alpha - 1 = \frac{(M_B - M_A)\omega^2}{RT} \frac{\int_0^{r_2} r\,dr \int_0^r r'cv\,dr'}{\int_0^{r_0} rcv\,dr} \tag{99}$$

From an inspection of the preceding three equations, it is evident that for the case of a given velocity profile in which v retains its functional dependence on r but is permitted to vary in magnitude by a factor, that is, $v = bf(r)$, the convective contribution to the stage length varies directly with the magnitude of L, whereas the diffusive contribution to the stage length varies inversely with the magnitude of the circulation rate L. Thus there exists a value of L for which the stage length for the separation process is a minimum. Designating this value of L by L_0, analysis of the expression for the stage lengths shows that (eq. 100):

$$L_0 = 2\pi \int_0^{r_0} rcv\,dr \left[\frac{\int_0^{r_2} rcD_{AB}\,dr}{\int_0^{r_2} \frac{dr}{rcD_{AB}} \left(\int_0^r r'cv\,dr'\right)^2}\right]^{1/2} \tag{100}$$

The corresponding minimum value of the stage length, designated by S_0, is given by (eq. 101):

$$S_0 = \frac{2\left[\int_0^{r_2} \frac{dr}{rcD_{AB}} \left(\int_0^r r'cv\,dr'\right)^2 \int_0^{r_2} rcD_{AB}\,dr\right]^{1/2}}{\int_0^{r_0} rcv\,dr} \tag{101}$$

The preceding two equations may be used (99) to write the gradient equation for the countercurrent gas centrifuge in an alternative form. If the ratio of the actual gas circulation rate in the centrifuge to the circulation rate which minimizes the stage length L/L_0 is designated by m, then equation 96 may be rewritten (eq. 102):

$$\left(\frac{1 + m^2}{2m}\right) S_0 \frac{dx}{dz} = (\alpha - 1)x(1 - x) - \frac{P}{mL_0}(x_P - x) \tag{102}$$

For the stripping section of the centrifuge, that is, the section between the point at which the feed is introduced and the end at which the waste stream is withdrawn, the gradient equation has the corresponding form (eq. 103):

$$\left(\frac{1+m^2}{2m}\right)S_0\frac{dx}{dz} = (\alpha - 1)x(1-x) - \frac{W}{mL_0}(x - x_W) \tag{103}$$

where W is the waste withdrawal rate, mol/s, and x_W is the concentration of component A in the waste material.

Maximum Separative Capacity and the Separative Efficiency. The separative efficiency of a gas centrifuge used for isotope separation is best defined in terms of separative work. Thus, the separative efficiency E is defined by (eq. 104):

$$E = \frac{\delta U(\text{experimental})}{\delta U(\text{max})} \tag{104}$$

where δU (experimental) is the actual separative work produced per unit time by the centrifuge under consideration and δU (max) is the maximum theoretical separative capacity of the machine. The maximum separative capacity of a gas centrifuge (99) is given by (eq. 105):

$$\delta U(\text{max}) = \frac{\pi Z c D_{AB}}{2}\left(\frac{\Delta M V^2}{2RT}\right)^2 \tag{105}$$

where δU is the separative capacity in moles per unit time, Z is the length of the rotor, ΔM is the difference in the molecular weights of the components being separated, and V is the peripheral velocity of the centrifuge ($V = \omega r_2$). The expression above for the maximum separative capacity of a centrifuge indicates a desirability for:

(1) Low-temperature operation since the theoretical maximum separative capacity of a centrifuge varies inversely as the temperature.

(2) Long centrifuge bowls since the theoretical maximum separative capacity varies directly as Z. It should also be noted that $\delta U(\text{max})$ is independent of the radius of the bowl.

(3) High peripheral velocity since the theoretical maximum separative capacity varies as the fourth power of the peripheral speed. It will be seen in the next section that at the higher speeds the predicted separative capacity increases with increasing pheripheral speed much more slowly than the fourth-power relationship. Nevertheless, over the entire range of speeds investigated there is still an appreciable gain in separative capacity to be realized from an increased in speed.

Theoretical Formulation of the Separative Efficiency. The separative efficiency E of a countercurrent gas centrifuge may be considered to be the product of four factors, all but one of which can be evaluated on the basis of theoretical considerations. In this formulation the separative efficiency is defined by the equation

$$E \equiv e_C e_I e_F e_E \tag{106}$$

where e_C designates the circulation efficiency, e_I designates the ideality efficiency, e_F designates the flow pattern efficiency, and e_E designates the experimental efficiency and includes all phenomena such as turbulence and end effects not taken into account by the preceding terms. The circulation efficiency for a countercurrent gas centrifuge is given by (eq. 107):

$$e_C = m^2/(1+m^2) \tag{107}$$

As has been previously noted, m is the ratio of the rate at which gas flows upward in a centrifuge to the quantity L_0 that depends on the geometry of the bowl, the physical properties of the gas, and the flow pattern. Thus, m is directly proportional to the upflow rate. It is evident from the definition of e_C that it approaches unity as m takes on increasingly larger values. This is understood when it is realized that the circulation efficiency is representative of the loss in separative capacity owing to axial diffusion against the axial concentration gradient established in the bowl, and is in fact, equal to the ratio of the convective contribution to the stage length of the sum of the convective and diffusive contributions, that is, to the total stage length. As m increases, the convective contribution of the stage length increases proportionally, and the diffusive contribution decreases as m^{-1}; at high circulation rates the diffusive transport becomes negligible with respect to the convective transport within the centrifuge.

The ideality efficiency takes into account the difference between the shape of a centrifuge which may be regarded as a square cascade and that of an ideal cascade. As has been pointed out in the section on cascade theory, the separative capacity of an element of length in a centrifuge is the greatest when the circulation rate L through the element bears a certain relationship to the withdrawal rate, withdrawal concentration, and the concentration in the centrifuge at that point, as has been indicated by equation 28. When this condition is satisfied the cascade is termed ideal. In a square cascade, however, this condition cannot be satisfied at more than a single point in the enricher and in the stripping sections. Thus, one can associate an efficiency with each point in the cascade that is a function of the departure of the actual flow from the ideal flow. The ideality efficiency may be regarded as the average of these point efficiencies over the entire cascade.

Analysis of the gradient equations for a countercurrent gas centrifuge shows that when the withdrawal rates are optimized, the ideal efficiency assumes a maximum value of 81%. Curves of the ideal efficiency for both a stripping and enriching section of five stages ($Z/S = 5$) are shown in Figure 25. The flow model assumed in these calculations is also shown. In order to achieve this maximum value of 0.81 for the ideality efficiency, it is necessary that in addition to operating at the optimum withdrawal rates there be no mixing of gas of unlike concentrations at the feed point. The difference in the behavior of the curves for the efficiency of the enriching and stripping sections is due primarily to the fact that in the model considered the feed is assumed to enter the downflowing stream, as indicated in the figure, and therefore the flows in the enriching and stripping sections are not symmetric.

The flow pattern efficiency e_F depends solely upon the shape of the velocity profile in the circulating gas. In terms of the integrals appearing in the gradient equation, the flow pattern efficiency is given by (eq. 108):

$$e_F = \frac{4 \left(\int_0^{r_2} r\, dr \int_0^r r'cv\, dr' \right)^2}{cD_{AB} r_2^4 \int_0^{r_2} \frac{dr}{rcD_{AB}} \left(\int_0^r r'cv\, dr' \right)^2} \tag{108}$$

Clearly, in order to evaluate the flow pattern efficiency, one must have a knowledge of the actual hydrodynamic behavior of the process gas circulating in the centrifuge. Primarily because of the lack of such knowledge, the flow pattern efficiency has been evaluated for a number of different assumed velocity profiles. The results of these studies are of interest and they are enumerated below. Isothermal centrifuge operation is assumed for all cases.

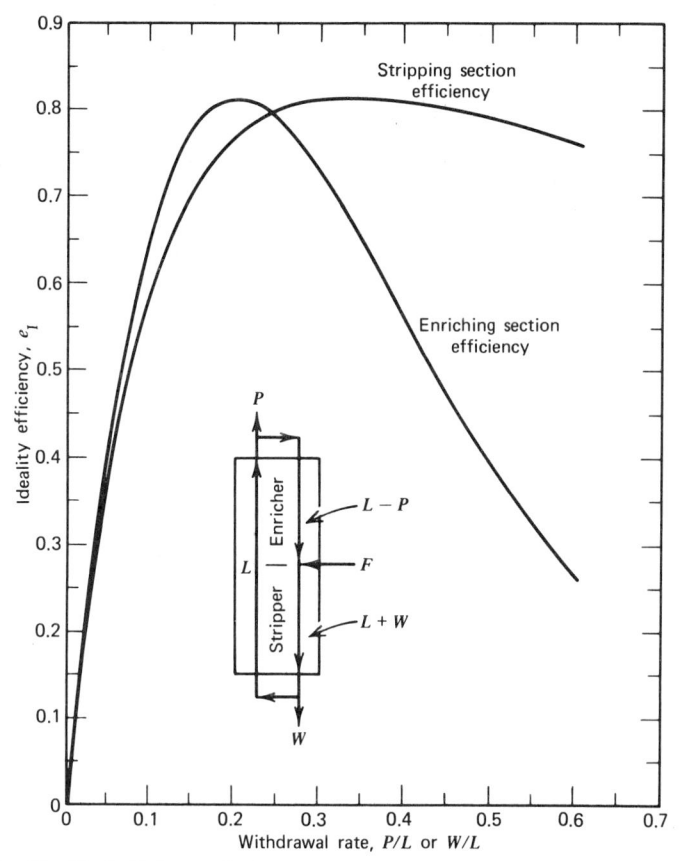

Figure 25. The ideal efficiency of a five-stage enricher and stripper as a function of the product or waste withdrawal rate.

The Optimum Velocity Profile. The optimum velocity profile (99), that is the velocity profile that yields the maximum value for the flow pattern efficiency, is one in which the mass velocity ρv is constant over the radius of the centrifuge except for a discontinuity at the wall of the centrifuge ($r = r_2$). This optimum velocity profile is shown in Figure 26a. For this case one obtains the following values for the separation parameters of the centrifuge (eqs. 109–111):

$$\alpha - 1 = \frac{1}{2} \frac{\Delta M V^2}{2 RT} \tag{109}$$

$$L_0 = 2\sqrt{2}\, \pi r_2 c D_{AB} \tag{110}$$

$$S_0 = r_2/\sqrt{2} \tag{111}$$

and $e_F = 1.0$.

The Two-Shell Velocity Profile. A second simple velocity profile (99) is shown in Figure 26b in which the flow consists of two thin streams, one situated at radius r, flowing upward, and the other situated at the wall ($r = r_2$), flowing downward. For this case the values of the separation parameters are (eqs. 112–114):

$$\alpha - 1 = \left[1 - \left(\frac{r_1}{r_2}\right)^2\right] \frac{\Delta M V^2}{2 RT} \tag{112}$$

708   DIFFUSION SEPARATION METHODS

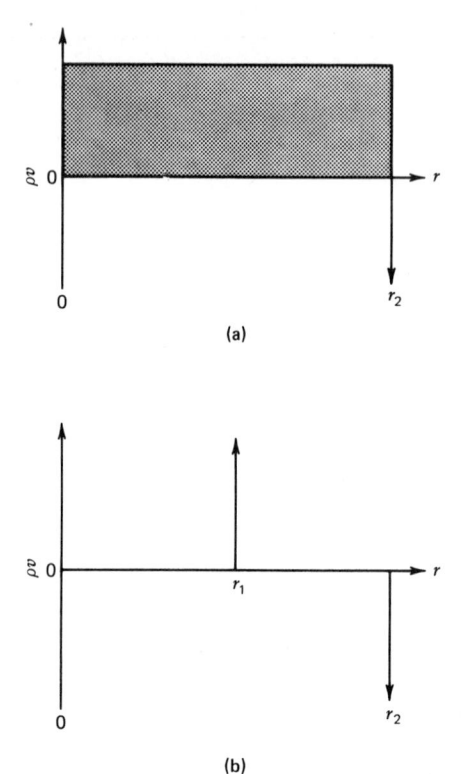

Figure 26. Hypothetical velocity profile models for a countercurrent-flow gas centrifuge. (a) The optimum velocity profile in a countercurrent gas centrifuge. (b) The two-shell velocity profile.

$$L_0 = \left[\frac{2}{\ln(r_2/r_1)}\right]^{1/2} \pi r_2 c D_{AB} \tag{113}$$

$$S_0 = [2 \ln(r_2/r_1)]^{1/2} r_2 \tag{114}$$

and (eq. 115):

$$e_F = \left[1 - \left(\frac{r_1}{r_2}\right)^2\right]^2 \bigg/ \ln\frac{r_2}{r_1} \tag{115}$$

The value of the flow pattern efficiency is shown as a function of the spacing between the streams in Figure 27. It can be seen that the flow pattern efficiency will be a maximum when the position of the upflowing stream is chosen such that r_1/r_2 is equal to 0.5335. For this particular case the flow pattern efficiency assumes the value $e_F = 0.8145$.

The simple velocity profiles considered above do not indicate directly any dependence of the flow pattern efficiency upon the rotational speed of the centrifuge. A dependence on speed is to be expected on the basis of the argument that at high speeds the gas in the centrifuge is crowded toward the periphery of the rotor and that the effective distance between the countercurrent streams is thereby reduced. It can be seen from the two-shell model that, as the position of upflowing stream approaches the periphery, the flow pattern efficiency drops off from its maximum value. This decrease of the flow pattern efficiency at high peripheral speeds is shown by calcula-

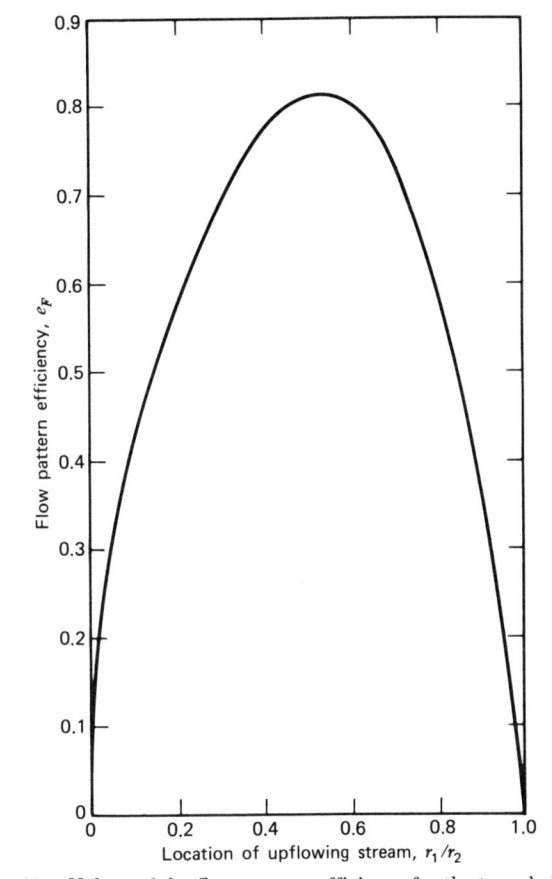

Figure 27. Values of the flow-pattern efficiency for the two-shell model.

tions of the flow pattern efficiency based upon more realistic velocity profiles than those considered above. As an example of this, flow pattern efficiencies calculated from a velocity profile hypothesized by Martin are shown below.

The Martin Velocity Profile. In a paper published in 1950, Martin (108) suggested that the velocity profile in a gas centrifuge in which the countercurrent flow is caused by a temperature difference between the circulating gas and the end caps is given by (eq. 116):

$$\int_0^r r'v dr' = \frac{\Delta T}{\omega} \left(\frac{\lambda^3}{\eta T}\right)^{1/4} \left(\frac{2\, rp(r_2)}{MRT}\right)^{1/2} e^{M\omega^2(r^2-r_2^2)/4RT} \qquad (116)$$

where λ is the thermal conductivity of the process gas, η is the viscosity of the process gas, and ΔT is the temperature difference between the gas and the end caps, one warmer, the other cooler than the gas. Equation 116 was derived by Martin by considering the flow along a heated plate in a strong gravity field. All other considerations such as coriolis forces and any resistance to the axial flow, were neglected.

The separation parameters have been calculated for a centrifuge in which the behavior of the circulationg gas is described by Martin's equation. The flow pattern efficiency is shown in Figure 28 as a function of the dimensionless parameter A, where

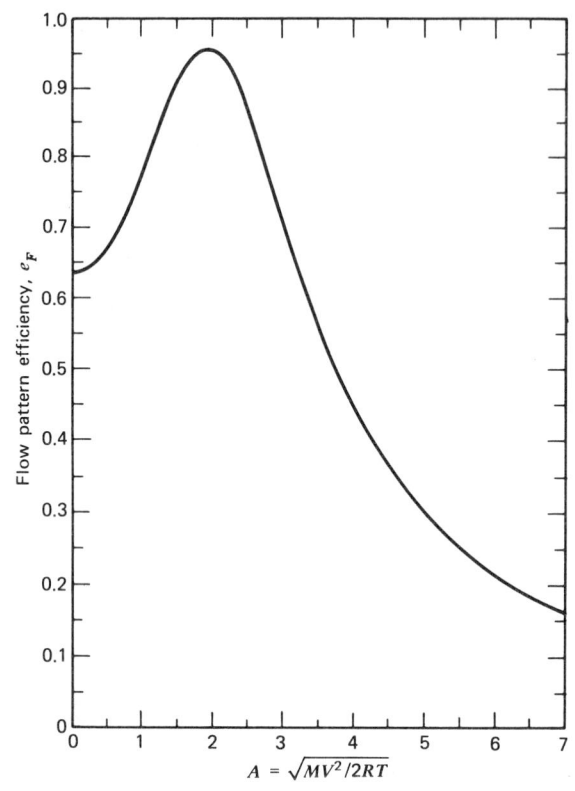

Figure 28. The dependence of the flow-pattern efficiency on the dimensionless parameter A for the Martin profile.

A is equal to $\sqrt{MV^2/2\,RT}$. In this case the maximum flow pattern efficiency attainable is 0.956.

Cascade Design. On the basis of the discussion of the separative efficiency of a single centrifuge presented above it is evident that the efficiency of a Zippe-type centrifuge, separating uranium isotopes with UF_6 as the process gas, operating at a peripheral speed of 350 m/s and at a temperature of 333 K (A = 2.8) would be expected to be (eq. 117):

$$E = \frac{m^2}{1+m^2}(0.81)(0.75) \tag{117}$$

The observed efficiency of 35% could be interpreted to mean that the circulation efficiency of this machine is about 60%, corresponding to an m value of 1.2. According to the theory presented, if the centrifuge could be operated with m = 3, it should be possible to obtain a separative efficiency of about 55%. With this assumption, the separative capacity of a single machine would be 1.68×10^{-3} kg uranium per day. A cascade for uranium isotope separation designed to produce 1 kg uranium per day of enriched uranium containing 90% ^{235}U from natural uranium would therefore require approximately 115,000 Zippe-type gas centrifuges. Table 13 shows the size of a cascade consisting of Zippe-type centrifuges required for the production of 1 kg of methane per day.

Table 13. Characteristics of an Ideal Gas Centrifuge Cascade[a]

Characteristics	Value	Source of value
centrifuge description		
length of rotor Z, cm	30.48	given[b]
diameter of rotor $2r_2$, cm	7.62	given[b]
peripheral speed $V = \omega r_2$, m/s	350	given
operating temperature T, K	333	assumed
centrifuge separative capacity		
maximum separative capacity $\delta U(\max)$, kg CH_4/d	3.4×10^{-4}	eq. 105
circulation efficiency e_C	0.90	eq. 107, $m = 3$
ideality efficiency e_I	0.81	maximum, Fig. 25
flow pattern efficiency e_F	0.70	Fig. 27, $A = 0.6$
overall efficiency E	0.51	$e_C e_I e_F$
actual separative capacity δU, kg CH_4/d	1.7×10^{-4}	$E \delta U(\max)$
cascade description		
product concentration y_P, mole fraction ^{13}C	0.90	given
feed concentration x_F, mole fraction ^{13}C	0.0106	given
waste concentration x_W, mole fraction ^{13}C	0.00955	given
product rate P, kg CH_4/d	1.0	given
feed rate F, kg CH_4/d	848	eqs. 34–35
separative capacity ΔU, kg CH_4/d	92.9	eq. 36
number of centrifuges required	550,000	$\Delta U/\delta U$

[a] Cascade to yield 1 kg methane per day enriched to 0.90 mol fraction ^{13}C.
[b] Predetermined by equipment or operation.

The Separation Nozzle Process

The separation nozzle process was developed by Becker at the Karlsruhe Nuclear Research Center in the Federal Republic of Germany for the enrichment of the light uranium isotope ^{235}U. It is also referred to as the jet diffusion method for the separation of gas mixtures. Isotopes were first separated (135) in a slit-type gas jet (136) at Oxford, in 1946. Attempts to improve the separation failed and the work was discontinued. A device for separating gaseous mixtures by jet diffusion was patented in the United States in 1952. Soon thereafter the separation effect associated with high-speed gas flow through a nozzle was investigated again by Becker and applied to the separation of isotopes. Subsequently he published a series of papers concerned with the development of the separation jet and presented experimental data on various gases and theoretical treatments of the separation mechanism, including comparisons of the process economics for separating uranium isotopes with those of alternative methods (137–139). More recent work by the German research group is described in many journal articles and KFK reports (140–159). Interest in the jet separation process also led to experimental and theoretical work in the United States (136,160–163), and in Japan (164–168).

Apparatus and Method of Operation. The separation nozzle process stage currently planned for use in a commercial implementation of the process differs appreciably from the stage used in early experimental investigations. The basic features of the separation nozzle method, incorporating the improvements developed in Germany, are illustrated in Figure 29 showing a cross-section of the separation nozzle system in current use. Gaseous uranium hexafluoride mixed with a light auxiliary gas is expanded through a nozzle into a curved flow channel. At the end of the curved flow path, after turning 180°, the stream is divided by a knife edge into two parts, an interior fraction which is enriched in ^{235}U and a wall fraction which is depleted with respect to the ^{235}U. The major differences between today's version of the process and its original form are the addition of the light auxiliary gas, substitution of the curved nozzle in place of a standard linear expansion, and the use of nozzles of smaller dimensions. The light auxiliary gas which is present in a large molar excess, increases the flow velocity of the UF_6 and, hence, it increases the centrifugal force determining the separation. In addition, the light gas delays the sedimentation of the two UF_6-isotopes in the centrifugal field slightly differently which also has a favorable effect upon the separation of the isotopes.

Usually, a mixture of 3–5 mol % UF_6 and 95–97 mol % H_2 is used as a process gas; the expansion ratios range from 1.8–2.5. According to the gas kinetic scaling relations the optimum operating pressure of the nozzle is inversely proportional to its characteristic dimensions; for example, the optimum inlet pressure of a commercial separation nozzle system with a radius of curvature of 0.1 mm is on the order of tens of kPa (tenths of atmospheres). Figure 30 illustrates the design of a commercial separation nozzle element manufactured by the Messerschmitt-Bölkow-Blohm Company, Munich, Germany. The ten slit-shaped separation nozzles are mounted on the periphery of an extruded aluminum tube. Feed gas is introduced into the segments marked F and expands through the nozzles. The heavy fraction is pumped off through the segments marked H and the light fraction is pumped off from the space around the element. Between 1968 and 1973, two prototype stages, one with 1620 m of slit length and a smaller one with 540 m of slit length have been built and successfully tested in

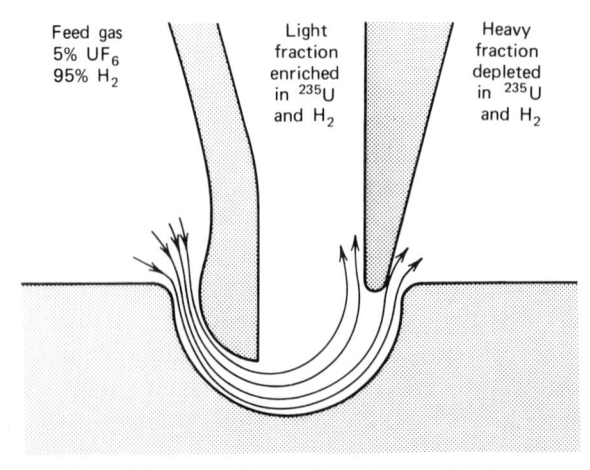

Figure 29. Cross-section of the separation nozzle system used in the commercial implementation of the separation nozzle process.

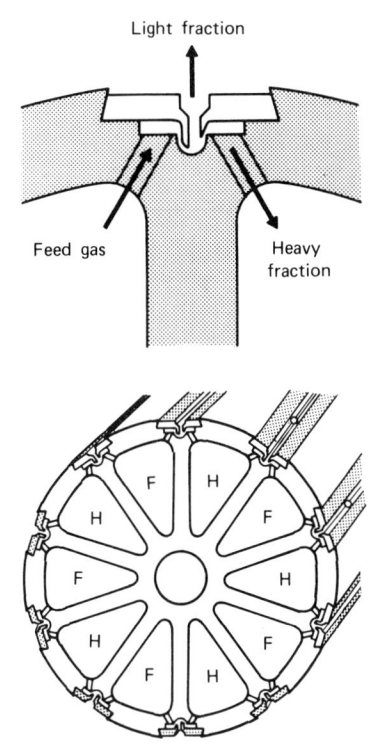

Figure 30. Schematic representation of a commercial separation element tube manufactured by the Messerschmitt-Bölkow-Blohm Company, Munich.

the Karlsruhe Center. The most important technical data of the small prototype separation stage are given in Table 14.

Theory. A good understanding of the separation phenomenon of the separation nozzle process is obtained from a very simple model that treats the separation nozzle as a gas centrifuge at steady state. It is assumed in this simplified process model that

Table 14. Characteristics of a Small Prototype Separation Nozzle Process Stage

Characteristics	Value
composition of gas, mol % H_2/UF_6	95/5
compressor inlet volume flow, m^3/s	9.6
inlet pressure, kPa[a]	19
pressure ratio	4.32
inlet temperature, °C	40
outlet temperature, °C	118
motor	
electrical power, kW	470
rotational speed, rpm	3,000
rotational speed of compressor rpm	14,000
number of separating elements	54
total slit length, m	540

[a] To convert kPa to mm Hg, multiply by 7.5.

714 DIFFUSION SEPARATION METHODS

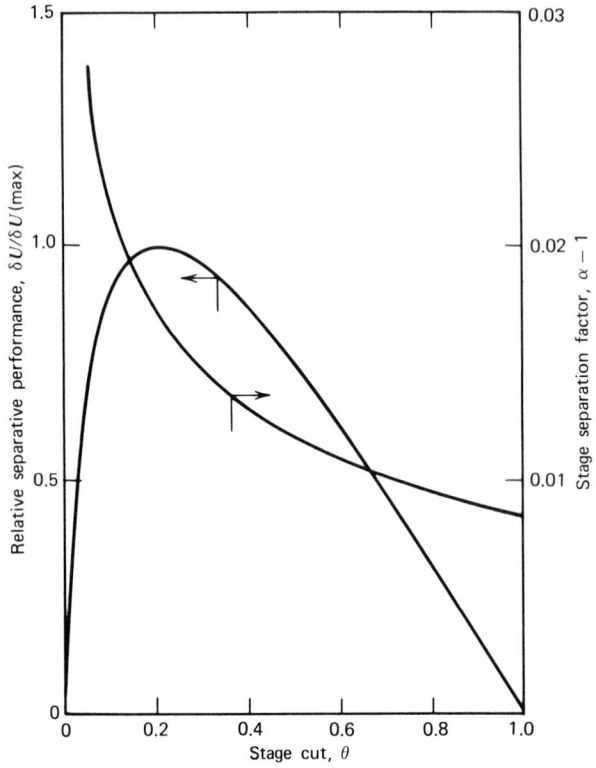

Figure 31. High-speed performance limit of an equilibrium separation nozzle.

the feed mixture traverses the circular flow path at a constant and uniform angular velocity (wheel flow) and that the peripheral velocity of the flow is equal to the sonic velocity of the entering feed gas. The separation of the isotopes is effected, as in the gas centrifuge, by pressure diffusion in the pressure gradient resulting from the curved streamlines and the associated centrifugal forces. One important function served by the light auxiliary gas in the feed is to increase the flow velocity of the mixture and hence the magnitude of the separation factor attained as shown in Figure 31.

When the gas speed is sufficiently high, the separation factor corresponding to a given value of the cut is essentially independent of the gas velocity and, hence, at high speeds, is given (146) to a good approximation as (eq. 118):

$$(\alpha - 1) \cong \left(\frac{1}{1-\theta} \ln \frac{1}{\theta}\right) \frac{\Delta M}{M} \qquad (118)$$

where θ is the stage cut defined as the fraction of the uranium in the feed stream to the separation nozzle that is withdrawn in the light or enriched product stream and $\Delta M/M$ is the fractional difference in the molecular weights of the isotopic species being separated. In the case of uranium isotopes with UF_6 as the process gas, $\Delta M/M = 3/352 = 0.0085$. The separative work produced by the stage is

$$\delta U = F \frac{\theta(1-\theta)}{2} (\alpha - 1)^2 \qquad (119)$$

where F is the feed rate of uranium to the separating unit and δU is the amount of separative work produced by the nozzle system per unit time. The separation performance of the separation nozzle in the limit of high gas speed, as described in equations 118 and 119, is shown in graphical form in Figure 31. It can be seen than an equilibrium separation nozzle produces its maximum separative work rate at a cut of about 0.2.

The secondary enrichment effects mentioned above, caused by ternary diffusion involving the light auxiliary gas, are treated in some detail in references 169 and 170.

Cascade for Uranium Enrichment. At the London Conference in 1975 (145), STEAG (Steinkohle-Electrizität AG) representatives who are collaborating on the development of the separation nozzle process in Germany presented the descriptions of a uranium isotope separation plant given in Table 15 and Figure 32. The design is for a 5 million SWU/yr plant producing uranium enriched to 3% ^{235}U and stripping the feed to 0.3% ^{235}U in the waste stream. The 570 stages and 2520 MW required by the 5 million SWU/yr nozzle plant should be compared with the 1180 stages (to span the range from 0.25–4.0% ^{235}U) and 2430 MW required by the 8.75 million SWU/yr gaseous diffusion plant shown in Figure 12.

Table 15. Conceptual Design Data For a 5 Million SWU/YR Plant

Characteristics	Value	
separating element[a]		
N_0 (mol % UF_6)	4.2	
expansion ratio, π	2.1	
uranium cut, ϑ_u	1/4	
elementary separation effect E_A (%)	1.48	
separation nozzle inlet pressure, kPa[b]	38.7	
separation nozzle outlet pressure, kPa[b]	18.4	
separation stage	*Large*	*Small*
suction flow of compressor, m³/h	1,030,600	312,000
rated power of compressor motor, kW	5,500	1,720
separating element slit length, m	26,660	8,050
separative work capacity, SWU/yr	12,800	3,900
cascade		
mass flow		
product[c], kg U/yr	1,585,000	
waste[d], kg U/yr	9,452,000	
feed[e], kg U/yr	11,037,000	
number of large stages	384	
number of small stages	186	
net separative work capacity, kg U/yr	5,045,000	
total energy requirement of plant, MW	2,520	

[a] Hydrogen is used as a carrier gas.
[b] To convert kPa to mm Hg, multiply by 7.5.
[c] 3% ^{235}U.
[d] 0.337% ^{235}U.
[e] Natural uranium.

Direction of Current Research. The discussion so far has described the separation nozzle process concept in which the flow is turned 180° during its passage through an expansion nozzle. This concept is illustrated in Figure 29. Current process research

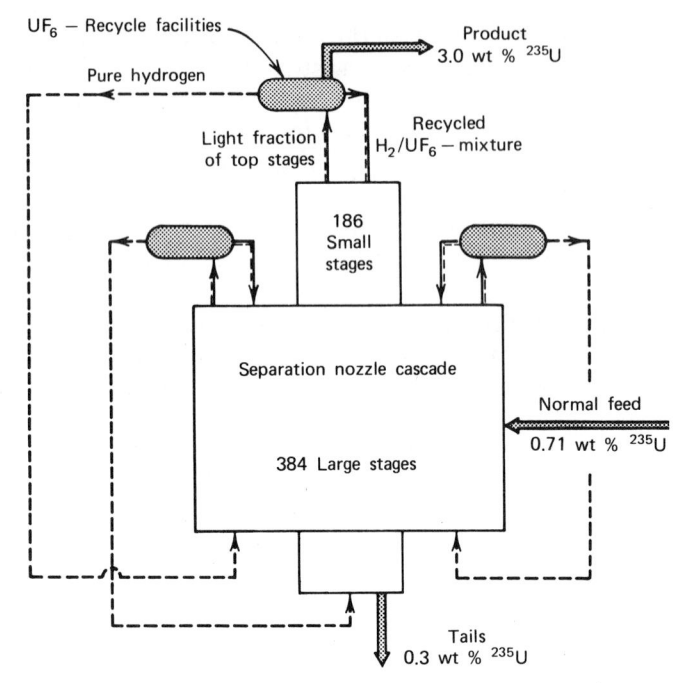

Figure 32. Separation nozzle cascade of 5 million SWU/yr.

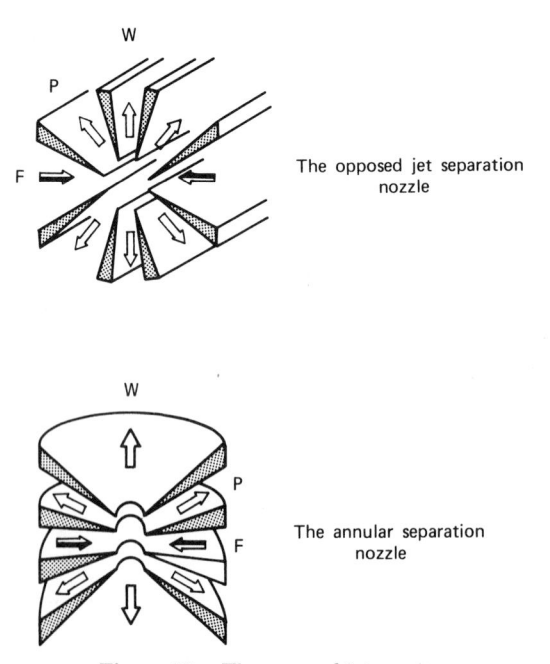

Figure 33. The opposed jet process.

is directed toward evaluation of an opposed jet concept which, it is hoped, will show an advantage over the 180° turning jet concept. The opposed jet concept is shown in Figure 33. An obvious problem that must be solved if the concept is to be useful is the tendency of the opposed jets to deflect each other and pass "shoulder to shoulder" rather than to impact each other squarely. The annular separation nozzle, also shown in Figure 33, is a newly developed configuration currently being investigated at the Karlsruhe Research Center; it does not exhibit the flow instability problem found with the original opposed jet configuration.

Nomenclature

Some symbols that appear in the text only once and are clearly defined at that time are not listed. Dimensions are given in terms of the mass, M, length, L, time, t, temperature, T, and moles.

α = thermal diffusion constant, eq. 71.
C_F = cost of feed material, eq. 14, \$/M or \$/mol
$C_{\Delta U}$ = cost of separative work, eq. 14, \$/M or \$/mol.
C_{total} = total cost of enriched product, eq. 14, \$/M or \$/mol.
c = total molar concentration, mol/L^3.
D_{AB} = binary diffusion coefficient for the pair $A - B$, L^2/t.
D_0 = mean diffusion coefficient of process gas mixture in sweep vapor, eq. 61, L^2/t.
E = overall efficiency of a separation process, eq. 104, dimensionless.
E_b = barrier efficiency in gaseous diffusion, eq. 51, dimensionless.
E_c = cut correction in gaseous diffusion, eq. 54, dimensionless.
E_M = mixing efficiency in gaseous diffusion, eq. 53, dimensionless.
E_p = back-pressure efficiency in gaseous diffusion, eq. 52, dimensionless.
e_C = circulation efficiency of a centrifuge, eq. 107, dimensionless.
e_E = experimental efficiency of a centrifuge, eq. 106, dimensionless.
e_F = flow pattern efficiency of a centrifuge, eq. 108, dimensionless.
e_I = ideality efficiency of a centrifuge, eq. 106, dimensionless.
F = feed flow rate to a unit or cascade, eq. 34, M/t or mol/t.
f = solubility fraction of process gas in sweep vapor, eq. 69, dimensionless.
H = steady-state enriched inventory of desired component in a cascade, eq. 42, M or moles.
H = transport coefficient in thermal diffusion, eq. 77, mol/t.
h = shape factor measuring thermal diffusion effect, eq. 77, dimensionless.
h_n = total material holdup of the nth stage, eq. 42, M or moles.
J_A = molar flux of component A in a binary mixture relative to the molar average velocity, eq. 71, mol/tL^2.
K_c = transport coefficient for convective remixing, eq. 78, mol L/t.
K_d = transport coefficient for back-diffusional remixing, eq. 79, mol L/t.
k_T = thermal diffusion ratio, eq. 72, dimensionless.
k_c = shape factor for convective remixing, eq. 78, dimensionless.
k_d = shape factor for back-diffusional remixing, eq. 79, dimensionless.
L = upflow rate of process gas in a stage, eq. 3, M/t or mol/t.
L_{ideal} = stage upflow of process gas in an ideal cascade, eq. 28, M/t or mol/t.
$L_{\min}$ = minimum stage upflow of process gas, eq. 22, M/t or mol/t.
L_0 = value of interstage flow rate which minimizes the stage length, eq. 100, M/t or mol/t.
l_f = thickness of effective stagnant film on gaseous diffusion barrier, eq. 53, L.
M_i = molecular weight of component i.
m = ratio of actual interstage flow rate to interstage flow rate which minimizes the stage length, eq. 102, dimensionless.
N_A = molar flux of component A with respect to stationary coordinates, mol/$L^2 t$.
N_{ideal} = number of stages required to span a given concentration span in an ideal cascade, eq. 27, dimensionless.
$N_{\min}$ = minimum number of stages required to span a given range of concentrations, eq. 21, dimensionless.
N_{sect} = number of stages required to span the concentration range of a square section, eq. 23, dimensionless.

718 DIFFUSION SEPARATION METHODS

n = stage counting index, eq. 17, dimensionless.
P = product flow rate of a unit or cascade, eq. 34, M/t or mol/t.
p = fluid pressure, $M/t^2 L$.
p_b = fluid pressure on the low-pressure side of the gaseous diffusion barrier, eq. 46, $M/t^2 L$.
p_c = core region pressure, eq. 118, $M/t^2 L$.
p_f = fluid pressure on the high-pressure side of the gaseous diffusion barrier, eq. 46, $M/t^2 L$.
p_p = peripheral pressure, eq. 118, $M/t^2 L$.
Q = power consumption of thermal diffusion column, eq. 83, ML^2/t^3.
q = thermal diffusion column separation factor, eq. 80, dimensionless.
R = gas constant, $ML^2/t^2 T$ mol.
R = sweep diffusion column wall spacing, eq. 62, L.
r = pressure ratio p_b/p_f across the gaseous diffusion barrier, eq. 52, dimensionless.
r_0 = transverse coordinate of the plane of zero axial velocity, eq. 97, L.
r_2 = radius of a centrifuge bowl, eq. 87, L.
S = stage length of a continuous process, L.
s = mole fraction solubility of process gas in condensed sweep vapor, eq. 69, dimensionless.
T = absolute temperature, T.
T_{eq} = equilibrium time of a cascade or separating unit, eq. 42, t.
δU = separative capacity of a unit, eq. 3, M/t or mol/t.
ΔU = separative capacity of a cascade, M/t or mol/t.
V = peripheral velocity of a centrifuge, eq. 105, L/t.
$\overline{V}_A$ = partial molal volume of a component A in a mixture, eq. 91, L^3/mol.
$v(x)$ = value function, eq. 12, dimensionless.
$v'(x)$ = first derivative of the value function with respect to concentration, eq. 13, dimensionless.
$v''(x)$ = second derivative of the value function with respect to concentration, eq. 11, dimensionless.
$v, v(r)$ = axial velocity in a column, L/t.
W = waste flow rate from a unit or cascade, eq. 34, M/t or mol/t.
x = mol fraction of desired component of a binary mixture in the downflow or depleted stream, dimensionless.
x_F = mol fraction of desired component of a binary mixture in the feed stream of a unit or cascade, dimensionless.
x_i = mol fraction of component i in a mixture, dimensionless.
x_W = mol fraction of desired component of a binary mixture in the waste stream of a unit or cascade, dimensionless.
y = mol fraction of desired component of a binary mixture in the upflow or enriched stream, dimensionless.
y_P = mol fraction of desired component of a binary mixture in the product stream of a unit or cascade, dimensionless.
Z = overall length of the separating column, L.
z = mol fraction of desired component of a binary mixture in the feed stream of a stage eq. 4, dimensionless.
z = axial distance or length coordinate in a column, L.
α = stage separation factor, eq. 1, dimensionless.
α = thermal diffusion constant, eq. 71, dimensionless.
α^* = ideal stage separation factor, eq. 49, dimensionless.
α_0 = equilibrium stage separation factor in a centrifuge, eq. 90, dimensionless.
γ = separability, eq. 61, dimensionless.
θ = fraction of stage feed which goes into the stage upflow, "cut," eq. 3, dimensionless.
μ = fluid viscosity, M/Lt.
ρ = mass density, M/L^3.
σ = Soret coefficient in thermal diffusion, eq. 72, $1/T$.
$\overline{\tau}$ = average net upward transport of desired material in a cascade approaching equilibrium, eq. 42, M/t or mol/t.
ω = angular velocity, eq. 84, radians/t.
ω_0, ω_R = mol fraction of process gas at the hot and cold walls, respectively, of a mass or sweep diffusion column, eq. 62, dimensionless.

BIBLIOGRAPHY

"Diffusion Separation Methods" in *ECT* 1st ed., Vol. 5, pp. 76–133, by Manson Benedict, Hydrocarbon Research, Inc.; "Diffusion Separation" in *ECT* 1st ed., 2nd Suppl., pp. 297–315, by K. B. McAfee, Jr., Bell

Telephone Labs.; "Diffusion Separation Methods" in *ECT* 2nd ed., Vol. 7, pp. 91–175, by J. Shacter, E. Von Halle, and R. L. Hoglund, Union Carbide Corporation.

1. M. Benedict and co-workers, *Report of Uranium Isotope Separation Review Ad-Hoc Committee*, ORO-694, U.S. Atomic Energy Commission, June 2, 1972.
2. *AEC Gaseous Diffusion Plant Operations*, ORO-684, U.S. Atomic Energy Commission, Jan. 1972.
3. U.S. Pat. 3,925,036; (March 23, 1951), J. Shacter (to U.S.A. as represented by the United States Energy Research & Development Administration).
4. E. Von Halle, *The Countercurrent Gas Centrifuge for the Enrichment of U-235*, K/OA-4058, Union Carbide Nuclear Div., Oak Ridge, Tenn., Nov., 1977.
5. P. R. Vanstrum and S. A. Levin, *New Process for Uranium Isotope Separation*, paper presented at IAEA meeting, Salzburg, Austria, 1977. Union Carbide Nuclear Div., Oak Ridge, Tenn., IAEE-CN-36/12 (II.3), 1977.
6. S. Blumkin, *Survey of Foreign Enrichment Capacity, Contracting and Technology, 1976*, K/OA-2547, Part 5; Union Carbide Nuclear Div.; Oak Ridge, Tenn., Oct. 30, 1978.
7. A. Kanagawa, I. Yamamoto, and Y. Mizuno, *J. Nucl. Sci. Tech.* **14,** 892 (1977).
8. G. Jansen and J. L. Robertson, *Analysis of Nonideal Asymmetric Cascades*, paper presented at American Chemical Society Meeting, Montreal, May 1976, to be published in *Can. J. Chem. Eng.*
9. G. A. Garrett and J. Shacter, *Proceedings of the International Symposium of Isotope Separation Amsterdam*, 1958, pp. 17–31.
10. G. R. H. Geoghegan in ref. 9, pp. 518–523.
11. H. Barwich, *Ann. Physik* **20,** 70 (1957).
12. E. Oliveri, *Energia Nucl. Milan* **8,** 453 (1961).
13. J. C. Guais, *BNES Intern. Conference on Uranium Isotope Separation*, Paper 21, London, March 5–7, 1975.
14. N. Ozaki and I. Harada, *BNES Intern. Conference on Uranium Isotope Separation*, Paper 22, London, March 5–7, 1975.
15. G. Hertz, *Z. Physik* **79,** 108 (1932).
16. H. Harmsen, G. Hertz, and W. Schütze, *Z. Physik* **90,** 703 (1934).
17. D. E. Wooldridge and W. R. Smythe, *Phys. Rev.* **50,** 233 (1936).
18. H. D. Smyth, *Atomic Energy for Military Purposes,* Princeton Univ. Press, Princeton, N.J., 1945.
19. P. C. Keith, *Chem. Eng.* **53,** 112 (1946).
20. J. F. Hogerton, *Chem. Eng.* **52,** 98 (1945).
21. K. E. B. Jay, *Britain's Atomic Factories,* H. M. Stationery Office, London, 1954.
22. H. Albert, *Proceedings of the U.N. International Conference on Peaceful Uses of Atomic Energy, 2nd Geneva,* Vol. 4, P/1268, 1958, pp. 412–417.
23. *Major Activities in the Atomic Energy Programs,* Jan.–Dec., 1962, U.S. Gov't. Printing Office, Washington, D.C., 1963.
24. *Electr. World* **149,** 38(1958).
25. D. Massignon in ref. 22, P/1266, pp. 388–394.
26. J. Charpin, P. Plurien, and S. Mommejac in ref. 22, P/1265, pp. 380–387.
27. C. Frejacques and co-workers in ref. 22, P/1262, pp. 418–421.
28. M. Martensson and co-workers in ref. 22, P/181, pp. 395–404.
29. O. Bilous and G. Counas in ref. 22, P/1263, pp. 405–411.
30. W. G. Pollard and R. D. Present, *Phys. Rev.* **73,** 762 (1948).
31. P. C. Carman, *Flow of Gases through Porous Media,* Butterworths Publications Ltd., London, 1956.
32. R. D. Present and A. J. de Bethune, *Phys. Rev.* **75,** 1050 (1949).
33. H. T. C. Pratt, *Countercurrent Separation Processes,* American Elsevier Publishing Company, Inc., New York, 1967.
34. C. Boorman in H. London, ed., *Separation of Isotopes,* George Newnes Ltd., London, 1961, Chapt. 8.
35. *AEC Gaseous Diffusion Plant Operations,* USAEC Report No. ORO-684, U.S. Atomic Energy Commission, Washington, D.C., Jan. 1972.
36. *AEC Data on New Gaseous Diffusion Plants,* USAEC Report No. ORO-685, U.S. Atomic Energy Commission, Washington, D.C., Apr. 1972.
37. Dept. of Energy, *Report ORO-685,* Oak Ridge Operations, Oak Ridge, Tenn., 1972.

38. *CEA Bulletin d'Informations Scientifiques et Techniques,* No. 206, 3-134, Sept. 1975 (in French).
39. G. Besse, "The Eurodif Program—Present Status of the Project," *BNES International Conference on Uranium Isotope Separation,* Paper 17, London, March 5–7, 1975.
40. J. P. Gougeau, *Developments in Uranium Enrichment, AIChE Symposium Series 169,* Vol. 73, 1977, pp. 12–14.
41. G. Hertz, *Z. Physik.* **23,** 433 (1922).
42. U.S. Pats. 1,486,521 (March 11, 1924), and 1,498,097 (June 17, 1924), G. L. Hertz (to Naamlooze Vennootschap Philips' Gloeilampenfabriken, Eindhoven, Netherlands).
43. G. Hertz, *Z. Physik.* **91,** 810–815 (1934); transl. as Brit. Rept. IGRL-T/CA **73,** (1958).
44. C. C. Maier, *Mechanical Concentration of Gases,* U.S. Bur. Mines Bull. 431, 1940.
45. C. H. Forsberg, *A Technical and Economic Study of Uranium Enrichment by Mass Diffusion, Ph.D. thesis,* Massachusetts Institute of Technology, 1973.
46. M. Benedict and A. Boas, *Chem. Eng. Progr.* **47,** 51, 111 (1951).
47. M. T. Cichelli, W. D. Weatherford, Jr., and J. R. Bowman, *Chem. Eng. Progr.* **47,** 63, 123 (1951).
48. D. Heymann and J. Kistemaker, *J. Chem. Phys.* **24,** 165 (1956).
49. L. G. Gverdtsiteli, R. Y. Kucherov, and V. K. Tskhakaya in ref. 22, P/2086, pp. 608–618.
50. R. S. Mayer, *Continuous Mass Diffusion in a Natural Convection Column, Ph.D. thesis,* Univ. of Michigan, 1958; Dissertation Abstr. 19, 496.
51. S. Chapman, *Phil. Trans. R. Soc. London Ser. A* **216,** 279 (1916); **217,** 115 (1917); *Proc. R. Soc. London Ser. A* **93,** 1 (1916).
52. D. Enskog, *Z. Physik.* **12,** 55, 533 (1911); D. Enskog, *The Kinetic Theory of Phenomena in Dilute Gases, Ph.D. thesis,* Upsala Univ., Sweden, 1917.
53. S. Chapman and F. W. Dootson, *Phil. Mag.* **33,** 248 (1917).
54. K. E. Grew and T. L. Ibbs, *Thermal Diffusion in Gases,* Cambridge University Press; Cambridge, Mass., 1952.
55. E. A. Mason, *J. Chem. Phys.* **27,** 75, 782 (1957).
56. S. C. Saxena, *Indian J. Phys.* **29,** 131, 453 (1955).
57. K. Clusius and G. Dickel, *Naturwissenschaften* **26,** 546 (1938).
58. P. H. Abelson and J. I. Hoover, *Proc. Intern. Symp. Isotope Separation Amsterdam, 1957,* 1958, pp. 483–508.
59. J. O. Hirschfelder, C. F. Curtiss, and R. B. Bird, *Molecular Theory of Gases and Liquids,* John Wiley & Sons, Inc., New York, 1954, Chapt. 8.
60. J. Kestin, S. T. Ro, and W. Wakeham, *Physica Utr.* **58,** 165 (1972).
61. J. Kestin and E. A. Mason, *AIP Conf. Proc.* **11,** 137 (1973).
62. W. L. Taylor, *J. Chem. Phys.* **62,** 3837 (1975); **64,** 3344 (1976).
63. H. R. Heath, T. L. Ibbs, and N. E. Wild, *Proc. R. Soc. London Ser. A* **178,** 380 (1941).
64. T. L. Ibbs, R. E. Grew, and A. A. Hirst, *Proc. R. Soc. London Ser.* **41,** 456 (1929).
65. R. E. Bastick, H. R. Heath, and T. L. Ibbs, *Proc. R. Soc. London Ser. A* **173,** 543 (1939).
66. B. E. Atkins, R. E. Bastick, and T. L. Ibbs, *Proc. R. Soc. London Ser. A* **172,** 142 (1939).
67. W. L. Taylor and co-workers, *J. Chem. Phys.* **50,** 4886 (1969).
68. K. P. Müller and A. Klemm, *Z. Naturforsch.* **27a,** 1755 (1972).
69. W. L. Taylor and S. Weissman, *J. Chem. Phys.* **55,** 4000 (1971).
70. W. L. Taylor, *J. Chem. Phys.* **62,** 3837 (1975).
71. W. L. Taylor, *J. Chem. Phys.* **64,** 3344 (1976).
72. E. Whalley and E. R. S. Winter, *Trans. Faraday Soc.* **47,** 1160 (1951).
73. S. Raman and co-workers, *J. Chem. Phys.* **49,** 4877 (1968).
74. A. O. Nier, *Phys. Rev.* **56,** 1009 (1930).
75. E. Von Halle, *A New Apparatus for Liquid Phase Thermal Diffusion,* Rept. K-1420, Union Carbide Corp., Nuclear Div., 1959 pp. 257–325.
76. G. D. Rabinovitch, *Inzh.-Fiz. Zh.* **15,** 1014 (1968).
77. W. M. Rutherford, *J. Chem. Phys.* **59,** 6061 (1973).
78. K. Clusius and M. Huber, *Z. Naturforsch.* **10a,** 230 (1955).
79. B. B. McInteer and M. J. Reisfeld, *Tabulated Values of the Thermal Diffusion Column Shape Factors for the Lennard-Jones (12-6) Potential,* Rept. LAMS-2517, Los Alamos Scientific Laboratory, Los Alamos, New Mexico, 1961; *J. Chem. Phys.* **33,** 570 (1960).
80. E. Von Halle, E. Greene, and R. L. Hoglund, *Thermal Diffusion Column Shape Factors Part I: Inverse Power Model,* Rept. K-1469, Union Carbide Corporation, Nuclear Div., 1965.

81. E. Von Halle and R. L. Hoglund, *Thermal Diffusion Column Shape Factors Part II: Shape Factors Based on the Lennard-Jones (6-12) Intermolecular Force Model,* Rept. K-1679, Union Carbide Corporation, Nuclear Div., 1966.
82. J. E. Powers in H. M. Schoen, ed., "Thermal Diffusion," *New Chemical Engineering Separation Techniques,* Interscience Publishers, a division of John Wiley & Sons, Inc., New York, Chapt. 1, 1962.
83. R. C. Jones and W. H. Furry, *Rev. Mod. Phys.* **18,** 152 (1946).
84. W. H. Furry, R. C. Jones, and L. Onsager, *Phys. Rev.* **55,** 1083 (1939).
85. L. Onsager and W. W. Watson, *Phys. Rev.* **56,** 474 (1939).
86. F. E. Frost, *Gaseous Thermal Diffusion. A Bibliography of the Report Literature 1941 Through 1957,* Rept. UCRL-5116, Part II, California Univ. Radiation Laboratory, Livermore, California, Jan. 1958.
87. C. McFadden, *Gaseous Thermal Diffusion: Part I, Bibliography of Open Literature, 1940–1957,* Rept. UCRL-5116, Part I, California Univ. Radiation Laboratory, Livermore, California, Jan. 1958.
88. M. Grazith, *Isotope Separation Methods: An Annotated Bibliography,* Rept. LS-101, Israel Atomic Energy Commission, Rehovoth, 1962.
89. G. R. Grove, *Thermal Diffusion: A Bibliography,* USAEC Research and Development Rept. MLM-1088, June 30, 1959.
90. G. Taniel, *Diffusion Thermique,* CEA Bibliographie No. 45, 1963.
91. G. Vasaru and co-workers, *The Thermal Diffusion Column,* VEB Deutscher Verlag der Wissenschaften, Berlin, 1969.
92. G. Vasaru, *Separation of Isotopes by Thermal Diffusion,* Publishing House of the Romanian Academy, Bucharest, 1972; (available in English as U.S. ERDA Rept. ERDA-tr-32).
93. H. J. Mürtz and H. G. Nöller, *Z. Naturforsch.* **16a,** 569 (1961).
94. Ger. Pat. 1,154,793 (Sept. 26, 1963), H. G. Nöller.
95. K. Bornkessel and J. Pilot, *Z. Physik. Chem.* **221,** 177 (1962).
96. F. A. Lindemann and F. W. Aston, *Phil. Mag.* **37,** 523 (1919).
97. W. Groth in H. London, ed., *Separation of Isotopes,* George Newnes Ltd., London, 1961, Chapt. 6.
98. J. W. Beams, L. B. Snoddy, and A. R. Kuhlthau in ref. 22, P/723, pp. 428–434.
99. K. Cohen, *The Theory of Isotope Separation as Applied to the Large-Scale Production of U-235,* Natl. Nuclear Energy Ser. Div. III, Vol. 1B, McGraw-Hill Book Co., N.Y., 1951, Chapt. 6.
100. W. Groth, E. Nann, and K. H. Welge, *Z. Naturforsch.* **12a,** 81 (1957).
101. W. Groth and K. H. Welge, *Z. Physik, Chem.* **19,** 1 (1959).
102. W. Bulang and co-workers, *Z. Physik. Chem. Frankfurt* **24,** 249 (1960).
103. W. E. Groth and co-workers in ref. 22, P/1807, pp. 439–446.
104. K. Beyerle and co-workers in ref. 9, pp. 667–694.
105. G. Zippe, *The Development of Short Bowl Ultracentrifuges,* Rept. EP-4420-101-60U, Research Laboratories for the Engineering Sciences, Univ. of Virginia, 1960.
106. A. Bramley, *Science* **92,** 427 (1940).
107. H. Martin and W. Kuhn, *Z. Physik. Chem.* **189,** 219 (1941).
108. H. Martin, *Z. Elektrochem.* **54,** 120 (1950).
109. M. Steenbeck, *Kernergie* **1,** 921 (1958).
110. J. Los and J. Kistemaker in ref. 9, pp. 695–700.
111. A. Kanagawa and Y. Oyama, *J. At. Energy Soc. Jpn.* **3,** 868 (1961).
112. A. Kanagawa and Y. Oyama, *Nippon Genshiryoku Gakkaishi* **3,** 918 (1961).
113. A. S. Berman, *A Theory of Isotope Separation in a Long Countercurrent Gas Centrifuge,* Rept. K-1536, Union Carbide Corp., Nuclear Div., 1962.
114. A. S. Berman, *A Simplified Model for the Axial Flow in a Long Countercurrent Gas Centrifuge,* Rept. K-1535, Union Carbide Corp., Nuclear Div., 1963.
115. H. M. Parker and T. T. Mayo, IV, *Countercurrent Flow in a Semi-Infinite Gas Centrifuge,* Rept. UVA-279-63U, Research Laboratories for the Engineering Sciences, Univ. of Virginia, 1963.
116. J. L. Ging, *Countercurrent Flow in a Semi-Infinite Gas Centrifuge: Axially Symmetric Solution in the Limit of High Angular Speed,* Rept. Ep-4422-198-62S, Research Laboratories for the Engineering Sciences, Univ. of Virginia, 1962.
117. J. L. Ging, *The Nonexistence of Pure Imaginary Eigenvalues and the Uniqueness Theorem for the Linearized Gas Flow Equations,* Rept. EP-4422-245-62S, Research Laboratories for the Engineering Sciences, Univ. of Virginia, 1962.

722 DIFFUSION SEPARATION METHODS

118. J. L. Ging, *Onsager Minimum Principle for Stationary Flow in Axially Symmetric Rotating Systems,* Rept. EP-3912-321-64U, Research Laboratories for the Engineering Sciences, Univ. of Virginia, 1964.
119. J. L. Ging, *Eigenvalue Problem—Limit of Low Angular Speed,* Rept. EP-3912-64U, Research Laboratories for the Engineering Sciences, Univ. of Virginia, 1964.
120. J. L. Ging, *Modified Minimum Principle for Stationary Flow in a Gas Centrifuge,* Rept. EP-3912-325-64U, Research Laboratories for the Engineering Science, Univ. of Virginia, 1964.
121. J. L. Ging, *Onsager Minimum Principle for Axially Decaying Eigenmodes,* Rept. EP-3912-326-64U, Research Laboratories for the Engineering Sciences, Univ. of Virginia, 1964.
122. G. F. Carrier and S. H. Maslen, *Flow Phenomena in Rapidly Rotating Systems,* Rept. TID 18065, U.S.-D.O.E., 1962.
123. G. F. Carrier in H. Görtler, ed., *Proceedings of the Eleventh International Congress of Applied Mechanics,* Springer, Berlin, 1964.
124. T. Sakurai and T. Matsuda, *J. Fluid Mech.* **62,** 727 (1974).
125. T. Sakurai, *J. Fluid Mech.* **72,** 321 (1975).
126. T. Matsuda, K. Hashimoto, and H. Takeda, *J. Fluid Mech.* **73,** 389 (1976).
127. K. Hashimoto, *J. Fluid Mech.* **76,** 289 (1976).
128. T. Matsuda and K. Hashimoto, *J. Fluid Mech.* **78,** 337 (1976).
129. E. Krause and E. H. Hirschel, eds., *DFVLR—Colloquium,* Proz-Wahn, West Germany, 1970.
130. J. J. H. Brouwers, *On the Motion of a Compressible Fluid in a Rotating Cylinder,* Doctoral Dissertation, The Technische Hogeschool, Twente, The Netherlands, June, 1976.
131. D. R. Olander, *Adv. Nucl. Sci. Tech.* **6,** 105 (1972).
132. D. G. Avery and E. Davies, *Uranium Enrichment by Gas Centrifuge,* Mills & Boon Limited, London, 1973.
133. S. Villani, *Isotope Separation,* American Nuclear Society, 1976.
134. W. H. Furry, R. C. Jones, and L. Onsager, *Phys. Rev.* **55,** 1083 (1939).
135. P. A. Tahourdin, *Final Report on the Jet Separation Methods,* Oxford Rept. No. 36, Br. 694, Clarendon Lab., Oxford, England, 1946.
136. S. A. Stern, P. C. Waterman, and T. F. Sinclair, *J. Chem. Phys.* 33, 805 (1960).
137. E. W. Becker and co-workers in ref. 22, P/1002, pp. 455–457.
138. E. W. Becker in ref. 9, pp. 560–578.
139. E. W. Becker in H. London, ed., *Separation of Isotopes,* George Newnes Ltd., London, 1961, Chapt. 9.
140. E. W. Becker and co-workers, *Angew. Chemie, Intern. Ed. (Engl.)* **6,** 507 (1967).
141. E. W. Becker and co-workers, *Atomwirtschaft* **18,** 524 (1973).
142. E. W. Becker and co-workers, *International Conference on Uranium Isotope Separation,* London, 1975.
143. E. W. Becker and co-workers, *European Nuclear Conference,* Paris, 1975.
144. E. W. Becker and co-workers, American Nuclear Society Meeting, *KFK-Bericht 2235,* Gesellschaft für Kernforschung, Karlsruhe, 1975.
145. H. Geppert and co-workers, *International Conference on Uranium Isotope Separation,* London, 1975.
146. E. W. Becker and co-workers, *Z. Naturforsch.* **26a,** 1377 (1971).
147. P. Bley and co-workers, *Z. Naturforsch.* **28a,** 1273 (1973).
148. K. Bier and co-workers, *KFK-Bericht 1440,* Gesellschaft für Kernforschung, Karlsruhe, 1971.
149. U. Ehrfeld and W. Ehrfeld, *KFK-Bericht 1634,* Gesellschaft für Kernforschung, Karlsruhe 1972.
150. W. Ehrfeld and E. Schmid, *KFK-Bericht 2004,* Gesellschaft für Kernforschung, Karlsruhe, 1974.
151. F.-J. Rosenbaum, *Thesis,* Univ. of Karlsruhe, 1975.
152. Ger. Pat. 1,096,875 (Jan. 12, 1961), E. W. Becker (to Deutsche Gold-und Silber-Scheideanstalt vorm. Roessler).
153. W. Ehrfeld and U. Knapp, *KFK-Bericht 2138,* Gesellschaft für Kernforschung, Karlsruhe, 1975.
154. E. W. Becker and co-workers, *4th United Nations International Conference on the Peaceful Uses of Atomic Energy,* Geneva, 1971, paper 383.
155. H. J. Fritsch and R. Schütte, *KFK-Bericht 1437,* Gesellschaft für Kernforschung, Karlsruhe, 1971.
156. R. Schütte and co-workers, *Chemie-Ing. Technik* **44,** 1099 (1972).
157. W. Fritz and co-workers, *Chemie-Ing. Technik* **45,** 590 (1973).
158. R. Schütte, *KFK-Bericht 1986,* Gesellschaft für Kernforschung, Karlsruhe, 1974.
159. P. Bley and co-workers, *KFK-Bericht 2092,* Gesellschaft für Kernforschung, Karlsruhe, 1975.

160. P. C. Waterman and S. A. Stern, *J. Chem. Phys.* **31,** 405 (1959).
161. R. R. Chow, *On the Separation Phenomenon of Binary Gas Mixture in an Axisymmetric Jet,* Rept. HE-150-175, Institute of Engineering Research, Univ. of California, Berkeley, 1959.
162. E. E. Gose, *Am. Inst. Chem. Engs. J.* **6,** 168 (1960).
163. V. H. Reis and J. B. Fenn, *J. Chem. Phys.* **39,** 3240 (1963).
164. H. Mikami, *J. Nucl. Sci. Tech.* **6,** 452 (1969).
165. H. Mikami, *I & EC Fundam.* **9,** 121 (1970).
166. H. Mikami and Y. Takashima, *Bull. Tokyo Inst. Tech.* **61,** 67 (1964).
167. H. Mikami and Y. Takashima, *J. Nucl. Sci. Tech.* **5,** 572 (1968).
168. H. Mikami and Y. Takashima, *Int. J. Heat Mass Transfer* **11,** 1597 (1968).
169. W. Berkahn, W. Ehrfeld, and G. Krieg, *Calculations of Uranium Isotope Separation in the Separation Nozzle for Small Mole Fractions of UF_6 in the Auxiliary Gas,* Institut für Kernverfahrenstechnik, Kernforschungszentrum Karlsruhe, West Germany, report KFK-2351, Nov. 1976.
170. G. F. Malling and E. Von Halle, *Aerodynamic Isotope Separation Processes for Uranium Enrichment: Process Requirement,* paper presented at the Symposium on New Advances in Isotope Separation, Div. of Nuclear Chemistry and Technology, American Chemical Society, San Francisco, Calif., Aug. 1976; *UCC-ND Report K/OA-2872,* Oak Ridge Gaseous Diffusion Plant, Oak Ridge, Tenn., Oct. 7, 1976.

General References

M. Benedict and T. H. Pigford, *Nuclear Chemical Engineering,* McGraw-Hill Book Company, Inc., New York, 1957, Chapt. 12.

K. P. Cohen, *The Theory of Isotope Separation as Applied to the Large-Scale Production of U-235,* National Nuclear Energy Series Division III, Vol. 1B, McGraw-Hill Book Company, Inc., New York, 1951.

H. London, ed., *Separation of Isotopes,* George Newnes Ltd., London, 1961.

H. R. C. Pratt, *Countercurrent Separation Processes,* American Elsevier Publishing Company, Inc., New York, 1967.

S. Villani, *Isotope Separation,* American Nuclear Society Monograph, ANS Publications, 1976.

J. Kistemaker, J. Bigeleisen, and A. O. Neir, eds., *Proceedings of the International Symposium on Isotope Separation,* Amsterdam, April 23–27, 1957, North-Holland Publishing Company, Amsterdam, and Interscience Publishers, Inc., New York, 1958.

Proceedings of the Second United Nations International Conference on the Peaceful Uses of Atomic Energy, Geneva, Sept. 1–3, 1958, United Nations publication, Geneva, 1958, particularly Vol. 4: *Production of Nuclear Materials and Isotopes.*

Proceedings of the Third International Conference on the Peaceful Uses of Atomic Energy, Geneva, Aug. 31–Sept. 9, 1964, United Nations publication, New York, 1965, particularly Vol. 12: *Nuclear Fuels—III, Raw Materials.*

Proceedings of the Fourth International Conference on the Peaceful Uses of Atomic Energy, Geneva, Sept. 6–16, 1971, United Nations and the International Atomic Energy Agency, 1972, particularly Vol. 9, *Isotope Enrichment, Fuel Cycles and Safeguards.*

Proceedings of the International Conference on Uranium Isotope Separation, London, March 5–7, 1975, British Nuclear Energy Society, 1975.

International Conference on Nuclear Power and Its Fuel Cycle, Salzburg, Austria, May 2–13, 1977, sponsored by the International Atomic Energy Agency; for this conference only a *Book of Abstracts* has been published.

M. Benedict, ed., *Development in Uranium Enrichment, AICHE Symposium Series,* Vol. 73, No. 169, American Institute of Chemical Engineers, New York, 1977.

The authors would like to acknowledge the contribution of William M. Rutherford, Monsanto Research Corporation, Mound Facility, Miamisburg, Ohio, to the section of thermal diffusion.

R. L. Hoglund
J. Shacter
E. Von Halle
Union Carbide Nuclear Division

DIGITAL DISPLAYS

The digital display is a device that indicates numerical information to a viewer. The data are usually generated by a semiconductor integrated circuit (IC) (see Semiconductors). The IC processes data entered by the viewer or other sources and reproduces the information in an electrical form. The display transforms the data from the electrical to the optical domain. The digital display is, therefore, one component in a system, eg, electronic watch, calculator, or microwave oven control. The rapid growth of digital displays over the past decade is a result of the utility of these systems, the introduction of large scale integrated circuits and the development of electrooptic technologies that are compatible with the system requirements.

There are considerable operational differences between light-emitting (active) and light-modulating (passive) displays. Light-emitting displays require a power source of sufficient capacity to create the required luminence. Light-modulating displays require much lower power since they do not have to generate photons. Passive displays are becoming dominant in applications that require battery operation. The power limitation and the requirement that many battery-powered displays be visible in bright sunlight give light-modulating displays a substantial advantage. In systems that can be operated from line power, however, light-emitting displays have an important advantage: people like bright lights and color. Since line power generally implies ambient home or office light levels, active displays remain dominant in this area.

The detailed information that is transmitted to the display is in an electrical form. There are a variety of electrooptic techniques for transforming these signals into visible numbers, but all rely on the local generation of electrical fields. It is well known that seven distinct segments can generate the ten Arabic numerals when arranged in the format shown in Figure 1. Consider each segment (a–g) to be an individual electrode; an eighth common electrode in the form of a block eight is placed adjacent to the seven-segment electrodes. The pattern is known as the seven-segment format. It gives numerals of equal size with pleasing, recognizable form. It is the dominant type of digital display in use today. Figure 1 also indicates a common geometrical feature of digital displays: flatness. This is in contrast to the bulkiness of the cathode ray tube

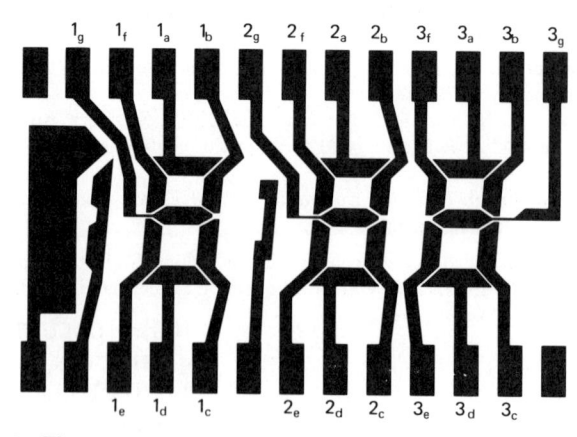

Figure 1. The seven-segment electrode pattern for a $3\frac{1}{2}$-digit watch display.

(CRT). The flat characteristic permits the design of small compact systems. This feature is in harmony with the semiconductor revolution. It is more important than the digital format, as recent analogue liquid crystal displays (LCDs) have demonstrated (see Liquid crystals).

Figure 1 explains how any single number can be formed by the seven-segment electrode pattern. It follows that any number with n digits can be formed by repeating the pattern n times and addressing $7n + 1$ electrodes. When n is large, eg, 12 as in a scientific calculator, the number of leads to the display for the LSI (large-scale integration) chip becomes too large to handle economically. Semiconductor technology has been able to advance as rapidly as it has owing in part to a serial in-serial out data format. Data are read in and out from the chip sequentially over a limited number of pins to the chip. The viewer, on the other hand, requiring a parallel data format, must be able to view the whole display simultaneously. The mismatch between the display data requirement and the chip data format becomes intolerable for n greater than four. The solution is to supply data to the display in a serial format and take advantage of the averaging properties of the display or the eye to produce the parallel format. Timesharing leads or multiplexing takes advantage of the fact that any point in an NXM matrix can be addressed by choosing one of N row electrodes and M column electrodes ($N + M$ leads). For an array of 12 digits in the seven-segment format, a 12 × 7 matrix requires 19 leads where the direct drive format requires $7 \times 12 + 1 = 85$ leads. Each point in a matrix is sequentially addressed by voltage pulses on the rows and columns, the repetition rate being larger than the flicker frequency of the eye (ca 30 Hz). The amplitudes of the pulses at row N, column M determine the contrast at point (N,M). Typically row N is selected ($N < M$) and data are read simultaneously to M columns for a duty cycle of $1/N$. The extent to which the multiplexing technique can be employed depends on the electrooptic response characteristics of individual effects. Specifically, the response must have a threshold characteristic so that a given point does not respond to the voltages being applied to the other points. Some are very easy to subject to timesharing, eg, vacuum fluorescence. Passive displays are traditionally difficult to multiplex. The matrix addressing capability is very important in deciding which display to use for higher information content systems.

Displays are evaluated in terms of two values: contrast and brightness. Color can be a third criterion, especially in light-emitting (active) displays. Contrast and brightness are defined differently for active and passive displays so that care must be used in interpreting the numerical values. For an active display, brightness is the absolute luminance given off by the active area. Luminance refers to the perception of brightness by the human eye. Contrast depends on the amount of emitted luminance relative to the amount of ambient light reflected by the background, and consequently can not be an absolute number (see Color). Generally light-emitting displays have absorbing backgrounds to reduce reflection, but viewing in direct sunlight may still be difficult.

Contrast for a passive display is defined as the ratio of luminance reflected from the display in its bright state to the luminance reflected in the absorbing state. Since light is reflected, the contrast ratio for a passive display can be expected to be insensitive to the ambient level. Brightness for a passive display is measured relative to some standard white diffuse reflector, eg, powdered MgO. A separate light source is required for dark viewing.

Liquid Crystal Displays

Liquid crystal displays (LCDs) are constructed to take advantage of the birefringence of the liquid crystal phase and the fact that the optic axis can be reoriented with an applied electric field. The devices are passive or light-modulating in character. The basic structure of the device is a sandwich, two pieces of glass with transparent conductors on the inside enclosing the liquid crystal. This structure is deceptively simple, and to understand why requires a description of the nematic liquid crystal phase.

Liquid Crystal Phase. The term liquid crystal refers to several thermodynamically stable phases of matter that occur over a limited temperature range for certain organic molecules (see Liquid crystals). The spatial structure of the nematic mesophase, the phase employed in commercial devices, is indicated schematically in Figure 2 which shows the molecules, represented by cigar-shaped ellipses, with all their axes parallel. This idealization ignores the certainty of significant Brownian motion. The figure does indicate, however, the key features that are necessary for device application.

The medium is anisotropic. One would expect different electrical susceptibilities for an electric field along the average molecular axis (optical axis) than for the field applied transverse to the axis. This is the case for both the low frequency susceptibility (dielectric constant) and the visible frequency response (index of refraction). The anisotropy of the dielectric response enables an applied electric field to reorient the optical axis of a liquid crystal. The optical anisotropy permits detection of this change via the difference in birefringence or scattering. These electrooptical effects are essential for the use of liquid crystals in digital display devices.

A second reason is that electrooptical switching can be accomplished with very little energy. Understanding the low-power-switching characteristics of liquid crystal

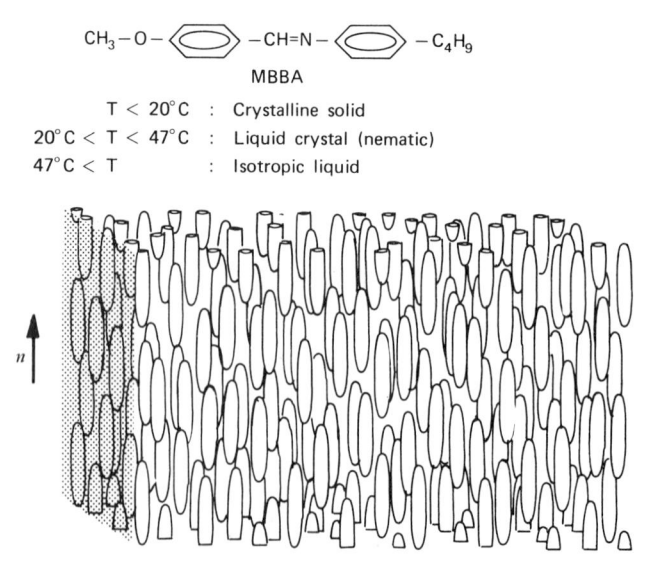

Figure 2. Schematic representation of the molecular order in the nematic phase and typical nematic characteristics. n = unit vector (= nematic director); MBBA = 4-butyl-N-[(4-methoxyphenyl)methylene]aniline.

materials requires knowledge of the electrical and mechanical properties of the phase. Consider a single crystal of nematic material as shown in Figure 3. The unit vector describing the local optical axis of the material varies in orientation through the sample owing to the electric field E_0 (V/m). The torque on the liquid crystal when it makes an angle θ with respect to the electrode plate is equal to $\epsilon_0(\epsilon_\| - \epsilon_\perp)E_0^2\theta$ where $\epsilon_\|$ and $\epsilon_\perp$ are the dielectric permittivities parallel and perpendicular to the liquid crystal director, respectively. ϵ_0, a constant, is the permittivity of free space. Distorting a liquid crystal in the manner shown in Figure 3 results in an elastic torque which opposes the distortion: $\pi^2 K\theta/d^2$ where K is an elastic constant of the liquid crystal and d is the thickness of the dielectric sandwich. It is assumed that the director is constrained by surface forces to lie in the plane of the surface at the surface. The distortion grows in amplitude when the dielectric torque is greater than the elastic torque. The condition for equilibrium results in a critical voltage.

$$V_c = E_0 d = \pi \sqrt{\frac{K}{\epsilon_0(\epsilon_\| - \epsilon_\perp)}}$$

The existence of a threshold voltage for reorienting the nematic director is characteristic of commercial LCDs. Ks are on the order of 10^{-12} newton (2.2×10^{-13} lbf) and $\epsilon_\| - \epsilon_\perp$ can be greater than 10. These values lead to threshold voltages of ca 1V, independent of the sample thickness.

The energy necessary to switch on the liquid crystal can be determined from the capacitance (C) of the sample and the voltage applied: $\frac{1}{2}$ CV^2. To turn on a display requires roughly twice the threshold voltage, and the sample has a capacitance determined by $\epsilon_\|$

$$\frac{\text{energy}}{\text{area}} = \frac{2(V_c)^2}{d} \epsilon_0 \epsilon_\|$$

For a typical commercial device $\epsilon_\|$ is 22.3×10^{-6} μF (20 esu) and d is 10 μm. The energy per unit area is thus ca 3 nJ/cm^2 (0.72×10^{-9} cal/cm^2). Since most LCDs are run with

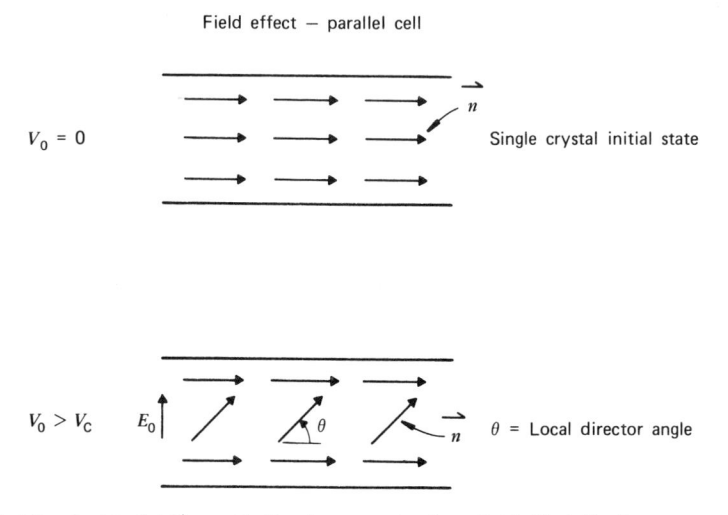

Figure 3. The electric field/nematic director geometry for a field effect display. n = unit vector (= nematic director); E_0 = electric field, V/m.

a-c voltages at frequencies of ≥32 Hz, the typical power to turn on a cm² of display is on the order of 1 µW. For properly made field effect devices, this corresponds to a capacitive current density on the order of 0.50 µA/cm². Ionic impurities add a resistive current component. Alternating current can be used because the electric torque is the product of the induced polarization times the applied field, ie, it is independent of the direction of the applied field. A-c operation prevents electrochemical corrosion, which can be a serious failure mechanism in d-c driven devices.

The low power operation results directly from the small elastic constant. The liquid crystal phase is very delicately structured and can be easily distorted. This ease of distortion generally means that liquid crystals are excellent transducers. Only electrooptical effects are discussed here, although other stimuli can cause optical changes in liquid crystals. Temperature can be sensed by color in cholesteric liquid crystals as demonstrated in the recently popular "Mood Rings."

The Twisted Nematic Display. The experimental geometry shown in Figure 3 does not produce a particularly effective display. It leads to dull colors that depend on the angle from which the display is viewed. The key to a high contrast LCD is to produce a twisted structure in the off-state as shown in Figure 4. The surfaces enclosing the liquid crystal layer are treated to promote alignment of the nematic director in one direction on the surface. The twisted structure is produced by orienting the surfaces so that the director at the top surface makes an angle of 90° with the bottom surface. In order to conform to these boundary conditions, the director in the bulk of the sample twists 90° in going from the top to the bottom surface. Thus the name twisted nematic.

The normal modes for electromagnetic waves in this structure are quite unique. Linearly polarized light entering the sample parallel to the optical axis on one side emerges on the other side still linearly polarized and twisted by 90°. If the nematic twist is only 45°, the light is twisted 45°. Twists of larger than 90° cannot be obtained for a pure nematic since the sample can attain a lower free energy state by twisting in the opposite helical sense by the supplement of the angle.

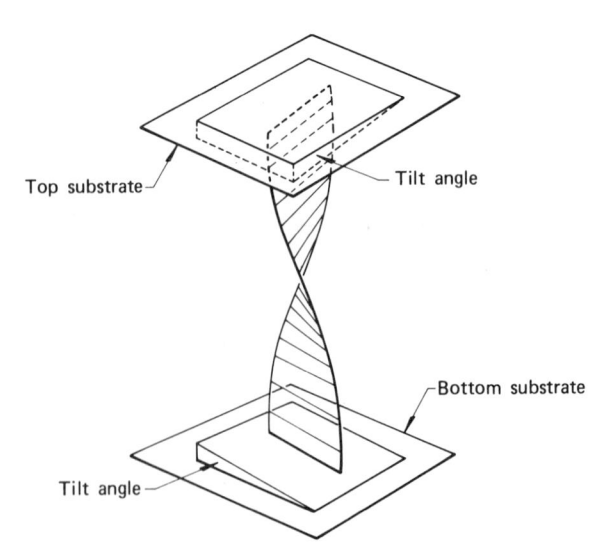

Figure 4. The geometry of a twisted nematic LED.

The 90° twist can be achieved by either a right-hand or left-hand helix. This leads to reverse-twist domains throughout a cell filled with pure nematic material. To avoid these domains and the poor device quality that arises because of them, it is standard practice to introduce a small amount of optically active material to the nematic host. The chiral additive, which may be a cholesterol derivative, makes the mixture cholesteric. This breaks the symmetry between right and left hand twist for the cell and produces a uniform, single-crystal twist structure.

In the early 1970s, Fergason (1) in the U.S. and Schadt and Helfrich (2) in Switzerland realized independently that the optical characteristics of the twisted nematic structure offered unique device characteristics. Between crossed polarizers oriented with the parallel director axes on the two plates the sample would appear transparent, allowing for the insertion loss of the polarizers. If the nematic/chiral host has a positive dielectric anisotropy ($\epsilon_\parallel > \epsilon_\perp$), the application of voltage above the threshold tilts the director parallel to the field. This eliminates the twist in the sample and the display appears dark. A schematic representation of the device operated in reflection is indicated in Figure 5. The electric field converts the sample from the twisted, transparent off-state to a uniaxial nematic, dark on-state when viewed between properly oriented crossed polarizers. This device is a high contrast display but the brightness is low owing to the polarizers.

Figure 5 indicates that the on-state has most of the molecules oriented parallel to the electric field. This is generally not the case in commercial devices because power supply considerations do not permit sufficient voltage to drive the director to this position throughout the sample. This fact and the birefringence of the material produces a pronounced viewing angle dependence to the darkness of the on-state. The optics of the situation are very difficult to understand, although significant progress has been made by computer modeling. The viewing characteristics can best be measured by convergent light observation with a polarizing microscope (conoscopic figure). The strong viewing directions correspond to the isogyres in the conoscopic Maltese Cross.

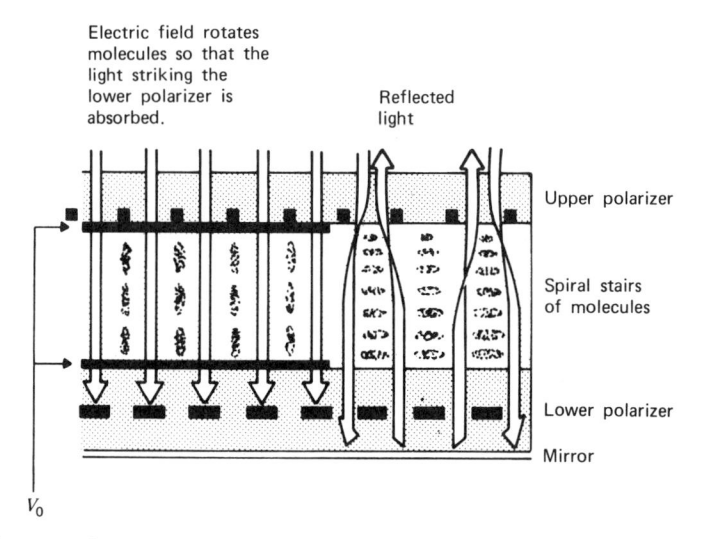

Figure 5. Cross section of the twisted nematic LCD in the on vs the off states.

The twisted nematic device has become the overwhelmingly favorite display for handheld calculators and is rapidly gaining ground on LEDs for watch applications. Volumes worldwide in 1977 reached 15 million devices. Projections indicate continued growth in the area of portable instrumentation.

The twisted nematic LCD was preceded in the marketplace by another liquid crystal display effect known as dynamic scattering. It was invented at RCA in the late 1960s (3). As the name implies, this effect scatters light in the on-state, and it can be seen without polarizers. Since the insertion loss of the polarizers is avoided, dynamic scattering displays appear bright relative to twisted nematic displays. Given the right viewing conditions, the display has the appearance of jewelry. Unfortunately the scattering characteristics of this display also leads to certain ambient lighting characteristics where the display cannot be read.

Device Fabrication. A liquid crystal display, whether twisted nematic or dynamic scattering is not simply a liquid crystal between two glass plates. The device actually employs a series of layers which are shown in cross section in Figure 6. Starting with the glass substrate and working towards the liquid crystal are a (1) transparent conductor, (2) a dielectric blocking layer and (3) an alignment layer. The transparent conductors are typically In_2O_3 and/or SnO_2 which have greater than 85% transmission in the visible and sheet resistivity of less than 1000 Ω/sq. Deposition techniques are sputtering, chemical vapor deposition, and evaporation (see Film deposition techniques). The uniform conducting sheets are patterned into seven-segment electrode digits by conventional photoresist or screened-ink techniques (see Photoreactive polymers; Reprography).

After the electrodes have been defined, the substrate surface alternates in composition from glass to transparent conductor. This heterogeneous structure is unsuitable for three reasons: (1) The liquid crystal, which is a good solvent, can leach impurities from the glass. (2) The electrodes in direct contact with the liquid crystal

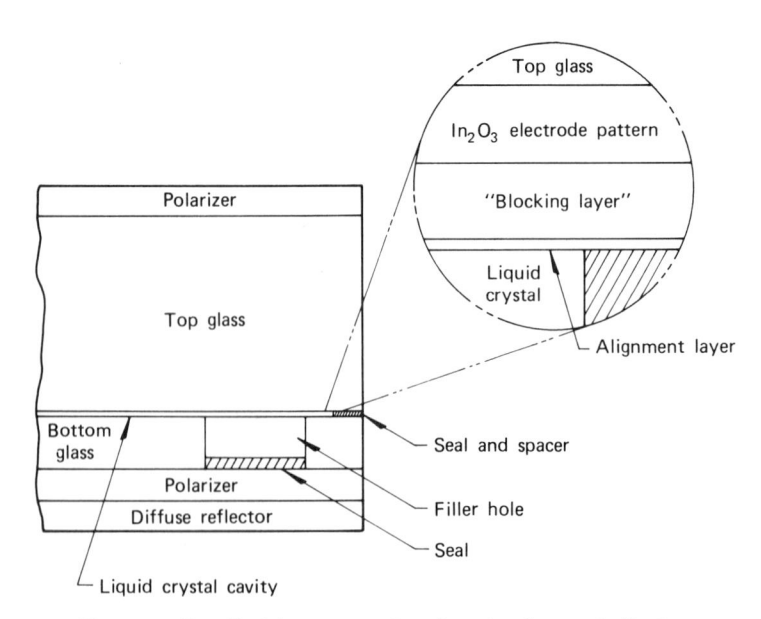

Figure 6. Detail of the construction of a twisted nematic display.

material can inject carriers, ie, electrochemical interactions are possible. (3) The surface energies of the two layers are different and can lead to different alignment textures of the liquid crystal in the filled cell. This is a failure mechanism since a "single-crystal" liquid crystal must be achieved over the whole viewing area of the display for good device performance. It is possible to avoid all three problems by depositing a dielectric coating over the patterned electrode substrate. This blocks the glass/transparent electrode surfaces from the liquid crystal and gives a uniform surface energy for the total active area. The dielectric layer typically is SiO_2 deposited by sputtering or evaporation techniques.

The layer between the blocking layer and liquid crystal is the alignment layer. This is the most important layer in the whole device because it is responsible for promoting a single-crystal liquid crystal. The techniques for how this alignment layer is achieved are regarded as proprietary by most LCD manufacturers. The alignment layer must have two characteristics: (1) The director must align more or less in the plane of the substrate, an alignment referred to as homogeneous in liquid crystal terminology. (2) There must be a preferred direction in the plane so that the orientation of the director in the plane is defined. Alignment layers with these characteristics constrain the director at the boundaries to a given direction. The elastic forces in the bulk liquid crystal cause the liquid crystal to align relative to these boundary conditions in the minimum energy configuration. The result for twisted boundary conditions is shown in Figure 4, which also indicates a slight tilt of the director out of the plane of the surface at the substrate. This third requirement for the alignment layer is necessary to ensure that the device turns on in the same direction throughout. The tilt bias breaks the symmetry that a zero-tilt alignment would give. The director turns in the direction of the bias when the field is applied.

The historical alignment technique is to rub the device surface with a cotton cloth several times in one direction. Most nematic liquid crystals align with the director parallel to the rubbing direction, but with a tilt of a few degrees out of the plane of the surface. This is an organic alignment method.

There are several inorganic alignment techniques, but only the most common is discussed here. When a dielectric such as SiO is deposited onto a glass substrate by evaporation such that the line between the source and the substrate makes a small angle (5°) with the substrate, a heterogeneous structure grows on the substrate for thicknesses on the order of 30 nm. This oriented dielectric layer aligns a nematic liquid crystal so that the director lies in the plane formed by the evaporation beam and the normal to the substrate (4). Typically, the director is tilted out of the plane by 30°. The evaporated alignment layers have been analyzed using the high resolution capabilities of the electron microscope. The pictures indicate a hill structure with the tip of the hills pointed toward the evaporation source.

It should be clear from the discussion above that cleanliness is an important requirement. Any spurious organic layer, eg, a fingerprint, will spoil the alignment. Dust particles can shadow the evaporated alignment surface and create alignment defects. Since the liquid crystal alignment is very sensitive to thin surface layers, there is the possibility of studying surfaces using liquid crystal alignment.

Package Sealing. The long term reliability of an LCD is a function of several process steps. One of the most important is the manner in which the liquid crystal is protected from the environment by the seals between the glass substrates. Generally there are two seals, a perimeter seal and a fill-hole seal. The perimeter seal attaches the two pieces of glass at the edges of the active area. The fill-hole seal closes the package after the liquid crystal has been loaded. A plastic seal is generally used in conjunction with the evaporated alignment technique since it is difficult to develop a rubbing process that will withstand the elevated temperature (500°C) associated with a frit seal.

There is considerable difference of opinion on which alignment or seal process is the best. The organic process is cheaper and is directly compatible with multiplex operation owing to the low tilt angle. The inorganic process is more resistant to accelerated life tests and provides a thin perimeter seal width. The width is important for watch applications because of limited space in many watch designs. Currently, the vast majority of calculator LCDs are of the organic type, and the watch LCDs are more equally divided between the organic and inorganic processes.

Liquid Crystal Materials. There are several types of liquid crystal materials currently in use for commercial LCDs. Only the generic types are listed (see Liquid crystals). The basic difference lies in the linkage between the aromatic rings, eg, the compound in Figure 2. Schiff-base materials are popular owing to their relatively low cost and easy alignment characteristics. Their sensitivity to moisture demands a frit-seal package. Biphenyls are tolerant to water but are expensive and tend to be difficult to align. Azoxy materials were used extensively in calculator/multiplex drive applications but required extra uv protection. Esters have been used more recently in calculator applications because of their relatively flat threshold voltage versus temperature characteristics. Slower response times are generally experienced with esters. Phenylcyclohexanes have recently been studied due in part to their low birefringence.

For twisted nematic displays all the basic molecular types discussed above have a $C{\equiv}N$ group on one end of the molecule to provide the strong positive dielectric anisotropy. The other end groups are alkyl or alkoxy groups of various lengths. An important practical matter is the achievement of desired temperature characteristics by mixing components with the same linkage group. Eutectic mixtures of wide temperature range are most desirable. Recent devices employ mixtures of different linkage types as well as dopants that are not liquid crystalline.

External Components. Polarizers, an essential component for the twisted nematic LCD, are generally stretched poly(vinyl alcohol) stained with iodine. The films are attached to the glass with an organic adhesive. For demanding environmental applications, a polarizer can be a weak element in the display system. LCDs can be illuminated at night with either an incandescent bulb (on demand) or with a tritium phosphor light emitter (see Luminescent materials). The light must travel through both polarizers, which means that the mirror in Figure 5 must be partly transmissive. The nature of this "transflector" surface is very important to successful backlighting.

Display Characteristics. It is difficult to specify the viewing characteristics of an LCD. Since the modulation mechanism involves birefringence, the contrast characteristics are angle dependent. This is especially true for the twisted nematic device. The viewing cone describes the solid angle over which the contrast ratio is greater than a specified value, typically seven to twelve depending on the type of reflector. For some applications, the viewing cone must be large, eg, watches. For other applications the viewing cone can be smaller with an increase in multiplex capability as a benefit, eg, calculators. Thus it is not very useful to specify contrast ratio without carefully understanding the optics under which the measurements are taken and the ultimate use intended for the display.

The brightness of a twisted nematic display on axis, relative to white scattering MgO, should be greater than 0.2 with a full reflector. When one employs a transflector for backlighting purposes, the brightness of the display in daylight conditions drops. This grey appearance of backlighted twisted nematic LCDs has led many groups to search for another passive display technology that has both brightness and contrast.

The dynamics of liquid crystal devices are determined by the viscous stresses that oppose rapid reorientation of the director. Turn-on times generally vary as $\eta d^2/V^2$ where η is an average viscosity (m^2/s or 10^4 St), V is the voltage above threshold, and d is the sample thickness. Turn-off times vary as $\eta d^2/K$. (K is an average elastic constant.) Clearly the more viscous the material the slower the response. Since viscosities depend on temperature, the dynamics of LCDs are significantly temperature dependent. The low-temperature operating limit is often determined by the long turn-on and turn-off times rather than by the crystalline/nematic transition temperature. Response times can be shortened by decreasing the thickness, but there are practical and theoretical reasons that keep the thickness greater than 5 μm.

Multiplex Drive. Liquid crystals have the threshold characteristic required for multiplexing, but the electrooptic response is angle dependent, temperature dependent and becomes saturated as the optical axis of the sample gets near the applied electric field. These latter characteristics make multiplexing of LCDs difficult. Operation at $1/3$ duty cycle ($n = 3$) has currently been achieved in commercial calculators. It is not at all clear, however, whether 100 lines can ever by scanned successfully. High-information-content LCDs probably will require drive schemes that employ a memory device, eg, a transistor circuit in series with each information position. This active matrix approach is currently under development for military applications.

Light-Emitting Diodes

The visible light-emitting diode (VLED) display is based on injection electroluminescence (see Light-emitting diodes and semiconductor lasers). The p-n junction represents the most effective way to inject minority charge carriers, and it is the recombination of these carriers with carriers of the opposite type, ie, electrons and holes, that produces the radiation. Nonradiative recombinations by several mechanisms are also possible, and dominate in many materials. The energy band structure of GaAs, described below, favors the radiative process.

Basically, in the physics of the p-n junction, the application of a forward bias lowers the potential barrier at the junction as shown in Figure 7, allowing electrons to diffuse from the n-region into the p-region, and holes from the p-region to diffuse to the n-region. The ratio of the electron current to hole current depends on the relative concentrations and electron-to-hole mobility ratio. This ratio is about 20 in GaAs; hence, most of the current is generally carried by electrons, ie, electrons are injected into the p-region, and the light is generated there. One of the important considerations in designing an efficient VLED is minimizing the internal absorption. To this end, the VLEDs or display segments are typically constructed as a thin p-region on an n-substrate.

Another very important consideration is the internal reflection of the light generated at the junction. When the index of refraction is large, as it is for the III–V compounds, most of the light is internally reflected. In GaAs, eg, the critical angle, given by

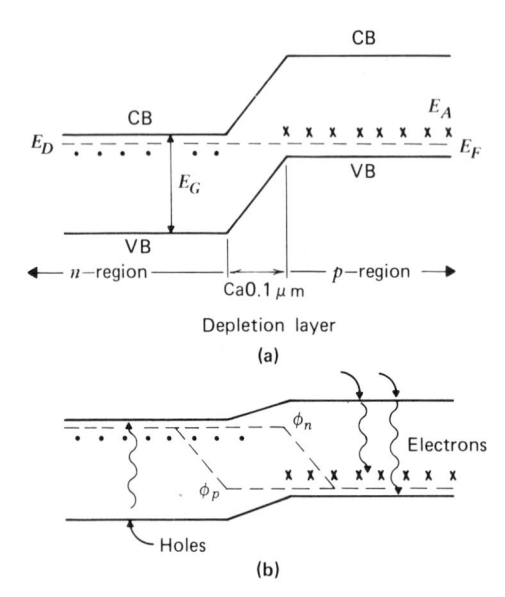

Figure 7. Schematic drawing of a p-n junction, showing the minority carrier injection and recombination processes. (**a**) Zero bias conditions. When p- and n-regions are in contact, the Fermi level S, E_f, in the two regions must line up. Electrons diffuse from the n-region and levels from the p-region until the holes' opposing electrostatic potential V_d is just sufficient to prevent further charge transport. (**b**) Under forward bias conditions, the built-in potential barrier V_d is lowered and current flows; electrons from n to p-region and holes from the p to n-region. The system is no longer in thermal equilibrium, and the quasi-Fermi levels ϕn, ϕp are bent. The injected electrons decay by combining with a hole, either at an acceptor center, or in the valence bend, producing radiation of wavelengths equivalent to the energy difference involved.

$$\sin \theta_c = \frac{n_{\text{air}}}{n_{\text{GaAs}}}$$

where θ_c = arcsin 1/3.6 = 16.1°. The fraction of light that lies within the critical angle and escapes from the front surface is $F_c = [1 - \cos\theta_c]/2$. For GaAs, this is only about 2%. If the wavelength of the light is not near the absorption edge of the material, multiple reflections will increase the percent of light that escapes. In the case of GaP red-emitting diodes the light is generated by recombination involving an impurity center quite far from the band edge; hence it is of longer wavelength, and therefore readily transmitted in multiple passes.

Materials for VLED Displays. In the late 1950s and early 1960s, soon after the successful development of the silicon transistor, an intensive search was underway for the next new material that would extend the operating temperature still higher (as silicon was an extension over germanium). It was that search which led, quite directly, even if unexpectedly, to the brilliant red, yellow and green displays so familiar in the calculators and watches of the 1970s. It happened because the search entailed a monumental program for the preparation, purification and characterization of a little-studied class of materials: the III–V compounds.

According to the Grimm-Sommerfeld rule (5), a binary compound formed from the elements on either side of a Group IV element is isoelectronic to the Group IV element and resembles it in many properties. Thus, boron nitride is similar to carbon (see Boron compounds, refractory) and, indeed, the cubic form of BN is harder than diamond; AlP relates to Si; and GaAs to Ge. Compounds adjacent to the semiconducting Group IV elements, Si, Ge, and Sn, exist in the tetrahedrally-bonded zinc blende lattice, and all are semiconductors (6).

The same trends occur in properties; eg, lower energy gaps and melting points with the heavier Group IV elements and the III–V compounds. From these considerations, the logical choice for a higher energy gap, and therefore, higher operating temperature, would be AlP, with an energy gap of 2.43 eV. It was quickly evident, however, that it was not to be the important candidate. AlP is difficult to prepare and unstable in moist air.

Instead, the workhorse of the III–V series became GaAs, the compound isoelectronic with Ge. With an energy gap of 1.43 eV giving theoretical transistor operating temperatures of >400°C, and a very high electron mobility of ca 9000 cm^2/Wb (9000 cm^2/(V·s)), it seemed the perfect semiconductor material. However, GaAs has not realized its potential as a transistor material. Transistor development has been hampered by technological difficulties (eg, an unexpected profusion of deep energy levels) and the high cost of gallium.

GaAs has a band structure as shown in Figure 8. The band gap is called direct since the minimum in the conduction band lies at the same position in wave vector space as the maximum in the valence band. During the 1960s, however, there were several remarkable discoveries that led to new applications of GaAs. One of these was that a forward-biased gallium arsenide p–n junction is a very efficient emitter of infrared radiation, owing in part to the direct band gap. It has been shown, in fact, that internal quantum efficiencies approaching 100% are achieved. However, the radiation is at 0.9 µm in the near infrared rather than the visible region.

In 1955, Folberth (7) predicted that ternary alloys could be formed among the III–V compounds. The second discovery made the visible displays possible with compounds such as GaAsP. The virtual crystal model of reference 8 holds well in the III–V system.

The first 7-segment red $GaAs_{0.6}P_{0.4}$ displays, offered about 1969, made possible

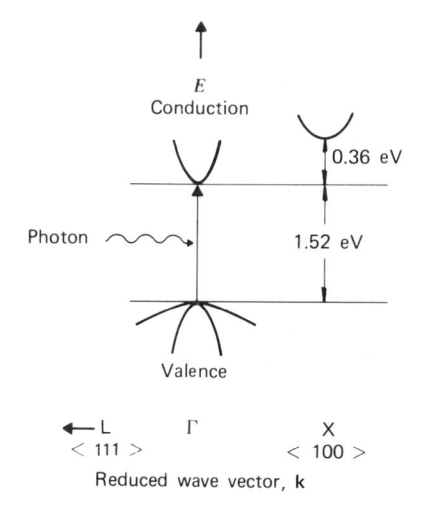

Figure 8. The band structure of III–V compound semiconductor showing a direct gap–GaAs.

inexpensive desk and pocket calculators. The calculators were followed about five years later by the digital wristwatch. At present, displays are available in formats ranging from single digits to arrays of ≥12 digits; in colors including red, amber, yellow and green; and sizes from 0.25 cm character height to >2.0 cm. They are also available in 7-segment, 9-segment, hexadecimal (using letters or other symbols to indicate the numbers 10 to 16), 5 by 7 dot-matrix alphanumeric, and other special fonts.

Although they cannot compete with liquid crystal displays where current drain is a major factor, VLED displays offer the advantage of appearance, readability, high speed multiplexing, low operating voltage (<2 V), and self-illumination.

Considerations for Designing VLED Displays. The important factors for a display are: brightness, color, hue, contrast, viewing angle, and size. In the case of battery-powered operations, the power efficiency is naturally preeminent.

Basic to the discussion of color and the materials that will be capable of producing a given color (qv) is the concept of the energy gap and, beyond that, the direct versus the indirect gap. Since the radiative decay times are long compared to the time for injected carriers to be scattered by the lattice vibrations, the free carriers are at the band edges, within about kT (the Boltzmann energy). For a band-to-band transition, then, the wavelength of the light generated is given by:

$$E = h\nu, \text{ or since } \nu = c/\lambda, E = hc/\lambda$$

where h = Planck's constant, and c = velocity of light. If E is in electron volts, the wavelength in μm is:

$$\lambda = 1.24/E$$

The longest wavelength the eye can detect with reasonable sensitivity is ca 0.72 μm, corresponding to an energy gap of ca 1.7 eV. As mentioned earlier, the gap in GaAs is ca 1.4 eV, much too small to produce visible light. Since the energy gap is higher for the lighter elements, the next candidate for a VLED is GaP because the aluminum compounds AlP and AlAs are unstable in moisture.

Direct vs Indirect Gap. GaP has an energy gap of 2.25 eV, corresponding to 0.55 μm, very near the green at the peak of the eye response curve. Gallium phosphide, however, is an indirect gap material like germanium and silicon. The lowest conduction band minimum is not the $k = 0$ in wave vector space as shown in Figure 9. This leads to significant device problems because, in optical transitions from conduction to valence band, both energy and momentum (or wave vector) must be conserved. If a transition does not conserve momentum, as is the case in indirect materials, without some additional mechanism, the probability of light emission will be very low.

The addition of As to GaP to form the ternary alloys $GaAs_{1-x}P_x$, produces a band structure that goes from direct ($x = 0$ to 0.45) to indirect ($x = 0.46$ to 1.0) as shown in Figure 10. The crossover from the direct gap being the lowest energy to the indirect gap being lowest in energy comes at $x = 0.45$ or 2.08 electron volts (626 nm).

Direct Gap Materials. The GaAsP system that was operated in the direct gap region provided the first commercial low-cost LEDs. Note that as x increases the emitted light moves toward the green on Figure 10. This means increased brightness as perceived by the human eye. As the crossover value of $x = 0.46$ is approached, however, the quantum efficiency of the direct gap composition begins to fall off sharply well before the crossover. This occurs because appreciable numbers of electrons are thermally excited into the higher indirect conduction band minimum (Fig. 9). The wave-

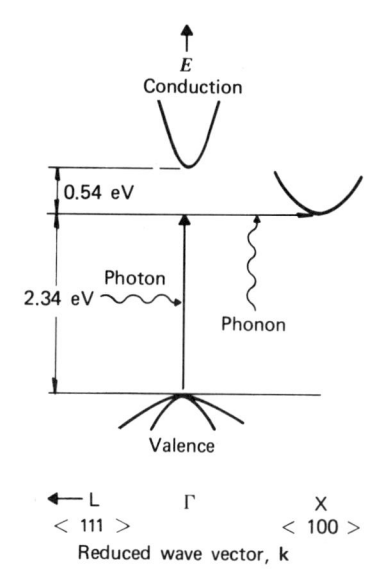

Figure 9. The band structure of III–V compound semiconductor showing an indirect gap, ie, GaP showing a phonon-assisted transition.

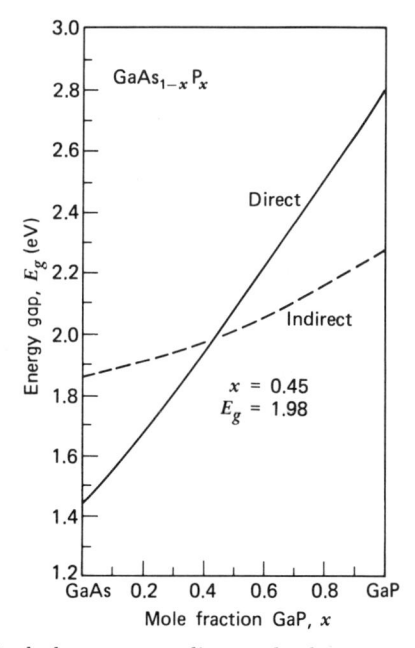

Figure 10. The change in the lowest energy direct and indirect gaps in the GaAs–GaP system.

length for optimum brightness (device efficiency × eye sensitivity) is about 650 nm which occurs at $x = 0.4$. This produces the familiar red LEDs in watches and calculators.

Other direct gap materials of interest for visible LEDs are: $Ga_{0.73}$, $Al_{0.27}$, As (red);

$In_{0.3}$, $Ga_{0.7}P$ (red to yellow); $In_{0.6}$, $Al_{0.4}$, P (red to green); and GaN (green to blue-violet) using transitions between different donor and acceptor levels.

The Indirect Gap Material. The most likely way to conserve momentum for a radiative combination in an indirect band-gap material is by the emission of a phonon. Although this occurs to some degree in indirect gap materials, it represents a three body event and is, therefore, orders of magnitude less probable than the direct radiative process taking place in direct band-gap material. GaP as grown is a very weak emitter of green light, and did not appear to offer any promise for displays. In the mid-1960s however, it was discovered that isoelectronic substituents induced a bound impurity state in semiconductors. These impurities provided a path by which a very efficient radiative recombination can take place (9). One such impurity in GaP is nitrogen, which gives rise to a green luminescence at around 550 nm. These transitions are possible with high probability even in indirect gap materials because momentum can be conserved for tightly bound electrons at the isoelectronic exciton traps. It is a consequence of the Heisenberg uncertainty principle that tightly bound electrons have values of momentum throughout the first Brillouin zone (10) (see Semiconductors).

A very significant point about the isoelectronic traps is that they do not act as normal dopants producing free carriers. Thus, it is possible to add relatively large amounts of the centers without limiting the internal quantum efficiency by Auger recombination or other problems associated with excessive free carriers. Fortunately, the solubility of nitrogen is high in GaP, and concentrations of 10^{19}–10^{20} atoms/cm^3 (0.02–0.17 M) can be used.

The other isoelectronic trap of importance in GaP is the zinc–oxygen pair. It lies considerably below the nitrogen level in the forbidden zone, and gives rise to a deep red electroluminescence at 690 nm. The host crystal is transparent to this wavelength with the result that the light generated is repeatedly internally reflected with negligible loss until most of it finally emerges, despite the very small critical angle. The highest external efficiencies observed have been achieved in GaP: Zn,O VLEDs, with values >12% (11). However, the red GaP:Zn,O VLEDs become saturated at low current densities (ca 20 A/cm^2), which is a serious disadvantage for multiplexing of the display segments.

Another important consideration in the use of a transmissive indirect gap material is the fact that the entire bar lights up, in contrast to the case of highly absorbing direct gap material. A direct gap material tends to absorb its own light whereas an indirect gap material does not. Since the light getting outside the device to the observer is paramount, design of the device geometry is influenced by the type of gap. For indirect materials, monolithic chips cannot readily be used. When the seven-segment bar display is produced by patterned diffusion onto one large chip, there is a tendency for unlighted segments to appear dimly lit as light propagates throughout the substrate.

Other indirect gap materials of commercial interest are nitrogen-doped $GaAs_{0.5}P_{0.5}$ to $GaA_{5.1}P_{0.9}$ (reddish-orange to orange), GaAlP (green), AlAs (yellow) and AlP (green). The tendency of the aluminum compounds to be unstable is mitigated to some extent in ternary or quaternary compounds with relatively low aluminum contact. Other considerations such as the lattice match to the substrate material and the chemistry of preparation tend to make the aluminum systems less desirable for commercial displays.

The hopes for blue VLEDs using GaN have not been realized. GaN resembles the II–VI compounds such as ZnS more than it does the other III–V compounds, and, as such, has not been susceptible to attempts to form a *p-n* junction. The II–VI compounds are strongly "self-compensating". That is, the solubility of a doping impurity in the host crystal tends to closely equal the concentration of the opposite type of impurity (or defect) centers already present. As a result, the strongly doped *p-n* junction needed for efficient minority carrier injection and radiative recombination cannot be achieved.

In the case of GaN, native defects, possibly nitrogen vacancies, cause the material to be very strongly *n*-type as grown, with free electron concentrations on the order of 10^{19} electrons/cm^3. Attempts to produce a *p*-type region by doping with the normal acceptor materials Zn, Cd, etc, have resulted in, at best, weakly *p*-type material of extremely high resistivity. The dopants normally used as acceptors are from Group II of the periodic table, which are one electron short when they substitute for the Group III element. Correspondingly, donor dopants are the Group VI elements, substituting for the Group V element, with one extra electron. The Group IV elements are amphoteric dopants in the III–V compounds, that is, they can be either donors or acceptors, depending on whether they substitute predominantly for the Group III or the Group V element.

The quantum efficiencies of these junctions can be exceptionally high, up to 1% (12–13). Quantum efficiency is the number of photons emitted per unit time, divided by the number of electrons flowing through the junction per unit time. Because of the large series resistance resulting from the high resistivity *p*-region, the power efficiencies achieved for blue emitters have only been of the order of 0.01%, too low for commerical displays.

In considering the optimum material for luminous efficacy (luminescence per unit current density) in direct gap material, it is necessary to determine the distribution of the electron population between the direct gap, Γ, and the indirect gap, X, and the radiative to nonradiative lifetimes, τ_R/τ_{NR}. The theoretical values in the equation below determine the electron distribution:

$$\frac{n_x}{n_\Gamma} = \frac{N_x}{N_\Gamma} \exp - [(E_x - E_\Gamma)/kT]$$

where N_x and N_Γ are the density of states at x and Γ, and $E_x - E_\Gamma$ is the energy difference in the two minima (14).

The external quantum efficiency is then given by:

$$\eta = f\left[1 + \frac{\tau_R}{\tau_{NR}}\left(1 + \frac{n_x}{n_\Gamma}\right)\right]^{-1}$$

where f = ratio of external to internal efficiency.

Both InGaP and InAlP are theoretically capable of well over an order of magnitude improvement over the widely used GaAsP (14). In practice, though, it has proved difficult to obtain either of these materials with a high degree of perfection and, since the nonradiative lifetime τ_{NR} is inversely proportional to the concentration of defect recombination centers, the theoretically expected improvements have not been realized.

The greatest gain in luminous efficacy has been accomplished by epitaxially depositing GaAsP on a transparent substrate of GaP, rather than the GaAs substrate originally used. This allows multiple passes, as discussed earlier in the case of red emitting GaP and, with a fairly thin epitaxial GaAsP layer, a much higher fraction of the internally generated light escapes.

Construction of VLED Displays. Most commercial VLED displays are constructed in a layer of GaAsP grown by vapor phase epitaxy. The growth is accomplished in a furnace of two or more temperature zones in which the gases H_2, HCl, AsH_3, PH_3, and a dopant gas such as H_2S are introduced at the high temperature end, and carefully polished single crystal GaAs (or GaP) substrates are positioned, usually in a rotating holder, at the other end. The $HCl-H_2$ stream is directed over a Ga reservoir where a volatile chloride is formed and swept by the flow of hydrogen, along with the other gases, across the substrates. There the GaAsP epitaxial layer is deposited according to a reaction (for the most generally used reactor conditions) of the form:

$$MCl_{(g)} = 1/z\, A_{z(g)} + \tfrac{1}{2} H_{2(g)} \rightarrow MA_{(s)} + HCl_{(g)},$$

where M is the Group III metal, A is As or P, and z is the number of atoms in the molecule.

There is a considerable lattice mismatch between GaAs and GaP; therefore, in the case of a GaAs substrate, the deposition is started with no PH_3 flow so as to deposit pure GaAs initially, and the PH_3 is gradually added to produce a graded transition region up to the desired GaAsP composition. In this way the defects due to strain caused by the lattice mismatch can be minimized.

To fabricate the displays, the GaAsP layer is coated with a diffusion masking layer, usually Si_3N_4, which is then patterned by photolithography to open windows for the p-type diffusant, usually Zn. After diffusion, the slice is lapped thin, metallized, scribed, and diced; then it is ready for assembly.

For very small digits, less than 2.5 mm high, it is practical to use monolithic chips, with all seven segments diffused on the same wafer. Generally a plastic lens is used over the digits to increase their apparent height to 2.5 mm or greater. This has the disadvantage, however, that the viewing angle is reduced, typically to about 35°.

In the case of intermediate sizes, a straight hybrid approach is used, in which the long, narrow bar segments are mounted in the familiar seven-segment pattern (usually with conductive epoxy) and viewed directly. For the large displays, eg, 13 mm characters, even this method consumes too much of the expensive GaAsP material, and small chips are used with reflectors or tapered light pipes (see Fiber optics) to achieve the larger size without reducing the viewing angle. Since the light from the small chip is, in effect, projected onto a larger slot, chips of high radiant intensity are required, and considerable care must be taken in the design to maintain a uniform appearance across the entire segment. Examples of each of the three types of assembly are shown in Figure 11.

Additional information about VLED displays can be obtained in several excellent survey articles such as refs 15–17. For textbooks on the physics of electroluminescence see, eg, refs. 18–20.

Gas Discharge Displays

Physics of the Discharge. Gas discharge displays emit light with an orange glow characteristic of neon. They employ a cold cathode from which electrons are emitted by positive ions striking the cathode. The electrons are accelerated by the electric potential (typically 150–250 V). In the immediate vicinity of the cathode, the electrons

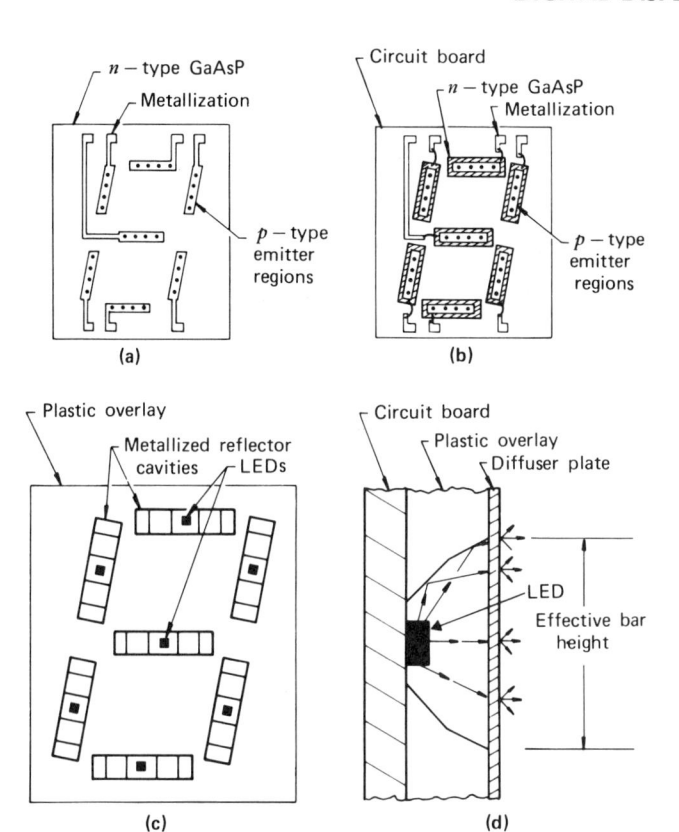

Figure 11. Methods of constructing VLED displays. (**a**) Monolithic chip; (**b**) hybrid chip; (**c**) and (**d**) use of a reflector cavity, front and side views, respectively.

acquire sufficient energy to ionize the gas. Some ions emit light and others strike the cathode. This negative glow is the most commonly used region for display purposes. The morphology of the discharge is shown in Figure 12. In most display applications, the cathode dark space is very small. It is possible to have a self-sustaining discharge without a positive column; in fact, the anode may even be located in the Faraday dark space.

The glow discharge has the peculiar property of running at a constant current density. The glow usually covers a portion of the cathode only, and its superficial area

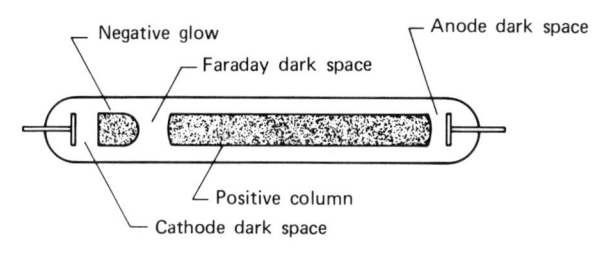

Figure 12. Morphology of the gas discharge tube. The figure identifies various zones in a typical d-c gas discharge.

expands or contracts in proportion to the current. The fundamental engineering relationships between firing voltage, steady state voltage, gap size and partial pressures of inert gases inside the tube are treated elsewhere in the Encyclopedia (see Helium-group gases) and in a text by Acton and Swift (21).

D-C and A-C Displays. There are two types of gas discharge tubes. In the d-c type, the display electrodes are in direct contact with the gas and the current flows only in one direction. In the a-c type the electrodes are coated with dielectric material which acts as a capacitor in series with the discharge. Once the a-c cell is fired with a high voltage pulse, charge stored on the dielectric aids the cell in firing at a lower applied alternating potential. This type of cell is said to have a memory as it will stay on once it is addressed. On the other hand, d-c displays must be refreshed at least every 33 ms to avoid flicker of the observed brightness.

Relative advantages and disadvantages of each display determine their applications. Presently, d-c gas discharge tubes are more commonly used on small to medium information displays (1–500 characters) whereas a-c plasma displays are more commonly used on large (ca 2000 characters) information displays. High voltage devices are required to drive a-c displays; multiplexing such a device is easier owing to its memory effect.

D-C Gas Discharge Display Fabrication. Since the fabrication materials and techniques are considerably different for these displays, fabrication of each display is discussed separately. There are five major steps in the manufacturing operation: (1) cathode fabrication, (2) anode fabrication, (3) assembly, (4) exhaust and backfill, and (5) burn-in and test.

Cathode Fabrication. In general, conductive thick film patterns are screen-printed on soda-lime glass to form the cathodes. Figure 13 shows a typical 9-digit cathode pattern where segments are interconnected for multiplexing purposes. Nickel is used as a conductor for its low work function as well as its superior resistance to sputtering. These conductors are fired in O_2/N_2 atmosphere at about 550°C for sintering purposes.

A thick-film dielectric layer is printed over the fired conductors leaving exposed only the part of the digit that needs to be visible. A black matte finish solder-glass layer is preferred to give a good contrast to orange glow. The firing temperature of this dielectric is about the same (ca 550°C) as for thick film conductors. Figure 14 shows the dielectric pattern.

In preparation for sealing the cathode substrate to the anode substrate, a seal ring of low temperature devitrified glass is screen-printed on the cathode substrate. The ring is 75–100 µm thick. The thermal expansion coefficient of the solder glass must

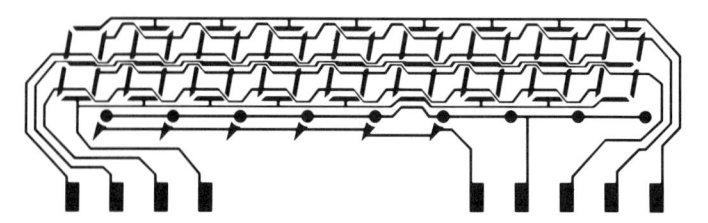

Figure 13. Cathode layout (7-segment digits). The figure shows pin layout for multiplexed 7-segment, 9-digit display.

Figure 14. Dielectric layer layout. The figure shows openings in the dielectric layer corresponding to a 7-segment, 9 digit-display.

match the soda-lime glass expansion coefficient. The seal ring is fired just below its glass-flow temperature so as not to cause any crystallization at this stage. At this stage the cathode substrate is ready for tube-sealing.

Anode Fabrication. Transparent conductors are deposited on a soda-lime glass substrate using CVD (chemical vapor deposition), spray or vacuum techniques (see Film deposition techniques). Tin oxide conductors are the most commonly used at present. A pattern of the type shown in Figure 15 is deposited. In general, sheet resistivity of 200–500 Ω/sq with 80–85% transparency is an acceptable specification. The same seal-glass is used as on the cathode substrate with a similar firing schedule.

Assembly. The cathode substrate, anode substrate and exhaust tubulation are assembled using a spacer preform and a tubulation attach preform, as shown in Figure 16. This assembly is done in a chain-driven furnace with a peak temperature of 475–500°C. The rate of temperature rise determines the degree of recrystallization of glass and the density of porosity in the seal ring. The spacer preform is generally fabricated to a thickness of ca 0.4–0.5 mm. An N_2 atmosphere is held inside of the furnace. The finished gap inside the display is held between 0.5–0.6 mm. The gap thickness and uniformity over the tube determine the firing characteristics of the cell.

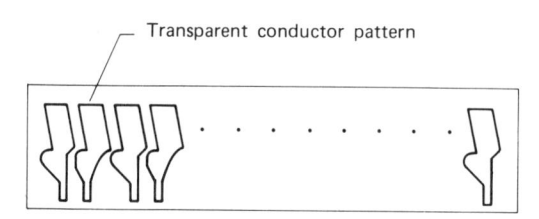

Figure 15. Anode substrate layout. The figure shows the anode pattern covering each digit.

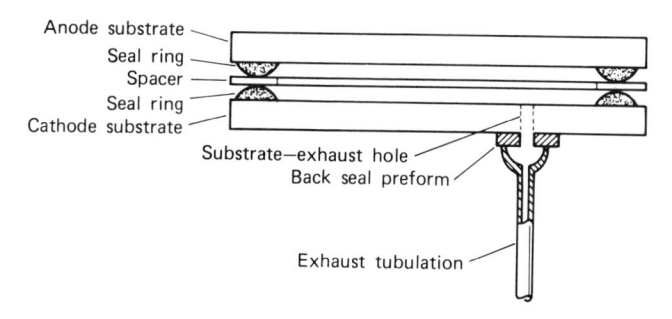

Figure 16. Assembly–d-c gas discharge display. The figure shows piece part assembly.

Exhaust and Backfill. In order to avoid sputtering of the cathode, mercury is dispensed inside the tube to protect the cathode surfaces. The mercury is encapsulated by an infrared-sensitive glass shell. The capsule is inserted in the back tubulation, and an extension is attached to the back tubulation so as to trap the mercury capsule. About 3–6 mg of mercury is used. The part is now ready for exhaust and backfill.

Conventional tube processing techniques are used to exhaust the tube at an elevated temperature of about 400°C at a vacuum of below 10^{-4} Pa (10^{-6} mm Hg). A Penning mixture of Ne + 0.3% Ar is filled in to 11–13 kPa (80–100 mm Hg) pressure. In most cases radioactive krypton-85 tracer gas is used to help initial ionization in the tube at a lower voltage. The tube is sealed (tipped-off) with mercury capsule intact in the back tubulation.

Using a laser beam or an ir source, the capsule is broken and mercury is distributed in a form of fine globules throughout the envelope of the display. In order to distribute the mercury evenly in the display, the tube is heated to ca 200°C through a chain furnace. This step assures an even mercury dispersion.

Burn-In and Test. Burn-in is generally done at a much higher voltage than the normal operating voltage. It is important to schedule the burn-in procedure in a manner that does not stress the segments. As a normal practice, individual segments are burned-in one at a time until the glow discharge seems to cover the segment area uniformly. This procedure may take 0.5–1 h. Then the tube may be completely turned on, in a multiplexed mode, for an additional 0.5–1 h to condition the cathode surfaces.

A-C Gas Discharge Display Fabrication. There are four major steps in the display fabrication: (*1*) front and back plate preparation, (*2*) assembly, (*3*) exhaust and backfill, and (*4*) burn-in.

Front and Back Plate Preparation. *Electrode fabrication.* For a 10-digit display, different electrode materials are used for front and back plates. Whereas front plates must use transparent conductors of low sheet-resistivity (20–30 Ω/sq), back plates can use any type of conductors as long as overall line resistance is not over 50 Ω.

Depending upon the size of the display, 1.5 mm-thick glass is used for both front and back plates. Conductive patterns are printed using silver, gold, etc, on the back plate. Thin-film metallization techniques are also used to deposit back plate patterns, though thick films are more cost-effective. The pattern layout is generally the same as for the d-c display devices.

Owing to the requirements of low resistance, front plate electrodes are not patterned in the d-c display manner. Instead, completely coated areas with a small gap between each electrode are patterned as shown in Figure 17.

Dielectric layer. The function of this dielectric layer is to provide a capacitive discharge, as discussed earlier. It therefore imposes very stringent requirements for holding the thickness of the fired layer within close tolerances. In general a fired thickness of 25 μm is used. The dielectric constant of this material is in the range of 12–16. It is also necessary for this layer to be free of pin holes and bubbles, and the surface of the dielectric needs to be clean and smooth. Both the front and back plates are completely coated with the dielectric and fired in a temperature range of 500–600°C.

Seal ring. Devitrified solder glasses are used to seal these displays, similar to d-c displays. A layer of solder glass in the form of a seal ring is printed on both plates and presintered to remove all binders and to achieve some adherence of glass particles to the substrates. The thickness of the deposit is 100–125 μm.

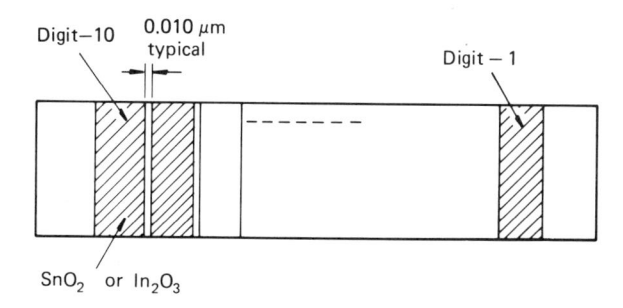

Figure 17. Front plate pattern–a-c plasma display. The transparent conductor pattern covers each digit.

Vacuum deposition of the emitter. As discussed earlier, lower work function surfaces are most desirable to drop the discharge voltage. MgO, Y_2O_3, and CeO_2 are the typical examples, though MgO is most commonly used. These barrier layers are generally evaporated or sputtered over the complete front and back plate surfaces. Owing to the nature of these layers, clean room environment and special handling precautions are taken so as not to poison the barrier layers. An advantage of such a layer is not only in lowering drive voltage but to provide a barrier layer between the dielectric layer and the discharge. Some tube reliability problems have been noted in the absence of any barrier layer. Lead oxide from the dielectric layer may decompose to contaminate the tube.

Assembly. Front plate, back plate, spacer and exhaust tubulation are combined as shown in Figure 16. In this display, fabrication of the spacer is unique as compared to d-c displays. The function of a spacer is not only to provide the correct size of the gap (ca 125 µm) but also to define discharge patterns more clearly. In the d-c display, the definition of the printed cathode layer defines the discharge. Since a dielectric coats the a-c electrode pattern, the discharge glow is considerably larger than the size of the electrode pattern. Generally luminance is bright at the center of the segment and of low intensity at the edge. Also, proper insulation is required between the front and back plates to keep the back plate bus bar from glowing. A dark blue spacer plate of ca 125 µm, is photoetched to a pattern as shown in Figure 14. This plate is fitted over the back plate and the assembly is sealed in a chain furnace using air or a nitrogen atmosphere.

Exhaust, Backfill and Burn-in. Conventional vacuum tube processing, similar to that used in d-c discharge display, is used to backfill and burn-in the display tube. The fill pressure of Penning gas mixture is ca 60–70 kPa (450–525 mm Hg). From a fabrication standpoint, the a-c plasma display might seem easier to build. The critical requirements on surface conditions, however, can lead to lower yields than for the d-c type display.

Operating Gas Discharge Displays. Although a firing voltage for a typical cathode-glow display may be 170–180 V, a swing of only 40–50 V is needed to turn the discharge on or off, so that lower voltage devices can be used to control the display. A typical 10-digit display is designed to be operated in a multiplexed mode where the cathode drive and decoder circuitry are timeshared among all of the digit positions. Corresponding cathode segments for all 10 character positions are bus-connected so that there are only nine cathode connections to the display. The conductor pattern layout is shown in Figure 13. Anodes, which are essentially transparent conductors, cover each character as shown in Figure 15. The anode is scanned fast enough to refresh information in less than 33 ms. Characters are displayed by successively applying positive signals to the anodes while negative signals are applied to the appropriate cathode bus line. Typical operating conditions and characteristics are shown in Table 1.

A sophisticated operating scheme has been employed in Burroughs "Self-Scan" alphanumeric dot matrix d-c displays (22). A priming technique is used to ensure that only the nearest neighbor to the existing discharge will fire when voltages are applied simultaneously. This technique makes possible the scanning of several hundred electrodes with only a few driving transistors at a great saving in drive circuitry costs.

As mentioned earlier, a-c plasma displays have an advantage of memory, but more expensive driver circuits are required (23).

Typical firing voltages for the a-c plasma are 120 V ac, with 90 V a-c sustaining voltages. Recent advances in solid state fabrication have made these operating voltages practical (24). The higher voltages mean that more expensive driver circuits are required for the a-c displays. This disadvantage is significant for small numeric displays, but it is not a factor when larger alphanumeric panels are required. Thus the popularity of the a-c plasma display has generally been in large information content displays.

Vacuum Fluorescent Displays

The vacuum fluorescent (VF) display emits green light which results when electrons from a hot cathode are accelerated into a phosphor (see Luminescent materials). Its extensive usage in handheld calculators is essentially due to multiplex capability at relatively low drive voltage (ca 25 V). This makes the electronic drivers cheaper for VF displays as compared to gas discharge displays.

Basic Principle and Materials. The display is built around the working principle of a triode tube with 3 basic parts to the display: (*1*) cathode filament (source of electrons); (*2*) metal grid (to accelerate and control the electrons); and (*3*) anode, coated with low voltage luminescent material.

In contrast to a vacuum tube configuration, the grid is located closer to the anode than to the cathode. In the on-state, the grid potential is kept at anode potential, 20–25

Table 1. Typical Operating Conditions

Operating conditions	Min	Max
cathode supply voltage, V_{KK}	−170	−210
cathode current, μA	−480	−750
anode off-state voltage, V	$V_{KA(on)}$[a] +125	−125
cathode off-state voltage, V_K (off), V	V_{KK} +125	−125
digit time period, μs	150	350
segment blanking interval, μs	25	55
luminance (1/10 duty cycle), cd/m^2	1.75	

[a] $V_{KA\ (on)}$ = segment on-state voltage: the voltage at the cathode terminal with respect to the anode terminal after segment is fired.

V d-c positive with respect to the cathode. Electrons are accelerated owing to the potential of the grid which is in the form of a stainless steel mesh, 90% optically transparent. The anode is coated with low voltage phosphor powder, such as ZnO:Zn. It emits in the visible (green) when struck by accelerated electrons. Standard texts on vacuum tubes may be consulted for detailed data on tube characteristics (25).

The most critical parts of the display are the filament cathode and the low voltage phosphor.

Filament Cathode. As a source of electrons, the cathode emission current density J_s is given by Richardson's equation (26):

$$J_s = 1.20 \times 10^6 \, T^2 \exp(-11605 \, \phi^1/T) \text{ A/m}^2,$$

where the work function ϕ^1 is in volts and cathode temperature T is in kelvins.

An important characteristic of a cathode is its relative efficiency, defined as the ratio of emission current to heater power; it is usually expressed in mA/w. The higher the work function of the emitting surface, the larger the cathode temperature needed to supply a given emission current density and the smaller the efficiency of the cathode. Oxide-coated cathodes are by far the most efficient as compared to metal emitters. Depending upon the activation process, materials used, etc, oxide coated filaments may emit 200–1000 mA/(cm^2·W).

These filament coatings consist of a mixture of barium, strontium, and calcium oxides, deposited on metal base. Since the oxides of the alkaline earth metals are not stable in air, they are generally applied in the form of the alkaline earth carbonates. The most efficient oxide coating composition reported to date is 50% $BaCO_3$, 43% $SrCO_3$, 7% $CaCO_3$. In a vacuum environment at a temperature of 1350–1500 K, the carbonates break down according to the process:

$$BaCO_3 \rightarrow BaO + CO_2$$

The oxide coating is left behind in the form of small crystallites of (Ba, Sr, Ca)O. The coating is not the mixture of BaO, SrO, or CaO grains, but each grain is a mixed crystal of (Ba, Sr, Ca)O. The difference in the size of each atom is relatively small, so that the atoms can take each other's place in the lattice. Usually the structure is quite porous. The cavities between the (Ba, Sr, Ca)O grains may be as large as the grains themselves.

To be a good emitter, the cathode coating should have a low work function. An activation process is performed to achieve this condition. Free barium is formed (26) by the reaction of the oxide with the tungsten metal base:

$$6 \, BaO + W \rightarrow Ba_3WO_6 + 3 \, Ba$$

The free barium thus obtained has a considerable vapor pressure. It flows slowly through the pores and migrates over the surface. At the operating temperature tungsten is normally covered by a monoatomic layer of oxygen. The barium is thus deposited on the top of this oxygen layer as well as partially covering the surface of the (Ba, Sr, Ca)O grains. Since the (Ba, Sr,Ca)O coating is a semiconductor, both its internal and its external work function is small.

In a typical 10-digit display, using an oxide coated cathode of ca 20 μm dia, the anode and grid currents are saturated when filament voltage is at ca 2.8 V RMS (root mean square) with the anode and grid held at +24 V dc with respect to cathode. The filament current is about 22 mA ac and the temperature of the filament is ca 700°C.

Low Voltage Phosphors. ZnO:Zn is the most commonly used phosphor powder in vacuum fluorescence displays. Light emission threshold is reported to be less than 4 eV and the emission intensity and the decay time characteristics make it a suitable material for low voltage displays. ZnO is typically an n-type semiconductor with donor levels at 0.04 to 0.05 eV. These levels result from an interstitial excess of Zn in concentrations of 10^{15}–10^{18} atoms/cm^3, depending upon the technique of preparation. Green luminescence appears if ZnO is prepared under reducing conditions, which produces further excess of Zn. The oxygen vacancies must be produced during the formation of the crystal. They cannot be obtained by a reducing agent acting on the surface. Based on these considerations, it has been proposed (28) that the green luminescence corresponds to a transition from Zn$^+$ to Zn^{2+} in the excess Zn ions or to a transition in an oxygen vacancy.

A theoretical expression for the threshold energy E_t is (29):

$$E_t = E_g[1 + m_e/(m_e + m_h)]$$

where E_g is the band gap energy (3.2 eV) and m_e and m_h are the electron and hole effective masses, respectively. The effective-mass ratio $m_e/m_h = 0.21 \pm 0.02$.

It has been shown that the ZnO:Zn has an adequate brightness and a long enough life potential for use in displays. However, the life is intimately tied to the proper choice of cathode filament and the careful preparation and aging of the device. It has also been shown that atomic hydrogen increases the luminescent yield and atomic oxygen decreases the luminescence yield. These effects have been explained by charging and nonradiative recombination at the surface. The effect on luminescence of small amounts of O$_2$ in the tube suggests that the slow evolution of oxygen by the tube structure may contribute to the observed brightness decay, as seen on many VF displays. It is almost certainly a factor in carelessly prepared devices.

Fabrication. There are six major steps in fabrication of VF displays: (*1*) anode fabrication, (*2*) cathode filament preparation, (*3*) mechanical assembly, (*4*) tube assembly, (*5*) exhaust and activation, and (*6*) getter flash and test.

Anode Fabrication. The basic conductor pattern is similar to the 10-digit gas discharge display cathode, as shown in Figure 13. Silver thick film on soda-lime glass is conventionally used and fired at 550°C in air.

A black dielectric layer of solder glass is screen-printed over silver, leaving small holes in the dielectric to connect to silver pattern. The part is now ready for ZnO:Zn phosphor deposition. Conventionally ZnO:Zn is not directly deposited over the dielectric for two reasons. If the underlayer is nonconductive, there will be accumulation of charge on the phosphor which affects its performance. The electrophoretically deposited phosphor tends to crack when deposited directly on the solder glass. Thus graphite in the pattern of the character to be displayed is screen-printed over the dielectric. Graphite powder in a screenable medium is generally used together with a low temperature fire to remove the solvent. Electrophoretic deposition of the phosphor is achieved in a solvent bath of acetone using Al(NO$_3$)$_3$ as a binder. With phosphor powder of 2–10 μm size a deposition of 40 μm is produced. The part is ready for mechanical assembly.

Cathode Filament Preparation. Pure tungsten or tungsten–rhenium wire of 5–10 μm dia is conventionally used. A triple carbonate of Ba–Sr–Ca, mixed with a dilute binder, is electrophoretically deposited on the tungsten wire at relatively low voltages. The deposition technique determines the density of deposit, a critical factor for an efficient electron emitter. The coated wire diameter is ca 25 μm.

Mechanical Assembly. The following parts are assembled at this stage: anode, cathode filament, pedestals (to support filament), Cu–Be spring (at one end, to keep wire in tension, when hot), grid (stainless steel), and getters. Mechanical assembly of display is shown in Figure 18. The grid is manufactured using 35-μm-thick stainless steel sheets and photo-etch pattern as shown. It is important that the grid is 85–90% transparent.

Tube Assembly. The front cover glass is in the form of a cup. It is sealed to the assembled anode substrate by a solder-glass seal. Exhaust tubulation is sealed at this time using the same glass. It has been found necessary to coat the inside of the front glass with transparent conductors (SnO_2, In_2O_3, etc), to avoid charge accumulation.

Exhaust and Filament Activation. The tube is evacuated and baked to $<10^{-4}$ Pa ($<10^{-6}$ mm Hg). The activation process involves conversion of carbonates to oxides in a vacuum. A small a-c voltage is applied to the heater and raised by small steps watching the ionization gage reading. The pressure should not be allowed to rise above 5×10^{-3} Pa (5×10^{-5} mm Hg). If the temperature of the cathode is raised too rapidly, there is a risk of blistering, flaking, or evaporation of the coating. The rise in pressure is due largely to the evolution of carbon dioxide with, perhaps, some water vapor and other gases. The normal operating temperature for an oxide-coated cathode is about 700–800°C. During the activation process, however, heater currents produce cathode temperatures to 1100°C for short periods (30 s) at intervals of several minutes (flashing). Maximum chemical change occurs at ca 900°C; the higher temperature is applied to ensure oxide conversion. If a very hard tube is required, the baking (both first and last) is extended many hours, with the cathode at normal operating temperature and a small anode current being drawn. After activation, the tube is sealed immediately after the tube has reached a vacuum of $\leq 3 \times 10^{-5}$ Pa (3×10^{-7} mm Hg).

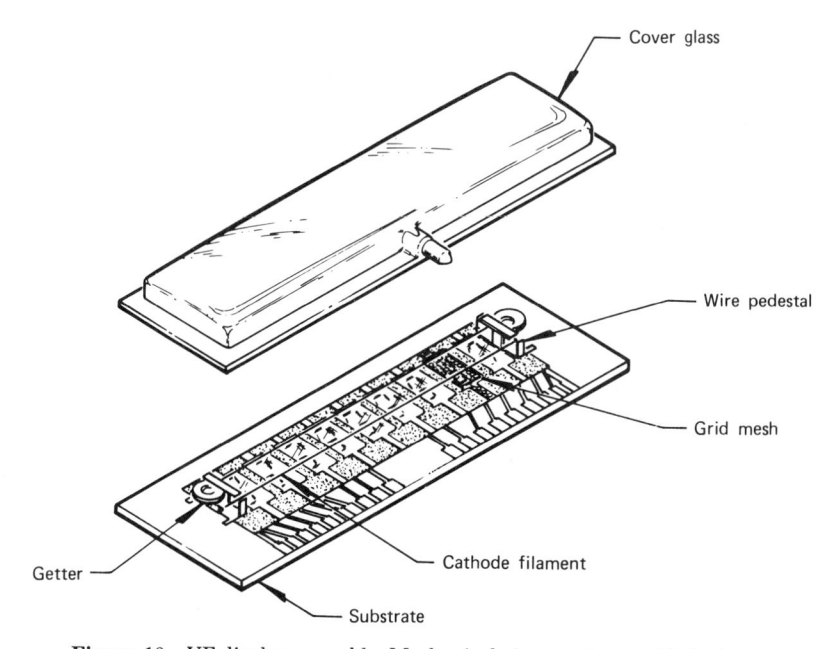

Figure 18. VF display assembly. Mechanical piece part assembly is shown.

Getter Flash and Test. As shown in Figure 18 a getter is located in the tube and shielded with nickel gates to prevent the sputtering of metal. The getter is in the form of a ring and contains barium which is flashed with an induction coil at this stage. Anode and grid potentials are then raised slowly to the tube operating conditions. The tube is left on for several hours and any change in brightness of phosphor is noted.

Tube Characteristics.

filament voltage	3.0 V
filament current	22 mA
anode, grid voltage (on voltage)	24 V
grid off voltage	−6 V
brightness ($1/9$ duty cycle)	514 cd/m^2
power consumption (per digit)	12 mW

For larger displays, requiring higher multiplexibility, anode and grid voltage are higher to get peak brightness. A typical 40-digit display is driven at 40–50 V.

The lifetime of the VF tube is a major problem. Although most manufacturers guarantee the minimum life to be 10,000 h, deterioration of the performance is generally seen much earlier. The display is, in fact, no different than a conventional vacuum tube. Further complications result from the aging of the ZnO–Zn phosphor. Mechanical design considerations are complex, restricting its use to relatively small displays.

Other Technologies

Four display technologies have been described in some detail: LCDs, LEDs, gas discharge displays and vacuum fluorescence displays. All four are in volume production for use in the vast majority of digital display applications. Light generation by a hot tungsten filament gives the brightest available display (as high as 41,100 cd/m^2). Power requirements, typically 5 V at 20 mA, and susceptibility to damage from shock have limited the market for these devices.

What are the technologies of the future? For light emitting displays, high field-injection electroluminescence has received a great deal of recent attention (30). The effect has been known since the 1930s, but lifetime problems had limited its usefulness. As opposed to LEDs, which use single crystal material with a *p-n* junction, electroluminescence can also be observed in thin films of polycrystalline, wide band-gap materials such as ZnS under large voltages. Protecting the active film with dielectric layers has led to greatly extended lifetimes, and work on graphics panels is underway in many laboratories.

In the passive display area, electrochromic displays (ECD) have received significant effort during the 1970s (see Chromogenic materials, electrochromic). The display mechanism is electrochemical in nature, ie, a battery. An absorbing layer is produced at the viewing electrode with a chemical change induced by electrons and ions. WO$_3$ can be changed from transparent to blue with voltages on the order of 1 V and charge densities on the order of 0.3 C/cm^2 (31). Another example in the same voltage/charge region is the deposition of viologen organic dyes (32). The main problem with these displays is lifetime. A device is needed that will completely charge and then discharge millions of times. The advantage is that the displays are bright with respect to the twisted nematic LCD, and do not require polarizers.

BIBLIOGRAPHY

1. U.S. Pat. 3,731,986 (May 8, 1973), J. Fergason (to International Liquid Xtal Co.)
2. M. Schadt and W. Hefrich, *Appl. Phys. Lett.* **18,** 127 (1971).
3. G. H. Heilmeier, L. A. Zanoni, and L. A. Barton, *Appl. Phys. Lett.* **13,** 46 (1968).
4. J. L. Janning, *Appl. Phys. Lett.* **21,** 173 (1972).
5. H. G. Grim and A. Sommerfeld, *Z. Physik* **36,** 36 (1926).
6. H. Welker, *Z. Naturforsch. Teil A* **7,** 744 (1952); **8,** 248 (1953).
7. O. G. Folberth, *Z. Naturforsch. Teil A* **10,** 502 (1955).
8. L. Nordheim, *Am. Phys.* **9,** 607, 641 (1931).
9. D. G. Thomas, J. J. Hopfield, and C. J. Frosch, *Phys. Rev. Lett.* **15,** 857 (1965).
10. N. W. Ashcroft and N. D. Mermin, *Solid State Physics,* Holt, Rinehart, and Winston, New York, 1976, Chapt. 8.
11. R. Solomon and D. DeFevere, *Appl. Phys. Lett.* **21,** 257 (1972).
12. J. I. Pankove, E. A. Miller, and J. E. Berkeyheiser, *J. Lumin.* **5,** 84 (1972).
13. J. I. Pankove, *RCA Rev.* **34,** 336 (1973).
14. R. J. Archer, *J. Electron. Mater.* **1,** 128 (1972).
15. A. A. Bergh and P. J. Dean, *Proc. Inst. Electr. Eng.* **60,** 156 (1972).
16. E. E. Loebner, *Proc. Inst. Electr. Eng.* **61,** 837 (1973).
17. M. G. Craford and W. O. Groves, *Proc. Inst. Electr. Eng.* **61,** 862 (1973).
18. A. A. Bergh and P. J. Dean, *Light-emitting Diodes,* Clarendon Press, Oxford, 1976.
19. J. I. Pankove, ed., *Topics in Applied Physics,* Vol. 17, *Electroluminescence,* Springer Verlag, Berlin, 1977.
20. J. I. Pankove, *Optical Processes in Semiconductors,* Prentice-Hall Inc., Engle Cliffs, N.J., 1971.
21. J. R. Acton, *Cold Cathode Discharge Tube,* Academic Press Inc., New York, 1963.
22. G. Chodil, *Proc. of the Soc. for Inf. Disp.* **17,** 14 (1976).
23. A. Sobel, *Proc. of the Soc. for Inf. Disp.* **18,** 51 (1977).
24. T. N. Criscimagna, *Proc. of the Soc. for Inf. Disp.* **17,** 124 (1976).
25. S. Spangenberg, *Vacuum Tubes,* McGraw-Hill, New York, Electrical and Electronics Engineering Series, 1948.
26. *Ibid.,* p. 30.
27. A. Van Der Ziel, *Solid State Physical Electronics,* Prentice-Hall, Electrical Engineering Series, 1957.
28. A. Pfahnl, *J. Electro. Chem. Soc.* **109,** 502 (1962).
29. K. Motizuki and M. Sparks, *J. Phys. Soc. Jpn.* **19,** 486 (1964).
30. T. Inoguchi and co-workers, *Digest 1974 Soc. Inf. Disp. Int. Symp.* 84 (1974).
31. S. K. Deb, *Phil. Mag.* **27,** 801 (1973).
32. C. J. Schoot and co-workers, *Digest 1973 Soc. Inf. Disp. Int. Symp.* 146 (1973).

P. Andrew Penz
Robert W. Haisty
Kishin H. Surtani
Texas Instruments Incorporated

DIMENSIONAL ANALYSIS

Dimensional analysis is a technique that treats the general forms of equations governing natural phenomena. It provides procedures of judicious grouping of variables associated with a physical phenomenon to form dimensionless products of these variables; therefore, without destroying the generality of the relationship, the equation describing the physical phenomenon may be more easily determined experimentally. It guides the experimenter in the selection of experiments capable of yielding significant information and in the avoidance of redundant experiments (see also Design of experiments). The method is particularly valuable when the problems involve a large number of variables. On such occasions, dimensional analysis may reveal that, whatever the form of the inaccessible final solution, certain features of it are obligatory. The technique has been utilized effectively in engineering modeling (1–7).

The method of dimensional analysis is not new. It can be traced to Newton who at the time was laying the foundation of mechanics as a fundamental branch of science. The validity of the method is based on the premise that any equation that correctly describes a physical phenomenon must be dimensionally homogeneous. This principle of dimensional homogeneity, which states in effect that quantities of different kinds cannot be added together, is of fundamental importance in dimensional analysis, and was first expressed by J. Fourier's classic work *La Théorie Analytique de la Chaleur*, published in 1822. He not only introduced the notion of dimensional homogeneity but also the conception of what is today termed the dimensional formula. In 1914, Buckingham (8) made a significant contribution with his famous Pi Theorem which possibly prompted Lord Rayleigh (9) shortly thereafter to observe, "It happens not infrequently that results in the form of 'laws' are put forward as novelties on the basis of elaborate experiments which might have been predicted *a priori* after a few minutes' consideration." Needless to say, Fourier's and Buckingham's works have been used, elaborated upon, and extended by many others (10–30). A measure of this interest is reflected in three comprehensive bibliographies owing to Higgins (10–11), Sloan, and Happ (12), in which more than 600 research contributions are referenced.

Units and Dimensions

The concepts used to describe natural phenomena are based on the precise measurement of quantities. The quantitative measure of anything is a number that is found by comparing one magnitude with another of the same type. It is necessary to specify the magnitude of the quantity used in making the comparison if the number is to be meaningful. The statement that "the length of a car is 6 meters" implies that we have chosen a length, namely, one meter, and that the ratio of the length of the car to the chosen length is 6. The chosen magnitudes, such as the meter, are called units of measurement. The result of a measurement is represented by a number followed by the name of the unit that was used in making the measurement (see Units). The statement that the area of a room is 30 square meters indicates that the unit of measurement is one square meter. Thus, to each kind of physical quantity there corresponds an appropriate kind of unit. The physical concepts such as length, area, and time are referred to as dimensions, which are different from units. The length of a car is 6 meters which is equivalent to 19.7 feet or 6.56 yards.

The classical physics is built on the foundation of the laws of motion. It was felt at the time that the entire subject could be based on the laws of classical mechanics, and further work would undoubtedly make electromagnetism another branch of mechanics. Under these circumstances, it was natural to regard length l, mass m, and time t as the fundamental, primary, or reference dimensions. However, such designations lead to dimensional ambiguity in that two distinct concepts may possess the same dimensions. The system works fairly well for mechanics. The most notable ambiguity occurs with energy and torque. However, in electromagnetism the situation is bad. The classical arrangement employs electrostatic and electromagnetic systems, and the same concept may lead to different dimensions. For example, in the electrostatic system capacitance and length have the same dimensions, and in the electromagnetic system inductance and length are not dimensionally distinguished. On the subject of heat, ambiguities occur between entropy and mass (13), or temperature and reciprocal length (14), depending upon the assumptions made about the dimensionality of some constants. This does not mean that dimensional ambiguity is a fact of nature; it merely shows an imperfection in our human scheme of assigning primary concepts.

Over the years the number of reference dimensions in physics has evolved from the original three, to four, to five, and then gradually downwards to an absolutely necessary one, and then upwards again through an understanding that, though only one is absolutely necessary, a considerable convenience can stem from using three, four, or five reference dimensions depending on the problem at hand (1,6–7,15–20). Bridgman (1) emphasizes the fact that there is nothing sacrosanct about the number of reference dimensions and that dimensional analysis is merely a tool that may be manipulated at will. This principle of free choice of the reference dimensions has been widely accepted, although one still finds references to "true dimensions". Thus, an important step in dimensional analysis is the selection of reference dimensions in such a way that the others, called the secondary or derived dimensions, can be expressed in terms of them. The relation between reference and derived dimensions is generally established either through the fundamental law or equation governing the phenomenon or through definitions. When length, mass, and time are taken to be the reference dimensions, the dimensions of velocity v, for example, are the dimensions of length divided by time, or expressed in symbols, $v = lt^{-1}$. Likewise, through Newton's law of motion which relates force, mass, and acceleration by

$$\text{force} = \text{constant} \cdot \text{mass} \cdot \text{acceleration}, \qquad (1)$$

the dimensions of force f must be mass-length/time^2 or $f = mlt^{-2}$. The expressions like $v = lt^{-1}$ and $f = mlt^{-2}$ are referred to as dimensional formulas. The exponents of dimensions of a physical quantity are the powers of the reference dimensions in which it is expressed. Thus, the exponents of the dimensions of force are 1 in mass, 1 in length, and -2 in time. If force, length, and time are chosen as the reference dimensions then mass becomes secondary. In either of these two choices, the constant in Newton's law is dimensionless. However if force, mass, length, and time are chosen as the reference dimensions, the constant is no longer dimensionless and the units are generally selected so that the constant is numerically equal to the standard acceleration of gravity. Table 1 lists the exponents of dimensions of some common variables in mechanics with respect to these three choices of reference dimensions which give rise to the absolute, gravitational, and engineering systems of dimensions, respectively.

754 DIMENSIONAL ANALYSIS

Table 1. Exponents of Dimensions for Mechanical Quantities

Quantity	m	l	t	f	l	t	f	m	l	t
acceleration	0	1	−2	0	1	−2	0	0	1	−2
angular acceleration	0	0	−2	0	0	−2	0	0	0	−2
angular velocity	0	0	−1	0	0	−1	0	0	0	−1
area	0	2	0	0	2	0	0	0	2	0
angular momentum	1	2	−1	1	1	1	0	1	2	−1
density	1	−3	0	1	−4	2	0	1	−3	0
energy, work	1	2	−2	1	1	0	1	0	1	0
force	1	1	−2	1	0	0	1	0	0	0
frequency	0	0	−1	0	0	−1	0	0	0	−1
length	0	1	0	0	1	0	0	0	1	0
linear acceleration	0	1	−2	0	1	−2	0	0	1	−2
linear momentum	1	1	−1	1	0	1	0	1	1	−1
linear velocity	0	1	−1	0	1	−1	0	0	1	−1
mass	1	0	0	1	−1	2	0	1	0	0
moment of inertia	1	2	0	1	1	2	0	1	2	0
power	1	2	−3	1	1	−1	1	0	1	−1
pressure	1	−1	−2	1	−2	0	1	0	−2	0
stress	1	−1	−2	1	−2	0	1	0	−2	0
surface tension	1	0	−2	1	−1	0	1	0	−1	0
time	0	0	1	0	0	1	0	0	0	1
viscosity, absolute	1	−1	−1	1	−2	1	1	0	−2	1
viscosity, kinematic	0	2	−1	0	2	−1	0	0	2	−1
volume	0	3	0	0	3	0	0	0	3	0

To eliminate the ambiguities in the subject of electricity and magnetism, it is convenient to add charge q to the traditional l, m, and t dimensions of mechanics to form the reference dimensions. In many situations permittivity ϵ or permeability μ is used in lieu of charge. For thermal problems temperature T is considered as a reference dimension. Tables 2 and 3 list the exponents of dimensions of some common variables in the fields of electromagnetism and heat.

Other dimensional systems have been developed for special applications which can be found in the technical literature. In fact, to increase the power of dimensional analysis it is advantageous to differentiate between the lengths in radial and tangential directions (13). In doing so ambiguities for the concepts of energy and torque, as well as for normal and shear stress are eliminated (see ref. 13).

Dimensional Matrix and Dimensionless Products

An appropriate set of independent reference dimensions may be chosen so that the dimensions of each of the variables involved in a physical phenomenon can be expressed in terms of these reference dimensions. In order to utilize the algebraic approach to dimensional analysis, it is convenient to display the dimensions of the variables by a matrix. The matrix is referred to as the dimensional matrix of the variables and is denoted by the symbol $\mathbf{D}$. Each column of $\mathbf{D}$ represents a variable under consideration, and each row of $\mathbf{D}$ represents a reference dimension. The ith row and jth column element of $\mathbf{D}$ denotes the exponent of the reference dimension corresponding to the ith row of $\mathbf{D}$ in the dimensional formula of the variable corresponding to the jth column. As an illustration, consider Newton's law of motion which relates force F, mass M, and acceleration A by (eq. 2):

$$F = \text{constant} \cdot MA \qquad (2)$$

Table 2. Exponents of Dimensions for Electromagnetic Quantities

Quantity	l	m	t	q	l	m	t	ε	l	m	t	μ
charge	0	0	0	1	3/2	1/2	−1	1/2	1/2	1/2	0	−1/2
capacitance	−2	−1	2	2	1	0	0	1	−1	0	2	−1
current	0	0	−1	1	3/2	1/2	−2	1/2	1/2	1/2	−1	−1/2
electric field intensity	1	1	−2	−1	−1/2	1/2	−1	−1/2	1/2	1/2	−2	1/2
electric potential difference	2	1	−2	−1	1/2	1/2	−1	−1/2	3/2	1/2	−2	1/2
electric flux	0	0	0	1	3/2	1/2	−1	1/2	1/2	1/2	0	−1/2
electric flux density	−2	0	0	1	−1/2	1/2	−1	1/2	−3/2	1/2	0	−1/2
inductance	2	1	0	−2	−1	0	2	−1	1	0	0	1
magnetic field intensity	−1	0	−1	1	1/2	1/2	−2	1/2	−1/2	1/2	−1	−1/2
magnetic flux	2	1	−1	−1	1/2	1/2	0	−1/2	3/2	1/2	−1	1/2
magnetic flux density	0	1	−1	−1	−3/2	1/2	0	−1/2	−1/2	1/2	−1	1/2
magnetomotive force	0	0	−1	1	3/2	1/2	−2	1/2	1/2	1/2	−1	−1/2
permeability	1	1	0	−2	−2	0	2	−1	0	0	0	1
permittivity	−3	−1	2	2	0	0	0	1	−2	0	2	−1
resistance	2	1	−1	−2	−1	0	1	−1	0	0	−1	1

Table 3. Exponents of Dimensions for Thermal Quantities

Quantity	l	m	t	T	f	l	t	T
coefficient of thermal expansion	0	0	0	−1	0	0	0	−1
entropy	2	1	−2	−1	1	1	0	−1
temperature	0	0	0	1	0	0	0	1
thermal energy (heat)	2	1	−2	0	1	1	0	0
thermal power	2	1	−3	0	1	1	−1	0
thermal conductivity	1	1	−3	−1	1	0	−1	−1
thermitivity	2	0	−2	−1	0	2	−2	−1

If length l, mass m, and time t are chosen as the reference dimensions, from Table 1 the dimensional formulas for the variables F, M, and A are shown below:

Variables	Dimensional formulas
F	$m^1 l^1 t^{-2}$
M	$m^1 l^0 t^0$
A	$m^0 l^1 t^{-2}$

The dimensional matrix associated with Newton's law of motion is obtained as (eq. 3):

$$\mathbf{D} = \begin{array}{c} \\ m \\ l \\ t \end{array} \begin{array}{c} F \quad M \quad A \\ \left[\begin{array}{ccc} 1 & 1 & 0 \\ 1 & 0 & 1 \\ -2 & 0 & -2 \end{array} \right] \end{array} \quad (3)$$

In the example, the exponents of dimensions in the dimensional formula of the variable F are 1, 1, and −2, and hence the first column is (1, 1, −2). Likewise, the second and

756 DIMENSIONAL ANALYSIS

third columns of D correspond to the exponents of dimensions in the dimensional formulas of the variables M and A.

As indicated earlier, the validity of the method of dimensional analysis is based on the premise that any equation that correctly describes a physical phenomenon must be dimensionally homogeneous. An equation is said to be dimensionally homogeneous if each term has the same exponents of dimensions. Such an equation is of course independent of the systems of units employed provided the units are compatible with the dimensional system of the equation. It is convenient to represent the exponents of dimensions of a variable by a column vector called dimensional vector represented by the column corresponding to the variable in the dimensional matrix. In equation 3, the dimensional vector of force F is $[1, 1, -2]'$ where the prime denotes the matrix transpose.

Suppose that there are n variables $Q_1, Q_2, \ldots, Q_n$ that are involved in a physical phenomenon whose dimensional vectors are $D_1, D_2, \ldots, D_n$, respectively. This phenomenon can generally be expressed by equation 4:

$$f(Q_1, Q_2, \ldots, Q_n) = 0 \tag{4}$$

When such a function is established or assumed, a function will still exist even after the variables are intermultiplied in any manner whatsoever. This means that each variable in the equation can be combined with other variables of the equation to form dimensionless products whose dimensional vectors are the zero vector. Equation 4 can then be transformed into the nondimensional form as (eq. 5):

$$f(\pi_1, \pi_2, \ldots, \pi_n) = 0, \tag{5}$$

where the dimensionless products π_i ($i = 1, 2, \ldots, n$) can generally be expressed as the power products of the form (eq. 6):

$$\pi_i = Q_1^{x_{1i}} Q_2^{x_{2i}} \ldots Q_n^{x_{ni}}. \tag{6}$$

Let $R_1, R_2, \ldots, R_m$ be a set of chosen reference dimensions. Then the dimensional formulas for the variables Q_i are given by (eq. 7):

$$R_1^{d_{1i}} R_2^{d_{2i}} \ldots R_m^{d_{mi}}, \tag{7}$$

where the exponents of dimensions are represented by the dimensional vectors as (eq. 8):

$$D_i' = [d_{1i}, d_{2i}, \ldots, d_{mi}] \quad i = 1, 2, \ldots, n. \tag{8}$$

Using (eq. 7) the dimensional formulas for π_i of equation 6 can be written to give (eq. 9):

$$(R_1^{d_{11}} R_2^{d_{21}} \ldots R_m^{d_{m1}})^{x_{1i}} (R_1^{d_{12}} R_2^{d_{22}} \ldots R_m^{d_{m2}})^{x_{2i}} \ldots (R_1^{d_{1n}} R_2^{d_{2n}} \ldots R_m^{d_{mn}})^{x_{ni}} \tag{9}$$

Since π_i are dimensionless products having dimensional vectors at the zero vector, the exponents of the R_j ($j = 1, 2, \ldots, m$) must add up to zero, giving (eq. 10):

$$\begin{aligned} d_{11} x_{1i} + d_{12} x_{2i} + \ldots + d_{1n} x_{ni} &= 0 \\ d_{21} x_{1i} + d_{22} x_{2i} + \ldots + d_{2n} x_{ni} &= 0 \\ &\vdots \\ d_{m1} x_{1i} + d_{m2} x_{2i} + \ldots + d_{mn} x_{ni} &= 0 \end{aligned} \tag{10}$$

In terms of the dimensional vectors of equation 8, equation 10 can be rewritten as (eqs. 11–13):

$$[\boldsymbol{D}_1, \boldsymbol{D}_2, \ldots, \boldsymbol{D}_n]\boldsymbol{X}_i = \boldsymbol{0}, \quad i = 1, 2, \ldots, n, \tag{11}$$

where

$$\boldsymbol{X}'_i = [x_{1i}, x_{2i}, \ldots, x_{ni}] \tag{12}$$

or more compactly

$$\boldsymbol{DX} = \boldsymbol{0}, \tag{13}$$

where $\boldsymbol{X} = \boldsymbol{X}_i, i = 1, 2, \ldots, n$. Thus, the product of a set of variables is dimensionless if, and only if, the exponents of these variables are a solution of the homogeneous linear algebraic equation (13). It is assumed here that $\boldsymbol{X}$ is a B-vector of $\boldsymbol{D}$ if it is a solution of equation 13. The corresponding dimensionless product associated with the variables of a B-vector is called a B-number (21–22). Frequently, the term pi-number is also used by many authors because it was first introduced by Buckingham (8) in 1914 who used the symbol π for a dimensionless product or group. In fact, the name was even attached to his contributions to dimensional analysis, and is known as Buckingham's Pi-Theorem. But the term is deprecated because of possible confusion with the universal constant of $\pi = 3.14159$ and his initial B is preferred to the term π.

The following example illustrates the above procedure.

Example 1. The problem is to find the period (P) of a simple pendulum swinging in a vacuum under the influence of gravity. To write an equation for the period, the first step is to consider what physical quantities affect the period. On this basis, it is clear that the period depends on the mass M of the bob, the length L of the string supporting the bob, and of course, the acceleration g owing to the force of gravity. As before, mass m, length l, and time t are chosen as the reference dimensions, ie, $R_1 = m$, $R_2 = l$, and $R_3 = t$. From Table 1 the dimensional formulas for the variables $M, L, P,$ and g together with their dimensional vectors are as shown below:

Variables	Dimensional formulas	Dimensional vectors
$Q_1 = M$	$m^1 l^0 t^0$	$\boldsymbol{D}'_1 = [1, 0, 0]$
$Q_2 = L$	$m^0 l^1 t^0$	$\boldsymbol{D}'_2 = [0, 1, 0]$
$Q_3 = P$	$m^0 l^0 t^1$	$\boldsymbol{D}'_3 = [0, 0, 1]$
$Q_4 = g$	$m^0 l^1 t^{-2}$	$\boldsymbol{D}'_4 = [0, 1, -2]$

Thus, the dimensional formulas for π_i of equation 6 can be expressed as (eq. 14):

$$(m^1 l^0 t^0)^{x_{1i}}(m^0 l^1 t^0)^{x_{2i}}(m^0 l^0 t^1)^{x_{3i}}(m^0 l^1 t^{-2})^{x_{4i}} \tag{14}$$

whose dimensional vector must be the zero vector requiring (eq. 15):

$$\begin{array}{c} \\ m \\ l \\ t \end{array} \begin{bmatrix} M & L & P & g \\ 1 & 0 & 0 & 0 \\ 0 & 1 & 0 & 1 \\ 0 & 0 & 1 & -2 \end{bmatrix} \begin{bmatrix} x_{1i} \\ x_{2i} \\ x_{3i} \\ x_{4i} \end{bmatrix} = \begin{bmatrix} 0 \\ 0 \\ 0 \end{bmatrix} \tag{15}$$

the coefficient matrix being identified as the dimensional matrix $\boldsymbol{D}$ associated with the pendulum problem, and $i = 1, 2, 3, 4$. Solving $x_{1i}, x_{2i},$ and x_{3i} in terms of x_{4i} yields the desired B-vectors of $\boldsymbol{D}$ as (eq. 16):

758   DIMENSIONAL ANALYSIS

$$X'_i = [0, -x_{4i}, 2x_{4i}, x_{4i}] \quad i = 1, 2, 3, 4, \tag{16}$$

where $x_{4i} \neq 0$ are arbitrary constants. The corresponding B-numbers become (eq. 17).

$$\pi_i = M^0 L^{-x_{4i}} P^{2x_{4i}} g^{x_{4i}} \tag{17}$$

Since the solutions X_i are related to one another by a multiplicative constant, there is only one linearly independent solution and hence only one independent B-number. Choose, for simplicity, $x_{4i} = 1$. Equation 17 can be rewritten as (eq. 18):

$$P = \text{constant } \sqrt{L/g} \tag{18}$$

since π_i is a constant. The value of this constant, which is known to be 2π, cannot be determined by the method of dimensional analysis, and must be evaluated experimentally or analytically.

The example demonstrates that not all the B-numbers π_i of equation 5 are linearly independent. A set of linearly independent B-numbers is said to be complete if every B-number of D is a product of powers of the B-numbers of the set. To determine the number of elements in a complete set of B-numbers, it is only necessary to determine the number of linearly independent solutions of equation 13. The solution to the latter is well known and can be found in any text on matrix algebra (see, for example, Hohn (31) and Bellman (32)). Thus, theorems can be stated.

Theorem 1.   The number of products in a complete set of B-numbers associated with a physical phenomenon is equal to $n - r$, where n is the number of variables that are involved in the phenomenon and r is the rank of the associated dimensional matrix.

This result was first discussed by Buckingham (8) and stated in its present form by Langhaar (23). It states in effect that an equation is dimensionally homogeneous if and only if it can be reduced to a relationship among a complete set of B-numbers. Buckingham's result (8) was originally stated as Theorem 2.

Theorem 2.   A necessary and sufficient condition for an equation $f(Q_1, Q_2, \ldots, Q_n) = 0$ to be dimensionally homogeneous is that it should be reducible to the form $g(\pi_1, \pi_2, \ldots, \pi_p) = 0$, where the πs are a complete set of B-numbers of the variables Qs.

This theorem does not specify how many products in a complete set of B-numbers can be expected from a given set of variables, but it does state that a physical phenomenon describable by n quantities can be rigorously and accurately described by a complete set of B-numbers. The number of products in a complete set of B-numbers may also be determined by another rule, which is equivalent to Theorem 1. The rule (Theorem 3) was first given by Van Driest (24).

Theorem 3.   The number of products in a complete set of B-numbers is equal to the total number of variables minus the maximum number of these variables that will not form a dimensionless product.

To show the equivalence of Theorems 1 and 3, it is only necessary to demonstrate that the maximum number of the variables that will not form a dimensionless product is equal to the rank of the dimensional matrix D.

In terms of linear vector space, Buckingham's theorem (Theorem 2) simply states that the null space of the dimensional matrix has a fixed dimension, and Van Driest's rule (Theorem 3) then specifies the nullity of the dimensional matrix. The problem of finding a complete set of B-numbers is equivalent to that of computing a fundamental system of solutions of equation 13 called a complete set of B-vectors. For simplicity, the matrix formed by a complete set of B-vectors will be called a complete B-matrix.

It can also be demonstrated that the choice of reference dimensions does not affect the B-numbers (22).

Theorem 4. The set of B-numbers associated with a physical phenomenon is invariant with respect to the choice of the reference dimensions provided that the reference dimensions are considered independent, and that the number of these reference dimensions is not altered.

The implication of this theorem is important in that in computing a complete set of dimensionless products or B-numbers associated with a physical phenomenon, it is immaterial as to which set of dimensions should be chosen as the reference dimensions as long as they are independent and their number is not altered.

In example 1, there are four variables that are involved in the pendulum problem. The associated dimensional matrix D is given in equation 15. Since the rank r of D is 3, according to Theorem 1 there is only $n - r = 4 - 3 = 1$ independent B-number, as expected.

Suppose now that force f, length l, and time t are chosen as the reference dimensions. From Table 1 the new dimensional matrix D'' becomes (eq. 19)

$$D'' = \begin{matrix} f \\ l \\ t \end{matrix} \begin{bmatrix} M & L & P & g \\ 1 & 0 & 0 & 0 \\ -1 & 1 & 0 & 1 \\ 2 & 0 & 1 & -2 \end{bmatrix} \quad (19)$$

The matrix D^* that will transform D'' to D is the dimensional matrix of the variables force, length, and time with respect to the reference dimensions m, l, and t. Again from Table 1 equation 20 is obtained;

$$D^* = \begin{matrix} m \\ l \\ t \end{matrix} \begin{bmatrix} f & l & t \\ 1 & 0 & 0 \\ 1 & 1 & 0 \\ -2 & 0 & 1 \end{bmatrix} \quad (20)$$

It is straightforward to confirm that $D = D^* D''$. Since D^* is nonsingular, $DX = 0$ and $D''X = 0$ are equivalent, possessing the same set of B-numbers.

It is evident that in applying dimensional analysis, it is first necessary to be able to identify the variables that govern a particular physical phenomenon. The naming of the governing variables requires some prior knowledge of the particular branch of physics involved, based on analytical studies, experimental observations, or both. Whatever the source, there must be some prior knowledge upon which the intuitive judgment or educated guess can be based.

Before proceeding to more complex problems to demonstrate the power of the method of dimensional analysis, a systematic procedure for computing a complete set of B-numbers is discussed.

Systematic Calculation of a Complete B-Matrix

Once the dimensional matrix has been set up and the number of products in a complete set of B-numbers determined, a complete set of B-vectors must be computed. In the following, a systematic procedure for this purpose is presented.

Let D be the dimensional matrix of order m by n associated with a set of variables of a physical phenomenon, m being the number of chosen reference dimensions and

760 DIMENSIONAL ANALYSIS

n the number of variables of the set. Without loss of generality, it may be assumed that $n \geq m$. Consider the augmented matrix (eq. 21):

$$[\boldsymbol{D'} \quad \boldsymbol{I}_n] \tag{21}$$

where, as before, the prime denotes the matrix transpose and $\boldsymbol{I}_n$ is the identity matrix of order n. Suppose the rank of $\boldsymbol{D}$ is r. Then a finite sequence of elementary row operations of equation 21 yields an equivalent matrix of the form (see, for example, Hohn (31)) (eq. 22):

$$\begin{bmatrix} \boldsymbol{D}_{11} & \boldsymbol{D}_{12} & \boldsymbol{C}_{13} \\ 0 & 0 & \boldsymbol{C}_{23} \end{bmatrix} \tag{22}$$

where $\boldsymbol{D}_{11}$ is a nonsingular upper triangular matrix of order r, and $\boldsymbol{D}_{12}$, $\boldsymbol{C}_{13}$, and $\boldsymbol{C}_{23}$ are matrices of orders $r \times (m - r)$, $r \times n$ and $(n - r) \times n$, respectively.

Theorem 5. The transpose $\boldsymbol{C}'_{23}$ of $\boldsymbol{C}_{23}$ is a complete B-matrix of equation 13.

As pointed out by Buckingham (8), it is advantageous if the dependent variables or the variables that can be regulated each occur in only one dimensionless product, so that a functional relationship among these dimensionless products may be most easily determined. For example, if a velocity is easily varied experimentally, then the velocity should occur in only one of the independent dimensionless variables (products). In other words, it is sometimes desirable to have certain specified variables each of which occurs in one and only one of the B-vectors. Otherwise, how is it known that there exists another one that possesses the above property for a given complete B-matrix? If a desired one indeed exists, then how can it be obtained without the necessity of exhausting all the possibilities by linear combinations? The solution to these problems is contained in the following theorem.

Theorem 6. Let $\boldsymbol{A}_1$ be a given complete B-matrix associated with a set of variables. Then there exists a complete B-matrix $\boldsymbol{A}_2$ of these variables such that certain specified variables each occur in only one of the B-vectors of $\boldsymbol{A}_2$ if, and only if, the rows corresponding to these specified variables in $\boldsymbol{A}_1$ are linearly independent.

The above procedures are illustrated by the following examples.

Example 2. A smooth spherical body of projected area A moves through a fluid of density ρ and viscosity μ with speed v. The total drag δ encountered by the sphere is to be determined. Clearly, the total drag δ is a function of v, A, ρ, and μ. As before, mass m, length l, and time t, are chosen as the reference dimensions. From Table 1 the dimensional matrix is (eq. 23):

$$\boldsymbol{D} = \begin{array}{c} \\ m \\ l \\ t \end{array} \begin{array}{c} \delta \quad\;\; v \quad\;\; A \quad\;\; \rho \quad\;\; \mu \\ \begin{bmatrix} 1 & 0 & 0 & 1 & 1 \\ 1 & 1 & 2 & -3 & -1 \\ -2 & -1 & 0 & 0 & -1 \end{bmatrix} \end{array} \tag{23}$$

To compute a complete B-matrix, the augmented matrix (eq. 24):

$$[\boldsymbol{D'} \quad \boldsymbol{I}_5] \tag{24}$$

is considered, which is given by (eq. 25):

$$\begin{bmatrix} 1 & 1 & -2 & \vdots & 1 & 0 & 0 & 0 & 0 \\ 0 & 1 & -1 & \vdots & 0 & 1 & 0 & 0 & 0 \\ 0 & 2 & 0 & \vdots & 0 & 0 & 1 & 0 & 0 \\ 1 & -3 & 0 & \vdots & 0 & 0 & 0 & 1 & 0 \\ 1 & -1 & -1 & \vdots & 0 & 0 & 0 & 0 & 1 \end{bmatrix} \qquad (25)$$

The objective is to apply a sequence of elementary row operations (see Hohn (31)) to equation 25 to bring it to the form of equation 22. Since the rank of D is 3, the order of the matrix C_{23} is $(n - r) \times n = 2 \times 5$. The following sequence of elementary row operations will result in the desired form:

new row 4 = row 4 − row 1 ≡ (designated as) row 4′;
new row 5 = row 5 − row 1 ≡ row 5′;
new row 3 = row 3 − 2 × row 2 ≡ row 3′;
new row 4 = row 4′ + 4 × row 2 ≡ row 4″;
new row 5 = row 5′ + 2 × row 2 ≡ row 5″;
new row 4 = row 3′ + row 4″ ≡ row 4‴;
new row 5 = row 5″ + ½ × row 3′ ≡ 5‴.

The corresponding matrix in partitioned form is given by (eq. 26):

$$\begin{bmatrix} 1 & 1 & -2 & \vdots & 1 & 0 & 0 & 0 & 0 \\ 0 & 1 & -1 & \vdots & 0 & 1 & 0 & 0 & 0 \\ 0 & 0 & 2 & \vdots & 0 & -2 & 1 & 0 & 0 \\ \hdashline 0 & 0 & 0 & \vdots & -1 & 2 & 1 & 1 & 0 \\ 0 & 0 & 0 & \vdots & -1 & 1 & ½ & 0 & 1 \end{bmatrix}$$

$$= \begin{bmatrix} D_{11} & C_{13} \\ 0 & C_{23} \end{bmatrix} \qquad (26)$$

where D_{12} is null. This gives (eq. 27):

$$C_{23} = \begin{array}{c} \\ \\ \end{array} \begin{matrix} \delta & v & A & \rho & \mu \\ \begin{bmatrix} -1 & 2 & 1 & 1 & 0 \\ -1 & 1 & ½ & 0 & 1 \end{bmatrix} \end{matrix} \qquad (27)$$

According to Theorem 5, the transpose of C_{23} is a complete B-matrix. Since there are five variables and since the rank of D is 3; Theorem 1 reveals that there are two dimensionless products in a complete set of B-numbers, each of which corresponds to a row of C_{23}. This yields a functional relation between the two B-numbers as (eq. 28):

$$f(v^2 A \rho / \delta, v \mu A^{1/2} / \delta) = 0, \qquad (28)$$

where $\pi_1 = v^2 A \rho / \delta$ and $\pi_2 = v \mu A^{1/2} / \delta$, or alternatively (eq. 29):

$$v^2 A \rho / \delta = f_1(v \mu A^{1/2} / \delta) \qquad (29)$$

This relation is not in the best form for the calculation of the drag since δ appears in both products. Hence it is necessary to change the two independent B-numbers by requiring that δ occur in only one of the B-numbers. To this end, we let M be a nonsingular submatrix of C'_{23} of order 2 containing the row corresponding to δ. Thus, row 1 and, eg, row 5 of C'_{23}, are chosen to give

$$M = \begin{bmatrix} -1 & -1 \\ 0 & 1 \end{bmatrix} \qquad (30)$$

whose adjoint matrix is given by (see Hohn (31)) (eq. 31):

$$M_a = \begin{bmatrix} 1 & 1 \\ 0 & -1 \end{bmatrix} \tag{31}$$

Then the matrix product (eq. 32):

$$C'_{23}M_a = \begin{bmatrix} -1 & -1 \\ 2 & 1 \\ 1 & 1/2 \\ 1 & 0 \\ 0 & 1 \end{bmatrix} \begin{bmatrix} 1 & 1 \\ 0 & -1 \end{bmatrix} = \begin{matrix} \delta \\ v \\ A \\ \rho \\ \mu \end{matrix} \begin{bmatrix} -1 & 0 \\ 2 & 1 \\ 1 & 1/2 \\ 1 & 1 \\ 0 & -1 \end{bmatrix} \tag{32}$$

is a desired B-matrix. The associated B-numbers are obtained as $\pi_1 = A\rho v^2/\delta$ and $\pi_2 = v\rho A^{1/2}/\mu$, yielding a functional relation (eq. 33):

$$\delta/A\rho v^2 = f_2(v\rho A^{1/2}/\mu) \tag{33}$$

Let d be the diameter of the sphere. Then $A = \pi d^2/4$ and $\pi_2 = \pi^{1/2}v\rho d/2\mu$. The dimensionless product $v\rho d/\mu$, which was first derived by Osborne Reynolds, is the familiar Reynolds number, and is denoted by Re. Equation 33 can now be expressed as (eq. 34):

$$\delta = (A\rho v^2/2)f_3(Re) \tag{34}$$

Defining a drag coefficient C_δ by (eq. 35):

$$\delta = C_\delta(A\rho v^2/2) \tag{35}$$

we have (eq. 36):

$$C_\delta = f_3(Re) \tag{36}$$

Thus, can be reduced the drag problem to an equation involving only two dimensionless products C_δ and Re. The plot of the drag coefficient C_δ as a function of the Reynolds number Re can be obtained from experimental data. Knowing the speed of the sphere, equation 34 together with the drag coefficient is now in the best form for the direct determination of the drag (see also Rheological measurements).

On the other hand, suppose that the speed is to be determined when the drag is given. Then equations 33 or 34 are not convenient, and the two independent B-numbers must be changed again, so that the speed v will occur only in one of the B-numbers. To this end, let M be a nonsingular submatrix of C'_{23} of order 2 containing the row corresponding to v. Thus, choosing row 2 and, again say, row 5 of the transpose of the matrix of equation (27) gives (eq. 37):

$$M = \begin{bmatrix} 2 & 1 \\ 0 & 1 \end{bmatrix} \tag{37}$$

Then the matrix product (eq. 38):

$$-C'_{23}M_a = -\begin{bmatrix} -1 & -1 \\ 2 & 1 \\ 1 & 1/2 \\ 1 & 0 \\ 0 & 1 \end{bmatrix} \begin{bmatrix} 1 & -1 \\ 0 & 2 \end{bmatrix} = \begin{matrix} \delta \\ v \\ A \\ \rho \\ \mu \end{matrix} \begin{bmatrix} 1 & 1 \\ -2 & 0 \\ -1 & 0 \\ -1 & 1 \\ 0 & -2 \end{bmatrix} \tag{38}$$

is a desired B-matrix, where M_a is the adjoint matrix of M. The associated B-numbers become $\pi_1 = \delta/A\rho v^2$ and $\pi_2 = \delta\rho/\mu^2$ yielding (eq. 39):

$$\delta/A\rho v^2 = f_4(\rho\delta/\mu^2) \tag{39}$$

From equation 35, it is simple to demonstrate that (eq. 40):

$$v\rho d/\mu = Re = (8\rho\delta/C_\delta\pi)^{1/2}/\mu, \qquad (40)$$

and that equation 39 can be expressed as (eq. 41):

$$C_\delta = f_5(\rho\delta/\mu^2) \qquad (41)$$

The drag coefficient C_δ can be plotted as a function of the dimensionless product $\rho\delta/\mu^2$. Thus, equations 40 and 41 are in proper form for direct determination of the speed once the drag is given.

Example 3. The mean free path P_m of electrons scattered by a crystal lattice is known to involve temperature θ, energy E, the elastic constant C, the Planck's constant h, the Boltzmann constant k, and the electron mass M (see, for example, Shockley (25)). The problem is to derive a general equation among these variables.

Again length l, mass m, time t, and temperature T are chosen as the reference dimensions. Then the associated dimensional matrix D is obtained as (eq. 42):

$$D = \begin{array}{c} \\ l \\ m \\ t \\ T \end{array} \begin{array}{c} P_m \quad C \quad E \quad h \quad k \quad \theta \quad M \\ \begin{bmatrix} 1 & -1 & 2 & 2 & 2 & 0 & 0 \\ 0 & 1 & 1 & 1 & 1 & 0 & 1 \\ 0 & -2 & -2 & -1 & -2 & 0 & 0 \\ 0 & 0 & 0 & 0 & -1 & 1 & 0 \end{bmatrix} \end{array} \qquad (42)$$

To compute a complete B-matrix, the augmented matrix (eq. 43):

$$[D' \quad I_7] \qquad (43)$$

is considered, which is given by (eq. 44):

$$\begin{bmatrix} 1 & 0 & 0 & 0 & | & 1 & 0 & 0 & 0 & 0 & 0 & 0 \\ -1 & 1 & -2 & 0 & | & 0 & 1 & 0 & 0 & 0 & 0 & 0 \\ 2 & 1 & -2 & 0 & | & 0 & 0 & 1 & 0 & 0 & 0 & 0 \\ 2 & 1 & -1 & 0 & | & 0 & 0 & 0 & 1 & 0 & 0 & 0 \\ 2 & 1 & -2 & -1 & | & 0 & 0 & 0 & 0 & 1 & 0 & 0 \\ 0 & 0 & 0 & 1 & | & 0 & 0 & 0 & 0 & 0 & 1 & 0 \\ 0 & 1 & 0 & 0 & | & 0 & 0 & 0 & 0 & 0 & 0 & 1 \end{bmatrix} \qquad (44)$$

and which may be put in the form of equation 22 by a sequence of elementary row operations yielding (eq. 45):

$$\begin{bmatrix} 1 & 0 & 0 & 0 & | & 1 & 0 & 0 & 0 & 0 & 0 & 0 \\ 0 & 1 & -2 & 0 & | & 1 & 1 & 0 & 0 & 0 & 0 & 0 \\ 0 & 0 & 0 & 0 & | & 2 & 0 & 1 & -2 & 0 & 0 & 1 \\ 0 & 0 & -1 & 0 & | & -2 & 0 & 0 & 1 & 0 & 0 & -1 \\ 0 & 0 & 0 & -1 & | & 2 & 0 & 0 & -2 & 1 & 0 & 1 \\ 0 & 0 & 0 & 0 & | & 2 & 0 & 0 & -2 & 1 & 1 & 1 \\ 0 & 0 & 0 & 0 & | & -5 & -1 & 0 & 2 & 0 & 0 & -1 \end{bmatrix} \qquad (45)$$

from which the appropriate submatrices can be identified: D_{12} is null and (eq. 46–48):

$$D_{11} = \begin{array}{c} (\text{row 1}) \\ (\text{row 2}) \\ (\text{row 4}) \\ (\text{row 5}) \end{array} \begin{bmatrix} 1 & 0 & 0 & 0 \\ 0 & 1 & -2 & 0 \\ 0 & 0 & -1 & 0 \\ 0 & 0 & 0 & -1 \end{bmatrix} \qquad (46)$$

$$C_{13} = \begin{array}{c}\text{(row 1)}\\ \text{(row 2)}\\ \text{(row 4)}\\ \text{(row 5)}\end{array} \begin{bmatrix} 1 & 0 & 0 & 0 & 0 & 0 & 0 \\ 1 & 1 & 0 & 0 & 0 & 0 & 0 \\ -2 & 0 & 0 & 1 & 0 & 0 & -1 \\ 2 & 0 & 0 & -2 & 1 & 0 & 1 \end{bmatrix} \quad (47)$$

$$C_{23} = \begin{array}{c} \\ \text{(row 3)}\\ \text{(row 6)}\\ \text{(row 7)}\end{array} \begin{array}{c}P_m \quad C \quad E \quad h \quad k \quad \theta \quad M \\ \begin{bmatrix} 2 & 0 & 1 & -2 & 0 & 0 & 1 \\ 2 & 0 & 0 & -2 & 1 & 1 & 1 \\ -5 & -1 & 0 & 2 & 0 & 0 & -1 \end{bmatrix}\end{array} \quad (48)$$

Thus, the transpose of C_{23} is a complete B-matrix. Since there are seven variables involved in the phenomenon and the rank of D is 4, from Theorem 1 there are three dimensionless products in a complete set of B-numbers, each of which corresponds to a row of C_{23}.

Suppose that the problem is to find a B-matrix of D such that the variables P_m, C, and E each occur in one and only one of the B-vectors. Since the submatrix M of C'_{23} consisting of the first three rows corresponding to the variables P_m, C, and E is nonsingular, according to Theorem 6 there exists a B-matrix with the desired property. Let M_a be the adjoint matrix of M. Then (eq. 49):

$$C'_{23} M_a = \begin{bmatrix} 2 & 2 & -5 \\ 0 & 0 & -1 \\ 1 & 0 & 0 \\ -2 & -2 & 2 \\ 0 & 1 & 0 \\ 0 & 1 & 0 \\ 1 & 1 & -1 \end{bmatrix} \begin{bmatrix} 0 & 0 & -2 \\ -1 & 5 & 2 \\ 0 & 2 & 0 \end{bmatrix} = \begin{bmatrix} -2 & 0 & 0 \\ 0 & -2 & 0 \\ 0 & 0 & -2 \\ 2 & -6 & 0 \\ -1 & 5 & 2 \\ -1 & 5 & 2 \\ -1 & 3 & 0 \end{bmatrix} \quad (49)$$

Hence, the right-hand side of equation 49 is a desired complete B-matrix. A functional relationship among the associated B-numbers can be obtained and is given by (eq. 50):

$$f(h^2/P_m^2 k\theta M, k^5\theta^5 M^3/C^2 h^6, k^2\theta^2/E^2) = 0 \quad (50)$$

Observe that the variables P_m, C, and E each occur in only one dimensionless product. Alternatively, can be written as: (eq. 51):

$$P_m^2 = (h^2/k\theta M) f_1(k^5\theta^5 M^3/C^2 h^6, k^2\theta^2/E^2) \quad (51)$$

The functional relation in equations 50 or 51 cannot be determined by dimensional analysis alone; it must be supplied by experiments. The significance is that the mean-free-path problem is reduced from an original relation involving seven variables to an equation involving only three dimensionless products, a considerable saving in terms of the number of experiments required in determining the governing equation.

Optimization of the Complete B-Matrices

In the foregoing, the computation of a complete B-matrix from a given dimensional matrix has been indicated. With the exception that each of certain variables may be required to occur in only one dimensionless product, the selection of a complete B-matrix is totally arbitrary. In order to simplify the formulas associated with a complete B-matrix and to provide a procedure for establishing an explicit set of B-numbers, it is necessary to impose additional constraints in the selection of these

B-vectors in forming a complete B-matrix. To avoid the fractional exponents of the formulas, the elements of the matrices are restricted to integers. In addition, Happ (21) proposes the following as criteria for the optimization of the B-matrices: (1) maximize the number of zeros in a complete B-matrix and (2) minimize the sum of the absolute values of all the integers of a complete B-matrix. These criteria are chosen so that the formulas associated with a physical phenomenon are in their simplest form, otherwise they are completely arbitrary. Evidently the order of the two optimization criteria is important. For the purpose of this review the sequence consisting of criterion (1) followed by criterion (2) is assumed.

Since the columns of any complete B-matrix are a basis for the null space of the dimensional matrix, it follows that any two complete B-matrices are related by a nonsingular transformation. In other words, a complete B-matrix itself contains enough information as to which linear combinations should be formed to obtain the optimized ones. Based on this observation, an efficient algorithm for the generation of an optimized complete B-matrix has been presented by Chen (22). No attempt is made here to demonstrate the algorithm. Instead, an example is being used to illustrate the results.

Example 4. For a given lattice, a relationship is to be found between the lattice resistivity and temperature using the following variables: mean free path L, the mass of electron M, particle density N, charge Q, Planck's constant h, Boltzmann constant k, temperature θ, velocity v, and resistivity ρ. Suppose that length l, mass m, time t, charge q, and temperature T are chosen as the reference dimensions. The dimensional matrix D of the variables is given by (eq. 52):

$$D = \begin{array}{c} \\ l \\ m \\ t \\ q \\ T \end{array} \begin{array}{c} \begin{matrix} L & M & N & Q & h & k & \theta & v & \rho \end{matrix} \\ \begin{bmatrix} 1 & 0 & -3 & 0 & 2 & 2 & 0 & 1 & 3 \\ 0 & 1 & 0 & 0 & 1 & 1 & 0 & 0 & 1 \\ 0 & 0 & 0 & 0 & -1 & -2 & 0 & -1 & -1 \\ 0 & 0 & 0 & 1 & 0 & 0 & 0 & 0 & -2 \\ 0 & 0 & 0 & 0 & 0 & -1 & 1 & 0 & 0 \end{bmatrix} \end{array} \quad (52)$$

Using the procedure outlined in the preceding section, a matrix similar to that of equation 22 is obtained as follows (eq. 53):

$$\begin{bmatrix} 1 & 0 & 0 & 0 & 0 & | & 1 & 0 & 0 & 0 & 0 & 0 & 0 & 0 & 0 \\ 0 & 1 & 0 & 0 & 0 & | & 0 & 1 & 0 & 0 & 0 & 0 & 0 & 0 & 0 \\ 0 & 0 & 0 & 0 & 0 & | & 3 & 0 & 1 & 0 & 0 & 0 & 0 & 0 & 0 \\ 0 & 0 & 0 & 1 & 0 & | & 0 & 0 & 0 & 1 & 0 & 0 & 0 & 0 & 0 \\ 0 & 0 & -1 & 0 & 0 & | & -2 & -1 & 0 & 0 & 1 & 0 & 0 & 0 & 0 \\ 0 & 0 & 0 & 0 & 0 & | & 2 & 1 & 0 & 0 & -2 & 1 & 1 & 0 & 0 \\ 0 & 0 & 0 & 0 & 1 & | & 0 & 0 & 0 & 0 & 0 & 0 & 1 & 0 & 0 \\ 0 & 0 & 0 & 0 & 0 & | & 1 & 1 & 0 & 0 & -1 & 0 & 0 & 1 & 0 \\ 0 & 0 & 0 & 0 & 0 & | & -1 & 0 & 0 & 2 & -1 & 0 & 0 & 0 & 1 \end{bmatrix} \quad (53)$$

Thus, the transpose of a complete B-matrix is given by (eq. 54):

$$C_{23} = \begin{array}{c} \\ (\text{row 3}) \\ (\text{row 6}) \\ (\text{row 8}) \\ (\text{row 9}) \end{array} \begin{array}{c} \begin{matrix} L & M & N & Q & h & k & \theta & v & \rho \end{matrix} \\ \begin{bmatrix} 3 & 0 & 1 & 0 & 0 & 0 & 0 & 0 & 0 \\ 2 & 1 & 0 & 0 & -2 & 1 & 1 & 0 & 0 \\ 1 & 1 & 0 & 0 & -1 & 0 & 0 & 1 & 0 \\ -1 & 0 & 0 & 2 & -1 & 0 & 0 & 0 & 1 \end{bmatrix} \end{array} \quad (54)$$

From this matrix, we can generate an optimized complete B-matrix by the algorithm proposed by Chen (22), which results in a matrix whose transpose is given by (eq. 55):

$$\begin{bmatrix} 3 & 0 & 1 & 0 & 0 & 0 & 0 & 0 & 0 \\ 1 & 1 & 0 & 0 & -1 & 0 & 0 & 1 & 0 \\ 0 & 1 & 0 & 0 & 0 & -1 & -1 & 2 & 0 \\ -1 & 0 & 0 & 2 & -1 & 0 & 0 & 0 & 1 \end{bmatrix} \qquad (55)$$

Thus, the general relationship between the lattice resistivity and temperature can be expressed as (eq. 56):

$$f(L^3N,\ LMv/h,\ Mv^2/k\theta,\ Q^2\rho/Lh) = 0 \qquad (56)$$

Observe that the number of independent variables is reduced from the original nine to four. This is a great saving in terms of the number of experiments required to determine the desired function. For example, suppose that a decision is made to test only four values for each variable. Then it would require $4^9 = 262144$ experiments to test all combinations of these values in the original equation. As a result of equation 56, only $4^4 = 256$ tests are now required for four values each, of each of the four B-numbers.

Nomenclature

A = area or acceleration
C = elastic constant
C_δ = drag coefficient
$\boldsymbol{D}$ = dimensional matrix of variables
E = energy
f, F = force
g = acceleration of gravity
h = Planck's constant
k = Boltzman constant
L = length or mean free path
l = length
m, M = mass
N = particle density
P = period
P_m = mean free path
Q = charge
q = charge
Re = Reynolds number
θ, T = temperature
t = time
v = velocity or speed
δ = drag
ϵ = permittivity
μ = permeability or viscosity
ρ = fluid density or resistivity

BIBLIOGRAPHY

"Dimensional Analysis" in *ECT* 1st ed., Vol. 5, pp. 133–141, by D. Q. Kern, The Patterson Foundry & Machine Co.; "Dimensional Analysis in *ECT* 2nd ed., Vol. 7, pp. 176–190, by I. H. Silberberg, Texas Petroleum Research Committee and John J. McKetta, The University of Texas.

1. P. W. Bridgman, *Dimensional Analysis,* Yale Univ. Press, New Haven, Conn., 1922.
2. G. Murphy, *Similitude in Engineering,* The Ronald Press Co., New York, 1950.
3. J. F. Douglas, *An Introduction to Dimensional Analysis for Engineers,* Sir Isaac Pitman & Sons, London, 1969.

4. H. L. Langhaar, *Dimensional Analysis and Theory of Models,* John Wiley & Sons, Inc., New York, 1951.
5. S. J. Kline, *Similitude and Approximation Theory,* McGraw-Hill Book Company, New York, 1965.
6. L. I. Sedov, *Similarity and Dimensional Methods in Mechanics,* Academic Press, Inc., New York, 1959.
7. H. E. Huntley, *Dimensional Analysis,* Dover Publications, Inc., New York, 1967.
8. E. Buckingham, *Phys. Rev.* **4,** 345 (1914).
9. Lord Rayleigh, *Nature* **95,** 66 (1915).
10. T. J. Higgins, *Appl. Mech. Rev.* **10,** 331 (1957).
11. *Ibid.,* p. 443.
12. A. D. Sloan and W. W. Happ, "Literature Search: Dimensional Analysis," *NASA Rept. ERC/CQD 68-631,* Aug. 1968.
13. P. Moon and D. E. Spencer, *J. Franklin Inst.* **248,** 495 (1949).
14. E. U. Condon, *Am. J. Phys.* **2,** 63 (1934).
15. R. C. Tolman, *Phys. Rev.* **9,** 237 (1917).
16. W. E. Duncanson, *Proc. Phys. Soc.* **53,** 432 (1941).
17. G. B. Brown, *Proc. Phys. Soc.* **53,** 418 (1941).
18. H. Dingle, *Phil. Mag.* **33,** 321 (1942).
19. E. A. Guggenheim, *Phil. Mag.* **33,** 479 (July 1942).
20. C. M. Focken, *Dimensional Methods and Their Applications,* Edward Arnold and Co., London, 1953.
21. W. W. Happ, *J. Appl. Phys.* **38,** 3918 (1967).
22. W. K. Chen, *J. Franklin Inst.* **292,** 403 (1971).
23. H. L. Langhaar, *J. Franklin Inst.* **242,** 459 (1946).
24. E. R. Van Driest, *J. Appl. Mech.* **13,** A-34 (1946).
25. W. Shockley, *Electrons and Holes in Semiconductors,* D. Van Nostrand Co., Princeton, N.J., 1950.
26. R. P. Kroon, *J. Franklin Inst.* **292,** 45 (1971).
27. A. Klinkenberg, and H. H. Mooy, *Chem. Eng. Progr.* **44,** 17 (1948).
28. S. Corrsin, *Am. J. Phys.* **19,** 180 (1951).
29. J. Geertsma, G. A. Croes, and N. Schwarz, *Trans. AIME* **207,** 118 (1956).
30. L. Brand, *Am. Math. Month.* **59,** 516 (Sept. 1952).
31. F. E. Hohn, *Elementary Matrix Algebra,* The Macmillan Co., New York, 1958.
32. R. Bellman, *Introduction to Matrix Analysis,* McGraw-Hill Book Co., New York, 1960.

WAI-KAI CHEN
Ohio University

DIMER ACIDS

The dimer acids [61788-89-4], C_{21} dicarboxylic acids [42763-46-2, 42763-47-3] and 9- and 10-carboxystearic acid [4165-02-0] and [4124-87-2] are products resulting from three different reactions of C_{18}, unsaturated fatty acids. These reactions are, respectively, self-condensation, Diels-Alder reaction with acrylic acid, and reaction with carbon monoxide followed by oxidation of the resulting 9- or 10-formylstearic acid (or, alternatively, by hydrocarboxylation of the unsaturated fatty acid).

The starting materials for these reactions have been almost exclusively tall oil fatty acids or oleic acid, although other unsaturated fatty acid feedstocks can be used (see Carboxylic acids, from tall oil; Tall oil).

The basic research that led to these products was done at the Northern Regional Research Center of the USDA: dimer acids research in the 1940s (1–3), C_{21} dicarboxylic acid work in 1957 (4), and recent carboxystearic acid synthetic studies (5–6).

Physical Properties

The physical properties of polymerized fatty acids are influenced by the basestock, by the dimerization conditions, and by the degree to which monomer, dimer, and higher oligomers are separated following the dimerization.

Dimer acids are relatively high molecular weight (ca 560) and yet are liquid at 25°C, a liquidity resulting from the many isomers present.

The products listed in Tables 1–3 are based on manufacture from tall oil fatty acids. Dimer acids based on other feedstocks (eg, oleic acid) have different properties.

Table 1. Properties of Dimer Acids[a]

Physical characteristics	High dimer	Intermediate dimer	High trimer
composition, %			
dimer	87	83	75
trimer	13	15.5	23.5
monobasic acids	trace	1.5	1.5
neutralization equivalent	284–295	285–295	285–297
acid number	190–198	190–197	189–197
saponification number	195–201	191–199	191–199
unsaponifiables, %, max	0.5	1.0	1.0
color, Gardner (1963), max	7	8	9
viscosity at 25°C, mm²/s (= cSt)	8000	8500	9000
specific gravity 25/25°C	0.95	0.95	0.96
100/25°C	0.91	0.91	0.91
density, kg/m³	952.4	953.6	953.6
refractive index, 25°C	1.484	1.484	1.484
pour point, °C	−10	−4	−4

[a] Hystrene series of dimer acids, Humko Sheffield Chemical.

Table 2. Properties of Distilled Dimer Acids [a]

Physical characteristics	Hydrogenated	Unhydrogenated
composition, %		
dimer	95	95
trimer	4	4
monomer	1	1
monobasic acids, max	1.5	1.5
neutralization equivalent	289–298	283–289
acid number	188–194	194–198
saponification number	190–197	198–202
unsaponifiables, %	0.1	0.5
color, Gardner (1963), max	1	5
iodine value	20 max	
viscosity at 25°C, mm^2/s (= cSt)	7000–8000	7000–8000
specific gravity, 25/25°C	0.98	0.95
100/25°C	0.90	0.91
density at 25°C, kg/m^3	951.5	951.2
refractive index, 25°C	1.478	1.483
pour point, °C	−11	−8

[a] Hystrene series of dimer acids, Humko Sheffield Chemical.

Table 3. Properties of Trimer Acids [a]

Physical characteristics	
composition, %	
dimer	40
trimer	60
monobasic acids, max	1.0
neutralization equivalent	295–308
acid number	182–190
saponification number	190–198
unsaponifiables, %	1.0
color, Gardner (1963) max	dark
viscosity at 25°C, mm^2/s (= cSt)	40,000
Gardner-Holdt	Z-6
density at 25°C, kg/m^3	8.0

[a] Hystrene 5460, Humko Sheffield Chemical.

Chemical Properties

Structure and Mechanism of Formation. Thermal dimerization of unsaturated fatty acids has been explained both by a Diels-Alder mechanism and by a free-radical route involving hydrogen transfer. The Diels-Alder reaction appears to apply to high-linoleic acid starting materials satisfactorily, and oleic acid polymerization is better rationalized by a free-radical reaction (7–8).

Clay-catalyzed dimerization of unsaturated fatty acids appears to be a carbonium ion reaction based on the observed double bond isomerization, acid catalysis, chain branching, and hydrogen transfer (7–8).

It has been shown (9) that different precursors for dimer acid preparation give quite different structures (see Table 4).

DIMER ACIDS

Table 4. Dimer Acids: Feedstock/Structure Relationship

	Dimer structure		
	Acyclic	Monocyclic	Polycyclic
oleic or elaidic acid	40	55	5
tall oil fatty acids	15	70	15
linoleic acid	5	55	40

Figure 1. Some possible dimer acids (methyl esters).

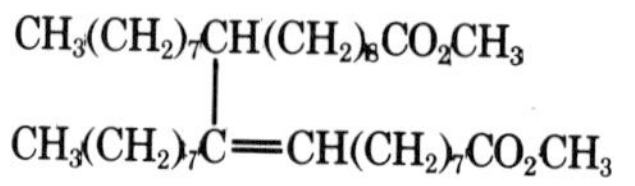

- Acyclic [28923-98-0] — C_{18} acid source — $\Delta 9$
- Monocyclic [26796-50-9, 56636-20-5] — $\Delta 9,11$
- Bicyclic [32733-04-3] — $\Delta 9,11,13$

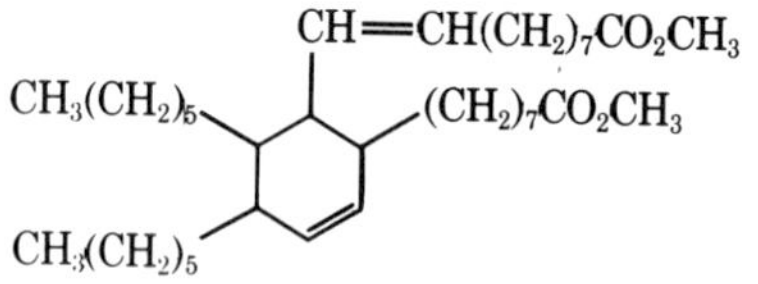

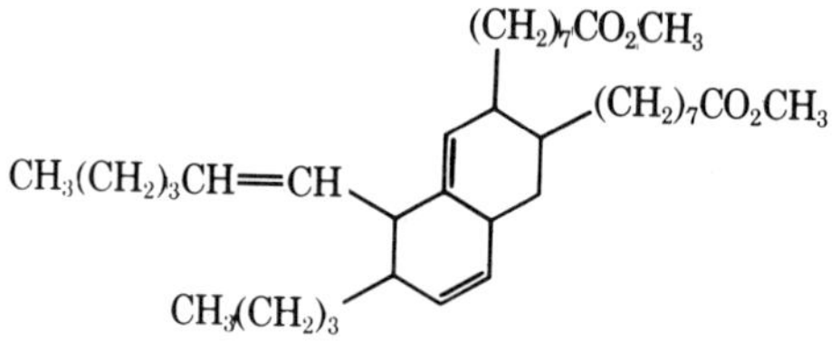

Figure 1 shows three possible structures of the methyl esters of dimer acids. There are a myriad of possible isomers, positional and geometrical isomers of the double bond, as well as structural isomers resulting from head-to-head or head-to-tail alignment of the starting materials.

Chemical Reactions. The reactions of dimer acids were reviewed in 1975 (10). The most important is polymerization, with the greatest tonnage of dimer acids incorporated into the non-nylon polyamides. Other reactions of dimer acids that are used commercially include polyesterification, hydrogenation, esterification, and conversion of the carboxy groups to various nitrogen-containing functional groups. Tables 5 and 6 summarize the chemical reactions of dimer acids and cite pertinent references.

Manufacture

The clay-catalyzed intermolecular condensation of oleic and/or linoleic acid on a commercial scale produces approximately a 60:40 mixture of dimer acids (C_{36} and higher polycarboxylic acids) and monomer acids (C_{18} cyclized, aromatized, and

Table 5. Nonpolymeric Chemical Reactions of Dimer Acids

Reactions at the double bond and at the α-carbon	Ref.	Reactions of the carboxyl group	Refs.
halogenation	11	salt formation	13
epoxidation	11	esterification	14–15
sulfation	11	hydrogenolysis	16–17
sulfurization	11	ethoxylation	18–21
hydrogenation	12	amidation	22–26
sulfonation	11	ammonolysis, reduction (dimer nitriles, dimer amines)	27–29
		isocyanate formation	30

Table 6. Polymerization Reactions of Dimer Acids

Reactions	Refs.
polyamidation	31–33
polyesterification	34–36
reactions resulting in polymeric nitrogen derivatives other than polyamides	37–39
reactions involving dimer diprimary amine	40–41

isomerized fatty acids). The polycarboxylic acid and monomer fractions are usually separated by wiped-film evaporation. The monomer fraction, after hydrogenation, can be fed to a solvent separative process that produces commercial isostearic acid, a complex mixture of saturated fatty acids that is liquid at 10°C.

Dimer acids can be further separated, also by wiped-film evaporation, into distilled dimer acids and trimer acids.

Thermal Polymerization. Commercial manufacture of dimer acids began in 1948 with Emery Industries' use of a thermal process involving steam pressure. Patents were issued in 1949 (42) and 1953 (43) that describe this process. Earlier references to fatty acid polymerization, antedating the USDA work of 1941–1948, occur in patents in 1918 and 1919 (44–45), and in papers written in 1929–1941 (46–48).

Clay-Catalyzed Polymerization. Emery Industries modified its commercial process in 1953 and began producing dimer acids by using a combination of thermal- and montmorillonite clay-catalyzed polymerization. Such a process has been described in patents (49–52). In general, in present-day commercial practice, 100 parts of the fatty acid, 4 parts of clay, and 2 parts of water are heated with agitation at approximately 230°C for 4 h. After cooling and removal of the clay by filtration, separation of monomeric and polymeric material can be carried out by wiped-film evaporation or molecular distillation.

In 1977, in addition to Emery, commercial producers of dimer acids were AZ Chemical Company, Crosby Chemical, Henkel, Humko Sheffield Chemical and Union Camp Corporation.

Process Modification. Dimer acid process modifications have fallen into three categories (7): those claiming higher dimer:trimer ratios, those utilizing varying types of clays, and those purporting to result in improved yields. Higher dimer:trimer ratios are said to be obtained through use of alkali (53), ammonia or amines (54), aryl sulfohalides (55), 1-mercapto-2-naphthol (56), and a "Texas natural acid clay" (57). Natural or synthetic clay catalysts have included lithium salt–acetic anhydride stabilized clay (58), synthetic magnesium silicate (59), and neutral Alabama bentonite (60). Improved yields of dimer acids are claimed to result from a number of two-step, clay-catalyzed procedures (61–62) and by the addition of low molecular weight, saturated alcohols to the reaction mixture (63) (see Clays).

Other Polymerization Methods. Although none has achieved commercial success, there are a number of experimental alternatives to clay-catalyzed or thermal polymerization of dimer acids. These include the use of peroxides (64), hydrogen fluoride (65), a sulfonic acid ion-exchange resin (66), and corona discharge (67) (see Initiators).

Energy Requirements. The production of dimer acids is quite energy intensive. A standard operating sequence at one manufacturing plant results in the expenditure of ca 18.6 MJ (17,600 Btu) (equiv to 0.67 kg coal or 0.33 kg natural gas or fuel oil) to produce each kilogram of crude dimer and to separate it into monomer, dimer and trimer. Of this energy it is estimated that 10% is electrical, consumed by the pumps and agitators of the system. The other 90% is fuel-derived and is consumed mainly for process thermal requirements with some usage for the steam necessary to obtain reduced pressures for certain operations. Energy requirements for the storage and handling of either the raw materials or the products are not included; thus the 18.6 MJ (17,600 Btu) used per kilogram represents only the energy consumption within the process battery limits (68).

Storage and Handling of Dimer Acids and Related Products

Since dimer acids, monomer acids, and trimer acids are unsaturated, they are susceptible to oxidative and thermal attack and will corrode metals. Special precautions are necessary, therefore, to prevent product color development and equipment deterioration.

Type 304 stainless steel is recommended for storage tanks for dimer acids. For heating coils and for agitators, 316 stainless steel is preferred (heating coils with ca 4¾ m² (50 ft²) of heat transfer surface in the form of a 5.1 cm schedule-10 U-bend scroll are recommended for a 37.9 m³ (10,000 gal) tank). Dimer acid storage tanks should have an inert gas blanket.

316 Stainless steel centrifugal pumps may be used to transfer dimer acid stocks. Pipe, valves, and fittings may be of 304 stainless steel if the liquid temperature is maintained below 107°C.

The recommended temperature ranges for transfer of dimer acids and related stocks are shown in Table 7. Specific gravities and viscosities for these chemicals, in the recommended temperature ranges for transfer, are shown in Table 8.

Table 7. Pumping-Temperature Ranges for Dimer Acid Stocks

Stock	Range, °C
monomer acids	46–49
dimer acids	54–77 (71 optimum)
trimer acids	77–82

Table 8. Specific Gravities and Viscosities at Pumping Temperatures

	Temperature, °C	Specific gravity, t/25°C	Viscosity[a], mm²/s
monomer acids	49	0.890	23
dimer acids	71	0.923	305
trimer acids	80	0.926	669

[a] mm²/s = cSt.

The temperature should never exceed 82°C. Even with an inert gas blanket, color deterioration of the products accelerates at higher temperatures. Dimer and trimer acids stored for a long time should be held at approximately 49°C. Tank agitators should be interlocked with the heating cycle of the steam coils so that the agitator will be moving the liquid to prevent discoloration of the stock near the coils during heating.

Stainless steel or epoxy-lined tank cars and tank trucks are recommended for shipping. Aluminum also has been used. The tank can be flushed with carbon dioxide before loading and blanketed with nitrogen after loading. Drum shipments are recommended in epoxy-lined open-head drums fitted with a bung. Dimer acids and their by-products contaminated with iron or copper show accelerated color deterioration. Exposure to these metals or their salts should be minimized.

Economic Aspects

The prices of dimer acids obviously depend on the price of the feedstock. Approximately 80% of the U.S. dimer acid production is based on tall oil fatty acids (qv), 14% on oleic acid (see Carboxylic acids), and 6% on soy acids (see Soybeans) (69). As of the end of 1977, dimer acids (ca 80% dimer) were priced at $1.15–1.21/kg. Tall oil fatty acids, depending on the grade, were selling at ca $0.60/kg. In the period 1973–1975, briefly characterized by a shortage of many commodities, including tall oil fatty acids, dimer acid prices were $0.62–1.39/kg and tall oil fatty acids $0.26–0.66/kg. The 1977 price of dimer acids reflects not only a lower price for raw material compared to 1974, but also an overcapacity in dimer acid production facilities.

Dimer (and trimer) acids production in the United States, based on the output of the six current producers, is approaching 18,000 metric tons per year, with a value of $20–30 million dollars. Sales of this product line began thirty years ago. No dibasic acid with more than twelve carbon atoms was commercially significant before the advent of dimer acids (see Dicarboxylic acids).

Analysis

The ASTM and the American Oil Chemists' Society (AOCS) provide standard methods for determining properties that are important in characterization of dimer acids. Characterization of dimer acids for acid and saponification values, unsaponifiables, and specific gravity are done by AOCS standard methods:

Property	Method
acid value	AOCS Te 1a-64
saponification number	AOCS Tl 1a-64
unsaponifiables, %	AOCS Tk 1a-64
specific gravity	AOCS To 1a-64

Flash and fire points, Gardner color, kinematic viscosity, and pour points are determined by ASTM methods:

Property	Method
flash and fire points	ASTM D92-72
Gardner color	ASTM D1544-63T
kinematic viscosity	ASTM D445-65
pour point	ASTM D97-66

The Gardner-Holdt method for viscosity determination has not been adopted as a standard by ASTM or AOCS (70).

The determination of iodine value (IV) is merely an approximation of unsaturation, because the tertiary allylic hydrogen in the compounds is capable of substitution by halogen atoms. Currently, there is no accepted, standardized method for the direct determination of monomer, dimer, and trimer acids. Urea adduction (of the methyl esters) has been suggested as a means of determining monomer in distilled dimer (71). The method is tedious, however, and the nonadducting branched-chain monomer is recovered with the polymeric fraction. A microsublimation procedure has been developed as an improvement on urea adduction for estimation of the polymer fraction. Incomplete removal of monomer esters or loss of dimer during distillation can lead to error (72).

Thin-layer chromatography (73–74) has been used for the estimation of the amounts of dimer acid methyl esters. Both this method and paper chromatography are characterized by lack of precision (75–76) (see Analytical methods).

Micromolecular distillation techniques have been used to determine monomer, dimer, and trimer in polymerized fatty acids. The most elegant of these is the "bale of hay" method in which the methyl esters of the polymerized fatty acids are dispersed in the still on a "bale" of glass wool to minimize thermal decomposition during the separation (77).

Various column chromatographic methods have been used for the analysis of dimer acids. These include liquid-partition chromatography (78) and gas-liquid chromatography (79–80). Gel-permeation chromatography has recently come into use for the quantitative analysis of dimer acids (81).

Once an accurate determination of monomeric, dimeric, and higher oligomeric dimer acids has been established for a specific feedstock's polymerization, subsequent polymerization extent can be determined very rapidly by a viscometric method (82).

Health and Safety Aspects

Toxicological Properties of Dimer Acids and Related Materials. The acute oral toxicity and the primary skin and acute eye irritative potentials of dimer acids, distilled dimer acids, trimer acids, and monomer acids have been evaluated based on the techniques specified in the Code of Federal Regulations (CFR) (83). The results of this evaluation are shown in Table 9. Based on these results, trimer acids, distilled

Table 9. Toxicity Data for Dimer Acids and Related Products

Sample	Oral LD$_{50}$	Primary irritation index	Eye irritation
trimer acids	>10.0 g/kg	0	very slight erythema in six rabbits
distilled dimer acids	>21.5 g/kg	0.50	very slight erythema in four rabbits
monomer acids	>21.5 g/kg	1.0	very slight erythema in four rabbits
dimer acids	>21.5 g/kg	0.75	very slight erythema in three rabbits

dimer acids, monomer acids, and dimer acids are classified as nontoxic by ingestion, are not primary skin irritants or corrosive materials, and are not eye irritants as these terms are defined in the Federal regulations.

Food Additive and Food Packaging Regulations. Federal regulations do not permit direct use of dimer acids in food products. The CFR, however, permits indirect use of dimer acids in packaging materials with incidental food contact. These permitted applications include use of dimer acids as components of polyamide, epoxy, and polyester resins for use in coating plastic films, paper, and paperboard. These regulations, Title 21, revised April 1, 1977, are: *21 CFR 177.1200*, dimer acids as a component of polyamide resins for coating cellophane; *21 CFR 175.300*, dimer acids as a component of epoxy, polyester, or polyamide resins for "resinous and polymeric coatings" that come into contact with food; *21 CFR 175.390*, dimer acids as a component of zinc–silicon dioxide matrix coating which is the food-contact surface of articles intended for use in producing, manufacturing, packing, processing, preparing, treating, packaging, transporting, or holding food; *21 CFR 177.1210*, dimer acids as components of "resinous and polymeric coatings" used in closures with sealing gaskets for food containers; *21 CFR 176.200*, dimer and trimer acids as defoaming agents in coatings ultimately destined for food use; *21 CFR 175.380*, dimer acids as components of "resinous and polymeric coatings" used as adjuvants for other resins used in food-contact coatings; *21 CFR 175.320*, dimer acids as components of polyamide resins used as coatings for food-contact polyolefin films; and *21 CFR 176.180*, dimer acids as components of "resinous and polymeric coatings" for paper and paperboard in contact with dry food.

Flammability. Dimer and trimer acids, as well as monomer acids derived from dimer acid processing, are neither flammable nor combustible as defined by the DOT and do not represent a fire hazard (see Table 10).

Table 10. Flash and Fire Points of Dimer Acids and Related Products[a]

Product	Flash point, °C Open cup	Flash point, °C Closed cup	Fire point, °C Open cup
monomer acids	193	154	216
dimer acids	279	246	318
trimer acids	329	299	352

[a] Ref. 84.

Uses

The Nonreactive Polyamide Resins. The non-nylon dimer-based polyamide resins are characterized by lack of crystallinity, relatively low softening points, adhesiveness, hydrophobicity and, generally, relatively low transition temperature ranges. Although these properties contrast sharply with the crystallinity and high melting temperatures of the nylon polyamides (qv) based on C_6 to C_{12} dibasic acids, they have been sufficiently unique to carve out large markets of their own (10).

Dimer-based polyamide resin markets are divided into those for reactive polyamides and those for nonreactive polyamides. The largest volume commercial application of dimer acids is in nonreactive polyamide resins. These resins, solids at 25°C, are manufactured by the reaction of dimer acids (or trimer acids) or their esters with aliphatic diamines. Polyamide resins with a broad spectrum of properties can be obtained, with the transition temperature (the phase change from a glass to a liquid) being determined by the diamine used, the stoichiometry of the reactants, and the amount and type of short-chain dibasic acids added to increase the melting range. Table 11 shows the melting ranges of typical neutral polyamide resins (31,33).

Table 12 (85–86) shows an assessment of the 1976 domestic market, compared to that in 1972, for nonreactive, dimer-based polyamide resins. Also shown is the estimate for the tonnage of dimer feedstock for these resins and a list of manufacturers of dimer-based, nonreactive polyamides.

Dimer acids impart flexibility, corrosion resistance, chemical resistance, moisture resistance, and adhesion to nonreactive polyamides.

Hot-melt adhesives (qv), the largest commercial application of nonreactive polyamide resins, are thermoplastics that have fairly sharp melting ranges. They are particularly useful in high-speed assembly operations such as packaging, can assembly, bookbinding, and shoe assembly because they can be applied in liquid form, eliminating the need for solvents.

Table 11. Melting Ranges for Typical Neutral Polyamides

Dimer:trimer mole ratio	Diamine	Polyamide melting range, °C
1.7:1	ethylenediamine	96–103
1.8:1	ethylenediamine	108–112
1.7:1	propylenediamine	53–59
1.7:1	hexamethylenediamine	70–80
1.7:1	dimer diprimary amine	liquid at 25

Table 12. U.S. Consumption of Dimer-Based, Nonreactive Polyamides, Metric Tons[a]

Application	1976 Resin	1976 Dimer feedstock for resin (est)	1972 Resin
hot-melt adhesives	4500	4050	3600
printing inks	3600	2700	3150
surface coatings	450	<450	900
other	450	<450	450
Total	9000	ca 7300	8100

[a] Producers are AZ Chemical Company, Cooper, Crosby, Emery, Henkel, Lawter, Sun, and Union Camp.

The dimer acid-based polyamide resins account for only 4% of the hot-melt adhesive market and are relatively expensive and limited to specialized uses requiring high performance. The dominant hot-melt adhesives are ethylene–vinyl acetate copolymers. The major application for the dimer acid-based polyamide hot-melt adhesives is in the shoe industry. Here good adhesion and high-temperature properties permit bonding of shoe soles to uppers without the need for stitching with thread.

Metal bonding (side-seam welding, eg) and plastic and metal film and foil lamination are other hot-melt adhesive applications of dimer acid-based polyamides.

Flexographic printing inks utilize nonreactive polyamides from dimer acids as resin binders. Polyamide resins are especially well suited for flexographic printing on plastic films and metallic foil laminates because the resins adhere very well to the printed surface, give high-gloss surfaces, and accommodate to substrate deformation (see Inks).

The most important coating application for the nonreactive polyamide resins is in producing thixotropy. Typical coating resins such as alkyds, modified alkyds, natural and synthetic ester oils, varnishes, and natural vegetable oils can be made thixotropic by the addition of dimer acid-based polyamide resins (see Alkyd resins).

As Table 12 shows, the estimated demand in nonreactive polyamide resins for dimer acids in 1976 was ca 7300 t, or approximately 42% of the total dimer acids marketed in that year.

The Reactive Polyamide Resins. The second largest commercial application of dimer acids is in reactive polyamide resins (10). These are formed by the reaction of dimer acids with polyamines such as diethylenetriamine to form polyamides containing reactive secondary amine groups (see Diamines). In contrast to nonreactive polyamides, these materials are generally liquids at 25°C.

They are used extensively to react with epoxy or phenolic resins, yielding adhesives that are useful in casting and laminating, in structural work, for patching and sealing compounds, and for protective coatings. The amount used in epoxy applications far exceeds the use with phenolic resins.

Table 13 (87) shows the market status in 1976 and 1972 of reactive polyamide resins.

The 5400 t of dimer acids estimated as feedstock for reactive polyamide resins represented about 32% of the total dimer acids market in 1976.

Table 13. U.S. Consumption of Dimer-Based, Reactive Polyamide Resins, Metric Tons[a]

Application	1976 Resin	1976 Dimer feedstock (est)	1972 Resin
surface coatings	4500	3600	4050
adhesives	1350	900	900
potting and casting	450	<450	450
other	900	<900	900
Total	7200	ca *5400*	6300

[a] Producers are AZ Chemical Company, Celanese, Crosby, Emery, Henkel, Humko Sheffield, Mobil, Nopko, Reichhold, and Schenectady Resins.

DIMER ACIDS

Miscellaneous Commercial Applications (10). Dimer acids are components of "downwell" corrosion inhibitors for oil-drilling equipment (see Petroleum). It has been estimated that this market consumes 1800–2000 t/yr of dimers (see Corrosion inhibitors).

The burgeoning synthetic lubricant field may offer a growth opportunity for dimer acids. The acids, alkyl esters, and polyoxyalkylene dimer esters are already used commercially as components of metal-working lubricants, with annual sales of the dimer stock in this application at 900–1800 t/yr (see Lubrication). It may be possible, however, for dimer esters to achieve larger use in specialty lubricant applications such as gear oils and compressor lubricants. The dimer esters, compared to polyol esters, poly-α-olefins, and dibasic acid esters, are higher in cost and of higher viscosity. The higher viscosity, however, is an advantage in some specialties and the dimer esters can be highly stable, oxidatively and thermally.

Other dimer-acid markets, probably accounting for usage of 900–1400 t/yr, include intermediates for nitriles, amines and diisocyanates. Dimers are also used in polyurethanes, corrosion inhibition uses other than for downwell equipment, as a "mildness" additive to metal-working lubricants, and in fiberglass manufacture.

9(10)-Carboxystearic Acid (C_{19} Dicarboxylic Acids)

The reaction of oleic acid with carbon monoxide can occur in several ways shown in Figure 2 (see Oxo process; Carbon monoxide).

The product of the hydroformylation reaction can be oxidized directly to the C_{19} dicarboxylic acids by a variety of oxidizing agents.

C_{19} dicarboxylic acids are liquid at 25°C. The products are usually mixtures of positional isomers, with the diacid produced from oleic acid by the Koch process showing skeletal rearrangement as well.

C_{19} dicarboxylic acids undergo the reactions of dimer acids with the difference

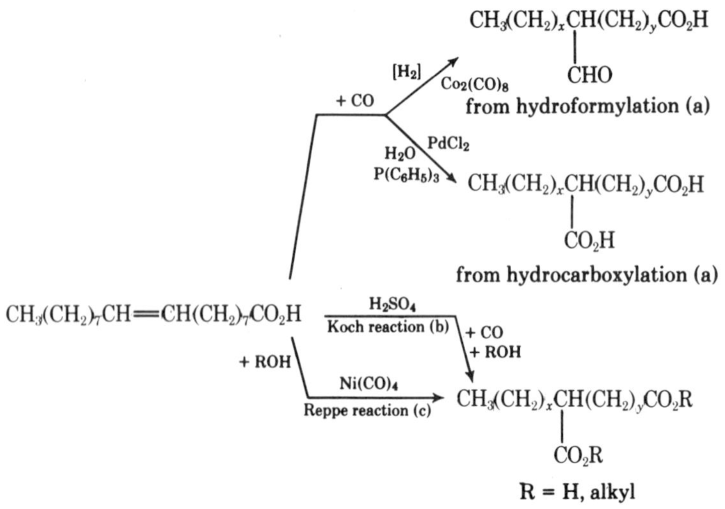

Figure 2. 9(10)-Carboxystearic acid formation, (a) Ref. 5; (b) Ref. 6; (c) Ref. 88.

that the internal carboxyl group is very much less reactive; in the dimer acids the two carboxyl groups have equivalent reactivities. The dissimilar carboxyl group reactivities permits selective transesterification of C_{19} dicarboxylic acids to mixed esters.

BASF and Union Camp Corporation have done developmental work on carboxystearic acid.

5(6)-Carboxy-4-Hexyl-2-Cyclohexene-1-Octanoic Acid (C_{21} Dicarboxylic Acids)

The C_{21} dicarboxylic acids (1) are produced by catalyzed condensation of acrylic acid with the diunsaturated (linoleic acid) fraction of tall oil fatty acids.

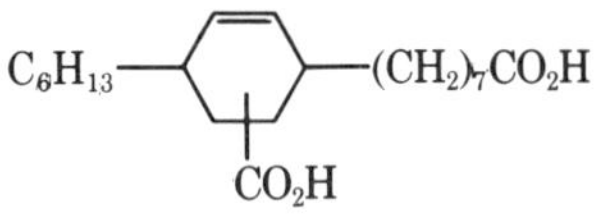

C_6H_{13}—⟨cyclohexene⟩—$(CH_2)_7CO_2H$
         |
        CO_2H

(1)
5-COOH [42763-47-3]
6-COOH [42763-46-2]

Typical properties of (1) are given in Table 14 (89).

Table 14. Typical Properties of C_{21} Dicarboxylic Acids (1)[a]

Property	Value
acid number	280
saponification number	305
dicarboxylic acid, %	92
monocarboxylic acid, %	7
unsaponifiables, %	1
iodine value	60
viscosity, 37.8°C, mm^2/s (= cSt)	5700
flash point, °C	236
density, 25°C	1.016
LD$_{50}$, acute oral, albino rats	6.176 g/kg

[a] Ref. 89.

BIBLIOGRAPHY

1. J. C. Cowan, W. C. Ault, and H. M. Teeter, *Ind. Eng. Chem.* **38,** 1138 (1946).
2. L. B. Falkenburg and co-workers, *Oil Soap* **22,** 143 (1945).
3. J. C. Cowan, A. J. Lewis, and L. B. Falkenburg, *Oil Soap* **21,** 101 (1944).
4. H. M. Teeter and co-workers, *J. Am. Oil Chem. Soc.* **22,** 512 (1957).
5. E. N. Frankel and E. H. Pryde, *J. Am. Oil Chem. Soc.* **54,** 873A (1977).
6. N. E. Lawson, T. T. Cheng, and F. B. Slezak, *J. Am. Oil Chem. Soc.* **54,** 215 (1977).
7. R. W. Johnson, "Dimerization and Polymerization," in E. H. Pryde, ed., *Fatty Acids,* American Oil Chemists' Society, Champaign, Illinois, in press.
8. M. J. A. M. den Otter, *Fette, Seifen, Anstrichm.* **72,** 667, 875, 1056 (1970).
9. D. H. McMahon and E. P. Crowell, *J. Am. Oil Chem. Soc.* **51,** 522 (1974).
10. E. C. Leonard, ed., *The Dimer Acids,* Humko Sheffield Chemical, Memphis, Tenn., 1975.
11. *Empol® Dimer and Trimer Acids,* technical bulletin Emery Industries, Inc., 1971, p. 8.
12. U.S. Pat. 3,595,887 (July 27, 1971), M. V. Kulkarni and R. L. Scheribel (to General Mills, Inc.).
13. J. Levy, "Utilization of Fatty Acids in Metallic Soaps and Greases," in E. Scott Pattison, ed., *Fatty Acids and Their Industrial Applications,* Marcel Dekker, Inc., New York, 1968, p. 209.

780 DIMER ACIDS

14. U.S. Pat. 2,673,184 (March 23, 1954), A. J. Morway, D. W. Young, and D. L. Cottle (to Standard Oil Development Company).
15. U.S. Pat. 2,849,399 (Aug. 26, 1958), A. H. Matuzak and W. J. Craven (to Esso Research and Engineering Company).
16. U.S. Pat. 2,347,562 (April 25, 1944), W. B. Johnston (to American Cyanamid Co.).
17. U.S. Pat. 2,413,612 (Dec. 31, 1946), E. W. Eckey and J. E. Taylor (to Proctor & Gamble Co.).
18. U.S. Pat. 3,173,887 (March 16, 1965), T. E. Yeates and C. M. Thierfelder (to the United States Department of Agriculture).
19. U.S. Pat. 2,473,798 (June 21, 1949), R. E. Kienle and G. P. Whitcomb (to American Cyanamid Co.).
20. U.S. Pat. 2,758,976 (Aug. 14, 1956), G. E. Barker (to Atlas Powder Co.).
21. U.S. Pat. 3,429,817 (Feb. 25, 1969), M. J. Furey and A. F. Turbak (to Esso Research & Engineering Co.).
22. U.S. Pat. 2,537,493 (Jan. 9, 1951) and U.S. Pat. 2,470,081 (May 10, 1949), J. T. Thurston and R. B. Warner (to American Cyanamid Co.).
23. U.S. Pat. 2,992,145 (July 11, 1961), C. Santangelo and B. H. Kress (to Quaker Chemical Products Corp.).
24. U.S. Pat. 3,256,182 (June 14, 1966), G. F. Scherer (to Rockwell Manufacturing Co.).
25. U.S. Pat. 2,965,591 (Dec. 20, 1960), J. Dazzi (to Monsanto Chemical Co.).
26. U.S. Pat. 3,219,612 (Nov. 23, 1965), E. L. Skau, R. R. Mod, and F. C. Magne (to the United States Department of Agriculture).
27. U.S. Pat. 2,526,044 (Oct. 17, 1950), A. W. Ralston, O. Turinsky, and C. W. Christensen (to Armour & Co.).
28. U.S. Pat. 3,223,631 (Dec. 14, 1965), A. J. Morway and A. J. Rutkowski (to Esso Research & Engineering Co.).
29. U.S. Pat. 3,010,782 (Nov. 28, 1961), K. E. McCaleb, L. Vertnik, and D. L. Anderson (to General Mills, Inc.).
30. U.S. Pat. 3,481,959 (Dec. 2, 1969), G. Egle (to Henkel).
31. D. E. Floyd, *Polyamide Resins,* 2nd ed., Reinhold Publishing, New York, 1966, p. 11.
32. L. B. Falkenburg and co-workers, *Oil Soap* **22,** 143 (1945).
33. U.S. Pat. 2,450,740 (Oct. 12, 1948), J. C. Cowan, L. B. Falkenburg, H. M. Teeter, and P. S. Skell (to the United States Department of Agriculture).
34. U.S. Pat. 2,411,178 (Nov. 19, 1946), D. W. Young and E. Lieber (to Standard Oil Development Co.); U.S. Pat. 2,424,588 (July 29, 1947), W. J. Sparks and D. W. Young (to Standard Oil Development Co.); U.S. Pat. 2,435,619 (Feb. 17, 1948), D. W. Young and W. J. Sparks (to Standard Oil Development Co.).
35. U.S. Pat. 3,492,232 (Jan. 27, 1970), M. Rosenberg (to The Cincinnati Milling Machine Co.).
36. U.S. Pat. 3,769,215 (Oct. 30, 1973), R. J. Sturwold and F. O. Barrett (to Emery Industries, Inc.).
37. U.S. Pat. 3,217,028 (Nov. 9, 1965), L. R. Vertnik (to General Mills, Inc.).
38. U.S. Pat. 3,281,470 (Oct. 25, 1966), L. R. Vertnik (to General Mills, Inc.).
39. U.S. Pat. 3,235,596 (Feb. 15, 1966), R. Nordgren, L. R. Vertnik, and H. Wittcoff (to General Mills, Inc.).
40. U.S. Pat. 3,231,545 (Jan. 25, 1966); U.S. Pat. 3,242,141 (March 22, 1966), L. R. Vertnik and H. Wittcoff (to General Mills, Inc.).
41. U.S. Pat. 3,483,237 (Dec. 9, 1969), D. E. Peerman and L. R. Vertnik (to General Mills, Inc.).
42. U.S. Pat. 2,483,761 (Sept. 27, 1949), C. G. Goebel (to Emery Industries, Inc.).
43. U.S. Pat. 2,664,429 (Dec. 29, 1953), C. G. Goebel (to Emery Industries, Inc.).
44. Brit. Pat. 121,777 (1918), J. Craven.
45. Brit. Pat. 127,814 (1919), (to De Nordiske Fabriker).
46. J. Scheiber, *Farbe Lack Anstrichst.* 585 (1929).
47. C. P. A. Kappelmeier, *Farben Ztg.* **38,** 1018, 1077 (1933).
48. T. F. Bradley and W. B. Johnston, *Ind. Eng. Chem.* **33,** 86 (1941); T. F. Bradley and D. Richardson, *Ind. Eng. Chem.* **32,** 963, 802 (1940).
49. U.S. Pat. 2,347,562 (April 25, 1944), W. B. Johnston (to American Cyanamid Co.).
50. U.S. Pat. 2,426,489 (Aug. 26, 1947), M. De Groote (to Petrolite Corp.).
51. U.S. Pat. 2,793,219 (May 21, 1957), F. O. Barrett, C. G. Goebel, and R. M. Peters (to Emery Industries, Inc.).
52. U.S. Pat. 2,793,220 (May 21, 1957), F. O. Barrett, C. G. Goebel, and R. M. Peters (to Emery Industries, Inc.).

53. U.S. Pat. 2,955,121 (Oct. 4, 1960), L. D. Myers, C. G. Goebel, and F. O. Barrett (to Emery Industries, Inc.).
54. U.S. Pat. 3,076,003 (Jan. 29, 1963), L. D. Myers, C. G. Goebel, and F. O. Barrett (to Emery Industries, Inc.).
55. U.S. Pat. 3,773,806 (Nov. 20, 1973), M. Morimoto, M. Saito, and A. Gouken (to Kao Soap Co.).
56. U.S. Pat. 3,925,342 (Dec. 9, 1975), R. P. F. Scharrer (to Arizona Chemical Co.).
57. U.S. Pat. 3,157,681 (Nov. 17, 1964), E. M. Fischer (to General Mills, Inc.).
58. U.S. Pat. 3,412,039 (Nov. 19, 1968), S. E. Miller (to General Mills, Inc.).
59. U.S. Pat. 3,444,220 (May 13, 1969), D. H. Wheeler (to General Mills, Inc.).
60. U.S. Pat. 3,732,263 (May 8, 1973), L. U. Berman (to Kraftco Corp.).
61. U.S. Pat. 3,110,784 (Aug. 13, 1963), C. G. Goebel (to Emery Industries, Inc.).
62. U.S. Pat. 3,632,822 (Jan. 4, 1972), N. H. Conroy (to Arizona Chemical Company).
63. U.S. Pat. 3,507,890 (April 21, 1970), G. Dieckelmann and H. Rutzen (to Henkel).
64. U.S. Pat. 2,964,545 (Dec. 13, 1960), S. A. Harrison (to General Mills, Inc.).
65. U.S. Pat. 2,670,361 (Feb. 23, 1954), C. B. Croston, H. B. Teeter, and J. C. Cowan (to the United States Department of Agriculture).
66. U.S. Pat, 3,367,952 (Feb. 6, 1968), H. G. Arlt (to Arizona Chemical Co.).
67. U.S. Pat. 3,533,932 (Oct. 13, 1970), J. A. Coffman and W. R. Browne (to General Electric Co.).
68. Private communication, E. E. Rice, Engineering Vice-President, Humko Sheffield Chemical, Sept. 14, 1978.
69. *Fatty Acids—North America, A Multiclient Market Survey,* Hull & Company, Bronxville, New York, 1975, p. 528.
70. H. A. Gardner and P. C. Holdt, *Circular #128, Paint Manufacturers Assoc. of the U.S.,* 1921.
71. D. Firestone and co-workers, *J. Am. Oil Chem. Soc.* **44,** 465 (1961); D. Firestone, S. Nesheim, and W. Horwitz, *J. Assoc. Off. Anal. Chem.* **38,** 253 (1961).
72. A. Huang and D. Firestone, *J. Assoc. Off. Anal. Chem.* **52,** 958 (1969).
73. A. K. Sen Gupta and H. Scharmann, *Fette Seifen Anstrichm.* **70,** 86 (1968).
74. G. Billek and O. Heisz, *Fette Seifen Anstrichm.* **71,** 189 (1969).
75. D. Firestone, *J. Am. Oil Chem. Soc.* **40,** 247 (1963).
76. H. E. Rost, *Fette Seifen Anstrichm.* **64,** 427 (1962); **65,** 463 (1963).
77. R. F. Paschke, J. R. Kerns, and D. H. Wheeler, *J. Am. Oil Chem. Soc.* **31,** 5 (1954).
78. E. N. Frankel and co-workers, *J. Org. Chem.* **26,** 4663 (1961); *J. Am. Oil Chem. Soc.* **38,** 130 (1961); C. D. Evans and co-workers, *ibid.,* **42,** 764 (1965).
79. J. H. Greaves and B. Laker, *Chem. Ind. (London),* 1709 (1961).
80. R. A. L. Paylor, R. Feinland and N. H. Conroy, *Anal. Chem.* **40,** 1358 (1968).
81. T. L. Chang, *Anal. Chem.* **40,** 989 (1968).
82. R. P. A. Sims, *Ind. Eng. Chem.* **47,** 1049 (1955).
83. *Regulations for the Enforcement of the Federal Hazardous Substances Act,* Title 16, Code of Federal Regulations, 1500.40, 1500.41, and 1500.42.
84. *Fed. Regist.* **39**(17), 2766, 2769, 2770, 2771 and 2772 (Jan. 24, 1974). Closed cup flash points and ranges are mandatory criteria for flammability determination after July 1, 1975.
85. E. C. Leonard, "Dimer Acids Applications," *American Oil Chemists' Society Convention, Higher Molecular Weight Dibasic Acids Symposium,* New York City, May 12, 1977.
86. A. Wolfe, "Polyamide Resins (Non-Nylon Types)," in *Chemical Economics Handbook,* Stanford Research Institute, Menlo Park, Calif., June, 1977, p. 580.1031L.
87. *Ibid.,* p. 580.1032M.
88. E. H. Pryde, E. N. Frankel, and J. C. Cowan, *J. Am. Oil Chem. Soc.* **49,** 451 (1972).
89. B. F. Ward, Jr. and co-workers, *J. Am. Oil Chem. Soc.* **52,** 219 (1975).

General References

References 7 and 10.
E. C. Leonard, "The Higher Aliphatic Di- and Polycarboxylic Acids," in E. H. Pryde, ed., *Fatty Acids,* American Oil Chemists' Society, Champaign, Ill., in press.
E. H. Pryde and J. C. Cowan, "Aliphatic Dibasic Acids," in *Condensation Monomers,* Wiley-Interscience, New York, 1972.

A. Wolfe, "The Non-Nylon Polyamides," in *Chemical Economics Handbook,* Stanford Research Institute, Menlo Park, Calif., 1977.

EDWARD C. LEONARD
Humko Sheffield Chemical

DIPHENYL ETHER, DIPHENYL OXIDE (C_6H_5)$_2$O. See Heat exchange technology—Heat transfer media other than water; Ethers.

DIPHENYL AND TERPHENYLS

Diphenyl (biphenyl, phenylbenzene) and terphenyl are the lowest members of a family of polyphenyls in which benzene rings are attached one to another in a chain-like manner, $C_6H_5(C_6H_4)_m C_6H_5$. Although many higher polyphenyls are known (1) only diphenyl and the terphenyls listed in Table 1 are of commercial significance.

Diphenyl was first reported in 1862 by Fittig, and then identified in 1867 by Berthelot as the main product obtained by passing benzene vapors through a hot tube. In 1874 Schultz, investigating the higher-boiling products of benzene pyrolysis, passed benzene vapors through a glowing iron pipe, and isolated, in addition to diphenyl, *m*- and *p*-terphenyl. The ortho isomer was identified in 1927 (2). Shortly thereafter, terphenyls became commercially available as by-products from the manufacture of diphenyl by benzene pyrolysis. For many years terphenyls were supplied by The Dow Chemical Company and Monsanto Company. Currently Monsanto is the sole U.S. producer of terphenyls. Other producers include Bayer A.G. (FRG), Rhone-Progil (France), Nippon Steel (Japan), and Monsanto, Ltd. (England).

Table 1. Diphenyl and the Terphenyls

Names	CAS Registry Number	Structure
diphenyl; 1,1'-biphenyl[a]	[92-52-4]	
terphenyl	[26140-60-3]	
o-terphenyl; 1,1':2',1"-terphenyl[a]	[84-15-1]	
m-terphenyl; 1,1':3',1"-terphenyl[a]	[92-06-8]	
p-terphenyl; 1,1':4',1"-terphenyl[a]	[92-94-4]	

[a] *Chemical Abstracts* name.

The principal uses of terphenyls are based on their outstanding thermal stability and high boiling points. Terphenyls and their partially hydrogenated derivatives also exhibit a high degree of radiation resistance. During the late 1950s and 1960s, these materials were considered to be the most promising organic coolants for nuclear reactors (see Heat exchange technology). Many studies and a large body of literature attests to the high level of interest in the nuclear application. During this period, Monsanto made available various individual terphenyl isomers and special isomer mixtures for development purposes. For composition and physical properties of some of these products see references 3 and 4. Eventually interest in organic-cooled reactors declined in favor of water-cooled designs and only minor quantities of terphenyls or hydrogenated terphenyls are now used in reactor applications (see Nuclear reactors).

Physical Properties

Diphenyl is a white or slightly yellow crystalline solid. It separates from solvents as plates or monoclinic prismatic crystals. As one of the most stable organic compounds, it is resistant to thermal decomposition and degradation by radiation (5–7). Its physical constants are listed in Table 2.

Pure terphenyls are white crystalline solids, whereas the commercial grades are yellow or tan. The physical properties of the three isomeric terphenyls are given in Table 3.

A mixture of terphenyls is currently marketed by Monsanto as a high-temperature

Table 2. Physical Properties of Diphenyl

Property	Value	Reference
melting point, °C	69.2	8
freezing or congealing point of commercial grade, °C	68.5–69.4	9
boiling point at 101.3 kPa[a], °C	255.2 ± 0.2	8
specific gravity, solid		
d_4^{20}	1.041	6
d_4^{15}	0.991	6
critical properties		
temperature, °C	515.7	10
pressure, kPa[a]	37.9	10
density, g/mL	0.314	10
flash point, °C	113	11
fire point, °C	123	11
ignition temperature of dust cloud	650	12

	Temperature, °C				
	100	200	300	350	
vapor pressure, kPa[a]		25.43	246.77	558.06	6
liquid density, g/mL	0.970	0.889	0.801	0.751	6
heat capacity, J/g[b]	1.786	2.129	2.129	2.640	6
heat of vaporization, J/g[b]	397	343	284.7	251	6
viscosity, mm²/s (= cSt)	0.98	0.43	0.240		11
thermal conductivity, liquid, W/(cm·K)[c]	13.39	11.92	10.46	9.75	13

[a] To convert kPa to mm Hg, multiply by 7.5.
[b] To convert J to cal, divide by 4.184.
[c] To convert W/(cm·K) to (cal·cm)/(s·cm²·°C), divide by 4.184.

Table 3. Physical Properties of Pure Terphenyl Isomers

Property	o-Terphenyl	m-Terphenyl	p-Terphenyl	References
melting point, °C	56.2	87.5	212.7	14
boiling point at 101.3 kPa[a], °C	332	365	376	15
heat of vaporization at 101.3 kPa[a] at bp, kJ/kg[b]	253	279	272	15
flash point, °C	171	206	210	16
fire point, °C	193	229	238	16
vapor pressure, kPa[a]				
93°C	0.01172	0.00165		15
204°C	2.834	0.8274		15
315.6°C	64.40	27.30		15
426.7°C	439.9	240.6		15
density of liquid g/L				
93°C	1022	1039	solid	17
204°C	935	958	solid	17
315.6°C	842	871	879	17
heat capacity of liquid, kJ/kg[b]				
93°C	1.007	0.970		15
315°C	1.300	1.298		15
398.9°C	1.400	1.379	1.116	15
viscosity of liquid, mPa·s (= cP)				
100°C	4.34	3.87	solid	18
225°C	0.66	0.78	0.74	18
300°C	0.30	0.40	0.43	18
350°C			0.32	18
thermal conductivity of liquid, W/(m·K)[c]				
100°C	0.1316	0.1347		18
150°C	0.1266	0.1306		18
210°C	0.1206	0.1256	0.1359	18
260°C			0.1339	18
heat of formation, MJ/kg[b]	68.35	68.33	68.36	19
heat of fusion, kJ/kg[b]	55.2	73.7	146.5	20
critical temp, K	891	925	926	18
critical pressure MPa[d]	3.903	3.503	3.330	21

[a] To convert kPa to mm Hg, multiply by 7.5.
[b] To convert J to cal, divide by 4.184.
[c] To convert W/(m·K) to (cal·cm)/(s·cm^2·°C), divide by 418.4.
[d] To convert MPa to psi, multiply by 145.

heat-transfer medium. Its composition is approximately 2–10% o-terphenyl, 45–49% m-terphenyl, and 25–35% p-terphenyl with trace amounts of diphenyl, plus higher polyphenyls normally in the 2–18% range. Typical physical properties of this commercial mixture are given in Table 4.

The terphenyls are essentially insoluble in water and dissolve only sparingly in the lower alcohols and glycols, whereas aromatic solvents readily dissolve o- and m-terphenyls. The p-isomer is much less soluble. The solubilities of terphenyls and technical-grade isomers formerly available are given in refs. 22 and 23, respectively.

Plots of weight percent terphenyls in various solvents as a function of temperature are given in ref. 24.

Table 4. Physical Properties of the Commercial Mixture[a] of Terphenyls

Property	Value
appearance	crumbly wax-like flakes at room temperature; light amber liquid above melt point
odor	faint, pleasant
av mol wt	230
melt point, °C	
begins to soften	60
completely liquid	145
flash point, COC[b], °C	191
fire point, COC, °C	238
autogenous ignition temperature (ASTM), °C	>540
coefficient of expansion per °C	0.00085
boiling range, °C	
10% over	364
90% over	418
density, g/L, 165°C	990
220°C	951
320°C	896
specific heat, kJ/(kg·K)[c], 140°C	1.891
220°C	2.079
320°C	2.305
thermal conductivity, W/(m·K)[d], 160°C	0.1295
220°C	0.1210
300°C	0.1112
kinematic viscosity, mm²/s (= cSt), 165°C	1.40
220°C	0.86
280°C	0.54
vapor pressure, kPa[d,e], 149°C	0.160
315°C	34.7
427°C	200

[a] *Monsanto Product Bulletin IC-/FF-58* and certain improved data to be included on forthcoming revisions.
[b] Cleveland open cup.
[c] To convert J to cal, multiply by 0.239.
[d] To convert W/(m·K) to (cal·cm)/(s·cm²·°C), divide by 418.4.
[e] To convert kPa to mm Hg, multiply by 7.5.

Chemical Properties

Diphenyl may be regarded as a substituted benzene, and thus exhibits most of the chemistry but it is less reactive than benzene. Partial reduction gives phenylcyclohexane, full hydrogenation gives dicyclohexyl.

Terphenyls, like diphenyl, undergo the usual organic reactions of aromatic hydrocarbons. The *o*- and *p*-isomers nitrate at the 4-position, whereas the *m*-isomer nitrates at the 4'-position (25). Halogenation apparently proceeds similarly. All three isomers are attacked most readily at the 4-position by acyl or sulfonyl halides in Friedel-Crafts reactions (qv) (26).

o- and *m*-Terphenyls can be isomerized to the *p*-isomer (27–28). The ortho isomer, refluxed for a short time with a small amount of aluminum chloride in benzene, gives 94% of *meta* isomer. More drastic isomerization conditions give conversions of up to

786 DIPHENYL AND TERPHENYLS

84% *para* isomer. Isomerization facilitates separation of diphenyl from *o*-terphenyl present in unrefined polyphenyls (29). Under strongly dehydrogenating conditions *o*-terphenyl cyclizes to triphenylene (30). The latter is a minor (1.5%) impurity present in crude polyphenyls prepared by pyrolytic dehydrocondensation of benzene.

o-terphenyl → triphenylene

In the presence of certain ether solvents, terphenyls can be metalated by sodium amalgam (31). The meta isomer reacts least rapidly and adds only one atom of sodium per molecule whereas *o*- and *p*-terphenyl give 2:1 adducts. Terphenyl can be reduced in liquid ammonia, and reductively methylated by alkali metals. The color of crude polyphenyls is improved by heating with small amounts of alkali metals (32–33).

Manufacture

Diphenyl has been produced commercially in the United States since 1926, mainly by The Dow Chemical Company, Monsanto Company, and Sun Oil Company. Before the early 1970s, the major process was the thermal dehydrogenation of benzene. This route also produces terphenyls.

Today, the main source of diphenyl is as a by-product in benzene production by hydrodealkylation of toluene (qv) (34–35). About 1 kg of diphenyl is obtained for 100 kg benzene produced (36). The by-product stream can be recycled into the process (37) or utilized to enrich fuel oil. For marketing, by-product diphenyl is refined to 93–97% purity by distillation.

toluene + H_2 $\xrightarrow{\Delta}$ benzene + CH_4 + diphenyl

When both diphenyl and terphenyls are desired, the thermal dehydrocondensation of benzene remains the commercial route of choice.

A number of other diphenyl processes have been reported in the literature involving a variety of hydrocarbon feedstocks and conditions. For example diphenyl can be produced by pyrolysis of coal tar in the presence of Mo-oxide/Al_2O_3 catalyst (38,39), hydrocarbon gas pyrolysis tar (40), crude benzene fractions (41–42), and CH_4 passed through silica gel at 1000°C (43). In addition, diphenyl can be prepared by heating benzene and ethylene to 130–160°C in the presence of Na- K- or KNH_2-containing catalysts on activated Al_2O_3 (44) or by the dehydrogenation of 2-cyclohexyl-cyclohexanone or 2-(1-cyclohexen-1-yl)cyclohexanone over Pt catalysts (45).

Terphenyls are obtained from the higher-boiling polyphenyl by-products formed in the pyrolysis process of diphenyl manufacture (46). After benzene and diphenyl are distilled from the pyrolysis mixture, the high-boiling residue consists of approximately 3–8% *o*-terphenyl, 44% *m*-terphenyl, 24% *p*-terphenyl, 1.5% triphenylene, and about 22–27% higher polyphenyls and tars. Distillation of this residue at reduced pressure yields the terphenyls. The mixture is noncorrosive and mild steel equipment suffices. Batchwise fractional distillation employing a suitable fractionating column gives relatively pure isomers. However, to obtain high purity *m*- or *p*-terphenyl, the appropriate distillation fraction (47) has to be further purified by recrystallizing, zone melting, or other refining technique (see Zone refining). Currently, little demand exists for the pure isomers, and only a mixture is produced.

Shipping

Terphenyls are sold as flaked solids in multi-wall bags or 125-kg fiber drums. They are available in carload lots. For shipping purposes, the terphenyl mixture is classified as Chemicals, NOIBN.

By-product diphenyl is usually sold as a dye carrier in the molten state in tank car or tank truck lots (see Dye carriers). Grades of higher purity are sold in the molten state or as flakes in 22.7-kg bags or 118-kg fiber drums.

Hydrogenated terphenyl products are shipped in steel cans, drums, tank trucks, or tank cars. The shipping classification is resin plasticizers.

Economic Aspects

Annual production of diphenyl in 1976 was estimated at 40,000 metric tons with approximately 70% obtained by refining dealkylated toluene by-product and 30% produced by thermal dehydrogenation of benzene (48). Annual consumption of diphenyl in textile dye carrier applications is estimated at 22,500 t. The actual capacity for diphenyl production is not reported, but considering the flexibility for recovery from hydrodealkylation of toluene and termination of polychlorinated diphenyl production, the potential for production far exceeds demand.

The price ranges from $0.35–0.80/kg, depending on purity. The cheapest grade assays at approximately 93–95% and is used for dye carriers. The highest grade is 99+% diphenyl and is often sold in a flake form. Prices doubled since 1974.

Currently, most of the terphenyl production is converted to a partially hydrogenated form. United States production of terphenyls has remained steady at several thousand metric tons per year over the past decade.

The 1978 U.S. price of terphenyl mixed isomers was $1.65/kg. It is a function not only of raw materials and capital costs, but also of the demand for and the price of high-purity pyrolysis diphenyl. Demand for this product has fallen after the phase-out of polychlorinated diphenyl production.

The U.S. production of hydrogenated terphenyls is on the order of several thousand metric tons per year at approximately $2/kg.

Health and Safety Factors

Because of its low vapor pressure and low toxicity, diphenyl does not present an industrial hazard (49–50). Dust explosions constitute a major hazard. The dust is usually caused by vapors originating at a hot diphenyl liquid surface and condensing in air (12,51). A TLV of 0.2 ppm time-weighted average (TWA) in air has been established by the American Conference of Governmental Industrial Hygienists

788   DIPHENYL AND TERPHENYLS

(ACGIH) (52) and adopted as a work place standard. This value is primarily based on worker's comfort; higher concentrations and overexposure can be irritating and may have toxic effects (53). Toxicological data for diphenyl are summarized in Table 5 (54–56).

Individual terphenyl isomers or the commercial mixture present a minimal hazard in a workplace operating in accordance with good industrial hygiene practice. Oral LD_{50} value in rats for mixed terphenyls is 4.6–4.7 g/kg (57). The 1977 threshold limit value–time weighted average (TLV–TWA) given for terphenyls is 1 ppm or 9 mg/m^3 at 25°C (52). A maximum permissible level in working areas of 5 mg/m^3 is recommended for both mixed terphenyls (58) and p-terphenyl (57). Various studies on the effect of terphenyls are reported in the literature (58–64). Toxicity of reduced terphenyls appears to be even lower than that of the parent compounds (57,65–68).

Environmental Considerations. Since much of the diphenyl released to the environment is in the wastewater effluent of textile mills, its environmental fate and impact on water quality is of prime concern (56,69–70). Laboratory data however, indicate that diphenyl is biodegradable (56). The perchlorination of sewage for disinfection could result in the formation of chlorinated diphenyls (47,69,71–72). However, the extent is open to question (47).

Terphenyls seem to be sufficiently biodegradable so as not to constitute an environmental problem (73–74).

Properties important for environmental considerations are given in Table 6.

Uses

In the past, the major use for diphenyl (75–76) and other polyphenyls was as heat-transfer fluids. However, its high freezing point necessitates elaborate precautions. The diphenyl oxide–diphenyl eutectic mixture, marketed under the trademarks

Table 5. Acute Toxicity of Diphenyl

Toxicity test	Value
inhalation, human, $TDL_0{}^a$, g/m^3	4400
oral, rat, LD_{50}, mg/kg	3280
oral, rabbit, LD_{50}, mg/kg	2400
skin, rabbit, LD_{50}, mg/kg	2500
fish, fathead minnow, $TL_m{}^b$, mg/L	1.5

^{a} TDL_0 = lowest published toxic dose.
^{b} 96 h.

Table 6. Properties of Diphenyl of Environmental Importance

Property	Value
vapor pressure at 25°C, Pa	1.3
water solubility, mg/L	7.5
aeration stripping, % COD removed in 4 h	89
BOD at 20°C	79
octanol/water partition coefficient	4.09
bioconcentration ratio	438 ± 48

Dowtherm A and Therminol VP-1, is widely used for heat transfer by the condensation of vapors in the temperature range of 250–360°C (see Heat transfer technology). Although a significant volume of diphenyl is still used in this application, currently the major use for biphenyl is in the formulation of dye carriers (qv) for textile dyeing (47).

Diphenyl is one of a variety of aromatic hydrocarbons that are used in dye carrier formulations. Typical formulations contain 60–90% hydrocarbon solvent and 10–40% emulsifier. The exact formulation of a dye carrier is often complicated and considered proprietary as demonstrated by the abundance of patented dye carrier formulations (77–81).

The consumption of diphenyl and all dye carriers is expected to remain stable or decrease in the future. Most of the diphenyl used as dye carrier is released to the environment either through stack emissions or in discharged wastewaters. Federal and local environmental restrictions tend to have a significant impact on the future use of diphenyl in textile applications (47,82).

Diphenyl is also used as a paper impregnant for citrus fruit wrappers where it acts as a mild fungicide to prevent blue mold (83–86). The diphenyl is bonded to the tissue paper by a paraffin wax-type solvent. Annual consumption of diphenyl in this application has been estimated at 270–360 t but has been decreasing (47).

Use of diphenyl in plasticizer applications for poly(vinyl chloride) (87), optical brighteners (qv) (88), and hydraulic fluid applications (89) is also reported in the literature (see Hydraulic fluids; Plasticizers).

Current use of terphenyls *per se* is confined almost exclusively to high-temperature heat-transfer applications. Despite excellent thermal stability and low vapor pressure, the relatively high melting points, even of the commercial mixture seriously restricts their utility in heat-transfer systems. This use of terphenyls amounts to only a few metric tons per year.

Derivatives

Diphenyls. Polychlorinated diphenyls were produced in large volumes and used in a variety of applications until the late 1960s (90). They came to be recognized as serious environmental contaminants and were restricted by governmental regulation. The Toxic Substances Control Act of 1976 essentially outlawed their manufacture and use in the United States (see Chlorocarbons).

Commercially important derivatives include *p*-phenylphenol, *o*-phenylphenol, and its sodium salt, 4-aminodiphenyl, and 4,4'-diaminodiphenyl. Diphenyl is partially reduced to cyclohexylbenzene or fully hydrogenated to dicyclohexyl. Paraphenylphenol reacts with formaldehyde to give a resin used in surface coatings. *p*-Phenylphenol is produced by the sulfonation of diphenyl followed by sodium hydroxide fusion or, together with *o*-phenylphenol, as a by-product from the hydrolysis of chlorobenzene with aqueous sodium hydroxide (91–94). *o*-Phenylphenol, particularly as its sodium salt, is used as a preservative, a germicide, or a fungicide. *p*-Dihydroxydiphenyl may also be produced by sulfonation of diphenyl followed by sodium hydroxide fusion (95).

Diphenyl is readily nitrated, and the nitrodiphenyls are reduced to give analogues of aniline. Benzidine (1,1-biphenyl-4,4'-diamine) is an important dye intermediate (96–98) (see Amines, aromatic).

A number of alkylated diphenyl products are available commercially. These are made primarily by Friedel-Crafts reaction of diphenyl with the appropriate alkene, or by-product formation during hydrodealkylation production of benzene from toluene (47,99–100).

Isopropyldiphenyl, the most important of the alkylated diphenyls, is primarily used in the production of carbonless copy paper. Methyldiphenyl is a by-product in benzene production and is used in dye carrier applications. Ethyl- and butyldiphenyl are produced in limited quantities for heat transfer fluid applications (47).

Terphenyls. For many years, most of the terphenyl production was converted to partially hydrogenated mixtures or polychlorinated terphenyl derivatives. In the United States manufacture of the latter was discontinued in 1972 (see Chlorocarbons).

However, demand for partially reduced terphenyls has increased. They are used, alone or in mixtures, as solvents for carbonless copy paper dyes, heat-transfer fluids, and plasticizers. A number of similar reduced terphenyl products are available in Japan, the USSR, and Europe.

Hydrogenated Terphenyls. Hydrogenated terphenyls are clear oils miscible at room temperature with common hydrocarbon and chlorinated hydrocarbon solvents and esters. These oils dissolve slowly in acetone and in ethanol only to the extent of 6%.

Terphenyls are hydrogenated over a variety of catalysts, with Raney nickel preferred for the commercial process.

Hydrogenated terphenyls are compatible with many plastic materials and solvate a variety of commercial polymers, rubbers, asphalts, and tars. Their main uses are as liquid-phase heat-transfer fluids and as dye solvents in carbonless copy paper technology.

Hydrogenated terphenyl is also used in certain plasticizer applications, primarily PVC. Minor amounts are sold for various other uses (lubricant, hydraulic, nuclear reactor coolant, etc.).

Quality control of various hydrogenated terphenyls is maintained by routine color, gravity, viscosity, moisture, and glc measurements. These are augmented by pour points, flash fire points, and trace analyses for various elements.

BIBLIOGRAPHY

"Diphenyl and Terphenyl" in *ECT* 1st ed., Vol. 5, pp. 145–155, by N. Poffenberger, The Dow Chemical Company, Midland, Michigan (Diphenyl), and C. F. Booth, Monsanto Co., St. Louis, Mo. (Terphenyls; Hydrogenated Derivatives); "Diphenyl and Terphenyls" in *ECT* 2nd ed., Vol. 7, pp. 191–204, by N. Poffenberger, The Dow Chemical Company, Midland, Mich. (Diphenyl), and H. L. Hubbard, Monsanto Co., St. Louis, Mo. (Terphenyls; Hydrogenated Derivatives).

1. W. Reid and D. Freitag, *Agnew. Chem. Inter. Ed.* **7,** 835 (1968).
2. W. E. Bachmann and H. T. Clarke, *J. Am. Chem. Soc.* **49,** 2089 (1927).
3. W. C. Taylor in R. F. Makens, ed., *Organic Coolant Summary Report, AEC Accession No. 15554, Report No. IDO-11401,* Idaho Operations Office, U.S. Atomic Energy Commission, avail. CFSTI, 1964, pp. 9–38.
4. H. Mandel, *Heavy Water Organic Cooled Reactor, Physical Properties of Some Polyphenyl Coolants AEC Report A1-CE-15,* April 15, 1966.
5. J. P. Stone, C. T. Ewing, and R. R. Miller, *J. Chem. Eng. Data* 7(1), 519 (1962).
6. M. McEwen, *Organic Coolant Handbook,* sponsored by U.S. Atomic Energy Commission, published by Monsanto Co., St. Louis, Mo., 1958.

7. J. P. Stone and co-workers, *Ind. Eng. Chem.* **50,** 895 (1958).
8. J. Chipman and S. B. Peltier, *Ind. Eng. Chem.* **21,** 1106 (1929).
9. *Monsanto Products 1964,* Monsanto Co., St. Louis, Mo., p. M5/22.
10. J. M. Cork, H. Mandel, and N. Ewbank (North American Aviation, Inc.), *U.S. Atomic Energy Comm. Sci. Rept. 5129, 1-28 (1960);* recalculated by A. N. Syverund, The Dow Chemical Company Thermal Lab, Midland, Mich.
11. M. McEwen and E. Wiederhold, *Nucl. Eng. Pt IV Symp. Ser.* **5**(22), 9 (1959).
12. I. Hartman, *U.S. Bur. Mines Rep.* 3489 (1955).
13. H. Ziebland and J. T. Burton, *J. Chem. Eng. Data* **6,** 579 (1961).
14. R. J. Good and co-workers, *J. Am. Chem. Soc.* **75,** 436 (1953).
15. J. A. Ellard and W. H. Yanko, *Thermodynamic Properties of Biphenyl and the Isomeric Terphenyls, Final Report, IDO-11,008,* Monsanto Res. Corp., U.S. Atomic Energy Comm. Contract AT (10-1)-1088, Oct. 31, 1963.
16. P. L. Geiringer, *Handbook of Heat Transfer Media,* Reinhold Publishing Company, New York, 1962, p. 173; G. Fritz and co-workers, *Atomkernenergie* **13**(1), 25 (1968) for tables of physical properties.
17. W. H. Hedley, M. V. Milnes, and W. H. Yanko, *Viscosity and Thermal Conductivity of Polyphenyls in Liquid and Vapor States, Final Report IDO-11,007,* Monsanto Res. Corp., U.S. Atomic Energy Comm., Contract-AT (10-1)-1088, July 31, 1963.
18. W. H. Hedley, M. V. Milnes and W. H. Yanko, *J. Chem. Eng. Data* **15,** 122 (1970).
19. A. K. Srivastava, S. C. Sharma, and B. Krishna, *Proc. Nat. Acad. Sci. India, Sect. A* **44**(1), 77 (1974).
20. H. L. Hubbard, *ECT* 2nd ed., Vol. 7, 1965, p. 20.
21. H. Mandel and N. Eubank, *Critical Constants of Diphenyl and the Terphenyl, Report No. NAA-SR-5129,* Atomics International, Dec. 1960, p. 18.
22. *Santowax O, Santowax M, and Santowax P, technical bulletin, IF-1,* Monsanto Co., St. Louis, Mo., July, 1964.
23. R. T. Keen, *U.S. At. Energy Comm. NAA-SR-Memo-2412,* 1958.
24. G. Mosselmans and J. Nienhaus, *Communuate,* EURATOM (Rapp.), EUR-4228, 1968; *Chem. Abstr.* **72,** 25672 (1970).
25. G. Descotes, J. Praly, and M. Lebaupain, *Bull. Soc. Chem. France,* 896, 901 (1976).
26. H. G. Goodman and A. Lowy, *J. Am. Chem. Soc.* **60,** 2155 (1938).
27. C. F. H. Allen and F. P. Pingert, *J. Am. Chem. Soc.* **64,** 1365 (1942).
28. U.S. Pat. 2,363,209 (Nov. 21, 1944), R. D. Swisher (to Monsanto Co.).
29. Czech. Pat. 123,389 (1967), K. Hyska and E. Kondrat.
30. C. Hansch and C. F. Geiger, *J. Org. Chem.* **23,** 477 (1958); T. Sato, Y. Goto, and K. Hata, *Bull. Chem. Soc. Jpn.* **40,** 1964 (1967).
31. M. Orchin, J. K. Blatchford, and S. I. Shupack, *Israel J. Chem.* 364 (1963).
32. R. G. Harvey, D. F. Lindlow, and P. W. Rabideau, *J. Am. Chem. Soc.* **94,** 5412 (1972).
33. U.S. Pat. 3,729,522 (April 24, 1973), G. A. Arnett and W. B. Dunlap (to Monsanto Co.).
34. *Oil, Paint Drug Rep.* 55 (Dec. 9, 1963).
35. *Hydrocarbon Process. Pet. Refiner.* **42**(11), 204 (1963).
36. L. C. Doelp, *ECT* 2nd ed., Vol. 11, 1966, pp. 453–462.
37. U.S. Pat. 3,700,566 (Oct. 24, 1972), R. N. Bellinger and J. T. Cabbage (to Phillips Petroleum Company).
38. USSR Pat. 215,911 (April 11, 1968), E. I. El'bert and Ya. R. Katsobashvili.
39. R. G. Ismailov, D. A. Aliev, and T. M. Ivanova, *Azerb. Neft. Khoz.* **44**(1), 30 (1965).
40. *Ibid.* **43**(5), 33 (1964).
41. *Ibid.* **42**(5), 33 (1963).
42. T. A. Bogdanova and co-workers, *Nefleperab. Neftekhim. (Moscow),* (2), 35 (1972).
43. J. Oro and J. Han, *Science* **153,** 1393 (1966).
44. U.S. Pat. 3,274,277 (Sept. 20, 1966), H. S. Block (to Universal Oil Products Company).
45. East Ger. Pat. 99,151 (July 20, 1973), H. Baltz, K. Becker, K. Kirschke, and H. Oberender.
46. E. Pajda and J. Rusin, *Chemik* 422 (1972).
47. USSR Pat. 570,584 (Aug. 1977), D. B. Orechkin and A. P. Bocharov.
48. W. M. Meylan and P. H. Howard, *Chemical Market Input/Output Analysis of Biphenyl and Diphenyl Oxide to Assess Sources of Environmental Contamination,* Syracuse Research Corp., Syracuse, NY, EPA Contract No. 68-01-3224, 1976.

49. H. W. Gerarde, *Toxicology and Biochemistry of Aromatic Hydrocarbons,* Elsevier Publishing Co., Amsterdam, 1960, p. 200.
50. H. W. Gerarde in F. A. Patty, ed., *Industrial Hygiene and Toxicology,* 2nd rev. ed., Interscience Publishers, Inc., New York, 1958, p. 1200.
51. M. McEwen, *Organic Coolant Handbook,* sponsored by U.S. Atomic Energy Comm., published by Monsanto Co., St. Louis, Mo., 1958.
52. *TLV's®—Threshold Limit Values for Chemical Substances in Workroom Air Adopted by ACGIH for 1977,* Am Conf. of Governmental Industrial Hygenists, Cincinnati, Ohio, p. 28.
53. I. Hakkinen and co-workers, *Arch. Environ. Health* **26**(2), 70 (1973).
54. H. E. Christensen and T. T. Luginbyhl, *Registry of Toxic Effects of Chemical Substances,* U.S. Dept. of H.E.W., U.S. Government Printing Office, Washington, D.C., 1975.
55. J. Opdyke, *Food Cosmet. Toxicol.* **12,** 707 (1974).
56. J. M. Haas, H. W. Earhart, and A. S. Todd, *Book of Papers, 1974 AATCC National Technical Conference,* American Association of Textile Chemists and Colorists, Research Triangle Park, N.C., p. 427.
57. Z. F. Khromenko, *Gig. Tr. Prof. Zabol.* 44 (1976).
58. Z. F. Khromenko, *Nauch. Tr. Irkutsk. Med. Inst.* (115), 121 (1972).
59. T. J. Haley and co-workers, *Toxicol. Appl. Pharmacol.* **1,** 575 (1953).
60. G. Young, A. Petkau, and J. Hoogstraten, *Am. Ind. Hyg. Assoc. J.* 7 (1969).
61. S. Kiriyama, M. Banjo, and H. Matsushima, *Nutr. Rep. Int.* 78 (1974).
62. H. Ott and D. Pirrwitz, *Eur. At. Energy Comm.,* EURATOM (Rep.) 1970, EUR-4558; *Chem. Abstr.* **75,** 86750 (1971).
63. P. Scoppa and K. Gerbault, *Boll. Soc. Ital. Biol. Sper.* 47(7), 194 (1971).
64. I. Y. R. Adamson, F. H. Frummond, and J. P. Wyatt, *Arch. Environ. Health* 499 (1969).
65. Y. R. Adamson and J. L. Weeks, *Arch. Environ. Health* **27**(8), 68 (1973).
66. Y. R. Adamson and J. M. Furlong, *Arch. Environ. Health* **28**(3), 155 (1974).
67. R. J. Hawkins and J. L. Weeks, *Report 1973,* AECL-4439; *Chem. Abstr.* **80,** 65963 (1974).
68. J. S. Henderson and J. L. Weeks, *Ind. Med.* **40,** 10 (1973).
69. R. G. Webb and co-workers, *Current Practice in GC-MS Analysis of Organics in Water,* PB-224-947, U.S. Nat. Tech. Inform. Service, Springfield, Va., 1973.
70. R. A. Hites, *J. Chromatogr. Sci.* **11,** 570 (1973).
71. P. Gaffney, *Science* **183,** 367 (1974).
72. R. M. Carlson and co-workers, *Environ. Sci. Technol.* **9,** 674 (1975).
73. D. Catelani and co-workers, *Experentia* 922 (1970).
74. T. Ohmori and co-workers, *Agr. Biol. Chem.* 1973, **37**(7), 1599 (1973).
75. U.S. Pat. 1,864,349 (June 21, 1932), F. X. Govers (to Indiana Refining Company).
76. G. H. Reid, *Refiner Nat. Gasoline Manuf.* **10,** 63 (1929).
77. J. W. Bullington, T. H. Guion, and T. S. Roberts, *J. Soc. Dyers Colour.* **82,** 405 (1966).
78. Belg. Pat. 644,371 (June 15, 1964), (to Cassella Farbwerke Mainkur A.G.).
79. H. Rath and P. Senner, *Z. Gesamte Text. Ind.* **69**(20), 71 (1967).
80. Fr. Pat. 1,505,175 (Dec. 8, 1967), (to Farbwerke Hoechst A.G.).
81. Ger. Pat. 2,225,565 (Dec. 7, 1972), H. P. Baumann (to Sandoz, Ltd.).
82. S. Cohen, *Am. Dyest. Rep.* **77**(9), 42 (1977).
83. U.S. Pat. 2,265,522 (Dec. 9, 1941), A. Farkas.
84. F. W. Hayward, W. Grierson and G. J. Edwards, *Citrus Ind.* **47**(10), 9 (1966).
85. A. I. Grimm, *Sb. Tr. Lenningr. Inst. Sov. Torgovli,* **23,** 116 (1964).
86. D. Leggo and J. A. Sebeny, *Aust. J. Exp. Agric. Anim. Husb.* 5(16), 91 (1965).
87. V. A. Zubchuk and S. V. Yukabovich, *Lakokras. Mater. Ikh Primen.* (1), 30 (1965).
88. Jpn. Pat. 29,450 (Dec. 18, 1964), Mitsui Chem. Industry Co., Ltd.
89. Brit. Pat. 1,075,571 (July 12, 1967), Mobil Oil Co., Ltd.
90. H. L. Hubbard, *ECT,* 2nd ed., Vol. 5, pp. 289–297.
91. U.S. Pat. 1,942,386 (Jan. 2, 1934), W. C. Stoesser and R. F. Marchner (to The Dow Chemical Company).
92. S. Billbrough, J. I. Jones, and F. M. Potter, *BIOS (British Intelligence Objectives Subcommittee) Report 1634,* 1964.
93. E. C. Britton and W. J. Hale, *Ind. Eng. Chem.* **20,** 114 (1928).
94. E. Hausmann, *Erdoel Kohle* **7,** 496 (1954).
95. U.S. Pat. 2,368,361 (Jan. 30, 1945), R. L. Jenkins (to Monsanto Co.).

96. U.S. Pat. 1,954,468 (April 10, 1934), C. F. Booth (to Monsanto Co.).
97. C. F. Booth, R. L. Jenkins and R. McCullough, *Ind. Eng. Chem.* **22,** 31 (1930).
98. U.S. Pat. 1,891,543 (Dec. 20, 1932), E. H. Huntress and C. R. McCullough (to Monsanto Co.).
99. P. Aeberli, *J. Med. Chem.* **18,** 177 (1975).
100. A. A. Kudinov and N. M. Indyukov, *Azerb. Khim. Zh.* **74**(5–6), 30 (1974).

W. C. WEAVER
P. B. SIMMONS
Dow Chemical U.S.A.

Q. E. THOMPSON
Monsanto Co.

DISINFECTANTS AND ANTISEPTICS

The procedures that are currently described as antiseptic or disinfectant began at the dawn of recorded history when food was preserved by smoking, on the application of essential oils and spices for embalming, or by burning of aromatic woods for protection against the plague. The Bible commands the burning of clothes worn by the diseased. Boiling of drinking water and its preservation by storage in silver or copper vessels was known to the Persians as well as to the army of Alexander the Great.

The bubonic plague was fought in medieval Europe by burning sulfur, cedar, or juniper wood in affected houses. Although a connection between pathogenic microorganisms and contagious disease was not established for another several centuries, several early students of disease, including Fracastorius in the 16th, and Robert Boyle in the 17th century clearly suspected the existence of pathogens.

Even before the publication of Pasteur's definitive work on transmission of disease (in 1860 and thereafter), Labarraque used a hypochlorite solution in the treatment of wounds (1825), Alcock recommended it for the purification of drinking water (1827), Lefevre extolled its virtues as a disinfectant (1843), and Semmelweis used it to inactivate the "cadaveric poisons" carried on the hands of physicians attending puerperal fever patients (1846).

Some twenty years later, Lister took up the use of phenol [*108-95-2*] in surgery. In this he was probably stimulated by Pasteur's demonstration of the ubiquity of germs and their involvement in putrefaction, which Lister assumed, resembled suppuration in wounds. At first he applied phenol in concentrated form, and later in an aqueous solution because of the corrosive action of the pure chemical upon tissue. In 1870, in an effort to decontaminate the ambient air, Lister introduced the carbolic acid (phenol) spray and used it for another fifteen years thereafter. It is noteworthy that both hypochlorite and phenol solutions were used for years on an essentially conjectural basis, that is, without any evidence of the actual causes of infection (see also Antibacterial agents; Antibiotics; Chemotherapeutics; Fungicides; Industrial antimicrobial agents; Sterile techniques).

Definitions

The Food and Drug Administration (FDA) has adopted the following definitions (1) for antimicrobial (active) ingredient and antimicrobial preservative (inactive) ingredient:

Antimicrobial (active) ingredient: A compound or substance that kills microorganisms or prevents or inhibits their growth and reproduction and that contributes to the claimed effect of the product in which it is included.

Antimicrobial preservative (inactive) ingredient: A compound or substance that kills microorganisms, or that prevents or inhibits their growth and reproduction, and that is included in a product formulation only at a concentration sufficient to prevent spoilage or prevent growth of inadvertently added microorganisms, but does not contribute to the claimed effects of the product to which it is added.

The Environmental Protection Agency (EPA) has adopted the following definitions (2) for antimicrobial agents:

Disinfectants destroy or irreversibly inactivate infectious or other undesirable bacteria, pathogenic fungi, or viruses on surfaces or inanimate objects.

Sanitizers reduce the number of living bacteria or viable virus particles on inanimate surfaces, in water, or in the air.

Bacteriostats inhibit the growth of bacteria in the presence of moisture.

Sterilizers destroy viruses and all living bacteria, fungi, and their spores on inanimate surfaces.

Fungicides and *fungistats* inhibit the growth of or destroy fungi (including yeasts) pathogenic to man or other animals on inanimate surfaces.

ANTISEPTICS

Antiseptics are generally understood to be agents applied to living tissue and are considered to be drugs. Therefore they may be inactivated by contact with body fluids, such as blood or serum.

The methods prescribed by Ruehle and Brewer for the examination of antiseptics, liquids, ointments, powders, oils, etc, enjoyed a quasi-official status for some time, having originated with a governmental authority (3). These tests are still useful for screening new preparations designed as antiseptics that are preliminary to more extensive clinical tests under conditions of intended use. However, the FDA no longer sponsors these methods; according to its present position, no standard or official tests for the evaluation of antiseptics exist, and no single *in vitro* test is considered sufficiently informative to serve as a criterion of practical performance.

To the extent that *Staphylococcus aureus* represents the most common cause of suppuration and exhibits marked resistance to both physical and chemical factors, its use as a test organism is logical in one of the several applicable screening tests. However, for antiseptics applied to prevent infection through a break in the skin, an *in vivo* method (4–6) yields more relevant results.

Klarmann and co-workers (7) reported a semimicro method that furnishes more information about the performance of liquid antiseptics likely to come in contact with tissue fluids. One of the features of this technique is control of the random-sampling error inherent in methods that depend upon loop transfers from the medication mixture to the subculture. In addition, this method incorporates the following premises: (1) The criterion of fitness of any antiseptic that is likely to come in contact with tissue fluids should be its demonstrated capacity for permanent suppression of bacterial activity under the conditions of use. (2) If an antiseptic that has prevented bacterial proliferation in a nutrient medium does not continue to prevent it upon contact with physiological material such as blood or serum, then its fitness for use as a preoperative or wound antiseptic is open to question.

Although the so-called degerming agents are not, strictly speaking, skin antiseptics, the *in vivo* testing method developed for the evaluation of degerming agents furnishes comparative information about their practical performance. This method has been devised by Price and is known as the serial basin test (8); later modifications were suggested to simplify the procedure (9–10). The procedure recognizes the fact that the skin cannot be disinfected in the true sense of the term, but that it is possible to reduce the "resident" potentially pathogenic bacterial flora for reasons of safety as required, for example, in surgery. With other factors being equal, the more valuable the degerming agent is, the better it depresses the bacterial count.

The glove juice test, ie, the procedure used to determine the effectiveness of surgical scrubs has, since 1973, been part of the guidelines of the Bureau of Drugs for evaluating compounds recommended for the surgical scrub. With some modification, this test was adopted by the OTC Antimicrobial I Panel and published in the Federal Register (11).

Since *in vivo* degerming tests are often cumbersome, an *in vitro* screening test such as the calf-skin disk test may be employed for the preliminary estimation of the effectiveness of antimicrobial soaps or detergents under actual conditions of use (12).

Although various test methods for the determination of tissue toxicity have been suggested in the practical screening of antiseptics (13–14), their results cannot be used as a rational basis of a comparative evaluation of such agents. The problem of tissue toxicity is actually of considerably greater importance in chemotherapy than in antisepsis. However, an antiseptic with lower tissue toxicity should not be given preference over a more toxic one.

DISINFECTANTS

Analytical and Test Methods

Disinfectants are generally understood to be agents suitable for application to inanimate objects. Environmental surfaces are deemed to be contaminated, not infected. Disinfectants kill the growing forms but not necessarily the resistant spore forms of microorganisms (2b).

Rideal and Walker were the first to recognize the need for standardizing a number of factors in the methodology of the evaluation of disinfectants (15). Perhaps their most fundamental requirement was that of determining the resistance of the test organisms to pure phenol at the time when the disinfectant sample was being tested with the same bacterial culture. Of course, this is not to minimize the importance of standardizing other details of the testing technique, without which reproducibility of results would be impossible.

The Rideal-Walker method has been revised and its application extended several times, both in Great Britain and the United States, during the 76 years since its introduction. The most important British revisions were those of the Lancet Commission in 1909 (16), by Rideal and Walker themselves in 1921 (17), and by the British Standards Institution in 1934 (18). In the United States revisions were undertaken by Anderson and McClintic on behalf of the Hygienic Laboratory in 1912 (19), by a committee of the American Public Health Association in 1918 (20), by Brewer and Reddish in 1929 (21), by Ruehle and Brewer in 1931 (3) on behalf of the FDA, and finally, by the Association of Official Agricultural Chemists (AOAC) in 1960 (22). This last method, designated the AOAC Phenol Coefficient Method, although patterned along the lines of the FDA method, provides for special formulas of subculture media designed to suspend bacteriostasis in order to distinguish between germicidal and inhibitory action. Thus, the addition of thioglycolate is specified for testing preparations containing heavy metals (eg, mercury), and lecithin (qv) is used in testing products containing quaternary ammonium compounds (22).

The phenol coefficient is a number obtained by dividing the greatest dilution of the disinfectant under test capable of killing *Salmonella typhosa* at room temperature by the greatest dilution of phenol showing the same result. According to the EPA, any phenol coefficient claim requires testing by the AOAC Phenol Coefficient Method from duplicate samples of one batch against *Salmonella typhosa*. The classic Ruehle and Brewer Circular No. 198 (3) added the requirement for the use of a second test organism, *Staphylococcus aureus* in 1931. This requirement serves to extend the method to testing antiseptics as well as disinfectants. The AOAC test for fungicidal action of environmental disinfectants call for the use of *Trichophyton interdigitale* as the test organism (22).

Methods such as the AOAC phenol coefficient method serve as screening procedures to yield a general idea as to whether or not a given product is of potential value as a germicide, but no single testing procedure is an adequate substitute for a comprehensive inquiry into the total antimicrobial spectrum of a given disinfectant used for pathogens of epidemiological and clinical significance.

The phenol coefficient test has lost some significance today and has been replaced by the AOAC Use-Dilution Test to measure the efficacy of disinfectants. The use-dilution procedure (23–24) differs from the phenol coefficient methods in that it employs bacteria deposited on carriers such as polished stainless steel rings (penicillin cups) rather than their suspensions in nutrient broth. Two test organisms are specified, *Staphylococcus aureus* and *Salmonella choleraesuis* which replaces *Salmonella typhosa* of the phenol coefficient method because the latter does not resist drying on the steel rings. The phenol resistance of a 48-h culture of *Salmonella choleraesuis* should fall within the range specified for a 24-h culture of *Salmonella typhosa*.

The use-dilution test distinguishes between two general classes of disinfectants, those for janitorial or household uses directed especially against enteric organisms, and those for medical, veterinary, hospital, and surgical uses directed against both enteric and pyogenic bacteria. The former must be germicidal for *Salmonella choleraesuis*, and the latter must be effective against both *Salmonella choleraesuis* and *Staphylococcus aureus* in their recommended use-dilutions under the conditions of the ring-carrier testing procedure (25). Supplementary confirmatory tests on floors and surgical instruments have been developed (25).

All disinfectant products must be formally registered with the EPA, according to the Federal Insecticide, Fungicide, and Rodenticide Act of 1947 and the Federal Environmental Pesticide Control Act of 1972 and registrations must be accompanied by efficacy data based on the use-dilution test.

Square Diluent Test. This test has acquired some importance in connection with a Federal specification for disinfectants (26). It recognizes the need for a broad spectrum antimicrobial (bactericidal and fungicidal) performance by requiring tests with three microorganisms representative of pathogenic types, that is *Staphylococcus aureus*, *Salmonella schottmuelleri*, and *Trichophyton interdigitale*. In this method test organisms are employed in mixture rather than individually.

Long-lasting disinfectants that persist in residual quantities on disinfected surfaces obviously have the advantage of inactivating infectious materials that may be later deposited on such surfaces (27).

Chemical Sterilization of Instruments. The tests described thus far relate to disinfection rather than to sterilization. That is, they assess the destruction of vegetative microorganisms but not of bacterial spores. Different chemicals have been claimed to possess sporicidal action. Disregarding those that would be too corrosive for practical use as instrument germicides, direct claims for sporicidal action at room temperature are being made today mostly for ethylene oxide [75-21-8] and aldehyde preparations. Hydrogen peroxide [7722-84-1], some iodophors and some quaternary ammonium compounds (qv) are also being credited with sporicidal potency, but several quaternary ammonium compounds tested are incapable of killing the spores of *Clostridium welchii*, *tetani* and *sporogenes*, and *Bacillus anthracis* even in 24 h (28).

Friedl (29) has reported a sporicidal test "applicable for use with germicides to determine presence or absence of sporicidal activity and potential effectiveness in disinfecting against specified spore-forming bacteria." The test is based on a procedure for obtaining spores of the desired resistance and exposing them to the action of disinfectants. The method provides for standardization by means of 20% hydrochloric acid of the spores of two test organisms, *Bacillus subtilis* and *Clostridium sporogenes*, although it may be carried out also with *Bacillus anthracis*, *Clostridium tetani*, and other species of *Bacilli* or *Clostridia* (30).

With a formaldehyde [50-00-0] germicide, for which sporicidal claims had been made, the spores of *Bacillus subtilis* were killed in 10–30 min, but the product did not kill the spores of *Clostridium sporogenes* in 2 h, the longest period of exposure employed in this study (31–32). These findings were subsequently extended to longer periods of exposure and the control of sporostasis in subcultures.

Steam under pressure, properly applied in an autoclave, is the sterilizing agent *par excellence* for instruments and appliances. Equivalent results, particularly with resistant bacterial spores, can be produced in a short period of time by exposure to boiling dilute aqueous solutions of certain disinfectants, preferably synthetic phenolics that exhibit low volatility with water vapors (31).

Proper sterilization of syringes and needles is important for the prevention of infection with hepatitis viruses A and B carried by human blood, serum, plasma, fibrinogen, etc, when administered by transfusion or injection (33) (see Blood). Since, at present, only man is known to be susceptible to infection by either virus, there is no laboratory method available for the evaluation of the sterilizing effectiveness of chemical disinfectants for these viruses. The Expert Committee on Hepatitis of the WHO considers the following procedures to be adequate: boiling in water for at least 10 min; autoclaving; or treatment in a hot-air oven for $\frac{1}{2}$ h at 170–180°C. No chemical disinfectants acting at room temperature are considered acceptable for this purpose (34).

In the cold sterilization of chemical thermometers and of instruments that cannot be subjected to heat treatment, tuberculocidal potency should be part of the nonspecific bactericidal spectrum of a disinfectant. The confirmatory *in vitro* tests require the use of a virulent culture of *Mycobacterium tuberculosis*; *in vivo* verification of the results obtained is carried out by inoculations in guinea pigs.

Virucidal Action. A nonspecific disinfectant is not necessarily a nonspecific virucide, some authoritative opinion to the contrary notwithstanding (35). Some of the most valuable germicides are not effective against important viruses, such as that of poliomyelitis, under practical conditions of use. Moreover, the differences in susceptibility of viruses to antiviral agents appear to be as great as, or even greater than, the corresponding differences in bacteria (see also Chemotherapeutics, antiviral).

Unlike bacteria, viruses are metabolically inert entities in the absence of a susceptible host cell and depend upon the living host cells to supply the wherewithal of viral multiplication. They accomplish this by altering the metabolism of the parasitized cell to force the production of new viral material. But here the uniformity among viruses comes to an end. The great chemical differences among them are illustrated by the finding that, in the case of influenza virus, protein material accounts for less than 50% of the dry weight, whereas of rabbit papilloma virus protein makes up over 90%. Although some 50% of influenza virus consists of lipids, less than 6% of the dry weight of vaccinia virus is of this character (36). Such differences in composition must be expected to find expression in the wide variations in susceptibility to chemical and physical agents (37).

Influenza virus may be readily inactivated by different substances, as shown by a variety of testing methods (38–40). This is in contrast to poliomyelitis virus whose inactivation presents a more difficult problem (41).

The EPA (42) has published *Methods for Evaluating the Virucidal Activity of Disinfectants;* however, this agency, in cooperation with other laboratories concerned, is continuing work on improved methods for testing virucides. Apparently, there is not yet a standard or official federal test for virucidal disinfection. Some of the difficulties encountered were described (43–44).

Role of Neutralizers. The testing methods mentioned occasionally call for the use of neutralizers. Their purpose is to inactivate the residual amounts of antimicrobial agent carried with the microorganisms into the subculture media. Otherwise, bacteriostasis might be mistaken for bactericidal action. These media must contain materials suppressing bacteriostasis by the disinfectant being tested. The character of these neutralizers depends on the antibacterial agent under test. For instance, Tween 80 with lecithin is a satisfactory neutralizer for quaternary ammonium compounds, sodium thiosulfate is effective for chlorine, and iodine compounds and sodium thioglycolate for mercurials.

Alcohols

The antibacterial effectiveness of several of the aliphatic alcohols has been known for a long time (45). In a study of their mode of action, the capacity of the monohydric alcohols from methyl to octyl to serve as substrates for *Escherichia coli* was investigated (46). Methyl alcohol [67-56-1] and the alcohols from butyl through octyl were unable to function as substrates; moreover, they inhibited the dehydrogenation of other substrates, the amount of inhibition increasing with the chain length of the alcohol. However, ethyl and propyl alcohols are capable of playing a dual role in that they act as substrates at low concentrations and as dehydrogenation inhibitors at high concentrations. By far the greatest amount of work has been done on the bactericidal action of ethyl alcohol [64-17-5] (47–50). In the proper concentration, ethyl alcohol is an effective germicide for vegetative pathogens as shown in Table 1.

The transmission of infectious or homologous serum hepatitis cannot be prevented by exposing virus-contaminated surgical instruments to the action of alcohol in any concentration or for any length of time. Only heat properly applied may be relied upon to produce inactivation of both types of hepatitis virus (34).

Isopropyl alcohol [67-63-0] appears to be somewhat more effective as an antimicrobial agent than ethyl alcohol. *Staphylococcus aureus* is reported to be killed by 50–91% isopropyl alcohol in 1 min at 20°C (58). However, earlier reports claim that 90% isopropyl alcohol fails to kill this organism in 2 hours (59). Isopropyl and ethyl alcohol are essentially similar in tuberculocidal performance when acting upon dried sputum smears (52).

In the homologous series of primary normal alcohols, the germicidal action increases with increasing molecular weight from methyl to octyl alcohol as shown in tests with *Salmonella typhosa* and *Staphylococcus aureus* (50). Moreover, there exists a fairly constant ratio between the molecular coefficients of successive members of the series as defined by the equation:

$$\text{molecular coefficient} = \frac{\text{mol wt alcohol}}{\text{mol wt phenol}} \times \frac{\text{phenol}}{\text{coefficient}}$$

Primary alcohols with 6 to 16 carbon atoms inhibit *Mycobacterium tuberculosis* and *Trichophyton gypseum*, but not *Staphylococcus aureus* (60). Ferguson's principle of relating thermodynamic activity to cytostatic activity accounts satisfactorily for many regularities encountered in homologous series of alcohols.

Vapors of propylene glycol [57-55-6] and triethylene glycol [112-27-6] can protect animals against airborne bacteria and influenza virus under controlled conditions of temperature and humidity (61–62) (see Alcohols; Glycols).

Halogens

Chlorine and Hypochlorites. Although the use of chlorinated lime as deodorant for sewage goes back to a report of the Royal Sewage Commission of Great Britain in 1854, it was R. Koch who in 1881 first referred to the bactericidal properties of hypochlorites. Calcium hypochlorite [7778-54-3] has now largely replaced the older chlorinated lime mostly for large-scale usage, and sodium hypochlorite [7681-52-9] is the

Table 1. Bactericidal and Virucidal Action of Ethanol

Organism	Effective conc, %	Time required	Reference
Salmonella typhosa	21.6	5 min	47
	17.3	10 min	47
	40–100	10 s	48
Staphylococcus aureus	34.5	5 min	47
	60–95	10 s	48
	100	50–90 s	48
Staphylococcus albus	70–80	5 min	49
	15	24 h	8
Streptococcus pyogenes	100	50–90 s	48
	50–100	30 s	48
	40	4 min	48
	30	45 min	48
	20	60 min	48
Escherichia coli	60	1–5 min	50
	70	40 s	51
	80	30 s	51
	60	5 min	51
	40–100	10 s	48
Pseudomonas aeruginosa	30–100	10 s	48
	20	30 min	48
Mycobacterium tuberculosis	95	15 s	52
aqueous suspension	100	30 s	52
	70	60 s	52
Mycobacterium tuberculosis			
in dry sputum, thin smears	70	1–5 min	52
thick smears	95	30 min	52
Trichophyton mentagrophytes	50	5 min	53
	25	20 min	53
Trichophyton rubrum	50	5 min	53
	25	20 min	53
Microsporum canis	50	5 min	53
	25	20 min	53
Microsporum audouini	70	24 h	53
fowl pox virus	70–95	10 min	54
	50	30 min	54
Newcastle virus	70–95	3 min	55
vaccinia virus	50	60 min	56
foot and mouth disease virus,	50	15–20 min	57
filtered (Berkefeld)	60	1–15 min	57
foot and mouth disease virus,			
unfiltered	20–60	6 h	57

active principle of many household products and of some hospital specialties (63). A number of organic chlorine compounds have been introduced; their antibacterial action is dependent upon their capacity for releasing "active" chlorine. In various areas of disinfection and sanitization, elementary chlorine [7782-50-5] is employed directly.

The bactericidal action of hypochlorites is caused primarily by the release of hypochlorous acid [7790-92-3], HOCl; however, the hypochlorite ion [13675-18-8], OCl⁻, may be a contributory factor since even distinctly alkaline hypochlorite solutions show some antibacterial potency (64). The pH of the hypochlorite solution has a marked effect on its germicidal activity in tests employing various test organisms (65–67).

The action of chlorine dioxide [10049-04-4], ClO$_2$, is more sporicidal than that of hypochlorous acid (68).

Although ordinarily all disinfectants suffer impairment of their activity in the presence of organic matter, the hypochlorites particularly lose much of their bactericidal potency under practical conditions of disinfection owing to the intense reactivity of both the hypochlorous acid and the hypochlorite ion. For this reason use of hypochlorites generally calls for prior removal of organic contamination (69).

The hypochlorites are subject to gradual deterioration over a period of time. The rate of deterioration depends upon several factors; the most important of which are the type of product, its pH, and the temperature of storage. The lower the pH, the more germicidal, but the less stable is the solution. A number of hypochlorite preparations available in dry form are quite stable if kept dry and cool (see Bleaching agents; Chlorine oxygen acids and salts).

Dakin's solution, which was used extensively during World War I for the irrigation of wounds, contained in its original form, sodium hypochlorite (0.45–0.5%) and boric acid [11113-50-1]. However, the success of the Carrel-Dakin surgical irrigation treatment is not caused by its antibacterial aspect, instead its action increases transudation from the wound which produces a drainage of stagnated fluid and causes replacement by fresh lymph (70).

N-Chloramines. This group comprises the derivatives of amines in which one or two valences of trivalent nitrogen are taken up by chlorine.

In comparison with hypochlorous acid, monochloramine [10599-90-3], NH$_2$Cl, requires more time and a higher concentration to produce sporicidal action. Table

Table 2. Available Chlorine Content of Some N-Chloramines

Name and CAS Registry No.	Structure	Available Cl, %
chloramine T [127-65-1]	CH$_3$—C$_6$H$_4$—SO$_2$NClNa·3H$_2$O	23–26
dichloramine T [473-34-7]	CH$_3$—C$_6$H$_4$—SO$_2$Cl$_2$N	56–60
halazone [80-13-7]	HO$_2$C—C$_6$H$_4$—SO$_2$NCl$_2$	48–52.5
succinchlorimide [128-09-6]	(succinimide N-Cl)	50–54
chloroazodin [502-98-7]	H$_2$NCN=NCNH$_2$, with NCl NCl	75–79
trichlorocyanuric acid [87-90-1]	(cyanuric acid tri-N-Cl)	88–90

2 lists several other chloramines and gives their available chlorine content (see Chloramines).

Chloramine-T (sodium N-chloro-p-toluenesulfonamide) was widely used during World War I for the treatment of infected wounds and subsequently for hygienic purposes such as mouthwashes, douches, etc. It can be used for sanitizing food-handling equipment, but its activity is considerably slower than that of hypochlorites.

Dichloramine-T (N,N-dichloro-p-toluenesulfonamide) is insoluble in water, but soluble in a number of organic solvents, including chlorinated paraffin. Its medical usage appears to have declined.

Halazone (N,N-dichloro-p-carboxybenzenesulfonamide) is suitable for the decontamination of water, as is also succinchlorimide (N-chlorosuccinimide).

Chlorinated cyanuric acid derivatives consist of dichloroisocyanuric acid [2782-57-2], their sodium [2893-78-9] and potassium [2244-21-5] salts, and trichloroisocyanuric acid [87-90-1]. These compounds can be used in sanitation as such or formulated into various products (see Bleaching agents; Cyanuric and isocyanuric acids).

Several commercial cleaner-sanitizer products contain 1,3-dichloro-5,5-dimethyl hydantoin [118-52-5] as antibacterial component. In a pH range of 5.8–7.0 it is found to have similar antimicrobial activity to hypochlorite; it is less effective under alkaline conditions.

Chloromelamine [7673-09-8] is a chlorination product of 1,3,5-triaminotriazine. Formulated with a suitable anionic surfactant, it has been considered as a bactericidal rinse for mess kits and also for the treatment of contaminated fruits and vegetables (71) (see Cyanamides).

Chloroazodin (azochloramide, N,N-dichloroazodicarbonamidine) is claimed to be relatively nontoxic to tissue. Applied to a wound it acts as a mild and slow oxidant (72).

N-Trichloromethylmercapto-4-cyclohexene-1,2-dicarboximide [133-06-2] (Captan) was recommended as degerming agent for incorporation in liquid or bar soap (73).

Iodine. Although bromine is of little significance as an antibacterial agent, iodine [7553-56-2] has held and even expanded its position, especially in the field of antisepsis. Like chlorine, iodine is a highly reactive substance combining with proteins partly by chemical reaction and partly by adsorption. Therefore its antimicrobial action is subject to substantial impairment in the presence of organic matter such as serum, blood, urine, milk, etc. However, where there is no such interference, nonselective microbicidal action is intense and rapid.

A saturated aqueous solution of iodine exhibits antibacterial properties. However, owing to the low solubility of iodine in water (33 mg/100 mL at 25°C), reaction with bacteria or with extraneous organic matter rapidly depletes the solution of its active content.

Iodide ion is often added to increase solubility of iodine in water. This increase takes place by the formation of triiodide, $I_2 + I^- = I_3^-$. An aqueous solution of iodine and iodide at a pH of less than 8 contains mainly free diatomic iodine I_2 and the triiodide [14900-04-0] I_3^-. Their ratio depends upon the concentration of iodide.

A quantitative investigation of the relative bactericidal activities of diatomic iodine and triiodide revealed a negligible effect of the latter upon the test organisms *Staphylococcus aureus* and *Escherichia coli* (74). Similar findings were reported for their sporicidal performance (75).

As a bactericidal antiseptic, iodine is used most frequently in the form of a 2% tincture. Stronger iodine preparations are employed occasionally, for example for preoperative antisepsis. The degerming capacity of iodine tinctures of different concentrations are compared to those of organomercuric and quaternary ammonium antiseptics by the serial basin test (76). The results of this test indicate that the iodine tinctures, within the range of 2–7% of iodine, can reduce the bacterial count of the skin by 97.5–100%; in this respect they appear to be markedly superior to the other antiseptics tested.

Because of the tuberculocidal action of iodine, its solutions are suitable for the disinfection of clinical thermometers (77–78) and for emergency sterilization of surgical appliances (79). Other uses include the sanitization of eating and drinking utensils (80) and the emergency treatment of contaminated drinking water (81–83). In most natural waters, 8 ppm of iodine will reduce a count of one million enteric bacteria per milliliter to less than 5/100 mL within 10 min. The same dosage of iodine will destroy thirty cysts of *Endamoeba histolytica* per milliliter in the same period of time.

Heliogen [127-65-1] contains potassium iodide [7681-11-0] and chloramine-T; free (diatomic) iodine is released upon addition of water (84).

Iodophors are combinations of iodine with suitable solubilizing organic compounds (usually nonionic surfactants) (41,85). Their effectiveness is caused primarily by the free, or the available iodine they contain (86).

Several disinfectant, antiseptic, and sanitizing agents have been formulated on the iodophor principle (63,87–89), among them Wescodyne, Iosan, and Iobac (all products of West Chemical Products, Inc.) (90).

An important solubilizing agent and carrier for iodine is polyvinylpyrrolidinone (PVP). Povidone–iodine (PVP–iodine), to be used externally on humans as an antiseptic, is marketed as Betadine and Isodine (The Purdue-Frederick Co.); other products are commercially available. The history of PVP–iodine, its toxicity and therapeutic uses have been described (91). PVP–iodine has been recommended as a topical antiseptic (92).

Iodine can also be solubilized by quaternary ammonium compounds with formation of complexes said to be nonirritating to the skin and mucous membranes (93–94). In contrast to the complexes produced with nonionic surfactants, those obtained with the cationic ammonium compounds are claimed to have all of their iodine available for antibacterial action. Surprisingly, such complexes do not precipitate with anionic surfactants, even when the original quaternary ammonium compounds would have reacted in such a manner; moreover, the bactericidal action of the iodine–quaternary complex is not abolished in the presence of the anionic surfactant (95–96).

Bisglycine hydriodide–iodine [69943-48-2], $(HO_2CCH_2NH_2)_2 \cdot 2HI \cdot I_2$ (Bursoline), contains 30.5–32.0% active iodine. It is used mostly for the disinfection of drinking water, reducing a count of 100 million of enteric pathogens per 100 mL of water to an average of 1–5 organisms per 100 mL in 5 min at normal temperatures.

Some other water-disinfecting preparations contain tetraglycine hydroperiodide [7097-60-1] (Globaline), potassium tetraglycine triiodide [55115-56-5] (Potadine), aluminum hexaurea sulfate triiodide [15304-14-0] (Hexadine-S), and aluminum hexaurea dinitrate triiodide [69943-58-4] (Hexadine-N) as active ingredients.

Organic Iodine Compounds. Iodoform [75-47-8], CHI_3, enjoyed considerable favor as an antiseptic wound powder for a long time; today its use is rather limited. Iodol (2,3,4,5-tetraiodopyrrole) [87-58-1] is still used.

804 DISINFECTANTS AND ANTISEPTICS

Aristol consists mostly of dithymol diiodide [552-22-7], in addition to other iodine compounds. It formerly served as a substitute for iodoform because it is free from odor.

Iodonium 235 is bis(*p*-chlorophenyl) iodonium chloride [34220-01-4]. In a suitable detergent base, it provides satisfactory degerming action because of its bacteriostatic effect upon both gram-negative and gram-positive microorganisms (97). It is ten times more inhibitory for *Salmonella typhosa* and 100 times more so for *Staphylococcus aureus* than for *Pseudomonas aeruginosa*.

Metals, Their Salts and Other Compounds

Inorganic Mercurials. Mercuric chloride [7487-94-7] (bichloride of mercury, $HgCl_2$) is one of the earliest antibacterial agents known. It was studied and credited by R. Koch with exceptional bactericidal and sporicidal potency. Only several years later did Geppert recognize the antibacterial action of mercuric chloride as primarily inhibitory rather than bactericidal in character, by observing that anthrax spores retained their infectiousness following exposure to mercuric chloride even though they did not germinate when subcultured. Moreover, he showed that ammonium sulfide eliminated the inhibitory effect of mercury with the result that a 0.1% solution of mercuric chloride was incapable of killing anthrax spores in 24 h. Similar findings were made subsequently with other pathogens such as *Streptococci* (98), *Staphylococci* (99), *Coli bacilli* (100), and influenza virus (101).

Organic Mercurials. Organic mercurials are mainly used as antiseptics rather than disinfectants. Here the mercury is linked directly with carbon and is not released in solution as an ion. If ionization takes place, organomercuric ions, RHg^+, are formed. Although the literature is replete with references to the antibacterial action of organic mercurials, most of the published information lacks the clear distinction between bacteriostatic and bactericidal action of the compounds studied. Pathogenic microorganisms in a state of bacteriostasis induced by organomercurials may continue to be infectious for the animal body (4,6,98,102). *In vitro* blood can reverse the action of inorganic and organic mercurials often after prolonged contact with the bacteria (7,103–104).

The inactivation of bacteria by mercurials may be caused by a blocking of cellular enzymatic thiol reception by formation of mercaptide bonds and without any other demonstrable cell injury (105–108). Because the mercaptides formed are dissociable, as a rule, the inhibition of enzyme activity by mercury compounds is reversible. The two reactions involved are illustrated below for a mercuric salt:

$$\text{enzyme}\begin{pmatrix}SH\\SH\end{pmatrix} + Hg^{2+} \longrightarrow \text{enzyme}\begin{pmatrix}S^-\\S^-\end{pmatrix}Hg^{2+} + 2H^+$$

$$\text{enzyme}\begin{pmatrix}S^-\\S^-\end{pmatrix}Hg^{2+} + H_2S \longrightarrow \text{enzyme}\begin{pmatrix}SH\\SH\end{pmatrix} + HgS$$

Over a period of time, a number of organic mercurials have been adopted for antiseptic usage. Some of the more important ones are listed in Table 3.

Many reports on the *in vitro* antiseptic performance of the aqueous organomer-

reduced salts applied individually (117). Substantial enhancement of the germicidal action of such antiseptics as iodine, phenol, cresol [1319-77-3], hexylresorcinol [136-77-6] is observed in the presence of such oxidation-reduction systems.

The germicidal action of the chlorides of several rare earth metals (Y, La, Ce, Pr, Nd, Sm, Eu, and Yb) increases generally with increasing atomic weight of the cation (118), except in the case of yttrium chloride [10361-92-9] which is more effective than the chlorides of the other rare earth metals in spite of its lower atomic weight. Of all the rare earth chlorides only ytterbium chloride [10361-91-8] is more bactericidal than copper sulfate [7758-98-7].

Peroxides and Other Oxidants

The antibacterial potency of some representatives of this group depends primarily upon the reactivity of free OH radicals. Although different microorganisms vary considerably in their susceptibility to oxidizing agents, the obligate anaerobic bacteria are most sensitive to oxidation. Of these, the species capable of forming hydrogen peroxide metabolically, but incapable of producing catalase to decompose it, are damaged and eventually killed by exposure to molecular oxygen (see Peroxides).

Hydrogen peroxide (qv) is used as an antiseptic in a 3% solution; concentrates of 30% or stronger are corrosive to skin and denuded tissue. Hydrogen peroxide solutions, containing as little as 0.1 and 0.25% H_2O_2, kill *Salmonella typhosa*, *Escherichia coli*, and *Staphylococcus aureus* in 1 h but not in 5 min (119). Addition of ferric or cupric ions, potassium dichromate [7778-50-9], cobaltous sulfate, or manganous sulfate (120), enhances the bactericidal action of hydrogen peroxide.

Synergism between ultrasonic waves and hydrogen peroxide in the killing of microorganisms has been reported (121) (see Ultrasonics).

With urea, hydrogen peroxide forms a solid compound capable of yielding over 35% of H_2O_2. A solution of this substance in anhydrous glycerol and stabilized with 8-hydroxyquinoline has been found effective in the treatment of certain eye and ear infections.

Zinc peroxide [1314-22-3], as used for medical purposes, consists of a mixture of this chemical with zinc carbonate [3486-35-9] and zinc hydroxide [20427-58-1]. Especially effective against anaerobic and microaerophilic bacteria (122), it is applied in the treatment of wound infections in the form of a thick suspension, diluted with talc in a dusting powder, or incorporated in an ointment vehicle.

Benzoyl peroxide [94-36-0], applied as an antiseptic dressing, is said to be particularly useful in treating wounds infected with gas-forming anaerobes (123). Peracetic acid [79-21-0] has been used both in aqueous solution and as an aerosol or vapor in germ-free or gnotobiotic research (124–125).

Potassium permanganate [7722-64-7] is no longer used for disinfection of drinking water because of the toxicity of the manganese residue. Different bacteria vary greatly in their resistance to permanganate and other oxidizing agents (126). Thus, *Staphylococcus aureus* and *Proteus vulgaris* are killed in 1 h by a dilution of 1:4000, whereas *Escherichia coli* is killed by 1:16,000 and *Pseudomonas aeruginosa* by 1:64,000 in the same period of time.

Ozone [10028-15-6] (qv) is actively germicidal in the presence of moisture (127). It has been used mainly for the disinfection of water in swimming pools, but chlorine is cheaper for this purpose. The ozone generators introduced for the purification of air do not provide efficient means of control of airborne pathogens; moreover, they may create a toxicological hazard.

Phenolic Compounds

As a disinfectant or antiseptic, phenol (carbolic acid) is mostly of historical interest; however, its extensive use continues in both investigative and analytical microbiology, eg, as in the AOAC phenol coefficient and use-dilution methods described above.

The bactericidal effectiveness of several phenol derivatives has been correlated with their respective partition ratios between oil and water; a high lipophilic character has been found to be associated with greater antibacterial activity (128–129). However, the surface activity of the phenol derivatives plays an important part in their total antibacterial performance by altering the permeability of the cell wall. In this respect their behavior resembles that of other antibacterial surfactants.

There are probably several steps involved in the mode of the antibacterial action of phenol, beginning with adsorption upon the bacterial cell, going through the inactivation of certain essential enzymes, and terminating with lysis and death of the cell. The distinctiveness of these steps depends upon the concentration of phenol present. In high concentrations, phenol acts as a gross protoplasmic poison, rapidly penetrating and rupturing the cell wall. Low concentrations act by causing the leakage of such cell constituents as glutamic acid and other metabolites (130–132) and inactivation of specific enzyme systems. Increased permeability is shown by enhanced penetration of certain dyes (133), and by a release of radioactivity from cells (*Escherichia coli*) incubated with carbon 14-labeled glutamate (134).

A 0.1% solution of phenol destroys the enzyme systems of *Staphylococcus aureus* that activate succinate, fumarate, pyruvate, and glutamate; the lactate, formate, glucose, and butanol systems suffer partial inactivation (135). In *Escherichia coli*, with sodium acetate providing the only source of energy in a synthetic medium, phenol at the retarding concentration of 0.075%, had little damaging effect upon dehydrogenase, oxidase, or catalase activity. At the inhibitory concentration of 0.15%, it inactivated oxidase but not dehydrogenase. At the lethal concentration of 1.2% both oxidase and dehydrogenase virtually ceased to function whereas catalase was inhibited only partially (136).

The effects of lethal and sublethal concentrations of phenol are irreversible by dilution with water. Moreover, unlike inhibitory agents such as sulfanilamide or bacteriostatic dyes, it was not possible to train or adapt microorganisms to grow in the inhibitory concentrations of several phenol derivatives. This is accepted as evidence that the phenols exert their bactericidal effect by a nonspecific mechanism (137–138). Certain phenol homologues, particularly those of higher molecular weight, display a selective (quasispecific) efficacy against different bacterial categories (139).

Coal Tar Disinfectants. Until recently, coal tar disinfectants constituted the most important category of disinfectants for environmental use. From raw materials obtained by distillation of coal tar, two groups of products, ie, the soluble and the emulsifiable types are produced. The soluble coal tar disinfectants are formulated with tar acids consisting essentially of low molecular weight phenol homologues (cresols, xylenols). The emulsifiable disinfectants are formulated with coal-tar fractions containing varying proportions of neutral oil, which consists mostly of methyl- and dimethylnaphthalenes and other hydrocarbons, organic bases, sulfur compounds, etc, in addition to phenol derivatives.

Technical cresol [1319-77-3], consisting of the three cresol isomers, is only slightly soluble in water; however, when combined with a soap in suitable proportions, it becomes completely and rapidly miscible, forming clear solutions in distilled or demineralized water. This principle is applied in preparing saponated cresol solution which contains 46–52% of cresol rendered soluble by means of a potassium or sodium soap.

Different soaps, including resin soaps, are used to formulate emulsifiable disinfectants which are characterized by the milky appearance of their dilutions. Because of the crude, unrefined character of the coal-tar fractions entering into their composition, these disinfectants emit a rather pungent, disagreeable odor (see Soap).

Phenol Homologues. In a homologous series of monoalkyl phenol derivatives, certain regularities are found in the dependence of bactericidal action upon chemical constitution (see Alkylphenols). As shown in Table 5, the potency against all four test organisms increases with the increase in molecular weight until the n-amyl derivative is reached. Further increase in molecular weight engenders a substantial increase of germicidal potency against three of the microorganisms, and a decline with respect to *Salmonella typhosa*. The term quasispecific has been proposed to describe the action of a disinfectant such as n-heptylphenol which is extremely effective against *Staphylococcus aureus*, but only slightly so against *Salmonella typhosa* (139).

The homologous series of alkylchlorophenol derivatives show a regularity like that of the phenol derivatives. The regularities noted in the series of o- [95-57-8] and p-chlorophenol [106-48-9] and o- [95-56-7] and p-bromophenol [106-41-2] derivatives (139–140) are as follows:

(1) Halogen substitution intensifies the microbicidal potency of phenol derivatives; halogen in the para position to the hydroxyl group is more effective than in the ortho position. (2) Introduction of aliphatic or aromatic groups into the aromatic

Table 5. Microbicidal Action of Phenol Derivatives (Phenol Coefficients, 37°C)

Substituting radical	CAS Registry No.	Salmonella typhosa	Staphylococcus aureus	Mycobacterium tuberculosis	Candida albicans
none	[108-95-7]	1.0	1.0	1.0	1.0
2-methyl	[95-48-7]	2.3	2.3	2.0	2.0
3-methyl	[108-39-4]	2.3	2.3	2.0	2.0
4-methyl	[106-44-5]	2.3	2.3	2.0	2.0
4-ethyl	[123-07-9]	6.3	6.3	6.7	7.8
2,4-dimethyl	[105-67-9]	5.0	4.4	4.0	5.0
2,5-dimethyl	[95-87-4]	5.0	4.4	4.0	4.0
3,4-dimethyl	[95-65-8]	5.0	3.8	4.0	4.0
2,6-dimethyl	[576-26-1]	3.8	4.4	4.0	3.5
4-n-propyl	[645-56-7]	18.3	16.3	17.8	17.8
4-n-butyl	[99-71-8]	46.7	43.7	44.4	44.4
4-n-amyl	[1322-06-1]	53.3	125.0	133.0	156.0
4-t-amyl	[80-46-6]	30.0	93.8	111.1	100.0
4-n-hexyl	[61902-50-9]	33.3	313.0	389.0	333.0
4-n-heptyl	[26997-02-4]	16.7[a]	625.0	667.0	556.0

[a] Approximate figure.

810   DISINFECTANTS AND ANTISEPTICS

nucleus of halogenated phenols increases the bactericidal potency up to certain limits, depending in the case of alkyl substitution upon the number of carbon atoms present in the substituting group or groups. (3) A normal aliphatic chain with a given number of carbon atoms generally exerts a greater intensifying effect upon the bactericidal potency than that of a branched chain or of two alkyl groups with the same total number of carbon atoms. (4) o-Alkyl derivatives of p-chlorophenol are more actively germicidal than the corresponding p-alkyl derivatives of o-chlorophenol.

In the case of the higher homologues, the germicidal action manifests the quasispecific character referred to before, as illustrated in Figure 1 which relates to the homologous series of o-alkyl derivatives of p-chlorophenol. Many instances of the same effect have been established also in the group of polyalkyl-, and of aryl- and aralkylchlorophenol derivatives (see also Chlorocarbons).

It is significant that the increase in molecular weight that accompanies the increase in the antimicrobial potential is also accompanied by a decrease in toxicity to animals (mice). Table 6 illustrates this condition for the three homologous series of alkyl derivatives of phenol, p-chlorophenol, and o-chlorophenol.

Several synthetic phenol derivatives have found use as disinfectants or antiseptics. 3,5-Dimethyl-4-chlorophenol [88-04-0] is used in Liquor Chloroxylenolis of the British Pharmacopoeia. p-Chloro-o-benzylphenol [120-32-1] is widely used as an ingredient of disinfectant formulas. o-Phenylphenol [90-43-7] plus xylenols or pine oil is present

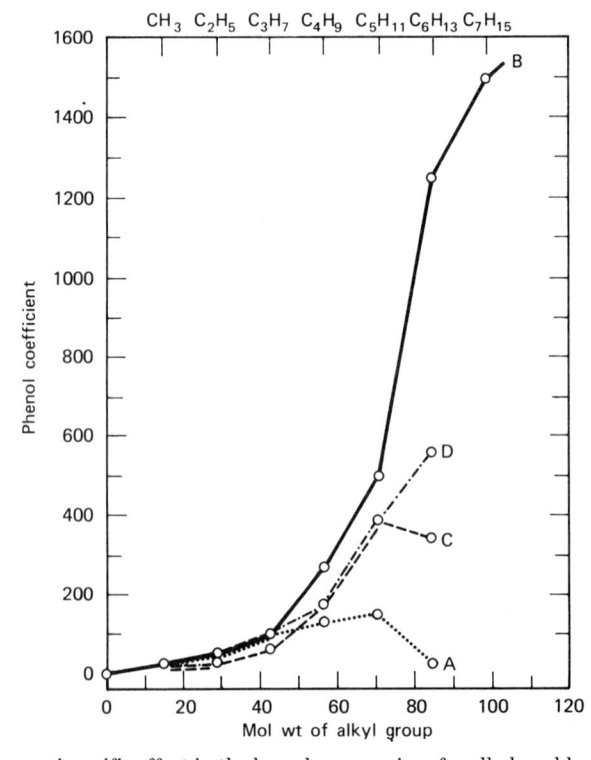

Figure 1. The quasispecific effect in the homologous series of o-alkyl-p-chlorophenol derivatives. A = *Salmonella typhosa*; B = *Staphylococcus aureus*; C = *Mycobacterium tuberculosis*; and D = *Candida albicans*.

Table 6. Toxicological Data for Phenols, mg/g, on Subcutaneous Injection in Mice

phenol	0.45		
p-chlorophenol		0.6	
o-chlorophenol			0.7
	p-Alkyl derivatives of phenol	o-Alkyl derivatives of p-chlorophenol	p-Alkyl derivatives of o-chlorophenol
methyl	0.5	2.0	1.5
ethyl	1.0	4.0	4.0
n-propyl	2.0	6.0	6.0
n-butyl	3.0	15.0	15.0
n-amyl	5.0	>20.0	20.0
n-hexyl	6.0	>20.0	>20.0
n-heptyl	10.0	>20.0	>20.0

in two versions of Lysol (Lehn & Fink); with p-tert-amylphenol [80-46-6] it is used in the composition of Amphyl (Lehn & Fink).

Properly formulated phenolic disinfectants are capable of yielding nonspecific or broad-spectrum microbicidal performance. These disinfectants offer considerable advantage over products such as the soap-containing solution of cresol in that they are virtually free of toxicity and corrosiveness to tissue; moreover, they are practically odorless when diluted for use.

Thymol [89-83-8], carvacrol [499-75-2], chlorothymol [89-68-9], and chlorocarvacrol [5665-94-1] have been used for antiseptic and disinfectant purposes (see Terpenoids), whereas polyhalogenophenol derivatives such as trichloro- or pentachlorophenol are useful as fungicides (qv).

A high phenol coefficient need not indicate the exceptional fitness of a given phenol derivative for all practical uses, particularly if protein is present to a significant degree. The microbicidal performance of phenol derivatives with high phenol coefficients is impaired by serum to a greater extent than that of analogues with low phenol coefficients. Some of these variables can be equalized, for practical screening purposes, by confirmatory testing methods, such as the AOAC use-dilution tests and the pertinent validation procedures (141).

Dihydric and Trihydric Phenols. Although resorcinol [108-46-3] is a comparatively weak bactericide, several of its nuclear-substituted alkyl derivatives are effective microbicidal agents. One of these, n-hexylresorcinol [136-77-6], has gained considerable reputation as a topical antiseptic. The study of the antibacterial properties of resorcinol derivatives dates back to the fundamental work of Johnson and co-workers (142–143) which was extended by others (144–145).

There exists a close quantitative resemblance of the bactericidal properties of the nuclear-substituted resorcinol derivatives and those of the corresponding monoethers. Apparently one free hydroxyl group of the resorcinol derivative is required for initiation of the antibacterial effect, and it matters little to the microbicidal potential whether the substituting radical is attached to an oxygen atom or to the carbon atom in the positions ortho–para to the two hydroxyl groups. The phenomenon of quasispecificity occurs in both series of resorcinol derivatives, as exemplified by the two n-octyl derivatives, 4-(n-octyl)resorcinol [6565-70-4] and 3-(n-octyloxy)phenol [34380-89-7], which combine high potency against Staphylococcus aureus with virtual impotence against Salmonella typhosa (146).

In the homologous series of alkyl monoethers of hydroquinone and catechol (47), a number of effective antibacterial agents have been found, some of which exhibit quasispecific character. However, they are less effective than the corresponding resorcinol monoethers. The same is true for several aromatic monoethers.

Several nuclear dialkylresorcinol derivatives have been studied bacteriologically (147). Although 4,6-diethylresorcinol [52959-32-7], with a phenol coefficient of 10, compared closely to the isomeric 4-n-butylresorcinol [18979-61-8] with a phenol coefficient of 8, the 4,6-di-n-propylresorcinol [69943-49-3] was less effective than the corresponding 4-n-hexylresorcinol [136-77-6], the phenol coefficients being 18 and 45, respectively.

As with alkylphenols, chlorine substitution in the nucleus of the alkylresorcinols intensifies the antibacterial potential, especially against *Staphylococcus aureus* (148–150). As with the higher alkylphenols and alkylchlorophenols, organic matter or soaps markedly depress the antibacterial performance of the higher alkylresorcinols (145).

2,4,4'-Trichloro-2'-hydroxydiphenyl ether [66943-50-6] (Irgasan DP300, Ciba-Geigy, S. A.) shows good activity against gram-positive and gram-negative microorganisms, also against yeasts (qv) and fungi. It had been recommended as an additive for soaps and washing products (151). A broad-spectrum soap contains a mixture of hexachlorophene (see below), triclocarban [101-20-2], and Irgasan DP300 (152).

Hydroxybenzoic Acids. The antibacterial and antifungal properties of the hydroxybenzoic acids depend primarily upon the reactivity of their phenolic hydroxy group. It is masked by intramolecular hydrogen bonding in o-hydroxybenzoic (salicylic) acid which is unimportant as an antimicrobial agent (see Salicylic acid and related compounds). On the other hand, salicylanilides with halogens as substituents have commercial use as antimicrobials. One of the first compounds to be employed as a soap additive was 3,3',4,5'-tetrachlorosalicylanilide [1322-37-8] (Anobial) (73); however, the initial success of this compound did not last when it was found to cause primary photodermatitis; 53 cases of photosensitization by the use of a soap containing tetrachlorosalicylanilide were confirmed (153–155). Brominated salicylanilides, such as 4,5'-dibromosalicylanilide [87-12-7] (158), 3,5-dibromo-3'-trifluoromethyl salicylanilide [4776-06-1] (fluorosalan) (157), and 3,4,5'-tribromosalicylanilide [87-10-5] (tribromsalan) (158–159) were also found to be photosensitizers (154) and withdrawn from use in soaps and cosmetics.

p-Hydroxybenzoate esters (parabens) play an important role in the preservation of different perishable organic materials against attack and destruction by airborne bacteria, fungi, and yeasts. Among the materials requiring such preservation are those based upon carbohydrates, gums, and proteins, and their many industrial, pharmaceutical, and cosmetic combinations with fats, oils, waxes, surfactants, etc.

The inhibitory capacity of the methyl [99-76-3], ethyl [120-47-8], propyl [94-13-3], and butyl p-hydroxybenzoates [94-26-8] is shown in Table 7 (160–161). The low order of toxicity of these esters (162–163), their lack of irritation, and their absorption and excretion characteristics both in men and animals indicate that the compounds approximate the ideal for a pharmaceutical preservative (164–166). Their application in antimycotic therapy has been considered by several investigators (167).

Bacteriological information on the activity of substituted salicylic and hydroxynaphthoic acids is also available (168).

Table 7. Percentage Inhibition of Bacterial and Fungal Growth by Esters of p-Hydroxybenzoic Acid

Microorganism	Methyl	Ethyl	Propyl	Butyl
Salmonella typhosa	0.2	0.1	0.1	0.1
Escherichia coli	0.4	0.1	0.1	0.4
Staphylococcus aureus	0.4	0.1	0.05	0.0125
Proteus vulgaris	0.2	0.1	0.05	0.05
Pseudomonas aeruginosa	0.4	0.4	0.8	0.8
Aspergillus niger	0.1	0.04	0.02	0.02
Rhizopus nigricans	0.05	0.025	0.0125	0.00625
Chaetomium globosum	0.05	0.025	0.00625	0.003125
Trichophyton interdigitale	0.008	0.008	0.004	0.002
Candida albicans	0.1	0.1	0.0125	0.0125
Saccharomyces cerevisiae	0.1	0.05	0.0125	0.00625

Bis(Hydroxyphenyl) Alkanes. Dichlorophene (G-4, Givaudan Corporation) is 2,2'-methylenebis(4-chlorophenol) [97-23-4]. Although this compound is active against bacteria, its special merit lies in the mildew proofing of textiles (see Textiles).

The most important member of the series is 2,2'-methylenebis(3,4,6-trichlorophenol) [70-30-4], also named bis(3,5,6-trichloro-2-hydroxyphenyl)methane (hexachlorophene, G-11, Givaudan Corporation) (169). It has found extensive use as an antibacterial soap additive as it does not lose activity in presence of soap, as most phenolic compounds do. Such soaps are able to reduce the bacterial skin flora to a small fraction of the original (170–171). Hexachlorophene has also been incorporated in soapless detergent bases (pHisoHex, Winthrop Laboratories, and other preparations). Such combinations of hexachlorophene and detergent or soaps are used for surgical scrub, in nurseries to prevent the spread of staphylococcal infections, and for other therapeutic purposes. Hexachlorophene acts as an inhibitory and slow germicidal agent since it is retained in very small quantities by skin that is repeatedly washed with a soap or detergent containing it (172). Extremely low amounts of hexachlorophene suffice to produce inhibition of *Staphylococcus aureus*, eg, dilutions from $1:10^6$ to $1:5 \times 10^6$, depending on the size of the inoculum (172–173). Hexachlorophene is considerably less active against *Escherichia coli;* the bacteriostatic concentration is in the range of 1:10,000 to 1:50,000 (174).

The action of hexachlorophene is reversed upon contact with blood; this is deemed significant for any prophylactic application to the broken skin against infection in injury, or for preoperative use. Skin that has been "degermed" with hexachlorophene is not proof against subsequent contamination with transient pathogens or against lesions of bacterial origin (174–175).

Hexachlorophene has been used successfully in deodorant soaps and cosmetics; here it controls the proliferation of cutaneous bacteria that cause perspiration malodor by decomposition of apocrine sweat (176) (see Soap; Cosmetics).

hexachlorophene

814   DISINFECTANTS AND ANTISEPTICS

Although the extensive use of hexachlorophene over the years has not shown toxic symptoms in humans (177–178), neurotoxicity has been demonstrated in rats with large doses (179–180), also in monkeys (181) and in premature infants (182). In view of these findings, the FDA (183) banned the over-the-counter sale of soaps, cosmetics, and drugs containing an amount of hexachlorophene exceeding 0.1%. All products with a higher percentage were put on a prescription basis.

Bacteriological data for hexachlorophene and its isomers (184) are presented in Table 8. A series of 2,2'-methylenebis(dichlorophenols) has also been described (185).

Although 2,2'-thiobis (4,6-dichlorophenol) [97-18-7], bithionol, is not a substituted diphenylalkane, it is structurally and functionally related to hexachlorophene. Like hexachlorophene, it shows a greater inhibitory effect upon *Staphylococcus aureus* than upon *Salmonella typhosa*, the respective concentrations being 1:10^6 and 1:10^3

Table 8. Antimicrobial Activity of Methylenebis (Trichlorophenol) Isomers[a]

Organism	\multicolumn{8}{c}{Inhibitory concentration, µg/mL (geometric mean, $n = 3$) isomers[b]}
	A	B	C	D	E	F	G	H
S. aureus	0.93	0.61	0.39	10	2.5	20	1.56	25
S. epidermidis	0.93	0.78	0.23	12.5	2.4	20	1.56	25
B. subtilis	0.19	0.39	0.19	3.9	1.9	9.9	1.56	15.6
B. ammoniagenes	0.39	0.48	0.19	15.6	2.4	40	1.56	31.5
P. vulgaris	3.9	1.9	3.9	25	c	63	10	63
E. coli	25	c	12.5	c	c	c	c	c
S. typhosa	40	c	10	c	c	c	c	c
Ps. aeruginosa	25	c	50	c	c	c	c	c
Ps. fluorescens	0.23	0.61	0.19	12.5	2.4	31.5	1.56	25
Sh. sonnei	40	c	9.9	c	c	c	c	c
K. pneumoniae	50	c	15.6	c	c	c	c	c
T. mentagrophytes	1.74	3.12	3.12	8.9	8.9	6.25	4.4	6.25
T. rubrum	4.4	3.12	8.9	8.9	3.1	1.56	1.74	9.9
M. audouini	3.12	1.74	3.12	3.12	4.4	3.12	8.9	4.4
C. albicans	c	c	c	c	c	c	c	c
Cl. tetani	0.19	0.39	0.09	1.56	0.27	1.74	0.19	4.4
Cl. perfringens	0.78	0.39	0.55	1.74	0.55	6.25	0.39	6.25
Cl. sporogenes	0.78	0.55	0.55	1.56	0.55	6.25	0.55	6.25
C. vulgaris	3.12	4.4	8.9	c	c	17.8	4.4	17.8

[a] Ref. 186.
[b] A 2,2'-methylenebis(3,4,6-trichlorophenol) [70-30-4], B 2,2'-methylenebis(4,5,6-trichlorophenol) [584-57-6], C 2,2'-methylenebis(3,4,5-trichlorophenol) [584-33-8], D 3,3'-methylenebis(2,4,6-trichlorophenol) [584-32-7], E 3,3'-methylenebis(2,4,5-trichlorophenol) [70495-29-3], F 3,3'-methylenebis(2,5,6-trichlorophenol) [70495-30-6], G 3,3'-methylenebis(4,5,6-trichlorophenol) [70495-31-7], H 4,4'-methylenebis(2,3,6-trichlorophenol) [70495-32-8].
[c] Denotes growth at 100 µg/mL.

bithionol

(187). It has been used as a soap additive (188) and for other purposes (189), but such use was discontinued when it showed photosensitivity in humans (190). In a series of eight thiobisphenols tested against *Trichophyton gypseum,* it exhibited the greatest fungistatic potency (191).

The antibacterial action of hexachlorophene and bithionol may be associated with their capacity to inactivate iron-containing enzyme systems of the microbial cell since iron and other metals are chelated by either compound (192).

Bacteriologic information is available for other bis(hydroxyphenyl)methane and sulfide derivatives (193–194) as well as for a number of bis(hydroxyphenyl)alkanes in which the two benzene rings are linked by aliphatic chains of greater length (129,195–196).

8-Hydroxyquinoline and Derivatives. Of the seven isomeric hydroxyquinolines, only 8-hydroxyquinoline [148-24-3] exhibits an antimicrobial character; this is ascribed to its capacity to chelate metals that constitute an integral part of some essential biological system of the microbial cell (197–199). 8-Hydroxyquinoline (8-quinolinol, oxine) used either by itself or as a salt such as the sulfate (chinosol) [134-31-6] or benzoate [86-75-9], is the active ingredient of several antiseptics whose effect is bacteriostatic and fungistatic rather than microbicidal. Inhibitory action upon gram-positive microorganisms is more pronounced than upon gram-negative ones, as illustrated by the following growth-preventing concentrations: for staphylococci 1:100,000; for streptococci 1:50,000; for *Salmonella typhosa* and *Escherichia coli* 1:10,000; for *Salmonella schottmuelleri* 1:5,000 (200–201).

However, 8-acetoxyquinoline [40245-26-9], which has no free hydroxyl group and is therefore incapable of chelation, also shows high antimicrobial activity (202). It is one of a group of 120 quinoline derivatives several of which were found to have comparatively high bacteriostatic potential.

Certain halogen derivatives of 8-hydroxyquinoline have a record of therapeutic efficacy in the treatment of cutaneous fungus infections and also of amebic dysentery. Among them are 5-chloro-7-iodo-8-quinolinol [130-26-7] (iodochlorhydroxyquin, Vioform), 5,7-diiodo-8-hydroxyquinoline [83-73-8] (diiodohydroxyquin), and sodium 7-iodo-8-hydroxyquinoline-5-sulfonate [885-04-1] (chiniofon) (203–205).

The copper compound of 8-hydroxyquinoline (copper 8-quinolinolate) [10380-28-6] is employed as an industrial preservative for a variety of purposes, including the protection of wood and textiles against fungus-caused rotting.

Quaternary Ammonium and Related Compounds

The antibacterial precursor of the quaternary ammonium compound (quat) is the primary aliphatic long-chain ammonium salt. In this category there are potent antibacterial agents that compare in effectiveness with some of the more active quaternary ammonium compounds. The primary ammonium salt may be regarded as the direct counterpart of soap. Both are surface-active substances; In soap, the anion contributes the hydrophobic principle, whereas in the primary ammonium salt the cation is hydrophobic. By way of illustration, the dissociation equations of potassium laurate (a type of soap) and of dodecylammonium chloride [929-73-7] (a type of primary ammonium salt) are given below:

$$C_{11}H_{23}CO_2K \rightarrow (C_{11}H_{23}CO_2)^- + K^+$$
$$C_{12}H_{25}NH_3Cl \rightarrow (C_{12}H_{25}NH_3)^+ + Cl^-$$

It is apparent that the designation invert soap could have been applied more properly to this type of primary ammonium salt although it was originally used for a quaternary ammonium compound (206) (see Ammonium compounds; Surfactants).

The primary long-chain ammonium salts are derived from the weakly basic aliphatic amines. Hence, their aqueous solutions require a pH low enough to counteract hydrolysis and partial liberation of the amine base. By contrast, because the quaternary ammonium compounds are salts of strong bases, they remain in solution in acidic as well as in basic media. Quaternary ammonium salts owe their surface activity and antibacterial quality primarily to the presence of certain aliphatic long-chain amino groups which, by themselves, or rather in the form of their soluble ammonium salts, display surface-active and antibacterial properties, often of a comparable order of magnitude (207–210).

Generally, the long-chain alkylammonium and the quaternary ammonium salts, like other cation-active compounds, are incompatible with soaps or other anion-active materials. Mutual precipitation usually occurs when aqueous solutions of the representatives of the cation-active and anion-active classes are brought into contact, except where the molecular weights of the cations and anions are sufficiently low.

The long-chain aliphatic amines combine readily with alkyl halides, sulfates, etc, to form quaternary ammonium compounds. In view of the advantages of the latter as outlined above, most of the cation-active antibacterials available today consists of quats, although some of the *N*-alkylammonium salts have also achieved importance.

The first quaternary ammonium compound to be tested for antiseptic activity was obtained by the reaction of hexamethylenetetramine [*100-97-0*] with chloroacetamidomethanol (211). Subsequently, other quaternary salts of hexamethylenetetramine were prepared and studied (212–213). These compounds are no longer of any practical importance and they bear only a remote structural resemblance to the important classes of modern quaternary ammonium compounds (206,214).

In the germicidal quaternary ammonium salts, at least one of the four organic radicals must be of such character as to impart surface activity to the compound. Since the activity of the members of this class resides in the cation, they are sometimes referred to as cationic germicides although in reality they constitute only a subgroup of the cationic group, the latter comprising "-onium" compounds other than ammonium-based ones.

The quaternary ammonium salts produce bacteriostasis in very high dilutions; this property is associated with the inhibition of certain bacterial enzymes, especially those involved in respiration and glycolysis (215–217). On the other hand, some microorganisms (*Serratia marcescens*, but not *Staphylococcus aureus* and *Escherichia coli*) may be adapted to grow in many times the originally inhibitory concentrations of the quats after a comparatively small number of transfers (218–220).

Although efforts have been made to explain the mode of antimicrobial activity of quats on the basis of inactivation of enzymes (221), their bactericidal effect appears to be owing to their ability to cause release of the bacterial cell content into the surrounding medium (222–224).

Of the numerous quaternary ammonium salts investigated (225), only a comparatively small number have retained importance as antibacterial agents. Among them are: benzalkonium chloride, ie, alkylbenzyldimethylammonium chloride, in which

$$C_6H_5CH_2\overset{+}{N}(CH_3)_2R \ Cl^-$$
benzalkonium chloride

$$(CH_3)_3CCH_2\underset{CH_3}{\overset{CH_3}{\underset{|}{\overset{|}{C}}}}-\underset{}{\bigcirc}-O(CH_2)_2O(CH_2)_2\underset{CH_3}{\overset{CH_3}{\underset{|}{\overset{|}{N^+}}}}-CH_2C_6H_5 \ Cl^-$$
benzethonium chloride

$$\underset{}{\bigcirc}\overset{Cl^-}{\underset{}{N^\pm(CH_2)_{15}CH_3 \cdot H_2O}}$$
hexadecylpyridinium chloride hydrate

R is a mixture of alkyls from C_8H_{17} to $C_{18}H_{37}$, with $C_{12}H_{25}$ predominating; benzethonium chloride [121-54-0], ie, benzyldimethyl [2-[2-(p-1,1,3,3-tetramethylbutylphenoxy)ethoxy]ethyl] ammonium chloride; methylbenzethonium chloride [106-45-0] with the cresoxy group replacing the phenoxy group of benzethonium chloride; hexadecylpyridinium chloride [25155-18-4], and alkylisoquinolinium bromide. Cetrimide (of the British Pharmacopoeia) is hexadecyltrimethylammonium bromide [57-09-0], $C_{16}H_{33}\overset{+}{N}(CH_3)_3Br^-$.

Although the quaternary ammonium compounds have an extensive record of satisfactory performance, pertinent quantitative bacteriological information is rather contradictory. A number of older papers require cautious reinterpretation in the light of more recent findings, since they tended to ascribe to the quats a disinfectant effectiveness and a broad microbicidal spectrum to which these compounds do not appear to be entitled. Therefore, the older phenol coefficient figures reported for the different quats must be regarded with reservations, especially if they have been obtained without regard to controlling the bacteriostatic action of the quaternary ions in tests for bactericidal potency (210,226–228). As to limitations of the microbicidal spectrum of the quats, *Mycobacterium tuberculosis, Pseudomonas aeruginosa, Trichophyton interdigitale* and *rubrum* are particularly resistant to these agents (225,229–234). However, in contrast to the quaternary ammonium salts, the primary tetradecylamine is credited with antituberculous properties, and N-dodecyl-1,3-propanediamine [5538-95-4], $C_{12}H_{25}$-NH(CH$_2$)$_3$NH$_2$, is held to offer possibilities as a surface disinfectant for tuberculosis hygiene (235).

Stuart and co-workers have shown that the phenol coefficient of the quaternary ammonium compounds as obtained, for example, by the AOAC method, cannot serve as a guide for the preparation of solutions that would provide adequate margins of safety for disinfection (228). The data that indicate the bacteriostatic action of the quats are more useful for this purpose. The inhibitory dilutions of benzalkonium chloride as observed with a variety of microorganisms are given in Table 9.

Soaps or other anionic surfactants must be removed by thorough rinsing from surfaces to be treated with quaternary ammonium germicides. The hardness of water can also affect the antibacterial performance of the quats (236–237). However, within these limitations the quaternaries render satisfactory service in the sanitization of eating and drinking utensils, the disinfection of equipment used in dairies and in other food processing plants, the cleaning of eggs, etc.

Mixtures of different quaternary ammonium compounds with hexachlorophene show a loss of antibacterial activity that is greatest at an equimolecular ratio (238).

An important application of quaternary ammonium compounds is the impregnation of fabrics to control the spread of infection from this source (239). Control of diaper rash, for example, is achieved by using a quat solution in the final rinse for the

Table 9. Bacteriostatic Dilutions of Benzalkonium Chloride

Microorganism	Dilution $\times 10^3$
Salmonella typhosa	1:256
Shigella dysenteriae	1:512
Escherichia coli	1:64
Aerobacter aerogenes	1:16
Salmonella paratyphi	1:64
Salmonella enteritidis	1:32
Proteus vulgaris	1:16
Pseudomonas aeruginosa	1:32
Vibrio cholerae	1:512
Staphylococcus aureus	1:800
Pneumococcus II	1:200
Streptococcus pyogenes	1:800
Clostridium welchii	1:200
Clostridium tetani	1:200
Clostridium histolyticum	1:200
Clostridium oedematiens	1:200

suppression of bacteria that attack the urinary urea with liberation of ammonia (240).

When used for strictly antiseptic purposes, such as the control of skin bacteria at the site of an operation, the quaternary ammonium compounds are more satisfactory in the form of tinctures (diluted with 50% alcohol–10% acetone), rather than in aqueous solution. This conclusion is based upon the results of the serial basin test, as employed in a comparative evaluation of the degerming effectiveness of several hospital antiseptics (76). A 1:1000 solution of benzalkonium chloride in water applied after thorough rinsing to remove soap reduced the bacterial count by 40%, whereas no significant reduction was observed after a superficial rinse. By contrast, the tincture of benzalkonium chloride produced reductions of 85 and 80%, respectively, under the corresponding test conditions. However, the solvent (50% alcohol–10% acetone) alone reduced the bacterial count by 70%, and 70% alcohol delivered an even superior performance of an 88% reduction. Similar results were obtained with cetrimide (241).

Amphoteric Surfactant Disinfectants

In Europe, a group of amphoteric surfactants are marketed as disinfectants under the trade name Tego (242–244). Chemically, they are amino acids, usually glycine, substituted with a long-chain alkylamine group. Table 10 gives the structure of some of the more widely used compounds.

The Tego disinfectants are claimed to be bactericidal, fungicidal, and virucidal, nontoxic and safe, and have been recommended for the surgical scrub, floor disinfection in hospital, and food plants, cleanup in dairies, breweries, and bottling plants, etc. However, contradictory reports on the activity of these compounds leave their effectiveness in doubt (see Surfactants).

Table 10. Composition of Tego Disinfectants[a]

Tego	Active ingredient	Composition	pH
103S	$RNH(CH_2)_2NH(CH_2)_2NHCH_2CO_2H \cdot HCl$	15% aqueous solution	7.7
103G or MHG	$R'NH(CH_2)_2NH(CH_2)_2NHCH_2CO_2H \cdot HCl$ + $[R'NH(CH_2)_2]_2NCH_2CO_2H \cdot HCl$	10% aqueous solution	7.7
51	$RNH(CH_2)_2NH(CH_2)_2NHCH_2CO_2H$ + $RNH(CH_2)_3NHCH_2CO_2H$	9% aqueous solution	8.1
51B	$R'NH(CH_2)_2NH(CH_2)_2NHCH_2CO_2H$ $R = C_{12}H_{25}$ $R' = C_8H_{17}$ to $C_{16}H_{33}$	22.5% aqueous solution	8.0

[a] Ref. 186.

Pine Oil Compounds

Pine oil is obtained from waste pine wood by destructive distillation or by distillation with superheated steam. Sometimes solvent extraction is added as a supplementary step, in which case a liquid hydrocarbon mixture is introduced into the retort after it has cooled and heat is applied again until extraction has been completed. The solvent extract is subjected to fractionation for rosin, pine oil, and recovery of the solvent. The volatile fraction obtained by any of these processes is separated into pine oil and turpentine. Although technical pine oil may vary considerably in composition depending upon its distillation range (from 170–350°C), a more uniform product is assured if it answers the distillation requirements of the *National Formulary* (NF), namely that 95% should distill between 200 and 225°C. In this case, the fraction consists mostly of isomeric terpineols, with the α-isomer predominating. The other constituents are dihydro-α-terpineol, borneol, and fenchyl alcohol (see Tall oil; Terpenoids).

Pine oil is insoluble in water but is readily emulsifiable when combined with a soap, a sulfonated oil, or other suitable dispersing agents. Most pine oil disinfectants contain from 60–65 vol % of pine oil. Diluted with water such preparations yield white, milky emulsions, with a characteristic, pungent odor.

A pine oil disinfectant prepared according to the hygienic laboratory formula may have a phenol coefficient against *Salmonella typhosa* of 3.5–4.5; higher coefficients (6 or more) may be obtained by using certain emulsifying agents other than the resin soap specified by the hygienic laboratory. However, the lack of ability to kill *Staphylococcus aureus* disqualifies the pine oil-based products for use as disinfectants for hospital, medical, veterinary, and related purposes where pyogenic cocci play a significant role. Thus, the use of pine oil disinfectants is limited to janitorial activities where a supplementary advantage is offered by their deodorant and odor-masking action (see Odor counteractants).

In recent years a serious problem has been created in the hospital field, owing to the emergence of antibiotic-resistant strains of staphylococci; however, these are susceptible, eg, to the action of most phenolic disinfectants (31). Combinations of pine oil or α-terpineol with phenolic broad-spectrum bactericides can be used for formulations with a range of antibacterial potential warranting a wider disinfectant use (245–246).

Carbamic Acid and Urea Derivatives

The class of dithiocarbamates was previously studied for its capacity to control fungus diseases of tomatoes, beans, potatoes, etc (see Fungicides). More recently some dithiocarbamates were found to exert powerful inhibitory action upon fungi pathogenic for man (*Trichophyton mentagrophytes*, *Epidermophyton floccosum*, *Microsporum audouini*, and *Torula histolytica*) (see Chemotherapeutics, antibacterial and antimycotic). Bacteria, especially those of the gram-positive variety, were also found to be susceptible, as illustrated by disodium bisdithiocarbamate [69943-51-7] whose inhibitory concentration for *Staphylococcus aureus* is 1:100,000. The corresponding dilutions for the gram-negative *Salmonella typhosa* and *Escherichia coli* are 1:5,000 and 1:10,000, respectively (247). In another investigation correlating the chemical structure of dithiocarbamic acid derivatives with their *in vitro* antibacterial and antifungal activity, it was found that within each category of dithiocarbamates, thiuram monosulfides, and methyl and ethyl esters were the most active whereas the higher alkyl derivatives were comparatively inactive (248). Tetramethylthiuram disulfide [137-26-8] (TMTD) was the most active of the compounds tested against pathogenic fungi as well as against bacteria; *Streptococcus pyogenes*, *Streptococcus faecalis*, and *Staphylococcus aureus* show a markedly greater susceptibility to its action than *Escherichia coli* and *Pseudomonas aeruginosa* (see also Rubber chemicals).

$$(CH_3)_2NCSSCN(CH_3)_2$$
$$\overset{S}{\|}\quad\overset{S}{\|}$$

tetramethylthiuram disulfide

Systematic investigation of over 200 urea and thiourea derivatives revealed several compounds with high bacteriostatic potency (249). Inhibitory dilutions for *Staphylococcus aureus* of the order of $1:3 \times 10^7$ have been established for 3,4,3'-trichloro- and 3,4,4'-trichlorocarbanilide [101-20-2], whereas inhibitory action in a dilution of $1:10^7$ was observed with 3,3',4,4'-tetrachloro- [1300-43-0] and 3,3',4,5,5'-pentachlorocarbanilides [69943-52-8], and with 3,4,4'-trichlorothiocarbanilide [5109-07-9], and *N*-formyl-3,4-dichloroaniline [5470-15-5].

3,4,4'-Trichlorocarbanilide (TCC, triclocarban, Monsanto Chemical Co.) has been introduced commercially as a soap additive for its effective skin degerming and deodorant action. In addition to *Staphylococcus aureus*, it inhibits *Diplococcus pneumoniae*, *Corynebacterium diphtheriae*, *Bacterium ammoniagenes*, and *Streptococcus viridans* in a dilution of $1:10^7$. Beta- and gamma-hemolytic streptococci require a dilution of $1:10^6$ for suppression of growth. However, most gram-negative bacteria and pathogenic fungi are not inhibited by a 1:1000 dilution. The combination of TCC with hexachlorophene exhibits antibacterial synergism (249).

Another commercial carbanilide is 3-(trifluoromethyl)-4,4'-dichlorocarbanilide [369-77-7] (Cloflucarban, Irgasan CF₃, Ciba-Geigy), also used as soap additive. The halogenated carbanilides continue to be approved by the FDA for use in bar soaps pending further studies on their toxicity.

1,6-Bis(4-chlorophenyl)diguanidinohexane [14007-07-9] (Chlorhexidine, Hibi-

3,4,4'-trichlorocarbanilide

tane, Imperial Chemical Industries Ltd.) (250) is a topical antiseptic. It reduces the bacterial flora of the hands when applied in a cream or other vehicle (251).

There are a number of substituted biguanides that are antibacterial and antimycotic and that are not inactivated by serum proteins (252) (see Cyanamides).

Nitrofuran Derivatives

Nitro-substituted furans possess antiseptic properties. The nitro group is essential for antibacterial action, as shown by a bacteriological comparison with the nonnitrated analogues (253–254). Of the many derivatives tested (58,255) a few have achieved practical importance. Among them is 5-nitro-2-furaldehyde-2-ethylsemicarbazone [5579-89-5] (nitrofurazone, Furacin, Eaton Laboratories) which is credited with antibacterial action upon both gram-positive and gram-negative microorganisms. Thus, *Salmonella typhosa* and *schottmuelleri*, and *Shigella dysenteriae* are inhibited in broth by a dilution of 1:10^5 and *Staphylococcus aureus* by a dilution of 1:80,000 of nitrofurazone. On the other hand, *Streptococcus pyogenes* required the much higher inhibitory concentration of 1:10,000, and the still higher concentration of 1:5000 (at the limit of solubility of nitrofurazone) is needed for the inhibition of *Streptococcus viridans* and of *Pseudomonas aeruginosa*.

$$O_2N \text{—furan—} CH=NNHCONH_2$$

nitrofurazone

Nitrofurazone is reduced by *Aerobacter aerogenes* to 5-amino-2-furaldehyde semicarbazone. The structure of this compound was confirmed by comparison with a sample made by catalytic hydrogenation of nitrofurazone. The bacterial reduction of nitrofuran semicarbazone derivatives in which the NH group of the semicarbazone grouping is blocked by an alkyl group or by involvement in a ring leads to reduction of the nitro group accompanied by furan ring cleavage (256) (see Furan derivatives).

Nitrofurazone has been found useful as a topical prophylactic and therapeutic agent in mixed infections common to contaminated wounds, burns, pyodermas, and the like, and also in the management of purulent otitis and conjunctivitis (257).

Furazolidone (Tricofuron, Eaton Laboratories) 3-(5-nitro-2-furfurylidene)amino-2-oxazolidone [67-45-8], is an antiprotozoan as well as an antibacterial agent. The range of its inhibitory effectiveness for a variety of pathogens extends from less than 0.5 mg/L for *Salmonella pullorum* to more than 99 mg/L for *Pseudomonas aeruginosa* (258). It is used mostly in treating the gynecological infection caused by *Trichomonas vaginalis*.

Another nitrofuran derivative *N*-(5-nitro-2-furfurylidene)-1-aminohydantoin [67-20-9] (nitrofurantoin, Furadantin, Eaton Laboratories) is also credited with a wide spectrum of antibacterial activity. It is administered orally in the treatment of bacterial infections of the urinary tract (see Chemotherapeutics; Antibacterial agents, synthetic-nitrofurans).

nitrofurantoin

Dyes

Those dyes that exhibit an antibacterial action are bacteriostatic rather than bactericidal, as a rule. Although it has been assumed that the basic dyes with their electropositive character are specific for gram-positive microoganisms, and the electronegative acid dyes, for gram-negative bacteria (259), the quantitative determination of the uptake of crystal violet by bacterial cells has failed to show any correlation with their gram character (260) (see Dyes; Triphenylmethane and related dyes; Azine dyes; Azo dyes).

The inhibitory action of basic dyes such as triphenylmethane [519-73-3] and acridine [260-94-6] derivatives is attributed to the tendency of their basic ions to form nonionizing or feebly ionizing complexes with the acidic groups of some cellular constituents of bacteria, probably a nucleoprotein in the cytoskeleton (261). Similarly in acid dyes such as fuchsin, the acidic ion is believed to react with some basic bacterial cell receptor, forming a nondissociating complex (262). These reactions are favored, respectively, by high and by low pH values. Bacteria become increasingly sensitive to basic dyes as the pH is increased, and to acidic dyes as it is decreased.

Thus, the bacteriostatic effectiveness of a dye should be a function of the strength of its ionization; the stronger the ionization, the greater should be the resistance to hydrolysis of the complexes formed with the reactive bacterial constituents. The basic strength of the isomeric aminoacridine dyes has been determined by potentiometric titration. Proceeding from the weakest members (1-amino- or 1,9-diaminoacridine [578-06-3]) to the strongest (5-amino- or 5-amino-1-methylacridine [23015-11-6]), there is a progressive increase in antibacterial potency for several test organisms (263). The relationship between ionization and antibacterial activity holds also in other series of related heterocyclic dyes, such as the benzacridines, benzoquinolines, and phenanthridines (264).

At one time the organic dyes were widely used for chemotherapy of infectious diseases. Several organic dyes are still in use, albeit to a minor extent; following is a selective listing:

Brilliant Green [633-03-4] is the quinonoid, bis(p-diethylamino)triphenylcarbinol anhydride. Tested by the semimicro method indicated a bactericidal capacity for *Staphylococcus aureus* in a dilution of 1:50,000. However, using 10% blood broth as reversing agent, the dye is not germicidal at 1:2000.

Gentian violet [548-62-9] (crystal violet) is essentially hexamethylpararosaniline hydrochloride. It suppresses the growth of *Lactobacillus acidophilus* and of *Micrococcus citreus* in a dilution of 1:10^6 or higher (265). It has been recommended as a topical antiseptic for the prevention of secondary infection of war wounds and burns (266).

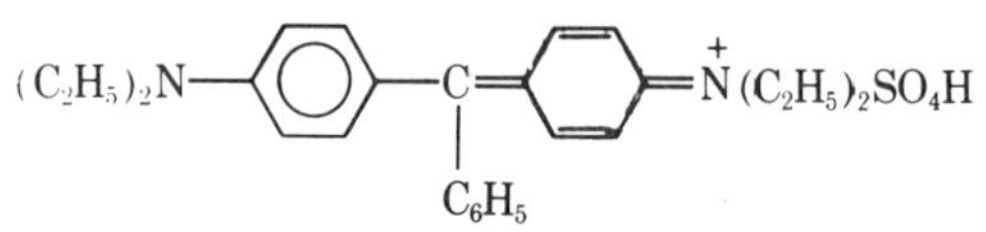

$(C_2H_5)_2N-\langle\bigcirc\rangle-\underset{C_6H_5}{C}=\langle\bigcirc\rangle=\overset{+}{N}(C_2H_5)_2 SO_4H$

Brilliant Green

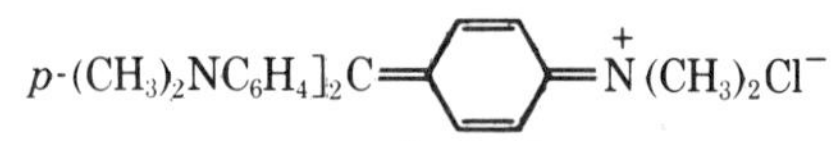

$p\text{-}(CH_3)_2NC_6H_4]_2C=\langle\bigcirc\rangle=\overset{+}{N}(CH_3)_2 Cl^-$

gentian violet

Methylene blue [61-73-4] stains plasmodia and is inhibitory for several pathogens including *Mycobacterium tuberculosis* (266).

Acriflavine is 3,6-diamino-10-methylacridinium chloride [6034-59-9]. Originally considered by Ehrlich for the treatment of trypanosomiasis, it gained importance subsequently as a topical antiseptic with bacteriostatic action. It enjoys limited use as a systemic drug (268). Rivanol, 2-ethoxy-6,9-diaminoacridine lactate [1837-57-6], exhibits antibacterial action *in vitro* that compares favorably with that of acriflavine (269).

Flavicid [525-12-2], 8-amino-2-dimethylamino-3,7,10-trimethylacridinium chloride, has been used as a surgical germicide for wound dressing as well as topically in gonorrhea (270).

Dimazon [83-63-6], diacetamidoazotoluene [83-63-6] is chemically related to scarlet red. It is applied in powder or ointment form to denuded areas as a means of controlling bacterial multiplication and of stimulating epithelization.

2,6-Diamino-3-phenylazopyridine hydrochloride [136-40-3], phenazopyridine hydrochloride, (Pyridium, Warner-Lambert Co.), is used primarily as a urinary antiseptic because of its claimed ability to exert bacteriostatic action in both acid and alkaline urine. It is inhibitory for *staphylococci, streptococci, gonococci,* and *coli bacilli* (271).

Two compounds in the sulfonamide series are dyes: sulfamidochrysoidine [103-12-8], *p*-[(2,4-diaminophenyl)azo] benzenesulfonamide (Prontosil rubrum) (272), and salicylazosulfapyridine [599-79-1], 5-[*p*-(2-pyridylsulfamoyl)phenylazo]salicylic acid (273).

The pigments produced by some chromogenic bacteria are inhibitory for other species. The antibiotic effect upon staphylococci and streptococci of pyocyanin [85-66-5], the pigment of *Pseudomonas aeruginosa,* has stimulated a systematic investigation of the series of phenazines to which pyocyanin belongs (274). Several of these dyes exhibited bacteriostatic effectiveness in the dilution range of $1:10^3$ to $1:10^4$ against *Staphylococcus aureus;* by contrast much higher dye concentrations, of the order of 5%, were required to inhibit *Escherichia coli* or *Proteus vulgaris.*

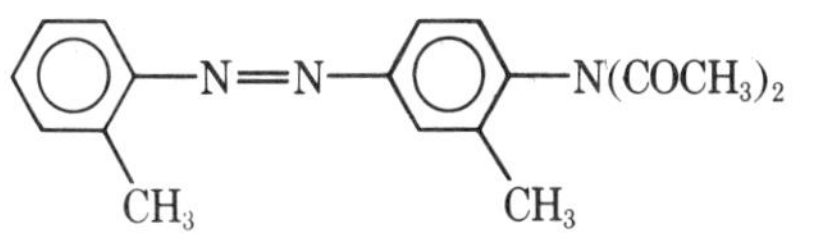

dimazon

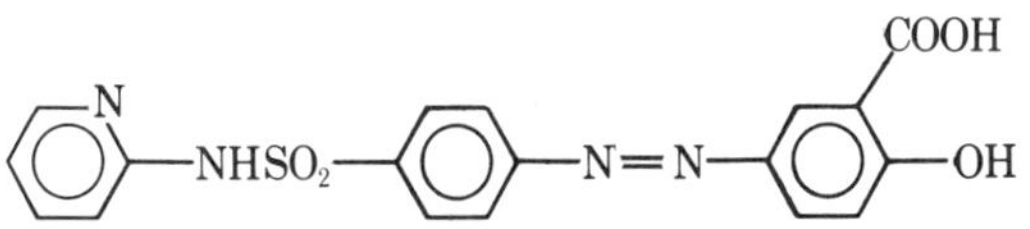

salicylazosulfapyridine

Formaldehyde and Other Aldehydes

At one time gaseous formaldehyde [50-00-0] was used to treat the premises, furniture, and objects exposed to patients with contagious illness. Fumigation with formaldehyde is now practiced only rarely since it has been found to be of little value as a disinfectant procedure. When used for sterilization of enclosed spaces, gaseous formaldehyde requires a very high humidity, approaching saturation (275) (see Sterile techniques).

Solutions of formaldehyde with soap are used as disinfectants, mostly outside the United States. Formaldehyde in hydroalcoholic solution is used widely for the "sterilization" of surgical instruments. Although it has been credited with sporicidal potency when employed for this purpose, its practical sporicidal effectiveness appears to be somewhat questionable (7,29,276). A mixture of formaldehyde and hexachlorophene has been recommended for cold sterilization of surgical instruments (276).

Hexamethylenetetramine [100-97-0], $C_6H_{12}N_4$ (methenamine), is still used as a urinary antiseptic, but to a lesser extent than prior to the advent of sulfonamides and antibiotics (qv). Its antibacterial action depends upon the slow liberation of formaldehyde in acid urine at a pH below 5.6 (see Antibacterial agents).

The slow release of formaldehyde from a new series of quaternary ammonium compounds, prepared by the reaction of hexamethylenetetramine with certain halohydrocarbons, furnishes the reason for their antimicrobial activity which renders them suitable for use in cutting-oil emulsions, latexes, and the like (277).

Methenamine mandelate [587-23-5] (Mandelamine, Warner-Chilcott Laboratories) utilizes the urinary antiseptic action of both methenamine and mandelic acid [90-64-2] ($C_6H_5CHOHCO_2H$). In concentrations of 35–50·g/100 liters of urine it is inhibitory to *Staphylococcus aureus, Bacillus proteus, Escherichia coli*, and *Aerobacter aerogenes* (278–279).

Recently, certain saturated dialdehydes and particularly glutaraldehyde [111-30-8], $OCH(CH_2)_3CHO$, have been found to possess a capacity for sterilizing (including sporicidal) action if employed in a hydroalcoholic solution, and in the presence of an alkalinizing agent, such as sodium bicarbonate. Sonacide is an aqueous 2% solution of glutaraldehyde with a nonionic ethoxylate of isomeric linear alcohols (280). The bactericidal effect upon vegetative pathogens (including *Mycobacterium tuberculosis* and *Pseudomonas aeruginosa*) is claimed to take place within 10 min, immersion for 3 h is required to destroy resistant bacterial spores. This type of preparation is suitable especially for the treatment of surgical instruments. The physical advantage of glutaraldehyde over formaldehyde is that in its use-dilution, the glutaraldehyde (unlike formaldehyde) has a mild odor and a low irritation potential for the skin and mucous membranes (281) (see Aldehydes).

dimethoxane

Dimethoxane (6-Acetoxyl-2,4-dimethyl-*m*-dioxane [828-00-2] (Giv-Gard DXN, Givaudan Corp.) is a bacteriostat and fungistat (282), to be used as preservative for aqueous systems and various types of industrial emulsions. It hydrolyzes in water to form active aldehydic compounds plus some acetic acid.

Ethylene Oxide and Other Alkylating Agents

As indicated above, true sterilization requires destruction of resistant bacterial spores and of all types of viruses, including those of infectious and serum hepatitis; it requires exposure to steam under pressure (autoclaving) for a sufficient period of time (283). Alternatively, such sterilization can be produced by suitable dilute solutions of certain phenolic disinfectants, eg, 2% aqueous Amphyl heated to the boiling point (284). Either treatment is suitable only for objects that tolerate heat (see Sterile techniques).

Important advances have been made in the cold sterilization of a variety of materials of low thermostability, based upon the discovery of the intensive microbicidal action of ethylene oxide [75-21-8] (qv) (285–286), which is extremely reactive in both the liquid and the gaseous states. Although there are a number of references to the bactericidal as well as the virucidal action of liquid ethylene oxide (287–288), the practical application of this chemical calls for its employment as a gas, in specially constructed autoclaves (289–293). Because of its flammability and explosiveness, gaseous ethylene oxide is not used as such; instead, it is diluted with an inert gas such as carbon dioxide or certain fluorohydrocarbons. Thus, the commercially available Carboxide R represents a mixture of 10% ethylene oxide and 90% carbon dioxide.

The bactericidal activity of ethylene oxide is thought to be the result of direct ethoxylation of functional groups of bacterial proteins, as illustrated below:

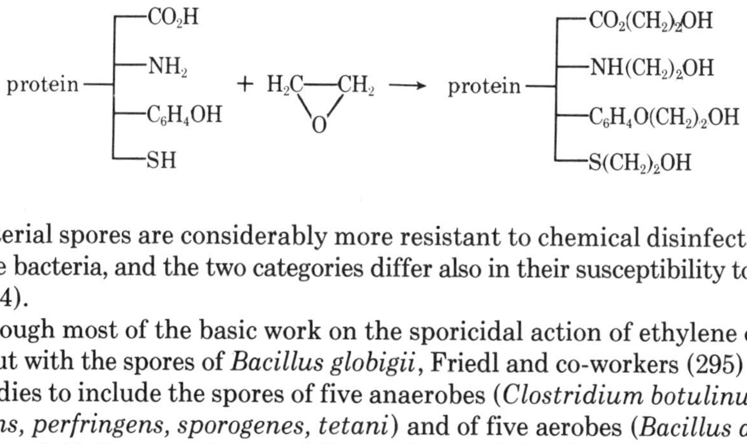

$$\text{protein} \begin{cases} -CO_2H \\ -NH_2 \\ -C_6H_4OH \\ -SH \end{cases} + H_2C\!\!-\!\!\!\overset{\diagdown\!O\!\diagup}{}\!\!\!-CH_2 \longrightarrow \text{protein} \begin{cases} -CO_2(CH_2)_2OH \\ -NH(CH_2)_2OH \\ -C_6H_4O(CH_2)_2OH \\ -S(CH_2)_2OH \end{cases}$$

Bacterial spores are considerably more resistant to chemical disinfectants than vegetative bacteria, and the two categories differ also in their susceptibility to ethylene oxide (294).

Although most of the basic work on the sporicidal action of ethylene oxide was carried out with the spores of *Bacillus globigii*, Friedl and co-workers (295) extended these studies to include the spores of five anaerobes (*Clostridium botulinum, lentoputrescens, perfringens, sporogenes, tetani*) and of five aerobes (*Bacillus anthracis, coagulans, globigii, stearothermophilus,* and *subtilis*). When exposed to ethylene oxide at room temperature, the dry spores of *Bacillus subtilis* and *Clostridium sporogenes* survived for several hours, whereas those of the other test organisms were killed after shorter periods of time. None survived an 18-h exposure. Spores of pathogenic fungi have also been studied (296). The sporicidal action of ethylene oxide is affected by the factors of concentration, temperature, and humidity (98). At a relative humidity of less than 25%, it is not reliably effective.

Vaccinia virus and Columbia-SK-encephalomyelitis virus are inactivated by ethylene oxide in 8 h at room temperature (297). Ethylene oxide also lends itself to mold control (298) and offers promise in certain areas of sterilization of foods (299–301) and drugs (302–303).

Propylene oxide [75-56-9] (qv) was found similar in properties to ethylene oxide, but less volatile and less active biologically (304). Interest in propylene oxide was revived when the FDA restricted use of ethylene oxide in the food industry because of the toxicity of ethylene glycol, a hydrolytic product found in small amounts (305). Propylene oxide hydrolyzes to nontoxic propylene glycol (see Glycols). Liquid propylene oxide has been suggested for sterilizing borated talc (306).

Other three-membered heterocyclic compounds also exhibit sporicidal action, in some instances superior to that of ethylene oxide (304). Among them are ethyleneimine [151-56-4], $(CH_2)_2NH$, its N-aminoethyl derivative, $(CH_2)_2$-$NCH_2CH_2NH_2$, and ethylene sulfide [420-12-2], $(CH_2)_2S$. Epichlorohydrin [106-89-8] and epibromohydrin [3132-64-7] are more effective than ethylene oxide but less so than either ethylene sulfide or ethylene imine (see Chlorohydrins). All of them are flammable as well as toxic, hence unsuitable for the sterilization of occupied premises.

β-Propiolactone [57-57-8] also exhibits sporicidal action. This chemical is applied usually in aqueous solution. Although stable in nonaqueous media, it hydrolyzes slowly in the presence of water to form 3-hydroxypropionic acid which is nontoxic. For this reason β-propiolactone is suitable for the sterilization of water and milk, of nutrient broth, and of biologicals such as vaccines. β-Propiolactone is active also against vegetative bacteria, as well as pathogenic fungi and viruses (307–312). It has been suggested for the sterilization of operating rooms and of enclosed spaces in general (313–315). As β-propiolactone is under suspicion as a carcinogen, it has been banned in interstate shipping by EPA for use as a pesticide.

$$\begin{array}{c} CH_2 \text{---} CH_2 \\ | \qquad | \\ O \text{---} CO \end{array}$$
β-propiolactone

BIBLIOGRAPHY

"Antiseptics, Disinfectants, and Fungicides" in *ECT* 1st ed., Vol. 2, "Survey," pp. 77–91, by W. C. Tobie, American Cyanamid Company; "Methods of Testing," pp. 91–105, by G. F. Reddish, Lambert Pharmacal Company; "Antiseptics and Disinfectants," in *ECT* 2nd ed., Vol. 2, pp. 604–648, by Emil G. Klarmann, Lehn & Fink, Inc.

1. *Fed. Reg.* **43,** 63771 (Jan. 6, 1978).
2. *Fed. Reg.* **40,** 28270 (July 3, 1975); **40,** 26808 (June 25, 1975).
3. G. L. A. Ruehle and C. M. Brewer, *U.S. Dept. Agric. Circ.* **198,** (1931).
4. W. J. Nungester and A. H. Kempf, *J. Infect. Dis.* **71,** 174 (1942).
5. W. J. Nungester and A. H. Kempf, *J. Am. Med. Assoc.* **121,** 593 (1945).
6. R. W. Sarber, *J. Pharmacol. Exp. Ther.* **75,** 277 (1942).
7. E. G. Klarmann, E. S. Wright, and V. A. Shternov, *Am. J. Pharm.* **122,** 5 (1950).
8. P. B. Price, *J. Infect. Dis.* **63,** 301 (1938).
9. A. R. Cade, *J. Soc. Cosmet. Chem.* **2,** 281 (1951).
10. A. R. Cade, *Am. Soc. Test. Mater., Spec. Tech. Publ.* **115,** 33 (1952).
11. *Fed. Reg.* **39,** 33103 (Sept. 13, 1974).
12. L. J. Vinson and co-workers, *J. Pharm. Sci.* **50,** 827 (1961).
13. A. J. Salle and co-workers, *J. Bacteriol.* **37,** 639 (1939).
14. E. H. Spaulding and J. A. Bondi, *J. Infect. Dis.* **80,** 194 (1947).
15. S. Rideal and J. T. A. Walker, *J. R. Sanit. Inst.* **24,** 424 (1903).
16. Lancet Commission, *Lancet* **177,** 1459, 1516, 1612 (1909).
17. S. Rideal and J. T. A. Walker, *Approved Technique of the Rideal-Walker Test,* Lewis, London, Eng., 1921.

18. *Br. Standards Inst. Bull.* **541,** 113 (1934).
19. J. F. Anderson and T. B. McClintic, *Hyg. Lab. Bull.* (*U.S. Treas. Dept.*) **82,** (1912).
20. American Public Health Association, *Am. J. Public Health* **8,** 506 (1918).
21. C. M. Brewer and G. F. Reddish, *J. Bacteriol.* **17,** 44 (1929).
22. *Official and Tentative Methods of Analysis,* 9th ed., Association of Official Agricultural Chemists, Washington, D.C., 1960.
23. Association of Official Analytical Chemists, *J. Assoc. Off. Anal. Chem.* **53,** 61 (1970).
24. L. S. Stuart, L. F. Ortenzio, and J. L. Friedl, *J. Assoc. Off. Agric. Chem.* **36,** 466 (1953).
25. E. G. Klarmann, *Am. J. Hosp. Pharm.* **15,** 795 (1958).
26. R. L. Steadman, E. Kravitz, and H. Bell, *Appl. Microbiol.* **2,** 119, 322 (1954); **3,** 71 (1955).
27. E. G. Klarmann, E. S. Wright, and V. A. Shternov, *Appl. Microbiol.* **1,** 19 (1953).
28. E. G. Klarmann and E. S. Wright, *Am. J. Pharm.* **122,** 330 (1950).
29. J. L. Friedl, *J. Assoc. Off. Agric. Chem.* **38,** 280 (1955).
30. L. F. Ortenzio, L. S. Stuart, and J. L. Friedl, *J. Assoc. Off. Agric. Chem.* **36,** 480 (1953).
31. E. G. Klarmann, *Am. J. Pharm.* **129,** 42 (1957).
32. *Ibid.* **131,** 86 (1959).
33. E. G. Klarmann, *Hosp. Top.* **35**(8), 90 (1957).
34. *W.H.O. Tech Rep. Ser.* **62,** (1953).
35. A. J. Rhodes and C. E. van Rooyen, *Textbook of Virology,* Williams and Wilkins, Baltimore, Md., 1953, p. 29.
36. C. L. Hoagland, *Ann. Rev. Biochem.* **12,** 615 (1943).
37. W. B. Dunham and W. J. MacNeal, *J. Immunol.* **49,** 123 (1944).
38. V. Groupe and co-workers, *Appl. Microbiol.* **3,** 333 (1955).
39. M. Klein and D. A. Stevens, *J. Immunol.* **50,** 265 (1945).
40. E. R. Parker, W. B. Dunham, and W. J. MacNeal, *J. Lab. Clin. Med.* **29,** 37 (1944).
41. N. A. Allawala and S. Riegelman, *J. Am. Pharm. Assoc. Sci. Ed.* **42,** 396 (1953).
42. *DIS-13,* Environmental Protection Agency, Washington, D.C., 1973.
43. H. S. Wright, *Appl. Microbiol.* **19,** 92 (1970).
44. J. H. Blackwell and J. H. Chen, *J. Assoc. Off. Anal. Chem.* **53,** 1229 (1970).
45. J. Christiansen, *Z. Physiol. Chem.* **102,** 275 (1918).
46. J. H. Quastel and M. D. Whetham, *Biochem. J.* **19,** 520 (1927).
47. E. G. Klarmann, L. W. Gates, and V. A. Shternov, *J. Am. Chem. Soc.* **54,** 3315 (1932).
48. H. E. Morton, *Ann. N.Y. Acad. Sci.* **53,** 191 (1950).
49. P. B. Price, *Arch. Surg.* **38,** 528 (1939).
50. F. W. Tilley and J. M. Schaffer, *J. Bacteriol.* **12,** 303 (1926).
51. P. B. Price, *Arch. Surg.* **38,** 528 (1929); **60,** 492 (1950).
52. C. R. Smith, *Public Health Rep.* (*U.S.*) **62,** 1285 (1947).
53. H. Neves, *Arch. Dermatol.* **84,** 132 (1961).
54. E. C. McCulloch, *Disinfection and Sterilization,* Lea and Febiger, Philadelphia, Pa., 1945, p. 319.
55. C. H. Cunningham, *Am. J. Vet. Res.* **9,** 195 (1948).
56. M. H. Gordon, *Med. Res. Council Spec. Rep. Ser.* **98,** (1925).
57. S. Stockman and F. C. Minett, *J. Comp. Pathol. Therap.* **39,** 1 (1926).
58. H. E. Paul and co-workers, *Proc. Soc. Exp. Biol. Med.* **79,** 199 (1952).
59. M. L. Tainter and co-workers, *J. Am. Dental Assoc.* **31,** 479 (1944).
60. G. Weitzel and E. Schraufstätter, *Z. Physiol. Chem.* **285,** 172 (1950).
61. O. H. Robertson and co-workers, *Science* **97,** 142 (1943).
62. O. H. Robertson and W. Lester, Jr., *Am. J. Hyg.* **53,** 69 (1951).
63. V. A. Chandler, R. E. Pepper, and L. E. Gordon, *J. Pharm. Sci.* **46,** 124 (1957).
64. C. K. Johns, *Sci. Agric.* **14,** 585 (1934).
65. S. M. Costigan, *J. Bacteriol.* **34,** 1 (1937).
66. M. Levine and A. S. Rudolph, *Iowa State Coll. Eng. Exp. Stn. Bull.* **150,** (1941).
67. H. C. Marks and F. B. Strandskov, *Ann. N.Y. Acad. Sci.* **53,** 163 (1950).
68. G. M. Ridenour, R. S. Ingols, and E. H. Armbruster, *Water Sewage Works* **96,** 279 (1949).
69. E. C. McCulloch and S. M. Costigan, *J. Infect. Dis.* **59,** 281 (1936).
70. A. Fleming, *Chem. Ind.* (*London*), 18 (1945).
71. S. L. Chang and G. Berg, *U.S. Armed Forces Med. J.* **10,** 33 (1959).
72. F. C. Schmelkes and E. S. Horning, *J. Bacteriol.* **29,** 323 (1935).
73. H. Lemaire, C. H. Schrammy, and A. Cahn, *J. Pharm. Sci.* **50,** 831 (1961).

74. B. Carroll, *J. Bacteriol.* **69,** 413 (1955).
75. O. Wyss and F. B. Strandskov, *Arch. Biochem.* **6,** 261 (1945).
76. P. B. Price, *Drug Stand.* **19,** 161 (1951).
77. M. Frobisher, Jr., L. Sommermeyer, and M. L. Blackwell, *Appl. Microbiol.* **1,** 187 (1953).
78. L. Gershenfeld, W. B. Flagg, and B. Witlin, *Mil. Surg.* **114,** 172 (1954).
79. L. Gershenfeld and B. Witlin, *J. Am. Pharm. Assoc. Sci. Ed.* **41,** 451 (1952).
80. L. Gershenfeld and B. Witlin, *Am. J. Pharm.* **123,** 87 (1951).
81. C. W. Chambers and co-workers, *Soap Sanit. Chem.* **28,** 149 (1952).
82. S. L. Chang and J. C. Morris, *Ind. Eng. Chem.* **45,** 1009 (1953).
83. J. C. Morris and co-workers, *Ind. Eng. Chem.* **45,** 1013 (1953).
84. L. Gershenfeld and B. Witlin, *Am. J. Pharm.* **125,** 129, 258 (1953).
85. P. G. Bartlett and W. Schmidt, *Appl. Microbiol.* **5,** 355 (1957).
86. R. Blatt and J. V. Maloney, Jr., *Surg. Gynecol. Obstet.* **113,** 699 (1961).
87. J. A. Boswick, E. Kissell, and W. I. Metzger, *J. Abdom. Surg.* **3,** 157 (1961).
88. L. Gershenfeld and B. Witlin, *Am. J. Pharm.* **128,** 335 (1956).
89. B. Witlin and L. Gershenfeld, *J. Milk Food Technol.* **32,** 155, 167 (1956).
90. C. K. Johns, *Can. J. Technol.* **32,** 71 (1954).
91. H. A. Shelanski and M. V. Shelanski, *J. Int. Coll. Surg.* **25,** 727 (1956).
92. L. Gershenfeld, *Am. J. Surg.* **94,** 938 (1957); *Am. J. Pharm.* **134,** 278 (1962).
93. A. W. Frisch, G. H. Davies, and W. Kreppachne, *Surg. Gynecol. Obstet.* **107,** 442 (1958).
94. C. A. Lawrence, *Am. J. Hosp. Pharm.* **17,** 100 (1960).
95. U.S. Pat. 2,679,533 (May 25, 1954), J. L. Darragh and R. House (to California Research Corp.).
96. U.S. Pat. 2,746,928 (May 22, 1956), J. L. Darragh and G. B. Johnson (to California Research Corp.).
97. W. E. Engelhard and A. W. Worton, *J. Am. Pharm. Assoc. Sci. Ed.* **45,** 402 (1956).
98. F. B. Engley, Jr., *Ann. N.Y. Acad. Sci.* **53,** 197 (1950).
99. H. Engelhardt, *Desinfektion* **7,** 63, 81 (1922).
100. J. M. McCalla, *J. Bacteriol.* **40,** 33 (1940).
101. M. Klein and co-workers, *J. Immunol.* **59,** 135 (1948).
102. H. E. Morton, L. L. North, and F. B. Engley, *J. Am. Med. Assoc.* **136,** 37 (1948).
103. L. Banti, *J. Am. Med. Assoc.* **140,** 404 (1948).
104. E. G. Klarmann, *Ann. N.Y. Acad. Sci.* **53,** 123 (1950).
105. E. S. G. Barron and G. Kalnitsky, *Biochem. J.* **41,** 346 (1947).
106. E. S. G. Barron and T. P. Singer, *J. Biol. Chem.* **157,** 221 (1945).
107. P. Fildes, *Br. J. Exp. Pathol.* **21,** 67 (1940).
108. J. A. de Loueiro and E. Lito, *J. Hyg.* **44,** 463 (1946).
109. A. E. Elkhouly and R. T. Yousef, *J. Pharm. Sci.* **63,** 681 (1974).
110. H. Kliewe, F. Steyskal, and K. Steyskal, *Z. Hyg. Infektionskrankh.* **127,** 110 (1947).
111. C. E. Hartford and S. E. Ziffren, *J. Trauma* **12,** 682 (1972).
112. H. S. Carr, T. J. Wodkowski, and H. S. Rosenkranz, *Antimicrob. Agents Chemotherap.* **4,** 585 (1973).
113. C. L. Fox, *Arch. Surg.* **96,** 184 (1968).
114. C. H. Brandes, *Ind. Eng. Chem.* **26,** 962 (1934).
115. R. K. Hoffman and co-workers, *Ind. Eng. Chem.* **45,** 287 (1953).
116. J. B. Sprowls and C. F. Poe, *J. Am. Pharm. Assoc.* **32,** 41 (1943).
117. H. L. Guest and A. J. Salle, *Proc. Soc. Exp. Biol. Med.* **51,** 572 (1942).
118. A. P. Muroma, *Ann. Med. Exp. Biol. Fenniae (Helsinki)* **31,** 432 (1953).
119. Th. Kunzman, *Fortschr. Med.* **52,** 357 (1934).
120. H. R. Dittmar, J. L. Baldwin, and S. B. Miller, *J. Bacteriol.* **19,** 203 (1930).
121. F. I. K. Ahmed and C. Russell, *J. Appl. Bacteriol.* **39,** 31 (1975).
122. F. L. Meleney, *J. Am. Med. Assoc.* **149,** 1450 (1952).
123. C. D. Leake, *J. Am. Med. Assoc.* **119,** 101 (1942).
124. J. P. Doll and co-workers, *Midl. Natural* **69,** 23 (1963).
125. F. P. Greenspan, M. A. Johnson, and P. C. Trexler, *Proc. 42nd Annual Meeting Chem. Spec. Manf. Assoc.*, 59 (1955).
126. S. Kojima, *J. Biochem. (Tokyo)* **14,** 95 (1931).
127. M. Ingram and R. B. Haines, *J. Hyg.* **47,** 146 (1949).
128. A. H. Fogg and R. M. Lodge, *Trans. Faraday Soc.* **41,** 309 (1945).

129. E. M. Richardson and E. E. Reid, *J. Am. Chem. Soc.* **62,** 413 (1940).
130. E. F. Gale and E. S. Taylor, *J. Gen. Microbiol.* **1,** 77 (1947).
131. D. A. Haydon, *Proc. R. Soc. (London) Ser. B* **145,** 583 (1956).
132. R. J. V. Pulvertaft and G. D. Lumb, *J. Hyg.* **46,** 62 (1948).
133. P. Maurice, *Proc. Soc. Appl. Bacteriol.* **15,** 144 (1952).
134. J. Judis, *J. Pharm. Sci.* **51,** 261 (1962).
135. D. Bach and J. Lambert, *Compt. Rend. Soc. Biol.* **126,** 298 (1937).
136. M. H. Roberts and O. Rahn, *J. Bacteriol.* **52,** 639 (1946).
137. J. H. Quastel and W. R. Wooldridge, *Biochem. J.* **21,** 148, 689 (1927).
138. G. Sykes, *J. Hyg.* **39,** 463 (1939).
139. E. G. Klarmann, V. A. Shternov, and L. W. Gates, *J. Lab. Clin. Med.* **19,** 835; **20,** 40 (1934).
140. E. G. Klarmann and co-workers, *J. Am. Chem. Soc.* **55,** 2576, 4657 (1933).
141. L. F. Ortenzio, C. D. Opalsky, and L. S. Stuart, *Appl. Microbiol.* **9,** 562 (1961).
142. T. B. Johnson and W. W. Hodge, *J. Am. Chem. Soc.* **35,** 1014 (1913).
143. T. B. Johnson and F. W. Lane, *J. Am. Chem. Soc.* **43,** 348 (1921).
144. A. R. L. Dohme, E. H. Cox, and E. Miller, *J. Am. Chem. Soc.* **48,** 1688 (1926).
145. B. Hampil, *J. Infect. Dis.* **43,** 25 (1928).
146. E. G. Klarmann, L. W. Gates, and V. A. Shternov, *J. Am. Chem. Soc.* **54,** 298, 1204 (1932).
147. E. G. Klarmann, *J. Am. Chem. Soc.* **48,** 2358 (1926).
148. E. G. Klarmann and J. Von Wowern, *J. Am. Chem. Soc.* **51,** 605 (1929).
149. M. L. Moore, A. A. Day, and C. M. Suter, *J. Am. Chem. Soc.* **56,** 2456 (1934).
150. R. R. Read, G. F. Reddish, and E. M. Burlingame, *J. Am. Chem. Soc.* **56,** 1377 (1934).
151. R. Zinkernagel and M. Koenig, *Seifen-Öle Fette-Wachse* **93,** 670 (1967).
152. E. Jungermann and D. Taber, *J. Am. Oil Chem. Soc.* **48,** 318 (1971).
153. P. S. Herman and W. M. Sams, *Soap Photodermatitis and Photosensitivity to Halogenated Salicylanilides,* Charles C Thomas, Springfield, Ill., 1972.
154. L. J. Vinson and R. S. Flatt, *J. Invest. Dermatol.* **38,** 327 (1962).
155. D. S. Wilkinson, *Br. J. Dermatol.* **73,** 213 (1961); **74,** 302 (1962).
156. A. Kraushaar, *Arzneim. Forsch.* **4,** 548 (1954).
157. U.S. Pat. 2,745,874 (May 15, 1956), G. Schetty, W. Stammbach, and R. Zinkernagel (to J. R. Geigy A.G.).
158. U.S. Pat. 2,906,711 (Sept. 29, 1959), H. C. Stecker.
159. U.S. Pat. 3,041,236 (June 26, 1962), H. C. Stecker.
160. T. R. Aalto, M. C. Firman, and N. E. Rigler, *J. Am. Pharm. Assoc. Sci. Ed.* **42,** 449 (1953).
161. H. Sokol, *Drug. Stand.* **20,** 89 (1952).
162. H. F. Cremer, *Z. Untersuch. Lebensm.* **70,** 136 (1935).
163. K. Schubel and I. Manger, Jr., *Münch. Med. Wochschr.* **77,** 13 (1929).
164. L. Gershenfeld and D. Perlstein, *Am. J. Pharm.* **111,** 227 (1939).
165. N. S. Gottfried, *Am. J. Hosp. Pharm.* **19,** 310 (1962).
166. C. Mathews and co-workers, *J. Am. Pharm. Assoc. Sci. Ed.* **45,** 260 (1956).
167. M. Huppert, *Antibiot. Chemother.* **7,** 29 (1957).
168. H. Gershon and R. Parmegiani, *Appl. Microbiol.* **10,** 348 (1962).
169. U.S. Pat. 2,250,408 (July 29, 1941), W. S. Gump (to Burton T. Bush, Inc.).
170. W. S. Gump, *Soap Sanit. Chem.* **21,** 36 (1945).
171. U.S. Pat. 2,535,077 (Dec. 26, 1950), E. C. Kunz and W. S. Gump (to Sindar Corp.).
172. C. V. Seastone, *Surg. Gynecol. Obstet.* **84,** 355 (1947).
173. P. B. Price and A. Bonnett, *Surgery* **24,** 542 (1948).
174. I. H. Blank and M. H. Coolidge, *J. Invest. Dermatol.* **15,** 257 (1950).
175. P. B. Price, *Ann. Surg.* **134,** 476 (1951).
176. W. B. Shelley, H. J. Hurley, and A. C. Nichols, *Arch. Dermatol. Syphilol.* **68,** 430 (1953).
177. W. S. Gump, *J. Soc. Cosmet. Chem.* **20,** 173 (1969).
178. B. P. Vaterlaus and J. J. Hostynek, *J. Soc. Cosmet. Chem.* **24,** 291 (1973).
179. R. D. Kimbrough and T. B. Gaines, *Arch. Environ. Health* **23,** 114 (1971).
180. T. B. Gaines, R. D. Kimbrough, and R. E. Linder, *Toxicol. Appl. Pharmacol.* **25,** 332 (1973).
181. J. A. Santolucito, *Toxicol. Appl. Pharmacol.* **22,** 276 (1972).
182. H. M. Powell, *J. Indiana State Med. Assoc.* **38,** 303 (1945).
183. *Fed. Reg.* **37,** 160, 219 (Jan. 7, 1972).
184. W. S. Gump, *J. Soc. Cosmet. Chem.* **14,** 269 (1963).

185. *Ibid.* **15,** 717 (1964).
186. S. S. Block, *Disinfection, Sterilization, and Preservation,* 2nd ed., Lea & Febiger, Philadelphia, Pa., 1977.
187. R. S. Shumard, D. J. Beaver, and M. C. Hunter, *Soap Sanit. Chem.* **29**(1), 34 (1953).
188. U.S. Pat. 2,353,735 (July 18, 1944), E. C. Kunz, M. Luthy, and W. S. Gump (to Burton T. Bush, Inc.).
189. K. M. Wood and S. H. Hopper, *J. Am. Pharm. Assoc. Sci. Ed.* **47,** 317 (1958).
190. O. F. Jillson and R. D. Baughma, *Arch. Dermatol.* **88,** 409 (1963).
191. R. Pfleger and co-workers, *Naturforsch.* **46,** 344 (1949).
192. J. B. Adams and M. Hobbs, *J. Pharm. Pharmacol.* **10,** 507 (1958).
193. W. S. Gump and G. R. Walter, *J. Soc. Cosmet. Chem.* **11,** 307 (1960).
194. G. R. Walter and W. S. Gump, *J. Soc. Cosmet. Chem.* **13,** 477 (1962).
195. W. C. Harden and E. E. Reid, *J. Am. Chem. Soc.* **54,** 4325 (1932).
196. B. Heinemann, *J. Lab. Clin. Med.* **29,** 254 (1944).
197. A. Albert, M. I. Gibson, and S. D. Rubbo, *Br. J. Exp. Pathol.* **34,** 119 (1953).
198. S. D. Rubbo, A. Albert, and M. I. Gibson, *Br. J. Exp. Path.* **31,** 425 (1950).
199. E. D. Weinberg, *Fed. Proc.* **20,** 132 (1961).
200. W. Liese, *Zentr. Bakteriol.* **105**(I), 137 (1927).
201. K. A. Oster and M. J. Golden, *J. Am. Pharm. Assoc. Sci. Ed.* **37,** 283 (1947).
202. S. M. Bahal, M. R. Baichwal, and M. I. Khorana, *J. Pharm. Sci.* **50,** 127 (1961).
203. W. Jadassohn and co-workers, *Schweiz. Med. Wochschr.* **74,** 168 (1944).
204. *Ibid.* **77,** 987 (1947).
205. K. Sigg, *Schweiz. Med. Wochschr.* **77,** 123 (1947).
206. M. Hartmann and H. Kägi, *Z. Angew. Chem.* **41,** 127 (1928).
207. P. M. Borick and M. Bratt, *Appl. Microbiol.* **9,** 475 (1961).
208. D. N. Eggenberger and co-workers, *Ann. N.Y. Acad. Sci.* **53,** 105 (1950).
209. P. H. H. Gray and L. J. Taylor, *Can. J. Botany* **30,** 674 (1952).
210. E. G. Klarmann and E. S. Wright, *Soap Sanit. Chem.* **22**(1), 125; **23**(7), 151 (1947).
211. A. Einhorn and M. Göttler, *Ann. Chem.* **343,** 207 (1908).
212. W. A. Jacobs, *J. Exp. Med.* **23,** 563 (1916).
213. W. A. Jacobs, M. Heidelberger, and C. G. Bull, *J. Exp. Med.* **23,** 577 (1916).
214. G. Domagk, *Dtsch. Med. Wochschr.* **61,** 829 (1935).
215. H. A. Krebs, *Biochem. J.* **43,** 51 (1948).
216. E. J. Ordal and A. F. Borg, *Proc. Soc. Exp. Biol. Med.* **50,** 332 (1942).
217. M. G. Sevag and O. A. Ross, *J. Bacteriol.* **48,** 677 (1944).
218. C. E. Chaplin, *Can. J. Bacteriol.* **29,** 373 (1951).
219. C. K. Crocker, *J. Milk Food Technol.* **14,** 138 (1951).
220. R. Fischer and P. Larose, *Nature* **170,** 175 (1952).
221. W. E. Knox and co-workers, *J. Bacteriol.* **58,** 443 (1949).
222. R. D. Hotchkiss, *Ann. N.Y. Acad. Sci.* **46,** 479 (1946).
223. M. R. J. Salton, *J. Gen. Microbiol.* **5,** 391 (1951).
224. M. R. J. Salton, R. W. Horne, and V. E. Cosslett, *J. Gen. Microbiol.* **5,** 405 (1951).
225. C. A. Lawrence, *Surface-Active Quaternary Ammonium Germicides,* Academic Press, New York, 1950.
226. G. R. Goetchius and H. Grinsfelder, *Appl. Microbiol.* **1,** 271 (1953).
227. E. G. Klarmann and E. S. Wright, *Am. J. Pharm.* **120,** 146 (1948).
228. L. S. Stuart, J. Bogusky, and J. L. Friedl, *Soap Sanit. Chem.* **26**(1), 121 (1950); **26**(2), 127 (1950).
229. W. A. Altemeier, *Ann. Surg.* **147,** 773 (1958).
230. H. Dold and R. Gust, *Arch. Hyg. Bakteriol.* **141,** 321 (1957).
231. J. G. Hirsch, *Am. Rev. Tuberc.* **70,** 312 (1954).
232. E. G. Klarmann, *Am. J. Pharm.* **126,** 267 (1954).
233. C. R. Smith, *Soap Sanit. Chem.* **27**(9–10), (1951).
234. C. R. Smith and co-workers, *Public Health Rep. (U.S.)* **65,** 588 (1950).
235. A. Hoyt, A. H. K. Djang, and C. R. Smith, *Public Health Rep. (U.S.)* **71,** 1097 (1956).
236. C. W. Chambers and co-workers, *Public Health Rep. (U.S.)* **70,** 545 (1955).
237. E. W. Dennis, *Soap Sanit. Chem.* **27,** 117 (1951).
238. G. R. Walter and W. S. Gump, *J. Pharm. Sci.* **51,** 707 (1962).
239. C. A. Lawrence and A. J. Maffia, *Bull. Am. Soc. Hosp. Pharm.* **14,** 164 (1957).

240. R. A. Benson and co-workers, *J. Pediat.* **31,** 369 (1947); **34,** 49 (1949).
241. P. Story, *Br. Med. J.,* 1128 (Nov. 22, 1952).
242. Th. Goldschmidt A. G., *Ned. Tijdschr. Geneeskd.* **3,** 234 (1967).
243. A. Schmitz and W. S. Harris, *Manuf. Chem.* **29,** 51 (1958).
244. A. Schmitz, *Milchwissenschaft* **7,** 250 (1952).
245. U.S. Pat. 2,253,182 (Aug. 19, 1941), E. G. Klarmann (to Lehn & Fink Products Corp.).
246. U.S. Pat. 2,359,241 (Sept. 26, 1944), A. M. Partansky (to The Dow Chemical Co.).
247. A. M. Kligman and W. Rosenzweig, *J. Invest. Dermatol.* **10,** 59 (1947).
248. C. R. Miller and W. O. Elson, *J. Bacteriol.* **57,** 47 (1949).
249. D. J. Beaver, D. P. Roman, and P. F. Stoffel, *J. Am. Chem. Soc.* **79,** 1236 (1957).
250. F. L. Rose and G. Swain, *J. Chem. Soc.,* 4422 (1956).
251. J. Murray and R. M. Calman, *Br. Med. J.* (I), 81 (1955).
252. E. D. Weinberg, *Antibiot. Chemotherapy* **11,** 572 (1961).
253. M. C. Dodd and W. B. Stillman, *J. Pharmacol. Exp. Therap.* **82,** 11 (1944).
254. U.S. Pat. 2,319,481 (May 18, 1943), W. B. Stilman, A. B. Scott, and J. M. Clampit (to Norwich Pharmacol Co.).
255. M. C. Dodd, D. L. Cramer, and W. C. Ward, *J. Am. Pharm. Assoc. Sci. Ed.* **39,** 313 (1950).
256. A. H. Beckett and A. E. Robinson, *Chem. Ind. (London),* 523 (1957).
257. M. C. Dodd, *J. Pharmacol. Exp. Therap.* **86,** 311 (1946).
258. J. A. Yourchenco, M. C. Yourchenco, and C. R. Piepoli, *Antibiot. Chemother.* **3,** 1035 (1953).
259. E. R. Kennedy and J. F. Barbaro, *J. Bacteriol.* **65,** 678 (1953).
260. J. W. Bartholomew and H. Finkelstein, *J. Bacteriol.* **67,** 689 (1954).
261. A. Albert, *Lancet,* 278 (1942).
262. T. M. McCalla, *Stain Technol.* **16,** 27 (1941).
263. A. Albert and co-workers, *Br. J. Exp. Pathol.* **26,** 60 (1945).
264. A. Albert, S. D. Rubbo, and M. Burvill, *Br. J. Exp. Pathol.* **30,** 159 (1949).
265. J. E. Weiss and L. F. Rettger, *J. Bacteriol.* **28,** 501 (1934).
266. C. P. G. Wakeley, *Practitioner* **146,** 27 (1941).
267. B. E. Greenberg and M. L. Brodney, *New Engl. J. Med.* **209,** 1153 (1933).
268. A. Albert and R. J. Goldacre, *Nature* **101,** 95 (1948).
269. G. R. Goetchius and C. A. Lawrence, *J. Lab. Clin. Med.* **29,** 134 (1944).
270. R. Wagner and A. Pohlner, *Z. Immunitätsforsch.* **94,** 171 (1938).
271. A. Goerner and F. L. Haley, *J. Lab. Clin. Med.* **16,** 957 (1931).
272. U.S. Pat. 2,085,037 (June 29, 1937), F. Mietzsch and J. Klarer (to Winthrop Chemical Co.).
273. U.S. Pat. 2,396,145 (Mar. 5, 1946), E. E. A. Askelöf, N. Svartz, and H. C. Willstaedt (to Akticbolaget Pharmacia).
274. G. Proske, *Arch. Hyg. Bakteriol.* **136,** 74 (1952).
275. G. Nordgren, *Acta Pathol. Microbiol. Scand.* **40**(Suppl. 1), (1939).
276. U.S. Pat. 2,519,565 (Aug. 22, 1950), H. T. Hallowell (to Bard-Parker Co., Inc.).
277. R. C. Scott and P. A. Wolf, *Appl. Microbiol.* **10,** 211 (1962).
278. J. V. Scudi and C. J. Duca, *J. Urol.* **61,** 459 (1949).
279. I. Simons, *J. Urol.* **64,** 586 (1950).
280. R. M. S. Boucher, *Am. J. Hosp. Pharm.* **29,** 660 (1972).
281. U.S. Pat. 3,016,328 (Jan. 9, 1962), R. E. Pepper and R. E. Lieberman (to Ethicon, Inc.).
282. U.S. Pat. 3,167,477 (Jan. 26, 1965), W. S. Gump and G. R. Walter (to Givaudan Corp.).
283. J. J. Perkins, *Principles and Methods of Sterilization,* Charles C Thomas, Springfield, Ill., 1956; *Sterilization of Surgical Materials, Symposium,* Pharmaceutical Press, London, Eng., 1961, p. 76.
284. E. G. Klarmann, *Am. J. Pharm.* **128,** 4 (1956).
285. C. W. Bruch, *Ann. Rev. Microbiol.* **15,** 245 (1961).
286. U.S. Pat. 2,037,439 (Apr. 14, 1936), H. Schrader and E. Bossert (to Union Carbide and Carbon Corp.).
287. H. S. Ginsberg and A. T. Wilson, *Proc. Soc. Exp. Biol. Med.* **73,** 614 (1950).
288. A. T. Wilson and P. Bruno, *J. Exp. Med.* **51,** 449 (1950).
289. S. Kaye, *J. Lab. Clin. Med.* **35,** 823 (1950).
290. S. Kaye and C. R. Phillips, *Am. J. Hyg.* **50,** 296 (1949).
291. C. R. Phillips, *Am. J. Hyg.* **50,** 280 (1949).
292. C. R. Phillips and S. Kaye, *Am. J. Hyg.* **50,** 270 (1949).
293. C. W. Walter, *Hosp. Top.* **35,** 104 (1957).

294. C. R. Phillips, *Bacteriol. Rev.* **16,** 135 (1952); *Sterilization of Surgical Materials, Symposium,* Pharmaceutical Press, London, Eng., 1961, p. 59.
295. J. L. Friedl, L. F. Ortenzio, and L. S. Stuart, *J. Assoc. Off. Agric. Chem.* **39,** 480 (1956).
296. J. D. Fulton and R. B. Mitchell, *U.S. Armed Forces Med. J.* **3,** 425 (1952).
297. A. Klarenbeck and H. A. E. Tongeren, *Reports on Virucidal Action of Gaseous Ethylene Oxide,* Netherlands Institute for Preventive Medicine, Leiden, The Netherlands, 1950.
298. G. W. Kirby, L. Atkin, and C. N. Frey, *Food Ind.* **8,** 450 (1936).
299. J. S. Barlow and H. L. House, *Science* **123,** 229 (1956).
300. E. A. Hawk and O. Mickelson, *Science* **121,** 442 (1955).
301. H. J. Pappas and L. A. Hall, *Food Technol.* **6,** 456 (1952).
302. S. Kaye, H. F. Irminger, and C. R. Phillips, *J. Lab. Clin. Med.* **40,** 67 (1952).
303. L. C. Miner, *Am. J. Hosp. Pharm.* **16,** 284 (1959).
304. S. Kaye, *Am. J. Hyg.* **50,** 289 (1949).
305. H. T. Gordon, W. W. Thornburg, and L. N. Werum, *Agric. Food Chem.* **7,** 196 (1959).
306. U.S. Pat. 2,809,879 (Oct. 15, 1957), J. N. Masci (to Johnson & Johnson Inc.).
307. F. Bernheim and G. R. Gale, *Proc. Soc. Exp. Biol. Med.* **80,** 162 (1952).
308. F. W. Dawson, H. J. Hearn, and R. K. Hoffman, *Appl. Microbiol.* **7,** 199 (1959).
309. F. W. Dawson, R. J. Janssen, and R. K. Hoffman, *Appl. Microbiol.* **8,** 39 (1960).
310. F. W. Hartman, S. L. Piepes, and A. M. Wallbank, *Fed. Proc.* **10,** 358 (1951).
311. A. R. Kelley and F. W. Hartmann, *Fed. Proc.* **10,** 361 (1951).
312. *Ibid.* **11,** 419 (1952).
313. G. H. Mangun and co-workers, *Fed. Proc.* **10,** 220 (1951).
314. C. W. Bruch, *Am. J. Hyg.* **73,** 1 (1961).
315. R. K. Hoffman and B. Warshowsky, *Appl. Microbiol.* **6,** 358 (1958).
316. D. R. Spiner and R. K. Hoffman, *Appl. Microbiol.* **8,** 152 (1960).

WILLIAM GUMP
Consultant

DISPERSANTS

Dispersants are a class of materials that are capable of bringing fine solid particles into a state of suspension so as to inhibit or prevent their agglomerating or settling in a fluid medium. Dispersants may break up agglomerates or aggregates of fine particles, bring fine particles into a colloidal solution, or solubilize supersaturated salts to leave a clear solution. In particular, dispersants are useful whenever one wishes to prevent deposition, precipitation, settling, agglomerating, adhering, caking, etc, of solid particles in a fluid medium. This article is concerned only with materials capable of dispersing completely aqueous systems. Dispersants can also refer to the large number of proprietary compositions that facilitate the formulation of paints, inks, dyes, etc, especially by the maintenance of uniform consistency of, eg, inorganic pigments in mixed organic–aqueous media. Often the term is applied to surface-active materials that stabilize oil–water emulsions in the manufacture of lotions, lubricating oils, resins, and latexes (see Latex technology). The latter groups of materials are primarily used in systems where a second, nonaqueous liquid phase is present (see Emulsions; Surfactants).

Dispersants can be contrasted with flocculants or coagulants which facilitate the aggregation of fine particles in aqueous media to improve separations. Chemically similar materials can often function as either a flocculant or dispersant for the same substrate simply by a change in dosage level or polymer molecular weight. However, dispersants are normally anionic materials of lower molecular weight and higher charge than flocculants (see Flocculating agents).

Within aqueous dispersants we can differentiate three major categories of materials: (1) deflocculants, primarily used to fluidize concentrated slurries, especially clay slurries, to reduce their bulk viscosity or stickiness in processing or handling as in drilling muds, pigments for paper coatings, and in mineral processing aids; (2) antinucleation agents, usually referred to by function as antiscalants or antiprecipitants, used to prevent the deposition of sparingly soluble salts or scale, such as $CaCO_3$ and $CaSO_4$ in boilers, recirculating cooling water systems, or desalination units; and (3) builders, the largest dispersants market, the additives in most cleaning and detergent compositions that facilitate cleaning action by sequestering water-hardness ions (see also Chelating agents).

Physico-Chemical Principles of Dispersancy

Virtually all dispersants must adsorb on the substrate in order to perform their function. The degree of adsorption is related to solubility, strength of surface interaction, and the presence or absence of coulombic or van der Waals forces. Once adsorbed, deflocculants have the ability to disaggregate insoluble or suspended matter in aqueous media. Scale inhibitors (or antiscalants) are agents that inhibit deposition of adherent, crystalline deposits at surfaces, especially heat transfer surfaces. Antiprecipitants are materials that hinder precipitation of solids or formation of turbidity in the bulk solution. Antiscalants and antiprecipitants are both classified as antinucleation agents.

The physico-chemical principle of antinucleation is based on the concept of threshold inhibition. This phenomenon was first described by Hatch and Rice in the early 1930s (1). The stabilizing substances (antinucleation agents) inhibit the development of microscopic nuclei to the critical size and then redisperse, freeing the stabilizing material to interact with other embryonic nucleii (2). Often less than 1 ppm of an agent can inhibit 1000 ppm of excess calcium beyond saturation levels without evident precipitation or scaling. Scanning electron micrographs confirm the ability of a threshold inhibitor to cause crystal distortion. Figures 1a and 1b (3) show the ability of very low levels of a proprietary low mol wt polyacrylate to produce growth distortion of $CaCO_3$ (aragonite-phase) crystals. Optimum antiprecipitants must strike a balance between low molecular weight for rapid adsorption and high molecular weight which promotes a high extent of adsorption (4). However, optimum antiprecipitants are not necessarily the best antiscalants. Homopolymers of methacrylic acid [79-41-4] (5) or polyacrylates (6) of very low molecular weight (200–2000) are the most effective $CaCO_3$ scale inhibitors (see Acrylic acid; Methacrylic acid). Increasing the percentage of higher molecular weight species or broadening the molecular weight distribution enhances inhibition of precipitation (7). Structurally, select phosphonates that are able to form strong, substitution-inert chelate bonds with the calcium ions present at kinks and dislocations on the crystal surface are superior calcium phosphate and calcium sulfate inhibitors (8–9).

All currently available commercial scale and precipitation inhibitors are threshold inhibitors, ie, they inhibit nucleation of macroscopic scale or precipitates. Applications of antinucleation agents include treatments for boiler and recirculating cooling water systems in electric power generation, paper making, oil refining, steel manufacture, and in desalination units and in mineral processing plants (see Water, industrial water treatment; Water, supply and desalination).

Deflocculation. Deflocculation has been termed: "the breaking up from a flocculant state: converting into very fine particles" (10). The key phenomenon of deflocculation is the breaking apart or tearing down of aggregates, ie, of reversing the tendency of particles in dense suspensions (not supersaturated solutions) to settle, stick, or adhere leading to inhibition of flow or increasing bulk viscosity. In pigment suspensions the basic function of a deflocculating agent is to provide a barrier around particles that will prevent their adhesion during random contact (11). A striking demonstration of deflocculation is the reduction of a putty-like glob of clay of about 70% solids to an easily poured watery mass by as little as 0.01% of a deflocculant. Specific physical interactions responsible for these deflocculation effects include: (1) alteration of the structure of the electric double layer around the substrate in aqueous solution; or (2) build-up of "protective barriers" which retard coagulation; or the (3) generation of coulombic repulsions, ie, like charges on all particles which foster repulsion and, thereby, prevent agglomeration.

The surfaces of particles in aqueous solutions (in particular polar inorganic solids such as oxides, carbonates, etc) interact with water molecules and thereby acquire a charge, ie, surfaces are positively charged in extremely acidic media and negatively charged in extremely alkaline media; somewhere between the two the point of zero charge is reached as a function of the affinity of the surface for hydroxyl or hydrogen ions. This isoelectric point corresponds to the pH where a particular surface has zero charge in pure water. Table 1 lists isoelectric points for typical pigments (11).

Adsorption of anionic dispersants tends to shift the isoelectric point towards lower pH, in effect increasing the negative potential at the same pH, and thus increasing coulombic repulsions between particles. In general, this charge repulsion reduces interparticle attraction and leads to greater freedom for particles to flow. Hence, the

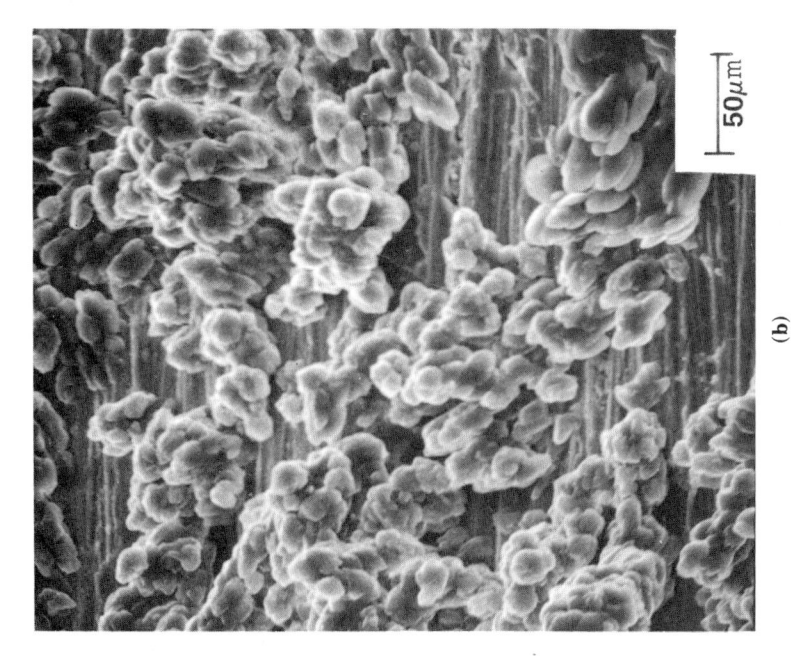

Figure 1. Scanning electron micrograph of (**a**) untreated calcium carbonate scale and (**b**) calcium carbonate scale treated by threshold inhibitor. Courtesy American Cyanamid Co.

Table 1. Isoelectric Point for Selected Pigments

Formula	Solid phase	Isoelectric point
$Al_2Si_2O_5(OH)_4$	kaolinite (clay)	4.8
$\alpha\text{-}Fe_2O_3$	hematite	5.2
PbO	litharge	10.3
SiO_2	quartz	2.2
TiO_2	rutile	4.7
ZnO	zincite	9.0

greater the affinity for the adsorbed ion and the more ionic it is, the greater its effectiveness as a deflocculant.

Sodium triphosphate [7758-29-4], or tripolyphosphate, STP, has been the predominant deflocculant of industry. A typical dosage vs viscosity curve is shown in Figure 2 (10). Recent work, however, shows striking synergistic improvement in deflocculation by mixtures of polyphosphates with polyacrylates (11). Polyacrylates being flexible polymers pucker in order to adsorb on the clay substrates; the small STP molecules fit between sites already occupied by the polyacrylates and provide further improvement in performance. In the deflocculation of finely ground pigment slimes, the smaller the average pigment particle size, the lower the requirement for the mol wt of the polymer (12). Molecular geometry is also important. For hydroxylated surfaces, such as in clays, the oxygen–oxygen distance for a typical octahedral arrangement is within 5% of the oxygen–oxygen distance of the phosphate tetrahedra (11).

Antinucleation and Deflocculation. Because many materials are active deflocculants as well as antinucleation agents, confusion in use can create an economic disadvantage. For example, insoluble particles, such as clay, settle as a deposit on the bottom of a container (Fig. 3a). If a pure deflocculant is added, a deflocculated medium

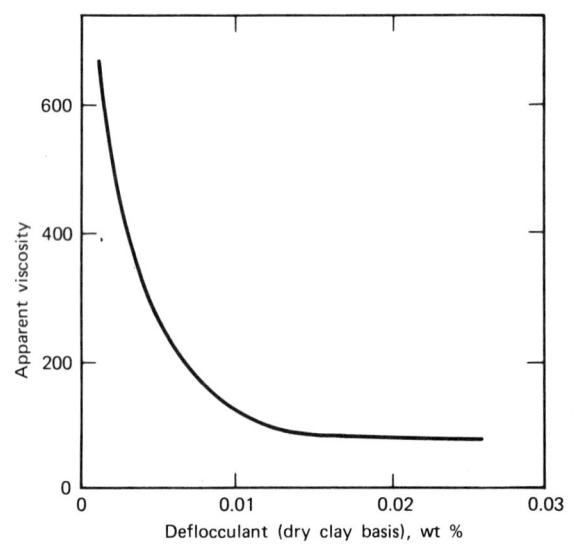

Figure 2. Reduction in apparent viscosity of a clay slurry as a function of the dose of sodium tripolyphosphate deflocculant added. Spindle G, Stormer viscometer, 300 rpm.

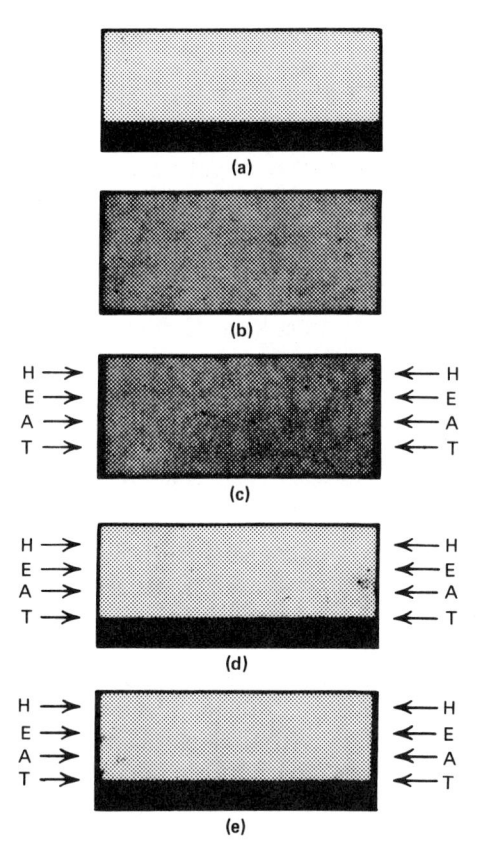

Figure 3. (a) Suspension of insoluble particles without dispersants. (b) Suspension treated with deflocculant. (c) Supersaturated scaling salt added to heated vessel containing suspension treated with deflocculant. (d) Supersaturated scaling salt and suspension treated with antiscalant. (e) Supersaturated scaling salt and suspension treated with antiprecipitant.

is obtained as shown in Figure 3**b**. In theory, if a pure antinucleation agent is added, the result is as in Figure 3**a** instead of as in Figure 3**b**. If heat is applied to the sides of the container, a scaling salt, eg, $CaCO_3$, could become supersaturated owing to its inverse solubility. If a pure deflocculant is added, the suspended particles would remain deflocculated as in Figure 3**c**, but adherent scale would form on the heat transfer surfaces. The water would also be turbid with precipitated, dispersed $CaCO_3$ particles. If only a pure scale inhibitor is added, ideally the suspended particles would settle to the bottom, as in Figure 3**d**, and the turbidity would remain, but no scale would form at the walls. If only a pure antiprecipitant is added (Fig. 3**e**), the haze is reduced, but suspended matter settles out and scale forms on the walls. In practical situations it is possible to use the wrong dispersant, ie, a deflocculant where an antiscalant is required, and never achieve adequate performance or only achieve performance at an exorbitant cost.

Detergent Builders. The major application of condensed phosphates (such as STP) is as detergent builders. Van Wazer (10) describes builders by function as affording peak surfactancy at lower surfactant concentrations and, with certain inorganic salts such as the polyphosphates, actually increasing a surfactant's peak performance even at reduced concentrations. Builders improve surfactant performance by acting as deflocculating agents, buffering agents, and water-softening agents, particularly as related to the removal of hardness which retards detergency (see Surfactants and detersive systems). Builders perform the following specific functions: (1) deflocculate soil aggregates coagulated by multivalent cations, especially Ca^{2+}, Mg^{2+}, Fe^{3+}, etc; (2) improve soil release from fabrics, ie, they strongly chelate with cations and hence break up the cation-enhanced attachment of soils to substrates; for example, sebum soil removal from fabrics dramatically decreases in waters having greater than 20 ppm calcium hardness unless sufficient builders are present (13); and (3) prevent spotting on hard shiny surfaces by preventing deposition of insoluble calcium soaps.

Dispersant Materials

Condensed Phosphates. Probably the most crucial development in the history of dispersants was the elucidation of the performance characteristics of the condensed phosphates (1). The term condensed phosphates refers to any dehydrated, condensed orthophosphate for which the $H_2O:P_2O_5$ ratio is less than 3:1 (10).

$$2\ Na_2HPO_4 \rightarrow Na_4P_2O_7 + H_2O$$

disodium-orthophosphate [7558-79-4]   tetrasodium-pyrophosphate [7722-88-5]

The most active threshold inhibitors for many applications (14) are amorphous polyphosphate salt glasses which contain long phosphate chains. Today, virtually all condensed phosphates are prepared by calcining the appropriate ratio of $Na_2O:P_2O_5:H_2O$ under carefully controlled conditions to form the desired compositions directly. They are used as oil field drilling mud thinners (15) (see Petroleum), detergent builders (16), and scale inhibitors for recirculating cooling water (17) (see Phosphoric acids and phosphates).

Despite the almost universal applicability of the condensed phosphates, their ease of manufacture and general economical availability, the total demand for them in the 1970s has waned. Total tripolyphosphate production, for example, peaked in 1970 at 1,100,000 metric tons and declined to less than 400,000 metric tons in 1977 (18). Probably the biggest single blow to phosphates was the discovery of their role in the eutrophication of enclosed bodies of water (19) (see Water pollution). Current EPA guidelines for industrial effluents restrict phosphorus to ≤0.05 ppm maximum at any time (20). However, polyphosphates as a group also suffer reversion or degradation to orthophosphates through hydrolysis in neutral or acidic solution. Orthophosphate is not an active dispersant and orthophosphate anions may actually favor coagulation or deposition under conditions of high calcium levels. Thus, as conditions of use become more severe, reversion of polyphosphates becomes a more significant problem. Further influencing the use of polyphosphates as dispersants is the large number of competing applications for phosphates, eg, in food products (in soft drinks, dentifrices, pet foods, etc), in flame retardants (21) for textiles, in insecticides and their biggest use in fertilizers (qv). For example, during 1974/1975, fertilizer demand skyrocketed carrying the price of phosphate rock from $6.25 per metric ton in 1973 to $25.20/t in 1975 (22). Though, as of 1978, phosphate prices are lower this sensitivity to fertilizer demand has stimulated interest in alternative dispersants. In addition, better products have frequently been developed for particular applications. These materials are discussed below.

Organic Polymers. Table 2 lists dispersant materials by types and typical applications and trademarked names for each class of materials. Organic phosphonates have significantly cut into the usage of condensed phosphates in antinucleation applications (recirculating cooling waters, boilers, etc) where, despite their high unit price, they show superior cost-performance in use because of superior thermal stability. Organic phosphonates are not particularly favored, however, as deflocculants. Both polyphosphates and phosphonates are preferred in applications where their performance as corrosion inhibitors is a useful adjunct property (see Corrosion).

Polyacrylates are probably the most flexible dispersant products, they are easily produced in a variety of molecular weights and degrees of anionic charge (see also Acrylamide polymers). They typically cost less than the phosphonates but more than polyphosphates. Polyacrylates show excellent stability and deflocculation. However, they provide no corrosion inhibition. Sales of all polyacrylate-based inhibitors remains well below that of the phosphorus-based materials.

Sulfonated polymers are particularly favored for many deflocculant applications, eg, ceramics, portland cements, clays, etc.

Polymaleates generally show similar properties to polyacrylates, but are particularly useful for evaporative technologies (see Maleic anhydride).

Natural product-derived dispersants, such as tannins, lignins, alginates, etc, are still widely used as drilling mud thinners (see Petroleum) or specialty dispersants where their low toxicity is a crucial property, eg, in shipboard boilers.

Builders. Because of the great economic impact of builders on the total dispersant market (see Economics below), alternatives to the condensed phosphates as builders merit detailed discussion.

Inorganics. Common inorganic salts, such as sodium sulfate, sodium carbonate, sodium silicate, etc, are used as builders, eg, to increase electrolyte levels and provide buffering; however, none of these materials sequester calcium or have any deflocculating strength and thus, strictly speaking, cannot function alone as complete builders. Insoluble sodium aluminum silicates (zeolites) act as *in situ* ion-exchange resins in cleaning formulations, ie, when added to detergents they remove calcium ions from the medium (see Molecular sieves). Further, they possess the ability to adsorb molecularly dispersed substances, eg, dyestuffs or iron oxides and prevent their redeposition, ie, they reduce staining of fabrics. However, since they possess no deflocculating or soil releasing power they cannot function alone as detergent builders. Zeolites, however, reduce the amount of deflocculants required in a detergent formulation and, in fact, zeolite-based detergents may become a significant factor in the future (18).

Organics. A possible organic STP replacement is trisodium nitrilotriacetate [5064-31-3] (NTA). It is an excellent chelating agent for calcium ions and is compatible with carbonate buffers; NTA is not a strong deflocculant (66). NTA has become a significant builder component in Canada. Though no studies have definitively shown carcinogenicity or teratogenicity, NTA has been reported to give a positive result in the Ames mutagenicity test and hence its use in the U.S. is restricted (67). Recently, oxygen-based analogues of NTA, such as 1-oxacyclopropane-*cis*-2,3-dicarboxylic acid [16533-72-5] (68), have been suggested as builders because of their strong chelation power for calcium as well as biodegradable ether linkages. Certain carboxylic polyelectrolytes identical to or analogous with deflocculants or antiprecipitants can function as detergent builders, eg, polyacrylates, polymaleates, polymethacrylates, etc. They are superior to STP, eg, in lime-soap redissolving power, sequestration of

Table 2. Dispersants (Table of Selected Examples)

Dispersant class with representative Manufacturers[a]	Trademarks[b]	Application	Reference
Condensed phosphates		recirculating cooling water	1
Calgon Chemical Corp.	Calgon	industrial boilers	23
FMC Corp.	[c]	evaporative desalination	24
Monsanto Chemical Corp.	[c]	desalination by reverse osmosis	25
		black liquor evaporators	26
		drilling mud thinner	27
		clay processing deflocculant	11
		pigment deflocculant	28
		minerals processing dispersant	29
		detergent builder	13
Polyacrylates—as $-[CH_2-\underset{\underset{Y}{\overset{\|}{C=O}}}{\overset{X}{\overset{\|}{C}}}]_x-$ where X = H, CH$_3$ Y = NH$_2$, OH, OCH$_3$, OC$_2$H$_5$, O$^-$Na$^+$, etc or copolymers with compatible monomers		industrial boilers	30
		marine boilers	31
		recirculating cooling water	3
		evaporative desalination	32
		industrial uses: reverse osmosis	33
		black liquor evaporators	34
		oil field: down-hole scale inhibition	35
		flue gas scrubbers	36
		drilling mud thinners	37
		clay processing deflocculants	11
		pigments deflocculants	38
		minerals processing dispersant	29
ALCO Chemical Corp.	Alco	leather finishing	39
Allied Colloid, Ltd.	Antiprex	detergent builder	40
American Cyanamid Co.	Cyanamer		
Goodrich Chemical Co.	Goodrite		
Aquaness Div.	Calnox		
Magna Chemical Co.			
Organic phosphonates, including methylenephosphonates and other structures possessing ionizable -PO$_3$H$_2$ structures		industrial boilers	41
		recirculating cooling water	42
		single stage evaporators	43
		oil field: down-hole scale inhibition	44
Monsanto Chemical Co.	Dequest	flue gas scrubbers	36
Petrolite Div.	[c]	drilling mud thinners	45
Tretolite Corp.		clay processing deflocculant	46
Textilana Div.	Fostex		
Henkel GmbH			
Polysulfonates[d] ie, linear polymeric structures having pendant -SO$_3$H (or -SO$_3$) groups, eg, lignosulfonates, petroleum sulfonates, poly(styrene sulfonates), etc		portland cement dispersants	47
		scale inhibition (oil field)	48
		black liquor transfer lines	49
		pigments dispersant	50
		nonwoven textiles	51
		lime-soap dispersant	52
National Starch Corp.	Versa-TL		
Sulfonated polycondensates, including naphthalene: formaldehyde sulfonated polycondensates and similar materials		ceramics dispersants	53
		pigments dispersant	53
		leather finishing	54
		clay processing	11
Diamond-Shamrock Corp.	Lomar	detergent builders	55

Table 2 (*continued*)

Dispersant class with representative Manufacturers[a]	Trademarks[b]	Application	Reference
Polymaleates, including		recirculating cooling water	56
$$\begin{array}{c}-\!\!\!\!\left[\mathrm{CH}\!-\!\!\!-\!\mathrm{CH}\!-\!\right]_{\!x}\\ \;\;\;\;\vert\quad\;\;\;\vert\\ \;\;\;\;\mathrm{C}\!\!=\!\!\mathrm{O}\;\;\;\mathrm{C}\!\!=\!\!\mathrm{O}\\ \;\;\;\;\vert\quad\;\;\;\vert\\ \;\;\;\;\mathrm{Y}\quad\;\;\mathrm{Y}\end{array}$$		evaporative desalination	57
		black liquor evaporators	26
		detergent builder	40
where Y = OH, ONa and copolymers with compatible monomer, including styrene, acrylic acid, etc			
Ciba Geigy Corp.	Belgard		
Tannins, lignins, glucosides,		industrial boilers	58
gluconates, alginates, etc		marine boilers	43
ie, polymeric materials		recirculating cooling water	59
derived from natural products	[d]	black liquor evaporators	26
Georgia Pacific Corp.	[c]	beet sugar evaporators	60
Kimberly Clark Corp.	[c]	oil field: down-hole scale inhibition	61
Marathon Paper Co.	Marasperse		
Westvaco Corp.	[c]	drilling mud thinners	62
		clay processing deflocculants	63
Other materials: phosphoester		recirculating cooling water	64
Nalco Chemical Co.	[c]	oil field: down-hole scale inhibition	65
Phospho glasses			
		clay processing deflocculants	10
		detergent builders	10

[a] Representative list only: some significant manufacturers may have been omitted. Many dispersants are sold as formulated products by local service organizations catering to specific applications.
[b] Representative list of typical names where manufacturers sell to end-users. Hundreds of names are extant for formulated products.
[c] Material sold either generically or by code number.
[d] Major product (in drilling muds) are heavy metal (ferrous or chrome) lignosulfonates which are polymers derived by reaction with waste lignins in paper manufacture.

calcium, or deflocculation of soils (40). They are compatible with sodium carbonate buffers. Mechanistically, the polyelectrolytes possess flexible backbones which allow the anionic groups to form stable calcium complexes. However, since the carbon–carbon backbones of the carboxylates are resistant to biodegradation, the potential ecological effects of large-scale usage are unknown and difficult to predict (66). Alpha-hydroxy acrylates possess a backbone that shows limited biodegradability (~10% that of low mol wt ether carboxylates) while preserving powerful polymeric builder strength (69). Generally, dispersancy of these molecules approaches that of STP. β-Naphthalenesulfonic acid–formaldehyde condensates can be used as organic builders in carbonate-based detergents (55). Lime-soap dispersing agents can also be used to produce phosphate-free soap-based detergents (52,70).

Economics of Dispersants

Because of the diversity of applications, the total worldwide market for dispersants is difficult to estimate. Builders represent the largest dispersant market: the 1978 U.S. market for all builders, ca 400,000 metric tons (40% polyphosphate, 40% silicates, 15% carbonates, 5% others), was in excess of $100,000,000. All other applications for dispersants in the U.S. (boilers, recirculating cooling water, clay and pigment processing aids, drilling mud thinners) can be approximated as $50,000,000 in 1979 based on implementation of EPA discharge standards (71). Worldwide total builder usage is approximately four times U.S. usage. Worldwide usage of dispersants (other than builders) is probably less than twice U.S. usage, with virtually no application outside Japan and Western Europe except in resource recovery (mining, desalination). The major dispersant commodity remains polyphosphate in all aspects of dispersant application. Phosphonates are significant in recirculating cooling waters (ca $15,000,000), natural products (tannins, lignosulfonates, etc) in drilling muds (ca $25,000,000), with polyacrylates and other polymers making inroads in all markets (ca $15,000,000).

Uses

Antinucleation Agents. *Boiler Treatments.* Current technology to reduce boiler scale has evolved along three main paths (72): (*1*) Phosphate cycle with polymer dispersants. The dispersants form nonadherent sludges which are easily removed during boiler blowdown. Typical dispersants used on phosphate cycle are polyacrylates or polymethacrylates of 2000–10,000 mol wt (30). Mixtures of higher mol wt natural product-based polyelectrolytes such as alginates, with very low mol wt (<1000) organic acid are recommended because they form highly peptized sludges (58). (*2*) Softened water with polymer–chelant combinations. Scale protection is maintained by very low levels of threshold inhibitors. A preferred treatment is a blend of a maleic anhydride copolymer with a chelant such as NTA (73). The use of chelants with organic phosphonate threshold inhibitors is also recommended (41). (*3*) Ultra-high purity water (no dispersants). Used at extremely high pressures where only volatile inhibitors are safe to use.

Recirculating Water. A typical recirculating cooling water system for a 1000 MW electric generating station as shown in Figure 4 (74) requires approximately 700 L/s (10,000 gal/min) of make-up water flow. At such a tremendous flow rate only the most rudimentary of pretreatments are economic, hence make-up water quality varies tremendously depending on its source. Historically, treatments for such recirculating cooling waters were first put on a sound footing with the use of the Langelier Index (75). The Langelier Index pinpoints the importance of high pH in encouraging $CaCO_3$ scaling and low pH in favoring iron corrosion. The electrochemical theory of corrosion (qv), ie, the importance of the passivating layer in preventing corrosion, provided the basis for the now traditional treatment regimen. Sulfuric acid is fed to lower the system pH, which reduces scaling, and chromate-zinc phosphate-based formulations are added to reduce corrosion. Depending on the exact effluent restrictions, there are many modifications of this technology. Where chromate and phosphorus discharges are delimited but zinc is not, the use of zinc with either organic phosphonate/organic polymer mixtures (59) or phosphoesters (64) or low mol wt polymaleates is suggested (56). At moderately alkaline pH (7.6–8.4) several metal-free treatments based on superior scale inhibitors are feasible, particularly organic phosphonates or blends of phosphonate and polyacrylate (76).

Desalination. Evaporative desalination technology has evolved to a high degree of sophistication; multistage flash evaporators can achieve high efficiencies. Desalination by reverse osmosis (RO) has also become a significant technology. Further,

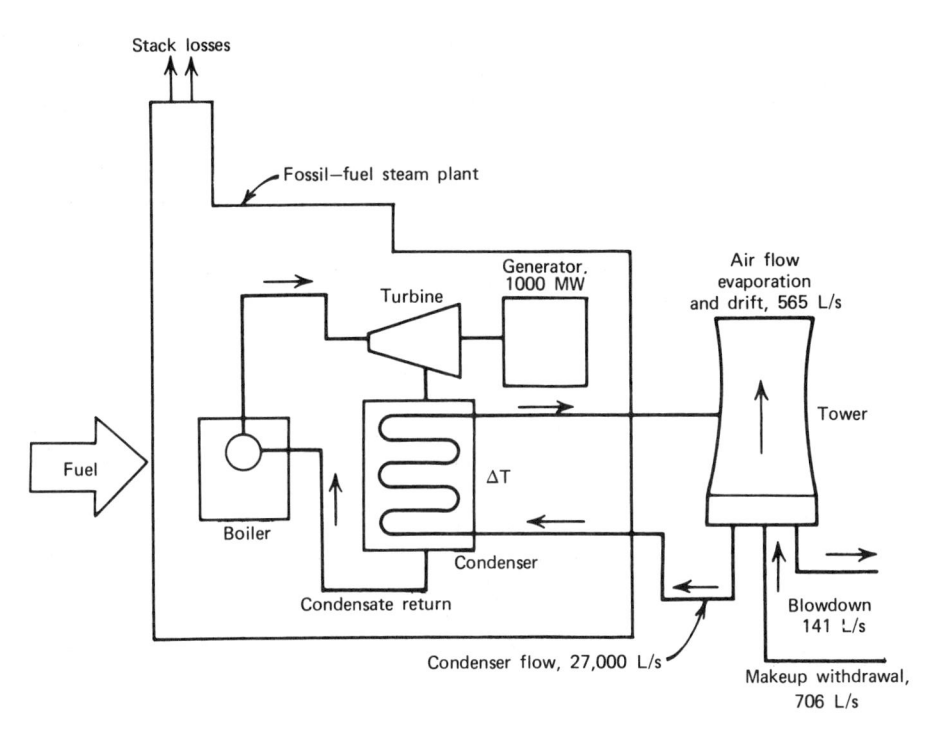

Figure 4. 1000 kW fossil-fuel Steam-Electric Plant: recirculating cooling water system. Courtesy *Power*.

RO is used currently in many industrial applications to purify waste liquors. Scaling is a major problem in all these technologies. Polyacrylates of low molecular weight are extremely effective inhibitors to reduce scale formation in operation of RO for pulp and paper wastes (33). Polyphosphates have also shown excellent performance (25) (see Water; Reverse osmosis).

Oil Field Uses. Scale or precipitation inhibitors are utilized in several aspects of oil-field operations. A problem unique to oil-field applications is the blockage of spaces in porous rock formations, usually with $CaCO_3$. $BaSO_4$ scaling can occur down-hole during secondary oil recovery (water-flooding). Preferred treatments include the traditional use of polyphosphates and natural polymers (61,77). More recently, phosphate glasses of certain composition have proven to be useful (65). Polyacrylates are superior as $BaSO_4$ inhibitors down-hole (35), but organic phosphonates are the best $CaCO_3$ inhibitors. Above 175°C, no inhibitors are reliable. One unusual technology occasionally tried is the generation of an *in situ* down-hole inhibitor (44). When deep well waters vent to the surface, the combination of rapid temperature and pressure change and high turbulence often causes spectacular impingement scale deposits (usually $CaCO_3$). Preferred treatments include a wide variety of threshold inhibitors (low mol wt polyacrylates, phosphonates, polyphosphates) in combination with petroleum sulfonates which help to emulsify the oil in the water and prevent the nucleation of scale (48).

Miscellaneous. In the manufacture of paper (qv) there is a large number of operations, from the preparation of the kraft pulps to the coating of the finished papers, where dispersants are used. Scaling can occur in pulp digester and liquor evaporators. Antinucleation agents have proven beneficial in reducing adherence of scale to digester walls and screens (by crystal distortion) and reducing maintenance requirements. However, to identify optimum antinucleation agents for each application in the paper mill usually requires extensive testing and evaluation on site (72).

In the production of drinking water aboard ship or in the concentration of edible juices, eg, beet sugar juices, evaporator scales may form (see Fruit juices; Sugar). However, the requirement that either the product water or the concentrate, or both, be safe to consume puts another constraint on antinucleation agents. Materials that have been shown to be nontoxic compositions include coconut oil fatty acid esters of methyl glucosides for beet sugar evaporators (60). Compositions of polyacrylates, organic phosphonates, tannins, and lignins (qv) provide safe distilled water from shipboard evaporators (31,43,78).

With the growth of air pollution controls, the use of scrubbers to remove sulfur dioxide from flue gases has become more significant (see Sulfur recovery). A key problem is deposition of $CaSO_4.2H_2O$ in these systems. Generally, equal levels of phosphonate–polyacrylate at up to 50 ppm dosage levels are effective; an optimum molecular weight exists for the polyacrylate, an optimum structure for the phosphonate (36).

Deflocculants. *Clay Deflocculation.* Clays are earthy materials which, when wet, develop plasticity (62). Their unique, flat, platelet structure is depicted in Figure 5. Effective clay deflocculants can absorb preferentially at the positively charged sites on the edges which destabilizes the ordered clay aggregates leading to random, dispersed structures, mutual repulsions of the individual clay particles, and greatly reduced slurry viscosity (79). For the polyphosphates, the longer the condensed phosphate chain, the greater is the degree of adsorption on clay substrates (see Clay).

Drilling Mud Thinners and Portland Cement Dispersants. The primary purpose of mud thinners is to maintain the clay-based drilling muds in a deflocculated state, decrease viscous drag on the drilling bit, provide for increased weighting of clay at constant viscosity, improve clay structure, and decrease fluid loss. Similarly, portland

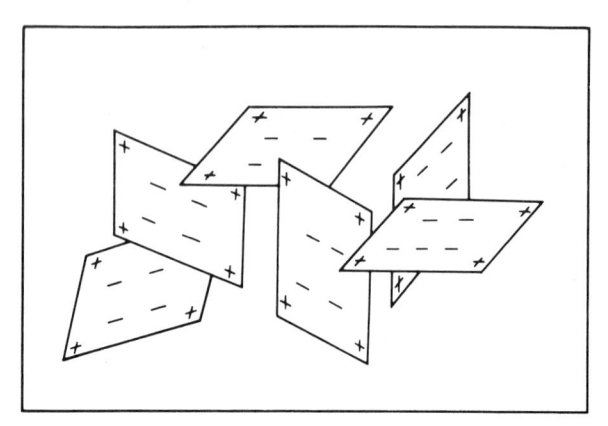

Figure 5. Simplified picture of kaolin clay platelets: house of cards structure.

cement dispersants allow increased weighting of the cement at lower water content, reduce viscosity during handling and improve strength and quality of the cement after drying. Many diverse deflocculants have been used in these applications (see Cement; Petroleum, drilling fluids).

Pigment Preparation. Various coatings and pigments used in paper and paint manufacture require the use of organic deflocculants during preparation. TiO_2, $CaCO_3$, $CaSO_4$, ZnO, and kaolin are the most frequently employed pigments. Kaolin is predispersed by use of silicates (46), phosphonates (80) and alkali alginates (63). Calcium carbonate is best pretreated with polysulfonates (50), or acrylic-based polymers with (28) or without (81) polyphosphates. Titanium dioxide is treated with potassium tripolyphosphate [13845-36-8]/polyacrylate combinations (38) (see Pigments).

Mineral Processing. For economic utilization of mined ores the mineral values must be liberated from the gangue. Liberation requires grinding of the ore to a sufficiently fine size so that each particle in the ore is chemically discrete and homogeneous, the mineral particles distinct from the gangue particles. Deflocculants reduce agglomeration of the particles during beneficiation processes. Dispersants (low mol wt polyacrylates) limit energy usage during grinding (82). Dispersants can optimize flotation (qv) separations (83). Selective flocculation usually requires dispersants. Caustic-silicate blends are the only dispersant systems presently employed commercially (84); however, so-called selective dispersants are reported to selectively interact with copper values facilitating their separation from the gangue (85). Similarly, a complex process for recovery of phosphate from Tennessee rock slimes features fine grinding in a mixed dispersant system of ammonia with tetrasodium pyrophosphate [7722-88-5] (29).

Recirculating Cooling Water. A key application of dispersants as deflocculating agents is in recirculating waters to prevent settling of suspended particles (clays, salts, sand, corrosion particles, etc) within distribution networks. The solids rarely exceed 1000 ppm and usually are ca 100 ppm; substrate chemical properties can often vary dramatically. Optimum reagents often are of much higher mol wt than in dense media chiefly because there is very little probability of flocculating particles in such dilute media and the higher adsorption and greater steric repulsion for the larger particles provides increased dispersion. Treatments are best selected by trained water treatment specialists (72).

Miscellaneous. Many industrial processes require the handling of high-solids slurries of unusual composition. For example, wet scrubbers produce slurries containing $CaSO_4$, $CaCO_3$, fly ash, etc. Fertilizer plants must handle slurries of high phosphate content. Deflocculants are widely used to facilitate their handling (86). Deflocculants are also employed in textile manufacture to facilitate improved formation of rayon fabric in the production of nonwoven textiles (qv) (51), or in tanning, coloring, and finishing (39,54).

Future Dispersant Markets (Ecological and Energy Considerations)

The future use pattern of dispersants will be determined by the following factors:

Environmental and energy considerations almost always increase demand for dispersants, eg, in removal of SO_2 from flue or stack gases and reuse of industrial process waters to conserve water and energy.

Detergent demand and hence demand for builders will rise as developing nations seek to improve sanitation.

Future energy shortages will put a premium on recovery of waste heat. All heat recovery systems require some form of heat exchange and favor use of dispersants in cooling water treatment. Energy shortages may also restrict chemical feedstocks to organic intermediates used for manufacture of many synthetic dispersants.

Toxic substances control legislation favors development of better solids-handling systems; many of these processes require dispersants. Likewise, dispersants themselves must meet toxic substance requirements. This is perhaps the biggest factor in designing replacements for sodium tripolyphosphate as a detergent builder.

Raw materials shortages, especially of high-grade metal ores and oil favor better recovery processes, eg, mineral beneficiating and tertiary oil recovery. These newer processes often require higher levels and better performance of dispersant products.

BIBLIOGRAPHY

1. G. B. Hatch and A. Rice, *Ind. Eng. Chem.* **31,** 51 (1939).
2. C. R. McCartney and A. G. Alexander, *J. Colloid Sci.* **13,** 383 (1958).
3. CYANAMER® P-70 and P-70D Antiprecipitant, American Cyanamid Co., 1976.
4. F. V. Williams and R. A. Ruehrwein, *J. Am. Chem. Soc.* **79,** 4898 (1957).
5. C. H. Nestler, J. Coll. *Interface Sci.* **26,** 10 (1968).
6. R. M. Goodman and A. M. Schiller, paper presented before Cooling Tower Institute Winter Meeting, Houston, Texas 1976; R. M. Goodman, paper presented before Division of Environmental Chemistry, American Chemical Society National Meeting, San Francisco, Calif., 1976.
7. U.S. Pat. 4,072,607 (Feb. 7, 1978), A. M. Schiller, R. M. Goodman, and R. E. Neff (to American Cyanamid Co.).
8. J. L. Meyer and G. H. Nancollas, *Calc. Tiss. Res.* **13,** 295 (1973).
9. A. E. Austin and co-workers, *Desalination* **16,** 345 (1975).
10. J. R. Van Wazer, *Phosphorus and Its Compounds,* in two volumes, Interscience Publishers, Inc., New York, 1958.
11. R. F. Conley, *J. Paint Technol.* **46**(594), 51 (1974).
12. A. S. Mallett and R. L. Craig, *Tappi* **60**(11), 101 (1977).
13. E. A. Matzner and co-workers, *Tenside Deterg.* **10**(3), 119 (1973).
14. U.S. Pat. 3,130,152 (Apr. 21, 1964), R. J. Fuchs (to FMC Corp.).
15. U.S. Pat. 2,361,760 (Oct. 31, 1944), A. D. Garrison (to the Texas Co.).
16. U.S. Pat. 1,956,515 (Apr. 24, 1934), R. E. Hall (to Hall Laboratories).
17. U.S. Pat. 3,130,002 (Apr. 21, 1964), R. J. Fuchs (to FMC Corp.).
18. *Chem. Eng. News,* 11 (May 22, 1978).
19. D. W. Schindler, *Science,* **184,** 897 (May 24, 1974).
20. *Fed. Regist.* **39**(196) Part III (Oct. 8, 1974).
21. Jpn. Kokai 75 45,097 (Apr. 22, 1975), S. Amane and co-workers (to Asaki Electro-Chemical Industry).
22. U.S. Bureau of the Census, *Statistical Abstract of the United States, 1977,* 98th ed. Washington, D.C., 1977, p. 744.
23. O. Rice and E. Partridge, *Ind. Eng. Chem.* **31,** 58 (1939).
24. M. H. Ali El-Saie, *3rd Int. Symp. Fresh Water, Sea Dubrovnik,* Vol. 1, A. Delyannis and E. Delyannis, eds., *Working Party on Fresh Water from the Sea, Athens, Greece,* 1970, pp. 405–416.
25. S. R. Redda, J. Glater, and J. W. McCutchon, *National Water Supply Improvement Assoc. J.* **3,** 11 (1976).
26. U.S. Pat. 3,298,734 (Dec. 6, 1966), R. S. Robertson (to Nalco Chem. Co.).
27. U.S. Pat. 2,368,823 (Feb. 6, 1945), A. D. Garrison (to the Texas Company).
28. Brit. Pat. 1,414,964 (Nov. 19, 1975), M. Hancock and D. G. Jeffs (to English Clays Lovering Pochin and Co., Ltd.).

29. U.S. Pat. 3,314,537 (Apr. 18, 1967), E. W. Greene and J. B. Duke (to Minerals and Chemicals Philipp Corp.).
30. U.S. Pat. 3,492,240 (Jan. 27, 1970), W. P. Hettinger, Jr. (to Nalco Chem. Co.).
31. U.S. Pat. 3,514,376 (May 26, 1970), M. L. Salutsky (to W. R. Grace and Co.).
32. U.S. Pat. 4,085,045 (Apr. 18, 1978), D. S. Song and co-workers (to American Cyanamid Co.).
33. U.S. Pat. 3,589,998 (June 29, 1971), H. L. Rice, A. Cizek, and M. D. Thaemor (to Milchem, Inc.).
34. U.S. Pat. 3,463,730 (Aug. 26, 1969), R. B. Booth and L. C. Mead (to American Cyanamid Co.).
35. O. J. Vetter, *J. Pet. Technol.* **997,** Aug. (1972).
36. W. L. Harpel and co-workers, "The Chemistry of Scrubbers," *Combustion* **47**(9), 33 (Mar. 1976).
37. U.S. Pat. 3,434,970 (Mar. 25, 1979), F. H. Siegele and co-workers (to American Cyanamid Co.).
38. Can. Pat. 935,255 (Feb. 12, 1971), E. L. Pyves (to Canadian Titanium Pigments, Ltd.).
39. Brit. Pat. 1,112,699 (May 8, 1968), (to BASF).
40. J. F. Schaffer and R. T. Woodhams, *Ind. Eng. Chem. Prod. Res. Dev.* **16**(1), 3 (1977).
41. U.S. Pat. 3,666,664 (May 30, 1972), W. F. Loreng and R. A. Beiner (to Nalco).
42. U.S. Pat. 3,434,969 (Mar. 25, 1969), P. H. Ralston (to Calgon Corp.).
43. U.S. Pat. 3,505,238 (Apr. 7, 1970), R. W. Liddell (to Calgon Corp.).
44. U.S. Pat. 3,633,672 (Jan. 11, 1972), C. F. Smith and T. J. Nolan (to Dow Chemical Co.).
45. U.S. Pat. 3,346,487 (Oct. 10, 1967), J. W. Lyons and R. R. Irani (to Monsanto Chemical Co.).
46. Fr. Demande 1,509,449 (Jan. 12, 1968), (to Henkel and Cie GmbH).
47. U.S. Pat. 3,448,096 (June 3, 1969), D. W. Reed (to Canadian International Paper, Inc.).
48. U.S. Pat. 2,592,511 (Apr. 8, 1952), J. F. Chittum (to California Research Corp.).
49. U.S. Pat. 3,518,204 (June 30, 1970), G. D. Hansen, Jr. and E. A. Guthrie (to Betz Laboratories, Inc.).
50. U.S. Pat. 3,324,063 (June 6, 1967), A. S. Teot (to The Dow Chemical Co.).
51. Ger. Offen. 2,213,980 (Nov. 2, 1972), R. H. Doggett and R. P. Tschirch (to Lubrizol).
52. W. M. Linfield, *J. Am. Oil. Chem. Soc.* **55**(1), 87 (1978).
53. LOMAR® PW, *Nopco Technical Bulletin GEN-10,* NOPCO Chemical Co., Div. Diamond Shamrock Corp., Morristown, New Jersey, 1976.
54. M. Urban and A. Dekwara, *Pr. Inst. Przem. Skorzanego* **14,** 73 (1970).
55. F. Tokiwa and T. Imamura, *J. Am. Oil Chem. Soc.* **47,** 422 (1970).
56. U.S. Pat. 3,963,636 (June 15, 1976), A. Harris, J. Burrows, and T. I. Jones (to Ciba-Geigy).
57. M. N. Elliot, T. D. Hodgson, and A. Harris, *Proc. 4th Int. Symp. Fresh Water, Sea, Heidelberg,* Vol. 2, 1973, pp. 97–110.
58. U.S. Pat. 3,520,813 (July 21, 1970), G. D. Hansen and E. A. Gurhrie (to Betz Laboratories, Inc.).
59. U.S. Pat. 3,518,203 (June 30, 1970), E. A. Savinelli and J. K. Rice (to Drew Chemical Corp.).
60. U.S. Pat. 3,061,478 (Oct. 30, 1962), S. G. Kent (to Hodag Chemical Corp.).
61. U.S. Pat. 2,777,818 (Jan. 15, 1957), M. Gambill (to United Chemical Corp. of N.M.).
62. W. F. Rogers, *Composition and Properties of Oil Well Drilling Fluids,* Gulf Publishing Company, Houston, Tex., 3rd ed., 1963.
63. Brit. Pat. 1,250,516 (Oct. 20, 1971), J. Cameron and R. C. D. Wylic (to ICI Ltd.).
64. U.S. Pat. 3,873,465 (Mar. 25, 1975), I. S. DiSimone (to Nalco Chem. Co.).
65. U.S. Pat. 3,288,217 (Nov. 29, 1966), P. H. Ralston (to Calgon Corp.).
66. E. A. Matzner and co-workers, *Tenside Deterg.* **10**(5), 239 (1973).
67. *Task Force Report to the Great Lakes Research Advisory Board of the International Joint Commission on the Health Implications of N.T.A.* P. D. Foley, Chairman, May 1977.
68. U.S. Pat. 3,796,223 (Oct. 30, 1973), T. H. Pearson and co-workers (to Ethyl Corp.).
69. U.S. Pat. 3,839,215 (Oct. 1, 1974), J. Mulders (to Solvay and Cie).
70. Ger. Offen. 2,048,066 (Apr. 22, 1971), D. G. S. Hirst (to Proctor and Gamble).
71. *Environmental Science and Technology,* A. C. Gross, May 1974, p. 414.
72. *Handbook of Industrial Water Conditions,* 7th ed., Betz Laboratories, Inc., Trevosa, Pa., 1976.
73. U.S. Pat. 3,549,538 (Dec. 22, 1970), C. Jacklin (to Nalco Chem. Co.).
74. "Cooling Towers," Power Special Report, (Mar. 1973).
75. W. F. Langelier, *J. Am. Water Works Assoc.* **28,** 1500 (1936).
76. U.S. Pat. 3,663,448 (May 16, 1972), P. H. Ralston and P. Hotchkiss (to Calgon Corp.).
77. U.S. Pat. 2,890,175 (June 9, 1959), H. J. Kipps (to Mobil Oil Co.).
78. U.S. Pat. 3,293,152 (Dec. 20, 1966), L. S. Herbert and U. J. Sterns (CSIRO, Australia).
79. A. G. Loomis, T. F. Ford, and J. F. Fidrian, *Trans. AIME* **142,** 86 (1941).
80. U.S. Pat. 3,713,859 (Jan. 30, 1973), F. Hoover and G. Di M. Sinkovitz (to Calgon).
81. Ger. Offen. 2,604,448 (Aug. 26, 1976), L. Plumet (to Solvay et Cie).

82. R. Klimpel and W. Mantroy, paper presented at 1977 SME Fall Meeting, St. Louis, Mo., AIME 1977.
83. U.S. Pat. 2,740,522 (Apr. 3, 1956), F. M. Aimone and R. B. Booth (to American Cyanamid Co.).
84. J. W. Villar and G. A. Dawe, *Min. Congr. J.* **40** (Oct. 1975).
85. Y. A. Attia, *Int. J. Miner. Process.* **4,** 209 (1977).
86. U.S. Pat. 3,657,134 (Apr. 18, 1972), T. M. King and H. L. Vandersall (to Monsanto Company).

General References

Handbook of Industrial Water Conditioning, 7th ed., Betz Laboratories, Inc., Trevose, Pa., 1976.

J. R. Van Wazer, *Phosphorus and Its Compounds,* in two vols., Interscience Publishers, Inc., New York, 1958.

W. F. Rogers, *Composition and Properties of Oil Well Drilling Fluids,* 3rd ed., Gulf Publishing Co., Houston, Tex., 1963.

E. E. Huebotter and G. R. Gray, "Drilling Fluids," A. Standen, ed., *ECT* 2nd ed., Vol. 7, 1965, pp. 287–306.

S. F. Adler and F. H. Siegele, "Drilling Mud Additives" in N. M. Bikales, ed., *Encyclopedia of Polymer Science and Technology,* Interscience Publishers, a division of John Wiley & Sons, Inc., Vol. 5, 1966, pp. 140–153.

A. M. Schwartz, "Detergency" in A. Standen, ed., *ECT,* 2nd ed., Vol. 6, 1965, pp. 853–845.

G. D. Parfitt, *Dispersion of Powders in Liquids,* Elsevier Publishing Co., Ltd., New York, 1969.

J. A. Szilard, *Descaling Agents and Methods,* Noyes Data Corp., Pack Ride, N.J., 1972.

J. Am. Oil Chem. Soc. **55**(1) (1978), for a complete review of detergents.

RICHARD M. GOODMAN
American Cyanamid Co.

DISPERSING AGENTS.   See Sulfonic acids.

DISTILLATION

Distillation is broadly defined as the separation of more volatile materials from less volatile materials by a process of vaporization and condensation. In engineering terminology, the separation of a liquid from a solid by vaporization is considered evaporation (qv), and the term distillation is reserved for the separation of two or more liquids by vaporization and condensation.

Distillation is the most widely used method of separating liquid mixtures and is at the heart of the separation processes in many chemical and petroleum plants. There are many types of distillations; some of them are discussed in this article. Azeotropic and extractive distillation (qv) are discussed under a separate heading in this Encyclopedia. A review of the historical development of distillation is given by Holland and Lindsey (1).

This article includes discussions of: vapor–liquid equilibria; binary and multicomponent distillation processes and calculational approaches; distillation columns (plate and packed); and distillation as a separational method (see also Separation systems synthesis). There are also brief discussions of steam distillation, molecular distillation, distillation equipment costs, column control, Fractionation Research, Inc., and distillation literature reviews.

Vapor–Liquid Equilibria (VLE)

Vapor–liquid equilibrium is the relationship of the composition of the vapor phase and the liquid phase when the phases are in physical equilibrium. The driving force for any distillation is a favorable VLE; if the VLE are unfavorable, the distillation is impossible. Reliable VLE data are essential for distillation column design and for most operations involving liquid–vapor-phase contacting. VLE data come from various sources such as in-house or contract experimental measurements or the literature. If the data are not available, they may be measured or estimated. Whatever their source, some evaluation should be made regarding accuracy. The VLE for the system at hand may be simple and easily represented by an equation or, in some systems, may be so complex that they cannot be adequately measured or represented. Excellent reviews of VLE are available (2–3).

Typical VLE are shown graphically in Figure 1. The diagrams are for binary systems. Figure 1a is a typical boiling point diagram; it has the liquid–vapor equilibrium as a function of temperature at a constant pressure. The lower line is the liquid bubble-point line—the point at which a liquid on heating forms the first bubble of vapor. The upper line is the vapor dew-point line—the point at which a vapor on cooling forms the first drop of condensed liquid. The liquid and vapor compositions are conventionally plotted in terms of the low boiler. In this discussion of the binary system, L–H, the low boiler is L. The system point A has a vapor composition of y_L^A in equilibrium with a liquid composition of x_L^A at a temperature of T^A. Figure 1b is a typical isobaric phase diagram (also called x-y diagram). See ref. 4 for further discussion.

Ideal Systems. Vapor–liquid equilibrium equations are of the general type:

$$y_i = f(x_i) \tag{1}$$

For ideal liquid and vapor systems, the following simple laws hold:

$$PV' = n'R'T \quad \text{ideal gas law} \tag{2}$$

850   DISTILLATION

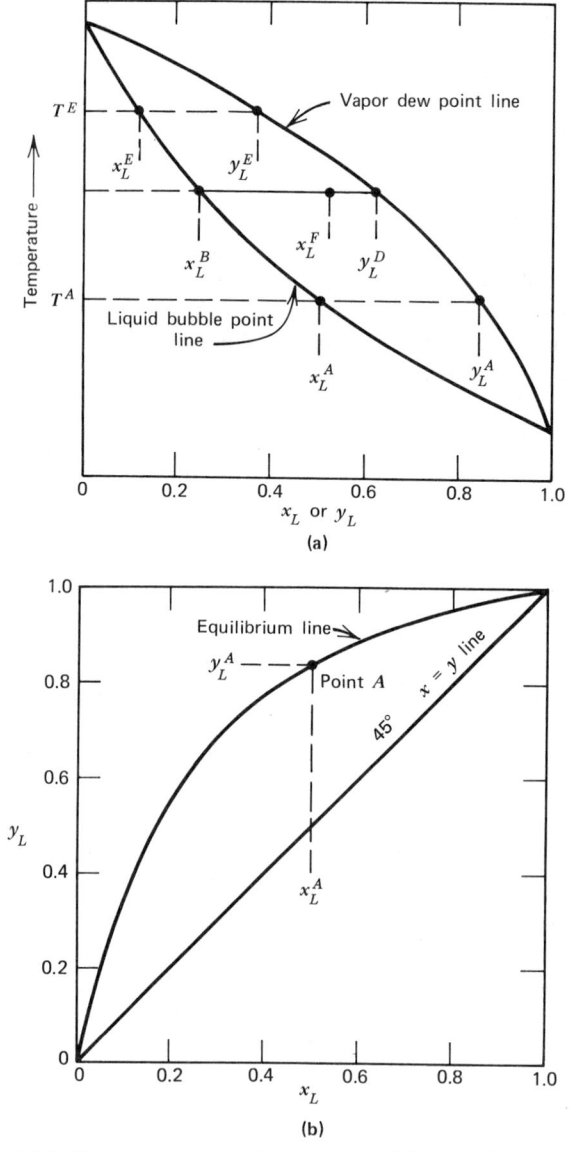

Figure 1. (a) Boiling point diagram (isobaric) and (b) phase diagram (isobaric).

$$p_i^L = x_i P_i^\circ \qquad \text{Raoult's law} \tag{3}$$

$$p_i^V = y_i P \qquad \text{Dalton's laws} \tag{4}$$

$$P = \sum_i^N p_i^V \tag{5}$$

At equilibrium between vapor and liquid:

$$p_i^V = p_i^L = y_i P = x_i P_i^\circ \tag{6}$$

$$y_i = \frac{x_i P_i^\circ}{P} \tag{7}$$

The system must also satisfy the general mass-balance relationships:

$$\Sigma x_i = 1 \text{ and } \Sigma y_i = 1 \tag{8}$$

For binary systems:

$$y_L = \frac{x_L P_L^\circ}{P} \tag{9}$$

$$y_H = \frac{x_H P_H^\circ}{P} = \frac{(1 - x_L) P_H^\circ}{P} \tag{10}$$

For ideal systems, the vapor–liquid equilibrium can be calculated from vapor pressures by means of these equations.

To facilitate distillation calculations VLE are frequently used in the form of relative volatility, α. For the binary system 1–2:

$$\alpha_{1/2} = \frac{(y/x)_1}{(y/x)_2} \tag{11}$$

$$y_1 = \frac{(\alpha_{1/2}) x_1}{1 + (1 - \alpha_{1/2}) x_i} \tag{12}$$

And for ideal systems:

$$\alpha_{1/2} = \frac{P_1^\circ}{P_2^\circ} \tag{13}$$

The relative volatility α is a direct measure of the ease of separation by distillation. If $\alpha = 1$, then separation is impossible because $y_1 = x_1$ (from eq. 12) and no phase equilibrium driving force exists. Separation by distillation becomes easier as the value of the relative volatility becomes increasingly greater than unity. Distillation separations with relative volatilities less than 1.2 are difficult, and those with relative volatilities above 2 are easy.

Real systems differ from ideal systems in two ways: Raoult's law does not hold in the liquid phase, and in the vapor phase the ideal gas law and Dalton's laws do not hold. Within engineering accuracy, Raoult's law applies for most compounds of the same homologous series and for some others. The ideal gas laws are approximated at less than 200 kPa (2 atm abs).

Nonideal Liquid-Phase Behavior. When the liquid phase is nonideal, Raoult's law (eq. 3) does not hold:

$$p_i^L \neq x_i P_i^\circ \tag{14}$$

A correction must be applied. Nonideal liquid-phase behavior is handled through activity coefficients γ, and the modified liquid-phase equation becomes:

$$p_i^L = \gamma_i x_i P_i^\circ \tag{15}$$

The development and thermodynamic significance of activity coefficients has been discussed (5–7). The liquid-phase activity coefficients are strong functions of liquid composition and temperature and, to a lesser degree, of pressure. Typical activity coefficient plots are shown in Figure 2. A system with positive deviation, ie, having activity coefficients larger than 1, is shown in Figure 2**a**; a system with negative deviation, ie, having activity coefficients smaller than 1, in Figure 2**b**.

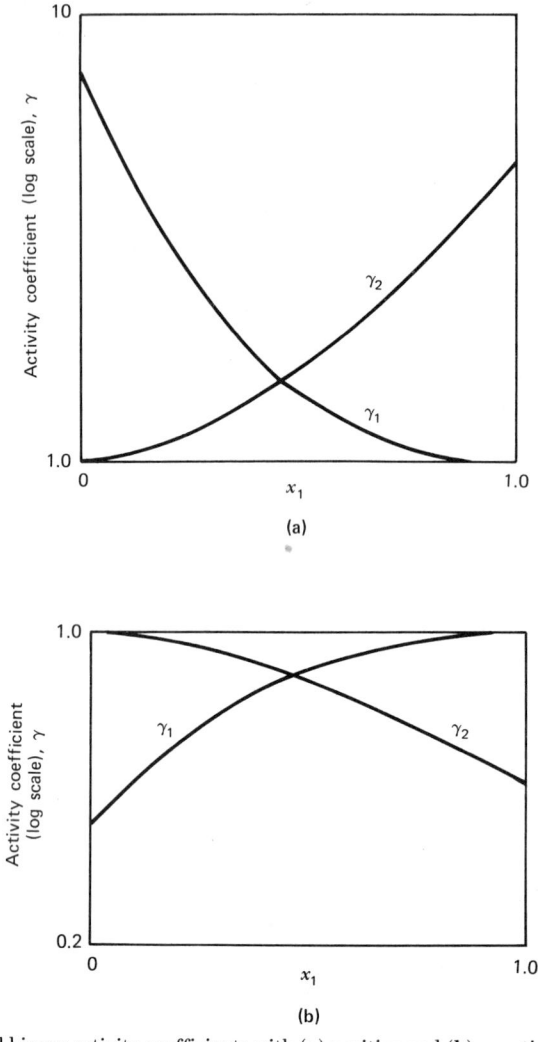

Figure 2. Typical binary activity coefficients with (**a**) positive and (**b**) negative deviations, in system 1–2, isobaric or isothermal.

Terminal activity coefficients (γ_i^∞) for some systems are given in Table 1. Systems that are of similar molecular species and size will tend to be ideal in the liquid state and to have activity coefficients close to 1, eg, hexane–heptane. As the molecular species become more dissimilar they are prone to repel each other, tend toward liquid immiscibility and have large positive activity coefficients.

If the molecular species in the liquid tend to form complexes, the system will have negative deviations and will have activity coefficients less than 1; eg, the system chloroform–ethyl acetate has terminal activity coefficients less than 0.5. Reviews of nonideal liquid behavior are available (8–12). In azeotropic and extractive distillation and in liquid–liquid extraction, nonideal liquid behavior is used to enhance component separation (see Extraction). An extensive discussion on the selection of nonideal addition agents is available (13).

Table 1. Approximate Terminal Activity Coefficients at Atmospheric Pressure[a]

Component 1	Component 2	γ_1^∞	γ_2^∞
chloroform	ethyl acetate	0.3	0.3
chloroform	benzene	0.9	0.7
n-hexane	n-heptane	1.0	1.0
ethyl acetate	ethanol	2.5	2.5
ethanol	toluene	6.0	6.0
benzene	methanol	9.0	9.0
ethanol	isooctane	11.0	8.0
methyl acetate	water	20.0	7.0
ethyl acetate	water	100.0	15.0
water	n-hexane	>100.0	>100.0

[a] Calculated from equilibria data from various sources.

The working equations of activity coefficients are based on solutions of the Gibbs-Duhem equation. The Gibbs-Duhem equation may be written:

$$x_i \left(\frac{\partial \ln \gamma_i}{\partial x_i}\right)_{T,P} + \ldots x_n \left(\frac{\partial \ln \gamma_n}{\partial x_i}\right)_{T,P} = 0 \tag{16}$$

Numerous activity coefficient models, which are based on integration of the Gibbs-Duhem equation with various assumptions, have been proposed. The Van Laar model (14) is a simple useful equation; the Margules model (15) is a simple equation which can deal with activity coefficients exhibiting maxima and minima. The Wilson equation (16) is probably the most widely used.

Rapid advances are being made toward the goal of predicting vapor–liquid equilibrium from molecular structure (17–21). Rao (22) briefly discusses current methods for preliminary estimation of activity coefficients from pure component properties.

The binary Van Laar equations are:

$$\ln \gamma_1 = \frac{A^*}{\left(1 + \frac{A^* x_1}{B^* x_2}\right)^2} \tag{17}$$

$$\ln \gamma_2 = \frac{B^*}{\left(1 + \frac{B^* x_2}{A^* x_1}\right)^2} \tag{18}$$

The Margules equations are:

$$\ln \gamma_1 = (x_2)^2 [A^* + 2x_1(B^* - A^*)] \tag{19}$$

$$\ln \gamma_2 = (x_1)^2 [B^* + 2x_2(A^* - B^*)] \tag{20}$$

The Wilson equations are:

$$\ln \gamma_1 = 1 - \ln(x_1 + \Lambda_{12} x_2) - \frac{x_1}{x_1 + \Lambda_{12} x_2} - \frac{\Lambda_{21} x_2}{x_2 + \Lambda_{21} x_1} \tag{21}$$

$$\ln \gamma_2 = 1 - \ln(x_2 + \Lambda_{21} x_1) - \frac{x_2}{x_2 + \Lambda_{21} x_1} - \frac{\Lambda_{12} x_1}{x_1 + \Lambda_{12} x_2} \tag{22}$$

854 DISTILLATION

An advantage of the Van Laar and Margules equations is their simplifying definition of the limiting coefficients at infinite dilution as the antilog of the A^* and B^* constants. In the Wilson equation, however, the limiting activity coefficients at infinite dilution are a more complex function of the binary constants and have no simple physical significance.

No single activity coefficient model is best for all systems. Null (23) compares the fit of these three models and concludes for systems exhibiting negative or slightly positive deviation or exhibiting moderate positive deviation from Raoult's Law that the three activity coefficient equations fit the systems equally well. In systems exhibiting large positive deviations and having complete liquid-phase miscibility, the Wilson equation is preferred followed by the Van Laar equation. Many other activity coefficient models are in current use (19,24–27). A summary of the more important equations being used is given in ref. 28.

Activity coefficients for multicomponent systems usually are generated from binary pair data. For example, for the ternary system A, B, and C, the binary pair data of AB, AC, and BC are used to define the ternary system. The Wilson multicomponent equation is of the form:

$$\ln \gamma_i = 1 - \ln \left[\sum_{j=1}^{N^*} x_j \Lambda_{ij} \right] - \sum_{k=1}^{N^*} \left[\frac{X_k \Lambda_{ki}}{\sum_{j=1}^{N^*} x_j \Lambda_{kj}} \right] \qquad (23)$$

Multicomponent activity equations are discussed (8,29–30).

The vapor–liquid equilibrium of dilute solutions is frequently expressed in terms of Henry's law which is:

$$p_i^L = H^* x_i \qquad (24)$$

where H^* is the Henry's law constant. From equation 15:

$$H^* = \gamma_i P_i^\circ \qquad (25)$$

Henry's law is useful for gas absorption problems. Henry's law data (eg, see ref. 31) are also useful for estimation of the terminal activity coefficients.

Nonideal Vapor-Phase Behavior. For most engineering purposes ideal gas behavior can be assumed up to pressures of 100–200 kPa (1–2 atm abs). At higher pressures the ideal gas law (eq. 2) and Dalton's laws (eqs. 4–5) may require modification to represent the VLE adequately. At these higher pressures, equation 2 is modified to the more general form:

$$PV' = zn'R'T \qquad (26)$$

At the higher pressures the basic equilibrium equation is usually written in terms of fugacities and the equilibrium partial pressure equation (eq. 6) is rewritten:

$$f_i^V = f_i^L \qquad (27)$$

For most calculations, equation 27 may be used in the form of:

$$y_i \phi_i^V P = x_i \gamma_i \phi_i^L P_i^\circ \qquad (28)$$

ϕ_i^V and ϕ_i^L are fugacity coefficients obtained from equations of state, such as the virial equation, the Redlich-Kwong equation (32), or from generalized correlations based on reduced properties and other parameters (33–34). A generalized fugacity coefficient plot is shown in Figure 3; in this plot the fugacity coefficient is a function of reduced temperature and pressure at a constant critical compressibility. Equilibrium based on fugacities has been discussed (8–9,12,36).

The VLE of light hydrocarbon systems are frequently used in the form of the equilibrium constant K.

$$K_i = \frac{y_i}{x_i} \qquad (29)$$

where K_i is a function of temperature and pressure. Sometimes K_i is also a function of a correlating parameter added to handle nonideal liquid behavior. These Ks then include the nonideal liquid and vapor corrections. The correlating third parameters include molar average boiling points from the Kellogg charts (37) and convergence

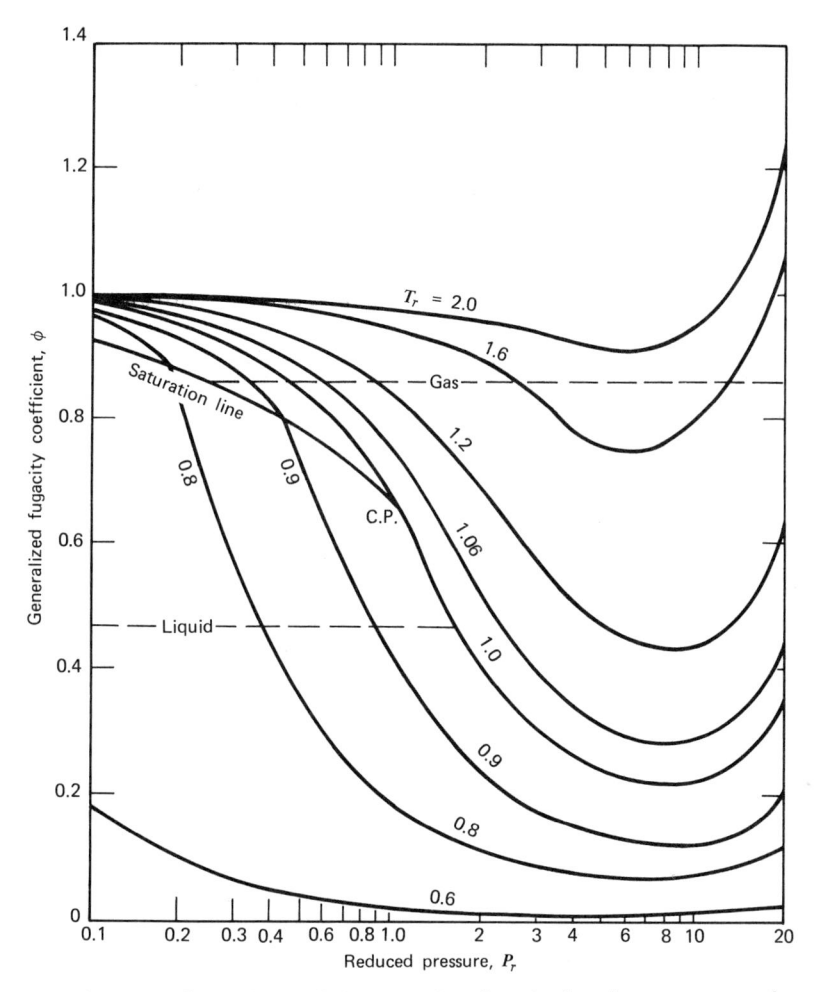

Figure 3. Generalized fugacity coefficients as a function of reduced temperature and pressure, at $z_c = 0.27$ (35).

856 DISTILLATION

pressure from the API *Technical Data Book* (38) as well as from others. These correlations have been discussed (39–41).

Azeotropic Systems. An azeotropic system is one that will vaporize without any change in composition. Figures 4 and 5 represent homogeneous azeotropic systems. Figure 4 depicts a minimum boiling azeotropic system such as ethanol–water. Figure 5 describes a maximum boiling azeotropic system such as acetone–chloroform. In these diagrams, the point Z defines the azeotropic composition; the azeotropic point is also called the constant boiling mixture (CBM). Positive activity coefficients tend to produce minimum boiling azeotropes, and negative coefficients tend to produce maximum boiling azeotropes.

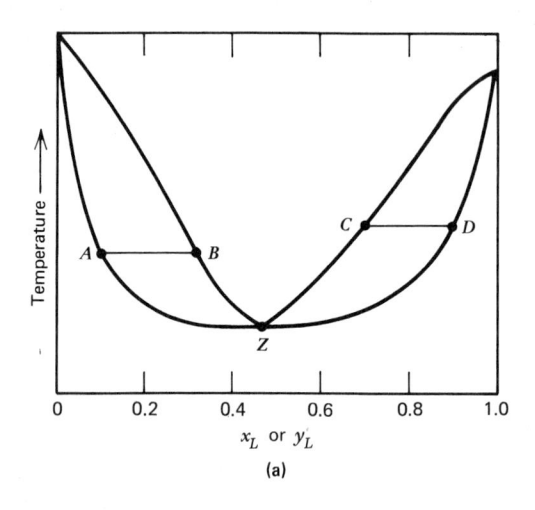

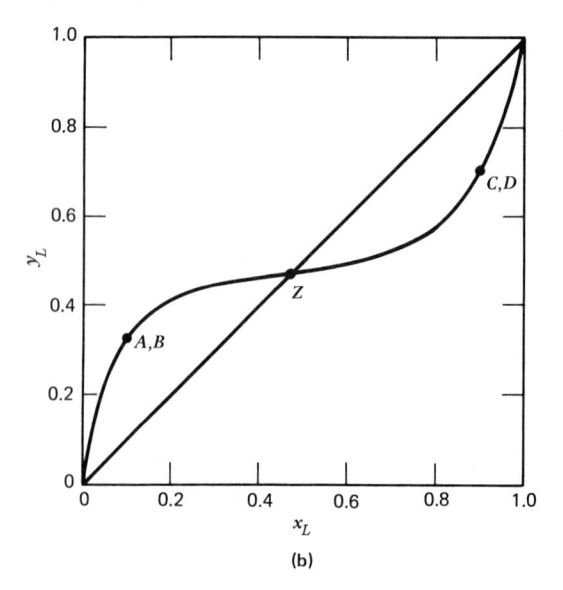

Figure 4. Typical boiling point (a) and (b) phase diagram for minimum boiling azeotropic system at constant pressure. A, B and C, D are representative equilibrium points; Z is the azeotropic point.

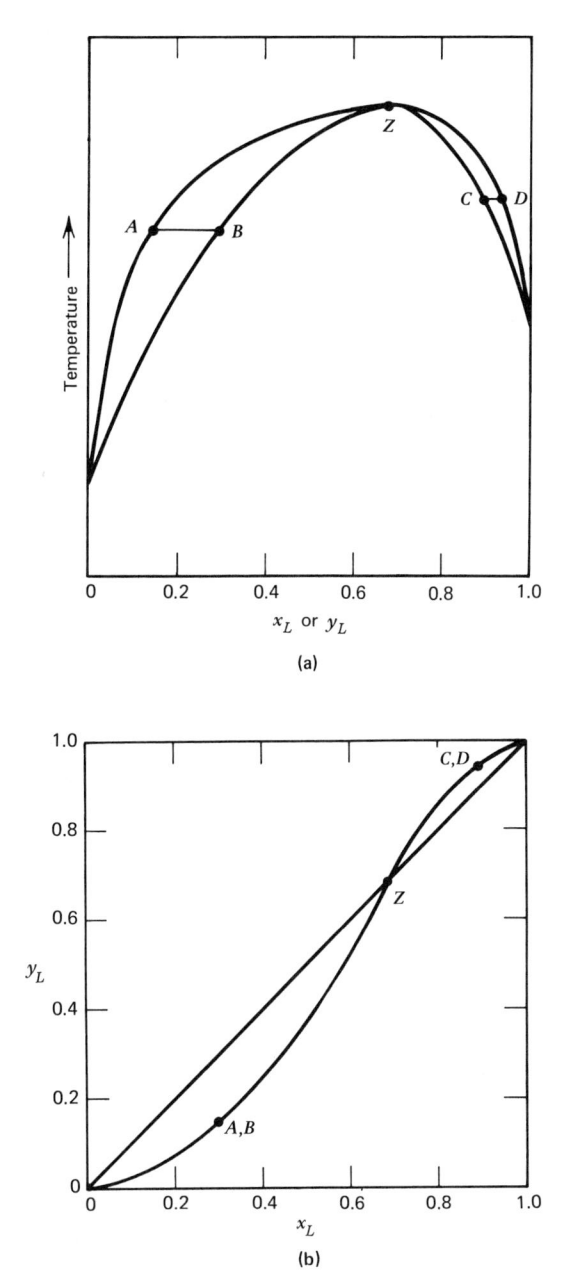

Figure 5. (a) Typical boiling point and (b) phase diagram for maximum boiling azeotropic system at constant pressure. A, B and C, D are representative equilibrium points; Z is the azeotropic point.

Heterogeneous azeotropes are formed when the positive activity coefficients are sufficiently large thereby producing two liquid phases which exist at the boiling point, and a constant boiling mixture which is formed at some composition, generally within the liquid immiscibility composition range. An example of a heterogeneous azeotropic system is the water–(1-butanol) system shown in Figure 6.

858   DISTILLATION

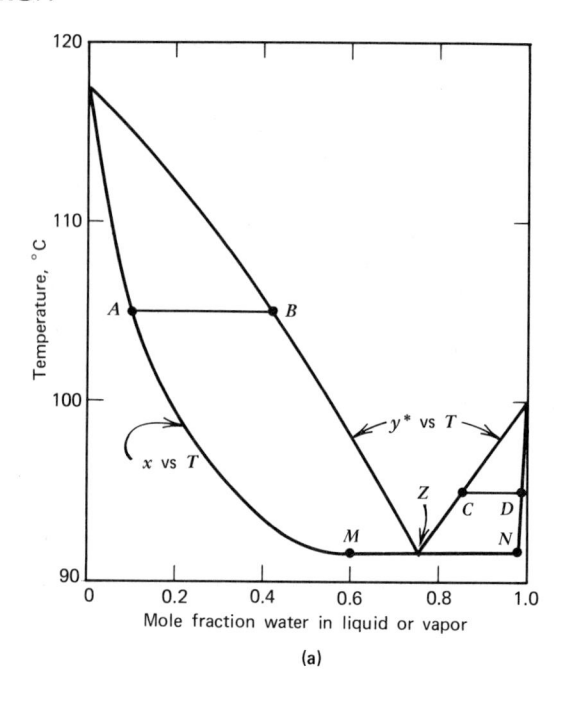

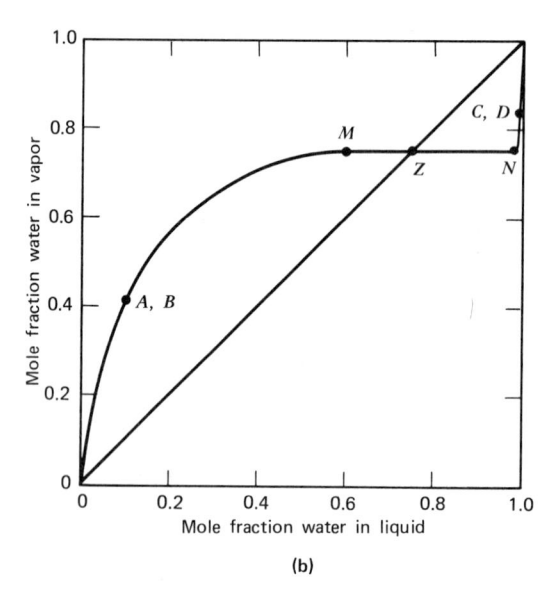

Figure 6. (a) Boiling point and (b) phase diagram for heterogeneous azeotropic system of water–(1-butanol) at atmospheric pressure. A, B, and C, D are representative equilibrium points; Z is the azeotropic point; M and N are liquid miscibility limits.

The abscissas of points M and N are the liquid miscibility limits. Within the immiscible range, M–N, the equilibrium vapor is the heterogeneous azeotrope (Z) of constant composition and the equilibrium temperature is constant. At liquid compositions lower in water than in the azeotrope, the relative volatility of water–(1-butanol) is greater than one; at liquid compositions higher in water than in the azeotrope, the relative volatility of water–(1-butanol) is less than one.

Sources of Vapor–Liquid Equilibrium Data. A VLE bibliography dating through 1965 (36) and one done through 1975 (42) are available. Collections of VLE data are contained in refs. 43–46. Over 400 sets of binary and ternary data have been fitted to the Van Laar, Margules, and relative volatility equations (46). Approximately 1000 binary systems have been adopted by computer to fit the Wilson equation (47). Gmehling and Onken (48) have started to release an excellent multivolume series, fitting moderate pressure VLE data, which are available up to 1975, to Van Laar, Margules, Wilson, NRTL, and UNIQUAC activity coefficient models, testing the data for thermodynamic consistency and graphically comparing the experimental data to the best-fit correlation model. The Gmehling and Onken compilation is exceptionally complete, eg, over 40 different sets of binary ethanol–water data are correlated. Horsley (49) presents 16,000 items of azeotropic data. Liquid–liquid solubility data (50) and gas–liquid solubility data (51–52) are useful for estimating vapor–liquid equilibrium. VLE data appear in many journals, particularly the *Journal of Chemical Engineering Data,* the *Journal of Applied Chemistry of the USSR,* and the *Collection of Czechoslovak Chemical Communications.*

Vapor–Liquid Equilibrium Measurement. The most common method to measure VLE data is direct vapor–liquid-phase analysis from recycle equilibrium stills and from total pressure measurements. Chromatographic methods are used extensively for initial equilibria screening. Equilibrium measurement techniques are discussed (53–54). VLE are sometimes deduced from overall column separations, principally at total reflux; the accuracy of such derived VLE depends on the accuracy of prediction of the plate efficiency of the test column.

Vapor–Liquid Equilibrium Data Evaluation. The accuracy of measured VLE data can be checked for scatter by empirical methods and, more significantly, by analysis of the thermodynamic consistency. Thermodynamic consistency testing has been discussed (55–58). Featherstone (59) considers the cost of distillation systems that should be added to allow for uncertainties in equilibrium data.

Distillation Processes

Basic distillation involves application of heat to a liquid mixture, vaporization of part of the mixture and removal of the heat from the vaporized portion. The resultant condensed liquid, the distillate, is richer in the more volatile components and the residual unvaporized bottoms are richer in the less volatile components. Most commercial distillations involve some form of multiple staging in order to obtain a greater enrichment than is possible by a single vaporization and condensation.

For ease of presentation and understanding, the initial discussion of distillation processes is based on binary systems. With reference to the binary boiling point diagram Figure 1**a** and the phase diagram Figure 1**b**, the enrichment from liquid composition x_L^A to vapor composition y_L^A represents a theoretical equilibrium stage or step.

Simple Distillations. Simple distillations utilize a single equilibrium stage to obtain separation. Simple distillations may be either batch or continuous. Simple batch distillation (also called differential distillation) may be represented on boiling point or phase diagrams. In Figure 1a, if the batch distillation started with a liquid of composition x_L^A, the initial distillate vapor composition would be y_L^A. As the distillate is removed, the remaining liquid becomes less rich in L (low boiler) and the boiling liquid composition moves to the left. If the distillation was continued until the liquid had a composition of x_L^E, the last vapor distillate would have had a composition of y_L^E. Simple batch distillation is not widely used in industry. Calculation methods are found in most standard distillation texts (60–61).

Simple continuous distillation (also called flash distillation) has a continuous feed to an equilibrium stage; the liquid and vapor leaving the stage are in phase equilibrium. A schematic representation is shown in Figure 7. On the boiling point diagram, Figure 1a, the feed is represented by x_L^F, the bottoms liquid by x_L^B, and the equilibrium vapor distillate by y_L^D. The mass balances are:

$$F = D + B \text{ (overall balance)} \tag{30}$$

$$x_L^F F = y_L^D D + x_L^B B \text{ (component balance)} \tag{31}$$

Flash distillations are widely used where a crude separation is adequate. Examples of flash multicomponent calculations are given in standard distillation texts (eg, ref. 62).

Multiple Equilibrium Staging. The component separation in simple distillation is limited to the composition difference between liquid and vapor in phase equilibrium; multiple equilibrium staging is used to increase the component separation. Figure 8 schematically represents a continuous distillation that employs multiple equilibrium stages stacked one upon another. The feed F enters the column at equilibrium stage $\bar{f}$. The heat $\bar{q}^S$ required for vaporization is added at the base of the column in a reboiler or calandria. The vapors V^T from the top of the column flow to a condenser from which heat $\bar{q}^C$ is removed. The liquid condensate from the condenser splits into two streams: the first, a distillate D which is the overhead product (also called heads or make) is withdrawn from the system and the second, a reflux R is returned to the top of the column. A bottoms stream B is withdrawn from the reboiler. The overall separation is the feed F separating into a distillate D and a bottoms B.

Above the feed a typical equilibrium stage is designated as n; the stage above n

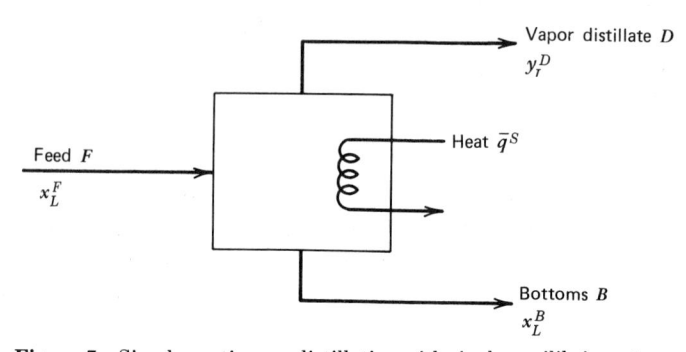

Figure 7. Simple continuous distillation with single equilibrium stage.

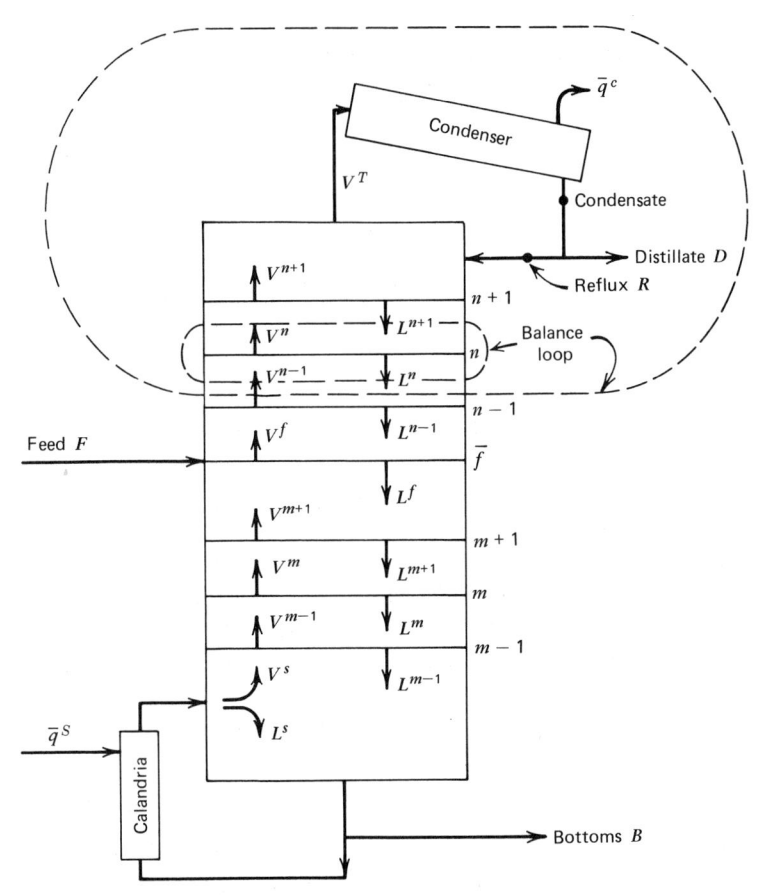

Figure 8. Distillation column with stacked multiple equilibrium stages.

is $n + 1$ and the stage below n is $n - 1$. The section of column above the feed is called the rectification section and the section below the feed is referred to as the stripping section.

The mass balance across stage n is: (1) Vapor (V^{n-1}) from the stage below ($n - 1$) flows up to stage n; (2) the liquid (L^{n+1}) from the stage above ($n + 1$) flows down to stage n; (3) On stage n the vapors leaving V^n are in equilibrium with the liquid leaving L^n.

The vapors moving up the column from equilibrium stage to equilibrium stage are increasingly enriched in the more volatile components. Similarly, the liquid streams moving down the column are increasingly diminished in the more volatile components.

The overall column mass balances are:
$$F = D + B \tag{32}$$

And for any component i:
$$Fx_i^F = Dx_i^D + Bx_i^B \tag{33}$$

The overall enthalpy balance is:
$$H^F F + \overline{H}^S = H^D D + H^B B + \overline{H}^C \tag{34}$$

A mass balance around plate n and the top of the column gives:
$$V^{n-1} = L^n + D \tag{35}$$
And for any component:
$$V^{n-1}y_i^{n-1} = L^n x_i^n + D x_i^D \tag{36}$$
$$\therefore y_i^{n-1} = \left(\frac{L^n}{V^{n-1}}\right)x_i^n + \left(\frac{D}{V^{n-1}}\right)x_i^D \tag{37}$$

Below the feed, a similar balance around plate m and the bottom of the column results in:
$$y_i^{m-1} = \left(\frac{L^m}{V^{m-1}}\right)x_i^m - \left(\frac{B}{V^{m-1}}\right)x_i^B \tag{38}$$

Equation 37 is the upper (or rectifying) operating line equation and equation 38 the lower (or stripping) operating line equation.

Graphical McCabe-Thiele Method. The graphical McCabe-Thiele (63) design method facilitates a visualization of distillation principles. In the following discussion of the McCabe-Thiele method, the subscripts L and H are not used and x and y refer to the low boiler of the binary system. A McCabe-Thiele diagram is given in Figure

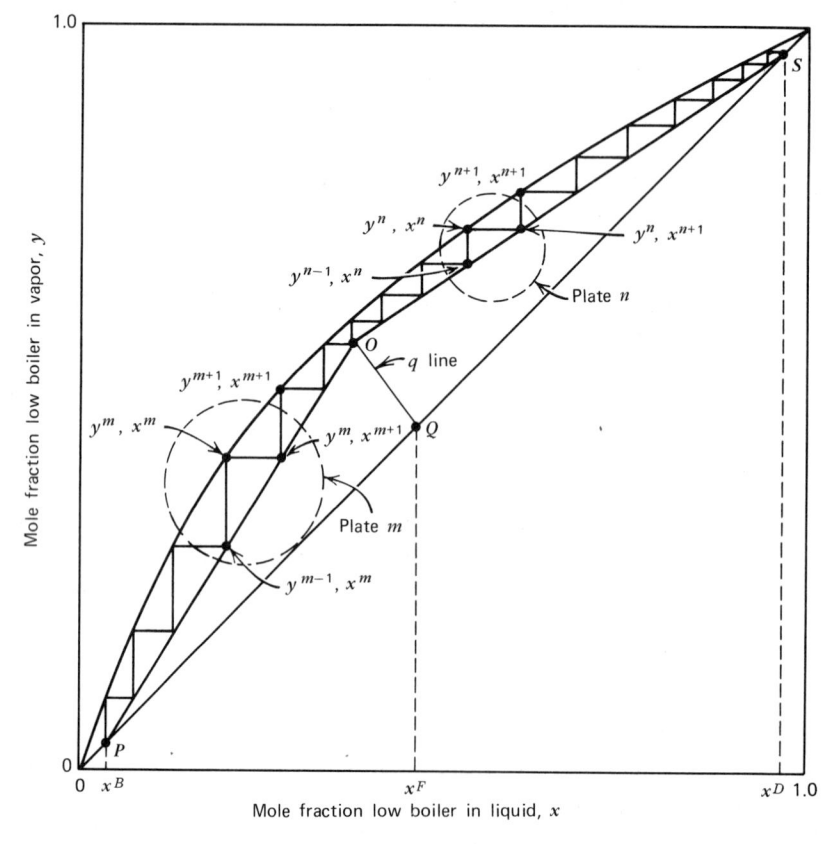

Figure 9. McCabe-Thiele diagram.

9. P, Q, and S are the x^B, x^F, and X^D compositions on the $x = y$, 45° construction line. Line OP is the stripping operating line and line OS is the rectifying operating line.

The McCabe-Thiele method makes the simplifying assumptions that the molal overflows in the stripping and the rectification sections are constant. These assumptions reduce the rectifying and stripping operating line equations to:

$$y^{n-1} = \left(\frac{\overline{L}}{\overline{V}}\right)_R x^n + \left(\frac{D}{\overline{V}_R}\right) x^D \tag{39}$$

$$y^{m-1} = \left(\frac{\overline{L}}{\overline{V}}\right)_S x^m - \left(\frac{B}{\overline{V}_S}\right) x^B \tag{40}$$

$\overline{L}$ and $\overline{V}$ designate the constant molal flow in each section. The McCabe-Thiele assumptions imply that the molal latent heats of the two components are identical, the sensible heat effects are negligible, and the heat of mixing and the heat losses are zero. These simplifying assumptions are closely approximated for many distillations. Equation 39 now represents the straight upper operating line OS and equation 40 represents the straight lower operating line OP. The upper operating line has the slope $(\overline{L}/\overline{V})_R$ and the intercept at x^D $(=y^T)$ on the $x = y$ line. (Note that this operating line slope is less than one.) Similarly, the lower operating line has a slope of $(\overline{L}/\overline{V})_S$ and the intercept is at x^B on the $y = x$ line. (Note that this operating line slope is greater than one.) The line QO from the feed intercept Q to the intersection of the operating lines at O is called the q line. The q line is discussed later.

The equilibrium line gives the vapor–liquid relationships of y^n and x^n above the feed and of y^m and x^m below the feed. The upper operating line gives the relationship of y^{n-1} and x^n and the lower operating line gives the relationship y^{m-1} and x^m. The graphical representation of theoretical equilibrium stages (or steps or plates) n and m is shown. The y^{m-1}, x^m to y^m, x^m to y^m, x_L^{m+1} represents the mass balance and phase equilibrium for theoretical stage m; similarly, y^{n-1}, x^n to y^n, x^n to y^n, x^{n+1} represents theoretical stage n. The total number of theoretical stages in the column can now be stepped off starting either at the x^B and stepping upward or starting at x^D and stepping downward.

Condition of Feed (q Line). The q line is determined by mass and enthalpy balances around the feed plate. The q line balances are detailed in distillation texts (eg, ref. 64).

The slope of the q line is $q/q - 1$ where:

$$q = \frac{\text{heat to vaporize one mole of feed}}{\text{molal latent heat of feed}}. \tag{41}$$

The q line, therefore, depends on the enthalpy condition of the feed. Types of q lines are shown in Figure 10 and are listed below.

Feed enthalpy condition	q	Slope of q line	q line coordinates
cold liquid	>1	+	Q–E
saturated liquid	1	∞	Q–D
partially vaporized	>0 to 1	−	Q–C
saturated vapor	0	0	Q–B
superheated vapor	<0	+	Q–A

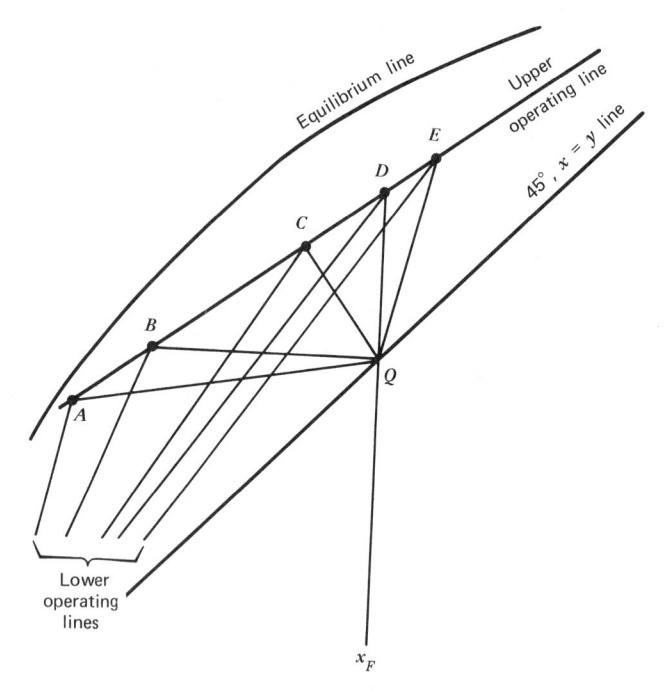

Figure 10. McCabe-Thiele q lines for various feed enthalpy conditions.

Reflux and Reflux Ratio. The liquid returned to the top of the column is called reflux. The ratio R/D is the external reflux ratio or simply reflux ratio. The ratio $(\bar{L}/v5V)_R$, which is the slope of the rectifying operating line, is the rectifying internal reflux ratio. Similarly, the ratio $(\bar{L}/\bar{V})_S$, which is the slope of the stripping operating line, is the stripping internal reflux ratio. As the reflux ratio R/D increases, the rectifying internal reflux ratio increases and numerically approaches 1; similarly, the stripping internal reflux ratio decreases and numerically approaches 1. In the McCabe-Thiele plot the two operating lines move away from the equilibrium line toward the $y = x$ diagonal as the reflux ratio increases; the individual theoretical stage steps become larger and fewer theoretical stages are required to make a given separation.

McCabe-Thiele Example. Assume a binary system L–H that has ideal vapor–liquid equilibria with a relative volatility of 2.0. The feed is 100 moles of $x^F = 0.6$; the required distillate is $x^D = 0.95$ and the bottoms $x^B = 0.05$. The feed is at the boiling point. Calculate the minimum reflux ratio, the minimum number of theoretical stages, the operating reflux ratio, and the number of theoretical stages. Assume the operating reflux ratio is one and a half times the minimum reflux ratio and there is no subcooling in reflux.

Calculate the vapor–liquid equilibria: an equation at $x = 0.60$ mole fraction from equation 12.

$$y^* = \frac{2(0.6)}{1 + (2-1)0.6} = 0.75 \text{ mole fraction} \tag{42}$$

Similarly, the whole equilibrium curve is calculated and it is plotted in Figure 11. The feed is at the boiling point so the q line is drawn with an infinite slope.

Calculate mass balances:

$$F = D + B = 100 \quad \text{(overall column balance)}$$

$$0.60\,F = 0.95\,D + 0.05\,B = 60 \text{ (component } L \text{ balance)}$$

$$D = 61.11 \text{ moles distillate} \tag{43}$$

$$B = 38.89 \text{ moles bottoms}$$

Calculate reflux ratios: The minimum internal reflux ratio is a line from the intercept of the q line with the equilibrium curve to the x^D point on the 45° line.

$$\text{slope} = \left(\frac{\overline{L}}{\overline{V}}\right)_R = \frac{0.95 - 0.75}{0.95 - 0.6} = 0.5714 \text{ minimum internal reflux ratio}$$

$$\therefore \overline{V} = \overline{L} + D = 0.5714\,\overline{V} + 61.11$$

$$\therefore \overline{V} = 142.58 \text{ moles (at minimum reflux)}$$

$$\therefore \overline{L} = 81.47 \text{ moles} = R \text{ (at minimum reflux)}$$

$$\therefore (R/D)_{\min} = \frac{81.47}{61.11} = 1.333 \text{ (minimum reflux ratio)} \tag{44}$$

$$\therefore (R/D)_{\text{operating}} = 1.5 \times 1.333 = 2.0 \text{ (operating reflux ratio)}$$

$$\therefore R = 2.0\,(61.11) = 122.22 \text{ moles reflux} = \overline{L} \text{ (at operating reflux ratio)}$$

$$\overline{V} = \overline{L} + D = 122.22 + 61.11 = 183.33 \text{ moles}$$

The upper operating line from equation 39 is:

$$y^{n-1} = \left(\frac{122.22}{183.33}\right)x^n + \left(\frac{61.11}{183.33}\right)0.95 = 0.667\,x^n + 0.317 \tag{45}$$

Since the feed is at the boiling point:

$$\overline{L}_S = \overline{L}_R + F = 122.22 + 100 = 222.22 \text{ moles} \tag{46}$$

$$\text{and } \overline{V}_S = \overline{V}_R = 183.33 \text{ moles} \tag{47}$$

the lower operating line (eq. 38) is:

$$y^{m-1} = \left(\frac{222.22}{183.33}\right)x^m + \left(\frac{38.89}{183.33}\right)(0.05) = 1.212\,x^m - 0.0106 \tag{48}$$

The complete McCabe-Thiele construction is shown in Figure 11. The theoretical plates were stepped off starting at the base; approximately 14.2 theoretical plates are required.

Unequal Molal Overflow. The McCabe-Thiele method is based on the simplifying assumption that the molal overflow is constant in both the rectifying and stripping sections. For many problems the assumption is not valid and more precise calculations are necessary. For the more general cases, detailed enthalpy balances are made around individual plates or sections. Standard distillation texts discuss the in-

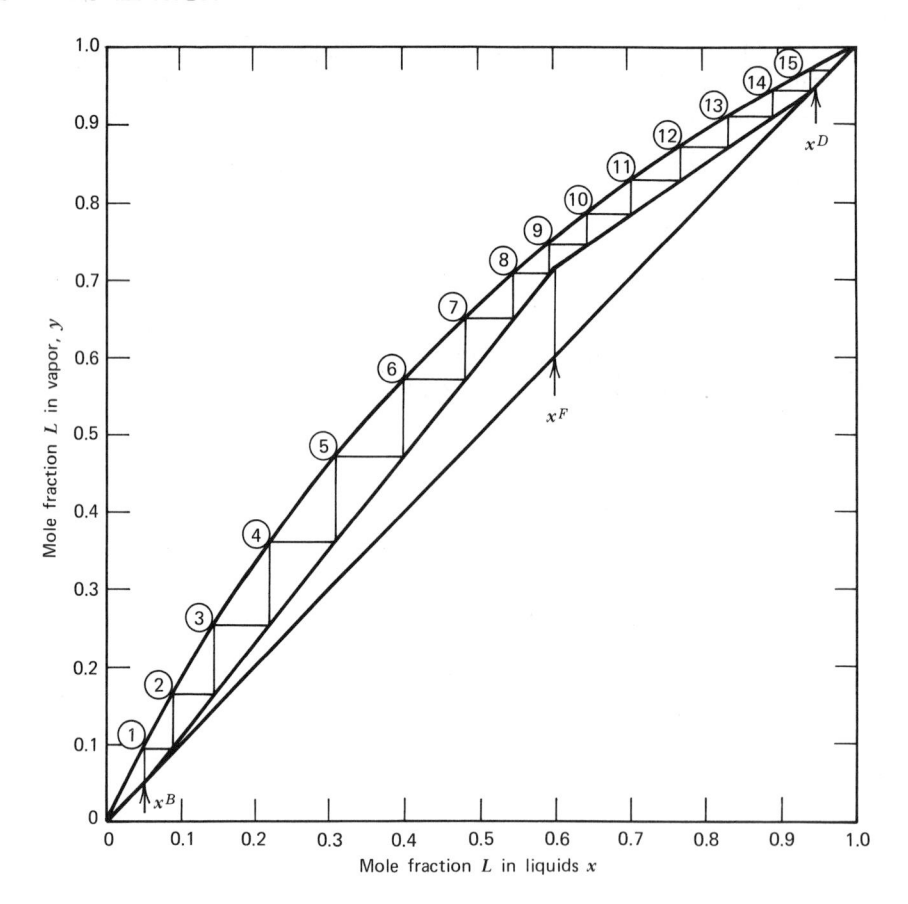

Figure 11. McCabe-Thiele example.

ternal enthalpy calculations by algebraic balances or by graphic procedures; eg, Van Winkle (65) details the plate-to-plate mass and enthalpy balances with equilibrium calculations and also by means of the graphical Ponchon-Savarit procedure (65–67). Hand algebraic and graphical methods requiring internal enthalpy calculations largely have been superseded by computer calculation (see Programmable pocket calculators).

Minimum Reflux Ratio and Minimum Number of Theoretical Stages. There are infinite combinations of reflux ratios and numbers of theoretical stages for any given distillation separation. The larger the reflux ratio, the fewer the theoretical stages required. For any distillation system with its given feed and its required distillate and bottoms composition, there are two constraints within which the variables of reflux ratio and number of theoretical stages must lie. The first constraint is the minimum number of theoretical stages, and the second is the minimum reflux ratio. The minimum reflux ratio occurs when the reflux ratio is reduced so that the upper and lower operating lines and the q line are coincident at a single point on the equilibrium line as shown in Figure 12a. When this condition exists, an infinite number of theoretical stages is required to make the separation. The minimum number of theoretical stages occurs when the system is at total reflux (no feed, distillate, or bottoms). This is illustrated in Figure 12b; the operating lines are coincident with the 45°, $Y = x$ line.

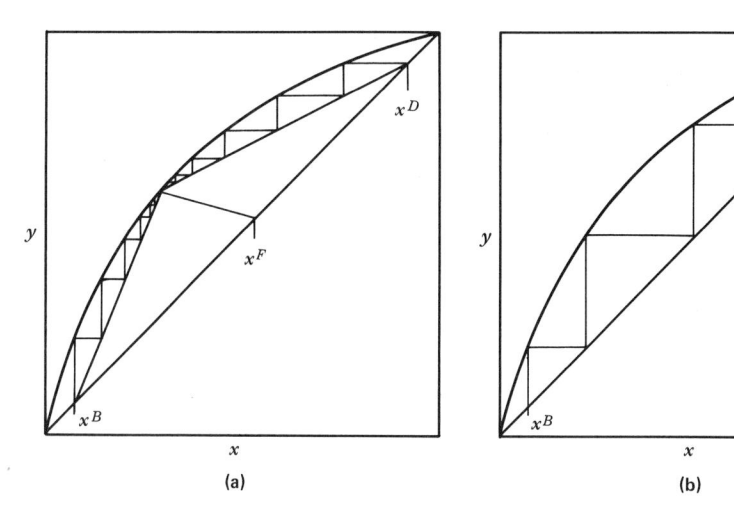

Figure 12. Limiting conditions in binary distillation. (**a**) Minimum reflux and infinite number of theoretical stages; (**b**) total reflux and minimum number of theoretical stages.

Both of these limits, the minimum number of stages and the minimum reflux ratio, are impractical for useful operation, but they are valuable guidelines within which the practical distillation must lie. As the reflux ratio decreases toward the minimum reflux, the required number of stages increases rapidly. Similarly, as the minimum number of stages is approached, the required reflux ratio increases rapidly. A plot of the number of theoretical stages vs reflux ratio for a typical distillation is shown in Figure 13. Both minimum limits may be calculated for any distillation, thereby bracketing the practical design. Actual operating reflux ratios for most commercial columns are in the range of 1.10–1.5 times the minimum reflux ratio.

The operating, fixed, and total costs of a distillation are functions of the relation

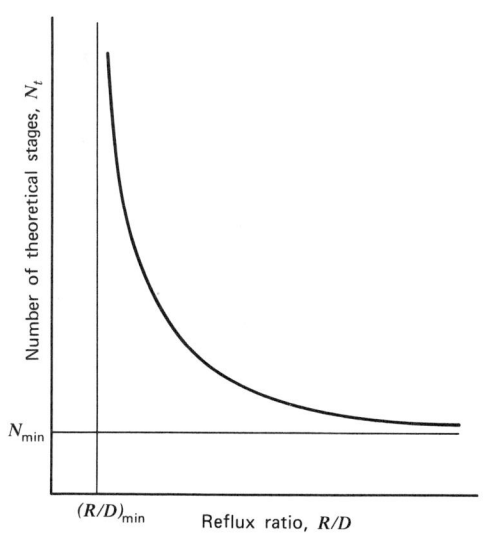

Figure 13. Reflux ratio vs number of theoretical stages for typical distillation.

of operating reflux ratio to minimum reflux ratio. Figure 14 shows a typical plot of costs; as the operating to minimum reflux ratio increases, the operating cost—principally energy costs for the boil-up—increases almost linearly. Similarly, the fixed costs at first decrease from the infinite number of stages, pass through a minimum, and then increase again as the diameter of column increases with increased reflux ratio. These costs for a typical distillation have been calculated (68).

False Minimum Reflux. There are some distillations where the minimum reflux does not occur at the intersection of the upper and lower operating lines and the q line. These cases arise when the equilibrium is skewed from positive activity coefficients (see Vapor–Liquid Equilibria section) and when the operating line intersects the equilibrium line in a zone of constant composition which is not at the q line intersection. Figure 15 illustrates such a case. Brown (69) provides an example of false minimum reflux in an ethanol–water column.

Batch Multistage Distillation. Batch multistage distillation is mainly limited to small lot, low volume distillations. The basics of batch multistage distillation are discussed (70–72).

Multicomponent Calculations. The calculations that determine the reflux and stage requirements are more difficult to make for multicomponent systems than for binary systems. When the concentration of a component in the distillate and in the bottoms is specified for the overall solution of a binary distillation, the component balance around the column also is completely specified. In the multicomponent case, only a single high boiling key component can be specified in the distillate and a single low boiling key component in the bottoms; the split of other components can be determined only by detailed calculations. These require a series of trial and error computations to obtain the solution at any given reflux ratio and number of stages. As the number of components and number of stages become large, the mathematical problem becomes formidable. Two approaches may be followed, one uses shortcut methods

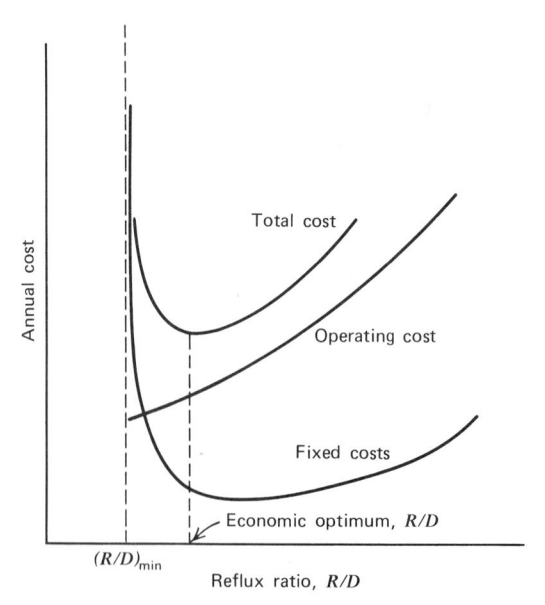

Figure 14. Fixed, operating, and total costs of typical distillation as a function of reflux ratio.

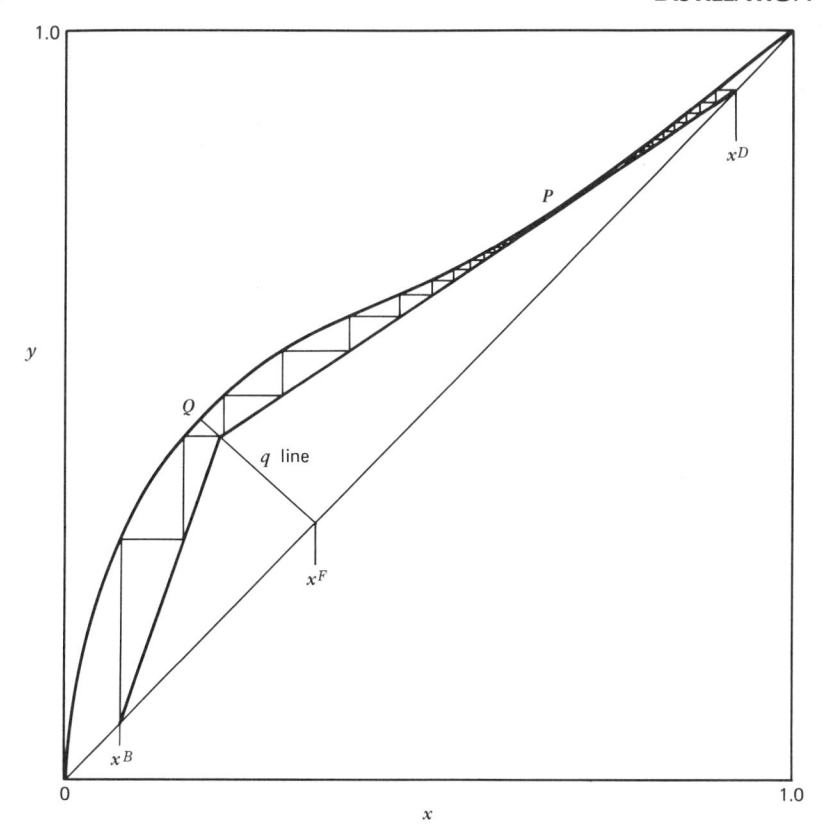

Figure 15. False minimum reflux for system of skewed equilibria. Minimum reflux occurring at intersection P of operating line and equilibrium line and not at intersection of q line and equilibrium line.

and the other a rigorous computer solution. The former are used when approximate solutions are adequate or when a computer is not available for a rigorous solution. Most shortcut methods involve: (*1*) estimating the minimum reflux; (*2*) estimating the minimum number of plates; and (*3*) estimating from empirical correlations the actual number of plates at an operating reflux. The minimum reflux is usually estimated from the Underwood equations (73) and the minimum number of plates from the Fenske equations (74). The relationship of operating to minimum reflux ratio and of operating to minimum number of plates is estimated from the Gilliland correlation (75) or from a more recent correlation such as that of Erbar and Maddox (76). An extensive discussion of shortcut methods is given in ref. 77, and two examples using different methods are given in ref. 78. The shortcut methods for multicomponent calculations necessarily are only approximate and, in some cases, as with highly nonideal equilibria, the short cut solutions may be grossly in error.

Rigorous computer solutions are used for complex distillations involving multiple stages, multicomponent, nonideal phase equilibria, multiple feeds and drawoffs, and heat addition or removal from intermediate stages. With the general availability of large computers and computer programs, which are able to handle the complex distillation problems, most distillation calculations are made by computer. The computer solution entails setting up component equilibrium and component mass and enthalpy balances around each theoretical stage and specifying the required design variables as well as solving the large number of simultaneous equations required for the solution. The explicit solution to these equations remains too complex for present methods. Studies to solve the mathematical problem by algorithm or iterational methods have been successful and, with a few exceptions, the most complex distillation problems can be solved (79).

If each component of a multicomponent distillation is to be essentially pure when recovered, the number of columns required for the distillation system is $N^* - 1$, where N^* is the number of components. Therefore in a six-component system, recovery of all six components as essentially pure products requires five separate columns.

The number of columns in a multicomponent train can be reduced from the $N^* - 1$ relationship if side-stream draw-offs are used for some of the component cuts. The feasibility of multicomponent separation by side-stream draw-offs depends on side draw-off purity requirements, feed compositions, and equilibrium relationships. In most cases, side-stream draw-off distillations are economically feasible only if component specifications in the side-stream draw-off are not tight. If a single component is to be recovered in an essentially pure state from a mixture containing both lower and higher boiling components, a minimum of two columns is required: one column to separate the lower boilers from the desired component and another column to separate the component from higher boilers.

The economics of the various methods that are employed to sequence multicomponent columns have been studied; eg, Freshwater and Henry (80) consider the separation of three-, four-, and five-component mixtures. They examine the heuristics (rules of thumb) developed by earlier investigators and make an economic analysis of various methods of sequencing the columns. The study of sequencing of multicomponent columns is part of a broader new field, process synthesis, which attempts to formalize and develop strategies for the optimum overall process (81) (see Separation systems synthesis).

Distillation Columns

Distillation columns (also called towers) provide the contact area between streams of descending liquid and ascending vapor and, thereby, furnish an approach toward physical vapor–liquid equilibrium. Depending upon the type of distillation column which is used, the contact may occur in discrete steps, called plates or trays, or in a continuous differential contacting on the surface of packing. The fundamental requirement of the column is to provide the mass transfer contacting at a required rate and at a minimum overall cost. Individual column requirements vary from high vacuum to high pressure, from low liquid to high liquid rates, from clean to dirty systems, etc. As a result, a large variety of equipment has been developed to fill these needs. The columns discussed in this section are used for both distillation and absorption; the internal column hardware providing the vapor–liquid contacting is identical for distillation and absorption. The principal operational difference is that in absorption, the gas flowing up the column is primarily a noncondensable phase at column conditions, whereas in distillation the gas phase is a condensable vapor (see Absorption).

Plate Columns. There are two general types of plates in use, the crossflow plate and the counterflow plate. The names refer to the direction of the liquid flow on the plate relative to the rising vapor flow. The liquid on the crossflow plate flows across the plate and from plate to plate via downcomers (Fig. 16). The liquid on the counterflow plate flows downward through the same orifices through which the vapor rises.

Crossflow Plates. The liquid enters the plate from the bottom of the downcomer of the plate above. The liquid flows across the bubbling area where it comes in contact with the vapors flowing through the bubbling area from the plate below. The aerated liquid flows over the exit weir into a downcomer. A vapor–liquid disengagement takes place in the downcomer and most of the trapped vapor escapes from the liquid and flows back to the interplate vapor space. The liquid, essentially free of entrapped vapor, leaves the plate by flowing under the downcomer to the inlet side of the next lower plate.

The vapor from the plate below rises to the bottom of the plate above and flows through the contacting orifices (see below) in the plate bubbling area. The vapor rises and makes vigorous contact with the liquid flowing across the bubbling area. The vapor–liquid in the bubbling area is a foaming, frothing, sometimes jetting mass where most of the mass transfer takes place. The vapor with the entrained liquid rises from the bubbling mass into the free space between the trays where the large entrained droplets settle out.

The pressure drop incurred by the vapor as it passes through the contacting orifices in the tray floor is fundamental to plate operation. In most plate designs, the pressure drop prevents the liquid that is flowing across the bubbling area from falling through the plate. In commercial crossflow plate columns, the column diameter is 0.3–15 m and the plate spacing is 0.15–1.2 m. The total pressure drop per plate is 0.3–1.6 kPa (2–12 mm Hg).

Three principal vapor–liquid contacting devices are used in current crossflow plate design: the sieve plate, the valve plate, and the bubble cap plate. These devices provide the mass transfer contacting area between the rising vapor and the crossflowing liquid.

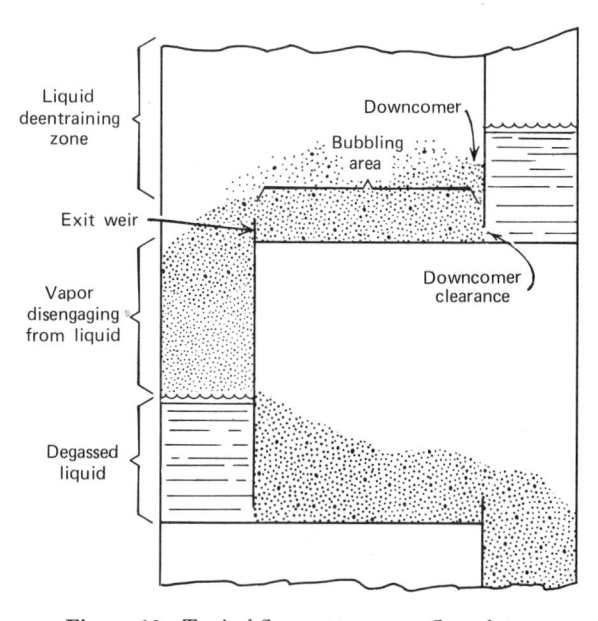

Figure 16. Typical flow pattern, crossflow plate.

Sieve Plates. The conventional sieve or perforated plate is inexpensive and widely used and is the simplest of the plates normally used. The contacting orifices in the conventional sieve plate (Fig. 17) are holes which measure 3.2–50 mm in diameter and which exhibit ratios of open perforated area to bubbling area ranging from 1:20 to 1:5. If the open area is too small, the pressure drop across the plate is excessive; if the open area is too large, the liquid will weep or dump through the plate.

Valve Plates. Valve plates are a relatively new proprietary development. The principal advantage of valve plates over sieve plates is their ability to maintain efficient operation over a wider operating range by using a variable orifice. The valve opens or closes to expose the holes for vapor passage. The most common valve units consist of flat disks with attached legs which allow the valve to open or close depending on the rate of vapor passage. Frequently, two weights of valves are used on a single plate to extend operating range and improve vapor distribution. The valve units usually have a tab or indentation that provides a minimum open area of vapor flow, even when the valve is closed, and also prevents the valve from sticking, which may result from corrosion or fouling.

The valve units may be round such as Glitsch's Ballast (82) or Koch's Flexitray valves (83), or rectangular such as the Nutter Float valves (84). These typical units are shown in Figure 18. Round valves usually have 39-mm diameter holes in the flat deck with valve lifts 6–12 mm. The rectangular valves have 25 by 127-mm deck holes with valve lifts similar in design to the round valves. Variations of the basic valves are available; these include venturi-throated decks for lower pressure drops, and single or double valve units within a retaining cage. Figure 19 shows a typical valve plate with round valves.

Bubble Cap Plates. Thirty years ago bubble caps were industry's standard design. Today their use in new installations is limited to very low liquid flow rate applications, or to those cases where the widest possible operating range is desired. A typical bubble cap is shown in Figure 20. The vapor flows through a hole in the plate floor, through the riser, reverses direction in the dome of the cap, flows downward in the annular area between the riser and the cap and exits through the slots in the cap.

Figure 17. Sieve tray. Courtesy of Glitsch Inc.

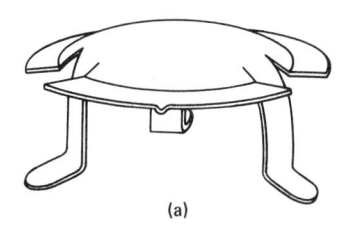

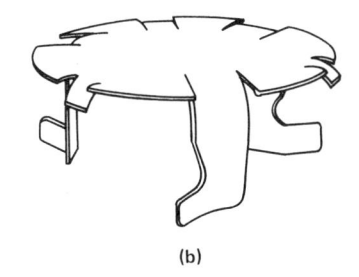

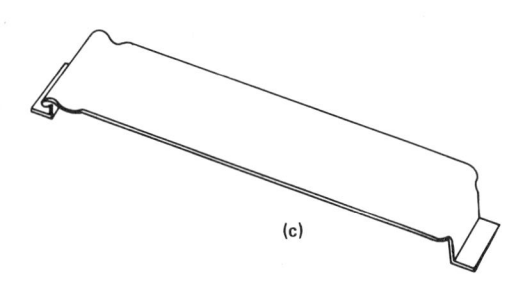

Figure 18. Representative individual valve units: (**a**) Koch Flexitray Valve (courtesy of Koch Engineering Co., Inc.); (**b**) Glitsch Ballast Valve (courtesy of Glitsch Inc.); (**c**) Nutter Float Valve (courtesy of Nutter Engineering Co.).

Commercial caps range from 50 to 150 mm in diameter; many slot design variations are used. Bubble cap trays are more expensive and have lower capacity than sieve or valve plates; therefore, their use has dropped to a small percentage of new column designs.

Multiple Liquid-Path Plates. As the liquid flow rate increases in large diameter crossflow plates (ca 4 m or larger), the crest heads on the overflow weirs and the hydraulic gradient of the liquid flowing across the plate become excessive. To obtain improved overall plate performance, multiple liquid-flow-path plates may be used. These designs are illustrated and discussed in standard distillation texts (eg, ref. 85).

Counterflow Plates. Counterflow plates are used less frequently than crossflow plates. The liquid flows downward and the vapor upward through the same orifices in a counterflow plate and the plate does not have downcomers. The openings are round holes (dual flow), slots (Turbogrids), or baffles that extend across a portion of the column diameter (86). Stone and Webster's Ripple Trays (Fig. 21) have round holes in a tray floor that has been formed in sinusoidal waves (87). Counterflow trays are used in high liquid flow rate or fouling systems. Counterflow plates generally have lower

Figure 19. Valve tray; tray shown is Koch Flexitray. Courtesy of Koch Engineering Co., Inc.

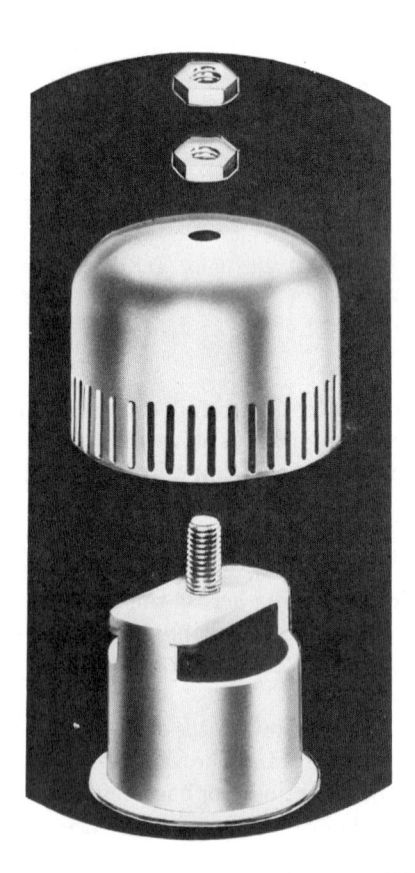

Figure 20. Expanded view of a typical bubble cap. Courtesy of Vulcan Manufacturing Co.

Figure 21. Ripple tray. Courtesy of Stone and Webster Engineering Corp.

plate efficiencies, narrower operating ranges and greater capacities than crossflow plates of the same diameter and plate spacing.

Vapor Capacity Parameters. The vapor capacity of a column is expressed in terms of vapor capacity parameters. The two common column characterizing parameters are:

$$C = V* \left(\frac{\rho_g}{\rho_l - \rho_g}\right)^{0.5} \quad (49)$$

and its simplification

$$F* = V*(\rho_g)^{0.5} \quad (50)$$

C in equation 49 is called a capacity factor and $F*$ in equation 50 is called an F-factor. For the equations to be meaningful, the column area to which they pertain must be specified. For example, the F-factor written for the column's superficial cross-sectional area is F_C^*, for the bubbling area F_B^*, and for the open hole area F_H^*. Typical ranges of operating F-factors for sieve plate columns are:

	kg$^{1/2}$/(m$^{1/2}$·s)	lb$^{1/2}$/(ft$^{1/2}$·s)
F_C^*	0.6–3.0	0.5–2.5
F_B^*	0.85–4.3	0.7–3.5
F_H^*	8.5–30	7–25

(Note, in distillation literature $F*$ and $V*$ are conventionally written without the asterisk.)

Flooding. The capacity of a column is limited by flooding. Flooding occurs when the liquid or vapor flow rates, or both, increase to the point where normal hydraulic functioning of the plate stops, which results in a precipitative drop in the plate efficiency. Three types of flooding occur in crossflow plates. They are: entrainment flooding; blow flooding; and downcomer flooding.

Entrainment flooding, the most common type, occurs when high velocity vapor in the column entrains massive amounts of liquid to the plate above. A typical entrainment flooding F_C^* plot is shown in Figure 22. Correlations of entrainment flooding are given in standard distillation texts (88–89). Also the proprietary valve-tray manufacturers publish design manuals that include entrainment flooding correlations (82–84).

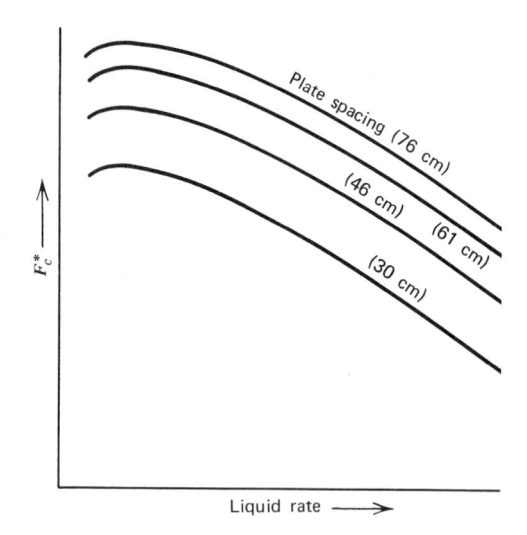

Figure 22. F_c^* for entrainment flooding of crossflow plates as a function of liquid rate and plate spacing. Absolute value of F_c^* is dependent on plate design and system factors.

Blow flooding, a process in which the liquid is literally blown off the plate, occurs at a combination of very low liquid flow rates and high vapor velocity through the vapor-contacting orifices.

Downcomer flooding occurs when the liquid backs up in the downcomer and cannot flow at the required rate to the plate below. The backup may be caused by insufficient deaeration time or by excessive velocity in the downcomer, or by excessive vapor pressure drop through the plate. The hydraulics of the liquid flow and the liquid deaeration in the downcomers is complex and no completely satisfactory fundamental correlation is available. Consequently, downcomer sizing is based on empirical approaches of either a maximum allowable liquid flow velocity or a minimum liquid downcomer residence time. Typical design parameters designate a maximum liquid velocity of 0.12 m/s or a minimum liquid residence time of 4 s, or both. A comprehensive study of downcomers (90) and an example of downcomer hydraulic calculations in a typical column have been reported (91).

Stable Operating Range. All plates have a stable operating envelope bounded by a range of liquid and vapor flow rates as shown in Figure 23. The size and shape of the stable area depends on the plate design and on the system properties. The line AD represents the minimum operable vapor flow rate at various liquid flow rates. Below AD, the vapor rate is too low to maintain the liquid on the plate and, as a result, the liquid will weep excessively or dump through the plate. Above line BC the column will flood by entrainment. To the right of CD the high liquid rate will cause downcomer flooding. The area to the left of AB represents high entrainment at low liquid flow rates, culminating in blow flooding.

Design procedures for bubble caps (92) and sieve trays (93) have been illustrated. Valve trays are designed by procedures developed by the valve vendors (82–84). Process and mechanical design of packed and plate columns have been discussed (94).

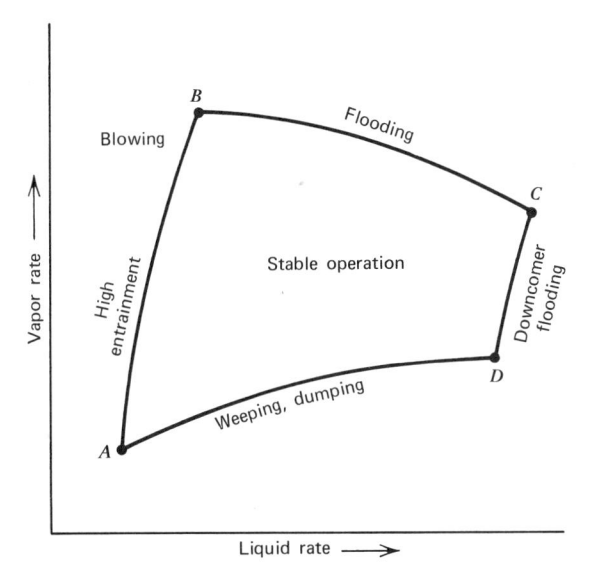

Figure 23. Stable operating range for crossflow plates.

Plate Efficiencies. Column requirements are calculated in terms of theoretical plates and are built and operated with actual plates. A plate efficiency, which is a measure of the rate of mass transfer, is used to relate actual and theoretical plates.

The simplest plate efficiency is the overall column efficiency which is the number of theoretical plates in a column divided by the number of actual plates.

$$E_o = \frac{N_t}{N_a} \qquad (51)$$

Therefore, the overall column efficiency is an averaged efficiency of the individual plates.

A more useful plate efficiency for theoretical prediction is the overall or Murphree plate efficiency (95):

$$E_{MV} = \frac{y^n - y^{n-1}}{y^{n,*} - y^{n-1}} \qquad (52)$$

where y^n and y^{n-1} are the vapor compositions from plate n and n-1 (the plate below n), and $y^{n,*}$ is the vapor in equilibrium with the liquid leaving plate n. Equation 52 is written in terms of vapor compositions; a similar equation can be written in terms of the liquid compositions. Other plate efficiency definitions are also used (96).

Theoretical Prediction of Plate Efficiency. The most comprehensive study of plate efficiency to date was made by an AIChE research committee (97). Unfortunately the AIChE model has been shown to be inadequate for many industrial distillations. Various studies have been and are being undertaken for improvement of the basic model. For example, Bell (98) discusses liquid mixing and residence time in commercial sieve plates, and Hughmark (99) uses a free-interface model to obtain improved fits to the AIChE data. A satisfactory method of theoretically predicting plate efficiency over the range of variables encountered does not exist and probably will not for some time. Haselden (100) states that it is still a brave designer who believes his tray efficiency predictions.

Recently there have been studies of the various types of flow regimes that occur on operating plates and of the effect of these regimes on tray performance, including plate efficiency. Pursuit of the flow regime studies (101–103) may lead to improved plate efficiency prediction methods.

Empirical Efficiency Prediction Methods. Numerous empirical methods of predicting plate efficiency have been proposed. Probably the most widely used empirical correlation correlates overall column efficiency as a function of feed viscosity and relative volatility (104). A statistical correlation of efficiency and system variables has been developed from numerous plate efficiency data (105).

General Comments on Plate Efficiency. The plate efficiencies of well-designed commercial bubble cap, sieve, and valve plates are approximately the same when the plates are operated within their normal design range.

The plate efficiency decreases both at the low end of the plate's operating range, where the liquid tends to leak through the plate, and at the high end of the operating range, where liquid entrainment becomes substantial.

Most distillation systems in commercial columns have Murphree plate efficiencies of 70% or higher. Lower efficiencies are found in systems of high λ (slope of equilibrium line/slope of operating line (97)), in systems of high liquid viscosity, and in absorbers. More complete discussion of plate efficiency is found in standard distillation texts (especially in refs. 106–107).

Packed Columns. In packed columns, the vapor–liquid contacting takes place in continuous beds of packing rather than in discrete individual plates as in plate columns. The contacting occurs in differential increments across the height of the packing. Mechanically, the packed column is a simple structure as compared to a plate column. The simplest packed column consists of a vertical shell with dumped packing on a packing support and a liquid distributor above the packing that distributes the liquid uniformly across the packed bed. A packed column with two packed beds and a midcolumn feed is shown in Figure 24. The vapor enters the column below the bottom packed bed and flows upward through the column. The liquid (reflux, or other liquid stream) enters at the top through the liquid distributor and flows downward through the packing countercurrently to the rising vapor. The height of the individual packed beds is limited to 2–9 m by the mechanical strength of the packing or by the need to redistribute the liquid so that good mass transfer efficiency can be maintained.

Packings. The majority of packed columns use randomly dumped packing; less common are beds made of regularly stacked packings or pads of woven or knitted wire. A variety of packings are in current use. Some of the common packings are shown in Figure 25. The Raschig ring, one of the oldest of packings, is an open cylinder of equal height and diameter. The Berl saddle and the more modern Intalox saddle (Norton Co.) have a higher capacity and efficiency than the Raschig ring. The Pall ring, one of the newer packings, is a modification of the Raschig ring. Pall rings have a higher capacity and efficiency than the Raschig rings. The dumped packings are made in ceramics, metals, and plastics; the approximate size range for individual packings is 12–75 mm.

Packed Column Operation. In the packed column, liquid flows downward concurrently with the upward flow of vapor through the same open space or interstices between the packing elements. At low liquid and gas flow rates, the descending liquid occupies only a small fraction of the interstices and, therefore, offers little hindrance to the rising vapor flow. Figure 26 shows a typical plot of pressure drop per unit of

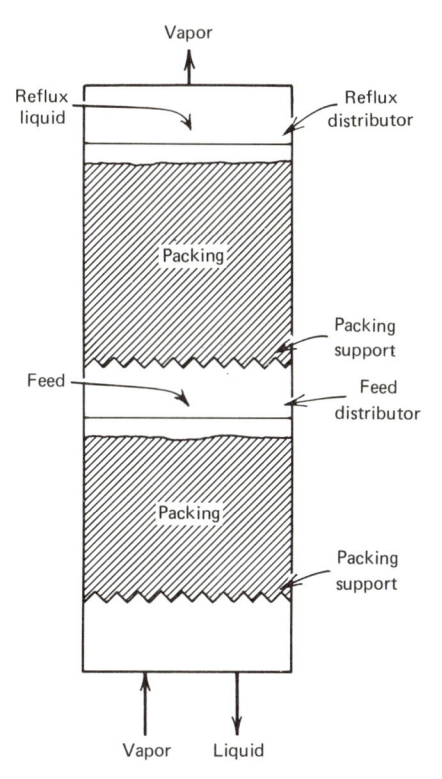

Figure 24. Typical packed column shell and internals. Column shown has single packed beds above and below the feed.

height ($\Delta P/h$) as a function of the gas rate G at low and high liquid flow rates. At a low rate of gas flow, the log slope of $\Delta P/(h \cdot G)$ is approximately 2. As the gas flow rate increases, there is an increasing tendency for the liquid to be held up in the void space, thereby decreasing the space available for the gas flow. As the gas flow rate increases further, more liquid is held up until at some high gas rate the packing floods. At this point, the liquid is essentially filling the interstices and can no longer flow downward. At flooding, the log slope $\Delta P/(h \cdot G)$ is infinite. The pressure drop at the inception of flooding ranges from 1.6 to 3.3 kPa/m (2–4 in. water/ft) of packing. More comprehensive discussion of packed column hydraulics is contained in distillation texts (eg, refs. 108–110).

Capacity of Packed Columns. Packed columns are usually designed to operate at some percentage of flooding, eg, 70–80%, or on a specific pressure drop per unit height of packing, eg, 0.8 kPa/m (1 in. water/ft) of packing. The most commonly used flooding correlation was initially proposed by Sherwood, Shipley, and Halloway (111). It has been revised by several authors; one revision (112) is shown in Figure 27. Experimentally-determined packing factors F_p from Table 2 are used in Figure 27 to obtain estimates of pressure drop and of flooding points for various packings.

Packing Mass Transfer Characteristics. The mass transfer contacting in a packed column takes place differentially along the length of the column. The mass transfer separation calculations can be made on a differential basis along the length of the column using mass transfer coefficients or transfer units (see Mass transfer; also Ad-

880   DISTILLATION

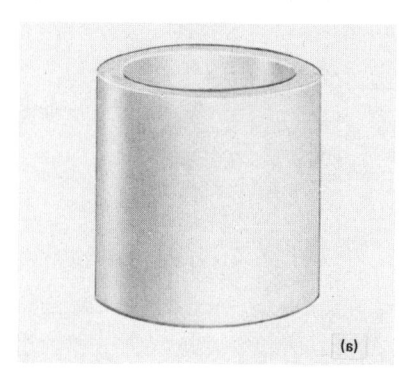

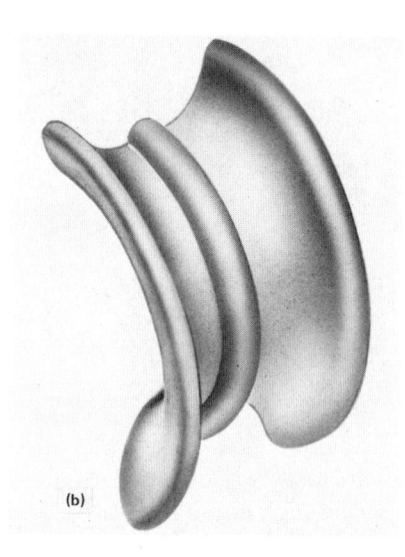

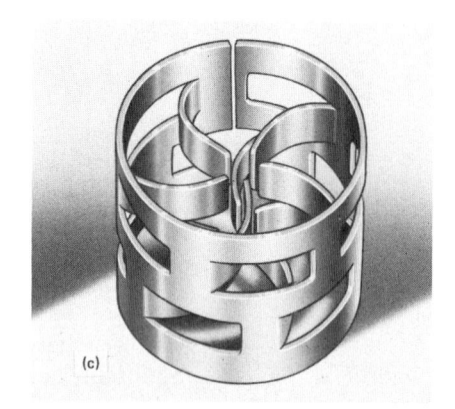

Figure 25. Typical packings: (**a**) Raschig ring; (**b**) Intalox saddle (Norton Co.); (**c**) Pall ring. Courtesy of Norton Chemical Process Products Division, Norton Company.

sorptive separation; Absorption). The calculations are somewhat imprecise because of the uncertainty in the fundamental mass transfer rates in commercial columns (114–115). A widely-used correlation for predicting packed column transfer units has been developed by Cornell, Knapp, and Fair (116–117). Examples of calculating the height of a packed distillation column by differential mass transfer are available (114, 118–119).

Instead of calculating the packed height on a differential basis, the required height frequently is calculated on the basis of theoretical plates and the height equivalent of a theoretical plate (HETP). The HETP of a packing is expressed in linear units of packing height. With the present inadequate understanding of packed column mass transfer fundamentals, there is frequently little incentive to use the more complex differential approach in packed distillation column calculations. Consequently, the efficiency of packed columns may be estimated using empirical HETP methods which have been developed by experience and are used within the range of variables covered by the empirical development. For example, Eckert (120) suggests the following guidelines for estimating HETPs for Pall rings, saddles, and Raschig rings when they are used over a range of specified pressure drop:

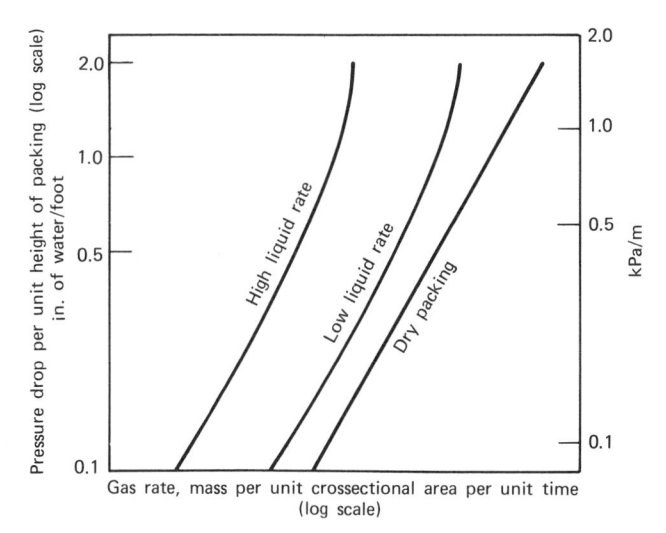

Figure 26. Pressure drop per unit height of typical packing at two liquid rates as a function of gas rate.

Packing size, mm (in.)	HETP, m (ft)
25 (1)	0.38–0.53 (1.25–1.75)
38 (1.5)	0.61–0.76 (2–2.5)
51 (2)	0.76–0.91 (2.5–3.0)

In order to obtain good packed-column mass transfer efficiency, the liquid must be distributed uniformly over the surface of the packing. Commercial columns using randomly dumped packing usually have a minimum of 11 distribution points per square meter (1/ft^2). The distributor design problem becomes severe at low liquid flow rates <700 cm^3/(s·m^2) [1 gal/(min·ft^2)] or in large (>3 m) diameter towers. Typical liquid distributor designs and typical packing supports are shown in ref. 121.

Specialty Packings. One type of specialty packing that is gaining increasing application is the metal gauze or cloth which is assembled in an oriented pattern, such as Koch-Sulzer packing. Billet (122) discusses the application of this type of packing in the high vacuum field. The combination of the characteristics of a low pressure drop per theoretical plate, high capacity, and improved mass transfer at low liquid rates is favorable for vacuum distillations. These oriented wire packings are also being used for debottlenecking other types of columns.

Packed vs Plate Towers. Relative to plate towers, packed towers are more useful for: general distillation in small towers (under 1.5 m) or for the following specific applications: severe corrosion environment—some corrosion-resistant materials (plastics, ceramics, and certain metallics) can easily be fabricated into packing but are difficult to fabricate into plates; vacuum operation where a low pressure drop per theoretical plate is a critical requirement; high liquid rates above 49,000 kg/(h·m^2) (10,000 lb/(h·ft^2)); foaming systems; or debottlenecking plate towers having relatively close plate spacing, say plate spacings under 0.5 m.

Steam Distillation

Steam distillation is used to lower the distillation temperatures of high boiling organic compounds that are essentially immiscible with water. If an organic compound

882 DISTILLATION

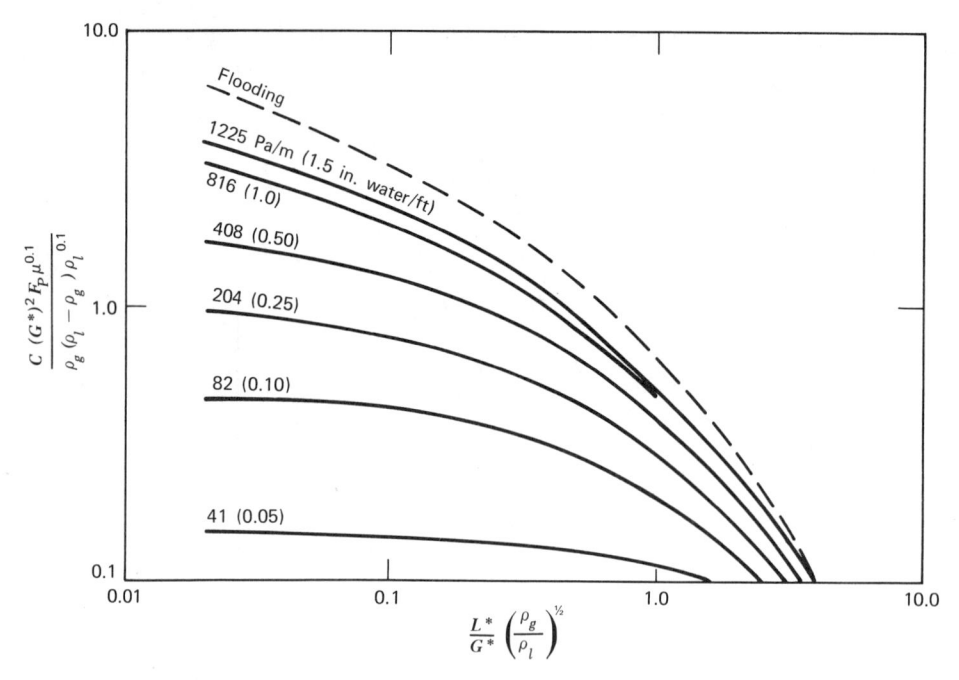

Figure 27. Generalized packing pressure drop correlation (112–113). Curve parameters are pressure drop per unit length of packed height, Pa/m (in. of water/ft).

Property	SI units	English units
C = conversion factor	42.81	1.512
F_p = packing factor		
G = gas rate	kg/(m²·s)	lb/(ft²·s)
L = liquid rate	kg/(m²·s)	lb/(ft²·s)
ρ_g = gas density	kg/m³	lb/ft³
ρ_l = liquid density	kg/m³	lb/ft³
μ_l = liquid viscosity	Pa·s	cP

is immiscible with water, both it and the water will exert their full vapor pressure upon vaporization from the immiscible two-component liquid. At a system pressure of P, the partial pressures would be:

$$P = p_{\text{water}} + p_{\text{organic}} \tag{53}$$

And since the water and organic compound are immiscible:

$$P = p_{\text{water}} + p_{\text{organic}} = P°_{\text{water}} + P°_{\text{organic}} \tag{54}$$

Treybal (123) gives the following example of the steam distillation of N-ethylaniline at atmospheric pressure. The vapor pressures at 99.15°C of water and N-ethylaniline are 98.27 and 3.04 kPa (737 and 22.8 mm Hg), respectively.

$$P = P°_{\text{water}} + P°_{\text{organic}} \tag{55}$$
$$= 98.27 + 3.04 = 101.3 \text{ kPa (760 mm Hg)} \tag{56}$$

The concentration of the N-ethylaniline in the vapor is:

$$y = \frac{3.04}{101.3} = 0.030 \text{ mole fraction} \tag{57}$$

Table 2. Typical Packing Factors[a], F_p

Type of packing	Material	6 (1/4)	13 (1/2)	16 (5/8)	25 (1)	38 (1 1/2)	51 (2)	76 (3)	89 (3 1/2)
Intalox saddles (Norton Co.)	ceramic	725	200		92	52	40	22	
Intalox saddles (Norton Co.)	plastic				33		21	16	
Raschig rings	ceramic	1600	580		155	95	65	37	
Raschig rings	metal[b]	700	300	170	115				
Raschig rings	metal[c]		410	290	137	83	57	32	
Pall rings	metal			70	48	33	20		16
Pall rings	plastic			97	52	40	24		16
Berl saddles	ceramic	900	240		110	65	45		

[a] Ref. 112–113. Courtesy of Norton Company.
[b] Thickness: 0.8 mm (1/32 in.).
[c] Thickness: 1.6 mm (1/16 in.).

The normal boiling point of N-ethylaniline is 204°C. Therefore, steam distillation makes possible the distillation of N-ethylaniline at atmospheric pressure at a temperature of 99.15°C instead of its normal boiling point of 204°C. Commercial application of steam distillation include the distilling of turpentine (see Terpenoids) and certain essential oils (see Oils, essential). A detailed calculation of steam distillation of turpentine has been reported (124).

Molecular Distillation

Molecular distillation occurs "where the vapor path is unobstructed and the condenser is separated from the evaporator by a distance less than the mean free path of the evaporating molecules" (125). This specialized branch of distillation is carried out at extremely low pressures ranging from 13–130 mPa (0.1–1.0 μm Hg). Molecular distillation is confined to applications where it is necessary to minimize component degradation by distilling at the lowest possible temperatures. Commercial usage includes the distillation of vitamins (qv) and fatty acid dimers (see Dimer acids).

Distillation as a Separation Method

Distillation is the most important industrial method of separation and purification of liquid components. Liquid separation methods of lesser importance include liquid–liquid extraction (see Extraction), membrane diffusion (see Dialysis; Membrane technology), ion exchange (qv), etc. At times distillation also competes indirectly with separation methods involving solid–liquid separation such as crystallization (qv), adsorption (see Adsorptive separation), etc. King (126) has an extensive discussion of the selection of alternative separation processes (see Separation systems synthesis).

The suitability and economics of a distillation separation depends on many factors, including: favorable vapor–liquid equilibria, feed composition, number of components to be separated, product purity requirements, the absolute pressure of the distillation, heat sensitivity, corrosivity, and continuous vs batch requirements.

Favorable Vapor–Liquid Equilibria. The suitability of distillation as a separation method is strongly dependent on favorable vapor–liquid equilibria. The absolute value of the key relative volatilities directly determines the ease and economics of a distillation. The energy requirements and the number of plates required for any given separation increase rapidly as the relative volatility becomes lower and approaches unity. This is illustrated by the following example.

Given an ideal binary mixture with a 50 mole % feed and a distillate and bottoms requirement of 99.8% purity each, the minimum reflux and minimum number of theoretical plates are shown below for assumed relative volatilities of 1.1, 1.5, and 4.

Relative volatility	Minimum reflux ratio, R/D	Minimum number of theoretical plates
1.1	20	130
1.5	4.0	31
4	0.66	9

In the example, the minimum reflux ratio and minimum number of theoretical plates increased 33- to 14-fold, respectively, when the relative volatility decreased from 4 to 1.1. Any other distillation system would have different specific reflux ratios and numbers of theoretical plates, but the trend would be the same. As the relative volatility approaches unity, distillation separations rapidly become more costly. Binary azeotropic systems are impossible to separate into pure components in a single column. Methods of breaking the azeotropes are discussed elsewhere in this Encyclopedia (see Azeotropic and extractive distillation).

The relative volatility of a given separation may be enhanced in order to make the distillation separation more favorable. Two enhancement methods are: (*1*) addition of a third component to improve the relative volatility (as is done in extractive or azeotropic distillation), or (*2*) reduction of the absolute pressure of the distillation to obtain more favorable equilibria.

Feed Composition. Feed composition has a substantial effect on the economics of a distillation. Distillations tend to become uneconomical as the feed becomes dilute. There are two types of dilute feed cases: one in which the valuable recovered component is a low boiler and the second when it is a high boiler. When the recovered component is the low boiler, the absolute distillate rate will be low but the reflux ratio and the number of plates will be high. An example is the recovery of methanol from a dilute solution in water. When the valuable recovered component is a high boiler, the distillate rate, the reflux relative to the high boiler, and the number of plates all are high. An example is the recovery of acetic acid from a dilute solution in water. Alternative recovery methods are usually more economical with dilute feeds.

Number of Components to be Separated. The number of components to be separated determines the overall distillation system design (see under Multicomponent Calculations).

Product Purity. Product purity requirements influence the choice between separation methods. With favorable equilibria, distillation energy requirements do not increase significantly as purity specifications become tighter. For example, in an ideal binary distillation of 50 mole % feed of A, the minimum and operating reflux ratio would be essentially the same as the required purity of A was 99 or 99.9999%. The number of plates would increase substantially as the purity requirements become more stringent. The shortcut methods of calculating minimum reflux ratio, minimum number of plates, operating reflux ratio, and number of operating plates (as discussed under Multicomponent Calculations) allow a rapid evaluation of the effect of changes in purity requirements on the key economic factor in distillation.

The Absolute Pressure. The absolute pressure of the distillation may have substantial economic impact. The temperature at which heat is supplied to the reboiler and is removed from the condenser determines the unit cost of the energy. The cost of removing heat in the condenser increases rapidly as the condensing temperature drops below the range of air or water cooling, eg, the cost of removing a unit quantity of heat at $-25°C$ may be one hundred times as high as removing it at $100°C$. Similarly, the cost of the energy required for the reboiler increases rapidly as the boiler temperature increases above some level determined by local conditions. For example, at a particular site low pressure waste steam at $110°C$ may be essentially without cost but if a temperature level of $200°C$ is required, the unit cost of the heat will be much higher. The relative cost of the heat being removed and supplied is the controlling factor determining the design of some distillations. Petterson and Wells (127) discuss the use of multiple interstage reboilers and condensers at different energy levels as well as the use of other operational modes used to optimize the overall economics.

The absolute pressure may have a significant effect on the vapor–liquid equilibrium. Generally, the lower the absolute pressure the more favorable the equilibrium. This effect has been discussed for the styrene–ethyl benzene system (128). In a given column, increased pressure increases the column capacity by reducing the capacity factor (see eqs. 49 and 50). Selection of the economic pressure can be facilitated by guidelines (129) that take into consideration the pressure effects on capacity and relative volatility. Low pressures are required for distillation involving heat-sensitive material (see below).

Heat Sensitivity. The heat sensitivity or polymerization tendencies of the materials being distilled influence the economics of distillation. Many materials cannot be distilled at their atmospheric boiling points because of high thermal degradation, polymerization, or other unfavorable reaction effects that are functions of temperature. These systems are distilled under high vacuum in order to lower distillation temperatures. In these systems, the pressure drop per theoretical plate is frequently the controlling factor in contactor selection. An excellent discussion of equipment requirements and characteristics of vacuum distillation is ref. 130.

Corrosivity. Corrosivity is an important factor in the economics of distillation. Corrosion rates increase rapidly with temperature, and in distillation, the separation is made at boiling temperatures. The boiling temperatures may require that expensive materials of construction be used for the distillation equipment; however, some of these corrosion-resistant materials are difficult to fabricate. Frequently, the easier-to-fabricate packed columns have an advantage over plate columns when there is a serious corrosion problem (see Corrosion).

Batch vs Continuous Distillation. The mode of operation also influences the economics of distillation. Batch distillation is generally limited to small-scale operations where the equipment serves several distillations.

Distillation Equipment Costs

Recent literature discussions of distillation equipment costs include presentations of British (96) and American (131) cost factors for plate and packed towers, consideration of cost optimization of packed and plate towers (132), a procedure for evaluating the installation cost of plate columns based on corporate experience (133), and a discussion of estimation of column costs (134).

Distillation Column Control

Distillation columns are controlled by hand or automatically. The parameters that must be controlled are: (1) the overall mass balance, (2) the overall enthalpy balance, and (3) the column operating pressure. Modern control systems are designed to control both the static and dynamic column and system variables. Some recent books discuss distillation column control (135–136) (see also Instrumentation and control).

Fractionation Research, Inc. (FRI)

Fractionation Research, Inc. is a nonprofit, industry-sponsored, research corporation with laboratories in South Pasadena, California. Much of the current research on commercial-size distillation equipment is being done by FRI. The industrial sponsors are fabricators, designers, and constructors, or users of distillation equipment. For the most part, the research is confidential to the members, however, some of the research results have been published in papers or motion pictures. Typical publications include: *Liquid Mixing on Sieve Plates* (137), *Bubble Cap Efficiency* (138), and *Motion Pictures of Downcomer Performance* (139).

Distillation Literature Reviews

The reviews of distillation literature in *Industrial and Engineering Chemistry*, which started in 1946 and have continued up to the present time in a modified form, are excellent sources for references to distillation literature (140–145).

Nomenclature

A^*, B^*	= end value constants in Van Laar or Margules activity coefficient equations 17–20
B	= bottoms from column, moles per unit time
C	= vapor capacity factor, equation 49
D	= distillate from column, moles per unit time
E_o	= overall column plate efficiency, equation 51
E_{MV}	= Murphree plate efficiency, equation 52
f	= fugacity
$\bar{f}$	= equilibrium stage at feed point
F	= feed, moles per unit time
F^*	= F-factor, equation 50
F_B^*	= F-factor, bubbling area
F_C^*	= F-factor, column cross-sectional area
F_H^*	= F-factor, column hole or slot area
F_p	= packing factor from Table 2
G	= gas rate, mass per unit time
G^*	= gas rate, mass per unit area per unit time
h	= unit height of packing
H	= enthalpy per mole
$\bar{H}$	= enthalpy per unit time

H^*	= Henry's law constant, equation 24
HETP	= height equivalent of theoretical plate, units of length
K	= y/x, vapor–liquid equilibrium constant, equation 29
L	= liquid rate, moles per unit time
L^*	= liquid rate, mass per unit area per unit time
$\overline{L}$	= constant molal liquid rate
m	= an equilibrium stage below the feed
n	= an equilibrium stage above the feed
n'	= number of moles in ideal gas law
N	= number of stages
N^*	= number of components
N_a	= number of actual stages
N_t	= number of theoretical stages
P	= total pressure of system
p	= partial pressure
P^o	= vapor pressure
q	= feed construction line on McCabe-Thiele diagram; also heat to vaporize one mole of feed divided by molal latent heat of feed, see equation 41
$\overline{q}$	= heat removed or added at column auxiliaries
R	= reflux, moles per unit time
R'	= gas law constant
T	= temperature, absolute
V	= vapor, moles per unit time
V^*	= vapor velocity
V'	= volume, per mole of gas
$\overline{V}$	= constant molal vapor rate
x	= mole fraction in liquid
y	= mole fraction in vapor
y^*	= mole fraction vapor in equilibrium with x
z	= compressibility factor in gas law, equation 26
z_c	= critical compressibility factor
Z	= azeotropic point
$\alpha_{1/2}$	= relative volatility of component 1 to 2 = $(y/x)_1/(y/x)_2$
γ	= activity coefficient, see equation 15
γ^∞	= terminal activity coefficient, ie, at infinite dilution of the specified component
λ	= slope of operating line/slope of equilibrium line
Λ	= constant in Wilson activity coefficient equations 21–22
ρ	= fluid-phase density
ϕ	= fugacity coefficient, equation 28

Superscripts

1,2...	= composition number
A	= a typical equilibrium point
B	= bottoms
C	= condenser
D	= distillate
E	= end
F	= feed
f	= feed stage
L	= liquid
m	= stage number m
$m-1$	= stage below m
$m+1$	= stage above m
n	= stage number n
$n-1$	= stage below n
$n+1$	= stage above n
N	= N^{th} component of components i to n
P	= pressure

888 DISTILLATION

S = calandria (S derives from the conventional reference to steam)
T = top column
V = vapor

Subscripts

$1,2,3 \ldots n$ = component numbers
B = bottoms
D = distillate
F = feed
g = gas
H = component H of binary system L–H, H is the high boiler
i = component i of components $i, j, k \ldots n$
L = component L of binary system L–H, L is the low boiler
l = liquid
MIN = minimum
P = pressure
r = reduced (temperature or pressure)
R = rectifying section
S = stripping section
T = temperature

BIBLIOGRAPHY

"Distillation" in *ECT* 1st ed., Vol. 5, pp. 156–187, by E. G. Scheibel, Hoffmann-LaRoche, Inc.; "Distillation" in *ECT* 2nd ed., Vol. 7, pp. 204–248, by C. D. Holland and J. D. Lindsay, Texas A&M University.

1. C. D. Holland and J. D. Lindsey, "Distillation" in A. Standen, ed., *Kirk-Othmer Encyclopedia of Chemical Technology,* 2nd ed., Vol. 7, Interscience Publishers, a division of John Wiley & Sons, Inc., New York, 1965, pp. 204–208.
2. R. C. Reid, J. M. Prausnitz, and T. K. Sherwood, *The Properties of Gases and Liquids,* 3rd ed., McGraw-Hill Book Co., New York, 1977, pp. 26, 288.
3. T. S. Storvick and S. I. Sandler, *Phase Equilibria and Fluid Properties in Chemical Industry,* ACS Symposium Series 60, American Chemical Society, Washington, D.C., 1977.
4. R. E. Treybal, *Mass Transfer Operations,* 2nd ed., McGraw-Hill Book Co., New York, 1968, p. 282.
5. G. N. Lewis and M. Randall, *Thermodyanmics,* revised by K. S. Pitzer and L. Brewer, 2nd ed., McGraw-Hill Book Co., New York, 1961.
6. J. M. Prausnitz, *Molecular Thermodynamics of Fluid-Phase Equilibria,* Prentice-Hall, Inc., Englewood Cliffs, N.J., 1968, pp. 20, 181.
7. E. Hala and co-workers, *Vapor–Liquid Equilibrium,* 2nd ed., Pergamon Press, Oxford, Eng., 1967, p. 24.
8. J. M. Prausnitz, *Molecular Thermodynamics of Fluid-Phase Equilibria,* Prentice-Hall, Inc., Englewood Cliffs, N.J., 1968.
9. H. R. Null, *Phase Equilibrium in Process Design,* Wiley-Interscience, New York, 1970.
10. R. Gilmont, "Vapor-Liquid Equilibria" in A. Standen, ed., *Kirk-Othmer Encyclopedia of Chemical Technology,* 2nd ed., Vol. 21, Interscience Publishers, a division of John Wiley & Sons, Inc., New York, 1965, pp. 196–240.
11. M. B. King, *Phase Equilibrium in Mixtures,* Pergamon Press, Oxford, Eng., 1969.
12. R. C. Reid, J. M. Prausnitz, and T. K. Sherwood, *The Properties of Gases and Liquids,* 3rd ed., McGraw-Hill Book Co., New York, 1977.
13. M. Van Winkle, *Distillation,* McGraw-Hill Book Co., New York, 1967, p. 390.
14. J. J. Van Laar, *Z. Phys. Chem.* **72,** 723 (1910); **83,** 599 (1913).
15. M. Margules, *Sitzungsber. Akad. Wiss. Wien Math. Naturwiss. Kl. Abt. 2* **104,** 1243 (1895).
16. G. M. Wilson, *J. Am. Chem. Soc.* **86,** 127 (1964).
17. E. L. Derr and C. N. Deal, *Proc. Int. Symp. Distill. Brighton, 1969* **3,** 40 (1970).
18. R. N Fleck and J. M. Prausnitz, *Chem. Eng.* **75**(11), 157 (1968).
19. A. Fredenslund, R. L. Jones, and J. M. Prausnitz, *AIChE J.* **21**(6), 1086 (1975).
20. A. Fredenslund, M. L. Michelsen, and J. M. Prausnitz, *Chem. Eng. Prog.* **72**(9), 67 (1976).

21. T. Nitta and co-workers, *AIChE. J.* **23**(2), 144 (1977).
22. A. K. Rao, *Chem. Eng.* **84**(10), 143 (1977).
23. Ref. 9, p. 121.
24. O. Redlich and A. T. Kister, *Ind. Eng. Chem.* **40,** 341 (1948).
25. B. Black, *Ind. Eng. Chem.* **50,** 403 (1958).
26. H. Renon and J. M. Prausnitz, *AIChE J.* **14,** 135 (1968).
27. A. Fredenslund and co-workers, *Ind. Eng. Chem. Proc. Des. Div.* **16,** 450 (1977).
28. Ref. 2, p. 300.
29. Ref. 9, p. 48.
30. Ref. 2, p. 288.
31. P. E. Liley and W. R. Gambill in R. H. Perry and C. H. Chilton, eds., *Chemical Engineers' Handbook,* 5th ed., McGraw-Hill Book Co., New York, 1973, p. 96.
32. O. Redlich and J. N. S. Kwong, *Chem. Rev.* **44,** 233, 1949.
33. B. D. Smith, B. Block, and K. C. D. Hickman in ref. 33, p. 4.
34. C. H. Holland, *Multicomponent Distillation,* Prentice-Hall, Inc., Englewood Cliffs, N.J., 1963, pp. 432, 446.
35. O. A. Hougen, K. M. Watson, and R. A. Ragatz, *Chemical Process Principles,* Part II, *Thermodynamics,* 2nd ed., John Wiley & Sons, Inc., New York, 1959, p. 600.
36. E. Hala and co-workers, *Vapour–Liquid Equilibrium,* 2nd ed., Pergamon Press, Oxford, Eng., 1967.
37. M. W. Kellogg Co., *Liquid–Vapor Equilibrium in Mixtures of Light Hydrocarbon,* M. W. Kellogg Co., Piscataway, N.J., 1970.
38. *Technical Data Book, Petroleum Refining,* Vols. I and II, 3rd ed., American Petroleum Institute, New York, 1976.
39. Ref. 13, p. 75.
40. W. C. Edmister, *Hydrocarbon Process.* **52**(3), 95, 1973.
41. B. D. Smith, *Design of Equilibrium Stage Processes,* McGraw-Hill Book Co., New York, 1963, p. 15.
42. I. Wichterle, J. Linek, and E. Hala, *Vapor–Liquid Equilibrium Data Bibliography,* Suppl. I, Elsevier Scientific Publishing Co., Amsterdam, The Netherlands, 1973, 1976.
43. J. C. Chu and co-workers, *Distillation Equilibrium Data,* Reinhold Publishing Corp., New York, 1950.
44. J. C. Chu and co-workers, *Vapor–Liquid Equilibrium Data,* J. W. Edwards, Publisher, Inc., Ann Arbor, Mich., 1956.
45. W. B. Kogan and W. M. Fridman, *Handbuch der Dampf-Flussigkeits-Gleichgewichte,* Veb Deutschere Verlag der Wissenschaften, Berlin, Ger., 1961.
46. E. Hala and co-workers, *Vapour Liquid Equilibrium Data at Normal Pressures,* Pergamon Press, Oxford, Eng., 1968.
47. M. Hirata, S. Ohe, and K. Nagahama, *Computer-Aided Data Book of Vapor-Liquid Equilibria,* Kodansha Limited, Elsevier Scientific Publishing Co., Tokyo, Japan, 1975.
48. J. Gmehling and U. Onken, *Vapor–Liquid Equilibrium Data Collection, Dechema Chemistry Data Series,* Deutsche Gesellschaft fur Chemisches Apparatewesen e.v., Frankfurt, FRG, 1977.
49. L. Horsley, *Azeotropic Data—III, Advances in Chemistry Series No. 116,* American Chemical Society, Washington, D.C., 1973.
50. R. E. Treybal in ref. 33, Sect. 15.
51. W. Gerrard, *Solubility of Gases and Liquids,* Plenum Press, New YOrk, 1976, p. 77.
52. Ref. 2, p. 356.
53. Ref. 36, p. 280.
54. Ref. 10, p. 230.
55. Ref. 9, p. 97.
56. Ref. 36, p. 329.
57. C. McDermott and S. R. M. Ellis, *Chem. Eng. Sci.* **20,** 293 (1965).
58. H. C. Van Ness, *Classical Thermodynamics of Non-Electrolyte Solutions,* The Macmillan Co., New York, 1964, p. 105.
59. W. Featherstone, *Processing* **21**(5), 21 (1975).
60. C. S. Robinson and E. R. Gilliland, *Element of Fractional Distillation,* 4th ed., McGraw-Hill Book Co., New York, 1950, p. 108.
61. Ref. 13, p. 170.

62. R. J. Hengstebeck, *Distillation Principles and Design Procedures,* Reinhold Publishing Corporation, New York, 1961, p. 69.
63. W. L. McCabe and E. W. Thiele, *Ind. Eng. Chem.* **17,** 605 (1925).
64. Ref. 13, p. 258.
65. Ref. 13, pp. 249, 266.
66. M. Ponchon, *Tech. Mod.* **13,** 20 (1921).
67. R. Savarit, *Arts Metiers* **65,** 142, 178, 241, 266, 307 (1922).
68. H. Sawistowski and W. Smith, *Mass Transfer Process Calculations,* Interscience Publishers, a division of John Wiley & Sons, Inc., New York, 1963, p. 262.
69. G. G. Brown, *Unit Operations,* John Wiley & Sons, Inc., New York, 1950, p. 368.
70. B. Block, *Chem. Eng.* **68**(3), 87 (1961).
71. R. W. Ellerbe, *Chem. Eng.* **80**(12), 110 (1973).
72. C. J. King, *Separation Processes,* McGraw-Hill Book Co., New York, 1971, p. 249.
73. A. J. V. Underwood, *Chem. Eng. Prog.* **44,** 603 (1948).
74. M. R. Fenske, *Ind. Eng. Chem.* **24,** 482 (1932).
75. E. R. Gilliland, *Ind. Eng. Chem.* **32,** 1101 (1940).
76. J. H. Erbar and R. M. Maddox, *Pet. Refiner* **40**(5), 183 (1961).
77. Ref. 41, p. 252.
78. Ref. 33, p. 26.
79. Ref. 33, p. 30.
80. D. C. Freshwater and B. D. Henry, *Chem. Eng. (London)* (301) 533 (1975).
81. J. E. Hendry, D. F. Rudd, and J. D. Seader, *AIChE J.* **19,** 1 (1973).
82. *Ballast Tray Design Manual, Bulletin 4900,* 3rd Ed., Glitsch, Inc., Dallas, Tex., 1974.
83. *Koch Flexitray Design Manual, Bulletin 960,* Koch Engineering Co., Wichita, Kansas, 1960.
84. *Float Valve Design Manual,* Nutter Engineering Co., Tulsa, Oklahoma, 1976.
85. Ref. 41, p. 500.
86. J. R. Fair and co-workers in ref. 33, Sect. 18. p. 5.
87. M. H. Hutchinson and R. F. Baddour, *Chem. Eng. Prog.* **52,** 503 (1956).
88. Ref. 13, p. 525.
89. Ref. 86, p. 10.
90. W. J. Thomas and A. N. Shah, *Trans. Inst. Chem. Eng.* **42,** T71 (1964).
91. Ref. 15, p. 578.
92. W. L. Bolles, *Pet. Process.* **11**(2), 65; (3), 82; (4), 72; (5), 109 (1956).
93. J. D. Chase, *Chem. Eng.* **74**(16), 105; (18), 139 (1967).
94. J. R. Barkhurst and J. H. Harker, *Process Plant Design,* American Elsevier Publishing Co., Inc., New York, 1973.
95. E. V. Murphree, *Ind. Eng. Chem.,* **17,** 747, 960 (1925).
96. Ref. 13, p. 533.
97. *Bubble Tray Design Manual—Prediction of Fractionation Efficiency,* American Institute of Chemical Engineers, New York, 1958.
98. R. L. Bell, *AIChE J.* **18**(3), 498 (1972).
99. G. A. Hughmark, *AIChE J.* **17,** 1295 (1971).
100. G. G. Haselden, *Chem. Eng. (London)* (299/300), 439 (1975).
101. W. V. Pinczewski, N. D. Benke, and C. J. D. Fell, *AIChe J.* **21,** 1210 (1975).
102. K. E. Porter, A. Safekourdi, and M. J. Lockett, *Trans. Inst. Chem. Eng.* **55,** 190 (1977).
103. M. A. da S. Jeronimo and H. Sawistowski, *Trans. Inst. Chem. Eng.* **51,** 265 (1973).
104. H. E. O'Connell, *Trans. AIChE* **42,** 741 (1946).
105. G. E. English and M. Van Winkle, *Chem. Eng.* **70**(23), 241 (1963).
106. Ref. 13, p. 550.
107. Ref. 86, p. 13.
108. W. S. Norman, *Absorption, Distillation, and Cooling Towers,* John Wiley & Sons, Inc., New York, 1961, p. 325.
109. Ref. 13, p. 604.
110. Ref. 86, p. 19.
111. T. K. Sherwood, G. H. Shipley, and F. A. L. Halloway, *Ind. Eng. Chem.* **30,** 765 (1938).
112. J. S. Eckert, *Chem. Eng. Prog.* **66**(3), 39 (1970).
113. *Engineering Data Charts GR-109 R6 and GR-122 R8,* Norton Company, Akron, Ohio, Dec. 1976.
114. Ref. 4, p. 365.

115. Ref. 86, p. 48.
116. D. Cornell, W. G. Knapp, and J. R. Fair, *Chem. Eng. Prog.* **56**(7), 68 (1960).
117. *Ibid.* **56**(8), 48 (1960).
118. Ref. 13, p. 627.
119. A. H. P. Skelland, *Diffusional Mass Transfer,* John Wiley & Sons, Inc., New York, 1974.
120. J. S. Eckert, *Chem. Eng.* **82**(8), 70 (1975).
121. Ref. 86, pp. 24, 37.
122. R. Billet, *Chem. Eng.* **79**(4), 68 (1972).
123. Ref. 4, p. 293.
124. W. L. McCabe and J. C. Smith, *Unit Operations of Chemical Engineering,* 3rd ed., McGraw-Hill Book Company, New York, 1976, p. 540.
125. Ref. 33, p. 55.
126. C. J. King, *Separation Processes,* McGraw-Hill Book Co., New York, 1971.
127. W. C. Petterson and T. A. Wells, *Chem. Eng.* **84**(20), 79 (1977).
128. J. C. Frank, G. R. Geyer, and H. Kehde, *Chem. Eng. Prog.* **65**(2), 79 (1969).
129. H. Z. Kister and I. D. Doig, *Hydrocarbon Process.* **56**(7), 132 (1977).
130. P. G. Nygren and G. K. S. Connolly, *Chem. Eng. Prog.* **67**(3), 49 (1971).
131. M. S. Peters and K. D. Timmerhaus, *Plant Design and Economics for Chemical Engineers,* McGraw-Hill Book Co., New York, 1968.
132. J. Happel and D. G. Jordon, *Chemical Process Economics,* 2nd ed., Marcel Dekker, Inc., New York, 1975.
133. J. S. Miller and W. A. Kappella, *Chem. Eng.* **84**(8), 129 (1977).
134. A. Pikulik and H. E. Diaz, *Chem. Eng.* **84**(21), 106 (1977).
135. F. G. Shinskey, *Distillation Control,* McGraw-Hill Book Co., New York, 1977.
136. O. Rademaker, J. E. Rijnsdorp, and A. Maarleveld, *Dynamics and Control of Continuous Distillation Units,* Elsevier Scientific Publishing Co., Amsterdam, The Netherlands, 1975.
137. T. Yanagi and B. D. Scott, *Chem. Eng. Prog.* **69**(10), 75 (1973).
138. B. D. Scott and H. S. Myers, *Chem. Eng. Prog.* **69**(10), 73 (1973).
139. T. Yanagi, "Performance of Downcomers in Distillation Columns"—motion picture, *AIChE Meeting, Atlanta, Ga., Feb. 1970.*
140. W. L. Bolles and J. R. Fair, *Ind. Eng. Chem.* **58**(11), 90 (1966); **59**(11), 86 (1967); **60**(12), **29**, (1968); **61**(11), 112, (1969); **62**(11), 81 (1970).
141. W. L. Bolles and J. R. Fair in V. W. Weekman, Jr., ed., *Annual Reviews of Industrial Engineering Chemistry, 1970,* American Chemical Society, Washington, D.C., 1972.
142. W. L. Bolles and J. R. Fair in V. W. Weekman, Jr., ed., *Annual Reviews of Industrial Engineering Chemistry,* Vol. II, American Chemical Society, Washington, D.C., 1974.
143. J. R. Fair, *Ind. Eng. Chem.* **54**(6), 53 (1962); **55**(5), 55 (1963); **56**(10), 61 (1964); **57**(11), 128 (1965).
144. T. J. Walsh, *Ind. Eng. Chem.* **38**(1), 8 (1946); **39**(1), 17 (1947); **40**(1), 13 (1948); **41**(1), 25 (1949); **42**(1), 32 (1950); **43**(1), 63 (1951); **44**(1), 45 (1952); **45**(1), 39 (1953); **46**(1), 79 (1954); **47**(3), 524 (1955); **48**(3), 492 (1956), **49**(3), 503 (1957); **51**(3), 370 (1959); **52**(3), 277 (1960), **53**(3), 248 (1961).
145. T. J. Walsh and S. Calvert, *Ind. Eng. Chem.* **50**(3), 453 (1958).

EARL R. HAFSLUND
E. I. du Pont de Nemours & Co., Inc.